COLLECTED WORKS OF HSUE-SHEN TSIEN

(1938~1956)

COLLECTED WORKS OF HSUE-SHEN TSIEN

(1938~1956)

HSUE-SHEN TSIEN

ELSEVIER

AMSTERDAM • BOSTON • HEIDELBERG • LONDON
NEW YORK • OXFORD PARIS • SAN DIEGO
SAN FRANCISCO • SINGAPORE • SYDNEY • TOKYO

Academic Press is an imprint of Elsevier

Academic Press is an imprint of Elsevier
The Boulevard, Langford Lane, Kidlington, Oxford OX5 1GB, UK
225 Wyman Street, Waltham, MA 02451, USA

First edition 2012

British Library Cataloguing in Publication Data
A catalogue record for this book is available from the British Library

Library of Congress Cataloging-in-Publication Data
A catalog record for this book is availabe from the Library of Congress

ISBN–13: 978-0-12-398277-3

For information on all Academic Press publications
visit our web site at books.elsevier.com

Printed and bound in the UK

12 13 14 15 16   10 9 8 7 6 5 4 3 2 1

Contents

ABOUT THIS BOOK

This volume collects the scientific works of Tsien Hsue-shen accomplished during his stay in the United States as a graduate student, scientist and professor and published in the period of 1938~1956, when the aeronautic exploration stepped from low-speed to high-speed regimes and astronautic technology entered its infant stage.

In these papers, he addressed and solved a series of key problems in aerodynamics, stability of shells, rocket ballistics and engine analyses, etc., some of which were path-breaking. Starting from 1946, with his strategic wisdom, Tsien Hsue-shen made pioneering contributions to some fields, such as jet propulsion, engineering cybernetics, physical mechanics and engineering sciences, and so on. All these works feature the unique methodology of turning basic theories in natural science into practical tools in tackling complicated engineering problems. It is worth noting that he first advocated the philosophy of engineering sciences, which has been elucidated and illustrated in the volume and proved to be the guideline of innovative industrial development.

The collected works might benefit to its extensive readers in getting deeper insight into the academic contributions, scientific thoughts and studying style of Tsien from various viewpoints.

HSUE-SHEN TSIEN'S BIOGRAPHY

HSUE-SHEN TSIEN
(11 December 1911 - 31 October 2009)

Professor Hsue-shen Tsien, an eminent Chinese scientist, was born in 1911 in Hangzhou, Zhejiang Province, China. He served as Director of the Institute of Mechanics of the Chinese Academy of Sciences, Director of the Fifth Institute of the Ministry of National Defense, Vice Minister of the Seventh Ministry of Machinery Industry, Vice Director of the Science and Technology Committee of the Commission of Science, Technology, and Industry for National Defense (COSTIND), and senior consultant of the General Armament Department of the Ministry of National Defense. He was also elected honorary Chairman of the Chinese Academy of Sciences, honorary Chairman of the China Science and Technology Association, and Vice Chairman of the CPPCC National Committee.

As a pioneer in engineering science and technology, his achievements in applied mechanics were considerable, and his contributions provided the theoretical bases for air vehicles to overcome sound and heat barriers to realize supersonic flight; his monograph *Engineering Cybernetics* has become a theoretical foundation for automation technology and his book *Lectures on Physical Mechanics* paved the way for the derivation of macroscopic characteristics from microscopic laws.

Starting in the mid-1950s, he took charge of key scientific projects and experiments for the Chinese space industry, making China one of the leading countries in space technology.

Even in later years, he continued to innovate in science: he introduced the concept of complex large-scale systems; he proposed the theory of Metasynthethic Engineering and subsequently achieved its implementation, thereby offering a methodological and effective approach to the identification and resolution of problems involving complex large-scale systems; he undertook extensive studies on Systematology, and generalized and extended project systems engineering into social systems engineering, thus transforming it into an effective social technology allowing people working in different social sectors to gain overall long-term optimal benefits.

He was the first, and remains the only, scientist awarded the honor of "National Outstanding Contribution Scientist" in China.

Boundary Layer in Compressible Fluids

Th. von Kármán and H. S. Tsien

(California Institute of Technology)

Summary

The first part of the paper is concerned with the theory of the laminar boundary layer in compressible fluids. The known solution for incompressible fluids is extended to large Mach's numbers by successive approximation. The compressibility effect on surface friction is discussed, and the results applied to estimate the ratio between wave resistance and frictional drag of projectiles and rockets. In the second part the heat transfer between a hot fluid and a cool surface, then between a hot body and a cool fluid is discussed. A general relation between drag and heat transfer as function of Mach's number is given. The limits where cooling becomes illusory because of the heat produced by friction are determined.

The solution of flow problems in which the density is variable is in general very difficult; hence, every case in which an exact or even an approximate solution of the equations of the motion of compressible fluids can be obtained has considerable theoretical interest. Several authors noticed that the theory of the laminar boundary layer can be extended to the case of compressible fluids moving with arbitrarily high velocities without encountering insurmountable mathematical difficulties. Busemann[1] established the equations and calculated the velocity profile for one speed ratio. (By speed ratio is understood the ratio of the airspeed to the velocity of sound.) Frankl[2] also made an analysis of the same problem, however, it is complicated and depends on several arbitrary approximations. The senior author[3] obtained a first approximation by a simple but apparently not sufficiently exact calculation. Hence, in the first part of the present paper, a better method for the solution of the problem is developed.

The boundary layer theory for very high velocities is not without practical interest. First, the statement can be found often in technical and semi-technical literature on rockets and similar high-speed devices that the skin friction becomes more and more insignificant at high speeds. Of course, it is known that with increasing Reynolds Number, the skin friction coefficient is decreasing, i. e., the skin friction becomes relatively small in comparison with the drag produced by wave formation or direct shock. Since high-speed flight will be performed mostly at high altitude where the air is of very low density, so that the kinematic viscosity is large, the resulting Reynolds Number will be relatively small in spite of the high speed.

Presented at the Aerodynamics Session, Sixth Annual Meeting, I. Ae. S. January 26, 1938.

Journal of the Aeronautical Sciences, vol. 5, pp. 227 – 232, 1938.

Another interesting point in the theory of the boundary layer in compressible fluids is the thermodynamic aspect of the problem. In the case of low speeds the influence of the heat produced in the boundary layer can be neglected both in the calculation of the drag and of the heat transferred to the wall. In the case of high speeds, however, the heat produced in the boundary layer is not negligible, but determines the direction of heat flow. In the second part of the paper a few simple examples of heat flow through the boundary layer are discussed.

It has been found necessary in most parts of this analysis to make the assumption of laminar flow. This assumption was found necessary because of the lamentable state of knowledge concerning the laws of turbulent flow of compressible fluids at high speeds. This assumption is somewhat justified by the fact that — as mentioned above — in many problems where the results of this paper can be applied, the Reynolds Number is relatively small, so that a considerable portion of the boundary layer is probably, *de facto*, laminar. Ackeret[4] called attention to the possibility that the stability conditions in supersonic flow might be quite different from those occurring in flow with low velocities. The authors also believe that the stability criteria as developed by Tollmien and others, cannot be applied without modification. Finally, some conclusions of the paper, as will be pointed out, are also applicable to turbulent flow. In other cases, as in the calculation of drag, the assumption of laminar flow surely gives at least the lower limit of its value.

I

If the x-axis is taken along the plate in the direction of the free stream, the y-axis perpendicular to the plate, and u and v indicate the x and y components of the velocity at any point, then the simplified equation of motion in the boundary layer is

$$\rho u \frac{\partial u}{\partial x} + \rho v \frac{\partial u}{\partial y} = \frac{\partial}{\partial y}\left(\mu \frac{\partial u}{\partial y}\right) \tag{1}$$

where both the density ρ and the viscosity μ are variables.

The equation of continuity in this case is

$$\frac{\partial}{\partial x}(\rho u) + \frac{\partial}{\partial y}(\rho v) = 0 \tag{2}$$

A third equation determines the energy balance between the heat produced by viscous dissipation and the heat transferred by conduction and convection. With the same simplification as used in Eqs. (1) and (2), one can write

$$\rho u \frac{\partial}{\partial x}(c_p \cdot T) + \rho v \frac{\partial}{\partial y}(c_p \cdot T) = \frac{\partial}{\partial y}\left(\lambda \frac{\partial T}{\partial y}\right) + \mu\left(\frac{\partial u}{\partial y}\right)^2 \tag{3}$$

where c_p is the specific heat at constant pressure, and λ is the coefficient of heat conduction. If Prandtl's number, $c_p \mu/\lambda$ is assumed to be equal to 1, then it can be easily shown that both Eqs. (1) and (3) can be satisfied by equating the temperature T to a certain parabolic function of the velocity u only. This relation between T and u is

$$\frac{T}{T_0}=\frac{T_w}{T_0}-\left(\frac{T_w}{T_0}-1\right)\frac{u}{U}+\frac{\kappa-1}{2}M^2\frac{u}{U}\left(1-\frac{u}{U}\right) \tag{4}$$

where

U = free stream velocity.

M = speed ratio, or Mach's number of the free stream.

T_0 = temperature of the free stream.

T_w = temperature at the wall of the plate.

Differentiating Eq. (4) one obtains

$$\frac{1}{T_0}\left(\frac{\partial T}{\partial y}\right)_w=\frac{1}{U}\left[\frac{\kappa-1}{2}M^2-\left(\frac{T_w}{T_0}-1\right)\right]\left(\frac{\partial u}{\partial y}\right)_w \tag{5}$$

where the subscript w refers to conditions existing at the surface of the plate. Now $(\partial u/\partial y)_w$ is always positive; therefore, if $[(\kappa-1)/2]M^2>(T_w/T_0)-1$ heat is transferred from the fluid to the wall, if $[(\kappa-1)/2]M^2=(T_w/T_0)-1$ there is no heat transfer between the fluid and the wall, and if $[(\kappa-1)/2]M^2<(T_w/T_0)-1$ heat is transferred from the wall to the fluid. If there is no heat transfer, the energy content per unit mass $(u^2/2)+c_pT$ is constant in the whole region of the boundary layer[5,6].

The pressure being constant the relation between ρ and T is,

$$\rho=\rho_0\frac{T_0}{T} \tag{6}$$

The expression for the viscosity based on the kinetic theory of gases is

$$\mu=\mu_0(T/T_0)^{1/2} \tag{7}$$

However, the following formula is in closer agreement with experimental data

$$\mu=\mu_0(T/T_0)^{0.76} \tag{7a}$$

Busemann[1] calculated the limiting case for which $[(\kappa-1)/2]M^2=(T_w/T_0)-1$ using Eq. (7) and found that for a high Mach's number, the velocity profile is approximately linear. The senior author[3], using this linear velocity profile, the integral relation between the friction and the momentum, and Eq. (7) found that

$$\begin{aligned}C_f&=\frac{\text{Frictional force per unit width of plate}}{(\rho_0U^2/2)\times\text{Length of plate}}\\&=\Theta\sqrt{\frac{\mu_0}{\rho_0Ux}}\left\{1+\frac{\kappa-1}{2}M^2\right\}^{-1/4}\end{aligned} \tag{8}$$

The dimensionless quantity Θ shown in Table 1 is a function of Mach's number only. However, if Eq. (7a) is used, then

$$C_f=\Theta\sqrt{\frac{\mu_0}{\rho_0Ux}}\left\{1+\frac{\kappa-1}{2}M^2\right\}^{-0.12} \tag{8a}$$

Table 1

M	0	1	2	5	10	∞
Θ	1.16	1.20	1.25	1.39	1.50	1.57

It is evident that this linear approximation is not satisfactory for small values of Mach's number. For $M=0$, the case is the same as the Blasius solution[7] for incompressible fluids for which Θ is 1.328.

To solve the problem more rigorously, one has to resort to Eqs. (1) and (2). By introducing the stream function ψ which is defined by

$$\frac{\rho}{\rho_0}u = \frac{\partial\psi}{\partial y},\quad -\frac{\rho}{\rho_0}v = \frac{\partial\psi}{\partial x}$$

the equation of continuity, Eq. (2), is satisfied automatically. Now, if in Eq. (1) ψ is introduced as the independent variable as was done by von Mises[8] in his simplification of the boundary layer equation for incompressible fluids, and all terms are reduced to non-dimensional form then

$$\frac{\partial u^*}{\partial n^*} = \frac{\partial}{\partial \psi^*}\left(u^*\rho^*\mu^*\frac{\partial u^*}{\partial \psi^*}\right) \tag{9}$$

where

$$\begin{aligned} u^* &= u/U \\ n^* &= n/L \\ \psi^* &= (\psi/UL)\sqrt{\rho_0 UL/\mu_0} \\ \rho^* &= \rho/\rho_0 \\ \mu^* &= \mu/\mu_0 \end{aligned} \tag{9a}$$

and L is a convenient length, say length of the plate.

Eq. (9) can be further simplified by introducing a new independent variable $\zeta = \psi^*/\sqrt{n^*}$, then

$$-\frac{\zeta}{2}\frac{du^*}{d\zeta} = \frac{d}{d\zeta}\left(u^*\rho^*\mu^*\frac{du^*}{d\zeta}\right) \tag{10}$$

This can be solved by the method of successive approximations. As ρ^* and μ^* are functions of temperature only as shown in Eqs. (6) and (7) or (7a) and the temperature is a function of u^* then by starting with the known Blasius' solution[6] the right-hand side of Eq. (10) can be expressed in terms of ζ. Therefore, one can write

$$u^*\rho^*\mu^* = f(\zeta)$$

Consequently, the solution of Eq. (10) is

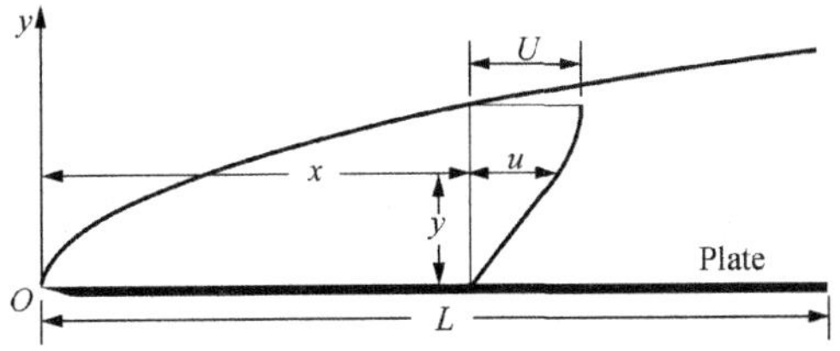

Fig. 1

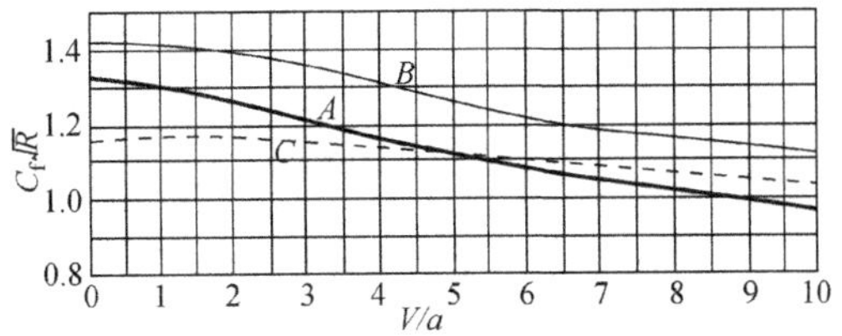

A: No heat transferred to wall. *B*: Wall temperature 1/4 of free stream temperature.

C: von Kármán's first approximation.

Fig. 2 Skin friction coefficients

$$u^* = C\int_0^\zeta \frac{F}{f}\mathrm{d}\zeta \tag{11}$$

where

$$F = \exp\left(-\int_0^\zeta \frac{\zeta\mathrm{d}\zeta}{f}\right)$$

and C is determined by the boundary condition,

$$\frac{1}{C} = \int_0^\infty \frac{F}{f}\mathrm{d}\zeta \tag{11a}$$

A second approximation can be made based upon the value of u^* obtained from Eq. (11). It has been found in the cases investigated that the third or fourth approximation gives sufficient accuracy.

Having computed the final u^*, the y corresponding to u^* can be calculated from

$$y\sqrt{U\rho_0/(\mu_0 x)} = \int_0^\zeta \mathrm{d}\zeta/(\rho^* u^*) \tag{12}$$

Also the skin friction can be computed by the momentum law,

$$C_f = \frac{F}{\frac{\rho_0}{2}U^2 L} = \frac{2\int_0^\infty (1-u^*)\mathrm{d}\zeta}{\sqrt{R}} \tag{13}$$

The velocity profile, the temperature distribution, and the frictional drag coefficient are calculated for different values of the Mach's number of the free stream, for the case $[(\kappa-1)/2]M^2 = (T_w/T_0) - 1$ using the approximate viscosity relation of Eq. (7a). The results are shown in Figs. 2 and 3. The velocity profiles for high speeds are very nearly linear, but it can be seen that the wall temperature for greater Mach's numbers is very high. If the free stream temperature is 40°F, then the wall temperature will be 1 600°F, 3 620°F, 6 540°F, and 10 170°F for Mach's numbers of 4, 6, 8, and 10, respectively. Therefore, there is no doubt that the law of viscosity as expressed by Eq. (7a) will not hold. Also at such high temperatures, the heat transfer due to radiation cannot be neglected. Therefore, the results for extreme Mach's numbers are qualitative only.

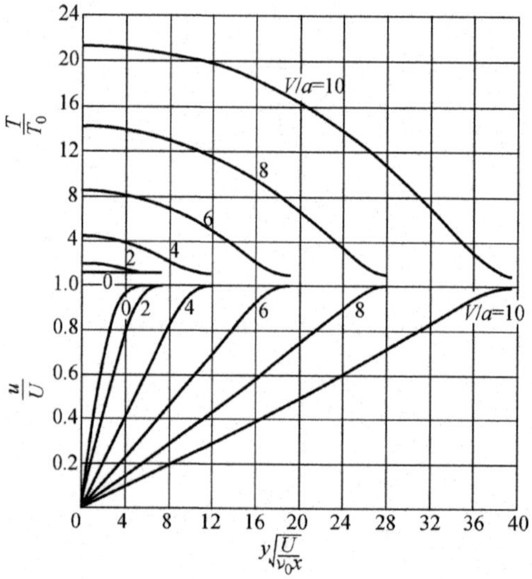

Fig. 3 Velocity and temperature distribution when no heat is transferred to wall

The change in the constant $C_f\sqrt{R}$ is appreciable, but not great. It decreases from 1.328 for $M=0$ to 0.975 for $M=10$, or about 30 percent. However, for $0<M<3$ the change of the constant is very small.

Fig. 2 also shows that Eq. (8a) which was obtained by using the linear approximation is fairly accurate for very high Mach's numbers.

As examples, consider first a projectile and second, a wingless sounding rocket. Taking the diameter of the projectile to be 6 in, the length 24 in, the velocity 1 500 ft/sec and the altitude 32 800 ft (10 km), then the Reynolds Number based on the total length is 7.86×10^6 and the speed ratio is 1.52. From Fig. 2 the skin friction coefficient is

$$C_f=(1.286\times10^{-3})/\sqrt{7.86}=0.000\,459$$

Changing the skin friction coefficient (based on the skin area) to the drag coefficient (based on the maximum cross-section), one obtains

$$C_{D_f}=0.005\,5$$

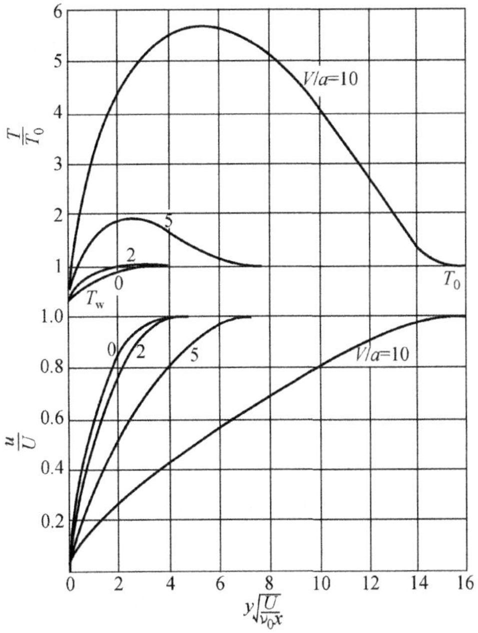

Fig. 4 Velocity and temperature distribution when the wall temperature is 1/4 of the free stream temperature

The drag coefficient due to wave formation taken from Kent's experiments[9] is

$$C_{D_w} = 0.190$$

Therefore, the ratio of skin friction to wave resistance is 0.005 5/0.190 = 0.029.

However, the ratio is greatly changed in the case of the rocket. Taking the diameter of the rocket to be 9 in, the length 8 ft, and the altitude of flight 50 km* (164 000 ft), the velocity 3 400 ft/sec, then the Reynolds Number based on a density ratio at that altitude of 0.000 67 and temperature 25 ℃. (deduced from data on meteors) is 6.14×10^5, and the speed ratio is 3.00. From Fig. 2, the skin friction coefficient is

$$C_f = (1.213 \times 10^{-2})/\sqrt{11.4} = 0.003\,60$$

Then

* The hydrodynamic equation holds so long as the mean free path of the molecules is small in comparison with the thickness of the boundary layer. For this case the thickness of the boundary layer is zero at the nose, however, at a distance 1/4 of the length of the rocket it already amounts to 3.2 cm, while the calculated mean free path of the air molecules at the altitude considered is about 1.1×10^{-2} cm Hence it appears that even for this case the theory can be safely applied. This conclusion is substantiated by the experimental results of H. Ebert in "*Darstellung der Strömungsvorgänge von Gasen bei neidrigen Drucken mittels Reynoldsscher Zahlen*," Zeitschrift für Physik, Bd. 85, S. 561 - 564, 1933.

$$C_{D_f} = 0.123$$

The drag coefficient due to wave formation from Kent's experiments[9] is

$$C_{D_w} = 0.100$$

Therefore, the ratio of skin friction and wave resistance is now 0.123/0.100 = 1.23. If the boundary layer is partly turbulent, the ratio will be even greater. This shows clearly the importance of skin friction in the case of a slender body moving with high speed in extremely rarefied air. It also disproves the belief that wave resistance would always be the predominating part in the total drag of a body moving with a velocity higher than that of sound. The reason underlying this fact can be easily understood when one recalls that the wave resistance of a body is approximately directly proportional to the velocity, while the skin friction is proportional to the velocity raised to a power between 1.5 and 2. Therefore, the ratio of skin friction to wave resistance increases with the speed. With very high velocities and high kinematic viscosity, the wave resistance may even be a negligible portion of the total drag of the body.

II

In order to point out the thermodynamic aspect of the problem two cases will be considered: the flow of a hot fluid along a surface which is kept at a constant temperature inferior to that of the fluid, and the case of a hot wall cooled by a fluid of lower temperature. The problems treated in this part have been discussed before in two very interesting papers by L. Crocco[5,6]. He especially gives an elegant treatment of the cooling problem in the case of very high velocities ("Hyperaviation"). The authors feel that their treatment is somewhat more general and extended than Crocco's previous analysis.

An interesting general relation between the heat transferred through the wall and the frictional drag can be obtained using the assumption that Prandtl's number, i.e., the ratio $c_p\mu/\lambda$, is equal to unity. The same assumption was used also in the previous calculations. It is remarkable that the relation holds also as well for laminar as for turbulent flow. The heat flow q per unit time and unit area of the wall surface is

$$q = \lambda_w (\partial T/\partial y)_w$$

and the frictional drag τ per unit area is

$$\tau = \mu_w (\partial u/\partial y)_w$$

Using Eq. (4) the ratio q/τ can be calculated from the relation

$$\frac{q}{\tau} = \frac{\lambda_w}{\mu_w}\frac{T_0}{U}\left[\left(1-\frac{T_w}{T_0}\right)+\frac{\kappa-1}{2}M^2\right] \tag{14}$$

where T_0 is the absolute temperature, and U is the velocity of the fluid in the free stream, T_w is the absolute temperature at the wall, λ_w and μ_w are the heat conduction and viscosity coefficients of the fluid corresponding to the wall temperature and M denotes Mach's number. Substituting

$M = 0$ one obtains from Eq. (14)

$$\frac{q}{\tau} = \frac{\lambda_w}{\mu_w}\frac{T_0 - T_w}{U} = \frac{c_p(T_0 - T_w)}{\rho_w U} \tag{15}$$

This is the relation known as Prandtl's or G. I. Taylor's formula, first discovered by O. Reynolds. Hence Eq. (14) gives the correction of this result for compressibility effects.

In the case $T_0 > T_w$, i. e., when the wall is colder than the free stream, the effect of compressibility is to increase the heat transferred through the wall. However, it would be erroneous to interpret this result as an "improvement" in cooling because at high speed the heat produced in the boundary layer is of the same order as the heat transferred through the wall. In order to determine the efficiency of the cooling a complete heat balance must be made. For this purpose Eq. (14) does not give sufficient information and the velocity and the temperature distribution in the boundary layer must be computed. Such calculations were carried out for the particular assumption $T_w = T_0/4$, i. e., for the particular case in which the absolute temperature of the wall is kept constant at a value equal to one-fourth of the temperature of the hot fluid. With the same assumption for the variation of μ as in Part Ⅰ, the results shown in Fig. 2 and Fig. 4 were obtained. The variation of $C_f\sqrt{R}$ with M is similar to that obtained in the case without heat conduction through the wall. Also the highest temperature in the boundary layer is very high for extreme Mach's numbers. However, the temperature maximum occurs some distance from the wall.

The heat transferred from the boundary layer to the wall can be calculated as follows:

The initial slope of the velocity profile is equal to

$$\left(\frac{\partial u}{\partial y}\right)_w = \frac{U}{L}\frac{\sqrt{R}}{\sqrt{n^*}}\left(\frac{\mu_0}{\mu_w}\right)\frac{\sqrt{R}C_f}{4} \tag{16}$$

By differentiation of Eq. (4) the relation between the velocity slope and the temperature gradient can be obtained. Using Eq. (7a) and substituting Eq. (16) into Eq. (5) then

$$(\partial T/\partial y)_w = K[T_0\sqrt{R}/(4L\sqrt{n^*})] \tag{17}$$

where

$$K = (4^{0.76}/2)(0.75 + [(\kappa - 1)/2]M^2)\sqrt{R}C_f$$

Therefore, the heat transferred to a strip of unit width of the wall of length L per unit time is equal to

$$Q = \int_0^L \left(\lambda\frac{\partial T}{\partial y}\right)_w \mathrm{d}x = \frac{K\lambda_w T_0\sqrt{R}}{2L}\int_0^L \frac{\mathrm{d}x}{\sqrt{n^*}}$$

or approximately

$$Q \approx K\lambda_w T_0\sqrt{R} \tag{18}$$

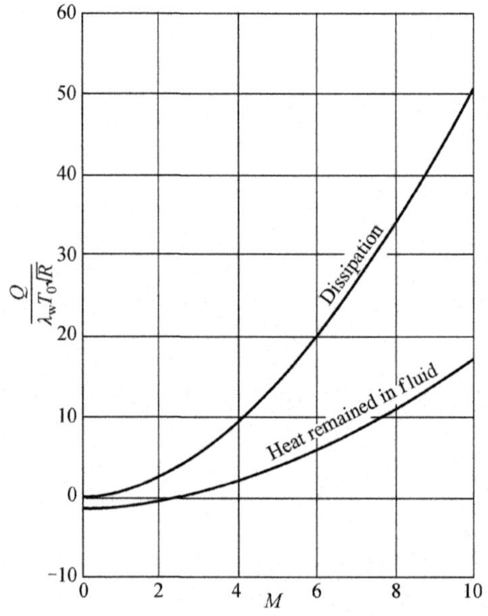

Fig. 5 Heat balance when the wall temperature is 1/4 of the free stream temperature

where K is given in Table 2.

Table 2

M	K
0	1.53
1	1.93
2	3.12
5	10.53
10	33.98

The total heat balance is shown in Fig. 5. The "dissipation" curve represents in dimensionless form the heat produced by friction per unit time and unit width of the plate. The lower curve shows the increase (or decrease) of the heat content per unit time and unit width. The difference of the ordinates corresponds to the heat transferred through the wall. It is seen that cooling takes place for $M < 2.6$. Beyond this limit more heat is produced by friction than the amount which can be transferred to the wall and, as a matter of fact, the fluid is heated.

In the case $T_w > T_0$, i.e., when the wall is hotter than the free stream, the ratio between the heat transfer and the drag decreases with increasing Mach's number. This is shown in Fig. 6 where the ordinate represents the ratio between q/τ with compressibility effect (according to Eq. (14)) to q/τ without compressibility effect (according to Eq. (15)). The calculation was carried out for a gas temperature of -55°F, and a wall temperature of 180°F, and 300°F. It is

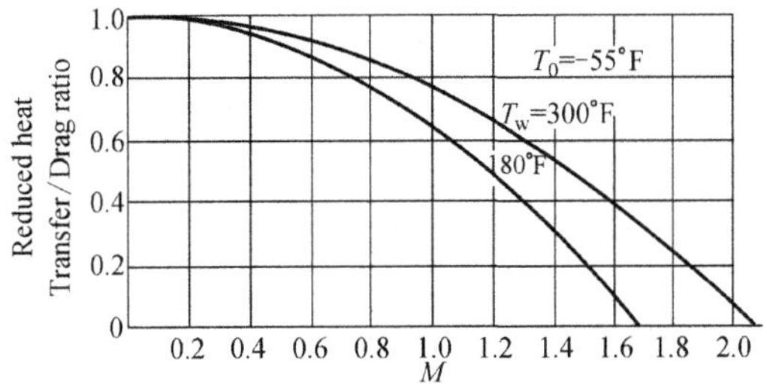

Fig. 6 The effect of high speed on cooling efficiency

seen that there is no cooling in the former case for $M = 1.69$ and in the latter case for $M = 2.08$. However, the decrease of cooling efficiency is appreciable even at much lower speeds. This emphasizes the benefit of the reduction of the speed of cooling air and the relatively poor efficiency of cooling surfaces exposed directly to a high-speed airstream. The curves in Fig. 6 being derived from Eq. (14) apply to laminar as well as to turbulent motion.

References

[1] Busemann A. Gas-strömung mit laminaren Grenzschicht entlang einer Platte. Z. A. M. M., 1935, 15:23.

[2] Frankl. Laminar Boundary Layer of Compressible Fluids. Trans. of the Joukowsky Central Aero-Hydrodynamical Institute, Moscow, 1934, (Russian).

[3] von Kármán Th. The Problem of Resistance in Compressible Fluids. V. Convengo della Foundazione Alessandro Volta (Tema: Le Alte Velocita in Aviazione), Reale Accademia D'Italia, Rome.

[4] Ackeret J. Über Luftkraft bei sehr grossen Geschwindigkeiten insbesondere bei ebenen Strömungen. Helvetica Physica Acta, 1928, 1: 301-322.

[5] Crocco L. Su di un valore massimo del coefficiente di transmissione del calore da una lamina piana a un fluido scorrente. Rendiconti R. Accademia dei Lincei, 1931, 14:490-496.

[6] Crocco L. Sulla Transmissione del calore da una lamina piana un fluido scorrente ad alta velocita. L'Aerotecnica, 1932, 12: 181-197.

[7] Blasius H. Grenzschichten in Flüssigkeiten mit kleiner Reibung. *Zeit*. F. Math. u. Phys., 1908, 56: 1.

[8] von Mises. Bemerkung zur Hydrodynamik. Z. A. M. M., 1927, 7: 425.

[9] Kent R. H. The Role of Model Experiment in Projectile Design. Mechanical Engineering, 1932, 54, 641-646.

Supersonic Flow over an Inclined Body of Revolution

Hsue-shen Tsien

(California Institute of Technology)

Summary

A first approximation is obtained for the side force or the lift of a body of revolution inclined in a supersonic flow from the linearized equation of motion of compressible fluids. It is shown that the lift at any fixed Mach's number is directly proportional to the angle of attack of the body. The case of the cone is calculated in detail and a general method using step-wise doublet distribution is developed for a pointed projectile.

The aerodynamic forces acting on a projectile can be divided into three parts: the resistance or drag in the direction of the axis of the body, the lift in the direction perpendicular to the axis of the body, and the forces due to the rotation of the body (Magnus effect). The first component, the resistance, is, of course, the most important one, because it is the predominating factor in determining the range of the projectile. However, in the case of an actual projectile, inclination and rotation are always present, and, therefore, accurate calculation of range is impossible without considering the second and third components of aerodynamic forces, i.e., the lift and the forces due to rotation of the body. von Kármán[2] suggested that his method of sources for the linearized hydrodynamical equation of axial flow[1] over a slender body of revolution can be generalized to the case in which the projectile is inclined to the flight path. This is carried out in the present paper. Strictly speaking, the solution is applicable only to a very slender body inclined at a small angle to the flight path, because second order quantities of the disturbance due to the presence of the body are neglected. However, for the case of axial flow over a cone, von Kármán-Moore's first approximation[1] differs very little from the exact solution of Taylor and Maccoll[3] for vertex angles up to 40°. Therefore, it is expected that the first approximation of the lift force as obtained in this paper can be applied to a pointed projectile with fair accuracy. This is substantiated by the example shown at the end of this paper.

If ϕ is the potential of the small disturbance velocity due to the presence of a body of revolution whose axis coincides with the x-axis, then the linearized equation of motion of compressible fluids in cylindrical coordinates x, r, and θ is

Received May 27, 1938.

Journal of the Aeronautical Sciences, vol. 5, pp. 480 – 483, 1938.

$$\left(1-\frac{V^2}{c^2}\right)\frac{\partial^2\phi}{\partial x^2}+\frac{1}{r}\frac{\partial\phi}{\partial r}+\frac{\partial^2\phi}{\partial r^2}+\frac{1}{r^2}\frac{\partial^2\phi}{\partial\theta^2}=0 \tag{1}$$

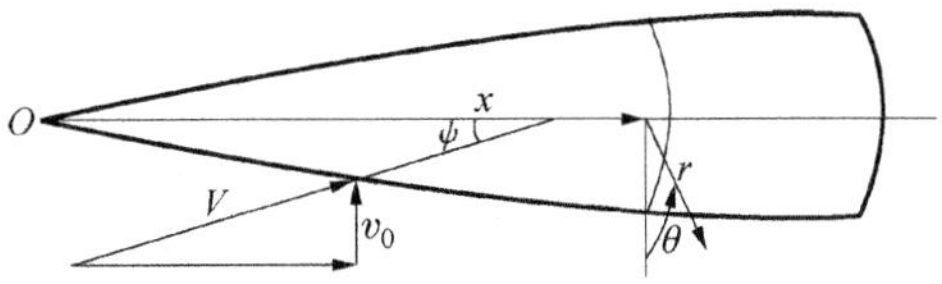

Fig. 1

In this equation, V is the velocity of the undisturbed flow for which the velocity of sound is c. If the direction of the undisturbed flow coincides with the axis of the body, then ϕ is independent of θ, and Eq. (1) reduces to

$$\left(1-\frac{V^2}{c^2}\right)\frac{\partial^2\phi}{\partial x^2}+\frac{1}{r}\frac{\partial\phi}{\partial r}+\frac{\partial^2\phi}{\partial r^2}=0 \tag{2}$$

The solution of this equation when the velocity of the undisturbed flow is greater than the velocity of sound, is the same as that for a two-dimensional wave diverging from a center. It was obtained by Levi-Civita[4] and by H. Lamb[5]. von Kármán and Moore[1] applied it to the present case and showed that it can be expressed as a source distribution given by the potential

$$\phi_1=\int_{\cosh^{-1}(x/ar)}^{0} f_1(x-ar\cosh u)\,du \tag{3}$$

where $\alpha=\sqrt{(V/c)^2-1}$. Analogy with a similar case of flow of an incompressible fluid leads one to expect the solution of Eq. (1) to be a doublet distribution given by the potential

$$\phi_2=-\alpha\cos\theta\int_{\cosh^{-1}(x/ar)}^{0} f(x-ar\cosh u)\cosh u\,du \tag{4}$$

This can be shown to be true, because, if the solution of Eq. (1) is of the form

$$\phi_2=\cos\theta F(x,r)$$

then Eq. (1) reduces to

$$\left(1-\frac{V^2}{c^2}\right)\frac{\partial^2 F}{\partial x}+\frac{1}{r}\frac{\partial F}{\partial r}+\frac{\partial^2 F}{\partial r^2}-\frac{F}{r^2}=0 \tag{1a}$$

Differentiation of Eq. (2) with respect to r gives

$$\left(1-\frac{V^2}{c^2}\right)\frac{\partial^2}{\partial x^2}\left(\frac{\partial\phi}{\partial r}\right)+\frac{1}{r}\frac{\partial}{\partial r}\left(\frac{\partial\phi}{\partial r}\right)+\frac{\partial^2}{\partial r^2}\left(\frac{\partial\phi}{\partial r}\right)-\frac{1}{r^2}\left(\frac{\partial\phi}{\partial r}\right)=0 \tag{2a}$$

Comparing Eq. (1a) with Eq. (2a), it is easily seen that Eq. (4) is a solution of Eq. (1). The function f has to be determined by the boundary condition

$$v_0=\frac{1}{\cos\theta}\left(\frac{\partial\phi}{\partial r}\right)_{r=R}=-\alpha^2\int_{\cosh^{-1}(x/ar)}^{0} f'(x-\alpha R\cosh u)\cosh^2 u\,du \tag{5}$$

where v_0 is the normal component of the velocity V of the undisturbed flow, and R is the radius of the body.

The complete solution of flow over an inclined body of revolution is then obtained by superimposing a crossflow upon an axial flow, i. e., $\phi = \phi_1 + \phi_2$. This solution was also suggested independently by C. Ferrari[6].

From the velocity potential ϕ, one can calculate the pressure distribution over the body and then the aerodynamic forces. However, since the theory is based upon the linearized equation, the cross product terms of derivatives of ϕ_1 and ϕ_2 in the pressure calculation can be neglected. Therefore the following simplification results: The resistance or drag can be calculated from the axial flow alone and the lift can be separately computed from the cross flow. Since the resistance was calculated before[1], the following treatment is concerned only with the lift force. Retaining only first order terms, the lift acting in a direction perpendicular to the axis of the body and the moment about the vertex are thus

$$L = \int_0^{\pi}\int_0^{\infty} \Delta p\, r\, \mathrm{d}\theta \cos\theta\, \mathrm{d}x \approx 2\rho V \int_0^{\pi}\int_0^{\infty} \frac{\partial \phi_2}{\partial x} r\cos\theta\, \mathrm{d}\theta\, \mathrm{d}x \tag{6}$$

$$M = \int_0^{\pi}\int_0^{\infty} \Delta p\, r\, \mathrm{d}\theta \cos\theta\, x\, \mathrm{d}x \approx 2\rho V \int_0^{\pi}\int_0^{\infty} \frac{\partial \phi_2}{\partial x} x r\cos\theta\, \mathrm{d}\theta\, \mathrm{d}x$$

where Δp is the difference between the pressure at the surface of the body and that of the undisturbed flow and ρ is the density of the fluid in the undisturbed flow.

Eq. (5) is a non-homogeneous linear integral equation in f which does not have a general solution of simple form. However, it is interesting to see how Eq. (5) simplifies in the limiting case when the radius of body approaches zero. It is convenient here to use $\xi = x - ar\cosh u$ as the independent variable, then Eq. (5) becomes

$$v_0 = \frac{1}{R^2}\int_0^{x-aR} \frac{f'(\xi)(x-\xi)^2\mathrm{d}\xi}{\sqrt{(x-\xi)^2 - a^2R^2}} \approx \frac{1}{R^2}\int_0^{x-aR} f'(\xi)(x-\xi)\mathrm{d}\xi = \frac{1}{R^2}\int_0^{x} f(x)\mathrm{d}x \tag{5a}$$

where $f(0)$ is put equal to zero, assuming that the projectile has a pointed nose. Since the cross-sectional area of the body of revolution is $S = \pi R^2$, Eq. (5a) gives

$$f(x) = (v_0/\pi)(\mathrm{d}S/\mathrm{d}x) \tag{7}$$

Substituting into Eq. (6), the lift force is then

$$L = 2\rho V \int_0^{\pi}\int_0^{\infty} \frac{v_0 \cos^2\theta}{\pi} \frac{\mathrm{d}S}{\mathrm{d}x} \mathrm{d}\theta\, \mathrm{d}x = \rho v_0 V A_b$$

where A_b = area of the base section of the body. Hence the lift coefficient can be evaluated as

$$C_L = \frac{L}{(\rho/2)V^2 A_b} = 2\frac{v_0}{V} \approx 2 \cdot \psi \tag{8}$$

in which ψ = angle of attack of the body.

The moment arm d, i. e., the distance between the point of application of the resultant lift force and the vertex can be obtained by dividing the moment computed from Eq. (6) by the lift force, and thus

$$d = [A_b - (A_m/A_b)]l \tag{9}$$

where A_m = area of the mean section of the body, i.e., the volume of the body divided by its length, l.

The results of Eq. (8) and Eq. (9) are identical to those found in Munk's theory of airships[7]. At first sight, this might be surprising. However, if the radius of the body approaches zero as assumed, the cross-flow pattern is the same as that for an infinitely long circular cylinder moving with its axes perpendicular to the flow. Therefore, in every plane perpendicular to the axis of the body, the flow can be considered as two dimensional, i.e., it is independent of the variable x. Hence Eq. (1) reduces simply to

$$\frac{\partial^2\phi}{\partial r^2} + \frac{1}{r}\frac{\partial\phi}{\partial r} + \frac{1}{r^2}\frac{\partial^2\phi}{\partial\theta^2} = 0 \tag{1b}$$

This is immediately recognized as the equation of motion for two-dimensional flow of incompressible fluids, which is the basis of Munk's theory.

Due to this two-dimensional character of the flow, the distribution of doublets is not effected by the change in Mach's number, which is only connected with the independent variable x, and, therefore, the lift coefficient and the moment arm are also independent of Mach's number as shown by Eq. (8) and Eq. (9). This can also be seen from the fact that when r approaches zero, the variable $\xi = (x - \alpha r\cosh u) - x$ and thus the effect of α, which is a function of Mach's number, is removed. To study the effect of Mach's number on the lift of the body, one has to go back to Eq. (5). To avoid the difficulty of solving this integral equation, the "indirect method" of solution can be employed, i.e., take a function f and determine the necessary shape of the body to comply with this function f.

Taking the simplest case

$$f(x - \alpha r\cosh u) = K(x - \alpha r\cosh u)$$

where K = a constant. Then

$$\phi_2 = -K\alpha\cos\theta\int_{\cosh^{-1}(x/\alpha r)}^{0}(x - \alpha r\cosh u)\cosh u\,du = K\alpha\cos\theta\left\{\frac{x}{2}\sqrt{\left(\frac{x}{\alpha r}\right)^2 - 1} - \frac{\alpha r}{2}\cosh^{-1}\frac{x}{\alpha r}\right\} \tag{4a}$$

And the boundary condition reduces to

$$v_0 = \frac{K\alpha^2}{2}\left\{\left(\frac{x}{\alpha R}\right)\sqrt{\left(\frac{x}{\alpha R}\right)^2 - 1} + \cosh^{-1}\left(\frac{x}{\alpha R}\right)\right\}$$

Therefore, the solution, Eq. (4a), is evidently a solution for a cone with half vertex angle ε, if $\cot\varepsilon = (x/R)$. By putting $(\cot\varepsilon/\alpha) = \zeta$, the boundary condition can be written in the form

$$v_0 = (\alpha^2K/2)\{\zeta\sqrt{\zeta^2 - 1} + \cosh^{-1}\zeta\} \tag{5b}$$

For any given value of vertex angle and Mach's number, the corresponding value of K can be obtained from Eq. (5b). Then the lift coefficient can be calculated by using Eq. (6) and it is found to be

$$C_L = K_1 \psi \tag{10}$$

where $K_1 = 2\zeta\sqrt{\zeta^2 - 1}/(\zeta\sqrt{\zeta^2 - 1} + \cosh^{-1}\zeta)$. In the limiting case when ε approaches zero, $K_1 \to 2$ which agrees with Eq. (8). Similarly from Eq. (6) the moment coefficient is

$$C_M = \frac{\text{Moment about Vertex}}{(\rho/2)V^2 A_b l} = \frac{2}{3} K_1 \psi \tag{11}$$

which satisfies Eq. (9).

Both Eq. (8) and Eq. (10) show that the lift at a given Mach's number is proportional to the angle of attack of the body. This is a general characteristic of flow around a body without separation. If the fluid separates from the body and creates a "dead water" region on the lee side of the body, then the lift will be proportional to the square of the angle of attack as was shown by W. Bollay[8]. The problem, whether the fluid separates or not, can only be answered by experiments. From the experimental data now available[9], it seems that the flow is continuous without separation, and, therefore, the lift is proportional to the angle of attack of the body.

Fig. 2 is the result of computation using Eq. (10). Calculations were carried out for values of $K_1 \geqslant 1$, because the value of $K_1 = 1$ corresponds to $\varepsilon = \beta$, where β is the wave angle. For $K_1 < 1$ it is found that $\beta < \varepsilon$. This means that the wave angle is smaller than the vertex angle which is, of course, impossible. Therefore, $K_1 = 1$ marks the limit of validity of this solution. In fact, even when K_1 is near to 1, the solution must be considered as qualitative only, since in this region the effect of the surface of the body on the shock wave cannot be neglected.

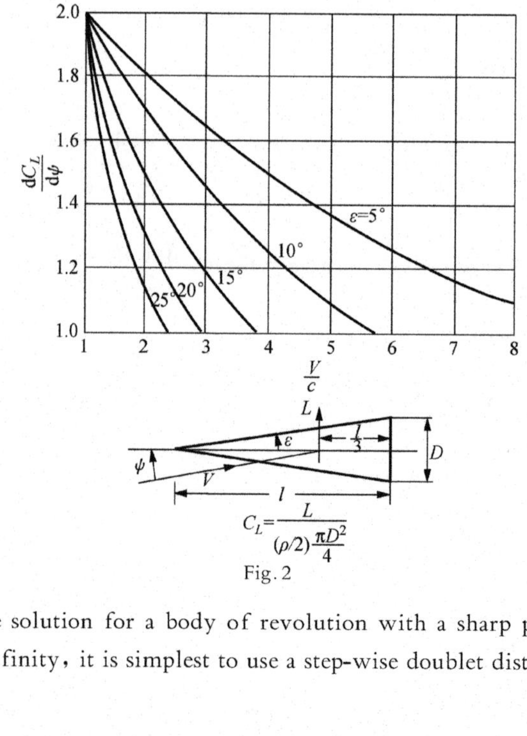

Fig. 2

To generalize the solution for a body of revolution with a sharp point at the origin and cylindrical shape at infinity, it is simplest to use a step-wise doublet distribution. Consider the points P_1, P_2, ..., P_n, ..., P_N of the meridian line of the body, and designate their coordinates by x_1, R_1; x_2, R_2; ...; x_n, R_n, ...; x_N, R_N and the corresponding values of $x-\alpha R$ by ξ_1, ξ_2, ..., ξ_n, ..., ξ_N. Then the boundary condition of Eq. (5) can be written as

$$v_0 = -\frac{\alpha^2}{2}\sum_{i=1}^{n} K_i\left\{\left(\cosh^{-1}\frac{x_n-\xi_{i-1}}{\alpha R_n} - \cosh^{-1}\frac{x_n-\xi_i}{\alpha R_n}\right) + \left(\frac{x_n-\xi_{i-1}}{\alpha R_n}\sqrt{\left(\frac{x_n-\xi_{i-1}}{\alpha R_n}\right)^2-1} - \frac{x_n-\xi_i}{\alpha R_n}\sqrt{\left(\frac{x_n-\xi}{\alpha R_n}\right)^2-1}\right)\right\} \tag{5c}$$

This condition actually gives a set of N equations to determine the N constants K_i. This set of equations can be solved rather easily because each following equation in the set only contains one more K_i which does not appear in the preceding one of the set. When K_is are determined, the lift can be calculated by using Eq. (6) The lift and moment coefficients are thus obtained as

$$\begin{aligned} C_L &= \frac{2\psi}{\alpha}\sum_{n=0}^{N}\frac{(x_{n+1}-x_n)}{R_N}\left(\frac{R_{n+1}+R_n}{R_N}\right)\left\{\sum_{i=0}^{n}K_i\left[\sqrt{\left(\frac{x_n-\xi_{i-1}}{\alpha R_n}\right)^2-1}-\sqrt{\left(\frac{x_n-\xi_i}{\alpha R_n}\right)^2-1}\right]\right\} \\ C_M &= \frac{2\psi}{\alpha l}\sum_{n=0}^{N}\left\{\frac{x_{n+1}+2x_n}{3}+\frac{(x_{n+1}-x_n)}{3}\left(\frac{R_{n+1}}{R_{n+1}+R_n}\right)\right\}\frac{(x_{n+1}-x_n)}{R_n}\left(\frac{R_{n+1}+R_n}{R_N}\right)\times \\ &\quad \left\{\sum_{i=0}^{n}K_i\left[\sqrt{\left(\frac{x_n-\xi_{i-1}}{\alpha R_n}\right)^2-1}-\sqrt{\left(\frac{x_n-\xi_i}{\alpha R_n}\right)^2-1}\right]\right\} \end{aligned} \tag{12}$$

where P_N is the last point on the meridian line, R_N is the base radius, and l is length of the body.

Fig. 3 is the result of the calculation using Eq. (12) for a body of revolution with "6-caliber head" and total length of 4.8 calibers, when it is traveling with a velocity 2.69 times that of sound ($\alpha = 2.5$). The lift coefficient is considerably higher than that of a cone at the same angle of attack and at the same Mach's number, evidently due to the cylindrical part of the body. The position of the resulting lift force is also shown in the figure. Since, as mentioned before, lift and drag are independent in the first order approximation, the calculated lift coefficient can be combined with the drag coefficient taken from experiments and thus give some information on the magnitude and direction of the resulting force. Fig. 4 shows the method applied to this projectile with the drag coefficient taken from Kent's experiment[10].

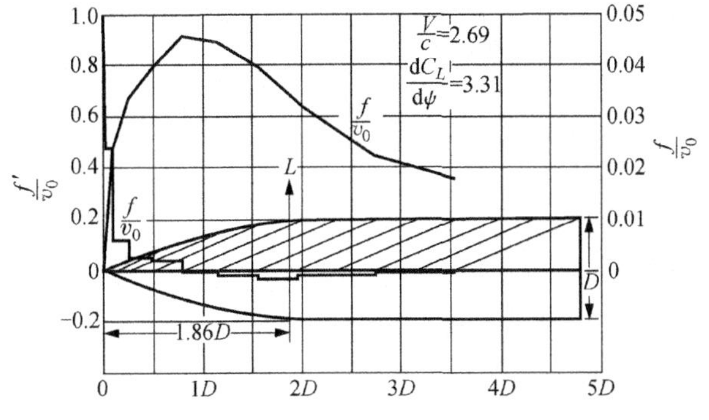

Fig. 3

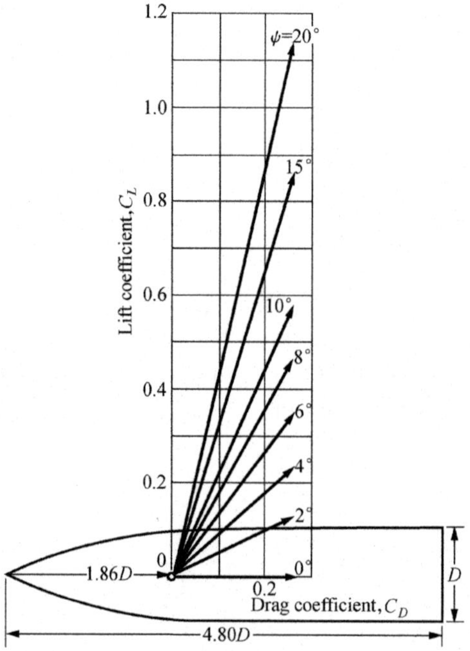

Fig. 4

If the projectile has a length of 4.34 diameters instead of 4.8 diameters and has the same nose shape, and, if its center of gravity is located at a point 2.68 diameters back of the nose, then the calculated moment about the center of gravity can be expressed as $M = \rho V^2 \times R_N^3 \psi f(V/c)$ where $f(V/c) = 9.35$ for $(V/c) = 2.69$. This compares closely with the value $f(V/c) = 10.7$ extrapolated from R. H. Fowler's experiment[9], for a projectile of the same proportions. This shows that the theory developed in this paper can be applied to a projectile with fair accuracy.

The author expresses his gratitude to Professor Th. von Kármán for suggesting the subject and for his frequent help during the course of the work.

References

[1] von Kármán Th, Moore N B. The Resistance of Slender Bodies Moving with Supersonic Velocities with Special Reference to Projectiles. Trans. A. S. M. E. 1932,54: 303 - 310.

[2] von Kármán Th. The Problem of Resistance in Compressible Fluids, Atti dei V Convegno "Volta": Le alte velocita in aviazione. Reale Accademia d'Italia, Rome. 1936:267.

[3] Taylor G I, Maccoll J W. The Air Pressure on a Cone Moving at High Speeds. Proc. Royal Society (A), 1933, 139: 278 - 298. Maccoll J W. The Conical Shock Wave Formed by a Cone Moving at a High Speed, Proc. Royal Society (A), 1937,159: 459 - 472. The comparison was mentioned in a paper by Taylor G I. Well Established Problems in High Speed Flow, Atti dei V Convegno "Volta":198 - 214.

[4] Levi-Civita, Nuovo Cimento (4). 1897, 6.

[5] Lamb H. On Wave-Propagation in Two Dimensions. Proc. London Math. Society (1), 1902, 35: 141.//See also Lamb H. Hydrodynamics, 6th Ed., Cambridge, 1932, 298.

[6] Ferrari C. Campi di corrente ipersonora attorno a solidi di rivoluzione. L'Aerotecnica, 1937, 17, 507-518.

[7] Munk M M. The Aerodynamic Forces on Airship Hulls. NACA Technical Report, 1923: 184.

[8] Bollay W. A Theory for Rectangular Wings of Small Aspect Ratio. Journal Aeronautical Sciences, 1937, 4: 294-296.

[9] Fowler R H, Gallop E G, Lock C N H, Richmond H W. The Aerodynamics of a Spinning Shell. Philosophical Trans. of Royal Society of London (A), 1921, 221: 295-389.

[10] Kent R H. The Rôle of Model Experiments in Projectile Design. Mechanical Engineering, 1932, 54: 641-646.

Problems in Motion of Compressible Fluids and Reaction Propulsion

Thesis

by

Hsue-shen Tsien

In Partial Fulfillment of the Requirements for
the Degree of Doctor of Philosophy

California Institute of Technology
Pasadena, California 1938

Acknowledgements

The author wants first to express his deep gratitude to Dr. Theodore von Kármán. Not only the subjects for PART (Ⅰ), (Ⅱ) and (Ⅲ) were suggested by Dr. von Karman but his constant guiding and help were also essential for carrying out these studies. His inspiration and his warm personality have won the author's highest respect and love.

The author also wants to thank Dr. Harry Bateman for his help in summing the series in PART (Ⅳ) and general discussions. During the preparation of PART (Ⅳ) the author was very much benefited by frequent discussions with Mr. Frank J. Malina who was also very kind in helping the author in numerical computations and figures in PART (Ⅳ).

Table of Contents

PART (I)
Boundary Layer in Compressible Fluids

The solution of flow problems in which the density is variable is in general very difficult; hence, every case in which an exact or even an approximate solution of the equations of the motion of compressible fluids can be obtained has considerable theoretical interest. Several authors noticed that the theory of the laminar boundary layer can be extended to the case of compressible fluids moving with arbitrarily high velocities without encountering insurmountable mathematical difficulties. Busemann[1] established the equations and calculated the velocity profile for one speed ratio (By speed ratio is understood the ratio of the airspeed to the velocity of sound). Frankl[2] also made an analysis of the same problem, however, it is complicated and depends on several arbitrary approximations. Von Karman[3] obtained a first approximation by a simple but apparently not sufficiently exact calculation. Hence, in Section (I), a better method for the solution of the problem is developed.

The boundary layer theory for very high velocities is not without practical interest. First, the statement can be found often in technical and semi-technical literature on rockets and similar high-speed devices that the skin friction becomes more and more significant at high speeds. Of course, it is known that with increasing Reynolds Number, the skin friction coefficient is decreasing, i. e., the skin friction becomes relatively small in comparison with the drag produced by wave formation or direct shock. Since high-speed flight will be performed mostly at high altitude where the air is of very low density, so that the kinematic viscosity is large, the resulting Reynolds Number will be relatively small in spite of the high speed.

Another interesting point in the theory of the boundary layer in compressible fluids is the thermodynamic aspect of the problem. In the case of low speeds the influence of the heat produced in the boundary layer can be neglected both in the calculation of the drag and of the heat transferred to the wall. In the case of high speeds, however, the heat produced in the boundary layer is not negligible, but determines the direction of heat flow. In Section (II) a few simple examples of heat flow through the boundary layer are discussed.

It has been found necessary in most parts of this analysis to make the assumption of laminar flow. This assumption was found necessary because of the lamentable state of knowledge concerning the laws of turbulent flow of compressible fluids at high speeds. This assumption is somewhat justified by the fact that — as mentioned above — in many problems where the results of this paper can be applied, the Reynolds Number is relatively small, so that a considerable portion of the boundary layer is probably, de facto, laminar. Ackeret[4] called attention to the possibility that the stability conditions in supersonic flow might be quite different from those occuring in flow with low velocities.

Recently Küchemann[5] studied the stability of a linear profile near a wall under small sinuous disturbance, and showed that as the velocity ratio increases, the flow is unstable at increasing wave length of disturbance. However, he assumed that the gas is non-viscous and the sound velocity is constant. Both assumptions tend to limit the usefulness of the theory, especially the latter one. Because the constancy of velocity of gas implies the constancy of gas temperature, which unfortunately is far from the truth for a perfect gas as will be shown in later calculations of this section. In spite of these uncertainties, some calculations of this section, as will be pointed out, are also applicable to turbulent flow. In other cases, as in the calculation of drag, the assumption of laminar flow surely gives at least the lower limit of its value.

Section (Ⅰ)

If the x-axis is taken along the plate in the direction of the free stream, the y-axis perpendicular to the plate, and u and v indicate the x and y components of the velocity at any point, then the simplified equation of motion in the boundary layer is for stationary flow

$$\rho u \frac{\partial u}{\partial x} + \rho v \frac{\partial u}{\partial y} = \frac{\partial}{\partial y}\left(\mu \frac{\partial u}{\partial y}\right) \tag{1.1}$$

where both the density ρ and the viscosity μ are variables.

The equation of continuity in this case is

$$\frac{\partial}{\partial x}(\rho u) + \frac{\partial}{\partial y}(\rho v) = 0 \tag{1.2}$$

A third equation determines the energy balance between the heat produced by viscous dissipation and the heat transferred by conduction and convection. With the same simplification as used in Eqs. (1.1) and (1.2), one can write

$$\rho u \frac{\partial}{\partial x}(c_p T) + \rho v \frac{\partial}{\partial y}(c_p T) = \frac{\partial}{\partial y}\left(\lambda \frac{\partial T}{\partial y}\right) + \mu\left(\frac{\partial u}{\partial y}\right)^2 \tag{1.3}$$

where c_p is the specific heat at constant pressure, and λ is the coefficient of heat conduction. If Prandtl's number, $c_p\mu/\lambda$ is assumed to be equal to 1, then it can be easily shown that both Eqs. (1.1) and (1.3) can be satisfied by equating the temperature T to a certain parabolic function of the velocity u only. Indeed, introducing $c_p T = f(u)$ into equation (1.3) and replacing λ by $c_p\mu$, one obtains

$$\left(\rho u \frac{\partial u}{\partial x} + \rho v \frac{\partial u}{\partial y}\right) f'(u) = \frac{\partial}{\partial y}\left(\mu \cdot \frac{\partial u}{\partial y}\right) f'(u) + \mu[f''(u) + 1]\left(\frac{\partial u}{\partial y}\right)^2$$

Hence Eq. (1.3) is reduced to Eq. (1.1) if $f''(u) = -1$

or

$$c_p T = C_1 + C_2 u - \frac{u^2}{2}$$

where C_1, C_2 are constants. Denoting the wall temperature ($u = 0$) by T_w and remembering that $T = T_0$ for $u = U$ where $U =$ free stream velocity, C_1, C_2 can be expressed in terms of T_w, T_0 and

$$\frac{T}{T_0} = \frac{T_w}{T_0} - \left(\frac{T_w}{T_0} - 1\right)\frac{u}{U} + \frac{\gamma - 1}{2}M^2\frac{u}{U}\left(1 - \frac{u}{U}\right) \tag{1.4}$$

Differentiating Eq. (1.4) one obtains

$$\frac{1}{T_0}\left(\frac{\partial T}{\partial y}\right)_w = \frac{1}{U}\left[\frac{\gamma - 1}{2}M^2 - \left(\frac{T_w}{T_0} - 1\right)\right]\left(\frac{\partial u}{\partial y}\right)_w \tag{1.5}$$

where the subscript w refers to conditions existing at the surface of the plate. Now $(\partial u/\partial y)_w$ is always positive; therefore, if $[(\gamma - 1)/2]M^2 > (T_w/T_0) - 1$, heat is transferred from the fluid to the wall; if

$$[(\gamma - 1)/2]M^2 = T_w/T_0 - 1$$

there is no heat transfer between the fluid and the wall; and if

$$[(\gamma - 1)/2]M^2 < T_w/T_0 - 1$$

heat is transferred from the wall to the fluid. If there is no heat transfer, the energy content per unit mass $(u^2/2) + c_p T$ is constant in the whole region of the boundary layer[6].

The pressure being constant the relation between ρ and T is,

$$\rho = \rho_0 \frac{T_0}{T} \tag{1.6}$$

The expression for the viscosity based on the kinetic theory of gases is

$$\mu = \mu_0 \left(\frac{T}{T_0}\right)^{\frac{1}{2}} \tag{1.7}$$

However, the following formula is in closer agreement with experimental data

$$\mu = \mu_0 \left(\frac{T}{T_0}\right)^{0.76} \tag{1.7a}$$

Busemann[1] calculated the limiting case for which $[(\gamma - 1)/2]M^2 = (T_w/T_0) - 1$ using Eq. (1.7) and found that for a high Mach's number, the velocity profile is approximately linear. von Kármán[3], using the linear velocity profile, the integral relation between the friction and the momentum, and Eq. (1.7) found that

$$\begin{aligned} C_f &= \frac{\text{Frictional drag per unit width of plate}}{(\rho_0 U^2/2) \times \text{Length of plate}} \\ &= \Theta\sqrt{\frac{\mu_0}{\rho_0 U x}}\left\{1 + \frac{\gamma - 1}{2}M^2\right\}^{-\frac{1}{4}} \end{aligned} \tag{1.8}$$

The dimensions quantity Θ shown in Table (1.1) is a function of Mach's number only. However, if Eq. (1.7a) is used, then

$$C_f = \Theta\sqrt{\frac{\mu_0}{\rho_0 U x}}\left\{1 + \frac{\gamma - 1}{2}M^2\right\}^{-0.12} \tag{1.8a}$$

Table 1.1

M	0	1	2	5	10	∞
Θ	1.16	1.20	1.25	1.39	1.50	1.57

It is evident that this linear approximation is not satisfactory for small values of Mach's number. For $M=0$, the case is the same as the Blasius solution[7] for incompressible fluids for which Θ is 1.328.

To solve the problem more rigorously, one has to resort to Eqs. (1.1) and (1.2). By introducing the stream function ψ which is defined by

$$\frac{\rho}{\rho_0}u=\frac{\partial\psi}{\partial y},\quad -\frac{\rho}{\rho_0}v=\frac{\partial\psi}{\partial x}$$

the equation of continuity, Eq. (1.2), is satisfied automatically. Now if in Eq. (1.1) ψ is introduced as the independent variable as was done by von Mises[8] in his simplification of the boundary layer equation for incompressible fluids then

$$\frac{\partial}{\partial y}=\frac{\rho}{\rho_0}u\frac{\partial}{\partial\psi},\quad \frac{\partial}{\partial x}=\frac{\partial}{\partial n}-\frac{\rho}{\rho_0}v\frac{\partial}{\partial\psi}$$

where n is a coordinate measured in normal direction to ψ coordinate. Using these new coordinates, the following relations exist

$$\rho u\frac{\partial u}{\partial x}+\rho v\frac{\partial u}{\partial y}=\rho u\frac{\partial u}{\partial n}$$

and

$$\mu\frac{\partial u}{\partial y}=\frac{\mu\rho u}{\rho_0}\frac{\partial u}{\partial\psi}$$

therefore Eq. (1.1) can be written as

$$\frac{\partial u}{\partial n}=\frac{\partial}{\partial\psi}\left[\mu\frac{\rho}{\rho_0}u\frac{\partial u}{\partial\psi}\right] \tag{1.9a}$$

Eq. (1.9a) can be put into non-dimensional form by introducing the following set of new quantities:

$$\left.\begin{aligned} u^* &= u/U \\ n^* &= n/L \\ \psi^* &= [\psi/(UL)]\sqrt{\rho_0 UL/\mu_0}=[\psi/(UL)]\sqrt{R} \\ \rho^* &= \rho/\rho_0 \\ \mu^* &= \mu/\mu_0 \end{aligned}\right\} \tag{1.9b}$$

where L is a convenient length, say length of the plate, and R is the corresponding Reynold's Number, then Eq. (1.9a) becomes

$$\frac{\partial u^*}{\partial n^*} = \frac{\partial}{\partial \psi^*}\left(u^* \rho^* \mu^* \frac{\partial u^*}{\partial \psi^*}\right) \tag{1.9}$$

Equation (1.9) can be further simplified by introducing a new dependent variable $\zeta = \psi^* / \sqrt{n^*}$, then

$$-\frac{\zeta}{2}\frac{du^*}{d\zeta} = \frac{d}{d\zeta}\left(u^* \rho^* \mu^* \frac{du^*}{d\zeta}\right) \tag{1.10}$$

This can be solved by the method of successive approximations. As ρ^* and μ^* are functions of temperature only as shown in Eqs. (1.6) and (1.7) or (1.7a) and the temperature is a function of u^* then by starting with the known Blasius' solution[7] the right-hand side of Eq. (1.10) can be expressed in terms of ζ. Therefore, one can write $u^* \rho^* \mu^* = f(\zeta)$ and Eq. (1.10) becomes $-\frac{\zeta}{2}\frac{du^*}{d\zeta} = \frac{d}{d\zeta}\left[f(\zeta)\frac{du^*}{d\zeta}\right]$ Consequently, the solution of Eq. (1.10) is

$$u^* = C\int_0^\zeta \frac{F}{f} d\zeta \quad \text{where} \quad F = e^{-\int_0^\zeta \frac{\zeta d\zeta}{f}} \tag{1.11}$$

and C is determined by the boundary condition that at $\zeta = \infty$

$$u^* = 1, \quad \frac{1}{C} = \int_0^\infty \frac{F}{f} d\zeta \tag{1.11a}$$

In the actual computation, two methods of evaluating the integrals in Eqs. (1.11) and (1.11a) are used. For small values of ζ, $\zeta < 0.2$, the function u^* and $f(\zeta)$ are expanded in a power series of ζ. Due to the uniform convergence of the power series for sufficiently small values of ζ, the integration is carried out term by term. For values of $\zeta > 0.2$, numerical integration is used.

A second approximation can be made based upon the value of u^* obtained from Eq. (1.11). It has been found in the cases investigated that the third and fourth approximation gives sufficient accuracy, if the velocity profile of next smaller Mach's number is used as the starting point of calculating the velocity profile of next larger Mach's number.

Having computed the final u^*, the y corresponding to u^* can be calculated in the following way:

It is known from Eq. (1.9b) that

$$\zeta = \frac{\psi^*}{\sqrt{n^*}} = \frac{\sqrt{R}\,\frac{\psi}{UL}}{\sqrt{n^*}}$$

Then remembering the definition of ψ, one has

$$\frac{\partial \zeta}{\partial y^*} = \frac{\sqrt{R}}{\sqrt{n^*}} \rho^* u^*$$

However due to the small slope of stream lines,

$$\frac{\partial \zeta}{\partial y^*} \approx \frac{d\zeta}{dy^*}$$

Hence

$$dy^* = \frac{\sqrt{n^*}}{\sqrt{R}} \frac{d\zeta}{\rho^* u^*} \tag{1.12a}$$

or

$$\frac{\sqrt{R}}{\sqrt{n^*}} y^* = \frac{y}{\sqrt{\dfrac{x\mu_0}{U\rho_0}}} = \int_0^{\zeta} \frac{d\zeta}{\rho^* u^*} \tag{1.12}$$

Here the expansion of $\rho^* u^*$ in a power series for small values of ζ is especially useful due to the singularity of integrand at $\zeta = 0$.

The skin friction can be computed by using momentum law, i.e.,

$$\text{Drag} = D = \left\{ \int_0^{\infty} \rho u (U - u) dy \right\}_{x=L}$$

Using Eq. (1.12a), on has

$$dy = L dy^* = \frac{\sqrt{n^*} L d\zeta}{\sqrt{R} \rho^* u^*}$$

thus

$$D = \left\{ L\rho_0 U^2 \frac{\sqrt{n^*}}{\sqrt{R}} \right\}_{x=L} \int_0^{\infty} (1 - u^*) d\zeta$$

But $n^* = \dfrac{n}{L} \approx \dfrac{x}{L}$, therefore

$$\{n^*\}_{x=L} = 1$$

Hence

$$D = \frac{L\rho_0 U^2}{\sqrt{R}} \int_0^{\infty} (1 - u^*) d\zeta$$

Thus the skin friction coefficient can be computed as

$$C_f = \frac{D}{\dfrac{\rho_0}{2} U^2 L} = \frac{2\displaystyle\int_0^{\infty} (1 - u^*) d\zeta}{\sqrt{R}} \tag{1.13}$$

The velocity profile, the temperature distribution, and the frictional drag coefficient are calculated for different values of the Mach's number of the free stream, for the case

$$[(\gamma-1)/2]M^2=(T_w/T_0)-1$$

using the approximate viscosity relation of Eq. (1.7a). The results are shown in Figs. 1.2 and 1.3. The velocity profiles for high speeds are very nearly linear, but it can be seen that the wall temperature for greater Mach's numbers is very high. If the free stream temperature is 40℉, then the wall temperature will be 1 600℉, 3 620℉, 6 540℉, and 10 170℉. for Mach's numbers of 4, 6, 8, and 10, respectively. Therefore, there is no doubt that the law of viscosity as expressed by Eq. (1.7a) will not hold. Also at such high temperatures, the heat transfer due to radiation cannot be neglected. The effect of radiation will be the equalization of gas temperature. In the extreme case of complete equalization, the temperature will be constant throughout the layer and due to the assumed constant pressure throughout the field, the density and viscosity of gas will be also constant throughout the field. Then the velocity profile will be again that calculated by Blasius for incompressible fluid. By this reasoning, the actual velocity profile for large Mach's number when radiation cannot be neglected is something between the Blasius profile and that shown in Fig. 1.3.

The change in the constant $C_f\sqrt{R}$ is appreciable, but not great. It decreases from 1.328 for $M=0$ to 0.975 for $M=10$, or about 30 percent. However, for $0<M<3$ the change of the constant is very small.

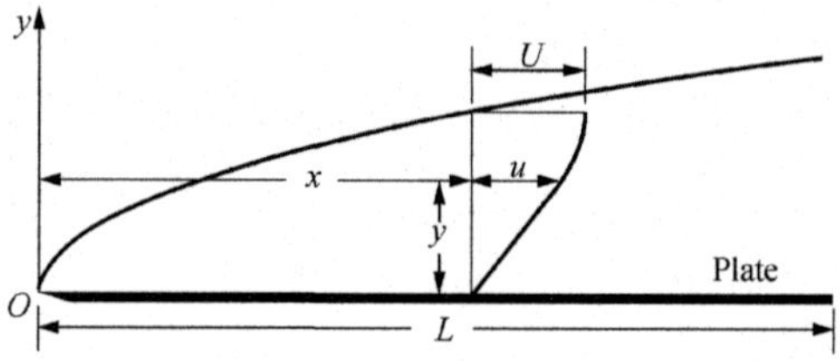

Fig. 1.1

Figure 1.2 also shows that Eq. (1.8a) which was obtained by using the linear approximation is fairly accurate for very high Mach's numbers.

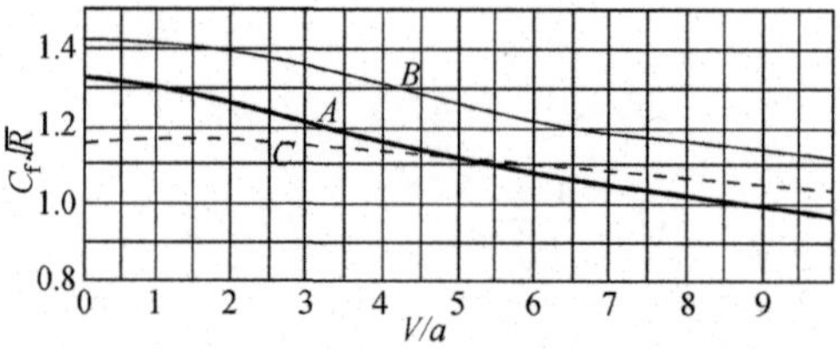

A: No heat transferred to wall B: Wall temperature 1/4 of free stream temperature

C: von Kármán's first approximation

Fig. 1.2 Skin friction coefficients

As examples, consider first a projectile and second, a wingless sounding rocket. Taking the diameter of the projectile to be 6 in, the length 24 in, the velocity 1 500 ft/sec and the altitude 32 800 ft (10 km), then the Reynold's Number based on the total length is 7.86×10^6

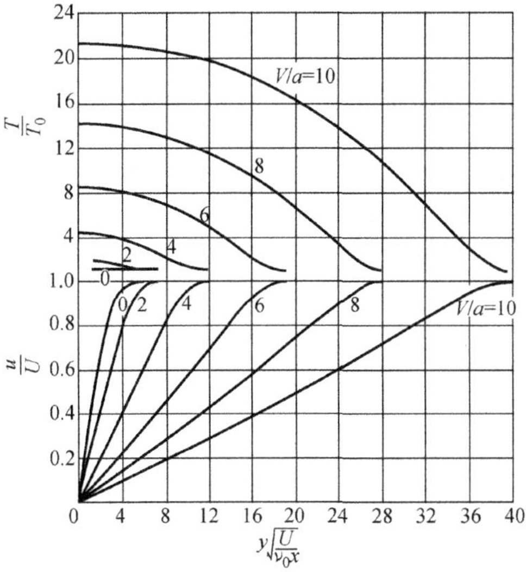

Fig. 1.3 Velocity and temperature distribution when no heat is transferred to wall

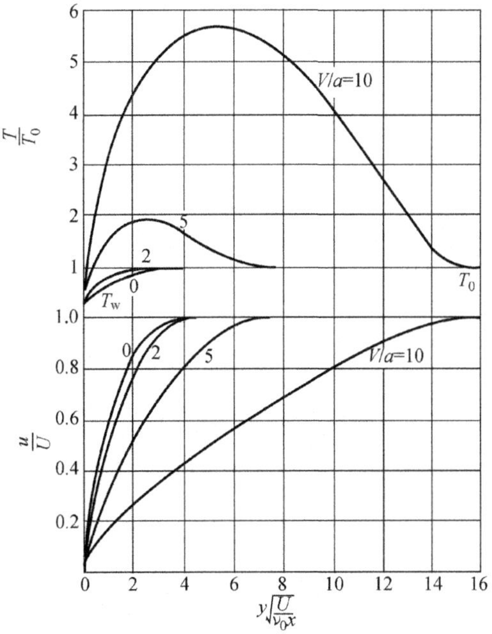

Fig. 1.4 Velocity and temperature distribution when wall temperature is 1/4 of the free stream temperature

and the speed ratio is 1.52. From Fig 1.2 the skin friction coefficient is

$$C_f = (1.286 \times 10^{-3})/\sqrt{7.86} = 0.000\,459$$

Changing the skin friction coefficient (based on the skin area) to the drag coefficient (based on the maximum cross-section), one obtains

$$C_{D_f} = 0.005\,5$$

The drag coefficient due to wave formation taken from Kent's experiments[9] is

$$C_{D_w} = 0.190$$

Therefore the ratio of skin friction to wave resistance is 0.005 5/0.190 = 0.029.

However, the ratio is greatly changed in the case of the rocket. Taking the diameter of the rocket to be 9 in., the length 8 ft., and the altitude of flight 50 km. (see Appendix) (164 000 ft.), the velocity 3 400 ft./sec., then the Reynold's Number based on a density ratio at that altitude of 0.000 67 and temperature 25℃. (deduced from data on meteors) is 6.14×10^5, and the speed ratio is 3.00. From Fig. 1.2, the skin friction coefficient is

$$C_f = (1.213 \times 10^{-2})/\sqrt{11.4} = 0.003\,60$$

Then

$$C_{D_f} = 0.123$$

The drag coefficient due to wave formation from Kent's experiment[9] is

$$C_{D_w} = 0.100$$

Therefore, the ratio of skin friction and wave resistance is now 0.123/0.100 = 1.23. If the boundary layer is partly turbulent, the ratio will be even greater. This shows clearly the importance of skin friction in the case of a slender body moving with high speed in extremely rarified air. It also disproves the belief that wave resistance would always be the predominating part in the total drag of a body moving with a velocity higher than that of sound. The reason underlying this fact can be easily understood when one recalls that the wave resistance of a body is approximately directly proportional to the velocity, while the skin friction is proportional to the velocity raised to a power between 1.5 and 2. Therefore, the ratio of skin friction to wave resistance increases with the speed. With very high velocities and high kinematic viscosity, the wave resistance may even be a negligible portion of the total drag of the body.

Section (Ⅱ)

In order to point out the thermodynamic aspect of the problem two cases will be considered: the flow of a hot fluid along a surface which is kept at a constant temperature inferior to that of the fluid, and the case of a hot wall cooled by a fluid of lower temperature. The problems treated in this part have been discussed before in two very interesting papers by L. Crocco[6]. He especially gives an elegant treatment of the cooling problem in the case of very high velocities ("Hyperaviation"). The author feels that his treatment is somewhat more general and extended than Crocco's previous analysis.

An interesting general relation between the heat transferred through the wall and the frictional drag can be obtained using the assumption that Prandtl's number, i. e., the ratio $c_p\mu/\lambda$, is equal to unity. The same assumption was used also in the previous calculations. It is remarkable that the relation holds also as well for laminar as for turbulent flow. The heat flow q per unit time and unit area of the wall surface is

$$q = \lambda_w (\partial T/\partial y)_w$$

and the frictional drag τ per unit area is

$$\tau = \mu_w (\partial u/\partial y)_w$$

Using Eq. (1.4) the ratio q/τ can be calculated from the relation

$$\frac{q}{\tau} = \frac{\lambda_w}{\mu_w}\frac{T_0}{U}\left[\left(1 - \frac{T_w}{T_0}\right) + \frac{\gamma - 1}{2}M^2\right] \tag{1.14}$$

where T_0 is the absolute temperature, and U the velocity of the fluid in the free stream, T_w the absolute temperature at the wall, λ_w and μ_w are the heat conduction and viscosity coefficients of the fluid corresponding to the wall temperature and M denotes Mach's number. Substituting $M = 0$ one obtains from Eq. (1.14)

$$\frac{q}{\tau} = \frac{\lambda_w}{\mu_w}\frac{T_0 - T_w}{U} = \frac{c_p(T_0 - T_w)}{U} \tag{1.15}$$

This is the relation known as Prandtl's or G. I. Taylor's formula, first discovered by O. Reynolds. Hence Eq. (1.14) gives the correction of this result for compressibility effects.

In the case $T_0 > T_w$ i. e., when the wall is colder than the free stream, the effect of compressibility is to increase the heat transferred through the wall. However, it would be erroneous to interpret this result as an "improvement" in cooling because at high speed the heat produced in the boundary layer is of the same order as the heat transferred through the wall. In order to determine the efficiency of the cooling a complete heat balance must be made. For this purpose Eq. (1.14) does not give sufficient information and the velocity and the temperature distribution in the boundary layer must be computed. Such calculations were carried out for the particular assumption $T_w = T_0/4$ i. e. , for the particular case in which the absolute temperature of the wall is kept constant at a value equal to one-fourth of the temperature of the hot fluid. With the same assumption for the variation of μ as in Section (I), the results shown in Fig. 1.2 and Fig. 1.4 were obtained. The variation of $C_f\sqrt{R}$ with M is similar to that obtained in the case without heat conduction through the wall. Also the highest temperature in the boundary layer is very high for extreme Mach numbers. However, the temperature maximum occurs some distance from the wall.

The heat transferred from the boundary layer to the wall can be calculated as follows:

By means of Eq. (1.12a), one has

$$\frac{\partial u}{\partial y}=\frac{U}{L}\frac{\partial u^*}{\partial y^*}=\frac{U}{L}\frac{\mathrm{d}u^*}{\mathrm{d}\zeta}\frac{\mathrm{d}\zeta}{\mathrm{d}y^*}=\frac{U\sqrt{R}}{L\sqrt{n^*}}(\rho^* u^*)\frac{\mathrm{d}u^*}{\mathrm{d}\zeta}$$

Hence

$$\begin{aligned}C_{\mathrm{f}}&=\frac{D}{\frac{\rho_0}{2}U^2L}=\frac{1}{\frac{\rho_0}{2}U^2L}\int_0^L\mu_{\mathrm{w}}\left(\frac{\partial u}{\partial y}\right)_{y=0}\mathrm{d}x\\&=\frac{1}{\frac{\rho_0}{2}U^2L}\frac{U\sqrt{R}\mu_0}{\sqrt{L}}\frac{\rho_{\mathrm{w}}}{\rho_0}\frac{\mu_{\mathrm{w}}}{\mu_0}\left\{u^*\frac{\mathrm{d}u^*}{\mathrm{d}\zeta}\right\}_{\mathrm{w}}\int_0^L\frac{\mathrm{d}x}{\sqrt{x}}\\&=\frac{4}{\sqrt{R}}\frac{\rho_{\mathrm{w}}}{\rho_0}\frac{\mu_{\mathrm{w}}}{\mu_0}\left\{u^*\frac{\mathrm{d}u^*}{\mathrm{d}\zeta}\right\}_{\mathrm{w}}\end{aligned}$$

Therefore, combining the above two equations,

$$\left(\frac{\partial u}{\partial y}\right)_{\mathrm{w}}=\frac{U}{L}\frac{\sqrt{R}}{\sqrt{n^*}}\left(\frac{\mu_0}{\mu_{\mathrm{w}}}\right)\frac{\sqrt{R}C_{\mathrm{f}}}{4}\tag{1.16}$$

Using Eq. (1.7a) and substituting Eq. (1.16) into Eq. (1.5), then

$$\left(\frac{\partial T}{\partial y}\right)_{\mathrm{w}}=K_1\frac{T_0\sqrt{R}}{2L\sqrt{n^*}}\tag{1.17}$$

where $K_1=(4^{0.76}/2)\left(0.75+\frac{\gamma-1}{2}M^2\right)\sqrt{R}C_{\mathrm{f}}$, as given in Table 1.2. Therefore, the heat transferred to a strip of unit width of the wall of length L per unit time is equal to

$$\begin{aligned}Q_1&=\int_0^L\left(\lambda\frac{\partial T}{\partial y}\right)_{\mathrm{w}}\mathrm{d}x\cong\frac{K_1\lambda_{\mathrm{w}}T_0\sqrt{R}}{2\sqrt{L}}\int_0^L\frac{\mathrm{d}x}{\sqrt{x}}\\&=K_1\lambda_{\mathrm{w}}T_0\sqrt{R}\end{aligned}\tag{1.18}$$

Now the increase in heat content of the gas per unit time by flowing from $x=0$ to $x=L$ can be calculated as

$$\begin{aligned}Q_2&=\int_0^\infty\{\rho u c_p(T_0-T)\mathrm{d}y\}_{x=L}=\rho_0Uc_pT_0\sqrt{\frac{\mu_0L}{U\rho_0}}\int_0^\infty(1-T^*)\mathrm{d}\zeta\\&=\lambda_0T_0\sqrt{R}\int_0^\infty(1-T^*)\mathrm{d}\zeta=\lambda_{\mathrm{w}}\left(\frac{\lambda_0}{\lambda_{\mathrm{w}}}\right)T_0\sqrt{R}\int_0^\infty(1-T^*)\mathrm{d}\zeta\\&=\lambda_{\mathrm{w}}\left(\frac{\mu_0}{\mu_{\mathrm{w}}}\right)T_0\sqrt{R}\int_0^\infty(1-T^*)\mathrm{d}\zeta=K_2\lambda_{\mathrm{w}}T_0\sqrt{R}\end{aligned}\tag{1.19}$$

The viscous dissipation of gas in the boundary layer of unit width plate per unit time is thus

$$Q=Q_1+Q_2\tag{1.20}$$

Table 1.2

M	K_1
0	1.53
1	1.93
2	3.12
5	10.53
10	33.98

The total heat balance at different Mach's numbers is shown in Fig. 1.5. The "dissipation" curve represents in dimensionless form the heat produced by friction per unit time and unit width of the plate. The lower curve shows the increase (or decrease) of the heat content per unit time and unit width. The difference of the ordinates corresponds to the heat transferred through the wall. It is seen that cooling takes place for $M < 2.6$. Beyond this limit more heat is produced by friction than the amount which can be transferred to the wall and, as a matter of fact, the fluid is heated.

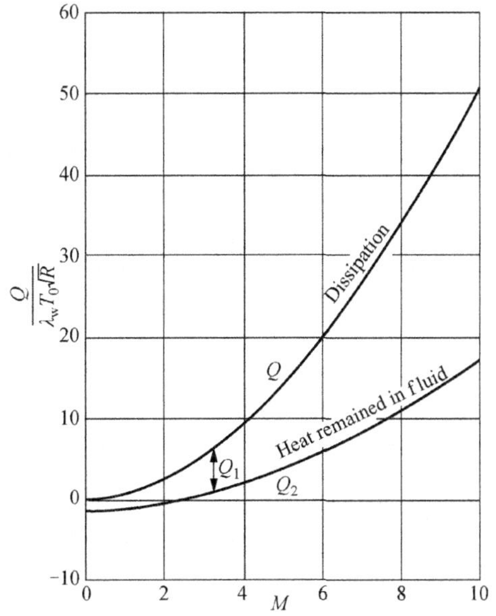

Fig. 1.5 Heat balance when the wall temperature is 1/4 of the free stream temperature

In the case $T_w > T_0$ i.e., when the wall is hotter than the free stream, the ratio between the heat transfer and the drag decreases with increasing Mach's number. This is shown in Fig. 1.6 where the ordinate represents the ratio between q/τ with compressibility effect (according to Eq. (1.14)) to q/τ without compressibility effect (according to Eq. (1.15)). The calculation was carried out for a gas temperature of -55℉. and a wall temperature of 180℉. and 300℉. It is seen that there is no cooling in the former case for $M > 1.69$ and in the latter case for $M > 2.08$. However, the decrease of cooling efficiency is appreciable even at much lower speeds. This emphasizes the benefit of the reduction of the speed of cooling air and the relatively poor efficiency of cooling surfaces exposed directly to a high-speed airstream. The curves in Fig. 1.6 being derived from Eq. (1.14) apply to laminar as well as to turbulent motion.

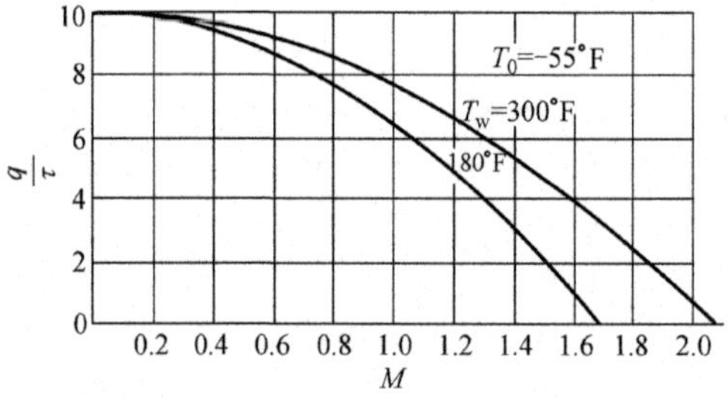

Fig. 1.6 The effect of high speed on cooling efficiency

Appendix to PART (I)
On the Validity of Theory in Very Rarefied Air

The hydrodynamic equation holds so long as the mean free path of the molecules is small in comparison with the thickness of the boundary layer. For this case the thickness of the boundary layer is zero at the nose, however, at a distance $\frac{1}{4}$ of the length of the rocket it already amounts to 3.2 cm., while the calculated mean free path of the air molecules at the altitude considered is about 1.1×10^{-2} cm. Hence it appears that even for this case the theory can be safely applied. This conclusion is substantiated by the experimental results of H. Ebert in "Darstellung der Strömungsvorgange von Gasen bei neidrigen Drucken mittels Reynoldsscher Zahlen", Zeitschrift für Physik, Bd. 85, S. 561-564, 1933.

References on PART (I)

[1] Busemann A. Gas-strömung mit laminaren Grenzschicht entlang einer Platte. Z.A.M.M., 1935,15: 23.

[2] Frankl. Laminar Boundary Layer of Compressible Fluids. Trans. of the Joukowsky Central Aero-Hydrodynamical Institute, Moscow, 1934, (Russian).

[3] von Kármán Th. The Problem of Resistance in Compressible Fluids. V. Convengo della Foundazione Alessandro Volta (Tema: Le Alte Velocita in Aviazione), Reale Accademia D'Italia, Rome.

[4] Ackeret J. Über Luftkraft bei sehr grossen Geschwindigkeiten insbesondere bei ebenen Strömungen. Helvetica Physica Acta, 1928,1: 301-322.

[5] Küchemann D. Störungsbewegungen in einer Gasströmung mit Grenzschicht. Z.A.M.M., 1938,18, S: 207-222.

[6] Crocco L. Su di un valore massimo del coefficient di transmissione del calore da una lamina piana a un fluido scorrente. Rendiconti R. Accademia dei Lincei, 1931, 14: 490 - 496. //Crocco, L., Sulla Transmissione del calore da una lamina piana un fluido scorrente ad alta velocita, L'Aerotecnica, 1932,

12:181 - 197.

[7] Blasius H. Grenzschichten in Flüssigkeiten mit kleiner Reibung. Zeit. F. Math. u. Phys., 1908,56: 1.

[8] von Mises. Bemerkung zur Hydrodynamik. Z.A.M.M., 1927,7: 425.

[9] Kent R H. The Rôle of Model Experiment in Projectile Design, Mechanical Engineering. 1932,54: 641 - 646.

PART (Ⅱ)

Supersonic Flow over an Inclined Body of Revolution

The aerodynamic forces acting on a projectile can be divided into three parts: the resistance or drag in the direction of the axis of the body, the lift in the direction perpendicular to the axis of the body, and the forces due to the rotation of the body (Magnus effect). The first component, the resistance, is, of course, the most important one, because it is the predominating factor in determining the range of the projectile. However, in the case of an actual projectile, inclination and rotation are always present, and therefore, accurate calculation of range is impossible without considering the second and third components of aerodynamic forces, i. e., the lift and the forces due to rotation of the body. It is found that the solution of von Kármán and Moore[1] for the linearized hydrodynamical equation of axial flow over a slender body of revolution can easily be generalized to the case in which the projectile is inclined to the flight path. Strictly speaking, the solution is applicable only to a very slender body inclined at a small angle to the flight path, because second order quantities of the disturbance due to the presence of the body are neglected. However, for the case of axial flow over a cone, von Kármán-Moore's first approximation[1] differs very little from the exact solution of Taylor and Maccoll[2] for vortex angles up to 40°. Therefore, it is expected that the first approximation of the lift force as obtained in this paper can be applied to a pointed projectile with fair accuracy.

If ϕ is the potential of the small disturbance v velocity due to the presence of a body of revolution whose axis coincides with the x-axis, then the linearized equation of motion of compressible fluids in cylindrical co-ordinates x, r, and θ is

$$\left(1-\frac{V^2}{c^2}\right)\frac{\partial^2\phi}{\partial x^2}+\frac{1}{r}\frac{\partial\phi}{\partial r}+\frac{\partial^2\phi}{\partial r^2}+\frac{1}{r^2}\frac{\partial^2\phi}{\partial\theta^2}=0 \qquad (2.1)$$

In this equation, V is the velocity of the undisturbed flow for which the velocity of sound is c. If the direction of the undisturbed flow coincides with the axis of the body, then ϕ is independent of θ, and Eq. (2.1) reduced to

$$\left(1-\frac{V^2}{c^2}\right)\frac{\partial^2\phi}{\partial x^2}+\frac{1}{r}\frac{\partial\phi}{\partial r}+\frac{\partial^2\phi}{\partial r^2}=0 \qquad (2.2)$$

The solution of this equation when the velocity of the undisturbed flow is greater than the velocity of sound, is the same as that for a two-dimensional wave diverging from a center. It was obtained by Levi-Civita[3] and by H. Lamb[4]. von Kármán and Moore[1] applied it to the present case and showed that it can be expressed as a source distribution given by the potential

$$\phi_1=\int_{\cosh^{-1}\frac{x}{\alpha r}}^{0} f_1(x-\alpha r\cosh u)\,du \qquad (2.3)$$

where $\alpha = \sqrt{(V/c)^2 - 1}$. Analogy with a similar case of flow of an incompressible fluid leads one to expect the solution of Eq. (2.1) to be a doublet distribution given by the potential

$$\phi_2 = -\alpha\cos\theta\int_{\cosh^{-1}\frac{x}{\alpha r}}^{0} f(x - \alpha r\cosh u)\cosh u\,du \tag{2.4}$$

This can be shown to be true, because, if the solution of Eq. (2.1) is of the form

$$\phi_2 = \cos\theta F(x, r)$$

then Eq. (2.1) reduces to

$$\left(1 - \frac{V^2}{c^2}\right)\frac{\partial^2 F}{\partial x^2} + \frac{1}{r}\frac{\partial F}{\partial r} + \frac{\partial^2 F}{\partial r^2} - \frac{F}{r^2} = 0 \tag{2.1a}$$

Differentiation of Eq. (2.2) with respect to r gives

$$\left(1 - \frac{V^2}{c^2}\right)\frac{\partial^2}{\partial x^2}\left(\frac{\partial \phi}{\partial r}\right) + \frac{1}{r}\frac{\partial}{\partial r}\left(\frac{\partial \phi}{\partial r}\right) + \frac{\partial^2}{\partial r^2}\left(\frac{\partial \phi}{\partial r}\right) - \frac{1}{r^2}\left(\frac{\partial \phi}{\partial r}\right) = 0 \tag{2.2a}$$

Comparing Eq. (2.1a) with Eq. (2.2a), it is easily seen that Eq. (2.4) is a solution of Eq. (2.1). The function f has to be determined by the boundary condition

$$v_0 = \frac{1}{\cos\theta}\left(\frac{\partial \phi_2}{\partial r}\right)_{r=R} = -\alpha^2\int_{\cosh^{-1}\frac{x}{\alpha R}}^{0} f'(x - \alpha R\cosh u)\cosh^2 u\,du \tag{2.5}$$

where v_1 is the normal component of the velocity V of the undisturbed flow, and R is the radius of the body.

The complete solution of flow over an inclined body of revolution is then obtained by superimposing a cross-flow upon an axial flow, i.e.

$$\phi = \phi_1 + \phi_2$$

This solution was also suggested independently by C. Ferrari[5].

From the velocity potential ϕ, one can calculate the pressure distribution over the body and then the aerodynamic forces. However, since the theory is based upon the linearized equation, the cross product terms of derivatives of ϕ_1 and ϕ_2 in the pressure calculation can be neglected. Therefore the following simplification results: the resistance or drag can be calculated from the axial flow alone and the lift can be separately computed from the cross flow. Since the resistance was calculated before[1], the following treatment is concerned only with the lift force. The lift acting in a direction perpendicular to the axis of the body and the moment about the vortex are thus

$$\begin{aligned} L &= \int_0^{\pi}\int_0^{\infty} \Delta p r\,d\theta\cos\theta\,dx \approx 2\rho V\int_0^{\pi}\int_0^{\infty} \frac{\partial \phi_2}{\partial x} r\cos\theta\,d\theta\,dx \\ M &= \int_0^{\pi}\int_0^{\infty} \Delta p r\,d\theta\cos\theta x\,dx \approx 2\rho V\int_0^{\pi}\int_0^{\infty} \frac{\partial \phi_2}{\partial x} x r\cos\theta\,d\theta\,dx \end{aligned} \tag{2.6}$$

where Δp is the difference between the pressure at the surface of the body and that of the undisturbed flow and ρ is the density of the fluid in the undisturbed flow.

Equation (2.5) is a non-homogeneous linear integral equation in f which does not have a general solution of simple form. However, it is interesting to see how Eq. (2.5) simplified in the limiting case when the radius of body approaches zero. It is convenient here to use $\xi = x - ar\cosh u$ as the independent variable, then Eq. (2.5) becomes

$$v_0 = \frac{1}{R^2}\int_0^{x-aR} \frac{f'(\xi)(x-\xi)^2 \mathrm{d}\xi}{\sqrt{(x-\xi)^2 - \alpha^2 R^2}} \approx \frac{1}{R^2}\int_0^{x-aR} f'(\xi)(x-\xi)\mathrm{d}\xi$$

Integrating by parts, one has

$$v_0 \approx \frac{1}{R^2}\left\{[f(\xi)(x-\xi)]_0^{x-aR} + \int_0^{x-aR} f(\xi)\mathrm{d}\xi\right\}$$

Now, if the projectile is pointed nose, the doublet strength must be zero at the nose $x = 0$, thus $f(0) = 0$. Let $R \to 0$, and writing x instead of ξ in the integrand the above equation reduces to

$$v_0 \approx \frac{1}{R^2}\int_0^x f(x)\mathrm{d}x \tag{2.5a}$$

Since the cross-sectional area of the body of revolution is $S = \pi R^2$, Eq. (2.5a) can be written as

$$v_0 = \frac{\pi}{S}\int_0^x f(x)\mathrm{d}x$$

or

$$\int_0^x f(x)\mathrm{d}x = \frac{v_0}{\pi}S$$

Differentiating, one arrives at

$$f(x) = \frac{v_0}{\pi}\frac{\mathrm{d}S}{\mathrm{d}x} \tag{2.7}$$

In order to calculate the lift, one has first to find the axial component of disturbance velocity. Thus

$$\begin{aligned}\left(\frac{\partial\phi_2}{\partial x}\right)_{r=R} &= -\alpha\cos\theta\int_{\cosh^{-1}\frac{x}{aR}}^{0} f'(x - aR\cosh u)\cosh u\,\mathrm{d}u \\ &= \frac{\cos\theta}{R}\int_0^{x-aR}\frac{f'(\xi)(x-\xi)\mathrm{d}\xi}{\sqrt{(x-\xi)^2-\alpha^2R^2}} \\ &\approx \frac{\cos\theta}{R}\int_0^x f'(\xi)\mathrm{d}\xi = \frac{\cos\theta}{R}f(x) = \frac{v_0\cos\theta}{\pi R}\frac{\mathrm{d}S}{\mathrm{d}x}\end{aligned}$$

Substituting into Eq. (2.6), the lift force is obtained as

$$L = 2\rho V\int_0^{\pi}\int_0^{\infty}\frac{v_0\cos^2\theta}{\pi}\frac{\mathrm{d}S}{\mathrm{d}x}\mathrm{d}\theta\,\mathrm{d}x = \rho_0 V v_0 A_b$$

where A_b = area of the base section of the body. Hence the lift coefficient can be evaluated as

$$C_L = \frac{L}{\frac{\rho}{2} V^2 A_b} = 2\frac{v_0}{V} \approx 2\psi \tag{2.8}$$

in which ψ = angle of attack of the body.

The moment arm d, i. e., the distance between the point of application of the resultant lift force and the vertex can be obtained by dividing the moment computed from Eq. (2.6) by the lift force, and thus

$$d = \left(\frac{A_b - A_m}{A_b}\right) l \tag{2.9}$$

where A_m = area of the mean section of the body, i. e., the volume of the body divided by its length, l.

The results of Eq. (2.8) and Eq. (2.9) are identical to those found in Munk's theory of airships[6]. At first sight, this might be surprising. However, if the radius of the body approaches zero as assumed, the cross-flow pattern is the same as that for an infinitely long circular cylinder moving with its axes perpendicular to the flow. Therefore, in every plane perpendicular to the axis of the body, the flow can be considered as two dimensional, i. e., it is independent of the variable x. Hence Eq. (2.1) reduces simply to

$$\frac{\partial^2 \phi}{\partial r^2} + \frac{1}{r}\frac{\partial \phi}{\partial r} + \frac{1}{r^2}\frac{\partial^2 \phi}{\partial \theta^2} = 0 \tag{2.1b}$$

This is immediately recognized as the equation of motion for two dimensional flow of incompressible fluids, which is the basis of Munk's theory.

Due to this two dimensional character of the flow, the distribution of doublets is not affected by the change in Mach's number, which is only connected with the independent variable x, and, therefore, the lift coefficient and the moment arm are also independent of Mach's number as shown by Eq. (2.8) and Eq. (2.9). This can also be seen from the fact that when r approaches zero, the variable $\xi = x - \alpha r \cosh u \to x$ and thus the effect of α, which is a function of Mach's number, is removed. To study the effect of Mach's number on the lift of the body, one has to go back to Eq. (2.5). To avoid the difficulty of solving this integral equation, the "indirect method" of solution can be employed, i. e., take a function f and determine the necessary shape of the body to comply with this function f.

Taking the simplest case

$$f(x - \alpha r \cosh u) = K(x - \alpha r \cosh u)$$

where K = a constant. Then

$$\begin{aligned}\phi_2 &= -K\alpha\cos\theta \int_{\cosh^{-1}\frac{x}{\alpha r}}^{0} (x - \alpha r\cosh u)\cosh u \, du \\ &= K\alpha\cos\theta \left\{ \frac{x}{2}\sqrt{\left(\frac{x}{\alpha r}\right)^2 - 1} - \frac{\alpha r}{2}\cosh^{-1}\frac{x}{\alpha r} \right\}\end{aligned}$$

And the boundary condition reduces to

$$v_0 = \frac{K\alpha^2}{2}\left\{\left(\frac{x}{\alpha R}\right)\sqrt{\left(\frac{x}{\alpha R}\right)^2 - 1} + \cosh^{-1}\left(\frac{x}{\alpha R}\right)\right\}$$

Therefore, the solution, Eq. (2.4a), is evidently a solution for a cone with half vertex angle ε, if $\cot\varepsilon = \frac{x}{R}$. By putting $\frac{\cot\varepsilon}{\alpha} = \zeta$ the boundary condition can be written in the form

$$v_0 = \frac{\alpha^2 K}{2}\{\zeta\sqrt{\zeta^2 - 1} + \cosh^{-1}\zeta\} \tag{2.5b}$$

For any given value of vertex angle and Mach's number, the corresponding value of K can be obtained from Eq. (2.5b)

In order to calculate the lift, one has first to find the axial component of disturbance velocity. Thus from Eq. (2.4a)

$$\left(\frac{\partial\phi_2}{\partial x}\right)_{r=R} = \frac{\alpha K\cos\theta}{2}\left[\sqrt{\left(\frac{x}{\alpha R}\right)^2 - 1} + \frac{\left(\frac{x}{\alpha R}\right)^2}{\sqrt{\left(\frac{x}{\alpha R}\right)^2 - 1}} - \frac{1}{\sqrt{\left(\frac{x}{\alpha R}\right)^2 - 1}}\right]$$

$$= \alpha K\cos\theta\sqrt{\zeta^2 - 1}$$

Substituting into Eq. (2.6), the lift force is found to be

$$L = 2\rho V v_0\int_0^{\pi}\int_0^{\infty}\frac{\partial\phi_2}{\partial x}r\cos\theta\,d\theta\,dx$$

$$= \frac{\alpha K\sqrt{\zeta^2 - 1}\rho V}{2}\int_0^{\infty}2\pi r\,dx$$

$$= \frac{\alpha K\sqrt{\zeta^2 - 1}\rho V\cos\varepsilon}{2}S$$

where S = lateral surface area of the cone. Therefore

$$C_L = \frac{L}{\frac{\rho}{2}V^2 A_b} = \frac{\alpha k\sqrt{\zeta^2 - 1}}{v_0}\cot\varepsilon\cdot\psi$$

But from Eq. (2.5b), k is obtained as

$$k = \frac{2v_0}{\alpha^2\{\zeta\sqrt{\zeta^2 - 1} + \cosh^{-1}\zeta\}}$$

Hence

$$C_L = K_1\psi \tag{2.10}$$

where

$$K_1 = \frac{2\zeta\sqrt{\zeta^2 - 1}}{\zeta\sqrt{\zeta^2 - 1} + \cosh^{-1}\zeta}$$

In the limiting case when ε approached zero,

$$K_1 \to 2$$

which agrees with Eq. (2.8). Similarly from Eq. (2.6) the moment coefficient is

$$C_M = \frac{\text{Moment about vertex}}{\frac{\rho}{2}V^2 A_b l} = \frac{2}{3}K_1\psi \tag{2.11}$$

which satisfies Eq. (2.9).

Both Eq. (2.8) and Eq. (2.10) show that the lift at a given Mach's number is proportional to the angle of attack of the body. This is a general characteristic of flow around a body without separation. If the fluid separates from the body and creates a "dead water" region on lee side of the body, then the lift will be proportional to the square of the angle of attack as was shown by *W*. Bollay[7]. The problem whether the fluid separates or not can only be answered by experiments. From the experimental data now available[8], it seems that the flow is continuous without separation, and, therefore, the lift is proportional to the angle of attack of the body.

Figure 2.2 is the result of computation using Eq. (2.10). Calculations were carried out for values of $K_1 \geqslant 1$, because the value of $K_1 = 1$ corresponds to $\varepsilon = \beta$, where β is the wave angle. For $K_1 < 1$ it is found that $\beta < \varepsilon$. This means that the wave angle is smaller than the vertex angle which is, of course, impossible. Therefore, $K_1 = 1$ marks the limit of validity of this solution. In fact, even when K_1 is near to 1, the solution must be considered as qualitative only, since in this region the effect of the surface of the body on the shock wave cannot be neglected.

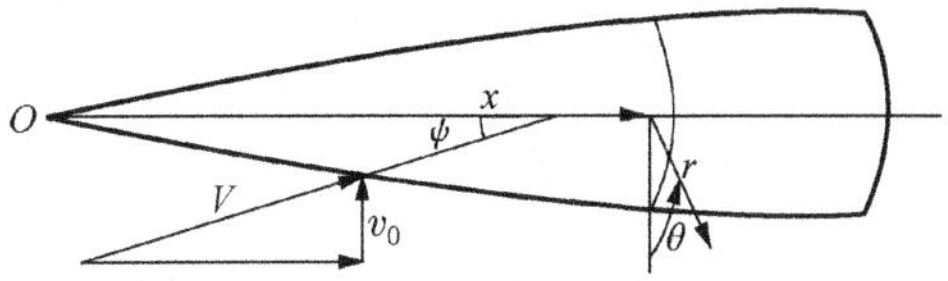

Fig. 2.1

To generalize the solution for a body of revolution with a sharp point at the origin and cylindrical shape at infinity, it is simplest to use a step-wise doublet distribution. Consider the points $P_1, P_2, \ldots, P_n, \ldots, P_N$ of the meridian line of the body, and designate their co-ordinates by $x_1, R_1; x_2, R_2; \ldots; x_n, R_n; \ldots; x_N, R_N$ and the corresponding values of $x - \alpha R$ by $\xi_1, \xi_2, \ldots, \xi_n, \ldots, \xi_N$ Then the boundary condition of Eq. (2.5) can be written as

$$v_0 = -\frac{\alpha^2}{2}\sum_{i=1}^{n} K_i\left\{\left(\cosh^{-1}\frac{x_n - \xi_{i-1}}{\alpha R_n} - \cosh^{-1}\frac{x_n - \xi_i}{\alpha R_n}\right) + \left(\frac{x_n - \xi_{i-1}}{\alpha R_n}\sqrt{\left(\frac{x_n - \xi_{i-1}}{\alpha R_n}\right)^2 - 1} - \frac{x_n - \xi_i}{\alpha R_n}\sqrt{\left(\frac{x_n - \xi_i}{\alpha R_n}\right)^2 - 1}\right)\right\} \tag{2.5c}$$

This condition actually gives a set of N equations to determine the N constants K_i. This set of equations can be solved rather easily because each following equation in the set only contains one more which does not appear in the preceding one of the set. When K_i's are determined,

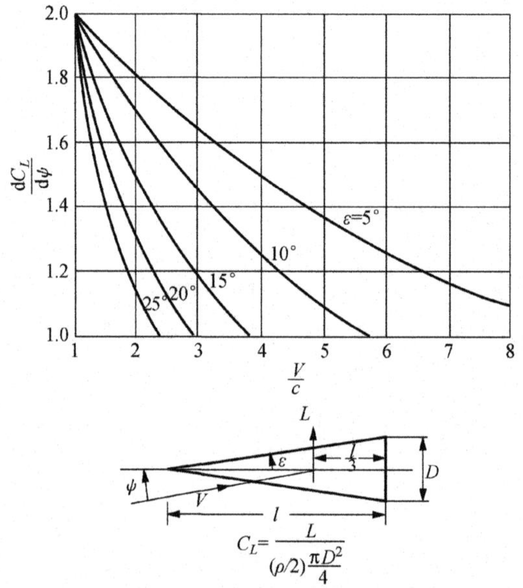

Fig. 2.2

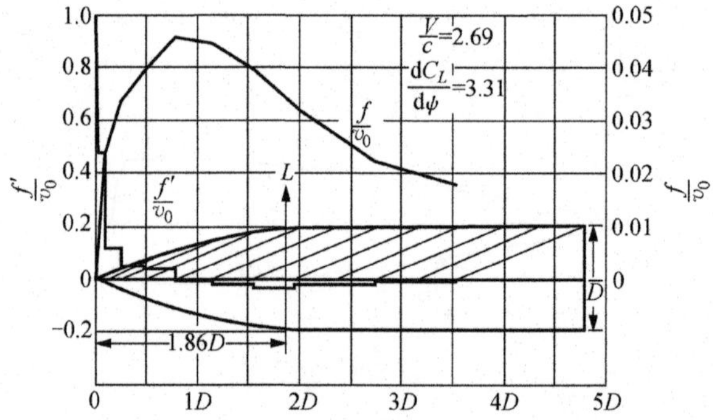

Fig. 2.3

the lift can be calculated by using Eq. (2.6). The pressure over each section of the conical surface is constant. The lift and moment coefficients are thus obtained as

$$
\begin{aligned}
C_L &= \frac{2\psi}{\alpha}\sum_{n=0}^{n=N}\frac{(x_{n+1}-x_n)}{R_N}\frac{(R_{n+1}+R_n)}{R_N}\left\{\sum_{i=0}^{i=n}K_i\left[\sqrt{\left(\frac{x_n-\xi_{i-1}}{\alpha R_n}\right)^2-1}-\sqrt{\left(\frac{x_n-\xi_i}{\alpha R_n}\right)^2-1}\right]\right\} \\
C_M &= \frac{2\psi}{\alpha l}\sum_{n=0}^{n=N}\left\{\frac{x_{n+1}+2x_n}{3}+\frac{(x_{n+1}-x_n)}{3}\left(\frac{R_{n+1}}{R_{n+1}+R_n}\right)\right\}\frac{(x_{n+1}-x_n)}{R_N}\left(\frac{R_{n+1}+R_n}{R_N}\right)\times \\
&\quad \left\{\sum_{i=0}^{i=n}K_i\left[\sqrt{\left(\frac{x_n-\xi_{i-1}}{\alpha R_n}\right)^2-1}-\sqrt{\left(\frac{x_n-\xi_i}{\alpha R_n}\right)^2-1}\right]\right\}
\end{aligned}
\tag{2.12}
$$

where P_N is the last point on the meridian line, R_N is the base radius and l is length of the body.

Figure 2.3 is the result of the calculation using Eq. (2.12) for a body of revolution with

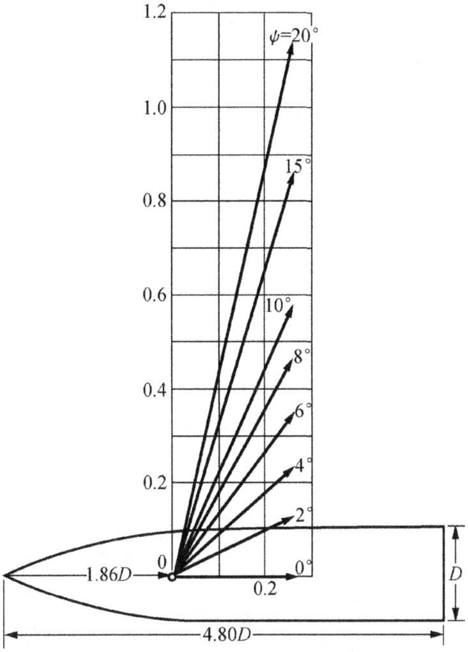

Fig. 2.4

"6-cliber head" and total length of 4.8 calibers, when it is travelling with a velocity 2.69 times that of sound ($\alpha = 2.5$). The lift coefficient is considerably higher than that of a cone at the same angle of attack and at the same Mach's number, evidently due to the cylindrical part of the body. The position of the resulting lift force is also shown in the figure. Since, as mentioned before, lift and drag are independent in first order approximation, the calculated lift coefficient can be combined with the drag coefficient taken from experiments and thus give some information on the magnitude and direction of the resulting force. Fig. 2. 4 shows the method applied to this projectile with the drag coefficient taken from Kent's experiment[9].

If the projectile has a length of 4.34 diameters instead of 4.8 diameters and has the same nose shape, and, if its center of gravity is located at a point 2.68 diameters back of the nose, then the calculated moment about the center of gravity can be expressed as $M = \rho V^2 R_N{}^3 \psi f\left(\frac{V}{c}\right)$ where $f\left(\frac{V}{c}\right) = 9.35$ for $\frac{V}{c} = 2.69$ This compares closely with the value $f\left(\frac{V}{c}\right) = 10.7$ extrapolated from R. H. Fowler's experiment[8], for a projectile of the same proportions. This shows that the theory developed in this paper can be applied to a projectile with fair accuracy.

References on PART (Ⅱ)

[1] von Kármán Th. Moore N B. The Resistance of Slender Bodies Moving with Supersonic Velocities with Special Reference to Projectiles. Trans. A.S.M.E. 1932,54: 303 - 310.

[2] Taylor G I, Maccoll J W, The Air Pressure on a Cone Moving at High Speeds. Proc. Royal Society (A),

1933,159: 278 - 298.//Maccoll J W.: "The Conical, Shock Wave Formed by a Cone Moving at a High Speed", Proc. Royal Society (A), 1937, 159: 459 - 472. The comparison was mentioned in a paper by Taylor G I.: Well Established Problems in High Speed Flow; Atti dei V Convegno "Volta," 1936,198 - 214. Reale Accademia d'Italia, Rome.

[3] Levi-Civita. Nuouo Cimento (4). 1897, 6.

[4] Lamb H. On Wave-Propagation in Two Dimensions. Proc, London Math. Society (I), 1902,35, 141.// See also Lamb H.: Hydrodynamics, 1932, 6th Ed: 298. Cambridge.

[5] Ferrari C. Campi di corrente ipersonora attorno a solidi di rivoluzione. L'Aerotecnica, 1937, 17, 507 - 518.

[6] Munk M M. The Aerodynamic Forces on Airship Hulls. NACA Technical Report No. 184, 1934.

[7] Bollay W. A Theory for Rectangular Wings of Small Aspect Ratio. Jour. Aeronautical Sciences, 1937, 4: 294 - 296.

[8] Fowler R H, Gallop E G, Lock C N H, Richmond H W. The Aerodynamics of a Spinning Shell. Philosophical Trans. of Royal Society of London (A), 1921,221: 295 - 389.

[9] Kent K M. The Rôle of Model Experiments in Projectile Design. Mechanical Engineering, 1932, 54: 641 - 646.

PART (Ⅲ)
Application of Tschapligin's Transformation to Two Dimensional Subsonic Flow

The equations of two dimensional irrotational motion of compressible fluids, assuming that the pressure is a single-valved function of density only, can be reduced to a single non-linear equation of the velocity potential. In the supersonic case, the problem is solved by Prandtl, Meyer and Busemann by means of the powerful method of characteristics. The essential difficulty of this problem lies in the subsonic case especially when the velocity is near to the velocity of sound. The first logical step is to linearize the equation based on the argument that the disturbance super-imposed on the parallel rectilinear flow due to the presence of a solid body is sufficiently small compared with parallel flow. This makes the second and higher order terms of disturbance potential to be negligible. An example of this method is the well-known theory of thin airfoil due to Prandtl and Glauert. But the presence of stagnation point at the nose of the airfoil makes the application of the linearized theory questionable, at least near this region; because there the disturbance due to the presence of the body is no longer small. On the same ground, the theory breaks down in case of bodies whose dimension across the stream is not small compared with the dimension parallel to the stream. The next method is that derived originally by Janzen and Lord Rayleigh. They solved the equation by successive approximations. However, the process is very tedious and the method converges very slowly if the velocity approaches that of sound.

Molenbroeck[1] and Tschapligin[2] suggested the use of the magnitude of velocity and inclination of velocity to the x-axis as independent variables and were able thus to reduce the equation of velocity potential to a linear equation. This equation was solved by Tschapligin[2] and recently put in a more convenient form by F. Clauser and M. Clauser[3]. The solution is essentially a series each term of which is a product of hypergeometric function and circular function. The chief difficulty in practical application of this solution is to obtain a proper set of boundary conditions in the transformed plane, or the hodograph plane.

Tschapligin[2] showed that a great simplification of the equation in hodograph plane results if the ratio of specific heats of the gas is equal to -1. Then the equation becomes the equation of minimal surface whose solution is well-known. However, at first, the hypothetical value of ratio of specific heats (all real gas has the value for this ratio ranging from 1.00 to 2.00) makes the practical application of Tschapligin's theory questionable. It was Demtschenko[4] and Busemann[5] who made the meaning of this special value of ratio clear. They found that this special value of ratio of specific heats really corresponds to take the tangent of $p-v$ curve of gas instead $p-v$ curve itself. However, they limit themselves to use the tangent at the state of rest of the gas. Thus their theory can only apply to velocities up to 0.5 times sound velocity. In this part, the theory is generalized to use the tangent at the state of gas corresponding to the undisturbed parallel flow. Therefore the range of usefulness of the theory is greatly extended.

In the first section, the general theory will be developed. In the second section, the theory will be applied to the case of symmetrical Joukowsky airfoil at zero angle of attack.

Section (I)

If p is the pressure and $1/v$ is the density of gas, the adiabatic process is expressed as a curve in the $p-v$ plane as shown in Fig. 3.1. Now conditions near to the point p_1, v_1 can be approximated by the tangent at this point. The equation of the tangent at this point can be

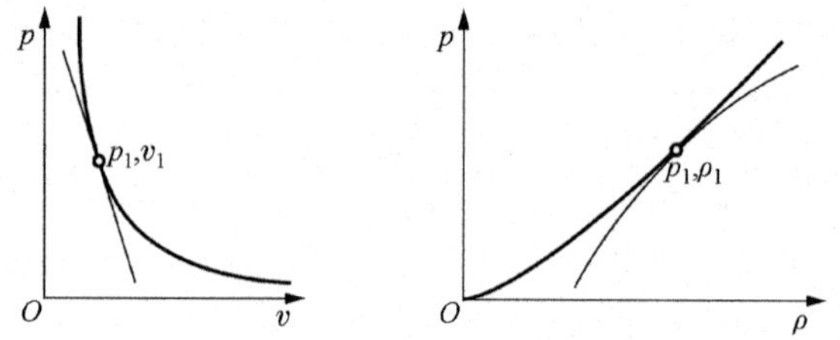

Fig. 3.1 The approximation of adiabatic curve by its tangent

written as

$$p_1 - p = C(v_1 - v) = C(\rho_1^{-1} - \rho^{-1}) \tag{3.1}$$

where ρ is the density of the gas. Now the slope C must be equal to the slope of the curve at the point p_1, v_1

$$C = \left(\frac{dp}{dv}\right)_1 = \left(\frac{dp}{d\rho}\frac{d\rho}{dv}\right)_1 = -\left(\frac{dp}{d\rho}\right)_1 \rho_1^2 = -a_1^2\rho_1^2$$

Therefore $\quad C = -a_1^2\rho_1^2$

Hence the approximate $p-\rho$ relation near p_1, ρ_1 can be written as

$$p_1 - p = a_1^2\rho_1^2\left(\frac{1}{\rho} - \frac{1}{\rho_1}\right) \tag{3.2}$$

From the generalized Bernoulli's theorem, the following relation is obtained

$$\frac{1}{2}w_2^2 - \frac{1}{2}w_3^2 = \int_2^3 \frac{dp}{\rho} \tag{3.3}$$

where w is the velocity of gas, the subscripts 2, 3 indicate two conditions. But from Eq. (3.2), p can be expressed as a function of ρ, thus

$$dp = \frac{a_1^2\rho_1^2}{\rho^2}d\rho \tag{3.4}$$

Substituting into the integrand in Eq. (3.3) and integrate, the following relation is obtained

$$\frac{1}{2}w_2^2 - \frac{1}{2}w_3^2 = \frac{1}{2}a_1^2\rho_1^2\left\{\frac{1}{\rho_3^2} - \frac{1}{\rho_2^2}\right\}$$

Now if $w_3 = 0$, $w_2 = w$, $\rho_3 = \rho_0$ and $\rho_2 = \rho$, then

$$\frac{a_1^2 \rho_1^2}{\rho_0^2} + w^2 = \frac{a_1^2 \rho_1^2}{\rho^2} \tag{3.5}$$

where the subscript 0 denotes the rest state of the gas. If the sound velocity a is defined as the derivative of p with respect to ρ, then Eq. (3.4) gives

$$\frac{dp}{d\rho}\rho^2 = a^2 \rho^2 = a_1^2 \rho_1^2 = \text{constant} \tag{3.6}$$

Therefore Eq. (3.5) can be written as

$$\left(\frac{\rho}{\rho_0}\right)^2 = 1 - \frac{w^2}{a^2}$$

or

$$\frac{\rho}{\rho_0} = \sqrt{1 - \frac{w^2}{a^2}} \tag{3.7}$$

Furthermore, from Eq. (3.6), $\rho_1^2 a_1^2 = \rho_0^2 a_0^2$, thus Eq. (3.7) can also be written as

$$\frac{\rho_0}{\rho} = \sqrt{1 + \frac{w^2}{a_0^2}} \tag{3.8}$$

It is interesting at this stage, to notice that from Eq. (3.8), the density decreases as velocity w increases, as it is expected. Thus from Eq. (3.7), the velocity of sound of the gas will increase as the velocity is increased. This is just opposite to real gas, because in the case of an adiabatic flow, it is well-known that the temperature of gas decreases as the velocity of gas is increased and thus the sound velocity also decreases. However from Eq. (3.7), the ratio $\frac{w}{a}$ or Mach's number increases as the velocity w increases. But this ratio only reaches the value unity when $w = \infty$, or from Eq. (3.8) when $\rho = 0$. It is thus seen that the entire region of flow is subsonic and thus the equation of motion is always of elliptic type. This may be considered as the physical reason why the complex representation of velocity potential and stream function is possible in all cases, as will be shown in following pages. However one should bear in mind that the portion of tangent that could be used as an approximation to the true adiabatic curve, must lie within the first quadrant. Thus the upper limit of velocity for practical application of the theory is when $p = 0$ By using Eqs. (3.2), (3.7) and (3.8), this upper limit is found to be

$$\left(\frac{w}{w_1}\right)_{\max} = \frac{1}{\left(\frac{w_1}{a_1}\right)} \sqrt{\left(\frac{p_1}{\rho_1 a_1^2} + 1\right)^2 - \left(1 - \frac{w_1^2}{a_1^2}\right)}$$

Or by putting $a_1^2 = \gamma \frac{p_1}{\rho_1}$, the above equation reduces to

$$\left(\frac{w}{w_1}\right)_{\max} = \frac{1}{\left(\frac{w_1}{a_1}\right)}\sqrt{(\gamma+1)^2 - \left\{1-\left(\frac{w_1}{a_1}\right)^2\right\}}$$

The values of $\left(\frac{w}{w_1}\right)_{\max}$ for different values of $\frac{w_1}{a_1}$ are shown in Table 3.1.

Table 3. 1

$\frac{w_1}{a_1}$	$\left(\frac{w}{w_1}\right)_{\max}$	$\left(\frac{w}{a_1}\right)_{\max}$
0	∞	2.186
0.2	10.91	2.195
0.4	5.56	2.225
0.6	3.78	2.265
0.8	2.92	2.335
1.0	2.405	2.405

It is thus seen that for most applications of this theory, p will remain positive. However due to large deviation from the true adiabatic process at high values of $\frac{w}{w_1}$, one has probably to limit the ratio $\left(\frac{w}{w_1}\right)$ to about 2.

Now if the flow is irrotational, there exists a velocity potential ϕ such that

$$\frac{\partial \phi}{\partial x} = u, \quad \frac{\partial \phi}{\partial y} = v \tag{3.9}$$

where u, v are the x, y components of the velocity w. To satisfy the equation of continuity, the stream function ψ is introduced. It is defined by

$$\frac{\rho}{\rho_0}u = \frac{\partial \psi}{\partial y}, \quad \frac{\rho}{\rho_0}v = -\frac{\partial \psi}{\partial x} \tag{3.10}$$

Now if the angle of inclination of the velocity w to the x-axis is β, then from Eqs. (3.9) and (3.10), one has

$$\begin{aligned} \mathrm{d}\phi &= w \cdot \cos\beta \,\mathrm{d}x + w \cdot \sin\beta \,\mathrm{d}y \\ \mathrm{d}\psi &= -w\frac{\rho}{\rho_0}\sin\beta \,\mathrm{d}x + w\frac{\rho}{\rho_0}\cos\beta \,\mathrm{d}y \end{aligned} \tag{3.11}$$

Solving for $\mathrm{d}x$ and $\mathrm{d}y$,

$$\begin{aligned} \mathrm{d}x &= \frac{\cos\beta}{w}\mathrm{d}\phi - \frac{\sin\beta}{w}\frac{\rho_0}{\rho}\mathrm{d}\psi \\ \mathrm{d}y &= \frac{\sin\beta}{w}\mathrm{d}\phi + \frac{\cos\beta}{w}\frac{\rho_0}{\rho}\mathrm{d}\psi \end{aligned} \tag{3.12}$$

So long as the correspondence between the physical plane and the hodoplane is one to one, or mathematically $\frac{\partial(x,y)}{\partial(u,v)} \neq 0$ one can express x, y as functions of w, β and so also ϕ and ψ as functions w and β. Thus

$$\begin{aligned} d\phi &= \phi'_w dw + \phi'_\beta d\beta \\ d\psi &= \psi'_w dw + \psi'_\beta d\beta \end{aligned} \tag{3.13}$$

where primes indicate the derivative, and subscript indicate the variables with respect to which the functions are differentiated. Now substitute Eq. (3.13) into Eq. (3.12), the following relations are obtained

$$\begin{aligned} dx &= \left(\frac{\cos\beta}{w}\phi'_w - \frac{\sin\beta}{w}\frac{\rho_0}{\rho}\psi'_w\right)dw + \left(\frac{\cos\beta}{w}\phi'_\beta - \frac{\sin\beta}{w}\frac{\rho_0}{\rho}\psi'_\beta\right)d\beta \\ dy &= \left(\frac{\sin\beta}{w}\phi'_w + \frac{\cos\beta}{w}\frac{\rho_0}{\rho}\psi'_w\right)dw + \left(\frac{\sin\beta}{w}\phi'_\beta + \frac{\cos\beta}{w}\frac{\rho_0}{\rho}\psi'_\beta\right)d\beta \end{aligned} \tag{3.14}$$

Since the left-hand sides of Eq. (3.14) are exact differential, one can apply the reciprocity relation and obtains

$$\begin{aligned} \frac{\partial}{\partial\beta}\left(\frac{\cos\beta}{w}\phi'_w - \frac{\sin\beta}{w}\frac{\rho_0}{\rho}\psi'_w\right) &= \frac{\partial}{\partial w}\left(\frac{\cos\beta}{w}\phi'_\beta - \frac{\sin\beta}{w}\frac{\rho_0}{\rho}\psi'_\beta\right) \\ \frac{\partial}{\partial\beta}\left(\frac{\sin\beta}{w}\phi'_w + \frac{\cos\beta}{w}\frac{\rho_0}{\rho}\psi'_w\right) &= \frac{\partial}{\partial w}\left(\frac{\sin\beta}{w}\phi'_\beta + \frac{\cos\beta}{w}\frac{\rho_0}{\rho}\psi'_\beta\right) \end{aligned} \tag{3.15}$$

Carrying out the differentiation, and cancelling identical terms in left-hand and right-hand side,

$$\begin{aligned} -\frac{\sin\beta}{w}\phi'_w - \frac{\cos\beta}{w}\frac{\rho_0}{\rho}\psi'_w &= -\frac{\cos\beta}{w^2}\phi'_\beta + \frac{\sin\beta}{w^2}\frac{\rho_0}{\rho}\left(1 - \frac{w^2}{a^2}\right)\psi'_\beta \\ \frac{\cos\beta}{w}\phi'_w - \frac{\sin\beta}{w}\frac{\rho_0}{\rho}\psi'_w &= -\frac{\sin\beta}{w^2}\phi'_\beta - \frac{\cos\beta}{w^2}\frac{\rho_0}{\rho}\left(1 - \frac{w^2}{a^2}\right)\psi'_\beta \end{aligned} \tag{3.16}$$

Using Eq. (3.7), Eq. (3.16) becomes

$$\begin{aligned} -\frac{\sin\beta}{w}\phi'_w - \frac{\cos\beta}{w}\frac{\rho_0}{\rho}\psi'_w &= -\frac{\cos\beta}{w^2}\phi'_\beta + \frac{\sin\beta}{w^2}\frac{\rho}{\rho_0}\psi'_\beta \\ \frac{\cos\beta}{w}\phi'_w - \frac{\sin\beta}{w}\frac{\rho_0}{\rho}\psi'_w &= -\frac{\sin\beta}{w^2}\phi'_\beta - \frac{\cos\beta}{w^2}\frac{\rho}{\rho_0}\psi'_\beta \end{aligned} \tag{3.17}$$

As in both equations ϕ'_w, ψ'_w, and ϕ'_β, ψ'_β are connected with a proportional factor, one can solve for them, and

$$\begin{aligned} \phi'_w &= -\frac{\rho}{\rho_0}\frac{1}{w}\psi'_\beta \\ \phi'_\beta &= \frac{\rho_0}{\rho}w\psi'_w \end{aligned} \tag{3.18}$$

Now Eq. (3.18) can be further reduced if a new variable ω is introduced. ω is defined as

$$d\omega = \frac{\rho}{\rho_0}\frac{dw}{w} \tag{3.19}$$

Then Eq. (3.18) becomes

$$\begin{aligned} \phi'_\omega &= -\psi'_\beta \\ \phi'_\beta &= \psi'_\omega \end{aligned} \tag{3.20}$$

This is the fundamental set of equations for the present theory. It can be easily recognized as the Cauchy-Riemann differential equation, and thus $\phi + i\psi$ must be an analytic function of $\beta + i\omega$. However for the convenience of numerical calculation, a new variable W is used instead of ω, such that

$$W = a_0 \cdot e^{\omega} \tag{3.21a}$$

Or by integrating Eq. (3.19),

$$W = \frac{2a_0 w}{\sqrt{a_0^2 + w^2} + a_0} \tag{3.21}$$

Hence by inverting,

$$w = \frac{4a_0^2 W}{4a_0^2 - W^2} \tag{3.22}$$

Thus by substituting into Eq. (3.8),

$$\frac{\rho_0}{\rho} = \frac{4a_0^2 + W^2}{4a_0^2 - W^2} \tag{3.23}$$

If another set of new variables $U = W\cos\beta$ and $V = W\sin\beta$ are used as independent variables, one has

$$\begin{aligned} \frac{\partial}{\partial\omega} &= \frac{\partial U}{\partial\omega}\frac{\partial}{\partial U} + \frac{\partial V}{\partial\omega}\frac{\partial}{\partial V} = W\left\{\cos\beta\frac{\partial}{\partial U} + \sin\beta\frac{\partial}{\partial V}\right\} \\ \frac{\partial}{\partial\beta} &= \frac{\partial U}{\partial\beta}\frac{\partial}{\partial U} + \frac{\partial V}{\partial\beta}\frac{\partial}{\partial V} = W\left\{-\sin\beta\frac{\partial}{\partial U} + \cos\beta\frac{\partial}{\partial V}\right\} \end{aligned} \tag{3.24}$$

Using Eq. (3.24), Eq. (3.20) can be written as

$$\begin{aligned} \cos\beta\frac{\partial\phi}{\partial U} + \sin\beta\frac{\partial\phi}{\partial V} &= \sin\beta\frac{\partial\psi}{\partial U} - \cos\beta\frac{\partial\psi}{\partial V} \\ -\sin\beta\frac{\partial\phi}{\partial U} + \cos\beta\frac{\partial\phi}{\partial V} &= \cos\beta\frac{\partial\psi}{\partial U} + \sin\beta\frac{\partial\psi}{\partial V} \end{aligned}$$

These equations are satisfied by

$$\frac{\partial\phi}{\partial U} = \frac{\partial\psi}{\partial(-V)}, \quad \frac{\partial\phi}{\partial(-V)} = -\frac{\partial\psi}{\partial U} \tag{3.25}$$

These are the Cauchy-Riemann differential equations, therefore the complex potential $F = \phi + i\psi$ is a function of $U - iV = \overline{W}$. Or.

$$\phi + i\psi = F(U - iV) = F(\overline{W})$$

Hence

$$\phi - i\psi = \overline{F}(U + iV) = \overline{F}(W) \tag{3.26}$$

To transform from hodograph plane back to physical plane, the expression of x and y in terms of U and V must be found. By using Eqs (3.22) and (3.23), Eq. (3.12) can be written as

$$dx = \frac{U \cdot d\phi}{W^2}\left\{1 - \frac{W^2}{4a_0^2}\right\} - \frac{V \cdot d\psi}{W^2}\left\{1 + \frac{W^2}{4a_0^2}\right\}$$

$$dy = \frac{V \cdot d\phi}{W^2}\left\{1 - \frac{W^2}{4a_0^2}\right\} + \frac{U \cdot d\psi}{W^2}\left\{1 + \frac{W^2}{4a_0^2}\right\}$$

where $W^2 = U^2 + V^2$. These equations can be combined into one equation by means of Eq. (3.26). Thus

$$dz = dx + idy = \frac{dF}{\overline{W}} - \frac{W \cdot d\overline{F}}{4a_0^2} \tag{3.27}$$

For practical application of the theory to the flow over an obstacle, the computation proceeds as follows: (1) Find the complex potential for the flow of incompressible fluid over the obstacle, say

$$w_1 G(\xi + i\eta) = w_1 G(\zeta)$$

where w_1, is the velocity of parallel rectilinear undisturbed flow, and ξ, η the space coordinate of the physical plane. (2) Now let $F = W_1 G(\zeta)$. Here W_1 is the transformed undisturbed velocity, to be interpreted as Eq. (3.21). But the complex variable ζ has no direct physical meaning. (3) Using the above value of F, Eq. (3.27) can be written as

$$dz = d\zeta - \frac{1}{4}\left(\frac{W_1}{a_0}\right)^2 \left(\frac{d\overline{F}}{d\overline{\zeta}}\right)^2 d\overline{\zeta}$$

Integrating,

$$z = \zeta - \frac{1}{4}\left(\frac{W_1}{a_0}\right)^2 \int \left(\frac{d\overline{F}}{d\overline{\zeta}}\right)^2 d\overline{\zeta} \tag{3.28}$$

Thus it is seen that the complex coordinate in the physical plane of compressible fluid is equal to the corresponding complex coordinate in the physical plane of incompressible fluid plus a correction term. The factor before the integral depends upon the Mach's number of the undisturbed flow only. By using Eqs. (3.7) and (3.8), and (3.21), one has

$$\frac{1}{4}\left(\frac{W_1}{a_0}\right)^2 = \frac{\left(\dfrac{w_1}{a_1}\right)^2}{\left\{1 + \sqrt{1 - \left(\dfrac{w_1}{a_1}\right)^2}\right\}^2} \tag{3.29}$$

The integration constant of the integral in Eq. (3.28) is not important, because it only means a translation of the whole z-plane. (4) The velocity w corresponds to z can be computed by starting with

$$\overline{W} = \frac{dF}{d\zeta} = W_1 \frac{dG}{d\zeta} = U - iV$$

By means of Eq. (3.29)

$$\frac{w}{w_1} = \frac{\dfrac{|W|}{W_1}}{\dfrac{w_1}{W_1}\left[1 - \dfrac{1}{4}\left(\dfrac{W_1}{a_0}\right)^2\left(\dfrac{|W|}{W_1}\right)^2\right]}$$

Thus by putting $w = w_1$, one obtains

$$\frac{w_1}{W_1} = \frac{1}{1 - \dfrac{1}{4}\left(\dfrac{W_1}{a_0}\right)^2}$$

Thus

$$\frac{w}{w_1} = \left\{1 - \frac{1}{4}\left(\frac{W_1}{a_0}\right)^2\right\}\frac{\dfrac{|W|}{W_1}}{1 - \dfrac{1}{4}\left(\dfrac{W_1}{a_0}\right)^2\left(\dfrac{|W|}{W_1}\right)^2} \tag{3.30}$$

Using Eqs. (3.30) and (3.29), the ratio $\dfrac{w}{w_1}$ can be calculated easily. (5) To determine the pressure acting on the surface of the body one has to use Eq. (3.2). With some manipulation, the following relation is obtained:

$$\frac{p - p_1}{\dfrac{1}{2}\rho_1 w_1^2} = \frac{2}{\left(\dfrac{w_1}{a_1}\right)^2}\left[1 - \frac{\rho_0}{\rho}\left(\frac{\rho_1}{\rho_0}\right)\right]$$

$$= \frac{2}{\left(\dfrac{w_1}{a_1}\right)^2}\left[1 - \sqrt{1 - \left(\frac{w_1}{a_1}\right)^2}\,\frac{\rho_0}{\rho}\right]$$

But

$$\frac{\rho_0}{\rho} = \sqrt{1 + \left(\frac{w}{w_1}\right)^2\left(\frac{w_1}{a_0}\right)^2}$$

$$= \sqrt{1 + \left(\frac{w}{w_1}\right)^2 \frac{\left(\dfrac{w_1}{a_1}\right)^2}{1 - \left(\dfrac{w_1}{a_1}\right)^2}} \tag{3.31}$$

Therefore

$$\frac{p - p_1}{\dfrac{1}{2}\rho_1 w_1^2} = \frac{2}{\left(\dfrac{w_1}{a_1}\right)^2}\left\{1 - \sqrt{1 + \left[\left(\frac{w}{w_1}\right)^2 - 1\right]\left(\frac{w_1}{a_1}\right)^2}\right\} \tag{3.32}$$

Section (Ⅱ)

In this section, the general theory developed in Section (Ⅰ) is applied to the simple case of flow over a symmetrical Joukowsky airfoil at zero angle of attack. The complex potential in

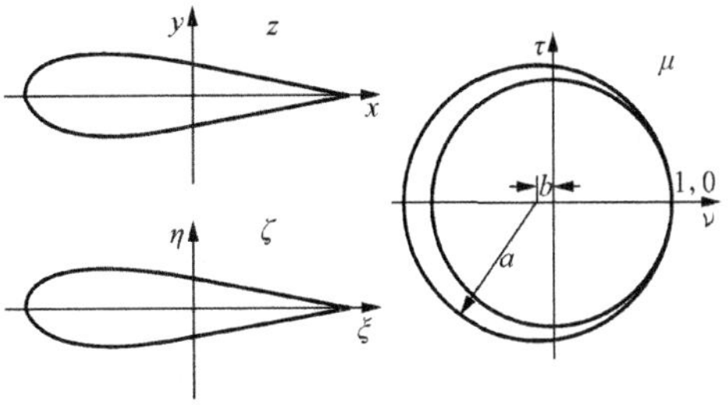

Fig. 3.2

the circle-plane (see Fig. 3.2) is known to be

$$w_1\left\{(\mu-b)+\frac{a^2}{\mu-b}\right\} \tag{3.33}$$

where a = radius of the airfoil circle, b = eccentricity of the airfoil circle. The relation between the airfoil plane and circle-plane is the well-known Joukowsky transformation

$$\zeta=\mu+\frac{1}{\mu} \tag{3.34}$$

if the radius of the transforming circle is unity.

Now the starting point of the calculation is the function to find $W\mathrm{d}\overline{F}$

$$W\mathrm{d}\overline{F}=\frac{\mathrm{d}\overline{F}}{\mathrm{d}\overline{\zeta}}\frac{\mathrm{d}\overline{F}}{\mathrm{d}\overline{\mu}}\mathrm{d}\overline{\mu}=\frac{\left(\frac{\mathrm{d}\overline{F}}{\mathrm{d}\overline{\mu}}\right)^2}{\frac{\mathrm{d}\overline{\zeta}}{\mathrm{d}\overline{\mu}}}\mathrm{d}\overline{\mu}$$

Therefore $$W\mathrm{d}\overline{F}=W_1^2\left\{1-\frac{a^2}{(\overline{\mu}-b)^2}\right\}^2\left\{1+\frac{1}{2}\left(\frac{1}{\overline{\mu}-1}-\frac{1}{\overline{\mu}+1}\right)\right\}$$

Thus the correction term in Eq. (3.28) is

$$\frac{1}{4a_0^2}\int W\mathrm{d}F=\frac{1}{4}\left(\frac{W_1}{a_0}\right)^2\{I_1+I_2-I_3\} \tag{3.35}$$

where
$$I_1=\int\left\{1-\frac{2a^2}{(\overline{\mu}-b)^2}+\frac{a^4}{(\overline{\mu}-b)^4}\right\}\mathrm{d}\overline{\mu}$$

$$I_2=\frac{1}{2}\int\left\{1-\frac{2a^2}{(\overline{\mu}-b)^2}+\frac{a^4}{(\overline{\mu}-b)^4}\right\}\frac{\mathrm{d}\overline{\mu}}{(\overline{\mu}-1)}$$

$$I_3=\frac{1}{2}\int\left\{1-\frac{2a^2}{(\overline{\mu}-b)^2}+\frac{a^4}{(\overline{\mu}-b)^4}\right\}\frac{\mathrm{d}\overline{\mu}}{\overline{\mu}+1}$$

These integrals can be easily computed and simplified, noting that $a-b=1$. If $(\overline{\mu}-b)=a\,\mathrm{e}^{\mathrm{i}\theta}$

and $\lambda = \dfrac{a}{1-b}$

$$
\begin{aligned}
I_1 &= a\left\{e^{i\theta} + 2e^{-i\theta} - \frac{1}{3}e^{-3i\theta}\right\} \\
I_2 &= \frac{1}{2}\left\{\frac{1}{6} - e^{-i\theta} + \frac{1}{2}e^{-2i\theta} + \frac{1}{3}e^{-3i\theta} + \log(a\, e^{i\theta})\right\} \\
I_3 &= \frac{1}{2}\left\{(1-\lambda^2)^2 \log a\left(e^{i\theta} + \frac{1}{\lambda}\right) + \lambda^2(2-\lambda^2)\log(a\, e^{i\theta}) + \right. \\
&\quad \left. \lambda(1-\lambda^2)e^{-i\theta} + \frac{\lambda^2}{2}e^{-2i\theta} - \frac{\lambda}{3}e^{-3i\theta}\right\}
\end{aligned}
\tag{3.36}
$$

Separating the real and imaginary parts, and adding,

$$
\begin{aligned}
\mathrm{Re}(I_1 + I_2 - I_3) &= a\left(3\cos\theta - \frac{1}{3}\cos 3\theta\right) + \\
&\quad \frac{1}{2}\left\{\frac{(1-\lambda^2)^2}{2}\log\left(1 + \frac{1}{\lambda^2} + \frac{2\cos\theta}{\lambda}\right) + (\lambda^3 - 2\lambda - 1)\cos\theta + \right. \\
&\quad \left. \frac{1}{2}(1-\lambda^2)\cos 2\theta + \frac{1}{3}(1+\lambda)\cos 3\theta\right\} \\
\mathrm{Im}(I_1 + I_2 - I_3) &= a\left(-\sin\theta + \frac{1}{3}\sin 3\theta\right) + \\
&\quad \frac{1}{2}\left\{(1 + 2\lambda - \lambda^3)\sin\theta - \frac{1}{2}(1-\lambda^2)\sin 2\theta - \right. \\
&\quad \left. \frac{1}{3}(1+\lambda)\sin 3\theta - (1-\lambda^2)^2 \tan^{-1}\frac{\sin\theta}{\cos\theta + \frac{1}{\lambda}} + (1-\lambda^2)^2\theta\right\}
\end{aligned}
\tag{3.37}
$$

These give the correction term to x and y coordinates.

The transformed velocity W over the surface of the airfoil can be easily found by means of graphical method[6]. Then the actual velocity and pressure can be computed by using Eqs. (3.22) and (3.32).

Figure 3.3 shows the result of calculation for the case $a = 1.20$ and $\dfrac{w_1}{a_1} = 0.550$. The nose of the airfoil is somewhat rounded by transforming into the case of compressible fluid. However, the pressure gradient is steeper, as would be expected. The main defect of this type of calculation is that during the transformation from incompressible flow to compressible flow, the shape of the body is also changed. To isolate the effect of compressibility of the fluid, it is necessary to bring back the original shape of the body. This is done by first deforming the original Joukowsky airfoil, such that the final profile after correction for compressibility is same as the original Joukowsky airfoil. The amount of deformation is obtained from the calculation assuming that the airfoil was a Joukowsky airfoil at start. That is the effect of deformation on the correction term of Eq. (3.28) is neglected. This is allowable because the quantity neglected is a second order quantity.

This deformation of Joukowsky airfoil can be carried out by using the method developed

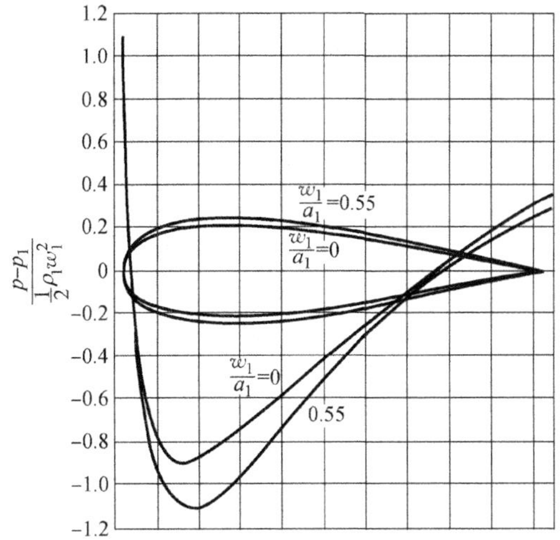

Fig. 3.3 Effect of transformation on Joukowsky airfoil with $b = 0.20$ at $w_1/a_1 = 0.55$

by von Kármán and Trefftz[7]. However, for some practical reasons, the Kármán-Trefftz method is somewhat modified:

Figure 3.4a shows two airfoils, both having the same chord, one is Joukowsky airfoil desired and the other is the airfoil resulted from the first step calculation. Now apply the Joukowsky transformation to this figure, then the Joukowsky airfoil will become a circle C_1 while the other airfoil a near-circular shape C, as shown in Fig. 3.4b. The desired deformed Joukowsky airfoil will appear like C_2 in this figure. The difference between C_2 and C_1 is just opposite and equal to that between C_1 and C. Now let C_2 be written as $\zeta_2 = re^{i\theta}$. Obviously,

$$r = 1 + g(\theta) \tag{3.38}$$

where $g(\theta)$ will be small compared with 1. The function which established the conformal transformation of the outside of this boundary C_2 into outside of the circle C_1 may be denoted by

$$\zeta_1 = \zeta_2[1 + f(\zeta_2)] \tag{3.39}$$

where ζ_1, ζ_2 have their origin at the center and the absolute value of $f(\zeta_2)$ is again small compared with 1. Then it is shown[7] that

$$g(\theta) + \mathrm{Re}[f(\zeta_2)] = 0 \tag{3.40}$$

In order to calculate $f(\zeta_2)$ we develop the function $g(\theta)$ in a Fourier series:

$$g(\theta) = \sum_0^\infty a_n \cos n\theta \tag{3.41}$$

Here only cosine terms appear because the airfoil is symmetrical about the chord. On the other hand, the complex function $f(\zeta_2)$ has the form, for $|\zeta_2| > 1$

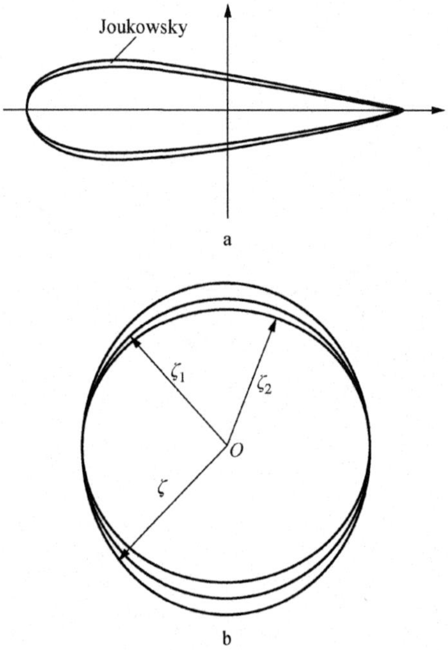

Fig. 3.4

$$f(\zeta_2) = \sum_0^{\infty} \frac{c_n}{\zeta_2^n} \tag{3.42}$$

Now put $\zeta_2 \approx e^{i\theta}$, then (3.40) is satisfied by

$$c_n = -a_n$$

Thus

$$f(\zeta_2) = -\sum_0^{\infty} \frac{a_n}{\zeta_2^n} \tag{3.43}$$

It can be easily seen that the velocity around the deformed Joukowsky airfoil can be calculated as

$$w = w_J \left| \frac{d\zeta_1}{d\zeta_2} \right| \tag{3.44}$$

where w_J = velocity around the Joukowsky airfoil. Now from Eqs. (3.39) and (3.43),

$$\begin{aligned} \frac{d\zeta_1}{d\zeta_2} &= 1 - \sum_0^{\infty} \frac{a_n}{\zeta_2^n} + \sum_0^{\infty} \frac{na_n}{\zeta_2^n} \\ &= \left\{1 + \sum_0^{\infty} (n-1)a_n \cos n\theta\right\} - i\sum_0^{\infty} (n-1)a_n \sin n\theta \end{aligned}$$

Neglecting small quantities of second order, and noting Eq. (3.41),

$$\begin{aligned}\left|\frac{d\zeta_1}{d\zeta_2}\right| &= 1 + \sum_0^\infty (n-1)a_n \cos n\theta \\ &= 1 + \sum_0^\infty na_n \cos n\theta - \sum_0^\infty a_n \cos n\theta \\ &= 1 + \frac{d}{d\theta}\sum_1^\infty a_n \sin n\theta - g(\theta)\end{aligned} \tag{3.45}$$

A trial calculation shows that the convergence of the coefficients a_n is not very good. Therefore, one must avoid manipulation on the Fourier series as required by Eq. (3.45). This is possible because $-\sum_0^\infty a_n \sin n\theta$ is known to mathematicians as the allied or conjugate series of $\sum_0^\infty a_n \cos n\theta$. It is also known[8] that if $g(\theta) = \sum_0^\infty a_n \cos n\theta$

$$-\sum_1^\infty a_n \sin n\theta = \frac{1}{2\pi}\int_0^\pi \frac{g(\theta+\xi)-g(\theta-\xi)}{\tan\frac{\xi}{2}}d\xi$$

Therefore

$$\frac{d}{d\theta}\sum_1^\infty a_n \sin n\theta = -\frac{1}{2\pi}\int_0^\pi \frac{g'(\theta+\xi)-g'(\theta-\xi)}{\tan\frac{\xi}{2}}d\xi$$

Integrating by parts,

$$-\frac{1}{2\pi}\left\{\left[\frac{g(\theta+\xi)+g(\theta-\xi)}{\tan\frac{\xi}{2}}\right]_0^\pi + \frac{1}{2}\int_0^\pi \frac{g(\theta+\xi)+g(\theta-\xi)}{\sin^2\frac{\xi}{2}}d\xi\right\}$$

$$= -\frac{1}{2\pi}\int_0^\pi \frac{\{g(\theta+\xi)-g(\theta)\}+\{g(\theta-\xi)-g(\theta)\}}{1-\cos\xi}d\xi$$

Hence Eq. (3.45) can be written as

$$\left|\frac{d\zeta_1}{d\zeta_2}\right| = 1 - \frac{1}{2\pi}\int_0^\pi \frac{\{g(\theta+\xi)-g(\theta)\}+\{g(\theta-\xi)-g(\theta)\}}{1-\cos\xi}d\xi - g(\theta) \tag{3.46}$$

The integral is evidently convergent for any continuous regular function $g(\theta)$, because then the integrand is always finite. Its evaluation can be done numerically.

Figure 3.5 is the result of calculation for a Joukowsky airfoil with the thickness parameter $b = 0.20$, at two speeds, $\frac{w_1}{a_1} = 0.450$ and 0.550. The suction peaks are considerably higher with higher speeds. Also the positions of pressure peaks tend to move backward with increasing speed. Both are in agreement with the experimental results obtained by J. Stack[9]. The values of $(p-p_1)\Big/\frac{1}{2}\rho_1 w_1^2$ for $\frac{w_1}{a_1} = 0.550$ and $\frac{w_1}{a_1} = 0.450$ at which real air will attain a velocity equal to the local sound velocity are equal to -1.653 and -2.755 respectively. It is thus seen

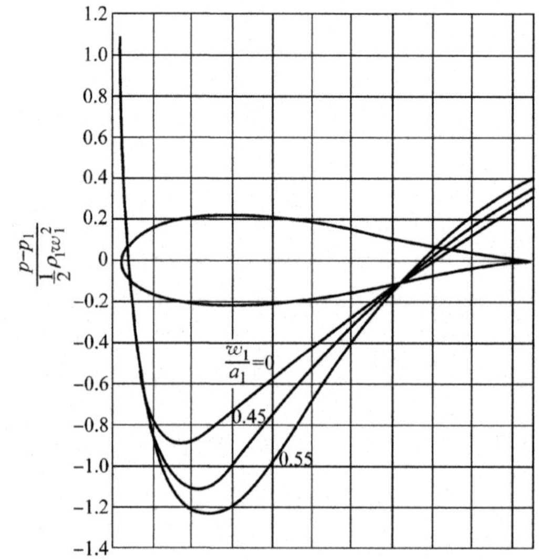

Fig. 3.5 Effect of compressibility on the pressure distribution of Joukowsky airfoil with $b = 0.20$

that the effect of compressibility on pressure distribution is appreciable, even when nowhere the local sound velocity is reached. One should, however, bear in mind that the effect on the force coefficient of the airfoil will probably not be so marked as with the pressure distribution, because the resultant force on the airfoil is the algebraic difference of pressure force acting on two sides of the section.

Appendix to PART (Ⅲ)
Comparison with Other Methods

In order to check the accuracy of the method developed in PART (Ⅲ), the flow over a finite circular cylinder with its axis perpendicular to the direction of undisturbed flow is studied. The method exposed in Section (Ⅱ) of PART (Ⅲ) for correction of shape of body is used. The following is the result of calculation for velocity at the top of the circular section, compared with results by other methods (collected by E. Pistolesi[10]).

$$\frac{w_1}{a_1} = 0.400$$

Method	$\frac{w}{w_1}$ at top of section
Part (Ⅲ)	2.268
Rayleigh	2.206
Poggi	2.194
Taylor's electric analogy	2.188

$\frac{w}{w_1} = 2.000$ for incompressible fluid. Thus the present method gives a higher value.

However, the flow over a cylinder is rather an extreme case. Because the difference between the velocities to be calculated and the undisturbed velocity is large, and thus this approximate method involves larger than usual error.

References on PART (Ⅲ)

[1] Molenbroek P. Über einige Bewegungen eines Gases bei Annahme eines Geschwindigkeitspotentials. Arch d. Mathem. u. Phys., Grunert Hoppe (1890), Reihe 2,9, 157.

[2] Tschapligin A. Scientific Memoirs of the Univ. Moscow (In Russian) (1902).

[3] Clauser F, Clauser M. New Methods of Solving the Equations for the Flow of a Compressible Fluid. Unpublished Ph.D. Thesis at C.I.T. 1937.

[4] Demtchenko B. Sur les mouvements lents des fluides compressibles. Comptes Rendus, 1932, 194: 1218.//Also, Variation de la résistance aux faibles vitesses sous l'influence de la compressibilité, Comptes Rendus, 194: 1720 (1932).

[5] Busemann A. Die Expansionsberichtigung der Kontraktionsziffer von Blenden. Forschung 1933, 4, S. 186-187.//Also, Hodographenmethode der Gasdynamik, Z.A.M.M. 1937,12,73-79.

[6] Durand W F. Aerodynamic Theory 1st Ed. Julius Springer, Berlin 1935. 2: 71-74.

[7] von Kármán Th., Trefftz E. Potentialströmung dem gegebene Tragflachenquerschnitte. Z.F.M. 1918, 9:111.//Also, W. F. Durand: Aerodynamic Theory. 1935,2:80-83.

[8] Hardy G H, Littlewood J E. The Allied Series of a Fourier Series. Proc. of London Math. Soc. (2), 1925,24, 211-246.

[9] Stack J. The Compressibility Burble. NACA Technical Note No. 543,1935.

[10] Pistolesi E. La portanza alle alte velocita inferiori a quella del suono. Atti dei V. Convengo "Volta", fasc 300, (1936) Reale Accademie. d'Italia, Rome.

PART (Ⅳ)

Flight Analysis of a Sounding Rocket with Special Reference to Propulsion by Successive Impulses

Introduction

In 1919 R. H. Goddard[1] published the historically important paper which suggested the use of nitrocellulose powder as a propellant for raising a sounding rocket to altitudes beyond the range of sounding balloons. To determine the feasibility of this propellant, a series of experiments had been carried out and it was found that thermal efficiency of 50% could be expected if the powder was exploded in a properly designed chamber and the resulting gases were allowed to escape at high velocity through an expanding nozzle. In 1931 R. Tilling used a mixture of potassium chlorate and naphthalene as propellant and actually reached an altitude of 6 600 feet. More recently, L. Damblanc[2] made static tests with a slow burning black powder and from these estimated that a height of 10 000 feet could be reached using a two-step arrangement. The results so far reported offer an incentive to further analysis.

The effect of decreasing gravitational acceleration on the maximum height reached by a rocket has been considered by A. Bartocci[3]. However, he assumes that the rocket itself has a constant acceleration during powered flight. L. Breguet and R. Devillers[4] also considered the effect of the variation of g. To simplify the analysis, they assumed that the acceleration of the rocket was equal to a constant multiple of g. Since the sounding rocket for practical reasons will be propelled by a nearly constant thrust or a uniform rate of successive impulses, in Section (Ⅱ) the author has studied the problem anew according to this mode of propulsion.

When the sounding rocket is ascending through the air the maximum height reached is less than that reached for flight in vacuo. Recently, studies have been made of the problem by W. Ley and H. Schaefer[5] and by F. J. Malina and A. M. O. Smith[6]. On the basis of the latter study a group of new performance parameters have been isolated from the general performance equation, and these are discussed in Section (Ⅲ).

Notation

Referring to Fig. (4.1), the following notation has been used throughout the paper:

w = weight of propellant and propellant container ejected per impulse, lbs

k = ratio of container weight to sum of container and propellant weight ejected per impulse

$\lambda = 1 - k$

W_0 = initial weight of rocket, lbs

M_0 = initial mass of the rocket, slugs

W_p = instantaneous weight of rocket, lbs

ζ = ratio of initial weight of propellants to initial total weight of a rocket propelled by constant thrust

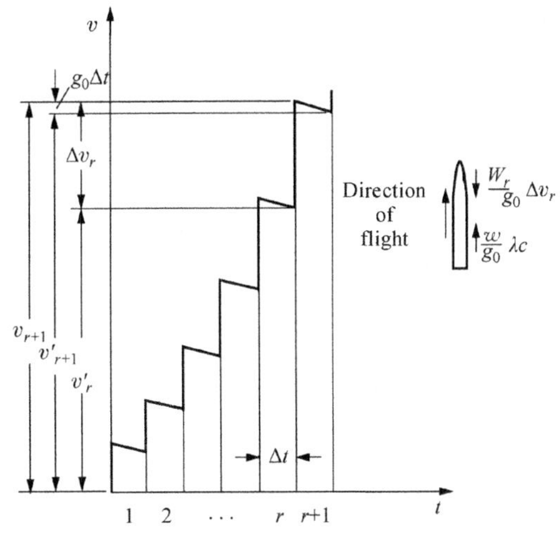

Fig. 4. 1

ζ_1 = ratio of initial weight of propellants to initial total weight of a rocket propelled by successive impulses

$\zeta_1' = \zeta_1 / \lambda$

n = number of impulses per second

N = total number of impulses occuring during powered flight

t = time, sec

Δt = interval between impulses, sec

a_0 = initial acceleration imparted to rocket, ft/sec^2

g_0 = acceleration of gravity at the starting point of flight, ft/sec^2

g = acceleration of gravity above the starting point of flight, ft/sec^2

c = effective exhaust velocity of ejected propellant, ft/sec

v = instantaneous velocity, ft/sec

Δv_r = velocity imparted to rocket by the rth impulse, ft/sec

v_r' = velocity at the end of the rth interval, ft/sec

v_{s_0} = velocity of sound corresponding to the atmospheric conditions at the starting point of the flight, ft/sec

v_s = velocity of sound corresponding to the atmospheric conditions at the height reached by the rocket at the time t, ft/sec

B = Mach's number = v/v_s

$V_{\max}$ = velocity of rocket at start of coasting flight, ft/sec

$V_{\max 0}$ = velocity of rocket as start of coasting flight if g is constant and equal to g_0, ft/sec

h = altitude above sea level, ft

h_r = height reached at the beginning of the rth interval, ft

h_r' = height reached at the end of the rth interval, ft

H_P = height travelled during powered flight, ft

H_{P_0} = height travelled during powered flight, if g is constant and equal to g_0, ft

H_C = height travelled during coasting flight, ft

H_{C_0} = height travelled during coasting flight, if g is constant and equal to g_0, ft

H_{max} = height travelled during powered flight and coasting flight, ft

H_{max0} = height travelled during powered flight and coasting flight, if g is constant and equal to g_0, ft

R = radius of earth, 2.088×10^8, ft

D = drag on rocket due to air resistance, lbs

C_D = drag coefficient of rocket shell

C_D^* = drag coefficient of rocket shell at the velocity of sound

Λ = drag-weight factor (discussed in the section on the effect of air resistance)

ρ_0 = mass density of air at the starting point of the flight, slugs/ft^3

σ = ratio of air densities at altitude and at the starting point of the flight

T = absolute temperature of the atmosphere at the height reached by the rocket at the time t, °F

T_0 = absolute temperature of atmosphere at the starting point of flight, °F

A = largest cross-sectional area of rocket shell, ft^2

d = largest diameter of rocket shell, ft

l = length of rocket shell, ft

Section (I)

An approximate method of calculating the maximum height reached by a rocket propelled by powder was developed by R. H. Goddard[1]. To simplify the analysis a continuous loss of mass was assumed and the problem was so stated that a minimum mass of propellant necessary to lift one pound of mass at the end of the flight to any desired height was determined. However, if high-powered powder is used, the rate of burning is so rapid that the propulsive action is instantaneous. The rocket is thus acted upon by an impulse rather than by a constant thrust.

In the following analysis, it has, therefore, been assumed that the propulsive force is an impulsive force, i.e., the force acts for such a brief interval of time that the rocket does not change its position during the application of the force, although its velocity and its momentum receive a finite change. If the combustion process of the propulsive unit takes place at constant volume this assumption is justified. Further, a study of interior ballistics of small arms reveals that the period between the ignition of the powder charge and the bullet's arrival at the end of a two-foot barrel is of the order of 14 ten-thousandths of a second. If the gases are not restrained and their travel through the burning chamber and the nozzle is of much shorter length, as is the case for the rocket motor, even shorter periods of duration of action can be expected.

Assuming that the propulsive force acts as an impulse, then the motion of the rocket can be calculated by Newton's third law, which states that impulses between two bodies are equal and opposite. Hence, equating the momentum of the exhaust gases to the momentum imparted to the rocket, using the quantities defined in the list of notation and referring to Fig. 4.1, the following relation can be written for flight in vacuo

$$\frac{\lambda w}{g_0}c = \frac{W_r}{g_0}\Delta v_r \tag{4.1}$$

where

$$\lambda = 1 - k$$

and

$$W_r = W_0 - rw \tag{4.2}$$

Or

$$\Delta v_r = \frac{w\lambda c}{W_0 - rw} = \frac{\zeta_1'\lambda c}{N}\left(\frac{1}{1 - r\frac{\zeta_1'}{N}}\right) \tag{4.3}$$

where

$$\zeta_1' = \frac{wN}{W_0} = \frac{\zeta_1}{\lambda}$$

During the interval between impulses, Δt, the velocity is reduced by the action of gravity so that at the end of the rth interval, the velocity of the rocket will be

$$\begin{aligned} v_r' &= v_r - g_0\Delta t \\ &= v_{r-1}' + \Delta v_r - g_0\Delta t \end{aligned} \tag{4.4}$$

Therefore

$$v_r' = \sum_{s=1}^{s=r}\Delta v_s - rg_0\Delta t \tag{4.5}$$

Substituting for Δv_s from Eq. (4.4)

$$v_r' = \frac{\zeta_1'\lambda c}{N}\sum_{s=1}^{s=r}\frac{1}{1 - s\left(\frac{\zeta_1'}{N}\right)} - rg_0\Delta t \tag{4.6}$$

Or

$$v_r' = \frac{\zeta_1'\lambda c}{N}S_1 - rg_0\Delta t$$

where

$$S_1 = \sum_{s=1}^{s=r}\frac{1}{1 - s\left(\frac{\zeta_1'}{N}\right)}$$

The height gained during each interval will be represented by the area under the velocity curve in the interval, or

$$h_r' - h_r = v_r'\Delta t + \frac{1}{2}g_0(\Delta t)^2 \tag{4.7}$$

Therefore, at the end of the Nth interval which is the end of the powered flight, the height will be

$$H_{P_0} = \sum_{r=1}^{r=N} v_r'\Delta t + \frac{N}{2}g_0(\Delta t)^2$$

Substituting for v_r' its value in Eq. (4.6)

$$H_{P_0} = \sum_{r=1}^{r=N} \Delta t \left\{ \frac{\zeta_1' \lambda c}{N} \sum_{s=1}^{s=r} \frac{1}{1 - s\left(\frac{\zeta_1'}{N}\right)} - r g_0 \Delta t \right\} + \frac{N}{2} g_0 (\Delta t)^2$$

$$= \frac{\zeta_1' \lambda c}{N} \Delta t \sum_{r=1}^{r=N} \frac{N+1-r}{1 - r\left(\frac{\zeta_1'}{N}\right)} - g_0 (\Delta t)^2 \sum_{r=1}^{r=N} \left(r - \frac{1}{2}\right)$$

$$= \frac{\zeta_1' \lambda c}{N} \Delta t S_2 - \frac{N^2}{2} g_0 (\Delta t)^2 \tag{4.8}$$

where
$$S_2 = \sum_{r=1}^{r=N} \frac{N+1-r}{1 - r\left(\frac{\zeta_1'}{N}\right)}$$

The maximum height reached will be the sum of the height at the end of powered flight and the height travelled during coasting or

$$H_{\max 0} = H_{P_0} + H_{C_0} = H_{P_0} + \frac{V_{\max 0}^2}{2 g_0} \tag{4.9}$$

To calculate the maximum height one has first to evaluate the sums S_1 and S_2. Noting that

$$\frac{1}{1 - s\left(\frac{\zeta_1'}{N}\right)} = \int_0^\infty e^{-x\left[1 - s\left(\frac{\zeta_1'}{N}\right)\right]} \mathrm{d}x$$

S_1 can be written in the form

$$S_1 = \int_0^\infty e^{-x} \sum_{s=1}^{s=r} \left(e^{x\frac{\zeta_1'}{N}}\right)^s \mathrm{d}x = \int_0^\infty e^{-x} \frac{e^{x\frac{\zeta_1'}{N}} - e^{(r+1)\frac{\zeta_1'}{N}}}{1 - e^{x\frac{\zeta_1'}{N}}} \mathrm{d}x$$

Putting $x = \frac{Ny}{\zeta_1'}$ the above integral becomes

$$S_1 = -\frac{N}{\zeta_1'} \int_0^\infty e^{-\frac{Ny}{\zeta_1'}} \left(\frac{1 - e^{ry}}{1 - e^{-y}}\right) \mathrm{d}y = \frac{N}{\zeta_1'} \left\{ \psi\left(\frac{N}{\zeta_1'}\right) - \psi\left(\frac{N}{\zeta_1'} - r\right) \right\} \tag{4.10}$$

where $\psi(z) = \frac{\mathrm{d}}{\mathrm{d}z}\{\log \Gamma(z)\}$, the so-called psi-function[7,8].

Similarly, S_2 can be summed as

$$S_2 = \frac{N}{\zeta_1'} \left\{ N - \left[\frac{N}{\zeta_1'} - (N+1)\right] \left[\psi\left(\frac{N}{\zeta_1'}\right) - \psi\left(\frac{N}{\zeta_1'} - N\right) \right] \right\} \tag{4.11}$$

Substituting Eqs. (4.10) and (4.11) into Eqs. (4.6) and (4.8), and then into Eq. (4.9) finally

$$H_{\max 0} = \frac{\lambda^2 c^2}{2 g_0} \Psi^2 - \frac{\lambda c}{n} \left\{ \left(\frac{N}{\zeta_1'} - 1\right) \Psi - N \right\} \tag{4.12}$$

where
$$n = \frac{1}{\Delta t} \text{ and } \Psi = \psi\left(\frac{N}{\zeta_1'}\right) - \psi\left(\frac{N}{\zeta_1'} - N\right)$$

For convenience of calculation in Fig. 4.2 the quantity Ψ is plotted against N for different values of ζ_1'.

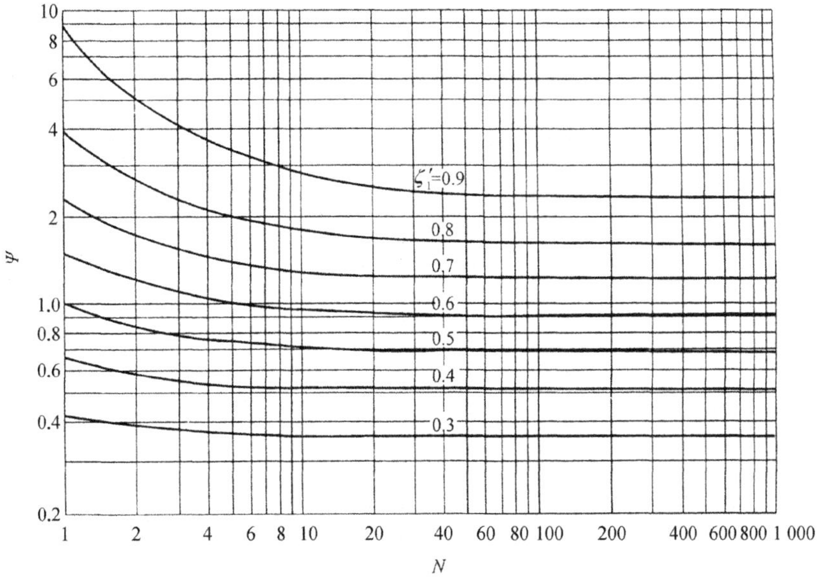

Fig. 4.2

It can easily be shown that when $N = 1$

$$\Psi = \frac{\zeta_1'}{1 - \zeta_1'}$$

so that Eq. (4.12) reduces to

$$H_{\max_0} = \frac{\lambda^2 c^2}{2g_0}\left(\frac{\zeta_1'}{1 - \zeta_1'}\right)^2 \tag{4.12a}$$

Also, as $N \to \infty$, $\Psi \to -\log(1 - \zeta_1')$ thus Eq. (4.12) reduces to

$$H_{\max_0} = \frac{\lambda^2 c^2}{2g_0}\left\{[\log(1 - \zeta_1')]^2 + \frac{\zeta_1' + \log(1 - \zeta_1')}{\frac{a_0}{g_0} + 1}\right\} \tag{4.12b}$$

where

$$a_0 = \frac{\zeta_1'}{N} n\lambda c - g_0 = \frac{nw\lambda c}{W_0} - g_0 \tag{4.12c}$$

The quantity a_0 can be considered as the initial acceleration of the rocket if $N \to \infty$. It is interesting to notice that Eq. (4.12b) is the equation obtained by Malina and Smith[6] for calculating the maximum height of a constant thrust rocket, as expected.

Figure 4.3 shows the variation of $H_{\max_0} g_0/(\lambda^2 c^2)$ with $n\lambda c/g_0$ for different values of ζ_1' and for four values of N. These curves show that when the total number of impulses, N, becomes larger than 100, the maximum height reached is imperceptibly changed by increasing the number.

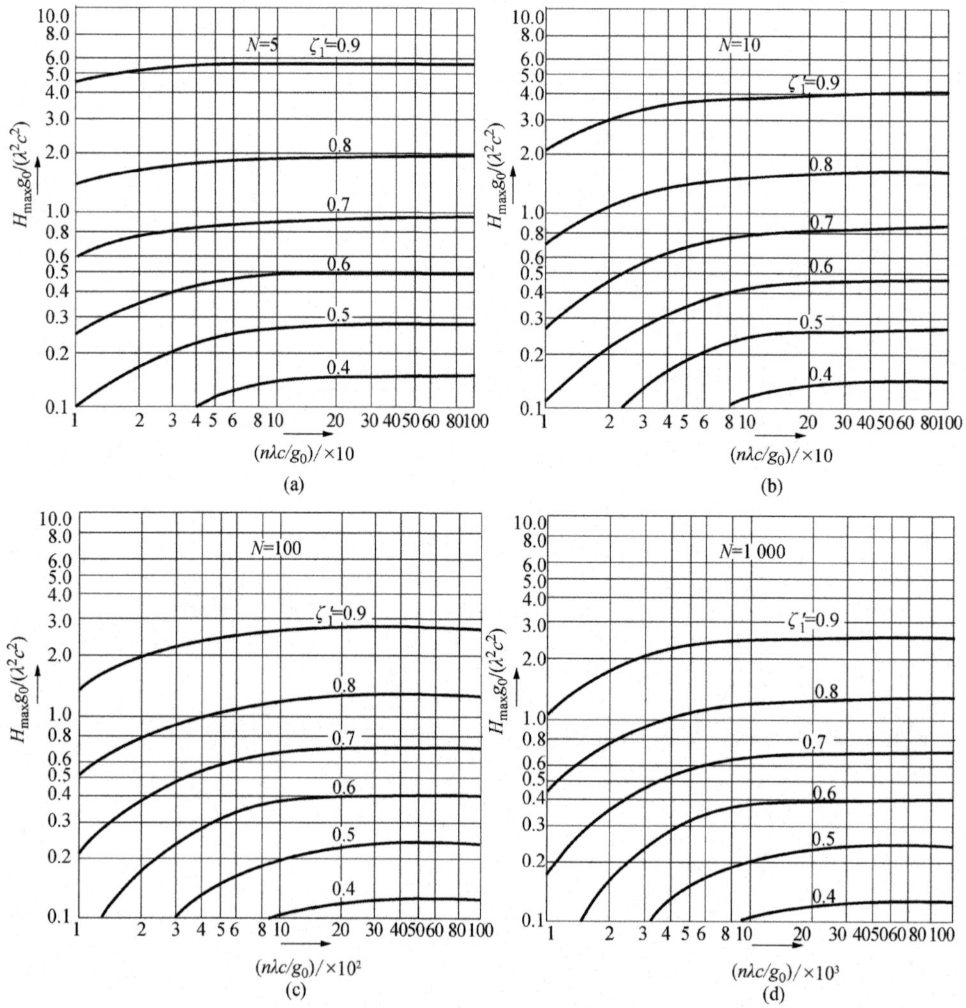

Fig. 4.3 $\dfrac{H_{max}g_0}{\lambda^2c^2}=\dfrac{1}{2}\Psi^2-\dfrac{1}{\frac{n\lambda c}{g_0}}\left[\left(\dfrac{N}{\zeta_1'}-1\right)\Psi-N\right]$

At this point it is necessary to discuss the similarity existing between a rocket propelled by successive impulses and a rocket propelled by constant thrust. The former loses not only the mass of the propellant, but also the containers for the individual charges. The difference in effect on the rocket between the propellant and its containers is that the propellant has an effective exhaust velocity, c, while the ejected containers leave the rocket without appreciable velocity. The propulsive action, however, will remain the same if the whole cartridge, that is, the propellant charge and its container, is considered wholly as propellant but leaving the motor at a reduced effective exhaust velocity λc. The rocket propelled by constant thrust loses only the mass equal to the propellant carried, therefore, it can be said to be equivalent to the "successive impulses" rocket if its effective exhaust velocity and its total mass of propellant are equal respectively to the reduced exhaust velocity and to the sum of the masses of all the containers of the "successive impulses" rocket. In other words, c is equal to λc and ζ is equal to ζ_1'.

In Table 4.1 the heights for four cases have been calculated to illustrate the effect of the exhuast gas velocity and the total number of impulses given to a rocket whose weight ratio, ζ_1' is 0.70. It will be noticed that for flight in vacuo a greater height will be reached if a smaller number of impulses is employed. The lower portion of the Table shows the maximum height reached by an equivalent "constant thrust" rocket for the same four cases with the initial acceleration given by Eq. (4.12c). The close agreement between the maximum height reached by use of successive impulses, when the total number of impulses exceeds 100, and that reached by the use of constant thrust simplifies the solution of the problem of decreasing acceleration of gravity with height, and enables prediction for flight with air resistance to be based on the results obtained for a rocket propelled by constant thrust (c. f. Ref. 6). These problems are considered in the following sections.

Table 4.1

Successive impulses

$$H_{\max_0} = \frac{\lambda^2 c^2}{g_0}\left\{\frac{1}{2}\Psi^2 - \frac{1}{\frac{n\lambda c}{g_0}}\left[\left(\frac{N}{\zeta_1'} - 1\right)\Psi - N\right]\right\}$$

Case	λc ft/sec	ζ_1'	N	n Impulses per sec.	$H_{\max 0}$ ft
1	10 000	0.70	326	3	1 472 000
2	10 000	0.70	10	0.092	1 686 000
3	7 000	0.70	326	3	560 000
4	7 000	0.70	10	0.092	676 000

Constant thrust

$$H_{\max 0} = \frac{c^2}{g_0}\left\{\frac{1}{2}[\log(1-\zeta)]^2 + \frac{1}{\frac{a_0}{g_0}+1}[\log(1-\zeta)+\zeta]\right\}$$

Case	ft/sec c	ζ	a_0	$H_{\max 0}$ ft
1	10 000	0.70	32.2	1 468 000
2	10 000	0.70	32.2	1 468 000
3	7 000	0.70	12.9	555 000
4	7 000	0.70	12.9	555 000

Section (Ⅱ)

It is well-known that the acceleration of gravity decreases with the height above the earth's surface according to the following relation

$$g = g_0 \left(\frac{R}{R+h} \right)^2 \tag{4.13}$$

At an altitude of 1 000 miles the acceleration is only 0.64 times that at sea level. Therefore, for flights up to such altitudes the assumption that g is approximately constant is no longer valid. It was shown by Malina and Smith[6] that a three-step rocket could theoretically reach such an altitude. Thus it is interesting to see how the decrease of g will increase the maximum height reached by the rocket.

First the effect on powered flight in vacuo will be considered and then on coasting flight in vacuo. For powered flight the analysis is based on the assumption that the thrust is constant. However, the results can be applied to the case of propulsion by successive impulses if the total number of impulses, N, exceeds 100 as was justified in the previous section.

The equivalent mass of gas flowing per second continuously for the case of successive impulses is

$$\frac{wn}{g_0} = m \tag{4.14}$$

Assuming that the rocket starts from rest at sea level the equation of motion in vacuo is

$$\frac{\mathrm{d}^2 h}{\mathrm{d}t^2} = - g_0 \left(1 + \frac{h}{R} \right)^{-2} + \frac{\frac{mc}{M_0}}{1 - \frac{m}{M_0} t} \tag{4.15}$$

This is a non-linear differential equation which can not be solved by usual means. However, for all practical purposes the ratio $\frac{h}{R}$ during powered flight is much smaller than 1, therefore, only first order terms in $\frac{h}{R}$ occurring in the expansions need to be retained. This approximation linearizes the equation to the form

$$\frac{\mathrm{d}^2 h}{\mathrm{d}t^2} = g_0 \left(\frac{2h}{R} - 1 \right) + \frac{\frac{mc}{M_0}}{1 - \frac{m}{M_0} t} \tag{4.16}$$

The solution of this equation with the initial condition that $h = 0$ and $\frac{\mathrm{d}h}{\mathrm{d}t} = 0$ when $t = 0$ is

$$h = \frac{R}{2} \left\{ 1 - \cosh \sqrt{\frac{2g_0}{R}} t \right\} + \frac{c}{2\sqrt{\frac{2g_0}{R}}} \left\{ \mathrm{e}^{\xi u} \int_{\xi}^{\xi u} \frac{\mathrm{e}^{-x} \mathrm{d}x}{x} - \mathrm{e}^{-\xi u} \int_{\xi}^{\xi u} \frac{\mathrm{e}^{x} \mathrm{d}x}{x} \right\} \tag{4.17}$$

where

$$\xi = \sqrt{\frac{2g_0}{R}} \frac{M_0}{m} \text{ and } u = 1 - \frac{m}{M_0} t$$

At the end of the powered flight, the time is

$$t = t_{\mathrm{P}} = \frac{M_0 \zeta}{m} \tag{4.18}$$

Therefore, the height at the end of the powered flight is

$$H_{\mathrm{P}} = \frac{R}{2}\left\{1 - \cosh\sqrt{\frac{2g_0}{R}}\frac{M_0\zeta}{m}\right\} + \frac{c}{2}\sqrt{\frac{R}{2g_0}}\left\{e^{\xi(1-\zeta)}\int_{\xi}^{\xi(1-\zeta)}\frac{e^{-x}\,dx}{x} - e^{-\xi(1-\zeta)}\int_{\xi}^{\xi(1-\zeta)}\frac{e^{x}\,dx}{x}\right\} \tag{4.19}$$

If the hyperbolic cosine term and the integrals are expanded and only first order terms in $\frac{1}{R}$ are retained in consistency with the linearization of Eq. (4.15), the equation becomes

$$\begin{aligned} H_{\mathrm{P}} &= -\left\{\frac{\zeta^2 g_0}{2}\left(\frac{W_0}{w}\right)^2 + \frac{\zeta^4 g_0^2}{12R}\left(\frac{W_0}{w}\right)^4\right\} + c\frac{W_0}{w}\{(1-\zeta)\log(1-\zeta)+\zeta\} + \\ &\quad \frac{cg_0}{18R}\left(\frac{W_0}{w}\right)^3\{6(1-\zeta)^3\log(1-\zeta)+\zeta(11\zeta^2-15\zeta+6)\} \\ &= H_{\mathrm{P}_0} + \frac{g_0}{6R}\left(\frac{W_0}{w}\right)^3\left\{\frac{c}{3}[6(1-\zeta)^3\log(1-\zeta)+\zeta(11\zeta^2-15\zeta+6)] - \frac{\zeta^4 g_0}{2}\left(\frac{W_0}{w}\right)\right\} \end{aligned} \tag{4.20}$$

Differentiating Eq. (4.17), and substituting the relation of Eq. (4.18), the maximum velocity at the end of powered flight is

$$V_{\max} = -\sqrt{\frac{Rg_0}{2}}\sinh\sqrt{\frac{2g_0}{R}}\frac{M_0\zeta}{m} - \frac{c}{2}\left\{e^{\xi(1-\zeta)}\int_{\xi}^{\xi(1-\zeta)}\frac{e^{-x}\,dx}{x} - e^{-\xi(1-\zeta)}\int_{\xi}^{\xi(1-\zeta)}\frac{e^{x}\,dx}{x}\right\} \tag{4.21}$$

Again expanding and retaining only first order terms in $\frac{1}{R}$, Eq. (4.21) becomes

$$\begin{aligned} V_{\max} &= -\left\{\left(\frac{W_0}{w}\right)\zeta + \frac{1}{3}\frac{\zeta^3 g_0^2}{R}\left(\frac{W_0}{w}\right)^3\right\} - \\ &\quad \left\{c\log(1-\zeta) + \frac{cg_0}{2R}\left(\frac{W_0}{w}\right)^3[2(1-\zeta)^2\log(1-\zeta)+2\zeta-3\zeta^2]\right\} \\ &= V_{\max 0} - \frac{g_0\left(\frac{W_0}{w}\right)^2}{R}\left\{\frac{\zeta^3 g_0\left(\frac{W_0}{w}\right)}{3} + \frac{c}{2}[2(1-\zeta)^2\log(1-\zeta)+2\zeta-3\zeta^2]\right\} \end{aligned} \tag{4.22}$$

It is seen that the second terms of Eqs. (4.20) and (4.22) are the correction to be applied to H_{P_0} and $V_{\max 0}$ to account for the variation of the acceleration of gravity. Since both corrections are first order approximations, they can be expected to apply approximately also to the case of successive impulses, even when the total number of impulses is less than 100.

The coasting height reached by the rocket due to its velocity at the end of powered flight can be obtained by equating the increase of potential energy during coasting flight to the kinetic energy at the end of powered flight. Thus

$$\frac{1}{2}V_{\max}^2 = g_0\int_{H_{\mathrm{P}}}^{H_{\mathrm{P}}+H_{\mathrm{C}}}\frac{dh}{\left(1+\frac{h}{R}\right)^2}$$

or

$$H_C = (H_P + R)\left[\frac{1}{1 - \dfrac{V_{max}^2}{2g_0\left(\dfrac{R}{H_P+R}\right)^2 (H_P+R)}} - 1\right] \tag{4.23}$$

Putting $V_{max}^2 \Big/ \left[2g_0\left(\dfrac{R}{R+H_P}\right)^2\right] = H_{C_0}$ which is coasting height obtained by assuming a constant gravitational acceleration of the value equal to that at the height H_C i.e., the height where coasting starts, then Eq. (4.23) can be written

$$H_C = (H_P + R)\left\{\frac{1}{1 - \dfrac{H_{C_0}}{H_P+R}} - 1\right\}$$

Upon expanding the second term this equation becomes,

$$H_C = H_{C_0}\left\{1 + \left(\frac{H_{C_0}}{H_P+R}\right) + \left(\frac{H_{C_0}}{H_P+R}\right)^2 + \left(\frac{H_{C_0}}{H_P+R}\right)^3 + \cdots\right\} \tag{4.24}$$

This equation shows that if the coasting flight starts from sea level, and if the maximum height reached is about 1 000 miles, the increase due to the decrease in g is over 25%, which is considerable.

Section (Ⅲ)

When the sounding rocket is ascending through the atmosphere instead of in vacuo, air resistance comes into play, causing the acceleration of the rocket to be reduced, which decreases the maximum height reached. Since air resistance increases with the air density and with the square of the flight velocity, it is desirable to keep the rocket from ascending too rapidly through the lower layers of the atmosphere where the air density is high. For this reason the optimum initial acceleration will no longer be infinite as shown by Eq. (4.12b). For the case of constant thrust Malina and Smith[6] have found that the optimum acceleration is around 30 ft/sec^2. For a total number of impulses greater than 100, the difference between propulsion by successive impulses and by constant thrust is very small, so one may expect the above optimum value of initial acceleration to hold for both cases of propulsion.

The actual amount of reduction in maximum height due to air resistance can be calculated by the method of step-by-step integration, if fair accuracy is desired. This integration is carried out by using the fundamental equation for vertical rocket flight which, as given in the previous paper[6] is

$$\frac{d^2h}{dt^2} = a = -g + \frac{a_0 + g_0}{1 - \dfrac{t(a_0+g_0)}{c}} - \frac{g_0\rho_0\sigma v^2}{2\left[1 - \dfrac{t(a_0+g_0)}{c}\right]}\frac{C_D A}{W_0} \tag{4.25}$$

The significance of the ratio $\frac{C_D A}{W_0}$ was discussed in that paper[6]. Greater significance can, however, be attached to the various terms in the equation if it is transformed into the non-dimensional form

$$\frac{a}{g_0} = -\frac{g}{g_0} + \frac{\frac{a_0}{g_0}+1}{1-\frac{g_0 t}{c}\left(\frac{a_0}{g_0}+1\right)} - \frac{\left(\sigma\frac{T}{T_0}\right)\left(\frac{C_D}{C_D^*}B^2\right)\Lambda}{1-\frac{g_0 t}{c}\left(\frac{a_0}{g_0}+1\right)} \tag{4.26}$$

where

$$\Lambda = \frac{\frac{\rho_0}{2}C_D^* A v_{s_0}^2}{W_0}$$

In Eq.(4.26) appear two types of significant quantities. First, quantities, called "factors", which are constant for any given family of rockets, and second, two quantities called "parameters", one of which is characteristic for a given family of rockets but changes in value along the flight path, and one which depends on the physical properties of the atmosphere. Thus there are the following factors:

$\frac{a_0}{g_0}$ = ratio of initial acceleration to g_0 "initial acceleration factor", a motor characteristic

c = exhaust velocity in ft/sec. ~ "exhaust velocity factor", a motor characteristic

Λ = "drag-weight factor"

ζ = ratio of weight of combustibles to total initial weight of the rocket ~ "loading factor"

The first two factors, i. e., the "initial acceleration factor" and the "exhaust velocity factor", determine the characteristics of the propelling unit for a given family of rockets while the "drag-weight factor" and the "loading factor" determine the physical dimensions of the rockets. The "drag-weight factor" is a ratio of the drag of the rocket at sea level when traveling with the velocity of sound to the initial weight of the rocket. Since for any given family of rocket shapes the only terms in the factor which can be varied are the maximum cross-sectional area A, and the initial weight W_0, it is clear that if the initial weight is doubled then the cross-sectional area must also be doubled to keep the factor the same. The "loading factor" needs to be discussed in some detail as it does not appear explicitly in Eq.(4.26). The Eq.(4.26) is a differential equation of the flight path which is satisfied at every point along the flight path. The loading factor ζ comes in only when this equation is integrated and the limits of integration are put in. For example, consider two rockets with identical performance factors and parameters, with the exception that one has a ζ of 0.90 and the other has a ζ of 0.50. The flight path of the two rockets will be identical up to the time that 0.50 times the initial weight of the rockets is used up as combustibles. At this point the rocket having a ζ of 0.50 will begin to decelerate while the one having a ζ of 0.90 will continue to accelerate until the remaining combustibles are used up. It is thus seen that the value of ζ controls the maximum height reached.

The two performance parameters are:

$\sigma \dfrac{T}{T_0}$ ~physical properties of the atmosphere called the "atmosphere parameter"

$\dfrac{C_D}{C_D^*}$ ~aerodynamic properties of the rocket shell called the "form parameter"

The "atmosphere parameter" for the earth's atmospheric lever will, of course, be the same for all rockets if standard conditions are assumed and its value depends only on the height the rocket has reached above the starting point of the flight. The "form parameter" is determined by the shape of the curve of C_D against B. This curve will be altered chiefly by the geometrical shape of the shell although it is also affected by the change in skin friction coefficient due to the change in Reynold's Number. As long as the rocket belongs to a family that has the same geometrical shape, which implies the same nose shape and the same l/d ratio, that is, the ratio of the length of the shell to the maximum diameter, the "form parameter" can be assumed to remain constant.

It is thus seen that the performance curves calculated for a typical rocket will also hold for a whole family of rockets determined by the values of the "factors" and of the "parameters" of the typical rocket and the design of a rocket to meet certain prescribed requirements is greatly simplified. Furthermore, for a good rocket form design the variation of $\dfrac{C_D}{C_D^*}B^2$, the form parameter, at the same values of B is small. Also, the deviation from standard atmospheric characteristics cannot be very large. Then, in view of the fairly accurate but not exact basic assumption of constant thrust, it is justified to use the same data for these two parameters for all cases. Thus, the performance problem is further simplified and depends only upon the four performance factors $\dfrac{a_0}{g_0}$, c, Λ and ζ.

Conclusion

This study shows that a sounding rocket propelled by successive impulses can theoretically reach heights of much use to those interested in obtaining data on the structure of the atmosphere and extra-terrestrial phenomena if a propelling unit gives an exhaust velocity of 7 000 ft per second or more.

The possibility of obtaining such exhaust velocities depends on two factors: first, the ability of the motor to transform efficiently the heat energy of the fuel into kinetic energy of the exhaust gases, and secondly, the amount of heat energy that can be liberated from the fuel. In an actual motor which burns its fuel at constant volume by igniting a powder charge in the combustion chamber the ratio of the chamber pressure to the outlet pressure drops from a maximum at the beginning of the expansion to zero at the end of the process. It is not possible to design a nozzle that will expand the products of combustion smoothly during the whole process. Therefore, the attainable efficiency must be less than that of a corresponding "constant pressure" motor which has a mixture of combustibles, e.g. gasoline and liquid oxygen, fed continuously into the combustion chamber at a constant pressure equal to the maximum pressure of the "constant volume" motor. However, very high maximum chamber pressures (up to 60 000 lbs per sq. in.) can be developed in a motor using constant volume burning, while the chamber pressure of a motor using constant pressure burning is limited to much lower pressures by the difficulty of feeding the combustibles. Therefore, the efficiency that can be obtained from motors using either of these processes should not be very different. As to the heat that can be liberated per unit mass of fuel, the present fuel, such as nitro-cellulose powder for a constant volume motor, is much lower than the liquid combustibles such as gasoline and oxygen for a constant pressure motor.

These considerations indicate that the attainable exhaust velocity of a "constant volume" motor for propulsion by successive impulses will probably be lower than that of a "constant pressure" motor for supplying a continuous thrust. This is the reason why many experimenters abandoned the "constant volume" motor and turned to the "constant pressure" motor, the so-called liquid propellant motor. Theoretically, this defect of the "constant volume" motor can be compensated if a small total number of impulses (c. f. Fig. 4. 3) is used. However, the use of few impulses is of doubtful practical value because the resulting extreme accelerations will be harmful to instruments carried and will necessitate a heavier construction of the rocket.

However, even with the lower exhaust velocities of the "constant volume" motor it is shown by the analysis in this paper that with the exhaust velocity of 7 000 feet/sec. obtained experimentally by R. H. Goddard[1] it should be possible to build a poder rocket capable of rising above 100 000 feet. Thus it seems to the author that a rocket propelled by successive impulses has useful possibilities and further experimental work is justified.

References on PART (Ⅳ)

[1] Goddard R H. A Method of Reaching Extreme Altitudes, Smithsonian Miscellaneous Collections. 1919, 71(2).

[2] Damblenc L. Les fusées autopropulsives a explosifs, L'Aerophile. 1935,43: 205 - 209, 241 - 247.

[3] Bartocci A. Le éscursioni in altezza col motore a reazione. L'Aerotecnica, 1933, 13:1646 - 1666.

[4] Breguet L, Devillers R. L'Aviation superatmosphérique les aérodynes propulsées par reaction directe. La Science Aerienne, 1936,5: 183 - 222.

[5] Ley W, Schaefer H. Les fusées volantes météorologiqus. L'Aérophile, 1936,44:228 - 232.

[6] Malina. F J, Smith A M O. Analysis of the Sounding Rocket. Journal of the Aeronautical Sciences, 1938, 5: 199 - 202.

[7] Whittaker, Watson. Modern Analysis. Cambridge, 4th Edition, 1927: 246 - 247.

[8] Davis H T. Tables of the Higher Mathematical Functions. Principia Press, 1st Edition, 1933:277 - 364.

Flight Analysis of a Sounding Rocket with Special Reference to Propulsion by Successive Impulses

Hsue-shen Tsien and Frank J. Malina
(California Institute of Technology)

Summary

In Part Ⅰ of this paper an exact solution of the problem of determining the height reached by a body in vertical flight in vacuo propelled by successive impulses is presented. On the basis of this analysis it is concluded that a rocket propelled by successive impulses — the impulses being obtained, for example, from rapidly burning powder — can theoretically reach much greater heights than is possible by sounding balloons and, therefore, further experimental research is justified. In Part Ⅱ the effect of the variation of the acceleration of gravity with height above sea level on the flight performance of a sounding rocket is analyzed. For a 1 000 mile sounding rocket the decrease in gravitational pull accounts for a 25 percent increase in the maximum height reached over that calculated on the basis of a constant gravitational acceleration. In Part Ⅲ the fundamental performance equation for flight of a sounding rocket in air is expressed in terms of dimensionless parameters and factors and their physical significance is discussed. Finally, in Part Ⅳ the theory of the preceding sections is applied to a specific case of a sounding rocket propelled by successive impulses which are supplied by a reloading type of powder rocket motor.

Introduction

In 1919, R. H. Goddard[1] published the historically important paper which suggested the use of nitro-cellulose powder as a propellant for raising a sounding rocket to altitudes beyond the range of sounding balloons. To determine the feasibility of this propellant, a series of experiments had been carried out and it was found that a thermal efficiency of 50 percent could be expected if the powder was exploded in a properly designed chamber and the resulting gases were allowed to escape at high velocity through an expanding nozzle. In 1931, R. Tilling used a mixture of potassium chlorate and naphthalene as propellant and actually reached an altitude of 6 600 feet. More recently, L. Damblanc[2] made static tests with a slow burning black powder and from these estimated that a height of 10 000 feet could be reached using a two-step arrangement. The results so far reported offer an incentive to further analysis.

The propulsion obtained by the use of powder charges in a rocket motor, which are

---

Received July 6, 1938.

Journal of the Aeronautical Sciences, vol. 6, pp. 50 – 58, 1939.

supplied by a reloading mechanism, is referred to in this paper as propulsion by successive impulses. This type of propulsion is essentially different from the type of propulsion made available by a rocket motor which continuously burns a combustible mixture at constant pressure. The thrust of the latter rocket motor is nearly constant, whereas in the former case, due to the rapidness of the combustion of the powder charges, the propulsion consists of a series of uniform impulses.

The effect of decreasing gravitational acceleration on the maximum height reached by a rocket has been considered by A. Bartocci[3]. However, he assumes that the rocket itself has a constant acceleration during powered flight. L. Breguet and R. Devillers[4] also considered the effect of the variation of g. To simplify the analysis, they assumed that the acceleration of the rocket was equal to a constant multiple of g. Since the sounding rocket will be propelled by a nearly constant thrust or a uniform rate of successive impulses, in Part II the authors have studied the problem anew according to this mode of propulsion.

When the sounding rocket is ascending through the air the maximum height reached is less than that reached for flight in vacuo. Recently, studies have been made of the problem by W. Ley and H. Schaefer[5] and by F. J. Malina and A. M. O. Smith[6]. On the basis of the latter study a group of new performance parameters and factors have been isolated from the general performance equation, and these are discussed in Part III.

Notation

Referring to Fig. 1, the following notation has been used throughout the paper:

w = weight of propellant and propellant container ejected per impulse, lbs

k = ratio of propellant container weight to sum of container and propellant weight ejected per impulse

$\lambda = 1 - k$

W_0 = initial weight of rocket, lbs

M_0 = initial mass of the rocket, slugs

W_r = instantaneous weight of rocket, lbs

ζ = ratio of initial weight of propellants to initial total weight of a rocket propelled by constant thrust

ζ_1 = ratio of initial weight of propellants to initial total weight of a rocket propelled by successive impulses

$\zeta_1' = \zeta_1/\lambda$

n = number of impulses per second

N = total number of impulses occurring during powered flight

t = time, sec

Δt = interval between impulses, sec

a_0 = initial acceleration imparted to rocket, ft/sec^2

g_0 = acceleration of gravity at the starting point of flight, ft/sec^2

g = acceleration of gravity above the starting point of flight, ft/sec^2

c = effective exhaust velocity of ejected propellant, ft/sec

v = instantaneous velocity, ft/sec

Δv_r = velocity imparted to rocket by the rth impulse, ft/sec

v_r' = velocity at end of the rth interval, ft/sec

v_{s_c} = velocity of sound corresponding to atmospheric conditions at the starting point of flight, ft/sec

v_s = velocity of sound corresponding to atmospheric conditions at height reached by the rocket at time t, ft/sec

B = Mach's number = v/v_s

$V_{max.}$ = velocity of rocket at start of coasting flight, ft/sec

$V_{max.0}$ = velocity of rocket at start of coasting flight if g is constant and equal to g_0, ft/sec

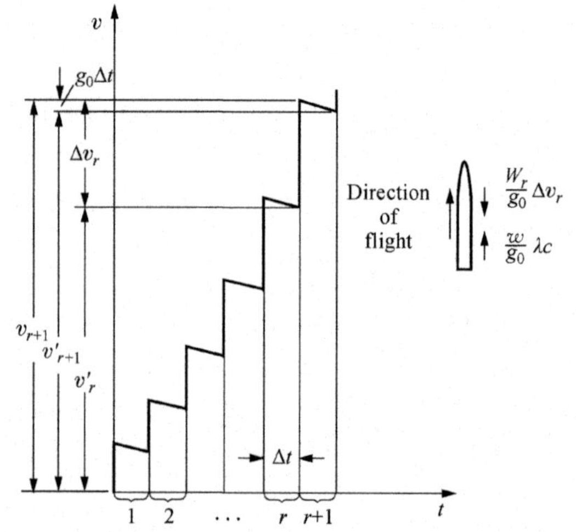

Fig. 1 Variation of the flight velocity v with the time t

h = altitude above sea level, ft

h_r = height reached at the beginning of the rth interval, ft

h'_r = height reached at the end of the rth interval, ft

H_P = height traveled during powered flight, ft

H_{P_0} = height traveled during powered flight, if g is constant and equal to g_0, ft

H_C = height traveled during coasting flight, ft

H_{C_0} = height traveled during coasting flight, if g is constant and equal to g_0, ft

$H_{max.}$ = height traveled during powered flight and coasting flight, ft

$H_{max.0}$ = height traveled during powered flight and coasting flight, if g is constant and equal to g_0, ft

R = radius of earth, 2.088×10^8, ft

D = drag on rocket due to air resistance, lbs

C_D = drag coefficient of rocket

C_D^* = drag coefficient of rocket at velocity of sound

Λ = drag-weight factor (discussed in the section on the effect of air resistance)

ρ_0 = mass density of air at the starting point of flight, slugs/ft^3

σ = ratio of air density at altitude to air density at the starting point of flight

T = absolute temperature of atmosphere at the height reached by the rocket at time t, °F

T_0 = absolute temperature of atmosphere at the starting point of flight, °F

A = largest cross-sectional area of rocket shell, ft^2

d = largest diameter of rocket shell, ft

l = length of rocket shell, ft

I

An approximate method of calculating the maximum height reached by a rocket propelled by powder was developed by R. H. Goddard[1]. To simplify the analysis a continuous loss of mass was assumed and the problem was so stated, that a minimum mass of propellant necessary to lift one pound of mass at the end of the flight to any desired height was determined. However, if high-powered powder is used, the rate of burning is so rapid that the propulsive action is instantaneous. The rocket is thus acted upon by an impulse rather than by a constant thrust.

In the following analysis, it has, therefore, been assumed that the propulsive force is an impulsive force, i. e., the force acts for such a brief interval of time that the rocket does not change its position during the application of the force, although its velocity and its momentum receive a finite change. If the combustion process of the propulsive unit takes place at constant volume this assumption is justified. Further, a study of interior ballistics of small arms reveals that the period between the ignition of the powder charge and the bullet's arrival at the end of a two-foot barrel is of the order of 14 ten-thousandths of a second. If the gases are not restrained and their travel through the burning chamber and the nozzle is of much shorter length, as is the case for the rocket motor, even shorter periods of duration of action can be expected.

Assuming that the propulsive force acts as an impulse, then the motion of the rocket can be calculated by Newton's third law, which states that impulses between two bodies are equal and opposite. Hence, equating the momentum of the exhaust gases to the momentum imparted to the rocket, using the quantities defined in the list of notation and referring to Fig. 1, the following relation can be written for flight in vacuo:

$$(\lambda w/g_0)c = (W_r/g_0)\Delta v_r \tag{1}$$

where

$$\lambda = 1 - k \text{ and } W_r = W_0 - rw \tag{2}$$

or

$$\Delta v_r = \frac{w\lambda c}{W_0 - rw} = \frac{\zeta_1'\lambda c}{N}\left(\frac{1}{1 - r\zeta_1'/N}\right) \tag{3}$$

where

$$\zeta_1' = wN/W_0 = \zeta_1/\lambda$$

During the interval between impulses, Δt, the velocity is reduced by the action of gravity so that at the end of the rth interval, the velocity of the rocket will be

$$v_r' = v_r - g_0\Delta t = v'_{r-1} + \Delta v_r - g_0\Delta t \tag{4}$$

Therefore

$$v_r' = \sum_{s=1}^{s=r} \Delta v_s - r g_0 \Delta t \tag{5}$$

Substituting for Δv_s from Eq. (3)

$$v_r' = \frac{\zeta_1' \lambda c}{N} \sum_{s=1}^{r} \frac{1}{1 - s(\zeta_1'/N)} - r g_0 \Delta t \tag{6}$$

or

$$v_r' = \frac{\zeta_1' \lambda c}{N} S_1 - r g_0 \Delta t$$

where

$$S_1 = \sum_{s=1}^{r} \frac{1}{1 - s(\zeta_1'/N)}$$

The height gained during each interval will be represented by the area under the velocity curve in the interval, or

$$h_r' - h_r = v_r' \Delta t + (1/2) g_0 (\Delta t)^2 \tag{7}$$

Therefore, at the end of the Nth interval which is the end of the powered flight, the height will be

$$H_{P_0} = \sum_{r=1}^{N} v_r' \Delta t + (N/2) g_0 (\Delta t)^2$$

Substituting for v_r' its value in Eq. (6)

$$\begin{aligned} H_{P_0} &= \sum_{r=1}^{N} \Delta t \left\{ \frac{\zeta_1' \lambda c}{N} \sum_{s=1}^{r} \frac{1}{1 - s(\zeta_1'/N)} - r g_0 \Delta t \right\} + \frac{N}{2} g_0 (\Delta t)^2 \\ &= \frac{\zeta_1' \lambda c}{N} \Delta t \sum_{r=1}^{N} \frac{N + 1 - r}{1 - r(\zeta_1'/N)} - g_0 (\Delta t)^2 \sum_{r=1}^{N} (r - 1/2) \\ &= \frac{\zeta_1' \lambda c}{N} \Delta t S_2 - \frac{N^2}{2} g_0 (\Delta t)^2 \end{aligned} \tag{8}$$

where

$$S_2 = \sum_{r=1}^{N} \frac{N + 1 - r}{1 - r(\zeta_1'/N)}$$

The maximum height reached will be the sum of the height at the end of powered flight and the height traveled during coasting or

$$H_{\max.0} = H_{P_0} + H_{C_0} = H_{P_0} + V_{\max.0}^2/(2g_0) \tag{9}$$

To calculate the maximum height one has first to evaluate the sums S_1 and S_2.* Noting that

* The authors wish to thank Prof. H. Bateman for his suggestion of this method of summing.

$$\frac{1}{1-s(\zeta_1'/N)} = \int_0^\infty \mathrm{e}^{-x\{1-s(\zeta_1'/N)\}}\,\mathrm{d}x$$

S_1 can be written in the form

$$S_1 = \int_0^\infty \mathrm{e}^{-x} \sum_{s=1}^{r} (\mathrm{e}^{x\zeta_1'/N})^s \mathrm{d}x = \int_0^\infty \mathrm{e}^{-x}\,\frac{\mathrm{e}^{x\zeta_1'/N} - \mathrm{e}^{(r+1)\zeta_1'/N}}{1-\mathrm{e}^{x\zeta_1'/N}}\mathrm{d}x$$

Putting

$x = Ny/\zeta_1'$, the above integral becomes

$$S_1 = -\frac{N}{\zeta_1'}\int_0^\infty \mathrm{e}^{-Ny/\zeta_1'}\left(\frac{1-\mathrm{e}^{ry}}{1-\mathrm{e}^{-y}}\right)\mathrm{d}y = \frac{N}{\zeta_1'}\left\{\psi\left(\frac{N}{\zeta_1'}\right) - \psi\left(\frac{N}{\zeta_1'} - r\right)\right\} \tag{10}$$

where $\psi(z) = (\mathrm{d}/\mathrm{d}z)\{\log \Gamma(z)\}$. (The reader is referred to references 7 and 8 for detailed information on this function.) Similarly, S_2 can be summed as

$$S_2 = \frac{N}{\zeta_1'}\left\{N - \left[\frac{N}{\zeta_1'} - (N+1)\right]\times\left[\psi\left(\frac{N}{\zeta_1'}\right) - \psi\left(\frac{N}{\zeta_1'} - N\right)\right]\right\} \tag{11}$$

Substituting Eqs. (10) and (11) into Eqs. (6) and (8), and then into Eq. (9)

$$H_{\max.0} = \frac{\lambda^2 c^2}{2g_0}\Psi^2 - \frac{\lambda c}{n}\left\{\left(\frac{N}{\zeta_1'} - 1\right)\Psi - N\right\} \tag{12}$$

where $n = 1/\Delta t$, and $\Psi = \psi(N/\zeta_1') - \psi[(N/\zeta_1') - N]$.

For convenience of calculation in Fig. 2 the quantity Ψ is plotted against N for different

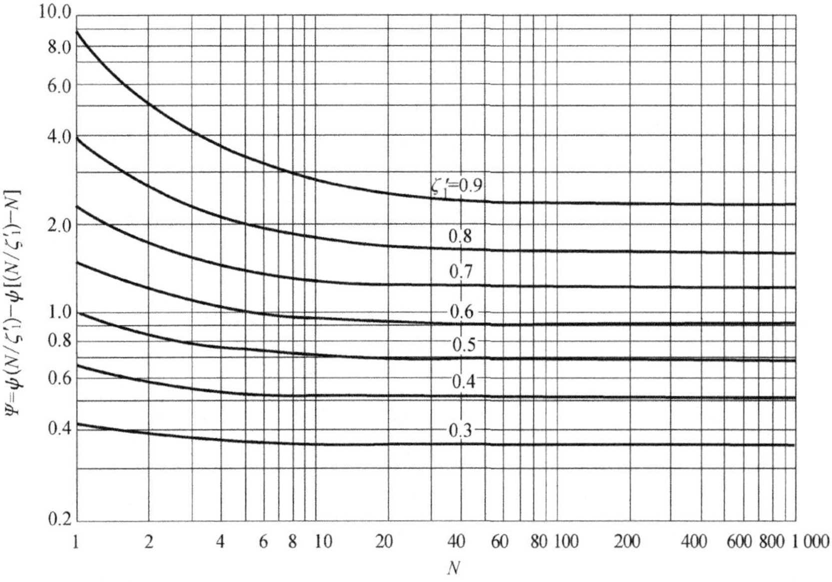

Fig. 2 Variation of Ψ with N for various values of ζ_1'

values of ζ_1'.

It can easily be shown that when $N = 1$

$$\Psi = \zeta_1'/(1 - \zeta_1')$$

so that Eq. (12) reduces to

$$H_{\max.0} = \frac{\lambda^2 c^2}{2g_0}\left(\frac{\zeta_1'}{1 - \zeta_1'}\right)^2 \tag{12a}$$

Also, as $N \to \infty$, $\Psi \to -\log(1 - \zeta_1')$, thus Eq. (12) reduces to

$$H_{\max.0} = \frac{\lambda^2 c^2}{g_0}\left\{\frac{[\log(1 - \zeta_1')]^2}{2} + \frac{\zeta_1' + \log(1 - \zeta_1')}{(a_0/g_0) + 1}\right\} \tag{12b}$$

where

$$a_0 = \frac{\zeta_1'}{N} n\lambda c - g_0$$

$$= \frac{nw\lambda c}{W_0} - g_0 \tag{12c}$$

The quantity a_0 can be considered as the initial acceleration of the rocket if $N \to \infty$. It is interesting to notice that Eq. (12b) is the equation obtained by Malina and Smith[6] for calculating the maximum height of a constant thrust rocket, as expected.

Figure 3 shows the variation of $H_{\max.0}\, g_0/(\lambda^2 c^2)$ with $n\lambda c/g_0$ for different values of ζ_1' and for four values of N. These curves show that when the total number of impulses, N, becomes larger than 100, the maximum height reached is imperceptibly changed by increasing the number.

At this point it is necessary to discuss the similarity existing between a rocket propelled by successive impulses and a rocket propelled by constant thrust. The former loses not only the mass of the propellant, but also the containers for the individual charges. The difference in effect on the rocket between the propellant and its containers is that the propellant has an effective exhaust velocity, c, while the ejected containers leave the rocket without appreciable velocity. The propulsive action, however, will remain the same if the whole cartridge, that is, the propellant charge and its container, is considered wholly as propellant, but leaving the motor at a reduced effective exhaust velocity λc. The rocket propelled by constant thrust loses only the mass equal to the propellant carried; therefore, it can be said to be equivalent to the "successive impulses" rocket if its effective exhaust velocity and its total mass of propellant are equal, respectively, to the reduced exhaust velocity and to the sum of the masses of all the containers of the "successive impulses" rocket. In other words, c is equal to λc and ζ is equal to ζ_1'.

$$\frac{H_{max}g_0}{\lambda^2c^2}=\frac{1}{2}\Psi'^2-\frac{g_0}{n\lambda c}\left[\left(\frac{N}{\zeta_1'}-1\right)\Psi-N\right]$$

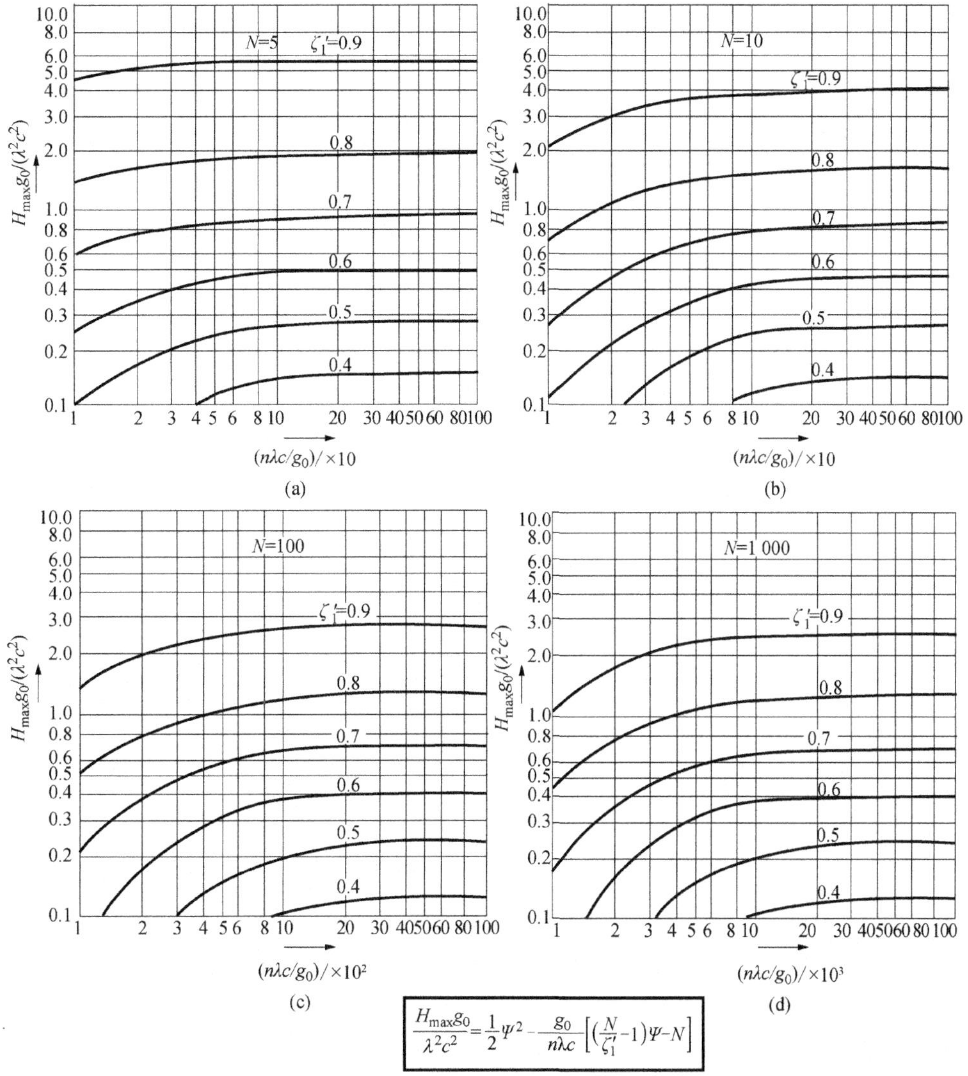

Fig. 3 $H_{max,0}\,g_0/(\lambda^2c^2)$ with nc/g_0 for various values of ζ_1'

In Table 1 the heights for four cases have been calculated to illustrate the effect of the exhaust gas velocity and the total number of impulses given to a rocket whose weight ratio. ζ_1' is 0.70. It will be noticed that for flight in vacuo a greater height will be reached if a smaller number of impulses is employed. The lower portion of the table shows the maximum height reached by an equivalent "constant thrust" rocket for the same four cases with the initial acceleration given by Eq. (12c). The close agreement between the maximum height reached by use of successive impulses, when the total number of impulses exceeds 100, and that reached by the use of constant thrust simplifies the solution of the problem of decreasing acceleration of gravity with height, and enables prediction for flight with air resistance to be based on the results obtained for a rocket propelled by constant thrust (cf. reference 6). Those problems are considered in the following sections.

Table 1

Successive impulses

$$H_{\max,0}=\frac{\lambda^2c^2}{g_0}\left\{\frac{1}{2}\Psi^2-\frac{1}{n\lambda c/g_0}\left[\left(\frac{N}{\zeta_1'}-1\right)\Psi-N\right]\right\}$$

Case	λc ft/sec	ζ_1'	N	n Impulses/sec	$H_{\max,0}$ ft
1	10 000	0.70	326	3	1 472 000
2	10 000	0.70	10	0.092	1 686 000
3	7 000	0.70	326	3	560 000
4	7 000	0.70	10	0.092	676 000

Constant thrust

$$H_{\max,0}=\frac{c^2}{g_0}\left\{\frac{1}{2}[\log(1-\zeta)]^2-\frac{1}{a_0/g_0+1}[\log(1-\zeta)+\zeta]\right\}$$

Case	c	ζ	a_0	$H_{\max,0}$ ft
1	10 000	0.70	32.2	1 468 000
2	10 000	0.70	32.2	1 468 000
3	7 000	0.70	12.9	555 000
4	7 000	0.70	12.9	555 000

Ⅱ

It is well known that the acceleration of gravity decreases with the height above the earth's surface according to the following relation:

$$g=g_0[R/(R+h)]^2 \tag{13}$$

At an altitude of 1 000 miles the acceleration is only 0.64 times that at sea level. Therefore, for flights up to such altitudes the assumption that g is approximately constant is no longer valid. It was shown by Malina and Smith[6] that a three-step rocket could theoretically reach such an altitude. Thus it is interesting to see how the decrease of g will increase the maximum height reached by the rocket.

First, the effect on powered flight in vacuo will be considered and then on coasting flight in vacuo. For powered flight the analysis is based on the assumption that the thrust is constant. However, the results can be applied to the case of propulsion by successive impulses if the total number of impulses, N, exceeds 100 as was justified in the previous section.

The equivalent mass of gas flowing per second continuously for the case of successive impulses is

$$wn/g_0=m \tag{14}$$

Assuming that the rocket starts from rest at sea level, the equation of motion in vacuo is

$$\frac{d^2h}{dt^2} = -g_0\left(1+\frac{h}{R}\right)^{-2} + \frac{mc/M_0}{1-(m/M_0)t} \tag{15}$$

This is a non-linear differential equation which cannot be solved by usual means. However, for all practical purposes the ratio h/R during powered flight is much smaller than 1, therefore, only first order terms in h/R occurring in the expansions need to be retained. This approximation linearizes the equation to the form

$$\frac{d^2h}{dt^2} = g_0\left(\frac{2h}{R}-1\right) + \frac{mc/M_0}{1-(m/M_0)t} \tag{16}$$

The solution of this equation with the initial condition that $h=0$ and $dh/dt=0$ when $t=0$ is

$$h = \frac{R}{2}\left\{1-\cosh\sqrt{\frac{2g_0}{R}}t\right\} + \frac{c}{2\sqrt{2g_0/R}}\left\{e^{\xi u}\int_{\xi}^{\xi u}\frac{e^{-x}dx}{x} - e^{-\xi u}\int_{\xi}^{\xi u}\frac{e^{x}dx}{x}\right\} \tag{17}$$

where $\xi = \sqrt{2g_0/R}M_0/m$ and $u = 1-(m/M_0)t$. At the end of the powered flight, the time is

$$t = t_P = M_0\zeta/m \tag{18}$$

Therefore, the height at the end of the powered flight is

$$H_P = \frac{R}{2}\left\{1-\cosh\sqrt{\frac{2g_0}{R}}\frac{M_0\zeta}{m}\right\} + \frac{c}{2}\sqrt{\frac{R}{2g_0}}\left\{e^{\xi(1-\zeta)}\int_{\xi}^{\xi(1-\zeta)}\frac{e^{-x}dx}{x} - e^{-\xi(1-\zeta)}\int_{\xi}^{\xi(1-\zeta)}\frac{e^{x}dx}{x}\right\} \tag{19}$$

If the hyperbolic cosine term and the integrals are expanded and only first order terms in $1/R$ are retained in consistency with the linearization of Eq. (15), the equation reduces to

$$\begin{aligned} H_P = &-\left\{\frac{\zeta^2 g_0}{2}\left(\frac{W_0}{w}\right)^2 + \frac{\zeta^4 g_0^2}{12R}\left(\frac{W_0}{w}\right)^4\right\} + c\frac{W_0}{w}\{(1-\zeta)\log(1-\zeta)+\zeta\} + \\ &\frac{cg_0}{18R}\left(\frac{W_0}{w}\right)^3\{6(1-\zeta)^3\log(1-\zeta)+\zeta(11\zeta^2-15\zeta+6)\} \\ = &H_{P_0} + \frac{g_0}{6R}\left(\frac{W_0}{w}\right)^3\left\{\frac{c}{3}[6(1-\zeta)^3\log(1-\zeta)+\zeta(11\zeta^2-15\zeta+6)] - \frac{\zeta^4 g_0}{2}\left(\frac{W_0}{w}\right)\right\} \end{aligned} \tag{20}$$

Differentiating Eq. (17), and substituting the relation of Eq. (18), the maximum velocity at the end of powered flight is

$$V_{\max.} = -\sqrt{\frac{Rg_0}{2}}\sinh\sqrt{\frac{2g_0}{R}}\frac{M_0\zeta}{m} - \frac{c}{2}\left\{e^{\xi(1-\zeta)}\int_{\xi}^{\xi(1-\zeta)}\frac{e^{-x}dx}{x} + e^{-\xi(1-\zeta)}\int_{\xi}^{\xi(1-\zeta)}\frac{e^{x}dx}{x}\right\} \tag{21}$$

Again expanding and retaining only first order terms in $1/R$, Eq. (21) reduces to

$$\begin{aligned} V_{\max.} = &-\left\{\left(\frac{W_0}{w}\right)\zeta + \frac{1}{3}\frac{\zeta^3 g_0^2}{R}\left(\frac{W_0}{w}\right)^3\right\} - \\ &\left\{c\log(1-\zeta) + \frac{cg_0}{2R}\left(\frac{W_0}{w}\right)^3[2(1-\zeta)^2\log(1-\zeta)+2\zeta-3\zeta^2]\right\} \\ = &V_{\max.0} - \frac{g_0(W_0/w)^2}{R}\left\{\frac{\zeta^3 g_0(W_0/w)}{3} + \frac{c}{2}[2(1-\zeta)^2\log(1-\zeta)+2\zeta-3\zeta^2]\right\} \end{aligned} \tag{22}$$

It is seen that the second terms of Eqs. (20) and (22) are the corrections to be applied to H_{P_0} and $V_{max.0}$ to account for the variation of the acceleration of gravity. Since both corrections are first order approximations they can be expected to apply approximately also to the case of successive impulses, even when the total number of impulses is less than 100.

The coasting height reached by the rocket due to its velocity at the end of powered flight can be obtained by equating the increase of potential energy during coasting flight to the kinetic energy at the end of powered flight. Thus,

$$\frac{1}{2}V_{max.}^2 = g_0\int_{H_P}^{H_P+H_C}\frac{dh}{[1+(h/R)]^2}$$

or

$$H_C = (H_P+R)\left\{\frac{1}{1-\dfrac{V_{max.}^2}{2g_0[R/(H_P+R)]^2(H_P+R)}}-1\right\} \tag{23}$$

Putting $V_{max.}^2/2g_0[R/(H_P+R)]^2 = H_{C_0}$, which is the coasting height obtained by assuming a constant gravitational acceleration of the value equal to that at the height H_P, i.e., the height where coasting starts, then Eq. (23) can be written

$$H_C = (H_P+R)\left\{\frac{1}{1-H_{C_0}/(H_P+R)}-1\right\}$$

Upon expanding the second term, this equation becomes,

$$H_C = H_{C_0}\left\{1+\left(\frac{H_{C_0}}{H_P+R}\right)+\left(\frac{H_{C_0}}{H_P+R}\right)^2+\left(\frac{H_{C_0}}{H_P+R}\right)^3+\cdots\right\} \tag{24}$$

This equation shows that if the coasting flight starts from sea level, and if the maximum height reached is about 1 000 miles, the increase due to the decrease in g is over 25 percent.

Ⅲ

When the sounding rocket is ascending through the atmosphere instead of in vacuo, air resistance comes into play, causing the acceleration of the rocket to be reduced, which decreases the maximum height reached. Since air resistance increases with the air density and with the square of the flight velocity, it is desirable to keep the rocket from ascending too rapidly through the lower layers of the atmosphere where the air density is high. For this reason the optimum initial acceleration will no longer be infinite as shown by Eq. (12b). For the case of constant thrust Malina and Smith[6] have found that the optimum acceleration is around 30 ft. per sec[2]. For a total number of impulses greater than 100, the difference between propulsion by successive impulses and by constant thrust is very small, so one may expect the above optimum value of initial acceleration to hold for both cases of propulsion.

The actual amount of reduction in maximum height due to air resistance can be calculated by the method of step-by-step integration, if fair accuracy is desired. This integration is carried out by using the fundamental equation for vertical rocket flight which, as given in the previous paper[6] is

$$\frac{d^2h}{dt^2}=a=-g+\frac{a_0+g_0}{1-t(a_0+g_0)/c}-\frac{g_0\rho_0\sigma v^2}{2\left[1-\frac{t(a_0+g_0)}{c}\right]}\frac{C_DA}{W_0} \tag{25}$$

The significance of the ratio C_DA/W_0 was discussed in that paper[6]. Greater significance can, however, be attached to the various terms in the equation if it is transformed into the non-dimensional form

$$\frac{a}{g_0}=-\frac{g}{g_0}+\frac{\frac{a_0}{g_0}+1}{1-\frac{g_0t}{c}\left(\frac{a_0}{g_0}+1\right)}-\frac{\left(\sigma\frac{T}{T_0}\right)\left(\frac{C_D}{C_D^*}B^2\right)\Lambda}{1-\frac{g_0t}{c}\left(\frac{a_0}{g_0}+1\right)} \tag{26}$$

where

$$\Lambda=\frac{\frac{\rho_0}{2}C_D^*Av_{s_0}^2}{W_0}$$

In Eq. (26) appear two types of significant quantities. First, quantities, called "factors," which are constant for any given family of rockets, and second, two quantities called "parameters," one of which is characteristic for a given family of rockets but changes in value along the flight path, and one which depends on the physical properties of the atmosphere. Thus there are the following factors:

a_0/g_0 = ratio of initial acceleration to g_0 ~ "initial acceleration factor," a motor characteristic.

c = exhaust velocity in ft/sec ~ "exhaust velocity factor," a motor characteristic.

Λ = "drag-weight factor."

ζ = ratio of weight of combustibles to total initial weight of the rocket ~ "loading factor."

The first two factors, i. e., the "initial acceleration factor" and the "exhaust velocity factor," determine the characteristics of the propelling unit for a given family of rockets while the "drag-weight factor" and the "loading factor" determine the physical dimensions of the rockets. The "drag-weight factor" is a ratio of the drag of the rocket at sea level when traveling with the velocity of sound to the initial weight of the rocket. Since for any given family of rocket shapes the only terms in the factor which can be varied are the maximum cross-sectional area, A, and the initial weight, W_0, it is clear that if the initial weight is doubled then the cross-sectional area must also be doubled to keep the factor the same. The "loading factor" needs to be discussed in some detail as it does not appear explicitly in Eq. (26). Eq. (26) is a differential equation of the flight path which is satisfied at every point along the flight path. The loading factor ζ comes in only when this equation is integrated and the limits of integration are put in. For example, consider two rockets with identical performance factors and parameters, with the exception that one has a ζ of 0.90 and the other has a ζ of 0.50. The flight path of the two rockets will be identical up to the time that 0.50 times the initial weight of the rockets is used up as combustibles. At this point the rocket having a ζ of 0.50 will begin to decelerate while the one having a ζ of 0.90 will continue to accelerate until the remaining combustibles are used up. It is thus seen that the value of ζ controls the maximum height reached.

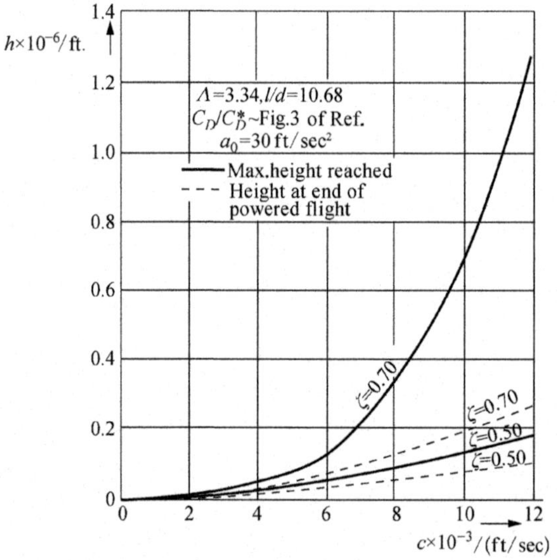

Fig. 4 Height reached in air at the end of powered flight and maximum height reached versus the exhaust velocity, c, for ζ of 0.50 and 0.70

The two performance parameters are:

$\sigma T/T_0$ ~physical properties of the atmosphere called the "atmosphere parameter."

C_D/C_D^* ~aerodynamic properties of the rocket called the "form parameter."

The "atmosphere parameter" for the earth's atmospheric layer will, of course, be the same for all rockets if standard conditions are assumed and its value depends only on the height the rocket has reached above the starting point of the flight. The "form parameter" is determined by the shape of the curve of C_D against B. This curve will be altered chiefly by the geometrical shape of the shell although it is also effected by the change in skin friction coefficient due to the change in Reynolds Number. As long as the rocket belongs to a family that has the same geometrical shape, which implies the same nose shape and the same l/d ratio, that is, the ratio of the length of the shell to the maximum diameter, the "form parameter" can be assumed to remain constant.

It is thus seen that the performance curves calculated for a typical rocket will also hold for a whole family of rockets determined by the values of the "factors" and of the "parameters" of the typical rocket and the design of a rocket to meet certain prescribed requirements is greatly simplified. Furthermore, for a good design of rocket form the variation of C_D/C_D^*, the form parameter, at the same values of B is small. Also, the deviation from standard atmospheric characteristics cannot be very large. Then, in view of the fairly accurate but not exact basic assumption of constant thrust, it is justified to use the same data for these two parameters for all cases. Thus, the performance problem is further simplified and depends only upon the four performance factors a_0/g_0, c, Λ, and ζ.

Ⅳ

The use of the results of the analysis developed in the preceding parts of this paper is illustrated in this section by the calculation of the performance of a rocket propelled by successive impulses, e.g., a powder rocket. The performance of the powder rocket can be predicted from the results obtained for an equivalent rocket propelled by constant thrust, provided the powder rocket is acted upon by more than 100 impulses.

In making use of the equivalence existing between the two methods of propulsion it is necessary that the following quantities discussed in Part Ⅰ and Part Ⅲ be identical for the two cases: initial acceleration factor, a_0/g_0; exhaust velocity factor, c, for constant thrust and λc for successive impulses; drag-weight factor, Λ; loading factor, ζ, for constant thrust, and ζ_1' for successive impulses; atmosphere parameter $\sigma T/T_0$; form parameter, C_D/C_D^*; and slenderness ratio l/d.

For the example the following characteristics of the powder rocket are assumed:

Powder Rocket

$W_0 = 85$ lbs	$c = 7\,000$ ft/sec
$\zeta_1 = 0.658$	$\lambda c = 6\,580$ ft/sec
weight of powder per shot = 0.108 lbs	$\zeta_1' = \zeta_1/\lambda = 0.70$
weight of cartridge = 0.007 lbs	$d = 0.75$ ft
$w = 0.115$ lbs/shot	$V = 3.34$
$k = 0.06$	C_D/C_D^* ~Figure 3 of reference 6
$\lambda = (l-k) = 0.94$	$l/d = 10.68$
$n = 7$	$\sigma T/T_0$ ~standard atmosphere, starting point — sea level
$N = 518$	

Equivalent Constant Thrust Rocket

$W_0 = 85$ lbs	$\Lambda = 3.34$
$\zeta = 0.70$	C_D/C_D^* ~Figure 3 of reference 6
$c = 6\,580$ ft/sec	$l/d = 10.68$
$a_0 = (\zeta_1' n\lambda c/N) - g_0 = 30.0$ ft/sec^2	$\sigma T/T_0$ ~standard atmosphere, starting point — sea level
$d = 0.75$ ft	

The data for the equivalent rocket are now complete and it is found from Fig. 4 that $H_{\max.} = 162\,000$ feet.

This is the maximum height reached by the rocket assuming that the acceleration of gravity does not vary with height. This assumption was shown in the preceding section to be practically valid for maximum heights up to about 800 000 feet.

Since the sounding rocket at the end of powered flight does not reach heights at which the acceleration of gravity is appreciably decreased, the heights calculated by Malina and Smith[6] can be used. In Fig. 4 the height at the end of powered flight is plotted against the exhaust velocity c for $\zeta = 0.70$ and 0.50 for flight with air resistance. If the height at the end of powered flight is subtracted from the maximum height reached, the coasting height is obtained. This height may be perceptibly affected by the decreasing acceleration of gravity. Using Eq. (24) the corrected coasting height can be calculated. For a sounding rocket propelled by constant thrust of the same dimensions as above but with an exhaust velocity of 12 000 ft/sec. it is found from Fig. 4 that

$$H_{\max.0} = 1\ 270\ 000 \text{ ft}$$

From Fig. 4

$$H_{\mathrm{P}} = 265\ 000 \text{ ft}$$

so that

$$H_{\mathrm{C}_0} = H_{\max.0} - H_{\mathrm{P}} = 1\ 005\ 000 \text{ ft}$$

Using Eq. (24) the corrected coasting height is

$$H_{\mathrm{C}} = 1\ 005\ 000\left[1 + \frac{1\ 005\ 000}{2.098 \times 10^8}\right] = 1\ 010\ 000 \text{ ft}$$

So that the maximum height with the coasting height corrected for decreasing acceleration of gravity is

$$H_{\max.0} = H_{\mathrm{P}} + H_{\mathrm{C}} = 265\ 000 + 1\ 010\ 000 = 1\ 275\ 000 \text{ ft}$$

Conclusion

This study shows that a sounding rocket propelled by successive impulses can theoretically reach heights of much use to those interested in obtaining data on the structure of the atmosphere and extra-terrestrial phenomena if a propelling unit gives an exhaust velocity of 7 000 ft per sec or more.

The possibility of obtaining such exhaust velocities depends upon two factors: first, the ability of the motor to transform efficiently the heat energy of the fuel into kinetic energy of the exhaust gases, and, second, the amount of heat energy that can be liberated from the fuel. In an actual motor which burns its fuel at constant volume by igniting a powder charge in the combustion chamber the ratio of the chamber pressure to the outlet pressure drops from a maximum at the beginning of the expansion to zero at the end of the process. It is not possible to design a nozzle that will expand the products of combustion smoothly during the whole process. Therefore, the attainable efficiency must be less than that of a corresponding "constant pressure" motor which has a mixture of combustibles, e.g., gasoline and liquid oxygen, fed continuously into the combustion chamber at a constant pressure equal to the maximum pressure of the "constant volume" motor. However, very high maximum chamber pressures (up to 60 000 lbs. per sq. in.) can be developed in a motor using constant volume burning, while the chamber pressure of a motor using constant pressure burning is limited to much lower pressures by the difficulty of feeding the combustibles. Therefore, the efficiency that can be obtained from motors using either of these processes should not be very different. As to the heat that can be liberated per unit mass of fuel, the present fuel, such as nitro-cellulose powder for a constant volume motor, is much lower than the liquid combustibles such as gasoline and oxygen for a constant pressure motor.

These considerations indicate that the attainable exhaust velocity of a "constant volume" motor for propulsion by successive impulses will probably be lower than that of a "constant pressure" motor for supplying a continuous thrust. This is the reason why many experimenters abandoned the "constant volume" motor and turned to the "constant pressure" motor, the so-called liquid propellant motor. Theoretically, this defect of the "constant volume" motor can be compensated if a small total number of impulses (cf. Fig. 3) is used. However, the use of few impulses is of doubtful practical value because the resulting extreme accelerations will be harmful to instruments carried and will necessitate a heavier construction of the rocket.

However, even with the lower exhaust velocities of the "constant volume" motor it is shown by the analysis in this paper that with the exhaust velocity of 7 000 ft. per sec. obtained experimentally by R. H. Goddard[1] it should be possible to build a powder rocket capable of rising above 100 000 feet. Thus it seems to the authors that a rocket propelled by successive impulses has useful possibilities and further experimental work is justified.

References

[1] Goddard R H. A Method of Reaching Extreme Altitudes. Smithsonian Miscellaneous Collections, 1919, 71(2).

[2] Damblanc L. Les fusées autopropulsives a explosifs. L'Aérophile, 1935,43: 205 - 209, 241 - 247.

[3] Bartocci A. Le éscursioni in altezza col motor a reazione. L'Aerotecnica, 1933, 13: 1646 - 1666.

[4] Breguet L, Devillers R. L' Aviation superatmosphérique les aérodynes propulsées par reaction directe. La Science Aérienne, 1936,5: 183 - 222.

[5] Ley W, Schaefer H. Les fusées volantes météorologiqus. L'Aérophile, 1936,44: 228 - 232.

[6] Malina F J, Smith A M O. Analysis of the Sounding Rocket. Journal of the Aeronautical Sciences, 1938, 5: 199 - 202.

[7] Whittaker, Watson. Modern Analysis. 4th Edition, Cambridge, 1927: 246 - 247.

[8] Davis H T. Tables of the Higher Mathematical Functions. Principia Press, 1st Edition, 1933: 277 - 364.

Two-Dimensional Subsonic Flow of Compressible Fluids

Hsue-shen Tsien

(*California Institute of Technology*)

Summary

The basic concept of the present paper is to use a tangent line to the adiabatic pressure-volume curve as an approximation to the curve itself. First, the general characteristics of such a fluid are shown. Then in Section I a theory is developed which can be applied to flows with velocities approaching that of sound, whereas the theory of Demtchenko and Busemann only give an approximation for flows with velocities smaller than one-half of the sound velocity. This is done by a generalization of the method of approximation to the adiabatic relation by a tangent line, conceived jointly by Th. von Kármán and the author. The theory is put into a form by which, knowing the incompressible flow over a body, the compressible flow over a similar body can be calculated. The theory is then applied to calculate the flow over elliptic cylinders. In Section II the work of *H.* Bateman is applied to this approximate adiabatic fluid and the results obtained are essentially the same as those obtained in Section I.

Introduction

Assuming that the pressure is a single-valued function of density only, the equations of two-dimensional irrotational motion of compressible fluids can be reduced to a single non-linear equation of the velocity potential. In the supersonic case, that is, in the case when flow velocity is everywhere greater than that of local sound velocity, the problem is solved by Meyer and Prandtl and Busemann using the method of characteristics. The essential difficulty of this problem lies in the subsonic case, that is, in the case when flow velocity is everywhere smaller than but near the local sound velocity, because then the method of characteristics cannot be used. Glauert and Prandtl[1] treated the case when the disturbance of parallel rectilinear flow, due to the presence of a solid body, is small. They were then able to linearize the differential equation for the velocity potential and obtain an equation very similar to that for incompressible fluids. But there are usually stagnation points either on the surface of the body or in the field of flow, where the disturbance is no longer small. Hence, it is doubtful whether the linear theory can be applied to the flow near a stagnation point. For the same reason, the theory breaks down in the case of bodies whose dimension across the stream is not small compared with the dimension parallel to the stream.

Received March 28, 1939.

Journal of the Aeronautical Sciences, vol.6, pp.399 - 407, 1939.

To treat these cases Janzen and Rayleigh developed the method of successive approximations. This method was put into a more convenient form by Poggi and Walther. Recently, Kaplan[2] treated the case of flow over Joukowsky air-foils and elliptic cylinders using Poggi's method. However, the method is rather tedious and the convergence very slow if the local velocity of sound is approached.

Molenbroek and Tschapligin suggested the use of the magnitude of velocity w and inclination β of velocity to a chosen axis as independent variables, and were thus able to reduce the equation of velocity potential to a linear equation. This equation was solved by Tschapligin. The solution is essentially a series, each term of which is a product of a hypergeometric function of w and a trigonometric function of β. The main difficulty in practical application of this solution is to obtain a proper set of boundary conditions in the plane of independent variables w, β and to put the solution in a closed form.

Tschapligin has shown that a great simplification of the equation in the hodograph plane results if the ratio of the specific heats of the gas is equal to -1. Since all real gases have their ratio of specific heats between 1 and 2, the value -1 seems without practical significance. Demtchenko[3] and Busemann[4] clarified the meaning of this specific value of -1. They found that this really means to take the tangent of the pressure-volume curve as an approximation to the curve itself. However, they limit themselves to the use of the tangent at the state of the gas corresponding to the stagnation point of flow. As a result their theory can only be applied to a flow with velocities up to about one-half the velocity of sound. Recently, during a discussion with Th. von Kármán he suggested to the author that the theory can be generalized by using the tangent at the state of the gas corresponding to undisturbed parallel flow. Thus the range of usefulness of the theory can be greatly extended. This is carried out in the first section of the present paper. This theory, based upon Demtchenko and Busemann's work, is then applied to the case of flow over elliptic cylinders and the results compared with those of Hooker[5] and Kaplan[2]. Furthermore, results calculated by Glauert-Prandtl linear theory are also included for comparison.

Recently, Bateman[6] demonstrated a remarkable reciprocity between two fields of flow, of two fluids related by a certain point transformation. It is shown in the second section of this paper that the flow of an incompressible fluid and the flow of a compressible fluid approximated by the use of the tangent to adiabatic pressure-volume curve can be interpreted as such a point transformation. It is therefore possible to obtain a solution for compressible flow whenever a solution of incompressible flow is known. This transformation from the flow of an incompressible fluid to the flow of a compressible fluid is found, however, to be essentially the same as that developed from Demtchenko and Busemann's work.

Approximation to the Adiabatic Relation

If p is the pressure, v is the specific volume and γ is the ratio of specific heats of a gas, the adiabatic relation $pv^{\gamma}=$ constant is a curve in the $p-v$ plane as shown in Fig. 1(a). The

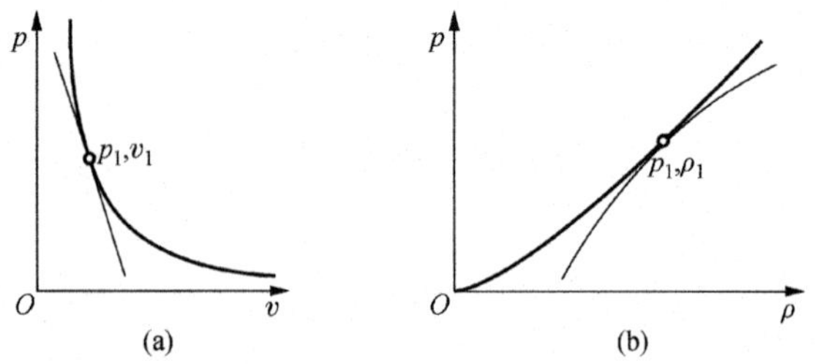

Fig. 1 Approximation to the adiabatic $p-v$ relation by means of a tangent

conditions near the point p_1, v_1 which correspond to a state of undisturbed flow can be approximated by the tangent to the curve at that point. The equation of the tangent at this point can be written

$$p_1 - p = C(v_1 - v) = C(\rho_1^{-1} - \rho^{-1}) \tag{1}$$

where C is the slope of the tangent and ρ is the density of the fluid. The slope C must be equal to the slope of the curve at the point p_1, v_1. Therefore,

$$C = \left(\frac{\mathrm{d}p}{\mathrm{d}v}\right)_1 = \left(\frac{\mathrm{d}p}{\mathrm{d}\rho}\frac{\mathrm{d}\rho}{\mathrm{d}v}\right)_1 = -\left(\frac{\mathrm{d}p}{\mathrm{d}\rho}\right)_1 \rho_1^2 = -a_1^2\rho_1^2$$

where a_1 is the sound velocity corresponding to the conditions p_1, v_1. Thus Eq. (1) can be written

$$p_1 - p = a_1^2\rho_1^2(\rho^{-1} - \rho_1^{-1}) \tag{2}$$

This is an approximation to the true adiabatic pressure density relation, and is shown in Fig. 1 (b) together with the true adiabatic relation.

The generalized Bernoulli theorem for compressible fluids is:

$$w_2^2 - w_3^2 = 2\int_2^3 \mathrm{d}p/\rho \tag{3}$$

where w is the velocity of the gas and the subscripts 2 and 3 denote two different states of the fluid. By substituting Eq. (2) into Eq. (3),

$$w_2^2 - w_3^2 = a_1^2\rho_1^2(\rho_2^{-2} - \rho_3^{-2}) \tag{4}$$

Now if $w_3 = 0$, $w_2 = w$, $\rho_3 = \rho_0$, and $\rho_2 = \rho$, with the subscript 0 denoting the state of the fluid corresponding to the stagnation point of flow, Eq. (4) gives

$$\frac{a_1^2\rho_1^2}{\rho_0^2} + w^2 = \frac{\rho_1^2 a_1^2}{\rho^2} \tag{5}$$

If the square of sound velocity a^2 is, as usual, defined as the derivative of p with respect to ρ, Eq. (2) gives

$$a^2\rho^2 = \rho^2\mathrm{d}p/\mathrm{d}\rho = a_1^2\rho_1^2 = \text{constant} \tag{6}$$

Therefore, Eq. (5) can be written as:

$$(\rho/\rho_0)^2 = 1 - (w/a)^2 \tag{7}$$

Similarly,

$$(\rho_0/\rho)^2 = 1 + (w/a_0)^2 \tag{8}$$

It is interesting to note that from Eq. (8) the density decreases as the velocity increases, as may be expected. Therefore Eq. (6) indicates that the local velocity of sound increases as the velocity increases. This behavior is opposite to that of a real gas, since in the case of an adiabatic flow of a real gas it is well known that the temperature of the gas decreases as the velocity increases; thus the local sound velocity, being proportional to the square root of the temperature, also decreases. However, in the present approximate theory, the ratio w/a or the Mach Number, still increases as the velocity increases, as can be seen from Eq. (7). But this ratio only reaches the value unity when $\rho = 0$, or from Eq. (8), when $w = \infty$. It is thus seen that the entire regime of flow is subsonic and the differential equation of the velocity potential is always of an elliptic type, i. e., always of the same type as the differential equation of the velocity potential of incompressible fluids. This is the reason why the complex representation of the velocity potential and the stream function is possible for all cases, as will be shown in the following paragraphs. However, one should realize that the portion of the tangent that could be used as an approximation to the true adiabatic relation is that portion which lies in the first quadrant. Thus the upper velocity limit for practical application of the theory occurs at $p = 0$. By using Eqs. (7) and (8) this upper limit is found to be

$$\left(\frac{w}{w_1}\right)_{\text{max.}} = \frac{1}{(w_1/a_1)}\sqrt{\left(\frac{p_1}{a_1^2\rho_1}+1\right)^2 - \left[1-\left(\frac{w_1}{a_1}\right)^2\right]} \tag{9}$$

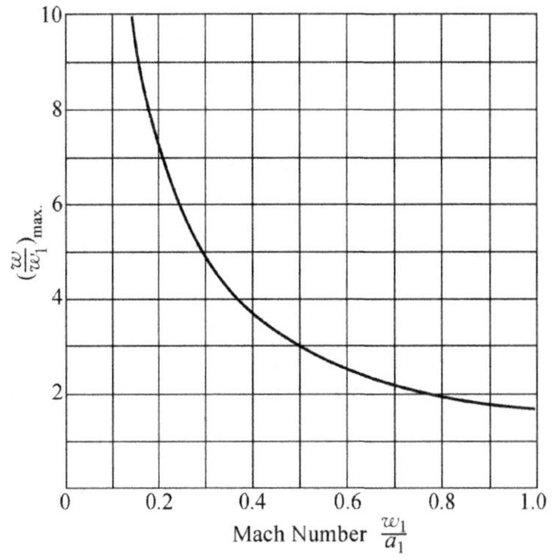

Fig. 2 Relation between the maximum velocity (w/w_1)$_{\text{max.}}$ (at which the pressure is zero) and the Mach Number (w_1/a_1)

Since the point p_1, ρ_1 being the tangent point to the true adiabatic curve, lies on the curve, the relation $a_1^2 = \gamma p_1/\rho_1$ which is true for the adiabatic relation $p\rho^{-\gamma} =$ constant can be used, and Eq. (9) becomes

$$\left(\frac{w}{w_1}\right)_{\max.} = \frac{1}{(w_1/a_1)}\sqrt{\left(\frac{1}{\gamma}+1\right)^2 - \left[1-\left(\frac{w_1}{a_1}\right)^2\right]} \tag{10}$$

This relation is plotted in Fig. 2 with $\gamma = 1.405$. Since for most practical cases it is not likely that the ratio (w/w_1) will rise to values much higher than 2, p will remain positive, and this theory will be sufficient to give an approximate solution.

Section I

Hodograph Method

If the flow is irrotational, there exists a velocity potential ϕ such that

$$\partial\phi/\partial x = u, \quad \partial\phi/\partial y = v \tag{11}$$

where u, v are the components of w in the x and y direction, respectively. The equation of continuity

$$\frac{\partial}{\partial x}\left(\frac{\rho}{\rho_0}u\right) + \frac{\partial}{\partial y}\left(\frac{\rho}{\rho_0}v\right) = 0$$

will be satisfied, if the stream function ψ is introduced such that:

$$u\rho/\rho_0 = \partial\psi/\partial y, \quad -v\rho/\rho_0 = \partial\psi/\partial x \tag{12}$$

If the angle of inclination of the velocity w to the x axis is β, Eqs. (11) and (12) give:

$$\begin{aligned} d\phi &= w\cos\beta dx + w\sin\beta dy \\ d\psi &= -w(\rho/\rho_0)\sin\beta dx + w(\rho/\rho_0)\cos\beta dy \end{aligned} \tag{13}$$

Solving for dx and dy,

$$\begin{aligned} dx &= \frac{\cos\beta}{w}d\phi - \frac{\sin\beta}{w}\frac{\rho_0}{\rho}d\psi \\ dy &= \frac{\sin\beta}{w}d\phi + \frac{\cos\beta}{w}\frac{\rho_0}{\rho}d\psi \end{aligned} \tag{14}$$

As long as the correspondence between the physical and hodograph plane is one to one, or mathematically $\partial(x,y)/\partial(u,v) \neq 0$; x and y can be expressed as functions of w, β, and ϕ and ψ as function of w, β. Thus,

$$\begin{aligned} d\phi &= \phi'_w dw + \phi'_\beta d\beta \\ d\psi &= \psi'_w dw + \psi'_\beta d\beta \end{aligned} \tag{15}$$

where primes indicate the derivative with respect to the independent variables indicated as subscripts. Substituting Eq. (15) into Eq. (14), the following expressions for dx and dy are obtained:

$$\begin{aligned} dx &= \left(\frac{\cos\beta}{w}\phi'_w - \frac{\sin\beta}{w}\frac{\rho_0}{\rho}\psi'_w\right)dw + \left(\frac{\cos\beta}{w}\phi'_\beta - \frac{\sin\beta}{w}\frac{\rho_0}{\rho}\psi'_\beta\right)d\beta \\ dy &= \left(\frac{\sin\beta}{w}\phi'_w + \frac{\cos\beta}{w}\frac{\rho_0}{\rho}\psi'_w\right)dw + \left(\frac{\sin\beta}{w}\phi'_\beta + \frac{\cos\beta}{w}\frac{\rho_0}{\rho}\psi'_\beta\right)d\beta \end{aligned} \tag{16}$$

Since the left-hand side of Eqs. (16) are exact differentials, the reciprocity relation can be applied, and, therefore,

$$\begin{aligned} \frac{\partial}{\partial\beta}\left(\frac{\cos\beta}{w}\phi'_w - \frac{\sin\beta}{w}\frac{\rho_0}{\rho}\psi'_w\right) &= \frac{\partial}{\partial w}\left(\frac{\cos\beta}{w}\phi'_\beta - \frac{\sin\beta}{w}\frac{\rho_0}{\rho}\psi'_\beta\right) \\ \frac{\partial}{\partial\beta}\left(\frac{\sin\beta}{w}\phi'_w + \frac{\cos\beta}{w}\frac{\rho_0}{\rho}\psi'_w\right) &= \frac{\partial}{\partial w}\left(\frac{\sin\beta}{w}\phi'_\beta + \frac{\cos\beta}{w}\frac{\rho_0}{\rho}\psi'_\beta\right) \end{aligned} \tag{17}$$

Carrying out these differentiations and simplifying with the aid of Eq. (7), Eq. (17),

$$\begin{aligned} -\frac{\sin\beta}{w}\phi'_w - \frac{\cos\beta}{w}\frac{\rho_0}{\rho}\psi'_w &= -\frac{\cos\beta}{w^2}\phi'_\beta + \frac{\sin\beta}{w^2}\frac{\rho}{\rho_0}\psi'_\beta \\ \frac{\cos\beta}{w}\phi'_w - \frac{\sin\beta}{w}\frac{\rho_0}{\rho}\psi'_w &= -\frac{\sin\beta}{w^2}\phi'_\beta - \frac{\cos\beta}{w^2}\frac{\rho}{\rho_0}\psi'_\beta \end{aligned} \tag{18}$$

Solving for ϕ'_w and ϕ'_β,

$$\begin{aligned} \phi'_w &= -\frac{\rho}{\rho_0}\frac{1}{w}\psi'_\beta \\ \phi'_\beta &= \frac{\rho_0}{\rho}w\psi'_w \end{aligned} \tag{19}$$

Eq. (19) can be further simplified by introducing a new variable ω, such that

$$d\omega = \frac{\rho}{\rho_0}\frac{dw}{w} \tag{20}$$

Then Eq. (19) becomes

$$\phi'_\omega = -\psi'_\beta, \quad \phi'_\beta = \psi'_\omega \tag{21}$$

This can be easily recognized as the Cauchy-Riemann differential equation, and thus $\phi + i\psi$ must be an analytic function of $\omega - i\beta$. However, for convenience of calculation, another new set of independent variables $U = W\cos\beta$, $V = W\sin\beta$ are introduced where $W = a_0 e^\omega$. Then Eq. (21) can be written as

$$\begin{aligned} \partial\phi/\partial U &= \partial\psi/\partial(-V) \\ \partial\phi/\partial(-V) &= -\partial\psi/\partial U \end{aligned} \tag{22}$$

Integrating Eq. (20),

$$W = 2a_0 w/\left(\sqrt{a_0^2 + w^2} + a_0\right) \tag{23}$$

and

$$w = 4a_0^2 W/(4a_0^2 - W^2) \tag{24}$$

Substituting into Eq. (8), the following expression for the density ratio ρ_0/ρ is obtained:

$$\rho_0/\rho = (4a_0^2 + W^2)/(4a_0^2 - W^2) \tag{25}$$

Eqs. (22), (23), (24), and (25) are the basic equations of the present theory. Eq. (22) is the Cauchy-Riemann differential equation, and thus the complex potential $F = \phi + \mathrm{i}\psi$ must be an analytic function of $\overline{W} = U - \mathrm{i}V$, or:

$$\begin{aligned} \phi + \mathrm{i}\psi &= F(U - \mathrm{i}V) = F(\overline{W}) \\ \phi - \mathrm{i}\psi &= \overline{F}(U + \mathrm{i}V) = \overline{F}(W) \end{aligned} \tag{26}$$

In Eq. (26), $\overline{W}$ and $\overline{F}$ are the complex conjugates of W and F, respectively.

It is now necessary to find the values of x and y corresponding to a given set of values of U and V, i. e., to find the transformation from the hodograph plane to the physical plane. By using Eqs. (24) and (25), Eq. (14) can be written

$$\begin{aligned} \mathrm{d}x &= \frac{U \cdot \mathrm{d}\phi}{W^2}\left(1 - \frac{W^2}{4a_0^2}\right) - \frac{V \cdot \mathrm{d}\psi}{W^2}\left(1 + \frac{W^2}{4a_0^2}\right) \\ \mathrm{d}y &= \frac{V \cdot \mathrm{d}\phi}{W^2}\left(1 - \frac{W^2}{4a_0^2}\right) + \frac{U \cdot \mathrm{d}\psi}{W^2}\left(1 + \frac{W^2}{4a_0^2}\right) \end{aligned} \tag{27}$$

where $W^2 = U^2 + V^2$. These equations can be combined into one equation by means of Eq. (26). Thus,

$$\mathrm{d}z = \mathrm{d}x + \mathrm{i}\mathrm{d}y = \frac{\mathrm{d}F}{\overline{W}} - \frac{W \cdot \mathrm{d}\overline{F}}{4a_0^2} \tag{28}$$

Hence, if an analytic function $F(\overline{W})$ is given for each value of W, the corresponding real velocity w can be calculated by Eq. (24). Then the coordinate of the point in the physical plane at which this velocity occurs can be calculated by integrating Eq. (28). The pressure at this point is then given by Eq. (2). However, using this procedure, it is not possible to predict whether the chosen function $F(\overline{W})$ will give the desired shape of the solid boundary and flow pattern. In other words, this procedure, in common with all hodograph methods, still suffers the difficulty of boundary conditions.

Transformation from Incompressible Flow to Compressible Flow

However, using the simple relation of Eq. (28) the resulting shape of the body can be ascertained approximately by starting with the function:

$$F(\overline{W}) = \phi + \mathrm{i}\psi = W_1 G(\zeta) \tag{29}$$

where W_1 is the transformed undisturbed velocity to be interpreted by Eq. (23), and ζ is the complex coordinate $\xi + \mathrm{i}\eta$. This function is so chosen as to give the flow of an incompressible fluid over the desired body shape in coordinates ξ and η. The real velocity in the ζ plane of the incompressible fluid is interpreted as the transformed velocity W in the hodograph plane for the compressible fluid. It is known that

$$\overline{W} = W_1 \, dG(\zeta)/d\zeta \tag{30}$$

Thus,

$$W = W_1 \, d\overline{G}(\overline{\zeta})/d\overline{\zeta} \tag{31}$$

where $\overline{G}$ and $\overline{\zeta}$ are the complex conjugates of G and ζ, respectively. With Eqs. (30) and (31), Eq. (28) gives

$$dz = d\zeta - \lambda [d\overline{G}/d\overline{\zeta}]^2 d\overline{\zeta}$$

where $\lambda = \frac{1}{4}(W_1/a_0)^2$. Integrating,

$$z = \zeta - \lambda \int [d\overline{G}/d\overline{\zeta}]^2 d\overline{\zeta} \tag{32}$$

Therefore, the complex coordinate in the physical plane of the compressible fluid is equal to the corresponding complex coordinate in the physical plane of the incompressible fluid plus a correction term. Since this correction term is usually small, the resulting shape of the body will be quite similar to the one in the incompressible fluid. The factor λ in the correction term depends upon the Mach Number of the undisturbed flow only. This can be shown by means of Eqs. (7), (8), and (23), because from those equations the following relation is obtained:

$$\lambda = \frac{1}{4}\left(\frac{W_1}{a_0}\right)^2 = \frac{(w_1/a_1)^2}{\left[1 + \sqrt{1 - (w_1/a_1)^2}\right]^2} \tag{33}$$

where w_1/a_1 is the Mach Number of the undisturbed flow. The values of λ for different Mach Numbers w_1/a_1 are plotted in Fig. 3.

To calculate the velocity in the physical plane, $\overline{W}$ is first obtained from Eq. (30) and then with Eq. (23):

$$\frac{w}{w_1} = \frac{|W|}{W_1} \frac{1-\lambda}{1 - \lambda(|W|/W_1)^2} \tag{34}$$

If the pressure coefficient $\tilde{\omega}$ at any point is defined as $\tilde{\omega} = (p - p_0)/\left(\frac{1}{2}\rho_1 w_1^2\right)$, then by using Eq. (2), the following relation is obtained:

$$\tilde{\omega} = (1 + \lambda) \frac{1 - (|W|/W_1)^2}{1 - \lambda(|W|/W_1)^2} \tag{35}$$

Flow over Elliptic Cylinders

The theory will now be applied to calculate the flow over an elliptic cylinder at zero angle of attack. The incompressible flow over an elliptic cylinder in the complex coordinate ζ can be obtained by applying Joukowsky's transformation to the flow over a circular cylinder in the complex coordinate ζ' with the center of the circle located at the origin of the ζ' plane. Therefore, the function $F(\overline{W})$ or $W_1 G(\zeta)$ can be written as:

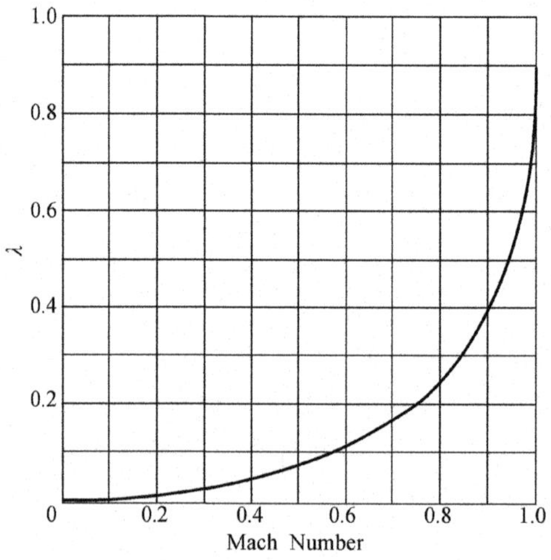

Fig. 3 Variation of the parameter λ of transformation from incompressible flow to compressible flow with the Mach Number w_1/a_1

$$\begin{aligned} F &= W_1(\zeta' + b^2/\zeta) \\ \bar{F} &= W_1(\bar{\zeta}' + b^2/\bar{\zeta}') \end{aligned} \tag{36}$$

where b is the radius of the circle in the ζ' plane. The Joukowsky transformation is:

$$\zeta = \zeta' + 1/\zeta' \tag{37}$$

It is convenient to carry out the computation by using the ζ' coordinates. Thus Eq. (32) is rewritten in the following form:

$$z = \left(\zeta' + \frac{1}{\zeta'}\right) - \lambda \int \left(\frac{d\bar{G}}{d\bar{\zeta}'}\right)^2 \frac{d\bar{\zeta}'}{d\bar{\zeta}/d\bar{\zeta}'} \tag{38}$$

If only the conditions over the surface of the elliptic cylinder are concerned, then:

$$\zeta' = b\,e^{i\theta}, \quad \bar{\zeta}' = b\,e^{-i\theta} \tag{39}$$

where θ is the argument as shown in Fig. 4 and b is the radius of the circular section in the ζ' plane which determines the thickness ratio of the elliptic section in the ζ plane. Substituting Eqs. (36), (37), and (39) into Eq. (38), and carrying out the integration, the following expressions for the x and y coordinates corresponding to ζ' are obtained by separating the real and imaginary parts:

$$\begin{aligned} x &= \left(b + \frac{1}{b}\right)\cos\theta - \lambda\left[b(1+b^2)\cos\theta + \frac{(b^2-1)^2}{4}\log\frac{(b^2-1)^2 + 4b^2\sin^2\theta}{(b^2 + 2b\cos\theta + 1)}\right] \\ y &= \left(b - \frac{1}{b}\right)\sin\theta + \lambda\left[b(1-b^2)\sin\theta + \frac{(b^2-1)^2}{2}\tan^{-1}\frac{2b\sin\theta}{b^2-1}\right] \end{aligned} \tag{40}$$

The horizontal and vertical semi-axis of the approximately elliptic section can then be calculated by substituting $\theta = 0$ and $\theta = \pi/2$, respectively, into Eq. (40). The thickness ratio δ is thus obtained as:

$$\delta = \left(\frac{b^2-1}{b^2+1}\right)\times\frac{1+\lambda\left[-b^2+\frac{b(b^2-1)}{2}\tan^{-1}\frac{2b}{b^2-1}\right]}{1-\lambda\left[b^2+\frac{b(b^2-1)}{2}\left(\frac{b^2-1}{b^2+1}\right)\log\left(\frac{b-1}{b+1}\right)\right]} \tag{41}$$

For a given thickness ratio and Mach Number for undisturbed flow, the value of λ is first computed by means of Eq. (33), and then Eq. (41) is solved graphically for b.

After b is obtained, the coordinate x and y for each value of θ can be computed by using Eq. (40). It is fortunate that the values of x, y so obtained lie very close to the true elliptic section. Hence, the velocity and the pressure distribution obtained by using Eqs. (34) and (35) are considered as those over the true elliptic sections.

Calculations for two thickness ratios, $\delta = 0.5$ and $\delta = 0.1$, are carried out and the results shown in Figs. 5 and 6, together with those of Kaplan[2]. Hooker's results[5] are very close to those of Kaplan. Computations are also carried out using the more simple theory of Glauert and Prandtl[1], and the results are included in Figs. 5 and 6 in order to compare with those of Kaplan and the present theory.

The difference between the various theories lies in the assumptions which are made to simplify the calculations. Glauert and Prandtl assumed that the disturbance introduced by the solid body to the parallel flow is small. In other words, they treated the flow over a body of small thickness ratio. On the other hand, Kaplan and Hooker assume that the Mach Number of the undisturbed flow is small, so that terms containing the third and higher powers of the Mach Number can be neglected. The present theory is essentially an improvement of the Glauert-Prandtl theory, so that the effect of large disturbances to the parallel flow is approximately taken into account. Therefore, for flow over thin sections at high Mach Numbers, the result of the present theory should agree well with the Glauert-Prandtl theory, especially at points not too close to the stagnation point. The results of Kaplan and Hooker should show smaller effect of the compressibility due to their second order approximation. For flow over thick sections at lower Mach Numbers, the situation is reversed. In this case results of the present theory should give better agreement with the results obtained by Kaplan and Hooker than with those obtained from the Glauert-Prandtl theory. The above reasoning is

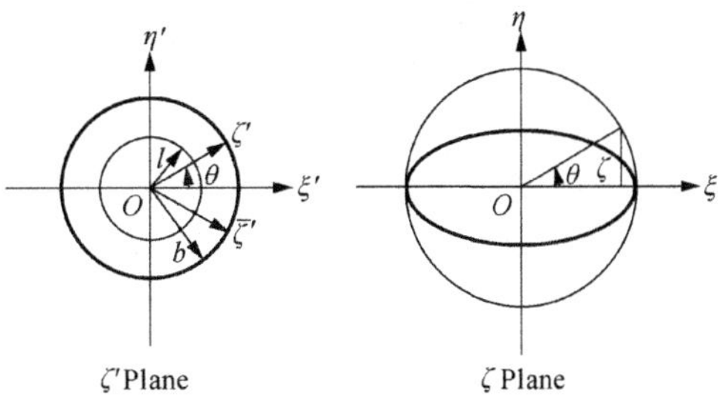

Fig. 4 Notations used in calculating the flow over an elliptical cylinder

substantiated by Figs. 5 and 6.

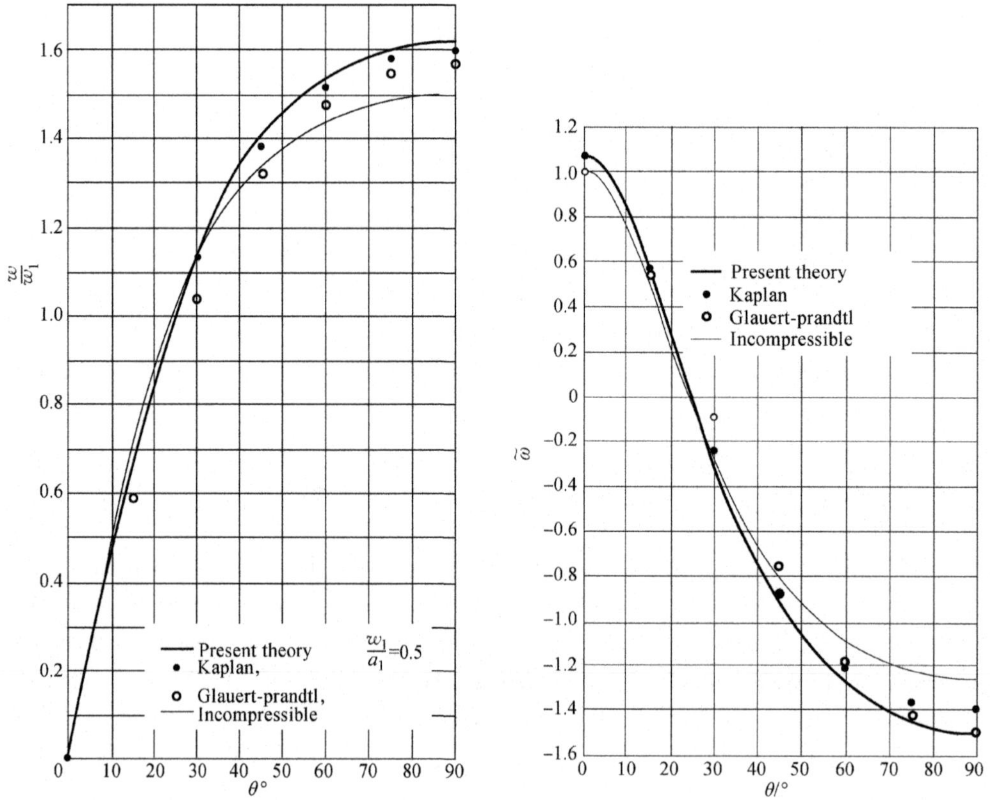

(a) (b)

Fig. 5 Flow over an elliptical cylinder with thickness ratio $\delta = 0.5$ at Mach Number $= 0.5$

(a) Velocity distribution; (b) Pressure distribution

Critical Velocities for Elliptic Cylinders

If the velocity of flow over a body is gradually increased, the maximum local velocity in the field will also be increased. When the maximum local velocity reaches the local velocity of sound, shock waves appear and the drag of the body suddenly increases. This velocity is, therefore, of considerable interest to practical engineers and is usually referred to as the critical velocity of the body. It is shown by Kaplan[2] and others that at this critical condition the ratio of maximum velocity of $w_{\max.}$ in the field to that of the undisturbed velocity w_1 is related to the Mach Number w_1/a_1 of the undisturbed flow in the following manner:

$$\frac{w_{\max.}}{w_1} = \left[\frac{2}{\gamma+1}\frac{1}{(w_1/a_1)^2} + \frac{\gamma-1}{\gamma+1}\right]^{1/2} \tag{42}$$

$w_{\max.}$ in the flow over an elliptic cylinder at zero angle of attack occurs at the top of the cylinder. Using Eqs. (34) and (33) the value of $w_{\max.}/w_1$ is found to be

$$\frac{w_{\max}}{w_1} = \frac{2b^2/(b^2+1)}{1+\dfrac{\left[1-\left(\dfrac{2b^2}{b^2+1}\right)^2\right]\left(\dfrac{w_1}{a_1}\right)^2}{2\left(1+\sqrt{1-\left(\dfrac{w_1}{a_1}\right)^2}\right)\sqrt{1-\left(\dfrac{w_1}{a_1}\right)^2}}} \tag{43}$$

Equating Eqs. (42) and (43) the equation for calculating the critical Mach Number (w_1/a_1) of the undisturbed flow for each value of b is

$$\left[\frac{2}{\gamma+1}\frac{1}{(w_1/a_1)_{\text{crit.}}}+\frac{\gamma-1}{\gamma+1}\right]^{1/2} = \frac{2b^2/(b^2+1)}{1+\dfrac{\left[1-\left(\dfrac{2b^2}{b^2+1}\right)^2\right]\left(\dfrac{w_1}{a_1}\right)^2_{\text{crit.}}}{2\left(1+\sqrt{1-\left(\dfrac{w_1}{a_1}\right)^2_{\text{crit.}}}\right)\sqrt{1-\left(\dfrac{w_1}{a_1}\right)^2_{\text{crit.}}}}} \tag{44}$$

Knowing (w_1/a_1)$_{\text{crit.}}$ for each value of b the corresponding value of δ can be calculated by means of Eqs. (34) and (41). Fig. 7 shows the result of this calculation with Kaplan's value included for comparison. It is seen that the critical Mach Number is lower than that obtained by Kaplan. This lower value of the critical Mach Number indicates a more pronounced effect of compressibility of a fluid and is consistent with the results shown in Figs. 5 and 6.

Section Ⅱ

The Use of Lift and Drag Functions

If two new functions X and Y are defined by:

$$\begin{aligned} p_0\,\mathrm{d}X &= p\mathrm{d}y + \rho_0 u\mathrm{d}\psi \\ p_0\,\mathrm{d}Y &= \rho_0 v\mathrm{d}\psi - p\mathrm{d}x \end{aligned} \tag{45}$$

Assuming that the flow is irrotational, it can be shown by means of Eqs. (11) and (12) that the following relations hold:

$$\begin{aligned} p_0\,\mathrm{d}X &= (p+\rho u^2)\mathrm{d}y - \rho uv\mathrm{d}x = (p+\rho w^2)\mathrm{d}y - \rho v\mathrm{d}\phi \\ p_0\,\mathrm{d}Y &= \rho uv\mathrm{d}y - (p+\rho v^2)\mathrm{d}x = \rho u\mathrm{d}\phi - (p+\rho w^2)\mathrm{d}x \end{aligned} \tag{46}$$

It is seen that by integrating Eq. (46) along any closed boundary, it will give the resultant of the pressure forces acting along the boundary and the rate of increase of momentum of the fluid passing out of the boundary. If there is a solid body in this boundary, then this integral will give the lift and the drag acting on the body. Therefore, X and Y are sometimes called the drag and lift functions. From Eq. (46) the following relations can be deduced:

$$p_0(v\mathrm{d}X - u\mathrm{d}Y) = p\mathrm{d}\phi \tag{47}$$

$$p_0(\rho/\rho_0)(u\mathrm{d}X + v\mathrm{d}Y) = (p+\rho w^2)\mathrm{d}\psi \tag{48}$$

Therefore, by writing $\partial\phi/\partial X = R$ and $\partial\phi/\partial Y = S$ Eq. (47) gives

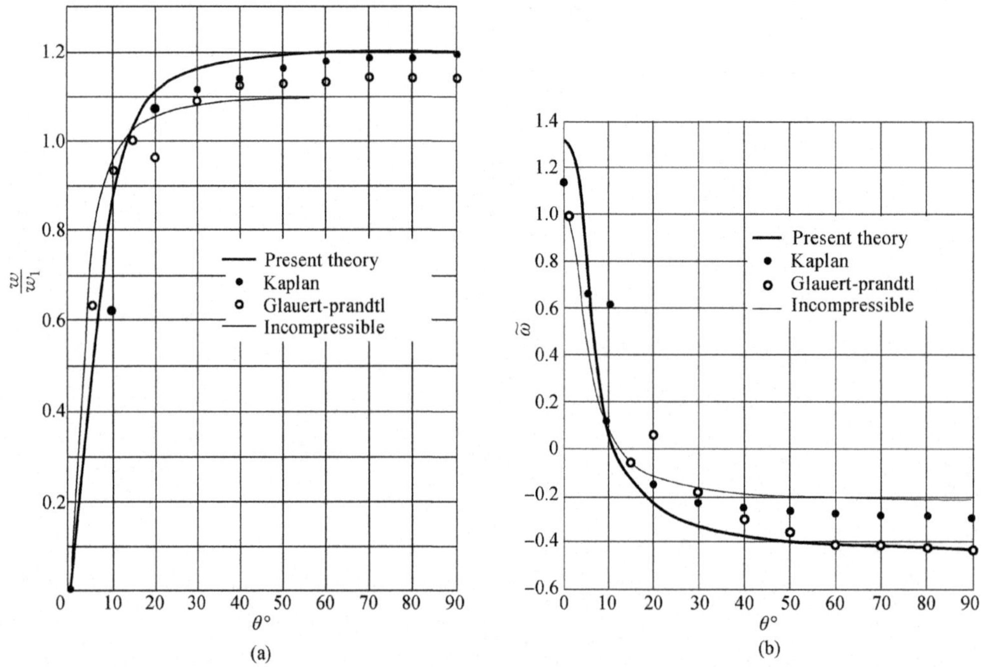

Fig. 6 Flow over an elliptical cylinder with thickness ratio $\delta = 0.1$ at Mach Number = 0.857

(a) Velocity distribution; (b) Pressure distribution

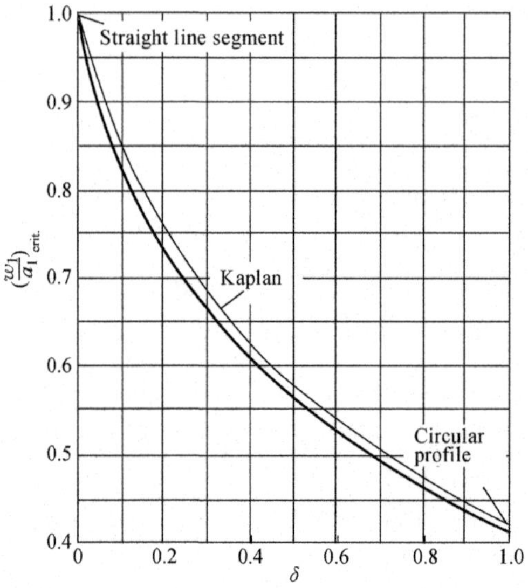

Fig. 7 Variation of critical Mach number $(w_1/a_1)_{\text{crit.}}$ of an elliptical cylinder with thickness ratio δ

$$R = \frac{\partial \phi}{\partial X} = \frac{p_0}{p}v, \quad S = \frac{\partial \phi}{\partial Y} = -\frac{p_0}{p}u \tag{49}$$

The quantities R and S have the dimension of a velocity and can be considered as components of a new velocity in the plane whose coordinates are denoted by X and Y. This relation between the xy plane and the XY plane is shown in Fig. 8. It is thus seen that if the undisturbed flow in the xy plane is in the positive x direction, the undisturbed flow in the XY plane will be in the negative Y direction. Furthermore, if σ is defined as

$$\sigma/\sigma_0 = p\rho/[\rho_0(p+\rho w^2)] \tag{50}$$

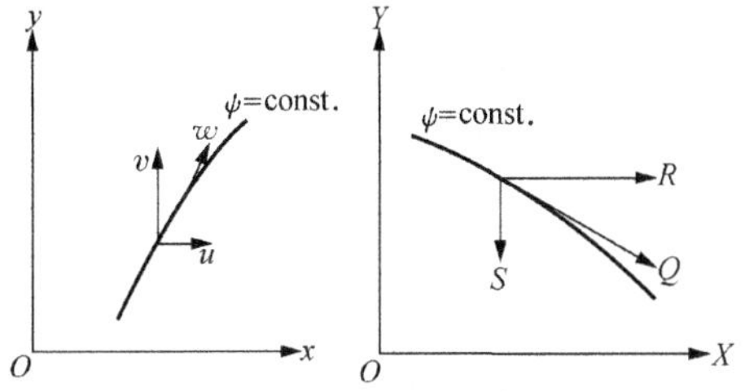

Fig. 8 Relation of the velocity components in the xy plane and XY plane

Eq. (48) gives:

$$\begin{aligned} \frac{\sigma}{\sigma_0}R &= -\frac{\partial \psi}{\partial X} = -\frac{u\rho/\rho_0}{(p+\rho w^2)/p_0} \\ \frac{\sigma}{\sigma_0}S &= \frac{\partial \psi}{\partial Y} = \frac{v\rho/\rho_0}{(p+\rho w^2)/p_0} \end{aligned} \tag{51}$$

Comparing Eq. (51) with Eq. (12), it is evident that σ can be considered as the density of a fluid in the XY plane. Therefore, there exists a complete reciprocity between the xy plane and the XY plane, as shown by Bateman[6].

Transformation Starting with Incompressible Flow

So far the relations obtained are general, i. e., they apply to fluids of arbitrary properties. However, since only the flow of incompressible fluids is well known, it would be interesting to find the properties of the fluid in the XY plane if the fluid in the xy plane is incompressible. If the fluid in the xy plane is incompressible, then $\rho/\rho_0 = 1$, and the Bernoulli theorem gives:

$$[p + (1/2)\rho w^2]/p_0 = 1 \tag{52}$$

Let P denote the pressure in the XY plane, and Q^2 denote $R^2 + S^2$; then Eq. (3), the generalized Bernoulli theorem, gives:

$$(1/2)Q^2 + \int \mathrm{d}p/\sigma = \text{constant} \tag{53}$$

In view of Eqs. (49), (50), and (52), Eq. (53) can be written in the following form:

$$\frac{1}{\sigma_0}\int \frac{\mathrm{d}\left(\frac{\sigma}{\sigma_0}\right)}{\frac{\sigma}{\sigma_0}} \frac{\mathrm{d}P}{\mathrm{d}\left(\frac{\sigma}{\sigma_0}\right)} + \frac{p_0}{\rho_0}\left(\frac{1}{4\frac{\sigma^2}{\sigma_0^2}} - \frac{1}{4}\right) = \text{constant} \tag{54}$$

By differentiating Eq. (54) with respect to σ/σ_0, multiplying the resulting expression by σ/σ_0 and then integrating with respect to σ/σ_0, the following relation connecting the pressure P and the density σ for the fluid in the XY plane is obtained:

$$P = C - \frac{1}{2}\frac{p_0}{\rho_0}\sigma_0^2\frac{1}{\sigma} \tag{55}$$

where C is the integration constant. Comparing Eq. (55) with the approximate adiabatic relation Eq. (2), also noting Eq. (6), it is evident that Eqs. (55) and (2) are identical, if

$$\frac{1}{2}\frac{p_0}{\rho_0} = A_0^2 = A_1^2\left[1 - \left(\frac{Q_1}{A_1}\right)^2\right] \tag{56}$$

and

$$C = P_1 + \frac{1}{2}\frac{p_0}{\rho_0}\sigma_0^2\frac{1}{\sigma_1}$$

In Eq. (56) A is the velocity of sound in the XY plane, and the subscript 1 refers to the conditions in the undisturbed flow. Hence, Q_1/A_1 is Mach Number of the undisturbed flow.

By using Eqs. (52) and (49) the components of velocity in the XY plane can be expressed as

$$\frac{R}{Q_1} = -\frac{v}{w_1}\frac{1 - \frac{1}{2}\frac{\rho_0}{p_0}w_1^2}{1 - \frac{1}{2}\frac{\rho_0}{p_0}w_1^2\left(\frac{w}{w_1}\right)^2} \tag{57}$$

$$\frac{S}{Q_1} = \frac{u}{w_1}\frac{1 - \frac{1}{2}\frac{\rho_0}{p_0}w_1^2}{1 - \frac{1}{2}\frac{\rho_0}{p_0}w_1^2\left(\frac{w}{w_1}\right)^2}$$

Hence,

$$\frac{Q}{Q_1} = \frac{w}{w_1}\frac{1 - \frac{1}{2}\frac{\rho_0}{p_0}w_1^2}{1 - \frac{1}{2}\frac{\rho_0}{p_0}w_1^2\left(\frac{w}{w_1}\right)^2} \tag{58}$$

The relation between w_1 and Q_1 can then be obtained from Eqs. (56) and (57), that is:

$$\frac{1}{2}\frac{\rho_0}{p_0}w_1^2 = \frac{(Q_1/A_1)^2}{\left[1+\sqrt{1-(Q_1/A_1)^2}\right]^2} = \lambda \tag{59}$$

Thus Eq. (58) can be rewritten as:

$$\frac{Q}{Q_1} = \frac{w}{w_1} \frac{1-\lambda}{1-\lambda(w/w_1)^2} \tag{60}$$

Using Eqs. (55), (56) and (50), the pressure coefficient Π in the XY plane, which is defined by $\Pi = (P - P_1)/[(1/2)\sigma_1 Q_1^2]$, can be expressed as:

$$\Pi = \left(1 + \frac{1}{2}\frac{p_0}{\rho_0}w_1^2\right)\frac{1-(w/w_1)^2}{1-\frac{1}{2}\frac{\rho_0}{p_0}w_1^2(w/w_1)^2}$$

Substituting the value of λ from Eq. (59) in the above expression, the following relation is obtained:

$$\Pi = (1+\lambda)\frac{1-(w/w_1)^2}{1-\lambda(w/w_1)^2} \tag{61}$$

To find the coordinates X and Y in terms of x, y, Eq. (46) must be integrated. It is convenient in this case to use the complex potential of the incompressible flow in the xy plane. If

$$\phi + i\psi = w_1 G(x + iy) = w_1 G(z) \tag{62}$$

then it can be shown with the aid of Eq. (52) that:

$$\bar{Z} = X - iY = i\bar{z} - \frac{1}{2}\frac{\rho_0}{p_0}w_1^2 i\int\left(\frac{dG}{dz}\right)^2 dz$$

where $\bar{z}$ is the complex conjugate of z. Or, writing Z and $\bar{G}$ as the complex conjugates of $\bar{Z}$ and G:

$$e^{i\pi/2}Z = z - \lambda\int\left(\frac{d\bar{G}}{d\bar{z}}\right)^2 d\bar{z} \tag{63}$$

where the factor $e^{i\pi/2}$ will rotate the Z plane through an angle equal to $\pi/2$ to make the directions of undisturbed flow in the Z plane and in the z plane coincide.

Comparing the set of Eqs. (59), (60), (61), and (63) with the previous set of Eqs. (33), (34), (35), and (32), it is evident that the two sets are identical except the change of notation. Therefore, Bateman's transformation does not give any new results as it leads to the same expressions as those obtained by the hodograph method.

Concluding Remarks

It is shown both in Section I and in Section II that starting from any solution of an incompressible fluid over a solid body, a solution of a nearly adiabatic flow over another approximately similar solid body can be calculated. The transformation from incompressible flow to compressible flow changes the shape of the body a small amount as represented by the correction terms in Eqs. (32) and (63). Thus, in order to investigate the effect of compressibility over the same body, it is necessary to use different functions $G(z)$ for different

Mach Numbers, as shown by the example given in Section I. This complicates the calculations to some extent, but the amount of labor involved is probably much less than the successive approximations devised by Janzen, Rayleigh, Poggi, and Walther, especially at higher Mach Numbers.

The main difficulty of the method lies in its application to flow involving circulation, e.g., the flow over a lifting airfoil. Then if the ordinary complex potential function $G(z)$ for the incompressible fluid is used, the correction terms in Eqs. (32) and (63) are no longer single-valued functions, that is, they do not return to their original value by increasing the argument of z by 2π. In other words, the boundary in compressible flow is no longer a closed curve. Therefore, in order to study this type of problem, it is necessary to use a function $G(z)$ which does not give a closed boundary in the incompressible flow, but will give a closed boundary in the compressible flow when the correction term is added. The problem is thus more difficult, and requires further study.

The author expresses his gratitude to Dr. Th. von Kármán for suggesting the subject and for his kindly criticism during the course of the work.

References

[1] Glauert H. The Effect of Compressibility on the Lift of an Airfoil. Proc. Roy. Soc. (A), 118:113, 1928; also. Reports and Memoranda No. 1135, British A.R.C., 1928.

[2] Kaplan C. Two-Dimensional Subsonic Compressible Flow Past Elliptic Cylinders. N.A.C.A. Technical Report No. 624, 1938.

[3] Demtchenko B. Sur les mouvements lents des fluides compressibles. Comptes Rendus, 1932, 194: 1218.//Variation de la résistance aux faibles vitesses sous l'influence de la compressibilite. Comptes Rendus, 1932, 194: 1720.

[4] Busemann A. Die Expansionsberichtigung der Kontraktionsziffer von Blenken. Forschung, 1933, 4, 186-187. Hodographenmethode der Gasdynamik, Z.A.M.M., 1937, 12: 73-79.

[5] Hooker S G. The Two-Dimensional Flow of Compressible Fluids at Subsonic Speeds Past Elliptic Cylinders. Reports and Memoranda No. 1684, British A.R.C., 1936.

[6] Bateman H. The Lift and Drag Functions for an Elastic Fluid in Two-Dimensional Irrotational Flow. Proc. Nat. Acad. Sci., 1938, 24: 246-251.

The Buckling of Spherical Shells by External Pressure*

Th. von Kármán and Hsue-shen Tsien
(*California Institute of Technology*)

General Considerations

The general theory of thin shells was developed by A. E. H. Love. He assumed small deflections and for that reason neglected in the energy expression all terms higher than quadratic, and thus obtained linear differential equations for the determination of the equilibrium position of a shell under given forces. The theory of buckling of thin shells is also based essentially on Love's equations. The buckling of cylindrical shells of uniform thickness under the action of a uniformly distributed axial load was calculated by R. Lorentz, R. V. Southwell, S. Timoshenko, W. Flügge, L. H. Donnell, and others. The same problem was also investigated experimentally by many authors, especially by E. E. Lundquist and L. H. Donnell. Unfortunately, a systematic discrepancy was found between the theoretically calculated and experimentally obtained buckling loads; the theoretical values are as much as 3 to 4 times higher than the experimental values. To remedy this situation W. Flügge[1] first considered the deviation of the assumed end conditions of the cylindrical shell from that realized in the laboratory. However, this effect is not sufficient to explain the discrepancy. The influence of the end conditions extends only to a distance approximately equal to $\sqrt{Rt}$, where R is the radius of the shell and t the thickness. The cylinders tested, however, usually have a length which is large compared to this value. Furthermore, Flügge's analysis would indicate a progressive increase of the wave amplitude until plastic deformation occurs, whereas the experimental evidence indicates that the failure of cylindrical shells under compression is not progressive but very rapid.

Another attempt was made by W. Flügge[1] and later by L. H. Donnell[2] to lower the theoretical buckling load by taking into account the initial deviation of the form of the shell from the exact cylindrical shape. According to their assumptions the buckling load, or rather the failing load would be determined by the plastic failure of the material. This explanation has

---

Received October 30. 1939.

Journal of the Aeronautical Sciences, vol. 7, pp. 43 – 50, 1940.

* The theoretical investigation reported in this paper was carried out by the authors in connection with the research project "General Instability Criteria for Stiffened Metal Cylinders," sponsored by the Civil Aeronautics Authority, Airworthiness Section.

several drawbacks: first, to obtain the low values of the buckling load found experimentally one has to assume deviations from the cylindrical shape as large as ten times the shell thickness. Such a large deviation in the shape of the specimens would be easily detected by visual observations. This is not substantiated by experience. Second, the failure of a cylindrical shell is not necessarily a plastic failure (yielding), especially when the wall of the cylinder is very thin. In many cases it has been observed that upon removing the load the buckling waves completely disappeared. Therefore, the phenomenon must be completely elastic, instead of being plastic as assumed in L. H. Donnell's analysis. Furthermore, the initial deviations from the exact cylindrical form would cause the deformation to increase gradually, which is again contrary to experimental observations.

A similar discrepancy between theory and experiment exists in the case of the buckling of spherical shells under uniform external pressure. The theoretical buckling load based upon Love's equations has been calculated by R. Zoelly, E. Schwerin, and A. Van der Neut[3]. If the buckling stress σ_{cr} is defined by

$$\sigma_{cr} = \frac{p_{cr}R}{2t} \tag{1}$$

where p_{cr} is the buckling pressure, t the thickness, and R the radius of the sphere, then the theory gives

$$\sigma_{cr} = \frac{Et/R}{\sqrt{3(1-\nu^2)}} \tag{2}$$

where ν is Poisson's ratio and E is Young's modulus of elasticity. While no systematic experimental work has been done on this problem — at least to the authors' knowledge — some tests made by E. E. Sechler and W. Bollay at the California Institute of Technology indicated that the experimental buckling load is only about 1/4 that of the theoretical value.

Besides these differences between the theoretically and experimentally obtained buckling loads the wave form predicted by the theory is also at variance with laboratory experience. According to theoretical calculations the same load would produce buckling either inward or outward; experiments have shown that the shell has a definite preference for buckling inward. For the case of a spherical shell it is observed that the buckling wave is restricted to a small dimple subtended by a solid angle of about 16°. The linear theory predicts a wave form extending over the whole spherical surface.

What may be the reason for these discrepancies between the prediction of the theory and the experimental evidence? It is very unlikely that there is an error in the fundamental equations of the theory of elasticity. For example, in the case of a flat plate not only the buckling load but also the behavior of the plate after buckling is predicted quite accurately by the theory. Hence, there must be an essential difference between the physical process of the buckling of a flat plate and a curved shell that is not embraced by the previous theory. The same opinion was expressed recently by H. L. Cox[4] in a lecture delivered to the Royal Aeronautical

Society.

In this paper a new conception of the mechanism of the collapse of curved sheets is presented.

It is advantageous to start the investigation with the case of a spherical shell loaded by a uniform external pressure, because the geometrical symmetry of the shell will considerably simplify the calculations.

Consider a segment of an extremely thin spherical shell as shown in Fig. 1, and assume that the bending stiffness, which is proportional to t^3 can be completely neglected. Under this assumption the strain energy consists only of the energy due to extension of the shell or compression of the median surface, and it is equal to zero in the deflected position ③ provided it is zero in the undeflected position ①, Fig. 1. In other words, neglecting the bending stiffness, the shell will be in equilibrium in the reflected position ③ without the aid of any external pressure applied to the shell surface.

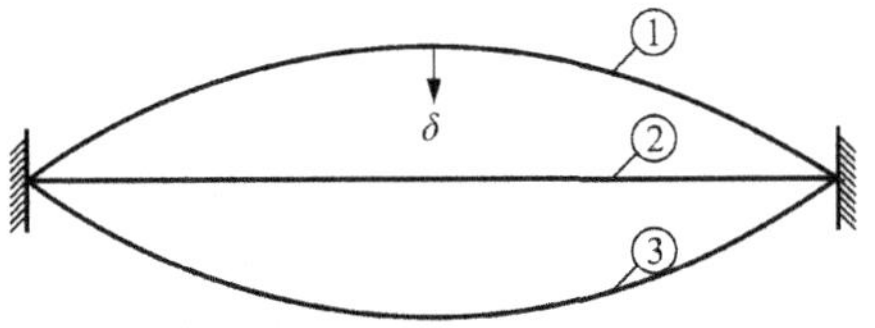

Fig. 1

On the other hand, the intermediate positions between ① and ③ do involve compression of the shell elements, and, therefore, the shell can be held in equilibrium in these positions only by external pressure. When the deflection δ of the shell is between the positions ② and ③, a negative external pressure is necessary to maintain equilibrium as the compressed elements tend to force the shell to take the equilibrium position ③. The pressure-deflection curve, under the assumption of negligible bending stiffness, is, therefore, of the form shown in Fig. 2a.

The effect of the bending stiffness is to increase the positive external pressure necessary to hold the shell in equilibrium. If the case of clamped edges is considered, then, for increasing values of the bending stiffness of the shell, the pressure-deflection curve will take the form shown by the curves 1, 2, etc., in Fig. 2b.

The next question concerns the load at which the shell, subjected to external pressure, will actually fail. The equilibrium positions represented by the portions A_1B_1, A_2B_2, etc. (Fig. 2b) are highly unstable; hence, if the load passes the peak A, which is determined by the stiffness of the particular shell, it immediately drops to the value represented by B. Now the ordinate of A and the form of the peak are extremely sensitive to initial imperfections of the shape of the shell to vibrations, etc. Furthermore, the curve in Fig. 2b is based upon the assumption of a symmetrical type of deflection; the peak in the curve may be lowered by an antisymmetric type of deflection, i.e., if one part of the shell is allowed to move out while the other part moves in. The somewhat analogous problem of a curved bar loaded by a concentrated force was recently investigated by K. Marguerre[5]; he demonstrated that the process indicated in Fig. 1 —

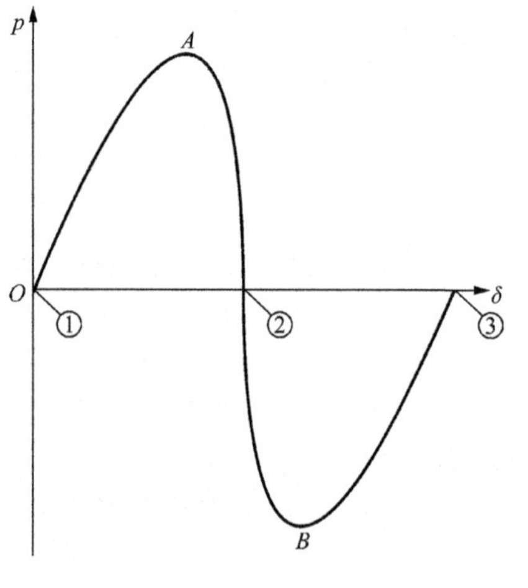

Fig. 2a

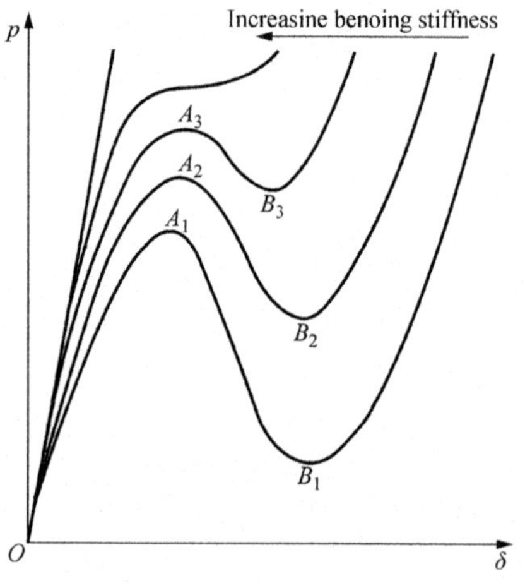

Fig. 2b

the "Durchschlag,"* according to German terminology — is actually precipitated by an anti-symmetric type of deflection. It seems, therefore, probable that if no special precaution is taken to reach the peak A, the failing load observed in the laboratory corresponds to the minimum load B.

It is believed that these simple considerations throw some light on the problem of buckling of curved shells. Consider, for example, a complete spherical shell under the action of uniform external pressure. The classical theory is correct in stating that until the buckling load obtained by the classical theory is reached, any infinitesimal deviation from the spherical form involves an increase of the potential energy of the shell, and, therefore, the spherical form is stable. However, the same classical theory fails to reveal that there are configurations not far away from the spherical form which involve a lower level of the potential energy, and, therefore, the shell will actually jump over into one of these configurations. Such configurations are clearly indicated by the foregoing considerations. Assume, for example, that a segment of the shell subtended by a solid angle 2β is deflected and takes the shape corresponding to the minimum load B, whereas the rest of the shell remains spherical. If it can be shown that the load corresponding to B is lower than the classical buckling load, then the discrepancy between the failing load predicted by the classical theory and found by experiment is explained. The problem is reduced to the determination of the value of the solid angle 2β, which gives the smallest value of the minimum load p_B.

The configuration indicated above is not an exact equilibrium position of the shell, since the curvature has a discontinuity at the boundary of the deflected region; in other words, it does not take into account the reaction moment at the clamped edge. This reaction moment

* There is a need for an English expression for this process; the most descriptive the authors can think of is the popular expression "oil canning."

must be taken by the rest of the shell. In the last section of the present paper (the calculation of the failing load of a spherical shell) the work done by this reaction moment is neglected. It is difficult to estimate the effect of this omission on the value of the failing load. Therefore, a more accurate computation would be highly desirable.

The calculations in this paper refer to the case of the spherical shell. The problem of buckling of a cylindrical shell under axial compression has been attacked by the authors by similar methods, but the investigations have not as yet been completed.

The Energy Expression and the Equation of Equilibrium for a Spherical Segment Under Uniform External Pressure

For the exact calculation of the load-deflection curve of a spherical segment under uniform external pressure the inclusion of non-linear terms in the equations of equilibrium is very complicated*. Therefore, the following simplifying assumptions were made:

(1) The solid angle of the segment is small.

(2) The deflection is rotationally symmetric.

(3) The deflection of any element of the shell is parallel to the axis of rotational symmetry.

(4) The effect of lateral contraction is neglected, i. e., Poisson's ratio is assumed to be zero.

Assumption (3) presumably increases the failing load, since it is a characteristic of the variational method used in the following calculation that any deviation from the exact equilibrium shape increases the load. Assumption (4) probably will not change the results to any appreciable amount. This assumption was made in this paper only because it materially reduces the numerical work.

Fig. 3 indicates the notations used in the following calculations. The choice of the inclination θ of the meridian line as the dependent variable is essential for the simplicity of the equations. As can be seen from Fig. 3, due to the assumed vertical deflection of the shell, stretching or compression of the shell occurs only in the meridian direction. An element whose initial length along the meridian is equal to $dr/\cos\alpha$ has the length $dr/\cos\theta$ after deflection. The strain is, therefore,

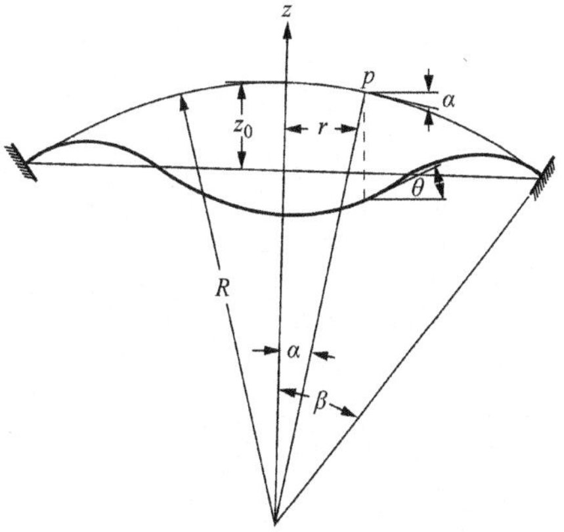

Fig. 3

* The "Durchschlag" of a spherical shell with small curvature under action of a concentrated force was calculated by C. B. Biezeno[6].

$$\varepsilon = \frac{\dfrac{\mathrm{d}r}{\cos\theta} - \dfrac{\mathrm{d}r}{\cos\alpha}}{\mathrm{d}r/\cos\alpha} = \frac{\cos\alpha}{\cos\theta} - 1$$

Hence, the strain energy due to the extension of the elements of the shell is given by

$$W_1 = \frac{Et}{2}\int_0^\beta \left(\frac{\cos\alpha}{\cos\theta} - 1\right)^2 2\pi r \frac{\mathrm{d}r}{\cos\alpha}$$

or, with $r = R\sin\alpha$ and $\mathrm{d}r = R\cos\alpha\mathrm{d}\alpha$,

$$W_1 = \frac{ER^3}{2}\left(\frac{t}{R}\right)2\pi\int_0^\beta \left(\frac{\cos\alpha}{\cos\theta} - 1\right)^2 \sin\alpha\,\mathrm{d}\alpha \tag{3}$$

The two components of the curvature of the shell at an arbitrary point P are equal to $1/R$ before deflection occurs. After deflection the curvature in a meridian plane is

$$\frac{\mathrm{d}\theta}{\mathrm{d}s} = \frac{\mathrm{d}\theta/\mathrm{d}\alpha}{\mathrm{d}s/\mathrm{d}\alpha} = \frac{\mathrm{d}\theta/\mathrm{d}\alpha}{R\cos\alpha/\cos\theta}$$

Hence, the change in curvature of a meridian section is

$$\frac{\mathrm{d}\theta/\mathrm{d}\alpha}{R\cos\alpha/\cos\theta} - \frac{1}{R} = \frac{1}{R}\left[\frac{\cos\theta}{\cos\alpha}\frac{\mathrm{d}\theta}{\mathrm{d}\alpha} - 1\right]$$

Similarly, the change in curvature of a section orthogonal to a meridian plane is $\frac{1}{R}\left[\frac{\sin\theta}{\sin\alpha} - 1\right]$

The strain energy due to bending is, therefore,

$$\begin{aligned} W_2 &= \frac{Et^3}{24}\int_0^\beta 2\pi R^2 \sin\alpha\mathrm{d}\alpha \frac{1}{R^2}\left[\left(\frac{\cos\theta}{\cos\alpha}\frac{\mathrm{d}\theta}{\mathrm{d}\alpha} - 1\right)^2 + \left(\frac{\sin\theta}{\sin\alpha} - 1\right)^2\right] \\ &= \frac{ER^3}{2}\left(\frac{t}{R}\right)^3 \frac{2\pi}{12}\int_0^\beta \sin\alpha\left[\left(\frac{\cos\theta}{\cos\alpha}\frac{\mathrm{d}\theta}{\mathrm{d}\alpha} - 1\right)^2 + \left(\frac{\sin\theta}{\sin\alpha} - 1\right)^2\right]\mathrm{d}\alpha \end{aligned} \tag{4}$$

The potential energy corresponding to the work done by the external pressure is equal to the pressure p times the volume included between the initial and the deflected surface of the shell. The volume enclosed between the deflected surface and the plane of its circular edge is equal to

$$\int_0^\beta 2\pi r z\,\mathrm{d}r = \left[2\pi\frac{r^2}{2}z\right]_0^\beta - \int_0^\beta \pi r^2 \frac{\mathrm{d}z}{\mathrm{d}r}\mathrm{d}r = \pi\int_0^\beta R^2\sin^2\alpha\tan\theta R\cos\alpha\mathrm{d}\alpha = R^3\pi\int_0^\beta \sin^2\alpha\tan\theta\cos\alpha\,\mathrm{d}\alpha$$

The volume enclosed by the initial surface and the plane is equal to $R^3\pi\int_0^\beta \sin^3\alpha\mathrm{d}\alpha$. Hence, the potential energy is given by

$$W_3 = R^3\pi\int_0^\beta \sin^2\alpha(\tan\theta - \tan\alpha)\cos\alpha\mathrm{d}\alpha \tag{5}$$

The total energy, W, of the system is the sum of the strain energy and the potential energy due to external pressure Thus, from Eqs. (3), (4), and (5),

$$\frac{W}{R^3\pi} = E\left(\frac{t}{R}\right)\int_0^\beta \left(\frac{\cos\alpha}{\cos\theta} - 1\right)^2 \sin\alpha d\alpha + \frac{E\left(\frac{t}{R}\right)^3}{12}\int_0^\beta \left[\left(\frac{\cos\theta}{\cos\alpha}\frac{d\theta}{d\alpha} - 1\right)^2 + \left(\frac{\sin\theta}{\sin\alpha} - 1\right)^2\right]\sin\alpha d\alpha +$$
$$p\int_0^\beta \sin^2\alpha \cos\alpha(\tan\theta - \tan\alpha)d\alpha \tag{6}$$

At the equilibrium position, the total energy must be a minimum, therefore, the equation of equilibrium can be obtained by finding the relation between θ and α which will make the integral (6) a minimum. Using the rules of the calculus of variations, the following equation is obtained:

$$2E\left(\frac{t}{R}\right)\left[\frac{\sin\alpha\cos\alpha}{\cos\theta}\tan\theta\left(\frac{\cos\alpha}{\cos\theta} - 1\right)\right] + \frac{E\left(\frac{t}{R}\right)^3}{6}\left[\cos\theta\left(\frac{\sin\theta}{\sin\alpha} + \tan^2\alpha\right) - \right.$$
$$\left.\frac{\cos^2\theta}{\cos\alpha}(2\tan^2\alpha + 1)\frac{d\theta}{d\alpha} + \frac{\sin\theta\cos\theta\tan\alpha}{\cos\alpha}\left(\frac{d\theta}{d\alpha}\right)^2 - \frac{\cos^2\theta\tan\alpha}{\cos\alpha}\frac{d^2\theta}{d\alpha^2}\right] + p\sin^2\alpha\cos\alpha\sec^2\theta = 0 \tag{7}$$

The boundary conditions for this equation are:

$$\theta = 0 \text{ at } \alpha = 0;\ \theta = \beta \text{ at } \alpha = \beta \tag{8}$$

Eqs. (6) and (7) are unwieldy. However, a great simplification results if β, the solid angle of the shell segment, is small. Then by expanding the sine and cosine functions into power series and neglecting terms of higher than the third order in α, θ, and the derivatives of θ, Eqs. (6) and (7) reduce to

$$\frac{W}{R^3\pi} = \frac{E\left(\frac{t}{R}\right)}{4}\int_0^\beta (\theta^2 - \alpha^2)^2\alpha d\alpha + \frac{E\left(\frac{t}{R}\right)^3}{12}\int_0^\beta \left[\left(\frac{d\theta}{d\alpha} - 1\right)^2 + \left(\frac{\theta}{\alpha} - 1\right)^2\right]\alpha d\alpha + p\int_0^\beta \alpha^2(\theta - \alpha)d\alpha \tag{9}$$

and

$$\alpha\frac{d^2\theta}{d\alpha^2} + \frac{d\theta}{d\alpha} - \frac{\theta}{\alpha} = \frac{6}{(t/R)^2}\alpha\theta(\theta^2 - \alpha^2) + \frac{6p}{E(t/R)^3}\alpha^2 \tag{10}$$

These are the simplified energy expression and the equation of equilibrium, respectively. It can be seen that the terms on the left side of Eq. (10) are linear in θ and in its derivatives. They appear also in the usual theory. The first term on the right side brings in the influence of finite deflection.

In order to calculate the maximum deflection δ at the center, first the ordinate z_0 (Fig. 3) at the center has to be computed. By means of the boundary condition $z = 0$ at $\alpha = \beta$, the following relation is obtained:

$$z_0 + \int_0^\beta (dz/dr)dr = 0$$

or

$$z_0 = R\int_0^\beta \tan\theta\cos\alpha d\alpha \tag{11}$$

Before deformation the ordinate at the center is equal to $R(1-\cos\beta)$. Therefore, the deflection δ at the center is given by:

$$\delta = R\int_0^\beta (\tan\alpha - \tan\theta)\cos\alpha\, d\alpha \tag{12}$$

If β is again assumed to be small, Eq. (12) is simplified to

$$\delta = R\int_0^\beta (\alpha - \theta)\, d\alpha \tag{13}$$

Approximate Solution by the Rayleigh-Ritz Method

To calculate the load-deflection curve, one can either solve the differential equation (10) or minimize the integral (9) directly by means of the Rayleigh-Ritz method. Due to the non-linear character of the Eq. (10), it is difficult, if not impossible, to solve it analytically. Therefore, in this paper the Rayleigh-Ritz method is used. To apply this method it is first necessary to find a plausible form of deflection satisfying the boundary conditions.

Due to the assumed symmetrical deflection it is evident that θ must be an odd function of α. The simplest form for a function $\theta(\alpha)$ which satisfies the boundary conditions is

$$\theta = \alpha[1 - K(1 - \alpha^2/\beta^2)] \tag{14}$$

where K is an undetermined parameter. Substituting the expression (14) in Eq. (13), the following relation between the parameter K and the deflection δ is obtained:

$$K = 4\delta/(R\beta^2) \tag{15}$$

If Eq. (14) is introduced into the energy expression, Eq. (9), and the integrations are carried out, the total energy is obtained in the form

$$\frac{W}{R^3\pi} = \frac{Et}{60R}\beta^6\left(K^2 - \frac{K^3}{2} + \frac{K^4}{14}\right) + \frac{Et^3}{18R^3}\beta^2K^2 - \frac{p\beta^4}{12}K \tag{16}$$

The equilibrium condition between the pressure p and the deflection is obtained by putting $\partial W/\partial K = 0$,

$$\frac{1}{R^3\pi}\frac{\partial W}{\partial K} = \frac{Et}{60R}\beta^6\left(2K - \frac{3K^2}{2} + \frac{2K^3}{7}\right) + \frac{Et^3}{9R^3}\beta^2K - \frac{p\beta^4}{12} = 0 \tag{17}$$

Writing $\sigma = pR/2t$, where σ is the uniform compression stress produced by the pressure p under the assumption of small deflections, the following relation is obtained from Eq. (14):

$$\frac{\sigma}{E} = \frac{\beta^2}{5}\left(K - \frac{3}{4}K^2 + \frac{1}{7}K^3\right) + \frac{2}{3}\left(\frac{t}{R}\right)^2\frac{K}{\beta^2} \tag{18}$$

Introducing the deflection δ by using Eq. (15), Eq. (18) can be written

$$\frac{\sigma}{E} = \frac{4}{5}\left(\frac{\delta}{R} - 3\frac{\delta^2}{R^2\beta^2} + \frac{16}{7}\frac{\delta^3}{R^3\beta^4}\right) + \frac{8t^2\delta}{3R^3\beta^4} \tag{19}$$

If σ/E is plotted as a function of δ/R, it is found that the load-deflection curve has the shape indicated in Fig. 2 when

$$\frac{t}{R} < \frac{1}{4}\sqrt{\frac{3}{2}}\beta^2 \tag{20}$$

If t/R is larger than the value on the right side of Eq. (20), the load increases with the deflection without having a maximum and a minimum; for $t/R = \frac{1}{4}\sqrt{\frac{3}{2}}\beta^2$ the curve for load *vs*. deflection has an inflection point with a horizontal tangent.

To find the smallest of the minimum values of the functions $\sigma/E = f(\delta/R)$, first vary β, and determine $(\sigma/E)_{\text{min.}}$ for given values of δ/R. By differentiation of Eq. (19) with respect to β^2, the value of β^2 which makes σ/E a minimum is obtained in the form:

$$\beta^2 = \frac{32}{21}\frac{\delta}{R} + \frac{20}{9}\frac{t^2}{\delta R} \tag{21}$$

Substituting this value of β^2 into Eq. (19), the following relation is obtained:

$$\frac{\sigma}{E} = \frac{4}{5}\frac{\delta}{R}\frac{1+\frac{3}{280}\left(\frac{\delta}{t}\right)^2}{1+\frac{24}{35}\left(\frac{\delta}{t}\right)^2} \tag{22}$$

or

$$\frac{\sigma R}{Et} = \frac{4}{5}\frac{\delta}{t}\frac{1+\frac{3}{280}\left(\frac{\delta}{t}\right)^2}{1+\frac{24}{35}\left(\frac{\delta}{t}\right)^2} \tag{23}$$

Therefore, $\sigma R/(Et)$ is a function of δ/t only. The function defined by Eq. (23) is represented in Fig. 4 by the curve labeled "envelope." The physical meaning of this curve is as follows: For values of $\sigma R/(Et)$ and δ/t corresponding to points that lie below the curve, no equilibrium position is possible, for any value of the solid angle of the segment 2β. This is particularly clear when Eq. (19) is rewritten in the following form:

$$\frac{\sigma R}{Et} = \frac{4}{105}\left(\frac{\delta}{t}\right)\left[21 - 63\left(\frac{\delta}{t}\right)\frac{t/R}{\beta^2} + \left(48\frac{\delta^2}{t^2}+70\right)\frac{(t/R)^2}{\beta^4}\right]$$

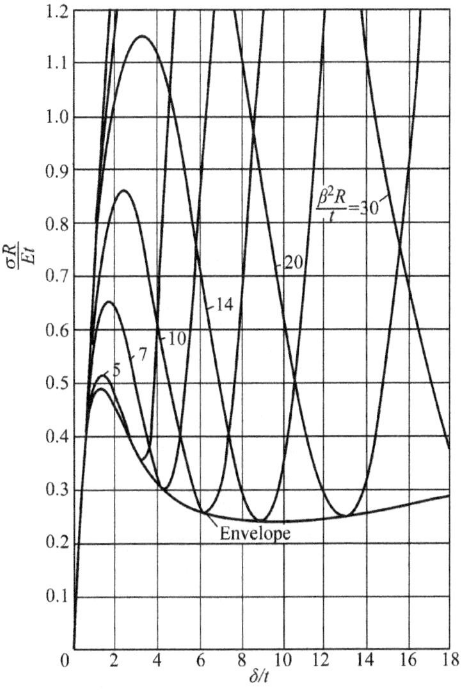

Fig. 4

Using $\beta^2/(t/R)$ as a parameter a family of curves expressing the relation between $\sigma R/(Et)$ and δ/t can be plotted (Fig. 4). Then the relation given by Eq. (23) represents the envelope of this family of curves. Hence, the maximum of the envelope curve $\sigma = 0.490\,8Et/R$ gives the smallest value for any peak through which the load has to pass before the shell collapses, provided the shell is of exact spherical shape and its deformation occurs with exact axial symmetry. It is noteworthy that the maximum corresponds to a deflection-thickness ratio as small as $\delta/t = 1.248$. The minimum value of the load which is able to keep the shell in a deflected shape corresponds to $\delta/t = 9.349$, and is given by $\sigma R/(Et) = 0.237\,7$ or

$$\sigma = 0.237\,7E\left(\frac{t}{R}\right) \tag{24}$$

Application to the Buckling Problem. Comparison with the Experiment

The results of the last section indicate that equilibrium positions involving finite deflections exist at much smaller loads than the buckling load given by the classical linear theory. To apply the method to the problem of buckling of a spherical shell under external pressure the energy expression, Eq. (9), must be modified to include the strain produced by uniform compression of the spherical shell before buckling. The total energy can be written as

$$\begin{aligned}\frac{W}{R^3\pi} = {}& \frac{Et}{R}\int_0^\beta\left[\frac{1}{2}(\theta^2-\alpha^2) - \frac{pR}{2Et}\right]^2\alpha\,\mathrm{d}\alpha + \\ & \frac{Et^3}{12R^3}\int_0^\beta\left[\left(\frac{\mathrm{d}\theta}{\mathrm{d}\alpha}-1\right)^2+\left(\frac{\theta}{\alpha}-1\right)^2\right]\alpha\,\mathrm{d}\alpha + p\int_0^\beta\alpha^2(\theta-\alpha)\,\mathrm{d}\alpha\end{aligned} \tag{25}$$

Using again Eq. (14) for θ, the equilibrium condition $\partial W/\partial K = 0$ leads in this case to

$$\frac{\sigma}{E} = \frac{1}{70}\beta^2[28-21K+4K^2] + \frac{4}{3}\left(\frac{t}{R}\right)^2\frac{1}{\beta^2} \tag{26}$$

Substituting K from Eq. (15), Eq. (26) becomes

$$\frac{\sigma}{E} = \frac{2}{5}\beta^2 - \frac{6}{5}\frac{\delta}{R} + \left[\frac{32}{35}\frac{\delta^2}{R^2} + \frac{4}{3}\frac{t^2}{R^2}\right]\frac{1}{\beta^2} \tag{27}$$

Now determine the minimum of σ/E for a given value of δ/R by varying β^2, thus

$$\beta^2 = \sqrt{\frac{16}{7}\left(\frac{\delta}{R}\right)^2 + \frac{10}{3}\left(\frac{t}{R}\right)^2}$$

and

$$\frac{\sigma R}{Et} = \frac{4}{5}\left\{\sqrt{\frac{16}{7}\left(\frac{\delta}{t}\right)^2+\frac{10}{3}} - \frac{3}{2}\left(\frac{\delta}{t}\right)\right\} \tag{28}$$

This relation is plotted in Fig. 5. For $\delta/t = 0$,

$$\sigma = 1.460\,6Et/R \tag{29}$$

Of course this value of the buckling stress is much higher than the value given by the linear theory. This is expected because the assumed buckling form is far away from that resulting from the linear theory, i.e., it is a very "unfavorable" shape for infinitesimal deflections. The minimum of $\sigma R/(Et)$ is equal to 0.182 58, i.e., the minimum load necessary to keep the shell in the buckled shape corresponds to:

$$\sigma_{\text{min.}} = 0.182\,58Et/R \tag{30}$$

This shows that the assumed shape is "favorable" for finite deflections.

The value of β corresponding to $\sigma_{\text{min.}}$ is

$$\beta = 3.821\,8\sqrt{t/R} \tag{31}$$

The value of β corresponding to the value given by Eq. (29) is equal to $\beta = 1.825\,7\sqrt{t/R}$. The deflection corresponding to $\sigma_{\text{min.}}$ is equal to about ten times the thickness t of the shell.

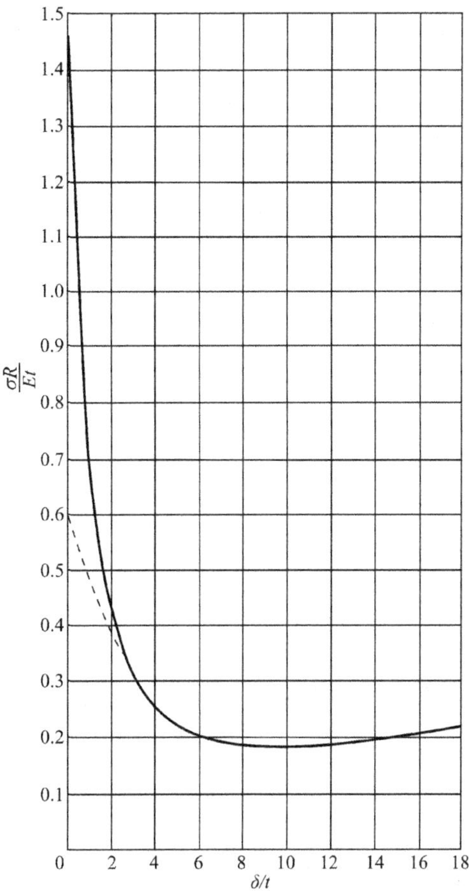

Fig. 5

E. E. Sechler and W. Bollay found, by subjecting a thin-walled copper hemisphere of 18 in. radius to external fluid pressure, a buckling stress of

$$\sigma = 2\,480 \text{ lbs. per sq. in.}$$

With $E = 14.5 \times 10^6$ lbs. per sq. in., $t = 0.020$ in., i.e., $R/t = 900$, this value corresponds to

$$\sigma = 0.154Et/R \tag{32}$$

The experimental result compares quite favorably with the theoretical value given by Eq. (30). The experimental values obtained for β and δ were: $\beta = 8°$ and $\delta \cong 0.25$ in., i.e., $\delta/t \cong 12.5$. The present theory gives: $\beta = 7.4°$ (cf. Eq. (31)) and $\delta/t \cong 10$. The theory, therefore, reproduces to a fair approximation the physical process. The linear theory gives (cf. Eq. (2)) $\sigma = 0.606Et/R$; the value of β (corresponding to the first nodal line) would be, according to the linear theory, 3.3°, and δ/t is undetermined.

It appears that in the case of curved shells it is necessary to introduce an "upper" buckling load given by the classical linear theory and a "lower" buckling load which is equal to the minimum load necessary to keep the shell in a buckled shape with finite deformations. In Fig. 5 the probable shape of the curve for load *vs.* deflection connecting both values is indicated by a dotted line. The essential feature of the present theory is that it determines a lower buckling load independently of the initial imperfectness of the specimen or the load arrangement, whereas all previous attempts to take into account finite deformations led to failing loads which depend on arbitrary assumptions concerning the magnitude of such imperfections or lack of symmetry. It seems that the upper buckling load can be approached experimentally only if extreme precaution is taken both in the manufacture of the specimen and in performing the test. With the amount of imperfections and tolerances in engineering practice, the buckling load obtained will invariably be very near the lower buckling load, and, of course, this lower value is to be specified for design.

The new theory also reveals the essential difference between the buckling of a flat and a curved plate. The finite deformations of a buckled flat plate were calculated by several authors, e.g., S. Timoshenko[7], H. L. Cox[8], M. Yamamoto and K. Kondo[9], and K. Marguerre[10]. The results of these authors do not agree completely due to the different simplifying assumptions introduced. However, all investigators agree that after buckling of the plate an increase in the load is necessary to increase the deflections. The load falls only if the elastic limit is passed. The senior author[11] has shown for the analogous case of a straight beam that, due to the decrease of the load after buckling, the experimental scatter is much larger if the buckling occurs in the plastic range than if it occurs in the elastic range. The rapid decrease of the load after buckling revealed in this paper for curved shells is a pure elastic phenomenon, and, therefore, changes the entire theoretical and practical aspects of the buckling problem as far as curved shells are concerned.

References

[1] Flügge W. Die Stabalität der Kreiszylinderschale. Ingenieur Archiv, 1932, 3: 463 - 506.

[2] Donnell L H. A New Theory for the Buckling of Thin Cylinders Under Axial Compression and Bending. Transactions A. S. M. E., 1934, 56: 795 - 806.

[3] Timoshenko S. Theory of Elastic Stability. McGraw-Hill, New York, 1936: 491 - 497 (A brief account of the classical theory of buckling of spherical shell under uniform external pressure, including references).

[4] Cox H L. Stress Analysis of Thin Metal Construction. Preprint of Royal Aeronautical Society, 1939.

[5] Marguerre K. Die Durchschlagskraft eines schwach gekrümmten Balkens. Sitzungsberichte der Berliner Mathematischen Gesellschaft, 1938, 37: 22 - 40.

[6] Biezeno C B. Über die Bestimmung der Durchschlagkraft einer schwach gekrümmten kreisförmigen Platte. Z. A. M. M., 1938, 15: 13 - 30.

[7] Timoshenko S. Theory of Elastic Stability. McGraw-Hill, New York, 1936: 390 - 393.

[8] Cox H L. The Buckling of Thin Plates in Compression. British A. R. C. Reports and Memoranda, No. 1554, 1933.

[9] Yamamoto M. Kondo K. Buckling and Failure of Thin Rectangular Plates in Compression. Rep. of

Aero. Res. Inst., Tokyo, No. 119, 1935.

[10] Marguerre K. Die mittragende Breite der gedrückten Platte. Luftfahrtforschung, 1937, 14: 121-128.

[11] von Kármán Th. Untersuchungen über Knickfestigkeit. Forschungsarbeiten, Berlin, 1910: 81.

The Influence of Curvature on the Buckling Characteristics of Structures

Th. von Kármán, Louis G. Dunn and Hsue-shen Tsien
(California Institute of Technology)

In the field of applied elasticity, one of the most perplexing problems is the prediction of the buckling load, or rather the failing load of a thin-walled structure with either simple or double curvature. Everyone who has contact with this subject will notice the gap between theory and experimental results. The designer, however, has to proceed with his work regardless of whether or not the theory of elasticity can give him the correction solution of his problem. Hence, in this case he has to resort to empirical relations determined by experimental methods. But such an empirical approach to a complex subject without solid physical basis has its definite limitations. Therefore, a correct picture of the interactions of the different factors which determine the failing load and the mechanism of the failing process will be always useful to the designer.

In this paper the authors do not present a new theory, but certain considerations which they believe bring out the crucial point of the subject. In Section I, a comparison is made between the buckling of one dimensional and two dimensional structures with and without curvature. Section II contains a critical examination of the discrepancies between the classical buckling theory of cylindrical shells and the experimental evidence together with a description of various investigations which have been made to reveal the true character of the mechanism of failure. In Section III, the buckling phenomena observed in the laboratory for different structures are discussed from the point of view developed in the previous sections.

Section I

Columns

One of the earliest problems of Applied Elasticity is that of a uniform column subjected to axial thrust, generally referred to as Euler's problem. This problem was first investigated by Euler in 1744 and has since occupied the attention of numerous investigators. The exact solution of the problem requires a solution of a non-linear differential equation which determines the shape of the deflected central line of the column. Using a rectangular coordinate system x, y, such that the x axis coincides with the line of thrust, then the differential equation for the displacement w in the y direction may be written in the form:

Presented at the Structures Session, Eighth Annual Meeting, I. Ae. S., New York, January 25, 1940.
Journal of the Aeronautical Sciences, vol. 7, pp. 276 – 289, 1940.

$$\frac{d^2 w}{dx^2}+\frac{Pw}{EI}\left[1+\left(\frac{dw}{dx}\right)^2\right]^{3/2}=0 \qquad (1)$$

An approximate solution can be readily obtained by assuming small deflections and neglecting the second order term (dw/dx)[2] in the bracket. This simplification leads to a simple second order linear differential equation of the form:

$$\frac{d^2 w}{dx^2}+\frac{P}{EI}w=0 \qquad (2)$$

The boundary conditions for a simple supported column are $w=0$ for $x=0$ and $x=l$, where l is the length of the column. The expression

$$w=A\sin\sqrt{P/(EI)}\,x \qquad (3)$$

is the solution of Eq. (2), which satisfies the boundary condition at $x=0$. In order to satisfy the boundary condition for $x=l$, $\sqrt{P/EI}l$ must be equal to $n\pi$, where n is an integer. Hence the "critical" values of P are

$$P_n=n^2\pi^2 EI/l^2 \quad (n=1,2,3,\cdots) \qquad (4)$$

Thus for a thrust corresponding to any one of the critical values of P, a state of equilibrium exists in which the deflection form is given by

$$w=A_n\sin n\pi x/l \qquad (5)$$

It is seen that the amplitude remains undetermined. The first critical load, i. e., $n=1$, corresponds to the "Euler buckling load."

If $P\neq P_n$, the linear theory gives $w=0$ as the only equilibrium shape. If it is desired to obtain the force-deflection relation beyond the first buckling load P_E it is necessary to integrate the non-linear Eq. (1). The integration of this equation by the use of elliptic integrals is well known and only the final results will be given here. A relation between the thrust P and the deflection δ at the center is obtained in the parametric form:

$$Pl^2/(EI)=4[K(\sin\alpha/2)]^2 \qquad (6)$$

$$\frac{\delta}{l}=\frac{\sin\alpha/2}{K(\sin\alpha/2)} \qquad (7)$$

Here K ($\sin\alpha/2$) denotes the complete elliptic integral of the first kind of the modulus $\sin\alpha/2$, where α is the angle between the line of thrust and the tangent to the centroidal axis at its end points. From these two equations it is a relatively simple matter to obtain the relation between P and δ illustrated in Fig. 1. It is seen that with a steadily increasing thrust the deflection δ will increase and then slowly decrease. The decrease in deflection is easily understood by considering the physical process illustrated in Fig. 2. It is seen that the deflection will increase to a maximum when the two end points are close together and as they pass each other a point will be reached at which the deflection starts to decrease. It is of course assumed in this discussion, which is generally referred to as the theory of the "Elastica," that no plastic deformation takes place. This would only apply to extremely slender columns.

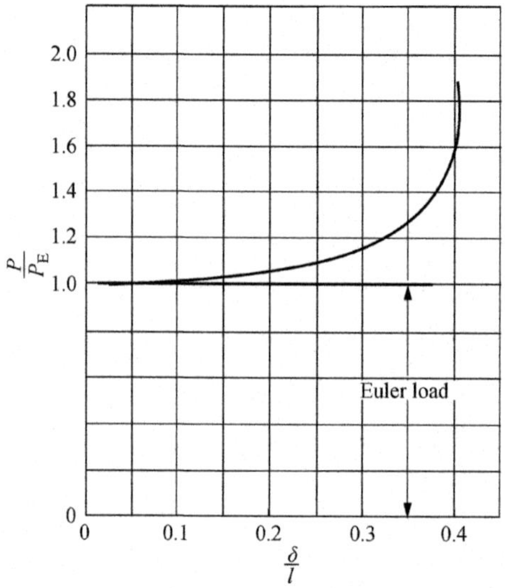

Fig. 1 Theoretical load-deflection curve for a slender column

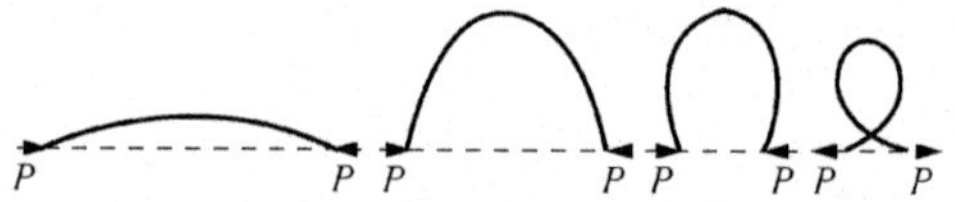

Fig. 2 Progressive deformations of an elastic strut

Another interesting aspect of the problem which seems to appeal particularly to the practical mind of the engineer, is the effect of small initial irregularities. This can be done quite readily by either assuming a small initial curvature, or a bending moment. If all deflections are considered small; the effect of an initial curvature or bending moment can be taken into account by an additional term in Eq. (2), which in this case also remains linear. The practical significance of the equation thus obtained is that it yields a solution in which the deflection δ appears explicitly, namely:

$$\frac{P}{P_E} = \frac{1}{1 + a_1/\delta} \tag{8}$$

where P is the column load, P_E the Euler buckling load and a_1 is a constant proportional to the initial deflection. It is evident that as δ becomes large compared to a_1 the load P always approaches Euler's load, as indicated by curve A in Fig. 3.

In engineering practice it is commonly considered that the approximate solution is sufficient to describe the collapsing of slender columns. A detailed consideration will show this to be generally true. It was shown that for the strut with a small initial displacement the thrust, P, according to the linear theory approaches asymptotically a horizontal line corresponding to the Euler load. The exact solution for this problem would yield a line B, Fig. 3, which follows in

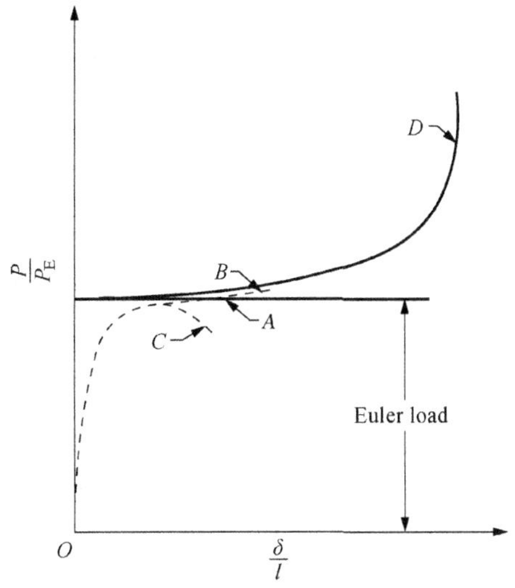

Fig. 3 Actual load-deflection curves of slender columns

general trend the curve D, obtained for an initially straight column, i. e., an increase in deflection is also accompanied by a steadily increasing thrust. However, there still remains to be considered the effect of the physical properties of the material. The column will, after deflection, be subjected to combined compression and flexural stresses. Since at first δ increases much faster than P, the flexural stress will predominate. For a slender column in which the compression stress is well below the yield stress it may then be concluded that during the loading process the deflection will follow the curve B until the combined compression and flexural stress reaches the yield stress. When this point, which will in general be close to the Euler load, is reached the load deflection curve follows a curve similar to C. Therefore, it can be said, that for slender columns in which the compression stress is well below the yield stress, the collapsing load will always be very near to the Euler buckling load.

However, if the column is short so that the stress in the column at buckling is very near the yield stress, the effect of plastic deformation on the load will be felt immediately after buckling starts. This would cause a drop in the load-deflection curve as shown in Fig. 4. This phenomenon was shown by the senior author in 1909[4]. An essential difference between this case and the case of a slender column is that the presence of initial deflection will make the failing load of a short column much lower than the Euler load (cf. Fig. 4).

Flat Plates

In the previous section it was noted, that in general the buckling load calculated by the linear theory was sufficient for determining the failing load of columns. However, the buckling load is in many cases not sufficient as a design criterion, as for instance, in the design of airplane structures of the semi-monocoque type where advantage is taken of the fact that the

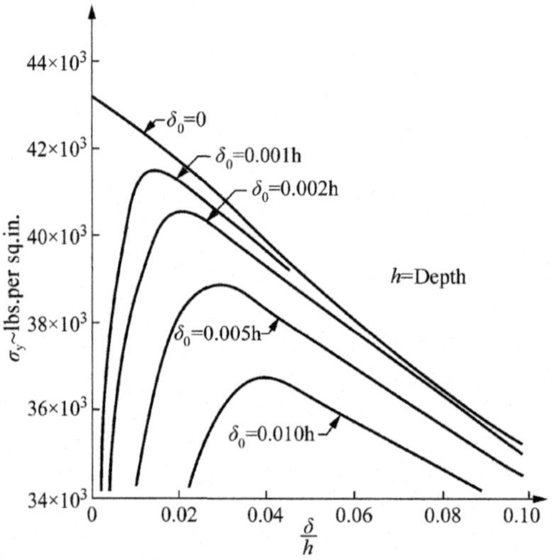

Fig. 4 Calculated load-deflection curves for short columns ($l/\rho = 75$) with various amounts of initial deflection for mild steel with a yield point = 45 000 lbs. per in² (cf. von Kármán[1])

thin metal sheet covering of wings, fuselages, etc., although buckling at a low load, has an ultimate strength or load-carrying ability which is in some cases a large multiple of the buckling load. For an efficient design it is then necessary that the designer be able to calculate with a fair degree of accuracy the amount of load which can be transferred through the metal sheet covering, apart from that which is carried by the longitudinal stiffening members.

For purposes of analysis the sheet between the longitudinal members may be considered as a flat plate. This problem is much more complicated than the analogous problem of the "Elastica" for, in the case of the deflected column the resultant of the normal stresses in any arbitrary cross-section is determined statically by the end thrust P, whereas in the case of the deflected plate the distribution of the resultant stresses acting in the median surface of the plate depends on the distribution of the normal deflection.

An exact solution of the problem is quite difficult as the amplitude is a non-linear function of the end thrust, and the wave form changes with increasing deflection. Approximate solutions for the case in which the edge stress is not very far beyond the buckling stress have been given by S. Timoshenko[2], K. Marguerre[3], E. Trefftz[4], H. L. Cox[5], and M. Yamamoto and K. Kondo[6]. They assumed either a wave form very near to the buckling wave form or an average extentional strain to simplify the differential equations. The former assumption precludes the possibility of multiple waves near the edges of the plate and the latter assumption oversimplifies the interaction between the bending and the extensional stiffnesses. Therefore, both tend to limit their usefulness for higher edge stresses. For loads far in excess of the buckling load, the senior author[7] of this paper developed an approximate method, based upon the following considerations: If a rectangular plate freely supported along two edges is compressed between two rigid plates, the distribution of stress will be uniform until buckling takes place. Further compression will cause buckling across the plate except at the edges. Hence, as the end plates are brought still further together the stress along the edges will increase to a large multiple of the buckling stress, whereas the stress in the center of the plate remains substantially of the order of the buckling stress.

To simplify the calculations it was then assumed that near failure the entire load is carried by two narrow strips adjacent to the edges. An analysis based on these simplifying assumptions leads to a simple and convenient expression for the total load carried by the plate, namely:

$$P = C\sqrt{E\sigma}t^2 \tag{9}$$

This equation is only valid below the elastic limit; beyond the elastic limit it is necessary to correct for the change in E. On the other hand if σ is replaced by σ_y, the yield point stress, the resulting value of P will differ but slightly from the maximum value of P. Hence for calculating the maximum load that can be sustained by the plate, Eq. (9) is replaced by

$$P_{\text{max.}} = C\sqrt{E\sigma_y}t^2 \tag{10}$$

where E = Young's modulus of elasticity in lbs. per sq. in.

σ_y = Yield point stress in lbs. per sq. in.

t = The plate thickness in inches.

The value of the constant could be either calculated in accordance with the assumptions or experimentally determined.

It is of interest to note that the experimental load deflection curve, Fig. 5, of a sheet attached to longitudinal stiffeners is, in the elastic range, analogous to that of the "Elastica." A non-dimensional plot is obtained by plotting the ratio of the applied load to the buckling load as a function of the ratio of the maximum amplitude to the half-wave length. As certain portions of the sheet are stressed beyond the elastic limit, the curve tends to flatten until finally the load decreases with increasing deflection. A characteristic behavior of the plate, of which a clear understanding is essential in design work, is, that the load carried by the plate is a non-linear function of the edge stress. This fact is clearly demonstrated by the experimental curve of Fig. 6.

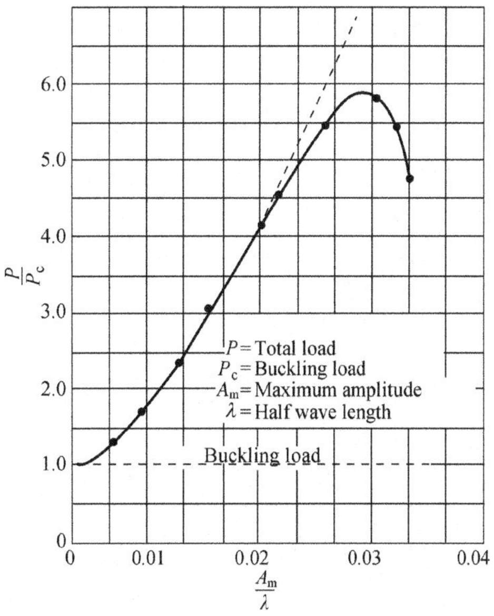

Fig. 5 Load-deflection curve for a thin dural sheet (t = 0.025 in.; b = 5 in.) riveted to strong stiffeners (I_{st} = 0.009 4 in[4].)

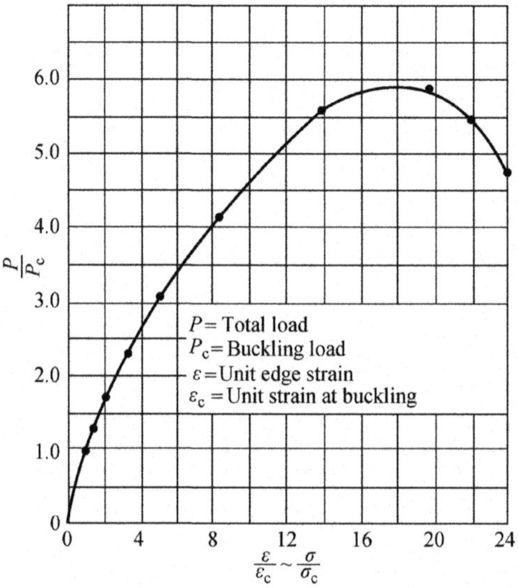

Fig. 6 The relation between the total load, P on the sheet and its edge strain ε (The same specimen as for Fig. 5.)

Curved Bars

So far, the discussion has dealt with structures in which there is no curvature in the undeformed state. If the undeformed structure is curved an entirely new phenomenon is revealed. This new phenomenon will be the central interest of the present paper.

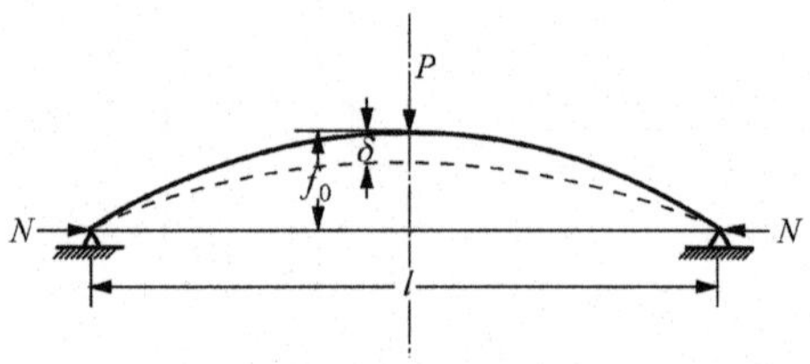

Fig. 7 A curved bar under a concentrated load P applied at the center of the bar

To simplify the problem, consider first the case of a curved bar under the action of a single concentrated load P at the center (Fig. 7). The ends of the bar are assumed to be laterally restrained. Now, starting from the undeformed position assume the load P is gradually increased and consider for the time being only the symmetrical type of deflection. During the initial stage of loading, the bar behaves in a manner similar to that of a straight beam under the action of a concentrated load, i.e., the load P increases with the deflection δ at the center of the bar. However, large deflections in a curved bar with fixed ends produce a shortening of the centroidal axis and consequently an increasing compression. This is contrary to the case of a straight beam with fixed ends where large deflections will produce tension. It is a well known fact that a beam under end compression is much weaker in sustaining a lateral load than one without end compression. In fact for the case of a straight beam, when the end compression reaches the Euler load, the beam loses all its ability to sustain a lateral load. This general property of beams applies also to the case of the curved bar under consideration. Thus, with increasing compression in the bar due to an increasing deflection δ at the center, the effective rigidity of the bar to sustain the lateral load P is gradually reduced. In other words, the slope of the P vs. δ curve decreases with increasing δ. Thus as the load P is increased, a point will be soon reached where the slope is decreased to zero. Therefore a maximum load P is reached.

Beyond this point, the load P will decrease with increasing deflection δ. That is, this part of the P vs. δ curve is highly unstable. This instability will continue until the actual shape of the bar has a curvature opposite to that of the undeformed bar. In other words, if the undeformed bar is curved downward, the bar has to be deformed so far as to curve upward before the decreasing process of the load P stops. If the deflection δ is further increased, then the load P again increases. The reason for this phenomenon can be again sought in the change of the end compression in the bar. Once the curvature of the bar is reversed, then the increase of deflection will decrease the compression in the bar, as shown in Fig. 8. This decrease in the end compression will, of course, increase its ability to carry lateral load. Finally the compression in the bar is changed into tension and its magnitude increases with further increase in the deflection δ. Thus increasing δ will further increase the ability of the bar to carry a lateral load P, i. e., the slope of the P vs. δ curve will now increase with an increase in δ.

The process described in the preceding paragraphs has been demonstrated very clearly by a

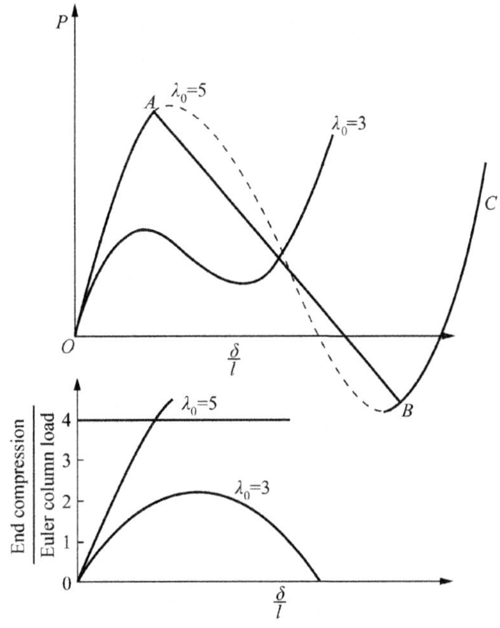

Fig. 8 Load-deflection relations for curved bars

($\lambda_0 = f_0/\rho$ where ρ = radius of gyration of the bar, cf. Fig. 7) (see reference 8) mathematical investigation of K. Marguerre[8]. Fig. 8 was taken from K. Marguerre's paper.

However, so far only the symmetrical type of deflection has been considered. Actually for slender bars, the situation is further complicated by the appearance of an anti-symmetrical type of deflection. It was shown by K. Marguerre that when the end compression in the bar reaches a value which is four times the Euler load, an anti-symmetrical type of deflection appears together with the symmetrical type of deflection discussed. This new component of deflection gives a P vs. δ curve as shown by the light straight line in Fig. 8. It is thus evident that the actual loading will start from the origin along the curve corresponding to a symmetrical deflection up to point A. Then, the anti-symmetrical component of deflection sets in and the bar will follow the light straight line to point B. From then on the deflection of the bar is again purely symmetrical and follows the heavy curve with increasing load. In a testing machine, the bar will actually "jump" from point A to some point C, depending upon the rigidity of the machine, with violent vibration due to a sudden release of elastic energy.

It is thus shown, in the case of a curved bar with laterally restrained ends and a load at the center, that the load-deflection relation is not linear but highly complicated. Moreover, the phenomenon occurs completely within the elastic range. Hence, the linear stress-strain relation gives a non-linear load-deflection relation involving a region where the load decreases with increase in deflection. This phenomenon is different from that of the plastic buckling of an Euler column in that it is purely elastic. It is again different from the case of a flat plate in that an increase in deflection can cause a decrease in load.

Spherical Shells

The buckling of a spherical shell under uniform external pressure is quite similar to the case of the curved bar just discussed. Consider a segment of an extremely thin spherical shell as shown in Fig. 9, and assume that the bending stiffness, which is proportional to t^3 can be completely neglected. Under this assumption the strain energy consists only of the energy due to extension or compression of the median surface of the shell, and it is equal to zero in the deflected position ③ provided it is zero in the undeflected position ①, Fig. 9. It is thus evident that it will be in equilibrium in position ③ without the aid of external pressure.

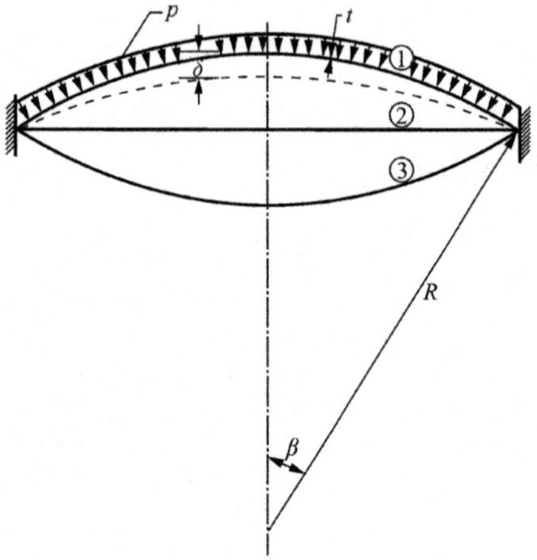

Fig. 9 Spherical shell segment under a uniform external pressure p

On the other hand, the intermediate positions between ① and ③ do involve compression of the shell elements, and, therefore, the shell can be held in equilibrium in these positions only by an external pressure. However, just as in the case of the curved bar, the compression in the shell elements tends to reduce its pressure-carrying ability. Thus, if only a rotationally symmetric type of deflection is considered, the initial part of the pressure p vs. maximum deflection δ curve again has a decreasing slope with increasing δ. When the deflection δ of the shell goes beyond the position ② and above position ③, a negative external pressure is necessary to maintain equilibrium as the compressed elements tend to force the shell to take the equilibrium position ③. The pressure-deflection curve under the assumption of negligible bending stiffness and symmetrical deflection is, therefore, of the form of curve A_1, shown in Fig. 10.

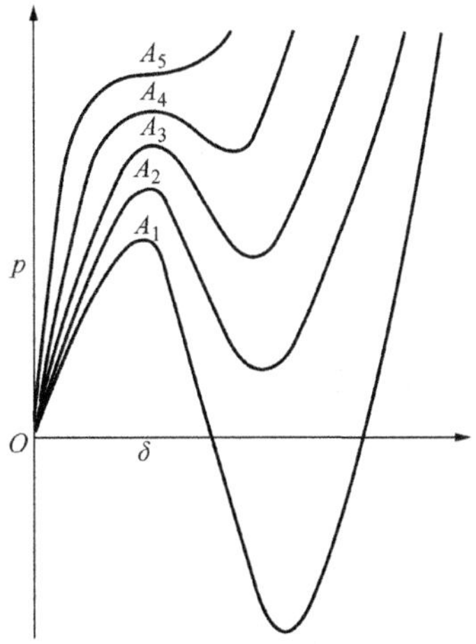

Fig. 10 Pressure-deflection curves for spherical shell segment with increasing thickness − radius ratios (A_1, A_2, A_3, ...)(cf. Fig. 9)

The effect of the bending stiffness is to increase the positive external pressure necessary to hold the shell in equilibrium. In other words, the pressure-deflection curves with increasing bending stiffnesses should be of the form of curves A_2, A_3, ... in Fig. 10.

The calculation of a spherical shell segment under uniform external pressure with clamped and laterally restrained edges was carried out by two of the present authors[9] under several simplifying assumptions. Fig. 11 is taken from this investigation. However, it should be noted that this load-deflection curve was obtained without the consideration of the anti-symmetrical type of deflection. Thus, it is possible that instability might occur before the point A is reached (Fig. 11) due to the anti-symmetrical type of deflection as in the case of a curved bar.

Hence, a segment of a spherical shell under a uniform external pressure behaves similarly to a curved bar under a lateral load. In both cases, the load-deflection curve is not a straight line and has an unstable portion within the elastic regime.

With this non-linear relation between the load and the deflection of a spherical segment the problem of buckling of spherical shells in general takes on an aspect quite different from that of the classical theory. The latter assumes that the shell buckles into a field of rotationally

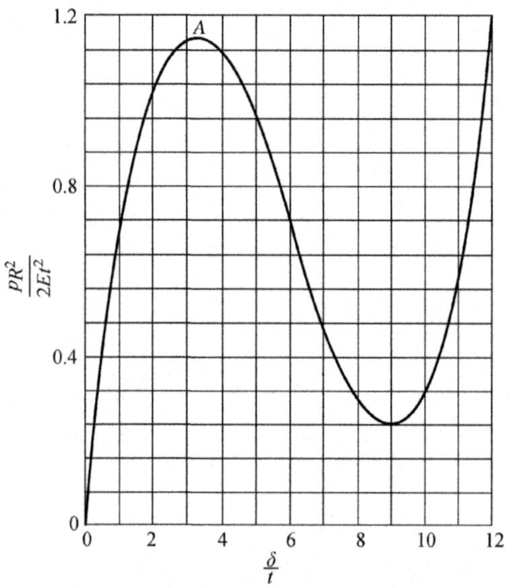

Fig. 11 Calculated load-deflection curve for a spherical shell segment with $\beta^2 R/t = 14$ where β is the semi-vertex angle of the segment (cf. Fig. 9) (The edge is clamped and laterally restrained.)

symmetrical waves of infinitesimal amplitude and gives a critical value for the external pressure as

$$pR/(2t) = 0.606E(t/R) \tag{11}$$

As this value is much higher than observed, the following calculations were carried out. It was assumed that the deflection is restricted to a small segment whose vertex angle is undetermined. The load-deflection curve was then calculated for various fixed values of the vertex angle. The minimum loads obtained in this way give an envelope as shown in Fig. 12. The significance of this envelope is that it shows corresponding values of minimum pressure and deflections, for which the deflected shell is in equilibrium. For very small deflections this calculation does not give the smallest possible loads due to the restrictions imposed by the assumed deflection shape. The correct location of the point of intersection between the envelope and the vertical axis ($\delta/t = 0$) is given by Eq. (11). The probable correct shape of the envelope for small values of δ/t is shown by the dotted line in Fig. 12. The most important point of this investigation is that the pressure which is necessary to keep the shell in equilibrium decreases once the shell starts to buckle. This is due to the decrease in effective stiffness of the shell with increasing deflection as in the case of curved bars discussed in the preceding paragraphs. The buckling load given by the new non-linear theory will be called the "minimum buckling load" and the point of intersection of the equilibrium curve with the vertical axis, given by the classical theory, the "initial buckling load."

The decrease in load after buckling appears superficially to be similar to the case of a short column (cf. Fig. 4). But there is one fundamental difference between these two cases: While

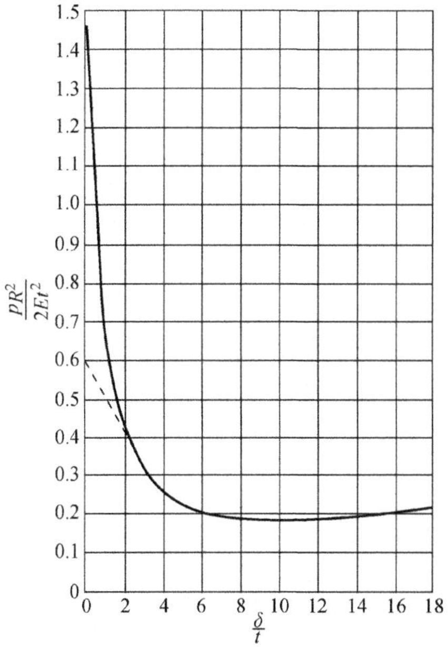

Fig. 12 Calculated buckling load-deflection curve for a spherical shell under uniform external pressure p

the decrease in load carried by a short column after buckling is due to plastic yielding of the material, the decrease of load on the spherical shell after buckling occurs completely within the elastic regime.

The only experimental data known to the authors are those obtained by E. E. Sechler and W. Bollay on a copper hemisphere of 18 inches radius and a thickness of 0.02 inch. Their results are compared with the theoretical values in the following table:

	σ	Maximum deflection	Angular extension of buckle
Experiments	$0.154E(t/R)$	$\approx 12.5t$	16°
Minimum buckling load (Theoretical)	$0.1826E(t/R)$	$\approx 10t$	14.8°
Initial buckling load (Theoretical)	$0.606E(t/R)$		6.6° (First nodal line)

It is seen that the theoretical "minimum buckling load" agrees closely with the experimental value whereas the "initial buckling load" does not. Thus, the shell actually does not reach the classical buckling load but "jumps" to an equilibrium position involving a large deflection at a considerably lower load.

Section Ⅱ

The Inadequacy of the Classical Theory of Cylindrical Shells

It appears from the results of the first section that any buckling theory of thin shells based upon considerations of small deflections alone will only give the "initial" buckling load and will not be able to explain the failing load experimentally obtained, which is much more important in practical applications. One example of this general inadequacy of the small deflection theory is the classical investigations of buckling of cylindrical shells.

The buckling load of cylindrical shells under axial compression, based on the classical theory, was calculated by R. Lorentz, R. V. Southwell, S. Timoshenko, W. Flügge, L. H. Donnell, and others. The buckling stress obtained by these calculations is given by the equation

$$\sigma_{cr} = 0.606E(t/R) \tag{12}$$

The same problem was also investigated experimentally by many authors, especially by E. E. Lundquist[10] and L. H. Donnell[11]. In comparing the theoretical with the experimental results, it is found that the theoretical buckling load is about 3 to 5 times higher than the experimental value, as shown in Fig. 13. To remedy this situation W. Flügge[12] first considered the deviation of the assumed end conditions of the cylindrical shell from that realized in the laboratory. However, as was indicated previously[9] neither this effect nor the assumption of initial deviations made by W. Flügge[12] and later by L. H. Donnell[11] are sufficient to explain the discrepancies.

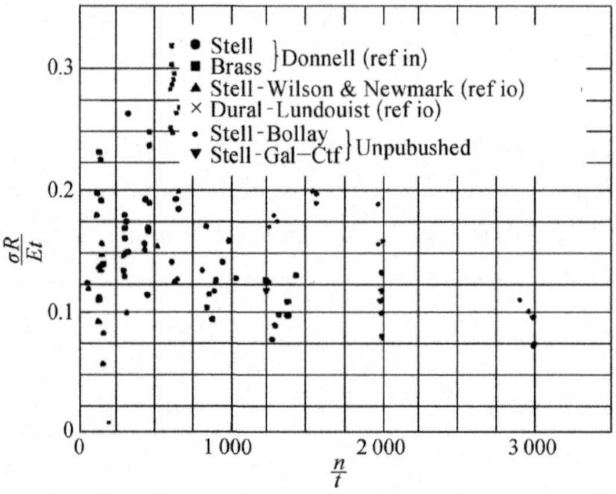

Fig. 13 Experimental buckling stress σ of cylindrical shells of different radius – thickness ratios, R/t $(L/R > 1.5)$

The classical theory also gives a wave form which differs from those experimentally observed. The theory indicates that the shell will buckle into a series of rectangular waves whose nodal lines are parallel and perpendicular to the axis of the cylinder and that the amplitudes of the waves which buckle outward and those which buckle inward are equal. However, experiments indicate that the wave pattern, instead of being rectangular, is diamond-shaped, as shown in

Fig. 17. Moreover, the outward radial deflections are much smaller than those inward. In other words, the shell definitely prefers to buckle inward.

If the exponent of (t/R) in Eq. (12) is correct, then when $\sigma_{cr}R/Et$ is plotted against (R/t) (Fig. 13) the points should lie on a horizontal line. This is evidently not the case. In fact by plotting the experimental buckling stress against (t/R) on a log-log-scale, it is found that the following expression is more correct:

$$\sigma_{cr} = \text{constant}\, E(t/R)^{1.4} \tag{13}$$

By considering each element of the shell as a small Euler column, the dimensional relation

$$\sigma_{cr} = \text{constant}\, E(t/l)^2$$

should be correct where l is the half wave length of this element. Comparing this relation with Eq. (12), it is evident that according to the theory the following relation should hold

$$l/t = \text{constant}(R/t)^{1/2} \tag{14}$$

Using the experimental data obtained by L. H. Donnell[11] it is found that

$$l/t = \text{constant}(R/t)^{0.7} \tag{15}$$

which is consistent with Eq. (13) but disagrees with Eq. (14) given by the classical theory. A strip of the cylindrical shell in an axial direction can also be considered as an elastically supported column, the supporting force being derived from the circumferential stress in the shell. It is well known that for such a column, the wave length decreases with increase in the elastic support, i. e., a stronger elastic support gives a shorter wave length. Therefore, by comparing Eqs. (14) and (15), it is evident that in the case of thin shells with large values of R/t the classical theory has overestimated the elastic support and thus arrived at a higher buckling load.

If the total length L of the cylinder is of the order of one wave length or less, then it can be expected that the length will have an effect on the buckling load. In this case the natural extension of the wave is restricted by the length of the shell and the buckling load will increase with decreasing length. As shown in the preceding paragraph, the classical theory has underestimated the natural wave length, and thus the theory would predict that the length effect is negligible at much lower values of L/R than experiments would indicate. This expectation is verified through tests made by N. Nojima and S. Kanemitsu at the California Institute of Technology under the direction of E. E. Sechler. Their results are shown in Fig. 14 together with the theoretical values based on small deflections. It appears that while the classical theory predicts a length effect only at $L/R \approx 0.1$, the test data show an increase of buckling load at a L/R ratio of about 15 times this value. Thus it is evident that the natural wave length of the classical theory is very much underestimated. However, the meaning of the large difference in the slopes of the theoretical and the experimental curves (Fig. 14) cannot be definitely stated; since, for most of the shorter specimens, failure is accompanied by a rotation around the axis of the cylindrical shell, due to the fact that the particular loading mechanism used is free to rotate in this manner. This type of failure is entirely different from that predicted by the classical theory.

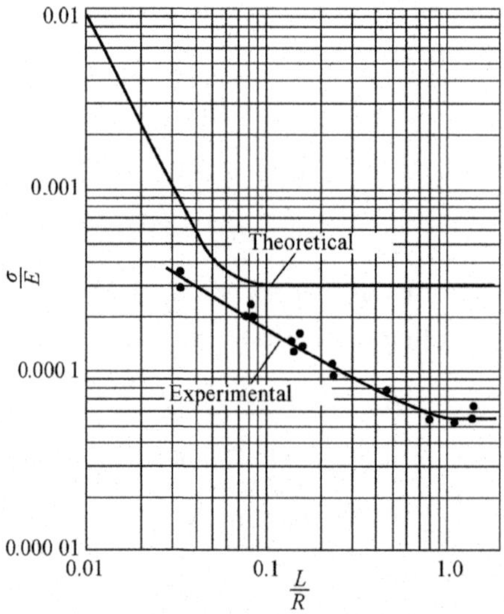

Fig. 14 The effect of the total length L of a cylindrical shell ($R/t = 2\ 000$) on its buckling stress

Visual Study of the Buckling Phenomena of Cylindrical Shells

To obtain a better understanding of the mechanism involved in the failure of thin wall cylinders it was felt that it would be desirable to determine the exact shape of the initial waves which actually appear on the cylindrical surface during a test. This was accomplished by restraining the loading mechanism so that every stage of the buckling process could be maintained for an arbitrary length of time.* Thus photographs could be taken which show the intermediate phases of the buckling process.

The test apparatus is shown in Figs. 15 and 16. The three upper set screws afford adjustment of the loading head and rest on a 3/4 inch plate. This plate in turn is held in position by three 1/2 inch screws resting on the base plate. These latter screws are turned by means of the gear system shown and lowers or raises the 3/4 inch plate as desired. The small 2 inch central gear, which turns the three 5 inch gears, can be externally operated. In this manner the motion of the loading head can be arrested at any desired position during loading of the specimen. The specimen was 9 inches long with a 0.003 4 inch wall thickness and a radius of 6.375 inches, clamped between the end plates.

* This was carried out by N. Nojima and S. Kanemitsu under the direction of E. E. Sechler at the California Institute of Technology.

Fig. 15 and Fig. 16 Loading mechanism for visual study of the buckling of a cylindrical shell

The progressive change in the wave shape and wave pattern is indicated in Fig. 17. It should be noted that the wave pattern does not agree with the uniform rectangular pattern which has been previously assumed for the theoretical solution. The initial wave form is elliptical in shape and scattered at random through the specimen. As the load is increased the waves tend toward a diamond shape and take on a more uniform configuration. This change in the wave form may be taken as an indication of the varying interaction between the bending

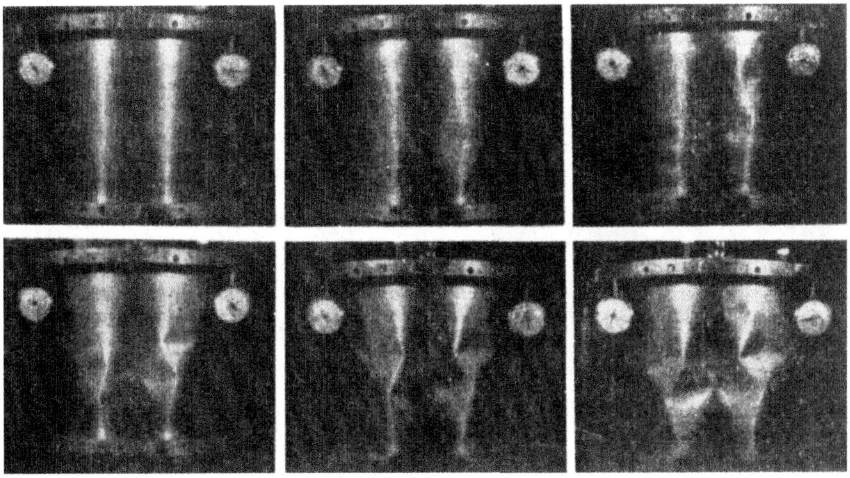

Fig. 17 Various stages during the buckling of a cylindrical shell under axial compression (R/t = 1 875)

and the extensional stiffnesses of the shell. If a longitudinal strip of the shell is considered as an elastically supported column, the increase in wave length indicates that the elastic support decreases as the deflection increases.

A Column Supported by a Non-Linear Elastic Element

The preceding discussion of the buckling of cylindrical shells suggested that a longitudinal strip of a cylindrical shell may be considered as an elastically supported column. However, it follows both from the physical considerations discussed in Section I and from the experimental observations of the different wave patterns described in the preceding paragraphs that the actual elastic support cannot be linear. Therefore, the explanation for the discrepancies observed between the theoretical and experimental values of the buckling load must be found in some characteristic property of a non-linear elastic support.

The problem of a column elastically supported has been discussed by H. Zimmerman for the case of concentrated supports, and by F. Engesser and others for the case of a uniformly distributed support[13]. In all cases the investigations have been confined to elastic supports exhibiting a linear force-deflection relation.

It was, therefore, felt that it would be of general interest to investigate the effect on the load-deflection relation of a column supported by a non-linear elastic element. Since rings are commonly used in structural design and are known to have the desired non-linear properties, a thin semi-circular steel ring, as shown in Fig. 18, was used as the elastic support. It may be of some interest to consider first the elastic behavior of a semi-circular ring. Designating the radial load by P and the corresponding radial deflection by δ, the curves of Fig. 19 indicate that if the load is applied radially inward, the value of P/δ, the elastic constant, decreases with increasing deflection. When the load is directed radially outward the value of P/δ increases with increasing deflection. Obviously then, if an initially straight column is supported by such an element, or elements, it may be expected that it would have a preference for buckling in the direction of decreasing P/δ, or, if due to an initial deformation in the direction of increasing P/δ, the buckling starts in that direction, then at some stage of the deflection a sudden "jump" may occur in the direction of decreasing P/δ.

Fig. 18 Test apparatus for columns supported by a semi-circular ring

Tests were conducted on columns 0.090 inch thick by 0.375 wide and 19 inches long. These columns were cut from 24SRT Alclad sheet stock. The steel rings were in all cases 8 inches in diameter with thicknesses of 0.008 and 0.015 inches; the width was varied between 1/2 to 1 inch. The test apparatus and method of testing is illustrated in Fig. 18.

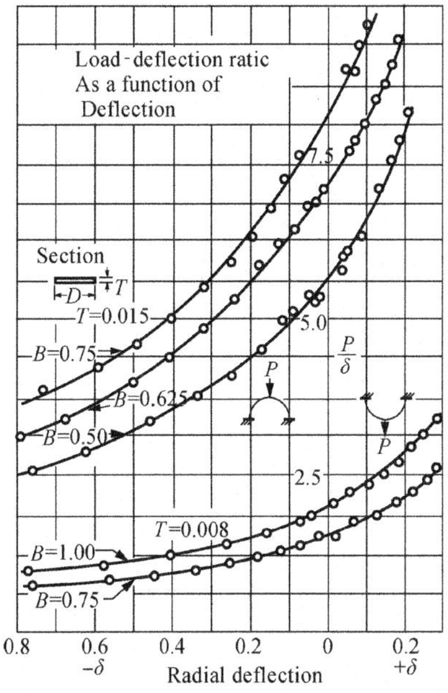

Fig. 19 Load-deflection ratio P/δ for semi-circular rings plotted as a function of deflection

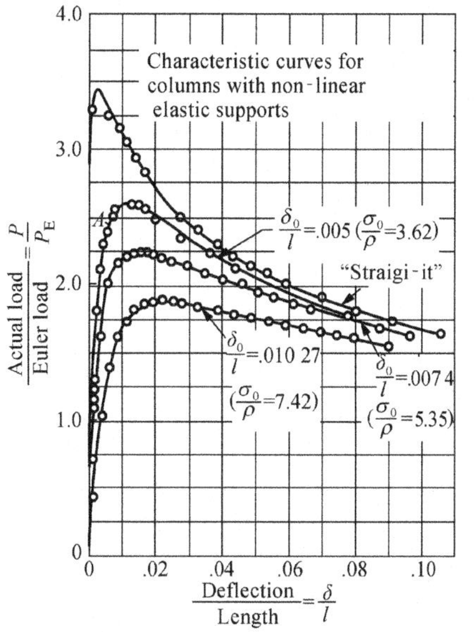

Fig. 20 Characteristic curves for columns with a non-linear elastic support (cf. Fig. 18) and different amounts of initial deflection

The results of these tests are indicated in Figs. 20 and 21, where the ratio of the column load to the Euler load is plotted as a function of the ratio of the normal deflection, δ, at the center of the column, to the column length l.

Considering first the results of Fig. 20, it is seen that the elastic support increases the buckling load of the "straight" column (upper curve) to nearly 3.5 times the Euler load. This load is reached at a relatively small deflection. Now, as the deflection δ increases, the decrease in load is at first quite rapid, then more gradual as the deflection becomes larger and may approach a minimum at large deflections. It was not possible to reach very large deflections because of plastic failure of both the rings and columns. The lower curves, in the same figure, indicate the effect of initial deflections, in which the column was rolled to approximately the form of a half-sine-wave, the maximum initial deflection being designated by δ_0. These curves show that with increasing initial deflection the maximum load decreases and occurs at increasingly larger deflections. In all cases the load sustained by the column tends to approach at large deflections the "minimum load" of the "straight" column. Thus in case of a non-linear support, any initial imperfections of the specimen will appreciably lower its buckling load. To illustrate the contrast between the column with a linear, and a column with non-linear elastic support, a number of tests were conducted on the same type of column, but with a linear elastic support (a coiled spring). As may be seen from the curves of Fig. 22 the columns with initial deflection in all cases approach the maximum load of the "straight" column.

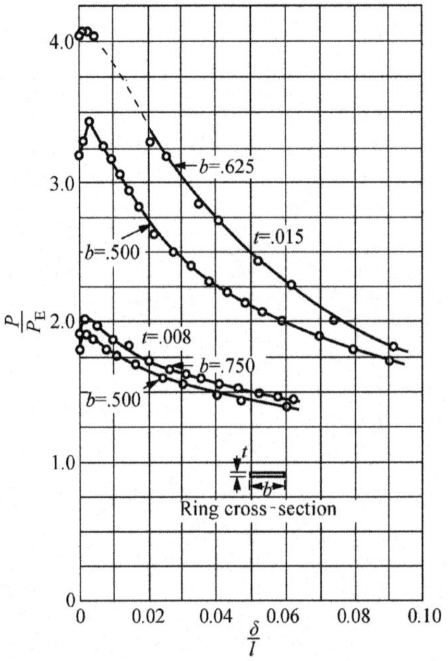

Fig. 21 Characteristic curves for columns with non-linear elastic supports of different stiffnesses

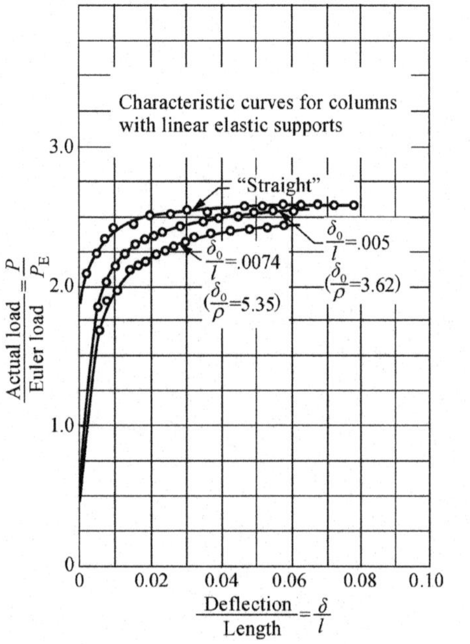

Fig. 22 Characteristic curves for columns with a linear elastic support and different amounts of initial deflection

A number of tests were made in which the width and thickness of the ring was varied. The results are shown in Fig. 21. As may be expected the maximum load increases as the ring stiffness increases, while the decrease in load at large deflections is correspondingly larger for the stiffer rings.

The most striking features of the column supported by a non-linear elastic element are, first, that as the deflection increases the load decreases, and secondly, that there are two or three possible configurations of equilibrium for the same load; one corresponds to $\delta = 0$, the others to $\delta \lesseqgtr 0$. Also in this case the decrease of the load with increasing deflection shows superficially a similarity to the behavior of a short column in the plastic state as discussed in Section I. However, in the case of the column with the non-linear elastic support this phenomenon is entirely elastic in character.

These tests were conducted for the purpose of illustrating the buckling characteristics of a structure involving a non-linear elastic-element. This particular experimental setup, as used, was chosen for the reason that the load P corresponding to any value of the deflection δ can be calculated analytically. Due to the complexity of the buckling phenomena of curved panels and cylindrical shells the authors are not yet able to give a thorough theoretical analysis. However, in the following section the phenomena will be discussed in the light of the above experimental observations.

Section Ⅲ

Curved Panels

A curved panel can be primarily considered as being intermediate between a flat panel and a cylindrical shell. From the previous discussion, it may be expected that for a panel with very small curvature, the load carried will increase even after buckling; while for a panel with large curvature, the load carried will decrease after buckling. This is verified by the experimental work of W. A. Wenzek[14]. His results are shown in Fig. 23, where the ratio of the actual load carried by the specimen to the observed buckling load is plotted as a function of the ratio of the axial shortening of the specimens to the shortening at buckling. These results indicate that for a panel with a b/R ratio of 0.4 the load carried will be constant after buckling. Hence, panels whose b/R ratio is less than 0.4 will fall in the flat plate category, i. e., the load increases after buckling, while those for which the b/R ratio is greater than 0.4 will fall in the curved shell category, i. e., the load decreases after buckling. Thus the value of $b/R = 0.4$ may be considered as a line of demarcation between these two categories of buckling phenomena.

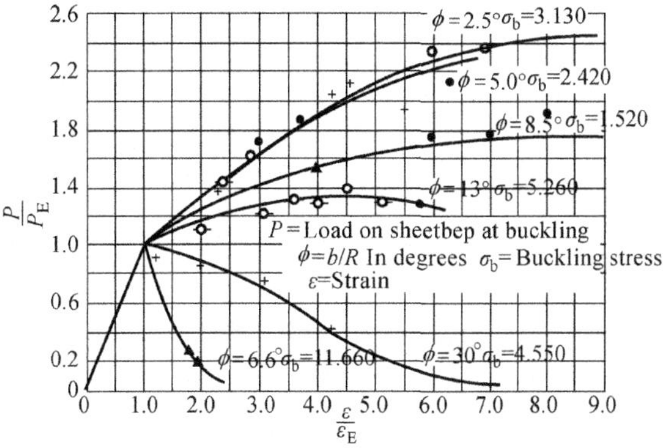

Fig. 23 The relation between the total load P and end strain ε for curved sheet of width b and radius R

However, a deeper understanding of these phenomena has to be sought in the concept of an elastically supported column developed in Section Ⅱ. It is found that the increase or decrease of load after buckling depends entirely on the characteristics of the supporting element. Thus, if an elementary strip in the direction of the axis of curvature is considered as a column elastically supported by strips of material in planes normal to the column axis, then the behavior of the panel after buckling depends entirely on the load-deflection characteristics of these supporting strips of material. These strips or elements can, of course, be considered as curved bars. If the panel is flat these bars are without curvature and the elastic support derived from them increases with deflection. In other words, the elastic support given to the elementary column concerned increases with an increase in wave amplitude. This results in a rising load-deflecting curve. By increasing the panel curvature, the curvature of the supporting bars is also increased resulting in a decrease of the load-carrying ability of these bars, i. e., the elastic support given to the elementary column concerned decreases. This is revealed by a drop in the rate of increase of the load carried by the buckled panel. It is thus clear that an increase in curvature of the panel will decrease its load-carrying ability after buckling. Hence, by increasing the curvature of the panel, one soon reaches a point where, the panel is no longer able to carry more load after buckling and a further increase in curvature of the panel causes the load carried by the panel to drop, just as the ring supported column discussed in Section Ⅱ. The panel has thus passed from the flat-plate category to the curved-shell category.

Stiffened Shell Structures

The elements of a stiffened shell such as used in metal aircraft construction may be divided into three distinct parts, namely, the sheet metal covering, longitudinal stiffeners and frames or bulkheads. A structure of this type if subjected to, say, compression loads parallel to the axis of the cylinder may fail in one of four distinct ways. The types of failure may be classified as material failure, local failure, panel failure, and general instability.

The first three types of failure are well known to the designer and are those for which present-day airplanes are analyzed. These types of failure may occur regardless of the size of the airplane. However, general instability, defined as a simultaneous buckling of both longitudinal stiffeners and frames, is a function of the stiffness of the structure as a whole. In small airplanes the frame sizes are determined by practical considerations rather than from a standpoint of stability. It is fortunate that these considerations have led to sufficiently rigid frames to minimize the danger of general instability. It is felt that general instability is more likely to occur in large airplanes, as it seems customary to increase the overall dimensions of the airplane while the frame dimensions are kept nearly constant. An investigation is now being conducted at the California Institute of Technology on stiffened circular cylinders for the purpose of determining when general instability may occur. *

At present very little is known about the phenomena of general instability. Considering the shell as a whole it is immediately evident that it is anisotropic and the influence of one member upon the other is extremely difficult to determine. The most elementary concept of the problem would be that of a column supported by continuous and concentrated elastic supports, a longitudinal stiffener being the column, the sheet covering providing the continuous elastic support and the frames the concentrated elastic supports. Since the frames and the sheet have the characteristics of a non-linear elastic support it may be expected that the longitudinal members under compression will behave in a manner similar to the column with a non-linear elastic support as discussed in Section Ⅱ. This expectation is verified by the results shown in Fig. 24. The abscissa is the deflection δ caused by a concentrated load P applied in a radial direction at the intersection of a longitudinal stiffener and frame (Fig. 24). The ordinate

---

* This investigation was made possible through a grant from the Civil Aeronautics Authority. The discussions presented in this paper have been stimulated to a large extent by this particular research project.

is the load P necessary for this deflection. It is thus evident that the stiffened shell is much less stiff when the load is applied radially inward than when applied radially outward. Furthermore, as the inward deflection is increased, the stiffness of the shell decreases. In this particular case, the slope of the P vs. δ curve at $\delta/R = -0.010$ is equal to only 1/3 of that at $\delta/R = 0$. Therefore, the buckling characteristics of a stiffened shell must be similar to that of the non-linearly supported column. Any theory which is based on the assumption of small deflections will probably give a much higher buckling load than that actually observed. This may be considered as an explanation of the inadequacy of the theory developed by D. D. Dschou[15], J. L. Taylor[16], and others. E. I. Ryder[17] has solved the difficulty by resorting to empirical coefficients to obtain agreement with experimental results.

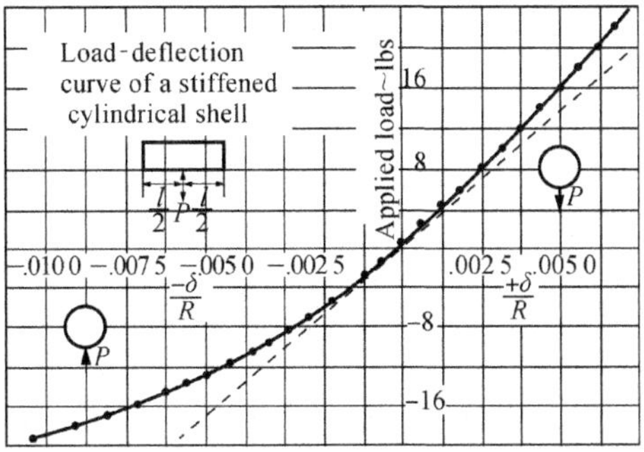

Fig. 24 Load-deflection curve of a stiffened cylindrical shell

Concluding Remarks

In Section Ⅰ, it was shown that while the load on thin-walled structures without curvature increases after buckling, the load on structures with curvature decreases after buckling. In Section Ⅱ, this fact is demonstrated by experiments on columns with a non-linear elastic support. It is thus clear that for the buckling of structures with curvature there are two important values of the applied load. First, the "initial" or "upper" buckling load given by the classical linear theory and then a "minimum" or "lower" buckling load which is equal to the minimum load necessary to keep the shell in a buckled shape with finite deformations. If the specimen is geometrically perfect, it will start to buckle only when the "initial" buckling load is reached. Once the shell starts to buckle, the load will decrease with increasing deflection. The elastic energy thus released will accelerate the buckling process until the "minimum" buckling load is passed. Actually the increase in deflection is so rapid that the shell appears to "jump" to a position involving large deflections. The kinetic energy associated with the rapid increase in deflection will cause the shell to vibrate around the equilibrium position corresponding to the "minimum" buckling load. This vibration will be rapidly damped by both the intrinsic friction of the material and the forces exerted by the testing machine.

It is a common engineering practice to record the maximum load which the structure can sustain as the failing load of the structure. Therefore, in the case of the perfect specimen, the "initial" buckling load will be the failing load. However, similar to the column with a non-linear elastic support shown in Section II, the maximum load which the shell can sustain is very sensitive to the amount of initial deflection of the shell. Larger deviations from the perfect geometrical shape will give a lower maximum load, without, perhaps, an appreciable effect on the "minimum" buckling load. Therefore, without extreme precautions, both in the manufacture of the specimen and in the testing, the failing load obtained will be invariably lower than the theoretical maximum given by the classical linear theory. But, this consideration also shows that by striving to make the test specimen geometrically perfect, as has been done by certain investigators, it is possible to obtain higher failing loads than usual, approaching the value given by the classical theory as an upper limit. The particular value of the failing load of a specimen is, therefore, determined by the degree of its geometrical perfection. This, perhaps, is one of the reasons for the bad scattering of the experimental points shown in Fig. 13. However, there seems to be a lower limit which corresponds to the "minimum" buckling load.

This explanation of the large difference between the failing load obtained by the classical theory and in experiments has a similarity with Donnell's theory for thin cylindrical shells[11] in that the initial deflection is a controlling factor in determining the failing load. However it should be clearly understood that whereas Donnell assumed that failure was caused by yielding of the material, the present explanation is based on the non-linear characteristics of certain elements of the structure without exceeding the elastic limit of the material.

Besides the effect of the initial deflection on the failing load of a thin shell, the effect of vibration occurring in the surroundings during the testing of the specimen should be considered. If a specimen with slight geometrical imperfections is loaded up to, say, the point A in Fig. 20, the vibrations occurring in the surroundings will impart a certain amount of kinetic energy to the specimen. This additional amount of kinetic energy when transformed into potential energy might be sufficient to help the specimen to pass the "hump" in the load-deflection curve and thus fail the specimen. The failing load recorded by the testing machine is, however, only that corresponding to the point A and not the maximum load. Therefore, the vibrations occurring in the testing surroundings can further reduce the failing load of a curved shell, besides the effect of initial deflections.

References

[1] von Kármán Th. Untersuchungen über Knickfestigkeit. Forschungsarbeiten, Berlin, 1910: 81.

[2] Timoshenko S. Theory of Elastic Stability. McGraw-Hill, New York, 1936: 390 - 393.

[3] Marguerre K. Die mittragende Breite der gedrückten Platte. Luftfahrtforschung, 1937, 14: 121 - 128.

[4] Marguerre K, Trefftz E. Über die Tragfähigkeit eines Plattenstreifens nach Überschreiten der Beullast. Z. A. M. M., 1937, 17: 85 - 100.

[5] Cox H L. The Buckling of Thin Plates in Compression. British A. R. C. Reports and Memoranda, No.

1554, 1933.

[6] Yamamoto M, Kondo K. Buckling and Failure of Thin Rectangular Plates in Compression. Rep. of Aero. Res. Inst., Tokyo, No. 119, 1935.

[7] von Kármán Th, Sechler E E, Donnell L H. The Strength of Thin Plates in Compression. Transactions A. S. M. E., 1932,54: 53-57.

[8] Marguerre K. Die Durchschlagskraft eines schwach gekrümmten Balkens. Sitzungsberichte der Berliner Mathematiscben Gesellschaft, 1938, 37: 22-40. See also Über die Anwendung der energetischen Methode auf Stabilitätsprobleme, Jahrbuch der Deutschen Versuchsanstalt für Luftfahrt, 1938: 252-262.

[9] von Kármán Th, Tsien Hsue-Shen. The Buckling of Spherical Shells by External Pressure. Journal of the Aeronautical Sciences, 1939,7: 43-50.

[10] Lundquist E E. Strength Tests of Thin-Walled Duralumin Cylinders in Compression. N. A. C. A. Technical Report, No. 473. 1933.

[11] Donnell L H. A New Theory for the Buckling of Thin Cylinders Under Axial Compression and Bending. Transactions A. S. M. E., 1934,56: 795-806.

[12] Flügge W. Die Stabilität der Kreiszylinderschale. Ingenieur Archiv, 1932,3: 463-506.

[13] Timoshenko S. Theory of Elastic Stability. McGraw-Hill, New York, 1936: 100-112 (A review of these works together with references).

[14] Wenzek W A. Die Mittragende Breite nach dem Ausknicken bei krummen Blechen. Luftfahrtforschung, 1938,15: 340-344.

[15] Dschou D D. Die Druckfestigkeit versteifter sydindrischer Schalen. Luftfahrtforschung, 1935, 11: 223-234.

[16] Taylor J L. Stability of a Monocoque in Compression. British A. R. C. Reports and Memoranda, No. 1679, 1935.

[17] Ryder E I. General Instability of Semi-monocoque Cylinders. Air Comm. Bulletin, 1938, 9: 241-246.

A Method for Predicting the Compressibility Burble

By Hsue-shen Tsien Ph. D.

(Dec. 7, 1940)

Summary*

In Part I of the present report, a method for calculating the two dimensional subsonic compressible flow is developed by using the tangent to the adiabatic pressure-volume curve as an approximation to the curve itself. In Part II, this method is applied to the case of flow over a circular cylinder with its axis perpendicular to the direction of undisturbed flow. The result is compared with that of perturbation method. In Part III, the conclusion drawn from Part I and Part II is used to develop a procedure which enables the prediction of the compressibility burble from test data obtained at low air speeds. Then the results of this method are compared with those obtained from Jacob's method and the experimental data. It is found that by using the new method, the predicted behavior agrees very well with the experiments.

Introduction

The experimental data obtained from high-speed wind tunnels show that when the maximum velocity of fluid over a solid body reaches approximately the local velocity of sound, the resistance experienced by the body suddenly increases. This phenomenon is generally called compressibility burble. The speed at which compressibility burble occurs is the critical speed of the body. Therefore, informations about the critical speed of different component parts of an airplane are very useful to designers of modern high-speed aircrafts. Unfortunately, to determine critical speed of a body experimentally requires a costly high-speed wind tunnel. Hence it is desirable to have a reliable method to calculate the critical speed either theoretically or from experimental data obtained from an ordinary low speed wind tunnel.

The calculation of the critical speed requires, of course, the solution of the problem of compressible flow over a given body. The known methods devised for this purpose are

(1) Glauert-Prandtl method

(2) Perturbation method

and (3) Hodograph method.

The theory of Glauert-Prandtl method is based on the assumption that the disturbance

Technical Report No. 2, The Aeronautical Research Institute, Chengtu, China, 1941.

* There was a summary in Chinese in the original version, which has been moved to the end in this version. — Noted by editor.

produced by the body placed in a parallel flow is small. They are thus able to linearize the partial differential equation for the velocity potential and obtain a very simple solution. It is evident that the theory can be applied only to a very thin airfoil or to a very slender body, because only then the disturbance produced by the body is small. But even in these cases, the theory breaks down in a region near the stagnation point. For bodies which are common in aircraft engineering, this method gives a higher critical speed than that experimentally observed. In other words, the Glauert-Prandtl theory is not conservative.

The perturbation method is developed by Lord Rayleigh, O. Janzen, L. Poggi and others. It consists essentially of expanding the velocity potential into a series of ascending powers of M_1^2 where M_1 is the ratio of the velocity of undisturbed parallel flow to the corresponding velocity of sound. M_1 is generally called the Mach number. This method is theoretically exact, provided that the convergence of the series can be established. However, the practical calculation is very tedious even for simple shaped bodies if one goes beyond the first approximation (terms involving M_1^2).

The hodograph method is first suggested by Molenbroek and Chaplygin. In this method, the inclination of the velocity vector to a fixed reference line and the magnitude of the velocity are used as the independent variables. It is a particular application of Legendre's contact transformation. The main drawback of this method is the difficulty of determining the solution by means of the boundary conditions. Therefore this method, although is exact and elegant, is applied till now only to a few isolated cases.

The method used in this report was first suggested by Dr. Th. von Kármán. Its general theory was discussed in a previous paper[1]. In the following a review of this theory will be first given.

Part Ⅰ

It is well-known that if the pressure in the fluid can be expressed as a function of density of the fluid only, the flow of a non-viscous compressible fluid is irrotational. If the flow is two-dimensional, the condition of zero rotation is expressed by

$$\frac{\partial u}{\partial y}-\frac{\partial v}{\partial x}=0 \tag{1}$$

where u and v are the components of velocity in x and y directions. Then there exists a velocity potential ϕ defined as

$$\mathrm{d}\phi = u\mathrm{d}x + v\mathrm{d}y. \tag{2}$$

The condition for the conservation of mass is expressed by the equation of continuity which in case of steady two-dimensional motion gives

$$\frac{\partial(\rho u)}{\partial x}+\frac{\partial(\rho v)}{\partial y}=0 \tag{3}$$

where ρ is the density of the fluid.

Eq. (3) can be automatically satisfied by introducing the so-called stream function ψ defined as

$$d\psi = -\frac{\rho}{\rho_0} v dx + \frac{\rho}{\rho_0} u dy \tag{4}$$

where ρ_0 indicates the value of the density when $u = v = 0$.

The velocity potential ϕ and stream function ψ are here introduced to satisfy the kinematical properties of the fluid field. The dynamical relation of the flow in this special case of irrotational steady flow is expressed by the generalized Bernoulli equation:

$$\frac{1}{2} w^2 + \int_{p_0}^{p} \frac{dp}{\rho} = 0 \tag{5}$$

where $w^2 = u^2 + v^2$, and p_0 is the pressure when $w = 0$, i.e., the stagnation pressure.

The Eqs. (1), (3) and (5) contain actually five unknowns u, v, p, ρ and T, where T is the temperature of the fluid. Therefore for complete solution of the problem, two more equations are necessary. They are the equation of state and the equation of conservation of energy. For perfect gas, the equation of state is

$$\frac{p}{\rho} = RT \tag{6}$$

where R is the gas constant. If no heat is added or removed from the fluid in the whole field, then the condition for conservation of energy can be expressed as the adiabatic pressure-density relation:

$$\frac{p}{\rho^\gamma} = \text{constant} \tag{7}$$

where γ is the ratio of specific heats of the fluid at constant pressure c_p and at constant volume c_V i.e.,

$$\gamma = c_p / c_V \tag{8}$$

The value of γ for a diatomic gas, such as air, is 1.405.

Now suppose that a solution of this system of equations is found, then the question arises: Whether this known solution can be utilized to construct another solution which satisfies the same type of equations as (1), (3) and (5) but perhaps a different equation of state and a different energy equation? The difference in the equation of state and the energy equation means only a difference in the properties of the fluids in the two solutions. To investigate this, two new functions X and Y are introduced such that

$$\begin{aligned} -p_0 dX &= \rho_0 v d\psi - p dx \\ p_0 dY &= p dy + \rho_0 u d\psi \end{aligned} \tag{9}$$

Substituting Eq. (4) into Eq. (9), the following relations are obtained:

$$\begin{aligned} -p_0 \mathrm{d}X &= \rho uv\mathrm{d}y - (p+\rho v^2)\mathrm{d}x \\ &= \rho u(v\mathrm{d}y + u\mathrm{d}x) - (p+\rho w^2)\mathrm{d}x \\ p_0\mathrm{d}Y &= (p+\rho u^2)\mathrm{d}y - \rho uv\mathrm{d}x \\ &= (p+\rho w^2)\mathrm{d}y - \rho v(v\mathrm{d}y + u\mathrm{d}x) \end{aligned} \tag{10}$$

To show that $\mathrm{d}X$ is an exact differential, the relation

$$\frac{\partial(\rho uv)}{\partial x} = -\frac{\partial(p+\rho v^2)}{\partial y} \tag{11}$$

has to be established. However,

$$\begin{aligned} \frac{\partial(\rho uv)}{\partial x} + \frac{\partial(p+\rho v^2)}{\partial y} &= v\left[\frac{\partial(\rho u)}{\partial x} + \frac{\partial(\rho v)}{\partial y}\right] + \rho\left[u\frac{\partial v}{\partial x} + v\frac{\partial v}{\partial y} + \frac{1}{\rho}\frac{\partial p}{\partial y}\right] \\ &= v\left[\frac{\partial(\rho u)}{\partial x} + \frac{\partial(\rho v)}{\partial y}\right] + \rho\left[u\frac{\partial u}{\partial y} + v\frac{\partial v}{\partial y} + \frac{1}{\rho}\frac{\partial p}{\partial y}\right] \end{aligned} \tag{12}$$

The third form is made possible by Eq. (1). Now the value in the first bracket on the right hand side of Eq. (12) is equal to zero due to Eq. (3). The value in the second bracket is also equal to zero, because they can be obtained by differentiating the left hand side of the Bernoulli equation, Eq. (5), with respect to y. Therefore Eq. (11) is satisfied and $\mathrm{d}X$ is an exact differential. Similarly $\mathrm{d}Y$ can be shown to be also an exact differential. By using Eq. (2), Eq. (10) can be reduced to

$$\begin{aligned} -p_0\mathrm{d}X &= \rho u\mathrm{d}\phi - (p+\rho w^2)\mathrm{d}x \\ p_0\mathrm{d}Y &= (p+\rho w^2)\mathrm{d}y - \rho v\mathrm{d}\phi \end{aligned} \tag{13}$$

Now by multiplying the first equation of Eq. (13) by u and the second by v and subtract, the following equation is obtained,

$$p_0(v\mathrm{d}Y + u\mathrm{d}X) = (p+\rho w^2)(u\mathrm{d}x + v\mathrm{d}y) - \rho w^2\mathrm{d}\phi \tag{14}$$

Using Eq. (2), Eq. (14) can be simplified to

$$p_0(u\mathrm{d}X + v\mathrm{d}Y) = p\mathrm{d}\phi \tag{15}$$

If two new variables are introduced such that

$$u\frac{p_0}{p} = U, \quad v\frac{p_0}{p} = V \tag{16}$$

then Eq. (15) gives

$$V\mathrm{d}\phi = U\mathrm{d}X + V\mathrm{d}Y \tag{17}$$

which is similar to Eq. (2) with U and V as the new velocity components in the $X-Y$ plane.

Similarly by multiplying the first equation of Eq. (9) by u and the second by v and adding, the following relation is obtained

$$p_0(udY - vdX) = p(udy - vdx) + \rho_0 w^2 d\psi \tag{18}$$

By means of Eq. (4), Eq. (18) can be simplified to

$$p_0(udY - vdX) = \frac{\rho_0}{\rho}(p + \rho w^2)d\psi \tag{19}$$

If another new variable σ is introduced such that

$$\frac{\sigma}{\sigma_0} = \frac{\rho}{\rho_0}\frac{p}{p + \rho w^2} \tag{20}$$

where σ_0 denotes the value of σ at $U = V = 0$. Then Eq. (19) with the aid of Eq. (16) can be rewritten as

$$d\psi = \frac{\sigma}{\sigma_0}(UdY - VdX) \tag{21}$$

which is similar to Eq. (4) with σ as the density of the fluid in $X-Y$ plane.

The relations expressed by Eqs. (16), (17), (20) and (21) indicate that from the original solution for flow of a compressible fluid in the $x-y$ plane with the velocity components u, v and the density ρ, a new solution for flow of a different compressible fluid in the $X-Y$ plane with the velocity components U, V and the density σ is obtained. Furthermore, a stream line in $x-y$ plane will be transformed by Eq. (10) into a stream line in $X-Y$ plane as shown in Fig. 1. This is evident from Eq. (21). This remarkable reciprocity between the $x-y$ plane and the $X-Y$ plane is first demonstrated by H. Bateman[2].

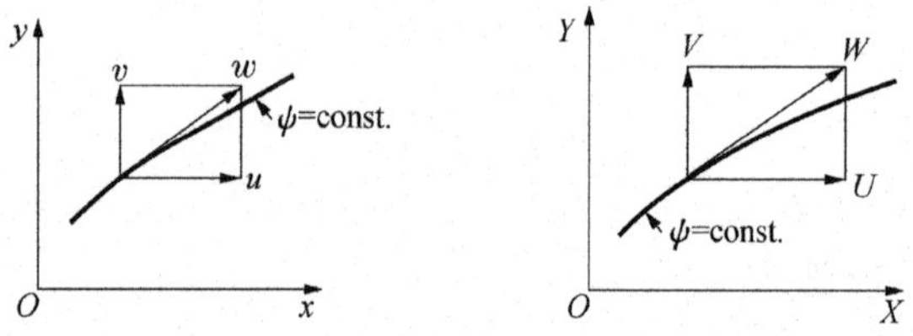

Fig. 1

To apply this theory, a solution of the flow around a certain body has to be found first. At first sight, the theory seems thus only of academic interest, because it is just as difficult to find this initial solution. However, the relations developed between the flows in the $x-y$ plane and the $X-Y$ plane are very general. No assumption about the specific properties of the fluids is made. The fluid in the $x-y$ plane can be incompressible. Then it is very easy to obtain a solution in that plane. The problem now is to investigate the properties of the fluid in the transformed $X-Y$ plane with incompressible flow in the $x-y$ plane.

If the flow in the $x-y$ plane is incompressible, then

$$\rho = \rho_0 = \text{constant}, \quad \text{or} \quad \frac{\rho}{\rho_0} = 1 \tag{22}$$

and Bernoulli theorem gives

$$p + \frac{1}{2}\rho w^2 = p_0 \quad \text{or} \quad \frac{p}{p_0} + \frac{1}{2}\frac{\rho}{p_0}w^2 = 1 \tag{23}$$

If P denotes the fluid pressure in the $X-Y$ plane, then the generalized Bernoulli theorem, Eq. (5), requires that

$$\frac{1}{2}W^2 + \int_{P_0}^{P}\frac{\mathrm{d}P}{\sigma} = 0 \tag{24}$$

where $W^2 = U^2 + V^2 = \left(\frac{p_0}{p}\right)^2 w^2$ by means of Eq. (16), and P_0 is the value of P when $W = 0$. However, from Eqs. (20) and (22), the following relation can be derived,

$$\frac{\rho w^2}{p} = \left(\frac{\sigma}{\sigma_0}\right)^{-1} - 1 \tag{25}$$

Thus

$$\begin{aligned} W^2 &= \left(\frac{p_0}{p}\right)^2 w^2 = \frac{p_0}{\rho_0}\frac{p_0}{p}\frac{\rho w^2}{p} \\ &= \frac{p_0}{\rho_0}\left(1 + \frac{1}{2}\frac{\rho w^2}{p}\right)\frac{\rho w^2}{p} = \frac{1}{2}\frac{p_0}{\rho_0}\left[\left(\frac{\sigma}{\sigma_0}\right)^{-2} - 1\right] \end{aligned} \tag{26}$$

Therefore Eq. (24) can be rewritten as

$$\frac{1}{\sigma_0}\int_{\sigma_0}^{\sigma}\frac{\mathrm{d}\left(\frac{\sigma}{\sigma_0}\right)}{\frac{\sigma}{\sigma_0}}\frac{\mathrm{d}P}{\mathrm{d}\left(\frac{\sigma}{\sigma_0}\right)} + \frac{1}{4}\frac{p_0}{\rho_0}\left[\left(\frac{\sigma}{\sigma_0}\right)^{-2} - 1\right] = 0 \tag{27}$$

Differentiate Eq. (27) with respect to $\frac{\sigma}{\sigma_0}$ and multiply by $\frac{\sigma}{\sigma_0}$,

$$\frac{1}{\sigma_0}\frac{\mathrm{d}P}{\mathrm{d}\left(\frac{\sigma}{\sigma_0}\right)} - \frac{1}{2}\frac{p_0}{\rho_0}\left(\frac{\sigma}{\sigma_0}\right)^{-2} = 0 \tag{28}$$

Integrating Eq. (28) the following relation is obtained,

$$P - P_1 = \frac{1}{2}\frac{p_0}{\rho_0}\sigma_0^2\left(\frac{1}{\sigma_1} - \frac{1}{\sigma}\right) \tag{29}$$

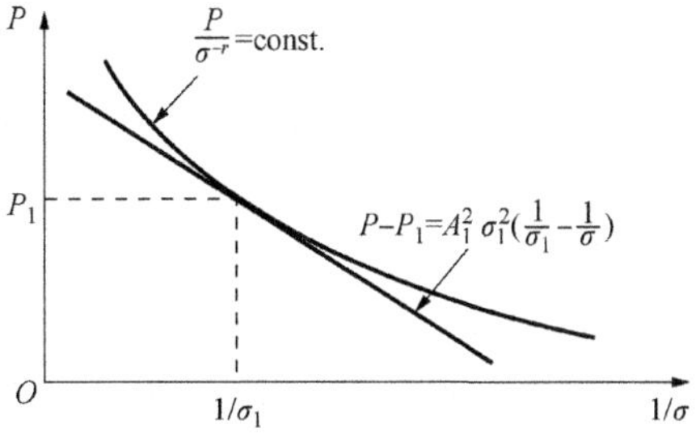

Fig. 2

where P_1 and σ_1 are constant. Since σ is the density of the fluid in the $X-Y$ plane, its reciprocal $1/\sigma$ is the specific volume. Using Eq. (29), P can be plotted against $1/\sigma$ as shown in Fig. 2. It is a straight line with negative slope and passes through the point (P_1, $1/\sigma_1$). Of course, there is no such fluid in nature that has this type of adiabatic pressure-volume relation. However, an ordinary adiabatic pressure-volume curve can be drawn through the point (P_1, $1/\sigma_1$) by means of Eq. (7) and then adjust the slope of the straight line given by Eq. (29) such that it is equal to that of the adiabatic curve. Then the relationship given by Eq. (29) can be considered as an approximation to the true adiabatic pressure-volume relation occuring during the actual flow of a compressible fluid. The slope of an adiabatic curve at the point (P_1, $1/\sigma_1$) is

$$\left[\frac{\mathrm{d}P}{\mathrm{d}\left(\frac{1}{\sigma}\right)}\right]_1 = \left[\frac{\mathrm{d}P}{\mathrm{d}\sigma}\Big/\frac{\mathrm{d}\left(\frac{1}{\sigma}\right)}{\mathrm{d}\sigma}\right]_1 = -A_1^2\sigma_1^2 \tag{30}$$

where $A_1^2 = \left(\frac{\mathrm{d}P}{\mathrm{d}\sigma}\right)_1$, A_1 being the velocity of sound in the fluid at the state $P = P_1$ and $\sigma = \sigma_1$. Therefore by equating the slope of the straight line and that of the adiabatic curve the following relation is obtained,

$$\frac{1}{2}\frac{p_0}{\rho_0} = A_1^2\left(\frac{\sigma_1}{\sigma_0}\right)^2 \tag{31}$$

Eq. (29) can then be rewritten as

$$P - P_1 = A_1^2\sigma_1^2\left(\frac{1}{\sigma_1} - \frac{1}{\sigma}\right) \tag{32}$$

Differentiating Eq. (32) with respect to σ, the velocity of sound, A, at any state can be calculated as

$$A^2 = A_1^2\left(\frac{\sigma_1}{\sigma}\right)^2$$

or

$$A^2\sigma^2 = A_1^2\sigma_1^2 = \text{constant} \tag{33}$$

By means of Eq. (32), the integral occuring in the generalized Bernoulli theorem, Eq. (24), can be calculated,

$$\int_{P_0}^{P}\frac{\mathrm{d}P}{\sigma} = A_1^2\sigma_1^2\int_{\sigma_0}^{\sigma}\frac{\mathrm{d}\sigma}{\sigma^3} = -\frac{A_1^2\sigma_1^2}{2}\left(\frac{1}{\sigma^2} - \frac{1}{\sigma_0^2}\right) \tag{34}$$

Substituting this value into Eq. (24), the following relation between the magnitude of velocity W and the density σ is obtained

$$\frac{1}{2}W^2 - \frac{A_1^2\sigma_1^2}{2}\left(\frac{1}{\sigma^2} - \frac{1}{\sigma_0^2}\right) = 0 \tag{35}$$

Or using Eq. (33)

$$\left(\frac{\sigma}{\sigma_0}\right)^2 = 1 - \left(\frac{W}{A}\right)^2 \tag{36}$$

Then Eq. (31) can be rewritten as

$$\frac{1}{2}\frac{p_0}{\rho_0} = A_1^2\left[1 - \left(\frac{W_1}{A_1}\right)^2\right] = A_1^2(1 - M_1^2) \tag{37}$$

where $M_1 = W_1/A_1$ is the Mach number corresponding to the state $P = P_1$ and $\sigma = \sigma_1$.

The point of tangency of the straight line of Eq. (32) and the adiabatic curve is not yet made to correspond to any particular state of the fluid. However, if the problem of a body in a parallel stream is concerned, then it seems natural to make this point of tangency correspond to the state of fluid in the undisturbed parallel flow. Thus W_1 is the velocity of the undisturbed parallel flow and A_1 the corresponding velocity of sound. M_1 will then be the Mach number of the undisturbed flow.

From Eqs. (16), (22) and (23), the magnitude of velocity W in the $X-Y$ plane can be expressed as

$$W = \frac{p_0}{p}w = w\frac{1}{1 - \frac{1}{2}\frac{\rho_0}{p_0}w^2} \tag{38}$$

Therefore

$$\frac{W}{W_1} = \frac{w}{w_1}\frac{1 - \frac{1}{2}\frac{\rho_0}{p_0}w_1^2}{1 - \frac{1}{2}\frac{\rho_0}{p_0}w_1^2\left(\frac{w}{w_1}\right)^2} \tag{39}$$

where w_1 is the magnitude of velocity in the $x-y$ plane corresponding to W_1 i. e., the undisturbed parallel flow in the $x-y$ plane. Eq. (39) relates the magnitude of velocity in the $X-Y$ plane where the fluid is compressible and approximately adiabatic to that in the $x-y$ plane where the fluid is incompressible. The relation can be put into a more useful form if the quantity $\frac{1}{2}\frac{\rho_0}{p_0}w^2$ can be expressed in terms of quantities relating to the velocity W_1 in the compressible flow. Using Eqs. (37) and (38)

$$\frac{1}{2}\frac{\rho_0}{p_0}w_1^2 = \frac{M_1^2}{4(1 - M_1^2)}\left(1 - \frac{1}{2}\frac{\rho_0}{p_0}w_1^2\right)^2 \tag{40}$$

Eq. (40) can be considered as a quadratic equation for $\frac{1}{2}\frac{\rho_0}{p_0}w_1^2$. Solving for $\frac{1}{2}\frac{\rho_0}{p_0}w_1^2$, it is found that the appropriate solution is

$$\frac{1}{2}\frac{\rho_0}{p_0}w_1^2 = \frac{M_1^2}{(1 + \sqrt{1 - M_1^2})^2} \tag{41}$$

Using this relation, Eq. (39) can be rewritten as

$$\frac{W}{W_1}=\frac{w}{w_1}\frac{1-\dfrac{M_1^2}{(1+\sqrt{1-M_1^2})^2}}{1-\dfrac{M_1^2}{(1+\sqrt{1-M_1^2})^2}\left(\dfrac{w}{w_1}\right)^2} \tag{42}$$

If the pressure coefficient k_p for incompressible flow is defined as

$$k_p=\frac{p-p_1}{(1/2)\rho_1 w_1^2} \tag{43}$$

then by means of Eq. (23)

$$k_p=1-\left(\frac{w}{w_1}\right)^2 \tag{44}$$

Similarly if the pressure coefficient K_p for the compressible flow is defined as

$$K_p=\frac{P-P_1}{(1/2)\sigma_1 W_1^2} \tag{45}$$

then by means of Eqs. (20), (23) and (29), the following relation is obtained,

$$K_p=\frac{\left(1+\dfrac{1}{2}\dfrac{\rho_0}{p_0}w_1^2\right)k_p}{1-\dfrac{1}{2}\dfrac{\rho_0}{p_0}w_1^2(1-k_p)} \tag{46}$$

Substituting the value of $\frac{1}{2}\frac{\rho_0}{p_0}w_1^2$ from Eq. (41)[①] into Eq. (46), a simple relation between the pressure coefficients in compressible flow and incompressible flow can be expressed as

$$K_p=\frac{k_p}{\sqrt{1-M_1^2}+\dfrac{M_1^2k_p}{2(1+\sqrt{1-M_1^2})}} \tag{47}$$

To find the coordinates X and Y in terms of x and y, Eq. (10) has to be integrated. Since the flow in the $x-y$ plane is incompressible, $\rho=\rho_0=$ constant, thus

$$\mathrm{d}X+\mathrm{i}\mathrm{d}Y=\frac{1}{p_0}[(p+\rho_0v^2)\mathrm{d}x-\rho_0uv\mathrm{d}y+\mathrm{i}(p+\rho_0u^2)\mathrm{d}y-\mathrm{i}\rho_0uv\mathrm{d}x] \tag{48}$$

Expressing p in terms of p_0 and w^2 as given by Eq. (23) and introducing the complex coordinates $Z=X+\mathrm{i}Y$ and $z=x+\mathrm{i}y$, Eq. (48) can be rewritten as

$$\begin{aligned}\mathrm{d}Z&=\mathrm{d}z-\frac{1}{2}\frac{\rho_0}{p_0}[(u^2-v^2)\mathrm{d}x+2uv\mathrm{d}y+\mathrm{i}(v^2-u^2)\mathrm{d}y+2\mathrm{i}uv\mathrm{d}x]\\&=\mathrm{d}z-\frac{1}{2}\frac{\rho_0}{p_0}(u+\mathrm{i}v)^2(\mathrm{d}x-\mathrm{i}\mathrm{d}y)\end{aligned} \tag{49}$$

It is well-known that if the flow is incompressible and irrotational then there is a complex

① Eq. (37) in the original paper has been corrected as Eq. (41). — Noted by editor.

potential $\phi + i\psi$ which is an analytic function of $z = x + iy$, and

$$\frac{d(\phi + i\psi)}{dz} = u - iv \tag{50}$$

Now introducing the complex potential function $G(z)$ such that

$$w_1 G(z) = \phi + i\psi \tag{51}$$

and denoting the complex conjugate of z as $\bar{z}$ and that of G as $\overline{G}$, Eq. (49) can be integrated and put into the form:

$$Z = z - \frac{1}{2}\frac{\rho_0}{p_0} w_1^2 \int \left(\frac{d\overline{G}}{d\bar{z}}\right)^2 d\bar{z} \tag{52}$$

or using Eq. (41), the relation between the coordinates in the $X-Y$ plane and the $x-y$ plane can be written as

$$Z = z - \frac{M_1^2}{(1+\sqrt{1-M_1^2})^2} \int \left(\frac{d\overline{G}}{d\bar{z}}\right)^2 d\bar{z} \tag{53}$$

The integration constant in Eq. (53) or (54) is immaterial, because its effect will be simply a translation of the whole $X-Y$ plane without changing the relative position of the points in that plane.

Equations (42), (47) and (53) are the three principal results of the present theory. If a solution of incompressible flow around a body is found, the velocity and pressure over an approximately similar body in compressible flow can be calculated by means Eqs. (42) and (47). The shape of the body in compressible flow is, however, not exactly the same as that of the body in incompressible flow. This distortion is given by the second term on the right hand side of Eq. (53) and depends upon the value of Mach number M_1 of the flow. To find the velocity and pressure distribution over a given body at different Mach number, it is necessary to start with different solutions corresponding to slightly different body shapes so that in the plane of compressible flow the given shape of the body can always be obtained. This introduces some complication, but, however, the labor involved is, perhaps, much less than that required by the perturbation method.

Part II

As stated previously, the approximation involved in the present theory is the substitution of the tangent to the adiabatic curve for the adiabatic curve itself. It thus becomes necessary to check the magnitude of error introduced by this approximation. The only two-dimensional subsonic flow calculated from the true adiabatic relation to sufficient accuracy is that past a circular cylinder with its axis placed perpendicular to the direction of undisturbed flow. It is calculated by means of the perturbation method up to terms of the order of M_1^4 by I. Imai[3], K. Tamada and Y. Saito[4]. The theory developed in Part I will be presently applied to this case and the result compared with those obtained by the authors mentioned.

The transformation from incompressible flow to compressible flow involves a slight distortion of the shape of the body as given by Eq. (53). The main effect is found to be a slight increase in the ratio of the maximum thickness to the length of the body. Therefore, in order to obtain a circular section after the transformation, it is necessary to start with an elliptic section with its major axis in the direction of the undisturbed flow. The incompressible flow over an elliptic cylinder can be obtained by applying the well-known Joukowsky transformation to the flow over a circular cylinder in complex coordinate ζ with the center of the circular section located at the origin of the ζ plane (see Fig. 3). Hence

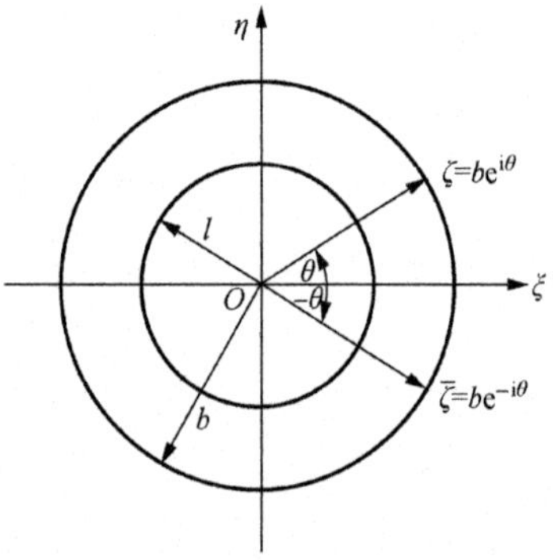

Fig. 3

$$G(\zeta) = \zeta + \frac{b^2}{\zeta}$$
$$\overline{G}(\overline{\zeta}) = \overline{\zeta} + \frac{b^2}{\overline{\zeta}} \qquad (54)$$

where b is the radius of the circle in the ζ plane. The Joukowsky transformations is

$$z = \zeta + \frac{1}{\zeta} \qquad (55)$$

The value of b controls the thickness ratio of the elliptic section in the $x-y$ plane. When $b \to \infty$, the thickness ratio becomes 1. When $b = 1$, the elliptic section is degenerated into a straight lamina. It is convenient to carry out the computation by using the ζ coordinate. Thus Eq. (53) can be written as

$$Z = z - \frac{M_1^2}{(1+\sqrt{1-M_1^2})^2} \int \frac{\left(\frac{d\overline{G}}{d\overline{\zeta}}\right)^2}{\left(\frac{d\overline{z}}{d\overline{\zeta}}\right)^2} d\overline{\zeta} \qquad (56)$$

Substituting Eq. (54) and (55) into Eq. (56), the complex coordinate in the plane of compressible flow is obtained as

$$Z = \zeta + \frac{1}{\zeta} - \frac{M_1^2}{(1+\sqrt{1-M_1^2})^2}\left[\overline{\zeta} + \frac{b^4}{\overline{\zeta}} + \frac{1}{2}(b^2-1)^2 \log\frac{\overline{\zeta}-1}{\overline{\zeta}+1}\right] \qquad (57)$$

In order to find the distortion of the shape of the original elliptic section by the transformation, the value of ζ corresponding to the surface of the elliptic section has to be substituted into Eq. (57). Thus

$$\zeta = b e^{i\theta}, \quad \overline{\zeta} = b e^{-i\theta} \qquad (58)$$

Then Eq. (57) gives by separating the real and the imaginary parts,

$$X = \left(b + \frac{1}{b}\right)\cos\theta - \frac{M_1^2}{(1+\sqrt{1-M_1^2})^2}\left[b(b^2+1)\cos\theta - \frac{(b^2-1)^2}{4}\log\frac{1+\dfrac{2b}{b^2+1}\cos\theta}{1-\dfrac{2b}{b^2+1}\cos\theta}\right] \tag{59}$$

$$Y = \left(b - \frac{1}{b}\right)\sin\theta - \frac{M_1^2}{(1+\sqrt{1-M_1^2})^2}\left[b(b^2-1)\sin\theta - \frac{(b^2-1)^2}{2}\tan^{-1}\frac{2b\sin\theta}{b^2-1}\right]$$

By setting $\theta = 0$ and $\frac{\pi}{2}$ in the first and second expressions of Eq. (59) respectively, the dimensions of the transformed section in the direction of the undisturbed flow and perpendicular to that direction can be calculated. The ratio formed by these two dimensions should be equal to unity if the section is nearly circular. Therefore

$$1 = \frac{b^2-1}{b^2+1}\,\frac{1 - \dfrac{M_1^2}{(1+\sqrt{1-M_1^2})^2}\left[b^2 - \dfrac{b(b^2-1)}{2}\tan^{-1}\dfrac{2b}{b^2-1}\right]}{1 - \dfrac{M_1^2}{(1+\sqrt{1-M_1^2})^2}\left[b^2 - \dfrac{b(b^2-1)^2}{2(b^2+1)}\log\dfrac{b+1}{b-1}\right]} \tag{60}$$

For a given Mach number M_1 of the parallel stream, the value of b can be solved numerically. For example, for $M_1 = 0.400$, b is found to be 5.782 7.

When the value of b is thus determined the detailed shape of the section in compressible flow can be calculated by means of Eq. (59). In the case of $M_1 = 0.400$, the section obtained by the transformation from incompressible flow with $b = 5.7827$ is shown in Table 1, where $r^2 = X^2 + Y^2$ and $a = \tan^{-1}\frac{Y}{X}$

Table 1

θ	r	a
0°	5.288 0	0°
10°	5.288 0	10.60°
20°	5.287 8	21.12°
30°	5.287 5	31.50°
40°	5.286 7	41.88°
50°	5.287 5	51.68°
60°	5.287 5	61.47°
70°	5.287 7	71.09°
80°	5.287 6	80.57°
90°	5.287 7	90°

If the section in compressible flow were a truely circular, r should be a constant. The variation of r in the above Table is less than 0.02%. This variation is, indeed, negligible. Therefore, the velocity and pressure distribution over this particular section can be considered as those over a truely circular section.

By putting $\zeta = i\eta$ in Eq. (57), the coordinate of a point along the Y axis can be obtained as

$$Y = \left(\eta - \frac{1}{\eta}\right) - \frac{M_1^2}{(1+\sqrt{1-M_1^2})^2}\left[\eta\left(\frac{b^4}{\eta^2} - 1\right) - \frac{1}{2}(b^2 - 1)^2 \tan^{-1}\frac{2\eta}{\eta^2 - 1}\right] \tag{61}$$

The velocity at the corresponding point in the incompressible flow is

$$w = w_1 \frac{dG}{dz} = w_1 \frac{\dfrac{dG}{d\zeta}}{\dfrac{dz}{d\zeta}} = w_1 \frac{\eta^2 + b^2}{\eta^2 + 1} \tag{62}$$

Hence using Eq. (42) the velocity in the compressible flow can be calculated by the following expression:

$$\frac{W}{W_1} = \frac{\eta^2 + b^2}{\eta^2 + 1} \frac{1 - \dfrac{M_1^2}{(1+\sqrt{1-M_1^2})^2}}{1 - \dfrac{M_1^2}{(1+\sqrt{1-M_1^2})^2}\left(\dfrac{\eta^2 + b^2}{\eta^2 + 1}\right)^2} \tag{63}$$

The results of this calculation for the case $M_1 = 0.400$, using the previously determined appropriate value for b is shown Fig. 4 as Curve Ⅲ. The ordinate is the increase in velocity ΔW due to compressibility divided by the velocity W_1 of the undisturbed flow. The abscissa is the ratio of the distance Y from the point concerned to the center of the circular section, to the radius of the section R.

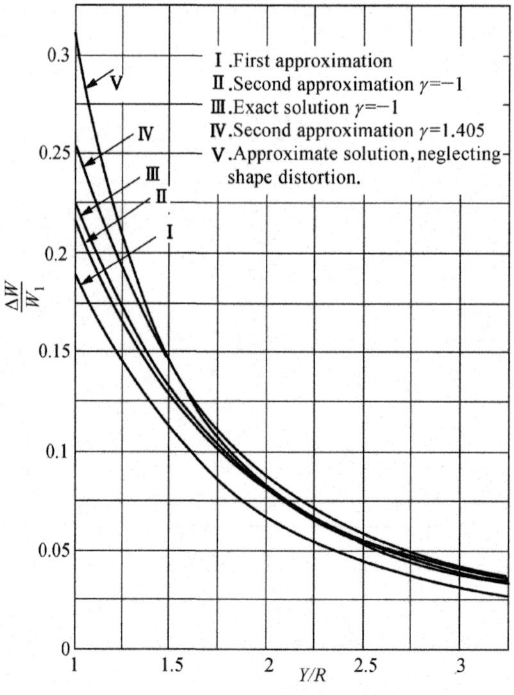

Fig. 4

According to Lord Rayleigh and O. Janzen, the perturbation method gives as first approximation:

$$\frac{\Delta W}{W_1}=\left[\frac{11}{6}\left(\frac{Y}{R}\right)^{-2}-\frac{3}{4}\left(\frac{Y}{R}\right)^{-4}+\frac{1}{12}\left(\frac{Y}{R}\right)^{-6}\right]M_1^2 \tag{64}$$

which is independent of the ratio of specific heats γ. The second approximation of perturbation method, carried out by I. Imai[3], K. Tamada, and Y. Saito[4], gives the increment of velocity as

$$\begin{aligned}\frac{\Delta W}{W_1}=&\left[\frac{11}{6}\left(\frac{Y}{R}\right)^{-2}-\frac{3}{4}\left(\frac{Y}{R}\right)^{-4}+\frac{1}{12}\left(\frac{Y}{R}\right)^{-6}\right]M_1^2+\\&(\gamma-1)\left[\frac{17}{60}\left(\frac{Y}{R}\right)^{-2}+\frac{19}{20}\left(\frac{Y}{R}\right)^{-4}-\frac{2}{3}\left(\frac{Y}{R}\right)^{-6}+\frac{1}{80}\left(\frac{Y}{R}\right)^{-8}+\frac{1}{80}\left(\frac{Y}{R}\right)^{-10}\right]M_1^4+\\&\left[\frac{257}{80}\left(\frac{Y}{R}\right)^{-2}-\frac{17}{24}\left(\frac{Y}{R}\right)^{-4}-\frac{3}{16}\left(\frac{Y}{R}\right)^{-8}+\frac{1}{40}\left(\frac{Y}{R}\right)^{-10}\right]M_1^4\end{aligned} \tag{65}$$

The value of $\Delta W/W_1$ calculated from both Eqs. (64) and (65)* with $\gamma=1.405$ are shown in Fig. 4 as Curve I and Curve IV respectively. It is seen that the theory developed in Part I of the present report gives a value of velocity increment $\Delta W/W_1$ between the first and the second approximation of the perturbation method with $\gamma=1.405$. Unfortunately, it is not certain whether the exact solution with $\gamma=1.405$ will further increase the value of $\Delta W/W_1$ or decrease it. Therefore, it is impossible at present to ascertain the exact magnitude of deviation introduced by using the tangent to the adiabatic curve instead of the adiabatic curve itself.

By comparing to approximate adiabatic relation given by Eq. (32) with the true adiabatic relation given by Eq. (8), it is seen that the relation given by Eq. (32) can be also considered as the exact adiabatic relation of a fictitious gas with $\gamma=-1$. Therefore the solution given by the present theory can be also considered as an exact solution of the flow of this fictitious gas. Hence by putting $\gamma=-1$ in Eq. (65) and comparing the value of $\Delta W/W_1$ so calculated with that obtained by transformation, the error involved in the second approximation can be determined. This comparison is shown in Fig. 4 as Curve II. The exact solution for $\gamma=-1$ is slightly higher than that of the second approximation of the perturbation method with $\gamma\doteq-1$.

If this tendency of increasing the value of $\Delta W/W_1$ also holds in the case of $\gamma=1.405$ then the value of velocity increment $\Delta W/W_1$ given by the straight line adiabatic relation will be definitely lower than that of exact solution. Using this lower value of velocity increment to calculate the critical speed of the section, the resulting value of the critical speed will be higher than it should be. In other words, the method is slightly unconservative.

Part III

The most difficult part of the calculation in applying the new theory to practical cases is

* Eqs. (63) and (64) in the original paper has been corrected as Eqs (64) and (65). — Noted by editor.

the correction for the shape of the body, because the correction formula, Eq. (53) involves the complex potential of a corresponding incompressible flow. Although the principle for finding this complex potential for any given shape of a body in a uniform flow is well-known, the actual calculation is generally very tedious except for a few simple geometrical shapes. The problem then arises: Whether it is possible to omit this correction without introducing serious error in the velocity and the pressure distribution? Furthermore, as shown by the calculation for elliptical cylinders, the correction for the shape of the body results a slight increase in the thickness ratio of the section. An increase in thickness ratio of the section generally raises the maximum local speed over the surface of the section. In view of the previously shown result that the method developed in Part I slightly underestimates the velocity in compressible flow, this increase in velocity due to neglect of the correction in shape of the body will tend to remedy this defect.

To test whether this reasoning is justified, the flow over a circular cylinder will be investigated anew*. However, now the slight change in shape of the body due to transformation from incompressible flow to compressible flow will be neglected. The complex potential for the incompressible flow over a circular cylinder is given by

$$\phi + i\psi = w_1\left(z + \frac{b^2}{z}\right) \tag{66}$$

where b is the radius of the cylinder. By differentiating Eq. (66), the velocity at a point corresponding to $z = iy$ can be found as

$$w = w_1\left(1 + \frac{b^2}{y^2}\right) \tag{67}$$

Therefore using Eq. (42), the velocity in compressible flow is

$$\frac{W}{W_1} = \left(1 + \frac{b^2}{y^2}\right)\frac{1 - \dfrac{M_1^2}{(1+\sqrt{1-M_1^2})^2}}{1 - \dfrac{M_1^2}{(1+\sqrt{1-M_1^2})^2}\left(1 + \dfrac{b^2}{y^2}\right)^2} \tag{68}$$

Substituting Eq. (66) into Eq. (53), the coordinate of a point in compressible flow can be expressed in terms of the coordinate of the corresponding point in incompressible flow. Thus

$$Z = z - \frac{M_1^2}{(1+\sqrt{1-M_1^2})^2}\left(\bar{z} + \frac{2b^2}{z} - \frac{b^4}{3\bar{z}^3}\right) \tag{69}$$

By putting $z = \bar{z} = b$ in Eq. (69), the length of the section parallel to the undisturbed flow is found to be

$$2b\left[1 - \frac{8}{3}\frac{M_1^2}{(1+\sqrt{1-M_1^2})^2}\right] \tag{70}$$

* This calculation is also carried out by K Tamada[5].

By putting $z = ib$ and $\bar{z} = -ib$ in Eq. (69), the thickness of the transformed section is found to be

$$2b\left[1-\frac{2}{3}\frac{M_1^2}{(1+\sqrt{1-M_1^2})^2}\right]=2R \tag{71}$$

where R is taken to be the radius of the circular section concerned. Therefore for $M_1 = 0.400$ the thickness ratio of the transformed section is equal to 1.098 6 instead of 1.000 0.

Setting $z = iy$ and $\bar{z} = -iy$ in Eq. (69), a point along the Y-axis can be calculated as

$$Y = y+\frac{M_1^2}{(1+\sqrt{1-M_1^2})^2}\left(y-\frac{2b^2}{y}-\frac{b^4}{3y^3}\right) \tag{72}$$

Using Eqs. (68) and (72), the velocity distribution along the Y-axis can be calculated. The result of the calculation is shown in Fig. 4.

By comparing with the curve for the second approximation of the perturbation method, it is clear that the new method gives higher velocity over the surface of the section. Therefore, if this value is used to predict the critical velocity of the section, the result will be slightly conservative which is very desirable from engineering standpoint. The circular section chosen for illustrating the method is, however, an extreme case in the sense that the velocity increment is very large. For airfoil sections generally used in design, the thickness ratio is smaller and therefore the velocity increment will be also smaller. Then the discrepancy between the present simplified theory and the exact solution will be much less than that shown in Fig. 4 as Curve V. Therefore, the neglect of the correction in shape of the body as given by Eq. (53) will not introduce serious error, but on the other hand, greatly reduce the amount of necessary calculations and make the method slightly conservative.

Summarizing, if the velocity and the pressure distributions over a body are measured at low speed in an ordinary wind tunnel, the velocity and the pressure distributions over the same body at high speeds can be calculated by means of Eqs. (42) and (47).* By integrating the pressure distribution, the lift coefficient and the moment coefficient of the airfoil at high speeds can be evaluated.

It is interesting to note that Eq. (47) is reduced to the relation of Glauert and Prandtl

$$K_p = \frac{k_p}{\sqrt{1-M_1^2}} \tag{73}$$

if k_p is very small so that k_p^2 can be neglected. However, for a finite suction pressure, the present method gives larger increase due to compressibility than Glauert-Prandtl method.

To calculate the critical speed of a body when its maximum suction pressure is given, the corresponding maximum velocity has to be determined first. Using Eq. (42) and (44), the

* It was mentioned by Dr. Th. von Kármán recently that the result of calculation by this method agrees well with recent N. A. C. A. tests.

following relation is obtained

$$\frac{W_{\max}}{W_1}=(1-k_{\text{pmin}})^{\frac{1}{2}}\frac{1-\dfrac{M_1^2}{(1+\sqrt{1-M_1^2})^2}}{1-\dfrac{M_1^2}{(1+\sqrt{1-M_1^2})^2}(1-k_{\text{pmin}})} \tag{74}$$

For calculating the critical speed, the straight line approximation to the adiabatic pressure-volume curve has to be abandoned. From Eq. (36) it is seen that at the minimum density $\sigma=0$, the maximum local Mach number is 1, i. e., the fluid velocity reaches the local sound velocity. But when $\sigma=0$, the fluid velocity W will be infinitely large according to Eq. (35). Therefore with the straight line approximation a finite fluid velocity will never reach the local sound velocity and there is no possibility of compressibility burble. However, if the velocity given by Eq. (74) is taken as a sufficiently close approximation to the fluid velocity in true adiabatic flow, the exact relation between the fluid velocity W and the velocity of sound A can then be used to predict the compressibility burble. According to Kaplan[6] and others, the maximum local velocity and critical speed are connected by the following relation in true adiabatic flow:

$$\left(\frac{W_{\max}}{W_1}\right)^2=\frac{2}{(\gamma+1)M_c^2}+\frac{\gamma-1}{\gamma+1} \tag{75}$$

where M_c is the critical Mach number, i. e., the Mach number of the undisturbed flow at which compressibility burble occurs. Equating the expressions of $W_{\max}/W_1$ given by Eqs. (74) and (75), the equation for computing the critical Mach number is

$$\frac{(1-k_{\text{pmin}})^{\frac{1}{2}}\left[1-\dfrac{M_c^2}{(1+\sqrt{1-M_c^2})^2}\right]}{1-\dfrac{M_c^2}{(1+\sqrt{1-M_c^2})^2}(1-k_{\text{pmin}})}=\left[\frac{2}{(\gamma-1)M_c^2}+\frac{\gamma-1}{\gamma+1}\right]^{\frac{1}{2}} \tag{76}$$

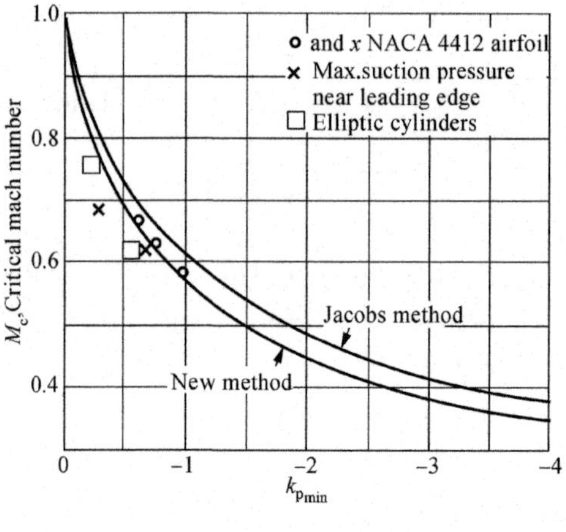

Fig. 5

The result of the calculation is shown in Fig. 5, where the ordinate is the critical Mach number M_c, and the abscissa is the maximum suction pressure coefficient k_{pmin} measured in a low speed wind tunnel. In the same figure a curve given by E. N. Jacobs[7] is also included. Jacobs based his calculation on the theory of Glauert and Prandtl. As already stated in connections with Eq. (73), this theory tends to underestimate the compressibility effect. Therefore the higher value of critical Mach number given by Jacobs is expected. The experimental data given by J. Stack, W. F. Lindsey and R. T. Littell[8] are also included in Fig. 5 for comparison. It is seen that while Jacobs' curve lies definitely above the experimental points, the curve obtained from Eq. (76) is a good representation of their average. Hence this method of predicting the compressibility burble, although not exact, is quite satisfactory for design calculations.

Appendix I

Functions of Mach Number Used in the Calculations of Velocity and Pressure

M_1	$\sqrt{1-M_1^2}$	$1-\frac{M_1^2}{(1+\sqrt{1-M_1^2})^2}$	$\frac{M_1^2}{2(1+\sqrt{1-M_1^2})}$
0.20	0.979 8	0.989 8	0.010 10
0.30	0.953 9	0.976 4	0.023 03
0.40	0.916 5	0.956 4	0.041 74
0.45	0.893 0	0.943 5	0.053 49
0.50	0.866 0	0.928 2	0.066 99
0.55	0.835 2	0.910 2	0.082 42
0.60	0.800 0	0.888 9	0.100 00
0.65	0.759 9	0.863 6	0.120 03
0.70	0.714 1	0.833 2	0.142 93
0.75	0.661 4	0.796 2	0.169 28
0.80	0.600 0	0.750 0	0.200 00
0.85	0.526 8	0.690 1	0.236 61
0.90	0.435 9	0.607 1	0.282 06
0.95	0.312 3	0.475 9	0.343 88
1.00	0	0	0.500 00

References

[1] Hsue-shen Tsien. Two-dimensional subsonic flow of compressible fluids. J. of Aero. Soc., 1939, 6: 399-407.

[2] Bateman H. The lift and drag functions for an elastic fluid in two-dimensional irrotational flow. Proceedings of the National Academy of Sciences, 1938, 24: 246-251.

[3] Imai I. On the flow of a compressible fluid past a circular cylinder. Proceedings of the Physico-Mathematical Society of Japan, 1938, 20(3): 635-645.

[4] Tamada K., Saito Y. Note on the flow of a compressible fluid past a circular cylinder. Ibid., 1939, 21

(3): 403 - 409.

[5] Tamada K. Application of the hodograph method to the flow of a compressible fluid past a circular cylinder. Ibid., 1940, 22(3): 208 - 219.

[6] Kaplan C. Two dimensional subsonic compressible flow past elliptic cylinders. N. A. C. A. T. R. No. 624, 1938.

[7] Jacobs E N. Method employed in America for the experimental investigation of aerodynamic phenomena at high speeds. Atti dei V Convergno "Volta", Rome: 1936, 369 - 467.

[8] Stack J, Lindsey W F, Littell R T. The compressibility burble and the effect of compressibility on pressure and forces acting on an airfoil. N. A. C. A. T. R. No. 646, 1939.

The Buckling of Thin Cylindrical Shells under Axial Compression

Theodore von Kármán and Hsue-shen Tsien
(California Institute of Technology)

In two previous papers[1,2] the authors have discussed in detail the inadequacy of the classical theory of thin shells in explaining the buckling phenomenon of cylindrical and spherical shells. It was shown that not only the calculated buckling load is 3 to 5 times higher than that found by experiments, but the observed wave pattern of the buckled shell is also different from that predicted. Furthermore, it was pointed out that the different explanations for this discrepancy advanced by L. H. Donnell[3] and W. Flügge[4] are untenable when certain conclusions drawn from these explanations are compared with the experimental facts. By a theoretical investigation on spherical shells[1] the authors were led to the belief that in general the buckling phenomenon of curved shells can only be explained by means of a non-linear large deflection theory. This point of view was substantiated by model experiments on slender columns with non-linear elastic support[2]. The non-linear characteristics of such structures cause the load necessary to keep the shell in equilibrium to drop very rapidly with increase in wave amplitude once the structure started to buckle. Thus, first of all, a part of the elastic energy stored in the shell is released once the buckling has started; this explains the observed rapidity of the buckling process. Furthermore, as it was shown in one of the previous papers[2] the buckling load itself can be materially reduced by slight imperfections in the test specimen and vibrations during the testing process.

In this paper, the same ideas are applied to the case of a thin uniform cylindrical shell under axial compression. First it is shown by an approximate calculation that again the load sustained by the shell drops with increasing deflection. Then the results of this calculation are used for a more detailed discussion of the buckling process as observed in an actual testing machine.

Stresses in the Median Surface and the expression for the Total Energy of the System

Let x and y be measured in the axial and the circumferential direction in the median surface of the undeformed cylindrical shell and u, v and w be the components of displacement of a

---

Received February 1, 1941.
Journal of the Aeronautical Sciences, vol. 8, pp. 303 – 312, 1941.

point on the median surface of the shell in the x-direction, the y-direction and the radial direction (Fig. 1). Then at an arbitrary point in the median surface the unit strains in the x and y-directions, ε_x, ε_y and the unit shear γ_{xy} can be expressed in the following forms, including

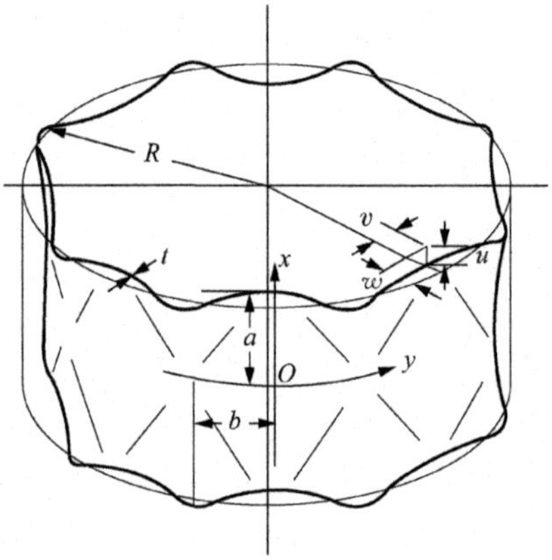

Fig. 1

terms up to second order:

$$\left.\begin{aligned}\varepsilon_x &= \frac{\partial u}{\partial x} + \frac{1}{2}\left(\frac{\partial w}{\partial x}\right)^2 \\ \varepsilon_y &= \frac{\partial v}{\partial y} + \frac{1}{2}\left(\frac{\partial w}{\partial y}\right)^2 - \frac{w}{R} \\ \gamma_{xy} &= \frac{\partial u}{\partial y} + \frac{\partial v}{\partial x} + \frac{\partial w}{\partial x}\frac{\partial w}{\partial y}\end{aligned}\right\} \tag{1}$$

R is the radius of the undeformed median surface of the shell. The stresses and the strains in the median surface of the shell are, however, related to each other by the following equations:

$$\left.\begin{aligned}\sigma_x &= \frac{E}{1-\nu^2}(\varepsilon_x + \nu\varepsilon_y) \\ \sigma_y &= \frac{E}{1-\nu^2}(\varepsilon_y + \nu\varepsilon_x) \\ \tau_{xy} &= \frac{E}{2(1+\nu)}\gamma_{xy}\end{aligned}\right\} \tag{2}$$

where E is Young's modulus of elasticity and ν is Poisson's ratio. Therefore, by substituting Eq. (1) into Eq. (2), the following connections between the components of stress in the median surface and the components of displacement of the median surface are obtained:

$$\left.\begin{aligned}\sigma_x &= \frac{E}{1-\nu^2}\left[\frac{\partial u}{\partial x} + \frac{1}{2}\left(\frac{\partial w}{\partial x}\right)^2 + \nu\left\{\frac{\partial v}{\partial y} + \frac{1}{2}\left(\frac{\partial w}{\partial y}\right)^2 - \frac{w}{R}\right\}\right] \\ \sigma_y &= \frac{E}{1-\nu^2}\left[\frac{\partial v}{\partial y} + \frac{1}{2}\left(\frac{\partial w}{\partial y}\right)^2 - \frac{w}{R} + \nu\left\{\frac{\partial u}{\partial x} + \frac{1}{2}\left(\frac{\partial w}{\partial y}\right)^2\right\}\right] \\ \tau_{xy} &= \frac{E}{2(1+\nu)}\left[\frac{\partial u}{\partial y} + \frac{\partial v}{\partial x} + \frac{\partial w}{\partial x}\frac{\partial w}{\partial y}\right]\end{aligned}\right\} \tag{3}$$

It is generally accepted that the conditions of equilibrium between the stresses acting in the median surface of a thin shell can be approximately expressed by the equations used for flat plates:

$$\left.\begin{aligned}\frac{\partial \sigma_x}{\partial x}+\frac{\partial \tau_{xy}}{\partial y}=0\\ \frac{\partial \tau_{xy}}{\partial x}+\frac{\partial \sigma_y}{\partial y}=0\end{aligned}\right\} \tag{4}$$

This pair of equations can be satisfied by introducing the well known Airy's stress function, $F(x, y)$, defined by the relations

$$\sigma_x=\frac{\partial^2 F}{\partial y^2}, \quad \tau_{xy}=-\frac{\partial^2 F}{\partial x \partial y}, \quad \sigma_y=\frac{\partial^2 F}{\partial x^2} \tag{5}$$

Eliminating the variables u and v in Eqs. (3) and (5) the following relation between Airy's stress function $F(x, y)$ and the radial component of the displacement, w is obtained.

$$\left(\frac{\partial^2}{\partial x^2}+\frac{\partial^2}{\partial y^2}\right)^2 F=E\left[\left(\frac{\partial^2 w}{\partial x \partial y}\right)^2-\frac{\partial^2 w}{\partial x^2}\frac{\partial^2 w}{\partial y^2}-\frac{1}{R}\frac{\partial^2 w}{\partial x^2}\right] \tag{6}$$

This equation expresses the condition of compatibility between stress and strain. When $R \to \infty$, it reduces to the corresponding equation for a flat plate derived by the senior author[5]. L. H. Donnell[3] first obtained Eq. (6) in its present form. With a given distribution of the radial component of the displacement, w, Eq. (6) gives the induced stresses in the median surface of the shell.

For one complete wave panel, the extensional elastic energy W_1 corresponding to these stresses can be written as

$$W_1=\frac{t}{2E}4\int_0^a\int_0^b\left[(\sigma_x+\sigma_y)^2-2(1+\nu)(\sigma_x\sigma_y-\tau_{xy}^2)\right]dx dy\ ^* \tag{7}$$

where a and b are the half wave lengths in the axial and the circumferential directions, respectively.

To calculate the elastic energy of bending, it is necessary to find the expressions for the change of curvatures and the unit twist of the median surface. In this paper, the following simplified expressions will be used:

$$\chi_x=\frac{\partial^2 w}{\partial x^2}, \quad \chi_{xy}=\frac{\partial^2 w}{\partial x \partial y}, \quad \chi_y=\frac{\partial^2 w}{\partial y^2} \tag{8}$$

In Eq. (8), certain additional terms in χ_y and χ_{xy} involving v are neglected. It was shown by L. H. Donnell[6] that if the terms retained in Eq. (8) are considered as of the order one, the neglected terms are of the order $1/n^2$, where n is the number of waves in the circumferential direction. For thin cylindrical shells, the value of n is around 10; therefore the neglection is justified. With these expressions for the change of curvatures and the unit twist of the median surface, the bending energy W_2 for one complete wave panel can be written as

* The original: $W_1=\frac{t}{2E}4\int_0^a\int_0^b\left[(\sigma_x+\sigma_y)^2-2(1+\nu)(\sigma_x\sigma_y-\tau_{xy}^2)\right]d\alpha dy.$

$$W_2 = \frac{t^3 E}{24(1-\nu^2)} 4 \int_0^a \int_0^b \left[\left\{ \frac{\partial^2 w}{\partial x^2} + \frac{\partial^2 w}{\partial y^2} \right\}^2 - 2(1-\nu) \left\{ \frac{\partial^2 w}{\partial x^2} \frac{\partial^2 w}{\partial y^2} - \left(\frac{\partial^2 w}{\partial x \partial y} \right)^2 \right\} \right] \mathrm{d}x \mathrm{d}y \tag{9}$$

The virtual work of the force applied on the end of the cylindrical shell can be calculated as the product of the applied force and the change in length of the shell. Therefore the following expression is obtained for one complete wave panel:

$$W_3 = 4t \int_0^b (\sigma_x)_{x=a} \mathrm{d}y \int_0^a \frac{\partial u}{\partial x} \mathrm{d}x \tag{10}$$

The equilibrium condition of the shell can be obtained either by equating the first variation of the difference between the sum of the energies W_1 and W_2 and the virtual work W_3 to zero, or by actually analyzing the moments and the stresses in the median surface of the shell. Using the approximations stated previously, Donnell[6] derived the equilibrium equation as

$$\frac{Et^3}{12(1-\nu^2)} \left(\frac{\partial^2}{\partial x^2} + \frac{\partial^2}{\partial y^2} \right)^4 w + \frac{Et}{R^2} \frac{\partial^4 w}{\partial x^4} = \left(\frac{\partial^2}{\partial x^2} + \frac{\partial^2}{\partial y^2} \right)^2 \left\{ p + t \left(\sigma_x \frac{\partial^2 w}{\partial x^2} + 2\tau_{xy} \frac{\partial^2 w}{\partial x \partial y} + \sigma_y \frac{\partial^2 w}{\partial x^2} \right) \right\} \tag{11}$$

where p is the external radial pressure on the surface of the shell. In the case concerned, $p = 0$, then using Eq. (5), the second equation connecting Airy's stress function $F(x, y)$ and the radial component of displacement w is obtained as

$$\frac{Et^2}{12(1-\nu^2)} \left(\frac{\partial^2}{\partial x^2} + \frac{\partial^2}{\partial y^2} \right)^4 w + \frac{E}{R^2} \frac{\partial^4 w}{\partial x^4} = \left(\frac{\partial^2}{\partial x^2} + \frac{\partial^2}{\partial y^2} \right)^2 \left[\frac{\partial^2 F}{\partial y^2} \frac{\partial^2 w}{\partial x^2} - 2 \frac{\partial^2 F}{\partial x \partial y} \frac{\partial^2 w}{\partial x \partial y} + \frac{\partial^2 F}{\partial x^2} \frac{\partial^2 w}{\partial y^2} \right] \tag{12}$$

When $R \to \infty$, Eq. (12) reduces to the corresponding equation for a flat plate.

There are two different ways to solve the problem of buckling of a thin uniform cylindrical shell under axial compression. The more exact method is to solve Eqs. (6) and (12) simultaneously, using appropriate boundary conditions. The approximate method is to first assume a plausible function for w, with undetermined parameters and then use Eq. (6) to determine the stresses in the median surface of the shell. The expressions W_1, W_2 and W_3 can then be calculated by means of Eqs. (7), (9) and (10). The undetermined parameters can be ascertained by the condition that $W_1 + W_2 - W_3$ must be a minimum. This approximate method will be used in the following calculations.

Calculation of the Total Energy

To obtain a plausible form for w, one has to resort to the experimental results. It is observed that, for large values of the wave amplitude, the waves show a so-called diamond shaped pattern. This particular wave shape can be approximately expressed by

$$\frac{w_1}{R} = \cos^2 \frac{(mx + ny)}{2R} \cos^2 \frac{(mx - ny)}{2R} \tag{13}$$

where the squares are introduced to account for the fact that the shell has a definite preference to buckle inward. Eq. (13) can be re-written as

$$\frac{w_1}{R} = \frac{1}{4} + \frac{1}{2}\Big[\cos\frac{mx}{R}\cos\frac{ny}{R} + \frac{1}{4}\cos\frac{2mx}{R} + \frac{1}{4}\cos\frac{2ny}{R}\Big] \tag{14}$$

On the other hand, the classical theory which is correct for infinitesimal values of the wave amplitude requires the wave to be of the form

$$\frac{w_2}{R} = \cos\frac{mx}{R}\cos\frac{ny}{R} \tag{15}$$

In order to satisfy this requirement, the wave form assumed in the following calculation is

$$\frac{w}{R} = \left(f_0 + \frac{f_1}{4}\right) + \frac{f_1}{2}\left(\cos\frac{mx}{R}\cos\frac{ny}{R} + \frac{1}{4}\cos\frac{2mx}{R} + \frac{1}{4}\cos\frac{2ny}{R}\right) + \frac{f_2}{4}\left(\cos\frac{2mx}{R} + \cos\frac{2ny}{R}\right) \tag{16}$$

where f_0, f_1, f_2 are unknowns to be determined by the minimum condition given above; f_0 is introduced in order to allow the shell to expand radially. The amplitude of the wave pattern defined as the maximum difference in the radial deflection w is evidently given by f_1. The wave lengths in the axial and the circumferential direction are $2\pi R/m$ and $2\pi R/n$, respectively. Hence the number of waves along the circumference of the shell is equal to n. It is evident that no end effect can be accounted for by this form of wave pattern, and therefore the following calculation really corresponds to the case of a very long cylindrical shell. This simplification is justified by the experimental findings of N. Nojima and S. Kanemitsu as reported in a previous paper[2]. It was found that there is no appreciable length effect when the length of the cylindrical is greater than 1.5 times the radius of the shell. Furthermore, it is seen that by setting $f_0 = f_2 = 0$, Eq. (16) is reduced to Eq. (14); while by setting $(f_1/4) + (f_2/2) = 0$ and $f_0 + f_1/4 = 0$, Eq. (16) is reduced to Eq. (15). With other values of these parameters, wave patterns intermediate between these two limits can be obtained.

Substituting Eq. (16) into Eq. (6), the differential equation for Airy's stress function $F(x, y)$ is obtained:

$$\left(\frac{\partial^2}{\partial x^2} + \frac{\partial^2}{\partial y^2}\right)^2 F = -E\mu^2\left(\frac{n}{R}\right)^2\Big[A\cos\frac{2mx}{R} + B\cos\frac{2ny}{R} + C\cos\frac{mx}{R}\cos\frac{ny}{R} + D\cos\frac{3mx}{R}\cos\frac{ny}{R} + G\cos\frac{mx}{R}\cos\frac{3ny}{R} + H\cos\frac{2mx}{R}\cos\frac{2ny}{R}\Big] \tag{17}$$

where $\mu = m/n$, the "aspect ratio" of the waves. If $\mu > 1$, the waves are longer in the circumferential direction; if $\mu < 1$, the waves are longer in the axial direction. The coefficients in Eq. (17) are given by the following relations:

$$\left.\begin{aligned} A &= \frac{1}{8} f_1^2 n^2 - \left(\frac{1}{2} f_1 + f_2\right) \\ B &= \frac{1}{8} f_1^2 n^2 \\ C &= \frac{1}{2} f_1 n^2 \left(\frac{1}{2} f_1 + f_2\right) - \frac{1}{2} f_1 \\ D &= \frac{1}{4} f_1 n^2 \left(\frac{1}{2} f_1 + f_2\right) \\ G &= \frac{1}{4} f_1 n^2 \left(\frac{1}{2} f_1 + f_2\right) \\ \text{and} \qquad H &= n^2 \left(\frac{1}{2} f_1 + f_2\right)^2 \end{aligned}\right\} \tag{18}$$

The solution of Eq. (17) can be easily obtained as

$$\begin{aligned} F = -E\mu^2 \left(\frac{R}{n}\right)^2 \Bigg[&\frac{A}{16\mu^4} \cos\frac{2mx}{R} + \frac{B}{16} \cos\frac{2ny}{R} + \frac{C}{(1+\mu^2)^2} \cos\frac{mx}{R} \cos\frac{ny}{R} + \\ &\frac{D}{(1+9\mu^2)^2} \cos\frac{3mx}{R} \cos\frac{ny}{R} + \frac{G}{(9+\mu^2)^2} \times \cos\frac{mx}{R} \cos\frac{3ny}{R} + \\ &\frac{H}{16(1+\mu^2)^2} \cos\frac{2mx}{R} \times \cos\frac{2ny}{R} \Bigg] + \frac{\alpha x^2}{2} + \frac{\beta y^2}{2} \end{aligned} \tag{19}$$

Using Eq. (5), the stress components in the median surface can be written as

$$\left.\begin{aligned} \sigma_x = \bar{E}\mu^2 \Bigg[&\frac{B}{4} \cos\frac{2ny}{R} + \frac{C}{(1+\mu^2)^2} \cos\frac{mx}{R} \cos\frac{ny}{R} + \frac{D}{(1+9\mu^2)^2} \cos\frac{3mx}{R} \cos\frac{ny}{R} + \\ &\frac{9G}{(9+\mu^2)^2} \cos\frac{mx}{R} \times \cos\frac{3ny}{R} + \frac{H}{4(1+\mu^2)^2} \cos\frac{2mx}{R} \cos\frac{2ny}{R} \Bigg] + \beta \\ \sigma_y = E\mu^2 \Bigg[&\frac{A}{4\mu^2} \cos\frac{2mx}{R} + \frac{\mu^2 C}{(1+\mu^2)^2} \cos\frac{mx}{R} \cos\frac{ny}{R} + \frac{9\mu^2}{(1+9\mu^2)^2} \cos\frac{3mx}{R} \cos\frac{ny}{R} + \\ &\frac{\mu^2 G}{(9+\mu^2)^2} \cos\frac{mx}{R} \times \cos\frac{3ny}{R} + \frac{\mu^2 H}{4(1+\mu^2)^2} \cos\frac{2mx}{R} \cos\frac{2ny}{R} \Bigg] + \alpha \\ t_{xy} = E\mu^2 \Bigg[&\frac{\mu C}{(1+\mu^2)^2} \sin\frac{mx}{R} \sin\frac{ny}{R} + \frac{3\mu D}{(1+9\mu^2)^2} \sin\frac{3mx}{R} \sin\frac{ny}{R} + \\ &\frac{3\mu G}{(9+\mu^2)^2} \sin\frac{mx}{R} \sin\frac{3ny}{R} + \frac{\mu H}{4(1+\mu^2)^2} \sin\frac{2mx}{R} \sin\frac{2ny}{R} \Bigg] \end{aligned}\right\} \tag{20}$$

In all experimental work, the data are usually expressed in terms of the average compression stress σ in the axial direction. It can be easily seen from Eq. (20) that

$$\beta = -\sigma \tag{21}$$

Using Eq. (3), the following expressions for $\partial u/\partial x$ and $\partial v/\partial y$ can be obtained:

$$\begin{aligned} \frac{\partial u}{\partial x} &= \frac{1}{E}(\sigma_x - \nu\sigma_y) - \frac{1}{2}\left(\frac{\partial w}{\partial x}\right)^2 \\ \frac{\partial v}{\partial y} &= \frac{1}{E}(\sigma_y - \nu\sigma_x) - \frac{1}{2}\left(\frac{\partial w}{\partial y}\right)^2 + \frac{w}{R} \end{aligned} \tag{22}$$

By substituting Eqs. (16) and (20) into Eq. (22), it is found that

$$\begin{aligned} \frac{\partial u}{\partial x} &= -\left[\left(\frac{\sigma}{E} + \nu\frac{\alpha}{E}\right) + \frac{1}{2}n^2\mu^2\left(\frac{3}{32}f_1^2 + \frac{1}{8}f_1 f_2 + \frac{1}{8}f_2^2\right)\right] + \\ &\quad \text{Terms of periodic functions} \\ \frac{\partial v}{\partial y} &= \frac{\alpha}{E} + \nu\frac{\sigma}{E} - \frac{1}{2}n^2\left(\frac{3}{32}f_1^2 + \frac{1}{8}f_1 f_2 + \frac{1}{8}f_2^2\right) + \left(f_0 + \frac{f_1}{4}\right) + \\ &\quad \text{Terms of periodic functions} \end{aligned} \tag{23}$$

Since y is measured along the circumference of the shell, v must be a periodic function of y; therefore, the constant term in $\frac{\partial v}{\partial y}$ must be equal to zero. Or

$$\frac{\alpha}{E} + \nu\frac{\sigma}{E} - \frac{1}{2}n^2\left(\frac{3}{32}f_1^2 + \frac{1}{8}f_1 f_2 + \frac{1}{8}f_2^2\right) + \left(f_0 + \frac{f_1}{4}\right) = 0 \tag{24}$$

This condition determines α.

Using Eqs. (7), (20) and (24), the extensional energy W_1 of the shell is obtained as

$$\begin{aligned} \frac{W_1}{\frac{1}{2}Etab} &= 4\Big[(1-\nu^2)\left(\frac{\sigma}{E}\right)^2 + n^4\left(\frac{3}{64}f_1^2 + \frac{1}{16}f_1 f_2 + \right. \\ &\left.\frac{1}{16}f_2^2\right)^2 + \left(f_0 + \frac{1}{4}f_1\right)^2 - 2n^2\left(\frac{3}{64}f_1^2 + \frac{1}{16}f_1 f_2 + \right. \\ &\left.\frac{1}{16}f_2^2\right)\left(f_0 + \frac{1}{4}f_1\right)\Big] + \Big[\frac{A^2}{8} + \frac{B^2\mu^4}{8} + \frac{\mu^4 C^2}{(1+\mu^2)^2} + \\ &\frac{\mu^4 D^2}{(1+9\mu^2)^2} + \frac{\mu^4 G^2}{(9+\mu^2)^2} + \frac{\mu^4 H^2}{16(1+\mu^2)^2}\Big] \end{aligned} \tag{25}$$

Using Eqs. (9) and (16), the energy of bending W_2 of the shell can be calculated as

$$\begin{aligned} \frac{W_2}{\frac{1}{2}Etab} &= \frac{1}{6(1-\nu^2)}\left(\frac{t}{R}\right)^2 n^4\left[f_1^2\left\{\frac{1}{8}(1+\mu^2)^2 + \frac{1}{4}(1+\mu^4)\right\} + \right. \\ &\left.(1+\mu^4)f_1 f_2 + (1+\mu^4)f_2^2\right] \end{aligned} \tag{26}$$

The virtual work of the applied force can be obtained by means of Eqs. (10), (20), (23) and (24). The result is

$$\begin{aligned} \frac{W_3}{\frac{1}{2}Etab} &= \left[2(1-\nu^2)\left(\frac{\sigma}{E}\right)^2 + n^2\frac{\sigma}{E}(\nu+\mu^2)\left\{\frac{3}{32}f_1^2 + \frac{1}{8}f_1 f_2 + \frac{1}{8}f_2^2\right\} - \right. \\ &\left. 2\nu\frac{\sigma}{E}\left(f_0 + \frac{1}{4}f_1\right)\right] \end{aligned} \tag{27}$$

Relation Between the compression Stress and the Amplitude of Waves

To find the relation between the average compression stress and the amplitude of the waves, the conditions which will make the expression $W_1 + W_2 - W_3$ a minimum have to be obtained. It was found that the calculations can be simplified to a certain extent by first using the condition that the sum of energies must be minimum with respect to f_0. Or

$$\frac{\partial}{\partial f_0}(W_1 + W_2 - W_3) = 0 \tag{28}$$

This condition determines a relation between f_0 and f_1, f_2, which can be written as:

$$f_0 + \frac{1}{4}f_1 = n^2\left(\frac{3}{64}f_1^2 + \frac{1}{16}f_1f_2 + \frac{1}{16}f_2^2\right) - \nu\frac{\sigma}{E} \tag{29}$$

Using this relation and Eq. (24), it is easily seen that $\alpha = 0$. In other words, the shell will expand radially to such an extent that the average of the circumferential stress σ_y is equal to zero. Substituting Eq. (29) into the expressions for W_1, W_2 and W_3 as given by Eqs. (25), (26) and (27) and using Eq. (18), the elastic energy of the system minus the virtual work is expressed finally in the following form:

$$\begin{aligned}\frac{W_1 + W_2 - W_3}{\frac{1}{2}Etab} = &-4\left(\frac{\sigma}{E}\right)^2 - \frac{\sigma}{E}n^2\mu^2\left(\frac{3}{8}f_1^2 + \frac{1}{2}f_1f_2 + \frac{1}{2}f_2^2\right) + \\ &n^4\left[\left\{\frac{1+\mu^4}{512} + \frac{17}{256}\frac{\mu^4}{(1+\mu^2)^2} + \frac{1}{64}\frac{\mu^4}{(1+9\mu^2)^2} + \frac{1}{64}\frac{\mu^4}{(9+\mu^2)^2}\right\}f_1^4 + \right. \\ &\left\{\frac{9}{32}\frac{\mu^4}{(1+\mu^2)^2} + \frac{1}{16}\frac{\mu^4}{(1+9\mu^2)^2} + \frac{1}{16}\frac{\mu^4}{(9+\mu^2)^2}\right\}f_1^3f_2 + \\ &\left\{\frac{11}{32}\frac{\mu^4}{(1+\mu^2)^2} + \frac{1}{16}\frac{\mu^4}{(1+9\mu^2)^2} + \frac{1}{16}\frac{\mu^4}{(9+\mu^2)^2}\right\}f_1^2f_2^2 + \\ &\left.\frac{1}{8}\frac{\mu^4}{(1+\mu^2)^2}f_1f_2^3 + \frac{1}{16}\frac{\mu^4}{(1+\mu^2)^2}f_2^4\right] - n^2\left[\left\{\frac{1}{64} + \frac{1}{4}\frac{\mu^4}{(1+\mu^2)^2}\right\}f_1^3 + \right. \\ &\left.\left\{\frac{1}{32} + \frac{1}{2}\frac{\mu^4}{(1+\mu^2)^2}\right\}f_1^2f_2\right] + \left[\left\{\frac{1}{32} + \frac{1}{4}\frac{\mu^4}{(1+\mu^2)^2}\right\}f_1^2 + \frac{1}{8}f_1f_2 + \right. \\ &\left.\frac{1}{8}f_2^2\right] + \frac{1}{6(1-\nu^2)}\left(\frac{t}{R}\right)^2 n^4\left[\left\{\frac{1}{8}(1+\mu^2)^2 + \frac{1}{4}(1+\mu^4)\right\}f_1^2 + \right. \\ &\left.(1+\mu^4)f_1f_2 + (1+\mu^4)f_2^2\right]\end{aligned} \tag{30}$$

The equilibrium conditions are then obtained by differentiating this expression with respect to f_1 and f_2, and then set those derivatives equal to zero. The results can be written in a simpler form by introducing the following parameters:

$$\rho = \frac{f_2}{f_1}, \quad \eta = n^2\frac{t}{R}, \quad \xi = f_1\frac{R}{t} = \frac{\delta}{t} \tag{31}$$

where δ is the wave amplitude of the buckled shape of the cylindrical shell. Then the equilibrium conditions are

$$
\left.\begin{aligned}
&\left(\frac{\sigma R}{Et}\eta\mu^2\left(\rho+\frac{3}{2}\right)=(\eta\xi)^2\left[\frac{\mu^4}{4(1+\mu^2)^2}\rho^3+\right.\right.\\
&\quad\left\{\frac{11}{8}\frac{\mu^4}{(1+\mu^2)^2}+\frac{1}{4}\frac{\mu^4}{(1+9\mu^2)^2}+\frac{1}{4}\frac{\mu^4}{(9+\mu^2)^2}\right\}\rho^2+\\
&\quad\left\{\frac{27}{16}\frac{\mu^4}{(1+\mu^2)^2}+\frac{3}{8}\frac{\mu^4}{(1+9\mu^2)^2}+\frac{3}{8}\frac{\mu^4}{(9+\mu^2)^2}\right\}\rho+\\
&\quad\left.\left\{\frac{1+\mu^4}{64}+\frac{17}{32}\frac{\mu^4}{(1+\mu^2)^2}+\frac{1}{8}\frac{\mu^4}{(1+9\mu^2)^2}+\frac{1}{8}\frac{\mu^4}{(9+\mu^2)^2}\right\}\right]-\\
&\quad(\eta\xi)\left[\left\{\frac{1}{8}+2\frac{\mu^4}{(1+\mu^2)^2}\right\}\rho+\left\{\frac{3}{32}+\frac{3}{2}\frac{\mu^4}{(1+\mu^2)^2}\right\}\right]+\\
&\quad\left[\frac{1}{4}\rho+\left\{\frac{1}{8}+\frac{\mu^4}{(1+\mu^2)^2}\right\}\right]+\frac{1}{3(1-\nu^2)}\eta^2\left[(1+\mu^4)\rho+\right.\\
&\quad\left.\left\{\frac{1}{4}(1+\mu^2)^2+\frac{1}{2}(1+\mu^4)\right\}\right]\\
&\frac{\sigma R}{Et}\eta\mu^2\left(\rho+\frac{1}{2}\right)=(\eta\xi)^2\left[\frac{1}{4}\frac{\mu^4}{(1+\mu^2)^2}\rho^3+\frac{3}{8}\frac{\mu^4}{(1+\mu^2)^2}\rho^2+\right.\\
&\quad\left\{\frac{11}{16}\frac{\mu^4}{(1+\mu^2)^2}+\frac{1}{8}\frac{\mu^4}{(1+9\mu^2)^2}+\frac{1}{8}\frac{\mu^4}{(9+\mu^2)^2}\right\}\rho+\\
&\quad\left.\left\{\frac{9}{32}\frac{\mu^4}{(1+\mu^2)^2}+\frac{1}{16}\frac{\mu^4}{(1+9\mu^2)^2}+\frac{1}{16}\frac{\mu^4}{(9+\mu^2)^2}\right\}\right]-\\
&\quad(\eta\xi)\left[\frac{1}{32}+\frac{1}{2}\frac{\mu^4}{(1+\mu^2)^2}\right]+\left[\frac{1}{4}\rho+\frac{1}{8}\right]+\\
&\quad\frac{1}{3(1-\nu^2)}\eta^2\left[(1+\mu^4)\rho+\frac{1}{2}(1+\mu^4)\right]
\end{aligned}\right\}\quad(32)
$$

Eliminating $\sigma R/(Et)$ from the above equations, the following equation for ρ is obtained

$$
A_3\rho^3+A_2\rho^2+A_1\rho+A_0=0 \quad (33)
$$

where the coefficients are

$$
\left.\begin{aligned}
A_3&=(\eta\xi)^2\left\{\frac{3\mu^4}{(1+\mu^2)^2}+\frac{\mu^4}{(1+9\mu^2)^2}+\frac{\mu^4}{(9+\mu^2)^2}\right\}\\
A_2&=(\eta\xi)^2\left\{\frac{9}{2}\frac{\mu^4}{(1+\mu^2)^2}+\frac{3}{2}\frac{\mu^4}{(1+9\mu^2)^2}+\frac{3}{2}\frac{\mu^4}{(9+\mu^2)^2}\right\}-(\eta\xi)\left\{\frac{1}{2}+\frac{8\mu^4}{(1+\mu^2)^2}\right\}\\
A_1&=(\eta\xi)^2\left\{\frac{1+\mu^4}{16}+\frac{1}{4}\frac{\mu^4}{(1+\mu^2)^2}+\frac{1}{4}\frac{\mu^4}{(1+9\mu^2)^2}+\frac{1}{4}\frac{\mu^4}{(9+\mu^2)^2}\right\}-\\
&\quad(\eta\xi)\left\{\frac{1}{2}+\frac{8\mu^4}{(1+\mu^2)^2}\right\}+\left\{\frac{4\mu^4}{(1+\mu^2)^2}-1\right\}-\frac{2}{3(1-\nu^2)}\eta^2\times\\
&\quad\left\{2(1+\mu^4)-\frac{1}{2}(1+\mu^2)^2\right\}\\
A_0&=(\eta\xi)^2\left\{\frac{1+\mu^4}{32}-\frac{5}{8}\frac{\mu^4}{(1+\mu^2)^2}-\frac{1}{8}\frac{\mu^4}{(1+9\mu^2)^2}-\frac{1}{8}\frac{\mu^4}{(9+\mu^2)^2}\right\}+\\
&\quad\left\{\frac{2\mu^4}{(1+\mu^2)^2}-\frac{1}{2}\right\}-\frac{2}{3(1-\nu^2)}\eta^2\left\{(1+\mu^4)-\frac{1}{4}(1+\mu^2)^2\right\}
\end{aligned}\right\}\quad(34)
$$

Therefore, when $\xi = 0$, i.e., when the wave amplitude approaches zero Eq. (32) gives $A_2 = A_3 = 0$; and

$$A_1 = -\frac{2}{3(1-\nu^2)}\eta^2\left\{2(1+\mu^4) - \frac{1}{2}(1+\mu^2)^2\right\}$$

$$A_0 = -\frac{2}{3(1-\nu^2)}\eta^2\left\{(1+\mu^4) - \frac{1}{4}(1+\mu^2)^2\right\}$$

$$= \frac{A_1}{2}$$

Substituting into Eq. (31) it is seen that $\rho = -1/2$, or $f_2 = -1/2f_1$. Putting this relation between f_1 and f_2 into Eq. (14), the wave pattern is reduced to that represented by Eq. (15)*, i.e., the wave pattern for infinitesimal wave amplitude given by the classical theory.

With a given value of μ and η, the coefficients for various values of the wave amplitude ξ can be first calculated by using Eq. (34). Then Eq. (33) can be solved for ρ corresponding to this particular set of values of μ and η at various wave amplitude ξ. When the value of ρ is known, Eq. (32) can be used to calculate the corresponding value of the "reduced compression stress," $\sigma R/(Et)$. It is found, however, that the following expression which is obtained from Eq. (30) by eliminating the third powers of ρ is more suitable for numerical computations:

$$\begin{aligned}\frac{\sigma R}{Et} = &\left\{\frac{1}{\eta}\frac{\mu^2}{(1+\mu^2)^2} + \frac{1}{12(1-\nu^2)}\frac{\eta(1+\mu^2)^2}{\mu^2}\right\} + \\ &\frac{1}{\eta\mu^2}\left[(\eta\xi)^2\left\{\frac{\mu^4}{(1+\mu^2)^2} + \frac{\mu^4}{4(1+9\mu^2)^2} + \frac{\mu^4}{4(9+\mu^2)^2}\right\}\rho^2 + \right. \\ &\left\{(\eta\xi)^2\left(\frac{\mu^4}{(1+\mu^2)^2} + \frac{1}{4}\frac{\mu^4}{(1+9\mu^2)^2} + \frac{1}{4}\frac{\mu^4}{(9+\mu^2)^2}\right) - \right. \\ &\left.(\eta\xi)\left(\frac{1}{8} + \frac{2\mu^4}{(1+\mu^2)^2}\right)\right\}\rho + \left\{(\eta\xi)^2\left(\frac{1+\mu^4}{64} + \frac{1}{4}\frac{\mu^4}{(1+\mu^2)^2} + \right.\right. \\ &\left.\left.\left.\frac{1}{16}\frac{\mu^4}{(1+9\mu^2)^2} + \frac{1}{16}\frac{\mu^4}{(9+\mu^2)^2}\right) - (\eta\xi)\left(\frac{1}{16} + \frac{\mu^4}{(1+\mu^2)^2}\right)\right\}\right]\end{aligned} \tag{35}$$

Therefore, when $\xi \to 0$, i.e., when the wave amplitude becomes very small, Eq. (35) reduces to

$$\left(\frac{\sigma R}{Et}\right)_{\xi\to 0} = \frac{1}{\eta}\frac{\mu^2}{(1+\mu^2)^2} + \frac{1}{12(1-\nu^2)}\frac{1}{\frac{1}{\eta}\frac{\mu^2}{(1+\mu^2)^2}} \tag{36}$$

The minimum value of the average compression stress σ is given by

$$\text{Min.}\left(\frac{\sigma R}{Et}\right)_{\xi\to 0} = \frac{1}{\sqrt{3(1-\nu^2)}} \tag{37}$$

---

* Eq. (15) in original paper has been corrected as Eq. (13). — Noted by editor

which is the well known result from the classical theory of infinitesimal deflections. This minimum value is obtained when

$$\eta \frac{(1+\mu^2)^2}{\mu^2} = 2\sqrt{3(1-\nu^2)} \tag{38}$$

It is interesting to notice that for infinitesimal wave amplitude, the minimum value of average compression stress occurs for an infinite number of pairs of the values of the parameters η and μ, for which the combined parameter shown in Eq. (38) has the same value.

Numerical computations were carried out for two values of the parameter μ, the ratio of the wave lengths in circumferential direction and in axial direction. These values of μ are 1 and 0.5. The value 1 was chosen because the experiments indicate that at large values of wave amplitude, the diamond waves have almost equal sides. The value 0.5 was chosen to investigate the possibility of occurrence of narrow waves. The results of these computations are shown in Fig. 2 and Fig. 3, where the reduced compression stress $\sigma R/(Et)$ is plotted against the wave amplitudes ξ. The parameter in the figures is η. The values written in the parenthesis after η are the actual number of waves n in circumferential direction for $R/t = 1000$. For given values of η and μ, i.e., a fixed side of the wave, the load sustained by the shell, $\sigma R/(Et)$ first decreases as the wave amplitude, ξ, is increased. After a minimum is reached, the load will rise with increase in wave amplitude. When the waves are larger, the initial buckling load, i.e., the value of $\sigma R/(Et)$ at $\xi = 0$, is higher. However, the minimum load reached tends to a lower value, except for $\eta < 0.169$ and $\mu = 1$. For $\mu = 0.5$, the lowest value of the minimum load is not yet reached at $\eta = 0.081$.

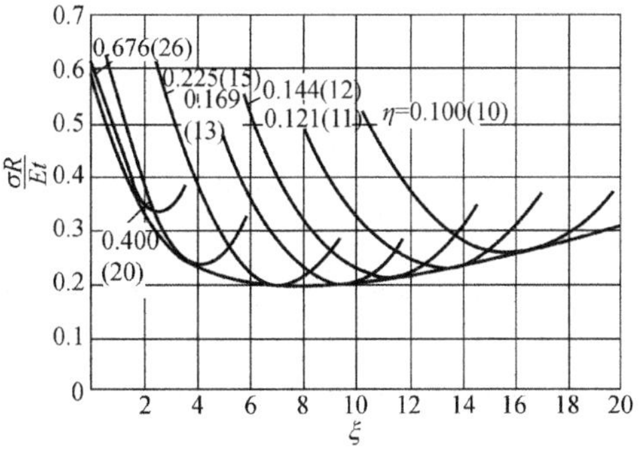

Fig. 2 Reduced compression stress $\sigma R/(Et)$ against amplitude of waves $\xi = \delta/t$, for $\mu = 1.00$ and different number of waves in circumferential direction

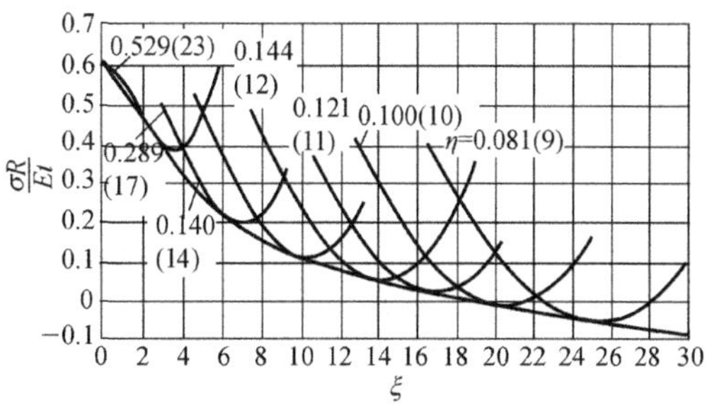

Fig. 3 Reduced compression stress $\sigma R/(Et)$ against amplitude of waves $\xi = \delta/t$, for $\mu = 0.50$ and different number of waves in circumferential direction

The Relation Between the Compression Stress and the Shortening of the Shell in the Axial Direction

Although the load characteristic of the cylindrical shell shown in the Figs. 2 and 3 gives the possible equilibrium relations between load and amplitude of the deflection wave, the actual behavior of a specimen in a testing machine cannot be directly seen from these figures. In a testing machine, the only factor under the control of the operator is the distance between the end plates; this is the geometrical restraint with which the specimen must conform. Therefore, in order to determine the behavior of the specimen the compression stress will have to be plotted as function of the end shortening. The unit end shortening, ε, i. e., the total shortening in one wave length of the shell in axial direction divided by the wave length, can be easily calculated from Eq. (23). It is found that

$$\frac{\varepsilon R}{t} = \frac{\sigma R}{Et} + \frac{\mu^2}{16}\xi(\eta\xi)\left(\rho^2 + \rho + \frac{3}{4}\right) \tag{39}$$

This equation for the unit end shortening contains only quantities already found such as the values of ρ and $\sigma R/(Et)$. In Figs. 4 and 5, $\sigma R/(Et)$ is plotted against $\varepsilon R/t$ for $\mu = 1$ and $\mu = 0.5$, respectively. It is immediately clear from these two figures that if the buckling process follows the curves drawn, after the shell starts to buckle the end shortening has to decrease. In other words, the end plates of the testing machine have to move apart. Therefore, the process of buckling in this region is highly unstable; as a matter of fact, before the operator has time to separate the end plates, the shell will jump to the point P (Fig. 4 or 5) which corresponds to the same end shortening as the starting point of the buckling process, but to a much lower compression stress. This jump in equilibrium position involves a release of elastic energy and thus explains the rapidity of the buckling process and the accompaning vibration.

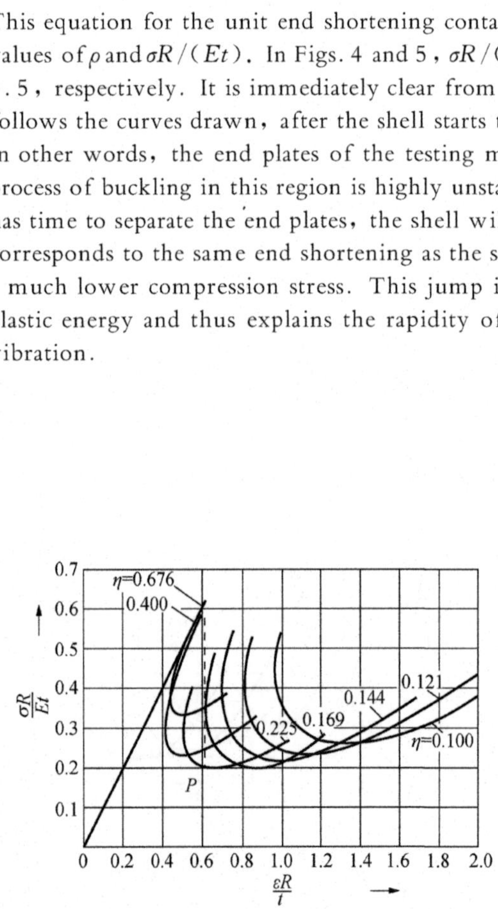

Fig. 4 Reduced compression stress $\sigma R/(Et)$ against unit end shortening $\varepsilon R/t$, for $\mu = 1.00$ and different number of waves in circumferential direction

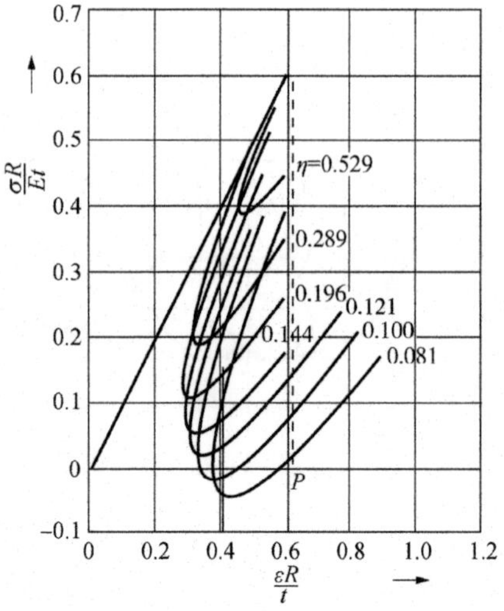

Fig. 5 Reduced compression stress $\sigma R/(Et)$ against unit end shortening $\varepsilon R/t$, for $\mu = 0.50$ and different number of waves in circumferential direction

It is tacitly assumed in the previous paragraph that if the system has an equilibrium position corresponding to the same end conditions, i.e., same end shortening, but involving a lower value of σ or a lower value of the stored elastic energy, a transition to one of these equilibrium positions will actually take place. This is not based on any vigorous application of the principles of mechanics. It is evident that if, for example, the straight equilibrium position is stable, i.e., all infinitesimally near configurations have higher energy, such a transition can only take place by application of an external impulse of finite magnitude. However, it can be assumed that such impulses will always be present in an experiment performed without extraordinary precaution in this respect, and of course if the structure is used in service. Then again, if it is assumed that the structure is helped over the "energy hump" by such external impulse, it cannot be proved that the jump has to end in the equilibrium position with the lowest energy level. However, if it is exposed to several random impulses, there is a certain probability that the position with the lowest energy level will be the "journey's end." An approximate calculation of the elastic energy of the shells shows that for large values of $\varepsilon R/t$ and $\eta < 0.121$, the elastic energy of the narrow waves is higher than that of the square waves at same value of the end shortening. Therefore, under such conditions the narrow waves appear to be less probable. However, for $\varepsilon R/t$ near 0.6, and $\eta \geqslant 0.121$ the elastic energy stored in the shell for the narrow waves is comparable to that for the square waves at the same value of the end shortening. This indicates the possibility of the appearance of narrow waves during the very initial stages of the buckling process.

In any case, it is certain that there are equilibrium positions of a buckled cylindrical shell which involve much lower average compression stress $\sigma R/(Et)$ than that at the beginning of buckling. For instance, in the case of square wave pattern, the lowest compression stress is given by

$$\frac{\sigma}{E} = 0.194\frac{t}{R} \tag{40}$$

Incidentally, this value corresponds closely to most of the experimental results obtained by L. H. Donnell[3] and E. E. Lundquist[7].

The corresponding value of the parameter η which determines the number of waves is equal to 0.225. In case of $R/t = 1\,000$, the number of waves, n, will be 15 which also agrees well with the experimental evidence. For this particular value of the radius to thickness ratio, the number of waves along the circumference decreases from the $n = 26$ at the beginning of the buckling to $n = 15$ at the calculated minimum buckling stress. This gradual increase in the size of waves with the unit shortening is also observed by the experiments reported in an earlier paper[2].

It is particularly interesting, however, to trace the gradual change in the wave pattern during the buckling process. Figs. 6 and 7 show the lines of equal deflection of the wave surfaces corresponding to different equilibrium states for two values of the aspect ratio of the wave pattern, $\mu = 1.0$ and $\mu = 0.5$. These particular equilibrium states are denoted in Figs. 2, 3, 4 and 5 by small circles in order to indicate their relative position during the buckling process. It is seen that there is a rapid shift from the rectangular waves bounded by lines, $x =$ const., and $y =$ const., as predicted by the classical theory for infinitesimal wave amplitudes, to staggered rows of circular or elliptical waves. Whereas, the rectangular waves are directed alternatively inward and outward, the circular or elliptical waves are all directed inward. The transition is practically completed for $\xi = 4$ or 6, i.e., when the wave amplitude is only 4 or 6

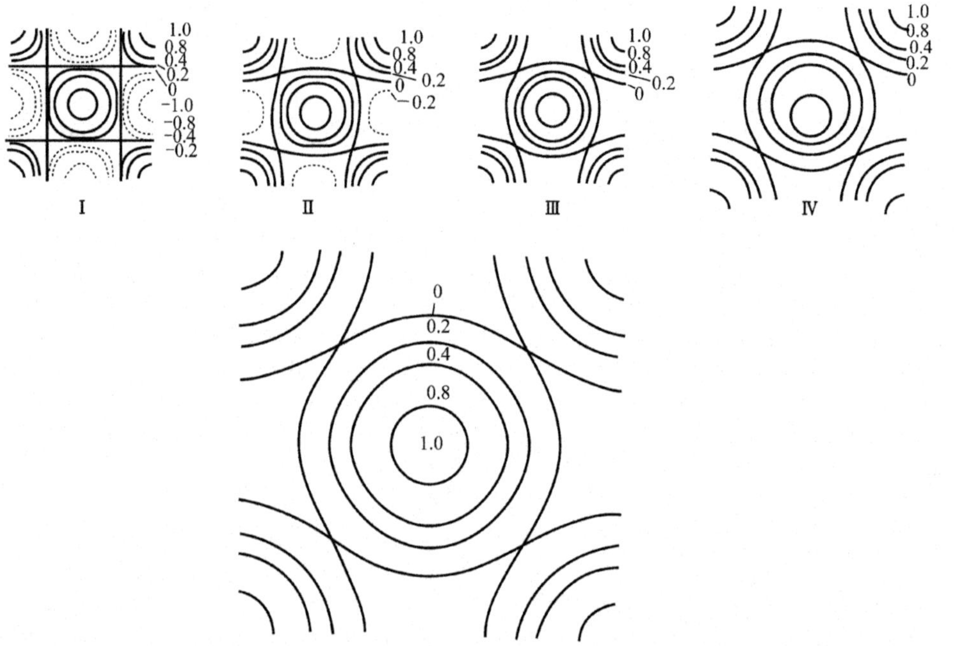

Fig. 6 Lines of equal deflection of wave surfaces. $+1.0$ = maximum inward deflection

Compression in vertical direction

	ξ	$\mu = 1.00$ η
I	0	0.676
II	1.00	0.676
III	2.00	0.676
IV	4.00	0.400
V	16.22	0.100

times the thickness of the shell. The occurrence of such inwardly directed circular and elliptical waves at this stage of the buckling process is in good agreement with the experimental observations[2]. If the experiment is continued to larger deflections ($\xi \sim 60$), these staggered waves acquire the characteristic diamond shape. The present approximate theory fails to give these sharp diamond shaped waves. It is obviously not sufficiently exact for such large deflections. Furthermore, when these diamond shaped waves occur, the load on the specimen actually falls to a very low value such as $\sigma R/(Et) \cong 0.06$, whereas the theory shows a slight increase of the stress at least for the case $\mu = 1.0$. Therefore, the present calculation can be only considered as a fair approximation to the earlier stages of buckling when the wave amplitude is only a few times the thickness of the shell. Nevertheless, it reproduces the characteristic features of the buckling process observed in the laboratory.

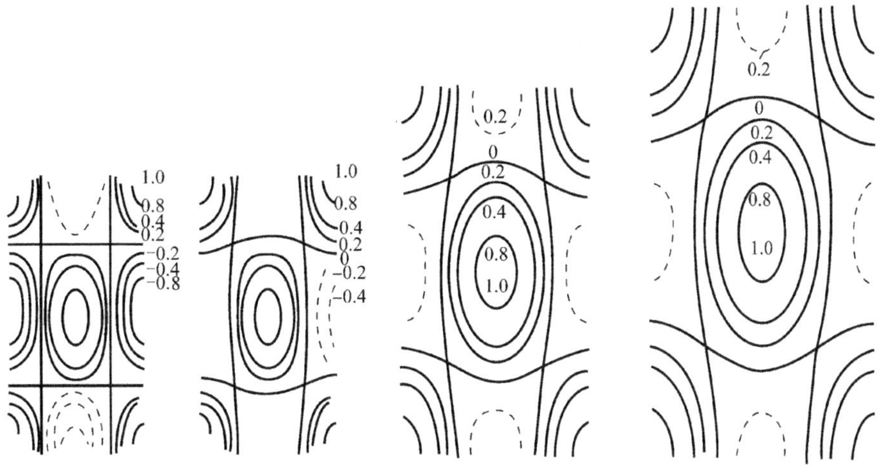

Fig. 7 Lines of equal deflection of wave surfaces. + 1. 0 = maximum inward deflection

Compression in vertical direction

	ξ	$\mu = 0.500$ η
I	0.00	0.529
II	2.50	0.529
III	5.50	0.289
IV	10.00	0.196

The Effect of the Elastic Characteristic of the testing Machine on the Buckling Phenomenon

It was stated in the previous paragraph that the state of the specimen is determined by the distance between the end plate and that this distance is the independent parameter controlled by the experimenter. This statement is correct only insofar the elasticity in the mechanism of the testing machine is neglected. There is always a certain amount of elastic deflection in the loading mechanism and this deflection is a function of the load. Hence, if, for example, the loading crank is held at a certain position, the compression force acting on the specimen will force the end plates apart and thus reduce the amount of end shortening of the specimen. The actual shortening is determined by the load-deflection characteristics of the specimen and the testing machine. Assuming that the testing machine has a linear elastic characteristic, the compression load is related to the end shortening by parallel straight lines, each line corresponding to, say, a fixed number of turns of the loading crank. If the loading crank of the machine is held at a fixed position, corresponding values of the compression load and end shortening of the specimen must lie on the straight line for this crank position. If the load-end shortening characteristics of the specimen itself are given, it is evident that the equilibrium positions of the entire system are determined by intersections of the curves representing the characteristics of the specimen with the straight lines representing the characteristics of the machine.

Fig. 8 shows representative curves for the characteristics of the specimen and two families of straight lines representing the characteristics of two different testing machines. It is evident that after the maximum or initial buckling load is reached, the shell will jump to a new equilibrium position involving much lower compression load. But this new equilibrium position is determined not only by the load-end shortening relationship and also by the elastic characteristic of the testing machine. A more elastic machine will give a set of characteristic straight lines with smaller slopes. Therefore, in case of curve A (Fig. 8), a more elastic machine will make the shell to jump to a higher load, while in case of curve B, a more elastic machine will make the shell to jump to a lower load. This influence of the elasticity of the testing machine has been discussed by the senior author in connection with the plastic buckling of columns[8].

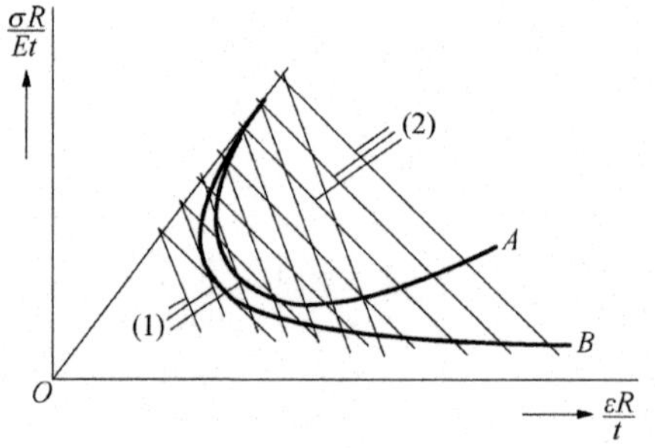

Fig. 8 Effect of rigidity of testing machine on behavior of specimen
(1) Represents a more rigid machine
(2) Represents a less rigid machine

Conclusions

In the previous paragraphs, the authors have shown that there are equilibrium positions with buckled shape involving much lower load than the buckling load predicted by the classical theory, and thus if the specimen is slightly imperfect, it is reasonable to expect much lower buckling loads. They have also pointed out that the elastic characteristic of the testing machine might have quite a large influence on the buckling process and this might be another cause of the large scattering of the data obtained by different experimenters. However, due to the complexity of the problem, the results given in this paper can be only considered as a rough approximation and most of the statements made are qualitative rather than quantitative. To put the new theory on a solid footing, a more accurate solution of the differential equations of equilibrium is necessary. Particular attention must be given to the calculation of the elastic energy stored in the shell, because it is found that the most probable equilibrium depends on the magnitude of the elastic energy stored in the various equilibrium positions compatible with the constraint exerted by the loading process.

Furthermore, an inquisitive mind will, perhaps, be pleased by a rigorous proof of the validity of all the large deflection equations. These equations are established by intuitive arguments, not by systematic reasoning. For instance, due to the appearance of sharp curvatures in the diamond shaped wave surfaces at large deflections, it is not certain whether the curvature of the shell has to be calculated more accurately by taking into account the second order terms, or the extensions of the median surface should be more accurately determined. It is the belief of the authors that an investigation of these problems by starting from the general non-linear theory of elasticity developed by G. Kirchhoff, J. Boussinesq and others is very desirable. The recent work by R. Kappus[9] is a noteworthy contribution in this field of investigation. The senior author has already expressed this opinion in his 1939 Gibbs Lecture[10] given before the American Mathematical Society.

References

[1] von Kármán Th., Tsien, Hsue-Shen. The Buckling of Spherical Shells by External Pressure. Journal of the Aeronautical Sciences, 1939,7(2): 43.

[2] von Kármán Th., Dunn Louis G., Tsien, Hsue-Shen. The Influence of Curvature on the Buckling Characteristics of Structures. Journal of the Aeronautical Sciences, 1940,7(7): 276.

[3] Donnell L H. A New Theory for the Buckling of Thin Cylinders Under Axial Compression and Bending. A.S.M.E. Transactions, 56:795 - 806,1934.

[4] Flügge W. Die Stabalität der Kreiszylinderschale. Ingenieur Archiv, 1932,(3): 463 - 506.

[5] von Kármán Th. Encyklopädie der Mathematischen Wissenschaften. 1910, Ⅳ.4: 349.

[6] Donnell L H. Stability of Thin-Walled Tubes Under Torsion. N.A.C.A. Technical Report No. 479, 1934.

[7] Lundquist E E. Strength Tests of Thin-Walled Duralumin Cylinders in Compression. N.A.C.A. Technical Report No. 473, 1933.

[8] von Kármán Th. Untersuchungen über Knickféstigkeit. Forschungsarbeiten, Berlin, 1910: 81.

[9] Kappus R. Zur Elastizitätstheorie endlicher Verschiebungen. ZAMM, 1939,19: 271 - 285. 344 - 361.

[10] von Kármán Th. The Engineer Grapples with Nonlinear Problems. Bulletin of the American Mathematical Society, 1940,46: 636 - 637.

Buckling of a Column with Non-Linear Lateral Supports

H. S. Tsien *

(*California Institute of Technology*)

Introduction

During the investigation of the buckling phenomenon of thin spherical shells[1] and thin cylindrical shells[2], it was found that for these structures the load sustained is not a linear function of the deflection even when the stresses are below the elastic limit and are proportional to the corresponding strains. This non-linear load vs. deflection relation gives a buckling phenomenon entirely different from that of the classical theory. However, the exact solution of these problems involves a pair of non-linear partial differential equations. It is difficult to obtain an exact solution. The method adopted in these investigations is the so-called energy method, where a plausible form of deflection of the shell is assumed, first with certain undetermined parameters, and then these parameters are determined by the condition that the first variation of the strain energy of the system must be zero. Although this method yields quite satisfactory results for the cases investigated, it is felt that due to the novel nature of the problem, an exact solution is very desirable. Experiments on a column with non-linear lateral supports[3] show that the essential characteristics of the buckling of curved shells can be reproduced by this structure. The problem of a column with non-linear lateral supports is, however, much simpler than the problem of curved shells, and an exact solution can be obtained without any mathematical difficulty. In the present paper, an exact solution of the column problem will be carried out.

During his investigation on the buckling of thin curved plates, Cox[4] has used the same analogy to demonstrate the non-linear character of the buckling phenomenon, and has also carried out the mathematical analysis of a column supported by two rods pin-jointed to the center of the column. However, Cox's calculation seems unnecessarily complicated. The method developed in this paper is believed to be more general, because it can be applied to supports with arbitrary elastic properties.

Received December 31, 1941.

* Research Fellow in Aeronautics.

Journal of the Aeronautical Sciences, vol. 9, pp. 119 – 132, 1942.

General Theory for Straight Columns

Using the variational method for the calculation of the equilibrium positions of the column, let the column be initially straight and its center line be taken as the x-axis. (Fig. 1) The lower end of the column is taken as the origin. The column is supported laterally at g- points x_1, x_2, ..., x_g by springs of non-linear load vs. deflection characteristics. The springs are assumed to be unloaded when the column is straight. Denote the deflections of the springs by δ_1, δ_2, ..., δ_g. Then the forces given by the springs when they are deflected can be expressed as $F_1(\delta_1)$, $F_2(\delta_2)$, ..., $F_g(\delta_g)$. The elastic energy S stored in these springs at the deflections δ_1, δ_2, ..., δ_g can be easily calculated as

$$S_1(\delta_1) = \int_0^{\delta_1} F_1(\xi)\mathrm{d}\xi$$

$$S_2(\delta_2) = \int_0^{\delta_2} F_2(\xi)\mathrm{d}\xi$$

or

$$S_g(\delta_g) = \int_0^{\delta_g} F_g(\xi)\mathrm{d}\xi \tag{1}$$

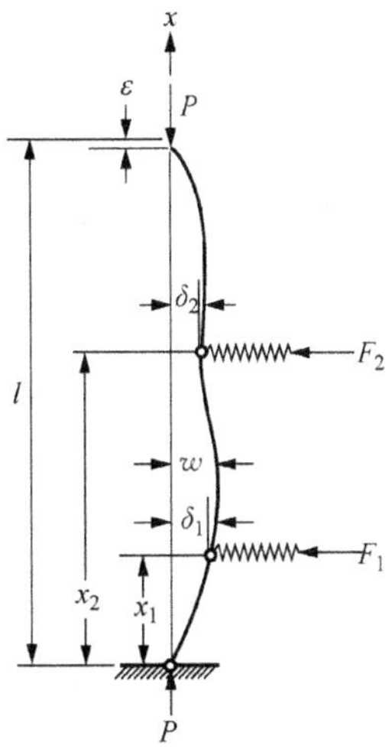

Fig. 1 Column with lateral supports

Let the deflection of the column normal to x-axis be denoted by $w(x)$. Then

$$\delta_1 = w(x_1)$$

$$\delta_2 = w(x_2)$$

or

$$\delta_g = w(x_g) \tag{2}$$

If it is assumed that the column is simply supported at the two ends, $w(x)$ can be expanded into a Fourier series such that

$$w(x) = \sum_{n=1}^{\infty} a_n \sin(n\pi x/l) \tag{3}$$

where l is the length of the column. Denoting the Young's modulus by E and the moment of inertia of the column section by I, the bending elastic energy involved in the deflection $w(x)$ is

$$W_1 = \frac{EI}{2}\int_0^l \left(\frac{\mathrm{d}^2 w}{\mathrm{d}x^2}\right)^2 \mathrm{d}x = \frac{EIl}{4}\sum_{n=1}^{\infty}\left(\frac{n\pi}{l}\right)^4 a_n^2 \tag{4}$$

The elastic energy stored in the springs can be calculated by means of Eq. (1)

$$W_2 = \sum_{m=1}^{g} S_m(\delta_m) = \sum_{m=1}^{g}\int_0^{\delta_m} F_m(\xi)\mathrm{d}\xi \tag{5}$$

The lowering of potential of the compression force P applied at the ends of the column due to the deflection $w(x)$ is

$$W_3 = -\frac{P}{2}\int_0^l \left(\frac{dw}{dx}\right)^2 dx = -\frac{Pl}{4}\sum_{n=1}^{\infty}\left(\frac{n\pi}{l}\right)^2 a_n^2 \tag{6}$$

The conditions of equilibrium can be found by putting the first variation of the sum of W_1, W_2, W_3 with respect to a_n equal to zero. Thus

$$\frac{EIl}{2}\left(\frac{n\pi}{l}\right)^4 a_n + \sum_{m=1}^{g} F_m(\delta_m)\frac{\partial \delta_m}{\partial a_n} - \frac{Pl}{2}\left(\frac{n\pi}{l}\right)^2 a_n = 0 \tag{7}$$

By means of Eqs. (2) and (3)

$$(\partial\delta_m/\partial a_n) = \sin(n\pi x_m/l) \tag{8}$$

Therefore by substituting this value of $\partial\delta_m/\partial a_n$ into Eq. (7), the following equation is obtained for the coefficients a_n.

$$a_n = \frac{\sum_{m=1}^{g}\sin(n\pi x_m/l)\cdot F_m(\delta_m)}{\frac{Pl}{2}\left(\frac{n\pi}{l}\right)^2 - \frac{EIl}{2}\left(\frac{n\pi}{l}\right)^4} \tag{9}$$

By introducing the Euler column load $P_E = \frac{EI\pi^2}{l^2}$, Eq. (9) can be rewritten in the following form

$$\frac{a_n}{l} = \frac{2}{\pi^2}\,\frac{\sum_{m=1}^{g}\sin\frac{n\pi x_m}{l}\cdot\frac{F_m(\delta_m)}{P_E}}{n^2\left[\frac{P}{P_E} - n^2\right]} \tag{10}$$

This equation allows one to calculate the coefficients a_n if the forces F_1, F_2, ..., F_g acting on the column are known. The known relations are the functions $F_1(\delta_1)$, $F_2(\delta_2)$, ..., $F_g(\delta_g)$, i.e., the spring forces as functions of their respective deflections. However, by using Eqs. (2) and (3), the following is obtained

$$(\delta_s/l) = \sum_{n=1}^{\infty}(a_n/l)\sin(n\pi x/l) \tag{11}$$

By substituting Eq. (10) into Eq. (11)

$$\frac{\delta_s}{l} = \frac{2}{\pi^2}\sum_{m=1}^{g}\frac{F_m(\delta_m)}{P_E}\sum_{n=1}^{\infty}\frac{\sin\frac{n\pi x_m}{l}\cdot\sin\frac{n\pi x_s}{l}}{n^2\left[\frac{P}{P_E} - n^2\right]},\ (s = 1,2,\ldots,g) \tag{12}$$

If

$$K_{sm} = \frac{2}{\pi^2}\sum_{n=1}^{\infty}\frac{\sin\frac{n\pi x_m}{l}\sin\frac{n\pi x_s}{l}}{n^2\left[\frac{P}{P_E} - n^2\right]} \tag{13}$$

Eq. (12) can be rewritten in the form

$$(\delta_s/l) = \sum_{m=1}^{g} K_{sm}(F_m/P_E), \quad (s = 1,2,\ldots,g) \tag{14}$$

With a given value of P/P_E, the coefficients K_{sm} can be calculated. These coefficients are symmetrical, i.e.,

$$K_{sm} = K_{ms} \tag{15}$$

which can be easily seen from Eq. (13). When these coefficients are calculated, Eq. (14) consists of a system of g equations for the g unknowns, δ_1, δ_2, ..., δ_g. Therefore the values of these deflections can be found. Then the spring forces F_1, F_2, ..., F_g can be calculated. The Fourier coefficients a_n/l can then be determined by means of Eq. (10). The problem is thus completely solved.

The merit of the present method is that the fundamental parameters K_{sm} are independent of the characteristics of the supporting springs. And therefore they are applicable to a wide class of problems.

Case of Two Equally Spaced Supports

For the case of two equally spaced supports, the coefficients K_{sm} can be immediately calculated as

$$\begin{aligned}
K_{11} &= \frac{2}{\pi^2}\sum_{n=1}^{\infty}\frac{\sin^2\frac{n\pi}{3}}{n^2\left[\frac{P}{P_E}-n^2\right]} \\
K_{12} = K_{21} &= \frac{2}{\pi^2}\sum_{n=1}^{\infty}\frac{\sin\frac{n\pi}{3}\sin\frac{2n\pi}{3}}{n^2\left[\frac{P}{P_E}-n^2\right]} \\
K_{22} &= \frac{2}{\pi^2}\sum_{n=1}^{\infty}\frac{\sin^2\frac{2n\pi}{3}}{n^2\left[\frac{P}{P_E}-n^2\right]}
\end{aligned} \tag{16}$$

These series can be easily summed by well-known relations as

$$\begin{aligned}
K_{11} = K_{22} &= \left[\frac{2}{9}+\frac{1}{4\theta}\left\{3\cot\theta-\cot\frac{\theta}{3}\right\}\right]\Big/\frac{P}{P_E} \\
K_{12} = K_{21} &= \left[\frac{1}{9}-\frac{1}{4\theta}\left\{3\csc\theta-\csc\frac{\theta}{3}\right\}\right]\Big/\frac{P}{P_E}
\end{aligned} \tag{17}$$

where

$$\theta = \pi\sqrt{P/P_E} \tag{18}$$

If the two spring supports have the same characteristics, and the deflection of the column is symmetrical, i.e., $\delta_1 = \delta_2 = \delta$, then $F_1 = F_2 = F$. Eq. (14) can be simplified into

$$(\delta/l) = (K_{11} + K_{22})(F/P_E) \tag{19}$$

$K_{11} + K_{12}$ can be obtained from Eq. (17), and written in the following form:

$$K_{11} + K_{12} = \left[\frac{1}{3} - \frac{1}{4\theta}\left\{3\tan\frac{\theta}{2} - \tan\frac{\theta}{6}\right\}\right] \Big/ \frac{P}{P_E} \tag{20}$$

For the degenerate case of springs with linear characteristics, i.e.,

$$F/P_E = \alpha \cdot (\delta/l) \tag{21}$$

from Eqs. (19) and (20) the following relation for symmetrical buckling is derived:

$$\alpha = \frac{P_1/P_E}{\frac{1}{3} - \frac{1}{4\theta}\left\{3\tan\frac{\theta}{2} - \tan\frac{\theta}{6}\right\}} \tag{22}$$

Eq. (22) can be also written in the form

$$\alpha = \frac{3(P_1/P_E)\theta}{\theta - 3\left\{\sin^2\frac{\theta}{3}\tan\frac{\theta}{2} + \sin\frac{\theta}{3}\cos\frac{\theta}{3}\right\}} \tag{23}$$

In case of anti-symmetric buckling with linear spring characteristics, $\delta_1 = -\delta_2$, and from Eq. (14)

$$(\delta_1/l) = K_{11}\alpha(\delta_1/l) - K_{12}\alpha(\delta_1/l) \tag{24}$$

Using the value of K_{11} and K_{12} from Eq. (17)

$$\alpha = \frac{9(P_2/P_E)\theta}{\theta - 9\left\{\cot\frac{\theta}{3} + \cot\frac{\theta}{6}\right\}} \tag{25}$$

Both Eqs. (23) and (25) check the calculations of Klemperer and Gibbons[5] which are limited to linear spring characteristics.

By means of Eq. (10), the shortening ε_1 of the column due to bending is

$$\frac{\varepsilon_1}{l} = \frac{\pi^2}{4}\sum_{n=1}^{\infty} n^2\left(\frac{a_n}{l}\right)^2 = C_{11}\left(\frac{F_1}{P_E}\right)^2 + C_{12}\frac{F_1F_2}{P_E^2} + C_{22}\left(\frac{F_2}{P_E}\right)^2 \tag{26}$$

The coefficients C_{11}, C_{12}, and C_{22} are functions of P/P_E and are first obtained from Eq. (10) as the following series

$$\begin{aligned} C_{11} &= \frac{1}{\pi^2}\sum_{n=1}^{\infty}\frac{\sin^2\frac{n\pi}{3}}{n^2\left[\frac{P}{P_E} - n^2\right]^2} \\ C_{12} &= \frac{2}{\pi^2}\sum_{n=1}^{\infty}\frac{\sin\frac{n\pi}{3}\sin\frac{2n\pi}{3}}{n^2\left[\frac{P}{P_E} - n^2\right]^2} \\ C_{22} &= \frac{1}{\pi^2}\sum_{n=1}^{\infty}\frac{\sin^2\frac{2n\pi}{3}}{n^2\left[\frac{P}{P_E} - n^2\right]^2} \end{aligned} \tag{27}$$

These series can be summed and give the following expressions for the coefficients

$$C_{11}=C_{22}=\left\{\frac{5}{18}+\frac{3}{16}\left(\cot^2\theta-\frac{1}{9}\cot^2\frac{\theta}{3}\right)+\frac{9}{16\theta}\left(\cot\theta-\frac{1}{3}\cot\frac{\theta}{3}\right)\right\}\Big/\left(\frac{P}{P_E}\right)^2$$

$$C_{12}=\left\{\frac{1}{9}-\frac{3}{8}\left(\frac{\cot\theta}{\sin\theta}-\frac{1}{9}\frac{\cot\frac{\theta}{3}}{\sin\frac{\theta}{3}}\right)-\frac{3}{8\theta}\left(\frac{3}{\sin\theta}-\frac{1}{\sin\frac{\theta}{3}}\right)\right\}\Big/\left(\frac{P}{P_E}\right)^2 \tag{28}$$

If the springs have same elastic properties and the deflection of the column is symmetrical, i.e., $F_1=F_2=F$, Eq. (26) reduces to

$$(\varepsilon_1/l)=(C_{11}+C_{12}+C_{22})\left(\frac{F}{P_E}\right)^2 \tag{29}$$

where

$$C_{11}+C_{12}+C_{22}=\left\{\frac{1}{2}+\frac{3}{16}\left(\tan^2\frac{\theta}{2}-\frac{1}{9}\tan^2\frac{\theta}{6}\right)-\frac{9}{8\theta}\left(\tan\frac{\theta}{2}-\frac{1}{3}\tan\frac{\theta}{6}\right)\right\}\Big/\left(\frac{P}{P_E}\right)^2 \tag{30}$$

The shortening of the column due to direct compression can be expressed as

$$(\varepsilon_2/l)=(P/A_E)=(\pi i/l)^2(P/P_E) \tag{31}$$

where A is the cross-sectional area of the column and $i^2=(I/A)$, i being the radius of gyration of the column section. Therefore, the total shortening of the column is

$$\varepsilon/l=(\varepsilon_1/l)+(\varepsilon_2/l) \tag{32}$$

The bending strain energy can be obtained by substituting the value of a_n from Eq. (10) into Eq. (4), thus

$$\frac{W_1}{P_E l}=B_{11}\left(\frac{F_1}{P_E}\right)^2+B_{12}\frac{F_1F_2}{P_E^2}+B_{22}\left(\frac{F_2}{P_E}\right)^2 \tag{33}$$

The coefficients are functions of P/P_E and can be expressed as the following series:

$$B_{11}=\frac{1}{\pi^2}\sum_{n=1}^{\infty}\frac{\sin^2(n\pi/3)}{[(P/P_E)-n^2]^2}$$

$$B_{12}=\frac{2}{\pi^2}\sum_{n=1}^{\infty}\frac{\sin(n\pi/3)\sin(2n\pi/3)}{[(P/P_E)-n^2]^2} \tag{34}$$

$$B_{22}=\frac{1}{\pi^2}\sum_{n=1}^{\infty}\frac{\sin^2(2n\pi/3)}{[(P/P_E)-n^2]^2}$$

These series can be summed into the following expressions

$$B_{11}=B_{22}=\left[\frac{1}{6}+\frac{3}{16}\left(\cot^2\theta-\frac{1}{9}\cot^2\frac{\theta}{3}\right)+\frac{3}{16\theta}\left(\cot\theta-\frac{1}{3}\cot\frac{\theta}{3}\right)\right]\Big/\left(\frac{P}{P_E}\right)$$

$$B_{12}=-\left[\frac{3}{16}\left(\frac{\cot\theta}{\sin\theta}-\frac{1}{9}\frac{\cot\frac{\theta}{3}}{\sin\frac{\theta}{3}}\right)+\frac{3}{16\theta}\left(\frac{1}{\sin\theta}-\frac{1}{3\sin\frac{\theta}{3}}\right)\right]\Big/\left(\frac{P}{P_E}\right) \tag{35}$$

If the buckling is symmetrical and springs of same properties are used, then Eq. (33) can be again reduced to

$$W_1/(P_E l) = (B_{11} + B_{12} + B_{22})(F/P_E)^2 \tag{36}$$

where

$$B_{11} + B_{12} + B_{22} = \frac{3}{16}\left\{\left(\sec^2\frac{\theta}{2} - \frac{1}{9}\sec^2\frac{\theta}{6}\right) - \frac{2}{\theta}\left(\tan\frac{\theta}{2} - \frac{1}{3}\tan\frac{\theta}{6}\right)\right\} \Big/ \left(\frac{P}{P_E}\right) \tag{37}$$

The strain energy due to the direct compression stress is evidently

$$W_4/(P_E l) = P\varepsilon_2/(2P_E l) \tag{38}$$

Or by means of Eq. (31),

$$\frac{W_4}{P_E l} = \frac{1}{2}\left(\frac{P}{P_E}\right)^2 \frac{P_E}{AE} = \frac{1}{2}\left(\frac{\pi i}{l}\right)^2 \left(\frac{P}{P_E}\right)^2 \tag{39}$$

The strain energy stored in the springs is given by Eq. (5). The total strain energy of the system is then given by

$$W/(P_E l) = (W_1 + W_2 + W_4)/(P_E l) \tag{40}$$

Different expressions for calculating the equilibrium positions, shortening of column and strain energies have been developed without any assumption as to the function $F(\delta)$. The elastic characteristics of a curved shell are found to correspond to a spring having the load-deflection relation shown in Fig. 2. This particular spring characteristic can be very well represented by the expression

$$F/P_E = a(\delta/l) + b(\delta/l)^2 + c(\delta/l)^3 \tag{41}$$

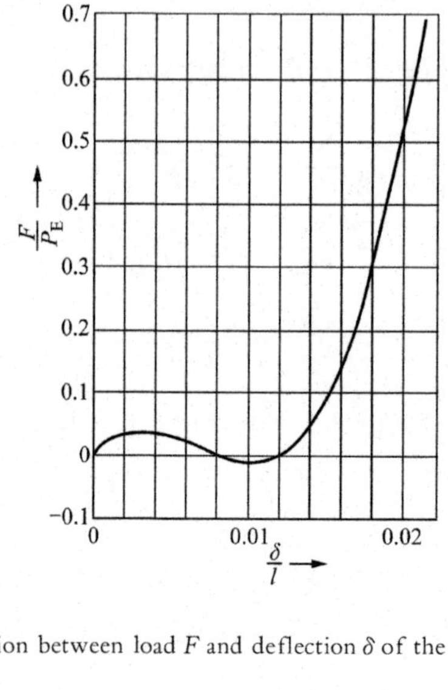

Fig. 2 Non-linear relation between load F and deflection δ of the lateral supporting spring

where a, b, c are constants. These constants can be determined by the initial slope, and the positions of the maximum and minimum of the F vs. δ curve.

For the case of symmetrical buckling with identical supporting springs, Eq. (19) can be combined with Eq. (41) and the following equation for determining δ with a given value of P/P_E obtained.

$$\frac{\delta}{l} = (K_{11} + K_{12})\left\{a\left(\frac{\delta}{l}\right) + b\left(\frac{\delta}{l}\right)^2 + c\left(\frac{\delta}{l}\right)^3\right\} \tag{42}$$

A simple solution of this equation is

$$\delta/l = 0 \tag{43}$$

which represents the unbuckled straight column. Other solutions are given by

$$1 = (K_{11} + K_{12})\left\{a + b\left(\frac{\delta}{l}\right) + c\left(\frac{\delta}{l}\right)^2\right\} \tag{44}$$

which is a quadratic equation for (δ/l), and can be easily solved. After the deflections δ of springs are obtained, other quantities can be calculated without difficulty by means of formulas already given.

For the unsymmetrical case, it is assumed that the support springs have the same elastic properties, i. e., the values of the constants a, b, c are the same for both springs. Then the combination of Eqs. (18) and (41) gives

$$\begin{aligned} \frac{\delta_1}{l} &= K_{11}\left\{a\left(\frac{\delta_1}{l}\right) + b\left(\frac{\delta_1}{l}\right)^2 + c\left(\frac{\delta_1}{l}\right)^3\right\} + K_{12}\left\{a\left(\frac{\delta_2}{l}\right) + b\left(\frac{\delta_2}{l}\right)^2 + c\left(\frac{\delta_2}{l}\right)^3\right\} \\ \frac{\delta_2}{l} &= K_{21}\left\{a\left(\frac{\delta_1}{l}\right) + b\left(\frac{\delta_1}{l}\right)^2 + c\left(\frac{\delta_1}{l}\right)^3\right\} + K_{22}\left\{a\left(\frac{\delta_2}{l}\right) + b\left(\frac{\delta_2}{l}\right)^2 + c\left(\frac{\delta_2}{l}\right)^3\right\} \end{aligned} \tag{45}$$

In trying to solve this system of equations by eliminating one variable, say δ_2, an algebraic equation of the 9th order is obtained for δ_1/l. Since already three possible solutions have been obtained by means of Eq. (42), only six more new solutions are possible. Thus we are faced with the task of deriving a 9th-order equation to obtain six new solutions. However, this lengthy calculation is unnecessary as can be seen in the following process. By adding and subtracting the two equations in Eq. (45),

$$\begin{aligned} \frac{\delta_1}{l} + \frac{\delta_2}{l} &= (K_{11} + K_{12})\left[a\left(\frac{\delta_1}{l} + \frac{\delta_2}{l}\right) + b\left(\frac{\delta_1^2}{l^2} + \frac{\delta_2^2}{l^2}\right) + c\left(\frac{\delta_1^3}{l^3} + \frac{\delta_2^3}{l^3}\right)\right] \\ \frac{\delta_1}{l} - \frac{\delta_2}{l} &= (K_{11} - K_{12})\left[a\left(\frac{\delta_1}{l} - \frac{\delta_2}{l}\right) + b\left(\frac{\delta_1^2}{l^2} - \frac{\delta_2^2}{l^2}\right) + c\left(\frac{\delta_1^3}{l^3} - \frac{\delta_2^3}{l^3}\right)\right] \end{aligned} \tag{46}$$

Putting

$$(\delta_1/l) + (\delta_2/l) = \xi_1 \qquad (\delta_1/l) - (\delta_2/l) = \eta \tag{47}$$

from Eq. (46)

$$\begin{aligned} \xi &= (K_{11} + K_{12})\left[a\xi + \frac{b}{2}(\xi^2 + \eta^2) + \frac{c}{4}(\xi^3 + 3\xi\eta^2)\right] \\ 1 &= (K_{11} - K_{12})\left[a + b\xi + \frac{c}{4}(3\xi^2 + \eta^2)\right] \end{aligned} \tag{48}$$

The second equation of Eq. (48) has already the root $\eta = 0$ eliminated, since it is contained in Eq. (42). By eliminating η^2, the following cubic equation is obtained for ξ,

$$0 = \xi^3 + 2\frac{b}{c}\xi^2 + \left\{\frac{ac + b^2}{c^2} - \frac{1}{2c} \times \left(\frac{3}{K_{11} - K_{12}} - \frac{1}{K_{11} + K_{12}}\right)\right\}\xi + \frac{b}{c^2}\left(a - \frac{1}{K_{11} - K_{12}}\right) \tag{49}$$

With a given value of P, this equation can be solved for ξ, and then η can be found from the second equation of Eq. (48). With these values of ξ and η, the values of δ_1/l and δ_2/l can be calculated by means of Eq. (47). With each value of ξ, there are two equal and opposite values of η, which correspond to an interchange of the values of δ_1 and δ_2. The shape of the column is not affected by this interchange and therefore η can be taken as always positive.

As a particular example,

$$a = 2F,\ b = -5\,640,\ c = 282\,000 \tag{50}$$

The corresponding relation between the spring force and the spring deflection is shown in Fig. 2. The initial slope of the load vs. deflection curve for the spring is so chosen in this particular case that the starting point of buckling occurs at a load $P = 9P_E$.

With these spring characteristics the equilibrium position is first calculated. The results are shown in Fig. 3 where the load P/P_E on the column is plotted against the deflection $\delta/\pi i$. Taking the slenderness ratio of the column to be

$$\pi i/l = 1/100 \tag{51}$$

Therefore the value $\delta/\pi i = 1.00$ corresponds to $\delta/l = 0.010$. Fig. 3 shows clearly the decrease in the load P with increase in deflection δ once the column starts to buckle. However, a minimum is soon reached and the load increases again at very large values of the deflection δ.

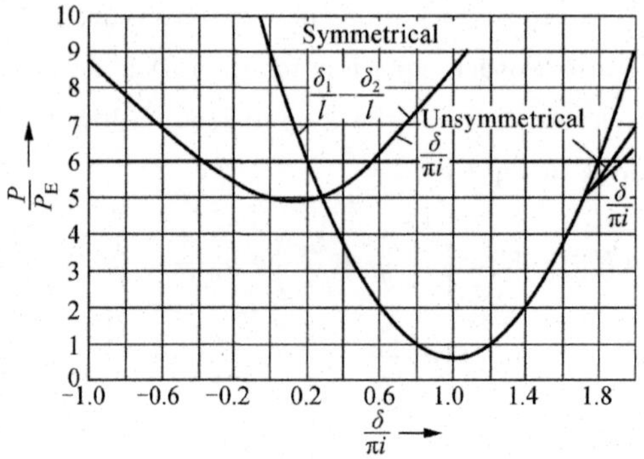

Fig. 3 Relation between the end load P on the column and the deflections δ_1, δ_2 of the two equally spaced lateral supports

However, in a testing machine the operator cannot control spring deflections δ_1 and δ_2 directly. On the other hand, he can increase or decrease the distance between the end plates of the testing machine. The distance between the end plates determines the end shortening ε of the column. Therefore, a better representation of the result of calculation is a diagram of the load plotted against the shortening ε. The shortening ε can be calculated from Eqs. (26), (28), (29), (30) and (31). The result is shown in Fig. 4. If the column keeps its straight form until the branch point of the solution for symmetrical equilibrium position is reached, it will be impossible to reach the equilibrium positions represented by the curve *A-B-C* in Fig. 4 in the loading process. This is the consequence of the fact that the corresponding end shortening ε for this part of the curve is smaller than the end shortening at the branch point. In the loading process, the shortening is always increased by diminishing the distance between the end plates of the testing machine. The actual behavior of the test specimen is probably a sudden jump from the point *A* to the point *C* (Fig. 4) with evident release of elastic energy. If the end plates of the testing machine are brought closer together, the load *P* will gradually increase and follow the curve *C-D* (Fig. 4). The equilibrium positions represented by the portion of the curve *B-C* can be reached by "unloading" the specimen after the buckling, i.e., by moving the end plates of the testing machine apart after it has reached the equilibrium position *C*. However, it is difficult to see how the equilibrium positions represented by the portion of curve *A-B* can be reached.

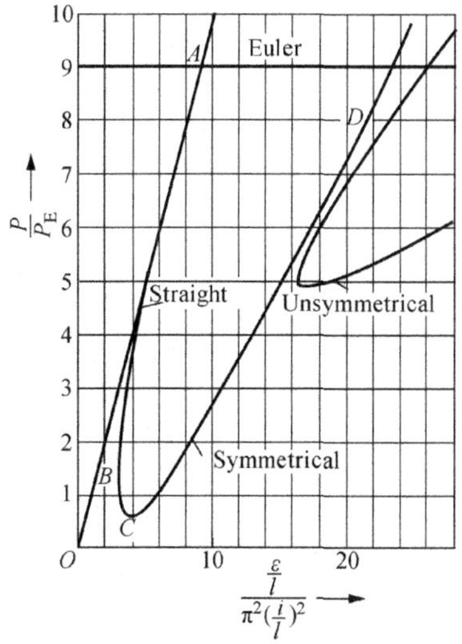

Fig. 4 Pelation between the end load *P* on the column with two equally spaced lateral supports and its end shortening ε

The three wave buckling of the column with the springs underformed is indicated by the horizontal line labeled "Euler" in Fig. 4. The unsymmetrical type of buckling gives a load vs. shortening relation quite similar to that of symmetrical type.

Under a given value of end shortening which is the geometrical restraint to which the specimen must conform, the condition for the most probable equilibrium state under actual testing circumstances is the minimum elastic energy stored in the system. This statement is based upon the assumption that even if the equilibrium state involving higher elastic energy is stable with respect to infinitesimal disturbances, a disturbance of finite magnitude will help the column to pass the "hump" in the energy and jump to an equilibrium state of lower elastic energy. When a series of such finite disturbances is present in the testing procedure, then the lowest energy state will be certainly the most probable one. Therefore, it is necessary to plot the elastic energy W for different equilibrium states against the end shortening ε. This is done in Fig. 5. Using the above mentioned criterion, the most probable equilibrium positions for the column are first, straight unbuckled state from 0 to F (Fig. 5), then symmetrical buckled state from F to G, and finally unsymmetrical buckled state from G to H. Whenever a change in the type of the equilibrium positions occurs, there is a sudden jump in the value of the load P on the column. Thus when the column changes from straight to symmetrically buckled shape, the value of the end load P jumps from $3.75\ P_E$ to $0.63\ P_E$. When the column changes from symmetrical to unsymmetrical shape, the value of the end load P jumps from $7.95\ P_E$ to $5.26\ P_E$. The "Euler" three wave buckling does not come into the picture at all in this particular case. It is especially interesting to notice that the branch point of the equilibrium position, A, which in the classical theory of infinitesimal deflections is the only equilibrium position other than the straight unbuckled shape, may not appear in the testing procedure. The column might buckle at a load P only $3.75/9.00 = 0.417$ times the "classical" buckling load of $9\ P_E$. In the

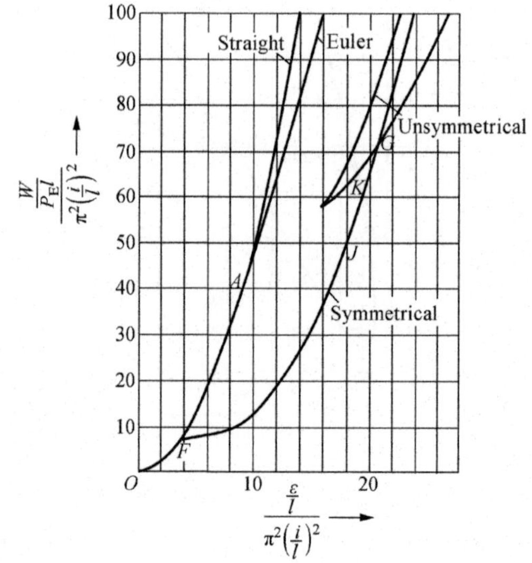

Fig. 5 Relation between the strain energy W stored in the system of a column with two equally spaced lateral supports and end shortening ε of the column

paper by Cox[4] and the author[2], the lower buckling load of curved shells observed in the testing machine is attributed to the slight imperfections of the test specimens. The present finding seems to indicate that even with perfect specimens, it is possible to get a lower buckling load than that predicted by theory of infinitesimal deflection. The second interesting point is that the magnitude of the end load P really does not determine whether the corresponding equilibrium position is a probable one. In fact the correct solution is usually opposite to what one would generally expect. One would generally expect that at a given value of the end shortening ε, the state involving lower load P is the probable one to occur. However, a point J (Fig. 5) having a higher load P than the point K (Fig. 5) as can be seen from Fig. 4, is the probable equilibrium state.

It has been stated in the previous paragraph that for the case shown the "Euler" three wave buckling is not a probable state. If, however, the supporting springs are much stiffer, it is possible to have a W vs. ε diagram as shown in Fig. 6. Then the "Euler" three wave buckling is actually the first buckled shape to appear. The symmetrical buckling involving deflections of the supporting springs appears only after the end shortening is increased beyond the point J (Fig. 6).

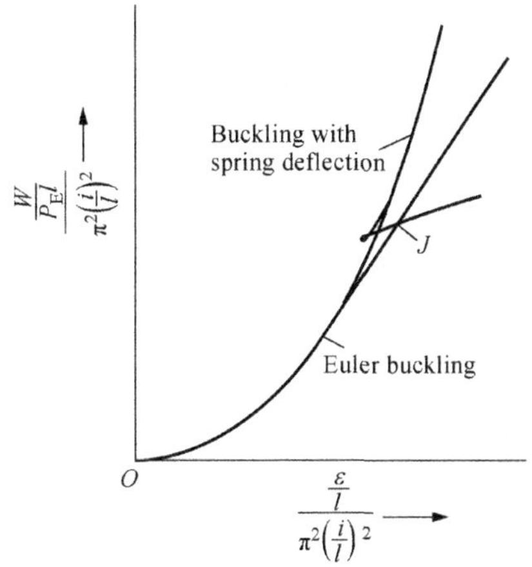

Fig. 6 Buckling of column started from Euler's type without deflection of the lateral supports

Case of a Single Lateral Support, Effect of Initial Deflection

To simplify calculation, the case of a single lateral support located at center of the span is considered. Assuming that the initial deflection of the column can be represented by

$$w^0 = a_1^0 \sin(\pi x/l) \tag{52}$$

and the buckled form of the column again by Eq. (3), it is also assumed that the supporting spring is undeflected when the column is at its original state given by Eq. (52). Then by variational method the relation between the Fourier coefficients and spring forces is obtained as

$$\frac{a_1}{l} = \frac{\dfrac{2}{\pi^2}\dfrac{F}{P_E} - \dfrac{a_1^0}{l}}{\dfrac{P}{P_E} - 1} \tag{53}$$

and

$$\frac{a_n}{l} = \frac{2(-1)^{\frac{n-1}{2}}\dfrac{F}{P_E}}{n^2\pi^2\left(\dfrac{P}{P_E} - n^2\right)} \quad (n = 3,5,7)$$

By a similar method to that used to obtain Eq. (12),

$$\frac{\delta}{l} = \frac{a_1^0}{l}\left[\frac{P/P_E}{1-(P/P_E)}\right] + K\frac{F}{P_E} \tag{54}$$

where δ is the deflection of the supporting spring, and using Eq. (18)

$$K = \frac{2}{\pi^2}\sum_{n=1,3,5}^{\infty}\frac{1}{n^2\left[\dfrac{P}{P_E} - n^2\right]} = \left\{\frac{1}{4} - \frac{1}{2\theta}\tan\frac{\theta}{2}\right\}\Big/\left(\frac{P}{P_E}\right) \tag{55}$$

The end shortening ε_1 due to buckling can be similarly obtained as

$$\frac{\varepsilon_1}{l} = \frac{\dfrac{\pi^2}{4}\left(\dfrac{a_1^0}{l}\right)^2\left[1 - \left(\dfrac{P}{P_E} - 1\right)^2\right] - \dfrac{a_1^0}{l}\dfrac{F}{P_E}}{\left(\dfrac{P}{P_E} - 1\right)^2} + C\left(\frac{F}{P_E}\right)^2 \tag{56}$$

where

$$C = \frac{1}{\pi^2}\sum_{n=1,3,5}^{\infty}\frac{1}{n^2\left[\dfrac{P}{P_E} - n^2\right]^2} = \left\{\frac{3}{16} + \frac{1}{16}\tan^2\frac{\theta}{2} - \frac{3}{8\theta}\tan\frac{\theta}{2}\right\}\Big/\left(\frac{P}{P_E}\right)^2 \tag{57}$$

The shortening due to direct compression stress can be calculated by Eq. (31).

The bending strain energy of the column is

$$\frac{W_1}{P_E l} = \frac{\pi^2}{4}\frac{\dfrac{a_1^0}{l}\dfrac{P}{P_E}\left(\dfrac{a_1^0}{l}\dfrac{P}{P_E} - \dfrac{4}{\pi^2}\dfrac{F}{P_E}\right)}{\left(\dfrac{P}{P_E} - 1\right)^2} + B\left(\frac{F}{P_E}\right) \tag{58}$$

where

$$B = \frac{1}{\pi^2}\sum_{n=1,3}^{\infty}\frac{1}{\left[\dfrac{P}{P_E} - n^2\right]^2} = \left\{\frac{1}{16} + \frac{1}{16}\tan^2\frac{\theta}{2} - \frac{1}{8\theta}\tan\frac{\theta}{2}\right\}\Big/\left(\frac{P}{P_E}\right) \tag{59}$$

The strain energy due to direct compression and the strain energy stored in the supporting spring can be calculated by Eqs. (38) and (5), respectively.

As an example,

$$\frac{F}{P_E} = 13.712\frac{\delta}{l}\left(1 - 208.89\frac{\delta}{l} + 10\,444\frac{\delta^2}{l^2}\right) \tag{60}$$

The shape of F vs. δ curve given by Eq. (60) is similar to that shown in Fig. 2, only the initial slope is made smaller. By substituting Eq. (60) into Eq. (54), a cubic equation for δ/l with a given value of P/P_E is obtained. When $a_1^0 = 0$, i.e., the column is initially straight, this cubic equation has a root $\delta/l = 0$ representing unbuckled position. The branch point of the solution occurs at $P/P_E = 3.61$. The shape of the P vs. δ curve is shown in Fig. 7 where P/P_E is plotted against $\delta/\pi i$ and with $\pi i/l = 1/100$. When $a_1^0 \neq 0$, the curves do not have a branch point. Furthermore, with larger values of a_1^0, the maximum possible end load becomes smaller. For $a_1^0/\pi i = 0.1$ or $a_1^0/l = 0.001/\pi$, the maximum value of P/P_E is reduced from 3.61 to 1.82. This effect is shown clearly in Fig. 8, where the maximum value of P/P_E is plotted as a function of $a_1^0\pi/l$.

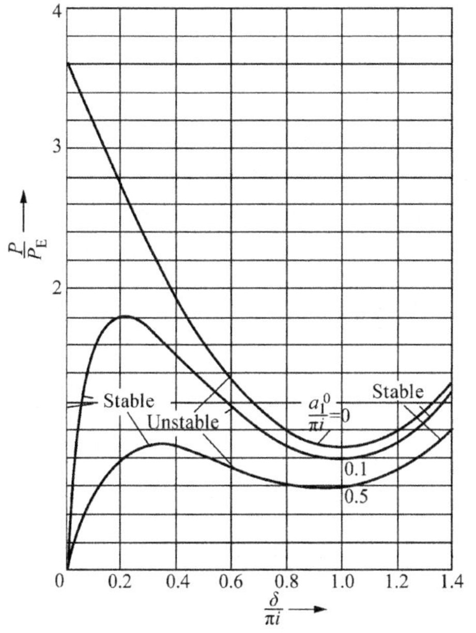

Fig. 7 Relation between the end load P on the column with various amounts of initial deflection a_1^0 and the deflection δ of a single lateral support at the middle

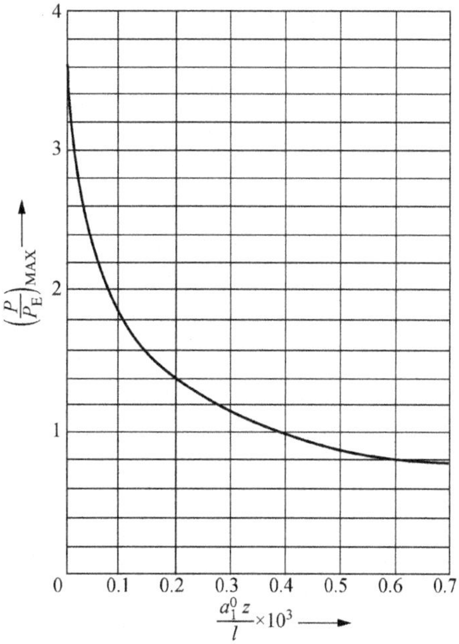

Fig. 8 The maximum end load P on the column as a function of its initial deflection a_1^0

In Fig. 9, the end load P is plotted against the shortening ε. Without initial deflection, there is a part of the buckling curve which gives decreasing value of ε, such as LM. Thus within this range of values of ε, three equilibrium positions are possible for a fixed ε. But with increasing value of initial deflection, this part of the buckling curve gradually disappears, and only one equilibrium position is possible for a given value of ε. At $a_1^0/\pi i = 0.1$, this change in the character of the curve is already completed.

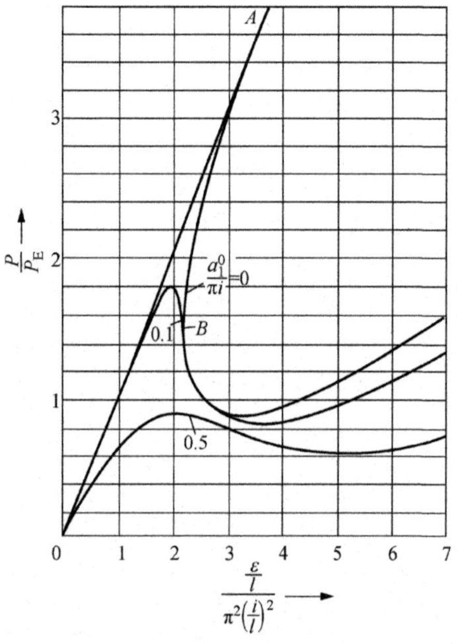

Fig. 9 Relation between the end load P on a column with a single lateral support at its middle and its end shortening ε

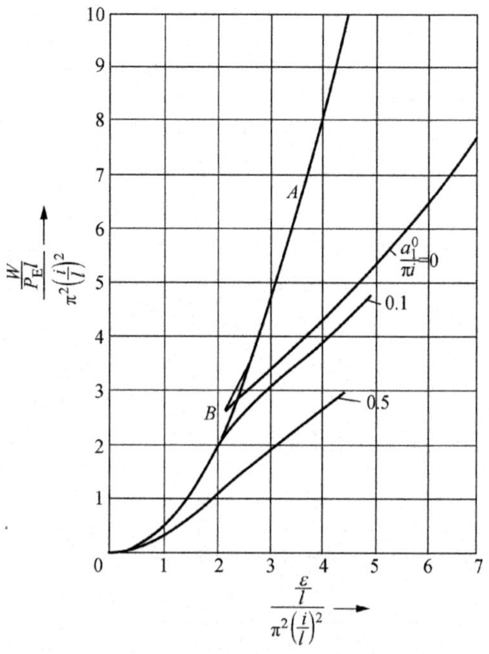

Fig. 10 Relation between the total strain energy W stored in the system of a column with a single lateral support and the end shortening ε of the column

In Fig. 10, where the elastic energy of the system is plotted as a function of unit end shortening ε, the above mentioned change in type of buckling curve removes the possibility of a jump in the value of P/P_E at a certain stage of the buckling process which is indicated by a node point such as C (Fig. 10). Therefore, by introducing a sufficient amount of initial deflection, there is an apparent change in the buckling characteristics of the column.

Case of a Single Lateral Support, Effect of Elasticity of the Testing Machine

In the foregoing paragraphs, it has been assumed that the testing machine is completely rigid. In other words, if the loading crank of the machine is held at a fixed position, the distance between the end plates will be constant and not affected by the magnitude of forces acting on them. Actually, this is not true. The machine always has a certain amount of "yield" or "give." This characteristic of the testing machine can be represented by a spring placed between the specimen and the supposedly rigid end plate (Fig. 11). The deflection ε_3 of this spring can be assumed to be a linear function of the applied load P, i.e.,

$$\varepsilon_3 = kP \tag{61}$$

Writing in terms of the Euler buckling load P_E,

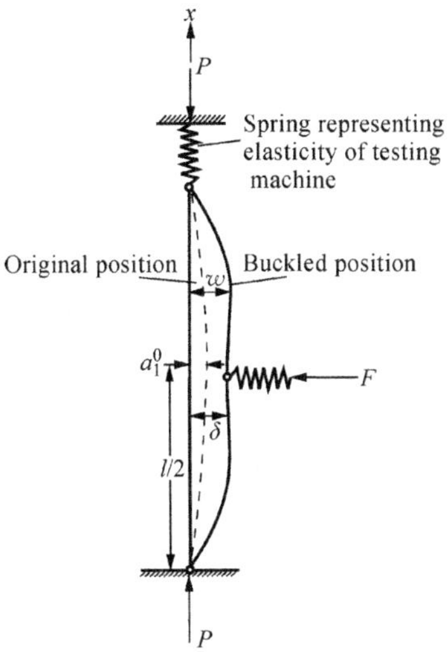

Fig. 11 Column with a single lateral support at its middle tested in a testing machine with certain amount of elasticits

$$\varepsilon_3/l = (k/l)P_E(P/P_E) = (kEA/l)(\pi i/l)^2(P/P_E) \tag{62}$$

The elastic energy stored in this spring is

$$\frac{W_5}{P_E l} = \frac{1}{2}\left(\frac{kEA}{l}\right)\left(\frac{\pi i}{l}\right)^2\left(\frac{P}{P_E}\right)^2 \tag{63}$$

The total shortening ε of the column can be obtained as the sum $\varepsilon_1 + \varepsilon_2 + \varepsilon_3$. For the case of a single lateral support, ε_1 and ε_2 are calculated by means of Eqs. (56) and (31). The total elastic energy is the sum $W_1 + W_2 + W_4 + W_5$ which can be calculated by means of Eqs. (58), (5), (39) and (63).

For the numerical example, $kEA/l = 3$, $\pi i/l = 1/100$ and a supporting spring with the property represented by Eq. (60). The results of the calculation are shown in Figs. 12 and 13. In Fig. 12, P/P_E is plotted against $(\varepsilon/l)/(\pi i/l)^2$. In Fig. 13, $(W/P_E l)/(\pi i/l)^2$ is plotted against $(\varepsilon/l)/(\pi i/l)^2$. For a straight column, the first possibility of buckling appears at a lower value of P/P_E due to the elasticity of the testing machine. If the machine is rigid, the jump occurs at $P/P_E = 2.40$ to $P/P_E = 1.09$. With $kEA/l = 3$, the jump occurs at $P/P_E = 1.62$ to $P/P_E = 0.89$. Therefore, the test buckling load of this type of column is profoundly affected by the elastic characteristics of the testing machine.

Furthermore, for $a_1^0/\pi i = 0.1$, a rigid testing machine excludes the possibility of a jump in equilibrium position, while for the case $kAE/l = 3$, Fig. 13 indicates the possibility of a jump at $(\varepsilon/l)/(\pi i/l)^2 = 6.1$. By referring back to Fig. 12, it will be noted that the jump involves a sudden change of the value of P/P_E from 1.52 to 0.80. Hence, even the apparent buckling characteristics of the column are affected by the elasticity of the testing machine.

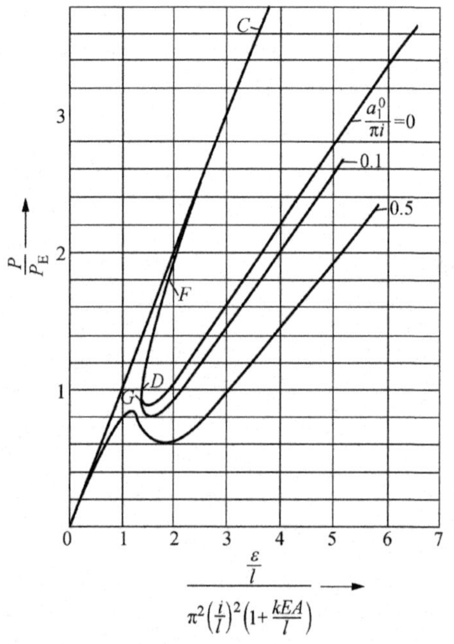

Fig. 12 Relation between the end load P on a column with a single lateral support at its middle and its end shortening ε, taking into account the elasticity of the testing machine (Compare with Fig. 9)

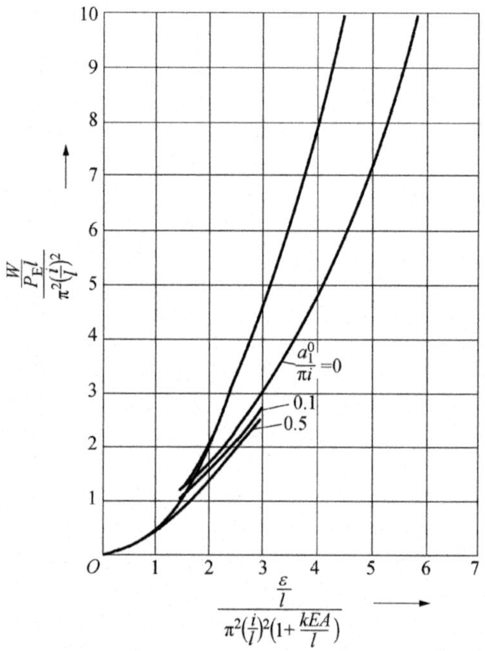

Fig. 13 Relation between the total strain energy W stored in the system of a column with a single lateral support and the end shortening ε of the column, taking into account the elasticity of the testing machine (Compare with Fig. 10)

Stability of the Equilibrium Positions Under Infinitesimal Disturbances

The problem of stability in connection with the criterion of the most probable equilibrium position has been mentioned. If the column is unstable under infinitesimal disturbances, it is certainly unstable under finite disturbances. Therefore, a first step in the stability investigation is the determination of the stability under infinitesimal disturbances. This calculation will be carried out for the case of a single central lateral support.

The elastic energy stored in the system is the sum of the direct compression energy in the column and the loading spring representing the testing machine, the elastic energy due to bending, and the elastic energy stored in the lateral support. By means of Eqs. (4), (5), (39) and (63), this total energy can be expressed as

$$\frac{W}{P_E l} = \frac{\pi^2}{4}\left[\left(\frac{a_1 - a_1^0}{l}\right)^2 + \sum_{n=3,5,F}^{\infty} n^4\left(\frac{a_n}{l}\right)^2\right] + \frac{1}{2}\left(\frac{P_E}{AE} + \frac{kP_E}{l}\right)\left(\frac{P}{P_E}\right)^2 + \int_0^{\delta/l}\frac{F}{P_E}\,d\xi \quad (64)$$

The end shortening due to bending can be calculated by

$$(1/2)\int_0^l \left(\frac{dw}{dx}\right)^2 dx$$

and thus the total end shortening ε is

$$\frac{\varepsilon}{l} = \frac{\pi^2}{4}\left[\sum_{n=1,3,5}^{\infty} n^2\left(\frac{a_n}{l}\right)^2 - \left(\frac{a_1^0}{l}\right)^2\right] + \left(\frac{P_E}{AE} + \frac{kP_E}{l}\right)\frac{P}{P_E} \tag{65}$$

The variations of $W/P_E l$ with respect to a_n have to conform to the geometrical constraint which can be expressed as

$$\varepsilon/l = \text{constant} \tag{66}$$

The equilibrium conditions are obtained by setting the first derivatives of $W/P_E l$ with respect to a_n equal to zero, taking into account the condition Eq. (66). These equilibrium conditions are, of course, the same as those obtained before. Since these first derivatives are zero at equilibrium position, the total elastic energy for configurations near the equilibrium position can be expanded into a Taylor series of a_n in the following manner:

$$\frac{W}{P_E l} = \left(\frac{W}{P_E l}\right)_{a_n = a_n^*} + \frac{1}{2}\sum_{n=1}^{\infty}\sum_{m=1}^{\infty} A_{nm}(a_n - a_n^*)(a_m - a_m^*) \tag{67}$$

where a_n^* is the value of a_n at equilibrium position, and

$$A_{nm} = \frac{\partial^2[W/(P_E l)]}{\partial a_n \partial a_m} \quad \text{at } a_n = a_n^*,\ a_m = a_m^* \tag{68}$$

Terms not explicitly written in Eq. (67) are terms involving $(a_n - a_n^*)$ in higher orders than second. For infinitesimal disturbances, or infinitesimal values of $(a_n - a_n^*)$, the higher order terms can be neglected. Then the elastic energy $W/P_E l$ for configurations near the equilibrium position will be larger or smaller than the elastic energy at the equilibrium position, depending upon whether the double sum in Eq. (67) is positive or negative. Therefore, if the double sum is always positive for any values of $(a_n - a_n^*)$ the equilibrium position is stable. In other words, the equilibrium is a stable one only when the double sum in Eq. (67) is positive definite. If

$$\Delta_1 = A_{11},\ \Delta_2 = \begin{vmatrix} A_{11} & A_{13} \\ A_{31} & A_{33} \end{vmatrix},\ \Delta_3 = \begin{vmatrix} A_{11} & A_{13} & A_{15} \\ A_{31} & A_{33} & A_{35} \\ A_{51} & A_{53} & A_{55} \end{vmatrix} \cdots$$

$$\Delta_g = \begin{vmatrix} A_{11} & A_{13} & \cdots & A_{1(2g-1)} \\ A_{31} & A_{33} & \cdots & A_{3(2g-1)} \\ \cdots & \cdots & \cdots & \cdots \\ A_{(2g-1)1} & A_{(2g-1)3} & \cdots & A_{(2g-1)(2g-1)} \end{vmatrix} \tag{69}$$

Then it is well-known that the conditions for the double sum in Eq. (67) to be positive definite are $\Delta_1 > 0$, $\Delta_2 > 0$, $\Delta_3 > 0$, ..., $\Delta_g > 0$, ...

From Eqs. (64) and (65)

$$
\left.\begin{aligned}
A_{11} &= \frac{\pi^2}{2}\left(1-\frac{P}{P_E}\right)+\frac{\mathrm{d}\left(\frac{F}{P_E}\right)}{\mathrm{d}\zeta}+\frac{1}{\left(\frac{P_E}{AE}+\frac{kP_E}{l}\right)}\frac{\left(\frac{F}{P_E}-\frac{\pi^2}{2}\frac{a_1^0}{l}\right)^2}{\left(\frac{P}{P_E}-1\right)^2}\\
A_{1n} &= A_{n1} = -(-1)^{\frac{1+n}{2}}\left\{\frac{\mathrm{d}\left(\frac{F}{P_E}\right)}{\mathrm{d}\zeta}+\frac{1}{\left(\frac{P_E}{AE}+\frac{kP_E}{l}\right)}\frac{\left(\frac{F}{P_E}-\frac{\pi^2}{2}\frac{a_1^0}{l}\right)\frac{F}{P_E}}{\left(\frac{P}{P_E}-1\right)\left(\frac{P}{P_E}-n^2\right)}\right\}\\
A_{nn} &= \frac{\pi^2}{2}n^2\left(n^2-\frac{P}{P_E}\right)+\left\{\frac{\mathrm{d}\frac{F}{P_E}}{\mathrm{d}\zeta}+\frac{1}{\left(\frac{P_E}{AE}+\frac{kP_E}{l}\right)}\frac{\left(\frac{F}{P_E}\right)^2}{\left(\frac{P}{P_E}-n^2\right)^2}\right\}\\
A_{nm} &= -(-1)^{\frac{m+n}{2}}\left\{\frac{\mathrm{d}\left(\frac{F}{P_E}\right)}{\mathrm{d}\zeta}+\frac{1}{\left(\frac{P_E}{AE}+\frac{kP_E}{l}\right)}\frac{\left(\frac{F}{P_E}\right)^2}{\left(\frac{P}{P_E}-n^2\right)\left(\frac{P}{P_E}-m^2\right)}\right\}
\end{aligned}\right\} \tag{70}
$$

where $\zeta = \delta/l$.

First the simpler case of a very elastic testing machine is investigated. This means that $k \to \infty$, as can be seen easily from Eq. (61). Then the last terms in each of the expressions in Eq. (70) drop out. The determinants defined by Eq. (69) are found to be as follows:

$$
\left.\begin{aligned}
\Delta_1 &= \left\{1+\frac{\mathrm{d}\left(\frac{F}{P_E}\right)}{\mathrm{d}\zeta}\left[\frac{2}{\pi^2}\frac{1}{\left(1-\frac{P}{P_E}\right)}\right]\right\}\frac{\pi^2}{2}\left(1-\frac{P}{P_E}\right)\\
\Delta_2 &= \left\{1+\frac{\mathrm{d}\left(\frac{F}{P_E}\right)}{\mathrm{d}\zeta}\left[\frac{2}{\pi^2}\frac{1}{\left(1-\frac{P}{P_E}\right)}+\frac{2}{\pi^2}\frac{1}{9\left(9-\frac{P}{P_E}\right)}\right]\right\}\frac{\pi^2}{2}\left(1-\frac{P}{P_E}\right)\frac{\pi^2}{2}9\left(9-\frac{P}{P_E}\right)
\end{aligned}\right\} \tag{71}
$$

or

$$
\Delta_g = \left\{1+\frac{\mathrm{d}\left(\frac{F}{P_E}\right)}{\mathrm{d}\zeta}\sum_{n=1,3,5}^{n=2g-1}\frac{2}{\pi^2}\frac{1}{n^2\left(n^2-\frac{P}{P_E}\right)}\right\}\prod_{n=1,3,5}^{n=2g-1}\frac{\pi^2}{2}n^2\left(n^2-\frac{P}{P_E}\right)
$$

Now take the case $\zeta = 0$ first, i.e., the unbuckled straight equilibrium positions. Here $\mathrm{d}(F/P_E)/\mathrm{d}\zeta$ is the initial slope of supporting force curve which is positive. For $0 < P/P_E < 1$, the multiplying factor before the curly bracket is also positive. Furthermore, each term under the summation sign is positive. Therefore, every quantity which appears in Δ_g is positive and hence Δ_g is positive. Consequently the equilibrium positions are stable. By direct computation, using the value of $\mathrm{d}(F/P_E)\mathrm{d}\zeta = 13.712$ as given by Eq. (60), it is found that the straight unbuckled equilibrium positions are stable under infinitesimal disturbances if $P/P_E < 3.610$, the branch point of equilibrium positions. For $\zeta \neq 0$, it is found that the stability is indicated by the sign of Δ_g with $g \to \infty$, i.e., all the determinant changes sign if $\mathrm{Lim}\ \Delta_g$ changes sign.

Now as stated above, the multiplying factor before the curly bracket is positive when $0 < P/P_E < 1$ and negative when $1 < P/P_E < 9$; therefore, the attention of the reader need only be fixed on the quantities within the curly bracket. When $g \to \infty$, the series in Eq. (60) can be summed and

$$1+\frac{d\left(\frac{F}{P_E}\right)}{d\zeta}\sum_{n=1,3,5}^{\infty}\frac{2}{\pi^2}\frac{1}{n^2\left(n^2-\frac{P}{P_E}\right)}=1-\frac{d\left(\frac{F}{P_E}\right)}{d\zeta}\frac{1}{\left(\frac{P}{P_E}\right)}\left(\frac{1}{4}-\frac{1}{2\theta}\tan\frac{\theta}{2}\right) \tag{72}$$

The limiting case between stability and instability is, of course, when the expression on the left of Eq. (72) is zero. Therefore, for the limiting case between stability and instability

$$\frac{d\zeta}{d(F/P_E)}=\frac{1}{(P/P_E)}\left(\frac{1}{4}-\frac{1}{2\theta}\tan\frac{\theta}{2}\right) \tag{73}$$

The meaning of this condition can be understood better by finding the corresponding value of $d\zeta/d(P/P_E)$ which is the reciprocal of the slope of P/P_E vs. δ/l curve as shown in Fig. 7. Eq. (73) can be rewritten as

$$\frac{d\zeta}{d(P/P_E)}=\frac{d(F/P_E)}{d(P/P_E)}\frac{1}{(P/P_E)}\left(\frac{1}{4}-\frac{1}{2\theta}\tan\frac{\theta}{2}\right) \tag{74}$$

Now differentiating Eq. (54) which is true for equilibrium positions,

$$\frac{d\zeta}{d\left(\frac{P}{P_E}\right)}=\frac{a_1^0}{l}\frac{1}{\left(1-\frac{P}{P_E}\right)^2}+\left\{\frac{d\left(\frac{F}{P_E}\right)}{d\left(\frac{P}{P_E}\right)}-\frac{\frac{F}{P_E}}{\frac{P}{P_E}}\right\}\frac{\frac{1}{4}-\frac{1}{2\theta}\tan\frac{\theta}{2}}{\left(\frac{P}{P_E}\right)}+\frac{\frac{F}{P_E}}{\left(\frac{P}{P_E}\right)^2}\times \left\{-\frac{1}{8}+\frac{1}{4\theta}\tan\frac{\theta}{2}-\frac{1}{8}\tan^2\frac{\theta}{2}\right\} \tag{75}$$

Substituting the value of $d(F/P_E)/(P/P_E)$ from Eq. (74) into Eq. (75),

$$\frac{d\zeta}{d\left(\frac{P}{P_E}\right)}=\frac{d\zeta}{d\left(\frac{P}{P_E}\right)}+\frac{a_1^0}{l}\frac{1}{\left(1-\frac{P}{P_E}\right)^2}-\frac{\frac{F}{P_E}}{\left(\frac{P}{P_E}\right)^2}\left\{\frac{3}{8}-\frac{3}{4\theta}\tan\frac{\theta}{2}+\frac{1}{8}\tan^2\frac{\theta}{2}\right\} \tag{76}$$

Eq. (76) can be satisfied by either

$$d\zeta/d(P/P_E)=\infty,\ \text{i.e.},\ d(P/P_E)/d\zeta=0 \tag{77}$$

or

$$0=\frac{a_1^0}{l}\frac{1}{\left(1-\frac{P}{P_E}\right)^2}-\frac{\frac{F}{P_E}}{\left(\frac{P}{P_E}\right)^2}\left\{\frac{3}{8}-\frac{3}{4\theta}\tan\frac{\theta}{2}+\frac{1}{8}\tan^2\frac{\theta}{2}\right\} \tag{78}$$

It is found that Eq. (78) cannot be satisfied with any of the equilibrium states. Therefore, for a very elastic testing machine the transition point from stable equilib ium positions to unstable equilibrium positions or *vice versa* is characterized by the condition Eq. (77), i.e., the slope of the P/P_E vs. δ/l curve must be horizontal. This condition is utilized to mark the stable and the unstable regions in Fig. 7.

With $k \to \infty$, the deflection of the loading spring representing the elasticity of the testing machine is very large compared with the shortening of the specimen. Therefore, the force in the spring, which is equal to load on the specimen, is not influenced by the small variation of shortening of the specimen. Similar situation exists also for a column loaded by a dead weight. Therefore, the stability criterion given by Eq. (77) also applies for such cases.

For the case $k \neq \infty$, the calculation for stability is more complicated. To simplify the calculation, let $a_1^0 = 0$, i.e., assuming the column is initially straight. Using the values of A_{nm} given by Eq. (70); it is found that g-th determinant is

$$
\begin{aligned}
\Delta_g = & \Bigg[1 + \left(\frac{\frac{F}{P_E}}{\frac{\pi i^*}{l}} \right)^2 \frac{2}{\pi^2} \sum_{n=1,3,5}^{2g-1} \frac{1}{n^2 \left(n^2 - \frac{P}{P_E} \right)^3} + \\
& \frac{d\left(\frac{F}{P_E}\right)}{d\zeta} \Bigg\{ \frac{2}{\pi^2} \sum_{n=1,3,5}^{2g-1} \frac{1}{n^2 \left(n^2 - \frac{P}{P_E} \right)} + \left(\frac{\frac{F}{P_E}}{\frac{\pi i^*}{i}} \right)^2 \left(\frac{2}{\pi^2} \right)^2 \times \\
& \Bigg[\sum_{n=1,3,5}^{2g-1} \frac{1}{n^2 \left(n^2 - \frac{P}{P_E} \right)^3} \sum_{m=1,3,5}^{2g-1} \frac{m^2}{\left(m^2 - \frac{P}{P_E} \right)^3} - \\
& \left(\sum_{n=1,3,5}^{2g-1} \frac{1}{\left(n^2 - \frac{P}{P_E} \right)^3} \right)^2 \Bigg] \Bigg\} \Bigg] \prod_{n=1,3,5}^{2g-1} \frac{\pi^2}{2} n^2 \left(n^2 - \frac{P}{P_E} \right)
\end{aligned}
\tag{79}
$$

where $(\pi i^*/l)^2 = [P_E/(AE)] + (kP_E/l)$.

All the straight unbuckled configurations are again found to be stable up to $P/P_E = 3.61$ for the particular case investigated. In other words, straight equilibrium states are stable under infinitesimal disturbances up to the branch point at which buckling begins. For equilibrium states involving buckling we again fix our attention on the quantities within the curly bracket in the expression for Δ_g, Eq. (79). It is again found that the sign of Δ_g when $g \to \infty$ is representative for the transition from stability to instability or *vice versa*. When $g \to \infty$, all the series in Eq. (79) can be summed and the condition of transition of stability is obtained

$$
\begin{aligned}
1 - & \left(\frac{\frac{F}{P_E}}{\frac{\pi i^*}{l}} \right)^2 \frac{1}{\left(\frac{P}{P_E} \right)^3} \left[\frac{1}{4} + \frac{7}{32} \sec^2 \frac{\theta}{2} - \frac{15}{16\theta} \tan \frac{\theta}{2} - \frac{1}{16} \frac{\theta}{2} \times \right. \\
& \left. \tan \frac{\theta}{2} \sec^2 \frac{\theta}{2} \right] - \frac{1}{\left(\frac{P}{P_E} \right)} \frac{d\left(\frac{F}{P_E} \right)}{d\zeta} \left[\left\{ \frac{1}{4} - \frac{1}{2\theta} \tan \frac{\theta}{2} \right\} + \right.
\end{aligned}
$$

$$\left(\frac{\dfrac{F}{P_E}}{\dfrac{\pi i^*}{l}}\right)^2 \frac{1}{128\left(\dfrac{P}{P_E}\right)^3}\left\{\frac{\tan\dfrac{\theta}{2}}{\dfrac{\theta}{2}}\left(3\frac{\tan\dfrac{\theta}{2}}{\dfrac{\theta}{2}} - 5\sec^2\frac{\theta}{2} - 1\right) + \sec^2\frac{\theta}{2}\left(3 + 2\frac{\theta}{2}\tan\frac{\theta}{2}\right)\right\}\Bigg] = 0 \tag{80}$$

Since

$$\frac{d\zeta}{d\left(\dfrac{P}{P_E}\right)} = \frac{d\left(\dfrac{F}{P_E}\right)\Big/ d\left(\dfrac{P}{P_E}\right)}{d\left(\dfrac{F}{P_E}\right)\Big/ d\zeta}$$

from (75),

$$\frac{1}{\left(\dfrac{P}{P_E}\right)}\frac{d\left(\dfrac{F}{P_E}\right)}{d\zeta} = \frac{\dfrac{d\left(\dfrac{F}{P_E}\right)}{d\left(\dfrac{P}{P_E}\right)}\Bigg/\dfrac{F}{P_E}}{\left\{\dfrac{\dfrac{d\left(\dfrac{F}{P_E}\right)}{d\left(\dfrac{P}{P_E}\right)}}{\dfrac{F}{P_E}} - \dfrac{1}{\dfrac{P}{P_E}}\right\}\left\{\dfrac{1}{4} - \dfrac{1}{4\dfrac{\theta}{2}}\tan\dfrac{\theta}{2}\right\} + \dfrac{1}{8\dfrac{P}{P_E}}\left\{\dfrac{1}{\dfrac{\theta}{2}}\tan\dfrac{\theta}{2} - \sec^2\dfrac{\theta}{2}\right\}} \tag{81}$$

By substituting Eq. (81) into Eq. (80), it is found that the condition for transition of stability is given by either

$$2 + \sec^2(\theta/2) - 3\frac{1}{(\theta/2)}\tan(\theta/2) = 0 \tag{82}$$

or

$$\left(\frac{\pi i^*}{l}\right)^2 + \frac{d}{d\left(\dfrac{P}{P_E}\right)}\left\{\left(\frac{\dfrac{F}{P_E}}{\dfrac{P}{P_E}}\right)^2 \times \left(\frac{3}{16} + \frac{1}{16}\tan^2\frac{\theta}{2} - \frac{3}{8\theta}\tan\frac{\theta}{2}\right)\right\} = 0 \tag{83}$$

Here again the condition represented by Eq. (82) cannot be satisfied by any of the equilibrium positions. Therefore, the only condition left for the transition of stability is that given by Eq. (83). By means of Eqs. (56), (31) and (62), the total end shortening including the shortening of the loading spring is

$$\frac{\varepsilon}{l} = \left(\frac{\pi i^*}{l}\right)^2 \frac{P}{P_E} + \left(\frac{\dfrac{F}{P_E}}{\dfrac{P}{P_E}}\right)^2 \times \left(\frac{3}{16} + \frac{1}{16}\tan^2\frac{\theta}{2} - \frac{3}{8\theta}\tan\frac{\theta}{2}\right) = 0 \tag{84}$$

Therefore, the condition given by Eq. (83) is simply

$$d(\varepsilon/l)/d(P/P_E) = 0, \text{ or } d(P/P_E)/d(\varepsilon/l) = \infty \tag{85}$$

Hence, the transition from stable equilibrium to unstable equilibrium or vice versa under infinitesimal disturbances occurs at points on the P/P_E vs. ε/l curves where the slope is vertical. Therefore, if the testing machine is rigid, Fig. 9 shows that for the case $a_1^0/\pi i = 0$, there is an unstable region between the points A and B (Fig. 9). However, for the cases $a_1^0/(\pi i) = 0.1$ and $a_1^0/(\pi i) = 0.5$, there is no unstable equilibrium position. If the testing machine has a certain amount of elasticity, e. g., $kEA/l = 3$, then Fig. 12 shows that for $a_1^0/(\pi i) = 0$ there is an unstable region between the points C and D (Fig. 12). Similarly for $a_1^0/(\pi i) = 0.1$, the equilibrium positions are unstable between the points F and G (Fig. 12). However, for $a_1^0/(\pi i) = 0.5$, all the equilibrium positions are stable.

The conditions found in this section for transition between stable equilibrium to unstable equilibrium under infinitesimal disturbances appear in very simple form and do not involve any parameter which is characteristic of the particular problem concerned. Therefore, it is possible that although these rules are derived from particular cases, they hold for all non-linear buckling problems of this category. This surmise has to be proved, of course, by further investigation.

Conclusions

In the previous papers[1~3], an explanation for the discrepancy between the classical theory of thin shells with infinitesimal deflections and the experimental result is suggested. It is stated then that there are equilibrium positions of the shell which involve much lower values of external load than the buckling load predicted by the classical theory. The initial imperfection of the shell tends to lower the maximum load the shell can sustain. The lower experimental buckling load is thus attributed to the slight imperfections always present in the test specimens. Furthermore, the elastic characteristics of the testing machine are shown to have profound effect on the buckling process. In the present paper these statements are subject to a quantitative verification not by attacking the shell problem itself which is too complicated, but by solving the similar problem of a column supported laterally by non-linear springs.

However, a new interesting fact is found. With supporting springs whose characteristic simulates that found in the case of a curved shell, a straight unbuckled column may have an elastic energy and end shortening equal to those of a buckled state with finite deflection even at a load far below the classical buckling load. If the end plates of the testing machine are brought slightly nearer, the elastic energy for the unbuckled state will be higher than the buckled state. Therefore, at a small disturbance, the column will jump to the buckled equilibrium state. This means that the recorded buckling load in a testing machine can be lower than that of classical theory even for a perfect specimen without initial deflection. This critical point can be further lowered by the elasticity of the testing machine. Hence, it seems that the buckling load calculated by means of classical theory really has very little bearing on the actual load carrying ability of this type of structure.

At the jumping point of equilibrium states, both the initial equilibrium and the final equilibrium states are stable under infinitesimal disturbances. The question then arises as to how large the external impulse has to be in order for the column to leave the initial stable state. Take, for instance, the point C in Fig. 10. This point represents two equilibrium states, namely, the straight unbuckled state and the buckled state. Both are stable under infinitesimal disturbances and have the same amount of elastic energy. At the same end shortening ε, there is another equilibrium state in the curve AB, which has a slightly higher elastic energy. For this particular value of end shortening, the elastic energy W given by Eq. (64) can be plotted as a surface over the plane with coordinates a_1, a_3, ..., a_∞. The load P is determined by the condition that the end shortening must be constant. Then the two stable equilibrium states will be represented by two minimum points. One of these points corresponding to the unstabled state is located at the origin of the a_1, a_3, ..., a_∞ plane. Between these two minimum points will be a saddle point representing the unstable equilibrium state with a slightly higher elastic energy. (The unstable equilibrium state being represented by a saddle point follows from the fact that Δ_1, Δ_2, ..., Δ_∞ are all negative.) Therefore, the necessary external impulse is equal to or larger than the difference between the elastic energy at the saddle point and the minimum points. For the case of point C in Fig. 10, this difference ΔW is

$$\Delta W \cong 0.15\pi^2 P_E (i^2/l)$$

For a 24SRT column of 0.09-inch thickness, 0.375-inch width and 19-inch length (tests were made on a column of this size[3]),

$$\Delta W = 3.26 \times 10^{-4} \text{inch-lbs}$$

which is very small indeed. This is, perhaps, sufficient to explain the "jumping" at point C.

References

[1] von Kármán Th., Tsien Hsue-shen, The Buckling of Spherical Shells by External Pressure. Journal of the Aeronautical Sciences, 1939, 7: 43.

[2] von Kármán Th., Tsien, Hsue-shen, The Buckling of Thin Cylindrical Shells under Axial Compression. Journal of the Aeronautical Sciences, 1941, 8: 303.

[3] von Kármán Th., Dunn Louis G., Tsien Hsue-shen. The Influences of Curvalure on the Buckling Characteristics of Structures. Journal of the Aeronautical Sciences, 1940, 7: 276.

[4] Cox H L. Stress Analysis of Thin Metal Construction. Journal Royal Aeronautical Society, 1940, 44: 231.

[5] Klemperer W B., Gibbons H B. Über die Knickfestigkeit eines auf elastischen Zwischenstützen gelagerten Balkens. ZAMM, 1933, 13: 251.

A Theory for the Buckling of Thin Shells

Hsue-shen Tsien*
(*California Institute of Technology*)

Summary

In a series of papers written by the present author in collaboration with Th. von Kármán and Louis G. Dunn, the effect of curvature of a structure on its buckling characteristics was investigated. The purpose of these investigations was to find an explanation for the discrepancy between the "classic" theory and the experiments. For the case of a thin spherical shell under external pressure and for the case of a thin cylindric shell under axial compression, equilibrium states involving large deflections are discovered which can be maintained by loads far less than the so-called buckling load calculated by the classic theory of infinitesimal deflections. It was felt then that since some of these newly found equilibrium states closely approach the observed phenomena the shell must "jump" suddenly from the unbuckled configuration to these equilibrium states and the structures fail as a result of this sudden change. However, the reason why the shell should jump to these particular equilibrium states and not others was not explained. In the present paper a new principle involving the energy level and the geometric restraint is developed to determine this sudden change in equilibrium states. By means of this new principle the buckling load of both the spherical shell and cylindric shell can be calculated. The agreement with experiments is good.

Introduction

Together with Theodore von Kármán and Louis G. Dunn, the author[1~3] has suggested an explanation for the apparent discrepancy between the observed facts and the well-known classic theory for certain cases of the buckling of thin shells. The essential result of these investigations is the discovery of equilibrium states of the shell which involve a much lower applied load than the buckling load predicted by the classic theory. The load actually decreases as the deflection of the shell is increased. Some of the equilibrium states found are close to the experimentally observed ones both in the applied load and in the wave pattern. The authors then concluded that the unbuckled shell will suddenly jump to those equilibrium positions during the loading process without reaching the classic buckling load. However, the question why the shell should jump to these particular equilibrium states and not others remains to be answered. This is essentially

---

Presented at the Structures Session, Tenth Annual Meeting, I. Ae. S., New York, January 28, 1942.
Journal of the Aeronautical Sciences, vol. 9, pp. 373 - 384, 1942.

* Research Fellow in Aeronautics.

the objection raised by K. O. Friedrichs[4] to the author's treatment of the buckling of spherical shells under external pressure. In the present paper the author will first try to clarify this point and then apply the theory to a number of special cases, for which experimental data are available, to test the validity of the theory. Finally, a more precise mathematical formulation of the buckling problem of spherical shells will be presented to indicate the possibilities of further investigations.

The Criterions for Buckling

In the author's paper on cylindric shells under axial compression[3], it is stated that under average laboratory and actual service conditions the most probable equilibrium state is the state with the lowest possible energy level. This statement is based upon the assumption that there are disturbances of sufficient magnitude so that the transitions from higher energy levels to lower energy levels are always possible. Although this assumption seems plausible, nevertheless it can only be verified by comparison with experiment. Since the available experimental data usually only give the buckling load of the specimen, the present investigation will be confined to buckling phenomena.

First of all the meaning of "possible energy levels," must be made clear. The two conditions for possible energy levels are: (1) the corresponding external forces and internal stresses must be in equilibrium; (2) the geometric restraint and loading conditions, if any, must be satisfied.

The second condition depends, of course, upon the manner in which the specimen is loaded. To facilitate the mathematical treatment, two limiting cases will be investigated first. If the specimen is loaded in axial compression, then one of these limiting loading processes is that of a rigid testing machine, and the other is that of "dead weight" loading.

In a testing machine the parameter over which the operator has direct control is the distance between the end plates, since this distance is determined by the position of the loading crank. The buckling of a thin shell usually occurs in a fraction of a second; therefore, the operator really has no time to turn the loading crank during the jump from the unbuckled state to the buckled state. In other words, the distance between the end plates is the same at the beginning and at the end of the buckling process. Hence, the geometric restraint in this case is the end shortening of the specimen. Thus, both the starting point of the jump and the end point of the jump must lie in a vertical line AB in the load vs. end-shortening diagram (Fig. 1). Since the end shortening of the specimen is constant, the potential energy of the loading force is not changed. The energy level of the system is given by the strain energy at a given value of the end shortening. Buckling will probably occur at a certain value of the end shortening which gives equal strain energy for both the unbuckled equilibrium state

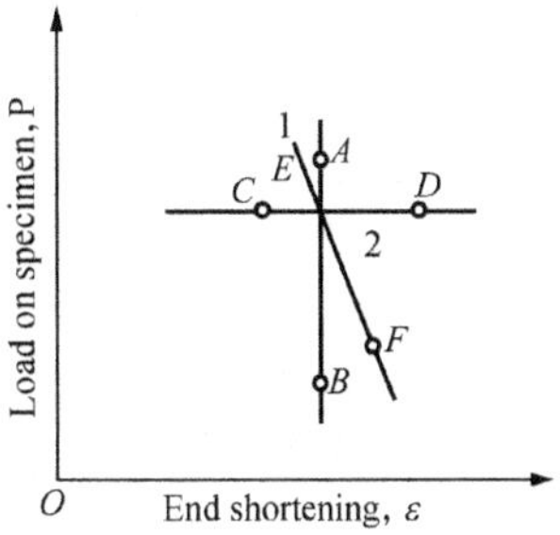

Fig. 1 Relation between the load and the end shortening for different types of loading processes

and the buckled equilibrium state. If these two particular equilibrium states are presented by points A and B, respectively, the load of the specimen will jump from a higher value to a lower value corresponding to the distance between A and B. The load given by A which corresponds to the unbuckled equilibrium state is generally recorded as the failing load of the specimen.

In case of dead weight loading, the parameter under the control of operator is the amount of weight on the test specimen. However, he has no control of the end shortening of the test specimen which is determined by the equilibrium of the external force and the internal stresses. Hence, the starting point and the end point of the buckling process must lie on a horizontal line CD in the load vs. end-shortening diagram (Fig. 1). Because of the movement of the loading force during the buckling process, the potential energy of the loading weight must be included in calculating the energy level of the system. Thus, the energy level is given by the total potential Φ of the system.

$$\Phi = S - P\varepsilon \tag{1}$$

where S is the strain energy, P the loading force and ε the end shortening of the specimen. Buckling will probably occur at a certain value of the load P which gives equal total potential Φ for both the unbuckled equilibrium state C and the buckled equilibrium state D. Therefore, referring to Fig. 1, the following relation exists

$$S_C - P\varepsilon_C = S_D - P\varepsilon_D \tag{2}$$

Or

$$S_D - S_C = P(\varepsilon_D - \varepsilon_C) \tag{3}$$

which means physically that the increase in strain energy of the specimen by jumping from the unbuckled to the buckled equilibrium states is provided by the decrease in potential energy of the load.

The limiting case of a perfectly rigid testing machine that will apply any load to the test specimen independent of the distance between the end plates really does not exist. In an actual testing machine there is always a certain amount of elasticity or "give." In other words, with the loading crank held fixed, the distance between the end plates will slightly increase if the load applied to the specimen is increased. Therefore, the starting point and end point of jumping of a specimen during buckling must lie on a straight line EF in Fig. 1, instead of the ideal case represented by the line AB. The more elastic the testing machine, the less will be the slope of the line EF. Therefore, the effect of the elasticity of the testing machine is to make the buckling phenomena more like that which would occur with dead weight loading. It will be shown later that the failing load of a specimen depends largely on the slope of the line EF. Hence, the elasticity of the testing machine has a strong influence on the failing load.

There are certain interesting relations connected with the important quantities, strain energy and total potential. Let the load P vs. end-shortening ε curve of a structure be given by Fig. 2. Then if the structure is brought up to the state P_1, ε_1, by following the equilibrium curve, the strain energy stored in the structure is certainly equal to the external work done by the loading force. Thus,

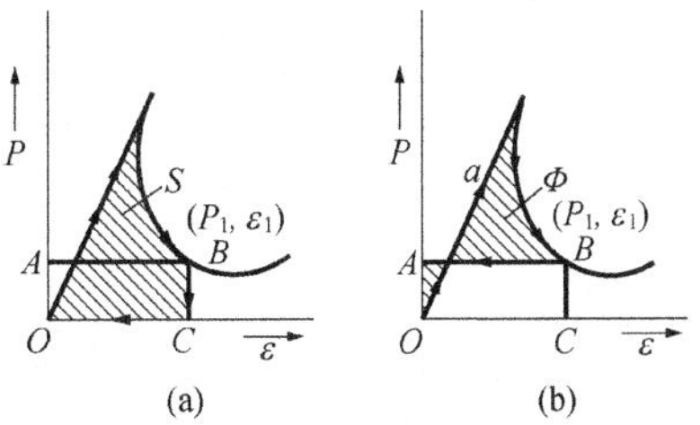

Fig. 2 Strain energy S and total potential Φ as integrals under the load P vs. end-shortening ε curve

$$S(\varepsilon_1) = \int_0^{\varepsilon_1} P\varepsilon \, d\varepsilon \tag{4}$$

which is represented by the shaded area in Fig. 2a. The direction followed around the contour of this area is as indicated and is according to the usual convention. From Eq. (4) it is evident that

$$dS/d\varepsilon = P \tag{5}$$

In other words, the slope of S vs. ε curve gives the load on the specimen.

Using Eq. (1), the total potential Φ is given by the difference of the shaded area in Fig. 2a and the rectangle OABC. In other words, it can be calculated as the algebraic sum of the shaded areas in Fig. 2b. Consequently,

$$\Phi(P_1) = -\int_0^{P_1} \varepsilon P \, dP \tag{6}$$

The integration is performed by following the contour as indicated in the figure. Eq. (6) can, of course, also be obtained from Eq. (4) by partial integration. From Eq. (6), it is evident that

$$d\Phi/dP = -\varepsilon \tag{7}$$

That is, the slope of Φ vs. P curve gives the end shortening of the specimen with a negative sign.

Both Eqs. (5) and (6) degenerate to rather trivial relations in case of linear load vs. end-shortening curves, but they are useful in nonlinear buckling problems. The relations between the quantities P and ε and the two potentials S and Φ are really equivalent to those existing between the pressure and volume and the internal energy and the heat content in thermodynamics.

Column with a Nonlinear Elastic Lateral Support

Using the results obtained in a previous investigation[3], a preliminary calculation of the strain energy in a cylindric shell under axial compression shows that the relation between this quantity and the geometric restraint — i. e., the end shortening — is rather complicated. It is felt that since the method of calculating the equilibrium positions in this particular investigation is only an approximate one, the complicated relation between the strain energy and end shortening may not be very reliable. However, an exact solution of this problem is difficult. During the investigation of the general character of the buckling of shells[2] it was found that the characteristics of this type of structure can be very well simulated by a column (Fig. 3) supported laterally by a spring whose load vs. deflection curve has a decreasing slope such as that shown in Fig. 4. This analogy was first suggested by H. L. Cox[5]. An exact solution of the column problem can be obtained without any mathematical difficulty[6].

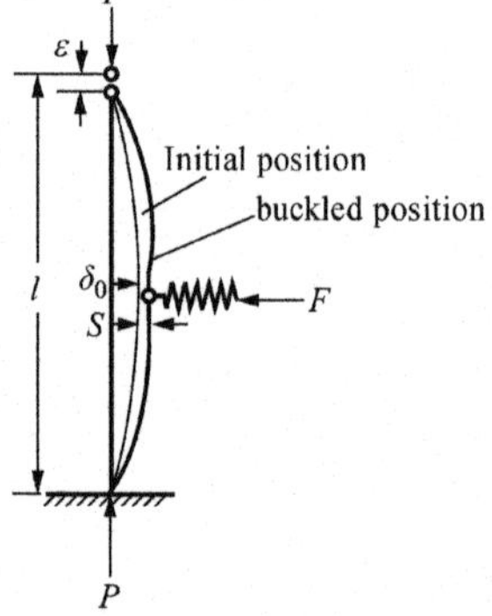

Fig. 3 Column supported laterally with a nonlinear spring

To test the validity of the theoretic analysis, the result of calculation is compared with experimental data. The experimental data were obtained from a column of 24RST of 0.090 in. thick by 0.375 in. wide and 19 in. long. The lateral support is a semicircular steel ring 0.015 in. thick by 0.50 in. wide. This steel ring gives a load F vs. deflection δ curve as indicated in Fig. 4. The columns were rolled to approximately a half sine wave. The amplitude of the initial deflection is denoted by δ_0. The test results are shown in Fig. 5 where $P_E = \pi^2 EI/l^2$ is the Euler column load, l is the length, I is the moment of inertia of the column section and E is Young's modulus of the material. The dotted, line is the theoretic curve for a perfectly straight column using the supporting spring characteristics shown in Fig. 4. One of the experimental curves is labeled "straight." However, it is practically impossible to obtain such a specimen; hence, the lower experimental values are to be expected. In Figs. 6 and 7 the maximum loads that the columns can sustain are plotted against the initial deflection ratio δ_0/l and against the deflection ratio δ/l at that maximum load, respectively. The agreement between the theoretically calculated curve and the experimental values taken from Fig. 5 is quite satisfactory.

With this verification of the theory, a further investigation is carried out in which the relation between the strain energy and the end shortening of the column is the primary consideration. The end shortening is the geometric restraint for a column in a rigid testing machine. For this calculation the load F vs. deflection δ relation for supporting spring is assumed to be represented by

$$\frac{F}{P_E} = 13.712 \frac{\delta}{l}\left(1 - 208.89 \frac{\delta}{l} + 10\,444 \frac{\delta^2}{l^2}\right) \tag{8}$$

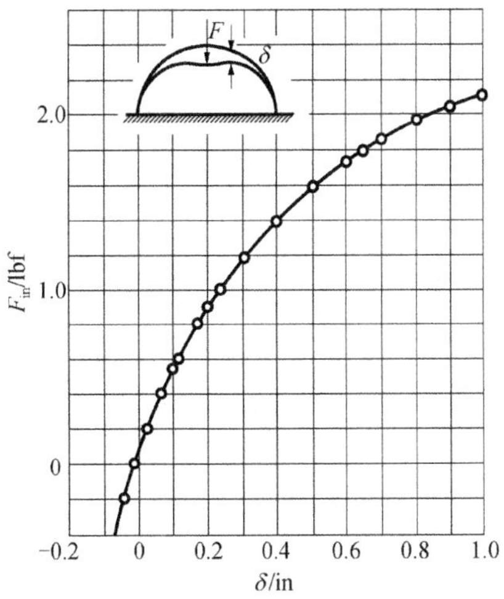

Fig. 4 Relation between the load F and the deflection δ for a semicircular ring

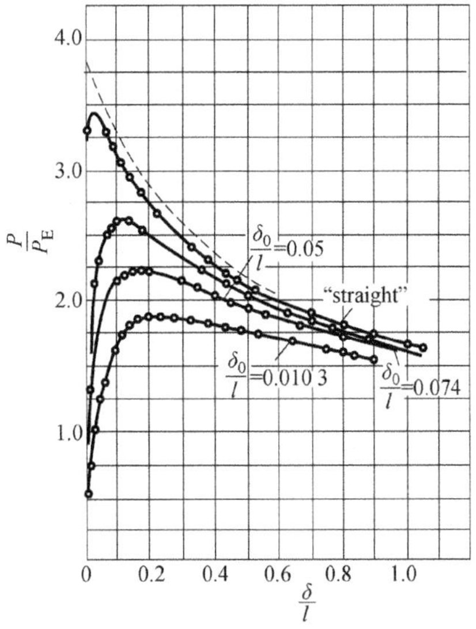

Fig. 5 Load P vs deflection δ curves for columns supported laterally with a semicircular ring and with various amounts of initial deflection δ_0

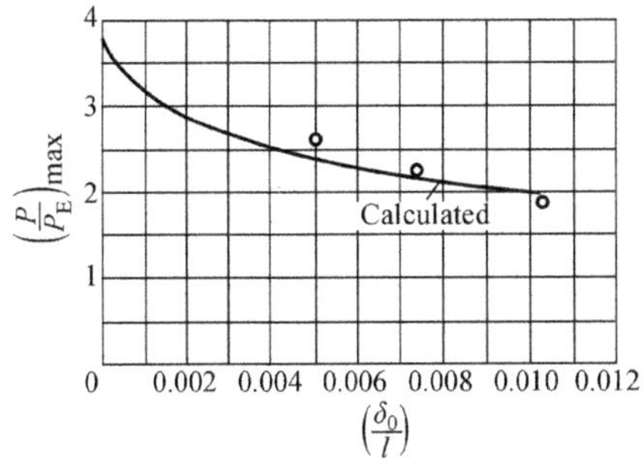

Fig. 6 Maximum load P_{max}, as a function of initial deflection δ_0 for columns supported laterally with a semicircular ring

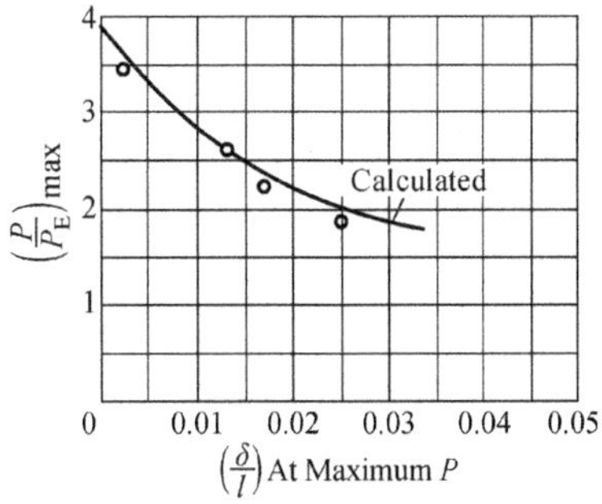

Fig. 7 Maximum load P_{max} as a function of deflection at the maximum load for columns supported laterally with a semicircular ring

This load vs. deflection relation is similar to that shown in Fig. 4. The results of this calculation are presented in Figs. 8 and 9, where the end load ratio P/P_E and the strain energy S are plotted against the end shortening ε, respectively. In these figures i denotes the radius of gyration of the column. For the case of a straight column, the load P/P_E first follows the straight line OA (Fig. 8) and then the curve AB. The point A represents the buckled state of the column with infinitesimal amplitude δ. All the other equilibrium positions the buckled state involve finite amplitudes δ. Since the classic theory of buckling only considers infinitesimal displacement, the point A can be said to represent the classic buckling load.

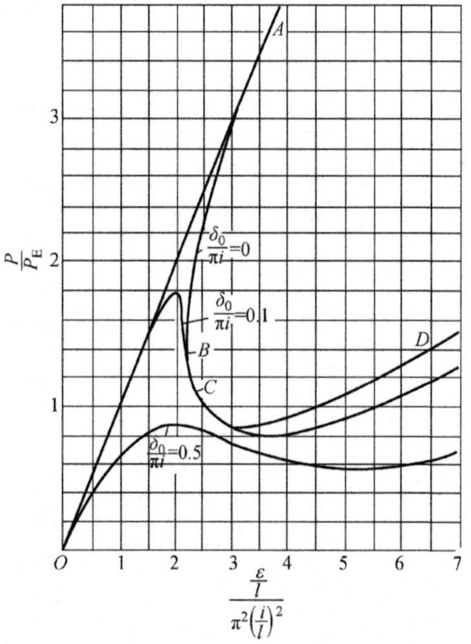

Fig. 8 Relation between the load P and the end shortening ε for columns with a nonlinear lateral support and various amounts of initial deflection δ_0

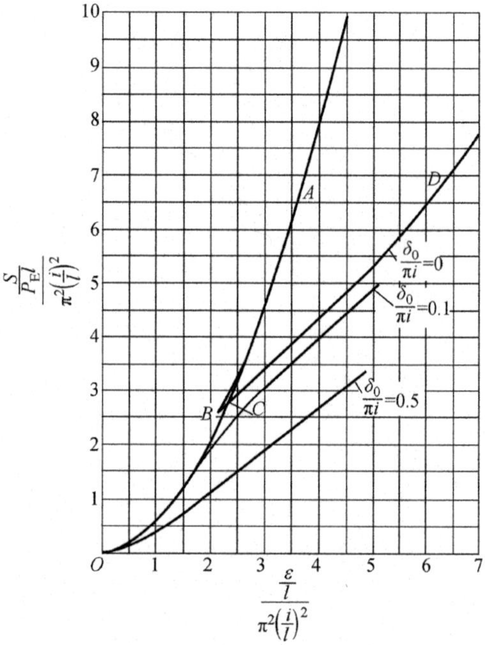

Fig. 9 Relation between the strain energy S and the end shortening ε for columns with a nonlinear lateral support and various amounts of initial deflection δ_0

In the strain energy diagram, Fig. 9, the corresponding points are denoted by the same letters as in Fig. 8. Thus, the straight equilibrium states of the column are represented by the parabola OA, and the buckled equilibrium states by the curve $ABCD$. It is interesting to notice that the curve AB closely follows the parabola OA — i. e., the strain energy is only slightly higher than that of straight column with the same end shortening. The curve turns sharply at B and meets the parabola at C. The branch CD has less strain energy for the same end shortening than either the straight column or the buckled state represented by AB. Therefore, according to the criterion of lowest strain energy, the actual behavior of the column in a testing machine when the end plates are gradually brought together will be, first, the straight unbuckled shape up to C. At C there will be a sudden change in the shape of the column to that which involves a finite amplitude δ and a jump in the magnitude of the load from $P/P_E = 2.40$ to $P/P_E = 1.09$. Since, according to Eq. (5), the slope of the S vs. ε curve is a measure of P/P_E, this jump in the magnitude of P/P_E is indicated in Fig. 8 by a sudden change in the slope of the two branches of the curve. From then on the column will behave according to the curve CD with increasing end shortening. Therefore, under ordinary conditions the classic buckling state does not appear in the buckling process at all, and the recorded failing load of the column will be

$P = 2.40\ P_E$ instead of the classic value of $P = 3.61\ P_E$ corresponding to the point A.

The initial deflection δ_0 of the column tends first to decrease the failing load. Finally, the sudden jump in the value of P disappears completely when, at any given value of end shortening ε, there is only one value of the load P. For this particular problem the small initial deflection of $\delta_0 = 0.1\ \pi i$ is already big enough to eliminate the jump in the value of P (Figs. 8 and 9).

To investigate the behavior of the columns when loaded by weights, the total potential Φ has to be plotted against the load P. This is done in Fig. 10, using the same characteristics of the lateral support as given by Eq. (8). For the straight column, $\delta_0/(\pi i) = 0$, the unbuckled equilibrium states are represented by the parabola OC_0A_0. The classic buckling load is indicated by the point A_0. The curve $A_0B_0C_0$ represents the buckled equilibrium states. Using the buckling criterion developed in the previous sections, the column will maintain its unbuckled straight shape up to the point C_0 with $P/P_E = 1.196$, while the loading is gradually increased. At this value of P/P_E the column will suddenly jump from the unbuckled equilibrium state to the buckled equilibrium state. At the same time the value of the end shortening will change from $\varepsilon/l = 1.19\ \pi^2 (i/l)^2$ to $\varepsilon/l = 5.60\ \pi^2 (i/l)^2$. According to Eq. (7) this jump in the magnitude of the end shortening is signified in Fig. 10 by the sudden change in the slope of the two branches of the curve. Here, again, the failing load will be $P = 1.196\ P_E$ instead of the classic buckling load of $P = 3.61\ P_E$.

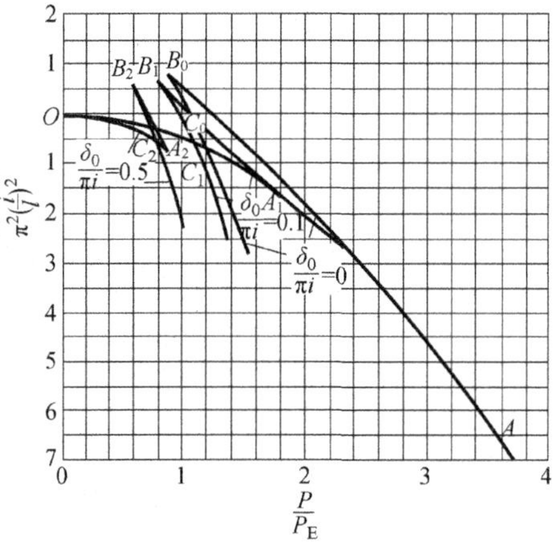

Fig. 10 Relation between the total potential Φ and the load P for columns with a nonlinear lateral support and various amounts of initial deflection δ_0

The initial deflection δ_0 of the column again tends first to decrease the failing load. For $\delta_0/\pi i = 0.1$ the failing load is $P = 1.088\ P_E$; for $\delta_0/(\pi i) = 0.5$ the failing load is $P = 0.778 P_E$. Finally, when at any given value of P there is only one equilibrium value of ε, the jump in the equilibrium states of the column during the loading process will disappear completely.

Summarizing the results, the failing load of a column with a nonlinear elastic support as given by Eq. (8) is given in Table 1. It is seen that under ordinary conditions the so-called classic buckling load really does not necessarily indicate the load carrying ability of the column.

Table 1

$\delta_0/(\pi i)$	P/P_E of Classic Theory	P/P_E of Testing machine	New Theory Dead weight
0	3.61	2.40	1.196
0.1	1.79	1.79	1.088
0.5	0.88	0.88	0.778

Thin Uniform Cylindric Shell Under Axial Compression

With the buckling criterions so clearly demonstrated by the failure of a column with nonlinear elastic support, the theory can now be applied with confidence to thin cylindric shells under axial compression. Using the results already obtained by the author in a previous paper[3], the strain energy S of the shell and the total potential Φ can be calculated. The results are shown in Figs. 11 to 14. In these figures R is the radius, t is the thickness and ε is the unit end shortening of the shell. Furthermore, s denotes the strain energy per unit area of the shell surface, and φ is the total potential per unit area of the shell surface. The number of waves n in the circumference of the shell is determined by the parameter η which is equal to $n^2(t/R)$. The ratio of the wave length in the circumferential direction to that in the axial direction is defined as μ. Thus, for $\mu = 1$ the waves are square; for $\mu = 0.5$, the waves are rectangular with long

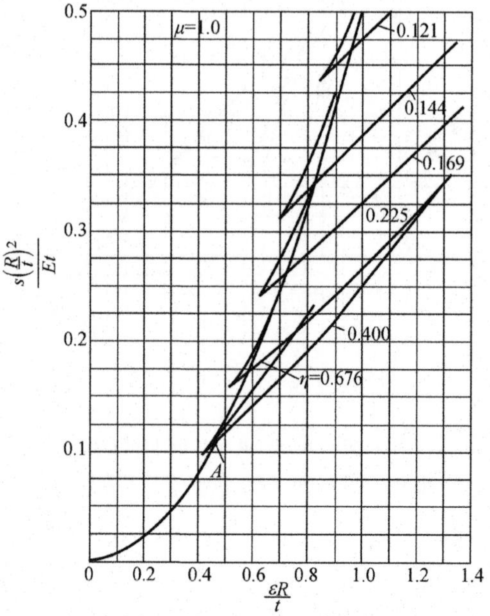

Fig. 11 Relation between the unit strain energy s and the unit end shortening ε for cylindric shells under axial compression with the aspect ratio μ of the waves equal to 1.0

side parallel to the axis of the cylindric shell. The value of Poisson's ratio is taken to be 0.3.

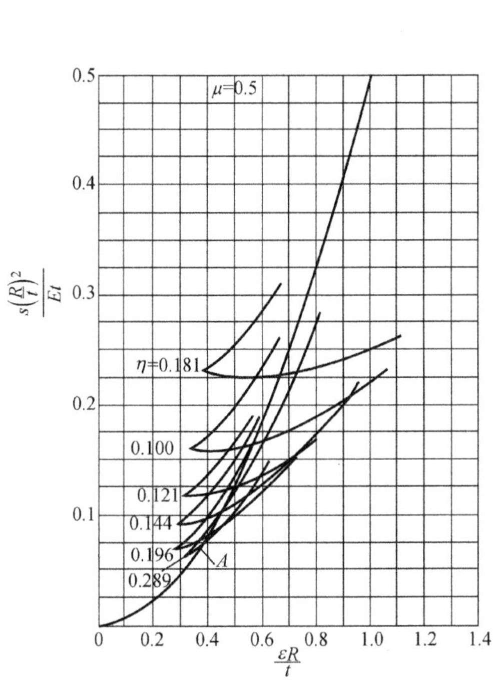

Fig. 12 Relation between the unit strain energy s and the unit end shortening ε for cylindric shells under axial compression with the aspect ratio μ of the waves equal to 0.5

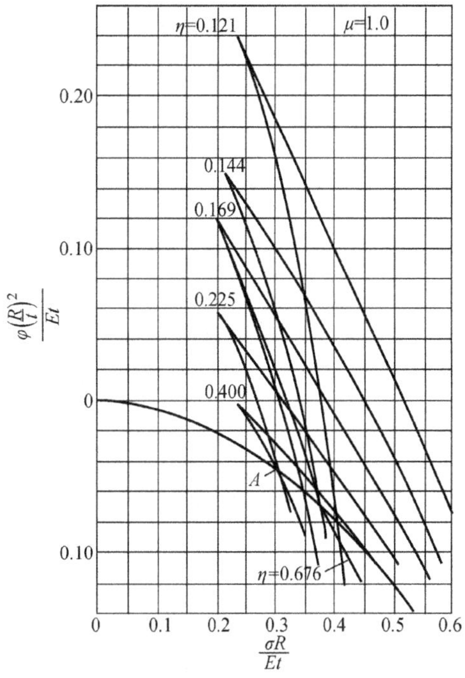

Fig. 13 Relation between the unit total potential φ and the compression stress σ for cylindric shells under axial compression with the aspect ratio μ of the waves equal to 1.0

To investigate the behavior of the shell under a rigid testing machine, the strain energy parameter $s(R/t)^2/(Et)$ is plotted against the end-shortening parameter $\varepsilon R/t$ as indicated in Figs. 11 and 12. According to the criterion developed in the previous sections, buckling will probably occur when the strain energy curve for the buckled equilibrium positions first crosses the curve for the unbuckled equilibrium positions. Thus, by referring to Fig. 11 for $\mu = 1.0$, the lowest probable buckling load is indicated by the point A. Then the corresponding buckling load can be calculated as $\sigma R/(Et) = 0.460$ where σ is the average axial compression stress. The value of η is 0.400. Thus if $R/t = 1\,000$, $n = 20$. At buckling, the stress will jump from $\sigma = 0.460\,Et/R$ to $\sigma = 0.239\,Et/R$. With $\mu = 0.5$, Fig. 12 shows that the lowest probable buckling stress is around

$$\sigma = 0.370\,Et/R \tag{9}$$

The corresponding value of η is 0.289. Thus if $R/t = 1\,000$, $n = 17$. At buckling, the stress will jump from the value given by Eq. (9) to $\sigma = 0.200\,0\,Et/R$.

The failing load of the cylindric shell under dead weight loading can be determined from

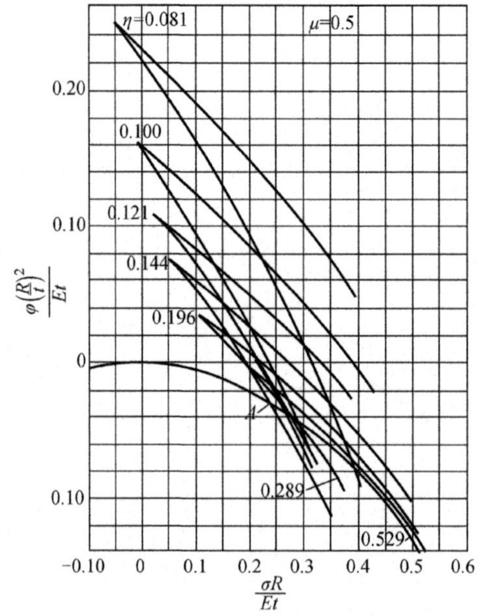

Fig. 14 Relation between the unit total potential φ and the compression stress σ for cylindric shells under axial compression with the aspect ratio μ of the waves equal to 0.3

Figs. 13 and 14, where the total potential parameter $\varphi(R/t)^2/(Et)$ is plotted against the stress parameter $\sigma R/(Et)$ for $\mu = 1.0$ and $\mu = 0.5$, respectively. Using the criterion that buckling will probably occur for dead weight loading when the total potential curve for the buckled equilibrium state crosses the curve for the unbuckled equilibrium state. For $\mu = 1.0$, this happens at point A (Fig. 13), which corresponds to $\sigma = 0.298\, Et/R$. Again, since there is a sudden change in slope of the two branches of the curve which the shell will follow, the buckling phenomena is accompanied by a sudden increase in end shortening. By Fig. 14, for $\mu = 0.5$, the lowest probable buckling stress is around

$$\sigma = 0.238\, Et/R \tag{10}$$

Again, there will be a sudden change in the end shortening at buckling.

Both Eqs. (9) and (10) give a buckling stress much less than the classic buckling stress of $\sigma = 0.606\, Et/R$. As shown in the previous sections, the elasticity of the testing machine tends to lower the buckling stress toward that which would be obtained by dead weight loading. However, even the dead weight failing load given by Eq. (10) is higher than the average of the reported experimental data which give $\sigma = 0.15\, Et/R$[2]. One of the reasons for this failure of quantitative agreement between the theory and experiment is, of course, the approximate character of the method used in reference 3 which is the basis of the present calculation. Furthermore, the aspect ratio μ of the wave is chosen rather arbitrarily. There may well be other values of μ which will give lower values of the buckling stress. These questions can only be clarified by investigations to obtain more accurate solutions of the problem.

Spherical Shells under External Pressure

The buckling of thin uniform spherical shells under external pressure was investigated by the author in collaboration with Th. von Karman[1] and by K. O. Friedrichs[4]. In the author's treatment there are several simplifying assumptions that were shown by Friedrichs to be unnecessary. Friedrichs, however, was unable to give a definite answer to the problem. Therefore, this problem will be treated now in the light of the present theory.

It was shown by Friedrichs that if the ratio of thickness to the radius of the shell and the angular extension of the buckle is small and if the additional radial and circumferential stresses due to buckling are zero at the boundary of the buckle then the difference between the sum of the additional strain energies of extension and bending and the virtual work of the external pressure force acting on the surface of the shell can be calculated by the following integral:

$$I = \frac{1}{2}\frac{(t/R)^2}{12(1-\nu^2)}\int_0^\beta \left(\frac{d\kappa}{d\alpha}\right)^2 \alpha^3 d\alpha + \frac{1}{2}\int_0^\beta \left[\int_\alpha^\beta \left(\kappa - \frac{1}{2}\kappa^2\right)\alpha d\alpha\right]^2 \alpha d\alpha - \frac{\sigma}{2E}\int_0^\beta \kappa^2 \alpha^3 d\alpha \tag{11}$$

Here, t is the thickness, R is the radius of the spherical shell, β is the angular extension of buckle as shown in Fig. 15 and σ is the stress in the unbuckled shell under an external pressure p, and is equal to

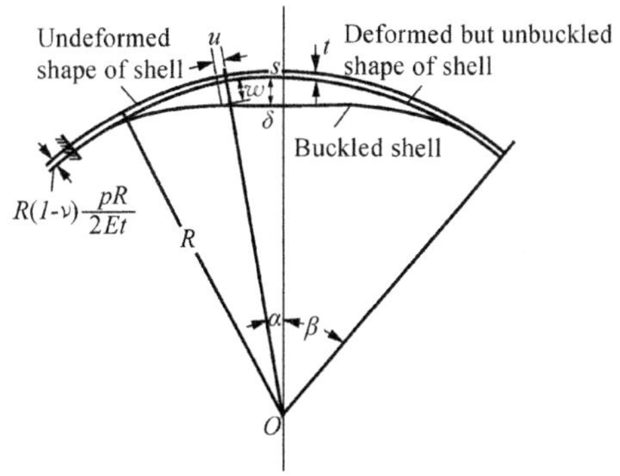

Fig. 15 Symmetric buckling of a spherical shell segment under external pressure

$$\sigma = PR/(2t) \tag{12}$$

If, due to buckling, the inward radial displacement of the shell element is w, the function $\kappa(\alpha)$ is given by

$$\kappa(\alpha) = -[1/(R\alpha)](dw/d\alpha) \tag{13}$$

To determine the equilibrium states the energy method can be used. For this purpose the radial displacement w is assumed to be

$$w/R = f[1-(\alpha/\beta)^2]^2 \tag{14}$$

where f is the ratio of the maximum radial displacement at the center of the buckle to the radius of the shell. This particular form of radial displacement has a zero slope at the boundary $\alpha = \beta$. However, the curvature is not zero at this boundary. Therefore, it requires a distribution of bending moment at the boundary to keep the buckled shell segment in equilibrium. Substituting Eq. (14) into Eq. (11), the integral I can be written as

$$I = \frac{4}{9}\frac{(t/R)^2}{(1-\nu^2)}\frac{f^2}{\beta^2} + \frac{\beta^2}{2}\left(\frac{1}{10}f^2 - \frac{2}{9}\frac{f^3}{\beta^2} + \frac{8}{63}\frac{f^4}{\beta^4}\right) - \frac{1}{3}f^2\frac{\sigma}{E} \tag{15}$$

To find the value of f at which the shell can be in equilibrium with a given value of σ, the derivative of I with respect to f has to be set equal to zero. Thus

$$\frac{\sigma}{E} = \frac{4}{3}\frac{(t/R)^2}{(1-\nu^2)}\frac{1}{\beta^2} + \beta^2\left(\frac{3}{20} - \frac{f}{2\beta^2} + \frac{8}{21}\frac{f^2}{\beta^4}\right) \tag{16}$$

Eq. (16) can be rewritten in a simpler form by using the following parameter:

$$\zeta = \beta^2 R/t, \quad \xi = fR/t \tag{17}$$

Then,

$$\frac{\sigma R}{Et} = \frac{4}{3}\frac{1}{(1-\nu^2)}\frac{1}{\zeta} + \frac{3}{20}\zeta - \frac{1}{2}\xi + \frac{8}{21}\frac{\xi^2}{\zeta} \tag{18}$$

When the amplitude of the buckle is very small, i.e., $\xi \to 0$, the minimum value of the buckling stress is given by

$$\sigma = 0.9376\, Et/R \tag{19}$$

with $\nu = 0.3$. This value is higher than the classic value

$$\sigma = 0.606\, Et/R \tag{20}$$

due to the special assumed deflection form given by Eq. (14). However, for the present calculation, the values of $\sigma R/(Et)$ for finite values of the amplitude ratio ξ are more interesting. Using Eq. (18), the value of $\sigma R/(Et)$ can be calculated as a function of ξ for various fixed values of ζ which determines the size of the buckle. The results are shown in Fig. 16.

The envelope of curves shown is Fig. 16 can be obtained by differentiating Eq. (18) with respect to ζ and then setting the resultant expression equal to zero. Thus,

$$-\frac{4}{3}\frac{1}{(1-\nu^2)}\frac{1}{\zeta^2} + \frac{3}{20} - \frac{8}{21}\frac{\xi^2}{\zeta^2} = 0 \tag{21}$$

Combining Eqs. (21) with (18), the equation for the envelope curve is obtained as

$$\left(\frac{\sigma R}{Et}\right)_{\text{Envelope}} = \frac{1}{2}\sqrt{\frac{16}{5}\frac{1}{(1-\nu^2)} + \frac{32}{35}\xi^2} - \frac{1}{2}\xi \tag{22}$$

For large values of ξ, the envelope value of $\sigma R/(Et)$ will become negative and decrease without

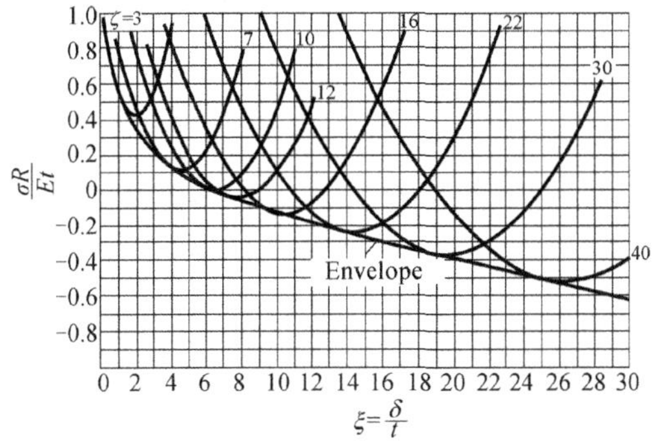

Fig. 16 Relation between the compression stress σ and the maximum deflection δ for a spherical shell segment under external pressure

limit. This is different from that which the author found in his previous investigation[1] where a minimum value of $\sigma R/(Et)$ exists. This minimum value was due to the particular displacement of the shell element used in that investigation. The negative values of $\sigma R/(Et)$ at large deflections, however, are not the failing stress of the shell, as will be shown presently.

When a segment of a spherical shell is loaded by external pressure, the experiments show that in general only one single small buckle appears, whereas the rest of the shell remains in approximately spherical shape. Therefore, in calculating the strain energy and total potential of the shell, both the buckled portion and the unbuckled portion have to be considered. In the present calculation the radial displacement of the shell element is assumed to be composed of two parts: the uniform inward displacement equal to $(1-\nu)R\sigma/E$ due to the external pressure and an additional displacement given by Eq. (14) within the boundary of the buckled portion. Furthermore, the additional extensional stresses due to buckling are assumed to be zero at the boundary of the buckle. These assumptions result in the following inaccuracies: (1) There is a discontinuity of curvature of the shell surface at the boundary of the buckle, (2) there is a discontinuity of the circumferential displacement of the shell elements at the boundary of the buckle; (3) the effective edge support at the boundary of the shell segment is neglected.

Since in all the calculations the ratio of the thickness to the radius of the shell is taken to be very small, it is probable that the errors introduced by the abovementioned inaccuracies will not be appreciable.

With these assumptions, the strain energy s per unit area of the shell surface can be calculated by using Eq. (14) and expressions similar to Eq. (11). The result is

$$\frac{sR^2}{(1-\nu)t^3E} = \left(\frac{\sigma R}{Et}\right)^2 + \lambda\frac{t}{R}\frac{\zeta\xi}{(1-\nu)}\Big[\frac{1}{3}\frac{\sigma R}{El} + \left(\frac{4}{9(1-\nu^2)}\frac{1}{\zeta^2} + \frac{1}{20} - \frac{1}{3}\frac{\sigma R}{Et}\frac{1}{\zeta}\right)\xi - \frac{\xi^2}{9\zeta} + \frac{4}{63}\frac{\xi^3}{\zeta^2}\Big] \tag{23}$$

where λ is the ratio of area of a hemisphere with radius R to the area of spherical segment. Therefore, if θ is the semiangular extension of the shell segment, λ can be written as

$$\lambda = 1/(1-\cos\theta) \tag{24}$$

If the shell is tested by means of external fluid pressure and the fluid is incompressible and enclosed in a chamber, the geometric restraint is the volume change of the shell due to its deflection. This volume change is caused by the uniform inward displacement due to the external pressure and the additional displacement given by Eq. (14) within the boundary of the buckled region. The change of volume ν per unit area of the shell surface can be shown to be

$$\frac{v}{(1-\nu)t} = \frac{\sigma R}{Et} + \lambda \frac{(t/R)\zeta\xi}{6(1-\nu)} \tag{25}$$

This type of closed chamber fluid pressure test is equivalent to the so-called rigid testing machine test previously treated. From the similarity to the previously treated cases, the probable failing load of the shell is determined by the condition that the unbuckled and buckled equilibrium states have the same strain energy and same volume change. That is, if the quantities corresponding to the starting point of buckling are denoted by the subscript 1 and those corresponding to the end point of buckling by subscript 2, the conditions are

$$\zeta_1 = \xi_1 = 0 \tag{26}$$

$$\frac{\sigma_1 R}{Et} = \frac{\sigma_2 R}{Et} + \lambda \frac{t}{R} \frac{\zeta_2 \xi_2}{6(1-\nu)} \tag{27}$$

$$\left(\frac{\sigma_1 R}{Et}\right)^2 = \left(\frac{\sigma_2 R}{Et}\right)^2 + \lambda \frac{t}{R} \frac{\zeta_2 \xi_2}{(1-\nu)} \left\{ \frac{1}{3} \frac{\sigma_2 R}{Et} + \left[\frac{4}{9(1-\nu^2)} \frac{1}{\zeta_2^2} + \frac{1}{20} - \frac{1}{3} \frac{\sigma_2 R}{Et} \frac{1}{\zeta_2} \right] \xi_2 - \frac{\xi_2^2}{9\zeta_2} + \frac{4}{63} \frac{\xi_2^3}{\zeta_2^2} \right\} \tag{28}$$

Eliminating σ_1 from Eqs. (27) and (28), the following relation is obtained.

$$\frac{4}{63} \frac{1}{\zeta_2^2} \xi_2^2 - \frac{1}{9\zeta_2} \xi_2 + \left[\frac{4}{9(1-\nu^2)} \frac{1}{\zeta_1^2} + \frac{1}{20} - \frac{1}{3} \frac{\sigma_2 R}{Et} \frac{1}{\zeta_2} - \frac{\lambda (t/R) \zeta_2}{36(1-\nu)} \right] = 0 \tag{29}$$

However, since at the end point of the buckling process the shell is at equilibrium, σ_2, ζ_2 and ξ_2 must also satisfy the equilibrium condition as given by Eq. (18). Thus,

$$\frac{8}{21} \frac{1}{\zeta_2} \xi_2^2 - \frac{1}{2} \xi_2 + \left[\frac{4}{3(1-\nu^2)} \frac{1}{\zeta_2} + \frac{3}{20} \zeta_2 - \frac{\sigma_2 R}{Et} \right] = 0 \tag{30}$$

Eqs. (29) and (30) give $\sigma_2 R/Et$ and ξ_2 as functions of ζ_2 for a given value of $\lambda t/R$. Then from Eq. (27) the expressions for $\sigma_1 R/(Et)$ and $\sigma_2 R/(Et)$ are,

$$\frac{\sigma_1 R}{Et} = \frac{4}{3(1-\nu^2)} \frac{1}{\zeta_2} + \frac{37}{480} \zeta_2 - \frac{3}{32} \frac{\lambda l/R}{(1-\nu)} \zeta_2^2 - \frac{\zeta_2}{24} \left[1 - \frac{\lambda t/R}{1-\nu} \zeta_2 \right] \sqrt{\left(\frac{7}{4}\right)^2 - \frac{7\lambda t/R}{1-\nu} \zeta_2} \tag{31}$$

$$\frac{\sigma_2 R}{Et} = \frac{4}{3(1-\nu^2)} \frac{1}{\zeta_2} + \frac{37}{480} \zeta_2 - \frac{1}{6} \frac{\lambda t/R}{(1-\nu)} \zeta_2^2 - \frac{\zeta_2}{24} \sqrt{\left(\frac{7}{4}\right)^2 - \frac{7\lambda t/R}{1-\nu} \zeta_2} \tag{32}$$

and

$$\xi_2 = \zeta_2\left[\frac{7}{16} + \frac{1}{4}\sqrt{\left(\frac{7}{4}\right)^2 - \frac{7\lambda t/R}{(1-\nu)}\zeta_2}\right] \tag{33}$$

Fig. 17 indicates the calculated results for $\lambda t/R = 0.021\ 01$. It is seen that as ζ_2 increases the value of $\sigma_1 R/(Et)$ first decreases and then increases again. The curves terminate at a value of ζ_2 which causes the expression under the radical in Eqs. (31), (32) and (33) to be equal to zero, because any further increase in ζ_2 will give imaginary values to the radical.

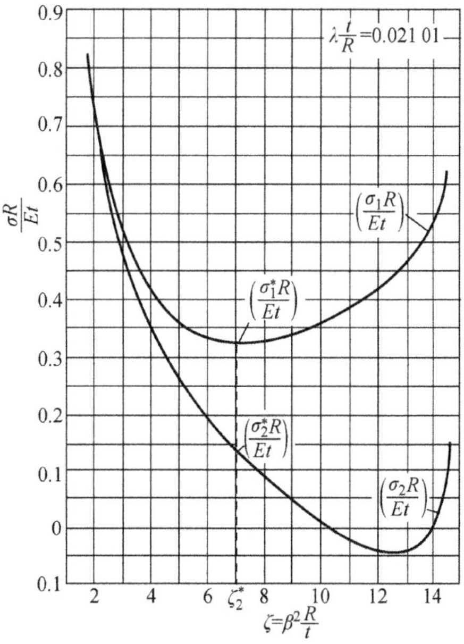

Fig. 17 Initial buckling stress σ_1 and final buckling stress σ_2 of a spherical shell segment under external pressure as a function of the angular extension β of the buckle

Since for a given shell segment only the value of $\lambda t/R$ is fixed, the size of the buckle or the parameter ζ_2 is free to vary, except that the size of the buckle cannot be larger than the size of the segment. Assuming for the moment that the last condition is not violated, then the actual probable failing load is determined by the minimum value of $\sigma_1 R/(Et)$. Denoting this minimum value by $\sigma_1^* R/(Et)$ and the corresponding $\sigma_2 R/(Et)$ by $\sigma_2^* R/(Et)$, then from Fig. 17, for $\lambda t/R = 0.021\ 01$, $\sigma_1^* R/(Et) = 0.320$ and $\sigma_2^* R/(Et) = 0.125$.

By carrying out a series of computations for different values of $\lambda t/R$, $\sigma_1^* R/(Et)$ and $\sigma_2^* R/(Et)$ can be plotted against the parameter $R/(t\lambda)$. This is shown in Fig. 18. The upper curve which represents $\sigma_1^* R/(Et)$ then gives the failing load of the shell segment under the previously stated conditions. This figure shows that the

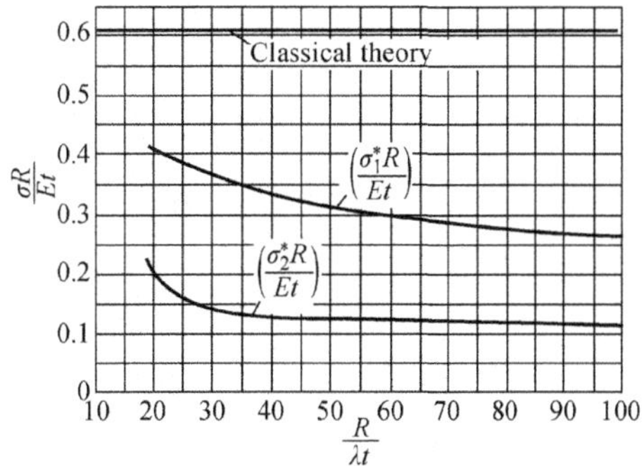

Fig. 18 Initial buckling stress σ_1^* and final buckling stress σ_2^* of spherical shell segments under external pressure

failing stress parameter $\sigma_1^* R/(Et)$ is no longer a constant as predicted by the classic theory but decreases slowly with increasing values of R/t if the solid angle of the segment — i.e., λ — is fixed. In fact, at a fixed value of λ it is found that

$$\sigma_1^* /E \sim (t/R)^{1.25} \tag{34}$$

for $20 < R/(\lambda t) < 100$. Furthermore, as shown in Fig. 19, the value of ζ_2^* at which $\sigma_1^* R/(Et)$ occurs also decreases with increasing R/t. It is found that with a fixed value of λ within the range stated ζ_2^* varies approximately as $(R/t)^{0.28}$. Then, according to Eq. (17), β_2^* varies approximately as $(R/t)^{-0.36}$. However, if l is the diameter or wave length of the buckle, $l/(2R) = \beta_2^*$. Then

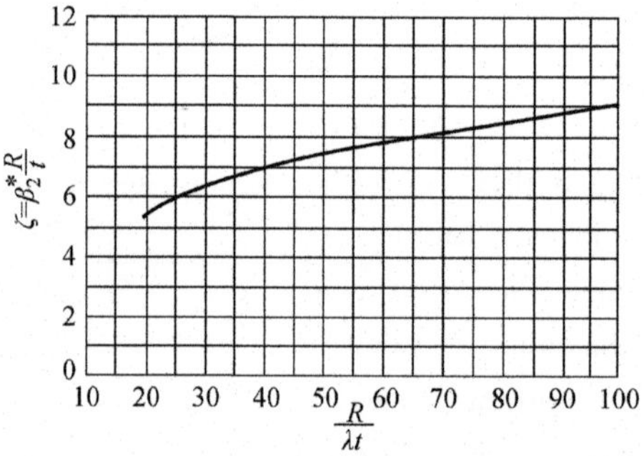

Fig. 19 Angular extension β_2^* of the buckle for spherical shell segments under external pressure

$$l/t \sim (t/R)^{-0.64} \tag{35}$$

The exponents of t/R for a buckling stress σ/E and a wave length l/R given by the classic theory are, of course, 1 and $-(1/2)$, respectively.

There is no experiment known to the author which is performed under the geometric restraint that the enclosed volume of the shell during the buckling must be constant. However, a number of experiments have been conducted at the California Institute of Technology on clamped spherical shell segments of $\theta = 17$ deg. 45 min. subjected to water pressure.* The water is slowly admitted to a closed chamber whose top cover is the shell segment. The water is immediately turned off as soon as the shell starts to buckle, so that the volume of water in the chamber is nearly constant during the buckling process. However, the pressure of the water is measured by a mercury manometer connected to the chamber by a piece of rubber tubing which is filled with air at the beginning of the test. Both the drop in height of the mercury column and the expansion of air trapped in the rubber tubing will effectively increase the volume of the chamber. Therefore, these tests are somewhat similar to that of a column loaded in a partially elastic testing machine. The results of these tests are shown in Fig. 20 where the values of

* The experiments were carried out by Mr. H. B. Crockett under Dr. L. G. Dunn's supervision. The author is deeply grateful to both of them.

$\sigma R/(Et)$ for the beginning of buckling and the end point of buckling are plotted against (R/t). It is seen that both values tend to drop off slightly as R/t increases, which is a feature also shown by the theoretic calculation for constant volume restraint. The dotted curves represent $\sigma_1^* R/(Et)$ and $\sigma_2^* R/(Et)$ previously calculated Although the test points do follow the curves very well, the agreement can only be regarded as qualitative because of the difference in the test conditions of these two cases.

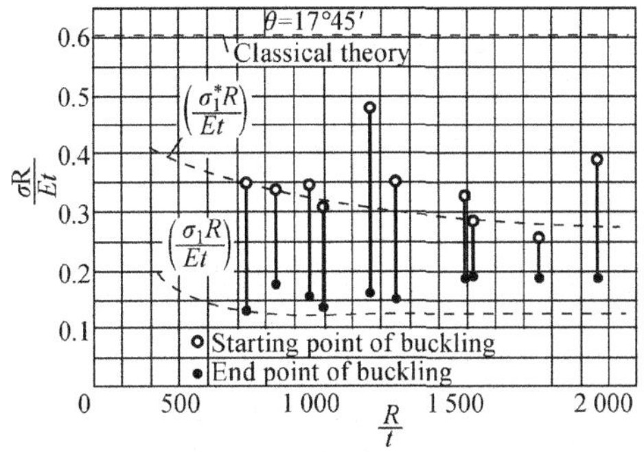

Fig. 20 Initial buckling stress and final buckling stress of spherical shell segments loaded with water pressure (experiments)

If the shell segment is loaded by means of fluid pressure and the fluid is drawn from a large reservoir with a free surface, then the small change of volume during the buckling process will not appreciably change the level of the fluid in the reservoir. Under such test conditions, the pressure acting on the shell at the starting point of buckling is practically the same as that at the end point of buckling. Let the volume change be ΔV ; then the corresponding work done by the fluid is given by $p\Delta V$. This amount of work done must be equal to the increase in strain energy of the shell. This will give a similar relation as Eq. (3). Therefore, if the total potential Φ of the system is defined as

$$\Phi = S - pV \tag{36}$$

the buckling will occur when, at equal values of $\sigma R/(Et)$, the total potential of the unbuckled equilibrium state is the same as that of the buckled equilibrium state. By means of Eqs. (36), (23) and (25), the total potential φ per unit area of the shell surface can be expressed as

$$\frac{\varphi R^2}{(1-\nu)t^3 E} = -\left(\frac{\sigma R}{Et}\right)^2 + \lambda \frac{t}{R}\frac{\zeta\xi^2}{(1-\nu)} \times \left\{\left[\frac{4}{9(1-\nu^2)}\frac{1}{\zeta^2} + \frac{1}{20} - \frac{1}{3}\frac{\sigma R}{Et}\frac{1}{\zeta}\right] - \frac{\xi}{9\zeta} + \frac{4}{63}\frac{\xi^2}{\zeta^2}\right\} \tag{37}$$

Let $\sigma_3 R/(Et)$ be the buckling stress parameter for this type of loading; then the buckling criterion stated above requires that

$$-\left(\frac{\sigma_3 R}{Et}\right)^2 = -\left(\frac{\sigma_3 R}{Et}\right)^2 + \lambda \frac{t}{R}\frac{\zeta_3\xi_3^2}{(1-\nu)} \times \left\{\left[\frac{4}{9(1-\nu^2)}\frac{1}{\zeta_3^2} + \frac{1}{20} - \frac{1}{3}\frac{\sigma_3 R}{Et}\frac{1}{\zeta_3}\right] - \frac{\xi_3}{9\zeta_3} + \frac{4}{63}\frac{\xi_3^2}{\zeta_3^2}\right\} \tag{38}$$

Thus,

$$\frac{4}{63}\frac{1}{\zeta_3^2}\xi_3^2 - \frac{1}{9}\frac{\xi_3}{\zeta_3} + \left[\frac{4}{9(1-\nu^2)}\frac{1}{\zeta_3^2} + \frac{1}{20} - \frac{1}{3}\frac{\sigma_3 R}{Et}\frac{1}{\zeta_3}\right] = 0 \tag{39}$$

On the other hand, since the buckled shell must be in equilibrium, Eq. (18) must also be satisfied. Therefore,

$$\frac{8}{21}\frac{1}{\zeta_3}\xi_3^2 - \frac{1}{2}\xi_3 + \left[\frac{4}{3(1-\nu^2)}\frac{1}{\zeta_3} + \frac{3}{20}\zeta_3 - \frac{\sigma_3 R}{Et}\right] = 0 \tag{40}$$

From Eqs. (39) and (40) $\sigma_3 R/Et$ and ξ_3 can be obtained as functions of ζ_3. Then,

$$\frac{\sigma_3 R}{Et} = \frac{1}{240}\zeta_3 + \frac{4}{3(1-\nu^2)}\frac{1}{\zeta_3} \tag{41}$$

and

$$\xi_3 = (7/8)\zeta_3 \tag{42}$$

If the shell segment is large enough, the actual buckling load will be determined by the minimum of $\sigma_3 R/(Et)$. Denoting quantities corresponding to this minimum by an asterisk, calculations show that

$$\sigma_3^* R/(Et) = 1/3\sqrt{5(1-\nu^2)} \tag{43}$$

$$\zeta_3^* = 8\sqrt{5/(1-\nu^2)} \tag{44}$$

and

$$\xi_3^* = 7\sqrt{5/(1-\nu^2)} \tag{45}$$

It should be noted that under this type of testing the buckling stress parameter $\sigma_3^* R/(Et)$ is again a constant as in the classic theory. However, the numerical value of this constant is only about one-fourth of the classic value.

The experiment on a hemispherical shell made by E. E. Sechler and W. Bollay was performed by immersing the shell in a mercury bath. Hence, the test conditions approximate the constant pressure type to which Eqs. (43), (44) and (45) applies. In Table 2 the values predicted by Eqs. (43), (44) and (45) are compared with those deduced from the test. The agreement between theory and experiment is truly remarkable.

Table 2

	Experiment	Theory
$\sigma_3 R/(Et)$	0.154	0.156 1
B_3^*	8 deg	8.3 deg
$\xi_3^* = \delta^*/t$	12.5	16.4

Concluding Remarks

In the previous sections a new principle for determining the buckling load of curved shell under ordinary test and service conditions has been developed. This principle of lowest energy level is verified by comparing the experimental data with the theoretic predictions. However, in view of the prerequisite that arbitrary disturbances of finite magnitude have to exist, the buckling load determined by this principle may be called the "lower buckling load." The classic buckling load that assumes only the existence of disturbances of infinitesimal magnitude may be called the "upper buckling load." Of course, by extreme care in avoiding all disturbances during the test, the upper buckling load can be approached. The lower buckling load, however, has to be used as the correct basis for design.

It may well be mentioned here that the possibility of the existence of the lower buckling load is connected with the elastic characteristics of the structure — i. e., the load on the structure must decrease as the deflection is increased. If this is not the case, the lower buckling load does not occur, and the buckling load of the structure can be closely calculated by the classic theory, as shown by the well-known case of a flat plate under edge compression. The author believes that when a cylindric shell is loaded by external lateral pressure, the load rises with the increase in the deflection of the shell surface after buckling. Therefore, for this case, also, the buckling load can be closely predicted by means of the classic theory of thin shells[7].

To indicate the possible future developments, the problem of buckling of a spherical shell of constant thickness under uniform dead weight pressures will be formulated more precisely. If the radial and circumferential displacements of the median surface of the shell from its spherical equilibrium positions are denoted by w and u, respectively, the strains and change in curvatures can be calculated as[4]

$$\begin{aligned}\varepsilon_r &= \frac{(du/d\alpha) - w}{R} + \frac{1}{2}\frac{[(dw/d\alpha) + u]^2}{R^2}\\ \varepsilon_c &= (u\cot\alpha - w)/R\\ \chi_r &= \frac{(d^2w/d\alpha^2) + w}{R^2} - \frac{1}{2}\frac{[(dw/d\alpha) + u]^2}{R^3}\\ \chi_c &= \frac{(dw/d\alpha)\cot\alpha + w}{R^2}\end{aligned} \tag{46}$$

α is the angle between the radius vector and the axis of symmetry. The difference in the total potential $\Phi - \Phi_0$ for the buckled and unbuckled state with same external pressure can then be expressed as

$$\begin{aligned}\frac{\Phi - \Phi_0}{\pi R^2 Et} &= \frac{1}{1-\nu^2}\int_0^\pi (\varepsilon_r^2 + \varepsilon_c^2 + 2\nu\varepsilon_r\varepsilon_c)\sin\alpha d\alpha + \\ &\frac{t^2}{12(1-\nu^2)}\int_0^\pi (\chi_c^2 + \chi_r^2 + 2\nu\chi_c\chi_r)\sin\alpha d\alpha - \\ &\frac{\sigma}{E}\int_0^\pi \Big[\Big(1 + \frac{1}{12}\frac{t^2}{R^2}\Big)\Big(\frac{dw}{d\alpha} + u\Big)^2 \Big/ R^2 - \frac{t^3}{3R^2}\frac{w}{R} + \\ &\frac{4w}{R}\Big(\frac{1}{R}\frac{du}{d\alpha} + \frac{u}{R}\cot\alpha\Big) - 4\frac{w^2}{R^2} - 8(1-\nu)\frac{\sigma}{E}\frac{w}{R}\Big]\sin\alpha d\alpha\end{aligned} \tag{47}$$

All equilibrium positions can be determined by the condition that the variations of the integral with respect to w and u must be zero. Or

$$\begin{aligned}\delta\,[(\Phi-\Phi_0)/(\pi R^2 Et)]_w &= 0\\ \delta\,[(\Phi-\Phi_0)/(\pi R^2 Et)]_u &= 0\end{aligned} \tag{48}$$

The upper buckling load is determined by the additional condition that

$$w = u \to 0 \tag{49}$$

The lower buckling load is determined by the additional condition that

$$(\Phi-\Phi_0)/(\pi R^2 Et) = 0 \tag{50}$$

The problem of determining the lower buckling load thus becomes a special case of the well-known isometric problem in variational calculus.

References

[1] von Kármán Th., Tsien Hsue-Shen. The Buckling of Spherical Shells by External Pressure. Journal of the Aeronautical Sciences, 1939, 7(2): 43 - 50.

[2] von Kármán Th., Dunn Louis G., Tsien Hsue-Shen. The Influence of Curvature on the Buckling Characteristics of Structures, Journal of the Aeronautical Sciences, 1940, 7(7): 276 - 289.

[3] von Kármán Th., Tsien Hsue-Shen. The Buckling of Thin Cylindrical Sheels Under Axial Compression. Journal of the Aeronautical Sciences, 1941, 8(8): 303 - 312.

[4] Friedrichs K O. On the Minimum Buckling Load for Spherical Shells. Theodore von Kármán Anniversary Volume, California Institute of Technology, Pasadena, 1941: 258 - 272.

[5] Cox H L. Stress Analysis of Thin Metal Construction. Journal of the Royal Aeronautical Society, 1940, 44(351): 231 - 282.

[6] Tsien Hsue-Shen. Buckling of a Column with Non-Linear Lateral Supports. Journal of the Aeronautical Sciences, 1942, 9(4): 119 - 132.

[7] A review of the theoretic investigations made by Lorentz R, Southwell R V, and Mises R v, together with references is given in: Timoshenko S, *Theory of Elastic Stability*, pp. 445 - 453, McGraw-Hill, New York, 1936. The comparison between the theory and the experiments is given by: Saunders H E, and Windenburg D F, Strength of Thin Cylindrical Shells under External Pressures, Trans. A. S. M. E., Vol. 53, pp. 207 - 218, 1931.

Heat Conduction across a Partially Insulated Wall

Hsue-shen Tsien

Recently, the author is interested in a problem of heat conduction across a partially insulated wall. The problem can be simplified to a two-dimensional one and stated in the following manner:

The upper surface of the wall with thickness t (Fig. 1) is in contact with the material to be cooled. The lower surface, however, is in contact with cooling medium only at regular intervals of length b. The distance between these cooled regions is a. The problem is to determine the heat conduction from a the upper surface to the lower surface.

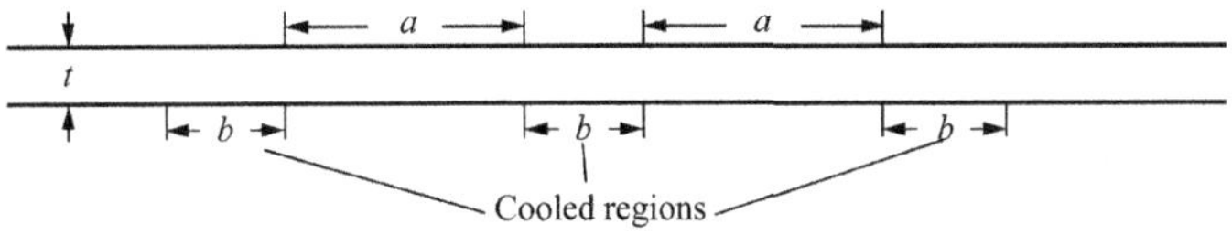

Fig. 1

The calculation can be further simplified by the circumstances in the engineering design problem which the author has to deal with. These are:

1) The temperature along the upper surface is constant because the material to be cooled is in constant agitation.

2) The lower surface at the uncooled regions is in contact with the atmosphere which is a very poor heat conductor. Therefor, it is justified to assume that the lower surface of the wall is insulated at these uncooled regions.

3) The thickness t of the wall is very small in comparison with the distances a & b. The temperature gradient across the wall will be constant except at the immediate neighborhood of the junction of the cooled and uncooled regions.

With these conditions in mind, the problem can be solved approximately in the following way:

First calculate the additional heat conducted due to the presence of an infinitely long uncooled wall attached to an infinitely long cooled wall. In other words, if the heat conducted per unit time for the problem shown in Fig. 2a is H_1, and the heat conducted per unit time for the problem shown in Fig. 2b is H_0, calculate first

---

Bulletin of the Chinese Natural Science Association, West Coast USA Chapter, vol. 1, pp. 7-11, 1942.

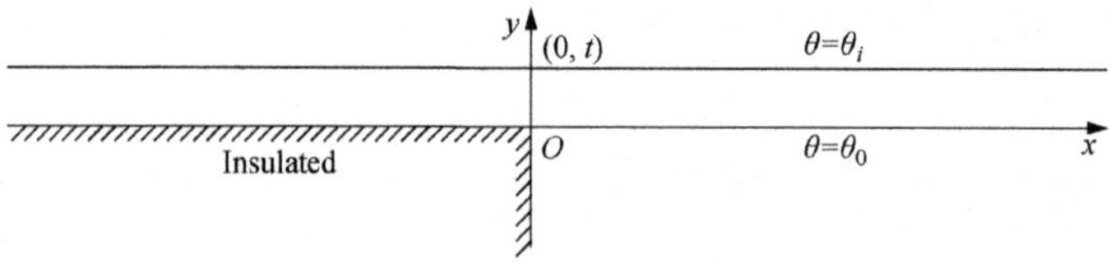

Fig. 2a

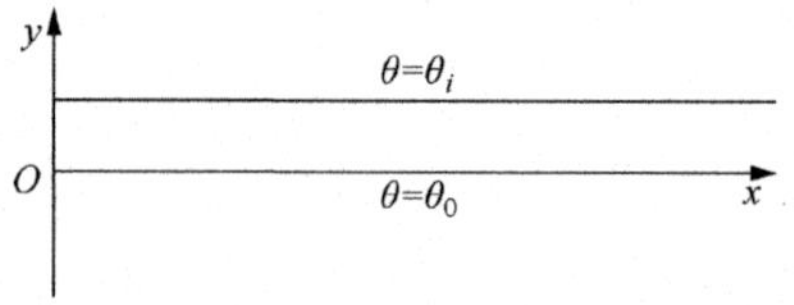

Fig. 2b

This value of ΔH can then be considered as the additional heat conducted due to the presence of one junction of a cooled and an uncooled wall. Then the total heat H conducted for one region of cooled wall per unit time per unit depth is

$$H = k\frac{\theta_i - \theta_0}{t} + 2\Delta H \tag{1}$$

To solve the problem shown in Fig. 2a, one observes first that since the wall surface for $x < 0$, $y = 0$, is insulated,

$$\frac{\partial \theta}{\partial y} = 0 \quad \text{at} \quad x < 0,\ y = 0$$

where θ is the temperature. This is evident because $k\frac{\partial \theta}{\partial y}$ is the heat conducted across a unit area parallel to x axis per unit time. Furthermore, for steady temperature field, it is well known that

$$\frac{\partial^2 \theta}{\partial x^2} + \frac{\partial^2 \theta}{\partial y^2} = 0 \tag{2}$$

Therefor, the method of conformal transformation can be used and the mathematical problem is to solve Eq. (2) together with the boundary conditions

$$\begin{aligned} &\theta = \theta_i,\ y = t \\ &\theta = \theta_0,\ y = 0 \quad x > 0 \\ &\frac{\partial \theta}{\partial y} = 0,\ y = 0 \quad x < 0 \end{aligned} \tag{3}$$

Then

$$\Delta H = k\int_0^\infty \left[\left(\frac{\partial \theta}{\partial y}\right)_{y=0} - \frac{\theta_i - \theta_0}{t}\right] \mathrm{d}x \tag{4}$$

By introducing a conjugate function φ and using the properties of the functions of a complex variable,

$$\theta + \mathrm{i}\varphi = F(x + \mathrm{i}y) \tag{5}$$

where$F$ is a function of $x + \mathrm{i}y$ It has to be determined in such a way as to satisfy the boundary conditions, Eq. (3). The lines of constant temperature θ can be sketched as shown in Fig. 3a.

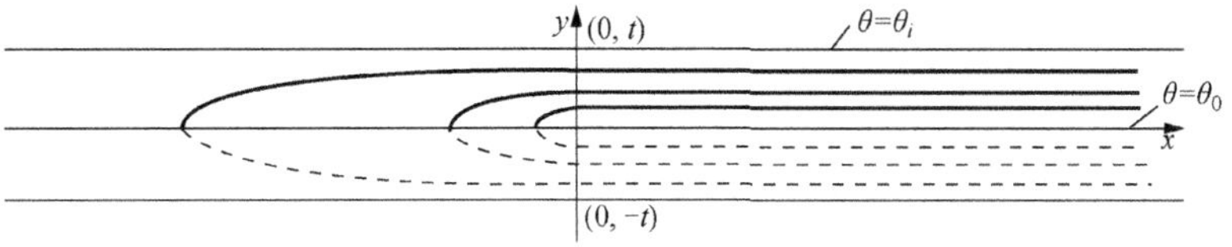

Fig. 3a

First use the transformation

$$\left.\begin{aligned} \xi &= \frac{\pi}{2}\frac{x}{t} \\ \eta &= \frac{\pi}{2}\frac{y}{t} + \frac{\pi}{2} \end{aligned}\right\} \tag{6}$$

Then the configuration of Fig. 3a in transformed into that shown in Fig. 3b.

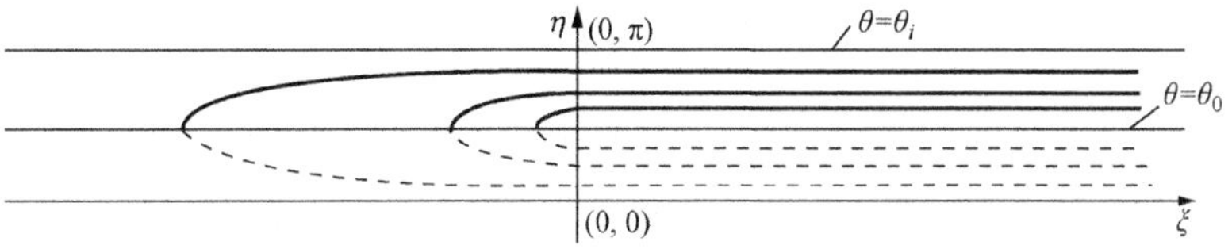

Fig. 3b

Now, by substituting

$$\xi' + \mathrm{i}\eta' = \mathrm{e}^{\xi + \mathrm{i}\eta} \tag{7}$$

the configuration of Fig. 3b is transformed into that shown in Fig. 3c.

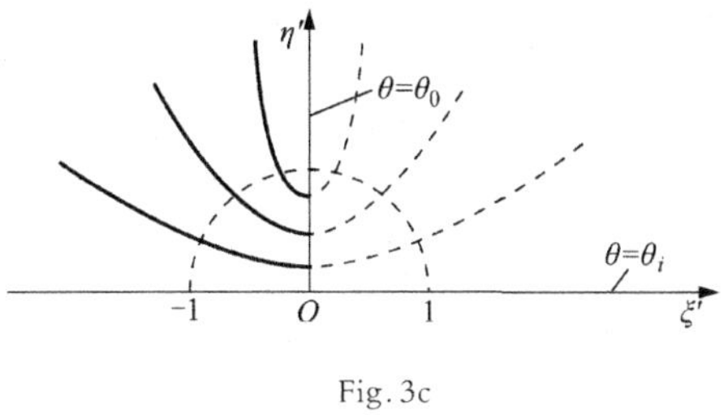

Fig. 3c

Consider the function

$$\cos(\alpha + i\beta) = -\eta' + i\xi' = \cos\alpha \cosh\beta - i\sin\alpha \sinh\beta \tag{8}$$

Thus

$$\left(\frac{\eta'}{\cos\alpha}\right)^2 - \left(\frac{\xi'}{\sin\alpha}\right)^2 = 1 \tag{9}$$

If $\alpha = 0$, the hyperbola is degenerated into the line $\xi' = 0$, $\eta' \geqslant 1$. If $\alpha = \frac{\pi}{2}$, the hyperbola is degenerated into the ξ'-axis. By putting $\alpha + i\beta = \frac{\pi}{2}\frac{(\theta + i\varphi - \theta_0)}{\theta_i - \theta_0}$, the curves of $\alpha =$ constant corresponds to the constant temperature lines required for the problem.

Since

$$\frac{\partial\alpha}{\partial y} = \frac{\partial\beta}{\partial x} = \frac{\pi/2}{\theta_i - \theta_0}\frac{\partial\theta}{\partial y} \tag{10}$$

the expression for ΔH, Eq. (4), can be written as

$$\Delta H = k \lim_{x\to\infty}\left[\frac{\theta_i - \theta_0}{\pi/2}\beta_{y=0} - \frac{\theta_i - \theta_0}{t}x\right] = k\frac{\theta_i - \theta_0}{\pi/2}\lim_{\xi\to\infty}[\beta_{y=0} - \xi] \tag{11}$$

By using Eqs. (6), (7), and (8)

$$\beta_{y=0} = \log\left[e^{\xi} + \sqrt{e^{2\xi} - 1}\right] \tag{12}$$

Then

$$\Delta H = k\frac{\theta_i - \theta_0}{\pi/2}\lim_{\xi\to\infty}\left[\log(1 + \sqrt{1 - e^{-2\xi}})\right] = k\frac{\theta_i - \theta_0}{t}\left(\frac{\log 4}{\pi}t\right) \tag{13}$$

Therefor the presence of the infinitely long wall with insulated surface is equivalent to an increase in the length of the cooled wall equal to $2\,[\log(2/\pi)] \cdot t$ or $0.441\,t$. Then Eq. (1) gives the total heat conducted for one region of the cooled wall per unit time per unit depth as

$$H = k\frac{\theta_i - \theta_0}{t}(b + 0.882\,t) \tag{14}$$

When $b \gg t$ as the case considered, the contribution of the uncooled portion of the wall to the heat conduction is very small indeed.

Pasadena, California

On the Design of the Contraction Cone for a Wind Tunnel

Hsue-shen Tsien *
(*California Institute of Technology*)

In desicning the contraction cone for a wind tunnel, the usual design condition is that the velocity at the end of the cone must befairly uniform. However, if the curvature of the wall along the flow direction is too large at certain points, local velocities at these points may exceed the uniform velocity at the end of the contraction cone. There are then regions of adverse pressure gradient and the boundary layer may separate from the wall. Furthermore, if the velocity at the end of the contraction cone is very high, say about 0.9 that of the sound, an additional factor enters — namely, the danger of compressibility shock. This danger can be avoided by keeping the velocity of flow below that of sound in the whole field of flow. In the particular case of a contraction cone, the highest velocity is reached at the wall of the cone. Therefore, if the velocity at the wall is made to increase monotonically† from the beginning of the cone to the end of cone the velocity in the cone will be always less than that of sound, provided that the velocity at the end of the contraction cone is less than the velocity of sound. The pressure along the wall will then be decreasing monotonically. Hence, the danger of boundary layer separation is also avoided.

Formulation of the Problem

To actually design such a contraction cone using the hydrodynamics of compressible fluids is rather complicated. However, it is substantiated by both theoretic and experimental investigations that the velocity of flow of a compressible fluid can be obtained by increasing the velocity of an incompressible flow with the same boundary by a certain factor. This multiplying factor is a function of the incompressible velocity itself and increases with it[1]. Hence, if a contraction cone is designed for incompressible fluid such that the velocity at the wall increases monotonically from the beginning to the end, the same cone with compressible flow will give a wall velocity that is the product of two monotonically increasing functions and

---

Received September 21, 1942.

Journal of the Aeronautical Sciences, vol. 10, pp. 68 – 70, 1943.

* Research Fellow in Aeronautics.

† A monotonic increasing function is one in which for every sequence of increasing values of the argument the corresponding values of the function always increase; similarly, a function is monotonic decreasing if its values decrease as the argument increases.

therefore is itself monotone.

To design such a contraction cone for incompressible flow is, however, much simpler. One can proceed in the following manner:

At the axis of the cone (Fig. 1), i.e., at $r = 0$, assume

$$u = f_0(x) \quad (r = 0) \tag{1}$$

where u is the velocity in the x-direction. If v is the velocity in the r-direction, by symmetry

$$v = 0 \text{ at } r = 0 \tag{2}$$

Values of u and v or the resultant velocity w can be found for other values of r by using the following fundamental equations of hydrodynamics:

$$\partial v/\partial x - \partial u/\partial r = 0 \tag{3}$$

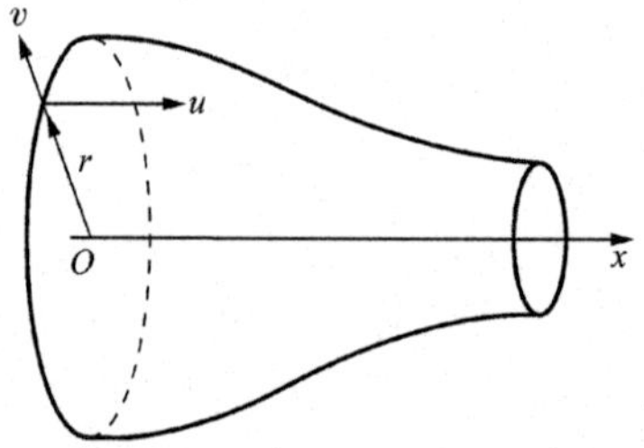

Fig. 1 Flow in a contraction cone

$$(\partial/\partial x)(ru) + (\partial/\partial r)(rv) = 0 \tag{4}$$

Eq. (3) is the condition for irrotational flow. Eq. (4) is the continuity equation.

By combining Eqs. (2) and (3), one obtains

$$\partial u/\partial r = 0 \text{ at } r = 0 \tag{5}$$

Hence, u is an even function of r. Similarly, v is shown to be an odd function of r. Therefore, it is appropriate to put

$$u = \sum_{n=0}^{\infty} r^{2n} f_{2n}(x) \tag{6}$$

$$v = \sum_{n=0}^{\infty} r^{2n+1} g_{2n+1}(x) \tag{7}$$

By substituting Eqs. (6) and (7) into Eq. (3) and equating equal powers of r,

$$g'_{2n-1}(x) = 2n f_{2n}(x) \tag{8}$$

where the prime indicates derivative with respect to x. By substituting Eqs. (6) and (7) into Eq. (4) and equating equal powers of r,

$$g_{2n-1}(x) = -\frac{1}{2n} f'_{2(n-1)}(x) \tag{9}$$

Eqs. (8) and (9) give the recurrence relation between the functions f_{2n} as follows

$$f_{2n}(x) = -\frac{1}{(2n)^2} f''_{2(n-1)}(x) \tag{10}$$

Therefore

$$f_{2n}(x) = \frac{(-1)^n}{2^{2n}(n!)^2} f_0^{(2n)}(x) \tag{11}$$

Hence, by substituting these expressions back into Eqs. (6) and (7),

$$u = \sum_{n=0}^{\infty} \frac{(-1)^n r^{2n}}{2^{2n}(n!)^2} f_0^{(2n)}(x) \tag{12}$$

$$v = \sum_{n=1}^{\infty} \frac{(-1)^n 2n r^{2n-1}}{2^{2n}(n!)^2} f_0^{(2n-1)}(x) \tag{13}$$

The resultant velocity w can then be calculated as

$$w = \sqrt{u^2 + v^2} \tag{14}$$

The stream lines are lines of constant values of the stream function ψ defined as

$$\psi(x, r) = \int_0^r r u(x, r) \mathrm{d}r \tag{15}$$

By starting from an assigned monotone velocity distribution $u = f_0(x)$ at $r = 0$, the resultant velocity w and the stream lines can be calculated by this procedure. The shape of the contraction cone is then determined by the stream line along which the velocity w still varies monotonically, but further out or for a further increase in the radius r this condition is no longer satisfied. In other words, the shape of the contraction eone is determined by the last monotone stream line.

It can be seen that the shape of the contraction cone depends upon the function $f_0(x)$ chosen at the beginning. However, with a plausible assumption of the function, it is believed that a satisfactory design for the contraction cone can be obtained.

It is interesting to notice that the solution in the form Eq. (12) can be easily identified with the well known Laplace expression for the symmetric potential function

$$u = \frac{1}{\pi} \int_0^{\pi} f_0(x + \mathrm{i} r \cos\theta) \mathrm{d}\theta \tag{16}$$

However, for practical numerical calculation the series form of Eq. (12) is more convenient.

Solution of the Problem

To carry out the computation, it is assumed that

$$f_0(x) = 0.55 + 0.90 \int_0^x \frac{\sqrt{2\pi}}{1} \mathrm{e}^{(x^2/2)} \mathrm{d}x \tag{17}$$

This distribution of velocity along the axis of the tunnel is shown in Fig. 2. The velocity at the

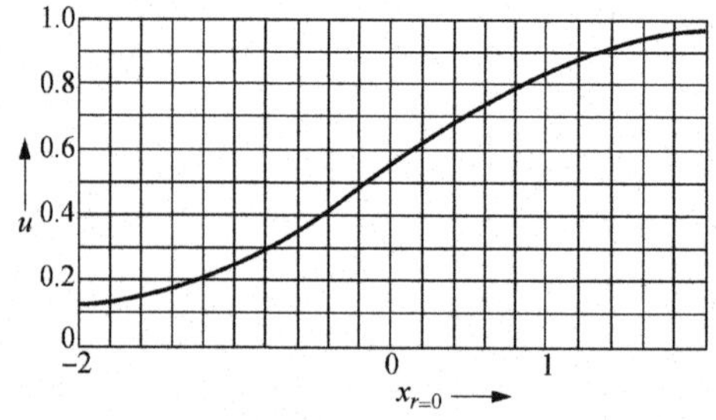

Fig. 2 Assumed velocity u along the axis of the cone as a function of the distance x on the axis

end of contraction cone is here taken as unity. It is seen that the chosen distribution of velocity is quite plausible. Furthermore the derivative of $f_0(x)$ can be easily calculated as follows: Put

$$\Phi(x) = \frac{1}{\sqrt{2\pi}} \mathrm{e}^{(x^2/2)} \tag{18}$$

then

$$f_0^{(m+1)}(x) = 0.90\, \Phi^{(m)}(x) \tag{19}$$

But

$$\Phi^{(m)}(x) = \frac{1}{\sqrt{2\pi}} \frac{\mathrm{d}^m}{\mathrm{d}x^m} (\mathrm{e}^{-(x^2/2)})$$

By using the substitutions, $x = \sqrt{2}z$, $x^2/2 = z^2$,

$$\Phi^{(m)}(x) = \frac{1}{\sqrt{2\pi}} \frac{1}{2^{m/2}} \frac{\mathrm{d}^m}{\mathrm{d}x^m} \mathrm{e}^{-z} = \frac{(-1)^m}{2^{m/2}} \Phi(x) H_m(z)$$

where $H_m(z)$ is the Hermite polynomial. By means of the recurrence relation between the Hermite polynomials,

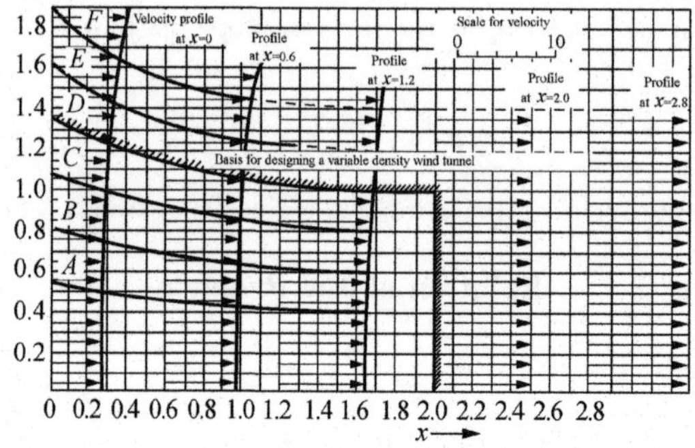

Fig. 3 Stream lines (A, B, C, D, E, F) in the meridian x_1, plane of the contraction cone together with the velocity profiles at different sections

$$H_m(z) = 2zH_{m-1}(z) - 2(m-1)H_{m-2}(z)$$

the following recurrence relation between $\Phi^{(m)}(x)$ is immediately obtained

$$\Phi^{(m)}(x) = -[x\Phi^{(m-1)}(x) + (m-1)\Phi^{(m-2)}(x)] \quad (20)$$

The values of $\int_{-1}^{1}(1/\sqrt{2\pi})e^{(-x^2/2)}dx$, $\Phi(x)$, $\Phi'(x)$, $\Phi''(x)$, $\Phi'''(x)$, $\Phi''''(x)$ are given in many mathematical tables[2]. The higher derivatives can be calculated by using the recurrence relation (20). Then Eq. (19) gives all the required functions for calculating u, and v is given by Eqs. (12) and (13). For each of the series (12) and (13), ten terms are used to give four significant figures. The stream function Ψ is obtained by numerical integration from Eq. (15).

The Results

The stream lines and velocity profiles are plotted in Fig. 3. The past of the stream lines where the velocity exceeds unity, the asymptotic velocity of the cone, is drawn as dotted lines. Therefore, theoretically the "last monotone" stream line must lie inside the line E. However, for the stream line F, the velocity at $x = 1.2$ is 1.032, while at $x = 2.0$ the value is 1.018. This slight excess of velocity will be easily removed by wall friction if this stream line is used as the wall of the contraction cone. In other words, the stream line F can be used as the shape for the contraction cone.

This result has been applied to the design of the contraction cone of a large variable density wind tunnel. The cross section of the contraction cone is not circular but partly circular and partly octagonal. Since this modification must tend to increase the local velocity at some points of the wall, it is thus thought safer to use the stream line D instead of stream line F as the basis for designing the contraction cone. Furthermore, the velocity change from $x = 2.0$ to $x = 2.8$ is negligible. Therefore, the final recommended shape of the contraction cone for this particular wind tunnel is based upon a boundary as marked in Fig. 3.

References

[1] von Kármán Th. Compressibility Effects in Aerodynamics. Journal of Aeronautical Sciences, 1941, 8: 337 - 356.

[2] Burington R S. Handbook of Mathematical Tables and Integrals. Handbook Publishers, Inc., Sancusky, O., 1940: 257 - 259.

Symmetrical Joukowsky Airfoils in Shear Flow

By Hsue-shen Tsien
(*California Institute of Technology*)

1. Problem. The usual two-dimensional theory of airfoils assumes a uniform velocity for points far from the airfoil. There are many applications where this condition is not satisfied. For example near the ground there is a large vertical velocity gradient, and therefore a first approximation to the problem can be obtained by assuming a linear velocity distribution. This has, in fact, been done by H. v. Sanden[1] in connection with O. Lilienthal's experiments in natural wind. However, v. Sanden used a numerical method of integrating the differential equation and carried out the calculation only for a wedge-shaped body. Th. von Kármán* suggested to the author to take up this problem again in order to develop a more complete theory. Hence in the first part of this paper, a generalization of the well-known Blasius theorem for calculating aerodynamic forces acting on an airfoil is given. Then the result is applied to the case of symmetrical Joukowsky airfoils and the final data are given in a number of tables and graphs.

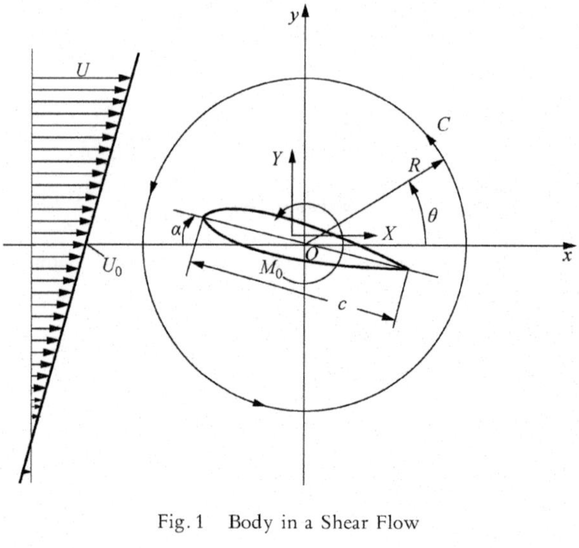

Fig. 1　Body in a Shear Flow

Received Dec. 27, 1942.

Quarterly of Applied Mathematics, vol. 1, No. 2, pp. 130－148, 1943.

* The author wishes to thank Dr. von Kármán for suggesting the problem and his kind interest during the course of the work.

2. Method of solution. By setting up the problem as shown in Fig. 1, the velocity distribution far from the airfoil is given by

$$u = U_0\left(1 + K\frac{y}{c}\right)$$
$$v = 0 \tag{1}$$

where U_0 is the undisturbed flow velocity along the x-axis and K is the non-dimensional velocity gradient of the undisturbed flow. c is some dimension of the body immersed in the stream, e.g., the chord of the airfoil. Then the vorticity at locations far from the airfoil can be calculated from Eq. (1) as

$$\frac{\partial v}{\partial x} - \frac{\partial u}{\partial y} = -U_0 K/c \tag{2}$$

which is a constant. However, in the flow of non-viscous incompressible fluid, the vorticity is associated with the fluid and maintains its strength. Consider the field of flow starting from the far left, where the vorticity is constant and equal to $-U_0K/c$. This value of vorticity is carried with the fluid over the whole field of flow. Therefore, the flow problem on hand is one with constant vorticity distribution.

To satisfy the equation of continuity,

$$\frac{\partial u}{\partial x} + \frac{\partial v}{\partial y} = 0 \tag{3}$$

the stream function ψ is introduced. It is defined by

$$u = \frac{\partial \psi}{\partial y}, \quad v = -\frac{\partial \psi}{\partial x} \tag{4}$$

u, v are the components of velocity in the x-direction and the y-direction. Due to constant vorticity distribution, the vorticity equation is

$$\frac{\partial v}{\partial x} - \frac{\partial u}{\partial y} = -U_0 K/c \tag{5}$$

By using Eq. (4), Eq. (5) can be written as

$$\frac{\partial^2 \psi}{\partial x^2} + \frac{\partial^2 \psi}{\partial y^2} = U_0 K/c \tag{6}$$

Therefore the flow problem is reduced to that of solving Eq. (6).

The stream function ψ_0 of the undisturbed flow given by Eq. (1) is

$$\psi_0 = U_0\left(y + \frac{K}{2}\frac{y^2}{c}\right) \tag{7}$$

which can be easily verified by means of Eq. (4). The mathematical problem can be considerably simplified by introducing the stream function ψ_1 due to the presence of the body defined as

$$\psi = \psi_0 + \psi_1 \tag{8}$$

By substituting this expression for ψ into Eq. (6), the equation for ψ_1 is simply

$$\frac{\partial^2 \psi_1}{\partial x^2} + \frac{\partial^2 \psi_1}{\partial y^2} = 0 \tag{9}$$

This is the Laplace equation. Therefore any solution of the Laplace equation combined with ψ_0

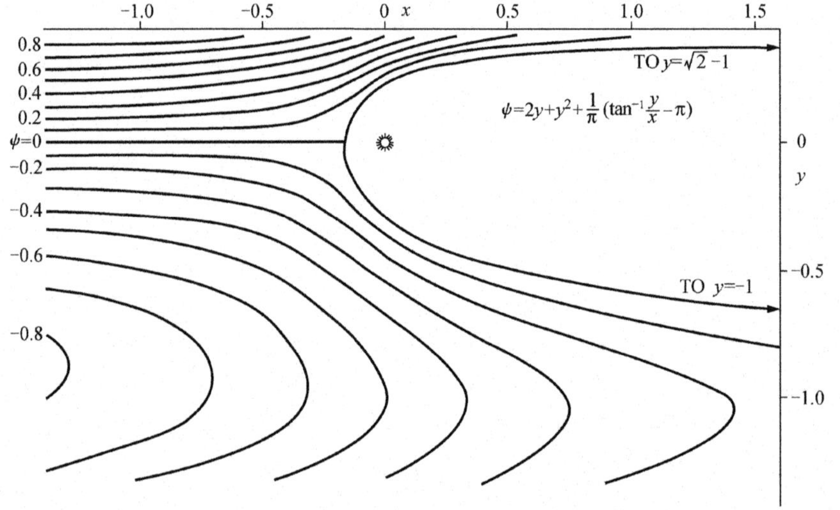

Fig. 2 Source in a Shear Flow

will satisfy Eq. (6). For example, we can combine ψ_0 with a source

$$\psi_1 = U_0 b_0 \theta \tag{10}$$

or a vortex

$$\psi_1 = U_0 a_0 \log r \tag{11}$$

as shown in Figs. 2, 3, and 4. Here

$$\theta = \tan^{-1}\frac{y}{x}, \quad r = \sqrt{x^2 + y^2} \tag{12}$$

It is interesting to notice that in the case of a source the zero stream line, which forms the walls of the "half body" when $K = 0$, is no longer symmetrical with respect to the flow direction. The velocity, and hence the pressure, along these two branches are also different. If the flow field within the zero stream line is replaced by a solid body, such lack of balance may cause the resultant pressure force to become infinitely large with the infinitely long solid boundary. This surmise will be verified later when the resultant force and moment are calculated.

If the boundary of the solid body is given, the flow problem indicated in Fig. 1 subjects ψ_1

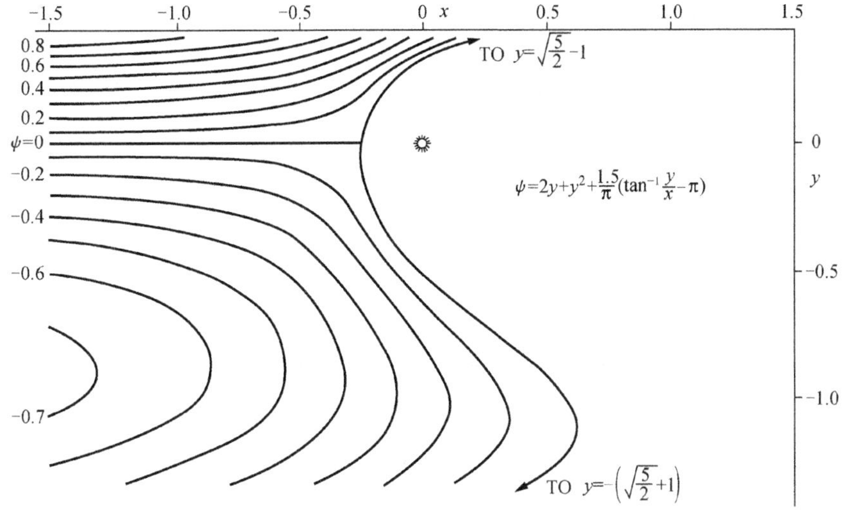

Fig. 3 Source in a Shear Flow

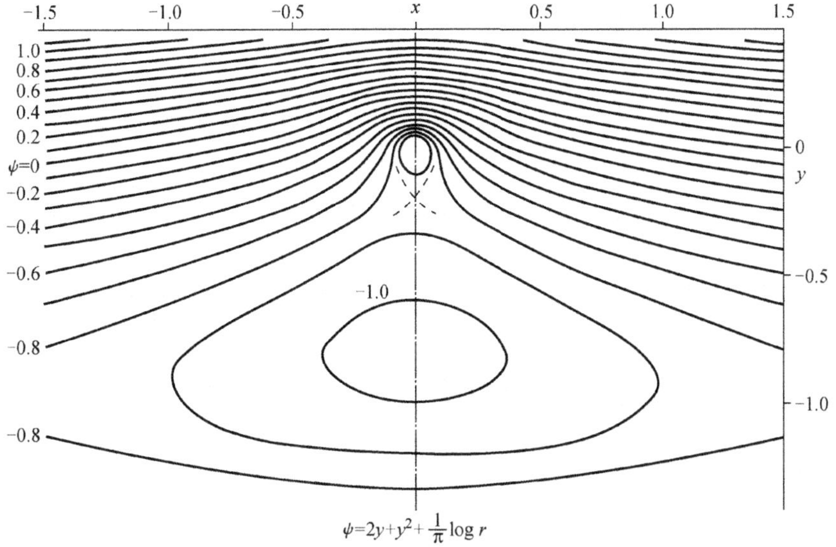

Fig. 4 Vortex in a Shear Flow

to following conditions:

(i) the disturbance velocity due to ψ_1 must vanish at points far from the body so that Eq. (1) is satisfied;

(ii) the normal component of the disturbance velocity at the surface of the body must equal the negative of that due to ψ_0 so that the resultant normal velocity is equal to zero.

3. Shear flow over a circular cylinder. To illustrate this method before solving the more complicated case of an airfoil, the flow over a circular cylinder will be investigated first. If the center of the circle is located at the origin, and the radius of the circle is $c/2$, the undisturbed velocity U due to ψ_0 at the surface of the cylinder is

$$U = U_0\left(1 + \frac{K}{2}\sin\theta\right) \tag{13}$$

where θ is given by Eq. (12). The normal component U_r of the velocity U is

$$U_r = U\cos\theta = U_0\left(\cos\theta + \frac{K}{4}\sin 2\theta\right) \tag{14}$$

Therefore, the normal component of the disturbance velocity due to ψ_1 at the surface of the circular cylinder must be equal to $-U_r$, or

$$\left(\frac{1}{r}\frac{\partial\psi_1}{\partial\theta}\right)_{r=c/2} = -U_0\left(\cos\theta + \frac{K}{4}\sin 2\theta\right) \tag{15}$$

where r is given by Eq. (12).

On the other hand, the solution of Eq. (9) that will give vanishing disturbance velocities at points far from the origin is

$$\psi_1 = U_0\left[a_0\log r + b_0\theta + \sum_{n=1}^{\infty}(a_n\cos n\theta + b_n\sin n\theta)\frac{1}{r^n}\right] \tag{16}$$

The a_n and b_n are undetermined coefficients. By substituting Eq. (16) into Eq. (15), one has immediately

$$\begin{aligned} b_1 &= -\left(\frac{c}{2}\right)^2 \\ a_2 &= \left(\frac{c}{2}\right)^3\frac{K}{8} \end{aligned} \tag{17}$$

All the other coefficients vanish. Therefore, the resultant stream function is

$$\begin{aligned} \psi = \psi_0 + \psi_1 = U_0\Bigg[&\left(r - \frac{c^2}{4r}\right)\sin\theta + \\ &\frac{K}{2}\left(\frac{r^2}{c}\sin^2\theta + \frac{c^3}{32r^2}\cos 2\theta\right)\Bigg] \end{aligned} \tag{18}$$

At the surface of the cylinder, $r = c/2$, Eq. (18) reduces to

$$\psi = U_0\frac{Kc}{16} \tag{19}$$

which is a constant, verifying the boundary condition of the problem. The stagnation point on the cylinder can be calculated by means of Eq. (18). The condition is $(\partial\psi/\partial r)_{r=c/2} = 0$. This is satisfied at the point where

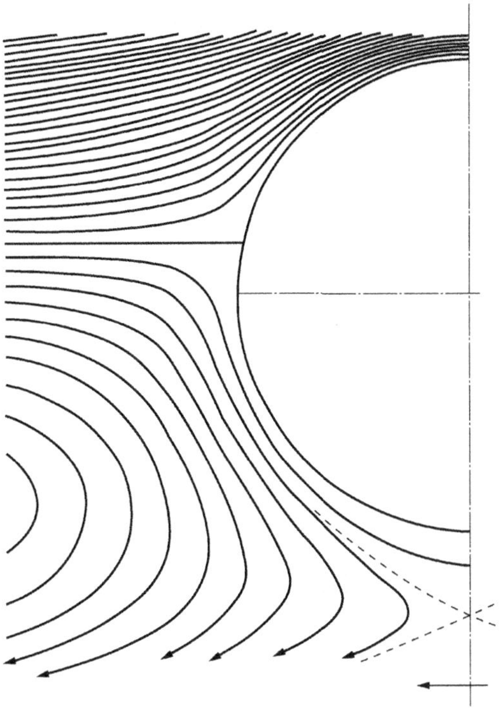

Fig. 5 Circular Cylinder in a Shear Flow with $K = 2$

$$\sin\theta = \frac{1}{K}\left[\pm\sqrt{1+\frac{K^2}{4}}-1\right]^{*} \tag{20}$$

Therefore, if $-8/3 < K < 8/3$, there will be two stagnation points on the cylinder located on that half of the surface where the velocity is higher. If K has a value beyond this range, there will be four stagnation points on the surface.

Fig. 5 shows the stream lines for the case of $K = 2$. It is seen that there is an additional stagnation point in the flow field on the negative y-axis.

4. Force and moment. In this section, the force and the moment acting on a body whose disturbance stream function ψ_1 is given by Eq. (16) will be calculated.

The pressure p in the fluid is related to the velocity components u, v and the density ρ by Euler's equations

$$\begin{aligned} \rho u\frac{\partial u}{\partial x}+\rho v\frac{\partial u}{\partial y} &= -\frac{\partial p}{\partial x} \\ \rho u\frac{\partial v}{\partial x}+\rho v\frac{\partial v}{\partial y} &= -\frac{\partial p}{\partial y} \end{aligned} \tag{21}$$

* "$\frac{1}{K}\left[\sqrt{1+\frac{K^2}{4}}\pm 1\right]$" in the original paper has been corrected as "$\frac{1}{K}\left[\pm\sqrt{1+\frac{K^2}{4}}-1\right]$". —Noted by editor.

By means of Eqs. (4), (7) and (16), the quantities on the left hand side of Eq. (21) can be calculated and then p can be obtained by integration. However in the final calculation of force and moment, the pressure will be integrated along a contour far away from the body; only terms up to $1/r^2$ in p need to be considered. Furthermore, the calculation can be simplified to a certain extent by differentiating the first of Eq. (21) with respect to x and the second with respect to y and adding the resultant. Then by using Eqs. (3) and (5),

$$\frac{1}{\rho}\nabla^2 p = 2\left[\frac{\partial^2\psi_1}{\partial x^2}\frac{\partial^2\psi_1}{\partial y^2} - \left(\frac{\partial^2\psi_1}{\partial x\partial y}\right)^2 + \frac{U_0 K}{c}\frac{\partial^2\psi_1}{\partial x^2}\right] \tag{22}$$

By substituting Eq. (16) into Eq. (22), the differential equation for p is then

$$\begin{aligned}\frac{1}{\rho}\nabla^2 p &= \frac{1}{\rho}\left[\frac{1}{r}\frac{\partial}{\partial r}\left(r\frac{\partial p}{\partial r}\right) + \frac{1}{r^2}\frac{\partial^2 p}{\partial\theta^2}\right] \\ &= 2U_0^2\Big[-\frac{1}{r^4}(a_0^2+b_0^2) + \frac{K}{c}\Big\{\frac{1}{r^2}(-a_0\cos 2\theta + b_0\sin 2\theta) + \\ &\quad \frac{1}{r^3}(2a_1\cos 3\theta + 2b_1\sin 3\theta) + \frac{1}{r^4}(6a_2\cos 4\theta + 6b_2\sin 4\theta) + \ldots\Big\}\Big]\end{aligned} \tag{23}$$

The appropriate solution for this non-homogeneous equation is evidently

$$\begin{aligned}\frac{1}{\rho}p &= 2U_0^2\Big[-\frac{1}{4r^2}(a_0^2+b_0^2) + \frac{K}{c}\Big\{-\frac{1}{4}(-a_0\cos 2\theta + b_0\sin 2\theta) - \\ &\quad \frac{1}{8r}(2a_1\cos 3\theta + 2b_1\sin 3\theta) - \frac{1}{12r^2}(6a_2\cos 4\theta + 6b_2\sin 4\theta)\Big\}\Big] + \\ &\quad A + A_0\log r + B_0\theta + \frac{1}{r}(A_1\cos\theta + B_1\sin\theta) + \frac{1}{r^2}(A_2\cos 2\theta + B_2\sin 2\theta) + \ldots\end{aligned} \tag{24}$$

where the A's and B's are undetermined constants. Either the first or the second equation (21) can be used to determine A_0, A_1, ... and B_0, B_1, B_2, ... The final result for the pressure p can be written as

$$\begin{aligned}\frac{p}{\rho} &= A + U_0^2\Big[\frac{K}{c}a_0\log r + \frac{K}{c}b_0\theta - \frac{K}{2c}(-a_0\cos 2\theta + b_0\sin 2\theta) - \\ &\quad \frac{K}{2cr}(a_1\cos 3\theta + b_1\sin 3\theta) + \frac{1}{r}\Big\{\left(\frac{3}{2}\frac{Ka_1}{c} - b_0\right)\cos\theta + \left(\frac{3}{2}\frac{Kb_1}{c} - a_0\right)\sin\theta\Big\} + \\ &\quad \frac{1}{r^2}\Big\{-\frac{a_0^2+b_0^2}{2} - \frac{K}{c}(a_2\cos 4\theta + b_2\sin 4\theta) + \\ &\quad \left(2\frac{Ka_2}{c} - b_1\right)\cos 2\theta + 2\left(\frac{Kb_2}{c} + a_1\right)\sin 2\theta\Big\} + \ldots\Big]\end{aligned} \tag{25}$$

Now by considering the pressure force and momentum of the fluid, the following relations can be obtained between the forces X, Y acting on the body and their moment M_0 about the origin[2] (Fig. 1).

$$
\begin{aligned}
X &= -\int_C p\,\mathrm{d}y - \int_C \rho u(u\mathrm{d}y - v\mathrm{d}x) \\
Y &= \int_C p\,\mathrm{d}x - \int_C \rho v(u\mathrm{d}y - v\mathrm{d}x) \\
M_0 &= \int_C p(x\mathrm{d}x + y\mathrm{d}y) - \int_C \rho(vx - uy)(u\mathrm{d}y - v\mathrm{d}x)
\end{aligned}
\tag{26}
$$

The integrals are taken along any closed curve C enclosing the body. If a circle with radius R is taken as the contour, then in the integrals of Eq. (26)

$$
\begin{aligned}
x &= R\cos\theta \\
y &= R\sin\theta
\end{aligned}
\tag{27}
$$

Therefore, by using Eq. (25),

$$
-\int_C p\,\mathrm{d}y = -R\int_0^{2\pi} p\cos\theta\mathrm{d}\theta = \pi\rho U_0^2\left[b_0 - \frac{3}{2}\frac{K}{c}a_1\right] \tag{28}
$$

Furthermore,

$$
\begin{aligned}
u\mathrm{d}y - v\mathrm{d}x = U_0\Big[&\frac{K}{2c}R^2\sin 2\theta + R\cos\theta + b_0 + \frac{1}{R}(-a_1\sin\theta + b_1\cos\theta) + \\
&\frac{1}{R^2}(-2a_2\sin 2\theta + 2b_2\cos 2\theta) + \ldots\Big]\mathrm{d}\theta
\end{aligned}
\tag{29}
$$

and

$$
u = U_0\left[\frac{K}{c}R\sin\theta + 1 + \frac{1}{R}(a_0\sin\theta + b_0\cos\theta) + \frac{1}{R^2}(-a_1\sin 2\theta + b_1\cos 2\theta) + \ldots\right] \tag{30}
$$

By combining Eqs. (29) and (30), the second term in the equation for the horizontal force X can be calculated as

$$
-\int_C \rho u(u\mathrm{d}y - v\mathrm{d}x) = -\pi\rho U_0^2\left[3b_0 - \frac{3K}{2c}a_1\right] \tag{31}
$$

Finally the horizontal force is expressed as

$$
X = -2\pi\rho U_0^2 b_0 \tag{32}
$$

Similarly, the vertical force Y or lift and the moment M_0 about the origin are obtained in the following forms:

$$
Y = 2\pi\rho U_0^2\left[a_0 + \frac{K}{c}(Rb_0 - b_1)\right] \tag{33}
$$

$$
M_0 = 2\pi\rho U_0^2\left[a_0 b_0 - a_1 + \frac{K}{c}\left(\frac{1}{2}R^2 b_0 - b_2\right)\right] \tag{34}
$$

Eqs. (32), (33) and (34) show that the force and moment on the body can be calculated in terms of the strength of the vortex, the source, the doublets and the quadruplets in the disturbance stream function ψ_1. These equations can be regarded as the generalization of the well-known Blasius formulas. They reduce to the latter formulas if $K = 0$. However, it should be noticed that if there is a source, i. e., $b_0 \neq 0$, both the lift Y and the moment M_0 grow to infinite magnitude with $R \to \infty$. This confirms the surmise stated previously. If the boundary of the body is closed, b_0 must vanish, and there can be no horizontal force or drag and the lift and moment will remain finite. v. Sanden[1] obtained a small drag force for his wedge shaped body. Evidently this is due to the unavoidable inaccuracies in his numerical method.

5. Symmetrical Joukowsky airfoils. For the flow over a symmetrical Joukowsky airfoil, it is difficult to determine the disturbance stream function ψ_1 directly. But, as seen from Eq. (9), ψ_1 satisfies Laplace's equation, and therefore conformal transformation can be used. It should be noticed, however, that ψ_0 does not satisfy Laplace's equation; and therefore it does not allow the use of this transformation in the ordinary sense.

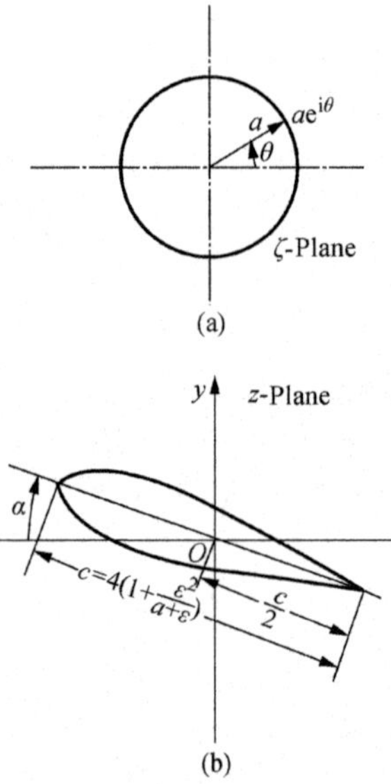

Fig. 6 Joukowsky Transformation of a Circle into an Airfoil

Consider a circle of radius a in the ζ plane (Fig. 6a). The transformation

$$z = e^{i\alpha}\left(\zeta - \varepsilon + \frac{1}{\zeta - \varepsilon} + \frac{2\varepsilon^2}{a + \varepsilon}\right) \tag{35}$$

will transform the circle into a symmetrical Joukowsky airfoil in the z plane (Fig. 6b) inclined at an angle α against the x -axis, ε is a positive quantity which increases with the thickness of the airfoil. ε is zero for a flat plate and is infinite for a circle. Furthermore

$$a = 1 + \varepsilon \tag{36}$$

The origin of the z -plane lies at the center of the airfoil (Fig. 6b). If we represent the circle in the ζ -plane by

$$\zeta = a\, e^{i\theta} \tag{37}$$

the trailing edge of the airfoil corresponds to the point where $\theta = 0$. The leading edge of the airfoil corresponds to the point where $\theta = \pi$. Therefore, the chord is

$$c = 4\left(1 + \frac{\varepsilon^2}{a + \varepsilon}\right) \tag{38}$$

With

$$\varepsilon / a = \lambda$$

the values of x and y corresponding to an arbitrary value of θ are given by Eq. (35) as

$$\begin{aligned} x &= a\left\{\cos(\theta - \alpha) - \lambda\cos\alpha + \frac{\cos(\theta + \alpha) - \lambda\cos\alpha}{a^2(1 - 2\lambda\cos\theta + \lambda^2)} + \frac{2\lambda^2\cos\alpha}{1 + \lambda}\right\} \\ y &= a\left\{\sin(\theta - \alpha) + \lambda\sin\alpha - \frac{\sin(\theta + \alpha) - \lambda\sin\theta}{a^2(1 - 2\lambda\cos\theta + \lambda^2)} - \frac{2\lambda^2\sin\alpha}{1 + \lambda}\right\} \end{aligned} \tag{39}$$

After Eq. (7), the velocity due to ψ_0 at a point in the z-plane corresponding to $\zeta = a\,\mathrm{e}^{\mathrm{i}\theta}$ is

$$U = U_0\left[1 + \frac{Ka}{c}\left\{\sin(\theta - \alpha) - \frac{\sin(\theta + \alpha) - \lambda\sin\alpha}{a^2(1 - 2\lambda\cos\theta + \lambda^2)} + \frac{\lambda\sin\alpha}{a(1 + \lambda)}\right\}\right] \tag{40}$$

This velocity is horizontal and has a tangential component in the counter-clockwise direction equal to

$$-U\left[\frac{\mathrm{d}x}{\sqrt{(\mathrm{d}x)^2 + (\mathrm{d}y)^2}}\right]_{\zeta = a\mathrm{e}^{\mathrm{i}\theta}} \tag{41}$$

and a normal component, directed along the outward normal, equal to

$$U\left[\frac{\mathrm{d}y}{\sqrt{(\mathrm{d}x)^2 + (\mathrm{d}y)^2}}\right]_{\zeta = a\mathrm{e}^{\mathrm{i}\theta}} \tag{42}$$

On the other hand if ψ_1 is the disturbance stream function, the velocity component normal to the circle $\zeta = a\,\mathrm{e}^{\mathrm{i}\theta}$ is

$$\left(\frac{1}{r}\frac{\partial\psi_1}{\partial\theta}\right)_{r=a} \tag{43}$$

Therefore, the corresponding velocity component in the z-plane, normal to the surface of the airfoil, is

$$\left(\frac{1}{r}\frac{\partial\psi_1}{\partial\theta}\right)_{r=a}\left|\frac{\mathrm{d}\zeta}{\mathrm{d}z}\right|_{\zeta = a\mathrm{e}^{\mathrm{i}\theta}} \tag{44}$$

Then the boundary condition at the airfoil surface requires that this normal component of the disturbance velocity be equal to the negative of the normal velocity component due to ψ_0. By means of Eqs. (42) and (44), this boundary condition is expressed by

$$\left(\frac{1}{r}\frac{\partial\psi_1}{\partial\theta}\right)_{r=a}\left|\frac{\mathrm{d}\zeta}{\mathrm{d}z}\right|_{\zeta = a\mathrm{e}^{\mathrm{i}\theta}} = -U\left[\frac{\mathrm{d}y}{\sqrt{(\mathrm{d}x)^2 + (\mathrm{d}y)^2}}\right]_{\zeta = a\mathrm{e}^{\mathrm{i}\theta}} \tag{45}$$

But

$$[\sqrt{(dx)^2+(dy)^2}]_{\zeta=ae^{i\theta}} = \frac{a\,d\theta}{\left|\frac{d\zeta}{dz}\right|_{\zeta=ae^{i\theta}}} \tag{46}$$

With the aid of Eqs. (39), (40), and (42), Eq. (45) can be re-written as follows

$$\frac{1}{U_0}\left(\frac{1}{r}\frac{\partial\psi_1}{\partial\theta}\right)_{r=a} = -\left[1+\frac{Ka}{c}\left\{\sin(\theta-\alpha)-\frac{\sin(\theta+\alpha)-\lambda\cos\alpha}{a^2(1-2\lambda\cos\theta+\lambda^2)}+\frac{\lambda\sin\alpha}{a(1+\lambda)}\right\}\right]\times \left[\cos(\theta-\alpha)-\frac{\cos(\theta+\alpha)+\lambda\cos(\theta-\alpha)+2\lambda\cos\alpha}{a^2(1-2\lambda\cos\theta+\lambda^2)^2}\right] \tag{47}$$

This equation alone would not determine the function ψ_1 completely but for the additional so-called Kutta-Joukowsky condition, which fixes the strength of the circulation over the airfoil.

6. Strength of circulation. The Kutta-Joukowsky condition states that the velocity at the trailing edge of the airfoil must be finite. The velocity at the trailing edge of the airfoil consists of two parts: one part is that due to ψ_0 and the other that due to ψ_1. Only tangential components need be considered because the normal components cancel each other as required by the boundary condition. The part due to ψ_1 in the counterclockwise direction is

$$-\frac{\partial\psi_1}{\partial r}\left|\frac{d\zeta}{dz}\right|_{\zeta=a} \tag{48}$$

Therefore, by means of Eq. (41), the resultant velocity at the trailing edge is

$$\left[-\frac{\partial\psi_1}{\partial r}-U\frac{dx}{\sqrt{(dx)^2+(dy)^2}}\frac{1}{\left|\frac{d\zeta}{dz}\right|_{\zeta=a}}\right]\left|\frac{d\zeta}{dz}\right|_{\zeta=a} \tag{49}$$

but $\zeta=a$ is the singular point of the transformation from ζ to z as can be easily verified. In other words, $|d\zeta/dz|$ tends to infinity at $\zeta=a$. Therefore, the resultant velocity at the trailing edge can only be finite if the quantity within the square bracket of Eq. (49) vanishes. By means of Eqs. (39), (40), (46), this condition can be written as

$$-\left(\frac{\partial\psi_1}{\partial r}\right)_{\substack{r=a\\ \theta=0}} - U_0\left[1+\frac{Ka}{c}\left\{\sin(\theta-\alpha)-\frac{\sin(\theta+\alpha)-\lambda\cos\alpha}{a^2(1-2\lambda\cos\theta+\lambda^2)}+\frac{\lambda\sin\alpha}{a(1+\lambda)}\right\}\right]\times \left[\sin(\theta-\alpha)+\frac{\sin(\theta+\alpha)-\lambda^2\sin(\theta-\alpha)-2\lambda\sin\alpha}{a^2(1-2\lambda\cos\theta+\lambda^2)}\right]_{\theta=0}=0 \tag{50}$$

The appropriate general solution ψ_1 in the ζ-plane is

$$\psi_1 = U_0\left[\alpha_0\log r+\beta_0\theta+\sum_{n=1}^{\infty}(\alpha_n\cos n\theta+\beta_n\sin n\theta)\frac{1}{r^n}\right] \tag{51}$$

Therefore,

$$\left.\begin{aligned}\frac{1}{U_0}\left(\frac{1}{r}\frac{\partial\psi_1}{\partial\theta}\right)_{r=a} &= \frac{\beta_0}{a}+\sum_{n=1}^{\infty}(-n\alpha_n\sin n\theta+n\beta_n\cos n\theta)\frac{1}{a^{n+1}}\\ -\frac{1}{U_0}\left(\frac{\partial\psi_1}{\partial r}\right)_{r=a} &= -\frac{\alpha_0}{a}+\sum_{n=1}^{\infty}(n\alpha_n\cos n\theta+n\beta_n\sin n\theta)\frac{1}{a^{n+1}}\end{aligned}\right\} \tag{52}$$

By expanding the right hand side of Eq. (47) into a trigonometric series, all the coefficients α_1, α_2, ... and β_0, β_1, ... in ψ_1 can be determined by the first of Eqs. (52). Then the second equation of Eqs. (52) together with Eq. (50) will determine the value of α_0. Actually, calculations are easier if each term on the right of Eq. (47) is taken separately. For example, the term $-U_0\cos(\theta-\alpha)$ can be expanded into $-U_0\cos\alpha\cos\theta - U_0\sin\alpha\sin\theta$. Then according to Eq. (52), the contribution to α_1 is $a^2\sin\alpha$; the contribution to. β_1 is $-a^2\cos\alpha$. Finally the contribution to $-(\partial\psi_1/\partial r)_{r=a}$ is $U_0(\sin\alpha\cos\theta - \cos\alpha\sin\theta) = -U_0\sin(\theta-\alpha)$. The other more complicated terms can be treated in a similar manner with the aid of the following expansions and their derivatives with respect to θ and λ:

$$\frac{\sin\theta}{1-2\lambda\cos\theta+\lambda^2} = \sum_{n=1}^{\infty}\lambda^{n-1}\sin n\theta$$

$$\frac{\cos\theta-\lambda}{1-2\lambda\cos\theta+\lambda^2} = \sum_{n=1}^{\infty}\lambda^{n-1}\cos n\theta$$

$$\frac{1}{1-2\lambda\cos\theta+\lambda^2} = \frac{1}{1-\lambda^2}\left(1+2\sum_{n=1}^{\infty}\lambda^n\cos n\theta\right)$$

The value of $-(\partial\psi_1/\partial r)_{r=a}$ thus found is

$$-\frac{1}{U_0}\left(\frac{\partial\psi_1}{\partial r}\right)_{r=a} = -\frac{\alpha_0}{a} - \left[1+\frac{K\lambda}{c(1+\lambda)}\right]\times$$
$$\left[\sin(\theta-\alpha) - \frac{\sin(\theta+\alpha)-\lambda^2\sin(\theta-\alpha)-2\lambda\sin\alpha}{a^2(1-2\lambda\cos\theta+\lambda^2)}\right]+$$
$$\frac{Ka}{c}\left[\frac{1}{2}\cos 2(\theta-\alpha) + \left\{\sin(\theta-\alpha) - \frac{\sin(\theta+\alpha)-\lambda\cos\alpha}{a^2(1-2\lambda\cos\theta+\lambda^2)}\right\}\times\right.$$
$$\frac{\sin(\theta+\alpha)-\lambda^2\sin(\theta-\alpha)-2\lambda\sin\alpha}{a^2(1-2\lambda\cos\theta+\lambda^2)^2} -$$
$$\frac{\cos(\theta-\alpha)[\cos(\theta+\alpha)-\lambda\cos\alpha]}{a^2(1-2\lambda\cos\theta+\lambda^2)} +$$
$$\left.\frac{1}{2a^4(1-\lambda^2)(1-2\lambda\cos\theta+\lambda^2)}\right] \tag{53}$$

The strength of circulation α_0 necessary to satisfy the Kutta-Joukowsky condition can then be calculated by means of Eq. (50). The result is

$$\alpha_0 = aU_0\left[2\sin\alpha + \frac{Ka}{c}\left\{\frac{2\lambda\sin^2\alpha}{a(1+\lambda)} - \frac{1}{2} + \cos 2\alpha - \frac{1}{a} + \frac{1}{2a(1+\lambda)}\right\}\right] \tag{54}$$

The strength of the source, β_0, is zero, as would be expected from the fact that the airfoil is represented by a closed contour.

7. Strengths of doublets and quadruplets. By collecting the contributions of the different terms of $[(1/r)\partial\psi_1/\partial\theta]_{r=a}$ in Eq. (47) to α_1, α_2, β_1 and β_2, the following values of the strength of doublets and quadruplets in the ζ-plane are obtained

$$\left.\begin{aligned}
\alpha_1 &= a^2 U_0\left[\left(1+\frac{1}{a^2}\right)\sin\alpha+\frac{Ka}{c}\left\{\frac{\lambda}{2a^2}\cos 2\alpha-\frac{\lambda}{2a^4(1-\lambda^2)}+\left(1+\frac{1}{a^2}\right)\frac{\lambda\sin^2\alpha}{a(1+\lambda)}\right\}\right]\\
\beta_1 &= a^2 U_0\left[\left(-1+\frac{1}{a^2}\right)\cos\alpha+\frac{Ka}{c}\sin 2\alpha\left\{-\frac{\lambda}{2a^2}+\left(-1+\frac{1}{a^2}\right)\frac{\lambda}{2a(1+\lambda)}\right\}\right]\\
\alpha_2 &= a^3 U_0\left[\frac{\lambda}{a^2}\sin\alpha+\frac{Ka}{c}\cos 2\alpha\left\{\frac{1}{4}-\frac{1}{2a^2}+\frac{\lambda^2}{2a^2}+\frac{1-3\lambda^2}{4a^4(1-\lambda^2)}\right\}\right]\\
\beta_2 &= a^3 U_0\left[\frac{\lambda}{a^2}\cos\alpha+\frac{Ka}{c}\sin 2\alpha\left\{\frac{1}{4}-\frac{\lambda^2}{2a^2}-\frac{1-3\lambda^2}{4a^4(1-\lambda^2)}\right\}\right]
\end{aligned}\right\}. \tag{55}$$

However, to calculate the lift and moment over the airfoil, it is not the strength of doublets and quadruplets in the ζ-plane that is needed but the strength of those in the z-plane, where the airfoil is located. From the known values of α_1, ... β_2 given by Eq. (35) and the transformation specified by Eq. (55), the desired quantities can be easily calculated in the following manner.

By introducing the conjugate function ϕ_1 of the stream function one can write

$$\phi_1 + i\psi_1 = i\alpha_0 \log\zeta + \frac{-\beta_1 + i\alpha_1}{\zeta} + \frac{-\beta_2 + i\alpha_2}{\zeta^2} + \ldots \tag{56}$$

With $\zeta = re^{i\theta}$, the real part of Eq. (56) is Eq. (51) since β_0 is found to be zero. Now Eq. (35) gives

$$\zeta = z e^{i\alpha}\left[1+\frac{\lambda}{(1+\lambda)z e^{i\alpha}}-\frac{1}{z^2 e^{2i\alpha}}+\ldots\right] \tag{57}$$

for sufficiently large values of z. Then $\phi_1 + i\psi_1$ can be expanded into a series in z by substituting Eq. (57) into Eq. (56). The result is

$$\begin{aligned}
\phi_1 + i\psi_1 = {}& i\alpha_0 \log z + \left[-\beta_1 + i\left(\alpha_1+\frac{\lambda\alpha_0}{1+\lambda}\right)\right] z^{-1} e^{-i\alpha} + \\
& \left[-\beta_2+\frac{\lambda\beta_1}{1+\lambda}+i\left(\alpha_2-\frac{\lambda\alpha_1}{1+\lambda}-\alpha_0-\frac{\lambda^2\alpha_0}{2(1+\lambda)^2}\right)\right] z^{-2} e^{-2i\alpha} + \ldots
\end{aligned} \tag{58}$$

If the strength of the circulation, the doublets, and the quadruplets in the z-plane are denoted by a_0, a_1, b_1, a_2 and b_2, respectively, the following relations are obtained by comparing Eq. (58) with Eq. (56):

$$\left.\begin{aligned}
a_0 &= \alpha_0, \quad b_0 = 0\\
a_1 &= \beta_1 \sin\alpha + \left(\alpha_1+\frac{\lambda\alpha_0}{1+\lambda}\right)\cos\alpha\\
b_1 &= \beta_1 \cos\alpha - \left(\alpha_1+\frac{\lambda\alpha_0}{1+\lambda}\right)\sin\alpha\\
a_2 &= \left(\beta_2-\frac{\lambda\beta_1}{1+\lambda}\right)\sin 2\alpha + \left(\alpha_2-\frac{\lambda\alpha_1}{1+\lambda}-\alpha_0-\frac{\lambda^2\alpha_0}{2(1+\lambda)^2}\right)\cos 2\alpha\\
b_2 &= \left(\beta_2-\frac{\lambda\beta_1}{1+\lambda}\right)\cos 2\alpha - \left(\alpha_2-\frac{\lambda\alpha_1}{1+\lambda}-\alpha_0-\frac{\lambda^2\alpha_0}{2(1+\lambda)^2}\right)\sin 2\alpha
\end{aligned}\right\} \tag{59}$$

Eqs. (54), (55), and (59) give all the necessary data to calculate the forces and moment acting on the airfoil.

8. Lift and moment coefficients. By means of Eqs. (32), (33), and (34), the drag, lift, and moment about the origin can be calculated with the values of a_0, a_1, a_2, b_0, b_1, b_2 given by Eq. (59). It is seen that the drag is also zero in the case of shear flow. The result of the computation can be conveniently represented by defining the functions l_0, l_1, l_2, l_3, l_4 and m_0, m_1, m_2, m_3, m_4 in connection with the lift coefficient C_L and moment coefficient C_{M_0} in the following manner:

$$\left.\begin{aligned} C_L &= \frac{Y}{\frac{1}{2}\rho U_0^2 c} = 2\pi\left[l_0 \sin\alpha + K(l_1 + l_2\cos 2\alpha) + K^2(l_3\sin\alpha + l_4\sin 3\alpha)\right] \\ C_{M_0} &= \frac{-M_0}{\frac{1}{2}\rho U_0^2 c^2} = \frac{\pi}{2}\left[\frac{m_0}{2}\sin 2\alpha + K(m_1\cos\alpha + m_2\cos 3\alpha) + K^2(m_3\sin\alpha + m_4\sin 4\alpha)\right] \end{aligned}\right\} \quad (60)$$

The negative sign in the moment coefficient is introduced in accordance with the usual convention of taking the stalling moment as positive. The functions l_0, l_1, l_2, l_3, l_4 and m_0, m_1, m_2, m_3, m_4 are given by the following equations:

$$\left.\begin{aligned} l_0 &= \frac{1}{h}, \quad h = \frac{1}{a} + \frac{\lambda^2}{1+\lambda} \\ l_1 &= \frac{\lambda(\lambda + 2/a)}{8h^2(1+\lambda)}, \quad l_2 = \frac{\lambda^2(2+1/a)}{8h^2(1+\lambda)} \\ l_3 &= -\frac{\lambda}{64h^3 a(1+\lambda)}\left[\frac{1}{a^2(1+\lambda)} - 2\lambda + \lambda^2\right] \\ l_4 &= \frac{\lambda}{64a^2h^3}\left[1 - \frac{\lambda}{(1+\lambda)^2}\right] \end{aligned}\right\} \quad (61)$$

$$\left.\begin{aligned} m_0 &= \frac{1}{ah^2}\left[\frac{1}{a} + \frac{\lambda}{1+\lambda}\right] \\ m_1 &= \frac{1}{8ah^3}\left[\frac{1}{a} + \frac{\lambda}{2(1+\lambda)}\left(1 + \lambda + \frac{\lambda}{a(1+\lambda)}\right)\right] \\ m_2 &= -\frac{1}{8ah^3}\left[\frac{1}{a}\left(1 - \frac{3}{2}\lambda\right) + \frac{\lambda}{2(1+\lambda)}\left(\frac{2}{a^2} - \lambda^2\frac{2+1/a}{1+\lambda}\right)\right] \\ m_3 &= -\frac{1}{64ah^4}\left[\frac{1}{1+\lambda}\left(\frac{1}{a^2} + \frac{\lambda^2}{2a^2(1+\lambda)^2}\right) - \frac{\lambda^2}{2a(1+\lambda)^2}\right] \\ m_4 &= \frac{1}{64ah^4}\left[1 + \lambda - \frac{\lambda}{a^2(1+\lambda)}\left(1 + \frac{\lambda}{a(1+\lambda)}\right) + \frac{\lambda^2(1+\lambda^2)}{2a(1+\lambda)(1-\lambda^2)} - \right. \\ &\qquad \left. \frac{1}{4a^2}\left(\frac{1}{a} + \frac{1}{1+\lambda}\right) + \frac{9\lambda^2}{4a^2(1+\lambda)}\right] \end{aligned}\right\} \quad (62)$$

Table I gives the numerical values of these functions together with the thickness ratio δ of the airfoil. It is seen that with increasing thickness the effect of the velocity gradient K becomes larger and larger. For example, if $\delta = 11.79\%$ the effect of K on C_L at small angles of attack is approximately given by the term $2\pi K(l_1 + l_2) = 2\pi K \times 0.026\,95$. In other words, it is equivalent to a shift in the angle of zero lift by 1.54 K degrees. For an airfoil of 21.50% thickness, this value is 3.21 K degrees.

Table I

ε	δ	l_0	l_1	l_2	l_3	l_4	m_0	m_1	m_2	m_3	m_4
0	0	1.000 0	0	0	0	0	1.000 0	0.125 0	−0.125 0	−0.015 6	0.007 8
0.05	0.061 8	1.047 6	0.012 2	0.000 9	−0.000 6	0.000 7	1.043 0	0.133 4	−0.126 7	−0.015 5	0.010 2
0.10	0.117 9	1.090 9	0.023 7	0.003 3	−0.000 9	0.001 4	1.073 7	0.141 3	−0.125 9	−0.015 2	0.012 8
0.15	0.168 7	1.130 4	0.034 5	0.006 9	−0.001 0	0.002 0	1.094 5	0.147 7	−0.123 1	−0.014 8	0.015 7
0.20	0.215 0	1.166 7	0.044 6	0.011 5	−0.000 9	0.002 5	1.107 3	0.153 0	−0.119 0	−0.014 3	0.018 8
0.30	0.295 8	1.230 8	0.062 8	0.022 7	−0.000 3	0.003 4	1.114 8	0.161 0	−0.108 0	−0.013 1	0.025 0
0.40	0.363 6	1.285 7	0.078 7	0.035 6	+0.000 5	0.004 0	1.105 8	0.166 0	−0.095 3	−0.011 9	0.031 0
0.50	0.421 0	1.333 3	0.092 6	0.049 4	+0.001 4	0.004 5	1.086 4	0.168 7	−0.082 3	−0.010 6	0.036 5
0.60	0.470 1	1.375 0	0.104 7	0.063 5	+0.002 3	0.004 8	1.060 8	0.169 7	−0.069 7	−0.009 5	0.041 6
0.80	0.548 9	1.444 4	0.124 8	0.091 1	+0.003 9	0.005 1	1.000 6	0.168 3	−0.047 4	−0.007 5	0.049 9
1.00	0.608 9	1.500 0	0.140 6	0.117 2	+0.005 1	0.005 1	0.937 5	0.164 1	−0.029 3	−0.005 9	0.056 0
∞	1.000 0	2.000 0	0.250 0	0.500 0	0	0	0	0	0	0	0

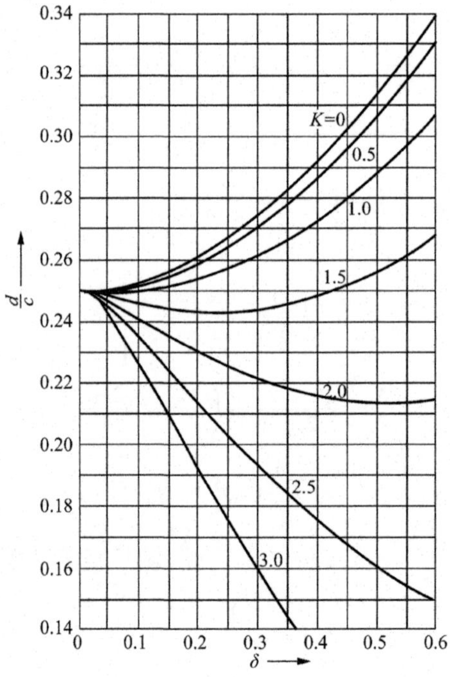

Fig. 7 Ratio of the distance d from leading edge to aerodynamic center (positive when it is behind the leading edge) to the chord c as a function of thickness ratio δ and non-dimensional velocity gradient K

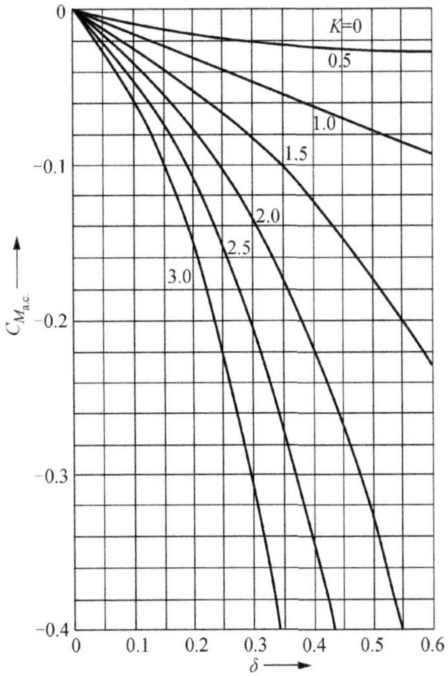

Fig. 8 Moment coefficient $C_{M_{a.c.}}$ about the aerodynamic center (positive for stalling moment) as a function of thickness ratio and non-dimensional velocity gradient K

9. Aerodynamic center. To demonstrate the effect of the velocity gradient on the moment more clearly, the aerodynamic center for small angles of attack will be calculated presently. For small values of α the expressions for C_L and C_{M_0} of Eq. (60) can be simplified to

$$\begin{aligned} C_L &= 2\pi\{[l_0 + K^2(l_3 + 3l_4)]\alpha + K(l_1 + l_2)\} \\ C_{M_0} &= \frac{\pi}{2}\{[m_0 + K^2(2m_3 + 4m_4)]\alpha + K(m_1 + m_2)\} \end{aligned} \tag{63}$$

The moment coefficient corresponds to the stalling moment about the center of the airfoil. If the moment is referred to a point on the chord at a distance d back of the leading edge, the corresponding moment coefficient C_M will be

$$\begin{aligned} C_M &= C_{M_0} - C_L\left(\frac{1}{2} - \frac{d}{c}\right) \\ &= \frac{\pi}{2}\{[m_0 + K^2(2m_3 + 4m_4)]\alpha + K(m_1 + m_2)\} - \\ &\quad \left(\frac{1}{2} - \frac{d}{c}\right)2\pi\{[l_0 + K^2(l_3 + 3l_4)]\alpha + K(l_1 + l_2)\} \end{aligned} \tag{64}$$

At the aerodynamic center, the corresponding moment coefficient should be independent of the angle of attack α. From Eq. (64), this condition gives the distance d of aerodynamic center back of the leading edge as

$$\frac{d}{c}=\frac{1}{2}-\frac{1}{4}\frac{m_0+K^2(2m_3+4m_4)}{l_0+K^2(l_3+3l_4)} \tag{65}$$

Then the moment about the aerodynamic center is given by the following coefficient:

$$C_{M_{a.c.}}=\frac{\pi K}{2}\left[(m_1+m_2)-\frac{m_0+K^2(2m_3+4m_4)}{l_0+K^2(l_3+3l_4)}(l_1+l_2)\right] \tag{66}$$

The numerical values of d/c and $C_{M_{a.c.}}$ calculated from Eqs. (65) and (66) are plotted in Figs. 7 and 8 against the thickness ratio δ for different values of K.

References

[1] v. Sanden H. Über den Auftrieb im natürlichen Winde. Zeitschrift f. Math. U. Phys., 61. 225, (1912).
[2] Glauert H. Aerofoil and airscrew theory. London: Cambridge University Press, 1930: 80.

NATIONAL ADVISORY COMMITTEE
FOR AERONAUTICS

TECHNICAL NOTE

No. 961

The "Limiting Line" in Mixed Subsonic and Supersonic Flow of Compressible Fluids

By Hsue - shen Tsien

California Institute of Technology

Washington
November 1944

The "Limiting Line" in Mixed Subsonic and Supersonic Flow of Compressible Fluids

By Hsue-shen Tsien

It is well known that the vorticity for any fluid element is constant if the fluid is non-viscous and the change of state of the fluid is isentropic. When a solid body is placed in a uniform stream, the flow far ahead of the body is irrotational. Then if the flow is further assumed to be isentropic, the vorticity will be zero over the whole field of flow. In other words, the flow is irrotational. For such flow over a solid body, it is shown by Theodorsen[1] that the solid body experiences no resistance. If the fluid has a small viscosity, its effect will be limited in the boundary layer over the solid body and the body will have a drag due to the skin friction. This type of essentially isentropic irrotational flow is generally observed for a streamlined body placed in a uniform stream, if the velocity of the stream is kept below the so-called "critical speed."

At the critical speed or rather at a certain value of the ratio of the velocity of the undisturbed flow and the corresponding velocity of sound, shock waves appear. This phenomenon is called the "compressibility burble." Along a shock wave, the change of state of the fluid is no longer isentropic, although still adiabatic. This results in an increase in entropy of the fluid and generally introduces vorticity in an originally irrotational flow. The increase in entropy of the fluid is, of course, the consequence of changing part of the mechanical energy into heat energy. In other words, the part of fluid affected by the shock wave has a reduced mechanical energy. Therefore, with the appearance of shock waves, the wake of the streamline body is very much widened, and the drag increases drastically. Furthermore, the accompanying change in the pressure distribution over the body changes the aerodynamic moment acting on it and in the case of an airfoil decreases the lift force.

All these consequences of the breakdown of isentropic irrotational flow are generally undesirable in applied aerodynamics. Its occurrence should be delayed as much as possible by modifying the shape or contour of the body. However, such endeavor will be very much facilitated if the cause or the criterion for the breakdown can be found first.

Criterion for the Breakdown of Isentropic Irrotational Flow

Taylor and Sharman[2] calculated the successive approximations to the flow around an airfoil by means of an electrolytic tank. They found that when the maximum velocity in the flow reaches the local velocity of sound, the convergence of the successive steps seems to break down. This fact led to the identification of critical speed or critical Mach number with the Mach number of the undisturbed flow for which the local velocity at some point reaches the local velocity of sound. However, there is no mathematical proof for the coincidence of the critical Mach number so defined and the breakdown of isentropic irrotational flow. Furthermore, such a definition for critical Mach number implies that a transition from a velocity less than that of sound, or subsonic velocity, to a velocity greater than that of sound, or supersonic velocity, does not occur in isentropic irrotational flow. On the other hand, Taylor[3] and others found solutions for which such a transition occurs. Furthermore, Binnie and Hooker[4] have shown that at least for the case of spiral flow the method of successive approximation is a convergent one even for supersonic velocities. With these facts in mind, it may be concluded that the identification of critical speed with local supersonic velocity cannot be correct.

Taylor's investigation on the spiral flow[3] indicates that there is a line in the flow field where the maximum velocity is reached and beyond which the flow cannot continue. Tollmien in a subsequent paper[5] called such lines limiting lines. The velocity at the limiting line is never subsonic. However, the true characteristics of such limiting lines and their significance were not investigated by Tollmien at that time. Recently Ringleb[6] obtained another particular solution of isentropic irrotational flow in which the maximum velocity reached is approximately twice the local sound velocity. For this flow also a limiting line appeared beyond which the flow cannot continue. Furthermore, he found the singular character of the limiting line, that is, infinite acceleration and infinite pressure gradient. Von Karman (reference 7, particularly pp. 351 – 356) demonstrated this fact for the general two-dimensional flow. He also suggested that the limiting line is the envelope of the Mach waves (Fig. 1) and thus can occur only in a supersonic region. He also took its appearance as the criterion for breakdown of isentropic irrotational flow. This general two-dimensional theory was established later by both Ringleb[8] and Tollmien[9]. Tollmien corrected some mistakes in Ringleb's paper and, in addition, showed that the flow definitely cannot continue beyond the limiting line. The later fact introduced a "forbidden region" in the flow bounded by the limiting line. This physical absurdity can be avoided only by relaxing the condition of irrotationality. But, as stated previously, for non-viscous fluids, the transition from a flow without vorticity to that with vorticity can be accomplished only by shock waves, which at the same time also cause an increase in the entropy.

However, before it can be concluded that the appearance of a limiting line, or the envelope of Mach waves, is the general condition for breakdown of isentropic irrotational flow, it must be proved that the singular behavior of limiting lines is general and not limited to two-dimensional flow. This is the purpose of the present paper. First the property of limiting line in axially symmetric flow will be investigated in detail. Then the general three-dimensional problem will be sketched. These investigations confirm the results of Ringleb, Von Karman, and Tollmien for these more general cases.

Therefore, by considering only the steady flow of non-viscous fluids, the criterion for

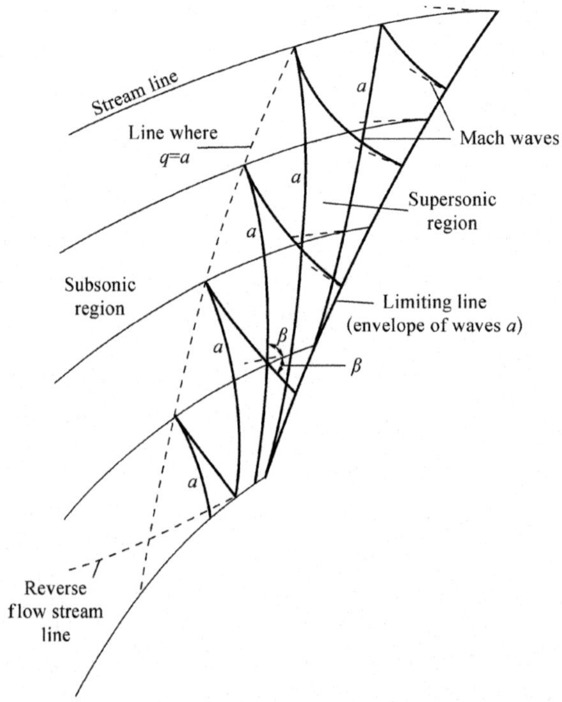

Fig. 1 Limiting line as the envelope of Mach waves

breakdown of isentropic irrotational flow is the appearance of a limiting line. For the actual motion of a solid body, however, the flow is neither steady nor non-viscous. Small disturbances always occur and almost all real fluids have appreciable viscosity. The small disturbances in the flow introduce the question of stability. In other words, the solution found for isentropic irrotational flow may be unstable even before the appearance of the limiting line, and tends to transform itself into a rotational flow involving shock waves at the slightest disturbance. If this is the case, the criterion concerns not the limiting line, but the stability limit. This problem has yet to be solved.

The effect of viscosity will be limited to the boundary layer if the pressure along the surface in the flow direction never increases too rapidly. Then outside the boundary layer the flow is isentropic and irrotational. If the gradient of pressure is too large, the boundary layer will separate from the surface. However, at low velocities such separation only widens the wake of the body and changes the pressure distribution over the body. But if the boundary layer separates at a point where the velocity outside the boundary layer is supersonic, additional effects may appear. The flow outside the boundary layer in this case can be regarded approximately as that of a solid body not of the original contour but of a new contour including the "dead water" region created by the separation. It is then immediately clear that the ideal isentropic irrotational flow around this new contour may have a limiting line. Hence, the actual flow then must involve shock waves. In other words, the separation of the boundary layer in the supersonic region may induce a shock wave and thus extend its influence far beyond the region of separation. Furthermore, the steep adverse pressure gradient across a shock wave may accentuate the separation. This interaction between the separation and the shock wave is frequently observed in experiments.

The above considerations indicate the possibility of the breakdown of isentropic irrotational flow outside the boundary layer even before the appearance of the limiting line. Therefore, the Mach number of the undisturbed flow at which the limiting line appears may be called the "upper critical Mach number." On the other hand, since shock waves can occur only in supersonic flow, the Mach number of the undisturbed flow at which the local velocity reaches the velocity of sound may be called the "lower critical Mach number." The actual critical Mach number for the appearance of shock waves and the compressibility burble must lie between these two limits. By carefully designing the contour of the body to avoid the crowding together or Mach waves to form an envelope and to eliminate adverse pressure gradients along the surface of the body, the compressibility burble can be delayed.

Axially Symmetric Flow

The solution of the exact differential equations for an axially symmetric isentropic irrotational flow was first given by Frankle[10]. The method was developed independently by Ferrari[11]. Their method applies particularly to the case of supersonic flow over a body of revolution with pointed nose. In this case, the flow at the nose can be approximated by the well-known solution for a cone. From this solution, the differential equation is solved step by step, using the net of characteristics which are real for supersonic velocities. In the following investigation, the chief concern is not the solution of the partial differential equation but rather the occurrence and the properties of the limiting line in an isentropic irrotational flow. The general plan of attack is that of Tollmien[9]. However, here the calculation is based on the Legendre transformation of velocity potential instead of the stream function.

If q is the magnitude of the velocity, a the corresponding velocity of sound assuming an isentropic process, p the pressure, and ρ the density of fluid, the Bernoulli equation gives

$$\frac{\rho}{\rho_0} = \left(1 - \frac{\gamma-1}{2}\frac{q^2}{a_0^2}\right)^{\frac{1}{\gamma-1}} = \left(1 + \frac{\gamma-1}{2}\frac{q^2}{a^2}\right)^{-\frac{1}{\gamma-1}} \tag{1}$$

$$\frac{a^2}{a_0^2} = 1 - \frac{\gamma-1}{2}\frac{q^2}{a_0^2} = \left(1 + \frac{\gamma-1}{2}\frac{q^2}{a^2}\right)^{-1} \tag{2}$$

$$\frac{p}{p_0} = \left(1 - \frac{\gamma-1}{2}\frac{q^2}{a_0^2}\right)^{\frac{\gamma}{\gamma-1}} = \left(1 + \frac{\gamma-1}{2}\frac{q^2}{a^2}\right)^{-\frac{\gamma}{\gamma-1}} \tag{3}$$

In these equations, the subscript o denotes quantities corresponding to $q = 0$, and γ is the ratio of specific heats of the fluid. Let the axis of symmetry be the x-axis, the distance normal to x- axis be denoted by y, and the velocity components along these two directions be denoted by u and v, respectively (Fig. 2). The $x - y$ plane is, therefore, a meridian plane. Then the

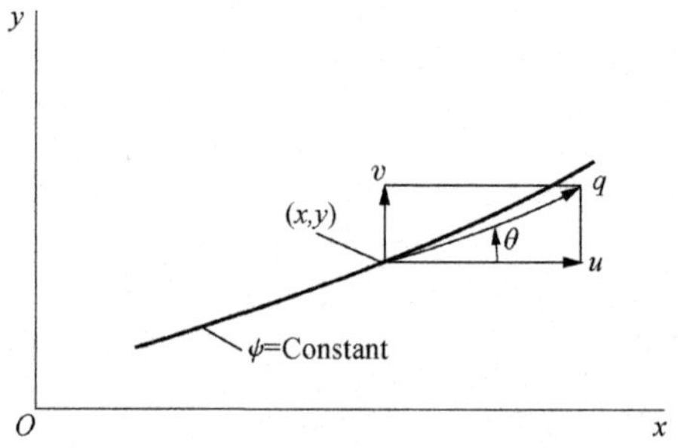

Fig. 2 Stream line and velocity components in an axially symmetric flow

kinematical relations of the flow are given by the vorticity equation

$$v_x - u_y = 0 \ ^* \tag{4}$$

and the continuity equation

$$\frac{\partial}{\partial x}\left(y\frac{\rho}{\rho_0}u\right)+\frac{\partial}{\partial y}\left(y\frac{\rho}{\rho_0}v\right)=0 \tag{5}$$

Equations (1) to (5), together with the relation $q^2 = u^2 + v^2$, specify the flow completely.

To simplify the problem, a velocity potential φ defined as follows is introduced:

$$u = \varphi_x, \quad v = \varphi_y \tag{6}$$

Then equation (4) is identically satisfied and equation (5), together with equations (1) and (2), gives the equation for φ.

$$\left(1-\frac{u^2}{a^2}\right)\varphi_{xx} - 2\frac{uv}{a^2}\varphi_{xy} + \left(1-\frac{v^2}{a^2}\right)\varphi_{yy} + \frac{v}{y} = 0 \tag{7}$$

The characteristics of this differential equation, to be called the characteristics in the physical plane, are given by $g(x,y) = 0$, where $g(x,y)$ is determined by the following equation

$$\left(1-\frac{u^2}{a^2}\right)g_x^2 - 2\frac{uv}{a^2}g_x g_y + \left(1-\frac{v^2}{a^2}\right)g_y^2 = 0 \tag{8}$$

It can be easily seen from this equation that g is real only when $q \geqslant a$. Therefore, the characteristics are real only in supersonic regions of the flow.

The meaning of characteristics in the physical plane is immediately clear if one calculates the relation between the slope of a characteristic and the slope of a stream line in the meridian or $x-y$ plane. By the definition of the function $g(x, y)$, the value of g is zero, or constant, along a characteristic. Therefore, by writing a quantity evaluated at a certain constant value of a parameter with that parameter as a subscript, the slope of the characteristic in the physical plane is

---

* Throughout this paper, partial derivatives are denoted by subscripts. Thus $v_x \equiv \frac{\partial v}{\partial x}$, $u_y \equiv \frac{\partial u}{\partial y}$

$$\left(\frac{dy}{dx}\right)_g = -\frac{g_x}{g_y} \tag{9}$$

Along a streamline, the stream function ψ defined by following equations is constant:

$$\psi_y = y\frac{\rho}{\rho_0}u, \quad \psi_x = -y\frac{\rho}{\rho_0}v \tag{10}$$

Therefore, the slope of a streamline is

$$\left(\frac{dy}{dx}\right)_\psi = \frac{v}{u} \tag{11}$$

Equations (8), (9), and (11) give

$$\left(\frac{dy}{dx}\right)_g = \frac{-\frac{uv}{a^2} \pm \sqrt{\frac{q^2}{a^2} - 1}}{1 - \frac{u^2}{a^2}} = \frac{\left(\frac{dy}{dx}\right)_\psi \pm \tan\beta}{1 \mp \left(\frac{dy}{dx}\right)_\psi \tan\beta} \tag{12}$$

where β is the Mach angle given by $\beta = \sin^{-1}\frac{a}{q}$. Therefore, equation (12) shows that the characteristics in the physical plane are inclined to the streamlines by an angle equal to the Mach angle. Such lines are the wave fronts of infinitesimal disturbances and are called Mach waves. In other words, characteristics in physical planes are the Mach waves in that plane. There are two families of Mach waves inclined symmetrically with respect to each streamline.

If to each pair of values of u and v, there is one pair of values of x, y, then x and y can be considered as functions of u, v. In other words, instead of taking x and y as independent variables u, v can be used as independent variables. The plane with u and v as coordinates is called the "hodograph plane." An equation in the hodograph plane corresponding to equation (7) can be obtained by means of Legendre's transformation. By writing

$$\chi = ux + vy - \varphi \tag{13}$$

it is seen that

$$\chi_u = x, \quad \chi_v = y \tag{14}$$

Then equation (7) can be written as

$$\left(1 - \frac{u^2}{a^2}\right)\chi_{vv} + 2\frac{uv}{a^2}\chi_{uv} + \left(1 - \frac{v^2}{a^2}\right)\chi_{uu} + \frac{v}{\chi_v}[\chi_{uu}\chi_{vv} - \chi_{uv}^2] = 0 \tag{15}$$

The characteristics of equation (15) are given by $f(u,v) = 0$ where f is the solution of the following differential equation

$$\left\{\left(1 - \frac{u^2}{a^2}\right) + \frac{v}{\chi_v}\chi_{uu}\right\}f_v^2 + 2\left(\frac{uv}{a^2} - \frac{v}{\chi_v}\chi_{uv}\right)f_u f_v + \left\{\left(1 - \frac{v^2}{a^2}\right) + \frac{v}{\chi_v}\chi_{vv}\right\}f_u^2 = 0 \tag{16}$$

Equation (16) shows that the characteristics in the hodograph plane depend upon the values of the derivatives of χ which must be obtained from equation (15). In other words, the characteristics in the hodograph plane change with the flow and are not a constant set of curves as are those in two-dimensional problems.

To obtain the relation between the characteristics in the physical plane and those in the hodograph plane, it is noticed that equation (9) can be rewritten as

$$(dy)_g : (dx)_g = -g_x : g_y \tag{17}$$

Then equation (8) is equivalent to

$$\left(1-\frac{u^2}{a^2}\right)(dy)_g^2 + 2\frac{uv}{a^2}(dy)_g(dx)_g + \left(1-\frac{v^2}{a^2}\right)(dx)_g^2 = 0 \tag{18}$$

However, in general, equation (14) gives the following relation between the differentials of x and y and those of u and v:

$$\begin{aligned} dx &= \chi_{uu}\,du + \chi_{uv}\,dv \\ dy &= \chi_{uv}\,du + \chi_{vv}\,dv \end{aligned} \tag{19}$$

By means of these relations, equation (18) can be transformed into an equation for $(du)_g$ and $(dv)_g$. This transformed equation can be simplified by using equation (15). The final relation is

$$\begin{aligned} &(\chi_{uu}\chi_{vv} - \chi_{uv}^2)\Big[\Big\{\left(1-\frac{u^2}{a^2}\right)+\frac{v}{\chi_v}\chi_{uu}\Big\}(du)_g^2 - \\ &2\left(\frac{uv}{a^2}-\frac{v}{\chi_v}\chi_{uv}\right)(du)_g(dv)_g + \Big\{\left(1-\frac{v^2}{a^2}\right)+\frac{v}{\chi_v}\chi_{vv}\Big\}(dv)_g^2\Big] = 0 \end{aligned} \tag{20}$$

Therefore, if the first factor of equation (20) is not zero, the variations $(du)_g$ and $(dv)_g$ along a characteristic in physical plane must satisfy the relation

$$\Big\{\left(1-\frac{u^2}{a^2}\right)+\frac{v}{\chi_v}\chi_{uu}\Big\}(du)_g^2 - 2\Big\{\frac{uv}{a^2}-\frac{v}{\chi_v}\chi_{uv}\Big\}(du)_g(dv)_g + \Big\{\left(1-\frac{v^2}{a^2}\right)+\frac{v}{\chi_v}\chi_{vv}\Big\}(dv)_g^2 = 0 \tag{21}$$

This is the same relation for the variations $(du)_f$ and $(dv)_f$ along a characteristic in the hodograph plane as can be seen from equation (16) and the following relation obtained from the definition of f

$$(dv)_f : (du)_f = -f_u : f_v \tag{22}$$

The transformed characteristics of the physical plane and the characteristics of the hodograph plane themselves satisfy then the same first order differential equation. Therefore, these two types of curves are the same. In other words, the characteristics of the hodograph plane are the representation of Mach waves in the $u-v$ plane.

The Limiting Line

Equation (20) shows that if

$$\chi_{uu}\chi_{vv} - \chi_{uv}^2 = 0 \tag{23}$$

then the transformed differential equation for the characteristics of the physical plane, or Mach waves, is satisfied. Therefore, if there is a line in the hodograph plane along which the values of the derivatives of χ are such that equation (23) is true, then this line when transferred to the physical plane will have its slope equal to that of one family of Mach waves. Such lines are called the limiting hodograph in $u-v$ plane and the limiting line in physical plane. Since Mach waves occur only in the supersonic regions it is then evident that the limiting line must appear in these regions. The significance of the adjective "limiting" will be made clear as other properties of such lines are investigated.

Now the question arises: Can the limiting hodograph be a characteristic in $u-v$ plane? Along a limiting hodograph, equation (23) gives

$$\left(\frac{dv}{du}\right)_l = -\frac{\chi_{uuu}\chi_{vv} - 2\chi_{uv}\chi_{uuv} + \chi_{uu}\chi_{uvv}}{\chi_{uuv}\chi_{vv} - 2\chi_{uv}\chi_{uvv} + \chi_{uu}\chi_{vvv}} \tag{24}$$

where the subscript l denotes the value along a limiting hodograph. Now the general differential equation for χ, equation (15), is true for the whole $u-v$ plane; therefore; the equation is still true by differentiating it with respect to u and v. The results can be simplified by using equation (15) itself and equation (23). Then at the limiting hodograph,

$$\left[\left(1-\frac{v^2}{a^2}\right)+\frac{v}{\chi_v}\chi_{vv}\right]\chi_{uuu} + 2\left[\frac{uv}{a^2}-\frac{v}{\chi_v}\chi_{uv}\right]\chi_{uuv} + \left[\left(1-\frac{u^2}{a^2}\right)+\frac{v}{\chi_v}\chi_{uu}\right]\chi_{uvv}$$
$$= (\gamma+1)\frac{u}{a^2}\chi_{vv} - 2\frac{v}{a^2}\chi_{uv} + (\gamma-1)\frac{u}{a^2}\chi_{uv} \tag{25a}$$

$$\left[\left(1-\frac{v^2}{a^2}\right)+\frac{v}{\chi_v}\chi_{vv}\right]\chi_{uuv} + 2\left[\frac{uv}{a^2}-\frac{v}{\chi_v}\chi_{uv}\right]\chi_{uvv} + \left[\left(1-\frac{u^2}{a^2}\right)+\frac{v}{\chi_v}\chi_{uu}\right]\chi_{vvv}$$
$$= (\gamma-1)\frac{v}{a^2}\chi_{vv} - 2\frac{u}{a^2}\chi_{uv} + (\gamma+1)\frac{v}{a^2}\chi_{uu} \tag{25b}$$

Equations (24), (25a), and (25b) are the only available equations involving no higher derivative than the third. On the other hand, the slope of a characteristic in the hodograph plane can be calculated by equation (22),

$$\left(\frac{dv}{du}\right)_f = -\frac{f_u}{f_v} \tag{26}$$

This equation together with equation (16) gives

$$\left\{\left(1-\frac{v^2}{a^2}\right)+\frac{v}{\chi_v}\chi_{vv}\right\}\left(\frac{dv}{du}\right)_f^2 - 2\left\{\frac{uv}{a^2}-\frac{v}{\chi_v}\chi_{uv}\right\}\left(\frac{dv}{du}\right)_f + \left\{\left(1-\frac{u^2}{a^2}\right)+\frac{v}{\chi_v}\chi_{uu}\right\} = 0 \tag{27}$$

Therefore, if the limiting hodograph is a characteristic, then $\left(\frac{dv}{du}\right)_l$ must satisfy equation (27). However, a simple calculation shows that it is not even possible to obtain a relation between $\left(\frac{dv}{du}\right)_l$ and other quantities not involving the third-order derivatives of χ. Hence, $\left(\frac{dv}{du}\right)_l$ does not satisfy equation (27). In other words, the limiting hodograph is not a characteristic. Transferred to the physical plane, this means that the limiting line is not a Mach wave. But as shown in previous paragraphs, the limiting line is everywhere tangent to one family of Mach waves. Consequently, the limiting line must be the envelope of a family of Mach waves. This property of the limiting line can be taken as its physical definition.

Limiting Hodograph and the Streamlines

At the limiting hodograph both equations (15) and (23) hold. By eliminating one of the second-order derivatives, say χ_{uu}, the following relation is obtained

$$(\chi_{vv})_l = \frac{-\frac{uv}{a^2} \pm \sqrt{\frac{q^2}{a^2} - 1}}{1 - \frac{u^2}{a^2}} (\chi_{uv})_l \tag{28}$$

The sign before the radical in equation (28) can be either positive or negative, but not both. This relation will be used presently to show that the streamlines and one family of characteristics are tangent in the $u-v$ plane.

From equation (10), the differential of the stream function can be calculated as

$$d\psi = -y\frac{\rho}{\rho_0}v\,dx + y\frac{\rho}{\rho_0}u\,dy \tag{29}$$

In this equation, y can be replaced by χ_v according to equation (14) and the differentials dx and dy replaced by the differential du and dv according to equation (19). Then

$$d\psi = \chi_v\frac{\rho}{\rho_0}[(-v\chi_{uu} + u\chi_{uv})du + (-v\chi_{uv} + u\chi_{vv})dv] \tag{30}$$

Along a streamline, $d\psi = 0$; therefore the slope of the streamline in the hodograph plane is given by

$$\left(\frac{dv}{du}\right)_\psi = \frac{v\chi_{uu} - u\chi_{uv}}{-v\chi_{uv} + u\chi_{vv}} \tag{31}$$

At the limiting hodograph, equation (23) holds; therefore, equation (31) together with equation (28) gives

$$\left(\frac{dv}{du}\right)_{\psi,l} = -\left(\frac{\chi_{uv}}{\chi_{vv}}\right)_l = \frac{1 - \frac{u^2}{a^2}}{\frac{uv}{a^2} \mp \sqrt{\frac{q^2}{a^2} - 1}} \tag{32}$$

where the sign before the radical can be either negative or positive corresponding to the sign in equation (28).

On the other hand, the slope of the characteristics in the hodograph plane is determined by equation (27). By solving for $\left(\frac{dv}{du}\right)_f$ and simplifying the result with the aid of equation (15).

$$\left(\frac{dv}{du}\right)_f = \frac{\frac{uv}{a^2} - \frac{v}{\chi_v}\chi_{uv} \pm \sqrt{\frac{q^2}{a^2} - 1}}{\left(1 - \frac{v^2}{a^2}\right) + \frac{v}{\chi_v}\chi_{vv}} \tag{33}$$

The sign before the radical is either positive or negative corresponding to the two families of characteristics. By using the positive sign in conjunction with the positive sign in equation (28), and similarly for the negative sign,

$$\left(\frac{dv}{du}\right)_{f,l} = \frac{1 - \frac{u^2}{a^2}}{\frac{uv}{a^2} \mp \sqrt{\frac{q^2}{a^2} - 1}} \tag{34}$$

Equations (32) and (34) show that the streamlines and one family of characteristics are tangent to each other at the limiting hodograph. This result is the same as that obtained for two-dimensional flow. (See references 7, 8, and 9.) These equations when compared with equation (12) for the slope of Mach waves in the physical plane yields the interesting result that the streamlines and one family of characteristics at the limiting hodograph are perpendicular to the corresponding Mach waves at the limiting line.

Since

$$\left(\frac{dv}{du}\right)_\psi = -\frac{\psi_u}{\psi_v} \tag{35}$$

Equation (32) gives the following equation which holds at the limiting hodograph

$$\left(1 - \frac{v^2}{a^2}\right)(\psi_u)_l^2 + 2\frac{uv}{a^2}(\psi_u)_l(\psi_v)_l + \left(1 - \frac{u^2}{a^2}\right)(\psi_v)_l^2 = 0 \tag{36}$$

This equation can be reduced to more familiar form by introducing the polar coordinates in the $u-v$ plane:

$$u = q\cos\theta, \quad v = q\sin\theta$$

where θ is the angle between the velocity vector and u-axis. Then equation (36) takes the form

$$(\psi_q)_l^2 + \left(\frac{1}{q^2} - \frac{1}{a^2}\right)(\psi_\theta)_l^2 = 0 \tag{37}$$

This can be regarded as the equivalent of equation (23) for defining the limiting hodograph. A similar relation exists for two-dimensional flow. (See references 7, 8, and 9.)

Along a streamline, the ratio between $(dv)_\psi$ and $(du)_\psi$ is given by equation (31). By substituting this ratio in equation (19), the differential $(dx)_\psi$ and $(dy)_\psi$ along a streamline is given as

$$\left.\begin{aligned}(dx)_\psi &= \frac{u[\chi_{uu}\chi_{vv}-\chi_{uv}^2]}{-v\chi_{uv}+u\chi_{vv}}(du)_\psi\\(dy)_\psi &= \frac{v[\chi_{uu}\chi_{vv}-\chi_{uv}^2]}{-v\chi_{uv}+u\chi_{vv}}(du)_\psi\end{aligned}\right\}\tag{38}$$

At the limiting line, equation (23) is satisfied. Then equation (38) shows that at the limiting line, the streamline has a singularity. Or, more plainly, $(dx)_\psi$ and $(dy)_\psi$ at these points are infinitesimals of higher order than $(du)_\psi$ and $(dv)_\psi$. By writing s for the distance measured along a streamline, equation (38) gives immediately

$$(u_s)_\psi = \frac{-v\chi_{uv}+u\chi_{vv}}{q[\chi_{uu}\chi_{vv}-\chi_{uv}^2]}\tag{39}$$

Similarly,

$$(v_s)_\psi = \frac{v\chi_{uu}-u\chi_{uv}}{q[\chi_{uu}\chi_{vv}-\chi_{uv}^2]}\tag{40}$$

Therefore, at the limiting line, the acceleration along a streamline is infinitely large. Furthermore, since the pressure gradient $(p_s)_\psi$ along a streamline is

$$(p_s)_\psi = -\rho q q_s = -\rho[u(u_s)_\psi + v(v_s)_\psi]\tag{41}$$

the pressure gradient at the limiting line is also infinitely large.

Such infinite acceleration and pressure gradient lead one to suspect that the fluid is thrown back at the limiting line. In other words, the streamlines are doubled back at this line of singularity. To investigate whether this is true, the character of the relation $\chi_{uu}\chi_{vv}-\chi_{uv}^2=0$ along a streamline has to be determined. If the derivative of this expression along a streamline is not zero, then $\chi_{uu}\chi_{vv}-\chi_{uv}^2$ has only a simple zero at the intersection of the limiting line and the streamline. Consequently, the differentials $(dx)_\psi$ and $(dy)_\psi$ will change sign by passing through the limiting hodograph in $u-v$ plane along a streamline. Hence, the streamlines will double back and form a cusp at the limiting line. The derivative of $\chi_{uu}\chi_{vv}-\chi_{uv}^2$ along the streamline can be calculated with the aid of equation (30)

$$\begin{aligned}\left[\frac{d}{du}(\chi_{uu}\chi_{vv}-\chi_{uv}^2)\right]_l &= \chi_{uuu}\chi_{vv}-2\chi_{uv}\chi_{uuv}+\chi_{uu}\chi_{uvv}+\\&\quad\frac{v\chi_{uu}-u\chi_{uv}}{-v\chi_{uv}+u\chi_{vv}}\{\chi_{uuv}\chi_{vv}-2\chi_{uv}\chi_{uvv}+\chi_{uu}\chi_{vvv}\}\end{aligned}\tag{42}$$

The expression on the right of equation (42) cannot be reduced to zero by the available relations, which consist of equation (23), equation (15), and differentiated forms of equation (15). Therefore, the expression concerned generally has only a simple zero at the limiting hodograph and the streamlines are doubled back at the limiting line. It will be shown later that there is no solution possible beyond the limiting line. Hence, the name limiting line.

Envelope of Characteristics in Hodograph Plane and Lines of Constant Velocity in Physical Plane

Since the limiting line is the envelope of the Mach waves in the physical plane, it is interesting to see whether there is also an envelope for the characteristics in the hodograph plane. The characteristics in the $u-v$ plane are determined by equation (26). The envelope to them can be found by eliminating $\left(\frac{dv}{du}\right)_f$ between equation (26) and the following equation

$$\left\{\left(1-\frac{v^2}{a^2}\right)+\frac{v}{\chi_v}\chi_{vv}\right\}\left(\frac{dv}{du}\right)_f-\left\{\frac{uv}{a^2}-\frac{v}{\chi_v}\chi_{uv}\right\}=0 \tag{43}$$

which is obtained by equating to zero the partial derivative of equation (27) with respect to $\left(\frac{dv}{du}\right)_f$. The result can be simplified by equation (15), and then it becomes simply

$$1-\frac{u^2+v^2}{a^2}+\frac{u^2v^2}{a^4}=\frac{u^2v^2}{a^4} \tag{44}$$

This is satisfied by either

$$a=0 \tag{45}$$

or

$$u^2+v^2=a^2 \tag{46}$$

The first, condition, equation (45), when substituted into equation (26) gives

$$\left(\frac{dv}{du}\right)_{f,a=0}=-\frac{u}{v} \tag{47}$$

which shows that the circle of maximum velocity corresponding to $a=0$, is the envelope of the characteristics in hodograph plane. The second condition, equation (46), is the spurious solution, since generally the characteristic at $q=a$ is not a tangent to the circle $q=a$. Hence $a=0$ is the only envelope.

The lines of constant velocity in the hodograph plane are simply circles. Therefore

$$\left(\frac{dv}{du}\right)_q=-\frac{u}{v} \tag{48}$$

By means of this relation and equation (19), the slope of the lines of constant velocity is given as

$$\left(\frac{dy}{dx}\right)_q=\frac{v\chi_{uv}-u\chi_{vv}}{v\chi_{uu}-u\chi_{uv}} \tag{49}$$

This equation together with equation (30) gives the following interesting relation

$$\left(\frac{dy}{dx}\right)_q=-\frac{1}{\left(\frac{dv}{du}\right)_\psi} \tag{50}$$

In other words, a line of constant velocity in the physical plane is perpendicular to the streamline in the hodograph plane at the corresponding points.

The Lost Solution

Throughout the previous calculation, the possibility of using the Legendre transformation is assumed. This requires that for each pair of values of u, v there is one and only one pair of values of x, y. However, it is not always true. It is possible to have a number of points in the physical plane having the same value of u and v. If this is the case, then evidently it is impossible to solve for x and y from the pair of functions $u = u(x,y)$, $v = v(x,y)$. Mathematically, the situation is expressed by saying that the Jacobian $\partial(u,v)/\partial(x,y)$ vanishes in the physical plane. Or

$$u_x v_y - u_y v_x = 0 \tag{51}$$

However, this is also the condition for a functional relation between u and v; for example, v can be expressed as a function of u. In other words, u and v are not independent. Hence if a solution is "lost" or not included in the family of solutions allowing Legendre transformation, then for that solution,

$$v = v(u) \tag{52}$$

It is seen that equation (51) is then identically satisfied.

By eliminating ρ from the continuity equation, there is obtained

$$\left(1-\frac{u^2}{a^2}\right)u_x - \frac{uv}{a^2}(u_y + v_x) + \left(1-\frac{v^2}{a^2}\right)v_y + \frac{v}{y} = 0 \tag{53}$$

This equation can be rewritten in the following form by using equation (52)

$$\left\{\left(1-\frac{u^2}{a^2}\right) - \frac{uv}{a^2}\frac{dv}{du}\right\}u_x + \left\{\left(1-\frac{v^2}{a^2}\right)\frac{dv}{du} - \frac{uv}{a^2}\right\}u_y + \frac{v}{y} = 0 \tag{54}$$

The vorticity equation, equation (4), can be expressed as

$$\frac{dv}{du}u_x - u_y = 0 \tag{55}$$

From equations (54) and (55), it is possible to solve for u_x and u_y. The result is

$$\left[\left(1-\frac{u^2}{a^2}\right) - 2\frac{uv}{a^2}\frac{dv}{du} + \left(1-\frac{v^2}{a^2}\right)\left(\frac{dv}{du}\right)^2\right]u_x = -\frac{v}{y} \tag{56a}$$

$$\left[\left(1-\frac{u^2}{a^2}\right) - 2\frac{uv}{a^2}\frac{dv}{du} + \left(1-\frac{v^2}{a^2}\right)\left(\frac{dv}{du}\right)^2\right]u_y = -\frac{v}{y}\frac{dv}{du} \tag{56b}$$

By differentiating the first of equation (56) with respect to y, the second with respect to x, the following relation is obtained by subtraction:

$$\frac{d^2v}{du^2}u_x + \frac{1}{y} = 0 \tag{57}$$

Therefore

$$\frac{dv}{du} = \frac{f(y) - x}{y} \tag{58}$$

or

$$y = \frac{f(y) - x}{\frac{dv}{du}}$$

where $f(y)$ is an undetermined function of y. However, equation (55) shows that for lines of constant values of u where $du = u_x(dx)_u + u_y(dy)_u = 0$,

$$\left(\frac{dy}{dx}\right)_u = -\frac{1}{\left(\frac{dv}{du}\right)_u} = \text{constant} \tag{59}$$

Hence, lines of constant values of u and v are straight lines. This restriction reduces the function $f(y)$ in equation (58) to a numerical constant. Put $f(y) = K$, equation (58) is then

$$y = \frac{K - x}{\frac{dv}{du}} \tag{60}$$

Therefore lines of constant value of u and v are radial lines passing through the point $x = K$. Thus the lost solution is nothing but the well-known solution for the flow over a conical surface.

From equation (59), it is seen that lines of constant velocity are perpendicular to the tangent of the $u - v$ curve at the corresponding points. By substituting the value of $\frac{1}{y}$ from equation (57) into equation (56a), a relation between u and v is obtained:

$$v\frac{d^2v}{du^2} - \left(1 - \frac{v^2}{a^2}\right)\left(\frac{dv}{du}\right)^2 + 2\frac{uv}{a^2}\frac{dv}{du} - \left(1 - \frac{u^2}{a^2}\right) = 0 \tag{61}$$

This is the differential equation for determining the hodograph representing the flow over a cone. Figure 3 shows the hodograph for a cone of 30° semivertex angle and with a velocity at the surface of cone equal to 0.35 c. The maximum velocity is c — that is, the value of q corresponding to $a = 0$. Figure 3 is drawn from data given by Taylor and Maccoll[12].

It may well be mentioned here that the lost solution for the axially symmetric flow is not limited to supersonic velocity as is the case for two-dimensional flow. In fact, Taylor and Maccoll show that for small forward velocity of the cone, supersonic velocities occur only just after the head shock wave. The velocity decreases as the surface of the cone is approached. Finally, it becomes subsonic for points near the surface of the cone. Figure 4 shows a few examples taken from their calculations[12]. The dotted curves in the figure are the Mach waves. The dotted straight lines are the boundaries between the supersonic and the subsonic regions. Furthermore, spark photographs of a conical shell in actual flight taken by Maccoll[13] do not

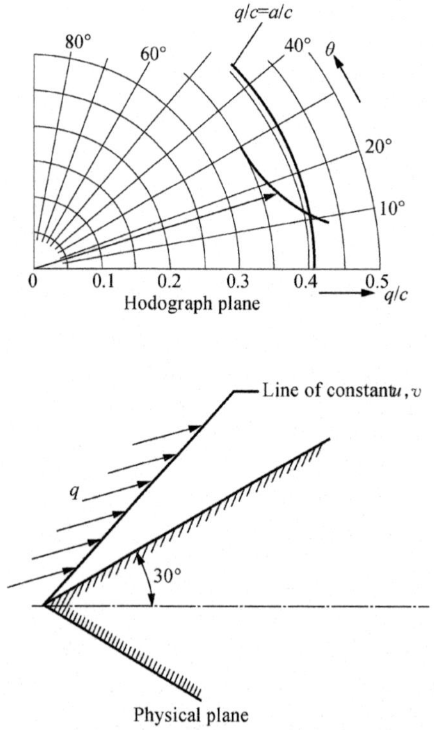

Fig. 3 Hodograph of the flow over a cone of 30° half vertex angle and a surface velocity q equal to $0.35\,c$

indicate the presence of shock waves in regions of flow where such transition from supersonic to subsonic velocities is expected. Therefore, at least for this particular type of flow, a smooth transition through sonic velocity actually takes place.

Continuation of Solution Beyond The Limiting Line

Since it is shown in a previous paragraph that the streamlines are generally turned back at the limiting line, the question arises: Is it possible to continue the solution beyond the limiting line? Of course, there are two ways of continuing the solution: The new solution is joined either smoothly to the given solution at the limiting line or with a discontinuity. As shown before, the limiting line is the envelope of one family of the Mach waves. Then at every point of this line its direction differs from that of the streamline by an angle equal to the Mach angle. But the Mach angle is not zero except at points where the velocity of fluid has reached the maximum velocity and the ratio $\frac{a}{q} = 0$. Therefore, the limiting line generally does not coincide with the streamline, and the discontinuity at the junction of the solution at the limiting line cannot be that of a vortex sheet. The only other type of discontinuity is the shock wave. However, the angle between the limiting line and the flow direction is equal to the

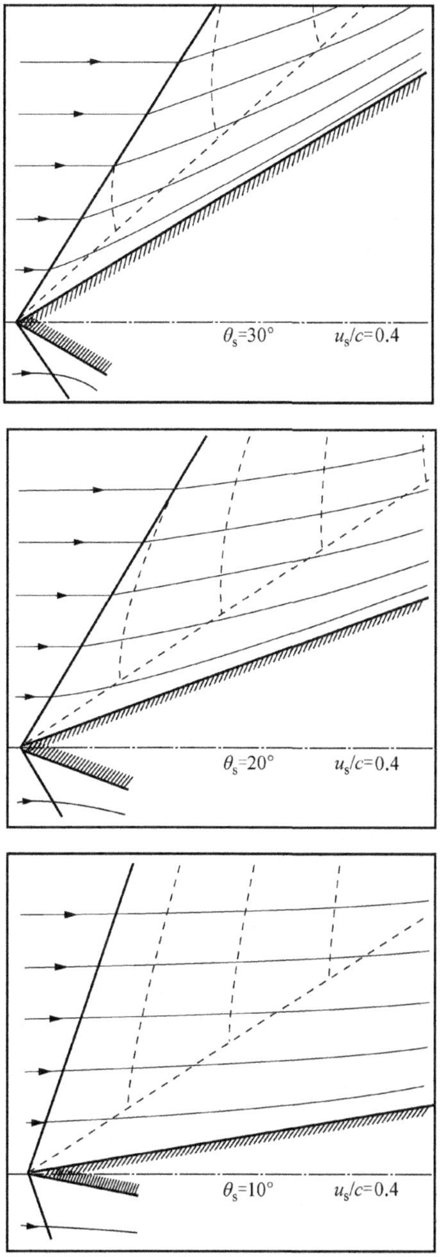

Fig. 4 Flow over cones of various vertex angles involving subsonic regions
θ_s = half vertex angle, u_s = velocity over the surface of cone

Mach angle. Then according to the result of the theory of shock waves, the discontinuity across such a line vanishes. In other words, there cannot be a discontinuity at the limiting line. Therefore, it is impossible to join a new solution at the limiting line with a discontinuity.

As to the second possibility of joining a new solution smoothly at the limiting line, it is seen that the flow beyond the limiting line must be irrotational and isentropic since the limiting line cannot be a shock wave. There are only two types of isentropic irrotational flow; namely, one that allows the Legendre transformation, and one that does not, the "lost solution." Investigate the second alternative first. If the solution beyond the limiting line belongs to the so-called lost solution, then since the junction at the limiting line must be smooth, the values of u and v at the limiting line must also satisfy equation (61). But the slope $\left(\frac{dv}{du}\right)_l$ at the limiting line is given by equation (24). The second derivative $\left(\frac{d^2v}{du^2}\right)_l$ will then involve the fourth order derivatives of χ. Besides these expressions, the available relations are equations (15), (23), (25a), (25b), and three more equations obtained by differentiating equations (25) with respect to u and v. However, it is still impossible for $\left(\frac{dv}{du}\right)_l$ to satisfy an equation like equation (61) where no derivative of χ appears. Hence, the limiting hodograph does not satisfy the equation for the lost solution. In other words, the lost solution cannot be used to continue the flow beyond the limiting line.

The only remaining possibility is to continue the flow smoothly by another solution obtainable by a Legendre transformation. Smooth continuation means that the values of u, v, and ρ must be the same at the junction, the limiting line. Since shock waves do not appear, isentropic relations still hold. The density ρ is determined by velocity only. The value of u and v are determined by the coordinates in the hodograph plane. The position of the limiting line in the physical plane is determined by χ_u, χ_v. Therefore, the problem can be stated as follows: At a certain given curve $u(\lambda)$, $v(\lambda)$ in the hodograph plane, the limiting hodograph, the values of χ_u, χ_v are given; λ is the parameter along the given curve. It is required to determine a new solution of the differential equation (equation (15)) with these initial values. First of all, it is seen that with the given data, the left-hand sides of the following equations are given:

$$\frac{d}{d\lambda}(\chi_u) = \chi_{uu}\frac{du}{d\lambda} + \chi_{uv}\frac{dv}{d\lambda} \tag{62a}$$

$$\frac{d}{d\lambda}(\chi_v) = \chi_{uv}\frac{du}{d\lambda} + \chi_{vv}\frac{dv}{d\lambda} \tag{62b}$$

Therefore

$$\chi_{uv} = \left[-\frac{dv}{d\lambda}\chi_{vv} + \frac{d}{d\lambda}(\chi_v)\right]\Big/\frac{du}{d\lambda} \tag{63a}$$

$$\chi_{uu} = \left[\left(\frac{dv}{d\lambda}\right)^2\chi_{vv} - \frac{dv}{d\lambda}\frac{d}{d\lambda}(\chi_v) + \frac{du}{d\lambda}\frac{d}{d\lambda}(\chi_u)\right]\Big/\left(\frac{du}{d\lambda}\right)^2 \tag{63b}$$

By substituting those values into equation (15), the second-degree terms reduce to

$$\chi_{uu}\chi_{vv} - \chi_{uv}^2 = \left[\frac{dv}{d\lambda}\frac{d}{d\lambda}(\chi_v) + \frac{du}{d\lambda}\frac{d}{d\lambda}(\chi_u)\right]\chi_{vv}\Big/\left(\frac{du}{d\lambda}\right)^2 + \left[\frac{d}{d\lambda}(\chi_v)\right]^2\Big/\left(\frac{du}{d\lambda}\right)^2 \tag{64}$$

which is linear in χ_{vv}. Therefore χ_{vv} can be uniquely determined by equation (15). In other words, with the given data, the second order derivatives of χ at the given curve $u(\lambda)$, $v(\lambda)$ can be determined uniquely, in spite of the fact that the differential equation (15) is of second degree. Friedrichs and Lewy[14]* have shown that under these circumstances, the function χ within a region R (Fig. 5) bounded by two characteristics and the given curve is uniquely determined except for an additional constant. Consequently there can be only one solution corresponding to the given data at the limiting hodograph. However, this solution is the very one which gives the reverse flow at limiting line. Therefore, it is impossible to continue the solution beyond the limiting line even by a Legendre transformation.

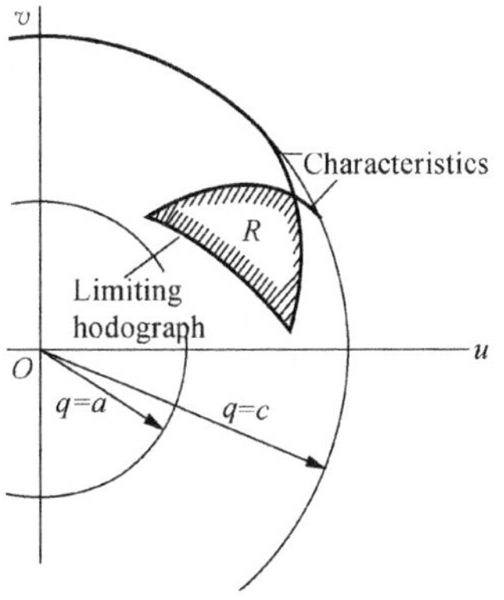

Fig. 5 Region R. where the solution is uniquely determined by given data at the limiting hodograph

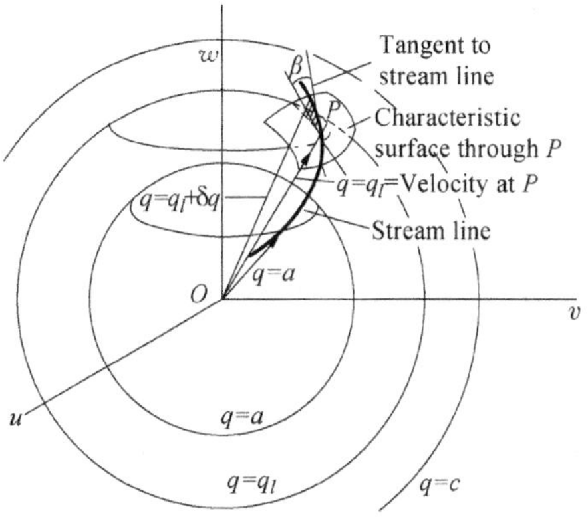

Fig. 6 Geometrical relations between stream line and characteristic surface at limiting surface in hodograph space

Since all three alternatives fail to offer a way of continuing the solution, the limiting line is truly an impossible boundary to cross. In other words, the region beyond the limiting line is a forbidden region. This physical absurdity can be resolved only by the breakdown of isentropic irrotational flow.

General Three-Dimensional Flow

The methods used in previous sections for investigating the axially symmetric flow can be easily extended to the general three-dimensional case. In the present section, this investigation will be sketched briefly and the results indicated.

Let the three components of velocity along the three coordinate axes x, y, and z be denoted by u, v, and w, respectively. Then by introducing a velocity potential φ defined by

$$u = \varphi_x, \quad v = \varphi_y, \quad w = \varphi_z \tag{65}$$

* "[13]" in the original paper has been corrected as "[14]"—Noted by editor.

the differential equation for φ of an isentropic irrotational flow can be written as[7]

$$a^2(\varphi_{xx}+\varphi_{yy}+\varphi_{zz})=u^2\varphi_{xx}+v^2\varphi_{yy}+w^2\varphi_{zz}+2vw\varphi_{yz}+2wu\varphi_{zx}+2uv\varphi_{xy} \tag{66}$$

If, for every triad of u, v, w, there is only one triad of x, y, z, then the Legendre transformation can be used. Thus

$$\chi = ux+vy+wz-\varphi \tag{67}$$

and

$$\chi_u = x, \quad \chi_v = y, \quad \chi_w = z \tag{68}$$

The differential equation for φ, equation (66), is then transformed into

$$\begin{aligned} a^2[BC-F^2+CA-G^2+AB-H^2] = {} & u^2(BC-F^2)+v^2(CA-G^2)+ \\ & w^2(AB-H^2)+2vw(GH-AF)+ \\ & 2wu(HF-BG)+2uv(FG-CH) \end{aligned} \tag{69}$$

where the following notations are used

$$A=\chi_{uu}, \quad B=\chi_{vv}, \quad C=\chi_{ww}, \quad F=\chi_{vw}, \quad G=\chi_{wu}, \quad H=\chi_{uv} \tag{70}$$

By analogy with the axially symmetric case, the limiting hodograph surface is defined as the surface in the u, v, w space, or hodograph space, where the following relation holds:

$$\Delta = \begin{vmatrix} A & H & G \\ H & B & F \\ G & F & C \end{vmatrix} = 0 \tag{71}$$

The properties of this limiting hodograph and the corresponding limiting surface can be found by considering the behavior of streamlines and characteristics at such surfaces.

From equation (68) the differentials of x, y, and z can be written as

$$dx = A du + H dv + G dw \tag{72a}$$
$$dy = H du + B dv + F dw \tag{72b}$$
$$dz = G du + F dv + C dw \tag{72c}$$

Along a streamline, the differentials dx, dy, and dz must be proportional to u, v, and w, respectively. Thus the equation of a streamline in physical space is

$$\frac{(dx)_\psi}{u} = \frac{(dy)_\psi}{v} = \frac{(dz)_\psi}{w} \tag{73}$$

where the subscript ψ indicates values taken along the streamline. The equation of a streamline in hodograph space is obtained by eliminating dx, dy, and dz from equation (73) by equation (72). The result is

$$\frac{(du)_\psi}{\bar{a}u+\bar{h}v+\bar{g}w} = \frac{(dv)_\psi}{\bar{h}u+\bar{b}v+\bar{f}w} = \frac{(dw)_\psi}{\bar{g}u+\bar{f}v+\bar{c}w} \tag{74}$$

where $\bar{a}$ is the co-factor of A in the determinant Δ of equation (71), $\bar{b}$ the co-factor of B, and so forth. Equation (74) can be used, in turn, to eliminate two of the three differentials du, dv, and dw in the right of equation (72). The result is

$$(dx)_\psi = \frac{u\Delta du}{\bar{a}u + \bar{h}v + \bar{g}w} \tag{75a}$$

$$(dy)_\psi = \frac{v\Delta dv}{\bar{h}u + \bar{b}v + \bar{f}w} \tag{75b}$$

$$(dz)_\psi = \frac{w\Delta dw}{\bar{g}u + \bar{f}v + \bar{c}w} \tag{75c}$$

At the limiting surface, $\Delta = 0$ as defined by equation (71); therefore the streamlines have a singularity there. Similar to the axially symmetric flow, the streamlines generally are turned back and form a cusp at this surface. The acceleration and the pressure gradient are, of course, infinitely large at such places.

The characteristic surface $g(x, y, z) = 0$ in physical space is determined by the equation

$$a^2[g_x^2 + g_y^2 + g_z^2] = u^2 g_x^2 + v^2 g_y^2 + w^2 g_z^2 + 2vwg_y g_z + 2wug_z g_x + 2uvg_x g_y \tag{76}$$

Since this equation is a second-degree equation, there are two families of surfaces passing through each point. These surfaces are the wave fronts of infinitesimal disturbances in the flow and can be called the Mach surfaces. The characteristic surface $f(u, v, w) = 0$ in the hodograph space is determined by the equation

$$\begin{aligned} &a^2[(B+C)f_u^2 + (C+A)f_v^2 + (A+B)f_w^2 - 2Ff_u f_w - 2Gf_w f_u - 2Hf_u f_v] \\ &= u^2[Cf_v^2 + Bf_w^2 - 2Ff_v f_w] + v^2[Cf_u^2 + Af_w^2 - 2Gf_w f_u] + \\ &\quad w^2[Bf_u^2 + Af_v^2 - 2Hf_u f_v] + 2vw[Hf_w f_u + Gf_u f_v - Ff_u^2 - Af_v f_w] + \\ &\quad 2wu[Hf_v f_w + Ff_u f_v - Gf_v^2 - Bf_w f_u] + 2uv[Gf_w f_v + Ff_w f_u - Hf_w^2 - Cf_u f_v] \end{aligned} \tag{77}$$

By transforming equation (76) for Mach surfaces to hodograph space, it can be shown that the transformed equation is satisfied either by the characteristics in hodograph space determined by equation (77) or by the limiting hodograph determined by equation (71). Therefore, here again the limiting surface is the envelope of a family of Mach surfaces.

By using equations (74) and (77), it is possible to show that the streamlines in the hodograph space are tangent to the characteristic surfaces at the limiting hodograph. Furthermore, by using equations (69), (71), and (74), the inclination of the streamlines at the limiting hodograph can be calculated. In fact, if $(ds)^2 = (du)^2 + (dv)^2 + (dw)^2$, $q^2 = u^2 + v^2 + w^2$, the following relation is obtained

$$\left(\frac{ds}{dq}\right)_{\psi,l} = \frac{q}{a} \quad \text{or} \quad -\frac{q}{a} \tag{78}$$

This relation is really equivalent to equation (32). In other words, at the limiting hodograph, the inclination of the streamlines and characteristics to the q = constant surface is equal to the

Mach angle (Fig.6). It thus seems the breakdown of general steady isentropic irrotational flow of nonviscous fluid is connected with the appearance of the envelope of Mach waves in physical space and the tangency of streamlines and characteristics in hodograph space.

California Institute of Technology,
Pasadena, Calif., August 24, 1943.

References

[1] Theodorsen, Theodore. The Reaction on a Body in a Compressible Fluid. Jour. of the Aero. Sci., 1937, 4: 239-240.

[2] Taylor G I., Sharman C F. A Mechanical Method for Solving Problems of Flow in Compressible Fluids. R. & M. No. 1195, British A.R.C., 1928.

[3] Taylor G I. Recent Works on the Flow of Compressible Fluids. Jour. London Math. Soc., 1930, 5: 224-240.

[4] Binnie A M, Hooker S G. The Radical and Spiral Flow of a Compressible Fluid. Phil. Mag., 1937, 23: 597-606.

[5] Tollmien W. Zum Übergang von Unterschall-in Überschall Strömungen. Z. f. a. M. M., 1937,17(2): 117-136.

[6] Ringleb, Friedrich: Exakte Lösungen der Differential-gleichungen einer adiabatischen Gasströmung. Z. f. a. M. M., 1940, 20(4):185-198.

[7] von Kármán, Th.: Compressibility Effects in Aerodynamics. Jour. of the Aero. Sci., 1941,8(9): 337-356.

[8] Ringleb F. Über die Differentialgleichungen einer adiabatischen Gasströmung und den Strömungsstoss. Deutsche Mathematik., 1940,5(5): 337-384.

[9] Tollmien W. Grenzlinien adiabatischer Potential-stromungen. Z. f. a. M. M., 1941,21(3): 140-152.

[10] Frankle F. Bulletin de lacademic des sciences. U. R. S. S. (7), 1934.//Frankle F, Aleksejeva R. Zwei Randwertaufgaben aus der Theorie der hyperbolischen partiellen Differentialgleichungen zweiter Ordnung mit Anwendungen auf Gasströmungen mit Überschallgeschwindigkeit. (Convergence proof.) Matematiceski Sbornik, 1935,41: 483-502.

[11] Ferrari Carlo. Campo aerodinamico a velocita iperacustica attorno a un solido di revoluzione a proraacuminata. L'Aerotecnica, 1936,16(2): 121-130.//also Determinazione della pressione sopra solidi di rivoluzione a prora acuminate disposti in deriva in corrente di fluido compressible a velocita ipersonora. Atti della R. Accad. Della Sci. Di Torino, 1937,72: 140-163.

[12] Taylor G I. Maccoll J W. The Air Pressure on a Cone Moving at High Speeds. Proc., Roy. Soc. of London, ser. A, 1937, 159: 278-298.

[13] Maccoll J W. The Conical Shock Wave Formed by a Cone Moving at a High Speed. Proc., Roy. Soc. of London, ser. A, 1937, 159: 459-472.

[14] Friedrichs K, Lewy H. Das Anfangswertproblem einer beliebigen nichtlinearen hyperbolischen Differentialeleichung beliebiger Ordnung in zwei Variablen. Math. Annalen, 1928, 99: 200-221.

Loss in Compressor or Turbine due to Twisted Blades

Hsue-shen Tsien*

Due to various design considerations, the gas stream leaving the guide wheel of an axial compressor or a turbine is sometimes made to have a twist along the radius of the wheel. The resulting non-uniform force distribution over the length of the rotating blades causes an induced velocity at the blade. The resulting loss can be calculated by a simple application of hydrodynamics. The author first develops the general theory and then examples are given to illustrate the use of the method.

Problem

The gas in the rotating wheel of an axial compressor or a turbine is subject to centrifugal forces. If there is no radial pressure gradient to counteract this centrifugal force field, a radial flow will take place. Similarly the rotation of the flow after the moving blades will cause a flow along the radial direction unless a radial pressure gradient is present to prevent it. Such lateral or radial flow has a detrimental effect on the boundary layers over the blades and may cause premature separation of the boundary layers with the accompanied bad results on the performance of the machine. To avoid these consequences, a twist is sometimes given to the moving blades or to the flow leaving the guide blades so that in case of a compressor the angle of attack of the rotating blades near the tip is larger than that near the root. This will make the lift force at the tip larger than that at the root of the blade. The result is a larger pressure rise across the rotating wheel at the tip than at the root. The pressure gradient so obtained will prevent the radial flow and thus insure a more satisfactory operation of the machine.

However, such expediency is not without a slight loss in the efficiency of the blading. The magnitude of velocity in the twisted flow behind the guide blades is approximately constant along the length of the blade, since the total twist is usually only a few degrees. The circulation, being the quotient obtained by dividing the lift per unit length of the blade by the product of density and velocity of the stream, increases towards the tip. This change in circulation will cause the "shedding" of vortices along the blade in the same manner as the shedding of vortices in an airplane wing. These vortices trail along the flow direction behind the blades and cause an induced velocity at the blade. This induced velocity will rotate the flow around the blade and change the effective local angle of attack of the blade section. The change

---

Journal of the Chinese Institute of Engineers/America Section/, vol.2, No.1, pp.40–53, 1944.

* Gugenheim Aeronautical Laboratory California Institute of Technology.

in effective angle of attack has two consequenses: First, the actual lift cannot be calculated directly from the geometrical angle of attack. Furthermore, the resultant change in the direction of action of the lift force of the blades will decrease the pressure rise across the blading. Both effects have to be considered for the proper design of the guide blades and rotating blades. The present paper is an attempt to calculate these effects by means of the well-known principles of hydrodynamics.

The Vortex System and the Induced Velocity

To make the discussion more definite, the following analysis is carried out with the axial compressor in mind, although the same procedure can be used for the turbine blading. As shown in Fig. 1, which represents the development of a cylindrical section of a compressor stage composed of a row of guide blades and a row of rotating blades, the blade sections are airfoil sections of small camber. Fig. 2 shows the velocity diagrams for the guide blades and the rotating blades. It is seen that by using the modern airfoil sections, the change in direction of flow in passing through the blade is quite small. Therefore, the trailing vortices can be considered as lying in the direction of the flow before the blade lattice. In other words, the trailing vortices of the guide blades are considered as following the direction of U_0, the inlet velocity to the guide blades. To treat the rotating blades in this case, one has to investigate the problem of blade lattice in a twisted flow. However, as an approximation the problem can be analyzed as that of twisted blades in a uniform velocity. In other words, the effect of twisting the flow and the effect of twisting the blades are approximately equivalent. With this simplification, both the guide blades and the rotating blades can be treated as a twisted blade lattice in the uniform flow with trailing vortices in the inflow direction.

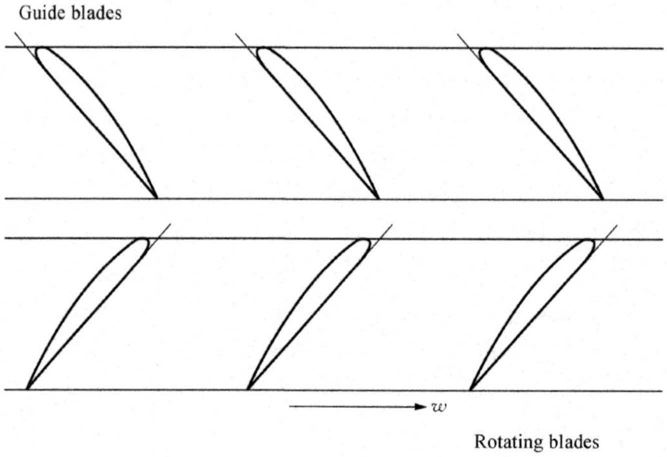

Fig. 1 Compressor stage

In an axial compressor or a turbine, the height of the blades is usually very small in comparison with the diameter of the wheel. Therefore the spacing of the blades can be

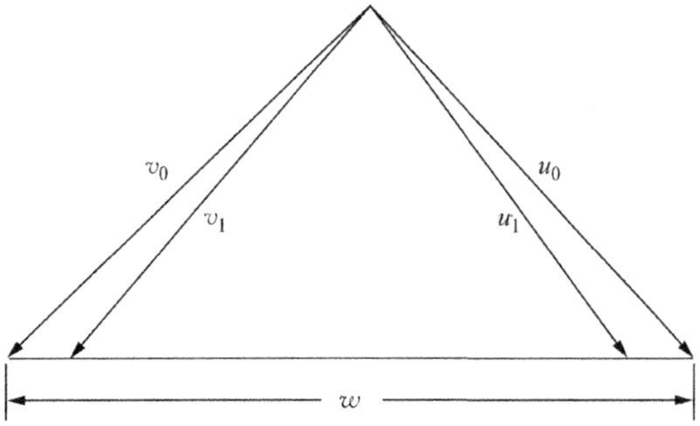

Fig. 2 Velocity diagram of a compressor stage

u_0 = Inlet velocity to the guide blades, absolute;

Or, Outlet velocity from the rotating blades, absolute

u_1 = Outlet velocity from the guide blades, absolute

v_0 = Outlet velocity from the guide blades; relative;

Or, Inlet velocity to the rotating blades, relative

v_1 = Outlet velocity from the rotating blades, relative

w = Circumferential velocity of the rotating blades

considered as uniform from root to tip. Hence, the problem on hand is finally reduced to that of a parallel blade lattice of length l between two walls (Fig. 3). One wall represents the surface of the hub. Another wall represents the casing of the machine. As in the usual wing theory of airplane, the blade will be replaced by a vortex of variable strength T. Now, it is well-known that the vortex between two walls can be calculated as an infinitely long vortex without walls, but with the additional condition of image symmetry. Let the positions of the walls be $y = 0$ and $y = 1$ (Fig. 4). Then the image symmetry requires

that $$\Gamma(y) = \Gamma(-y)$$

and $$\Gamma(y) = \Gamma(l + l - y) \tag{1}$$

It is easily verified that Eq. (1) is satisfied by putting

$$\Gamma(y) = \sum_0^\infty a_n \cos\frac{n\pi y}{l} \tag{2}$$

where a_n is the Fourier coefficient of the circulation distribution.

The rate of decrease of circulation in the y-direction is $-\,\mathrm{d}\Gamma/\mathrm{d}y$. Therefore, between y and $y + \mathrm{d}y$, the amount of vortex shed off is $(-\,\mathrm{d}\Gamma/\mathrm{d}y)\mathrm{d}y$. This is the strength of the trailing vortex at y. The direction of the trailing vortex is, according to the simplifying assumption that of the inflow velocity U or the x-axis. The induced velocity at a point y' due to the trailing vortices on the x-y plane is then[1]:

$$\frac{1}{4\pi}\int_{-\infty}^{\infty} -\frac{\mathrm{d}\Gamma}{\mathrm{d}y}\frac{1}{y - y'}\mathrm{d}y \tag{3}$$

However, there are other vortices above and below the y-axis representing other blades in

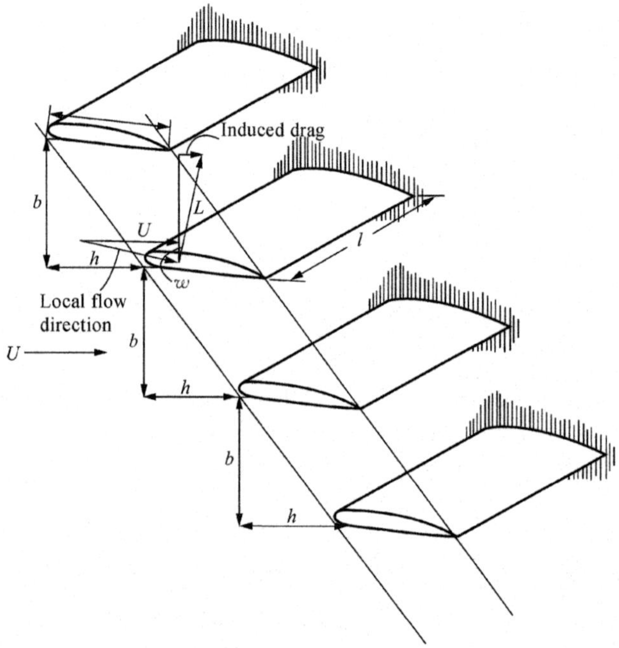

Fig. 3 Airfoil lattice representing compressor blades

the lattice. Let the spacing or gap of the blades be b and the stagger be h. The stagger is considered as positive if the upper blades are ahead of the lower blades, as shown in Fig. 4. The induced velocity at apoint y' due to trailing vortices shed by the vortex at $x = h$, $Z = b$ is then

$$\frac{1}{4\pi}\int_{-\infty}^{\infty} -\frac{\mathrm{d}\Gamma}{\mathrm{d}y}\frac{1}{\sqrt{(y-y')^2+b^2}}\left\{1+\frac{h}{\sqrt{(y-y')^2+b^2+h^2}}\right\}\mathrm{d}y \tag{4}$$

This velocity is in a direction perpendicular to the line joining $x = 0$, $y = y'$, $Z = 0$ and $x = 0$, $y = y'$, $Z = b$ and lies in the y-Z plane. The vertical component of this velocity is directed downward and is equal to

$$\frac{1}{4\pi}\int_{-\infty}^{\infty} -\frac{\mathrm{d}\Gamma}{\mathrm{d}y}\frac{y-y'}{(y-y')^2+b^2}\left\{1+\frac{h}{\sqrt{(y-y')^2+b^2+h^2}}\right\}\mathrm{d}y \tag{5}$$

Similarly, the downward vertical component of the induced velocity due to the trailing vortices from the blade at $x = -h$, $Z = -b$ is

$$\frac{1}{4\pi}\int_{-\infty}^{\infty} -\frac{\mathrm{d}\Gamma}{\mathrm{d}y}\frac{y-y'}{(y-y')^2+b^2}\left\{1-\frac{h}{\sqrt{(y-y')^2+b^2+h^2}}\right\}\mathrm{d}y \tag{6}$$

By adding Eqs. (5) and (6), the vertical component of the induced velocity due to the pair of blades is

$$\frac{1}{4\pi}\int_{-\infty}^{\infty} -\frac{\mathrm{d}\Gamma}{\mathrm{d}y}\frac{2(y-y')}{(y-y')^2+b^2}\mathrm{d}y \tag{7}$$

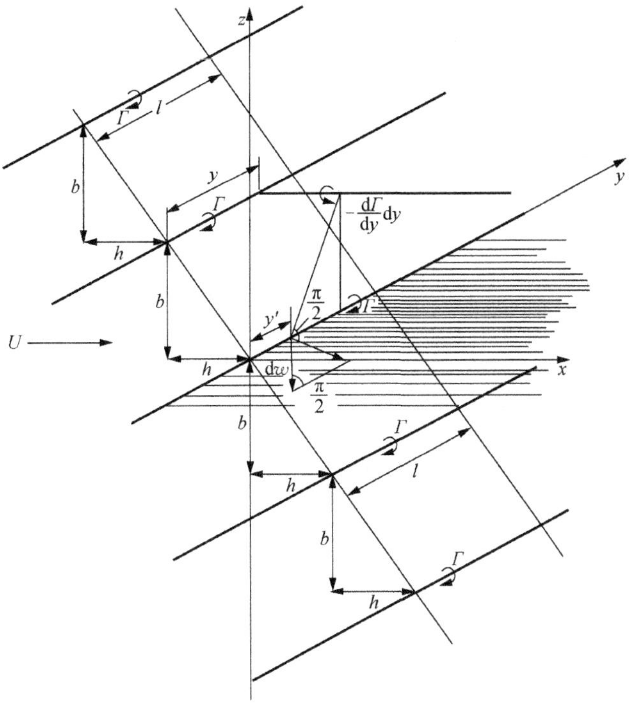

Fig. 4 Vortex system representing compressor blades

For the calculation of the change in angle of attack due to induced velocity, one needs only to consider the vertical component of the induced velocity. Then as shown in Eq. (7), there is no effect due to stagger.

There are blades at $Z = \pm 2b, \pm 3b, \pm 4b, \ldots$ Therefore, the total induced downwash at $x = 0, y = y', Z = 0$ is

$$w(y') = \frac{1}{4\pi}\int_{-\infty}^{\infty} -\frac{d\Gamma}{dy}\left\{\frac{1}{y-y'} + 2\sum_{m=1}^{\infty}\frac{y-y'}{(y-y')^2+(mb)^2}\right\}dy \tag{8}$$

The downwash at other blades is, of course, the same as that at y-axis. By using the formula

$$\coth x = \frac{1}{x} + 2\sum_{m=1}^{\infty}\frac{x}{x^2+m^2\pi^2}$$

Eq. (8) can be re-written as

$$w(y') = \frac{1}{4b}\int_{-\infty}^{\infty} -\frac{d\Gamma}{dy}\coth\frac{\pi(y-y')}{b}dy \tag{9}$$

This is the general equation for the downwash due to an infinite lattice of airfoils of gap b.

The General Blade Equation

By substituting Eq. (2) into Eq. (9), $w(y')$ can be expressed as

$$w(y') = \sum_{n=1}^{\infty} a_n w_n(y') \tag{10}$$

where
$$w_n(y') = \frac{n}{4b}\frac{\pi}{l}\int_{-\infty}^{\infty} \sin\frac{n\pi y}{l}\coth\frac{\pi(y-y')}{b}dy \tag{11}$$

Eq. (11) can be greatly simplified by introducing the new variable $\xi = y - y'$. Then

$$w_n(y') = \frac{n}{4b}\frac{\pi}{l}\int_{-\infty}^{\infty}\left\{\sin\frac{n\pi\xi}{l}\cos\frac{n\pi y'}{l} + \cos\frac{n\pi\xi}{l}\sin\frac{n\pi y'}{l}\right\}\coth\frac{\pi\xi}{b}d\xi$$

However, due to the fact that $\cos\frac{n\pi\xi}{l}$ is an even function of ξ while $\coth\frac{\pi\xi}{b}$ is an odd function of ξ,

$$\int_{-\infty}^{\infty}\cos\frac{n\pi\xi}{l}\coth\frac{\pi\xi}{b}d\xi = 0$$

Hence,

$$w_n(y') = \frac{n\pi}{l}\cos\frac{n\pi y'}{l}I_n \tag{12}$$

where

$$\begin{aligned} I_n &= \frac{1}{4b}\int_{-\infty}^{\infty}\sin\frac{n\pi\xi}{l}\coth\frac{\pi\xi}{b}d\xi \\ &= \frac{1}{2b}\int_{0}^{\infty}\sin\frac{n\pi\xi}{l}\coth\frac{\pi\xi}{b}d\xi \end{aligned} \tag{13}$$

This integral is not difficult to evaluate. The result is*

$$I_n = \frac{1}{4}\coth\frac{n\pi b}{2l} \tag{14}$$

Therefore Eq. (12) can be re-written as

$$w_n(y') = \frac{1}{4}\frac{n\pi}{l}\coth\frac{n\pi b}{2l}\cos\frac{n\pi y'}{l} \tag{15}$$

By combining Eqs. (10) and (15), the induced downwash due to trailing vortices is then

$$w(y) = \frac{\pi}{4l}\sum_{n=1}^{\infty} na_n\coth\frac{n\pi b}{2l}\cos\frac{n\pi y}{l} \tag{16}$$

Of course, the vortices parallel to y-axis on the "bound" vortices also give an induced velocity at y-axis where the typical blade is located. However, such induced velocity also exists in two-dimensional flow without trailing vortices and is considered in calculating the two-dimensional or section characteristics of an airfoil lattice. Therefore in accordance with the usual wing

* See Appendix.

theory, the induced downwash angle is taken as w/U with w calculated from trailing vortices only. The effect of the bound is incorporated into the section characteristic of the lattice.

If Θ is the local geometrical angle of attack of the airfoil, i.e., angle between the zero lift direction of the airfoil and the direction of U, then the effect local angle of attack is $\Theta - w/U$. Let k be the slope of the lift coefficient for the two-dimensional flow over a lattice of the same profile and the same gap and stagger. Then the local lift coefficient of the blade is $k(\Theta - w/U)$. The lift per unit length of blade is then

$$\frac{\rho}{2}U^2 ck\left(\Theta - \frac{w}{U}\right)$$

where ρ is the density of the fluid and c is the chord or width of the blade. This quantity must be equal to $\rho U\Gamma$. Hence, the general blade equation is

$$\frac{\rho}{2}U^2 ck\left\{\Theta - \frac{\pi}{4lU}\sum_1^\infty na_n \coth\frac{n\pi b}{2l}\cos\frac{n\pi y}{l}\right\} = \rho U\sum_0^\infty a_n \cos\frac{n\pi y}{l} \tag{17}$$

Let c_0 be the mean chord, then Eq. (17) can be put into non-dimensional form by introducing α_n defined by

$$a_n = \frac{1}{2}Uc_0 x_n \tag{18}$$

The resultant general blade equation is then

$$k(y)\frac{c}{c_0}(y)\left\{\Theta - \frac{\pi c_0}{8l}\sum_1^\infty n\alpha_n \coth\frac{n\pi b}{2l}\cos\frac{n\pi y}{l}\right\} = \sum_0^\infty \alpha_n \cos\frac{n\pi y}{l} \tag{19}$$

This equation determines the relation between the geometrical properties of the blade and its aerodynamic behavior. It can be used to calculate the geometrical angle at attack Θ when the lift distribution is given, or to calculate the lift distribution from given Θ. Since for constant gap and stagger the life slope k is a function of chord c, both k and c/c_0 are variables along the blade. In order to justify the use of two-dimensional section properties of the lattice, the rate of change of k and c along the blade should be small. This condition is fortunately satisfied for the practical cases where the chord of the blades is approximately constant.

Application I

As an example for the application the results obtained in the previous sections, take the case of constant chord and linearly increasing lift from root to tip of the blade. Thus $c/c_0 = 1$, and

$$\Theta = \frac{\alpha_0}{k} + \sum_1^\infty\left\{\frac{1}{k} + \frac{\pi c_0}{8l}n\coth\frac{n\pi b}{2l}\right\}\alpha_n \cos\frac{n\pi y}{l} \tag{20}$$

However, since the lift coefficient c_l is defined as the lift per unit length or $\rho U\Gamma$ divided by $\frac{1}{2}\rho U^2 c_0$, Eqs. (2) and (18) give

$$c_l = \sum_{0}^{\infty} \alpha_n \cos \frac{n\pi y}{l} \tag{21}$$

Eq. (20) can be written as

$$\Theta = \frac{c_l}{k} + \frac{\pi c_0}{8l} \sum_{1}^{\infty} n \coth \frac{n\pi b}{2l} \alpha_n \cos \frac{n\pi y}{l} \tag{22}$$

If the value of c_l at the root is c_{l_0} and that at the tip is c_{l_1} then at any point y, the lift coefficient is expressed as

$$c_l = c_{l_0} + (c_{l_1} - c_{l_0}) \frac{y}{l} \tag{23}$$

By substituting Eq. (23) into Eq. (21), the values of α_n's can be easly determined as

$$\left.\begin{aligned} \alpha_0 &= \frac{c_{l_0} + c_{l_1}}{2} \\ \alpha_n &= 0 \quad \text{for } n \text{ even} \\ \alpha_n &= -\frac{4}{\pi^2} \frac{c_{l_1} - c_{l_0}}{n^2} \quad \text{for } n \text{ odd} \end{aligned}\right\} \tag{24}$$

Therefore Eq. (22) gives the geometrical angle of attack Θ as

$$\begin{aligned} \Theta &= \frac{c_l}{k} - \frac{1}{2\pi} \frac{c_0}{l} (c_{l_1} - c_{l_0}) \sum_{1,3,5}^{\infty} \frac{\coth \frac{n\pi b}{2l}}{n} \cos \frac{n\pi y}{l} \\ &= \frac{c_l}{k} - \frac{1}{2\pi} \frac{c_0}{l} (c_{l_1} - c_{l_0}) \left\{ \sum_{1,3,5}^{\infty} \frac{1}{n} \cos \frac{n\pi y}{l} + \sum_{1,3,5}^{\infty} \frac{\left(\coth \frac{n\pi b}{2l} - 1\right)}{n} \cos \frac{n\pi y}{l} \right\} \end{aligned} \tag{25}$$

But the first infinite series can be summed, i.e.,

$$\sum_{1,3,5}^{\infty} \frac{1}{n} \cos \frac{n\pi y}{l} = \frac{1}{2} \log\left(\cot \frac{\pi y}{2l} \right)$$

Then the final form for the geometrical angle of attack is given by

$$\left(\Theta - \frac{c_l}{k}\right) = -\frac{1}{2\pi} \frac{c_0}{l} (c_{l_1} - c_{l_0}) \left[\frac{1}{2} \log\left(\cot \frac{\pi y}{2l} \right) + \sum_{1,3,5}^{\infty} \frac{\coth \frac{n\pi b}{2l} - 1}{n} \cos \frac{n\pi y}{l} \right] \tag{26}$$

The series in Eq. (26) is rapidly convergent and can be calculated without any difficulty. It represents the effect of the lattice, because the value of the series vanishes when the gap b approaches infinity. The result of calculation for the case of $l/b = 3$ is shown in Fig. 5, where the additional twist necessary to obtain the linearly increasing lift distribution is plotted against y/l, the distance along the blade. At points near the root and the tip, the additional twist becomes very large. This is in contradiction to the simplifying assumptions made and the result is only quantitative at those regions.

The "induced drag" or the component of the lift force in the direction of the inflow velocity U for the blade from y to $y + \mathrm{d}y$ is (Fig. 3):

$$\frac{w}{U}\rho U\Gamma \mathrm{d}y$$

Let D be the total induced drag of one blade, then by using Eqs. (2), (16) and (18)

$$D = \frac{\rho}{2}U^2 c_0 \frac{\pi}{8}\frac{c_0}{l}\int_0^l \sum_0^\infty \alpha_n \cos\frac{n\pi y}{l}\sum_1^\infty n\alpha_n \coth\frac{n\pi b}{2l}\cos\frac{n\pi y}{l}\mathrm{d}y$$

$$= \frac{\rho}{2}U^2 c_0 \frac{\pi}{8}\frac{c_0}{l}\frac{l}{2}\sum_1^\infty n\alpha_n^2 \coth\frac{n\pi b}{2l} \tag{27}$$

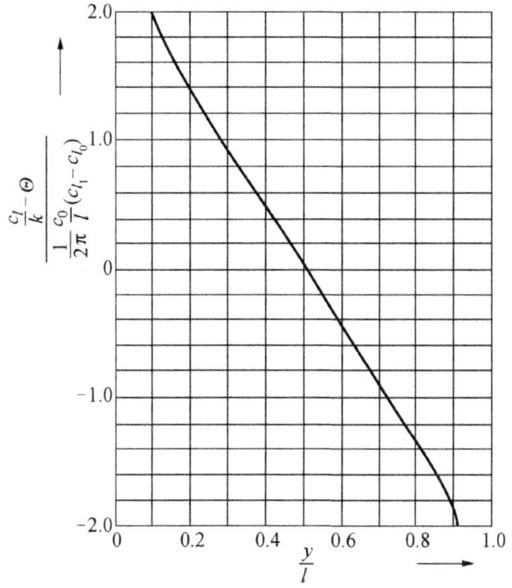

Fig. 5 Additional twist due to trailing vortices for linear lift distribution, $l/b = 3$

For the case under consideration, α_n's are given by Eq. (24). Therefore

$$D = \frac{\rho}{2}U^2 c_0 \frac{\pi}{8}\frac{c_0}{l}\frac{l}{2}\frac{16}{\pi^4}(c_{l_1} - c_{l_0})^2 \sum_{1,3,5}^\infty \frac{\coth\frac{n\pi b}{2l}}{n^3}$$

$$= \frac{\rho}{2}U^2 c_0 \left(\frac{c_0}{l}\right)\frac{(c_{l_1} - c_{l_0})^2}{\pi^3} l\left\{1.051\,80 + \sum_{1,3,5}^\infty \frac{\coth\frac{n\pi b}{2l} - 1}{n^3}\right\} \tag{28}$$

The energy loss associated with this drag is DU. The total kinetic energy in the inflow streams between two blades is $\rho Ubl \cdot U^2/2$. The ratio ε of loss due to trailing vortices and the inflow kinetic energy is thus

$$\varepsilon = \frac{c_0}{b}\frac{c_0}{l}\frac{(c_{l_1} - c_{l_0})^2}{\pi^3}\left\{1.051\,80 + \sum_{1,3,5}^\infty \frac{\coth\frac{n\pi b}{2l} - 1}{n^3}\right\} \tag{29}$$

The series is again rapidly convergent. It can be seen from Eq. (19) that the loss due to trailing vortices is quite small as c_l is limited to values less than unity for airfoil blade sections.

Application Ⅱ

In the previous section, the problem of calculating the geometrical angle of attack Θ from given lift distribution is considered. The inverse problem of finding the lift distribution from given Θ will be now investigated. As before, k and c are assumed to be constants, and Θ is taken as linearly increasing from Θ_0 at the root to Θ_1 at the tip. Then

$$\Theta = \Theta_0 + (\Theta_1 - \Theta_0)\frac{y}{l} \tag{30}$$

The first step to solve this problem is to expand Θ into a Fourier series. Thus

$$\Theta = \sum_0^\infty \theta_n \cos\frac{n\pi y}{l} \tag{31}$$

where

$$\left.\begin{aligned} \theta_0 &= \frac{\Theta_0 + \Theta_1}{2} \\ \theta_n &= 0 \quad \text{for } n \text{ even} \\ \theta_n &= -\frac{4}{\pi^2}\frac{\Theta_1 - \Theta_0}{n^2} \quad \text{for } n \text{ odd} \end{aligned}\right\} \tag{32}$$

By substituting Eqs. (31) and (32) into Eq. (19), the lift coefficient distribution is determined. The result is

$$c_l = k\frac{\Theta_0 + \Theta_1}{2} - \frac{4k}{\pi^2}(\Theta_1 - \Theta_0)\sum_{1,3,5}^\infty \frac{\cos\frac{n\pi y}{l}}{\left(1 + \frac{\pi}{8}k\frac{c_0}{l}n\coth\frac{n\pi b}{2l}\right)n^2} \tag{33}$$

The series converges rapidly and can be easily calculated.

The fraction ε of the kinetic energy lost to the trailing vortices is obtained by using Eq. (27). The value is given by:

$$\varepsilon = \frac{c_0}{b}\frac{c_0}{l}\frac{k^2(\Theta_1 - \Theta_0)^2}{\pi^3}\sum_{1,3,5}^\infty \frac{\coth\frac{n\pi b}{2l}}{n^3\left[1 + \frac{\pi}{8}k\frac{c_0}{l}n\coth\frac{n\pi b}{2l}\right]^2} \tag{34}$$

Since $k\Theta$ is of the order of lift coefficient, the magnitude of loss is therefore again quite small. Hence, the method of twisting the blade is a very efficient device to avoid the radial flow in the axial compressor or turbine.

Appendix

The integral I_n of Eq. (13) can be evaluated in the following way

$$I_n = \frac{1}{2b}\int_0^\infty \sin\frac{n\pi\xi}{l}\coth\frac{\pi\xi}{b}d\xi = \frac{1}{2\pi}\int_0^\infty \sin\left(\frac{nb}{l}t\right)\coth t dt$$

where $t = \dfrac{\pi\xi}{b}$. However $\coth t = 1 + 2\sum_1^\infty e^{-2mt}$. This series is uniformly convergent except at the lower limit of the integral. Hence it can be substituted into the integrand and the integral evaluated term by term, provided the resultant series is convergent. Thus

$$\begin{aligned}
I_n &= \frac{1}{2\pi}\left[\int_0^\infty \sin\left(\frac{nb}{l}\cdot t\right)dt + 2\sum_{m=1}^\infty\int_0^\infty \sin\left(\frac{nb}{l}\cdot t\right)e^{-2mt}\cdot dt\right] \\
&= C + \frac{1}{2\pi}\left[\frac{1}{\frac{nb}{l}} + 2\sum_{m=1}^\infty \frac{\frac{nb}{l}}{\left(\frac{nb}{l}\right)^2 + (2m)^2}\right] \\
&= C + \frac{1}{4}\left[\frac{1}{\frac{n\pi b}{2l}} + 2\sum_{m=1}^\infty \frac{\frac{n\pi b}{2l}}{\left(\frac{n\pi b}{2l}\right)^2 + m^2\pi^2}\right] \\
&= C + \frac{1}{4}\coth\frac{n\pi b}{2l}
\end{aligned}$$

The constant C is not determined due to the first integral in the above equation. However, when $b \to \infty$, $I_n = C + \dfrac{1}{4}$ But then

$$\begin{aligned}
\lim_{b\to\infty} I_n &= \lim_{b\to\infty}\frac{1}{2b}\int_0^\infty \sin\frac{n\pi\xi}{l}\frac{b}{\pi\xi}d\xi \\
&= \frac{1}{2\pi}\int_0^\infty \sin\frac{n\pi\xi}{l}\frac{d\xi}{\xi} = \frac{1}{4}
\end{aligned}$$

Therefore, $C = 0$. Consequently the integral I_n is given by Eq. (14).

Reference

[1] Glauert H. Aerofoil and Aircrew Theory. Cambridge University Press, 1930: 127 - 128.

Lifting-Line Theory for a Wing in Non-uniform Flow *

By Theodore von Kármán and Hsue-shen Tsien
(*California Institute of Technology*)

1. Introduction. Prandtl's theory of the lifting line gave the answer to most of the questions in the aerodynamic design of airplane wings. Thus the three-dimensional wing theory became a standard tool of airplane designers. One restriction involved in the conventional wing theory is the uniformity of the undisturbed flow in which the wing is placed. Now there are many important cases which do not satisfy this condition. For instance, in the case of a wing spanning an open jet wind tunnel, the velocity of the air stream has a maximum at the center of the jet and drops to zero outside of the jet. Another example is the problem of the influence of the propeller slip-stream on the characteristics of the wing. Here the higher velocity of the propeller slip-stream makes the application of the Prandtl wing theory difficult. Such cases led several authors to investigate the problem of a wing in non-uniform flow. Some investigators found a satisfactory solution of the problem for the case of "stepwise" velocity distribution. In this case the flow in regions of uniform velocity can be determined by using Prandtl's concepts with additional continuity conditions at the boundaries between such regions. On the other hand, the problem of a continuously varying velocity field seems to need an appropriate treatment. K. Bausch[1] has tried to modify the Prandtl theory for the case of small inhomogeneity in the air stream; however, besides the restriction of slight deviation from uniform flow, his method encounters a further difficulty in estimating the error introduced by the approximations. The seriousness of this difficulty becomes evident when one tries to compare the results of Bausch with that of F. Vandrey[2]. Vandrey considers the problem with variable velocity as the limiting case of a wing in a stepwise velocity field, and his result seems to differ from that of Bausch. Recently R. P. Isaacs[3] has investigated the same problem, but the authors have not yet had the opportunity to study his work.

It seems to the authors that a general and more satisfactory solution for the flow of a wing in a non-uniform stream can be obtained by studying the three-dimensional problem anew in this generalized case, introducing the modifications of Prandtl's fundamental concepts. The first fundamental concept is the following: the span of the wing is sufficiently large compared with the chord so that the variation of the velocities in the spanwise direction is small when

---

Quarterly of Applied Mathematics, vol. 3, pp. 1 – 11, 1945.

* Received September 27, 1944.

compared with the variation of the velocities in a plane normal to the span; then the flow at each sectional plane perpendicular to the span can be considered as a two-dimensional flow around an airfoil. The only additional feature for the flow in this sectional plane is the modification of the geometrical angle of attack, as defined by the undisturbed flow, on account of the so-called induced velocity. The second fundamental concept of Prandtl is the replacement of the wing by a lifting line having the same distribution of lifting forces along the span as the wing. This concept, with the additional assumption that the disturbance caused by the lifting line is small, i. e., that the wing is lightly loaded, makes the calculation of the induced velocity relatively simple. In this paper the authors will study the flow around a lightly loaded lifting line placed in a parallel stream whose velocity is perpendicular to the span (Fig. 1) and is assumed to vary in both directions normal to the flow. Due to the rather complicated character of the flow, the usual concept of the picturesque system of trailing vortices encountered in Prandtl's wing theory is not very useful here. A method, which is mathematically more convenient, has to be adopted. This method has already been used by the senior author[4] in explaining the similarity between Prandtl's wing theory and the theory of planning surfaces. After the general theory is formulated, the problem of minimum induced drag will be considered. Finally a general expression for calculating the induced drag of a wing in a stream of varying velocity will be presented.

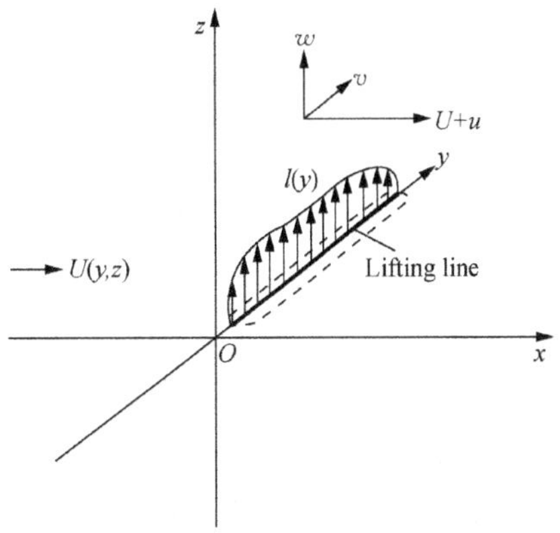

Fig. 1 Lifting line in a non-uniform flow

Of course, the complete solution of the problem of a wing in a non-uniform stream requires a knowledge of the "section characteristic" or the two-dimensional properties of the airfoil sections of the wing. If the velocity of the main stream is varying only in the direction of the span, the required section characteristics are those of an airfoil in a two-dimensional uniform flow, and are common knowledge in applied aerodynamics. However, if the velocity of the main stream is also varying in a direction perpendicular to the span and to the velocity itself, the required section characteristics are those of an airfoil in a two-dimensional non-uniform flow. Such flow problems have not yet been studied extensively[5].

2. General theory of a lifting line. Let the x-axis be parallel to the direction of the main flow, the y-axis coincide with the lifting line and the z-axis be normal to the lifting line (Fig. 1). If p is the pressure, ρ the density, and v_1, v_2, v_3 the components of the velocity, the dynamical equations for the steady motion of an inviscid, incompressible fluid without external forces are

$$v_1 \frac{\partial v_1}{\partial x} + v_2 \frac{\partial v_1}{\partial y} + v_3 \frac{\partial v_1}{\partial z} = -\frac{1}{\rho} \frac{\partial p}{\partial x} \tag{1}$$

$$v_1 \frac{\partial v_2}{\partial x} + v_2 \frac{\partial v_2}{\partial y} + v_3 \frac{\partial v_2}{\partial z} = -\frac{1}{\rho} \frac{\partial p}{\partial y} \tag{2}$$

$$v_1 \frac{\partial v_3}{\partial x} + v_2 \frac{\partial v_3}{\partial y} + v_3 \frac{\partial v_3}{\partial z} = -\frac{1}{\rho} \frac{\partial p}{\partial z} \tag{3}$$

The equation of continuity is

$$\frac{\partial v_1}{\partial x} + \frac{\partial v_2}{\partial y} + \frac{\partial v_3}{\partial z} = 0 \tag{4}$$

Equations (1) to (4) constitute a system of four simultaneous equations for the four unknowns v_1, v_2, v_3 and p.

For the particular problem of a lightly loaded lifting line, the velocity components can be expressed in the following forms:

$$v_1 = U + u \quad (5); \qquad v_2 = v \quad (6); \qquad v_3 = w \quad (7)$$

Here u, v, w are the velocity components due to the presence of the lifting line and U is the main stream velocity assumed to be a function of y and z but independent of x. Since the lifting line is assumed to be lightly loaded, u, v and w are small compared with the main velocity U. By substituting Eqs. (5) to (7) into the dynamical equations and neglecting higher order terms, a set of linear equations for u, v and w is obtained. Thus

$$U \frac{\partial U}{\partial x} + v \frac{\partial U}{\partial y} + w \frac{\partial U}{\partial z} = -\frac{1}{\rho} \frac{\partial p}{\partial x} \tag{8}$$

$$U \frac{\partial v}{\partial x} = -\frac{1}{\rho} \frac{\partial p}{\partial y} \quad (9); \qquad U \frac{\partial w}{\partial x} = -\frac{1}{\rho} \frac{\partial p}{\partial z} \quad (10)$$

Then the equation of continuity becomes

$$\frac{\partial u}{\partial x} + \frac{\partial v}{\partial y} + \frac{\partial w}{\partial z} = 0 \tag{11}$$

If Eqs. (8), (9) and (10) are differentiated with respect to x, y and z respectively and the results added, the sum can be simplified by using Eq. (11) and can, finally, be written in the form

$$\frac{1}{U^2} \frac{\partial^2 p}{\partial x^2} + \frac{\partial}{\partial y}\left(\frac{1}{U^2} \frac{\partial p}{\partial y}\right) + \frac{\partial}{\partial z}\left(\frac{1}{U^2} \frac{\partial p}{\partial z}\right) = 0 \tag{12}$$

This is now an equation for the pressure p only and can be used conveniently as the starting point of the solution. If the pressure of the undisturbed main flow is chosen as the reference pressure and set equal to zero, one of the boundary conditions to be satisfied by p is

$$p = 0, \quad \text{for} \quad |x| \to \infty, \quad |y| \to \infty, \quad \text{or} \quad |z| \to \infty \tag{13}$$

The condition at the lifting line, or y-axis, is that the lifting force is represented by a suction force on the "upper surface" of the lifting line and a pressure force of equal magnitude on the "lower surface" (Fig. 2). Hence the pressure p must satisfy the following expressions

$$\int_{-\varepsilon}^{\varepsilon} p\,dx = -\frac{1}{2}l(y), \quad \text{for} \quad z = +0 \tag{14}$$

and

$$\int_{-\varepsilon}^{\varepsilon} p\,dx = \frac{1}{2}l(y), \quad \text{for} \quad z = -0 \tag{15}$$

where $l(y)$ is the lift per unit length of the lifting line at the point y. Furthermore, on account of the symmetry of the flow,

$$p = 0 \quad \text{for} \quad z = 0, \quad |x| > \varepsilon \tag{16}$$

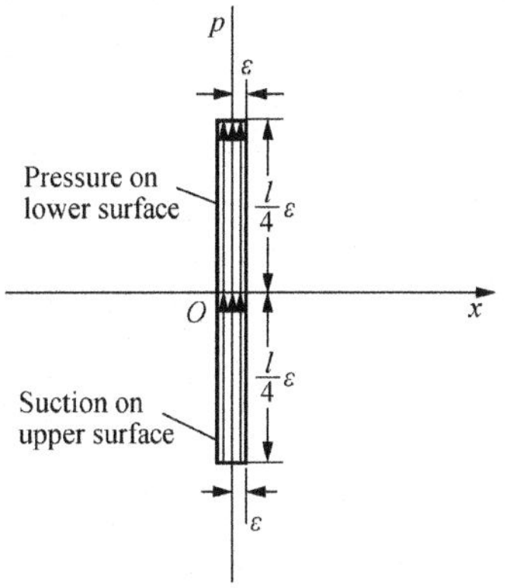

Fig. 2 Representation of lift as pressure forces acting on the two "surfaces" of the lifting line

To solve Eq. (12) together with the boundary conditions given by Eqs. (13) to (16), the Fourier integral theorem can be used to build up the solution of the problem from the elementary solutions of Eq. (12) of the form

$$P(y, z, \lambda)\cos\lambda x$$

The equation to be satisfied by P is

$$U^2 \frac{\partial}{\partial y}\left(\frac{1}{U^2}\frac{\partial P}{\partial y}\right) + U^2 \frac{\partial}{\partial z}\left(\frac{1}{U^2}\frac{\partial P}{\partial z}\right) - \lambda^2 P = 0 \tag{17}$$

To determine P uniquely, it is convenient to impose the following conditions

$$P = 0, \quad \text{for} \quad |y| \to \infty, \quad |z| \to \infty \tag{18}$$

$$P = -\frac{1}{2}l(y) \quad \text{for} \quad z = +0 \tag{19}$$

$$P = \frac{1}{2}l(y) \quad \text{for} \quad z = -0 \tag{20}$$

The required solution for p can then be written as

$$p = \frac{1}{\pi}\int_0^{\infty} \cos\lambda x P(y, z, \lambda)\,d\lambda \tag{21}$$

By substituting Eq. (21) into Eqs. (9) and (10), the "induced velocities" v and w are obtained;

$$v(x, y, z) = v(0, y, z) - \frac{1}{\rho U}\frac{1}{\pi}\int_0^{\infty} \frac{\sin\lambda x}{\lambda}\frac{\partial}{\partial y}P(y, z, \lambda)\,d\lambda \tag{22}$$

$$w(x, y, z) = w(0, y, z) - \frac{1}{\rho U}\frac{1}{\pi}\int_0^{\infty} \frac{\sin\lambda x}{\lambda}\frac{\partial}{\partial z}P(y, z, \lambda)\,d\lambda \tag{23}$$

Because the integrals are odd functions of x, the following relations hold for velocities far ahead of the lifting line and far behind the lifting line:

$$\frac{1}{2}[v(-\infty,y,z)+v(\infty,y,z)]=v(0,y,z),\frac{1}{2}[w(-\infty,y,z)+w(\infty,y,z)]=w(0,y,z)$$

However, it is evident that the induced velocities far ahead of the lifting lines must be zero. Hence

$$v(0,y,z)=\frac{1}{2}v(\infty,y,z) \qquad (24); \qquad w(0,y,z)=\frac{1}{2}w(\infty,y,z) \qquad (25)$$

The induced velocities v and w at the lifting line are then one-half of those far downstream. This is in accordance with the usual wing theory based upon the concept of trailing vortices.

One meets an apparent difficulty if the x component of the induced velocity is calculated; integration of Eq. (8) with respect to x furnishes the x-component of the induced velocity:

$$u=-\frac{1}{\rho U}p-\frac{1}{U}\frac{\partial U}{\partial y}\int_{-\infty}^{x}v\,dx-\frac{1}{U}\frac{\partial U}{\partial z}\int_{-\infty}^{x}w\,dx \qquad (26)$$

Since p tends to zero, v and w tend to finite quantities as x tends to infinity, and u increases indefinitely as x tends to infinity. This is in contradiction to the assumption of small disturbances introduced at the beginning of the present investigation. However, it is believed that this difficulty does not prevent the application of the theory to practical cases, since the apparent large value of the u component is due to the distortion of the variable main stream by the induced cross flow and the infinite value for $x\to\infty$ is due to the linearization of the differential equations. Some further remarks on this point are given in Section 4.

3. Conditions far downstream. For the application of the lifting-line theory to the wing problem, the quantity of primary interest is the z component of the induced velocity at the lifting line. The simple relations given by Eqs. (24) and (25) suggest a possible simplification of the calculation by considering conditions far downstream, or the "Trefftz plane" according to the terminology of the conventional wing theory. To abbreviate the notation, we let

$$\left.\begin{aligned} v_0&=v(0,y,z), & w_0&=w(0,y,z)\\ v_1&=v(\infty,y,z), & w_1&=w(\infty,y,z)\end{aligned}\right\} \qquad (27)$$

Then, according to (24) and (25), $v_0=\frac{1}{2}v_1$, $w_0=\frac{1}{2}w_1$. Therefore, Eqs. (22) and (23) give

$$v_1=-\frac{1}{\rho U}\lim_{x\to\infty}\frac{1}{\pi}\int_0^{\infty}\frac{\sin\lambda x}{\lambda}\frac{\partial}{\partial y}P(y,z,\lambda)\,d\lambda$$

$$w_1=-\frac{1}{\rho U}\lim_{x\to\infty}\frac{2}{\pi}\int_0^{\infty}\frac{\sin\lambda x}{\lambda}\frac{\partial}{\partial z}P(y,z,\lambda)\,d\lambda$$

Let us consider $P(y, z, \lambda)$ as a regular function of λ; then

$$P(y,z,\lambda)=P(y,z,0)+\lambda\left[\frac{\partial P}{\partial\lambda}\right]_{\lambda=0}+\ldots$$

By using the variable $t=\lambda x$, the expressions for v_1 and w_1 can be rewritten,

$$v_1 = -\frac{1}{\rho U}\lim_{x\to\infty}\frac{2}{\pi}\int_0^\infty \frac{\sin t}{t}\frac{\partial}{\partial y}\left[P(y,z,0)+\frac{t}{x}\left(\frac{\partial P}{\partial \lambda}\right)_{\lambda=0}+\ldots\right]\mathrm{d}t$$

$$w_1 = -\frac{1}{\rho U}\lim_{x\to\infty}\frac{2}{\pi}\int_0^\infty \frac{\sin t}{t}\frac{\partial}{\partial z}\left[P(y,z,0)+\frac{t}{x}\left(\frac{\partial P}{\partial \lambda}\right)_{\lambda=0}+\ldots\right]\mathrm{d}t$$

At the limit, only the first terms of the integrands are significant, and furthermore

$$\frac{2}{\pi}\int_0^\infty \frac{\sin t}{t}\mathrm{d}t = 1$$

Hence

$$v_1 = -\frac{1}{\rho U}\frac{\partial}{\partial y}P(y,z,0) \qquad (28); \qquad w_1 = -\frac{1}{\rho U}\frac{\partial}{\partial z}P(y,z,0) \qquad (29)$$

Equations (28) and (29) simplify the problem of calculating the induced velocities at the Trefftz plane considerably. In fact, by introducing a "potential function" ϕ defined by the relation

$$\phi(y,z) = -P(y,z,0) \qquad (30)$$

the problem can be formulated as follows: the differential equation to be satisfied by ϕ can be deduced from Eq. (17) by setting $\lambda = 0$; thus

$$\frac{\partial}{\partial y}\left(\frac{1}{U^2}\frac{\partial\phi}{\partial y}\right)+\frac{\partial}{\partial z}\left(\frac{1}{U^2}\frac{\partial\phi}{\partial z}\right)=0 \qquad (31)$$

The boundary conditions to be satisfied by ϕ are

$$\phi = 0 \quad \text{for} \quad |y|\to\infty, \quad |z|\to\infty \qquad (32)$$

$$\phi = l(y)/2 \quad \text{for} \quad z=+0 \qquad (33)$$

$$\phi = -l(y)/2 \quad \text{for} \quad z=-0 \qquad (34)$$

Then

$$v_1 = \frac{1}{\rho U}\frac{\partial\phi}{\partial y} \qquad (35); \qquad w_1 = \frac{1}{\rho U}\frac{\partial\phi}{\partial z} \qquad (36)$$

By substituting Eqs. (35) and (36) into Eq. (31), one has

$$\frac{\partial}{\partial y}\left(\frac{v_1}{U}\right)+\frac{\partial}{\partial z}\left(\frac{w_1}{U}\right)=0 \qquad (37)$$

This equation has a very simple physical meaning. Since v_1 and w_1 are considered to be small quantities, the ratios v_1/U and w_1/U are the angles of inclination, β and γ, of the stream lines with respect to the zx and xy planes. Consider parallel planes perpendicular to the x-axis and $\mathrm{d}x$ apart (Fig. 3). If the width of the stream tube at the section x is δ_y, then at the section $x+\mathrm{d}x$, the width of the stream tube is $\delta_y[1+\mathrm{d}x\partial\beta/\partial y]$. If the height of the stream tube at the section x is δ_z, then at the section $x+\mathrm{d}x$, the height of the stream tube is $\delta_z[1+\mathrm{d}x\partial\gamma/\partial z]$. The total increase in the cross-sectional area of the stream tube from x to $x+\mathrm{d}x$ is then

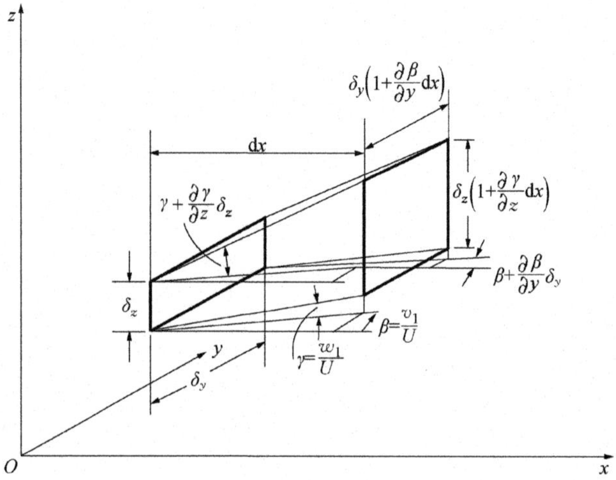

Fig. 3 Stream tube far downstream from the lifting line

approximately

$$\delta_y\delta_z\left(\frac{\partial\beta}{\partial y}+\frac{\partial\gamma}{\partial z}\right)dx$$

Now at the Trefftz plane, the flow field can be considered as settled into a uniform condition; i.e., the pressure is constant in the x-direction. Hence, the velocity of the flow along any stream tube is constant. Then the cross-sectional area of the stream tube must be also constant. Therefore,

$$\frac{\partial\beta}{\partial y}+\frac{\partial\gamma}{\partial z}=0$$

which is simply Eq. (37). From this point of view, Eq. (37) is really the equation of continuity, simplified under the conditions prevailing at the Trefftz plane.

On the other hand, ϕ can be eliminated from Eqs. (35) and (36). The result is

$$\frac{\partial}{\partial z}(Uv_1)-\frac{\partial}{\partial y}(Uw_1)=0 \tag{38}$$

This equation can be considered as the modified vorticity equation. It actually holds for all values of x under the approximation assumed in the present investigation. This can be seen in the following way: since U is a function of y and z but independent of x, Eqs. (9) and (10) can be written in the form

$$\frac{\partial}{\partial x}Uv=-\frac{1}{\rho}\frac{\partial p}{\partial y},\quad \frac{\partial}{\partial x}Uw=-\frac{1}{\rho}\frac{\partial p}{\partial z}$$

By differentiating the first equation with respect to z and the second equation with respect to y and then subtracting, the result is

$$\frac{\partial}{\partial x}\left[\frac{\partial}{\partial z}(Uv)-\frac{\partial}{\partial y}(Uw)\right]=0$$

Thus

$$\frac{\partial}{\partial z}(Uv) - \frac{\partial}{\partial y}(Uw) = \text{a function of } y \text{ and } z$$

But for points far upstream, or for $x = -\infty$, v and w vanish; therefore the function of y and z on the right of above equation must be identically zero. Hence for all values of x,

$$\frac{\partial}{\partial z}(Uv) - \frac{\partial}{\partial y}(Uw) = 0 \tag{39}$$

It should be noted here that Eqs. (37), (38) and (39) are obtained without any reference to the lifting line and hence they are true for more general cases. However, the complete determination of v_1 and w_1 requires a knowledge of the relation between the induced velocities and the lift on the wing. This relation depends upon the type of lift distribution. For the particular case of a lifting line, this relation is supplied by Eqs. (33) and (34).

Equation (37) can be identically satisfied by introducing the "stream function" ψ defined by

$$v_1 = U\frac{\partial \psi}{\partial z}, \quad w_1 = -U\frac{\partial \psi}{\partial y} \tag{40}$$

Then Eq. (38) gives the differential equation for ψ :

$$\frac{\partial}{\partial y}\left(U^2\frac{\partial \psi}{\partial y}\right) + \frac{\partial}{\partial z}\left(U^2\frac{\partial \psi}{\partial z}\right) = 0 \tag{41}$$

Both Eq. (31) and Eq. (41) reduce to the Laplace equation for the conventional wing theory when U is a constant.

4. Minimum induced drag. The induced downwash angle at the lifting line is equal to w_0/U or $\frac{1}{2}w_1/U$, according to Eq. (25). Therefore, Eq. (36) gives the downwash angle at the lifting line as $[1/2\rho U^2](\partial\phi/\partial z)_{z=0}$, and the induced drag D_i can then be expressed as

$$D_i = -\frac{1}{2\rho}\int[\phi(y, +0) - \phi(y, -0)]\frac{1}{U^2}\left(\frac{\partial \phi}{\partial z}\right)_{z=0} dy = \frac{1}{2\rho}\int_C \frac{\phi}{U^2}\frac{\partial \phi}{\partial z} ds \tag{42}$$

The first integral is evaluated across the span of the lifting line. The second integral is calculated along a contour following the upper and lower "surface" of the horizontal strip shown in Fig. 4. Since $\phi \to 0$ for points far from the lifting line, the contour integral can be transformed into an area integral by Green's theorem, and

$$D_i = \frac{1}{2\rho}\iint\left\{\frac{\partial}{\partial y}\left(\frac{1}{U^2}\phi\frac{\partial \phi}{\partial y}\right) + \frac{\partial}{\partial z}\left(\frac{1}{U^2}\phi\frac{\partial \phi}{\partial z}\right)\right\} dydz \tag{43}$$

This integral extends throughout the region outside of the lifting line. Since ϕ satisfies the differential equation (31), Eq. (43) reduces to

$$D_i = \frac{\rho}{2}\iint\left\{\left(\frac{1}{\rho U}\frac{\partial \phi}{\partial y}\right)^2 + \left(\frac{1}{\rho U}\frac{\partial \phi}{\partial z}\right)^2\right\} dydz \tag{44}$$

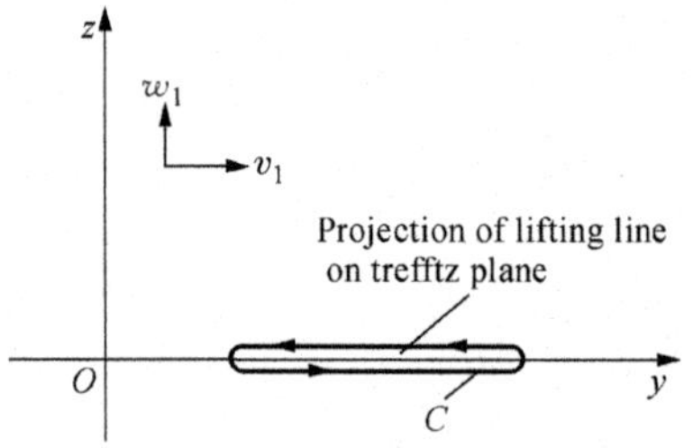

Fig. 4 Contour integration in the Trefftz plane

Therefore, the induced drag is represented by the kinetic energy corresponding to the velocity components v_1 and w_1 at the Trefftz plane. It is seen that the u component of the velocity does not appear in the expression for the induced drag. This is due to the fact that the increase of u with increasing x does not represent a real acceleration of a fluid element in the x direction. Rather, it is due to the fact that the cross flow transports fluid elements from regions of lower main velocity to regions of higher main velocity and vice versa. This is in accordance with the modified continuity equation (37) which clearly indicates that the cross section of the individual stream tubes has a definite limiting value for $x \to \infty$, and therefore the velocity component in the direction of the stream tube tends to a finite value.

The problem of minimum induced drag requires the determination of the minimum of D_i as given by Eq. (44) together with the condition that the total lift L remains fixed. Thus

$$L = \int l \mathrm{d}y = \int [\phi(y, +0) - \phi(y, -0)] \mathrm{d}y = -\int_C \phi \mathrm{d}s = \text{constant} \tag{45}$$

By using the method of Lagrange's multiplier, the above problem can be reduced to that of finding the minimum of $D_i + K/\rho L$, where K is a constant. Hence,

$$\delta D_i + \frac{K}{\rho}\delta L = 0 \tag{46}$$

The variation of the induced drag can be obtained from Eq. (44),

$$\delta D_i = \frac{1}{\rho}\iint\left\{\frac{1}{U}\frac{\partial\phi}{\partial y}\frac{1}{U}\frac{\partial\delta\phi}{\partial y} + \frac{1}{U}\frac{\partial\phi}{\partial z}\frac{1}{U}\frac{\partial\delta\phi}{\partial z}\right\}\mathrm{d}y\mathrm{d}z$$

However, ϕ must satisfy the differential equation (31); thus

$$\delta D_i = \frac{1}{\rho}\iint\left\{\frac{\partial}{\partial y}\left(\frac{1}{U^2}\frac{\partial\phi}{\partial y}\delta\phi\right) + \frac{\partial}{\partial z}\left(\frac{1}{U^2}\frac{\partial\phi}{\partial z}\delta\phi\right)\right\}\mathrm{d}y\mathrm{d}z = \frac{1}{\rho}\int_C \frac{1}{U^2}\frac{\partial\phi}{\partial z}\delta\phi\mathrm{d}s$$

On the other hand,

$$\delta L = -\int_C \delta\phi \mathrm{d}s$$

By substituting these results into Eq. (46), the condition of minimum induced drag is obtained in the form

$$\frac{1}{\rho}\int_C\left(\frac{1}{U^2}\frac{\partial\phi}{\partial z}-K\right)\delta\phi\, ds=0 \tag{47}$$

The variation of $\delta\phi$ on the lifting line is arbitrary; therefore the minimum induced drag is given by the condition that the induced downwash angle must be constant along the span. If the main stream velocity U is constant, the above condition is reduced to the requirement of constant downwash. This is in agreement with the well-known result of Prandtl's wing theory.

5. Flow with velocity varying in the direction of span only. If the stream velocity varies only in the y direction, i. e., in the direction of the wing span, the calculation of induced velocity and induced drag can be simplified with the aid of characteristic functions connected with the differential equation for the potential function ϕ. In this case Eq. (31) becomes

$$\frac{\partial^2\phi}{\partial y^2}+\frac{\partial^2\phi}{\partial z^2}-2\frac{\frac{dU}{dy}}{U}\frac{\partial\phi}{\partial y}=0 \tag{48}$$

To satisfy the boundary condition given by Eq. (32), ϕ is expressed by the following integral

$$\phi(y,z)=\int_0^\infty f(\lambda)e^{-\lambda z}Y_\lambda(y)\,d\lambda \tag{49}$$

for $z>0$. $f(\lambda)$ is an unknown function to be determined. For $z<0$,

$$\phi(y,z)=-\phi(y,-z) \tag{50}$$

By substituting Eq. (49) into Eq. (48), the differential equation for $Y_\lambda(y)$ is obtained,

$$\frac{d^2Y_\lambda}{dy^2}-2\frac{\frac{dU}{dy}}{U}\frac{dY_\lambda}{dy}+\lambda^2Y_\lambda=0 \tag{51}$$

This equation will determine $Y_\lambda(y)$ uniquely if proper normalizing and boundary conditions are imposed.

At the span, the condition (33) must be satisfied. Thus

$$\frac{l(y)}{2}=\int_0^\infty f(\lambda)Y_\lambda(y)\,d\lambda \tag{52}$$

This relation can be considered as the equation for determining $f(\lambda)$ with the given lift distribution $l(y)$. For example, in the case of constant stream velocity U or Prandtl's case, $Y_\lambda(y)$ is a trigonometric function and therefore $f(\lambda)$ can be determined easily by means of Fourier's inversion theorem. Equation (50) shows that with $f(\lambda)$ so determined, the condition (34) will be automatically satisfied.

The downwash velocity w_0 at the wing can then be easily calculated by using Eqs. (25), (36) and (49). The result is

$$w_0(y,0)=-\frac{1}{2\rho U}\int_0^\infty \lambda f(\lambda)Y_\lambda(y)\,d\lambda \tag{53}$$

The induced drag D_i is given by

$$D_i = -\int_{-\infty}^{\infty} l(y) \frac{w_0(y,0)}{U} \mathrm{d}y$$

Therefore, in terms of $Y_\lambda(y)$, the following general expression for the induced drag is obtained:

$$D_i = \int_{-\infty}^{\infty} \frac{1}{\rho U^2} \mathrm{d}y \int_0^{\infty} f(\lambda) Y_\lambda(y) \mathrm{d}\lambda \int_0^{\infty} \eta f(\eta) Y_\eta(y) \mathrm{d}\eta \tag{54}$$

Thus the problem of calculating the induced drag with a given distribution of lift $l(y)$ is reduced to the problem of solving the integral equation (52) for $f(\lambda)$ and then evaluating the integral given by Eq. (54).

If the chord c, the geometrical angle of attack α and the slope k of the lift coefficient are given instead of the lift distribution $l(y)$, then

$$l(y) = \frac{1}{2} \rho U^2 ck \left\{ \alpha + \frac{w_0(y,0)}{U} \right\} \tag{55}$$

Thus Eq. (52) is replaced by the following equation

$$\frac{1}{4} \rho U^2 ck \left\{ \alpha - \frac{1}{2\rho U^2} \int_0^{\infty} \lambda f(\lambda) Y_\lambda(y) \mathrm{d}\lambda \right\} = \int_0^{\infty} f(\lambda) Y_\lambda(y) \mathrm{d}\lambda$$

or

$$\frac{1}{4} \rho U^2 ck\alpha = \int_0^{\infty} \left(1 + \frac{ck}{8} \lambda \right) f(\lambda) Y_\lambda(y) \mathrm{d}\lambda \tag{56}$$

This is now the integral equation for $f(\lambda)$. When $f(\lambda)$ is determined, the induced drag D_i can be again calculated by using Eq. (54).

References

[1] Bausch K. Auftriebsverteilung und daraus abgeleitete Grössen für Tragflügel in schwach inhomogenen Strömungen. Luftfahrtforschung, 1939, 16:129 – 134.

[2] Vandrey F. Beitrag zur Theorie des Tragflügels in schwach inhomogener Parallelstromung. Zeitschrift f. angew. Math. u. Mech. 1940, 20: 148 – 152.

[3] Isaacs R P. Airfoil theory for. flows of variable velocity. abstract in Bulletin of the American Mathematical Society, 1944, 50:186.

[4] von Kármán Th. Neue Darstellung der Tragflügeltheorie. Zeitschrift f. angew. Math. u. Mech., 1935, 15:56 –61.

[5] Tsien H S. Symmetrical Joukowsky airfoils in shear flow. Quart. Appl. Math., 1943, 1: 130 – 148.

Atomic Energy

Hsue-shen Tsien*
(*California Institute of Technology*)

Introduction

The spectacular results achieved by the use of atomic bombs toward the end of World War II have greatly stimulated the interest in the possibilities of atomic energy in other fields of engineering applications. The very fact that the energy release by nuclear reactions is approximately a million times that of the more conventional chemical reactions seems to stagger one's imagination. Although there are wide differences in opinion as to the time required to develop this branch of new-found knowledge to practical power-plant engineering, there is no diversity in the belief that a new era of technological evolution has begun. To keep abreast with the expected rapid advancement in this field, the engineers should have a clear understanding of the fundamental concepts involved as advocated by von Kármán in a recent article[1].

The aeronautical engineers have a further interest in the matter due to the belief that the aeronautical power plant will probably be the first prime mover to utilize the atomic energy. This belief is based upon the following fact: For stationary power plants, criterion for economic operation is the cost to produce one kilowatt-hour and thus the cost of the fuel, be it chemical or atomic, is of prime importance instead of the weight of the fuel. For automotive applications and especially aeronautical applications, the weight of the fuel is extremely important. In case of the projected supersonic flight at extreme speeds, it seems that only the use of atomic energy will make such flight economically feasible by reducing the fuel load and increasing the pay load.

This article is a digest of the basic concepts in this branch of knowledge as an introductory study of this new field. The enormous energy release of nuclear reaction is explained through the binding energy of the atomic nuclei, with the energy generation in the stars and the atomic bombs as typical examples. Frequent comparison between molecular reaction and nuclear reaction will be made to guide the exploration of an unfamiliar land.

Received September 29, 1945.
Journal of the Aeronautical Sciences, vol. 13, pp. 171–180, 1946.
* Daniel Guggenheim Aeronautical Laboratory.

The Equivalence of Energy and Mass

The astonishing amount of energy released by the nuclear reactions can be easily understood through the still greater energy potential in the atom itself represented by its mass. Therefore, one of the fundamental concepts of the atomic energy is the concept of the equivalence of energy and mass. This concept was first announced by Einstein as a conclusion of his special theory of relativity. This theory deals with the comparison of measurements made by observers who are assumed to be in unaccelerated relative motion and in a region of space and time where the action of gravitation can be neglected. The basis of the theory can be conveniently regarded as the two following fundamental postulates[2]:

(a) It is impossible to measure or detect the unaccelerated translatory motion of a system through free space or through any ether-like medium that might be assumed to pervade it.

(b) The velocity of light in free space is the same for all observers, independent of the relative velocity of the source of light and the observers.

The first postulate is derived from the failure to detect the ether drift of the motion of the earth through the space, and the second postulate can be derived from both astronomical evidences and laboratory experiments. Therefore, although Einstein's theory is often disseminated by popular science writers as an abstruse mathematical theory, it really is solidly founded on observational facts. In this sense, the relativity theory is just as empirical as a stress-strain curve drawn through test points obtained on a steel specimen.

The first postulate of special relativity is already contained in the Newtonian dynamics and can be accepted as part of everyday experience. The second postulate, although familiar in wave theory of light, is somewhat beyond the daily experience. In fact, one would generally expect the velocity of propagation and the velocity of the source to be additive. The combination of these two postulates of widely different natures would certainly lead to entirely new results. This expectation is, of course, fully substantiated. One of the results of deduction is that, if the mass of a particle is m_0 at rest, then its mass m at velocity V is

$$m = m_0 / \sqrt{1 - (V^2/c^2)} \tag{1}$$

where c is the constant velocity of light. Since the velocity of light is about a million times the velocity of sound at ordinary conditions, for all velocities realized in the terrestrial macroscopic world the mass m is practically the rest mass m_0. The kinetic energy calculated by the relativity theory is

$$E = c^2 (m - m_0) \tag{2}$$

For small velocities — i.e., $V \ll c$ — the preceding equation reduces to the familiar form $E = 1/2 m_0 V^2$.

This close relationship between energy and mass, as well as other evidences, prompted Einstein to take a revolutionary step. He says that there is no difference between energy and mass. One is just a form of the other. They are closely associated with each other. The energy E associated with any mass m is $E = mc^2$. Any energy E has a mass $m = E/c^2$ associated with it. In other words, a complete equivalence of energy and mass is achieved here. The old concept of separate conservation of energy and mass is thus merged into one single concept of conservation of "energy-mass."

Since the velocity of light is generally given as approximately 3×10^{10} cm per sec, 1 lb. of mass is equivalent to

$$\left.\begin{aligned} \left(\frac{1}{g}\right)c^2 &= \left(\frac{1}{32.2}\right)\left(\frac{1}{778}\right)\frac{(3 \times 10^1)_0}{10^2} \times (3.280)^2 \\ &= 3.86 \times 10^{13}\ \text{Btu} \\ &= 1.130 \times 10^{10}\ \text{kW} \cdot \text{h} \end{aligned}\right\} \qquad (3)$$

The problem is then how to transform the mass into energy or how to release the energy intrinsically bounded with the mass.

Atomic Structure

The problem is similar to that of releasing the chemical energy, intrinsically bounded with a mixture of hydrogen and oxygen. Until a device for igniting this gaseous mixture is discovered, the chemical reaction, although known to be possible, cannot be started and no chemical energy of this mixture will be released. Thus, the central problems of converting mass into energy are first to discover the type of reactions which can perform this mass-energy transformation and then to find the means of "igniting" or starting such reactions. The only known reactions that will transform mass into energy are nuclear reaction or the reactions with the atomic nuclei as the participants. Before entering into the discussion of the nuclear reaction, it is thus profitable to study the structure of atoms first.

The atom is composed of the central core or the nucleus and the outer collection of electrons. The mass of the atom is mostly concentrated in the nucleus. The electrons are extremely light. The mass of an electron at rest is 9.035×10^{-28} g, and in the physical atomic weight* scale it is 0.000 549. The total mass of the electrons is only several thousandths that of the nucleus. The charge of the nucleus is positive and that of the electrons is negative. For a normal atom the positive charge of the nucleus is just equal to the negative charge of the electrons and thus gives a neutral atom.

The electrons are arranged in "shells" around the nucleus. In terms of atomic scales there is an enormous distance between the nucleus and the first shell. This is indicated by the fact that the diameter of the nucleus is less than 1/10 000 of the diameter of the atom. The shells are

---

* In the physical atomic weight scale, the atomic weight of O^{16}, or the oxygen atom with 8 protons and 8 neutrons, is exactly 16. Because of a slight amount of heavier atoms in natural oxygen, the usual atomic weight scale or the chemical atomic weight scale is slightly larger (by about 1 part in 4 000).

named as K, L, M, N, O, P, Q. The states of motion the electron might occupy are designated by three-positional or orbitual quantum numbers and a "nonclassical" spin quantum number. For each orbitual quantum number there are two spin quantum numbers. These quantum numbers, together with the energy of the corresponding states of motion, can be determined either experimentally by measuring the spectrum of the element or theoretically by calculation with the rules of quantum mechanics assuming Coulomb forces between the electrons and the nucleus. In fact, the excellent agreement between the theory and the experiment is one of the major triumphs of modern physics. It is also found that no two electrons can occupy a state with the same four quantum numbers, according to the "Pauli exclusion principle." The orbitual states are listed in Table 1. The number in the parentheses is the number of orbitual states. Since there are two spin numbers with each orbitual state, the number of electron positions is twice the number shown in the parentheses.

Table 1 Electronic quantum states

Shell	Orbitual quantum states						
K	1s(1)						
L	2s(1)	2p(3)					
M	3s(1)	3p(3)	3d(5)				
N	4s(1)	4p(3)	4d(5)	4f(7)			
O	5s(1)	5p(3)	5d(5)	5f(7)	5g(9)		
P	6s(1)	6p(3)	6d(5)	6f(7)	6g(9)	6h(11)	
Q	7s(1)	7p(3)	7d(5)	7f(7)	7g(9)	7h(11)	7i(13)

With increasing atomic weight, the nucleus gets heavier, and the charge of the nucleus is increased with increase in the number of electrons attached to it. In other words, the shells of electrons gradually fill up. The states or "positions" with the lowest potential is filled up first. The potential ladder is shown in Fig. 1. Hydrogen uses one of the 1s electron positions. Helium uses both of the 1s electron positions and completes the K. shell. Hence, helium shows little tendency to draw electrons of other atoms to form chemical compounds and is an inert gas. Neon completes the L shell and becomes an inert gas. Now, the chemical property of the atom is mainly determined by the outmost shell. Since the Fe, Pd, Pt, transition elements, and the rare earth elements are made by filling up the inner shells with 3d, 4d, 5d, and 4f positions, respectively, and without changing the outer shell, their respective similarity in chemical property is expected[3]. For the same reason, the similarity between the elements U, Np,* Pu,* indicate that they may be the elements made by filling up the 5f positions in the O shell with the two electrons in the Q shell unchanged. These elements then might constitute the

* Np and Pu stand for neptunium and plutonium, two new elements beyond uranium with 93 and 94 electrons, respectively (atomic numbers 93 and 94).

beginning of a new series similar to the rare earth elements[4].

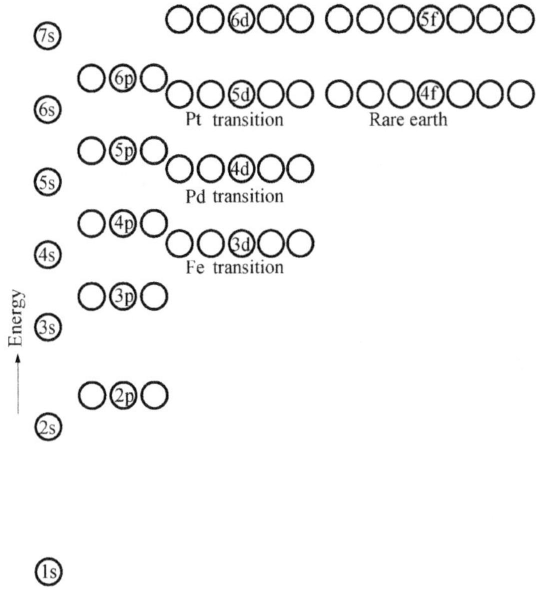

Fig. 1 Energy levels of electronic quantum states. 3d states are first occupied by the iron transition elements. 4d states are first occupied by the palladium transition elements. 5d states are first occupied by the platinum transition elements. 4f states are first occupied by the rare earth elements

The nucleus is now generally considered as composed of neutrons and protons. They have about the equal mass, with the neutron being slightly heavier, and are approximately 1 840 times heavier than electrons. In the physical atomic weight scale, a neutron is 1.008 93 and a proton is 1.007 57. Similar to the electron, their states are determined by the orbitual numbers and the two spin numbers. Since only the proton has a unit positive charge, the number of protons in the nucleus must be equal to the number of electrons. This number is the atomic number Z. Hydrogen has the atomic number 1 and uranium 92. The total number of neutrons and protons in the nucleus is the mass number A. Thus $N = A - Z$ is the number of neutrons in the nucleus. Since the mass of neutrons or protons is approximately unity in the scale of atomic weight, A is also roughly the atomic weight. Therefore, the nucleus of the hydrogen is a proton. The nucleus of deuterium is deuteron, or a proton and a neutron. The nucleus of helium is usually called the α-particle and is composed of two neutrons and two protons. Elements with equal number Z are called "isotopes." Thus, hydrogen and deuterium are isotopes. Similarly, the uranium isotopes U-235 and U-238 are composed of 92 protons and 143 and 146* neutrons, respectively.

The neutrons and protons in the nucleus are not a loose assemblage but are tightly bound

* "133 and 136" in the original have been corrected as "143 and 146". — Noted by editor

together. In fact, their bounds are much stronger than the bounds between the nucleus and the electrons in the outer shells. For instance, while it takes only 13.53 volts to remove the electron from the hydrogen atom — i.e., a binding energy of 13.53 eV.† — the binding energy between a proton and a neutron is approximately 2.15×10^6 eV. The average binding energy per particle for heavier elements is even higher, being approximately 8.5×10^6 eV. The ratio for these two types of bounds is thus roughly $10^{[6]}$. The ordinary chemical energy released by molecular reactions is linked with the rearrangement of electrons in the molecules. For instance, the electrons in hydrogen molecules and in oxygen molecules are bound to the hydrogen nuclei and the oxygen nuclei, respectively. By rearranging the electrons so that they are bound to both the hydrogen nucleus and the oxygen nucleus at the same time, water molecules are formed. The closer binding between the electrons and the nuclei in the water molecule corresponds to the chemical energy released during the combustion of hydrogen with oxygen. Therefore, if nuclear reaction involving rearrangements of the particles in nuclear structure is realized, then the associated energy released must be of the order of the nuclear binding energy — i.e., approximately a million times that of chemical or molecular reaction. This difference in the order of magnitude between the nuclear reactions and the molecular reactions is basic and will occur frequently in the subsequent discussions.

Nuclear Reactions

The nuclear reactions known up until 1939 are reactions produced by the bombardment of the atoms with various kinds of projectiles, such as α-particle, proton, neutron, and deuteron. The result of this bombardment is either absorption of the bombarding particle with emission of short wave-length electromagnetic waves or γ-ray, or disintegration of the atom with emission of other particles. The reactions can be classified as in Table 2. For instance, in 1932 J. D. Cockcroft and E. T. S. Walton of Rutherford's laboratory, bombarded a target of lithium with protons of 700 kV. energy and found that α-particles were ejected from the target as a result of the bombardment. The nuclear reaction that occurred can be written symbolically as

$$ {}_3\mathrm{Li}^7 + {}_1\mathrm{H}^1 \rightarrow {}_2\mathrm{He}^4 + {}_2\mathrm{He}^4 \tag{4} $$

where the subscript represents the atomic number Z and the superscript the mass number A. In a chemical equation, the number of atoms of any element entering into the reaction must be equal to the number of atoms of that element in the product. In a nuclear reaction as shown in Eq. (4), the number of the nuclear particles, protons, and neutrons must be the same on both sides. Thus the sum of subscripts is four and the sum of superscripts eight on each side.

Neither mass nor energy has been included in this equation. In general, the sum of the

† eV. is the abbreviation for electron-volt — i.e., the energy spent in moving a unit electron charge through a potential of 1 volt. 1 eV. $= 4.45 \times 10^{-26}$ kW · h.

masses of the incident proton and the lithium atom will not be precisely the same as the sum of the masses of the α-particles produced. According to Einstein's law of conservation of energy-mass, the sums of mass and energy taken together should be the same before and after the reaction. The masses were known from mass spectra. On the left ($Li^7 + H^1$) they totaled 8.024 1; on the right ($2He^4$), 8.005 6, so that 0.018 5 unit of mass had disappeared in the reaction. The experimentally determined energies of the α-particles were approximately 8.5 MeV.* each, a figure compared to which the kinetic energy of the incident proton could be neglected. Thus, 0.018 5 units of mass are 3.07×10^{-26} g. or 17.2 MeV. according to Einstein's equation. Therefore the experimental results have completely proved Einstein's mass energy equivalence.

Table 2 Types of nuclear reaction

	Incident particle	Emitted particle	Z becomes	A becomes
Reactions by particle bombardments	α	Proton	$Z+1$	A** $+3$
	α	Neutron	$Z+2$	$A+3$
	Proton	...	$Z+1$	$A+1$
	Proton	α	$Z-1$	$A-3$
	Proton	Neutron	$Z+1$	A
	Proton	Deuteron	Z	$A-1$
	Deuteron	Proton	Z	$A+1$
	Deuteron	Neutron	$Z+1$	$A+1$
	Deuteron	α	$Z-1$	A*** -2
	Neutron	...	Z	$A+1$
	Neutron	Proton	$Z-1$	A
	Neutron	α	$Z-2$	$A-3$
Radioactivity	...	α	$Z-2$	$A-4$
	...	Proton	$Z-1$	$A-1$
	...	Electron	$Z+1$	A
	...	Positron	$Z-1$	A

As stated, the protons for the lithium reaction are accelerated by 700 kV. or have 0.7 MeV. energy per particle. Such an amount of kinetic energy gives the proton an appreciable probability of penetrating into a lithium nucleus and causing a reaction. However, out of a million accelerated protons there still is only one that will make a successful hit; the others will be slowed down by collisions with atoms without producing the desired reaction, and their kinetic energy will be lost by transforming into heat. Thus 0.7×10^6 MeV. energy is required to accelerate a million protons to produce one reaction giving 17 MeV. energy. Hence, 40 000 times more energy is spent than recovered. It is obviously impracticable to use

* 1 MeV. is one million electron-volts or 10^6 eV. and is equivalent to 4.45×10^{-20} kW • h.

** Z in the original paper has been corrected as A. — Noted by editor

*** Z in the original paper has been corrected as A. — Noted by editor this nuclear process to produce energy. The situation could be compared to that of a hypothetical molecular reaction where an inefficient igniting device has to be used for every pair of molecules. Then the chemical reaction, even if it were an energetic one, cannot be practically utilized for energy production.

There is a special class of nuclear reactions which does not involve incident particles. This means that the nuclei automatically disintegrate with the emission of particles. This phenomenon is called the radioactivity and is discovered actually earlier than the usual nuclear reactions described above. The emitted particles could be the α-particles, the protons, the electrons, and the positrons, positive electrons, as shown in Table 2. The emission of electrons and positrons are found, however, to give continuous spectra of the kinetic energy of the emitted particles, while the energy states of the nuclei are found to be discrete. This is in contradiction to the law of conservation energy-mass, which is observed to hold in all other cases. To preserve this law, a neutral particle with spin but of practically zero mass called neutrino is assumed by W. Pauli. The neutrino is, so far, not directly detected. However, because of the assumed existence of neutrinos, part of the energy could be transferred to them, and the continuous spectra of electron energy in radioactivity is explained. The kinetic energy of the neutrinos is "lost" because it cannot be detected.

Nuclear Structure — Binding Energy

An entirely new type of nuclear reactions is the breaking up of a heavy nucleus into two approximately equal fragments. This is called the nuclear fission reaction. It was discovered by L. Meitner, O. Hahn, and F. Strassmann toward the end of 1938. To understand the reaction one must look into the structure of the nucleus in more detail[5].

In studying nuclear structure one is, however, faced with the difficulty of not knowing the character of the forces between the nuclear particles. The desired information must be deduced from the scanty experimental data. First, by plotting the atomic number Z against the mass number A for all known elements, there is an obvious relation between these two quantities. This relation is, of course, not precise because of the variations by isotopes. For instance, with the same atomic number 1 for hydrogen and deuterium, the mass numbers are 1 and 2, respectively. However, for light elements, the value of A is approximately twice that of Z. This is exactly true for ${}_6C^{12}$, ${}_7N^{14}$, ${}_8O^{16}$, etc. This means that the stable light nuclei are composed of equal numbers of protons and neutrons. Then, the binding force between the nuclear particles must be the greatest between the proton and the neutron. If the force between protons were stronger, the stable nucleus would contain more protons than neutrons. If the force between the neutrons were stronger, the stable nucleus would contain more neutrons. However, the fact that the force between the neutron and the proton is the strongest does not exclude the possibility of forces between neutrons themselves and protons themselves. But if such binding force between the like particles exists, they cannot be widely different, because, if the binding force between a pair of neutrons were much higher than that between a pair of protons, then the stable nucleus would again have a preponderance of neutron instead of the observed equality in numbers of these two kinds of particles. Of course, there will be Coulomb repulsion between two protons. But this repulsion can be shown to be small in comparison with the specifically nuclear forces. In fact, it is generally assumed that aside from the Coulomb force, the force between a pair of neutrons is equal to the force between a pair of protons.

The α-particle, or the nucleus of the helium atom, consists of two protons and two neutrons. The binding energy per particle of this nucleus is found to be much higher than the deuteron. Furthermore, a nucleus of five particles, by adding either a proton or a neutron to the α-particle, is never observed. This means that a nucleus of five particles is highly unstable and its lifetime is so short that its detection by experiment is highly unlikely. This great stability of the helium nucleus and the difficulty of adding another particle to it show the saturation of the binding forces in helium nucleus. The situation is similar to that of the helium atom with the two electrons in the same orbitual quantum state but different spin states. The resultant closed K-shell structure refuses to combine with additional electrons from other atoms to form molecules. Thus, similarly, the two neutrons must be in the same orbitual quantum state with different spin states, and the two protons must be in the same orbitual quantum state with different spin states. But the large binding energy of the helium nucleus requires that there is binding force between every pair of the particles in this nucleus. Hence, the binding force between a proton and a neutron and the binding force between the like particles cannot depend to any considerable extent upon the relative spin directions of the interacting particles.

If there is binding energy between *every* pair of interacting particles, then, since the number of pairs in a nucleus is $(1/2)A(A-1)$, the total binding energy of a nucleus must be proportional to the square of A. However, experimental data show a practically linear relation between the binding energy and the mass number A. This is the same as the relation between molecular binding energy of a liquid drop containing many molecules. It is known that the force between the molecules in liquid or solid shows saturation. In other words, a molecule is attracted only to the other molecules in its immediate neighborhood. Thus, the intermolecular forces can be said to be short ranged, speaking in terms of the dimension of the drop. This character of the binding force greatly limits the number of effectively interacting pairs of particles and makes the total binding energy proportional to the number of particles. The nuclear forces then must show a similar saturation character. Here, a neutron or a proton only interacts strongest toward another particle of nearest orbitual quantum states. As the number of the particles increases with increasing mass number A, the number of particles having nearly a given orbitual quantum state does not increase, since only two particles of the same kind can occupy the same orbitual quantum states, according to Pauli's exclusion principle. Then the nuclear forces can also be said to be short-ranged in terms of the dimension of the nucleus. Thus, the number of effectively interacting pairs is limited in a similar manner as in the liquid drop, and the total binding energy is only proportional to the total number of particles instead of proportional to the square of the number of particles.

This liquid drop analogy is also useful in two other connections. First, since the volume of the liquid drop is proportional to the number of molecules, the volume of a nucleus must also be proportional to the number of nuclear particles, since they are similarly held together. This means that the radius of a nucleus is proportional to $A^{1/3}$. This is found to be true experimentally. Secondly, a liquid drop exhibits the well-known surface tension phenomenon. This is due to the existence of a free surface where the molecules are only attracted by the molecules within the drop. Therefore, for the molecules on the surface, only half of their binding force is satisfied, and their contribution to the total binding energy of the drop is not so large as the molecules within the drop. A similar situation exists for the nucleus, where the protons or neutrons on the surface of the nucleus contribute much less binding energy than the particles within the nucleus. For a heavy nucleus, the ratio of the particle at the surface to the total number of particles is reduced and the binding energy per particle should be increased. In other words, the surface tension effect indicates an increase in binding energy per particle with increasing mass number A.

So far, the effect of Coulomb repulsion between protons is neglected. A simple calculation shows that the average energy due to this effect is 1/4 MeV. per pair of protons. The average binding energy due to nuclear forces is $8^{1/2}$ MeV. per particle. Thus the Coulomb repulsion is negligible if compared on the basis of a single pair of protons. However, the nuclear forces show saturation, and the binding energy due to nuclear forces is approximately proportional to A. The number of protons in the nucleus and, thus, the total proton charge are also approximately proportional to A. The radius of the nucleus is proportional to $A^{1/3}$. Hence, the total repulsion energy is proportional to $A \times A/A^{1/3}$ or $A^{5/3}$. This difference of the Coulomb repulsion energy from the energy due to nuclear forces is, or course, the result of nonsaturation character of the Coulomb forces. The ratio of these two energies is thus proportional to $A^{2/3}$. Therefore, for a large nucleus, the Coulomb repulsion energy is important in spite of its smallness for one pair of protons. Then the total binding energy of the nucleus will be increased and the nucleus made more stable by substituting some of the protons by neutrons. This is actually found to be the case: The uranium nucleus has a neutron-proton ratio N/Z equal to 1.6, while for light nuclei this ratio is 1 as stated before. Furthermore, the effect of the Coulomb repulsion will also make the total binding energy increase less rapidly as the increase in number of particles A. This effect is thus opposite to the surface tension effect. However, for small nuclei, the Coulomb repulsion is unimportant. Hence, the binding energy per particle will increase with A in this range. For large nuclei, or large A, the binding energy per particle will again decrease as the surface tension effect is overcome by the Coulomb repulsion.

To actually calculate the nuclear binding energy from the experimental data, one can proceed as follows: An atom of atomic number Z and mass number A contains Z protons and Z electrons together with $N = A - Z$ neutrons. Because of the negligible binding energy between electrons and the nucleus, the mass of Z protons and Z electrons is equal to Z times the mass of hydrogen atom. The atomic mass of hydrogen and the mass of neutrons are 1.008 13 and 1.008 93, respectively, in the physical scale. Therefore, the mass of components that go into the structure of the atom is $1.008\ 13Z+1.008\ 93\ N$. If the actually measured atomic weight of the element is M, then the mass defect is $1.008\ 13\ Z+1.008\ 93\ N-M$. This mass defect must represent the binding energy of the atom according to Einstein's law of equivalence of energy and mass. Since the binding energy of the electrons to the nucleus is negligible, the mass defect is practically the binding energy of the nucleus. Generally, a different parameter is introduced. This parameter is called the packing fraction f defined as

$$f=(M-A)/A \tag{6}$$ *

The binding energy per particle is then

$$\begin{aligned}\frac{1.008\ 13Z+1.008\ 93N-M}{A} &= -f+0.008\ 13\frac{Z}{A}+0.008\ 93\frac{N}{A}\\ &= -f+0.017\ 06\frac{Z}{A}-0.008\ 93\end{aligned} \tag{7}$$

By neglecting the small variation of the ratio Z/A, the binding energy per particle is given by the negative of the packing fraction f. Therefore, for increasing binding energy per particle, the packing fraction increases. According to the discussion given in the previous paragraph, the binding energy per particle must show a maximum for medium heavy nuclei. Thus, the packing fraction must show a minimum for the medium values of the mass number A. This is seen in Fig. 2[5].

By examining the graph for the packing fractions (Fig. 2), it is seen that very large energy can be released by either of the two methods: manufacture of the medium heavy elements out of light elements, or breakdown of an extremely heavy element into the medium heavy elements. In other words, by going from the two ends of the periodic table toward the middle, closer packing of the nucleus can be achieved with the release of the enormous binding energy. Nature provides an example for each case. The manufacture of helium from hydrogen is of the first type and

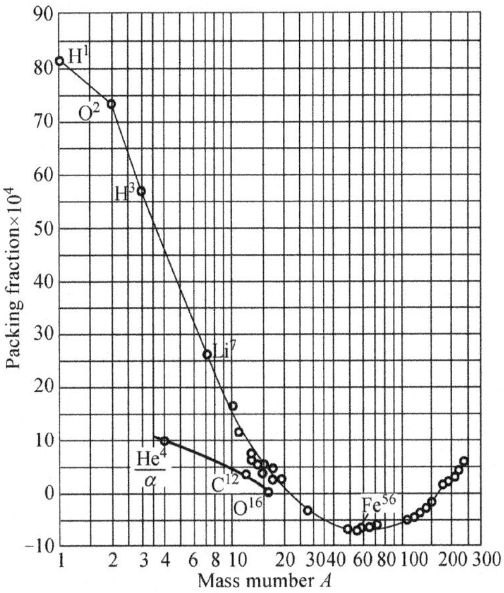

Fig. 2 Packing fraction f as a function of mass number A

* Eq. (5) in the original paper has been missed. — Noted by editor.

is carried out in the center of the main sequence stars such as the sun. The fission of uranium is of the second type. This reaction is used in the atomic bomb. These two typical examples will be examined in more detail presently.

Energy Production in the Stars

In the laboratory nuclear transmutations are produced by accelerating a few particles to high kinetic energy. The resultant poor efficiency is explained under "Nuclear Reactions." In stars, different conditions prevail: Because of the high temperature in the interior of stars, all protons in the stars have high kinetic energies. Moreover, they are not slowed down by collisions with other atoms because all atoms have equally high energies. Therefore, unlike the laboratory nuclear reaction, no energy need be spent in accelerating the bombarding particles such as protons, and any energy released by the nuclear reaction is a net gain. On the other hand, the kinetic energies of the nuclei in stars are small compared with those used in the laboratory. It is true that the temperature is about 20 000 000℃. at the center of sun, and the atomic nuclei are therefore traveling with much greater speed than the molecules in the terrestrial atmosphere. However, their average kinetic energy is still only about 0.003 MeV — 200 times smaller than the energies commonly used in the laboratory. Therefore, their probability of penetrating into the nucleus of the bombarded atom is much smaller. This is, however, remedied by the presence of enormous numbers of bombarding protons, and the net reaction rate is not small.

Through a careful examination of the probable reaction, Bethe[7] picked out a series of the nuclear reactions as follows:

$$ {}_6C^{12} + {}_1H^1 \rightarrow {}_7N^{13} + \gamma \tag{8}$$

$$ {}_7N^{13} \rightarrow {}_6C^{13} + {}_1e^0 \tag{9}$$

$$ {}_6C^{13} + {}_1H^1 \rightarrow {}_7N^{14} + \gamma \tag{10}$$

$$ {}_7N^{14} + {}_1H^1 \rightarrow {}_8O^{15} + \gamma \tag{11}$$

$$ {}_8O^{15} \rightarrow {}_7N^{15} + {}_1e^0 \tag{12}$$

$$ {}_7N^{15} + {}_1H^1 \rightarrow {}_6C^{12} + {}_2He^4 \tag{13}$$

Here γ represents the emission of γ-rays and ${}_1e^0$ represents the positron that is emitted together with neutrinos, are explained previously. This series of reactions is remarkable in that it reproduces the carbon atom after six reactions. This is important, since the content of carbon in the stars is not high while hydrogen is plentiful. The net result is the manufacture of an α-particle and two positive electrons from four protons. The two positive electrons will react with two ordinary electrons to disappear into γ-ray, the so-called annihilation of electron pairs. Then the net energy released per cycle is that given by the difference in mass of four protons plus two electrons, and an α-particle. Since the binding energy between the electron and nuclei particles is negligible, this difference is the same as the difference between four hydrogen atoms (four protons plus four electrons) and the helium atom (α-particle plus two electrons).

This difference is equivalent to 27 MeV. A small fraction 2 MeV of this energy goes into neutrinos and is lost. The remainder goes into radiation, which is finally emitted from the sun.

Thus about 25 MeV are set free for each helium nucleus produced in the stars, or 6 MeV for each proton destroyed. One gram of sun material has been calculated to contain about 2×10^{23} protons. Therefore, if all the protons can be converted into helium, the available energy supply is 1.2×10^{24} MeV per g At present, the sun radiates about 1.2×10^{6} MeV per g per sec At this rate, the energy supply will last about 30 000 000 000years. The rate of energy release can also be approximated by calculations based upon the reactions (8) to (13) and laboratory measured constants. Thus Bethe's "carbon cycle" is indeed the energy process in the sun. In fact, the "carbon cycle" is found to be the energy process for all the "main-sequence stars" from "red dwarfs" to "blue giants."

On the physical atomic weight scale, four protons and two electrons total 4.021 5 units. This means that 4.021 5 g of the reacting material in stars will carry out 6.064×10^{23} processes, each of which generates 25 MeV energy. The energy generated by 1 lb of the reacting substance is then

$$[1/(0.004\,021\,5 \times 2.205)] \times 25 \times 6.064 \times 10^{23} \times 4.45 \times 10^{-20} \times 3\,413 = 2.59 \times 10^{11}\ \text{Btu} \tag{14}$$

By comparing this value with that given by Eq. (3), it is seen that the helium manufacturing process is able to transform 0.67 per cent of the original reacting mass into energy. While this percentage may seem to be small, because of an enormous conversion factor from mass to energy, the "heat value" per pound of reacting material is still tremendous. One of the most energetic molecular reactions is the combustion of hydrogen with oxygen. The heat value of the stoichiometric mixture is 6 850 Btu per lb. The heat value of the nuclear reaction given in Eq. (14) is thus 3.78×10^{7} times larger than that of the molecular reaction. This confirms the previous surmise that the ratio of energy releases for the nuclear reaction and the molecular reaction must be of the order of a million.

Nuclear Fission — Chain Reaction

The second method of energy release by breaking down the extremely heavy elements depends upon the instability of the heavy nuclei under external excitation. The liquid drop model of the nucleus is also useful in the stability considerations for this nuclear fission reaction. In fact, by assuming that the drop is composed of an incompressible fluid of radius R and of volume $(4\pi/3)R^3 = (4\pi/3)r_0^3A$, uniformly electrified to a charge Ze and possessing a surface tension O, the critical excitation energy, E_f, for separating into two equal nuclei is shown by Bohr and Wheeler[8] to be expressed by

$$E_f = 4\pi r_0^2 O A^{2/3} f[(Z^2/A)/(Z^2/A)_{\text{limiting}}] \tag{15}$$

r_0 is the radius of a single nuclear particle and $(Z^2/A)_{\text{limiting}} = 47.8$, limiting value for Z^2/A giving instability for infinitesimal deformation of the "drop." E_f is actually the height of the barrier which leads to instability. In other words, if a certain excitation has an energy of less than E_f, the nucleus is stable against splitting into two fragments. If a certain excitation has an energy in excess of E_f, then the nucleus can be deformed to such an extent that it can be further deformed into two fragment nuclei without additional energy expenditure. The situation is similar to that of rolling a ball against a hump of ground. In order to pass the hump, one must apply sufficient energy for the ball to reach the hump, then further motion of the ball is automatic.

Consider now what would happen if a neutron were added to such a near unstable nucleus. The value of E_f is calculated by Bohr and Wheeler and the result included in Table 3. After the capture of a "zero" kinetic energy neutron (thermal neutron), the equilibrium of the collection of neutrons and protons is disturbed. In general, a readjustment is possible with the subsequent release of energy to excite the product nucleus. The energy of excitation E_c of the compound nucleus is then the binding energy between the initial nucleus and the neutron. This is also calculated by Bohr and Wheeler and the result listed in Table 3. Therefore, the threshold energy of neutrons required for producing the splitting of fission of a given compound nucleus is obtained by subtracting the excitation energy of zero kinetic energy neutron from the critical energy for fission. The result for this energy of the bombarding neutrons is listed in Table 3 as $E_f - E_c$ for different bombarded nuclei. The negative values for U^{235}, U^{234}, and Pu^{239} mean that these nuclei will undergo fission after capture of thermal neutrons. These theoretical results are in full agreement with experiments.

Table 3 Excitation of heavy nuclei by neutron capture

Initial nucleus	Produce nucleus	Z^2/A	x	E_f	E_c	$E_f - E_c$
${}_{92}U^{234}$	${}_{92}U^{235}$ *	36.00	0.753	5.0MeV**	5.4MeV**	−0.4MeV**
${}_{92}U^{235}$	${}_{92}U^{236}$	35.86	0.750	5.2	6.4	−1.2
${}_{92}U^{238}$	${}_{92}U^{239}$	35.40	0.741	5.9	5.2	+0.7
${}_{94}Pu^{239}$	${}_{94}Pu^{240}$	36.80	0.770	4.0	6.2	−2.2
${}_{90}Th^{232}$	${}_{90}Th^{233}$	34.76	0.727	6.9	5.2	+1.7
${}_{91}Pa^{231}$	${}_{91}Pa^{232}$	35.70	0.747	5.5	5.4	+0.1

However, there are other competitive reactions that could happen to an excited nucleus. These reactions are: (a) emission of γ-ray, (b) emission of α-particle, and (c) re-emission of the neutron. To estimate the relative probability of occurrence, one must turn to the statistical mechanics of the excited nucleus. It is found that, for energies of excited nucleus below, or not far above, the critical value, radiation of γ-ray will predominate. With increasing energy

* ${}_{92}U^{234}$ in the original paper has been corrected as ${}_{92}U^{235}$. — Noted by editor.

** Mv in the original paper has been corrected as MeV. — Noted by editor.

the probability of fission will increase rapidly. The probability of fission will not increase indefinitely, however, because at still higher energies the re-emission of neutrons becomes the predominant process.

Now there still remains the problem of making the capture possible. Of course, the nuclei will be bombarded with neutrons, but the probability of actual absorption depends upon the kinetic energy of the bombarding neutron and the character of the bombarded nucleus. This probability is generally expressed as a "cross section" σ defined by

$$N = n\nu\sigma \tag{16}$$

N is the number of occurrences of the kind in question (fission) per sq cm of bombarded surface per sec; n is the number of bombarding particles per sq cm per sec. (normal incidence around); and ν is the number of atoms of the kind involved per sq cm of surface. The situation is as if each atom presented a target area σ. If the target is struck, the effect in question occurs; if not, nothing happens. It is apparent that σ gives a sort of averaged probability of occurrence of an effect for a random spatial distribution of impinging particles.

It is thus evident for the successful fission process that two conditions must be satisfied: (a) a large cross section of capture of neutrons of proper energy; (b) the energy of the neutron must be such that the excitation energy of the compound nuclei, after capture of neutrons, is above the critical barrier energy. It is found that $_{92}U^{235}$ and $_{94}Pu^{239}$ can be split by either thermal neutrons or fast neutrons. Then, as explained before, a very large energy release is expected. In fact, the measured energy release per fission for $_{92}U^{235}$ is approximately 177 MeV[9]. Since 1 gram mole or 235 g. of $_{92}U^{235}$ contains 6.064×10^{23} atoms, the energy generated by 1 lb of $_{92}U^{235}$ with complete fission is

$$\begin{aligned}&[1/(0.235 \times 2.205)] \times 177 \times 6.064 \times 10^{23} \times 4.45 \times \\ &10^{-20} \times 3\,413 = 3.14 \times 10^{10}\ \text{Btu}\end{aligned} \tag{17}$$

By comparing with the value given by Eq. (3), it is seen that the fission of U-235 is able to convert 0.081 percent of the original mass into energy. Thus, the reaction is not so efficient as that of helium manufacture from hydrogen, treated in the previous section. This is, of course, expected from the diagram for the packing fraction (Fig. 2), which shows a much greater change in binding energy per nuclear particle from hydrogen to helium than from uranium to medium heavy elements. Nevertheless, the "heat value" of the U-235 is still approximately 1.5×10^{6} times the heat value of gasoline, which is the most powerful fuel used today.

However, to make the fission process a really going process, a continuous supply of neutrons of proper energy is necessary. This is actually accomplished by slowing down the fast neutrons released by the fission fragments themselves. For $_{92}U^{235}$, the compound nucleus $_{92}U^{236}$ first breaks down into $_{56}Ba$ and $_{36}Kr$ of unequal mass, since this splitting involves smaller potential barriers. The product nuclei would be highly excited because of the energy released and would contain more neutrons than stable equilibrium requires, because the ratio of neutron to proton is much higher in uranium than for atoms of medium atomic weight. These extra neutrons will be "vaporated" or emitted with high energy and a small time delay of about 1 sec. For one fission, more than one neutron is thus generated. If these neutrons can be slowed down to proper energy for fission of new $_{92}U^{235}$ atoms without absorbing so many of them as to reduce the desired neutron to less than one per fission, the reaction becomes a chain reaction. The chain carrier is the neutron of proper energy. Then the reaction will tend to build up to explosive violence.

The agents for slowing down the fast neutrons are called moderators. They must not absorb the neutrons above the desired energy; otherwise the useful yield will be extremely small. They must also be light elements so that they can effectively share the kinetic energy of fast neutrons by collision and thus slow down the neutrons with but few collisions. The thickness and the bulk of the moderator can thus be reduced. Hydrogen is good because it is the lightest element, but it is undesirable for slow neutron fission because of its large capture cross section of slow neutrons. Deuterium is best from this aspect, and heavy water is a good moderator.

Besides the absorption of neutrons by the moderator, there is another loss due to escape of the neutrons out of the reacting mass. This is proportional to the surface of the mass. Therefore, by increasing the size of the reacting mass, this loss can be cut down relative to the production, which is proportional to volume. Hence, if a small quantity of fission material and moderator cannot carry on the chain reaction, a large quantity of the same material might be able to do so. For any given geometrical arrangement and effective composition, there is thus a critical size. Generally, single fission can be produced automatically or by cosmic ray; therefore, once the critical size is reached, the mass will react by itself or explode. The atomic bomb is based essentially on the principle of bringing parts of subcritical size together quickly to build up the reaction. The rate of reaction and particularly the rate of building up the reaction are controlled by the diffusion of the neutrons through the moderator to the fission nuclei. For fast build-up in explosion, fast neutrons are preferred.

Engineering Approach to the Nuclear Reaction

The previous discussion clearly shows that the fundamental concepts involved in a nuclear reaction — such as collision, capture, excitation, and energy barrier — are all familiar to students of molecular reactions. The study of the molecular reactions or chemical reaction from kinetic point of view is, however, only a recent development. Its consideration by engineers is even more so. For instance, the power-plant engineers learned to design large boiler installations burning tons of coal per hour without any concept of molecular physics. The design of internal combustion engines was no exception, until the difficulty caused by detonation became serious. Then the investigation of the slow oxidation of the hydrocarbons by chain reactions is helpful. However, the study of nuclear reactions took a different road. Here, the effort was concentrated immediately on the detailed dynamic processes and on the development of a fundamental theory for the nuclear dynamics to explain these processes. This, of course, is the natural result of the emphasis by physicists on the understanding of the fundamentals.

For an engineer, his assignment is not so much to understand a particular phenomenon as to learn to utilize it. Electrical engineers knew how to design long-distance transmission lines before the development of satisfactory physical theory for the conduction of electricity. A similar situation will certainly appear in the engineering utilization of atomic energy. A satisfactory fundamental theory for nuclear structure will no doubt be a great help in "atomic engineering," but to wait for such a theory is certainly unwise and unnecessary. Then shall the ad hoc experiments without coordination be encouraged? Such an approach would not only be uneconomical but extremely dangerous because of possible uncontrollable explosion with greater violence. A better approach would be a semi-empirical one. In other words, first the performance of each elementary process should be determined experimentally. Then the kinetic theory of reaction developed for the chemical kinetics can be used to predict the performance of the overall reaction involving these elementary processes.

For instance, the basic knowledge required for predicting the fission reaction is the fission cross section; the elastic scattering cross section; the inelastic scattering cross section of the atoms U-234, U-235, U-238, and Pu-239 as a function of the energy of the neutrons; the number and energy of neutrons emitted by fission and its dependence on the energy of incident neutron; etc. With these empirical data, the fission reaction in an atomic bomb can be calculated. To those familiar with the theory of chemical kinetics, the extreme complexity of the treatment due to the great variety of possible molecules is well known. The application of kinetic theory to nuclear reactions, however, may be much simpler, since the total number of possible nuclei is only several hundred, while the number of known molecular compounds may be hundreds of thousands. The reacting components that need to be considered in a nuclear reaction will certainly be fewer than for the case of molecular reactions.

According to H. D. Smyth's report[4] on the development of the atomic bomb, this semi-empirical approach is actually followed. Similar methods must be used for the development of the general utilization of atomic energy to determine the optimum process or processes and then finally test the validity of calculation by experiments. This approach would certainly eliminate, as far as humanly possible, the danger and the waste that through improper handling might be associated with the release of energy a million times as great as the conventional combustion process of molecular reaction. It is seen then that the concept of nuclear power engineering is necessarily somewhat different from its counterpart in pure science, the nuclear physics.

References

[1] von Kármán Th. Atomic Engineering. Mechanical Engineering, 1945, 67: 672.

[2] Tolman R C. Relativity, Thermodynamics and Cosmology. Chapter II, Part I, and Chapter III, Part

Ⅰ, Oxford, 1934.

[3] For a complete tabulation of electron positions in various atoms, see The handbook of Chemistry and Physics. 25th ed., 1941: 275 - 276.

[4] Smyth H D. A General Account of the Development of Methods of Using Atomic Energy for Military Purposes under the Auspices of the United States Government. 1940 - 45, 1945.

[5] Bethe H A. Nuclear Physics-A stationary state of Nuclei. Reviews of Modern Physics, 1936, 8: 92 - 97.

[6] Oliphant M L. Masses of Light Atoms. Nature, 1936, 137: 396 - 397//Dempster A J. The Energy Content of the Heavy Nulcei. Physical Review, 1938, 53: 869 - 874.

[7] Bethe H A. Energy Production in Stars. American Scientist, 1942,30:243 - 264.//Energy Production in Stars, Physical Review, 1939,55: 434 - 456.

[8] Bohr N, Wheeler J A. The Mechanism of Nuclear Fission. Physical Review, 1939, 56:426 - 450.

[9] Henderson M C. The Heat of Fission of Uranium. Physical Review, 1940, 58:774 - 780.

NATIONAL ADVISORY COMMITTEE FOR AERONAUTICS

TECHNICAL NOTE

No. 995

Two-Dimensional Irrotational Mixed Subsonic and Supersonic Flow of a Compressible Fluid and the Upper Critical Mach Number

By Hsue-shen Tsien and Yung-huai Kuo
California Institute of Technology

Washington
May 1946

Two-Dimensional Irrotational Mixed Subsonic and Supersonic Flow of a Compressible Fluid and the Upper Critical Mach Number

By Hsue-shen Tsien and Yung-huai Kuo
(California Institute of Technology)

Summary

The problem of flow of a compressible fluid past a body with subsonic flow at infinity is formulated by the hodograph method. The solution in the hodograph plane is first constructed about the origin by superposition of the particular integrals of the transformed equations of motion with a set of constants which would determine, in the limiting case, a known incompressible flow. This solution is then extended outside the circle of convergence by analytic continuation.

The previous difficulty of the Chaplygin method of slow convergence of the series has been overcome by using the asymptotic properties of the hypergeometric functions so that numerical solutions can be obtained without difficulty. It is emphasized that, for a solution covering the whole domain of the field of flow, both fundamental solutions of the hyper-geometrical differential equation are required.

Explicit formulas for numerical calculations are given for the flow about a body, such as an elliptic cylinder, and for the periodic flow such as would exist over a wavy surface.

Numerical examples based on the incompressible flow solution of an elliptic cylinder of thickness ratio of 0.6 are computed for free-stream Mach numbers of 0.6 and 0.7.

The results of this investigation indicate an appreciable distortion in the shape of the bodies in compressible flow from that of incompressible flow, which necessitates a series of computations with various values of the geometric parameter in order that the desired body shapes can be selected for a given Mach number. It also is shown that the breakdown of irrotational flow depends solely upon the occurrence of limiting lines, which, in turn, are dependent on the boundary conditions.

The numerical calculations show that at a free-stream Mach number of 0.6, irrotational supersonic flow exists up to a local Mach number of 1.25; whereas breakdown occurs at 1.22 for a Mach number of 0.7.

Introduction

When a flow of nonviscous incompressible fluid is irrotational, it is well known that the problem can be reduced to either the problem of Dirichlet or that of Neumann, and that there exists a unique solution for any given boundary conditions. When the fluid is nonviscous but compressible, the variation of density makes the mathematical problem very difficult and complex. In this case, a pure potential flow throughout the region is not always possible for a given body; this depends very much upon the condition at infinity. If a certain speed of the flow at infinity is reached, regions within the field of flow will be created in which the irrotational flow does not exist owing to the appearance of "limiting lines." Such regions were picturesquely designated as "forbidden regions" by Th. von Kármán[1], and they appear when the local speed of the flow considerably exceeds the local speed of sound. It has been shown that the occurrence of limiting lines is directly connected with the breakdown of irrotational flow and with the resultant increase in drag of the body due to shock waves. In other words, if there is a limiting line in the field of flow, the isentropic irrotational flow must break down. However, the irrotational flow may break down before the appearance of limiting line due to the instability of the velocity field. On the other hand, shock waves can occur only in supersonic flow. Therefore, there is no danger of breakdown of isentropic flow if the whole field of flow is subsonic. Consequently, the Mach number corresponding to the first appearance of local speed equal to that of sound can be designated as the "lower critical Mach number"; and the Mach number corresponding to the first appearance of limiting lines can be designated as the "upper critical Mach number." The actual critical Mach number for a given body will be influenced by the boundary layer and hence the Beynolds number. However, it must lie between these two limiting critical values. (See reference 2.) Thus, knowledge of these critical speeds of the flow are essential for the design of efficient aerodynamic bodies.

To determine the critical Mach numbers, the general problem of flow of a compressible fluid about a given body must be solved. The often-used methods treating such a problem are Janzen-Rayleigh's method of successive approximations and Glauert-Prandtl's method of small perturbation. The latter method has been extended recently by both Hantzsche and Wendt[3] and C. Kaplan[4]. Indeed, both methods yield valuable information regarding the effects of compressibility and are useful for many practical design problems, particularly the determination of the lower critical Mach number of a given body. But, so far as the general problem of limiting line and upper critical number is concerned, none seems to be adequate, owing to the doubtful convergence of such successive approximations at the required high Mach numbers.

An entirely different approach first was made by Molenbroek[5] and Chaplygin[6] by introducing the velocity components instead of the usual space coordinates as independent variables. The advantage of the method is that, instead of a nonlinear differential equation as is the case in the physical plane, it leads to a linear one in the velocity or hodograph plane. The particular solutions of this linear equation are found to be products of trigonometric functions of the angle of inclination of velocity vector and hypergeometric functions of the magnitude of the velocity vector. It is then possible to construct a general solution from the particular solutions of the differential equation. The difficulty, however, is that the character of the field in the physical plane to which the solution in the hodograph plane corresponds cannot be determined beforehand. This difficulty prevents the exact formulation of the boundary value problem in the hodograph plane. Chaplygin has overcome this handicap by first choosing a "suitable solution" in the hodograph plane and then proceeding to find the corresponding flow in the physical plane. The suitable solution is one which, in the limiting case of zero Mach number at infinity, becomes identical with the incompressible flow over a body similar to the body concerned. This will ensure the satisfaction of the proper boundary conditions in the physical plane. Furthermore, such a solution would be exact both for the subsonic and for the supersonic regions, as no approximation is introduced. Therefore, it is particularly suitable for the problem of determining the upper critical Mach number for a given body, as limiting lines occur only in mixed subsonic and supersonic flows. This method is followed in the present report, except for the introduction of the transformed potential function χ, for easy calculation of the space coordinates.

For the flow around a body, Chaplygin's procedure will lead to a solution in the form of an infinite series, each term of which is a product of a trigonometric function and a hypergeometric function. To put the method on a firm foundation, it is necessary to establish the convergence of the infinite series. Chaplygin himself has done this for the subsonic region. Thus, only the extension to include the supersonic region remains to be completed. In part I of this report, the general properties of hypergeometric functions of large order are investigated in preparation for the proof of the convergence given in part II. The essential point in these parts is to establish the upper and lower bounds for the hypergeometric functions so that the sum of the infinite series can be discussed. It is appropriate to mention here that for the proper representation of the general solution in the hodograph plane, both fundamental solutions of the hypergeometric differential equation are required. This fact has not been considered by many of the previous investigators in this field. In other cases[7] the investigator has chosen to work with only the first solution.

The general solution constructed by the Chaplygin method is really an existence theorem. The extremely slow convergence of the series makes numerical calculation very difficult, if not impossible. This, in fact, constitutes the main difficulty of the method. In part III of the present report, this difficulty is overcome by using the asymptotic properties of the hypergeometric functions. The result is the separation of the solution in the hodograph plane into two parts. One part is of closed form and is the product of a universal function of the velocity and the same solution as for incompressible flow but with a velocity distortion, or velocity correction. For instance, the first part of the stream function for the compressible flow is equal to the product of the universal function of velocity and the stream function for the incompressible flow with the magnitude of velocity modified by a given rule. The other part is an infinite series which converges rapidly everywhere except in a small region on both sides of a critical circle with a radius equal to $q = c$ in the hodograph plane. In practice, by using only a few terms of the infinite series, this zone of slow convergence can be limited to such a small interval that it is of no consequence. Thus the Chaplygin procedure is improved to a point where actual numerical calculations can be made without difficulty.

As a result of this part of the study it becomes clear that by the mere substitution of a different speed scale, or velocity distortion, in the solution for an incompressible fluid, an accurate enough solution for the compressible flow cannot be obtained. For if this were the case, then not only the second part of the solution (the rapidly convergent series given by the present method) would be negligible, but also the value of the multiplying universal function of velocity in the first part of the solution would be unity. However, the value of the second part of the solution is not small compared with that of the first part for a speed near that of sound, and the value of the multiplying function of velocity is far from unity. In other words, the usual so-called hodograph method[8] cannot, in general, yield satisfactory results, for mixed subsonic and supersonic flow. On the other hand, the present method does show that the second part of the solution is zero and the multiplying function in the first part takes the constant value of unity, if the isentropic exponent is equal to -1. This means that for this particular case, a simple speed distortion is sufficient. This is, of course, in accordance with the previous investigation of von Kármán[1] and Tsien[9] and L. Bers[10].

Furthermore, the present method also shows that the rules of speed distortion for the first part of the solution can be used only for subsonic flow and that there is a singularity at the local sonic speed. For regions of supersonic flow, the first part of the solution involves both the incompressible stream function and the incompressible potential function. Thus even without considering the second part of the solution, there is no possibility of making the compressible stream lines coincide with those for incompressible flow in the hodograph plane by a simple stretching of the speed scale. The mathematical basis of this fact is the change in character of the differential equation from elliptic to hyperbolic in the transition from subsonic to supersonic flow. For the supersonic regions, it is not possible to use a real transformation of the velocity variable to convert the differential equation of flow to the Laplace equation, and thus make a simple connection between the compressible and the incompressible flows. This is one of the difficulties of the previously proposed hodograph method. In fact, writers using this method must generally limit their calculation to subsonic speeds. (See references 9, 10.) Now this limit is removed, and the whole field of mixed subsonic and supersonic flows can be treated at once with ease.

For the purely subsonic flow, the second part of the solution is small compared with the first part and may be neglected. Furthermore, if only the zero streamline representing the body is considered, the universal multiplying function of velocity is of no importance. In other words, for this case, a simple speed distortion from the solution of incompressible flow is sufficient to give accurate enough results. However, the subject of the "best" velocity distortion rule in subsonic regions has been the subject of many discussions. (See references 1 and 8.) The present analysis is considered to settle this question. This is due to the fact that the present velocity distortion rule is obtained from the asymptotic properties of the hypergeometric functions, and that such properties are difinite and unique. Therefore, the resultant velocity distortion rule is not the result of uncertain speculation. Furthermore, it is also the best rule, because the analysis implies that this rule will make the second part of the solution, or the correction terms, the smallest. This distortion rule is found to coincide with that of Temple and Yarwood (See reference 11.).

For the purely supersonic flow, the second part of the solution is again small compared with the first part and may be neglected. In fact, the solution then can be reduced to that of the simple wave equation with the inclination of the velocity vector and the distorted velocity as independent variables. This is, of course, the counterpart of the fact that by a simple distortion in velocity, the differential equation for subsonic flows can be reduced to the Laplace equation. The usefulness of this new result for purely supersonic flow has yet to be exploited.

Once the general problem of mixed subsonic and supersonic flow around a body is solved, the determination of the upper critical Mach number or the Mach number for the first appearance of the limiting lines is a simple matter. This problem is discussed in part Ⅳ of the report. A simple method is developed, based on the properties of the limiting line as given by von Kármán[1], Ringleb[12], Tollmien[13], and Tsien[2].

To test the practicability of the method developed, two numerical examples are worked out in detail. However, in order to reduce the amount of computational work and in view of the limited time available, a slightly different procedure actually is used. This procedure is only approximate but is believed to be sufficiently accurate in the supersonic region to give a satisfactory description of the most interesting features of such flows. The examples chosen are derived from the incompressible solution of an elliptic cylinder of thickness ratio 0.6. The free-stream Mach numbers of the compressible flow are 0.6 and 0.7 for these two exampleo. The first case gives a smooth flow over an "elliptic" cylinder of thickness ratio 0.42. The maximum local Mach number is approximately 1.25. Thus a considerable supersonic region exists. The second case gives a flow with limiting line.

Finally, it must be said that owing to the limitation of time, only the case of flow without circulation is investigated in detail. The explicit formulas for numerical calculation are given for two cases: (a) Flow around a body such as an ellipse, (b) periodic flow pattern such as that over a wavy surface. However, it is believed that more general cases can be studied by a slight extension of the present results and use of the same method of approach.

This investigation, conducted at the Guggenheim Aeronautics Laboratory, California Institute of Technology, was sponsored by and conducted with the financial assistance of the National Advisory Committee for Aeronautics.

Notations

The symbols used in this report are classified according to the following groups:

A. Physical Quantities

x, y Cartesian coordinates

u, v the velocity components

q the absolute value of the velocity vector

θ the inclination of the velocity vector with x-axis

ρ density of the fluid

ρ_0 density of the fluid at $q = 0$

p pressure within the fluid corresponding to ρ

p_0 pressure at $q = 0$

γ ratio of the specific heats

c the local speed of sound

c_0 the speed of sound at $q = 0$

U the value of q at infinity, assuming parallel to the x-axis. With subscript, however, it may be a function of τ.

B. Hydrodynamic Functions in the Physical Plane

$z = x + \mathrm{i}y$

$W_0(z) = \varphi_0(x, y) + \mathrm{i}\psi_0(x, y)$ complex potential for incompressible flow in z

φ_0 velocity potential for incompressible flow

ψ_0 stream function for incompressible flow

φ velocity potential for compressible flow

ψ stream function for compressible flow

C. Hydrodynamic Functions in the Hodograph Plane

$w = u - \mathrm{i}v$

$W_0(w) = \varphi_0(u, v) + \mathrm{i}\psi_0(u, v)$ complex potential for incompressible flow in w

$\varphi_0(u, v)$ velocity potential for incompressible flow

$\psi_0(u, v)$ stream function for incompressible flow

$\Lambda_0(w) = zw - W_0(w) = \chi_0(u, v) - \mathrm{i}\sigma_0(u, v)$ transformed complex potential function

$\chi_0(u, v) = ux + vy - \varphi_0(x, y)$; $x = \dfrac{\partial \chi_0}{\partial u}$, $y = \dfrac{\partial \chi_0}{\partial v}$ transformed potential function

$W(w;\tau)$ the complex potential function for compressible flow

$\psi(u, v) = \mathrm{Im}\{W(w;\tau)\}$ stream function for compressible flow

$\Lambda(w;\tau)$ transformed complex potential function for compressible flow

$\chi(u, v) = ux + vy - \varphi(x, y) = \mathrm{Re}\{\Lambda(w;\tau)\}$ transformed potential function for compressible flow

$$\Theta_0(u, v) = \frac{\partial \chi_0}{\partial \theta}$$

$$\Omega_0(u, v) = \frac{\partial \sigma_0}{\partial \theta}$$

$\psi(q, \theta) = \psi_1(q, \theta) + \psi_2^{(l)}(q, \theta)$; $\psi_1(q, \theta)$ represents the contribution by the velocity distorsion; $\psi_2^{(l)}(q, \theta)$ stands for the transformed infinite series, where the superscript l may either mean i the inner, or o the outer solution. In the case of coordinates, the notation is exactly the same.

$$G_{\nu}^{(a)}(\tau) = \underline{F}_{\nu}(\tau)\Delta B_n^{(a)} + \frac{B_n \Delta \underline{F}_{\nu}(\tau)}{f(\tau_1)T^{\nu}(\tau_1)}$$

$$\widetilde{G}_{\nu}^{(a)}(\tau) = \underline{\widetilde{F}}_{\nu}(\tau)\Delta \widetilde{B}_n^{(a)} + \frac{\widetilde{B}_n \Delta \underline{\widetilde{F}}_{\nu}(\tau)}{f(\tau_1)T^{\nu}(\tau_1)}$$

$$\widetilde{G}_{\nu,1}^{(a)}(\tau) = \frac{\nu-1}{\nu+1}\underline{\widetilde{F}}_{\nu,1}(\tau)\Delta \widetilde{B}_n^{(a)} + \frac{\widetilde{B}_n \Delta \widetilde{F}_{\nu,1}(\tau)}{f(\tau_1)T^{\nu}(\tau_1)}$$

D. Parameters and Variables

ν positive rational numbers

m, n positive integers

α denotes 1 or 2 when used as superscript with a bracket or $\alpha = \sqrt{\frac{\gamma+1}{\gamma-1}}$

β denotes the dependence on β when used as subscript or $\beta = \frac{1}{\gamma-1}$

$\lambda = \frac{2(2\beta)^{\alpha/2}}{(1+\alpha)^{\alpha}} \frac{1}{\sqrt{2\beta\tau_1}} \frac{1}{T(\tau_1)}$ the ratio of the distorted speed to that at infinity

$\tau = \frac{1}{2\beta}\frac{q^2}{c_0^2}$

$\mu = \cos^{-1}\sqrt{\frac{\alpha^2\tau-1}{2\beta\tau}}$

ξ, η With superscript or subscript they denote some functions of τ or stand for the two families of the characteristic parameters $\theta + \omega(\tau)$, $\theta - \omega(\tau)$ or the partial differential equations for $\psi(q, \theta)$ or $\chi(q, \theta)$.

ζ complex variable or $\zeta(\tau)$ a function of τ

$M_1 = \frac{U}{c_1}$ the Mach number at infinity

$\tau_1 = \frac{1}{2\beta}\frac{U^2}{c_0^2}$

ε geometrical parameter of the body

Δ Laplacian or difference between exact and approximate values of a function or a constant

E. Hypergeometric Functions

a, b, c parameters of the hypergeometric functions. In particular, a_ν, b_ν, c_ν are defined by (29).

$\underline{F}_{\nu}(\tau) = F(a_\nu, b_\nu; c_\nu; \tau)$ first integral of the hypergeometric equation associated with the stream function

$\underline{F}_{-\nu}(\tau) = F(1+a_\nu - c_\nu, 1+b_\nu - c_\nu; 2-c_\nu; \tau)$

$\underline{F}_\nu(\tau) = \dfrac{\pi\tau^{-\nu}}{(2\beta c_0^2)^\nu \Gamma(c_\nu - 1)\Gamma(c_\nu)}\left[\dfrac{\Gamma(a_\nu)\Gamma(b_\nu)}{\Gamma(1+a_\nu - c_\nu)\Gamma(1+b_\nu - c_\nu)}\tau^\nu \underline{F}_\nu(\tau) - \dfrac{\Gamma(c_\nu)}{\Gamma(2-c_\nu)}\underline{F}_{-\nu}(\tau)\right]$ **

second integral of the same equation

$\underline{G}_\nu(\tau) = q^{2\nu}F_\nu(\tau)$

$\underline{F}_{\nu,1}(\tau) = F(1+a_\nu, 1+b_\nu; 1+c_\nu; \tau)$

$\underline{F}_\nu^{(r)}(\tau) = \underline{F}_\nu(\tau)/\underline{F}_\nu(\tau_1)$

$\underline{F}_{\nu,1}^{(r)}(\tau) = \underline{F}_{\nu,1}(\tau)/\underline{F}_\nu(\tau_1)$

$\underline{F}_\nu^*(\tau) = \underline{F}_\nu(\tau) + i\underline{F}_\nu(\tau)$

$R_\nu(\tau) = |\underline{F}_\nu^*(\tau)|$

$\phi_\nu(\tau) = \arg \underline{F}_\nu^*(\tau)$

If any function or a constant is associated with $\chi(q, \theta)$, it will be marked on top by a symbol $\sim$, such as $\widetilde{\underline{F}}_\nu(\tau)$.

** $F_\nu(\tau)$ in the original paper has been corrected as $\underline{F}_\nu(\tau)$ and all T's in the original expression have been corrected as Γ. — Noted by editor

Part I
Differential Equations of Compressible Flow and Properties of Their Particular Solutions

1. Equations of Motion

It is proposed to study the irrotational steady motion of an inviscid nonconducting compressible fluid in an infinitely extended domain containing a cylindrical body with its axis perpendicular to the constant velocity at infinity. The flow is then two-dimensional. Let x and y be the Cartesian coordinates and u and v the velocity components parallel to the x- and the y- axis. The dynamical equations governing such a motion, in the absence of body force, are

$$\rho u \frac{\partial u}{\partial x} + \rho v \frac{\partial u}{\partial y} = -\frac{\partial p}{\partial x} \tag{1}$$

$$\rho u \frac{\partial v}{\partial x} + \rho v \frac{\partial v}{\partial y} = -\frac{\partial p}{\partial y} \tag{2}$$

Here p is the pressure and ρ the density of the fluid, both being continuous functions of x and y. In addition, the following equation of continuity must be satisfied:

$$\frac{\partial}{\partial x}(\rho u) + \frac{\partial}{\partial y}(\rho v) = 0 \tag{3}$$

Furthermore, since the velocity is constant at infinity, the flow is irrotational there. Then, according to Thomson's theorem, if the pressure is a function of the density alone, the flow will remain irrotational; that is,

$$\frac{\partial v}{\partial x} - \frac{\partial u}{\partial y} = 0 \tag{4}$$

In the case of flow of an inviscid nonconducting gas, the thermodynamic change of state of the gas is adiabatic. If the flow is assumed to be continuous, excluding shock waves, then the relation between p and ρ must be that of an isentropic process:

$$p = \text{constant} \cdot \rho^{\gamma} \tag{5}$$

where γ is the ratio of the specific heats.

As in the case of incompressible flow, there are more equations than the number of the variables. However, by virtue of equations (4) and (5), the dynamical equations (1) and (2) reduce to a single differential equation and can be integrated easily to give a relation between the pressure and the magnitude q of the velocity: namely,

$$p = p_0 \left\{1 - \frac{\gamma - 1}{2} \frac{q^2}{c_0^2}\right\}^{\frac{\gamma}{\gamma-1}}, \quad \text{with} \quad q^2 = u^2 + v^2 \tag{6}$$

Here p_0 and c_0 are respectively the pressure and the speed of sound at the stagnation point $q = 0$ and $c = \sqrt{\frac{dp}{d\rho}}$. It is possible to obtain a similar relation between ρ and q by means of equation (5):

$$\rho = \rho_0 \left\{ 1 - \frac{\gamma - 1}{2} \frac{q^2}{c_0^2} \right\}^{\frac{1}{\gamma-1}} \tag{7}$$

where ρ_0 denotes the value of ρ at $q = 0$.

After integrating the dynamical equations, the velocities u and v can be determined from the kinematic conditions specified by equations (3) and (4). By eliminating ρ from equation (3), the result is

$$\left(1 - \frac{u^2}{c^2}\right)\frac{\partial u}{\partial x} - \frac{2uv}{c^2}\frac{\partial u}{\partial y} + \left(1 - \frac{v^2}{c^2}\right)\frac{\partial v}{\partial y} = 0 \tag{8}$$

where $c^2 = \gamma p/\rho$ and thus can be calculated in terms of the speed by equations (6) and (7). It is of interest to note that the equation of continuity (8) now, unlike the case of incompressible flow, becomes dependent on the dynamical equations and, consequently, is nonlinear. This change in the character of the fundamental equation makes the direct solution of the problem in space coordinates very difficult.

2. Transformation of the Differential Equations

The assumption of irrotationality implies the existence of a velocity-potential for such a flow. If this function is introduced to eliminate u and v, equations (4) and (8) would give rise to a nonlinear partial differential equation of the second order. The problem is further complicated by the possible appearance of supersonic regions, or regions where the speed of flow is larger than the local sonic speed. This means that for some part of the domain, the equation is of the elliptic type; while in the other part, it is of the hyperbolic type. Thus the equation not only is nonlinear but also is of mixed type, and there is as yet no successful method to deal with it directly in the physical plane. Molenbroek[5] and Chaplygin[6] made some progress in solving the problem by transforming the equations from the physical to the hodograph plane in which u and v are taken as the independent variables. If this is done, the differential equations become linear and thus can be solved by well-known methods.

Let the transformation be defined by

$$u = u(x, y) \tag{9}$$

$$v = v(x, y) \tag{10}$$

If u and v are continuous functions of x and y with continuous partial derivatives, and if the Jacobian $\left(\frac{\partial(x, y)}{\partial(u, v)}\right)$ is finite and nonvanishing, a unique inverse transformation exists. Under these conditions, equations (8) and (4) are easily transformed into

$$\left(1-\frac{u^2}{c^2}\right)\frac{\partial y}{\partial v}+\frac{2uv}{c^2}\frac{\partial x}{\partial v}+\left(1-\frac{v^2}{c^2}\right)\frac{\partial x}{\partial u}=0 \tag{11}$$

$$\frac{\partial x}{\partial v}-\frac{\partial y}{\partial u}=0 \tag{12}$$

Corresponding to $\varphi(x, y)$ in the physical plane, there is introduced here a function $\chi(u, v)$ defined by

$$\chi = xu + yv - \varphi; \quad x=\frac{\partial \chi}{\partial u}, \quad y=\frac{\partial \chi}{\partial v} \tag{13}$$

While equation (12) is satisfied identically, equation (11) becomes

$$\left(1-\frac{u^2}{c^2}\right)\frac{\partial^2 \chi}{\partial v^2}+\frac{2vu}{c^2}\frac{\partial^2 \chi}{\partial v \partial u}+\left(1-\frac{v^2}{c^2}\right)\frac{\partial^2 \chi}{\partial u^2}=0 \tag{14}$$

As c is a function of q alone, the equation for $\chi(u, v)$ is then linear. From equation (13) it is recognized that if $\chi(u, v)$ is known, a one-to-one correspondence between the space coordinates and the velocity components can be easily established.

However, it is also clear that this function is inconvenient for obtaining the streamlines and the flow in the physical plane. To solve this part of the problem, a plan may be adopted similar to Chaplygin's by introducing both the potential function $\varphi(x, y)$ and the stream function $\psi(x, y)$ defined by:

$$u=\frac{\partial \varphi}{\partial x}, \quad v=\frac{\partial \varphi}{\partial y} \tag{15}$$

$$\rho u = \rho_0 \frac{\partial \psi}{\partial y}, \quad \rho v = -\rho_0 \frac{\partial \psi}{\partial x} \tag{16}$$

From these definitions are obtained immediately the following equivalent relations:

$$d\varphi = u\,dx + v\,dy \tag{17}$$

$$\rho_0\, d\psi = -\rho v\,dx + \rho u\,dy \tag{18}$$

For the subsequent calculations, it was found convenient to introduce the polar coordinates in the hodograph plane defined by:

$$u = q\cos\theta, \quad v = q\sin\theta \tag{19}$$

where θ is the inclination of the velocity vector to the x-axis. Functions dx and dy can be solved for from equations (17) and (18). As dx and dy are exact differentials, the conditions of integrability then give:

$$\frac{\partial \varphi}{\partial q} = -\frac{\rho_0}{\rho}\left(1-\frac{q^2}{c^2}\right)\frac{1}{q}\frac{\partial \psi}{\partial \theta} \tag{20}$$

$$\frac{1}{q}\frac{\partial \varphi}{\partial \theta} = \frac{\rho_0}{\rho}\frac{\partial \psi}{\partial q} \tag{21}$$

By eliminating φ between equations (20) and (21), an equation for ψ is obtained:

$$q^2 \frac{\partial^2 \psi}{\partial q^2} + \left(1 + \frac{q^2}{c^2}\right) q \frac{\partial \psi}{\partial q} + \left(1 - \frac{q^2}{c^2}\right) \frac{\partial^2 \psi}{\partial \theta^2} = 0 \tag{22}$$

Equation (14) can also be transformed in polar coordinates. The procedure is straightforward and yields

$$q^2 \frac{\partial^2 \chi}{\partial q^2} + \left(1 - \frac{q^2}{c^2}\right) q \frac{\partial \chi}{\partial q} + \left(1 - \frac{q^2}{c^2}\right) \frac{\partial^2 \chi}{\partial \theta^2} = 0 \tag{23}$$

There is an additional relation between χ and φ derived from equation (13):

$$\varphi = q\chi_q - \chi \tag{24}$$

Since φ is connected with ψ, this relation ensures that ψ and χ are properly connected and represent the same flow pattern in the physical plane. It can be thus considered as the equation of compatibility. Equations (22), (23), and (24) are the three fundamental equations in the present problem dealing with the two-dimensional flow of a compressible fluid.

3. The Particular Solutions of the Differential Equations

As the differential equations for $\psi(q, \theta)$ and $\chi(q, \theta)$ are linear, a general solution can certainly be built by superimposing the particular integrals of the equations. To obtain the particular integrals, let $\psi(q, \theta)$ and $\chi(q, \theta)$ be of the following forms:

$$\psi(q, \theta) = q^{\nu}\psi_{\nu}(q)e^{i\nu\theta}$$
$$\chi(q, \theta) = q^{\nu}\chi_{\nu}(q)e^{i\nu\theta}$$

where ν is any real number. By substituting in equations (22) and (23), the equations satisfied by $\psi_\nu(q)$ and $\chi_\nu(q)$ are:

$$q^2 \frac{d^2\psi_\nu}{dq^2} + \left(2\nu + 1 + \frac{q^2}{c^2}\right) q \frac{d\psi_\nu}{dq} + \nu(\nu+1)\frac{q^2}{c^2}\psi_\nu = 0 \tag{25}$$

$$q^2 \frac{d^2\chi_\nu}{dq^2} + \left(2\nu + 1 - \frac{q^2}{c^2}\right) q \frac{d\chi_\nu}{dq} + \nu(\nu-1)\frac{q^2}{c^2}\chi_\nu = 0 \tag{26}$$

Now each of these equations can be further reduced by changing the independent variable. The appropriate transformation is found to be

$$\tau = \frac{1}{2\beta}\frac{q^2}{c_0^2}, \quad \text{with} \quad \beta = \frac{1}{\gamma - 1}$$

By expanding the gas to zero pressure, or vacuum, the maximum velocity is obtained. Equation (6) shows that the maximum speed is $q_{\max} = \sqrt{\frac{2}{\gamma - 1}}c_0$. Therefore, the maximum value of τ is unity. Similarly, it is found that for the speed of the flow equal to the local sonic speed, $\tau = \frac{1}{2\beta + 1}$, equations (25) and (26) then become

$$\tau(1-\tau)\psi''_\nu(\tau) + [c_\nu - (a_\nu + b_\nu + 1)\tau]\psi'_\nu(\tau) - a_\nu b_\nu \psi_\nu(\tau) = 0 \tag{27}$$

$$\tau(1-\tau)\chi_\nu''(\tau)+[c_\nu-(a_\nu+\beta+b_\nu+\beta+1)\tau]\chi_\nu'(\tau)-(a_\nu+\beta)(b_\nu+\beta)\chi_\nu(\tau)=0 \quad (28)$$

where

$$a_\nu+b_\nu=\nu-\beta,\quad a_\nu b_\nu=-\frac{1}{2}\beta\nu(\nu+1),\quad \text{and}\quad c_\nu=\nu+1 \quad (29)$$

These are the hypergeometric equations, of which equation (27) was first obtained by Chaplygin in 1904 (See reference 6). The differential equation of this type has three regular singularities at 0, 1, and $+\infty$. If the differences of the two exponents at the respective singularities: namely, $c-1$, $a-b$, $a+b-c$, are not integers or zero, the two fundamental independent solutions are $F(a,b;\ c;\ \tau)$ and $\tau^{1-c}F(1+a-c,\ 1+b-c;\ 2-c;\ \tau)$. They are single-valued and regular in the whole plane with a cut from $+1$ to $+\infty$. The function $F(a,\ b;\ c;\ \tau)$ known as the hypergeometric function of general parameters a, b, and c, is defined by the hypergeometric series which is absolutely and uniformly convergent when $|\tau|<1$, provided $\mathrm{Re}(c-a-b)>0$. For $|\tau|>1$, analytic continuation has to be used. Furthermore, it is normalized so that at $\tau=0$

$$F(a,b;\ c;\ 0)=1 \quad (30)$$

Hence, the particular solutions of equation (27) are

$$F(a_\nu,b_\nu;\ c_\nu;\ \tau),\quad \tau^{1-c_\nu}F(1+a_\nu-c_\nu,\ 1+b_\nu-c_\nu;\ 2-c_\nu;\ \tau) \quad (31)$$

The particular solutions of equations (28) are

$$F(a_\nu+\beta;\ b_\nu+\beta;\ c_\nu;\ \tau),\quad \tau^{1-c_\nu}F(1+a_\nu+\beta-c_\nu,\ 1+b_\nu+\beta-c_\nu;\ 2-c_\nu;\ \tau) \quad (32)$$

Here a_ν, b_ν, and c_ν are parameters defined by equation (29).

When ν is a positive integer while a_ν and b_ν remain as they are, the second integral will reduce to a constant multiple of the first one. This case was first studied by Gauss[14], who found a second integral involving a logarithmic term by considering the limiting value of the integrals given as ν tends to an integral value. The method has been further developed by Tannery[15] and Goursat[16]. However, the form regarded as conventional nowadays was that obtained by Frobenius' general method. According to this method, the pair of fundamental solutions of a hypergeometric equation are

$$F(a,b;n+1;\tau),\quad K_n\tau^{-n}\{\tau^nF(a,b;n+1;\tau)\log\tau+\tau^nQ_n^{(1)}(a,b;\tau)+P_{n-1}^{(1)}(\tau)\} \quad (33)$$

when $c_n=n+1$, n being a positive integer; and

$$Q_n^{(1)}(a,b;\tau)=\frac{\Gamma(n+1)}{\Gamma(a)\Gamma(b)}\sum_0^\infty\frac{\Gamma(a+m)\Gamma(b+m)}{\Gamma(m+1)\Gamma(n+1+m)}\psi(a,b;m)\tau^m$$

$$P_{n-1}^{(1)}(\tau)=(-1)^{n+1}\frac{\Gamma(n+1)}{\Gamma(a)\Gamma(b)}\sum_0^{n-1}(-1)^m\frac{\Gamma(a-n+m)\Gamma(b-n+m)\Gamma(n-m)}{\Gamma(m+1)}\tau^m \quad (34)$$

$$\Psi(a,b;m)=\sum_{r=0}^{m-1}\left[\frac{1}{a+r}+\frac{1}{b+r}-\frac{1}{n+1+r}\right]-\sum_{r=1}^{m}\frac{1}{r}$$

Here a, b may be either a_n, b_n or $a_n+\beta$, $b_n+\beta$ defined in equation (29) according to whether the system (33) is referred to as solutions of equation (27) or (28). And K_n can be determined so that the product of the second integral and q^{2n} satisfies the condition (30).

In view of the fact that the second integral in (33) does not constitute a family of solutions with the second integral given in (31) or (32), it is very desirable to define a new function as second integral which will be continuous in ν as well as in τ. Let $\underline{F}_\nu(\tau)$ denote the first integral $F(a,b;c_\nu;\tau)$. As a second integral, take the linear combination of the solutions:

$$F_\nu(\tau) = K_\nu\{\Gamma(1-c_\nu)\Gamma(a)\Gamma(b)\,\underline{F}_\nu(\tau)+\Gamma(1-c_\nu)\Gamma(1+a-c_\nu)\Gamma(1+b-c_\nu)\tau^{1-c_\nu}\underline{F}_{-\nu}(\tau)\} \quad (35)$$

where

$$\underline{F}_{-\nu}(\tau) = F(1+a-c_\nu, 1+b-c_\nu; 2-c_\nu;\tau)$$

This is evidently a solution and valid for all values of ν. The constant K_ν is determined subject to the following condition:

$$q^{2\nu}F_\nu(\tau) = 1 \quad \text{for} \quad \tau = 0 \quad (36)$$

The value of K_ν then is found to be

$$K^{-1} = (2\beta c_0)^\nu\Gamma(c_\nu-1)\Gamma(1+a-c_\nu)\Gamma(1+b-c_\nu)$$

Using the relation

$$\Gamma(c_\nu)\Gamma(1-c_\nu) = \pi\csc c_\nu\pi$$

equation (35), when multiplied by $q^{2\nu}$, will define a new function $\underline{G}_\nu(\tau)$: $a,b\neq -n$

$$\underline{G}_\nu(\tau) = \frac{\pi}{\sin c_\nu\pi}\left[\frac{\Gamma(a)\Gamma(b)\tau^\nu\underline{F}_\nu(\tau)}{\Gamma(c_\nu)\Gamma(c_\nu-1)\Gamma(1+a-c_\nu)\Gamma(1+b-c_\nu)}-\frac{\underline{F}_{-\nu}(\tau)}{\Gamma(c_\nu-1)\Gamma(2-c_\nu)}\right] \quad (37)$$

When ν takes integral values, the expression in the bracket vanishes; however, the limit of the ratio exists:

$$\underline{G}_n(\tau) = \lim_{\nu\to n}\underline{G}_\nu(\tau) \quad (38)$$

The usual definition of the limit of a quotient gives

$$\underline{G}_n(\tau) = (-1)^{n+1}\left[\frac{\partial}{\partial\nu}\frac{\Gamma(a)\Gamma(b)\tau^\nu\underline{F}_\nu(\tau)}{\Gamma(\nu+1)\Gamma(\nu)\Gamma(a-\nu)\Gamma(b-\nu)}-\frac{\partial}{\partial\nu}\frac{\underline{F}_{-\nu}(\tau)}{\Gamma(\nu)\Gamma(1-\nu)}\right]_{\nu=n}$$

By considering separatoly the first n terms in $\underline{F}_{-\nu}(\tau)$, as $\Gamma(1-\nu)$ has poles at $\nu=n$, a straightforward reduction yields:

$$\underline{G}_n(\tau) = \underline{c}_n\tau^n\log\tau\underline{F}_n(\tau)+\tau^nQ_n^{(2)}(\tau)+P_{n-1}^{(2)}(\tau) \quad (39)$$

where

$$Q_n^{(2)}(\tau) = \frac{(-1)^{n+1}}{\Gamma(n)\Gamma(-n+a)\Gamma(-n+b)} \sum_{m=0}^{\infty} [\psi(a+m) + \psi(b+m) -$$

$$\psi(c_n + m) - \psi(m+1)] \frac{\Gamma(a+m)\Gamma(b+m)}{\Gamma(c_n+m)\Gamma(m+1)} \tau^m$$

$$P_{n-1}^{(2)}(\tau) = \frac{1}{\Gamma(n)\Gamma(a-n)\Gamma(b-n)} \sum_{m=0}^{n-1} (-1)^m \frac{\Gamma(a-n+m)\Gamma(b-n+m)\Gamma(n-m)}{\Gamma(m+1)} \tau^m$$

$$\underline{c}_n = \frac{(-1)^{n+1}\Gamma(a)\Gamma(b)}{\Gamma(n)\Gamma(n+1)\Gamma(a-n)\Gamma(b-n)}$$

and $\psi(\xi)$ denotes the derivative of $\log \Gamma(\xi)$. It can be seen that the difference between (33) and (39) lies only in a constant multiple of the first integral which has been absorbed in $Q_n^{(2)}(\tau)$.

In the following discussions, the two fundamental solutions of the hypergeometric differential equation will be taken as $\underline{F}_\nu(\tau)$ and $q^{-2\nu}\underline{G}_\nu(\tau)$. The normalization conditions given by (30) and (36) are chosen for the continuous passage of a compressible to an incompressible flow. Ultimately, these functions are again defined in terms of power series which are absolutely and uniformly convergent within the domain $|\tau| < 1$. However, since the maximum value of τ attainable by the fluid is unity, the continuation of the solutions beyond the unit circle will not be discussed here.

Thus $\underline{F}_\nu(\tau)$ and $q^{-2\nu}\underline{G}_\nu(\tau)$ denote the two independent integrals of equation (27) where ν is any positive number. The particular solutions of equation (22) are then:

$$q^\nu \underline{F}_\nu(\tau)[A_\nu^{(1)} \cos\nu\theta + A_\nu^{(2)} \sin\nu\theta], \quad q^{-\nu}\underline{G}_\nu(\tau)[B_\nu^{(1)}\cos\nu\theta + B_\nu^{(2)}\sin\nu\theta] \tag{40}$$

where $A_\nu^{(1)}$, $A_\nu^{(2)}$, $B_\nu^{(1)}$, and $B_\nu^{(2)}$ are constants. Similarly, those of equation (23) are

$$q^\nu \underline{\tilde{F}}_\nu(\tau)[\tilde{A}_\nu^{(1)} \cos\nu\theta + \tilde{A}_\nu^{(2)} \sin\nu\theta], \quad q^{-\nu}\underline{\tilde{G}}_\nu(\tau)[\tilde{B}_\nu^{(1)}\cos\nu\theta + \tilde{B}_\nu^{(2)}\sin\nu\theta] \tag{41}$$

where $\underline{\tilde{F}}_\nu(\tau)$ and $q^{-2\nu}\underline{\tilde{G}}_\nu(\tau)$ are the two independent integrals of equation (28) and $\tilde{A}_\nu^{(1)}$, $\tilde{A}_\nu^{(2)}$, $\tilde{B}_\nu^{(1)}$, and $\tilde{B}_\nu^{(2)}$ are constants.

In addition to these solutions, there are two other integrals each of which is a function of one variable only. Assuming $\psi = \psi(q)$ or $\psi(\theta)$, then equations (22) and (23) yield respectively:

$$c_1\theta \quad \text{and} \quad c_2\int (1-\tau)^\beta \frac{d\tau}{\tau} \tag{42}$$

$$\tilde{c}_1\theta \quad \text{and} \quad \tilde{c}_2\int (1-\tau)^{-\beta} \frac{d\tau}{\tau} \tag{43}$$

which correspond to the fundamental solution of the Laplace equation.

As c_0 approaches infinity, all these particular solutions reduce to the familiar harmonic functions: namely,

$$q^\nu [A_\nu^{(1)}\cos\nu\theta + A_\nu^{(2)}\sin\nu\theta], \quad q^{-\nu}\{B_\nu^{(1)}\cos\nu\theta + B_\nu^{(2)}\sin\nu\theta\} \tag{44}$$

and

$$c_1\theta, \quad c_2\log q \tag{45}$$

This property which is the consequence of (30) and (36) is essential in the method presented in this report for connecting a compressible flow with the incompressible flow of similar configuration.

In the subsequent calculations, another integral will be encountered for the function $\chi(q, \theta)$ which corresponds to the imaginary part of $w\log we^{i\pi}$ or $q\log q\sin\theta - q(\pi-\theta)\cos\theta$ of the incompressible flow. Suppose the solution possesses the form:

$$\chi(q, \theta) = \chi_1(q)\sin\theta - \chi_2(q)(\pi-\theta)\cos\theta \tag{46}$$

By substituting the expression in equation (23), χ_1 and χ_2 are found to satisfy simultaneously the following differential equations:

$$q^2\chi''_1(q) + \left(1-\frac{q^2}{c^2}\right)(q\chi'_1-\chi_1) = 2\left(1-\frac{q^2}{c^2}\right)\chi_2 \tag{47}$$

$$q^2\chi''_2 + \left(1-\frac{q^2}{c^2}\right)(q\chi'_2-\chi_2) = 0 \tag{48}$$

Equation (48) can be easily integrated by putting $\chi_2 = qk_2(q)$. The condition that $\chi_2 \to q$ as $c_0 \to \infty$ requires $k_2(q)$ to be a constant. The second integral of equation (48) is just the second of (43) which, in the limit, tends to $\log q$. Thus $\chi_2 = q$ is the appropriate solution. With this solution, it is possible to proceed to solve equation (47) by assuming $\chi_1 = qk_1(q)$. The equation for $k_1(q)$ is again integrable by quadrature, and the result is

$$k_1(q) = \frac{1}{2(\beta+1)}\left[(2\beta+1)\log\tau - \frac{1}{\tau} + K_1\int^{\tau}(1-\tau)^{-\beta}\frac{d\tau}{\tau^2}\right] + K_2 \tag{49}$$

where K_1 and K_2 are the constants of integration. Hence, the desired particular integral is

$$\chi(q, \theta) = qk_1(\tau)\sin\theta - q(\pi-\theta)\cos\theta \tag{50}$$

The correspondence between solutions for compressible flow and for incompressible flow is summarized in table 1.

4. The Properties of the Hypergeometric Functions of Large Order

The behavior of $F(a_\nu, b_\nu; c_\nu; \tau)$ for large positive values of ν has been discussed by Chaplygin in connection with the question of convergence of his series solution for the flow of a gas jet. However, his discussions are limited to the subsonic flow and, for this reason, the value of τ is restricted to the interval $0 \leqslant \tau \leqslant \frac{1}{2\beta+1}$. In the more general problem where both subsonic and supersonic flow may exist, the whole interval $0 \leqslant \tau \leqslant 1$ has to be considered. Furthermore, both integrals of the hypergeometric equation are involved, as will be shown in part II. As a preparation for the proof of the convergence of the solutions, the properties of the hypergeometric functions of large order in the extended interval will be discussed presently.

Chaplygin[6] introduced a new function $\frac{\nu}{2\tau}\xi_\nu(\tau)$ defined as the logarithmic derivative of $\tau^{\frac{\nu}{2}}\underline{F}_\nu(\tau)$: *namely*,

$$\nu\xi_\nu(\tau) = 2\tau\frac{\mathrm{d}}{\mathrm{d}\tau}\log\tau^{\frac{\nu}{2}}\underline{F}_\nu(\tau), \quad \nu\neq 0 \tag{51}$$

where $\underline{F}_\nu(\tau)$ denotes the first integral of the hypergeometric equation (27) or (28) and ν can be either an integer or not an integer. Then in the place of equation (27) or (28), the corresponding differential equation for ξ_ν is a Riccati equation:

$$X(\xi_\nu)\equiv\xi'_\nu\pm\frac{\beta}{1-\tau}\xi_\nu+\frac{\nu}{2\tau}\left[\xi_\nu^2-\frac{1-(2\beta+1)}{1-\tau}\right]=0 \tag{52}$$

where the lower sign corresponds to equation (28). As shown by Chaplygin, $\underline{F}_\nu(\tau)$, although an oscillatory function, can have no root in $0\leqslant\tau\leqslant\frac{1}{2\beta+1}$ and, consequently, $\xi_\nu(\tau)$ is finite and continuous in the same interval. Moreover, it can be deduced also that $\xi_\nu(0)=1$ and $\xi'_\nu(0)=-\beta$. Since $\xi'_\nu(\tau)$ does not change sign in $0\leqslant\tau\leqslant\frac{1}{2\beta+1}$; $\xi_\nu(\tau)$ is monotonic decreasing and eventually vanishes at $\tau_0\leqslant\tau^*$, τ^* being the first root of the hypergeometric function for $\nu>0$. Since τ^* is a decreasing function of ν, when ν becomes large, τ^* and consequently τ_0 will differ from $\frac{1}{2\beta+1}$ by a small quantity.

Chaplygin's theorem. In $0\leqslant\tau\leqslant\frac{1}{2\beta+1}$, if a monotonic continuous function $\eta_\nu(\tau)$ satisfies (i) $\eta_\nu(0)=1$ and (ii) $X(\eta_\nu)\gtreqless 0$, then

$$\eta_\nu(\tau)\gtreqless\xi_\nu(\tau), \quad \nu>1 \tag{53}$$

The proof is given in Chaplygin's paper[6]. In the case of the second integral $F_\nu(\tau)$, the theorem is still true with the signs of inequalities reversed because it can be verified that $X(\xi_{-\nu})=0$, where $\xi_{-\nu}(\tau)$ corresponds to the case of $F_\nu(\tau)$ instead of $\underline{F}_\nu(\tau)$ in (51), and $\xi_{-\nu}(0)=-1$; therefore $\xi_{-\nu}(\tau)$ is negative in $0\leqslant\tau\leqslant\frac{1}{2\beta+1}$.

Corollary (51). In $0\leqslant\tau\leqslant\frac{1}{2\beta+1}$, the functions $\underline{F}_\nu(\tau)$ and $\underline{G}_\nu(\tau)$ fall respectively between the limits:

(i) $$T_1^\nu(\tau)<\underline{F}_\nu(\tau)<T_2^\nu(\tau) \tag{54}$$

(ii) $$T_1^{-\nu}(\tau)>\underline{G}_\nu(\tau)>T_2^{-\nu}(\tau), \quad \nu>1 \tag{55}$$

where

$$T_1(\tau)=\exp\left\{-\int_0^\tau\left[1-\sqrt{\frac{1-(2\beta+1)\tau}{1-\tau}}\right]\frac{\mathrm{d}\tau}{2\tau}\right\} \tag{56}$$

$$T_2(\tau)=\exp\left\{-\int_0^\tau[1-(1-\tau)^\beta]\frac{\mathrm{d}\tau}{2\tau}\right\} \tag{57}$$

This can be verified easily by choosing η_ν to be $\sqrt{\dfrac{1-(2\beta+1)\tau}{1-\tau}}$ or $(1-\tau)^\beta$. As, evidently, in $0 \leqslant \tau \leqslant \dfrac{1}{2\beta+1}$ when

$$\nu > 1, \quad \sqrt{\frac{1-(2\beta+1)\tau}{1-\tau}} < \xi_\nu < (1-\tau)^\beta \tag{58}$$

and

$$-\left(1+O\left(\frac{1}{\nu}\right)\right)\sqrt{\frac{1-(2\beta+1)\tau}{1-\tau}} > \xi_{-\nu} > -(1-\tau)^\beta \tag{59}$$

and furthermore, $X(\eta_\nu) \gtreqless 0$ are satisfied, consequently, it follows the results.

Corollary (52). In $0 \leqslant \tau \leqslant \dfrac{1}{2\beta+1}$, the absolute value of the logarithmic derivative of $F(a_\nu, b_\nu; c_\nu; \tau)$ divided by ν, is bounded both above and below, that is,

$$M_1(\tau) \leqslant \frac{F(a_\nu+1, b_\nu+1; c_\nu+1; \tau)}{F(a_\nu, b_\nu; c_\nu; \tau)} \leqslant M_2(\tau) \tag{60}$$

where $M_1(\tau)$ and $M_2(\tau)$ are independent of ν. This really is a consequence of (58) and (59).

It shall be noted that the results established in the foregoing are applicable to $\underset{\sim}{\widetilde{F}}_\nu(\tau) = F(a_\nu+\beta, b_\nu+\beta; c_\nu; \tau)$, provided ν is large, because then the two equations (27) and (28) tend to be the same.

Obviously, Chaplygin's theorem ceases to be true when $\tau > \dfrac{1}{2\beta+1}$. For in the interval $\dfrac{1}{2\beta+1} < \tau < 1$, the solutions of the hypergeometric equation are oscillatory and, hence, within any closed interval in $\dfrac{1}{2\beta+1} < \tau < 1$ the number of roots of $\underline{F}_\nu(\tau)$ will be proportional to ν. (See reference 17) When ν is large, there will be a large number of roots in the interval. As a consequence, the function $\xi_\nu(\tau)$ will have there an ever increasing number of simple poles, and a finite interval in which $\xi_\nu(\tau)$ remains finite for all ν does not exist.

To carry the investigation over into $\dfrac{1}{2\beta+1} < \tau < 1$, the method is modified. Let $\underline{F}_\nu(\tau)$ and $F_\nu(\tau)$ be two independent solutions of equation (27) or (28); and let the linear combination be denoted by

$$F_\nu^*(\tau) = \underline{F}_\nu(\tau) + iF_\nu(\tau) \tag{61}$$

The complex function is, of course, a solution of the same differential equation. In terms of its modulus $R_\nu(\tau)$ and argument $\phi_\nu(\tau)$, the function may also be expressed as

$$F_\nu^*(\tau) = R_\nu(\tau)e^{i\phi_\nu(\tau)} \tag{62}$$

where both $R_\nu(\tau)$ and $\phi_\nu(\tau)$ are continuous functions with continuous derivatives. By comparing with (61), it is necessary to have

$$\underline{F}_\nu(\tau) = R_\nu(\tau)\cos\phi_\nu(\tau) \tag{63}$$

$$F_\nu(\tau) = R_\nu(\tau)\sin\phi_\nu(\tau) \tag{64}$$

According to the Sturm separation theorem, $\underline{F}_\nu(\tau)$ and $F_\nu(\tau)$ never vanish simultaneously in any closed interval and $R_\nu(\tau)$ never vanishes in $\frac{1}{2\beta+1} < \tau < 1$ and remains positive in the whole interval. Then corresponding to (51), a complex function $\xi_\nu^*(\tau)$ can be defined as follows:

$$\nu\xi_\nu^*(\tau) = 2\tau\frac{d}{d\tau}\log\tau^{\frac{\nu}{2}}F_\nu^*(\tau) \tag{65}$$

which satisfies the same equation (52). On separating into real and imaginary parts, the Riccati equation for $\xi_\nu^*(\tau)$ becomes

$$X_1(\xi_\nu^{(1)},\xi_\nu^{(2)}) \equiv \xi'^{(1)}_\nu + \frac{\beta}{1-\tau}\xi_\nu^{(1)} + \frac{\nu}{2\tau}\left[\xi_\nu^{2(1)} - \xi_\nu^{2(2)} + \frac{(2\beta+1)\tau-1}{1-\tau}\right] = 0 \tag{66}$$

$$X_2(\xi_\nu^{(2)},\xi_\nu^{(1)}) \equiv \xi'^{(2)}_\nu + \frac{\beta}{1-\tau}\xi_\nu^{(2)} + \frac{\nu}{\tau}\xi_\nu^{(2)}\xi_\nu^{(1)} = 0 \tag{67}$$

where $\xi_\nu^{(1)}$ and $\xi_\nu^{(2)}$ are real continuous functions of τ defined as

$$\xi_\nu^*(\tau) = \xi_\nu^{(1)}(\tau) + i\xi_\nu^{(2)}(\tau) \tag{68}$$

Their connection with $R_\nu(\tau)$ and $\phi_\nu(\tau)$ separately are given by means of (65): namely,

$$\nu\xi_\nu^{(1)}(\tau) = 2\tau\frac{d}{d\tau}\log\tau^{\frac{\nu}{2}}R_\nu(\tau) \tag{69}$$

$$\nu\xi_\nu^{(2)}(\tau) = 2\tau\frac{d}{d\tau}\phi_\nu(\tau) \tag{70}$$

Now equation (67) can be integrated in terms of $\xi_\nu^{(1)}(\tau)$ and whence $\xi_\nu^{(2)}(\tau)$ can be eliminated from equation (66). Then the equations for $\xi_\nu^{(1)}$ and $\xi_\nu^{(2)}$ are

$$X_1(\xi_\nu^{(1)}) \equiv \xi'^{(1)}_\nu + \frac{\beta}{1-\tau}\xi_\nu^{(1)} + \frac{\nu}{2\tau}\left[\xi_\nu^{2(1)} - \xi_0^2(1-\tau)^{2\beta}e^{-2\nu}\int_{\tau_0}^{\tau}\xi_\nu^{(1)}\frac{d\tau}{\tau} + \frac{(2\beta+1)\tau-1}{1-\tau}\right] = 0 \tag{71}$$

$$\xi_\nu^{(2)}(\tau) = -\xi_0(1-\tau)^{\beta}e^{-\nu}\int_{\tau_0}^{\tau}\xi_\nu^{(1)}\frac{d\tau}{\tau}, \quad \xi_0 = \frac{2}{(\tau_0^{\frac{1}{2}}R_\nu(\tau_0))^2} \tag{72}$$

Equation (71) together with the condition $\xi_\nu^{(1)}(0) = -1$ determines uniquely the solution $\xi_\nu^{(1)}(\tau)$. The actual value of $\xi_\nu^{(1)}(\tau)$ can be expressed, of course, in terms of the known hypergeometric functions. But the problem on hand is to determine the properties of $\xi_\nu^{(1)}(\tau)$ for large ν which are given by the following theorem.

Theorem (52). If $\eta_\nu^{(1)}(\tau)$ is continuous and monotonic in $\tau_0 < \tau < 1$ and satisfies $X_1(\eta_\nu^{(1)}) \gtreqless 0$, then for all $\nu > N$

$$\eta_\nu^{(1)}(\tau) \gtreqless \xi_\nu^{(1)}(\tau) \tag{73}$$

The proof is given in appendix A.

Corollary (53). In $\tau_0 < \tau < 1$, the following inequality holds for the modulus of $\underline{F}_\nu^*(\tau)$:

$$R_\nu(\tau)/R_\nu(\tau_0) < \left(\frac{\tau_0}{\tau}\right)^{\nu/2}, \quad \nu > N \tag{74}$$

where

$$(2\beta + 1)\tau_0 - 1 \geqslant 0$$

For in $\tau_0 < \tau < 1$, $\xi_\nu^{(1)} < 0$; and hence $\eta_\nu^{(1)}(\tau) = 0$ satisfies the condition $0 > \xi_\nu^{(1)}(\tau)$, which gives (74) by integration.

Now, since $\xi_\nu^{(1)}(\tau)$ is bounded by zero for all $\nu \neq 0$ in $\tau_0 < \tau < 1$, it is implied also that

$$R_\nu(\tau) < T_3^\nu(\tau) \tag{75}$$

where $T_3(\tau) = \frac{t_0}{\tau^{1/2}}$. Here the constant t_0 can be determined by joining T_3 at $\tau = \tau_0$ with T_1 or T_2 defined by equations (56) and (57). Then from equations (63) and (64) it follows that for $\nu > N$

$$|\underline{F}_\nu(\tau)| < T_3^\nu(\tau) \tag{76}$$

$$|\underline{G}_\nu(\tau)| < T_3^{-\nu}(\tau), \quad \tau_0 < \tau < 1 \tag{77}$$

Part Ⅱ
Construction of the Solutions for Compressible Flow Around a Body

5. Chaplygin's Procedure

In the previous sections, the particular solutions of the differential equations in the hodograph plane are obtained. Since the differential equations in the hodograph plane are linear, superposition of solutions is allowed. In other words, if these particular solutions are multiplied by different constants and then added together, the sum is again a solution of the differential equations. By this procedure, general solutions can be constructed from the particular solutions.

However, there is a difficulty in such a method of constructing the general solution — the difficulty of making a proper choice of the multiplying constants for the particular solutions so that the resultant solution will give a flow satisfying the boundary conditions specified in the physical plane. This can be seen from the fact that the space coordinates x and y are obtained from χ which is not explicitly connected with ψ, the stream function. In fact, to obtain the coordinate x and y directly from ψ would involve an integration in the hodograph plane. Thus the linearization of differential equations in the hodograph plane is obtained at the expense of the simplicity in boundary value problem. To guarantee that ψ and χ do actually belong to the same flow in the physical plane, an additional condition besides the differential equations for ψ and χ has to be satisfied. This condition will be discussed in section 11.

Chaplygin[6] suggested an ingenious method of solving this difficulty by using the well-known solutions of the incompressible flow. The first step in this method is to find the incompressible flow around a body "similar" to the body concerned. (The meaning of the word "similar" will be made clear in the following paragraph.)

The stream function ψ_0, for instance, is then expressed in terms of the speed q and the inclination θ. The function $\psi_0(q, \theta)$ can be expanded into an infinite series each term of which is of the form $q^n \cos n\theta$ or $q^n \sin n\theta$. For the flow around a body with constant velocity U at infinity, the function $\psi_0(q, \theta)$ has a singularity at the point $q = U$, $\theta = 0$ in the hodograph plane, since there all the streamlines, or lines of constant ψ_0 originate. Thus, there are two forms of the series expansion of ψ_0 : One is convergent within the circle $q = U$; while the other is convergent outside of the circle $q = U$. The first, or "inside," series must be regular at the origin of the hodograph plane. Therefore, only positive values of the integers n can occur. The second, or "outside," series can have both positive and negative ν. Chaplygin's method is to use the inside series for ψ_0 as the starting point for obtaining the desired solution ψ for the compressible fluid. He suggested choosing the multiplying coefficient of the particular solutions for the compressible flow by the condition that for the limiting case of infinite sonic speed, or incompressible fluid, the series will degenerate to the inside series of the incompressible flow already obtained. The series for the compressible stream function ψ so constructed can be called as the inside series of ψ. The outside series for ψ then can be obtained by the method of analytical continuation with the aid of the "outside series" of the incompressible flow.

These solutions so constructed for the compressible flow contain the Mach number of the undisturbed flow as a parameter. They constitute a family of singly infinite solutions. Included in this family of solutions is the limiting case of zero Mach number of the free stream. This limiting case will give the incompressible flow around a body used as the starting point of this method. For other values of the free-stream Mach number, the body contour is generally different from that corresponding to zero Mach number. Thus, if the compressible flow around a given body is desired, the body shape for the initial incompressible flow must be slightly different from the given body shape. However, if a geometric parameter is included in the solution, such an adjustment is not difficult to make.

It may be stated here that owing to the regularity of the solution at the origin of the hodograph plane, only the first solution of the hypergeometric differential equation appears in the inside series. For the outside series, both the first and the second solution of the hypergeometric differential equation are necessary. This is in direct analogy with the appearance of both positive and negative exponents of q in the incompressible outside series. This fact is particularly important, since previous investigators seem to be unaware of it. Chaplygin himself did not use the second solution of the hypergeometric differential equation, but that is simply because, for his problem, there is no singularity in the hodograph plane and hence only the inside series is needed.

6. The Functions for Incompressible Flow

Following the procedure outlined in the previous section, the analysis starts with the functions required in defining an irrotational incompressible flow. For this case, the sonic speed c_0 tends to infinity, and the equations for the velocity potential $\varphi_0(x, y)$ and the stream function $\psi_0(x, y)$ all became harmonic:

$$\Delta\varphi_0 = 0 \tag{78}$$

$$\Delta\psi_0 = 0 \tag{79}$$

where Δ stands for the Laplacian operator. If $W_0(z)$ is the complex potential, it can be shown that

$$W_0(z) = \varphi_0 + \mathrm{i}\psi_0 \tag{80}$$

where

$$z = x + \mathrm{i}y$$

If w denotes the complex velocity $u - \mathrm{i}v$, it is connected with $W_0(z)$ by

$$w = \frac{dW_0}{dz} \equiv w(z) \tag{81}$$

If $w'(z) \neq 0$, it always is possible to solve for z in terms of w; namely,

$$z = z_0(w) \tag{82}$$

In general, this solution is not single-valued and will be discussed later. By introducing this relation into equation (80), the complex potential function in the hodograph plane can be obtained

$$W_0(w) = \varphi_0(u, v) + i\psi_0(u, v) \tag{83}$$

In case equation (82) is many-valued, this would correspond to one branch of the function.

It is clear that in this case $\chi_0(u, v)$ is also a harmonic function. Let $\sigma_0(u, v)$ be the conjugate function defined by

$$\frac{\partial \chi_0}{\partial u} = -\frac{\partial \sigma_0}{\partial v} \tag{84}$$

$$\frac{\partial \chi_0}{\partial v} = \frac{\partial \sigma_0}{\partial u} \tag{85}$$

Hence

$$\Lambda_0(w) = \chi_0 - i\sigma_0 \tag{86}$$

where

$$w = u - iv$$

Thus $\Lambda_0(w)$ is an analytic function of w. From equation (13) the derivative of $\Lambda_0(w)$ with respect to w must be z. That is,

$$\frac{d\Lambda_0}{dw} = z_0(w)$$

But $z_0(w)$ already has been found from equation (82). Therefore,

$$\Lambda_0(w) = \int z_0(w)dw + \text{constant} \tag{87}$$

The real part of $\Lambda_0(w)$ gives $\chi_0(u, v)$ as required, according to (86).

7. Conformal Mapping of Incompressible Flow on the Hodograph Plane

Before the construction of solutions for the compressible flow, the general character of the solutions in the hodograph plane should be examined. This can be done easily by investigating the behavior of the transition function $z_0(w)$ for an incompressible fluid. To start with the simplest case first, consider a steady irrotational flow in an infinite, simply connected domain D bounded by a curve C in the z-plane, with a parallel flow at infinity (Fig. 1). At every point z of D there is one, and only one, velocity vector $\vec{q}$. If the curve C is mapped into $\underline{C}$ and infinity corresponds to a point $\underline{P}$ on the axis of reals of w within $\underline{C}$, then the domain D is mapped into $\underline{D}$ by a mapping function

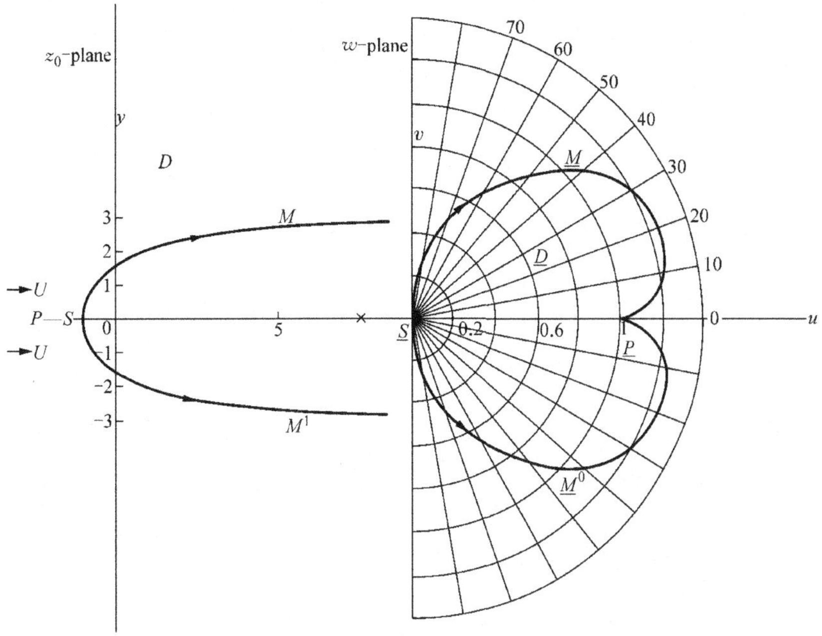

Fig. 1 The mapping of the whole plane D

$$w = w(z)$$

defined in (81), where $w(z)$ is an analytic function of z. The inverse function

$$z = z_0(w)$$

will set up a continuous one-to-one correspondence between w- and z-plane, provided the mapping is conformal. This requires that $w(z)$ is analytic, simple within D, and $w'(z) \neq 0$.

However, for most problems these conditions cannot be satisfied throughout the field of flow. In the first place, the function $w(z)$ is generally nonsimple, for example, in the case of a uniform flow, $w(z) =$ constant, thus $w'(z) \equiv 0$ and the whole z-plane corresponds only to a point in the w-plane. Furthermore, the complex velocity for a two-dimensional boundary-value problem generally can be put in the following form:

$$w = w_\infty + w^*(z)$$

where w_∞ is a constant. The boundary condition requires that $w^*(z) = 0$ and, as a consequence, $w'^*(z) = 0$ as z becomes infinite. Therefore, in all cases, the point $\underline{P}$ in the w-plane, is a singular point. It is a branch point at w_∞ if $z_0(w)$ is many-valued; or a pole, if otherwise. In practice, there are two kinds of singularities that play a dominant role in the problem of two-dimensional flow. These singularities will be investigated presently.

Branch point of order 1.① — It may be recalled that, when a closed body is present in a uniform flow, there always exist two stagnation points both of which correspond to the origin of the w-plane. If a streamline PS is followed, for instance, (see Fig. 2) from $+\infty$ to S, the portion SMS' and then to $-\infty$, a curve $\underline{PS}$ in w-plane would be described twice. This indicates that the function $z_0(w)$ possesses two branches of Riemann surfaces joining together about the branch point $\underline{P}$. In order to make the domain $\underline{D}$ single-valued, a cut is put along the axis of reals from the branch point to $+\infty$. Then one portion of the z-plane is mapped into a definite branch of the Riemann surfaces in the w-plane, and this will be defined as the domain D. If the body is symmetrical with respect to the coordinate axes with parallel flow at infinity, then the domain D : $\text{Re}z \leqslant 0$ will be mapped conformally into $\underline{D}$ on one branch of the Riemann surfaces and D' : $\text{Re}z > 0$ on the other, where the region within C is excluded.

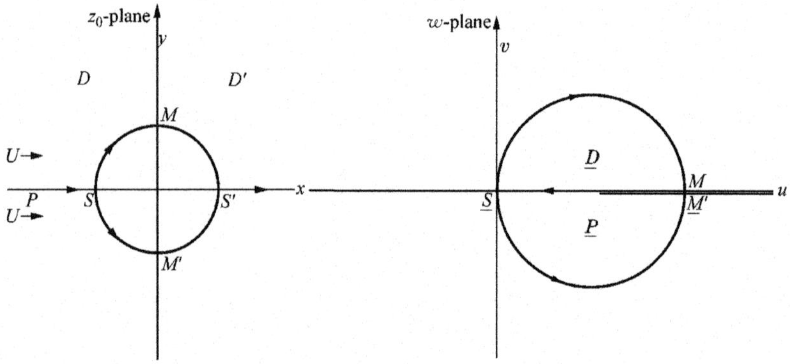

Fig. 2 The mapping of the half plane D

Logarithmic singularity. — The flow over a wavy surface, for instance, placed parallel to a uniform stream has a periodic nature. For such flows there are infinitely many points in the physical plane that have the same velocity. Hence, there are an infinite number of branches in the w-plane, each of which corresponds to a definite portion of the z-plane. The function $z_0(w)$ must have a term $\log\left(1-\frac{w}{U}\right)$ and the point $\underline{P}$ now is a logarithmic singularity. If, however, a cut is introduced from the branch point to $+\infty$ and $-\pi < \arg\left(1-\frac{w}{U}\right) < \pi$, then the domain $\underline{D}$ is again made single-valued.

8. Construction of a Solution about the Origin

Stream function. — From the considerations of the last section, the domain within a circle with radius $|w| = \underline{PS} = U$, where U is the absolute value of w at infinity in z-plane, is in all cases single-valued. If a function $W_0(w)$ is associated with a definite flow in z-plane,

① The function $z_0(w)$ is said to have a branch point of order k at $w = w_\infty$ if its inverse $w(z)$ contains the part w^* which has a zero of order $k+1$ at $z = \infty$.

from section 6 it is an analytic function of w and regular within the circle $|w| = U$. Consequently, the following Taylor expansion exists:

$$W_0(w) = \sum_{n=0}^{\infty} A_n w^n, \quad |w| < U \tag{88}$$

where $A_n's$ are, in general, complex. Since $w = qe^{-i\theta}$ and by (80) the imaginary part of $W_0(w)$ is equal to incompressible stream function ψ_0, it can be written as

$$\psi_0(q, \theta) = \mathrm{Im}\{W_0(w)\} = \sum_{n=0}^{\infty} q^n \{A_n^{(1)} \cos n\theta + A_n^{(2)} \sin n\theta\} \tag{89}$$

According to Chaplygin's procedure, the corresponding compressible solution can be obtained by simply replacing the function q^n in equation (89) by the corresponding $q^n \underline{F}_n^{(r)}(\tau)$ as shown by (40). The second integral is excluded by the regularity requirement at $q = 0$. However, in order to preserve the proper singularity at the point $(U, 0)$ in the hodograph plane, the compressible stream function ψ is written as

$$\psi(q, \theta) = \sum_{n=0}^{\infty} q^n \underline{F}_n^{(r)}(\tau)\{A_n^{(1)} \cos n\theta + A_n^{(2)} \sin n\theta\} \tag{90}$$

where

$$\underline{F}_n^{(r)}(\tau) = \frac{\underline{F}_n(\tau)}{\underline{F}_n(\tau_1)} = \frac{F(a_n, b_n; c_n; \tau)}{F(a_n, b_n; c_n; \tau_1)}, \quad q < U \tag{91}$$

and $\tau_1 = \dfrac{1}{2\beta}\dfrac{U^2}{c_0^2}$, the value of τ_1 corresponding to the free-stream velocity U. It is seen that if $c_0 \to \infty$, then $\tau = \tau_1 \to 0$, and $\underline{F}_n^{(r)}(\tau) \to 1$ due to the normalizing condition (30). Thus the solution is reduced to the incompressible form. Furthermore, if $q \to U$ the character of the solution is exactly like that of the incompressible solution. Hence, all the specified conditions are satisfied. Of course, for the mixed subsonic and supersonic flow, the free-stream Mach number is always less than unity. Thus $\tau_1 < 1/(2\beta + 1)$.

For later analysis as given in part Ⅲ, it is convenient to write ψ in a different form. Since $\underline{F}_n^{(r)}(\tau)$ is a purely real quantity, a complex function $W(w;\tau)$ can be constructed as

$$W(w;\tau) = \sum_{n=0}^{\infty} A_n \underline{F}_n^{(r)}(\tau) w^n, \quad |w| < U \tag{92}$$

Then, similar to the relation between equations (88) and (89), $\psi(q, \theta)$ can be taken as the imaginary part of the new function $W(w;\tau)$. Thus,

$$\psi(q, \theta) = \mathrm{Im}\{W(w;\tau)\} \tag{93}$$

Transformed potential function. — Similarly, it is possible to construct another function $\Lambda(w;\tau)$ defined by

$$\Lambda(w;\tau) = \sum_{n=0}^{\infty} \widetilde{A}_n \underline{\widetilde{F}}_n^{(r)}(\tau) W^n, \quad q < U \tag{94}$$

In this expression, the coefficients $\tilde{A}_n$ are obtained from the expansion of $\Lambda_0(w)$ for the incompressible flow (87):

$$\Lambda_0(w) = \sum_{n=0}^{\infty} \tilde{A}_n w^n, \quad |w| < U \tag{95}$$

and

$$\underline{\tilde{F}}_n^{(r)}(\tau) = \frac{\underline{\tilde{F}}_n(\tau)}{\underline{F}_n(\tau_1)} \tag{96}$$

Equation (96) is the result of equation (91) and the equation of compatibility given by equation (24). Then the function $\chi(q, \theta)$ for the compressible flow is given by

$$\chi(q, \theta) = \mathrm{Re}\{\Lambda(w;\tau)\} \tag{97}$$

The functions $W(w;\tau)$ and $\Lambda(w;\tau)$ are actually regular at the origin and satisfy the imposed conditions. However, the following question may be raised: Do the series (92) and (94) converge and represent the functions $\psi(q, \theta)$ and $\chi(q, \theta)$ in the domain of validity? To settle this question, it is necessary to prove the following theorem:

Theorem (88). If the constants A_n and $\tilde{A}_n$ are given in equations (88) and (95), while $\underline{F}_n^{(r)}(\tau)$ and $\underline{\tilde{F}}_n^{(r)}(\tau)$ are defined respectively by equations (91) and (96), the series (92) and (94) are uniformly and absolutely convergent in the same domain as those of (88) and (95). The proof is given in appendix B.

9. Analytic Continuation of the Solution Branch Point or Order 1

Stream function. — As proved in the appendix B, the series (92) is absolutely and uniformly convergent and does represent a regular function $W(w;\tau)$ for every τ in $0 \leqslant \tau \leqslant \tau_1$ and on the circle of convergence it agrees with $W_0(Ue^{-i\theta})$, of which the Fourier expansion exists:

$$W_0(Ue^{-i\theta}) = \sum_{n=0}^{\infty} A_n U^n e^{-in\theta} \tag{98}$$

In the present section, it is proposed to continue the solution (92) analytically outside the domain $|w| \leqslant U$ with the initial value given by equation (98). The domain outside $|w| \leqslant U$ is generally many-valued. To fix ideas, discuss first the case of a branch point of order 1. Generally, the function $W_0(w)$ has other singularities besides the one at $w = U$.① However,

① For instance, in the problem, of the flow around an elliptic cylinder treated in part V, there are two singularities of the W_0 function given by equation (280): namely, $w = 1$ and $w = 1/\varepsilon^2$. The first singularity corresponds to the flow at infinity and is the singularity under discussion. The second singularity corresponds to a point inside the circle of the ζ-plane, the plane of the circular section. Since only the flow outside the circle of the ζ-plane is of interest here, the singularity $w = 1/\varepsilon^2$ need not be investigated. In other words, it is necessary only to expand the W_0 function in the annular region $1 < \left|\frac{w}{U}\right| < \frac{1}{\varepsilon^2}$.

such singularities lie outside the region of interest and thus need not be investigated. Let the nearest singularity be given by $|w| = V > U$. Then, the domain to be considered outside $|w| = U$ is an annulus with a cut joining the two singularities. The proper representation of $W_0(w)$ in such a region which has a branch point of order 1 at $w = U$, is

$$W_0(w) = \mathrm{i}w^{-\frac{1}{2}}W_0^*(w) \tag{99}$$

where $W_0^*(w)$ is single-valued and regular within the open annulus $U < |w| < V$. Hence, in any closed domain

$$U+\delta \leqslant |w| \leqslant V-\delta, \delta$$

being a small number, there exists a uniformly and absolutely convergent series:

$$W_0^*(w) = \sum_{n=0}^{\infty}[B_n w^n + C_n w^{-n}] \tag{100}$$

which, on substituting in (99), will give the continuation of the Taylor series (88).

The solution for a compressible fluid, which has the same character of singularities as $W_0(w)$ and is valid in the annulus $U < |w| < V$, can be obtained from (100) by introducing the proper hypergeometric functions corresponding to each exponent of w. That is:

$$W^{(0)}(w;\tau) = \mathrm{i}\sum_{n=0}^{\infty}[B_n^* \underline{F}_\nu(\tau)w^\nu + C_n^* \underline{G}_\nu(\tau)w^{-\nu}] \tag{101}$$

which is the continuation of $W^{(i)}(w;\tau)$. Here $\nu = n+\dfrac{1}{2}$, n being a positive integer; $\underline{F}_\nu(\tau)$ and $q^{-2\nu}\underline{G}_\nu(\tau)$ are the first and second integrals of the hypergeometric equation; and B_n^* and C_n^* are constants. It should be noticed that the coefficients B_n^* and C_n^* in the outside series for the compressible flow are not equal to B_n and C_n in equation (100) for the outside series of the incompressible flow. The outside series of the incompressible flow is used only to give the proper form of $W^{(0)}(w;\tau)$ for the desired branch point characteristics; while the exact determination of $W^{(0)}(w;\tau)$ has to be made by the conditions of continuity, which will be discussed presently.

Since the partial differential equation considered here is of the second order, to ensure that $W^{(0)}(w;\tau)$ is the analytic continuation of $W^{(i)}(w;\tau)$, two conditions have to be imposed at the boundary of the respective regions of convergence; that is, the circle $q = U$. These two conditions are the following:

$$W^{(i)}(Ue^{-\mathrm{i}\theta};\tau_1) = W^{(0)}(Ue^{-\mathrm{i}\theta};\tau_1) \tag{102}$$

$$\left[\frac{\partial}{\partial q}W^{(i)}(w;\tau)\right]_{\tau=\tau_1} = \left[\frac{\partial}{\partial q}W^{(0)}(w;\tau)\right]_{\tau=\tau_1} \tag{103}$$

On account of equations (102) and (103), there are two relations which have the imaginary parts:

$$\sum_{n=0}^{\infty}[B_n^* \underline{F}_\nu(\tau_1)U^\nu + C_n^* \underline{G}\nu(\tau_1)U^{-\nu}]\cos\nu\theta = -\sum_{n=0}^{\infty}A_n U^n \sin n\theta \quad 0 \leqslant \theta < 2\pi$$

$$\sum_{n=0}^{\infty}[B_n^* U^\nu(\nu\underline{F}_\nu(\tau)+2\tau_1\underline{F}'_\nu(\tau_1)) + C_n^* U^{-\nu}(-\nu\underline{G}_\nu(\tau_1)+2\tau_1\underline{G}'_\nu(\tau_1))]\cos\nu\theta$$

$$= -\sum_{n=0}^{\infty}A_n U^n\left(n+2\tau_1\frac{\underline{F}'_n(\tau_1)}{\underline{F}_n(\tau_1)}\right)\sin n\theta$$

Here the prime denotes differentiation with respect to τ. Evidently, the coefficients on the left-hand side can be solved for in terms of the known constants A_n. They are:

$$B_n^* \underline{F}_\nu(\tau_1)U^\nu + C_n^* \underline{G}_\nu(\tau_1)U^{-\nu} = -\frac{1}{\pi}\sum_{m=0}^{\infty}A_m U^m\left(\frac{1}{m+\nu}+\frac{1}{m-\nu}\right) \tag{104}$$

$$B_n^* U^\nu(\nu\underline{F}_\nu(\tau_1)+2\tau_1\underline{F}'_\nu(\tau_1)) + C_n^* U^{-\nu}(-\nu\underline{G}_\nu(\tau_1)+2\tau_1\underline{G}'_\nu(\tau_1))$$
$$= -\frac{1}{\pi}\sum_{m=0}^{\infty}mA_m U^m\xi_m(\tau_1)\left(\frac{1}{m+\nu}+\frac{1}{m-\nu}\right) \tag{105}$$

From these two equations, the constants B_n^* and C_n^* can be uniquely determined, provided the determinant $\Delta(\underline{F}_\nu, F_\nu)$ does not vanish. These results are:

$$B_n^* U^\nu = -\frac{\underline{G}_\nu(\tau_1)}{2\nu\pi(1-\tau_1)^\beta}\sum_{m=0}^{\infty}A_m U^m\left(\frac{1}{m+\nu}+\frac{1}{m-\nu}\right)(m\xi_m(\tau_1)-\nu\xi_{-\nu}(\tau_1)) \tag{106}$$

$$C_n^* U^{-\nu} = +\frac{\underline{F}_\nu(\tau_1)}{2\nu\pi(1-\tau_1)^\beta}\sum_{m=0}^{\infty}A_m U^m\left(\frac{1}{m+\nu}+\frac{1}{m-\nu}\right)(m\xi_m(\tau_1)-\nu\xi_\nu(\tau_1)) \tag{107}$$

as the Wronskian $\Delta(\underline{F}_\nu, F_\nu) = -\frac{\nu}{\tau}q^{-2\nu}(1-\tau)^\beta \neq 0$ and $\xi_\nu(\tau)$ is defined in (51).

The solution is again formal. To prove that the function $W(w;\tau)$ is a regular function in the annulus region, the truth of the following theorem must be first demonstrated. (See appendix C)

Theorem (98). If the constants B_n^* and C_n^* are determined according to (102) and (103) and if the series (100) converges uniformly and absolutely in a closed domain $U+\delta \leqslant |w| \leqslant V-\delta$, then the series (101) will converge uniformly and absolutely in the domain $U+\delta \leqslant |w| \leqslant V-\delta$, $\delta > 0$.

Transformed potential function. — By a similar procedure, the continuation of (94) is

$$\Lambda^{(0)}(w;\tau) = \mathrm{i}\sum_{n=0}^{\infty}[\tilde{B}_n^* \tilde{\underline{F}}_\nu(\tau)w^\nu + \tilde{C}_n^* \tilde{\underline{G}}_\nu(\tau)w^{-\nu}] \tag{108}$$

where $\tilde{\underline{F}}_\nu(\tau)$ and $\tilde{\underline{G}}_\nu(\tau)$ are the first and second integrals of equation (28) and the constants $\tilde{B}_n^*$ and $\tilde{C}_n^*$ can be similarly determined; namely,

$$\tilde{B}_n^* U^\nu = \frac{\tilde{\underline{G}}_\nu(\tau_1)}{2\pi\nu(1-\tau_1)^{-\beta}}\sum_{m=0}^{\infty}\tilde{A}_m U^m\left(\frac{1}{m+\nu}-\frac{1}{m-\nu}\right)[m\tilde{\xi}_m(\tau_1)-\nu\tilde{\xi}_{-\nu}(\tau_1)]\tilde{\underline{F}}_m^{(r)}(\tau_1) \tag{109}$$

$$\tilde{C}_n^* U^{-\nu} = \frac{\tilde{\underline{F}}_\nu(\tau_1)}{2\pi\nu(1-\tau_1)^{-\beta}}\sum_{m=0}^{\infty}A_m U^m\left(\frac{1}{m+\nu}-\frac{1}{m-\nu}\right)[m\tilde{\xi}_m(\tau_1)-\nu\tilde{\xi}_\nu(\tau_1)]\tilde{\underline{F}}_m^{(r)}(\tau_1) \tag{110}$$

The solution determined so far is understood to be the principal branch of the function $W(w;\tau)$. It was assumed that the flow at infinity is parallel to the x-axis. If, in addition, the body is symmetrical with respect to the coordinate axes, the expression for the second branch of $W^{(0)}(w;\tau)$ will be identical. However, in a more general case where asymmetry exists, the two branches will require separate consideration.

10. Continuation — Logarithmic Singularity

Stream function. — Consider now the second important type of singularity: it is assumed here that the only singularity possessed by the function $W_0(w)$ in the finite part of the w-plane is a logarithmic branch point at $w = U$ about which infinitely many Riemann surfaces are joined. By analogy with (99), $W_0(w)$ now can be conveniently written as

$$W_0(w) = W_0^*(w) + \widetilde{W}_0(w) \tag{111}$$

where $W_0^*(w)$ is a regular function in the entire domain with possibly an essential singularity at infinity, and hence generally is given by a Taylor series or a polynomial in w, and $\widetilde{W}_0(w) = \varphi_0(q,\theta) + i\,\widetilde{\psi}_0(q,\theta)$ is an analytic function which characterizes the singularity of $W_0(w)$. Thus, aside from a constant factor,

$$\widetilde{W}_0(w) = \frac{1}{i}\log\left(1-\frac{w}{U}\right) \tag{112}$$

If a cut is laid from $+U$ to $+\infty$ and the argument of $\left(1-\frac{w}{U}\right)$ is restricted in $-\pi < \arg\left(1-\frac{w}{U}\right) < \pi$, then the function $\widetilde{W}_0(w)$ will be single-valued in the whole cut plane.

The question of constructing a solution for the compressible fluid consists, therefore: of two parts: $W_0^*(w)$ and $\widetilde{W}_0(w)$. However, the construction for $W_0^*(w)$ is, in principle, exactly the same as that of (92) and hence only $\widetilde{W}_0(w)$ will be considered. First, let $\widetilde{W}_0(w)$ be developed into power series in the respective domains of validity. The imaginary parts are:

$$\widetilde{\psi}_0^{(i)}(q,\theta) = \sum_{n=1}^{\infty}\frac{1}{n}\left(\frac{q}{U}\right)^n\cos n\theta, \quad q<U \tag{113}$$

$$\widetilde{\psi}_0^{(0)}(q,\theta) = -\log\frac{q}{U} + \sum_{n=1}^{\infty}\frac{1}{n}\left(\frac{q}{U}\right)^{-n}\cos n\theta, \quad q>U \tag{114}$$

The corresponding expression for $\widetilde{\psi}(q,\theta)$, accordingly, will be:

$$\widetilde{\psi}^{(i)}(q,\theta) = \sum_{n=1}^{\infty}A_n\underline{F}_n(\tau)\left(\frac{q}{U}\right)^n\cos n\theta, \quad q<U \tag{115}$$

$$\widetilde{\psi}^{(0)}(q,\theta) = -B\int_{\tau_1}^{\tau}(1-\tau)^{\beta}\frac{d\tau}{\tau} + \sum_{n=1}^{\infty}C_n\underline{G}_n(\tau)\left(\frac{q}{U}\right)^{-n}\cos n\theta, \quad q>U \tag{116}$$

where $\underline{F}_n(\tau)$ stands for $F(a_n, b_n; c_n; \tau)$ and $\underline{G}_n(\tau)$ is defined by (39).

The function $\widetilde{W}_0(w)$ may be regarded as the complex potential of a complex source situated at $w=U$. It is known that in this case the normal derivative of $\tilde{\psi}_0(q, \theta)$ on $|w|=U$ is a constant, except at $w=U$, where it becomes infinite. This boundary value can be expanded uniquely:

$$-\sum_{n=1}^{\infty}\cos n\theta = \frac{1}{2}, \quad \theta \neq 0 \tag{117}$$

The corresponding problem in the case of compressible flow can be put in an analogous manner: to find a function $\tilde{\psi}(q, \theta)$ which is continuous together with continuous partial derivatives and the normal derivative of which on $|w|=U$ is constant. Thus, the conditions (102) and (103) in conjuction with equation (117) demand:

$$\underline{F}_n(\tau_1)A_n - \underline{G}_n(\tau_1)C_n = 0 \tag{118}$$

$$[n\underline{F}_n(\tau_1) + 2\tau_1\underline{F}'_n(\tau_1)]A_n + [n\underline{G}_n(\tau_1) - 2\tau_1\underline{G}'_n(\tau_1)]C_n = 4B(1-\tau_1)^{\beta} \tag{119}$$

where the constant B can be determined when the normal derivative $\psi_q(q, \theta)$ on $|w|=U$ is assigned. By solving equations (118) and (119) and using the relation of the Wronskian of the two independent integrals of equation (27), there is obtained

$$A_n = \frac{2}{n}B\underline{G}_n(\tau_1) \tag{120}$$

$$C_n = \frac{2}{n}B\underline{F}_n(\tau_1) \tag{121}$$

Thus the function $\tilde{\psi}(q, \theta)$ is completely determined.

<u>Transformed potential function</u>. The associated function $\chi(q, \theta)$ can be similarly constructed. As $\Lambda_0(w)$ is derived from (87) by integration of the inverse mapping function, it must involve a term $\left(1-\frac{w}{U}\right)\log\left(1-\frac{w}{U}\right)$ which represents the singularity of the function $\Lambda_0(w)$. As in equation (111), $\Lambda_0(w)$ is split again into two parts:

$$\Lambda_0(w) = \Lambda_0^*(w) + \tilde{\Lambda}_0(w) \tag{122}$$

where $\Lambda_0^*(w)$ is an entire function and $\tilde{\Lambda}_0(w)$ is

$$\tilde{\Lambda}_0(w) = \frac{1}{i}\left(1-\frac{w}{U}\right)\log\left(1-\frac{w}{U}\right) \tag{123}$$

Now the solution corresponding to $\log\left(1-\frac{w}{U}\right)$ can be determined in exactly the same manner except that the hypergeometric functions involved are $\tilde{\underline{F}}_n(\tau)$ and $\tilde{\underline{G}}_n(\tau)$ instead of $\underline{F}_n(\tau)$ and $\underline{G}_n(\tau)$. The part that will require special consideration is the term $\frac{w}{U}\log\left(1-\frac{w}{U}\right)$. Let it be denoted by $\tilde{\lambda}_0(w) = \tilde{\chi}_0 - i\tilde{\sigma}_0$:

$$\tilde{\lambda}_0(w) = -\frac{1}{i}\frac{w}{U}\log\left(1-\frac{w}{U}\right) \tag{124}$$

This function is also multiple-valued. Let the argument of $\left(1-\frac{w}{U}\right)$ again be restricted in $-\pi<\arg\left(1-\frac{w}{U}\right)<\pi$; then in the cut plane the result will be

$$\tilde{\lambda}_0^{(i)}=\frac{1}{i}\sum_{n=1}^{\infty}\frac{1}{n}\left(\frac{w}{U}\right)^{n+1},\quad |w|<U \tag{125}$$

$$\tilde{\lambda}_0^{(0)}=\frac{1}{i}\left[-\frac{w}{U}\log\frac{w}{U}e^{i\pi}+\sum_{n=1}^{\infty}\frac{1}{n}\left(\frac{w}{U}\right)^{-n+1}\right],\quad |w|>U \tag{126}$$

According to equation (86), the function $\tilde{\chi}_0(q, \theta)$ is defined as the real part of $\Lambda_0(w)$. That part represented by equations (125) and (126) is then

$$\tilde{\chi}_0^{(i)}(q, \theta)=-\sum_{n=2}^{\infty}\frac{1}{n-1}\left(\frac{q}{U}\right)^n\sin n\theta \tag{127}$$

$$\tilde{\chi}_0^{(0)}(q, \theta)=\frac{q}{U}\log\frac{q}{U}\sin\theta-\frac{q}{U}(\pi-\theta)\cos\theta+\sum_{n=1}^{\infty}\frac{1}{n+1}\left(\frac{q}{U}\right)^{-n}\sin n\theta \tag{128}$$

The particular solution corresponding to

$$\frac{q}{U}\log\frac{q}{U}\sin\theta-\frac{q}{U}(\pi-\theta)\cos\theta$$

already has been given in equation (50). Hence the solution for the compressible flow is

$$\tilde{\chi}^{(i)}(q, \theta)=-\sum_{n=2}^{\infty}\tilde{A}_n\underline{\tilde{F}}_n(\tau)\left(\frac{q}{U}\right)^n\sin n\theta \tag{129}$$

$$\tilde{\chi}^{(0)}(q, \theta)=\frac{q}{U}k(\tau)\sin\theta-\frac{q}{U}(\pi-\theta)\cos\theta+\sum_{n=1}^{\infty}\tilde{C}_n\underline{\tilde{G}}_n(\tau)\left(\frac{q}{U}\right)^{-n}\sin n\theta \tag{130}$$

where

$$k(\tau)=\frac{1}{2(\beta+1)}\left[(2\beta+1)\log\frac{\tau}{\tau_1}-\left(\frac{1}{\tau}-\frac{1}{\tau_1}\right)+K_1\int_{\tau_1}^{\tau}(1-\tau)^{-\beta}\frac{d\tau}{\tau^2}\right] \tag{131}$$

The conditions (102) and (103) together with an expansion

$$\frac{1}{2}\sin\theta+\sum_{n=2}^{\infty}\left(\frac{1}{n+1}+\frac{1}{n-1}\right)\sin n\theta=(\pi-\theta)\cos\theta,\quad 0<\theta<2\pi$$

require that:

$$\underline{\tilde{F}}_n(\tau_1)\tilde{A}_n+\underline{G}_n(\tau_1)\tilde{C}_n=\frac{1}{n+1}+\frac{1}{n-1} \tag{132}$$

$$[n\underline{\tilde{F}}_n(\tau_1)+2\tau_1\underline{\tilde{F}}'_n(\tau_1)]\tilde{A}_n+[-n\underline{\tilde{G}}_n(\tau_1)+2\tau_1\underline{\tilde{G}}'_n(\tau_1)]\tilde{C}_n=\frac{1}{n+1}+\frac{1}{n-1};\quad n\neq 1 \tag{133}$$

and

$$\underline{\tilde{G}}_1(\tau_1)\tilde{C}_1=\frac{1}{2} \tag{134}$$

$$[-\underset{\sim}{\widetilde{G}}_1(\tau_1)+2\tau_1\underset{\sim}{\widetilde{G}}'_1(\tau_1)]\widetilde{C}_1+2\tau_1 k'(\tau_1)=\frac{1}{2}, \quad n=1 \tag{135}$$

By solving (132) and (133) for $\widetilde{A}_n$ and $\widetilde{C}_n$, there is obtained:

$$\widetilde{A}_n=\frac{(1-\tau_1)^\beta}{n^2-1}(1-n\widetilde{\xi}_{-n}(\tau_1))\underset{\sim}{\widetilde{G}}_n(\tau_1) \tag{136}$$

$$\widetilde{C}_n=\frac{(1-\tau_1)^\beta}{n^2-1}(1-n\widetilde{\xi}_n(\tau_1))\underset{\sim}{\widetilde{F}}_n(\tau_1), \quad n\neq 1 \tag{137}$$

by using the Wronskian of the independent integrals of equation (28). With C_1 given by (134), the constant K_1 can be solved for from (135); it is

$$K_1=-(1-\tau_1)^\beta\left[1+\beta\tau_1+(\beta+1)\tau_1^2\frac{\underset{\sim}{\widetilde{G}}'_1(\tau_1)}{\underset{\sim}{\widetilde{G}}_1(\tau_1)}\right] \tag{138}$$

The solutions $\widetilde{\psi}(q, \theta)$ and $\widetilde{\chi}(q, \theta)$ in the whole domain under consideration are uniquely determined. Since the dominant properties of the hypergeometric functions discussed in section 4 hold, in general, the equation of convergence can be similarly settled.

11. Transition to Physical Plane

In the previous sections, it has been proved that, for a given incompressible flow for which two associated functions $\psi_0(q, \theta)$ and $\chi(q, \theta)$ are defined, there exist two associated functions $\psi(q, \theta)$ and $\chi(q, \theta)$ for the corresponding compressible flow, depending upon two parameters γ and τ_1. The question is whether the associated functions $\psi(q, \theta)$ and $\chi(q, \theta)$ belong to the same flow pattern in the physical plane. To answer this question it is necessary to fall back once more on the equation of compatibility (24); since when $\psi(q, \theta)$ is given, $\varphi(q, \theta)$ is known by solving equations (20) and (21). Hence, if $\chi(q, \theta)$, satisfying equation (23) and approaching χ_0 as $C_0 \to \infty$ is to be associated with $\psi(q, \theta)$ for the same flow, then it is necessary that the equation of compatibility be satisfied. Except in the case of logarithmic singularity in section 10 where the complete $\psi(q, \theta)$ function was not discussed, this condition has been properly considered.

Once the relationship between $\psi(q, \theta)$ and $\chi(q, \theta)$ is established, the next object is to calculate the flow pattern $\psi(x, y)=$ constant in the physical plane corresponding to $\psi(q, \theta)$ and $\chi(q, \theta)$. In the first place, the fact that the transformation defined by equations (9) and (10) is generally one-to-one must be recalled. Suppose that in the hodograph plane there is a line defined by

$$\psi(q, \theta)=\text{constant}=K \tag{139}$$

which will correspond to a definite streamline in the physical plane or a definite part of it. The streamline can be obtained by eliminating one of the two variables in $x(q, \theta)$ and $y(q, \theta)$. To do this, first the equation (139) is solved for θ; namely,

$$\theta=\theta(q,K) \tag{140}$$

provided $\psi_\theta(q, \theta) \neq 0$. Introducing this relation into equation (13) which, when transformed into polar coordinates, are

$$x = \cos\theta \frac{\partial\chi}{\partial q} - \frac{\sin\theta}{q}\frac{\partial\chi}{\partial\theta} \tag{141}$$

$$y = \sin\theta \frac{\partial\chi}{\partial q} + \frac{\cos\theta}{q}\frac{\partial\chi}{\partial\theta} \tag{142}$$

gives a parametric representation of this particular streamline corresponding to $\psi(q, \theta) = K$ in the hodograph plane.

Part Ⅲ
Improvement of the Convergence of Solution by the Asymptotic Properties of Hypergeometric Functions

12. General Concepts

The significance of the general solutions constructed in part Ⅱ of the present report when viewed from the practical point, rests in the fact that they constitute an existence theorem. It shows that an irrotational isentropic flow about a body can be obtained from the corresponding problem of an incompressible fluid, if the free-stream Mach number is not too high. However, the solution in the form of a slowly convergent infinite series cannot be conveniently used to obtain numerical values, as the labor of computation would be prohibitive.

By examining the infinite series obtained in part Ⅱ, the essential difference between the compressible flow solution and the incompressible flow solution is seen to be associated with the fact that, while in incompressible flow solution the individual terms of the series are of the forms

$$q^{\nu}\begin{matrix}\cos\nu\theta\\ \sin\nu\theta\end{matrix} \qquad q^{-\nu}\begin{matrix}\cos\nu\theta\\ \sin\nu\theta\end{matrix}$$

in compressible flow solution the individual terms of the series are of the forms

$$q^{\nu}\underline{F}_{\nu}(\tau)\begin{matrix}\cos\nu\theta\\ \sin\nu\theta\end{matrix} \qquad q^{-\nu}\underline{G}_{\nu}(\tau)\begin{matrix}\cos\nu\theta\\ \sin\nu\theta\end{matrix}$$

If it were possible to write

$$q^{\nu}\underline{F}_{\nu}(\tau) = [Q(q)]^{\nu}, \quad q^{-\nu}\underline{G}_{\nu}(\tau) = [Q(q)]^{-\nu}$$

there would be no difference between the incompressible flow solution and the compressible flow solution except the "distortion of the speed" q by the new scale Q. In fact, this possibility is realized under the special condition of $\gamma = -1$ as shown by von Kármán[1] and Tsien[9].

For the case of isentropic flow with the general exponent γ there is no such scale factor Q. However, if ν is assumed to be very large, then there is such a function Q, at least to a first approximation. In other words, the leading term in the asymptotic representations of $\underline{F}_{\nu}(\tau)$ and $\underline{G}_{\nu}(\tau)$ does give the desired form. On the other hand, the use of asymptotic representation necessarily implies an approximation. But this defect is not difficult to remedy as the difference between an exact hypergeometric function and its asymptotic form can be added to the approximate solution as a correction term. Since there are an infinite number of terms in the series form of the solution and each gives a correction term, the correction terms also constitute an infinite series. Therefore, the original infinite series is now transformed into a closed function plus another infinite series of correction terms. At first sight, such a transformation seems unable to give a result that will avoid the difficulty of prohibitive computational work. But actually, owing to the good approximation given by the asymptotic representation even for moderate values of ν, the correction series converges very rapidly. A few terms seem to be all that are necessary. Thus, for all practical purposes, the original infinite series is now converted into a closed function with "speed distortion" plus a few correction terms. The fundamentally interesting point is that for the case of a general exponent γ, the simple method of speed distortion will not give an accurate enough solution. (Cf. reference 8)

The change in type of the differential equation at the sonic speed also introduces a singularity in the speed distortion function Q. However, by using the correction terms, the effect of the singularity can be limited to a very narrow range in the neighborhood of sonic speed, and no practical difficulty is experienced. This will be made clear by the numerical example given in part V of this report.

13. Asymptotic Solutions of the Hypergeometric Equations

Let $U_\nu(\tau)$ and $V_\nu(\tau)$ be two new dependent variables defined by

$$\psi_\nu(\tau) = \tau^{-\frac{\nu+1}{2}}(1-\tau)^{\frac{\beta}{2}}U_\nu(\tau) \tag{143}$$

$$\chi_\nu(\tau) = \tau^{-\frac{\nu+1}{2}}(1-\tau)^{-\frac{\beta}{2}}V_\nu(\tau) \tag{144}$$

The differential equations (27) and (28) reduce respectively to

$$U''_\nu(\tau) - [\nu^2\varphi(\tau) + \rho_\beta(\tau)]U_\nu(\tau) = 0 \tag{145}$$

$$V''_\nu(\tau) - [\nu^2\varphi(\tau) + \rho_{-\beta}(\tau)]V_\nu(\tau) = 0 \tag{146}$$

where

$$\varphi(\tau) = \frac{1-(2\beta+1)\tau}{4\tau^2(1-\tau)}$$

$$\rho_{\pm\beta}(\tau) = \frac{\beta\tau(\beta\tau\pm 2)-(1-\tau)^2}{4\tau^2(1-\tau)^2}$$

Both equations (145) and (146) involve a constant parameter ν which is real and positive but otherwise arbitrary for any fixed constant β. In the interval $0<\tau<1$ in which the flow takes place, the functions $\varphi(\tau)$ and $\rho_{\pm\beta}(\tau)$ are finite and continuous except at the extremities $\tau=0$ and $\tau=1$. To avoid the repetition, let equations (145) and (146) be replaced by

$$U''_{\alpha,\nu}(\tau) - [\nu^2\varphi(\tau) + \rho_\alpha(\tau)]U_{\alpha,\nu}(\tau) = 0 \tag{147}$$

where $U_{\beta,\nu}(\tau) = U_\nu(\tau)$ when $\alpha=\beta$; and $U_{-\beta,\nu}(\tau) = V_\nu(\tau)$ when $\alpha=-\beta$. In the interval $\delta\leqslant\tau\leqslant\frac{1}{2\beta+1}-\delta$, $\delta>0$, $\varphi(\tau)$ is bounded from zero and is positive. F. Horn[18] showed that when ν is a large positive number, a pair of solutions of the following forms exist in the interval concerned:

$$U_{a,\nu}^{(1)}(\tau) \sim e^{\nu K}\left[\varphi^{-\frac{1}{4}} + \frac{f_{11}(\tau)}{\nu} + \frac{f_{12}(\tau)}{\nu^2} + \ldots + \frac{f_{1s}(\tau)}{\nu^s}\right] \tag{148}$$

$$U_{a,\nu}^{(2)}(\tau) \sim e^{-\nu K}\left[\varphi^{-\frac{1}{4}} + \frac{f_{21}(\tau)}{\nu} + \frac{f_{22}(\tau)}{\nu^2} + \ldots + \frac{f_{2s}(\tau)}{\nu^s}\right] \tag{149}$$

where

$$K(\tau) = \int^{\tau} \varphi^{\frac{1}{2}}(\tau)\mathrm{d}\tau, \quad 0 < \tau < \frac{1}{2\beta+1} \tag{150}$$

A constant in equation (150) was left out, as it can be absorbed in the constant factor in equations (148) and (149). This representation can be shown to be unique as long as ν remains greater than a large positive number N. By substituting $U_{a,\nu}^{(1)}(\tau)$ and $U_{a,\nu}^{(2)}(\tau)$ in equation (147) and choosing the coefficients $f_{r,s}(\tau)$ ($r = 1$ and 2; and $s = 1, 2, 3, \ldots$) to make the individual terms vanish, equation (147) reduces to

$$2K'f'_{1,s+1} + K''f_{1,s+1} = \rho_a f_{1,s} - f''_{1,s} \tag{151}$$

$$2K'f'_{2,s+1} + K''f_{2,s+1} = -\rho_a f_{2,s} + f''_{2,s}, \quad s = 0, 1, 2, \ldots \tag{152}$$

where $f_{1,0}(\tau) = f_{2,0}(\tau) = \varphi^{-\frac{1}{4}}$. The coefficients $f_{r,s}(\tau)$ then are given successively by a first order ordinary differential equation and their determination does not involve any difficulty. The problem is thus formally solved.

Obviously, the solution is of the exponential type when $\varphi(\tau)$ is positive in the range concerned and of an oscillatory type when $\varphi(\tau)$ is negative. Now in the interval $\delta \leqslant \tau \leqslant 1-\delta$, $\delta > 0$ where $\varphi(\tau) \gtreqless 0$ when $\tau \lesseqgtr \frac{1}{2\beta+1}$, both types of solution exist. It is evident that in the neighborhood of $\tau = \frac{1}{2\beta+1}$ a change of character of the solutions must take place, but the manner in which the transition occurs cannot be deduced from equations (148) and (149) because of the failure of the representation of the solutions in the neighborhood $\tau = \frac{1}{2\beta+1}$. This is closely related to the Stokes phenomenon.

The method was extended by Jeffreys[19] to include the case where $\varphi(\tau)$ has a simple root in an interval under consideration and can be applied suitably to the first order of approximation. The general problem has been treated by Langer[20] in a series of papers, considering both the case where ν and τ are real and that where ν and τ are complex. Attention was given especially to the Stokes phenomenon, and a law of connection of the solution valid on each side of the critical point was explicitly stated. In the present case, however, only the first approximation is used and Jeffreys' method is adopted for convenience.

It is seen from ecuations (148) and (149) that the first approximation depends only on $\varphi(\tau)$, and the effect of $\rho_a(\tau)$ is felt only by the higher order terms. Hence, for the first approximation only, equation (147) can be written as

$$U''_{\nu}(\tau) - \nu^2\varphi(\tau)U_{\nu}(\tau) = 0 \tag{153}$$

where $U_{\beta,\nu} = U_{-\beta,\nu} = U_\nu$. Thus, when $\nu > N$, the dominant terms of the asymptotic solutions are

$$U_\nu^{(1)}(\tau) \sim \varphi^{-\frac{1}{4}} e^{\nu K}\left[1 + O\left(\frac{1}{\nu}\right)\right] \qquad 0 < \tau < \frac{1}{2\beta+1} \tag{154}$$

$$U_\nu^{(2)}(\tau) \sim \varphi^{-\frac{1}{4}} e^{-\nu K}\left[1 + O\left(\frac{1}{\nu}\right)\right] \tag{155}$$

Here $O\left(\frac{1}{\nu}\right)$, in each case, denotes the fact that the term is uniformly of the order ν^{-1} when ν is sufficiently large in an interval $\delta \leqslant \tau \leqslant \frac{1}{2\beta+1} - \delta$, $\delta > 0$ and is a function of ν^{-1}.

On the other hand, in the interval $\frac{1}{2\beta+1} + \delta \leqslant \tau \leqslant 1 - \delta$ where $\varphi(\tau) < 0$ and K is a pure imaginary quantity $i\omega$ where ω is real, the dominant terms of the asymptotic solutions must be a linear combination of equations (148) and (149) and must be of the forms:

$$U_\nu^{(1)}(\tau) \sim \frac{c_1}{\varphi^{\frac{1}{4}}}\cos(\nu\omega + \varepsilon_\nu) \tag{156}$$

$$U_\nu^{(2)}(\tau) \sim \frac{c_2}{\varphi^{\frac{1}{4}}}\sin(\nu\omega + \varepsilon_\nu); \qquad \frac{1}{2\beta+1} < \tau < 1 \tag{157}$$

where c_1, c_2, and ε_ν are constants to be determined.

The question of determination of these constants is actually the same as that of determining the more of continuation of the asymptotic representation of the solutions in the range $\frac{1}{2\beta+1} + \delta \leqslant \tau \leqslant 1 - \delta$. This can be done, according to Jeffreys, by considering the solutions valid in the neighborhood of $\tau = \frac{1}{2\beta+1}$ Let $\xi = \tau - \frac{1}{2\beta+1}$. When ξ is sufficiently small and ν is large, equation (153) can be written approximately as

$$U_\nu''(\xi) + \nu^2\varphi'(0)\xi U_\nu(\xi) = 0 \tag{158}$$

provided $\frac{\varphi^{(n)}(0)}{n!\ \varphi'(0)} \sim 1$. This is known as Stokes equation. The independent integrals are

$$\xi^{\frac{1}{2}} H_{\frac{1}{3}}^{(1)}(\zeta), \quad \xi^{\frac{1}{2}} H_{\frac{1}{3}}^{(2)}(\zeta); \quad \text{with} \quad \zeta = \frac{2}{3}\nu\varphi'^{\frac{1}{2}}(0)\xi^{\frac{3}{2}} \tag{159}$$

where $H_{\frac{1}{3}}^{(1)}(\zeta)$ and $H_{\frac{1}{3}}^{(2)}(\zeta)$ are the Hankel functions of order $\frac{1}{3}$. Consider as two independent solutions the following linear combinations:

$$U_\nu^{(1)}(\xi) = \xi^{\frac{1}{2}} H_{\frac{1}{3}}^{(1)}(\zeta) + \xi^{\frac{1}{2}} H_{\frac{1}{3}}^{(2)}(\zeta) \tag{160}$$

$$U_\nu^{(2)}(\xi) = \xi^{\frac{1}{2}} H_{\frac{1}{3}}^{(1)}(\zeta) - \xi^{\frac{1}{2}} H_{\frac{1}{3}}^{(2)}(\zeta) \tag{161}$$

As $H_{\frac{1}{3}}^{(1)}(\zeta)$ and $H_{\frac{1}{3}}^{(2)}(\zeta)$ are analytic functions in the whole ζ-plane, this immediately provides a means of identifying the asymptotic forms that represent the same function.

Suppose first that for $\arg\xi = 0$, the solutions are given in equations (160) and (161). The same solutions for which $\arg\xi = \pi$ and $\arg\zeta = \frac{3}{2}\pi$ are

$$U_\nu^{(1)}(\xi) = \xi^{\frac{1}{2}} e^{\frac{\pi i}{2}} H_{\frac{1}{3}}^{(1)}\left(\zeta e^{\frac{3\pi i}{2}}\right) + \xi^{\frac{1}{2}} e^{\frac{\pi i}{2}} H_{\frac{1}{3}}^{(2)}\left(\zeta e^{\frac{3\pi i}{2}}\right) \tag{162}$$

$$U_\nu^{(2)}(\xi) = \xi^{\frac{1}{2}} e^{\frac{\pi i}{2}} H_{\frac{1}{3}}^{(1)}\left(\zeta e^{\frac{3\pi i}{2}}\right) - \xi^{\frac{1}{2}} e^{\frac{\pi i}{2}} H_{\frac{1}{3}}^{(2)}\left(\zeta e^{\frac{3\pi i}{2}}\right) \tag{163}$$

Now

$$H_{\frac{1}{3}}^{(1)}\left(\zeta e^{\frac{\pi i}{2}} e^{\pi i}\right) = -e^{-\frac{\pi i}{2}} H_{\frac{1}{3}}^{(2)}\left(\zeta e^{\frac{\pi i}{2}}\right)$$

$$H_{\frac{1}{3}}^{(2)}\left(\zeta e^{\frac{\pi i}{2}} e^{\pi i}\right) = 2\cos\frac{\pi}{3} H_{\frac{1}{3}}^{(2)}\left(\zeta e^{\frac{\pi i}{2}}\right) + e^{\frac{\pi i}{3}} H_{\frac{1}{3}}^{(1)}\left(\zeta e^{\frac{\pi i}{2}}\right)$$

and when ζ is large and $-\pi < \arg\zeta e^{\frac{\pi i}{2}} < \pi$, the dominant terms of the asymptotic expansions of $H_{\frac{1}{3}}^{(1)}\left(\zeta e^{\frac{\pi i}{2}}\right)$ and $H_{\frac{1}{3}}^{(2)}\left(\zeta e^{\frac{\pi i}{2}}\right)$ are

$$H_{\frac{1}{3}}^{(1)}\left(\zeta e^{\frac{\pi i}{2}}\right) \sim \sqrt{\frac{2}{\pi\zeta}}\, e^{i\left(\zeta e^{\frac{\pi i}{2}} - \frac{\pi}{6} - \frac{\pi}{2}\right)} \left\{1 + O\left(\frac{1}{\zeta}\right)\right\}$$

$$H_{\frac{1}{3}}^{(2)}\left(\zeta e^{\frac{\pi i}{2}}\right) \sim \sqrt{\frac{2}{\pi\zeta}}\, e^{-i\left(\zeta e^{\frac{\pi i}{2}} - \frac{\pi}{6}\right)} \left\{1 + O\left(\frac{1}{\zeta}\right)\right\}$$

By substituting in equations (162) and (163) and neglecting the term of lower order in ζ, there is obtained by expanding at the same time equations (160) and (161):

$$2\xi^{-\frac{1}{4}} \cos\left(\zeta - \frac{\pi}{4}\right) \to \xi^{-\frac{1}{4}} e^{-\zeta} \tag{164}$$

$$\xi^{-\frac{1}{4}} \sin\left(\zeta - \frac{\pi}{4}\right) \to \xi^{-\frac{1}{4}} e^{\zeta} \tag{165}$$

Here the arrow is used to indicate the transition of the asymptotic representation of the same function from the lefthand to the right-hand member. For small ξ, $\xi^{-\frac{1}{4}} \sim \varphi^{-\frac{1}{4}}$, and $\zeta \sim -\nu\omega$; (156) and (157) finally become

$$U_\nu^{(1)}(\tau) \sim \frac{2}{\varphi^{\frac{1}{4}}} \cos\left(\nu\omega - \frac{\pi}{4}\right)\left\{1 + O\left(\frac{1}{\nu}\right)\right\} \tag{166}$$

$$\frac{1}{2\beta+1} < \tau < 1$$

$$U_\nu^{(2)}(\tau) \sim \frac{1}{\varphi^{\frac{1}{4}}} \cos\left(\nu\omega + \frac{\pi}{4}\right)\left\{1 + O\left(\frac{1}{\nu}\right)\right\} \tag{167}$$

with $c_1 = 2$, $c_2 = -1$, and $\varepsilon_\nu = -\frac{\pi}{4}$. Under the hypothesis just made, the pair of expressions (154), (166) and (155), (167) actually represent respectively the dominant terms of the two asymptotic expansions of the solutions $U_\nu^{(1)}(\tau)$ and $U_\nu^{(2)}(\tau)$ for a ν which may be any positive but large number.

14. The Asymptotic Representation of $F(a_\nu, b_\nu; c_\nu; \tau)$ and $F(a_\nu+\beta, b_\nu+\beta; c_\nu; \tau)$

The dominant terms of the asymptotic expansion of $U_\nu^{(1)}(\tau)$ and $U_\nu^{(2)}(\tau)$ are given respectively by (154), (166) and (155), (167). By evaluating the simple integrals in (154) and (166), the explicit expressions for the first approximation of $U_\nu^{(1)}(\tau)$ and $U_\nu^{(2)}(\tau)$ are

$$U_\nu^{(1)}(\tau) \sim (2\beta)^{\frac{-\nu(\alpha-1)}{2}} \left\{\frac{4(1-\tau)}{1-\alpha^2\tau}\right\}^{\frac{1}{4}} \tau^{\frac{\nu+1}{2}} T^{\nu *}(\tau) \tag{168}$$

$$0<\tau<\frac{1}{2\beta+1}$$

$$U_\nu^{(2)}(\tau) \sim (2\beta)^{\frac{\nu(\alpha-1)}{2}} \left\{\frac{4(1-\tau)}{1-\alpha^2\tau}\right\}^{\frac{1}{4}} \tau^{-\frac{\nu+1}{2}} T^{-\nu *}(\tau) \tag{169}$$

$$U_\nu^{(1)}(\tau) \sim 2\left\{\frac{4(1-\tau)}{\alpha^2\tau-1}\right\}^{\frac{1}{4}} \tau^{\frac{1}{2}} \cos\left(\nu\omega-\frac{\pi}{4}\right) \tag{170}$$

$$\frac{1}{2\beta+1}<\tau<1$$

$$U_\nu^{(2)}(\tau) \sim \left\{\frac{4(1-\tau)}{\alpha^2\tau-1}\right\}^{\frac{1}{4}} \tau^{\frac{1}{2}} \cos\left(\nu\omega+\frac{\pi}{4}\right) \tag{171}$$

where

$$T^*(\tau) = \frac{[\alpha(1-\tau)^{\frac{1}{2}}+(1-\alpha^2\tau)^{\frac{1}{2}}]^\alpha}{(1-\tau)^{\frac{1}{2}}+(1-\alpha^2\tau)^{\frac{1}{2}}}, \quad \alpha = \left[\frac{\gamma+1}{\gamma-1}\right]^{\frac{1}{2}} \tag{172}$$

$$\omega(\tau) = \alpha\tan^{-1}\sqrt{\frac{\alpha^2\tau-1}{\alpha^2(1-\tau)}} - \tan^{-1}\sqrt{\frac{\alpha^2\tau-1}{1-\tau}}$$

The values of the function $\omega(\tau)$ are given in Fig. 3 together with the function $\mu(\tau)$, defined by $\cos\mu = 1/M$. In the respective ranges of validity, each pair of expressions differs from the exact solution only by a constant factor which can be determined to satisfy the normalization conditions (30) and (36). By substituting equation (168) into equation (143), these were found to be

$$c_{\pm\nu} = \frac{1}{\sqrt{2}}(2\beta)^{\pm\frac{\alpha-1}{2}\nu}\left\{\frac{2}{(1+\alpha)^2}\right\}^{\pm\nu}$$

Thus, the expressions for the desired asymptotic forms, when $\nu > N$, are, for the interval $0 \leqslant \tau < \dfrac{1}{2\beta+1}$,

$$\underline{F}_\nu(\tau) \sim f(\tau)T^\nu(\tau) \tag{173}$$

$$\underline{G}_\nu(\tau) \sim f(\tau)T^{-\nu}(\tau) \tag{174}$$

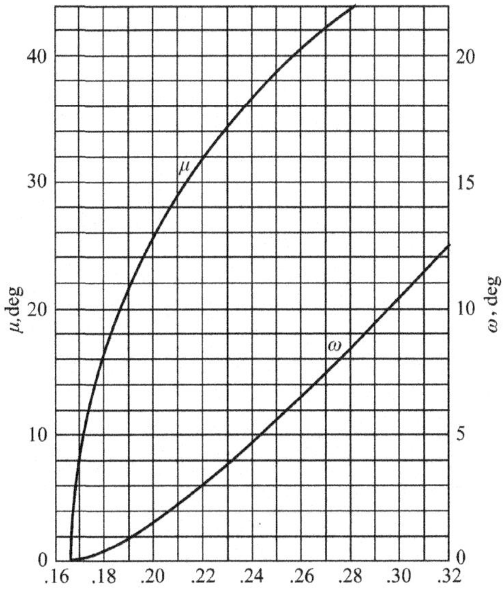

Fig. 3 $\omega(\tau); \mu(\tau); \gamma = 1.405$

where

$$f(\tau) = \frac{(1-\tau)^{\frac{\beta}{2}+\frac{1}{4}}}{(1-\alpha^2\tau)^{\frac{1}{4}}}, \quad T(\tau) = \frac{2}{(1+\alpha)^\alpha}\frac{[\alpha(1-\tau)^{\frac{1}{2}}+(1-\alpha^2\tau)^{\frac{1}{2}}]^\alpha}{(1-\tau)^{\frac{1}{2}}+(1-\alpha^2\tau)^{\frac{1}{2}}} \tag{175}$$

For the interval $\frac{1}{2\beta+1} < \tau < 1$, they are

$$\underline{F}_\nu(\tau) \sim f(\tau) T^{\nu}(\tau) \cos\left(\nu\omega - \frac{\pi}{4}\right) \tag{176}$$

$$\underline{G}_\nu(\tau) \sim \frac{1}{2} f(\tau) T^{-\nu}(\tau) \cos\left(\nu\omega + \frac{\pi}{4}\right) \tag{177}$$

where

$$\left.\begin{aligned} f(\tau) &= 2\,\frac{(1-\tau)^{\frac{\beta}{2}+\frac{1}{4}}}{(\alpha^2\tau-1)^{\frac{1}{4}}}, \\ T(\tau) &= 2\,\frac{(2\beta)^{\frac{\alpha}{2}}}{(1+\alpha)^{\alpha}}\,\frac{1}{\sqrt{2\beta\tau}} \end{aligned}\right\} \tag{178}$$

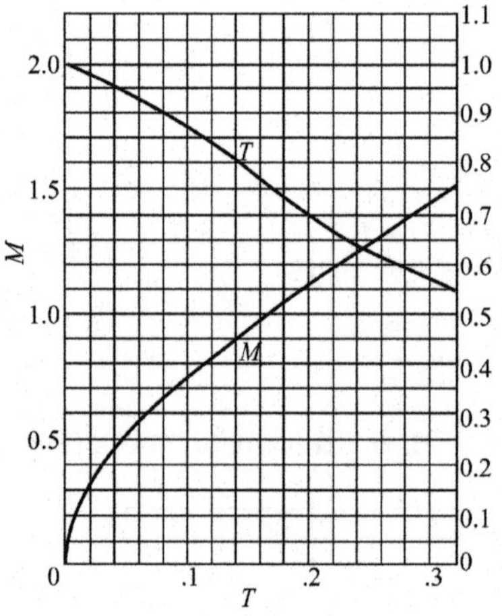

Fig. 4 $T(\tau); M(\tau); \gamma = 1.405$

The values of $T(\tau)$ are given (Fig. 4) as a function of τ together with the local Mach number M.

Similarly, as from (153) $U_\nu(\tau) \sim V_\nu(\tau)$, corresponding expressions for $F(a_\nu + \beta, b_\nu + \beta; c_\nu; \tau)$ are:

$$\left.\begin{aligned} \underline{\tilde{F}}_\nu(\tau) &\sim g(\tau) T^{\nu}(\tau) \\ \underline{\tilde{G}}_\nu(\tau) &\sim g(\tau) T^{-\nu}(\tau) \end{aligned}\right\} \quad 0 \leqslant \tau < \frac{1}{2\beta+1} \qquad \begin{aligned} &(179) \\ &(180) \end{aligned}$$

where

$$g(\tau) = \frac{(1-\tau)^{-\frac{\beta}{2}+\frac{1}{4}}}{(1-\alpha^2\tau)^{\frac{1}{4}}} \tag{181}$$

and

$$\left.\begin{aligned} \underline{\tilde{F}}_\nu(\tau) &\sim g(\tau) T^{\nu}(\tau) \cos\left(\nu\omega - \frac{\pi}{4}\right) \\ \underline{\tilde{G}}_\nu(\tau) &\sim \frac{1}{2} g(\tau) T^{-\nu}(\tau) \cos\left(\nu\omega + \frac{\pi}{4}\right) \end{aligned}\right\} \quad \frac{1}{2\beta+1} < \tau < 1 \qquad \begin{aligned} &(182) \\ &(183) \end{aligned}$$

where

$$g(\tau) = 2\,\frac{(1-\tau)^{-\frac{\beta}{2}+\frac{1}{4}}}{(\alpha^2\tau-1)^{\frac{1}{4}}} \tag{184}$$

Here $\underline{F}_\nu(\tau)$ denotes invariahly the first integral $F(a_\nu, b_\nu; c_\nu; \tau)$ while $\underline{G}_\nu(\tau)$, when multiplied by $q^{-2\nu}$, denotes the second integral $F_\nu(\tau)$, defined by equation (37) when ν is not an integer or by equation (39) when ν is an integer, since the asymptotic expansions are valid for both integral and nonintegral values of ν, provided $\nu > N$.

In the domains of validity, the asymptotic expansions may be differentiated with respect to τ with the same order of approximation. Hence, for $\nu > N$, it can be shown that for $0 \leqslant \tau <$

$\frac{1}{2\beta+1}$

$$\underline{\tilde{F}}_{\nu,1}(\tau) \sim h(\tau)T^{\nu}(\tau)\left\{1+O\left(\frac{1}{\nu}\right)\right\} \tag{185}$$

$$\underline{\tilde{G}}_{\nu,1}(\tau) \sim h(\tau)T^{-\nu}(\tau)\left\{1+O\left(\frac{1}{\nu}\right)\right\} \tag{186}$$

where

$$h(\tau) = 2(1-\tau)^{-\frac{\alpha^2}{4}}(1-\alpha^2\tau)^{-\frac{1}{4}}\left[(1-\tau)^{\frac{1}{2}}+(1-\alpha^2\tau)^{\frac{1}{2}}\right]^{-1} \tag{187}$$

and for $\frac{1}{2\beta+1} < \tau < 1$

$$\underline{\tilde{F}}_{\nu,1}(\tau) \sim h(\tau)T^{2}(\tau)\cos\left(\nu\omega-\mu-\frac{\pi}{4}\right)\left\{1+O\left(\frac{1}{\nu}\right)\right\} \tag{188}$$

$$\underline{\tilde{G}}_{\nu,1}(\tau) \sim \frac{1}{2}h(\tau)T^{-\nu}(\tau)\cos\left(\nu\omega+\mu+\frac{\pi}{4}\right)\left\{1+O\left(\frac{1}{\nu}\right)\right\} \tag{189}$$

where

$$h(\tau) = 4(1-\tau)^{-\frac{\alpha^2}{4}}(\alpha^2\tau-1)^{-\frac{1}{4}}(2\beta\tau)^{-\frac{1}{2}}, \quad \mu(\tau) = \cos^{-1}\sqrt{\frac{1-\tau}{2\beta\tau}} \tag{190}$$

The values of the functions $g(\tau)$ and $h(\tau)$ are given in Fig. 5.

It is interesting to note that when $\gamma = -1$ the constant α vanishes and only the exponential type of solutions exist. In the case of $\psi_0(\tau)$ the solution is exact, namely, for $\beta = -\frac{1}{2}$

$$\underline{F}_{\nu}(\tau) = \left[\frac{2}{1+\sqrt{1+\frac{q^2}{c_0^2}}}\right]^{\nu} \tag{191}$$

$$\underline{F}_{-\nu}(\tau) = \left[\frac{2}{1+\sqrt{1+\frac{q^2}{c_0^2}}}\right]^{-\nu} \tag{192}$$

of which the first is in agreement with the result obtained by Tsien[9], while for $\chi_{\nu}(\tau)$ the solutions which are not exact reduce to

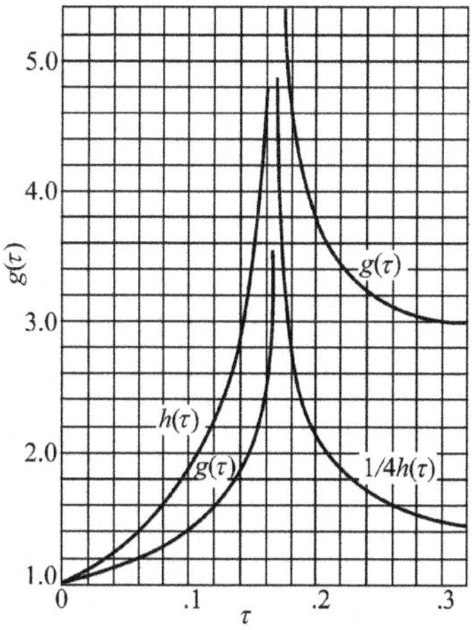

Fig. 5 $g(\tau);h(\tau);\gamma = 1.405$

$$\underline{\tilde{F}}_{\nu}(\tau) \sim \left[1+\frac{q^2}{c_0^2}\right]^{\frac{1}{2}}\left[\frac{2}{1+\left(1+\frac{q^2}{c_0^2}\right)^{\frac{1}{2}}}\right]^{\nu}\left\{1+O\left(\frac{1}{\nu}\right)\right\} \tag{193}$$

$$\nu \geqslant N$$

$$\underline{\tilde{G}}_{\nu}(\tau) \sim \left[1+\frac{q^2}{c_0^2}\right]^{\frac{1}{2}}\left[\frac{2}{1+\left(1+\frac{q^2}{c_0^2}\right)^{\frac{1}{2}}}\right]^{-\nu}\left\{1+O\left(\frac{1}{\nu}\right)\right\} \tag{194}$$

This may be the cause that destroys the analogy between the coordinates of the corresponding compressible flows and the incompressible flows.

For $\gamma = 1.405$ and $\nu = n + \dfrac{1}{2}$, n being a positive integer, the three groups of functions $\underline{F}_\nu(\tau)$, $\underline{F}_{-\nu}(\tau)$; $\underline{\tilde{F}}_\nu(\tau)$, $\underline{\tilde{F}}_{-\nu}(\tau)$; and $\underline{\tilde{F}}_{\nu,1}(\tau)$, $\underline{\tilde{F}}_{-\nu,1}(\tau)$, together with their asymptotic expressions were calculated for τ varying from 0 to 0.5 and n from 0 to 10. The results are presented in tables 2 to 13. The behavior of the approximation is illustrated in Figs. 6 to 11. It

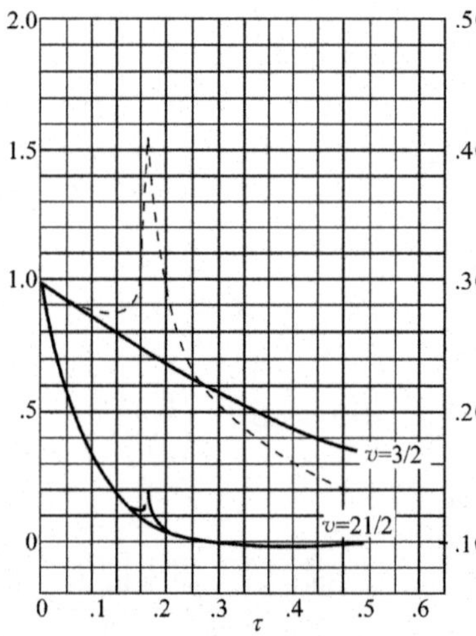

Fig. 6 $\underline{F}_v(\tau)$; $\gamma = 1.405$. The dash line denotes the approximate values of $\underline{F}_v(\tau)$

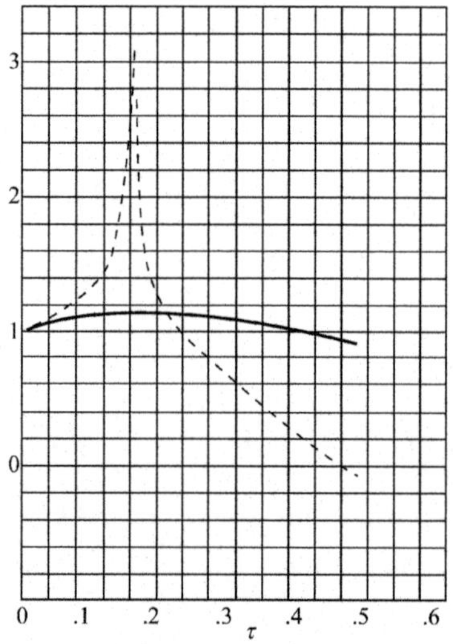

Fig. 7 $\underline{F} - 3/2(\tau)$; $\gamma = 1.405$

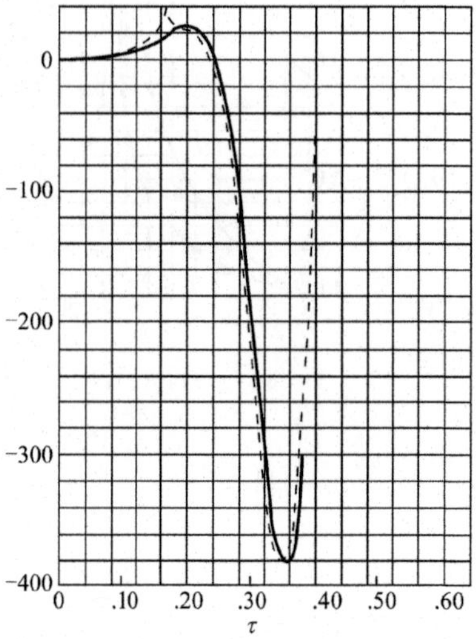

Fig. 8 $\underline{F} - 21/2(\tau)$; $\gamma = 1.405$

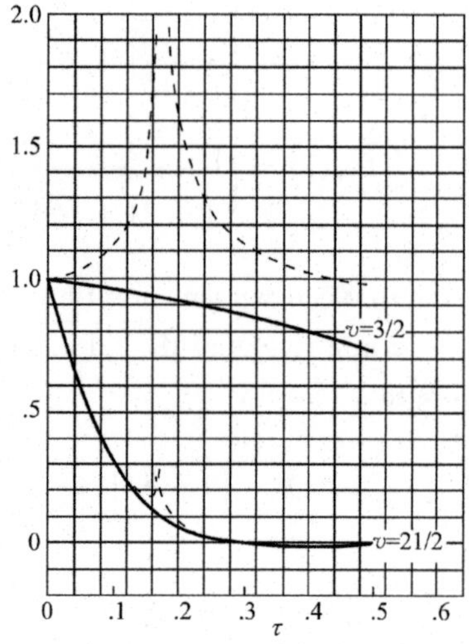

Fig. 9 $\underline{\tilde{F}}_v(\tau)$; $\gamma = 1.405$

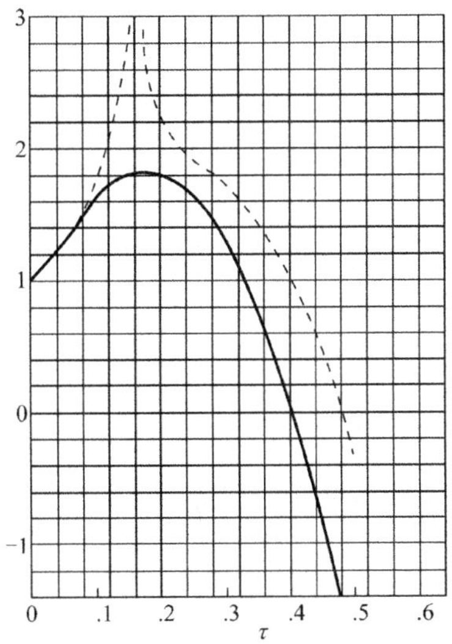

Fig. 10 $\underline{\tilde{F}} - 3/2(\tau)$; $\gamma = 1.405$

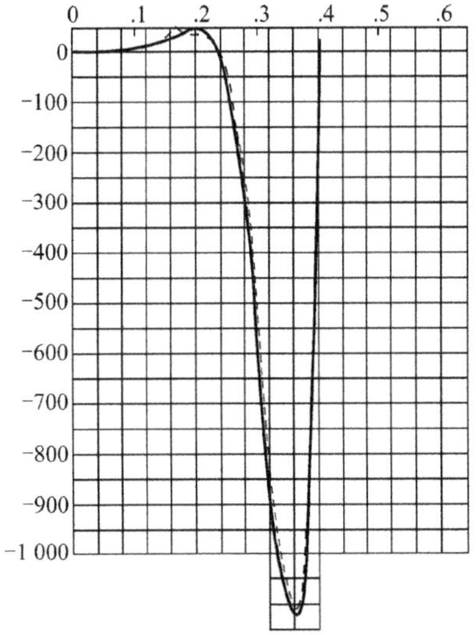

Fig. 11 $\underline{\tilde{F}} - 21/2(\tau)$; $\gamma = 1.405$

can be observed that the degree of approximation of the functions increases, on the one hand, with ν for any fixed τ. On the other hand, for any fixed n, the approximation becomes worse as τ approaches the critical point $\tau = \frac{1}{2\beta + 1}$, corresponding to the local sonic speed. On the whole, if the critical point $\tau = \frac{1}{2\beta + 1}$ is not reached, the agreement can generally be regarded as excellent, especially for larger values of n.

15. Transformation of the Function $W(w;\tau)$ Branch Point of Order 1

The function $W(w;\tau)$ for a flow that possesses a branch point of order 1 was given in sections 8 and 9. The forms of representation, as can be seen, are not, in general, suitable for practical calculation. The difficulty is twofold: First, the series involves infinitely many hypergeometric functions which themselves are, in turn, defined as infinite series. The convergence of hypergeometric series generally decreases with an increase of the parameter ν_0. This means that it is very difficult to compute the value of the later terms of the series for $W(w;\tau)$. Secondly, the convergence of the power series defining the function $W(w;\tau)$ itself is, as expected, very slow in the neighborhood of the singularity. To increase the convergence, the following method is used:

Observe that the corresponding function for the incompressible flow that has the same character of singularity is

$$W_0^{(i)}(w) = \sum_{n=0}^{\infty} A_n w^n, \quad |w| < U$$

which is absolutely and uniformly convergent in any closed domain in $|w| < U$. Now, if in

(92) $\underline{\tilde{F}}_n^{(r)}(\tau)$ is replaced by

$$\underline{F}_n^{(r)}(\tau) \cong \frac{f(\tau)}{f(\tau_1)} t^n(\tau), \quad 0 \leqslant \tau < \frac{1}{2\beta+1} \tag{195}$$

where $t(\tau) = \dfrac{T(\tau)}{T(\tau_1)}$, as by hypothesis, $0 < \tau_1 \ll \dfrac{1}{2\beta+1}$; then it is clear that

$$W^{(i)}(w;\tau) \cong \frac{f(\tau)}{f(\tau_1)} \sum_{n=0}^{\infty} A_n (tw)^n, \quad |tw| < U \tag{196}$$

which is also absolutely and uniformly convergent in the same domain as $W_0(w)$ and, consequently, (196) will be denoted by $\dfrac{f(\tau)}{f(\tau_1)} W_0(tw)$. In doing this, however, the restriction that (195) holds only when n is greater than a large number N is violated. The error can be removed by adding to and subtracting from (91) the quantity given in (196); then it follows immediately that

$$W^{(i)}(w;\tau) = W_1(w;\tau) + W_2^{(i)}(w;\tau) \tag{197}$$

where

$$W_1(w;\tau) = \frac{f(\tau)}{f(\tau_1)} W_0(tw) \tag{198}$$

$$W_2^{(i)}(w;\tau) = \sum_{n=0}^{\infty} A_n G_n(\tau) w^n, \quad |w| < U \tag{199}$$

with

$$G_n(\tau) = \underline{F}_n^{(r)}(\tau) - \frac{f(\tau)}{f(\tau_1)} t^n(\tau)$$

Here n is a positive integer. The function $W(w;\tau)$ then is represented by the sum of two functions $W_1(w;\tau)$, which is of closed form, and $W_2^{(i)}(w;\tau)$, which is the difference of two convergent power series and hence is also convergent. But, according to the theory of asymptotic expansion, $G_n(\tau)$ tends to zero as n approaches infinity. In fact, $G_n(\tau)$ is of order n^{-1}; therefore, the convergence of $W(w;\tau)$ is increased by the order of n^{-1}. This actually is the gist of the whole problem.

As the form of the representation of the hypergeometric function given in equation (195) is valid for all τ in $0 \leqslant \tau < \dfrac{1}{2\beta+1}$, $W_1(w;\tau)$ given by equation (198) holds automatically even outside the circle $|w| = U$. For this reason, $W_1(w;\tau)$ should be identical in form with that derived from equation (101). That this is the case can be seen from the following consideration. For, in addition to equation (195), if it is assumed that

$$\underline{G}_\nu(\tau) \cong f(\tau) T^{-\nu}(\tau) \tag{200}$$

it follows that

$$\xi_{\nu}(\tau_1) = -\xi_{-\nu}(\tau_1) = \sqrt{\frac{1-\alpha^2\tau_1}{1-\tau_1}} \tag{201}$$

then equations (106) and (107) yield, by equations (108) and (109),

$$B_n^* \cong \frac{B_n}{f(\tau_1)} T^{-\nu}(\tau_1), \quad C_n^* \cong \frac{C_n}{f(\tau_1)} T^{\nu}(\tau_1) \tag{202}$$

By using these sets of approximate coefficients and replacing $\underline{F}_{\nu}(\tau)$ and $\underline{G}_{\nu}(\tau)$ by their respective asymptotic expression, the following relation is obtained with the aid of equation (100)

$$W^{(0)}(w;\tau) = W_1(w;\tau) + W_2^{(0)}(w;\tau) \tag{203}$$

where

$$W_2^{(0)}(w;\tau) = \mathrm{i}\sum_{n=0}^{\infty}\{G_{\nu}^{(1)}(\tau)w^{\nu} + G_{\nu}^{(2)}(\tau)w^{-\nu}\} \tag{204}$$

In this case the coefficients B_n^* and C_n^*, as well as the functions $\underline{F}_{\nu}(\tau)$ and $\underline{G}_{\nu}(\tau)$ used in deriving $W_1(w;\tau)$, are approximate. Hence, if both are corrected, $G_{\nu}^{(1)}(\tau)$ and $G_{\nu}^{(2)}(\tau)$ should be of the forms

$$\left.\begin{aligned} G_{\nu}^{(1)}(\tau) &= \Delta B_n^* \underline{F}_{\nu}(\tau) + \frac{B_n}{f(\tau_1)} T^{-\nu}(\tau_1)\Delta\underline{F}_{\nu}(\tau) \\ G_{\nu}^{(2)}(\tau) &= \Delta C_n^* \underline{G}_{\nu}(\tau) + \frac{C_n}{f(\tau_1)} T^{\nu}(\tau_1)\Delta\underline{G}_{\nu}(\tau) \end{aligned}\right\} \tag{205}$$

where

$$\left.\begin{aligned} \Delta B_n^* &= B_n^* - \frac{B_n}{f(\tau_1)} T^{-\nu}(\tau_1), \quad \Delta\underline{F}_{\nu}(\tau) = \underline{F}_{\nu}(\tau) - f(\tau)T^{\nu}(\tau) \\ \Delta C_n^* &= C_n^* - \frac{C_n}{f(\tau_1)} T^{\nu}(\tau_1), \quad \Delta\underline{G}_{\nu}(\tau) = \underline{G}_{\nu}(\tau) - f(\tau)T^{-\nu}(\tau) \end{aligned}\right\} \tag{206}$$

Here the differences ΔB_n^* and ΔC_n^* depend upon the condition at infinity for any sets of constants B_n and C_n, while those of $\Delta\underline{F}_{\nu}(\tau)$ and $\Delta\underline{G}_{\nu}(\tau)$ are functions of τ only and, for this reason, can be tabulated once for all. It also can be shown that the order of ΔB_n^* is at least of n^{-1} and therefore the convergence of (204) is again increased by n^{-1}.

Consequently, if $\psi(q, \theta) = \psi_1(q, \theta) + \psi_2^{(l)}(q, \theta)$ where the superscript (l) denotes either (i) or (0), and if the coefficients are real, the stream function for the subsonic flow is according to (93) given by

$$\psi_1(q, \theta) = \frac{f(\tau)}{f(\tau_1)}\psi_0(tq, \theta), \quad 0 \leqslant \tau \leqslant \frac{1}{2\beta+1} \tag{207}$$

$$\psi_2^{(i)}(q, \theta) = -\sum_{n=0}^{\infty} A_n G_n(\tau) q^n \sin n\theta, \quad q < U \tag{208}$$

and in $U < q < V$

$$\psi_2^{(0)}(q,\theta) = \sum_{n=0}^{\infty}[G_\nu^{(1)}(\tau)q^\nu + G_\nu^{(2)}(\tau)q^{-\nu}]\cos\nu\theta \tag{209}$$

with θ restricted by $0 \leqslant \theta \leqslant 2\pi$. This result is striking in that for $\tau = \tau_1$, $\psi(U,\theta) \equiv \psi_1(U,\theta)$ as $G_\nu(\tau_1) = 0$; that is, the function $\psi_1(q,\theta)$ represents the correct singularity of the exact function. Far away from the singularity the term $\psi_2^{(l)}(q,\theta)$ ($l = i$ or 0) gradually comes into prominence, especially near $\tau = \dfrac{1}{2\beta+1}$; but the convergence there is already so rapid that a small number of terms is enough to secure a high accuracy in $\psi(q,\theta)$.

It is interesting to estimate the magnitude of the second part of the stream function. By noting the fact that $G_n(\tau_1) = 0$, $G_\nu(\tau_1) = 0$, the expansions of the $G_n(\tau)$ and $G_\nu(\tau)$ are

$$G_n(\tau) = G'_n(\tau_1)(\tau-\tau_1)+\ldots, \quad 0<\tau<\tau_1$$

$$G_\nu(\tau) = G'_\nu(\tau_1)(\tau-\tau_1)+\ldots, \quad \tau_1<\tau<\frac{1}{2\beta+1}$$

Then from corollary (52), it is shown that for $\theta \neq 0$

$$\psi_2^{(l)}(q,\theta) \sim \left(\frac{\partial\varphi_0}{\partial\theta}\right)_{q=U}(\tau-\tau_1)+\ldots$$

In other words, the second part of the solution is of the order of magnitude of $(\tau-\tau_1)$. However, the magnitude of $(\tau-\tau_1)$ depends essentially upon τ_1 for a given incompressible flow. If τ_1 is not small, then when $\tau = 0$, $|\tau-\tau_1|$ will be large. Thus for large free-stream Mach numbers, the second part of the solution ψ_2 cannot be neglected. This means that for high free-stream Mach numbers the correct solution for compressible flow is considerably more complicated than the usually assumed simple speed distortion rule would lead one to believe. Thus, any theory based upon such a simple rule cannot be accurate enough for transonic flows.

On the other hand, if τ_1 is small, or $\tau_1 \ll \dfrac{1}{2\beta+1}$, then the value of $|\tau-\tau_1|$ for $\tau = 0$ is small. Furthermore, if the maximum velocity of the flow is well below the sonic velocity, then the maximum value of τ also is small, thus $|\tau-\tau_1|$ for the whole flow is small. Then the second part of the solution ψ_2 is negligible. However, even then the solution for the compressible flow cannot be expressed as the solution of the incompressible solution by a simple distortion of the velocity scale, as is generally assumed in the so-called hodograph method. For this would be the case only if the multiplying factor $f(\tau)/f(\tau_1)$ is identically equal to 1. Since the multiplying factor is a function of the magnitude of velocity, the streamlines of the compressible flow and the streamlines of the incompressible flow cannot be made to correspond to each other. On the other hand, equation (207) shows that if ψ_0 is zero, then ψ_1 is also zero. Thus there is this one streamline, the zero streamline, in both flows satisfying the requirement of direct mapping. Since the zero streamline generally is chosen to represent the contour of the body, on the surface of the body in purely subsonic flows, the velocity of the compressible flow can be calculated from the incompressible flow by a simple "correction formula." This formula is given by equating the velocity q of the incompressible fluid to the velocity function tq of the compressible flow. Thus

$$\left(\frac{q}{U}\right)_0 = \sqrt{\frac{\tau}{\tau_1}}t = \sqrt{\frac{\tau}{\tau_1}}\frac{T(\tau)}{T(\tau_1)}$$

where the subscript 0 denotes the quantity for incompressible flow and $T(\tau)$ is given by equation (175). This formula is the same as that suggested by G. Temple and J. Yarwood[11]. This coincidence of Temple's theory with the present investigation can be considered as a further substantiation of the method.

For the supersonic regions, $\underline{F}_\nu(\tau)$ and $\underline{G}_\nu(\tau)$ in (101) should be replaced by

$$\left.\begin{aligned} \underline{F}_\nu(\tau) &\cong f(\tau)T^\nu(\tau)\cos\left(\nu\omega - \frac{\pi}{4}\right) \qquad (210)\\ \underline{G}_\nu(\tau) &\cong \frac{1}{2}f(\tau)T^{-\nu}(\tau)\cos\left(\nu\omega + \frac{\pi}{4}\right) \qquad (211)\end{aligned}\right\} \quad \frac{1}{2\beta+1} < \tau < 1$$

where $f(\tau)$, $T(\tau)$ and $\omega(\tau)$ are given in (178) and (172); then by writing

$$\underline{F}_\nu(\tau) \cong \frac{1}{2}f(\tau)\{e^{i\left(\nu\omega-\frac{\pi}{4}\right)} + e^{-i\left(\nu\omega-\frac{\pi}{4}\right)}\}$$

and substituting as before in equation (101), it leads again to the sum of $W_1(w;\tau)$ and $W_2^{(0)}(w;\tau)$, where

$$W_1(w;\tau) = \frac{f(\tau)}{4f(\tau_1)}\Big[e^{-\frac{\pi i}{4}}i\sum_{n=0}^{\infty}\{B_n(twe^{i\omega})^\nu + C_n(twe^{i\omega})^{-\nu}\} + e^{\frac{\pi i}{4}}i\sum_{n=0}^{\infty}\{B_n(twe^{-i\omega})^\nu + C_n(twe^{-i\omega})^{-\nu}\}\Big]$$

and

$$W_2(w;\tau) = i\sum_{n=0}^{\infty}\{G_\nu^{(1)}(\tau)w^\nu + G_\nu^{(2)}(\tau)w^{-\nu}\}, \quad \frac{1}{2\beta+1} < \tau < 1$$

According to equation (100), $W_1(w;\tau)$ also can be summed:

$$W_1(w;\tau) = \frac{1}{4}\frac{f(\tau)}{f(\tau_1)}[e^{-\frac{\pi i}{4}}W_0(twe^{\omega i}) + e^{\frac{\pi i}{4}}W_0(twe^{-\omega i})] \qquad (212)$$

Furthermore, from (178) it can be seen that $|tw| = \lambda U$, λ being a constant given by

$$\lambda = \frac{2(2\beta)^{\frac{\alpha}{2}}}{(1+\alpha)^\alpha(2\beta\tau_1)^{\frac{1}{2}}}\frac{1}{T(\tau_1)} > 1 \qquad (213)$$

which is a function of the Mach number and the characteristic constant β of the gas but independent of the shape of the boundary. The value of this function λ is given in table 14 and figure 12 for $\gamma = 1.405$. As a consequence, the functions constituting the stream function for the supersonic flow are

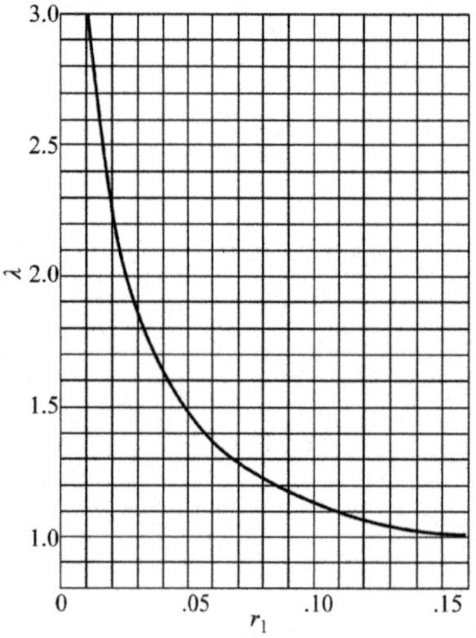

Fig. 12 $\lambda(\tau_1); \gamma = 1.405$

$$\psi_1(q, \theta) = 2^{-\frac{5}{2}} \frac{f(\tau)}{f(\tau_1)} \Big[\psi_0(\theta + \omega) + \psi_0(\theta - \omega) + \Phi_0(\theta + \omega) - \Phi_0(\theta - \omega) \Big] \quad \theta - \omega \geqslant 0 \tag{214}$$

$$\psi_2(q, \theta) = \sum_{n=0}^{\infty} \{ G_\nu^{(1)}(\tau) q^\nu + G_\nu^{(2)}(\tau) q^{-\nu} \} \cos \nu\theta, \quad U < q < V \tag{215}$$

Here the functions ψ_0 and Φ_0 are defined, on account of (213), by

$$\begin{aligned} \Psi_0(\theta \pm \omega) &\equiv \psi_0(\lambda U, \theta \pm \omega), \\ \Phi_0(\theta \pm \omega) &\equiv \varphi_0(\lambda U, \theta \pm \omega) \end{aligned} \tag{216}$$

where φ_0 and ψ_0 are the velocity potential and the stream function, respectively, of the corresponding incompressible flow. The functions $G_\nu^{(1)}(\tau)$ and $G_\nu^{(2)}(\tau)$ are the same as defined in (205) except that $\Delta \underline{F}_\nu(\tau)$ and $\Delta \underline{G}_\nu(\tau)$ now are given by

$$\left. \begin{aligned} \Delta \underline{F}_\nu(\tau) &= \underline{F}_\nu(\tau) - \frac{f(\tau)}{2} T^\nu \cos\left(\nu\omega - \frac{\pi}{4}\right) \\ \Delta \underline{G}_\nu(\tau) &= \underline{G}_\nu(\tau) - \frac{f(\tau)}{2} T^{-\nu} \cos\left(\nu\omega + \frac{\pi}{4}\right) \end{aligned} \right\} \tag{217}$$

Unlike the previous calculations, $G_\nu^{(1)}(\tau)$ in (211) is not of the order of ν^{-1} due to the presence of 1/2 in front of $f(\tau) T^\nu \cos\left(\nu\omega - \frac{\pi}{4}\right)$. This, however, does not offer a serious objection, as the series in which it appears already converges with $(tq)^\nu$, t being less than unity.

It is worth noting, moreover, that in the hyperbolic domain the function $\psi_1(q, \theta)$ depends, aside from a factor $f(\tau)$, only on the two independent families of characteristics defined by

$$\xi = \theta + \omega, \quad \eta = \theta - \omega \tag{218}$$

This result is most striking, as it shows that the main part of the solution satisfies the simple wave equation and thus clearly demonstrates its hyperbolic character. With both the incompressible stream function ψ_0 and the incompressible potential function φ_0 appearing in the solution, it in impossible to establish a simple relation between the incompressible streamlines and the compressible streamlines. Since such a simple relation is the foundation of the so-called speed correction formula for a quick estimation of velocity distribution in compressible flow from that of incompressible flow over the same body, this idea cannot be extended to supersonic regions. On the other hand, this also indicates that although the differential equation for $\psi(q, \theta)$ is hyperbolic in the supersonic range, it cannot be reduced to the simple wave equation by a mere distortion of the speed scale as given by the function $w(\tau)$. For if this were the case, then $\psi_1(q, \theta)$ would constitute an exact solution without the additional $\psi_2^{(0)}(q, \theta)$. *This fact is all the more important as the additional* $\psi_2^{(0)}(q, \theta)$ is not small in comparison with $\psi_1(q, \theta)$ for the mixed subsonic and supersonic flows, especially for the transitional region near sonic velocity. However, in the case of pure supersonic flow, $\psi_2^{(0)}(q, \theta)$ might be small; then $\psi_1(q, \theta)$ alone may be used as a satisfactory approximation. Of course, when $\gamma = -1$, then, as in the corresponding case in subsonic flow, the exact differential equation for the stream function can be reduced to the simple wave equation. In this case, the appropriate form for the speed function ω is

$$\omega(q) = -\tan^{-1}\sqrt{\frac{1}{\frac{q^2}{q_1^2 - c_1^2} - 1}} \tag{219}$$

where the subscript 1 denotes the conditions at the point of tangency of the true isentropic curve and the approximating tangent. This agrees with the result obtained by *N*. Coburn. (See reference 21.)

16. Continuation: Logarithmic Singularity

In the case of the logarithmic singularity the function $W(w;\tau)$ was broken up into two parts of which only the one that characterizes the singularity was given in equations (115) and (116). As an example, it is proposed to show that this problem can be treated by the same method. If the same approximation is introduced as in equations (195) and (201), then the coefficients defined in equations (121) and (122) become approximately:

$$A_n \cong \frac{1}{n}\frac{T^{-n}(\tau_1)}{f(\tau_1)}, \quad C_n \cong \frac{1}{n}\frac{T^n(\tau_1)}{f(\tau_1)} \tag{220}$$

with $Bf^2(\tau_1) = \frac{1}{2}$, so chosen that the form of equation (207) is again preserved. With these coefficients and if there is written for the function inside the circle $q = U$:

$$\tilde{\psi}^{(i)}(q, \theta) = \tilde{\psi}_1(q, \theta) + \tilde{\psi}_2^{(i)}(q, \theta)$$

Equation (115) reduces to the sum of

$$\tilde{\psi}_1(q, \theta) = \frac{f(\tau)}{f(\tau_1)}\tilde{\psi}_0(tq, \theta), \quad 0 \leqslant \tau < \frac{1}{2\beta + 1} \tag{221}$$

$$\tilde{\psi}_2^{(i)}(q, \theta) = \sum_{n=1}^{\infty}\frac{1}{n}\tilde{G}_n(\tau)\left(\frac{q}{U}\right)^n \cos n\theta, \quad q < U \tag{222}$$

where

$$\tilde{G}_n(\tau) = \underline{F}_n(\tau)\Delta\underline{G}_n(\tau_1) + \frac{\Delta\underline{F}_n(\tau)}{f(\tau_1)T^n(\tau_1)} \tag{223}$$

with

$$\left.\begin{aligned}\Delta \underline{F}_n(\tau) &= \underline{F}_n(\tau) - f(\tau)T^n(\tau) \\ \Delta \underline{G}_n(\tau_1) &= \frac{\underline{G}_n(\tau_1)}{f^2(\tau_1)} - f^{-1}(\tau_1)T^{-n}(\tau_1)\end{aligned}\right\} \tag{224}$$

Similarly, in the case of equation (116) it reduces to

$$\tilde{\psi}^{(0)}(q, \theta) = \tilde{\psi}_2(q, \theta) + \tilde{\psi}_2^{(0)}(q, \theta)$$

Here $\tilde{\psi}_1(q, \theta)$ is again the same as (221); while $\tilde{\psi}_2^{(0)}(q, \theta)$ is

$$\tilde{\psi}_2^{(0)}(q, \theta) = -\frac{1}{2f^2(\tau_1)}\int_{\tau_1}^{\tau}(1-\tau)^{\beta}\frac{d\tau}{\tau} + \frac{f(\tau)}{f(\tau_1)}\log\frac{tq}{U} + \sum_{n=1}^{\infty}\frac{1}{n}\tilde{G}_n^{(0)}\left(\frac{q}{U}\right)^{-n}\cos n\theta \tag{225}$$

where

$$\tilde{G}_n^{(0)}(\tau) = \underline{G}_n(\tau)\Delta\underline{F}_n(\tau_1) + f^{-1}(\tau_1)T^n(\tau_1)\Delta\underline{G}_n(\tau) \tag{226}$$

with

$$\Delta\underline{F}_n(\tau_1) = \frac{\underline{F}_n(\tau_1)}{f^2(\tau_1)} - \frac{T^n(\tau_1)}{f(\tau_1)}, \quad \Delta\underline{G}_n(\tau) = \underline{G}_n(\tau) - f(\tau)T^{-n}(\tau) \tag{227}$$

Unlike the previous case, $\tilde{\psi}(q, \theta) = \tilde{\psi}_0(q, \theta)$ when, and only when, c_0 tends to infinity. Because of (221), however, the singularity of $\tilde{\psi}(q, \theta)$ remains unchanged.

Again, if in (116)

$$\underline{G}_n(\tau) \cong \frac{1}{2}f(\tau)T^{-n}(\tau)\cos\left(n\omega + \frac{\pi}{4}\right)$$

is substituted for $\underline{G}_n(\tau)$, it can similarly be shown that

$$\begin{aligned}\tilde{\psi}_1(q, \theta) = 2^{-\frac{5}{2}}\frac{f(\tau)}{f(\tau_1)}[&\tilde{\Psi}_0(\theta+\omega) + \tilde{\Psi}_0(\theta-\omega) \\ &- \tilde{\Phi}_0(\theta+\omega) + \tilde{\Phi}_0(\theta-\omega)] \quad \theta-\omega \geqslant 0\end{aligned} \tag{228}$$

$$\tilde{\psi}_2^{(0)}(q, \theta) = -\frac{1}{2f^2(\tau_1)}\int_{\tau_1}^{\tau}(1-\tau)^{\beta}\frac{d\tau}{\tau} + 2^{-\frac{3}{2}}\frac{f(\tau)}{f(\tau_1)}(\log\lambda - \omega) + \sum_{n\pm 1}^{\infty}\frac{\tilde{G}_n^{(0)}}{n}\left(\frac{q}{U}\right)^{-n}\cos n\theta \tag{229}$$

where $\tilde{\Psi}_0(\theta\pm\omega)$ and $\tilde{\Phi}_0(\theta\pm\omega)$ are defined analogously to (216), and $\Delta\underline{G}_n(\tau)$ in $\tilde{G}_n^{(0)}(\tau)$ is now given by

$$\Delta\underline{G}_n(\tau) = \underline{G}_n(\tau) - \frac{1}{2}f(\tau)T^{-n}(\tau)\cos\left(n\omega + \frac{\pi}{4}\right) \tag{230}$$

This seems to indicate that the results obtained so far for $\psi_1(q, \theta)$ are quite general: It may differ for different cases, at most, by a constant factor. The general property, however, is not shared by $\psi_2(q, \theta)$, the character of which changes radically with the nature of the singularity and the shape of the boundary. Hence, its importance in the present problem is evident.

17. The Coordinate Functions $x(q, \theta)$ and $y(q, \theta)$

Whenever the function $\chi(q, \theta)$ for a boundary problem is determined, the coordinate functions $x(q, \theta)$ and $y(q, \theta)$ can be calculated according to equations (141) and (142). Suppose, for instance, a boundary is assigned with the property that the function $\Lambda(w;\tau)$ is truly described by (94) and (110), of which the real part $\chi(q, \theta)$, defined within the circle $|w|=U$, is

$$\chi(q, \theta) = \sum_{n=0}^{\infty} \widetilde{A}_n \underline{\widetilde{F}}_n^{(r)}(\tau) q^n \cos n\theta, \quad q < U \tag{231}$$

where the constants $\widetilde{A}_n$ in (94) are again real and are regarded as known, and $\underline{\widetilde{F}}_n^{(r)}(\tau) = \underline{\widetilde{F}}_n(\tau)/\underline{F}_n(\tau_1)$.

As the series is absolutely and uniformly convergent in $q < U$, it can be differentiated partially term by term with respect to q and θ. When the differential coefficients $\chi_q(q, \theta)$ and $\chi_\theta(q, \theta)$ are calculated and are substituted in equations (141) and (142), there results:

$$x(q, \theta) = \sum_{n=1}^{\infty} n\widetilde{A}_n \underline{\widetilde{F}}_n^{(r)} q^{n-1} \cos(n-1)\theta - \beta\tau \sum_{n=1}^{\infty} n\widetilde{A}_n \frac{n-1}{n+1} \underline{\widetilde{F}}_{n,1}^{(r)}(\tau) q^{n-1} \cos n\theta \cos\theta$$

$$q < U \tag{232}$$

$$y(q, \theta) = -\sum_{n=1}^{\infty} n\widetilde{A}_n \underline{\widetilde{F}}_n^{(r)} q^{n-1} \sin(n-1)\theta - \beta\tau \sum_{n=1}^{\infty} n\widetilde{A}_n \frac{n-1}{n+1} \underline{\widetilde{F}}_{n,1}^{(r)}(\tau) q^{n-1} \cos n\theta \sin\theta \tag{233}$$

where

$$\underline{F}_{n,1}^{(r)}(\tau) = \frac{F(a_n+\beta+1, b_n+\beta+1; c_n+1; \tau)}{F(a_n, b_n; c_n; \tau_1)} \tag{234}$$

Now, since

$$x_0(q, \theta) = \sum_{n=1}^{\infty} n\widetilde{A}_n q^{n-1} \cos(n-1)\theta$$

and

$$y_0(q, \theta) = -\sum_{n=1}^{\infty} n\widetilde{A}_n q^{n-1} \sin(n-1)\theta$$

$$\sigma_0(q, \theta) = \sum_{n=1}^{\infty} \widetilde{A}_n q^n \sin n\theta$$

by introducing the approximation given by equations (179) and (185), that is

$$\left.\begin{aligned} \underline{\widetilde{F}}_n^{(r)}(\tau) &\cong \frac{g(\tau)}{f(\tau_1)} t^n(\tau) \\ \underline{\widetilde{F}}_{n,1}^{(r)}(\tau) &\cong \frac{h(\tau)}{f(\tau_1)} t^n(\tau) \end{aligned}\right\} \quad 0 \leqslant \tau < \frac{1}{2\beta+1}$$

by defining

$$x(q, \theta) = x_1(q, \theta) + x_2^{(l)}(q, \theta) \tag{235}$$

$$y(q, \theta) = y_1(q, \theta) + y_2^{(l)}(q, \theta) \tag{236}$$

it can be shown by the same manner that

$$x_1(q,\theta) = \frac{g(\tau)}{f(\tau_1)}t(\tau)x_0(tq,\theta) - \frac{\beta\tau}{q}\frac{h(\tau)}{f(\tau_1)}\Omega_0(tq,\theta)\cos\theta \tag{237}$$

$$0 \leqslant \tau < \frac{1}{2\beta+1}$$

$$y_1(q,\theta) = \frac{g(\tau)}{f(\tau_1)}t(\tau)y_0(tq,\theta) - \frac{\beta\tau}{q}\frac{h(\tau)}{f(\tau_1)}\Omega_0(tq,\theta)\sin\theta \tag{238}$$

and

$$x_2^{(i)}(q,\theta) = \sum_{n=1}^{\infty} n\widetilde{A}_n\widetilde{G}_n(\tau)q^{n-1}\cos(n-1)\theta - \beta\tau\sum_{n=1}^{\infty} n\widetilde{A}_n\widetilde{G}_{n,1}(\tau)q^{n-1}\cos n\theta\cos\theta \tag{239}$$

$$q < U$$

$$y_2^{(i)}(q,\theta) = -\sum_{n=1}^{\infty} n\widetilde{A}_n\widetilde{G}_n(\tau)q^{n-1}\sin(n-1)\theta - \beta\tau\sum_{n=1}^{\infty} n\widetilde{A}_n\widetilde{G}_{n,1}(\tau)q^{n-1}\cos n\theta\sin\theta \tag{240}$$

where

$$\widetilde{G}_n(\tau) = \frac{F(a_n+\beta, b_n+\beta; c_n; \tau)}{F(a_n, b_n; c_n; \tau_1)} - \frac{g(\tau)}{f(\tau_1)}t^n(\tau) \tag{241}$$

$$\widetilde{G}_{n,1}(\tau) = \frac{n-1}{n+1}\frac{F(a_n+\beta+1, b_n+\beta+1; c_n+1; \tau)}{F(a_n, b_n; c_n; \tau_1)} - \frac{h(\tau)}{f(\tau_1)}t^n(\tau) \tag{242}$$

$$\Omega_0(q,\theta) = \frac{\partial\sigma_0}{\partial\theta} \tag{243}$$

On the other hand, the expression for $\chi(q,\theta)$ valid outside the circle of convergence is

$$\chi(q,\theta) = \sum_{n=0}^{\infty}[\widetilde{B}_n^*\underline{\widetilde{F}}_\nu(\tau)q^\nu - \widetilde{C}_n^*\underline{\widetilde{G}}_\nu(\tau)q^{-\nu}]\sin\nu\theta \tag{244}$$

provided the coefficients $\widetilde{B}_n^*$ and $\widetilde{C}_n^*$ in (110) are real. The functions $x(q,\theta)$ and $y(q,\theta)$ corresponding to (244) can be found similarly. These are:

$$x(q,\theta) = \sum_{n=0}^{\infty}\{\nu\widetilde{B}_n^*\underline{\widetilde{F}}_\nu(\tau)q^{\nu-1}\sin(\nu-1)\theta + \nu\widetilde{C}_n^*\underline{\widetilde{G}}_\nu(\tau)q^{-\nu-1}\sin(\nu+1)\theta\} -$$
$$\beta\tau\sum_{n=0}^{\infty}\left\{\nu\widetilde{B}_n^*\frac{\nu-1}{\nu+1}\underline{\widetilde{F}}_{\nu,1}(\tau)q^{\nu-1} + \nu\widetilde{C}_n^*\frac{\nu+1}{\nu-1}\underline{\widetilde{G}}_{\nu,1}(\tau)q^{-\nu-1}\right\}\sin\nu\theta\cos\theta \tag{245}$$

$$U < q < V$$

$$y(q,\theta) = \sum_{n=0}^{\infty}\{\nu\widetilde{B}_n^*\underline{\widetilde{F}}_\nu(\tau)q^{\nu-1}\cos(\nu-1)\theta - \nu\widetilde{C}_n^*\underline{\widetilde{G}}_\nu(\tau)q^{-\nu-1}\cos(\nu+1)\theta\} -$$
$$\beta\tau\sum_{n=0}^{\infty}\left\{\nu\widetilde{B}_n^*\frac{\nu-1}{\nu+1}\underline{\widetilde{F}}_{\nu,1}(\tau)q^{\nu-1} + \nu\widetilde{C}_n^*\frac{\nu+1}{\nu-1}\underline{\widetilde{G}}_{\nu,1}(\tau)q^{-\nu-1}\right\}\sin\nu\theta\sin\theta \tag{246}$$

Here the constants $\widetilde{B}_n^*$ and $\widetilde{C}_n^*$ satisfy the relations (109) and (110) and can be reduced to

$$\widetilde{B}_n^* \cong \frac{\widetilde{B}_n}{f(\tau_1)}T^{-\nu}(\tau_1), \quad \widetilde{C}_n^* \cong \frac{\widetilde{C}_n}{f(\tau_1)}T^{\nu}(\tau_1) \tag{247}$$

provided the same approximation is made as in (202). Furthermore,

$$x_0(q,\theta) = \sum_{n=0}^{\infty}\{\nu\widetilde{B}_nq^{\nu-1}\sin(\nu-1)\theta + \nu\widetilde{C}_nq^{-\nu-1}\sin(\nu+1)\theta\}$$

$$y_0(q,\theta) = \sum_{n=0}^{\infty}\{\nu\widetilde{B}_nq^{\nu-1}\cos(\nu-1)\theta - \nu\widetilde{C}_nq^{-\nu-1}\cos(\nu+1)\theta\}$$

and if $\underline{\widetilde{F}}_{\nu}(\tau)$ and $\underline{\widetilde{F}}_{\nu,1}(\tau)$ for the high-order terms are substituted by the asymptotic forms: namely,

$$\underline{\widetilde{F}}_{\nu}(\tau) \cong g(\tau)T^{\nu}(\tau), \quad \underline{\widetilde{F}}_{\nu,1}(\tau) \cong h(\tau)T^{-\nu}(\tau); \quad 0 \leqslant \tau < \frac{1}{2\beta+1}$$

then in like manner (245) and (246) can be transformed and can each be represented by the sum of two functions $x_1(q, \theta)$, $y_1(q, \theta)$, and $x_2(q, \theta)$, $y_2(q, \theta)$, where x_1 and y_1 are the same as (237) and (238); while x_2 and y_2 are:

$$\left.\begin{aligned} x_2^{(0)}(q, \theta) = & \sum_{n=0}^{\infty} \nu\{\widetilde{G}_{\nu}^{(1)}(\tau)q^{\nu-1}\sin(\nu-1)\theta + \widetilde{G}_{\nu}^{(2)}(\tau)_q^{-\nu-1}\sin(\nu+1)\theta\} - \\ & \beta\tau\sum_{n=0}^{\infty}\nu\{\widetilde{G}_{\nu,1}^{(1)}(\tau)q^{\nu-1} + \widetilde{G}_{\nu,1}^{(2)}(\tau)q^{-\nu-1}\}\sin\nu\theta\cos\theta \end{aligned}\right. \tag{248}$$

$$\tau_1 \leqslant \tau < \frac{1}{2\beta+1}$$

$$\left.\begin{aligned} y_2^{(0)}(q, \theta) = & \sum_{n=0}^{\infty} \nu\{\widetilde{G}_{\nu}^{(1)}(\tau)q^{\nu-1}\cos(\nu-1)\theta - \widetilde{G}_{\nu}^{(2)}(\tau)q^{-\nu-1}\cos(\nu+1)\theta\} - \\ & \beta\tau\sum_{n=0}^{\infty}\nu\{\widetilde{G}_{\nu,1}^{(1)}(\tau)q^{\nu-1} + \widetilde{G}_{\nu,1}^{(2)}(\tau)q^{-\nu-1}\}\sin\nu\theta\sin\theta \end{aligned}\right. \tag{249}$$

where $\widetilde{G}_{\nu}^{(a)}$ and $\widetilde{G}_{\nu,1}^{(a)}$ are defined by:

$$\left.\begin{aligned} \widetilde{G}_{\nu}^{(1)}(\tau) &= \Delta\widetilde{B}_n^*\underline{\widetilde{F}}_{\nu}(\tau) + \frac{\widetilde{B}_n}{f(\tau_1)}T^{-\nu}(\tau_1)\Delta\underline{\widetilde{F}}_{\nu}(\tau) \\ \widetilde{G}_{\nu}^{(2)}(\tau) &= \Delta C_n^*\underline{\widetilde{G}}_{\nu}(\tau) + \frac{\widetilde{C}_n}{f(\tau_1)}T^{\nu}(\tau_1)\Delta\underline{\widetilde{G}}_{\nu} \end{aligned}\right\} \tag{250}$$

$$\left.\begin{aligned} \widetilde{G}_{\nu,1}^{(1)}(\tau) &= \Delta\widetilde{B}_n^*\frac{\nu-1}{\nu+1}\underline{\widetilde{F}}_{\nu,1}(\tau) + \frac{\widetilde{B}_n}{f(\tau_1)}T^{-\nu}(\tau_1)\Delta\underline{\widetilde{F}}_{\nu,1}(\tau) \\ \widetilde{G}_{\nu,1}^{(2)}(\tau) &= \Delta\widetilde{C}_n^*\frac{\nu-1}{\nu+1}\underline{\widetilde{G}}_{\nu,1} + \frac{\widetilde{C}_n}{f(\tau_1)}T^{\nu}(\tau_1)\Delta\underline{\widetilde{G}}_{\nu,1} \end{aligned}\right\} \tag{251}$$

with

$$\left.\begin{aligned} \Delta\underline{\widetilde{F}}_{\nu,1}(\tau) &= \frac{\nu-1}{\nu+1}\underline{\widetilde{F}}_{\nu,1}(\tau) - h(\tau)T^{\nu}(\tau) \\ \Delta\underline{\widetilde{G}}_{\nu,1}(\tau) &= \frac{\nu+1}{\nu-1}\underline{\widetilde{G}}_{\nu,1}(\tau) - h(\tau)T^{-\nu}(\tau) \end{aligned}\right\} \tag{252}$$

while $\Delta\widetilde{B}_n^*$ and $\Delta\underline{\widetilde{F}}_{\nu}(\tau)$ are defined just the same as those given in equation (206).

Similarly, if the hypergeometric functions involved in the high-order terms are substituted by

$$\underline{\widetilde{F}}_{\nu}(\tau) \cong g(\tau)T^{\nu}\cos\left(\nu\omega - \frac{\pi}{4}\right), \quad \underline{\widetilde{F}}_{\nu,1}(\tau) \cong h(\tau)T^{\nu}\cos\left(\nu\omega - \mu - \frac{\pi}{4}\right)$$

$$\underline{\widetilde{G}}_{\nu}(\tau) \cong \frac{1}{2}g(\tau)T^{-\nu}\cos\left(\nu\omega + \frac{\pi}{4}\right), \quad \underline{\widetilde{G}}_{\nu,1}(\tau) \cong \frac{1}{2}h(\tau)T^{-\nu}\cos\left(\nu\omega + \mu + \frac{\pi}{4}\right)$$

and by resolving the products of the trigonometric functions into sums: for instance,

$$2\sin(\nu-1)\theta\cos\left(\nu\omega-\frac{\pi}{4}\right)=\sin\left[(\nu-1)(\theta+\omega)+\left(\omega-\frac{\pi}{4}\right)\right]+\sin\left[(\nu-1)(\theta-\omega)-\left(\omega-\frac{\pi}{4}\right)\right]$$

$$2\sin(\nu+1)\theta\cos\left(\nu\omega+\frac{\pi}{4}\right)=\sin\left[(\nu+1)(\theta+\omega)-\left(\omega-\frac{\pi}{4}\right)\right]+\sin\left[(\nu+1)(\theta-\omega)+\left(\omega-\frac{\pi}{4}\right)\right]$$

a brief reduction gives when $\frac{1}{2\beta+1}<\tau<1$,

$$\begin{aligned}x_1(q,\theta)=&\frac{t(\tau)}{4}\frac{g(\tau)}{f(\tau_1)}\left\{[X_0(\theta+\omega)+X_0(\theta-\omega)]\cos\left(\frac{\pi}{4}-\omega\right)-\right.\\&\left.[Y_0(\theta+\omega)-Y_0(\theta-\omega)]\sin\left(\frac{\pi}{4}-\omega\right)\right\}-\\&\frac{\beta\tau}{4q}\frac{h(\tau)}{f(\tau_1)}\left\{[\Omega_0(\theta+\omega)+\Omega_0(\theta-\omega)]\cos\left(\mu+\frac{\pi}{4}\right)-\right.\\&\left.[\Theta_0(\theta+\omega)-\Theta_0(\theta-\omega)]\sin\left(\frac{\pi}{4}+\mu\right)\right\}\cos\theta\end{aligned}\tag{253}$$

$$\begin{aligned}y_1(q,\theta)=&\frac{t(\tau)}{4}\frac{g(\tau)}{f(\tau_1)}\left\{[Y_0(\theta+\omega)+Y_0(\theta-\omega)]\cos\left(\frac{\pi}{4}-\omega\right)+\right.\\&\left.[X_0(\theta+\omega)-X_0(\theta-\omega)]\sin\left(\frac{\pi}{4}-\omega\right)\right\}-\\&\frac{\beta\tau}{4q}\frac{h(\tau)}{f(\tau_1)}\left\{[\Omega_0(\theta+\omega)+\Omega_0(\theta-\omega)]\cos\left(\mu+\frac{\pi}{4}\right)-\right.\\&\left.[\Theta_0(\theta+\omega)-\Theta_0(\theta-\omega)]\sin\left(\frac{\pi}{4}-\omega\right)\right\}\sin\theta\end{aligned}\tag{254}$$

by the fact that $qt=\lambda U$ in the interval under consideration. Here

$$X_0(\theta\pm\omega)=x_0(\lambda U,\theta\pm\omega),\quad Y_0(\theta\pm\omega)=y_0(\lambda U,\theta\pm\omega)$$
$$\Theta_0(\theta\pm\omega)\equiv\Theta_0(\lambda U,\theta\pm\omega),\quad \Omega_0(\theta\pm\omega)\equiv\Omega_0(\lambda U,\theta\pm\omega)$$

where

$$\Theta_0(q,\theta)=\frac{\partial\chi_0}{\partial\theta}$$

and

$$\begin{aligned}x_2^{(0)}(q,\theta)=&\sum_{n=0}^{\infty}\nu\{\tilde{G}_\nu^{(1)}(\tau)q^{\nu-1}\sin(\nu-1)\theta+\tilde{G}_\nu^{(2)}(\tau)q^{-\nu-1}\sin(\nu+1)\theta\}-\\&\beta\tau\sum_{n=0}^{\infty}\nu\{\tilde{G}_{\nu,1}^{(1)}(\tau)q^{\nu-1}+\tilde{G}_{\nu,1}^{(2)}(\tau)q^{-\nu-1}\}\sin\nu\theta\cos\theta\end{aligned}\tag{255}$$

$$\frac{1}{2\beta+1}<\tau<1$$

$$\begin{aligned}y_2^{(0)}(q,\theta)=&\sum_{n=0}^{\infty}\nu\{\tilde{G}_\nu^{(1)}(\tau)q^{\nu-1}\cos(\nu-1)\theta-\tilde{G}_\nu^{(2)}(\tau)q^{-\nu-1}\cos(\nu+1)\theta\}-\\&\beta\tau\sum_{n=0}^{\infty}\nu\{\tilde{G}_{\nu,1}^{(1)}(\tau)q^{\nu-1}+\tilde{G}_{\nu,1}^{(2)}(\tau)q^{-\nu-1}\}\sin\nu\theta\sin\theta\end{aligned}\tag{256}$$

where $\widetilde{G}_{\nu}^{(a)}(\tau)$ and $\widetilde{G}_{\nu,1}^{(a)}(\tau)$ retain the definitions given in (250) and (251) except that $\Delta\underline{\widetilde{F}}_{\nu}(\tau)$, $\Delta\underline{\widetilde{F}}_{\nu,1}(\tau)$, $\Delta\underline{\widetilde{G}}_{\nu}(\tau)$, and $\Delta\underline{\widetilde{G}}_{\nu,1}(\tau)$ are replaced by

$$\left.\begin{aligned}
\Delta\underline{\widetilde{F}}_{\nu}(\tau) &= \underline{\widetilde{F}}_{\nu}(\tau) - \frac{1}{2}g(\tau)T^{\nu}\cos\left(\nu\omega - \frac{\pi}{4}\right)\\
\Delta\underline{\widetilde{F}}_{\nu,1}(\tau) &= \frac{\nu-1}{\nu+1}\underline{\widetilde{F}}_{\nu,1} - \frac{h(\tau)}{2}T^{\nu}\cos\left(\nu\omega - \mu - \frac{\pi}{4}\right)\\
\Delta\underline{\widetilde{G}}_{\nu}(\tau) &= \underline{\widetilde{G}}_{\nu}(\tau) - \frac{g(\tau)}{2}T^{-\nu}\cos\left(\nu\omega + \frac{\pi}{4}\right)\\
\Delta\underline{\widetilde{G}}_{\nu,1}(\tau) &= \frac{\nu+1}{\nu-1}\underline{\widetilde{G}}_{\nu,1} - \frac{h(\tau)}{2}T^{-\nu}\cos\left(\nu\omega + \mu + \frac{\pi}{4}\right)
\end{aligned}\right\} \qquad (257)$$

respectively. It must be noted again that the orders of $\widetilde{G}_{\nu}^{(1)}(\tau)$ and $\widetilde{G}_{\nu,1}^{(1)}(\tau)$ are the same as those of $\Delta\underline{\widetilde{F}}_{\nu}(\tau)$ and $\Delta\underline{\widetilde{F}}_{\nu,1}(\tau)$, respectively, because of the way they are defined in (257). For the same reason as stated in section 15, this again cannot jeopardize the basic assumption of convergence of the series.

Part Ⅳ
Criteria for the Upper Critical Mach Number

18. Limiting Line and the Breakdown of Isentropic Flow

The solutions constructed in the previous sections are known to be regular in the hodograph plane except at a few singular points. It is also known that for the limiting case of infinite sonic speed, or $c_0 \to \infty$, the solution will give the desired flow pattern in the physical plane. When the sonic speed is finite or when the Mach number of the free stream is different from zero, there is no guarantee as to the behavior of the solution in the physical plane except the probable continuity of the flow pattern with respect to the free-stream Mach number. It is found that such continuity in the flow pattern actually exists up to a certain Mach number. In other words, the pattern of the compressible flow is only slightly different from that of the incompressible flow up to a certain Mach number at which the so-called limiting lines appear. At the limiting line, the acceleration of the flow is infinite and the flow is reversed. It was shown by Tollmein[13] and Tsien[2] that, without considering viscosity, the flow cannot be continued across the limiting lines, and a forbidden region is created in the space where no fluid can enter. In other words, continuity of flow pattern exists up to a critical Mach number beyond which no isentropic flow is possible with the imposed physical boundary conditions.

The breakdown of isentropic flow, or the compressibility burble, can be effected in two ways. First of all, the acceleration in the neighborhood of the limiting line is very large. Thus each one of the following factors gives appreciable alterations in the dynamic relations:

(a) Viscous stress due to ordinary viscosity of the fluid[22].

(b) Stress due to expansion or compression of the fluid or viscous stress due to the second viscosity coefficient (reference 23, pp. 351 and 358).

(c) Small but appreciable relaxation time required for the vibrational modes of the molecules to reach equilibrium state[24].

(d) Heat conduction from fluid element to fluid element Secondly, the isentropic flow also can break down through the appearance of shock waves. The breakdown of isentropic flow is associated with the introduction of vorticity to the flow. Thus the flow becomes rotational with part of the mechanical energy of the fluid converted into heat energy. All these factors tend to increase the entropy of the fluid and finally to increase the drag of the body. Thus the critical Mach number so defined is of great physical importance to the aerodynamic characteristics of the body concerned.

Of course, the isentropic flow might break down due to the instability of flow fluid with the final appearance of shock waves. Furthermore, the action of boundary layer and possible condensation of one component of the fluid① on the flow might lead also to the premature

① The phenomenon of condensation shocks due to water vapor in the air flow around an airfoil was first brought to the attention of the authors by Kate Liepmann, who observed them in wind-tunnel experiments.

destruction of the isentropic flow. On the other hand, shock waves can appear only in supersonic flow; thus, if the speed of the fluid is everywhere subsonic, there is no danger of the compressibility burble. Hence, the free-stream Mach number for the first appearance of sonic speed in the field is called the "lower critical Mach number"; while the free-stream Mach number for the first appearance of limiting lines is called the "upper critical Mach number." (See reference 2). The latter is always higher than the former, due to the fact that limiting lines appear only in supersonic flow. The actual critical Mach number for the compressibility burble must lie between these two limits and depends, among other parameters, upon the Reynolds number of the flow.

19. The Condition for the Limiting Line

At the limiting hodograph, or the hodograph of the limiting line, it was shown (references 1, 2, 12, and 13) that

$$\frac{\partial(x, y)}{\partial(u, v)} \equiv -\left(\frac{\rho_0}{\rho_q}\right)^2\left[\psi_q^2-\left(\frac{1}{c^2}-\frac{1}{q^2}\right)\psi_\theta^2\right]=0 \tag{258}$$

Since the factor before the term ψ_θ^2 is positive for supersonic regions only, $c<q$, where $\rho\neq 0$, the limiting line can appear only when the local speed exceeds that of sound. It should be noted that the vanishing of the Jacobian is the condition for the failure of the hodograph method, as the transformation (9) and (10) would no longer be one-to-one and continuous. Thus, the appearance of the limiting lines is then the physical counterpart of the singularity of the transformation.

As $\psi(\tau,\theta)$ is known, equation (258) defines two lines in the hodograph plane:

$$2\tau\left[\frac{1-\tau}{\alpha^2\tau-1}\right]^{\frac{1}{2}}\psi_\tau-\psi_\theta=0 \tag{259}$$

$$2\tau\left[\frac{1-\tau}{\alpha^2\tau-1}\right]^{\frac{1}{2}}\psi_\tau+\psi_\theta=0, \quad \tau\geqslant\frac{1}{2\beta+1} \tag{260}$$

Geometrically, this expresses the fact that the streamline $\psi(q, \theta)=$ constant and a characteristic curve belonging to either family has a common tangent[1]. The problem can then be formulated based on this property: the necessary and sufficient condition for the existence of a limiting line is that there exists a solution between the two simultaneous equations

$$2\tau\left[\frac{1-\tau}{\alpha^2\tau-1}\right]^{\frac{1}{2}}\psi_\tau-\psi_\theta=0 \tag{261}$$

$$\psi=0 \tag{262}$$

or

$$2\tau\left[\frac{1-\tau}{\alpha^2\tau-1}\right]^{\frac{1}{2}}\psi_\tau+\psi_\theta=0 \tag{263}$$

$$\psi=0 \tag{264}$$

where $\psi(\tau,\theta)$ is a definite branch associated with the largest possible τ for a given boundary and a free-stream Mach number. The zero streamline is chosen, as it generally gives the highest velocity and is the place for the earliest appearance of the limiting line.

Generally, these equations may not possess a solution for a known function $\psi(\tau,\theta)$ when the parameter M_1 is assigned. This means that there will be a system of boundaries corresponding to a sequence of values of M_1, for which the limiting line does not occur. The first Mach number for which equations (261) and (262) have a solution will be defined as the upper critical Mach number and the corresponding boundary as the critical boundary.

The actual solution of the equation is, in general, difficult owing to the fact that $\psi(\tau,\theta)$ is, in most cases, represented by an infinite series. However, if the streamlines are determined in the hodograph plane for the calculation of the shape of the body, a simple graphical test of whether there is a point of tangency between the zero streamline and the characteristic can be easily made. On the other hand, if the form (214) and (215), for instance, is used, an approximate analytic solution can be obtained without involving much labor.

20. The Approximate Determination of the Upper Critical Mach Number

As can be seen from section 15, the importance of $\psi_2^{(0)}(\tau,\theta)$ relative to $\psi_1(\tau,\theta)$ will decrease as τ recedes from the critical circle $\tau=\dfrac{1}{2\beta+1}$ toward the supersonic region. For the first appearance of the limiting line, τ is almost always high, especially when the boundary is a slender closed body. Let this be the case; then $\psi_2^{(0)}(\tau,\theta)$ can be neglected in comparison with $\psi_1(\tau,\theta)$ and a great simplification is possible. The zero streamline then can be represented approximately by

$$\psi(\tau,\theta)\equiv\psi_1(\tau,\theta)=0$$

Furthermore, a simple reduction shows that the two pairs of equations, (261), (262) and (263), (264) reduce respectively to

$$\Phi'_0(\eta)+\Psi'_0(\eta)=0 \tag{265}$$

$$\Phi_0(\xi)+\Psi_0(\xi)=\Phi_0(\eta)-\Psi_0(\eta) \tag{266}$$

or

$$\Phi'_0(\xi)+\Psi'_0(\xi)=0 \tag{267}$$

$$\Phi_0(\eta)-\Psi_0(\eta)=\Phi_0(\xi)+\Psi_0(\xi) \tag{268}$$

where ξ and η are the characteristic parameters defined in equation (218). This reduction is made possible by the fact that $f(\tau)$ never vanishes in the interval $\dfrac{1}{2\beta+1}<\tau<1$.

Whenever the stream function ψ_0 and the potential function φ_0 of the incompressible flow are given, the functions Ψ_0 and Φ_0 can be easily obtained by substituting λU for q according to equation (216). Then, since λ decreases with an increase in the free-stream Mach number M, as shown in table 14 and figure 12, the upper critical Mach number will be given by the largest value of λ that gives a solution either of equations (265) and (266) or equations (267) and (268). An analytical solution can be made, as the functions Ψ_0 and Φ_0 are quite simple.

There is, however, an interesting direct geometrical interpretation of these sets of equations in the physical plane of the incompressible flow as shown by figure 13. According to equations (216), the functions Ψ_0 and Φ_0 are the stream function ψ_0 and the potential function φ_0 at the constant value of the speed λU. Since $\lambda > 1$, for the body shown in figure 13, the constant speed λU curve C_λ forms a loop symmetrical with respect to the y-axis. The variables are really the angle of inclination of the incompressible velocity vector. Along the constant speed curve C_λ from the point S_2 to P, the angle of inclination of the velocity vector is monotonically decreasing. Therefore, the parameter of the angle of inclination can be replaced by the distances along the curve C_λ. Let equation (267) be satisfied at the point $S = S_2$; then

$$\Phi'_0(S_2) = -\psi'_0(S_2) \qquad (269)$$

This means that, at the point $S = S_2$, the rate of change of the potential function φ_0 along C_λ is equal to the negative of the rate of change of the stream function ψ_0. Since potential lines and streamlines in incompressible flow form an infinitesimal square mesh, this condition requires that the angle between the tangent to the curve C_λ at $S = S_2$ be 45°, as shown in figure 13. This is easily seen by remembering that from S_2 to P, the value of the stream function increases while the value of the potential function decreases, because of the indicated flow direction. Thus the point S_2 can be easily determined by this graphical condition. Equation (268) can then be written as

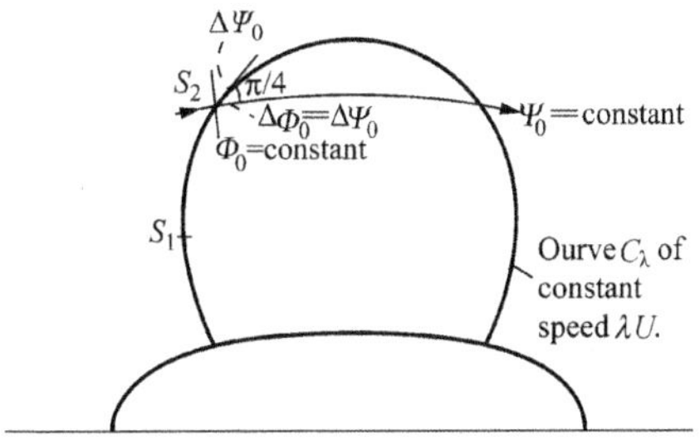

Fig. 13 Geometrical condition of eq. (269)*

$$\Phi_0(s) - \Psi_0(S) = \Phi_0(S_2) + \Psi_0(S_2) \qquad (270)$$

If this condition is satisfied at a point S_1, then the condition for a limiting line is completely satisfied. A similar graphical interpretation for the equations (265) and (266) can be worked out for the side of the constant speed curve lying to the right of the y-axis. From these considerations, it is clear that the upper critical Mach number is the lowest free-stream Mach number which gives a constant speed C_λ containing two points, S_1 and S_2, defined by equations (269) and (270).

* (20.5) in the original paper has been corrected as (269). — Noted by editor.

Part V
Application — Elliptic Cylinders

21. Preliminary Discussions

This part of the report is devoted to the application of the general method, developed in part III, to the study of the flow of a compressible fluid around an elliptic cylinder. According to sections 8 and 9, if a solution were constructed about the stagnation point, the continuation of this solution would require that conditions (102) and (103) and, hence, (106) and (107) be satisfied. These equations involve two sets of hypergeometric functions with parameters m and $m+1/2$, as well as their derivatives. To shorten the lengthy calculations, in view of the limited amount of time available, the following approximate procedure was adopted.

Given the domain $\underline{D}$, the solution valid in the annulus region, rather than that about the stagnation point, was first constructed. The constants which determine the Laurent expansion of the solution, B_n^* and C_n^*, for example, are now assigned and, consequently, the set of hypergeometric functions with integral parameters is not immediately required. The difficulty, however, is the question of whether it is possible to continue the solution within the circle of convergence. This continuation may not be possible owing to the stringent continuity conditions given by equations (102) and (103), and to the requirement that the function must be regular within the circle $q=U$.

This, however, does not offer a serious objection from the practical point of view. In the first place, the summed function ($\psi_1(q, \theta)$, for instance) actually holds even within the circle of convergence $q<U$, and the correction function $\psi_2(q, \theta)$, is generally small compared with $\psi_1(q, \theta)$ due to the close asymptotic approximation of the hypergeometric functions in the elliptic domain. In other words, although the solution within the circle of convergence strictly represents a different flow, numerically it approximates very closely that defined in the annulus region. In the second place, since this region $q<U$ is relatively unimportant in the case of mixed flow, where τ_1 is very much less than $\frac{1}{2\beta+1}$ — that is, for free-stream Mach number considerably less than unity — the inaccuracy of the solution is limited to a small region in the hodograph plane. Furthermore, the most interesting phenomena of such a flow, such as the appearance of limiting lines, always take place in the annulus region. Therefore, this modified procedure, although unsatisfactory from the general view point, is an expedient capable of yielding an interesting result and furnishing a test of the practicability of the proposed solution.

The situation also may be considered from another angle. The procedure used in this section can be derived by replacing the functions $\xi_\nu(\tau)$ and $\xi_{-\nu}(\tau)$ with the approximate values given in equation (201) in the expressions for the coefficients involved in the solution within the annulus region, that is, (106) and (107). Thus the procedure may be regarded as an appropriate method of approximation. The error introduced is generally negligible if $\tau_1 \ll \frac{1}{2\beta+1}$. This is indicated by the fact that the correction function $\psi_2(q,\tau)$, for instance, is very small in comparison with $\psi_1(q, \theta)$ when $q \leqslant U$.

Another simplification is made by using the elementary integral $q^{-2\nu}\underline{F}_{-\nu}(\tau)$ instead of $q^{-2\nu}\underline{G}_\nu(\tau)$ in the continued solution, as, in this case, $\underline{F}_{-\nu}(\tau)$ is a well-defined function. In doing so, the asymptotic behavior of the second solution remains unchanged because the first term in $\underline{G}_\nu(\tau)$ is always small in comparison with the second.

If, however, all the required hypergeometric functions are computed, there is no difficulty in carrying out the exact method developed in part Ⅲ of the report for any accurate study of two-dimensional flow. For this reason, the expressions for the hydrodynamic functions derived for both the exact and approximate procedures for the problem at hand are given.

In the numerical example, detailed calculations are made for the flow of air about a cylindrical body derived from the incompressible flow about an elliptic section with a ratio of the minor and major axes equal to 0.6. The calculations were carried out for two different free-stream Mach numbers, 0.6 and 0.7.

22. The Functions $z_0(w)$, $W_0(w)$ and $\Lambda_0(w)$

The irrotational flow of an incompressible fluid about an elliptic cylinder with the velocity at infinity parallel to the major axis is represented by the complex potential $W_0(z_0)$:

$$W_0(z_0) = \zeta + \frac{1}{\zeta} \tag{271}$$

with

$$z_0 = \zeta + \frac{\varepsilon^2}{\zeta} \tag{272}$$

For convenience in practical calculation, all the physical quantities z_0, q, and ρ, will be normalized consistently throughout the present part. The major and minor axes of the section are respectively $1+\varepsilon^2$ and $1-\varepsilon^2$, where $\varepsilon < 1$; $q=1$ at infinity and $\rho = 1$ when $q=0$. This will automatically render the hydrodynamic functions dimensionless and the constants U and ρ_0 will be eliminated from the formulas in the succeeding sections.

By differentiating (271) with respect to z_0, the dimensionless complex velocity of the flow is

$$w = \frac{\zeta^2 - 1}{\zeta^2 - \varepsilon^2}$$

Thus

$$\zeta = -\left[\frac{1-\varepsilon^2 w}{1-w}\right]^{1/2}, \quad |1-\varepsilon^2 w| \neq 0 \tag{273}$$

This function is two-valued with two branch points at $w = 1$ and $w = \varepsilon^{-2}$. In order to make $z_0(w)$ a single-valued function of w, the expression (273) is supposed to be the principal value so that $|\arg(1-w)| < \pi$ and $1 < |w| < \varepsilon^{-2}$. The condition $|\varepsilon^2 w| < 1$ must be satisfied, for $w = \varepsilon^{-2}$ corresponds to $\zeta = 0$, which is another singularity. With the principal value so defined, if the negative sign in (273) is taken, then the domain $\underline{D}$ corresponds to the half plane $Rl\zeta \leqslant 0$ and $|\zeta| \geqslant 1$. On the other hand, since the transformation (272) is one-to-one when $|\zeta| \geqslant 1$, then the domain D, which is $Rlz_0 \leqslant 0$ with the region inside the section excluded, corresponds uniquely to $\underline{D}$.

Consequently, the inverse mapping function $z_0(w)$ is

$$z_0(w) = -\left\{\left[\frac{1-\varepsilon^2 w}{1-w}\right]^{1/2} + \varepsilon^2\left[\frac{1-w}{1-\varepsilon^2 w}\right]^{1/2}\right\} \tag{274}$$

which will be single-valued, provided a cut is introduced to join the branch points in such a way that the argument of $(1-w)$ is restricted to $-\pi < \arg(1-w) < \pi$ and $|\varepsilon^2 w| < 1$. On separating into real and imaginary parts, it is found that as $0 \leqslant \theta < 2\pi$

$$x_0(q, \theta) = -\frac{1}{2^{\frac{1}{2}}}[\{I(q, \theta) + J(q, \theta)\}^{1/2} + \varepsilon^2\{I_\varepsilon(q, \theta) + J^{-1}(q, \theta)\}^{1/2}] \tag{275}$$

$$y_0(q, \theta) = \frac{1}{2^{\frac{1}{2}}}[\{-I(q, \theta) + J(q, \theta)\}^{1/2} - \varepsilon^2\{-I_\varepsilon(q, \theta) + J^{-1}(q, \theta)\}^{1/2}] \tag{276}$$

with $w = qe^{-i\theta}$, where the functions $I(q, \theta)$, $I_\varepsilon(q, \theta)$, and $J(q, \theta)$ stand for:

$$I(q, \theta) = \frac{1-(1+\varepsilon^2)q\cos\theta + \varepsilon^2 q^2}{1-2q\cos\theta + q^2} \tag{277}$$

$$I_\varepsilon(q, \theta) = \frac{1-(1+\varepsilon^2)q\cos\theta + \varepsilon^2 q^2}{1-2\varepsilon^2 q\cos\theta + \varepsilon^4 q^2} \tag{278}$$

$$J(q, \theta) = \left[\frac{1-2\varepsilon^2 q\cos\theta + \varepsilon^4 q^2}{1-2q\cos\theta + q^2}\right]^{1/2} \tag{279}$$

On the other hand, substituting equation (273) in equation (271), the function $W_0(z_0)$ is carried over into D; namely

$$W_0(w) = -\left\{\left[\frac{1-\varepsilon^2 w}{1-w}\right]^{\frac{1}{2}} + \left[\frac{1-w}{1-\varepsilon^2 w}\right]^{\frac{1}{2}}\right\} \tag{280}$$

Now $W_0(w) = \varphi_0(q, \theta) + i\psi_0(q, \theta)$, and similarly

$$\varphi_0(q, \theta) = \frac{1}{2^{\frac{1}{2}}}[\{I(q, \theta) + J(q, \theta)\}^{\frac{1}{2}} + \{I_\varepsilon(q, \theta) + J^{-1}(q, \theta)\}^{\frac{1}{2}}] \tag{281}$$

$$\psi_0(q, \theta) = \frac{1}{2^{\frac{1}{2}}}[\{-I(q, \theta) + J(q, \theta)\}^{\frac{1}{2}} - \{-I_\varepsilon(q, \theta) + J^{-1}(q, \theta)\}^{\frac{1}{2}}] \tag{282}$$

By integrating $z_0(w)$, according to (87), the transformed potential function $\Lambda_0(w)$, aside from a constant, takes the form:

$$\Lambda_0(w) = 2(1-w)^{\frac{1}{2}}(1-\varepsilon^2 w)^{\frac{1}{2}} \tag{283}$$

The principal value of this function is again defined by restricting the argument of $(1-w)$ to $-\pi < \arg(1-w) < \pi$ and $|w| < \varepsilon^{-2}$. Within this domain D, the real and imaginary parts are:

$$\chi_0(q, \theta) = 2^{\frac{1}{2}}[K(q, \theta) + L(q, \theta)]^{\frac{1}{2}} \tag{284}$$

$$0 \leqslant \theta < 2\pi$$

$$\sigma_0(q, \theta) = -2^{\frac{1}{2}}[-K(q, \theta) + L(q, \theta)]^{\frac{1}{2}} \tag{285}$$

as $\Lambda_0(w) = \chi_0(q, \theta) - i\sigma_0(q, \theta)$, where the functions $K(q, \theta)$ and $L(q, \theta)$ are defined by:

$$K(q, \theta) = 1 - (1+\varepsilon^2)q\cos\theta + \varepsilon^2 q^2 \cos 2\theta \tag{286}$$

$$L(q, \theta) = [1 - 2q\cos\theta + q^2]^{\frac{1}{2}}[1 - 2\varepsilon^2 q\cos\theta + \varepsilon^4 q^2]^{\frac{1}{2}} \tag{287}$$

23. Expansions of $W_0(w)$ and $\Lambda_0(w)$

The function $W_0(w)$ defined in (280) is single-valued and regular everywhere in $|w| < 1$ and, hence, possesses the following expansion:

$$W_0(w) = -\sum_{n=0}^{\infty} A_n w^n, \quad |w| < 1 \tag{288}$$

where the coefficients A_n are real and given by

$$A_n = 2S_n^{(i)} - (1+\varepsilon^2)S_{n-1}^{(i)} \quad n \geqslant 1 \tag{289}$$

$$A_0 = 2S_0^{(i)} = 2$$

with

$$S_n^{(i)}(\varepsilon^2) = \frac{1}{\pi}\sum_{m=0}^{n} \frac{\Gamma\left(n-m+\frac{1}{2}\right)\Gamma\left(m+\frac{1}{2}\right)}{\Gamma(n-m+1)\Gamma(m+1)}\varepsilon^{2m}$$

However, in the region outside $|w| < 1$ the function $W_0(w)$ is doubled-valued; and when a cut is put between the branch points $w = 1$ and $w = \varepsilon^{-2}$, the principal value is discontinuous along the positive axis of reals within the annulus region. To obtain the desired expansion, the function is written in the following form

$$W_0(w) = \frac{1}{w^{\frac{1}{2}}} \frac{2-(1+\varepsilon^2)w}{(1-w^{-1})^{\frac{1}{2}}(1-\varepsilon^2 w)^{\frac{1}{2}}} \tag{290}$$

Now $(1-w^{-1})^{-\frac{1}{2}}(1-\varepsilon^2 w)^{-\frac{1}{2}}$ is single-valued and continuous within the annulus region; its Laurent expansion is

$$(1-w^{-1})^{-\frac{1}{2}}(1-\varepsilon^2 w)^{-\frac{1}{2}} = S_0^{(0)} + \sum_{n=1}^{\infty} S_n^{(0)}[\varepsilon^{2n}w^n + w^{-n}], \quad 1 < |w| < \varepsilon^{-2} \tag{291}$$

where

$$S_n^{(0)}(\varepsilon^2) = \frac{1}{\pi}\sum_{m=0}^{\infty}\frac{\Gamma\left(n+m+\frac{1}{2}\right)\Gamma\left(m+\frac{1}{2}\right)}{\Gamma(n+m+1)\Gamma(m+1)}\varepsilon^{2m} \tag{292}$$

Substituting $(1-w)^{-\frac{1}{2}}(1-\varepsilon^2 w)^{-\frac{1}{2}}$ from (291) in (290), the expansion for $W_0(w)$ in the annulus region is

$$W_0(w) = \mathrm{i}\sum_{n=0}^{\infty}[B_n\varepsilon^{2n}w^{\nu} + C_n w^{-\nu}], \quad 1 < |w| < \varepsilon^{-2} \tag{293}$$

when the constants B_n, C_n and the exponent ν are defined by:

$$\left.\begin{aligned} B_n &= 2\varepsilon^2 S_{n+1}^{(0)} - (1+\varepsilon^2)S_n^{(0)} \\ C_n &= 2S_n^{(0)} - (1+\varepsilon^2)S_{n+1}^{(0)} \\ \nu &= n+\frac{1}{2} \end{aligned}\right\} \tag{294}$$

Similarly, the transformed potential function $\Lambda_0(w)$ can be expanded and is:

$$\Lambda_0(w) = 2\sum_{n=0}^{\infty}\widetilde{A}_n w^n, \quad |w| < 1 \tag{295}$$

when the constants $\widetilde{A}_n$ are

$$\left.\begin{aligned} \widetilde{A}_n &= S_n^{(i)} - (1+\varepsilon^2)S_{n-1}^{(i)} + \varepsilon^2 S_{n-2}^{(i)} \\ \widetilde{A}_1 &= -\frac{1}{2}(1+\varepsilon^2), \quad \widetilde{A}_0 = 1 \end{aligned}\right\} \tag{296}$$

and $S_n^{(i)}$ is given in (289).

On the other hand, in the annulus region the expansion is

$$\Lambda_0(w) = -2\mathrm{i}\sum_{n=0}^{\infty}[\widetilde{B}_n\varepsilon^{2n}w^n + \widetilde{C}_n w^{-n}], \quad 1 < |w| < \varepsilon^{-2} \tag{297}$$

with the constants $\widetilde{B}_n$ and $\widetilde{C}_n$ defined as

$$\left.\begin{aligned} \widetilde{B}_n &= S_{n-1}^{(0)} - (1+\varepsilon^2)S_n^{(0)} + \varepsilon^2 S_{n+1}^{(0)}, \quad n \geqslant 1 \\ \widetilde{B}_0 &= 2\varepsilon^2 S_1^{(0)} - (1+\varepsilon^2)S_0^{(0)} \\ \widetilde{C}_n &= S_n^{(0)} - (1+\varepsilon^2)S_{n+1}^{(0)} + \varepsilon^2 S_{n+2}^{(0)} \end{aligned}\right\} \tag{298}$$

where $S_n^{(0)}(\varepsilon^2)$ is defined in (292).

24. The Stream Function $\psi(q, \theta)$

The relationship between the domain $\underline{D}$ and D is thus fully established and the functions corresponding to such domains are also given. From the general scheme developed in sections 8 and 9 the solutions for the similar motion of a compressible fluid can be constructed. First of all, the stream function $\psi(q, \theta)$ governing the subsonic flow is the sum of $\psi_1(q, \theta)$ and

$\psi_2(q, \theta)$. According to (207), (208), and (209), for $0 \leqslant \tau < \frac{1}{2\beta+1}$

$$\psi_1(q, \theta) = \frac{1}{2^{\frac{1}{2}}} \frac{f(\tau)}{f(\tau_1)} \{[-I(tq, \theta) + J(tq, \theta)]^{\frac{1}{2}} - [-I_\varepsilon(tq, \theta) + J^{-1}(tq, \theta)]^{\frac{1}{2}}\} \tag{299}$$

where the functions $I(tq, \theta)$, $I_\varepsilon(tq, \theta)$, and $J(tq, \theta)$ are obtained from I, I_ε, and J in (272) to (279) by replacing q by tq, t being defined in (195). For $q < 1$, the function $\psi_2(q, \theta)$ is

$$\psi_2^{(i)}(q, \theta) = \sum_{n=0}^{\infty} A_n G_n(\tau) q^n \sin n\theta \tag{300}$$

where A_n is defined in (289) and $G_n(\tau)$ in (199). For $q > 1$ and in subsonic region the function $\psi_2^{(0)}(q, \theta)$:

$$\psi_2^{(0)}(q, \theta) = \sum_{n=0}^{\infty} [G_\nu^{(1)}(\tau)\varepsilon^{2n} q^\nu + G_\nu^{(2)}(\tau) q^{-\nu}] \cos \nu\theta, \quad 0 \leqslant \theta < 2\pi \tag{301}$$

where $G_\nu^{(1)}(\tau)$ and $G_\nu^{(2)}(\tau)$ are defined by (205) with the constants B_n and C_n defined in (294).

When the motion becomes supersonic, the continuation of $\psi_1(q, \theta)$ defined in (299) gives

$$\begin{aligned}\psi_1(q, \theta) = \frac{1}{8} \frac{f(\tau)}{f(\tau_1)} \{ & [-I(\lambda, \xi) + J(\lambda, \xi)]^{\frac{1}{2}} - [-I_\varepsilon(\lambda, \xi) + J^{-1}(\lambda, \xi)]^{\frac{1}{2}} + \\ & [-I(\lambda, \eta) + J(\lambda, \eta)]^{\frac{1}{2}} - [-I_\varepsilon(\lambda, \eta) + J^{-1}(\lambda, \eta)]^{\frac{1}{2}} - \\ & [+I(\lambda, \xi) + J(\lambda, \xi)]^{\frac{1}{2}} - [-I_\varepsilon(\lambda, \xi) + J^{-1}(\lambda, \xi)]^{\frac{1}{2}} \pm \\ & [I(\lambda, \eta) + J(\lambda, \eta)]^{\frac{1}{2}} \pm [I_\varepsilon(\lambda, \eta) + J^{-1}(\lambda, \eta)]^{\frac{1}{2}}\} \quad \frac{1}{2\beta+1} < \tau < 1\end{aligned} \tag{302}$$

according to (214). Here ξ and η are the characteristic parameters defined in (218). The upper sign in the last two terms corresponds to $\eta > 0$ while the lower one, to $\eta < 0$. The accompanying function $\psi_2^{(0)}(q, \theta)$ is

$$\psi_2^{(0)}(q, \theta) = \sum_{n=0}^{\infty} [G_\nu^{(1)}(\tau)\varepsilon^{2n} q^\nu + G_\nu^{(2)}(\tau) q^{-\nu}] \cos \nu\theta, \quad \frac{1}{2\beta+1} < \tau < 1 \tag{303}$$

Here the functions $G_\nu^{(1)}(\tau)$ and $G_\nu^{(2)}(\tau)$ are defined by (205) in conjunction with (217) in such a way that (303) will be the continuation of (301). It also should be noticed that the variable is restricted to $\frac{1}{2\beta+1} < \tau < 1$ instead of $\frac{1}{2\beta+1} < \tau < \tau_1 \varepsilon^{-4}$, as $\tau_1 \varepsilon^{-4}$ is generally greater than unity, which is impossible for the actual gas.

It should be remembered that $\psi_2^{(i)}(q, \theta)$ is always negligible compared with $\psi_1(q, \theta)$ within and on the unit circle $q = 1$ when τ_1 is small in comparison with $\frac{1}{2\beta+1}$; $\psi(q, \theta)$ can be approximately represented by $\psi_1(q, \theta)$ alone throughout the interior of the unit circle. As a consequence, the calculation can be simplified considerably by constructing first a solution for the annulus region by using $\underline{F}_{-\nu}(\tau)$ instead of $\underline{G}_\nu(\tau)$ and making an approximate connection across the unit circle. In that event, the stream function will be reduced to

$$\psi(q,\theta) \cong \psi_1(q,\theta) \tag{304}$$

when $0 \leqslant q \leqslant 1$; here $\psi_1(q,\theta)$ is again defined in (299). On the other hand, when $\tau_1 < \tau < \frac{1}{2\beta+1}$,

$$\psi(q,\theta) = \psi_1(q,\theta) + \psi_2^{(0)}(q,\theta) \tag{305}$$

where the function $\psi_2^{(0)}(q,\theta)$ which is small on $q=1$ is given by

$$\psi_2^{(0)}(q,\theta) = \sum_{n=0}^{\infty}[B_n G_\nu(\tau)\varepsilon^{2n}q^\nu + C_n G_{-\nu}(\tau)q^{-\nu}]\cos\nu\theta \tag{306}$$

Here the functions $G_\nu(\tau)$ and $G_{-\nu}(\tau)$ can be shown to be

$$G_\nu(\tau) = \underline{F}_\nu^{(r)}(\tau) - \frac{f(\tau)}{f(\tau_1)}t^\nu, \quad G_{-\nu}(\tau) = \underline{F}_{-\nu}^{(r)}(\tau) - \frac{f(\tau)}{f(\tau_1)}t^{-\nu} \tag{307}$$

and the coefficients B_n and C_n are defined in (294).

The continuation of $\psi_1(q,\theta)$ is naturally the expression given in (302) while that of (306) differs only in the definition of $G_\nu(\tau)$ and $G_{-\nu}(\tau)$ which are

$$\left.\begin{aligned} G_\nu(\tau) &= \underline{F}_\nu^{(r)}(\tau) - \frac{1}{2}\frac{f(\tau)}{f(\tau_1)}t^\nu\cos\left(\nu\omega - \frac{\pi}{4}\right) \\ G_{-\nu}(\tau) &= \underline{F}_{-\nu}^{(r)}(\tau) - \frac{1}{2}\frac{f(\tau)}{f(\tau_1)}t^{-\nu}\cos\left(\nu\omega + \frac{\pi}{4}\right) \end{aligned}\right\} \quad \frac{1}{2\beta+1} < \tau < 1 \tag{308}$$

25. The Coordinate Functions $x(q,\theta)$ and $y(q,\theta)$

With the functions $z_0(w)$ and $\Lambda_0(w)$ defined in sections 22 and 23, the corresponding functions $\Lambda(w;\tau)$ and consequently $z(w;\tau)$ for the motion of a compressible fluid can be constructed. The coordinate functions derived from $\Lambda(w;\tau)$ are given respectively by the sum of two functions $x_1(q,\theta)$ and $y_1(q,\theta)$ which, according to equations (237) to (238), are

$$\begin{aligned} x_1(q,\theta) = &-\frac{t(\tau)}{2^{\frac{1}{2}}}\frac{g(\tau)}{f(\tau_1)}\{[I(tq,\theta)+J(tq,\theta)]^{\frac{1}{2}} + \varepsilon^2[I_\varepsilon(tq,\theta)+J^{-1}(tq,\theta)]^{\frac{1}{2}}\} - \\ &\frac{\beta\tau}{2}\frac{h(\tau)}{f(\tau_1)}\frac{t\sin 2\theta}{\sigma_0(tq,\theta)}\{-1+4\varepsilon^2 tq\cos\theta - \varepsilon^2 + J(tq,\theta) + \varepsilon^2 J^{-1}(tq,\theta)\} \end{aligned} \tag{309}$$

$$\begin{aligned} y_1(q,\theta) = &\frac{t(\tau)}{2^{\frac{1}{2}}}\frac{g(\tau)}{f(\tau_1)}\{[-I(tq,\theta)+J(tq,\theta)]^{\frac{1}{2}} - \varepsilon^2[-I_\varepsilon(tq,\theta)+J^{-1}(tq,\theta)]^{\frac{1}{2}}\} - \\ &\beta\tau\frac{h(\tau)}{f(\tau_1)}\frac{t\sin^2\theta}{\sigma_0(tq,\theta)}\{-1+4\varepsilon^2 tq\cos\theta - \varepsilon^2 + J(tq,\theta) + \varepsilon^2 J^{-1}(tq,\theta)\} \end{aligned} \tag{310}$$

where $\sigma_0(tq,\theta)$ is obtained from $\sigma_0(q,\theta)$ in (285) by replacing q by tq. The functions

$x_2^{(i)}(q, \theta)$ and $y_2^{(i)}(q, \theta)$, according to equations (239) and (240), are

$$x_2^{(i)}(q, \theta) = 2\sum_{n=1}^{\infty} n\widetilde{A}_n\widetilde{G}_n(\tau)q^{n-1}\cos(n-1)\theta - 2\beta\tau\sum_{n=1}^{\infty} n\widetilde{A}_n\widetilde{G}_{n,1}(\tau)q^{n-1}\cos n\theta\cos\theta \tag{311}$$

$$q<1$$

$$y_2^{(i)}(q, \theta) = -2\sum_{n=1}^{\infty} n\widetilde{A}_n\widetilde{G}_n(\tau)q^{n-1}\sin(n-1)\theta - 2\beta\tau\sum_{n=1}^{\infty} n\widetilde{A}_n\widetilde{G}_{n,1}(\tau)q^{n-1}\cos n\theta\sin\theta \tag{312}$$

Here the functions $\widetilde{G}_n(\tau)$ and $\widetilde{G}_{n,1}(\tau)$ are defined by equations (241) and (242) and the constants $\widetilde{A}_n$ by (296).

The same functions valid in the annulus region are again represented by the sums $x_1(q, \theta) + x_2^{(0)}(q, \theta)$ and $y_1(q, \theta) + y_2^{(0)}(q, \theta)$, where $x_1(q, \theta)$ and $y_1(q, \theta)$ are defined by equations (309) and (310), respectively. When $\tau_1 \leqslant \tau < \dfrac{1}{2\beta+1}$, $x_2^{(0)}(q, \theta)$, and $y_2^{(0)}(q, \theta)$ are

$$x_2^{(0)}(q, \theta) = -2\sum_{n=0}^{\infty}\nu[\widetilde{G}_\nu^{(1)}(\tau)\varepsilon^{2n}q^{\nu-1}\sin(\nu-1)\theta + \widetilde{G}_\nu^{(2)}(\tau)q^{-\nu-1}\sin(\nu+1)\theta] +$$
$$2\beta\tau\sum_{n=0}^{\infty}\nu[\widetilde{G}_{\nu,1}^{(1)}(\tau)\varepsilon^{2n}q^{\nu-1} + \widetilde{G}_{\nu,1}^{(2)}(\tau)q^{-\nu-1}]\sin\nu\theta\cos\theta \tag{313}$$

$$y_2^{(0)}(q, \theta) = -2\sum_{n=0}^{\infty}\nu[\widetilde{G}_\nu^{(1)}(\tau)\varepsilon^{2n}q^{\nu-1}\cos(\nu-1)\theta + \widetilde{G}_\nu^{(2)}(\tau)q^{-\nu-1}\cos(\nu+1)\theta] +$$
$$2\beta\tau\sum_{n=0}^{\infty}\nu[\widetilde{G}_{\nu,1}^{(1)}(\tau)\varepsilon^{2n}q^{\nu-1} + \widetilde{G}_{\nu,1}^{(2)}(\tau)q^{-\nu-1}]\sin\nu\theta\sin\theta \tag{314}$$

The functions $\widetilde{G}_\nu^{(a)}(\tau)$, $\widetilde{G}_{\nu,1}^{(a)}(\tau)$ are defined in equations (250) and (251) together with equations (252) with the constants $\widetilde{B}_n$ and $\widetilde{C}_n$ defined in equations (298).

On the other hand, when $\dfrac{1}{2\beta+1} < \tau < 1$, the continued expressions of $x_1(q, \theta)$, $y_1(q, \theta)$ across the critical circle $\tau = \dfrac{1}{2\beta+1}$ are, according to equations (253) and (254),

$$x_1(q, \theta) = -\frac{t(\tau)}{2^{5/2}}\frac{g(\tau)}{f(\tau_1)}\{[I(\lambda,\xi) + J(\lambda,\xi)]^{\frac{1}{2}} + \varepsilon^2[I_\varepsilon(\lambda,\xi) + J^{-1}(\lambda,\xi)]^{\frac{1}{2}} +$$
$$[I(\lambda,\eta) + J(\lambda,\eta)]^{\frac{1}{2}} + \varepsilon^2[I_\varepsilon(\lambda,\eta) + J^{-1}(\lambda,\eta)]^{\frac{1}{2}}\}\cos\left(\frac{\pi}{4} - \omega\right) -$$
$$\frac{t(\tau)}{2^{5/2}}\frac{g(\tau)}{f(\tau_1)}\{[-I(\lambda,\xi) + J(\lambda,\xi)]^{\frac{1}{2}} - \varepsilon^2[-I_\varepsilon(\lambda,\xi) + J^{-1}(\lambda,\xi)]^{\frac{1}{2}} -$$
$$[-I(\lambda,\eta) + J(\lambda,\eta)]^{\frac{1}{2}} + \varepsilon^2[-I_\varepsilon(\lambda,\eta) + J^{-1}(\lambda,\eta)]^{\frac{1}{2}}\}\sin\left(\frac{\pi}{4} - \omega\right) -$$
$$\frac{\beta\tau h(\tau)\cos\theta}{4qf(\tau_1)}\left\{\left[\frac{\lambda\sin\xi}{\sigma_0(\lambda,\xi)} \times (-1 + 4\varepsilon^2\lambda\cos\xi - \varepsilon^2 + J(\lambda,\xi) + \varepsilon^2 J^{-1}(\lambda,\xi)) +\right.\right.$$
$$\left.\frac{\lambda\sin\eta}{\sigma_0(\lambda,\eta)}(-1 + 4\varepsilon^2\lambda\cos\eta - \varepsilon^2 + J(\lambda,\eta) + \varepsilon^2 J^{-1}(\lambda,\eta))\right]\cos\left(\mu + \frac{\pi}{4}\right) -$$
$$\left[\frac{\lambda\sin\xi}{\chi_0(\lambda,\xi)} \times (1 - 4\varepsilon^2\lambda\cos\xi + \varepsilon^2 + J(\lambda,\xi) + \varepsilon^2 J^{-1}(\lambda,\xi)) -\right.$$

$$\frac{\lambda\sin\eta}{\chi_0(\lambda,\eta)}(1-4\varepsilon^2\lambda\cos\eta+\varepsilon^2+J(\lambda,\eta)+\varepsilon^2J^{-1}(\lambda,\eta))\Big]\sin\left(\mu+\frac{\pi}{4}\right)\Big\} \tag{315}$$

$$y_1(q,\theta)=\frac{t(\tau)}{2^{5/2}}\frac{g(\tau)}{f(\tau_1)}\{[-I(\lambda,\xi)+J(\lambda,\xi)]^{\frac{1}{2}}-\varepsilon^2[-I_\varepsilon(\lambda,\xi)+J^{-1}(\lambda,\xi)]^{\frac{1}{2}}+$$

$$[-I(\lambda,\eta)+J(\lambda,\eta)]^{\frac{1}{2}}-\varepsilon^2[-I_\varepsilon(\lambda,\eta)+J^{-1}(\lambda,\eta)]^{\frac{1}{2}}\}\cos\left(\frac{\pi}{4}-\omega\right)+$$

$$\frac{t(\tau)}{2^{5/2}}\frac{g(\tau)}{f(\tau_1)}\{-[I(\lambda,\xi)+J(\lambda,\xi)]^{\frac{1}{2}}-\varepsilon^2[I_\varepsilon(\lambda,\xi)+J^{-1}(\lambda,\xi)]^{\frac{1}{2}}+$$

$$[I(\lambda,\eta)+J(\lambda,\eta)]^{\frac{1}{2}}+\varepsilon^2[I_\varepsilon(\lambda,\eta)+J^{-1}(\lambda,\eta)]^{\frac{1}{2}}\}\times\sin\left(\frac{\pi}{4}-\omega\right)-$$

$$\frac{\beta\tau h(\tau)\sin\theta}{4qf(\tau_1)}\Big\{\Big[\frac{\lambda\sin\xi}{\sigma_0(\lambda,\xi)}\times(-1+4\varepsilon^2\lambda\cos\xi-\varepsilon^2+J(\lambda,\xi)+\varepsilon^2J^{-1}(\lambda,\xi))+$$

$$\frac{\lambda\sin\eta}{\sigma_0(\lambda,\eta)}(-1+4\varepsilon^2\lambda\cos\eta-\varepsilon^2+J(\lambda,\eta)+\varepsilon^2J^{-1}(\lambda,\eta))\Big]\cos\left(\mu+\frac{\pi}{4}\right)-$$

$$\Big[\frac{\lambda\sin\xi}{\chi_0(\lambda,\xi)}(1-4\varepsilon^2\lambda\cos\xi+\varepsilon^2+J(\lambda,\xi)+\varepsilon^2J^{-1}(\lambda,\xi))-$$

$$\frac{\lambda\sin\eta}{\chi_0(\lambda,\eta)}(1-4\varepsilon^2\lambda\cos\eta+\varepsilon^2+J(\lambda,\eta)+\varepsilon^2J^{-1}(\lambda,\eta))\Big]\sin\left(\mu+\frac{\pi}{4}\right)\Big\} \tag{316}$$

While $x_2(q,\theta)$ and $y_2(q,\theta)$ remain to be defined by equations (313) and (314) except the functions $\widetilde{G}_\nu^{(a)}(\tau)$ and $\widetilde{G}_{\nu,1}^{(a)}(\tau)$ are replaced by those given in equations (250), (251) together with equations (257).

By the same argument as that used for the stream function, the practical calculation of $x(q,\theta)$ and $y(q,\theta)$ can be simplified by neglecting $x_2^{(i)}(q,\theta)$ and $y_2^{(i)}(q,\theta)$ when $q<1$; namely,

$$x(q,\theta)\cong x_1(q,\theta) \tag{317}$$

$$y(q,\theta)\cong y_1(q,\theta),\quad 0\leqslant q\leqslant 1 \tag{318}$$

where $x_1(q,\theta)$ and $y_1(q,\theta)$ are defined in equations (309) and (310); and in the annulus region

$$x(q,\theta)=x_1(q,\theta)+x_2^{(0)}(q,\theta) \tag{319}$$

$$y(q,\theta)=y_1(q,\theta)+y_2^{(0)}(q,\theta)\qquad \tau_1<\tau<1 \tag{320}$$

Here $x_1(q,\theta)$ and $y_1(q,\theta)$ are either given by equations (309), (310) or (315), (316). The terms $x_2^{(0)}(q,\theta)$ and $y_2^{(0)}(q,\theta)$, on the other hand, become

$$x_2^{(0)}(q,\theta)=-2\sum_{n=0}^{\infty}\nu[\widetilde{B}_n\widetilde{G}_\nu(\tau)\varepsilon^{2n}q^{\nu-1}\sin(\nu-1)\theta+\widetilde{C}_n\widetilde{G}_{-\nu}(\tau)q^{-\nu-1}\sin(\nu+1)\theta]+$$

$$2\beta\tau\sum_{n=0}^{\infty}\nu[\widetilde{B}_n\widetilde{G}_{\nu,1}(\tau)\varepsilon^{2n}\times q^{\nu-1}+\widetilde{C}_n\widetilde{G}_{\nu,1}(\tau)q^{-\nu-1}]\sin\nu\theta\cos\theta \tag{321}$$

$$y_2^{(0)}(q,\theta)=-2\sum_{n=0}^{\infty}\nu[\widetilde{B}_n\widetilde{G}_\nu(\tau)\varepsilon^{2n}q^{\nu-1}\cos(\nu-1)\theta-\widetilde{C}_n\widetilde{G}_{-\nu}(\tau)q^{-\nu-1}\cos(\nu+1)\theta]+$$

$$2\beta\tau\sum_{n=0}^{\infty}\nu[\widetilde{B}_n\widetilde{G}_{\nu,1}(\tau)\varepsilon^{2n}\times q^{\nu-1}+\widetilde{C}_n\widetilde{G}_{-\nu,1}q^{-\nu-1}]\sin\nu\theta\sin\theta \tag{322}$$

For $\tau_1 \leqslant \tau < \dfrac{1}{2\beta+1}$, the functions $\widetilde{G}_\nu(\tau)$, $\widetilde{G}_{\nu,1}(\tau)$ are defined as

$$\widetilde{G}_\nu(\tau) = \underline{\widetilde{F}}_\nu^{(r)}(\tau) - \frac{g(\tau)}{f(\tau_1)}t^\nu, \quad \widetilde{G}_\nu(\tau) = \underline{\widetilde{F}}_{-\nu}^{(r)}(\tau) - \frac{g(\tau)}{f(\tau_1)}t^{-\nu} \tag{323}$$

$$\left.\begin{aligned} \widetilde{G}_{\nu,1}(\tau) &= \frac{\nu-1}{\nu+1}\underline{\widetilde{F}}_{\nu,1}^{(r)}(\tau) - \frac{h(\tau)}{f(\tau_1)}t^\nu \\ \widetilde{G}_{-\nu,1}(\tau) &= \frac{\nu+1}{\nu-1}\underline{\widetilde{F}}_{-\nu,1}^{(r)}(\tau) - \frac{h(\tau)}{f(\tau_1)}t^{-\nu} \end{aligned}\right\} \tag{324}$$

For $\dfrac{1}{2\beta+1} < \tau < 1$

$$\left.\begin{aligned} \widetilde{G}_\nu(\tau) &= \underline{\widetilde{F}}_\nu^{(r)}(\tau) - \frac{1}{2}\frac{g(\tau)}{f(\tau_1)}t^\nu\cos\left(\nu\omega - \frac{\pi}{4}\right) \\ \widetilde{G}_{-\nu}(\tau) &= \underline{\widetilde{F}}_{-\nu}^{(r)}(\tau) - \frac{1}{2}\frac{g(\tau)}{f(\tau_1)}t^{-\nu}\cos\left(\nu\omega + \frac{\pi}{4}\right) \end{aligned}\right\} \tag{325}$$

$$\left.\begin{aligned} \widetilde{G}_{\nu,1}(\tau) &= \frac{\nu-1}{\nu+1}\underline{\widetilde{F}}_{\nu,1}^{(r)}(\tau) - \frac{1}{2}\frac{h(\tau)}{f(\tau_1)}t^\nu\cos\left(\nu\omega - \mu - \frac{\pi}{4}\right) \\ \widetilde{G}_{-\nu,1}(\tau) &= \frac{\nu+1}{\nu-1}\underline{\widetilde{F}}_{-\nu,1}^{(r)}(\tau) - \frac{h(\tau)}{2f(\tau_1)}t^{-\nu}\cos\left(\nu\omega + \mu + \frac{\pi}{4}\right) \end{aligned}\right\} \tag{326}$$

Conclusions

As an example, the motion of air past a cylindrical body was considered by taking $\varepsilon = \dfrac{1}{2}$. The flow patterns in the τ, θ-plane for two free-stream Mach numbers $M_1 = 0.6$ and 0.7 have been calculated and were given in Figs. 14 and 15. It should be noticed that there is considerable

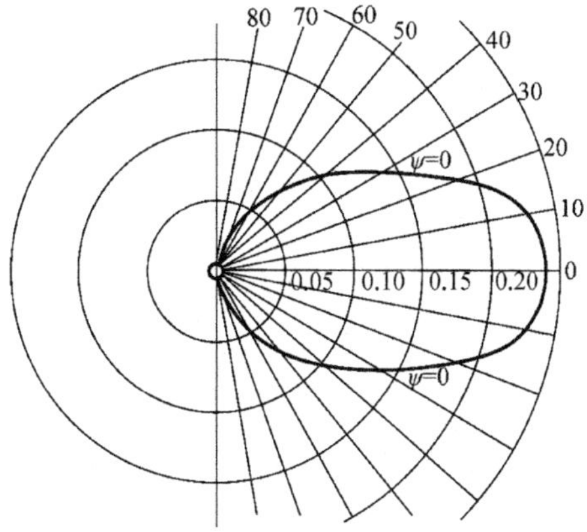

Fig. 14 The compressible flow in τ, θ-plane
$\varepsilon = 1/2$; $M_1 = 0.6$

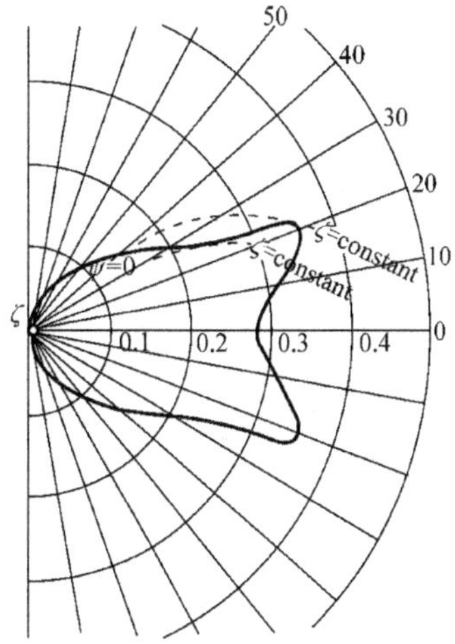

Fig. 15 The compressible flow in τ
θ-plane. $\varepsilon = 1/2$; $M = 0.7$

distortion in the shape of the bodies in the compressible flow from that in the incompressible flow. If the compressible flow around a given body is desired, a series of computations should be made with various geometric parameters ε, so that the desired body shape at a definite Mach number M_1 could be picked out.

These computations definitely demonstrate the practicability of the proposed method. They also show that, in the case of two-dimensional motion of a compressible fluid, the mixed subsonic and supersonic flows exist within the field of an irrotational isentropic flow about a suitable body, and the transition from one to the other is continuous and reversible. Furthermore, the breakdown of the irrotational isentropic flow depends solely upon the occurrence of limiting lines which, in turn, is determined by the condition at infinity or the shape of the boundary, while the magnitude of the local speed attained is immaterial. In the case of $M_1 = 0.6$, the irrotational supersonic flow continues to exist up to the local Mach number $M = 1.25$; whereas for $M_1 = 0.7$ it breaks down as soon as $M = 1.22$ is reached. The singular behavior of the streamline is marked by the point of tangency of $\psi = 0$ with a characteristic at $M = 1.22$.

The calculation of the flow pattern in the physical plane is yet to be completed. When this is done, the pressure distribution can be compared with that over the same body of the incompressible flow.

Guggenheim Aeronautical Laboratory,
California Institute of Technology,
Pasadena, Calif., April 17, 1945.

References

[1] von Kármán Th. Compressibility Effects in Aerodynamics. Jour. Aero. Sci., 1941, 8(9): 337 – 356.

[2] Tsien Hsue-Shen: The "Limiting Line" in Mixed Subsonic and Supersonic Flow of Compressible Fluids. NACA TN No. 961, 1944.

[3] Hantzsche W, Wendt H. Der Kompressiblitätseinfluss für dünne wenig gekrümmte profile bei Unterschallgeschwindigkeit. Z. f. a. M. M., 1942, 22: 72 – 86.

[4] Kaplan Carl. The Flow of a Compressibic Fluid past a Curved Surface. NACA ARR No. 3K02, 1943. The Flow of a Compressible Fluid past a Circular profile. NACA ARR. No. L4G15, 1944.

[5] Molenbroek P. Über einige Bewegungen eines Gases mit Annahme eines Geschwindigke tspotentials. Archiv der Math. und phys. (Grunert-Hoppe), 1890, 9(2): 157.

[6] Chaplygin S. Gas Jets. NACA TM No. 1063, 1944.

[7] Bergman Stefan A. On Two-Dimensional Flows of Compressible Fluids. NACA TN No. 972, 1945.

[8] Garrick I E, Kaplan Carl. On the Flow of a Compressible Fluid by the Hodograph Method. I-Unification and Extension of Present-Day Results. NACA ACR No. L4C24, 1944. (Classification changed to "Restricted," oct. 11, 1944)

[9] Tsien Hsuo-Shen. Two-Dimensional Subsonic Flow of Compressible Fluids. Jour. Aero. Sci., 1939, 6 (10): 399 – 407.

[10] Bers Lipman. On a Method of Constructing Two-Dimensionl Subsonic Compressible Flows around Closed Profiles. NACA TN No. 966, 1945.

[11] Temple G, Yarwood J. The Approximate Solution of the Hodograph Equations for Compressible Flow. Rep. No. S. M. E. 3201, R. A. E., 1942.

[12] Ringleb, Friedrich Über die Differentialgleichungen einer adiabatischen Gasströmung und den Strömungs-Stoss. Deutsche Math., 1940, 5(5): 337 - 384.

[13] Tollmien W. Grenzlinien adiabatischer potential-Strömungen. Z. f. a. M. M. 1941, 21: 140 - 152.

[14] Gauss C F. Werke, 3: 207 - 229.

[15] Tannery J. Proprigtes des intégrales des éguations différentielles linéaires. Ann. Sci. de e'Ecole Normale Superieure. 1875, 4(2): 113 - 182.

[16] Goursat E. Sur l'équations différentielle linéaire qui admet pour intégrale hypergéométrique. Ann. Sci. de e Ecole Normale Superieure, 1881, 10(2): 3 - 142.

[17] Ince E L. Ordinary Differential Equations. Longmans. Green and Co., 1927: 227.

[18] Horn F. Über eine lineare Differentialgleichung zweiter Ordnung mit einem willkürlich en parameter. Math. Ann., 1899, 52: 271 - 292.

[19] Jeffreys H. on Certain Approximate. Solutions of Linear Differential Equations of the Second Order. Proc. London Math. Soc., 23(2): 428 - 436.

[20] Langer R E. On the Asymptotic Solutions of Ordinary Differential Equations. Trans. Am. Math. Soc., 1931, 33: 23 - 64; 1932, 34: 447 - 480.

[21] Coburn N. The Kármán-Tsien Pressure-Volume Relation in the Two-Dimensional Supersonic Flow of Compressible Fluids. Quarterterly of Appl. Math., 1945.

[22] Knaft Hans, Dibble Charles G. Some Two Dimen-Sional Adiabatic Compressible Flow Patterns. Jour. Aero.: Sci., 1944, 11: 283 - 298.

[23] Busemann A. Gasdynamik. Handbuch der Experimentalphysik. Akademische Verlags Leipzig, Sec. 1, pt. Ⅳ, 1931: 341 - 460.

[24] Kantrowitz A. Effects of Heat-Capacity Lag in Gas Dynamics. NACA ARR No. 4A22, 1944.

Appendix A
Proof of Theorem (52)

To facilitate the discussion, equation (71) is first written in the form:

$$X_1(\xi_\nu^{(1)}) \equiv \xi'^{(1)}_\nu(\tau) + \frac{\nu}{2\tau}\zeta_1\zeta_2 = 0$$

where

$$\zeta_1(\tau) = \xi_\nu^{(1)}(\tau) + \frac{\beta\tau}{\nu(1-\tau)} + \gamma_\nu(\tau)$$

$$\zeta_2(\tau) = \xi_\nu^{(1)}(\tau) + \frac{\beta\tau}{\nu(1-\tau)} - \gamma_\nu(\tau)$$

and

$$\gamma_\nu(\tau) = \left\{\frac{1-(2\beta+1)\tau}{1-\tau} + \frac{\beta^2\tau^2}{\nu^2(1-\tau)^2} + \frac{4(1-\tau)^{2\beta}}{\tau^\nu R_\nu^2(\tau)}\right\}^{\frac{1}{2}}$$

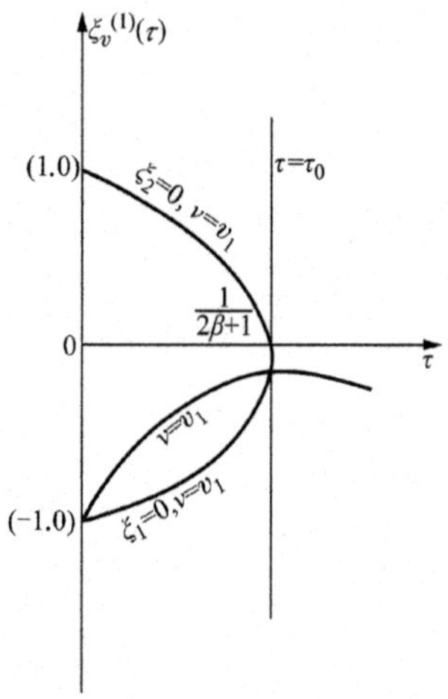

Fig. 16 The behavior of the integral-curve; $\xi_\nu^{(1)}(\tau)$

when ν is large, the character of the functions ζ_1 and ζ_2 can be easily studied in the τ, $\xi_\nu^{(1)}$-plane (fig. 16) by neglecting the third term under the radical sign. This can be justified in the following manner: Consider the case when ν is positive and large but not an integer. In the interval $0 \leqslant \tau \leqslant \frac{1}{2\beta+1}$, $\underline{F}_\nu(\tau) \ll F_\nu$ because $F_\nu(\tau) \sim \tau^{-\nu}\underline{F}_{-\nu}(\tau)$ by equations (35) and (55). Then $\tau^{\frac{\nu}{2}}R_\nu(\tau) \sim \tau^{-\frac{\nu}{2}}T_1^{-\nu}$. Therefore, $\tau^{\frac{\nu}{2}}R_\nu(\tau) \gg 1$ when ν is large. But both $\underline{F}_\nu(\tau)$ and $F_\nu(\tau)$ are continuous with respect to ν; so the foregoing result applies equally to the case of integral ν. Hence, the third term in the radical for $\gamma_\nu(\tau)$ can be neglected for large ν.

Owing to the manner in which γ_ν is defined, corresponding to each ν there is a line $\tau = \tau_0 > \frac{1}{2\beta+1}$ such that $\gamma_\nu^2(\tau) \gtreqless 0$ when $\tau \lesseqgtr \tau_0$. As a consequence ζ_1 and ζ_2 are real or complex conjugate according as $\tau \lesseqgtr \tau_0$. In $0 \leqslant \tau \leqslant \tau_0$, $\zeta_1 = 0$ and $\zeta_2 = 0$ will give two 1-parameter families radiating from $(0, -1)$ and $(0, 1)$, respectively, and joining together at a point where $\gamma_\nu^2 = 0$. If $0 \leqslant \tau \leqslant \tau_0$, the product $\zeta_1\zeta_2$ may be negative or positive according to whether the point lies to the left or the right of the curve $\zeta_1 = 0$ and $\zeta_2 = 0$. On the other hand, if $\tau > \tau_0$, $\zeta_1\zeta_2$ is always positive.

Now $\xi'^{(1)}_\nu(0) = \beta$, while the initial slope of $\zeta_1 = 0$ is $\beta\left(1 - \frac{1}{\nu}\right)$, the integral curve must lie above $\zeta_1 = 0$, and below $\zeta_2 = 0$. If it were not, the integral curve would cross the curve

$\zeta_1 = 0$, $\zeta_2 = 0$, where $\xi'^{(1)}_\nu(\tau) = 0$, and $\xi'^{(1)}_\nu(\tau)$ would be negative somewhere in $0 \leqslant \tau \leqslant \frac{1}{2\beta+1}$. This is not possible, for $\xi^{(1)}_\nu \sim \xi_{-\nu}$ by an argument similar to that used for determining the magnitude of $\tau^{\frac{\nu}{2}} R_\nu(\tau)$ and according to (55) $-\sqrt{\frac{1-(2\beta+1)\tau}{1-\tau}} > \xi_{-\nu} > -(1-\tau)^\beta$ in $0 \leqslant \tau \leqslant \frac{1}{2\beta+1}$. Hence $\xi'^{(1)}_\nu > 0$ in $0 \leqslant \tau \leqslant \frac{1}{2\beta+1}$ and $\xi^{(1)}_\nu$ continues to increase until it intersects with $\zeta_1 = 0$. After it crosses the curve $\zeta_1 = 0$, $\xi'^{(1)}_\nu < 0$ and never changes sign as $\zeta_1 \zeta_2 > 0$ in $\tau_0 < \tau < 1$. Consequently, $\xi^{(1)}_\nu(\tau)$ is monotonic and decreasing in the interval $\tau_0 < \tau < 1$. When ν is sufficiently large, τ_0 will approach very rapidly to $\frac{1}{2\beta+1}$ and $\tau_0 = \frac{1}{2\beta+1}$ when ν becomes infinite.

Proof of theorem (52). — Form the following identity:

$$X_1(\eta^{(1)}_\nu) \equiv (\eta'^{(1)}_\nu - \xi'^{(1)}_\nu) + (\eta^{(1)}_\nu - \xi^{(1)}_\nu)\left[\frac{\beta}{1-\tau} + \frac{\nu}{2\tau}(\eta^{(1)}_\nu + \xi^{(1)}_\nu)\right] + \frac{\nu}{\tau}\xi_0^2(1-\tau)^{2\beta} e^{-\nu\int_{\tau_0}^{\tau}(\eta^{(1)}_\nu + \xi^{(2)}_\nu)\frac{d\tau}{\tau}} \sinh \nu \int_{\tau_0}^{\tau} (\eta^{(1)}_\nu - \xi^{(1)}_\nu)\frac{d\tau}{\tau} \gtreqless 0 \quad \text{(A1)}$$

It can be shown that the differential expression possesses an integration factor

$$(\eta^{(1)}_\nu - \xi^{(1)}_\nu)\tau^{2\nu}(1-\tau)^{-2\beta} R_\nu^2 S_\nu^2 \quad \text{(A2)}$$

where

$$R_\nu = R_\nu(\tau_0)\exp\left\{\nu\int_{\tau_0}^{\tau}(\xi^{(1)}_\nu - 1)\frac{d\tau}{2\tau}\right\}$$

$$S_\nu = S_\nu(\tau_0)\exp\left\{\nu\int_{\tau_0}^{\tau}(\eta^{(1)}_\nu - 1)\frac{d\tau}{2\tau}\right\}$$

It will be noticed that the sign of (A2) is determined by the first factor $(\eta^{(1)}_\nu - \xi^{(1)}_\nu)$ only. On multiplying (A1) by (A2) and integrating the resulting total differential from τ_0 to τ, with a suitably chosen initial value $\eta^{(1)}_\nu(\tau_0) = \xi^{(1)}_\nu(\tau_0)$ it is found that

$$\frac{1}{2}(\eta^{(1)}_\nu - \xi^{(1)}_\nu)^2\tau^{2\nu}(1-\tau)^{-2\beta}R_\nu^2 S_\nu^2 + \xi_0^2 R_\nu^2(\tau_0)S_\nu^2(\tau_0) \times \left[\cos h\nu\int_{\tau_0}^{\tau}(\eta^{(1)}_\nu - \xi^{(1)}_\nu)\frac{d\tau}{\tau} - 1\right] > 0$$

which is positive if and only if $\eta^{(1)}_\nu - \xi^{(1)}_\nu \gtreqless 0$ everywhere in $\tau_0 < \tau < 1$. Since both $\xi^{(1)}_\nu$ and $\xi^{(1)}_\nu$ are continuous and monotonic, the condition is both necessary and sufficient. Furthermore, it should be noticed that the condition $\eta^{(1)}_\nu(\tau_0) = \xi^{(1)}_\nu(\tau_0)$ is purely a convenience. If $\eta^{(1)}_\nu(\tau_0) \neq \xi^{(1)}_\nu(\tau_0)$, the validity of the theorem is not in the least impaired.

Appendix B
Proof of Theorem (88)

Consider the first series: Multiplying throughout the inequality (58), namely,

$$\xi_n(\tau) > \sqrt{\frac{1-(2\beta+1)\tau}{1-\tau}}, \quad 0<\tau<\frac{1}{2\beta+1}$$

by $\frac{u}{2\tau}$ and integrating both sides from τ to τ_1 shows that

$$\underline{F}_n^{(r)}(\tau) < t_1^n(\tau)$$

where $t_1(\tau) = \dfrac{T_1(\tau)}{T_1(\tau_1)} \geqslant 1$. Then it follows that

$$|A_n \underline{F}_n^{(r)}(\tau) w^n| < |A_n(t_1 w)^n|$$

Now $\sum_{n=0}^{\infty} |A_n(t_1 w)^n|$ converges when $|t_1 w| < U$ due to equation (88). By Weirstrass's theorem, the series (92) is uniformly and absolutely convergent if $|t_1 w| = t_1 q < U$. Now $t_1(\tau_1) = 1$; thus $t_1 q$ is equal to U when $q = U$ and $= 1$. The term $t_1 q$ is zero if $q = 0$ and remains positive for $0 < q < U$. By the definition of $T_1(\tau)$ given by equation (56), it can be easily shown that

$$\frac{\mathrm{d}}{\mathrm{d}q} t_1 q > 0$$

for $0 < \tau < \tau_1$. Thus $t_1 q$ increases monotonically from zero to U in the interval $0 \leqslant \tau \leqslant \tau_1$. Therefore, the series (92) is uniformly and absolutely convergent in any closed domain in $|w| < U$.

Similarly, the convergence of the series (94) can be established.

Appendix C
Proof of Theorem (98)

It is observed that the following identities exist among the constants involved in (98) and (99):

$$B_n U^{\nu} = -\frac{1}{2\nu\pi}\sum_{m=0}^{\infty} A_m U^m \left(\frac{1}{m+\nu}+\frac{1}{m-\nu}\right)(m+\nu)$$

$$C_n U^{-\nu} = \frac{1}{2\nu\pi}\sum_{m=0}^{\infty} A_m U^m \left(\frac{1}{m+\nu}+\frac{1}{m-\nu}\right)(m-\nu)$$

Now, by the inequalities (58) and (59), the functions $\xi_\nu(\tau_1)$, $\xi_{-\nu}(\tau_1)$ can be bounded both above and below for all $\nu \neq 0$, when $0 \leqslant \tau \leqslant \frac{1}{2\beta+1}$. And if a smaller value of $\Delta(\underline{F}_\nu, F_\nu)$ is taken, it can be deduced that

$$|B_n^*| \leqslant M_1 \frac{|B_n|}{\underline{F}_\nu(\tau_1)}$$

$$|C_n^*| \leqslant M_2 \frac{|C_n|}{\underline{G}_\nu(\tau_1)}$$

where M_1 and M_2 are constants independent of n. On the other hand, from the inequality (58)

$$\xi_\nu(\tau) < (1-\tau)^{\beta}, \quad 0 \leqslant \tau \leqslant \frac{1}{2\beta+1}$$

it follows that

$$\frac{\underline{F}_\nu(\tau)}{\underline{F}_\nu(\tau_1)} < t_2^{\nu}(\tau), \quad \tau_1 \leqslant \tau \leqslant \frac{1}{2\beta+1}$$

Consequently, the first part of (101) can be dominated:

$$|B_n^* \underline{F}_\nu(\tau) w^{\nu}| < |B_n (t_2 w)^{\nu}|$$

where $t_2(\tau) = \frac{T_2(\tau)}{T_2(\tau_1)}$. The continuation of this inequality for $\tau > \frac{1}{2\beta+1}$ can be easily done by defining a new $t_2(\tau)$. By hypothesis, $\sum_{n=0}^{\infty} |B_n (t_2 w)^{\nu}|$ converges if $|t_2 w| < V$. Since $t_2(\tau) \leqslant t_2(\tau_1)$ for $\tau_1 \leqslant \tau < 1$, the inequality $|t_2 w| < V$ is uniformly bounded.

Similarly, it can be shown that

$$|C_n^* \underline{G}_\nu(\tau) w^{-\nu}| < |C_n (t_1 w)^{-\nu}|$$

But $\sum_{n=0}^{\infty} |C_n (t_1 w)^{-\nu}|$ converges if $|t_1 w| > U$. Since on $|w| = U$ $t_1(\tau_1) = 1$ and $\frac{d}{dq}\log|t_1 w| > 0$ when $0 < \tau < \frac{1}{2\beta+1}$ or $\frac{d}{dq}|t_1 w| = 0$ when $\frac{1}{2\beta+1} < \tau < 1$, the condition $|t_1 w| > U$ holds for all τ in $\tau_1 \leqslant \tau < 1$. Hence, by Weierstrass's theorem the series (101) converges uniformly and absolutely in $U+\delta \leqslant |w| \leqslant V-\delta$.

Tables of the Hypergeometric Functions

The values of the hypergeometric functions given in tables 1 to 5 are calculated from power series for $\gamma = 1.405$. The function $\tilde{\underline{F}}_{-\nu,1}(\tau)$ in table 6 is connected with $\tilde{\underline{F}}_{\nu}(\tau)$, $\tilde{\underline{F}}_{-\nu}(\tau)$, and $\tilde{\underline{F}}_{\nu,1}(\tau)$ through the following equation:

$$\frac{\beta(\nu+1)}{2(\nu-1)}\tau\tilde{\underline{F}}_{\nu}(\tau)\tilde{\underline{F}}_{-\nu,1}(\tau) = \tilde{\underline{F}}_{\nu}(\tau)\tilde{\underline{F}}_{-\nu}(\tau) - \frac{\beta(\nu-1)}{2(\nu+1)}\tau\tilde{\underline{F}}_{\nu,1}(\tau)\tilde{\underline{F}}_{-\nu}(\tau) - (1-\tau)^{-\beta}$$

This is simply the Wronskian of the two independent integrals of the hypergeometric equation and it holds everywhere except at the singularities $\tau = 0$ and $\tau = 1$. Tables 7 to 12 contain the corresponding approximate functions as indicated.

The numbers in these tables are expressed in terms of appropriate powers of 10. However, a notation was devised in which only the powers are given while the base "10" is omitted. Thus, $3.141\,59 \times 10^{m} = 3.141\,59, m$. Here m may be either a positive or negative integer, or zero. Unless indicated by the sign † on the heading, accidental errors were detected and eliminated by the difference method.

Table 1 — Corresponding Particular Integrals for the Solutions of Compressible Flow and Incompressible Flow

	Compressible	Incompressible
$\psi(q,\theta)$	$q^{\nu}\underline{F}_{\nu}(\tau)$ $\cos\nu\theta$ $\sin\nu\theta$	q^{ν} $\cos\nu\theta$ $\sin\nu\theta$
	$q^{-\nu}\underline{G}_{\nu}(\tau)$ $\cos\nu\theta$ $\sin\theta$	$q^{-\nu}$ $\cos\nu\theta$ $\sin\nu\theta$
	$\int(1-\tau)^{\beta}\frac{d\tau}{\tau}$	$\log q$
	θ	θ
$\chi(q,\theta)$	$q^{\nu}\tilde{\underline{F}}_{\nu}(\tau)$ $\cos\nu\theta$ $\sin\nu\theta$	q^{ν} $\cos\nu\theta$ $\sin\theta$
	$q^{-\nu}\tilde{\underline{G}}_{\nu}(\tau)$ $\cos\nu\theta$ $\sin\theta$	$q^{-\nu}$ $\cos\nu\theta$ $\sin\theta$
	$\int(1-\tau)^{-\beta}\frac{d\tau}{\tau}$	$\log q$
	θ	θ

The functions $\underline{F}_{\nu}(\tau)$, $q^{-2\nu}\underline{G}_{\nu}(\tau)$ and $\tilde{\underline{F}}_{\nu}(\tau)$, $q^{-2\nu}\tilde{\underline{G}}_{\nu}(\tau)$ are respectively the two independent integrals of equations (27) and (28).

Table 2

τ	$F_{1/2}(\tau)$	$F_{3/2}(\tau)$	$F_{5/2}(\tau)$	$F_{7/2}(\tau)$	$F_{9/2}(\tau)$	$F_{11/2}(\tau)$	$F_{13/2}(\tau)$	$F_{15/2}(\tau)$	$F_{17/2}(\tau)$	$F_{19/2}(\tau)$	$F_{21/2}(\tau)$
0	1.000 00, 0	1.000 00, 0	1.000 00, 0	1.000 00, 0	1.000 00, 0	1.000 00, 0	1.000 00, 0	1.000 00, 0	1.000 00, 0	1.000 00, 0	1.000 00, 0
.10	9.405 92, −1	8.267 48, −1	7.235 08, −1	6.317 26, −1	5.508 40, −1	4.798 86, −1	4.178 17, −1	3.636 14, −1	3.163 39, −1	2.751 37, −1	2.395 25, −1
.12	9.292 61, −1	7.948 94, −1	6.755 36, −1	5.721 09, −1	4.834 87, −1	4.080 15, −1	3.439 78, −1	2.897 78, −1	2.439 80, −1	2.053 32, −1	1.727 39, −1
.14	9.181 12, −1	7.639 45, −1	6.298 60, −1	5.166 93, −1	4.225 13, −1	3.447 61, −1	2.808 76, −1	2.285 61, −1	1.858 21, −1	1.509 67, −1	1.225 75, −1
.15	9.126 05, −1	7.488 07, −1	6.078 73, −1	4.904 98, −1	3.942 90, −1	3.161 20, −1	2.529 60, −1	2.021 24, −1	1.613 22, −1	1.286 41, −1	1.024 95, −1
.16	9.071 43, −1	7.338 92, −1	5.864 10, −1	4.652 82, −1	3.674 96, −1	2.893 41, −1	2.272 71, −1	1.781 96, −1	1.395 15, −1	1.091 10, −1	8.524 37, −2
.165	9.044 29, −1	7.265 16, −1	5.759 10, −1	4.530 33, −1	3.546 21, −1	2.766 21, −1	2.152 19, −1	1.671 09, −1	1.295 59, −1	1.003 03, −1	7.756 97, −2
.17	9.017 26, −1	7.191 96, −1	5.655 03, −1	4.410 18, −1	3.420 87, −1	2.643 33, −1	2.036 72, −1	1.565 81, −1	1.201 63, −1	9.208 55, −2	7.047 64, −2
.175	8.990 35, −1	7.119 30, −1	5.552 49, −1	4.292 35, −1	3.298 81, −1	2.524 68, −1	1.926 11, −1	1.465 85, −1	1.133 35, −1	8.442 75, −2	6.392 85, −2
.18	8.963 55, −1	7.047 19, −1	5.451 30, −1	4.176 82, −1	3.180 04, −1	2.410 12, −1	1.820 27, −1	1.371 02, −1	1.030 35, −1	7.729 69, −2	5.789 04, −2
.185	8.936 84, −1	6.975 63, −1	5.351 40, −1	4.063 56, −1	3.064 48, −1	2.299 56, −1	1.718 98, −1	1.281 11, −1	9.524 08, −2	7.066 36, −2	5.232 90, −2
.19	8.912 17, −1	6.904 61, −1	5.252 81, −1	3.952 53, −1	2.952 08, −1	2.192 88, −1	1.622 13, −1	1.195 92, −1	8.792 55, −2	6.449 91, −2	4.721 28, −2
.195	8.883 80, −1	6.834 11, −1	5.155 53, −1	3.843 70, −1	2.842 74, −1	2.090 03, −1	1.529 55, −1	1.115 23, −1	8.106 47, −2	5.964 99, −2	4.251 22, −2
.20	8.857 43, −1	6.764 17, −1	5.059 52, −1	3.737 07, −1	2.736 42, −1	1.990 83, −1	1.441 09, −1	1.038 90, −1	7.463 65, −2	5.346 93, −2	3.819 90, −2
.21	8.807 15, −1	6.625 89, −1	4.878 34, −1	3.530 20, −1	2.532 58, −1	1.803 15, −1	1.276 01, −1	8.984 54, −2	6.299 06, −2	4.400 49, −2	3.063 07, −2
.22	8.753 12, −1	6.489 74, −1	4.688 17, −1	3.331 72, −1	2.340 26, −1	1.629 11, −1	1.125 82, −1	7.732 25, −2	5.282 29, −2	3.592 19, −2	2.431 31, −2
.23	8.701 61, −1	6.355 71, −1	4.509 95, −1	3.141 41, −1	2.158 82, −1	1.468 39, −1	1.083 52, −1	6.618 95, −2	4.398 13, −2	2.905 44, −2	1.907 36, −2
.24	8.650 54, −1	6.223 79, −1	4.336 62, −1	2.959 01, −1	1.987 92, −1	1.318 93, −1	8.659 39, −2	5.632 45, −2	3.632 54, −2	2.325 24, −2	1.475 90, −2
.25	8.599 91, −1	6.093 96, −1	4.168 07, −1	2.784 36, −1	1.827 08, −1	1.181 45, −1	7.543 63, −2	4.761 40, −2	2.973 02, −2	1.837 80, −2	1.126 07, −2
.26	8.549 73, −1	5.966 18, −1	4.004 25, −1	2.617 21, −1	1.675 91, −1	1.054 80, −1	6.538 35, −2	3.995 19, −2	2.407 58, −2	1.431 92, −2	8.380 40, −3
.27	8.499 98, −1	5.840 48, −1	3.845 07, −1	2.457 34, −1	1.534 01, −1	9.384 18, −2	5.635 24, −2	3.323 99, −2	1.925 62, −2	1.095 83, −2	6.093 34, −3
.28	8.450 66, −1	5.716 82, −1	3.690 46, −1	2.304 57, −1	1.400 94, −1	8.316 00, −2	4.826 37, −2	2.738 64, −2	1.517 53, −2	8.201 53, −3	4.282 02, −3
.29	8.401 77, −1	5.595 19, −1	3.540 34, −1	2.158 69, −1	1.276 36, −1	7.338 11, −2	4.104 44, −2	2.230 67, −2	1.174 06, −2	5.962 81, −3	2.867 31, −3
.30	8.353 32, −1	5.475 57, −1	3.394 63, −1	2.019 51, −1	1.159 89, −1	6.444 93, −2	3.462 31, −2	1.792 23, −2	8.875 99, −3	4.165 79, −3	1.780 66, −3
.32	8.257 71, −1	5.242 32, −1	3.116 19, −1	1.760 35, −1	9.497 54, −2	4.891 40, −2	2.391 54, −2	1.095 49, −2	4.569 47, −3	1.634 91, −3	3.639 55, −4
.34	8.163 81, −1	5.016 92, −1	2.854 51, −1	1.525 58, −1	7.677 39, −2	3.615 99, −2	1.565 92, −2	5.970 58, −3	1.756 36, −3	1.617 65, −4	−3.446 49, −4
.36	8.071 62, −1	4.799 26, −1	2.608 99, −1	1.313 70, −1	6.111 84, −2	2.581 57, −2	9.430 50, −3	2.536 32, −3	3.658 67, −5	−5.940 75, −4	−6.151 56, −4
.38	7.981 11, −1	4.589 23, −1	2.379 08, −1	1.123 20, −1	4.775 56, −2	1.755 08, −2	4.858 85, −3	2.909 12, −4	−9.064 62, −4	−8.863 34, −4	−6.354 43, −4
.40	7.892 28, −1	4.386 70, −1	2.164 18, −1	9.526 80, −2	3.646 02, −2	1.106 41, −2	1.622 80, −3	−1.063 71, −3	−1.319 17, −3	−8.993 0?, −4	−5.311 85, −4
.42	7.805 12, −1	4.195 14, −1	1.963 69, −1	8.007 64, −2	2.700 13, −2	6.069 27, −3	−5.543 79, −4	−1.770 29, −3	−1.389 33, −3	−7.625 1?, −4	−3.820 33, −4
.44	7.719 62, −1	4.003 62, −1	1.777 06, −1	6.661 19, −2	1.911 78, −2	2.366 21, −3	−1.908 72, −3	−2.023 54, −3	−1.256 58, −3	−5.631 3?, −4	−2.346 60, −4
.46	7.635 77, −1	3.822 84, −1	1.603 68, −1	5.474 62, −2	1.279 36, −2	1.500 59, −4	−2.639 49, −3	−1.076 81, −3	−1.021 27, −3	−3.562 3?, −4	−1.127 18, −4
.48	7.553 55, −1	3.648 99, −1	1.443 04, −1	4.435 54, −2	7.674 04, −3	−2.125 46, −3	−2.902 84, −3	−1.748 15, −3	−7.525 60, −4	−1.733 3?, −4	−2.481 3?, −5
.50	7.472 96, −1	3.482 02, −1	1.294 54, −1	3.532 04, −2	3.652 70, −3	−4.823 62, −3	−2.864 66, −3	−1.425 91, −3	−4.951 52, −4	−2.923?, −5	+2.968 27, −5

Table 3

τ	$\underline{F}_{-1/2}(\tau)$	$\underline{F}_{-3/2}(\tau)$	$\underline{F}_{-5/2}(\tau)$	$\underline{F}_{-7/2}(\tau)$	$\underline{F}_{-9/2}(\tau)$	$\underline{F}_{-11/2}(\tau)$	$\underline{F}_{-13/2}(\tau)$	$\underline{F}_{-15/2}(\tau)$	$\underline{F}_{-17/2}(\tau)$	$\underline{F}_{-19/2}(\tau)$	$\underline{F}_{-21/2}(\tau)$
0	1.000 00, 0	1.000 00, 0	1.000 00, 0	1.000 00, 0	1.000 00, 0	1.000 00, 0	1.000 00, 0	1.000 00, 0	1.000 00, 0	1.000 00, 0	1.000 00, 0
.10	1.058 39, 0	1.120 55, 0	1.417 71, 0	1.726 08, 0	2.051 19, 0	2.403 70, 0	2.793 48, 0	3.230 17, 0	3.723 68, 0	4.284 71, 0	4.924 93, 0
.12	1.069 28, 0	1.131 21, 0	1.492 11, 0	1.908 84, 0	2.385 96, 0	2.938 35, 0	3.583 02, 0	4.339 40, 0	5.229 92, 0	6.280 99, 0	7.523 51, 0
.14	1.079 93, 0	1.138 14, 0	1.553 69, 0	2.077 68, 0	2.724 58, 0	3.525 56, 0	4.519 62, 0	5.754 83, 0	7.289 97, 0	9.197 94, 0	1.156 85, 1
.15	1.085 16, 0	1.140 30, 0	1.578 87, 0	2.152 67, 0	2.884 80, 0	3.819 70, 0	5.014 70, 0	6.542 53, 0	8.495 24, 0	1.098 96, 1	1.417 35, 1
.16	1.090 32, 0	1.141 63, 0	1.599 95, 0	2.219 20, 0	3.032 64, 0	4.100 61, 0	5.503 01, 0	7.344 28, 0	9.760 40, 0	1.292 87, 1	1.707 96, 1
.165	1.092 89, 0	1.142 01, 0	1.608 89, 0	2.248 83, 0	3.100 39, 0	4.232 37, 0	5.737 21, 0	7.737 04, 0	1.039 32, 1	1.391 81, 1	1.859 44, 1
.17	1.095 43, 0	1.142 19, 0	1.616 75, 0	2.275 77, 0	3.163 14, 0	4.356 28, 0	5.960 26, 0	8.115 76, 0	1.101 04, 1	1.489 61, 1	2.010 70, 1
.175	1.097 96, 0	1.142 18, 0	1.623 48, 0	2.299 84, 0	3.220 29, 0	4.470 61, 0	6.168 44, 0	8.472 96, 0	1.159 90, 1	1.583 67, 1	2.157 72, 1
.18	1.100 48, 0	1.141 99, 0	1.629 10, 0	2.320 87, 0	3.271 19, 0	4.573 66, 0	6.357 82, 0	8.800 50, 0	1.214 25, 1	1.671 16, 1	2.295 39, 1
.185	1.102 98, 0	1.141 62, 0	1.633 56, 0	2.338 75, 0	3.315 24, 0	4.663 69, 0	6.524 26, 0	9.089 55, 0	1.262 35, 1	1.748 78, 1	2.417 73, 1
.19	1.105 46, 0	1.141 08, 0	1.636 87, 0	2.353 25, 0	3.351 82, 0	4.738 92, 0	6.663 43, 0	9.330 66, 0	1.302 29, 1	1.812 89, 1	2.517 78, 1
.195	1.107 93, 0	1.140 36, 0	1.639 03, 0	2.364 30, 0	3.380 32, 0	4.797 56, 0	6.770 89, 0	9.513 83, 0	1.331 97, 1	1.859 06, 1	2.587 66, 1
.20	1.110 39, 0	1.139 49, 0	1.640 00, 0	2.371 72, 0	3.400 14, 0	4.837 84, 0	6.841 98, 0	9.622 82, 0	1.348 16, 1	1.883 07, 1	2.618 58, 1
.21	1.115 25, 0	1.137 24, 0	1.638 42, 0	2.375 25, 0	3.411 59, 0	4.856 25, 0	6.857 11, 0	9.609 26, 0	1.336 53, 1	1.844 50, 1	2.524 19, 1
.22	1.120 06, 0	1.134 39, 0	1.632 10, 0	2.362 93, 0	3.381 94, 0	4.780 45, 0	6.671 56, 0	9.158 10, 0	1.244 49, 1	1.655 04, 1	2.149 34, 1
.23	1.124 81, 0	1.130 95, 0	1.621 06, 0	2.334 11, 0	3.307 45, 0	4.597 64, 0	6.249 57, 0	8.262 38, 0	1.052 58, 1	1.270 29, 1	1.402 74, 1
.24	1.129 50, 0	1.126 95, 0	1.605 36, 0	2.288 16, 0	3.184 93, 0	4.296 34, 0	5.557 98, 0	6.761 10, 0	7.409 95, 0	6.465 95, 0	1.903 68, 0
.25	1.134 14, 0	1.122 44, 0	1.585 03, 0	2.224 77, 0	3.011 81, 0	3.866 85, 0	4.567 82, 0	4.604 97, 0	2.919 33, 0	−2.551 13, 0	−1.567 29, 1
.26	1.138 71, 0	1.117 43, 0	1.560 18, 0	2.143 73, 0	2.786 19, 0	3.301 52, 0	3.256 17, 0	1.733 52, 0	−3.084 69, 0	−1.465 18, 1	−3.931 95, 1
.27	1.143 23, 0	1.111 96, 0	1.530 92, 0	2.045 04, 0	2.506 93, 0	2.595 26, 0	1.607 07, 0	−1.892 68, 0	−1.069 08, 1	−3.000 24, 1	−6.931 67, 1
.28	1.147 70, 0	1.106 05, 0	1.497 38, 0	1.928 84, 0	2.173 66, 0	1.745 70, 0	−3.865 33, −1	−6.288 54, 0	−1.991 69, 1	−4.859 17, 1	−1.054 66, 2
.29	1.152 11, 0	1.099 73, 0	1.459 73, 0	1.795 45, 0	1.786 83, 0	8.534 85, −1	−3.821 93, 0	−1.143 93, 1	−3.069 82, 1	−7.018 48, 1	−1.470 11, 2
.30	1.156 46, 0	1.093 04, 0	1.418 12, 0	1.645 35, 0	1.347 69, 0	−3.775 64, −1	−5.385 66, 0	−1.729 70, 1	−4.287 51, 1	−9.427 83, 1	−1.925 79, 2
.32	1.165 00, 0	1.078 61, 0	1.323 81, 0	1.297 80, 0	3.215 87, −1	−3.024 35, 0	−1.158 68, 1	−3.076 18, 1	−7.025 22, 1	−1.465 95, 2	−2.863 12, 2
.34	1.173 33, 0	1.062 96, 0	1.216 09, 0	8.930 74, −1	−8.784 44, −1	−6.098 00, 0	−1.865 50, 1	−4.657 05, 1	−9.866 07, 1	−1.959 22, 2	−3.630 51, 2
.36	1.181 45, 0	1.046 30, 0	1.096 82, 0	4.402 08, −1	−2.214 29, 0	−9.451 45, 0	−2.605 88, 1	−6.005 62, 1	−1.231 74, 2	−2.287 98, 2	−3.826 23, 2
.38	1.189 35, 0	1.028 81, 0	9.679 76, −1	−5.002 66, −2	−3.637 55, 0	−1.289 41, 1	−3.321 51, 1	−7.212 57, 1	−1.378 46, 2	−2.299 03, 2	−3.153 71, 2
.40	1.197 06, 0	1.010 64, 0	8.316 23, −1	−5.654 87, −1	−5.091 75, 0	−1.620 38, 1	−3.912 43, 1	−7.968 83, 1	−1.360 28, 2	−1.856 09, 2	−1.328 8?, 2
.42	1.204 56, 0	9.919 68, −1	6.898 72, −1	−1.093 09, 0	−6.514 10, 0	−1.913 85, 1	−4.327 31, 1	−7.993 67, 1	−1.144 10, 2	−8.633 6?, 1	1.703 8?, 2
.44	1.211 87, 0	9.729 48, −1	5.448 47, −1	−1.619 23, 0	−7.843 57, 0	−2.145 84, 1	−4.481 51, 1	−7.172 73, 1	−6.873 9?, 1	+6.634??, 1	5.671 0?, 2
.46	1.218 99, 0	9.537 07, −1	3.986 39, −1	−2.130 27, 0	−9.014 12, 0	−2.294 47, 1	−4.312 90, 1	−5.395 05, 1	4.000 07, −1	2.595??, 2	9.941??, 2
.48	1.225 91, 0	9.344 08, −1	2.532 86, −1	−2.612 93, 0	−9.968 14, 0	−.234 099, 1	−3.792 01, 1	−2.671 17, 1	8.747 0?, 1	4.674 6?, 2	1.354 2?, 3
.50	1.232 65, 0	9.151 43, −1	1.107 27, −1	−3.054 72, 0	−1.065 45, 1	−2.272 61, 1	−2.911 57, 1	−8.728 6, 0	1.835 16, 2	6.529 3?, 2	1.535 5?, 3

Table 4

τ	$\widetilde{F}_{1/2}(\tau)$	$\widetilde{F}_{3/2}(\tau)$	$\widetilde{F}_{5/2}(\tau)$	$\widetilde{F}_{7/2}(\tau)$	$\widetilde{F}_{9/2}(\tau)$	$\widetilde{F}_{11/2}(\tau)$	$\widetilde{F}_{13/2}(\tau)$	$\widetilde{F}_{15/2}(\tau)$	$\widetilde{F}_{17/2}(\tau)$	$\widetilde{F}_{19/2}(\tau)$	$\widetilde{F}_{21/2}(\tau)$
0	1.000 00, 0	1.000 00, 0	1.000 00, 0	1.000 00, 0	1.000 00, 0	1.000 00, 0	1.000 00, 0	1.000 00, 0	1.000 00, 0	1.000 00, 0	1.000 00, 0
.10	1.022 53, 0	9.606 32, −1	8.656 36, −1	7.685 44, −1	6.774 56, −1	5.947 30, −1	5.207 66, −1	4.552 15, −1	3.974 26, −1	3.466 62, −1	3.021 75, −1
.12	1.027 56, 0	9.521 36, −1	8.382 24, −1	7.243 88, −1	6.203 21, −1	5.284 08, −1	4.486 02, −1	3.799 76, −1	3.213 18, −1	2.713 83, −1	2.289 94, −1
.14	1.032 81, 0	9.434 06, −1	8.106 16, −1	6.809 74, −1	5.656 17, −1	4.666 75, −1	3.833 76, −1	3.140 03, −1	2.566 17, −1	2.093 74, −1	1.706 10, −1
.15	1.035 51, 0	9.389 48, −1	7.967 35, −1	6.595 51, −1	5.391 73, −1	4.374 84, −1	3.532 40, −1	2.842 46, −1	2.281 46, −1	1.827 70, −1	1.461 99, −1
.16	1.038 27, 0	9.344 27, −1	7.828 03, −1	6.383 17, −1	5.133 32, −1	4.197 73, −1	3.246 41, −1	2.565 17, −1	2.020 45, −1	1.588 15, −1	1.246 04, −1
.165	1.039 67, 0	9.321 41, −1	7.758 16, −1	6.277 73, −1	5.006 38, −1	3.957 46, −1	3.109 98, −1	2.433 85, −1	1.898 50, −1	1.477 68, −1	1.147 81, −1
.17	1.041 09, 0	9.298 39, −1	7.688 17, −1	6.172 78, −1	4.880 93, −1	3.823 70, −1	2.976 82, −1	2.307 25, −1	1.781 96, −1	1.373 09, −1	1.055 68, −1
.175	1.042 52, 0	9.275 20, −1	7.618 04, −1	6.068 30, −1	4.756 99, −1	3.692 58, −1	2.847 39, −1	2.185 27, −1	1.670 69, −1	1.274 15, −1	9.693 63, −2
.18	1.043 97, 0	9.251 83, −1	7.547 77, −1	5.964 32, −1	4.634 55, −1	3.565 74, −1	2.721 64, −1	2.067 81, −1	1.564 50, −1	1.180 63, −1	8.885 60, −2
.185	1.045 43, 0	9.228 29, −1	7.477 36, −1	5.860 83, −1	4.513 60, −1	3.438 22, −1	2.599 49, −1	1.954 75, −1	1.463 23, −1	1.092 32, −1	8.130 07, −2
.19	1.046 91, 0	9.204 57, −1	7.406 81, −1	5.757 84, −1	4.394 15, −1	3.314 91, −1	2.480 89, −1	1.845 98, −1	1.366 73, −1	1.009 00, −1	7.424 42, −2
.195	1.048 41, 0	9.180 68, −1	7.336 11, −1	5.655 35, −1	4.276 19, −1	3.194 16, −1	2.365 79, −1	1.741 40, −1	1.274 83, −1	9.304 51, −2	6.766 13, −2
.20	1.049 92, 0	9.156 59, −1	7.265 28, −1	5.553 35, −1	4.159 72, −1	3.075 95, −1	2.254 12, −1	1.640 91, −1	1.187 37, −1	8.565 88, −2	6.152 77, −2
.21	1.052 30, 0	9.107 86, −1	7.123 17, −1	5.350 87, −1	3.931 24, −1	2.847 26, −1	2.040 85, −1	1.451 77, −1	1.025 21, −1	7.215 22, −2	5.051 58, −2
.22	1.056 14, 0	9.058 35, −1	6.980 47, −1	5.150 41, −1	3.708 70, −1	2.628 03, −1	1.840 65, −1	1.277 78, −1	8.790 50, −2	6.025 99, −2	4.103 22, −2
.23	1.059 36, 0	9.008 06, −1	6.837 15, −1	4.952 03, −1	3.492 08, −1	2.418 70, −1	1.653 04, −1	1.118 13, −1	7.477 76, −2	4.983 07, −2	3.291 42, −2
.24	1.062 66, 0	8.956 94, −1	6.693 22, −1	4.755 71, −1	3.281 36, −1	2.218 89, −1	1.477 59, −1	9.720 75, −2	6.303 01, −2	4.073 23, −2	2.601 14, −2
.25	1.066 04, 0	8.904 96, −1	6.548 64, −1	4.561 50, −1	3.076 52, −1	2.028 41, −1	1.313 86, −1	8.388 76, −2	5.255 89, −2	3.284 09, −2	2.018 63, −2
.26	1.069 50, 0	8.852 10, −1	6.403 42, −1	4.369 42, −1	2.877 55, −1	1.847 08, −1	1.161 40, −1	7.178 13, −2	4.326 51, −2	2.604 03, −2	1.531 24, −2
.27	1.073 05, 0	8.798 33, −1	6.257 52, −1	4.179 49, −1	2.684 43, −1	1.674 73, −1	1.019 79, −1	6.081 89, −2	3.505 38, −2	2.022 19, −2	1.127 40, −2
.28	1.076 69, 0	8.743 62, −1	6.110 79, −1	3.991 74, −1	2.497 14, −1	1.511 15, −1	8.885 98, −2	5.093 33, −2	2.783 47, −2	1.528 52, −2	7.965 72, −3
.29	1.080 42, 0	8.687 92, −1	5.963 14, −1	3.806 21, −1	2.315 66, −1	1.356 18, −1	7.674 13, −2	4.205 88, −2	2.152 15, −2	1.113 58, −2	5.291 81, −3
.30	1.084 25, 0	8.631 20, −1	5.815 62, −1	3.622 89, −1	2.139 97, −1	1.209 63, −1	6.558 17, −2	3.413 21, −2	1.603 20, −2	7.686 58, −3	3.165 44, −3
.32	1.092 21, 0	8.514 57, −1	5.517 34, −1	3.263 10, −1	1.805 88, −1	9.410 32, −2	4.597 60, −2	2.087 96, −2	7.214 12, −3	2.572 58, −3	2.497 54, −4
.34	1.100 63, 0	8.393 38, −1	5.215 91, −1	2.912 62, −1	1.494 70, −1	7.038 61, −2	2.972 27, −2	1.071 05, −2	7.989 90, −4	−6.296 83, −4	−1.316 70, −3
.36	1.109 54, 0	8.267 26, −1	4.911 16, −1	2.871 70, −1	1.206 28, −1	4.966 11, −2	1.651 04, −2	3.192 66, −3	−3.739 23, −3	−2.402 70, −3	−1.959 31, −3
.38	1.118 98, 0	8.137 92, −1	4.602 88, −1	2.240 64, −1	9.404 21, −2	3.177 71, −2	6.037 59, −3	−2.074 04, −3	−6.873 87, −3	−3.147 16, −3	−2.008 87, −3
.40	1.129 02, 0	7.998 50, −1	4.290 84, −1	1.919 64, −1	6.969 46, −2	1.657 72, −2	−1.987 61, −3	−5.457 76, −3	−9.029 91, −3	−3.190 32, −3	−1.715 58, −3
.42	1.139 72, 0	7.858 22, −1	3.974 80, −1	1.609 33, −1	4.756 56, −2	3.909 22, −3	−7.847 25, −3	−7.295 13, −3	−1.058 81, −2	−2.793 84, −3	−1.262 03, −3
.44	1.151 14, 0	7.704 25, −1	3.654 50, −1	1.309 76, −1	2.763 46, −2	−6.384 82, −3	−1.181 31, −2	−7.891 79, −3	−1.188 92, −2	−2.161 46, −3	−7.749 04, −4
.46	1.163 38, 0	7.545 98, −1	3.329 55, −1	1.021 45, −1	9.879 62, −3	−1.446 37, −2	−1.414 65, −2	−7.522 73, −3	−1.323 83, −2	−1.446 23, −3	−3.356 4?, −4
.48	1.176 53, 0	7.379 24, −1	2.999 68, −1	7.448 10, −2	−5.722 30, −3	−2.363 44, −2	−1.509 77, −2	−6.432 88, −3	−1.490 87, −2	−7.574 47, −4	+1.003??, −5
.50	1.190 71, 0	7.203 12, −1	2.664 46, −1	4.803 25, −2	−1.919 57, −2	−2.462 14, −2	−1.490 59, −2	−4.837 56, −3	−1.714 71, −2	−1.672 92, −4	2.435 9?, −4

Table 5

τ	$\widetilde{F}_{-1/2}(\tau)$	$\widetilde{F}_{-3/2}(\tau)$	$\widetilde{F}_{-5/2}(\tau)$	$\widetilde{F}_{-7/2}(\tau)$	$\widetilde{F}_{-9/2}(\tau)$	$\widetilde{F}_{-11/2}(\tau)$	$\widetilde{F}_{-13/2}(\tau)$	$\widetilde{F}_{-15/2}(\tau)$	$\widetilde{F}_{-17/2}(\tau)$	$\widetilde{F}_{-19/2}(\tau)$	$\widetilde{F}_{-21/2}(\tau)$
0	1.000 00, 0	1.000 00, 0	1.000 00, 0	1.000 00, 0	1.000 00, 0	1.000 00, 0	1.000 00, 0	1.000 00, 0	1.000 00, 0	1.000 00, 0	1.000 00, 0
.10	8.008 92, −1	1.671 85, 0	2.115 94, 0	2.513 48, 0	2.926 85, 0	3.376 77, 0	3.877 57, 0	4.442 05, 0	5.082 74, 0	5.813 60, 0	6.649 72, 0
.12	7.573 00, −1	1.742 79, 0	2.346 32, 0	2.952 10, 0	3.628 62, 0	4.407 40, 0	5.314 93, 0	6.379 10, 0	7.631 49, 0	9.108 73, 0	1.085 44, 1
.14	7.122 57, −1	1.791 53, 0	2.548 79, 0	3.380 58, 0	4.379 11, 0	5.604 13, 0	7.117 64, 0	8.992 43, 0	1.131 69, 1	1.419 97, 1	1.777 43, 1
.15	6.891 60, −1	1.807 40, 0	2.635 22, 0	3.601 50, 0	4.749 77, 0	6.229 64, 0	8.111 57, 0	1.050 92, 1	1.356 51, 1	1.745 91, 1	2.241 92, 1
.16	6.656 63, −1	1.817 52, 0	2.709 76, 0	3.760 74, 0	5.101 84, 0	6.844 54, 0	9.118 76, 0	1.209 84, 1	1.599 28, 1	2.110 94, 1	2.675 21, 1
.165	6.537 60, −1	1.820 41, 0	2.742 09, 0	3.843 32, 0	5.266 76, 0	7.726 87, 0	9.615 56, 0	1.289 42, 1	1.723 61, 1	2.298 49, 1	3.058 82, 1
.17	6.417 51, −1	1.821 82, 0	2.770 90, 0	3.919 53, 0	5.421 81, 0	7.420 84, 0	1.009 38, 1	1.367 35, 1	1.846 60, 1	2.488 35, 1	3.347 24, 1
.175	6.296 36, −1	1.821 77, 0	2.796 00, 0	3.988 68, 0	5.565 12, 0	7.684 38, 0	1.054 69, 1	1.441 83, 1	1.965 54, 1	2.673 80, 1	3.631 28, 1
.18	6.174 10, −1	1.820 24, 0	2.817 20, 0	4.050 05, 0	5.694 71, 0	7.925 51, 0	1.096 52, 1	1.511 21, 1	2.077 05, 1	2.848 88, 1	3.901 36, 1
.185	6.050 740, −1	1.817 20, 0	2.834 34, 0	4.102 91, 0	5.808 52, 0	8.139 35, 0	1.133 84, 1	1.573 37, 1	2.177 24, 1	3.006 55, 1	4.144 78, 1
.19	5.926 25, −1	1.812 67, 0	2.847 24, 0	4.146 55, 0	5.904 45, 0	8.320 74, 0	1.165 52, 1	1.626 01, 1	2.261 69, 1	3.138 54, 1	4.346 94, 1
.195	5.800 60, −1	1.806 62, 0	2.855 74, 0	4.180 23, 0	5.980 32, 0	8.464 31, 0	1.190 36, 1	1.666 60, 1	2.325 40, 1	3.235 40, 1	4.490 29, 1
.20	5.673 78, −1	1.799 04, 0	2.859 65, 0	4.203 22, 0	6.033 94, 0	8.564 41, 0	1.207 06, 1	1.692 41, 1	2.362 84, 1	3.286 38, 1	4.554 60, 1
.21	5.416 53, −1	1.779 28, 0	2.853 09, 0	4.214 25, 0	6.065 34, 0	8.610 69, 0	1.210 56, 1	1.687 74, 1	2.334 25, 1	3.201 72, 1	4.351 96, 1
.22	5.154 36, −1	1.753 30, 0	2.826 24, 0	4.173 83, 0	5.980 32, 0	8.410 95, 0	1.164 41, 1	1.586 26, 1	2.121 76, 1	2.775 20, 1	3.525 12, 1
.23	4.886 98, −1	1.721 00, 0	2.777 89, 0	4.076 23, 0	5.760 31, 0	7.914 61, 0	1.056 26, 1	1.359 85, 1	1.664 90, 1	1.951 77, 1	1.824 95, 1
.24	4.614 29, −1	1.682 29, 0	2.706 81, 0	3.915 86, 0	5.386 85, 0	7.070 39, 0	8.733 18, 0	9.790 24, 1	8.992 49, 0	3.874 23, 0	−1.020 29, 1
.25	4.336 05, −1	1.637 07, 0	2.611 82, 0	3.687 33, 0	4.841 83, 0	5.827 52, 0	6.028 26, 0	4.141 23, 0	−2.402 21, 0	−1.844 21, 1	−5.283 08, 1
.26	4.052 02, −1	1.585 25, 0	2.491 82, 0	3.385 43, 0	4.107 92, 0	4.137 14, 0	2.325 52, 0	−3.631 81, 0	−1.814 66, 1	−4.938 64, 1	−1.120 90, 2
.27	3.762 00, −1	1.526 71, 0	2.345 74, 0	3.005 28, 0	3.168 80, 0	1.953 94, 0	−2.487 09, 0	−1.378 02, 1	−3.876 54, 1	−8.997 03, 1	−1.897 92, 2
.28	3.465 71, −1	1.461 34, 0	2.172 55, 0	2.542 32, 0	2.009 60, 0	−7.623 41, −1	−8.504 72, 0	−2.650 47, 1	−6.563 40, 1	−1.407 99, 2	−2.876 66, 2
.29	3.162 91, −1	1.389 03, 0	1.971 29, 0	1.992 36, 0	6.172 18, −1	−4.045 21, 0	−1.660 05, 1	−4.293 31, 1	−9.591 51, 1	−2.019 87, 2	−4.028 93, 2
.30	2.853 32, −1	1.309 64, 0	1.741 05, 0	1.351 68, 0	−1.019 31, 0	−7.919 65, 0	−2.441 35, 1	−6.009 89, 1	−1.324 92, 2	−2.724 72, 2	−5.326 52, 2
.32	2.212 58, −1	1.129 15, 0	1.190 21, 0	−2.143 14, −1	−5.055 71, 0	−1.749 00, 1	−4.558 01, 1	−1.041 67, 2	−2.192 61, 2	−4.341 21, 2	−9.162 99, 2
.34	1.540 90, −1	9.187 60, −1	5.139 02, −1	−2.174 79, 0	−1.013 05, 1	−2.943 68, 1	−7.149 52, 1	−1.562 52, 2	−3.160 41, 2	−5.977 82, 2	−1.059 87, 3
.36	8.354 13, −2	6.772 47, −1	−2.930 08, −1	−4.538 49, 0	−1.621 71, 1	−4.347 90, 1	−1.007 72, 2	−2.111 16, 2	−4.168 18, 2	−7.149 23, 2	−1.132 99, 3
.38	9.281 31, −3	4.032 56, −1	−1.234 54, 0	−7.302 06, 0	−2.322 05, 1	−5.945 45, 1	−1.310 03, 2	−2.604 30, 2	−4.636 63, 2	−7.182 56, 2	−8.755 30, 2
.40	−6.906 36, −2	9.527 28, −2	−2.313 48, 0	−1.044 83, 1	−3.096 75, 1	−7.535 60, 1	−1.596 94, 2	−2.930 13, 2	−4.571 59, 2	−5.303 41, 2	−1.238 27, 2
.42	−1.519 22, −1	−2.483 82, −1	−3.531 26, 0	−1.394 44, 1	−3.919 85, 1	−9.083 02, 1	−1.793 35, 2	−2.947 89, 2	−3.586 61, 2	−7.817 4?, 1	1.231 62, 3
.44	−2.397 85, −1	−6.296 46, −1	−4.887 82, 0	−1.774 01, 1	−4.756 05, 1	−1.042 38, 2	−1.875 49, 2	−2.513 83, 2	−1.159 12, 2	+6.792 17, 2	3.159 28, 3
.46	−3.332 23, −1	−1.050 64, 0	−6.381 42, 0	−2.176 61, 1	−5.560 41, 1	−1.135 86, 2	−1.774 39, 2	−1.492 08, 2	+2.623 99, 2	1.725 53, 3	5.420 30, 3
.48	−4.328 96, −1	−1.514 24, 0	−8.008 40, 0	−2.593 18, 1	−6.312 98, 1	−1.167 51, 2	−1.431 78, 2	+2.211 92, 1	7.931 97, 2	2.957 75, 3	7.508 3?, 3
.50	−5.395 75, −1	−2.021 98, 0	−9.762 94, 0	−3.012 37, 1	−6.845 33, 1	−1.114 66, 2	−7.963 15, 1	2.657 68, 2	2.435 21, 3	4.164 43, 3	8.653 0?, 3

Table 6

τ	$\widetilde{F}_{1/2,1}(\tau)$	$\widetilde{F}_{3/2,1}(\tau)$	$\widetilde{F}_{5/2,1}(\tau)$	$\widetilde{F}_{7/2,1}(\tau)$	$\widetilde{F}_{9/2,1}(\tau)$	$\widetilde{F}_{11/2,1}(\tau)$	$\widetilde{F}_{13/2,1}(\tau)$	$\widetilde{F}_{15/2,1}(\tau)$	$\widetilde{F}_{17/2,1}(\tau)$	$\widetilde{F}_{19/2,1}(\tau)$	$\widetilde{F}_{21/2,1}(\tau)$
0	1.000 00, 0	1.000 00, 0	1.000 00, 0	1.000 00, 0	1.000 00, 0	1.000 00, 0	1.000 00, 0	1.000 00, 0	1.000 00, 0	1.000 00, 0	1.000 00, 0
.10	1.200 03, 0	1.131 60, 0	1.032 56, 0	9.273 34, −1	8.253 07, −1	7.303 12, −1	6.437 52, −1	5.658 90, −1	4.964 34, −1	4.348 27, −1	3.804 05, −1
.12	1.248 30, 0	1.162 43, 0	1.039 84, 0	9.120 19, −1	7.908 37, −1	6.807 72, −1	5.831 06, −1	4.976 60, −1	4.235 97, −1	3.598 13, −1	3.051 38, −1
.14	1.299 93, 0	1.195 07, 0	1.047 41, 0	8.964 12, −1	7.565 34, −1	6.327 18, −1	5.258 48, −1	4.350 30, −1	3.586 52, −1	2.948 89, −1	2.419 42, −1
.15	1.327 12, 0	1.212 11, 0	1.051 31, 0	8.884 95, −1	7.394 46, −1	6.092 52, −1	4.984 69, −1	4.057 36, −1	3.289 63, −1	2.659 01, −1	2.144 00, −1
.16	1.355 28, 0	1.229 67, 0	1.055 30, 0	8.804 99, −1	7.224 02, −1	5.861 59, −1	4.719 13, −1	3.777 47, −1	3.010 38, −1	2.390 71, −1	1.893 26, −1
.165	1.369 73, 0	1.238 65, 0	1.057 33, 0	8.764 71, −1	7.138 96, −1	5.747 54, −1	4.589 40, −1	3.642 23, −1	2.877 15, −1	2.264 29, −1	1.776 63, −1
.17	1.384 45, 0	1.247 77, 0	1.059 37, 0	8.724 22, −1	7.054 02, −1	5.634 43, −1	4.401 71, −1	3.510 33, −1	2.748 09, −1	2.142 86, −1	1.665 55, −1
.175	1.399 44, 0	1.257 02, 0	1.061 44, 0	8.683 53, −1	6.969 19, −1	5.522 26, −1	4.336 02, −1	3.381 45, −1	2.623 10, −1	2.026 25, −1	1.559 83, −1
.18	1.414 70, 0	1.266 42, 0	1.063 53, 0	8.642 62, −1	6.884 47, −1	5.411 04, −1	4.212 35, −1	3.255 63, −1	2.502 11, −1	1.914 36, −1	1.459 27, −1
.185	1.430 24, 0	1.275 97, 0	1.065 65, 0	8.601 50, −1	6.799 87, −1	5.300 77, −1	4.090 66, −1	3.132 85, −1	2.385 03, −1	1.807 03, −1	1.363 70, −1
.19	1.446 07, 0	1.285 67, 0	1.067 78, 0	8.560 17, −1	6.715 38, −1	5.191 44, −1	3.970 96, −1	3.013 05, −1	2.271 79, −1	1.704 14, −1	1.272 92, −1
.195	1.462 20, 0	1.295 52, 0	1.069 94, 0	8.518 61, −1	6.631 01, −1	5.083 06, −1	3.853 23, −1	2.896 22, −1	2.162 30, −1	1.605 57, −1	1.186 77, −1
.20	1.478 63, 0	1.305 53, 0	1.072 13, 0	8.476 84, −1	6.546 75, −1	4.975 63, −1	3.737 47, −1	2.782 30, −1	2.056 50, −1	1.511 18, −1	1.105 07, −1
.21	1.512 43, 0	1.326 03, 0	1.076 58, 0	8.392 60, −1	6.378 59, −1	4.763 63, −1	3.511 78, −1	2.563 05, −1	1.841 02, −1	1.334 50, −1	9.543 30, −2
.22	1.547 55, 0	1.347 20, 0	1.081 12, 0	8.307 44, −1	6.210 92, −1	4.555 46, −1	3.293 81, −1	2.355 02, −1	1.668 50, −1	1.173 11, −1	8.194 21, −2
.23	1.584 05, 0	1.369 09, 0	1.085 78, 0	8.221 32, −1	6.043 73, −1	4.351 12, −1	3.083 46, −1	2.157 89, −1	1.494 59, −1	1.026 12, −1	6.991 08, −2
.24	1.622 00, 0	1.391 71, 0	1.090 55, 0	8.134 21, −1	5.877 04, −1	4.150 63, −1	2.880 67, −1	1.971 36, −1	1.333 29, −1	8.926 18, −2	5.922 27, −2
.25	1.661 50, 0	1.415 12, 0	1.095 43, 0	8.046 09, −1	5.710 85, −1	3.954 01, −1	2.685 32, −1	1.795 14, −1	1.184 02, −1	7.717 56, −2	4.976 77, −2
.26	1.702 62, 0	1.439 35, 0	1.100 44, 0	7.956 92, −1	5.545 18, −1	3.761 26, −1	2.497 34, −1	1.628 93, −1	1.046 22, −1	6.627 07, −2	4.144 15, −2
.27	1.745 46, 0	1.464 43, 0	1.105 57, 0	7.866 66, −1	5.380 04, −1	3.572 41, −1	2.316 63, −1	1.472 42, −1	9.193 44, −2	5.646 81, −2	3.414 30, −2
.28	1.790 11, 0	1.490 43, 0	1.110 84, 0	7.775 29, −1	5.215 43, −1	3.387 47, −1	2.143 10, −1	1.325 33, −1	8.028 49, −2	4.769 17, −2	2.778 98, −2
.29	1.836 70, 0	1.517 38, 0	1.116 24, 0	7.682 76, −1	5.051 36, −1	3.206 45, −1	1.976 66, −1	1.187 37, −1	6.962 12, −2	3.986 86, −2	2.228 50, −2
.30	1.885 32, 0	1.545 34, 0	1.121 79, 0	7.589 04, −1	4.887 85, −1	3.029 38, −1	1.817 21, −1	1.058 23, −1	5.989 20, −2	3.292 90, −2	1.755 05, −2
.32	1.989 18, 0	1.604 52, 0	1.133 34, 0	7.397 86, −1	4.562 55, −1	2.687 12, −1	1.518 91, −1	8.252 83, −2	4.303 85, −2	2.143 62, −2	1.009 26, −2
.34	2.102 82, 0	1.668 45, 0	1.145 56, 0	7.201 38, −1	4.239 61, −1	2.360 82, −1	1.247 44, −1	6.241 99, −2	2.933 87, −2	1.271 44, −2	4.865 32, −3
.36	2.227 52, 0	1.737 71, 0	1.158 50, 0	6.999 20, −1	3.919 14, −1	2.050 64, −1	1.002 00, −1	4.527 05, −2	1.842 72, −2	6.310 80, −3	1.391 98, −3
.38	2.364 81, 0	1.812 98, 0	1.172 26, 0	6.790 89, −1	3.601 25, −1	1.756 73, −1	7.817 76, −2	3.085 61, −2	9.957 91, −3	1.816 85, −3	−7.354 94, −4
.40	2.516 46, 0	1.895 02, 0	1.186 91, 0	6.575 94, −1	3.286 08, −1	1.479 24, −1	5.859 41, −2	1.895 58, −2	3.605 10, −3	−1.133 27, −3	−1.862 16, −3
.42	2.684 60, 0	1.984 76, 0	1.202 55, 0	6.353 78, −1	2.973 74, −1	1.218 36, −1	4.136 37, −2	9.351 88, −3	−9.371 88, −4	−2.864 47, −3	−2.275 52, −3
.44	2.871 73, 0	2.083 28, 0	1.219 31, 0	6.123 78, −1	2.664 41, −1	9.742 75, −2	2.639 83, −2	1.829 97, −3	−3.954 79, −3	−3.662 43, −3	−2.210 80, −3
.46	3.080 88, 0	2.191 85, 0	1.237 32, 0	5.885 18, −1	2.358 25, −1	7.471 81, −2	1.360 70, −2	−3.820 48, −3	−5.713 41, −3	−3.775 45, −3	−1.856 20, −3
.48	3.315 71, 0	2.312 02, 0	1.256 75, 0	5.637 15, −1	2.055 44, −1	5.372 97, −2	2.895 93, −3	−7.806 48, −3	−6.458 57, −3	−3.416 26, −3	−1.357 90, −3
.50	3.580 63, 0	2.445 63, 0	1.277 79, 0	5.378 70, −1	1.756 21, −1	3.448 58, −2	−5.832 18, −3	−1.033 10, −2	−6.415 47, −3	−2.763 89, −3	−8.250 06, −4

Table 7

τ	$\widetilde{\underline{F}}_{-1/2,1}(\tau)$	$\widetilde{\underline{F}}_{-3/2,1}(\tau)$	$\widetilde{\underline{F}}_{-5/2,1}(\tau)$	$\widetilde{\underline{F}}_{-7/2,1}(\tau)$	$\widetilde{\underline{F}}_{-9/2,1}(\tau)$	$\widetilde{\underline{F}}_{-11/2,1}(\tau)$	$\widetilde{\underline{F}}_{-13/2,1}(\tau)$	$\widetilde{\underline{F}}_{-15/2,1}(\tau)$	$\widetilde{\underline{F}}_{-17/2,1}(\tau)$	$\widetilde{\underline{F}}_{-19/2,1}(\tau)$	$\widetilde{\underline{F}}_{-21/2,1}(\tau)$
0	1.000 00, 0	1.000 00, 0	1.000 00, 0	1.000 00, 0	1.000 00, 0	1.000 00, 0	1.000 00, 0	1.000 00, 0	1.000 00, 0	1.000 00, 0	1.000 00, 0
.10	1.158 23, 0	4.421 63, −1	1.676 70, 0	2.779 63, 0	3.773 24, 0	4.717 87, 0	5.659 71, 0	6.634 94, 0	7.693 22, 0	8.806 58, 0	7.878 90, 0
.12	1.196 15, 0	3.235 81, −1	1.520 93, 0	2.825 19, 0	4.218 70, 0	5.748 67, 0	7.464 18, 0	9.415 64, 0	1.175 33, 1	1.425 13, 1	1.311 98, 1
.14	1.236 63, 0	2.022 70, −1	1.275 91, 0	2.642 75, 0	4.302 63, 0	6.349 74, 0	8.888 12, 0	1.205 20, 1	1.633 76, 1	2.093 70, 1	2.097 48, 1
.15	1.257 90, 0	1.405 31, −1	1.121 11, 0	2.455 30, 0	4.163 85, 0	6.366 61, 0	9.220 60, 0	1.272 40, 1	1.829 64, 1	2.394 92, 1	2.563 53, 1
.16	1.279 91, 0	7.964 50, −2	9.453 88, −1	2.199 25, 0	3.876 20, 0	6.124 55, 0	9.146 21, 0	1.320 59, 1	1.957 39, 1	2.593 75, 1	3.030 61, 1
.165	1.291 20, 0	4.649 67, −2	8.498 44, −1	2.044 65, 0	3.672 04, 0	5.889 94, 0	8.917 03, 0	1.304 47, 1	1.981 70, 1	2.628 34, 1	3.244 52, 1
.17	1.302 69, 0	1.475 56, −2	7.492 60, −1	1.872 03, 0	3.425 10, 0	5.571 06, 0	8.538 24, 0	1.263 52, 1	1.966 91, 1	2.602 92, 1	3.430 95, 1
.175	1.314 39, 0	−1.718 97, −2	6.436 98, −1	1.681 12, 0	3.133 63, 0	5.161 66, 0	7.993 46, 0	1.194 07, 1	1.955 42, 1	2.503 34, 1	3.576 45, 1
.18	1.326 28, 0	−4.934 30, −2	5.332 27, −1	1.471 77, 0	2.795 97, 0	4.655 77, 0	7.266 47, 0	1.092 37, 1	1.818 67, 1	2.314 29, 1	3.665 46, 1
.185	1.338 40, 0	−8.170 85, −2	4.179 12, −1	1.243 83, 0	2.410 65, 0	4.047 65, 0	6.341 40, 0	9.546 61, 0	1.661 76, 1	2.019 75, 1	3.680 30, 1
.19	1.350 73, 0	−1.142 94, −1	2.978 19, −1	9.972 06, −1	1.976 32, 0	3.331 98, 0	5.202 79, 0	7.771 85, 0	1.441 18, 1	1.602 85, 1	3.601 06, 1
.195	1.363 28, 0	−1.470 92, −1	1.730 16, −1	7.318 47, −1	1.491 82, 0	2.503 83, 0	3.836 03, 0	5.562 83, 0	1.149 47, 1	9.382 76, 0	3.405 87, 1
.20	1.376 06, 0	−1.801 20, −1	4.357 29, −2	4.477 34, −1	9.561 01, −1	1.558 72, 0	2.227 31, 0	2.884 30, 0	7.793 62, 0	3.335 07, 0	3.071 04, 1
.21	1.402 34, 0	−2.468 67, −1	−2.289 49, −1	−1.766 09, −1	−2.717 91, −1	−6.976 55, −1	−1.765 06, 0	−4.010 80, 0	−2.232 85, 0	−1.628 44, 1	1.879 86, 1
.22	1.429 61, 0	−3.145 70, −1	−5.191 78, −1	−8.751 35, −1	−1.711 54, 0	−3.461 62, 0	−6.857 60, 0	−1.314 05, 1	−1.614 73, 1	−4.403 72, 1	−1.885 91, 0
.23	1.457 93, 0	−3.832 68, −1	−8.265 16, −1	−1.646 50, 0	−3.363 98, 0	−6.746 85, 0	−1.310 49, 1	−2.466 41, 1	−3.528 10, 1	−8.082 64, 1	−3.347 72, 1
.24	1.487 34, 0	−4.530 05, −1	−1.150 35, 0	−2.488 72, 0	−5.236 40, 0	−1.055 48, 1	−2.052 60, 1	−3.865 01, 1	−5.675 02, 1	−1.270 63, 2	−7.802 58, 1
.25	1.517 92, 0	−5.238 25, −1	−1.490 03, 0	−3.399 13, 0	−7.292 48, 0	−1.487 29, 1	−2.909 76, 1	−5.505 23, 1	−8.338 37, 1	−1.834 72, 2	−1.370 56, 2
.26	1.549 72, 0	−5.957 76, −1	−1.844 93, 0	−4.374 43, 0	−9.552 01, 0	−1.967 38, 1	−3.874 93, 1	−7.368 74, 1	−1.136 64, 2	−2.458 81, 2	−2.113 39, 2
.27	1.582 81, 0	−6.689 08, −1	−2.214 34, 0	−5.410 65, 0	−1.199 08, 1	−2.491 38, 1	−5.035 64, 1	−9.421 90, 1	−1.466 56, 2	−3.150 74, 2	−3.005 05, 2
.28	1.617 28, 0	−7.432 77, −1	−2.597 56, 0	−6.503 14, 0	−1.459 05, 1	−3.053 29, 1	−6.073 80, 1	−1.161 38, 2	−1.809 65, 2	−3.866 20, 2	−4.026 93, 2
.29	1.653 20, 0	−8.189 40, −1	−2.993 85, 0	−7.645 61, 0	−1.732 86, 1	−3.645 28, 1	−7.265 25, 1	−1.387 54, 2	−2.147 06, 2	−4.558 10, 2	−5.141 94, 2
.30	1.690 65, 0	−8.959 59, −1	−3.402 46, 0	−8.835 08, 0	−2.017 88, 1	−4.257 81, 1	−8.479 73, 1	−1.611 90, 2	−2.454 93, 2	−5.166 24, 2	−6.291 34, 2
.32	1.770 54, 0	−1.054 33, 0	−4.253 39, 0	−1.131 98, 1	−2.608 70, 1	−5.497 28, 1	−1.082 59, 2	−2.011 11, 2	−2.862 67, 2	−5.832 20, 2	−8.340 89, 2
.34	1.857 78, 0	−1.218 99, 0	−5.143 57, 0	−1.389 59, 1	−3.201 82, 1	−6.630 00, 1	−1.275 20, 2	−2.254 68, 2	−2.755 16, 2	−5.184 08, 2	−9.236 27, 2
.36	1.953 35, 0	−1.390 57, 0	−6.065 62, 0	−1.649 12, 1	−3.761 12, 1	−7.609 01, 1	−1.381 94, 2	−2.219 06, 2	−1.817 91, 2	−2.509 24, 2	−7.737 39, 2
.38	2.058 37, 0	−1.569 90, 0	−7.011 43, 0	−1.902 22, 1	−4.244 50, 1	−8.186 55, 1	−1.354 08, 2	−1.775 35, 2	+2.467 45, 1	2.739 28, 2	−2.529 34, 2
.40	2.174 18, 0	−1.757 84, 0	−7.972 08, 0	−2.139 45, 1	−4.604 61, 1	−8.221 40, 1	−1.142 30, 2	−8.081 23, 1	3.629 99, 2	1.073 53, 3	+7.370 65, 2
.42	2.302 37, 0	−1.955 46, 0	−8.937 68, 0	−2.350 18, 1	−4.789 83, 1	−7.538 55, 1	−7.024 19, 1	+7.536 41, 1	8.408 63, 2	2.075 30, 3	2.213 36, 3
.44	2.444 82, 0	−2.164 00, 0	−9.897 19, 0	−2.522 66, 1	−4.745 56, 1	−5.971 72, 1	+2.232 0?, −1	2.918 40, 2	1.430 76, 3	3.141 85, 3	4.041 31, 3
.46	2.603 78, 0	−2.384 91, 0	−1.083 82, 1	−2.643 93, 1	−4.415 75, 1	−3.377 87, 1	9.705 06, 1	5.597 80, 2	2.074 04, 3	4.020 83, 3	5.836 12, 3
.48	2.782 01, 0	−2.619 92, 0	−1.174 68, 1	−2.699 82, 1	−3.845 13, 1	+3.449 03, 0	2.200 14, 2	8.576 13, 2	2.678 61, 3	4.362 93, 3	7.231 71, 3
.50	2.982 80, 0	−2.871 14, 0	−1.260 69, 1	−2.674 89, 1	−2.681 33, 1	5.238 18, 1	3.634 60, 2	1.150 05, 3	3.125 06, 3	3.779 55, 3	7.329 7?, 3

Table 8 $-\underline{F}_{\nu}^{(0)}(\tau)=f(\tau)T'(\tau)\to f(\tau)T'(\tau)\cos\left(\nu\omega-\frac{\pi}{4}\right)$

τ	$\underline{F}_{1/2}^{(0)}(\tau)$	$\underline{F}_{3/2}^{(0)}(\tau)$	$\underline{F}_{5/2}^{(0)}(\tau)$	$\underline{F}_{7/2}^{(0)}(\tau)$	$\underline{F}_{9/2}^{(0)}(\tau)$	$\underline{F}_{11/2}^{(0)}(\tau)$	$\underline{F}_{13/2}^{(0)}(\tau)$	$\underline{F}_{15/2}^{(0)}(\tau)$	$\underline{F}_{17/2}^{(0)}(\tau)$	$\underline{F}_{19/2}^{(0)}(\tau)$	$\underline{F}_{21/2}^{(0)}(\tau)$
.02	9.890 43, −1	9.643 66, −1	9.403 05, −1	9.168 45, −1	8.939 70, −1	8.716 65, −1	8.499 17, −1	8.287 12, −1	8.080 35, −1	7.878 75, −1	7.682 17, −1
.04	9.814 77, −1	9.319 47, −1	8.849 17, −1	8.402 59, −1	7.978 56, −1	7.575 92, −1	7.193 60, −1	6.830 58, −1	6.485 87, −1	6.158 56, −1	5.847 77, −1
.06	9.786 34, −1	9.036 32, −1	8.343 78, −1	7.704 31, −1	7.113 85, −1	6.568 65, −1	6.065 23, −1	5.600 39, −1	5.171 18, −1	4.774 86, −1	4.408 91, −1
.08	9.827 97, −1	8.809 76, −1	7.897 04, −1	7.078 89, −1	6.345 49, −1	5.688 08, −1	5.098 78, −1	4.570 53, −1	4.097 01, −1	3.672 55, −1	3.292 06, −1
.10	9.983 16, −1	8.670 01, −1	7.529 59, −1	6.539 18, −1	5.679 04, −1	4.932 04, −1	4.283 30, −1	3.719 89, −1	3.230 59, −1	2.805 65, −1	2.436 61, −1
.12	1.034 92, 0	8.685 73, −1	7.289 64, −1	6.117 96, −1	5.134 60, −1	4.309 30, −1	3.616 65, −1	3.035 34, −1	2.547 46, −1	2.138 00, −1	1.794 35, −1
.14	1.121 50, 0	9.064 99, −1	7.327 13, −1	5.922 44, −1	4.787 04, −1	3.869 31, −1	3.127 52, −1	2.527 94, −1	2.043 31, −1	1.651 58, −1	1.334 96, −1
.15	1.215 94, 0	9.628 70, −1	7.624 05, −1	6.036 26, −1	4.779 54, −1	3.784 47, −1	2.996 56, −1	2.372 69, −1	1.878 71, −1	1.487 57, −1	1.177 87, −1
.16	1.437 33, 0	1.112 86, 0	8.616 44, −1	6.671 36, −1	5.165 36, −1	3.990 13, −1	3.096 51, −1	2.397 50, −1	1.856 29, −1	1.437 25, −1	1.111 80, −1
.165	1.775 27, 0	1.357 77, 0	1.038 45, 0	7.942 35, −1	6.074 50, −1	4.645 92, −1	3.553 31, −1	2.717 66, −1	2.078 53, −1	1.589 71, −1	1.215 85, −1
.17	2.983 46, 0	2.251 11, 0	1.698 54, 0	1.281 60, 0	9.670 06, −1	7.296 35, −1	5.505 31, −1	4.153 92, −1	3.134 26, −1	2.364 83, −1	1.784 38, −1
.175	2.062 47, 0	1.537 51, 0	1.146 14, 0	8.543 87, −1	6.368 92, −1	4.747 54, −1	3.538 87, −1	2.519 16, −1	1.966 26, −1	1.465 61, −1	1.092 42, −1
.18	1.765 92, 0	1.302 66, 0	9.608 44, −1	7.086 67, −1	5.226 37, −1	3.854 09, −1	2.841 92, −1	2.095 41, −1	1.544 88, −1	1.138 91, −1	8.395 55, −2
.185	1.592 44, 0	1.163 51, 0	8.499 24, −1	6.207 21, −1	4.532 32, −1	3.308 68, −1	2.414 91, −1	1.762 23, −1	1.285 69, −1	9.378 40, −2	6.839 72, −2
.19	1.471 26, 0	1.065 72, 0	7.716 81, −1	5.584 86, −1	4.040 13, −1	2.921 40, −1	2.111 58, −1	1.525 62, −1	1.101 82, −1	7.954 42, −2	5.740 38, −2
.195	1.378 64, 0	9.908 78, −1	7.117 64, −1	5.105 89, −1	3.660 84, −1	2.622 72, −1	1.877 71, −1	1.343 34, −1	9.603 88, −2	6.861 50, −2	4.899 04, −2
.20	1.303 90, 0	9.303 66, −1	6.629 38, −1	4.717 69, −1	3.353 10, −1	2.380 40, −1	1.687 95, −1	1.195 62, −1	8.459 94, −2	5.980 00, −2	4.222 85, −2
.21	1.187 53, 0	8.361 16, −1	5.870 26, −1	4.110 51, −1	2.871 15, −1	2.000 77, −1	1.391 14, −1	9.652 16, −2	6.683 36, −2	4.618 70, −2	3.185 87, −2
.22	1.098 62, 0	7.643 85, −1	5.292 04, −1	3.647 20, −1	2.503 05, −1	1.711 09, −1	1.165 39, −1	7.909 55, −2	5.350 33, −2	3.607 57, −2	2.424 94, −2
.23	1.026 75, 0	7.066 81, −1	4.826 31, −1	3.273 78, −1	2.205 73, −1	1.477 63, −1	9.843 94, −2	6.523 68, −2	4.301 63, −2	2.822 68, −2	1.840 94, −2
.24	9.664 05, −1	6.584 45, −1	4.436 16, −1	2.959 65, −1	1.956 25, −1	1.282 53, −1	8.343 20, −2	5.387 36, −2	3.453 89, −2	2.198 82, −2	1.390 07, −2
.25	9.143 89, −1	6.170 21, −1	4.100 28, −1	2.688 44, −1	1.741 59, −1	1.115 74, −1	7.073 45, −2	4.439 35, −2	2.758 71, −2	1.697 30, −2	1.033 85, −2
.26	8.686 46, −1	5.807 16, −1	3.805 11, −1	2.450 35, −1	1.553 56, −1	9.708 98, −2	5.985 02, −2	3.640 24, −2	2.184 40, −2	1.292 64, −2	7.536 83, −3
.27	8.277 90, −1	5.483 63, −1	3.541 41, −1	2.237 72, −1	1.386 67, −1	8.437 58, −2	5.044 44, −2	2.963 05, −2	1.708 84, −2	9.662 25, −3	5.343 48, −3
.28	7.908 40, −1	5.191 56, −1	3.302 82, −1	2.045 84, −1	1.237 14, −1	7.313 73, −2	4.227 85, −2	2.388 02, −2	1.315 38, −2	7.040 83, −3	3.640 46, −3
.29	7.570 68, −1	4.924 79, −1	3.084 61, −1	1.871 10, −1	1.102 37, −1	6.316 35, −2	3.517 91, −2	1.900 52, −2	9.914 62, −3	4.954 55, −3	2.336 74, −3
.30	7.259 75, −1	4.679 50, −1	2.883 76, −1	1.710 84, −1	9.802 91, −2	5.428 46, −2	2.900 00, −2	1.487 48, −2	7.256 98, −3	3.306 19, −3	1.351 46, −3
.32	6.701 51, −1	4.239 17, −1	2.523 64, −1	1.426 89, −1	7.684 66, −2	3.933 80, −2	1.898 78, −2	8.486 27, −3	3.367 74, −3	1.049 00, −3	1.081 78, −3
.34	6.209 92, −1	3.851 59, −1	2.208 17, −1	1.183 10, −1	5.927 61, −2	2.751 51, −2	1.155 89, −2	4.073 67, −3	9.272 34, −4	−2.061 00, −4	−4.753 33, −4
.36	5.769 27, −1	3.504 41, −1	1.928 35, −1	9.724 85, −2	4.471 72, −2	1.825 67, −2	6.106 91, −3	1.156 48, −3	−4.928 92, −4	−7.999 40, −4	−6.615 75, −4
.38	5.369 04, −1	3.189 76, −1	1.678 29, −1	7.901 76, −2	3.271 09, −2	1.111 27, −2	2.282 04, −3	−6.501 67, −4	−1.201 50, −3	−9.814 36, −4	−6.310 46, −4
.40	5.001 58, −1	2.901 93, −1	1.453 83, −1	6.325 97, −2	2.290 20, −2	5.721 87, −3	−2.850 08, −4	−1.646 55, −3	−1.441 86, −3	−9.251 38, −4	−5.004 05, −4
.42	4.661 39, −1	2.636 99, −1	1.252 02, −1	4.968 53, −2	1.498 33, −2	1.771 23, −3	−1.888 52, −3	−2.481 34, −3	−1.393 53, −3	−7.514 45, −4	−3.417 63, −4
.44	4.344 24, −1	2.391 99, −1	1.070 48, −1	3.806 19, −2	8.697 61, −3	−1.002 15, −3	−2.670 53, −3	−2.117 22, −3	−1.185 95, −3	−5.387 97, −4	−2.000 86, −4
.46	4.046 96, −1	2.164 73, −1	9.073 89, −2	2.818 67, −2	3.815 61, −3	−2.827 46, −3	−3.098 94, −3	−1.922 64, −3	−9.100 79, −4	−3.349 57, −4	−8.071 36, −5
.48	3.762 33, −1	1.953 51, −1	6.740 98, −2	1.988 03, −2	1.321 00, −4	−3.902 15, −3	−3.332 62, −3	−1.598 26, −3	−6.263 81, −4	−1.654 24, −4	−2.522 59, −6
.50	3.502 30, −1	1.756 84, −1	5.432 18, −2	1.297 82, −2	−2.535 21, −3	−4.397 83, −3	−2.784 13, −3	−1.222 35, −3	−3.716 28, −4	−4.027 91, −5	+4.188 85, −5

Table 9 $-\underline{F}^{(0)}_{-\nu}(\tau)=f(\tau)T^{-\nu}(\tau)\rightarrow\frac{1}{2}f(\tau)T^{-\nu}(\tau)\cos\left(\nu\omega+\frac{\pi}{4}\right)$

τ	$\underline{F}^{(0)}_{-1/2}(\tau)$	$\underline{F}^{(0)}_{-3/2}(\tau)$	$\underline{F}^{(0)}_{-5/2}(\tau)$	$\underline{F}^{(0)}_{-7/2}(\tau)$	$\underline{F}^{(0)}_{-9/2}(\tau)$	$\underline{F}^{(0)}_{-11/2}(\tau)$	$\underline{F}^{(0)}_{-13/2}(\tau)$	$\underline{F}^{(0)}_{-15/2}(\tau)$	$\underline{F}^{(0)}_{-17/2}(\tau)$	$\underline{F}^{(0)}_{-19/2}(\tau)$	$\underline{F}^{(0)}_{-21/2}(\tau)$
0.02	1.014 35, 0	1.040 31, 0	1.066 93, 0	1.094 23, 0	1.122 23, 0	1.150 94, 0	1.180 39, 0	1.210 60, 0	1.241 57, 0	1.273 34, 0	1.305 93, 0
0.04	1.033 64, 0	1.088 57, 0	1.146 43, 0	1.207 36, 0	1.271 53, 0	1.339 10, 0	1.410 27, 0	1.485 23, 0	1.564 16, 0	1.647 29, 0	1.734 84, 0
0.06	1.059 86, 0	1.147 83, 0	1.243 10, 0	1.346 28, 0	1.458 02, 0	1.579 04, 0	1.710 10, 0	1.852 05, 0	2.005 77, 0	2.172 25, 0	2.352 55, 0
0.08	1.096 39, 0	1.223 10, 0	1.364 43, 0	1.522 17, 0	1.698 09, 0	1.894 35, 0	2.113 30, 0	2.357 55, 0	2.630 03, 0	2.934 00, 0	3.273 10, 0
0.10	1.149 52, 0	1.323 62, 0	1.524 10, 0	1.754 93, 0	2.020 73, 0	2.326 79, 0	2.679 20, 0	3.084 99, 0	3.552 24, 0	4.090 25, 0	4.709 75, 0
0.12	1.233 22, 0	1.469 25, 0	1.750 68, 0	2.085 96, 0	2.485 45, 0	2.961 46, 0	3.528 62, 0	4.204 41, 0	5.009 62, 0	5.969 05, 0	7.112 21, 0
0.14	1.387 50, 0	1.716 59, 0	2.123 73, 0	2.627 44, 0	3.250 62, 0	4.021 61, 0	4.975 46, 0	6.155 55, 0	7.615 53, 0	9.421 79, 0	1.165 65, 1
0.15	1.535 66, 0	1.939 44, 0	2.449 39, 0	3.093 43, 0	3.906 80, 0	4.934 46, 0	6.231 39, 0	7.869 86, 0	9.939 13, 0	1.255 25, 1	1.585 30, 1
0.16	1.856 39, 0	2.397 64, 0	3.096 69, 0	3.999 55, 0	5.165 65, 0	6.671 73, 0	8.616 92, 0	1.112 93, 1	1.437 41, 1	1.856 48, 1	2.397 77, 1
0.165	2.321 15, 0	3.034 88, 0	3.968 07, 0	5.188 22, 0	6.783 55, 0	8.869 42, 0	1.159 67, 1	1.516 26, 1	1.982 49, 1	2.592 09, 1	3.389 13, 1
0.17	1.977 04, 0	2.620 22, 0	3.476 96, 0	4.602 37, 0	6.099 65, 0	8.084 00, 0	1.071 40, 1	1.419 95, 1	1.881 89, 1	2.494 12, 1	3.305 51, 1
0.175	1.383 31, 0	1.855 58, 0	2.489 04, 0	3.338 70, 0	4.478 31, 0	6.006 85, 0	8.056 95, 0	1.080 66, 1	1.449 44, 1	1.944 03, 1	2.607 37, 1
0.18	1.196 87, 0	1.622 26, 0	2.198 64, 0	2.979 57, 0	4.037 52, 0	5.470 65, 0	7.411 85, 0	1.004 10, 1	1.360 14, 1	1.842 29, 1	2.495 08, 1
0.185	1.089 51, 0	1.490 49, 0	2.038 56, 0	2.787 51, 0	3.810 68, 0	5.208 10, 0	7.116 15, 0	9.720 65, 0	1.327 48, 1	1.812 36, 1	2.473 62, 1
0.19	1.015 02, 0	1.399 82, 0	1.929 51, 0	2.658 22, 0	3.660 14, 0	5.036 80, 0	6.927 15, 0	9.521 15, 0	1.307 82, 1	1.795 21, 1	2.462 53, 1
0.195	9.582 26, −1	1.330 82, 0	1.846 51, 0	2.559 47, 0	3.543 96, 0	4.901 70, 0	6.771 65, 0	9.343 45, 0	1.287 52, 1	1.771 74, 1	2.434 45, 1
0.20	9.123 86, −1	1.274 94, 0	1.778 69, 0	2.477 27, 0	3.443 96, 0	4.778 57, 0	6.616 55, 0	9.140 90, 0	1.259 76, 1	1.731 52, 1	2.373 06, 1
0.21	8.407 56, −1	1.186 56, 0	1.667 88, 0	2.337 14, 0	3.259 28, 0	4.502 31, 0	6.242 45, 0	8.561 35, 0	1.165 69, 1	1.573 71, 1	2.103 11, 1
0.22	7.870 16, −1	1.115 60, 0	1.574 19, 0	2.203 97, 0	3.057 75, 0	4.196 90, 0	5.686 05, 0	7.580 25, 0	9.904 10, 0	1.257 25, 1	1.535 38, 1
0.23	7.393 96, −1	1.054 31, 0	1.485 79, 0	2.064 59, 0	2.819 52, 0	3.766 01, 0	4.883 19, 0	6.069 90, 0	7.064 75, 0	7.303 95, 0	7.107 40, 0
0.24	7.000 90, −1	9.989 20, −1	1.398 93, 0	1.913 22, 0	2.535 30, 0	3.212 18, 0	3.793 53, 0	3.939 31, 0	2.954 15, 0	−4.999 69, −1	−2.814 51, 0
0.25	6.654 02, −1	9.472 80, −1	1.310 99, 0	1.746 12, 0	2.198 31, 0	2.520 35, 0	2.382 06, 0	1.107 02, 0	−2.614 75, 0	−1.122 76, 1	−2.916 50, 1
0.26	6.341 58, −1	8.980 63, −1	1.220 48, 0	1.561 28, 0	1.804 48, 0	1.680 68, 0	6.237 65, 0	−2.486 11, 0	−9.774 45, 0	−2.515 42, 1	−5.565 70, 1
0.27	6.055 84, −1	8.505 27, −1	1.126 75, 0	1.357 93, 0	1.352 67, 0	6.898 15, −1	−1.490 96, 0	−6.864 30, 0	−1.857 50, 1	−4.236 45, 1	−8.847 45, 1
0.28	5.791 47, −1	8.059 93, −1	1.029 38, 0	1.135 98, 0	8.426 60, −1	−4.529 03, −1	−3.963 52, 0	−1.202 62, 1	−2.899 43, 1	−6.275 15, 1	−1.272 25, 2
0.29	5.544 75, −1	7.586 97, −1	9.286 25, −1	8.966 65, −1	2.781 21, −1	−1.738 90, 0	−6.772 25, 0	−1.791 49, 1	−4.087 78, 1	−8.589 20, 1	−1.707 75, 2
0.30	5.312 44, −1	7.137 34, −1	8.238 00, −1	6.388 75, −1	−3.428 30, −1	−3.168 72, 0	−9.909 20, 0	−2.447 50, 1	−5.405 35, 1	−1.112 32, 2	−2.175 16, 2
0.32	4.884 04, −1	6.269 10, −1	6.047 95, −1	7.797 00, −2	−1.724 09, 0	−6.379 15, 0	−1.694 45, 1	−3.905 49, 1	−8.260 15, 1	−1.640 58, 2	−3.091 89, 2
0.34	4.494 10, −1	5.388 27, −1	3.749 12, −1	−5.340 45, −1	−3.255 21, 0	−9.931 45, 0	−2.347 83, 1	−5.436 95, 1	−1.104 63, 2	−2.094 52, 2	−3.739 29, 2
0.36	4.135 15, −1	4.540 18, −1	1.382 04, −1	−1.179 27, 0	−4.871 07, 0	−1.360 61, 1	−3.213 58, 1	−6.807 90, 1	−1.318 99, 2	−2.338 93, 2	−3.731 54, 2
0.38	3.801 99, −1	3.711 85, −1	−1.016 42, −1	−1.839 46, 0	−6.499 30, 0	−1.714 78, 1	−3.877 58, 1	−7.805 25, 1	−1.403 83, 2	−2.198 33, 2	−2.728 73, 2
0.40	3.491 07, −1	2.908 07, −1	−3.400 12, −1	−2.493 11, 0	−8.056 30, 0	−2.026 70, 1	−4.360 49, 1	−8.168 40, 1	−1.296 98, 2	−1.544 54, 2	−4.985 21, 1
0.42	3.199 80, −1	2.132 67, −1	−5.727 70, −1	−3.119 42, 0	−9.458 30, 0	−2.267 40, 1	−4.575 67, 1	−6.467 65, 1	−9.526 30, 1	−3.147 25, 1	2.931 40, 2
0.44	2.926 30, −1	1.390 91, −1	−7.954 20, −1	−3.696 68, 0	−1.062 11, 1	−2.409 59, 1	−4.450 07, 1	−6.180 20, 1	−3.562 33, 0	−1.439 21, 2	7.164 35, 2
0.46	2.669 12, −1	6.874 72, −2	−1.003 82, 0	−4.204 72, 0	−1.146 89, 1	−2.430 71, 1	−3.935 44, 1	−3.642 54, 1	4.645 26, 1	3.526 80, 2	1.141 48, 3
0.48	2.427 13, −1	2.690 75, −3	−1.538 58, 0	−4.625 20, 0	−1.193 75, 1	−2.315 34, 1	−2.086 25, 1	−1.656 51, 0	1.433 91, 2	5.619 60, 2	1.458 80, 3
0.50	2.199 55, −1	−5.864 93, −2	−1.675 10, 0	−4.943 66, 0	−1.198 40, 1	−2.057 34, 1	−1.723 85, 1	−4.007 40, 1	2.431 37, 2	7.291 25, 2	1.678 58, 3

Table 10 $-\widetilde{\underline{F}}_\nu^{(0)}(\tau)=g(\tau)T^\nu(\tau)\to g(\tau)T^\nu(\tau)\cos\left(\nu\omega-\frac{\pi}{4}\right)$

τ	$\widetilde{\underline{F}}^{(0)}_{1/2}(\tau)$	$\widetilde{\underline{F}}^{(0)}_{3/2}(\tau)$	$\widetilde{\underline{F}}^{(0)}_{5/2}(\tau)$	$\widetilde{\underline{F}}^{(0)}_{7/2}(\tau)$	$\widetilde{\underline{F}}^{(0)}_{9/2}(\tau)$	$\widetilde{\underline{F}}^{(0)}_{11/2}(\tau)$	$\widetilde{\underline{F}}^{(0)}_{13/2}(\tau)$	$\widetilde{\underline{F}}^{(0)}_{15/2}(\tau)$	$\widetilde{\underline{F}}^{(0)}_{17/2}(\tau)$	$\widetilde{\underline{F}}^{(0)}_{19/2}(\tau)$	$\widetilde{\underline{F}}^{(0)}_{21/2}(\tau)$
.02	1.039 63, 0	1.013 69, 0	9.884 02, −1	9.637 42, −1	9.396 97, −1	9.162 51, −1	8.933 90, −1	8.711 01, −1	8.493 66, −1	8.281 75, −1	8.075 11, −1
.04	1.085 56, 0	1.030 78, 0	9.787 62, −1	9.293 68, −1	8.824 69, −1	8.379 35, −1	7.956 48, −1	7.554 96, −1	7.173 70, −1	6.811 68, −1	6.467 93, −1
.06	1.140 18, 0	1.052 79, 0	9.721 09, −1	8.976 06, −1	8.288 13, −1	7.652 94, −1	7.066 42, −1	6.524 85, −1	6.024 79, −1	5.563 05, −1	5.136 69, −1
.08	1.207 46, 0	1.082 37, 0	9.702 30, −1	8.967 12, −1	7.796 07, −1	6.988 38, −1	6.264 36, −1	5.615 35, −1	5.033 59, −1	4.512 09, −1	4.044 62, −1
.10	1.294 95, 0	1.124 61, 0	9.766 86, −1	8.482 17, −1	7.366 45, −1	6.397 50, −1	5.556 00, −1	4.825 18, −1	4.190 50, −1	3.639 29, −1	3.160 60, −1
.12	1.419 02, 0	1.190 94, 0	9.995 12, −1	8.388 58, −1	7.040 26, −1	5.908 65, −1	4.958 93, −1	4.161 88, −1	3.492 92, −1	2.931 50, −1	2.460 31, −1
.14	1.627 54, 0	1.315 53, 0	1.063 33, 0	8.594 76, −1	6.947 05, −1	5.615 22, −1	4.538 72, −1	3.668 60, −1	2.965 29, −1	2.396 81, −1	1.937 32, −1
.15	1.816 30, 0	1.438 28, 0	1.138 83, 0	9.016 60, −1	7.139 39, −1	5.653 01, −1	4.476 08, −1	3.544 18, −1	2.806 30, −1	2.222 04, −1	1.759 43, −1
.16	2.210 66, 0	1.711 61, 0	1.325 40, 0	1.026 08, 0	7.944 48, −1	6.136 94, −1	4.762 53, −1	3.687 43, −1	2.855 03, −1	2.210 53, −1	1.709 98, −1
.165	2.770 97, 0	2.119 30, 0	1.620 89, 0	1.239 70, 0	9.481 50, −1	7.251 68, −1	5.546 25, −1	4.241 91, −1	3.244 32, −1	2.481 33, −1	1.897 78, −1
.17	4.726 37, 0	3.550 34, 0	2.690 81, 0	2.030 30, 0	1.531 93, 0	1.155 87, 0	8.721 44, −1	6.582 17, −1	4.965 33, −1	3.746 29, −1	2.826 83, −1
.175	3.316 47, 0	2.472 33, 0	1.843 00, 0	1.373 86, 0	1.024 13, 0	7.634 03, −1	5.690 59, −1	4.050 90, −1	3.161 83, −1	2.356 70, −1	1.756 59, −1
.18	2.882 55, 0	2.126 36, 0	1.568 40, 0	1.156 77, 0	8.531 16, −1	6.291 12, −1	4.638 89, −1	3.420 36, −1	2.521 77, −1	1.859 05, −1	1.370 41, −1
.185	2.638 93, 0	1.928 12, 0	1.408 45, 0	1.028 63, 0	7.510 75, −1	5.483 05, −1	4.001 88, −1	2.920 25, −1	2.130 61, −1	1.554 15, −1	1.133 45, −1
.19	2.475 44, 0	1.793 11, 0	1.298 37, 0	9.396 69, −1	6.797 59, −1	4.915 34, −1	3.552 83, −1	2.566 87, −1	1.853 81, −1	1.338 35, −1	9.658 40, −2
.195	2.355 34, 0	1.692 87, 0	1.216 01, 0	8.723 16, −1	6.254 36, −1	4.480 75, −1	3.207 96, −1	2.294 96, −1	1.640 78, −1	1.172 25, −1	8.369 70, −2
.20	2.262 19, 0	1.614 13, 0	1.150 30, 0	8.184 91, −1	5.817 43, −1	4.129 85, −1	2.928 41, −1	2.074 29, −1	1.467 74, −1	1.037 49, −1	7.326 30, −2
.21	2.125 29, 0	1.496 37, 0	1.050 58, 0	7.356 46, −1	5.138 41, −1	3.580 72, −1	2.489 68, −1	1.727 42, −1	1.196 10, −1	8.265 95, −2	5.701 66, −2
.22	2.029 00, 0	1.411 71, 0	9.773 66, −1	6.735 87, −1	4.622 78, −1	3.160 14, −1	2.152 31, −1	1.460 78, −1	9.881 31, −2	6.662 68, −2	4.478 52, −2
.23	1.957 65, 0	1.347 39, 0	9.202 08, −1	6.241 95, −1	4.205 56, −1	2.817 32, −1	1.876 89, −1	1.243 84, −1	8.201 70, −2	5.381 86, −2	3.510 03, −2
.24	1.903 04, 0	1.296 61, 0	8.735 69, −1	5.821 84, −1	3.852 25, −1	2.525 56, −1	1.642 94, −1	1.060 88, −1	6.801 40, −2	4.329 92, −2	2.737 33, −2
.25	1.860 46, 0	1.255 42, 0	8.342 63, −1	5.470 03, −1	3.543 53, −1	2.270 14, −1	1.439 20, −1	9.032 52, −2	5.613 01, −2	3.453 41, −2	2.103 52, −2
.26	1.826 95, 0	1.221 37, 0	8.002 98, −1	5.153 63, −1	3.267 48, −1	2.042 01, −1	1.258 78, −1	7.656 23, −2	4.594 27, −2	2.718 71, −2	1.585 16, −2
.27	1.801 55, 0	1.193 42, 0	7.707 31, −1	4.870 04, −1	3.017 87, −1	1.836 30, −1	1.097 84, −1	6.448 60, −2	3.719 02, −2	2.102 83, −2	1.162 92, −2
.28	1.779 72, 0	1.168 32, 0	7.432 73, −1	4.604 00, −1	2.784 08, −1	1.645 90, −1	9.514 44, −2	5.374 04, −2	2.960 16, −2	1.584 48, −2	8.192 56, −3
.29	1.703 72, 0	1.108 28, 0	6.941 67, −1	4.210 76, −1	2.480 80, −1	1.421 44, −1	7.916 78, −2	4.276 97, −2	2.231 21, −2	1.114 98, −2	5.258 65, −3
.30	1.751 44, 0	1.128 95, 0	6.957 19, −1	4.127 47, −1	2.364 99, −1	1.309 64, −1	6.996 37, −2	3.588 60, −2	1.750 78, −2	7.976 32, −3	3.260 45, −3
.32	1.736 73, 0	1.098 60, 0	6.540 14, −1	3.697 86, −1	1.991 52, −1	1.019 46, −1	4.920 78, −2	2.199 26, −2	8.727 67, −3	2.718 54, −3	2.803 49, −3
.34	1.732 44, 0	1.074 52, 0	6.160 35, −1	3.300 61, −1	1.653 68, −1	7.676 16, −2	3.224 70, −2	1.136 47, −2	2.575 64, −3	−5.749 78, −3	−1.326 08, −3
.36	1.736 56, 0	1.054 83, 0	5.804 35, −1	2.927 19, −1	1.345 99, −1	5.495 28, −2	1.838 19, −2	3.481 02, −3	−1.483 61, −3	−2.407 83, −3	−1.991 35, −3
.38	1.747 87, 0	1.038 42, 0	5.463 62, −1	2.572 39, −1	1.064 89, −1	3.617 71, −2	7.429 11, −3	−2.116 60, −3	−3.911 45, −3	−3.195 04, −3	−2.054 35, −3
.40	1.765 56, 0	1.024 38, 0	5.132 02, −1	2.233 07, −1	8.084 41, −2	2.019 80, −2	−1.006 08, −3	−5.812 32, −3	−5.089 77, −3	−3.265 74, −3	−1.766 43, −3
.42	1.789 14, 0	1.012 13, 0	4.805 53, −1	1.907 03, −1	5.750 92, −2	6.798 37, −3	−7.248 56, −3	−9.523 93, −3	−5.348 67, −3	−2.884 21, −3	−1.311 76, −3
.44	1.818 33, 0	1.001 19, 0	4.480 61, −1	1.593 12, −1	3.640 48, −2	−4.194 61, −3	−1.117 78, −2	−8.861 86, −3	−4.963 92, −3	−2.255 19, −3	−8.374 82, −4
.46	1.853 05, 0	9.912 00, −1	4.154 81, −1	1.290 63, −1	1.747 11, −2	−1.294 65, −2	−1.418 96, −2	−8.803 50, −3	−4.167 12, −3	−1.533 72, −3	−3.695 76, −4
.48	1.890 97, 0	9.818 46, −1	3.388 06, −1	9.991 96, −2	6.639 43, −4	−1.961 24, −2	−1.674 99, −2	−8.032 95, −3	−3.148 23, −3	−8.314 31, −4	−1.267 87, −5
.50	1.939 27, 0	9.727 85, −1	3.007 87, −1	7.186 20, −2	−1.403 78, −2	−2.435 14, −2	−1.541 61, −2	−6.768 31, −3	−2.057 75, −3	−2.230 31, −4	−2.319 421, −4

Table 11 $-\widetilde{\underline{F}}^{(0)}_{-\nu}(\tau)=g(\tau)T^{-\nu}(\tau)\to\frac{1}{2}g(\tau)T^{-\nu}(\tau)\cos\left(\nu\omega+\frac{\pi}{4}\right)$

τ	$\widetilde{\underline{F}}^{(0)}_{-1/2}(\tau)$	$\widetilde{\underline{F}}^{(0)}_{-3/2}(\tau)$	$\widetilde{\underline{F}}^{(0)}_{-5/2}(\tau)$	$\widetilde{\underline{F}}^{(0)}_{-7/2}(\tau)$	$\widetilde{\underline{F}}^{(0)}_{-9/2}(\tau)$	$\widetilde{\underline{F}}^{(0)}_{-11/2}(\tau)$	$\widetilde{\underline{F}}^{(0)}_{-13/2}(\tau)$	$\widetilde{\underline{F}}^{(0)}_{-15/2}(\tau)$	$\widetilde{\underline{F}}^{(0)}_{-17/2}(\tau)$	$\widetilde{\underline{F}}^{(0)}_{-19/2}(\tau)$	$\widetilde{\underline{F}}^{(0)}_{-21/2}(\tau)$
.02	1.066 23, 0	1.093 52, 0	1.121 50, 0	1.150 20, 0	1.179 63, 0	1.209 81, 0	1.240 77, 0	1.272 52, 0	1.305 08, 0	1.338 47, 0	1.372 73, 0
.04	1.143 26, 0	1.204 01, 0	1.268 01, 0	1.335 40, 0	1.406 38, 0	1.481 11, 0	1.559 83, 0	1.642 74, 0	1.730 04, 0	1.821 99, 0	1.918 82, 0
.06	1.234 81, 0	1.337 30, 0	1.448 30, 0	1.568 51, 0	1.698 70, 0	1.839 69, 0	1.992 39, 0	2.157 77, 0	2.336 86, 0	2.530 82, 0	2.740 89, 0
.08	1.347 02, 0	1.502 70, 0	1.676 34, 0	1.870 14, 0	2.086 27, 0	2.327 40, 0	2.596 40, 0	2.896 49, 0	3.231 25, 0	3.604 71, 0	4.021 33, 0
.10	1.491 08, 0	1.716 91, 0	1.976 96, 0	2.276 37, 0	2.621 15, 0	8.018 15, 0	3.475 27, 0	4.001 63, 0	4.607 72, 0	5.305 59, 0	6.109 16, 0
.12	1.690 92, 0	2.014 55, 0	2.400 43, 0	2.860 14, 0	3.407 90, 0	4.060 58, 0	4.838 23, 0	5.764 83, 0	6.868 89, 0	8.184 40, 0	9.751 84, 0
.14	2.013 57, 0	2.491 15, 0	3.082 00, 0	3.812 89, 0	4.717 36, 0	5.836 24, 0	7.220 49, 0	8.933 06, 0	1.105 18, 1	1.367 31, 1	1.691 61, 1
.15	2.293 88, 0	2.897 02, 0	3.658 75, 0	4.620 78, 0	5.835 74, 0	7.370 80, 0	9.308 08, 0	1.175 55, 1	1.484 65, 1	1.875 02, 1	3.368 03, 1
.16	2.855 18, 0	3.687 64, 0	4.762 80, 0	6.151 43, 0	7.944 92, 0	1.026 13, 1	1.325 31, 1	1.711 72, 1	2.210 78, 1	2.855 32, 1	3.687 84, 1
.165	3.623 01, 0	4.737 05, 0	6.193 64, 0	8.098 14, 0	1.058 82, 1	1.384 40, 1	1.810 09, 1	2.366 68, 1	3.094 41, 1	4.045 92, 1	5.289 99, 1
.17	3.132 01, 0	4.150 93, 0	5.508 17, 0	7.291 03, 0	9.663 00, 0	1.280 66, 1	1.697 29, 1	2.249 46, 1	2.981 27, 1	3.951 16, 1	5.237 76, 1
.175	2.224 38, 0	2.983 79, 0	4.002 40, 0	5.368 66, 0	7.201 17, 0	9.659 07, 0	1.295 57, 1	1.737 71, 1	2.330 71, 1	3.126 04, 1	4.192 68, 1
.18	1.953 67, 0	2.648 05, 0	3.588 88, 0	4.863 61, 0	6.590 52, 0	8.929 85, 0	1.209 85, 1	1.639 01, 1	2.220 18, 1	3.007 21, 1	4.072 77, 1
.185	1.805 49, 0	2.469 98, 0	3.378 22, 0	4.619 35, 0	6.314 91, 0	8.630 65, 0	1.179 26, 1	1.610 87, 1	2.199 85, 1	3.003 37, 1	4.099 18, 1
.19	1.707 80, 0	2.355 24, 0	3.246 46, 0	4.472 53, 0	6.158 30, 0	8.474 57, 0	1.165 51, 1	1.601 96, 1	2.200 45, 1	3.020 49, 1	4.143 28, 1
.195	1.637 09, 0	2.273 64, 0	3.154 67, 0	4.372 73, 0	6.054 68, 0	8.374 31, 0	1.156 90, 1	1.596 28, 1	2.199 66, 1	3.026 93, 1	4.159 14, 1
.20	1.582 94, 0	2.211 94, 0	3.085 92, 0	4.297 91, 0	5.975 06, 0	8.290 53, 0	1.147 93, 1	1.585 89, 1	2.185 61, 1	3.004 08, 1	4.117 12, 1
.21	1.504 68, 0	2.123 55, 0	2.984 95, 0	4.182 71, 0	5.833 04, 0	8.057 65, 0	1.117 19, 1	1.532 70, 1	2.086 20, 1	2.816 42, 1	3.763 87, 1
.22	1.453 52, 0	2.060 36, 0	2.907 31, 0	4.070 42, 0	5.647 24, 0	7.751 09, 0	1.050 13, 1	1.399 97, 1	1.829 15, 1	2.321 96, 1	2.835 63, 1
.23	1.409 78, 0	2.010 20, 0	2.832 88, 0	3.936 45, 0	5.375 84, 0	7.180 46, 0	9.310 53, 0	1.157 32, 1	1.347 00, 1	1.392 61, 1	1.355 13, 1
.24	1.378 62, 0	1.967 07, 0	2.754 77, 0	3.767 51, 0	4.992 51, 0	6.325 42, 0	7.470 22, 0	7.757 29, 0	5.817 31, 0	−9.845 41, −1	−5.542 33, 0
.25	1.353 86, 0	1.927 38, 0	2.667 41, 0	3.552 74, 0	4.472 79, 0	5.128 03, 0	4.846 66, 0	2.252 40, 0	−5.320 10, 0	−2.284 44, 1	−5.934 06, 1
.26	1.333 78, 0	1.888 84, 0	2.566 94, 0	3.283 72, 0	3.795 22, 0	3.534 84, 0	1.311 93, 0	−5.228 84, 0	−2.055 78, 1	−5.290 48, 1	−1.170 59, 2
.27	1.317 97, 0	1.851 00, 0	2.452 19, 0	2.955 32, 0	2.943 86, 0	1.501 28, 0	−3.244 84, 0	−1.493 90, +1	−4.042 55, 1	−9.219 96, 1	−1.925 50, 2
.28	1.303 33, 0	1.813 84, 0	2.316 54, 0	2.556 43, 0	1.896 34, 0	−1.019 22, 0	−8.919 58, 0	−2.706 40, 1	−6.524 94, 1	−1.412 17, 2	−2.863 10, 2
.29	1.291 67, 0	1.767 40, 0	2.163 25, 0	2.088 80, 0	6.478 83, −1	−4.050 81, 0	−1.577 60, 1	−4.173 29, 1	−9.522 52, 1	−2.000 86, 2	−3.978 22, 2
.30	1.281 57, 0	1.721 93, 0	1.987 45, 0	1.541 32, 0	−8.270 91, −1	−7.644 62, 0	−2.390 63, 1	−5.904 69, 1	−1.304 06, 2	−2.683 52, 2	−5.247 66, 2
.32	1.265 72, 0	1.624 67, 0	1.567 37, 0	2.020 63, −1	−4.468 07, 0	−1.653 19, 1	−4.391 25, 1	−1.012 13, 2	−2.140 66, 2	−4.251 65, 2	−8.012 79, 2
.34	1.253 76, 0	1.503 23, 0	1.045 93, 0	−1.489 88, 0	−9.081 38, 0	−2.770 68, 1	−6.549 98, 1	−1.516 80, 2	−3.081 70, 2	−5.843 29, 2	−1.043 19, 3
.36	1.244 68, 0	1.366 60, 0	4.159 95, −1	−3.549 61, 0	−1.466 20, 1	−4.095 45, 1	−9.672 91, 1	−2.049 18, 2	−3.970 17, 2	−7.040 20, 2	−1.123 20, 3
.38	1.237 73, 0	1.208 38, 0	−3.308 92, −1	−5.988 31, 0	−2.115 83, 1	−5.582 41, 1	−1.262 33, 2	−2.540 98, 2	−4.570 13, 2	−7.156 60, 2	−8.883 30, 2
.40	1.232 35, 0	1.026 55, 0	−1.200 24, 0	−8.800 68, 0	−2.843 87, 1	−7.154 25, 1	−1.539 25, 2	−2.883 45, 2	−4.578 37, 2	−5.452 23, 2	−1.759 78, 2
.42	1.228 15, 0	8.185 66, −1	−2.198 44, 0	−1.197 30, 1	−3.630 30, 1	−8.702 78, 1	−1.756 24, 2	−2.482 43, 2	−3.656 40, 2	−1.207 98, 2	1.125 14, 3
.44	1.224 84, 0	5.821 81, −1	−3.329 32, 0	−1.547 29, 1	−4.445 58, 1	−1.008 56, 2	−1.862 63, 2	−2.586 79, 2	−1.491 05, 1	6.023 97, 2	2.998 72, 3
.46	1.222 15, 0	3.147 83, −1	−4.596 35, 0	−1.925 28, 1	−5.251 45, 1	−1.112 99, 2	−1.801 98, 2	−1.667 87, 2	2.127 00, 2	1.614 87, 3	5.226 68, 3
.48	1.219 89, 0	1.352 41, −2	−7.733 00, 0	−2.324 65, 1	−5.999 86, 1	−1.163 70, 2	−1.048 56, 2	−8.325 72, 0	7.206 92, 2	2.824 44, 3	7.332 02, 3
.50	1.217 92, 0	−3.247 53, −1	−9.275 25, 0	−2.737 37, 1	−6.635 70, 1	−1.139 18, 2	−9.545 18, 1	−2.218 95, 2	1.346 28, 3	4.037 26, 3	9.294 52, 3

Table 12 $-\widetilde{\underline{F}}^{(0)}_{\nu,1}(\tau)=h(\tau)T^{\nu}(\tau)\to h(\tau)T^{\nu}(\tau)\cos\left(\nu\omega-\mu-\frac{\pi}{4}\right)$

τ	$\widetilde{\underline{F}}^{(0)}_{1/2,1}(\tau)$	$\widetilde{\underline{F}}^{(0)}_{3/2,1}(\tau)$	$\widetilde{\underline{F}}^{(0)}_{5/2,1}(\tau)$	$\widetilde{\underline{F}}^{(0)}_{7/2,1}(\tau)$	$\widetilde{\underline{F}}^{(0)}_{9/2,1}(\tau)$	$\widetilde{\underline{F}}^{(0)}_{11/2,1}(\tau)$	$\widetilde{\underline{F}}^{(0)}_{13/2,1}(\tau)$	$\widetilde{\underline{F}}^{(0)}_{15/2,1}(\tau)$	$\widetilde{\underline{F}}^{(0)}_{17/2,1}(\tau)$	$\widetilde{\underline{F}}^{(0)}_{19/2,1}(\tau)$	$\widetilde{\underline{F}}^{(0)}_{21/2,1}(\tau)$
.02	1.089 0, 0	1.061 8, 0	1.035 4, 0	1.009 5, 0	9.843 3, −1	9.597 7, −1	9.358 3, −1	9.124 8, −1	8.897 1, −1	8.675 1, −1	8.458 7, −1
.04	1.195 8, 0	1.135 5, 0	1.078 2, 0	1.023 8, 0	9.721 2, −1	9.230 6, −1	8.764 8, −1	8.322 4, −1	7.902 4, −1	7.503 6, −1	7.125 0, −1
.06	1.327 4, 0	1.225 7, 0	1.131 8, 0	1.045 0, 0	9.649 3, −1	8.909 8, −1	8.226 9, −1	7.596 4, −1	7.014 2, −1	6.476 7, −1	5.980 3, −1
.08	1.495 4, 0	1.340 4, 0	1.201 6, 0	1.077 1, 0	9.654 9, −1	8.649 3, −1	7.758 0, −1	6.954 2, −1	6.233 8, −1	5.587 9, −1	5.009 0, −1
.10	1.721 3, 0	1.494 9, 0	1.298 3, 0	1.127 5, 0	9.791 9, −1	8.503 9, −1	7.385 4, −1	6.413 9, −1	5.570 2, −1	4.837 6, −1	4.201 3, −1
.12	2.052 2, 0	1.722 4, 0	1.445 5, 0	1.213 2, 0	1.018 2, 0	8.545 3, −1	7.171 8, −1	6.019 0, −1	5.051 6, −1	4.239 6, −1	3.558 2, −1
.14	2.623 3, 0	2.120 4, 0	1.713 9, 0	1.385 3, 0	1.119 7, 0	9.050 7, −1	7.315 6, −1	5.913 1, −1	4.779 5, −1	3.863 2, −1	3.122 6, −1
.15	3.145 8, 0	2.491 1, 0	1.972 5, 0	1.561 7, 0	1.236 5, 0	9.791 0, −1	7.752 5, −1	6.138 5, −1	4.860 5, −1	3.848 5, −1	3.047 3, −1
.16	4.232 2, 0	3.276 8, 0	2.537 1, 0	1.964 4, 0	1.520 9, 0	1.174 9, 0	9.117 5, −1	7.059 3, −1	5.465 8, −1	4.231 9, −1	3.273 6, −1
.165	5.744 0, 0	4.393 2, 0	3.360 0, 0	2.569 8, 0	1.965 5, 0	1.503 2, 0	1.149 7, 0	8.793 2, −1	6.725 2, −1	5.143 6, −1	3.934 0, −1
.17	1.005 5, 0	7.587 1, 0	5.725 2, 0	4.320 3, 0	3.260 0, 0	2.460 0, 0	1.856 2, 0	1.400 7, 0	1.055 9, 0	7.975 5, −1	6.018 2, −1
.175	6.006 5, 0	4.423 2, 0	3.348 6, 0	2.499 6, 0	1.866 0, 0	1.391 2, 0	1.039 9, 0	7.763 0, −1	5.796 4, −1	4.325 5, −1	3.228 7, −1
.18	4.617 2, 0	3.423 3, 0	2.537 8, 0	1.881 0, 0	1.394 0, 0	1.032 9, 0	7.561 6, −1	5.666 4, −1	4.199 1, −1	3.108 5, −1	2.301 6, −1
.185	3.789 1, 0	2.799 1, 0	2.066 6, 0	1.525 0, 0	1.124 7, 0	8.291 8, −1	6.108 7, −1	4.499 0, −1	3.312 0, −1	2.437 1, −1	1.792 5, −1
.19	3.209 3, 0	2.371 0, 0	1.749 2, 0	1.289 3, 0	9.487 0, −1	6.970 3, −1	5.117 4, −1	3.753 5, −1	2.750 3, −1	2.014 0, −1	1.472 5, −1
.195	2.770 4, 0	2.054 5, 0	1.558 6, 0	1.120 2, 0	8.240 4, −1	6.048 4, −1	4.429 8, −1	3.303 1, −1	2.363 4, −1	1.721 8, −1	1.252 5, −1
.20	2.421 7, 0	1.809 4, 0	1.344 2, 0	9.943 2, −1	7.322 0, −1	5.371 5, −1	4.008 2, −1	2.945 4, −1	2.080 2, −1	1.508 0, −1	1.090 6, −1
.21	1.896 2, 0	1.454 2, 0	1.099 4, 0	8.205 2, −1	5.849 6, −1	4.450 2, −1	3.240 3, −1	2.345 3, −1	1.688 9, −1	1.209 6, −1	8.629 2, −2
.22	1.516 3, 0	1.212 8, 0	9.393 8, −1	7.107 5, −1	5.281 8, −1	3.869 2, −1	2.801 8, −1	2.009 0, −1	1.428 6, −1	1.008 5, −1	7.074 1, −2
.23	1.228 0, 0	1.041 9, 0	8.317 4, −1	6.385 0, −1	4.758 7, −1	3.474 3, −1	2.494 8, −1	1.772 7, −1	1.236 8, −1	8.620 7, −2	5.611 4, −2
.24	1.033 2, 0	9.466 5, −1	7.571 0, −1	5.885 2, −1	4.412 4, −1	3.211 5, −1	2.291 5, −1	1.575 6, −1	1.082 8, −1	7.278 3, −2	4.936 1, −2
.25	8.208 4, −1	8.349 8, −1	7.047 4, −1	5.528 4, −1	4.117 0, −1	2.980 0, −1	2.093 7, −1	1.415 9, −1	9.525 9, −2	6.313 1, −2	4.130 0, −2
.26	6.729 8, −1	7.578 0, −1	6.678 8, −1	5.332 2, −1	3.905 5, −1	2.770 5, −1	1.904 1, −1	1.265 6, −1	8.385 2, −2	5.383 9, −2	3.437 3, −2
.27	5.505 7, −1	7.072 8, −1	6.420 3, −1	5.084 5, −1	3.733 2, −1	2.608 4, −1	1.741 3, −1	1.141 7, −1	7.327 4, −2	4.693 9, −2	2.832 1, −2
.28	4.482 4, −1	6.701 9, −1	6.241 7, −1	4.940 4, −1	3.586 8, −1	2.462 4, −1	1.623 4, −1	1.036 4, −1	6.439 0, −2	3.905 2, −2	2.314 9, −2
.29	3.615 8, −1	6.431 4, −1	6.119 9, −1	4.825 9, −1	3.455 8, −1	2.305 6, −1	1.497 7, −1	9.262 4, −2	5.597 6, −2	3.274 2, −2	1.858 7, −2
.30	2.883 3, −1	6.247 9, −1	6.045 8, −1	4.735 4, −1	3.337 5, −1	2.199 4, −1	1.379 3, −1	8.309 1, −2	4.827 5, −2	2.708 0, −2	1.417 1, −2
.32	1.713 3, −1	6.059 0, −1	5.984 4, −1	4.591 0, −1	3.117 6, −1	1.958 1, −1	1.158 9, −1	6.518 9, −2	3.492 5, −2	1.777 6, −2	8.520 9, −3
.34	8.409 7, −2	6.047 4, −1	6.001 4, −1	4.473 0, −1	2.908 8, −1	1.729 4, −1	9.571 9, −2	4.958 6, −2	2.393 8, −2	1.060 1, −2	4.127 7, −3
.36	1.813 3, −2	6.156 8, −1	6.062 8, −1	4.363 4, −1	2.702 3, −1	1.509 8, −1	7.725 6, −2	3.611 9, −2	1.509 4, −2	5.277 7, −3	1.183 1, −3
.38	−3.207 5, −2	6.362 2, −1	6.154 7, −1	4.218 0, −1	2.496 1, −1	1.299 4, −1	6.005 0, −2	2.470 5, −2	8.162 9, −3	1.508 9, −3	−6.372 6, −4
.40	−7.024 5, −2	6.629 7, −1	6.252 3, −1	4.136 2, −1	2.284 3, −1	1.096 1, −1	4.536 9, −2	1.512 8, −2	2.914 3, −3	−9.833 7, −4	−1.603 9, −3
.42	−9.907 1, −2	6.959 7, −1	6.365 6, −1	4.015 4, −1	2.071 3, −1	9.031 2, −2	3.192 4, −2	7.360 5, −3	−8.608 8, −4	−2.454 3, −3	−1.966 1, −3
.44	−1.204 3, −1	7.341 1, −1	6.480 8, −1	3.873 9, −1	1.855 7, −1	7.197 4, −2	2.015 1, −2	1.242 0, −3	−3.378 0, −3	−3.133 5, −3	−1.910 7, −3
.46	−1.356 3, −1	7.769 7, −1	6.596 5, −1	3.724 2, −1	1.638 1, −1	5.474 3, −2	1.002 5, −2	−3.653 1, −3	−4.844 1, −3	−3.226 3, −3	−1.602 1, −3
.48	−1.456 6, −1	8.245 4, −1	7.345 6, −1	3.561 0, −1	1.419 1, −1	3.869 7, −2	−4.100 7, −3	−6.618 4, −3	−5.456 0, −3	−2.912 3, −3	−1.167 2, −3
.50	−1.511 9, −1	8.767 6, −1	7.342 4, −1	3.382 5, −1	1.200 8, −1	2.387 4, −2	−5.421 2, −3	−8.653 9, −3	−5.064 8, −3	−2.342 8, −3	−7.010 9, −4

Table 13 $-\underline{\widetilde{F}}^{(0)}_{-\nu,1}(\tau)=h(\tau)T^{-\nu}(\tau)\to\frac{1}{2}h(\tau)T^{-\nu}(\tau)\cos\left(\nu\omega+\mu+\frac{\pi}{4}\right)$

τ	$\underline{\widetilde{F}}^{(0)}_{-1/2,1}(\tau)$	$\underline{\widetilde{F}}^{(0)}_{-3/2,1}(\tau)$	$\underline{\widetilde{F}}^{(0)}_{-5/2,1}(\tau)$	$\underline{\widetilde{F}}^{(0)}_{-7/2,1}(\tau)$	$\underline{\widetilde{F}}^{(0)}_{-9/2,1}(\tau)$	$\underline{\widetilde{F}}^{(0)}_{-11/2,1}(\tau)$	$\underline{\widetilde{F}}^{(0)}_{-13/2,1}(\tau)$	$\underline{\widetilde{F}}^{(0)}_{-15/2,1}(\tau)$	$\underline{\widetilde{F}}^{(0)}_{-17/2,1}(\tau)$	$\underline{\widetilde{F}}^{(0)}_{-19/2,1}(\tau)$	$\underline{\widetilde{F}}^{(0)}_{-21/2,1}(\tau)$
.02	1.116 9, 0	1.145 5, 0	1.174 8, 0	1.204 8, 0	1.235 7, 0	1.267 3, 0	1.299 7, 0	1.333 0, 0	1.367 1, 0	1.402 1, 0	1.437 9, 0
.04	1.259 4, 0	1.326 3, 0	1.396 8, 0	1.471 1, 0	1.549 2, 0	1.631 6, 0	1.718 3, 0	1.809 6, 0	1.905 8, 0	2.007 1, 0	2.113 8, 0
.06	1.437 6, 0	1.556 9, 0	1.686 2, 0	1.826 1, 0	1.977 7, 0	2.141 8, 0	2.319 6, 0	2.512 1, 0	2.720 7, 0	2.946 5, 0	3.191 0, 0
.08	1.668 2, 0	1.861 0, 0	2.076 0, 0	2.316 0, 0	2.583 7, 0	2.882 3, 0	3.215 5, 0	3.587 1, 0	4.001 7, 0	4.464 2, 0	4.980 2, 0
.10	1.982 0, 0	2.282 2, 0	2.527 9, 0	3.025 9, 0	3.484 2, 0	4.011 9, 0	4.619 5, 0	5.319 2, 0	6.124 8, 0	7.052 5, 0	8.120 6, 0
.12	2.445 3, 0	2.913 5, 0	3.471 6, 0	4.136 4, 0	4.928 6, 0	5.872 6, 0	6.997 2, 0	8.337 3, 0	9.934 0, 0	1.183 7, 1	1.410 3, 1
.14	3.245 5, 0	4.015 3, 0	4.967 6, 0	6.145 8, 0	7.603 5, 0	9.407 0, 0	1.163 8, 1	1.439 8, 1	1.781 3, 1	2.203 9, 1	2.726 6, 1
.15	3.973 0, 0	5.017 6, 0	6.336 9, 0	8.003 1, 0	1.010 7, 1	1.276 6, 1	1.612 1, 1	2.036 0, 1	2.571 4, 1	3.247 5, 1	4.101 4, 1
.16	5.466 1, 0	7.059 7, 0	9.118 1, 0	1.177 6, 1	1.521 0, 1	1.964 5, 1	2.537 2, 1	3.277 0, 1	4.232 4, 1	5.466 3, 1	7.060 1, 1
.165	7.510 3, 0	9.819 6, 0	1.283 9, 1	1.678 7, 1	2.194 9, 1	2.869 8, 1	3.752 2, 1	4.906 0, 1	6.414 5, 1	8.386 9, 1	1.096 6, 2
.17	6.662 4, 0	8.829 5, 0	1.170 1, 1	1.550 6, 1	2.054 8, 1	2.723 2, 1	3.608 9, 1	4.782 5, 1	6.338 2, 1	8.399 4, 1	1.113 1, 2
.175	4.022 6, 0	5.387 5, 0	7.215 6, 0	9.656 9, 1	1.294 2, 1	1.733 7, 1	2.321 0, 1	3.108 0, 1	4.161 8, 1	5.573 0, 1	7.465 0, 1
.18	3.114 3, 0	4.199 0, 0	5.658 1, 0	7.625 9, 0	1.027 6, 1	1.384 4, 1	1.864 8, 1	2.511 2, 1	3.381 2, 1	4.553 6, 1	6.125 8, 1
.185	2.563 1, 0	3.465 5, 0	4.682 6, 0	6.322 9, 0	8.531 8, 0	1.150 4, 1	1.550 0, 1	2.086 6, 1	2.806 5, 1	3.771 4, 1	5.063 1, 1
.19	2.169 9, 0	2.927 9, 0	3.941 9, 0	5.298 1, 0	7.110 2, 0	9.523 7, 0	1.271 8, 1	1.692 6, 1	2.247 1, 1	2.973 5, 1	3.922 9, 1
.195	1.861 7, 0	2.575 7, 0	3.324 6, 0	4.412 9, 0	5.808 2, 0	7.644 5, 0	1.045 1, 1	1.286 0, 1	1.643 6, 1	2.073 7, 1	2.573 7, 1
.20	1.610 3, 0	2.125 7, 0	2.899 6, 0	3.600 6, 0	4.848 3, 0	5.780 0, 0	7.105 0, 0	8.473 5, 0	9.669 7, 0	1.027 7, 1	9.551 1, 0
.21	1.213 3, 0	1.514 0, 0	1.824 2, 0	2.087 4, 0	2.177 4, 0	1.885 4, 0	8.251 6, −1	−1.664 2, 0	−6.691 4, 0	−1.606 9, 1	−3.285 4, 1
.22	9.042 7, −1	9.990 9, −1	9.526 3, −1	6.124 4, −1	−3.694 1, −1	−2.470 1, 0	−6.599 4, 0	−1.425 9, 1	−2.793 8, 1	−5.171 6, 1	−9.221 5, 1
.23	6.505 0, −1	5.411 4, −1	1.191 1, −1	−9.210 6, −1	−3.117 1, 0	−7.362 6, 0	−1.535 9, 1	−2.966 3, 1	−5.474 5, 1	−9.791 0, 1	−1.660 3, 2
.24	4.345 6, −1	1.210 6, −1	−5.745 3, −1	−2.492 4, 0	−6.092 1, 0	−1.294 2, 1	−2.551 4, 1	−4.691 9, 1	−8.734 6, 1	−1.551 6, 2	−2.702 9, 2
.25	2.904 9, −1	−2.741 9, −1	−1.512 8, 0	−3.934 3, 0	−9.308 8, 0	−1.910 5, 1	−3.706 7, 1	−6.921 8, 1	−1.256 6, 2	−2.231 4, 2	−3.891 5, 2
.26	7.745 7, −2	−6.522 0, −1	−2.332 5, 0	−5.826 3, 0	−1.276 6, 1	−2.587 0, 1	−4.995 1, 1	−8.820 0, 1	−1.691 4, 2	−3.005 3, 2	−5.242 0, 2
.27	−7.533 7, −2	−1.018 0, 0	−3.161 6, 0	−7.630 6, 0	−1.645 0, 1	−3.318 8, 1	−6.400 7, 1	−1.193 7, 2	−2.167 5, 2	−3.846 3, 2	−6.690 6, 2
.28	−2.158 6, −1	−1.373 9, 0	−4.002 6, 0	−9.286 1, 0	−2.030 6, 1	−4.097 7, 1	−7.900 8, 1	−1.472 4, 2	−2.667 4, 2	−4.713 3, 2	−8.138 8, 2
.29	−3.467 4, −1	−1.726 2, 0	−4.856 0, 0	−1.140 8, 1	−2.440 1, 1	−4.913 1, 1	−9.452 1, 1	−1.758 7, 2	−3.169 1, 2	−5.549 9, 2	−9.452 5, 2
.30	−4.657 8, −1	−2.073 7, 0	−5.723 7, 0	−1.338 8, 1	−2.861 0, 1	−5.754 0, 1	−1.151 6, 2	−2.042 3, 2	−3.644 1, 2	−6.284 9, 2	−1.045 9, 3
.32	−6.981 6, −1	−2.762 1, 0	−7.498 7, 0	−1.747 8, 1	−3.725 4, 1	−7.443 9, 1	−1.410 4, 2	−2.544 6, 2	−4.368 4, 2	−7.090 7, 2	−1.072 5, 3
.34	−9.103 2, −1	−3.450 1, 0	−9.315 9, 0	−2.166 7, 1	−4.585 0, 1	−9.016 6, 1	−1.659 1, 2	−2.849 2, 2	−4.503 8, 2	−4.707 9, 2	−7.103 8, 2
.36	−1.114 4, 0	−4.143 3, 0	−1.116 7, 1	−2.583 3, 1	−5.387 0, 1	−1.028 7, 2	−1.795 3, 2	−2.803 7, 2	−3.675 9, 2	−3.144 5, 2	2.063 1, 2
.38	−1.309 9, 0	−4.849 6, 0	−1.304 1, 1	−2.985 7, 1	−6.073 5, 1	−1.092 7, 2	−1.757 2, 2	−2.248 4, 2	−1.533 2, 2	3.088 0, 2	1.760 5, 3
.40	−1.504 2, 0	−5.563 6, 0	−1.489 6, 1	−3.352 8, 1	−6.565 1, 1	−1.105 2, 2	−1.480 0, 2	−1.045 3, 2	2.151 7, 2	1.249 0, 3	3.850 5, 3
.42	−1.700 7, 0	−6.297 6, 0	−1.673 1, 1	−3.672 4, 1	−6.797 7, 1	−1.008 2, 2	−9.099 4, 1	2.076 4, 2	7.376 1, 2	2.436 3, 3	6.126 6, 3
.44	−1.901 7, 0	−7.051 2, 0	−1.851 1, 1	−3.922 6, 1	−6.689 1, 1	−7.921 9, 1	−8.690 8, −1	3.570 4, 2	1.379 3, 3	3.688 7, 3	7.934 4, 3
.46	−2.109 6, 0	−7.826 8, 0	−2.020 1, 1	−4.081 7, 1	−6.160 0, 1	−4.384 1, 1	1.236 1, 2	6.859 8, 2	2.062 1, 3	4.703 5, 3	8.383 2, 3
.48	−2.327 2, 0	−8.626 7, 0	−2.131 7, 1	−4.126 4, 1	−5.133 8, 1	6.467 5, 0	3.550 9, 2	1.048 7, 3	2.661 2, 3	5.074 1, 3	6.390 1, 3
.50	−2.557 5, 0	−9.526 8, 0	−2.220 6, 1	−4.032 2, 1	−3.541 7, 1	7.213 1, 1	4.608 0, 2	1.400 9, 3	3.010 5, 3	4.346 3, 3	1.165 3, 3

Table 14

τ	λ	M	τ	M	τ	M
0	∞	0	0.17	1.005 7	0.28	1.385 8
0.02	2.255 4	0.100 78	0.18	1.041 2	0.29	1.420 2
0.04	1.637 6	0.205 76	0.19	1.076 3	0.30	1.454 8
0.06	1.375 1	0.315 21	0.20	1.111 1	0.32	1.524 4
0.08	1.226 7	0.429 41	0.21	1.145 7	0.34	1.595 0
0.10	1.132 2	0.548 70	0.22	1.180 2	0.36	1.666 7
0.12	1.069 7	0.673 40	0.23	1.214 5	0.38	1.739 8
0.14	1.028 3	0.803 91	0.24	1.249 8	0.40	1.814 0
0.15	1.014 1		0.25	1.283 0	0.42	1.891 0
0.16	1.004 1	0.940 62	0.26	1.317 2	0.44	1.969 8
0.165	1.001 1		0.27	1.351 5	0.46	2.051 0

Superaerodynamics, Mechanics of Rarefied Gases

Hsue-shen Tsien*
(*California Institute of Technology*)

Introduction

Zahm[1] in 1934 published an article on the aerodynamics of highly rarefied gases, a branch of fluid mechanics which he called superaerodynamics. At that time, however, with the means of propulsion then available, flight at extreme altitudes did not seem to be realizable. Therefore, superaerodynamics has been considered as a subject of academic interest rather than one of practical engineering importance. With the recent perfection of the rocket as a propulsive power plant, the situation is radically changed and there should be no limit to the altitude that can be reached by an aircraft. There are even indications that the optimum flight altitude of long-range rocket airplanes is approximately 60 miles. At these and higher altitudes, the air density is so low that the fluid must be thought of as one having a coarse structure and not as a continuous medium of conventional fluid mechanics. Then concepts of superaerodynamics are needed to guide the design of such an aircraft.

Besides its applications to high-altitude flights, superaerodynamics should have many applications to industrial processes in which low density gases are involved. The knowledge of superaerodynamics should be of invaluable help in improving the efficiency of these processes. For instance, the pumping of low density gases will be of more and more importance because of the increasingly frequent use of high vacuum in distillation and other chemical engineering processes. The improvement in the design of such gas pumps definitely needs the understanding of the principles of mechanics of low density gases. It is the purpose of the present paper to discuss the fundamental concepts of this new branch of fluid mechanics and to indicate some of the results already obtained. The field of superaerodynamics is still relatively uncultivated. Here and there the ground has been broken by a few physicists. However, further effort is required to develop this branch of fluid mechanics into an aid in engineering design and research.

Received May 20, 1946.

Journal of the Aeronautical Sciences, vol. 13, pp. 653 – 664, 1946.

* Associate Professor of Aeronautics, Daniel Guggenheim Aeronautical Laboratory. Now Associate Professor of Aeronautical Engineering, Massachusetts Institute of Technology.

Mean Free Path and Realms of Fluid Mechanics

As a first approximation, the gas may be considered as an aggregate of rapidly moving particles which are constantly colliding with each other. The influence of the particles on each other can be conveniently neglected until they are so close together that a "collision" takes place. Then the coarseness of the structure of the gaseous medium can be expressed by the parameter l which is the distance the particles or molecules travel between collisions. Since the instantaneous velocity distribution and density distribution in the gas are far from uniform, one can only conveniently use the statistical average of a quantity instead of the instantaneous value of the quantity. The distance l is then the statistical average over the billions and billions of molecules concerned. This average l is called the mean free path of the gas. If the mean free path is small in comparison with the dimension of the flow field or the dimension of the body in the flow field, then the gas can be considered as a continuum and the ordinary gas dynamics is sufficient for the analysis of these flows. If the means free path is not negligible when compared with the dimension of the body, then the effects of the discrete character of the gas must be taken into account in the calculation. Then if L is the linear dimension of the body, superaerodynamics can be defined as the aerodynamics of flows where the ratio l/L is not negligible.

The free mean path l can be calculated from the mean velocity of the molecules, the density of the gas, and the "effective radius of the molecules." However, there is no direct way of measuring the effective radius of the molecules; therefore, it is more fruitful to express the quantity l in terms of a measurable quantity such as the viscosity of the gas. This is easily done by the following consideration. If the gas flows in the x-direction with the macroscopic velocity $u(y)$ (Fig. 1), the gradient of velocity is then $\partial u/\partial y$. Now the gas molecules in a lower layer move into an upper layer with the average velocity $\bar{v}$ of the molecules. The distance they will travel before they lose their identity by collision with other molecules is l. The difference in macroscopic velocity of the layer where the molecules originated and the layer where the molecules are mixed is then $l(\partial u/\partial y)$. The mass of molecules crossing a unit area of the layer proportional to $\rho\bar{v}$ where ρ is the density of the fluid. Therefore, momentum exchange between layers is proportional to $\rho\bar{v}l(\partial u/\partial y)$. This is the viscous shearing stress. Since the coefficient of viscosity is defined as

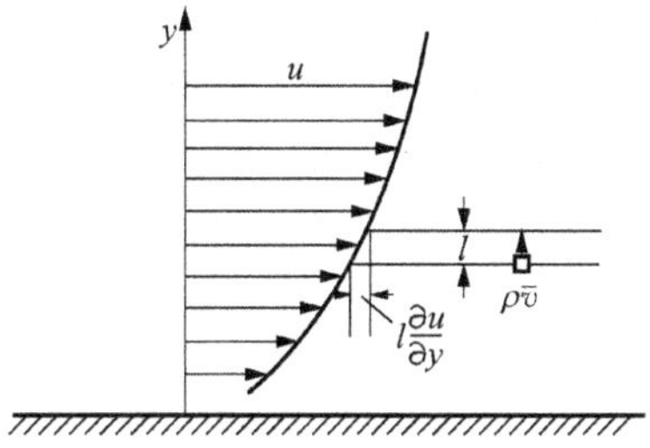

Fig. 1 Shearing action in parallel flows

$$\tau = \mu(\partial u/\partial y)$$

μ is proportional to $\rho\bar{v}l$. The proportionality constant can be calculated by the kinetic theory. The most accurate calculation is that due to Chapman[2] who gives

$$\mu = 0.499\,\rho\bar{v}l \tag{1}$$

The average velocity $\bar{v}$ is closely connected with the "velocity of sound" a by the equation

$$\bar{v} = \sqrt{\frac{8}{\pi}}\sqrt{\frac{p}{\rho}} = \sqrt{\frac{8}{\pi\gamma}}a \tag{2}$$

where γ is the ratio of specific heats. From Eqs. (1) and (2), the mean free path is given in terms of the kinematic viscosity $\nu = \mu/\rho$ and the velocity of sound a by

$$l = 1.255\sqrt{\gamma}(\nu/a) \tag{3}$$

For air, the quantity $l(p/p_0)$ in inches is tabulated in Table 1, where p and p_0 are the pressure and standard pressure (one atmosphere), respectively. Thus, for ordinary pressures, the mean free path is truly negligible in comparison with the body dimensions, and therefore gas dynamics is sufficient for aerodynamic calculations. However, at extremely high altitudes where the pressure may be only one-millionth of the pressure at the surface of the earth, the mean free path will be comparable to the body dimensions. A clear demonstration of this fact is perhaps that of Fig. 2 where the mean free path calculated by Maris[3] is plotted against the altitude. Maris assumed that the air temperature above 80 km (50 miles) is constant and has the value 97℃, −43℃, −23℃, and −53℃ for summer day, summer night, winter day, and winter night, respectively. At 50 miles altitude, the mean free path is seen to be nearly 1 in. The air there is certainly not a continuum.

Table 1 Mean free path for air

Temperature/℃	$l(p/p_0)$ /in	Temperature/℃	$l(p/p_0)$ /in
0	2.32×10^{-6}	260	5.33×10^{-6}
20	2.55×10^{-6}	280	5.57×10^{-6}
40	2.78×10^{-6}	300	5.85×10^{-6}
60	3.00×10^{-6}	320	6.05×10^{-6}
80	3.23×10^{-6}	340	6.28×10^{-6}
100	3.46×10^{-6}	360	6.52×10^{-6}
120	3.69×10^{-6}	380	6.77×10^{-6}
140	3.92×10^{-6}	400	7.00×10^{-6}
160	4.16×10^{-6}	420	7.24×10^{-6}
180	4.40×10^{-6}	440	7.49×10^{-6}
200	4.62×10^{-6}	460	7.73×10^{-6}
220	4.85×10^{-6}	480	7.96×10^{-6}
240	5.10×10^{-6}	500	8.20×10^{-6}

p_0 = Standard pressure, one atmosphere.

The flow conditions near the wall for the case in which the mean free path l is small but not negligible compared with the body dimension L or the thickness δ of the boundary layer were first investigated theoretically by J. C. Maxwell in 1879. It was found that the gas no longer sticks to the surface but slips over the surface with a definite velocity. This phenomenon

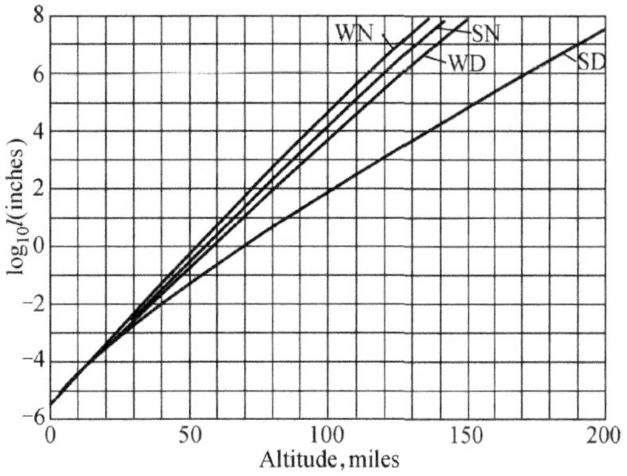

Fig. 2 Mean free path at different altitudes; SD = summer day, SN = summer night, WD = winter day, and WN = winter night (H. B. Maris)

was also suggested by the experiments of Kundt and Warburg in 1875, on the damping of a vibrating disc by gas at low pressure. Later experiments and theoretical study completely verified this conclusion. This type of flow can be called the slip flow. Because of Eq. (3),

$$\frac{l}{\delta}=\frac{l}{L}\frac{L}{\delta}\sim\frac{L}{\delta}\frac{\nu}{a}\frac{1}{L}\sim\frac{L}{\delta}\frac{M}{Re} \tag{4}$$

where M is the Mach Number of the free stream and Re is the Reynolds Number of the flow referred to the length L of the body. For extremely small Reynolds Number, $L/\delta\sim 1$, then

$$l/\delta\sim M/Re,\quad Re\ll 1 \tag{5}$$

For large Reynolds Number, it is well known that $L/\delta\sim\sqrt{Re}$. Then,

$$l/\delta\sim M/\sqrt{Re},\quad Re\gg 1 \tag{6}$$

Therefore, if the interval $1/100 < l/\delta < 1$ is considered as the proper range for, the slip flow, then in the plane of M and Re, this realm of fluid mechanics occupies a region as shown in Fig. 3. The region below this slip-flow region belongs to the realm of the conventional gas dynamics with the usual boundary conditions at the wall.

If the mean free path is much larger compared with the body dimensions, one enters an entirely new realm of fluid mechanics. Here the chances for the collision of molecules among themselves are much smaller than the chances for the collision of molecules with the wall or the surface of the body. Therefore, for the calculation of forces, one need only consider the impact of a stream of molecules with a velocity and energy distribution determined by the thermal equilibrium in the free stream, the Maxwellian distribution. The re-emission of the molecules from the surface will be governed by the accommodation coefficient. But the greatest simplification comes from the fact that one need not consider the distortion of the Maxwellian distribution due to the collision of the re-emitted molecules with the molecules in the stream. The realm of fluid mechanics can thus be called the free molecule flow.

If one takes the limiting ratio of l/L to be 10 for the free molecule flow, then this realm of fluid mechanics occupies a region of the $M-Re$ plan given by

$$l/L \sim M/Re > 10$$

This is shown in Fig. 3. The region between the free molecule flow and the slip flow is the realm of fluid mechanics where the collision between molecules and the collision of molecules with the wall are of equal importance. The problem is extremely complicated and no satisfactory theoretical solution can yet be offered. However, it is certain that as far as the fluid itself is concerned, the characteristic parameters are still the Mach Number M and the Reynolds Number Re as used in gas dynamics.

Stresses and Boundary Conditions in Slip Flow

The fundamental equations governing the fluid flows are the continuity equation, the dynamic equations, and the energy equation. If u_1, u_2, u_3 are the components of velocity in the x_1, x_2, x_3 directions, and p, ρ, T the pressure, density, and temperature of the gas, respectively, these equations are*

$$(\partial\rho/\partial t) + (\partial\rho u_i/\partial x_i) = 0 \tag{7}$$

$$\rho\frac{\partial u_i}{\partial t} + \rho u_j\frac{\partial u_i}{\partial x_j} = -\frac{\partial p}{\partial x_i} - \frac{\partial \tau_{ij}}{\partial x_j} \tag{8}$$

$$\left(\rho\frac{\partial}{\partial t} + \rho u_i\frac{\partial}{\partial x_i}\right)\left(\frac{u_i u_i}{2} + c_p T\right) = \frac{\partial p}{\partial t} - \frac{\partial q_i}{\partial x_i} - \frac{\partial}{\partial x_i}(T_{ij}u_j) \tag{9}$$

In these equations, the τ's are the components of the stress tensor, the q's are the components of the heat flux vector, and c_p is the specific heat at constant pressure.

For gases under ordinary pressures, the stresses are the products of the coefficient of viscosity and linear combinations of the first order derivatives of the velocity components. The components of heat flux are simply the products of coefficient of heat conduction and the components of the gradient of temperature.

These relations between the stresses and the heat flux with the flow quantities are confirmed by the kinetic theory of nonuniform gases developed by Maxwell, Boltzmann, Chapman, D. Enskog, and others. In this theory, the distribution function of the molecules in the velocity space on the probability of molecules having a particular range of velocities plays the predominating role.

If, in the macroscopic sense, the gas is at rest with uniform temperature, the probability function is the well-known Maxwellian distribution. In a nonuniform gas, the first order correction to this Maxwellian distribution gives the ordinary viscous stresses and the heat flux

---

* The summation convention is used here. For instance,

$$\frac{\partial u_K}{\partial x_k} = \frac{\partial u_1}{\partial x_1} + \frac{\partial u_2}{\partial x_2} + \frac{\partial u_3}{\partial x_3}.$$

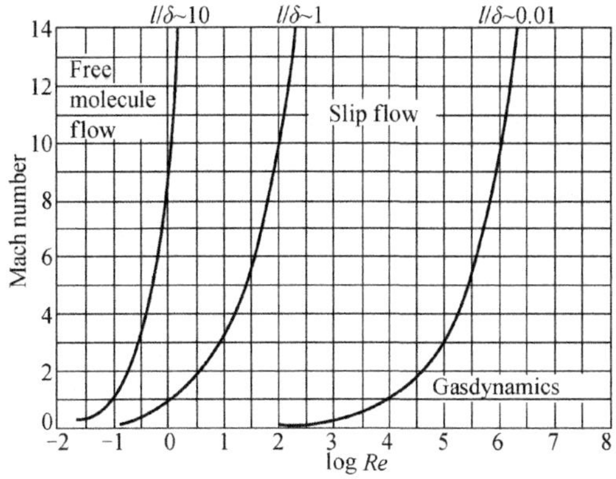

Fig. 3 Realms of fluid mechanics

by conduction, as observed by experiments.

However, the theory also gives a second order correction to the Maxwellian distribution which introduces highly complicated additional terms to the stresses and the heat flux. These additional terms, while negligible for gases under ordinary pressure, are important for rarefied gases. In fact, they form the basis for interpretation of many experimentally observed phenomena, such as the operation of a radiometer.

The investigation of Burnett[4] gives the following forms of the heat flux and the stresses:

$$q_i = -\lambda \frac{\partial T}{\partial x_i} + \theta_1 \frac{\mu^2}{\rho T}\frac{\partial u_j}{\partial x_j}\frac{\partial T}{\partial x_i} + \theta_2 \frac{\mu^2}{\rho T}\left[\frac{2}{3}\frac{\partial}{\partial x_i}\left(T\frac{\partial u_j}{\partial x_j}\right) + 2\frac{\partial u_j}{\partial x_i}\frac{\partial T}{\partial x_j}\right] + \left(\theta_3 \frac{\mu^2}{\rho p}\frac{\partial p}{\partial x_j} + \theta_4 \frac{\mu^2}{\rho}\frac{\partial}{\partial x_j} + \theta_5 \frac{\mu^2}{\rho T}\frac{\partial T}{\partial x_j}\right) e_{ji} \tag{10}$$

$$\tau_{ij} = -2\mu e_{ij} + K_1 \frac{\mu^2}{p}\frac{\partial u_k}{\partial x_k} e_{ij} + K_2 \frac{\mu^2}{p}\left[-\overline{\frac{\partial}{\partial x_i}\left(\frac{1}{\rho}\frac{\partial p}{\partial x_j}\right)} - \overline{\frac{\partial u_k}{\partial x_i}\frac{\partial u_j}{\partial x_k}} - 2\overline{e_{ik}\frac{\partial u_j}{\partial x_k}}\right] + K_3 \frac{\mu^2}{\rho T}\overline{\frac{\partial^2 T}{\partial x_i \partial x_j}} + K_4 \frac{\mu^2}{\rho p T}\overline{\frac{\partial p}{\partial x_i}\frac{\partial T}{\partial x_j}} + K_5 \frac{\mu^2}{\rho T^2}\overline{\frac{\partial T}{\partial x_i}\frac{\partial T}{\partial x_j}} + K_6 \frac{\mu^2}{p}\overline{e_{ik} e_{kj}} \tag{11}$$

In these equations, λ is the coefficient of heat conduction and μ is the coefficient of viscosity. The first terms of the equations are thus the ordinary heat flux and viscous stresses. θ_1, θ_2, θ_3, θ_4, θ_5 and K_1, K_2, K_3, K_4, K_5, K_6 are pure numbers. Furthermore,

$$e_{ij} = 1/2\left(\frac{\partial u_i}{\partial x_j} + \frac{\partial u_j}{\partial x_i}\right) - 1/3\left(\frac{\partial u_k}{\partial x_k}\delta_{ij}\right) \tag{12}$$

where $\delta_{ij} = 1$ if $i = j$ and $\delta_{ij} = 0$ if $i \neq j$. Any tensor A_{ij} with a bar over it has the following meaning:

$$\overline{A_{ij}} = 1/2(A_{ij} + A_{ji}) - 1/3(A_{kk}\delta_{ij}) \tag{13}$$

The approximate values for θ's and K's are

$$\theta_1 = \frac{15}{4}\left(\frac{7}{2} - \frac{T}{\mu}\frac{d\mu}{dT}\right), \theta_2 = \frac{45}{8}, \theta_3 = -3, \theta_4 = 3, \theta_5 = \frac{3}{2}\left(5 - \frac{T}{\mu}\frac{d\mu}{dT}\right) \tag{14}$$

$$K_1 = \frac{4}{3}\left(\frac{7}{2} - \frac{T}{\mu}\frac{d\mu}{dT}\right), K_2 = 2, K_3 = 3, K_4 = 0, K_5 = \frac{3T}{\mu}\frac{d\mu}{dT}, K_6 = 8 \tag{15}$$

More accurate values for K's are given by Burnett[4] for molecules of rigid elastic spheres. They differ little from those of Eq. (15).

It is immediately clear by examining Eqs. (10) and (11) that the additional terms in the heat flux and the stress tensor are only of importance in rarefied gases. As an example, the magnitude of the ratio of the second terms to the first terms of these equations will be estimated presently. Let U be the free stream velocity and T_0 the free stream density and temperature, c_p the specific heat at constant pressure. For gases, the Prandtl number $c_p\mu/\lambda$ is rarely unity. Then these ratios have the same order of magnitude given by

$$\frac{U}{L}\frac{\mu}{\rho_0 c_p T_0} = \frac{\mu}{\rho_0 UL}\frac{U^2}{c_p T_0} = (\gamma - 1)\frac{M^2}{Re}$$

where M is the free stream Mach Number and Re the Reynolds Number. According to Eq. (4), the order of magnitude of the additional heat flux or stresses to the ordinary heat flux and stresses is then $M(l/L)$. That is, the additional terms are only of importance if the product of Mach Number and the ratio of mean free path to body dimension is large. Therefore, for the gas dynamical flows with small mean free path, the ordinary heat conduction terms and viscous stress terms suffice. This is also true for the slip flows if the Mach Number is small. For slip flows at high Mach Number, the conventional Navier-Stokes equations are no longer a true description of the physical relations and the complicated forms of heat flux q_i and the stresses τ_{ij} given by Eqs. (10) and (11) must be used.

Since the additional terms in q_i and τ_{ij} contain higher derivatives than the first order, they will raise the order of the system of partial differential equations given by Eqs. (7), (8), and (9). Hence, to solve these equations, one needs more boundary conditions than in the case of gas dynamics. In fact, the question of proper boundary conditions for slip flows of large Mach Number where such additional terms are required, has not yet been answered. This is then one of the fundamental problems in superaerodynamics. The theoretical study of the boundary conditions must, of course, be based on the kinetic point of view. Parallel to such theoretical investigation, it seems advisable to obtain also an experimental check on the validity of Eqs. (10) and (11). One of the possible experiments would be to observe the flow of rarefied gas between two concentric cylinders, one rotating and the other held fixed. If the length of the cylinders is large in comparison with the gap between the cylinders, the flow will be two-dimensional with the radial distance as the only parameter. The physical situation is thus the simplest, and amenable to detailed examination. Indeed, the same experiment has been performed by Millikan and his collaborators[5] at low rotative speeds for the same type of investigation. The high rotative speeds necessary to obtain high flow velocities will certainly introduce difficulties in the technique but these do not seem to be insurmountable.

The results of the kinetic theory discussed in the previous paragraphs are based upon the assumption of simple molecules with the only intrinsic energy due to translation. Generally, however, the gas molecules have three kinds of intrinsic energies: the translational energy, the rotational energy, and the vibrational energy. When the gas is in equilibrium, the distribution of the total energy among the three forms is uniquely determined by the properties of the gas molecules concerned. If this equilibrium is destroyed by a sudden change in the external conditions, such as an expansion, new equilibrium will be reached by numerous collisions of the molecules under these new conditions. However, it is known through ultrasonic dispersion, etc., that the process of reaching equilibrium in the vibrational energy is a slow one. In other words, a large number of collisions is necessary to excite the vibration degrees of freedom of the molecules properly. Since the average velocity of the molecules is $\bar{v}$, the average distance traveled by the molecules per unit time is $\bar{v}\,l$. The number of collisions made by the molecules per unit time is then $\bar{v}/l$. Therefore, the time t, generally called the relaxation time, necessary to excite the vibrational energy is proportional to the mean free path at a given temperature. In other words, the relaxation time is inversely proportional to the pressure of the gas. The measured value of t for several typical gases is given in Table 2[6]. Therefore, at extremely low pressures, the relaxation time may be of the order of 1 sec. If this is coupled with high flow velocity, the gas must have passed the body before the vibrational degrees of freedom are excited. Therefore, in low density gas flows, the vibrational energy of the gas can be considered as fixed at the "free stream" value. This decrease in the effective number of degrees of freedom of the gas molecules tends to raise the value of γ, the ratio of specific heats. It is fortunate that for air at ordinary temperatures, the equilibrium vibrational energy is small, therefore this freezing of vibrational degrees will not greatly alter the value of γ. However, at higher temperatures and for other polyatomic gases, the effect may be appreciable.

Table 2 Relaxation time t for the excitation of vibrational energy at one atmosphere[6]

Pure CO_2		O_2	
Temperature/℃	$t \times 10^6$/sec	Temperature/℃	$t \times 10^6$/sec
19.2	10.8	40.7	4.4
47.1	10.6	49.7	2.9
58.1	8.1	77.9	4.0
88.7	7.9	100.2	3.6
		138.0	3.1
		140.6	3.7
		141.4	3.3

Boundary Conditions for Slip Flows of Small Mach Number

As concluded in the preceding section, the slip flows of small Mach Number can be calculated by using the Navier-Stokes equations. The proper boundary conditions for such flows were first investigated by Maxwell. Millikan[5] has reduced Maxwell's consideration to a simple form. A more detailed theoretical treatment of the problem by Boltzmann's H-theorem was given by Epstein[7]. The result of these investigations shows that if the Mach Number of the flow is small, then the flow velocity along the wall, the normal gradient $\partial u/\partial y$ of u, and the temperature gradient $\partial T/\partial x$ along the wall are connected by the relation[8]

$$u = \frac{\mu}{\beta}\frac{\partial u}{\partial y} + \frac{3}{4}\frac{\mu}{\rho T}\frac{\partial T}{\partial x} \tag{16}$$

at $y = 0$, where β is the coefficient of sliding or external friction and μ the coefficient of viscosity. The second term in Eq. (16) is the velocity due to temperature gradient and is called the creep velocity. The ratio of μ and β is given by

$$\mu/\beta = 0.998\ [(2/f) - 1]l \tag{17}$$

f is the fraction of the tangential momentum of the impinging molecules transmitted to the wall. Maxwell himself interpreted the fractional value of f as meaning that a fraction f of the surface reflects diffusely and the remainder specularly. The value of f is listed in Table 3 as given by Millikan[5]. It is seen from Eq. (17) that if l is small compared with the scale of the velocity gradient or the thickness δ of the boundary layer, then β is large and u at the wall is small. This is the reason why in gas dynamics one generally imposes the boundary condition $u = 0$ at the wall. If l/δ is not negligible, then there is a slip of the gas at the wall.

Table 3 Values of f[5]

	f
Air or CO_2 on machined brass or old shellac	1.00
Air on mercury	1.00
Air on oil	0.895
CO_2 on oil	0.92
Hydrogen on oil	0.925
Air on glass	0.89
Helium on oil	0.874
Air on fresh shellac	0.79

Table 3 shows that the value of f is very nearly unity. Therefore, the molecules, after striking the surface, have a strong tendency to reflect diffusely. In the scale of molecular dimensions, the surface of a solid body must be extremely jagged even if it is highly polished in the macroscopic sense. Then one cannot expect the molecules to reflect in any uniform direction. The tendency toward diffusive reflection is thus, perhaps, anticipated. Furthermore, a closer study of the behavior of the molecules colliding with the solid surface seems to indicate the temporary adsorption of the molecules on the surface. This combined structure of gas molecule and the surface materials then breaks up with a certain average rate determined by the "lifetime" of the combined structure. The molecules thus freed are re-emitted from the surface. With this procedure, the molecules re-emerge from the surface with all their prior history obliterated. In particular, the direction of re-emission is completely dependent upon the direction of impinging. In other words, the molecules re-meit diffusely.

Similar to this velocity discontinuity, there is a temperature discontinuity at the wall first found by Smoluchowski in 1898. Namely,

$$\kappa(T - T_w) = \lambda(\partial T/\partial y), y = 0 \tag{18}$$

where T is temperature of the gas, T_w is the temperature of the wall, and κ is the coefficient of temperature discontinuity. This coefficient κ can be expressed in terms of the mean free path by a consideration similar to Maxwell's[9]. The result is

$$\frac{1}{\kappa} = 1.996\left(\frac{2-\alpha}{\alpha}\right)\frac{\gamma}{\gamma+1}\frac{1}{\mu C_p}l \tag{19}$$

where α is the accommodation coefficient introduced by Knudsen[10]. α can be defined as the fractional extent to which those molecules that fall on the surface and are reflected or re-emitted from it, have their mean energy adjusted or "accommodated" toward what it would be if the returning molecules were issuing as a stream out of the mass of gas at the temperature of the wall. This adaptation of the impinging molecules to the wall conditions is, of course, expected from the temporary adsorption of molecules on the surface, as mentioned before. Thus, if E_i is the energy brought up to unit area of the wall per second by the incident molecules, and E_r the energy carried away by the re-emitted molecules, and E_w the energy that would be carried away if the gas is at the wall temperature, then

$$E_i - E_r = \alpha(E_i - E_w) \tag{20}$$

Of course, strictly speaking, α is different for the three different types of molecular energies: translational, rotational, and vibrational. However, experimental evidence seems to indicate that one coefficient is generally sufficient. The older measured values of α are given in many books on kinetic theory of gases[11]. Wiedmann[12] recently determined the value of α for air on metals. The results are given in Table 4. Wiedmann concluded that the value of the accommodation coefficient is independent of the nature of the technical surface of the metal. This again supports the concept of the microscopic character of the surface collision.

Table 4 Accommodation coefficient α for air[12]

	Accommodation coefficient	
Surface description	Minimum	Maximum
Flat black lacquer on bronze	0.881	0.894
Bronze, polished	0.91	0.94
Bronze, machined	0.89	0.93
Bronze, etched	0.93	0.95
Cast iron, polished	0.87	0.93
Cast iron, machined	0.87	0.88
Cast iron, etched	0.89	0.96
Aluminum, polished	0.87	0.95
Aluminum, machined	0.95	0.97
Aluminum, etched	(0.89)	(0.97)

Slip Flows at Small Mach Numbers

If the Mach Number is extremely small, the effect of the compressibility of the gas can be neglected and the fluid can be treated as incompressible. In other that the mean free path will be comparable with body dimension, the Reynolds Number must also be small. The flow can then be calculated by the well-known method of Stokes[13] who neglected the inertia terms, which are small in comparison with the viscous and pressure terms. The resistance of a sphere of radius R moving with a velocity U through a fluid of viscosity and coefficient β of sliding friction is given by Basset[14]. If the drag coefficient C_D is defined as the drag divided by $(\rho/2)U^2\pi R^2$ then

$$C_D = \frac{12}{Re}\left\{\frac{1+2[\mu/(\beta R)]}{1+3[\mu/(\beta R)]}\right\} \tag{21}$$

where Re is the Reynolds Number UR/ν. According to Eqs. (3) and (17),

$$\frac{\mu}{\beta R} = 0.998\left(\frac{2}{f}-1\right)\frac{l}{R} = 1.253\sqrt{\gamma}\left(\frac{2}{f}-1\right)\frac{M}{Re} \tag{22}$$

Hence, Eq. (21) can be rewritten as

$$C_D = \frac{12}{Re}\left\{\frac{1+2.506\sqrt{\gamma}[(2/f)-1](M/Re)}{1+3.759\sqrt{\gamma}[(2/f)-1](M/Re)}\right\}(\text{sphere}) \tag{23}$$

Fig. 4 shows the drag coefficient as function of M and Re with $\gamma = 1.405$ and $f = 1$. It is seen that there is an appreciable reduction in the drag coefficient due to the slip phenomenon. It also introduces a large effect of Mach Number, although this is not the compressibility effect in the ordinary sense.

The similar case of the uniform two-dimensional motion of a circular cylinder can be obtained by a slight modification of Lamb's solution[15]. If the drag coefficient is defined as the drag per unit length divided by $(\rho/2)U^2(2R)$ where R is the radius of the cylinder,

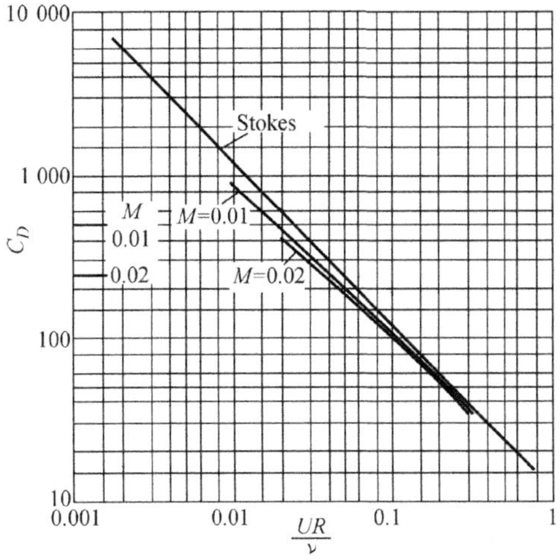

Fig. 4 Drag coefficient C_D as functions of Mach Number M and Reynolds Number for spheres of radius R, $f=1$, $\gamma=1.405$

$$C_D = 4\pi/Re\left[\log\frac{4}{Re} - 1.281\ 07 + 1.253\sqrt{\gamma}\left(\frac{2}{f}-1\right)\frac{M}{Re}\right]\text{(cylinder)} \tag{24}$$

where Re is again the Reynolds Number UR/ν. The results of calculation are again plotted in terms of M and Re in Fig. 5. The reduction of drag due to slip is again evident.

For the fully established laminar flow through a circular tube of radius R, the velocity u as a function of the radial distance from the axis of the tube is determined by

$$\frac{1}{\mu}\frac{dp}{dx} = \frac{1}{r}\frac{d}{dr}\left(r\frac{du}{dr}\right) \tag{25}$$

The pressure gradient along the tube is a constant. Therefore, Eq. (25) yields

$$du/dr = [1/(2\mu)](dp/dx)r \tag{26}$$

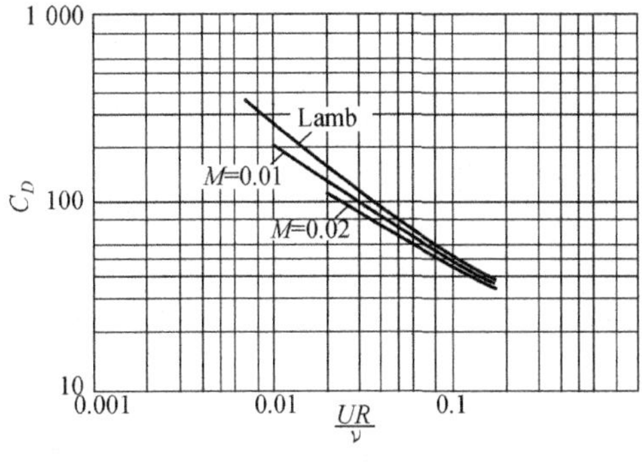

Fig. 5 Drag coefficient C_D as functions of Mach Number M and Reynolds Number for cylinders of radius R, $f=1$, $\gamma=1.405$

and

$$u=(1/4\mu)(\mathrm{d}p/\mathrm{d}x)r^2+u_0 \tag{27}$$

where u_0 is the velocity at the center of the tube. The boundary condition that

$$\beta u_{r=R}=-\mu(\mathrm{d}u/\mathrm{d}r)_{r=R}$$

then gives

$$-\mathrm{d}p/\mathrm{d}x=(4\mu/R^2)/[1+2\mu/(\beta R)]u_0 \tag{28}$$

The relation between the average velocity U and the maximum velocity u_0 can be easily obtained by integrating Eq. (27). Thus

$$u_0=U\{[1+2\mu/(\beta R)]/[1/2+2\mu/(\beta R)]\} \tag{29}$$

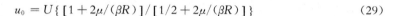

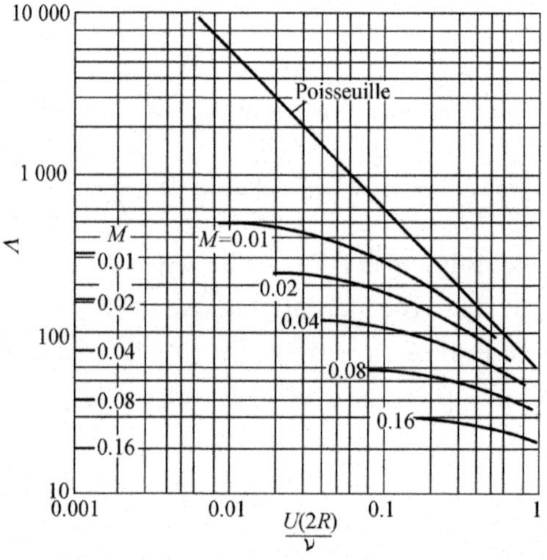

Fig. 6 Pressure drop coefficient Λ as functions of Mach Number M and Reynolds Number Re for flow in a tube of radius R, $f=1$, $\gamma=1.405$

As the ratio of mean free path to the radius increases, the value of $\mu/\beta R$ will increase according to Eq. (22). Then Eq. (29) shows that the average velocity U approaches the maximum velocity u_0. In other words, the velocity distribution across the tube becomes more uniform. This is, of course, expected from the "slip" at the wall.

If the pressure drop coefficient Λ is defined as

$$\Lambda=[4R/(\rho U^2)](-\mathrm{d}p/\mathrm{d}x) \tag{30}$$

then Eqs. (19) and (20) give immediately

$$\Lambda=\frac{64}{Re}\frac{1}{1+10.024\sqrt{\gamma}[(2/f)-1](M/Re)} \tag{31}$$

where Re is the Reynolds Number based upon the diameter, $U(2R)/\nu$. Here again the effect of slip is to reduce the resistance coefficient. For extremely small mean free path the value of the second fraction is unity and the equation reduces to the well-known formula for Poiseuille flow. Fig. 6 is a graph for Λ as function of M and Re, assuming as before $f = 1$ and $\gamma = 1.405$. Experiments on the slip flow in a tube were made by Knudsen[16,17], Gaede[18], and Ebert[19,20]. The correctness of Eq. (31) is fully verified. In fact, it can be used to determine the reflection factor f for the wall materials.

Free Molecule Flows at Small Mach Numbers

Free molecule flow at low velocities has been studied for two cases: the flow around a sphere, and the flow in a tube. The calculation made by Epstein[7] has shown that the drag coefficient C_D can be expressed

$$C_D = 8/3\sqrt{8/(\pi\gamma)}\,C(1/M) \tag{32}$$

where C is a constant. Epstein determined C with diffuse reflection of molecules to be 1.442 for spheres that do not conduct heat and 1.393 for spheres that conduct heat perfectly. The experimental value of C for charged oil drop through air found by Millikan[21] is 1.365. The lower experimental value can be accounted for by a small tendency toward specular reflection of the molecules at the surface. By using the value of $C = 1.393$ and $\gamma = 1.405$, the drag coefficients are indicated in Fig. 4 for small Reynolds Number Re. In this flow régime, the C_D is independent of Re. Fig. 4 shows that in the transition region between slip flow and free molecule flow, the drag coefficient actually decreases with decrease in Reynolds Number.

The free molecule flow through a circular tube was first investigated by Knudsen[16] and Smoluchowski[22]. Here the macroscopic velocity of the gas is constant across the section of the tube as could be expected from the large slip in the present case. The pressure drop can be expressed in terms of the pressure drop coefficient Λ defined by Eq. (30) as

$$\Lambda = [3\sqrt{\pi}/(\sqrt{2}\gamma)](f/2 - f)(1/M) \tag{33}$$

where f is fraction of diffuse reflection as defined previously. By assuming $f = 1$, and $\gamma = 1.405$, values of Λ are calculated from Eq. (33). They are indicated in Fig. 6 at the left side of the diagram. It is seen that, similar to the drag coefficient for the sphere, Λ actually decreases for a decrease of Reynolds Number in the transition region from slip flow to free molecule flow.

The absence of Reynolds Number in Eqs. (32) and (33) simply means that the mean free path does not enter into this type of flow because of the smallness of the body dimension when compared with it. Thus, for both gas dynamical flow and for free molecule flow, viscosity or Reynolds Number is a secondary effect. The primary effect is given by the Mach Number. In the slip flow region, both the Reynolds Number and the Mach Number are of the same importance.

Free Molecule Flow at Large Mach Numbers

Epstein and Smoluchowski assumed the velocity of macroscopic motion to be small in comparison with the microscopic molecular velocity in their investigations of free molecule flows. This means, of course, that the Mach Number of the flow is much smaller than unity. For large Mach Numbers, this restriction must be removed. The calculation should be simple because of the inherent simplicity of the physical situation. Sänger[23,24] has solved this problem but his calculation[23] is rather complicated. With the belief that a simpler derivation of the fundamental formulas will be helpful for an understanding of his results, the author will try to reproduce his basic equations by following Epstein's general method for free molecule flows[7].

If ξ', η', ζ' are the components of velocity of a molecule in the directions x', y', z' of a coordinate system in which the macroscopic velocity of the gas is zero, then since the velocity distribution is not modified by the impact with the body because of absence of collision between molecules in the free stream and re-emitted molecules from the surface of the body, the distribution is Maxwellian, or

$$N_{\xi'\eta'\zeta'} = N(h/\pi)^{3/2} e^{-h(\xi'^2+\eta'^2+\zeta'^2)} \tag{34}$$

$N_{\xi'\eta'\zeta'}$ is the number of molecules per unit volume in the range of velocities ξ' to $\xi' + d\xi'$, η' to $\eta' + d\eta'$, and ζ' to $\zeta' + d\zeta'$ divided by $d\xi'\, d\eta'\, d\zeta'$. N is total of molecules per unit volume. h is related to the most probable velocity c_i of the molecules in the free stream by the expression

$$h = 1/c_i^2 \tag{35}$$

where $c_i^2 = 2RT = 2(p/\rho)$.

Now, if the observer moves with a velocity e_1U, e_2U, e_3U in the directions x', y', z', then in the relative coordinate system x, y, z where the observer is considered at rest, the velocities ξ, η, ζ of the molecule in the directions x, y, z are

$$\begin{aligned} \xi &= \xi' - e_1U \\ \eta &= \eta' - e_2U \\ \zeta &= \zeta' - e_3U \end{aligned} \tag{36}$$

Therefore, in the new coordinate system, the number of molecules having velocity component between ξ, η, ζ and $\xi + d\xi$, $\eta + d\eta$, $\zeta + d\zeta$ is

$$N_{\xi\eta\zeta}\, d\xi d\eta d\zeta = N\left(\frac{h}{\eta}\right)^{3/2} e^{-h[(\xi+e_1U)^2+(\eta+e_2U)^2+(\zeta+e_3U)^2]} \tag{37}$$

Suppose there is a surface dS whose normal is the x-axis. During 1 sec., the molecules, having velocity components between ξ, η, ζ and $\xi + d\xi$, $\eta + d\eta$, $\zeta + d\zeta$, and striking the surface dS, will be contained at a given moment in the cylinder with dS as base and the length $\sqrt{\xi^2 + \eta^2 + \zeta^2}$ in the direction ξ, η, ζ. The volume of the cylinder is equal to the area of the base dS multiplied by the heights $-\xi$. The number of this kind of molecules is then $N_{\xi\eta\zeta}\mathrm{d}\xi\mathrm{d}\eta\mathrm{d}\zeta(-\xi)\mathrm{d}S$. The number of molecules between ξ, η, ζ and $\xi+\mathrm{d}\xi$, $\eta+\mathrm{d}\eta$, $\zeta+\mathrm{d}\zeta$ that will strike a unit area with x-axis as normal is then $-\xi N_{\xi\eta\zeta}\mathrm{d}\xi\mathrm{d}\eta\mathrm{d}\zeta$.

To find the total number n of molecules striking this unit area, one has to integrate $-\xi N_{\xi\eta\zeta}\mathrm{d}\xi\mathrm{d}\eta\mathrm{d}\zeta$ from $\xi=-\infty$, $\eta=-\infty$, $\zeta=-\infty$ to $\xi=0$, $\eta=\infty$, $\zeta=\infty$. The upper limit for ξ is easily understood since no molecules with positive velocity in the x- direction can strike this area. Thus

$$n=-N\left(\frac{h}{\pi}\right)^{3/2}\int_{-\infty}^{0}\mathrm{d}\xi\int_{-\infty}^{\infty}\mathrm{d}\eta\int_{-\infty}^{\infty}\zeta\,\mathrm{e}^{-h[(\xi+e_1U)^2+(\eta+e_2U)^2+(\zeta+e_3U)^2]}\mathrm{d}\zeta \tag{38}$$

The integration can easily be carried out; the result is

$$n=N\left\{\frac{1}{2\sqrt{\pi h}}\mathrm{e}^{-h(e_1U)^2}+\frac{e_1U}{2}\left[1+\mathrm{erf}(e_1U\sqrt{h})\right]\right\} \tag{39}$$

where erf(t) is the error function[25] defined as

$$\mathrm{erf}(t)=\frac{2}{\sqrt{\pi}}\int_0^t\mathrm{e}^{-s^2}\mathrm{d}s \tag{40}$$

If the velocity U of the motion of the area concerned is inclined with the angle θ with y- axis and in the $x-y$ plane (Fig. 7), then

$$e_1=\sin\theta, e_2=\cos\theta, e_3=0 \tag{41}$$

Hence, the number of molecules per second striking a unit area of the plate inclined at an angle θ to the stream with velocity U is

$$n=N\left\{\frac{1}{2\sqrt{\pi h}}\mathrm{e}^{-h(U^2)\sin^2\theta}+\frac{U\sin\theta}{2}\left[1+\mathrm{erf}(U\sqrt{h}\sin\theta)\right]\right\}$$

By multiplying the above equation by the mass of the molecule, the mass m_i of the stream per second striking a unit area of the plate is obtained,

$$m_i=\rho\frac{c_i}{2\sqrt{\pi}}\left\{\mathrm{e}^{-(U^2/c_i^2)\sin^2\theta}+\sqrt{\pi}\frac{U}{c_i}\sin\theta\left[1+\mathrm{erf}\left(\frac{U}{c_i}\sin\theta\right)\right]\right\} \tag{42}$$

The component of velocity of the molecule in a direction, having directional cosines e'_1, e'_2, e'_3 with the axes is $e'_1\xi+e'_2\eta+e'_3\zeta$. If this molecule is absorbed by the surface after striking it, the corresponding momentum will be transferred to the surface. For the surface with x-axis as normal as considered previously, the total momentum $M_{e'_1,e'_2,e'_3}$ per second per unit area is

$$M_{e'_1,e'_2,e'_3}=-\rho\left(\frac{h}{\pi}\right)^{3/2}\int_{-\infty}^{0}\mathrm{d}\xi\int_{-\infty}^{\infty}\mathrm{d}\eta\int_{-\infty}^{\infty}\xi(e'_1\xi+e'_2\eta+e'_3\zeta)\mathrm{e}^{-h[(\xi+e_1U)^2+(\eta+e_2U)^2+(\zeta e_3U)^2]}\mathrm{d}\zeta \tag{43}$$

The result of integration is

$$M_{e_1', e_2', e_3'} = -\frac{\rho}{2}U^2\left\{\frac{1}{\sqrt{\pi}hU}(e_1e_1' + e_2e_2' + e_3e_3')e^{-h(e_1U)^2} + \left[\frac{e_1'}{2hU^2} + e_1(e_1e_1' + e_2e_2' + e_3e_3')\right][1 + \mathrm{erf}(U\sqrt{h}e_1)]\right\} \tag{44}$$

For calculating the pressure p_i due to impact of molecules on a plate inclined at an angle θ to the stream of velocity U,

$$\begin{aligned} e_1 &= \sin\theta & e_1' &= -1 \\ e_2 &= \cos\theta & e_2' &= 0 \\ e_3 &= 0 & e_3' &= 0 \end{aligned} \tag{45}$$

By substituting Eq. (45) into Eq. (43), the impact pressure p_i is calculated by

$$\frac{p_i}{\frac{1}{2}\rho U^2} = \sin\theta\frac{c_i}{U}e^{-(U^2/c_i^2)\sin^2\theta} + \left(\frac{1}{2}\frac{c_i^2}{U^2} + \sin^2\theta\right)\left[1 + \mathrm{erf}\left(\frac{U}{c_i}\sin\theta\right)\right] \tag{46}$$

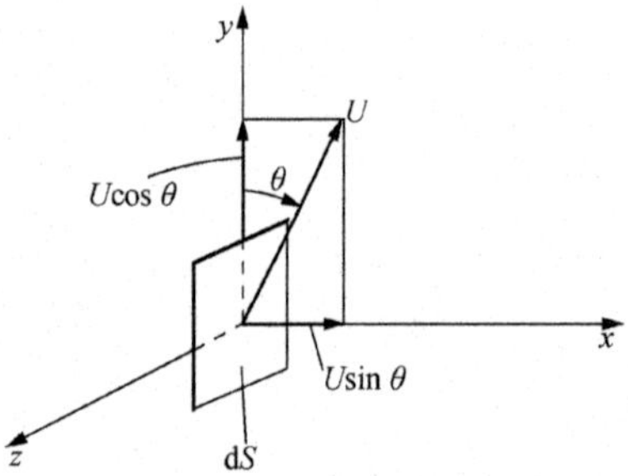

Fig. 7 Coordinate system for calculation free molecule flow

Eq. (46) reduces to particularly simple forms for the two limiting cases of $U = 0$ and $U \gg c_i$. When $U = 0$, the impinging pressure is

$$p_i = (1/4)\rho c_i^2$$

This is one-half of the pressure acting on the surface calculated by the kinetic theory as the other half is the result of the reflection of molecules. For $U \gg c_i$ and $\theta = 90°$, the impinging pressure is

$$p_i = \rho[(1/2)c_i^2 + U^2]$$

This checks with the result of Zahm[1] except for a factor of 2 which is accounted for by the fact that here only pressure due to impinging molecules is calculated, not the total pressure. Zahm obtained the result by simply assuming that $U \geqslant c_i$ and the molecules all having the same speed c_i but with random distribution in their directions of motion. The later assumption is, of course, an oversimplification. However, if $U \gg c_i$, the spread of the speeds of the molecule is relatively unimportant. Then the present more accurate calculation should rightly give the same result as Zahm.

To calculate the shearing stress τ_i due to impact of molecules on the plate, the direction cosines are

$$\begin{aligned} e_1 &= \sin\theta & e_1' &= 0 \\ e_2 &= \cos\theta & e_2' &= -1 \\ e_3 &= 0 & e_3' &= 0 \end{aligned} \tag{47}$$

By substituting Eq. (47) into Eq. (43),

$$\frac{\tau_i}{(1/2)\rho U^2} = \cos\theta\frac{c_i}{U}e^{-(U^2/c_i^2)\sin\theta} + \sin\theta\cos\theta\left[1 + \mathrm{erf}\left(\frac{U}{c_i}\sin\theta\right)\right] \tag{48}$$

If the molecules reflect specularly from the surface, then motion of the impinging molecules normal to the surface is simply reversed in re-emission, while the tangential velocity is maintained. Then it is easy to see that the pressure due to re-emission is $p_r = p_i$ and the shearing stress due to re-emission is $\tau_r = -\tau_i$. On the other hand, if the molecules re-emit diffusely; $\tau_r = 0$, since no preferred direction exists. To calculate the pressure p_r due to diffuse re-emission, one observes that according to the kinetic theory,

$$p_r = (\text{mass emitted per second per unit area})\left(\sqrt{\pi}/2\right)c_r \tag{49}$$

where c_r is the most probable velocity of the molecule in thermal equilibrium at a temperature T_r of the re-emitted gas. Thus

$$c_r^2 = 2RT_r \tag{50}$$

But the mass re-emitted must be equal to the mass m_i striking the surface, as given by Eq. (40). Thus,

$$\frac{p_r}{(1/2)\rho U^2} = \frac{\sqrt{\pi}}{2}\frac{c_i}{U}\frac{c_r}{U}\frac{e^{-(U^2/c_i^2)\sin^2\theta}}{\sqrt{\pi}} + \frac{\sqrt{\pi}}{2}\frac{c_r}{U}\sin\theta\left[1 + \operatorname{erf}\left(\frac{U}{c_i}\right)\sin\theta\right] \tag{51}$$

(diffuse reflection).

The total pressure due to both impinging and re-emission, with f fraction of the molecules reflected diffusely and $(1-f)$ fraction of the molecule reflected specularly is

$$\frac{p}{(1/2)\rho U^2} = (2-f)\frac{p_i}{(1/2)\rho U^2} + f\frac{p_r}{(1/2)\rho U^2} \tag{52}$$

The corresponding shearing stress is

$$\tau/(1/2)\rho U^2 = f[\tau_i/(1/2)\rho U^2] \tag{53}$$

These equations can be also written in terms of Mach Numbers by the substitution

$$\frac{U}{c_i} = \sqrt{\frac{\gamma}{2}}M, \frac{U}{c_r} = \sqrt{\frac{\gamma}{2}}M\left(\frac{c_i}{c_r}\right) \tag{54}$$

where M is the free stream Mach Number.

The temperature T_r has to be determined by a consideration of the accommodation coefficient α. The energy content of the molecules from the wall is proportional $(1-f)T_0^\circ + fT_r$ where T_0° is the stagnation temperature of the free stream. Therefore, according to the definition given in Eq. (20),

$$\alpha = \frac{T_0^\circ - [(1-f)T_0^\circ + fT_r]}{T_0^\circ - T_w} = \frac{f(T_0^\circ - T_r)}{T_0^\circ - T_w} \tag{55}$$

where T_w is the temperature of the surface. If α and T_w are given, T_r can be calculated from this equation.

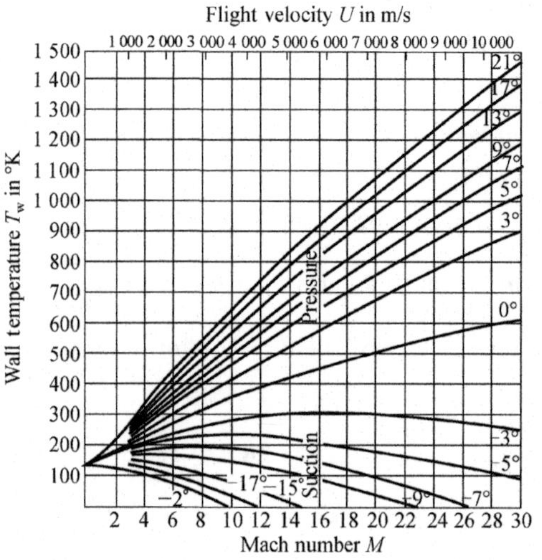

Fig. 8 Surface temperature T_w for various angles θ to the wind in an atmosphere of 86 per cent N_2 and 14 per cent O_2 at the density 1.9×10^{-9} slugs per cu ft. Air temperature = 320°K

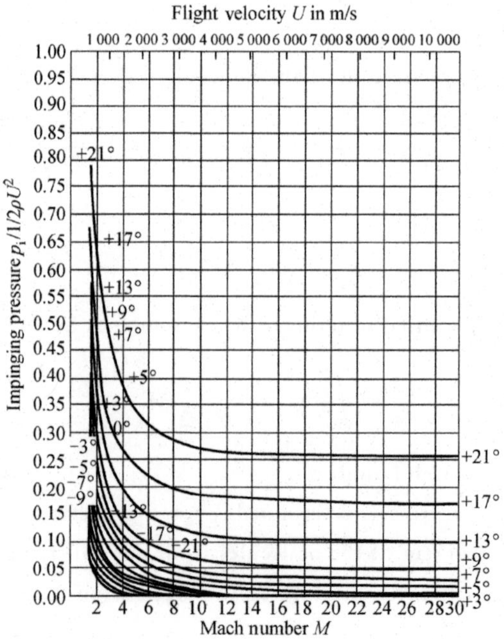

Fig. 9 Impact pressure ratio $p_i \Big/ \left(\frac{1}{2}\rho U^2\right)$ for various angle θ to the wind in an atmosphere of 86 per cent N_2 and 14 per cent O_2 at the density 1.9×10^{-9} slugs per cu ft

Free Molecule Flow Over an Inclined Plate

To determine T_r for the case of an inclined plate moving through air, Sänger[24] utilizes the condition of balance between energy lost through radiation and energy brought to the surface by impact of molecules. If E_s is the energy radiated from the surface per second per unit area, then it is related to the energy E_i brought to the plate and the energy E_r carried away from the plate by

$$E_s = E_i - E_r \tag{56}$$

Let $\varepsilon_{trans.}$, $\varepsilon_{rot.}$, $\varepsilon_{vib.}$ be the translational energy, the rotational energy, and the vibrational energy per until mass of the gas. Since experiments and theoretical considerations indicate that the re-emission is mostly diffuse, Sänger takes $f = 1$. Then, if m_i is the mass brought to and re-emitted from the surface per unit per unit time,

$$E_i = m_i\left(\frac{1}{2}U^2 + \varepsilon_{itrans.} + \varepsilon_{irot.} + \varepsilon_{ivib.}\right) \tag{57}$$

$$E_r = m_i(\varepsilon_{rtrans.} + \varepsilon_{rrot.} + \varepsilon_{rvib.}) \tag{58}$$

The corresponding energies for a gas temperature T_w is $\varepsilon_{wtrans.}$, $\varepsilon_{wrot.}$ and $\varepsilon_{wvib.}$. Sänger then assumes that the accommodation coefficients for translational and rotational energies are unity, while the accommodation coefficient for vibrational energy is zero because of its slow rate of adjustment.

Then

$$\begin{aligned} \varepsilon_{itrans.} - \varepsilon_{rtrans.} &= \varepsilon_{itrans.} - \varepsilon_{wtrans.} \\ \varepsilon_{irot.} - \varepsilon_{rrot.} &= \varepsilon_{irot.} - \varepsilon_{wrot.} \\ \varepsilon_{ivib.} - \varepsilon_{rvib.} &= 0 \end{aligned} \tag{59}$$

Hence, $\varepsilon_{rtrans.} = \varepsilon_{wtrans.}$ $\varepsilon_{rrot.} = \varepsilon_{wrot.}$ $\varepsilon_{rvib.} = \varepsilon_{ivib.}$ and

$$c_r = \sqrt{2RT_w} \tag{60}$$

Therefore, Eq. (58) can be written as

$$E_r = m_i(\varepsilon_{wtrans.} + \varepsilon_{wrot.} + \varepsilon_{ivib.}) \tag{61}$$

By substituting Eqs. (57) and (61) into Eq. (56), one has immediately

$$E_s = m_i\left[\frac{1}{2}U^2 + (\varepsilon_{itrans.} - \varepsilon_{wtrans.}) + (\varepsilon_{irot.} - \varepsilon_{wrot.})\right] \tag{62}$$

For nitrogen, and oxygen the equipartition value of the rotational energy is reached at even exceedingly low temperatures. Therefore,

$$\varepsilon_{trans.} = \frac{3}{2}RT$$

$$\varepsilon_{rot.} = \frac{2}{2}RT$$

Hence Eq. (61) gives

$$E_s = m_i\left[\frac{1}{2}U^2 + \frac{5}{2}R(T_i - T_w)\right] \tag{63}$$

where m_i is given by Eq. (42). T_i is the free stream temperature T_0.

As an example, Sänger assumes that the air temperature is equal to $T_i = T_0 = 320$K. (47℃). The air density is 1.904×10^{-9} slugs per cu ft. The density ratio referred standard conditions is thus 0.802×10^{-6}, corresponding to an altitude of approximately 55 miles. The mean free path according to Table 1 is thus approximately 3.5 in. Sänger actually assumed the air to be composed of 14 per cent oxygen and 86 per cent nitrogen without considering ionization and dissociation. Neglecting the radiation from the air to the surface, he assumed the emissivity of the surface to be 0.80. Thus

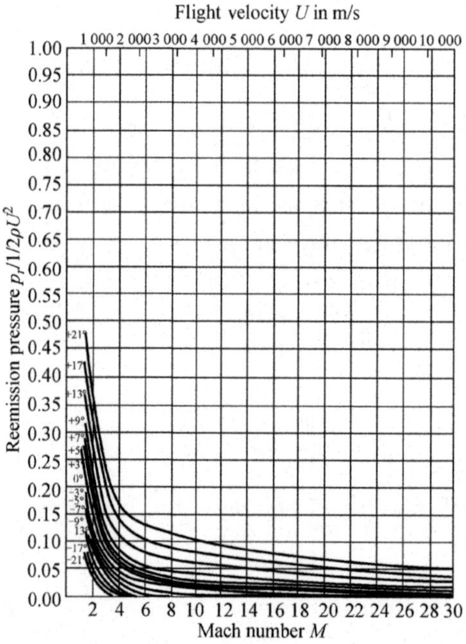

Fig. 10 Re-emission pressure ratio $p_r\Big/\left(\frac{1}{2}\rho U^2\right)$ for various angle θ to the wind in an atmosphere of 86 per cent N_2 and 14 per cent O_2 at the density 1.9×10^{-9} slugs/ft^3

$$E_s = 0.80\,\sigma T_w^4$$

where σ is the Stefan-Boltzmann constant for radiation. The results of calculation of the wall temperature T_w are shown in Fig. 8. It is interesting to note that because of radiation, the wall temperature is considerably lower than the air temperature for low air velocities and suction side. The value of c_r can be calculated by Eq. (60). The stresses acting on the surface can then be computed with Eqs. (52) and (53). The pressure ratios $p_i\Big/\left(\frac{1}{2}\rho U^2\right)$ and $p_r\Big/\left(\frac{1}{2}\rho U^2\right)$ for impinging and diffuse re-emission ($f = 1$) are given in Figs. 9 and 10, also taken from Sänger's paper. The shearing stress is given in Fig. 11. The extremely high shearing force is evident.

If the dimension of a flat plate airfoil is much less than the mean free path, Eqs. (52) and (53) can be used to calculate the lift and drag. Thus, for diffuse reflection ($f = 1$), and for an angle of attack α, the lift coefficient C_L and the drag coefficient C_D are

$$C_L = \sqrt{\frac{2\pi}{\gamma}} \frac{1}{M}\left(\frac{c_r}{c_i}\right)\sin\alpha\cos\alpha + \frac{2}{\gamma}\frac{1}{M^2}\cos\alpha\,\mathrm{erf}\left(\sqrt{\frac{\gamma}{2}}M\sin\alpha\right) \tag{64}$$

$$C_D = \sqrt{\frac{8}{\gamma}}\frac{1}{M}\mathrm{e}^{-(\gamma/2)M^2\sin^2\alpha} + \sqrt{\frac{2\pi}{\gamma}}\frac{1}{M}\left(\frac{c_r}{c_i}\right)\sin^2\alpha + 2\left(\frac{1}{\gamma M^2}+1\right)\sin\alpha\,\mathrm{erf}\left(\sqrt{\frac{\gamma}{2}}M\sin\alpha\right) \tag{65}$$

Sänger's result for the flat plate airfoil is given in Fig. 12. The lift-drag ratio is less than unity, showing the poor efficiency of the surface under free molecule flow conditions. Of course, the calculated values hold true only for wing dimensions much less than the mean free path, i.e., less than 3.5 in for the case under consideration.

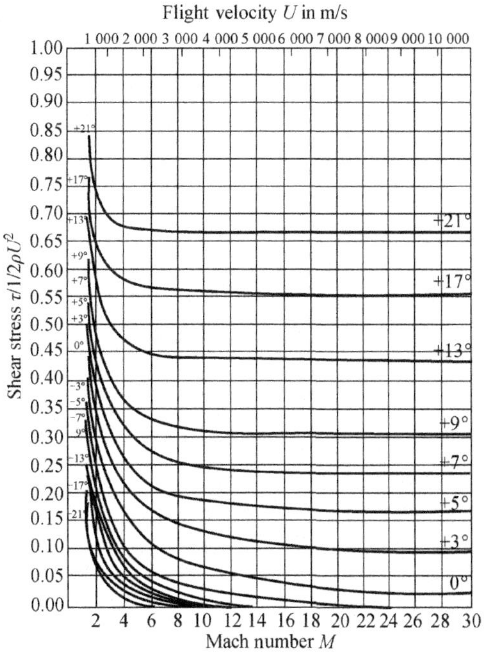

Fig. 11 Shear stress ratio $\tau/\left(\frac{1}{2}\rho U^2\right)$ for various angle θ to the wind in an atmosphere of 86 per cent N_2 and 14 per cent O_2 at the density 1.9×10^{-9} slugs/ft^3

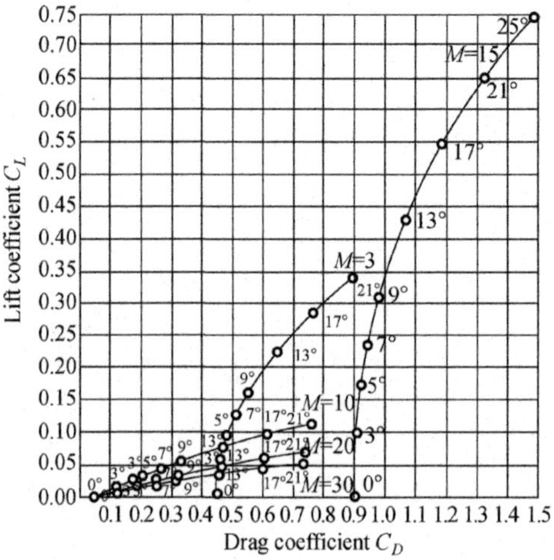

Fig. 12 Polar diagram for a flat plate at various angles of attack in an atmosphere of 86 per cent N_2 and 14 per cent O_2 at the density 1.9×10^{-9} slugs/ft^3

References

[1] Zahm A F. Superaerodynamics. Journal of the Franklin Institute, 1934, 217: 153 - 166.

[2] Chapman S. On the Law of Distribution of Molecular Velocities, and on the Theory of Viscosity and Thermal Conduction, in a Non-Uniform Simple Monatomic Gas. Phil. Trans. Roy. Soc., 1915, 216 (A): 279 - 348.

[3] Maris H B. The Upper Atmosphere. Terr. Mag, 1928, 33: 233 - 255; 1929, 34: 45 - 53.

[4] Burnett D. The Distribution of Molecular Velocities and the Mean Motion in a Non-Uniform Gas. Proc. Lond. Math. Soc., 1935, 40: 382// See also Chapman S, and Cowling TG. The Mathematical Theory of Non-Uniform Gases, 1st Ed., Chap 15; Cambridge University Press, London, 1939.

[5] Millikan R A. Coefficients of Slip in Gases and the Law of Reflection of Molecules from the Surfaces of Solids and Liquids. Physical Review, 1923, 21: 217 - 238.

[6] van Itterbeck A, Mariens P. Measurements on the Absorption of Sound in O_2 Gas and in CO_2 Gas Containing Small Quantities of H_2O, D_2O and Ne. Relaxation Times for Vibrational Energy, Physica, 1940, 7: 125 - 130.

[7] Epstein P S. On the Resistance Experienced by Spheres in Their Motion through Gases. Physical Review, 1924, 23: 710 - 733.

[8] See, for instance, Kennard EH. Kinetic Theory of Gases. McGraw-Hill Book Co., Inc. New York, 1938, 328 - 332.

[9] Kennard. Ibid. p.312 - 315.

[10] Knudsen M. Die molekulare Warmeleitung der Gase und der Akkommodationskoefficient. Ann. Physik, 1911, 34: 593 - 656.

[11] See, for instance, Kennard, op. cit., p.323.

[12] Wiedmann ML. Thermal Accommodation Coefficient. Trans. of ASME, 1946, 68: 57 - 64.

[13] See, for example, Lamb H, Hydrodynamics, 6th Ed. Cambridge University Press, 1932, 594 - 617.

[14] Basset A B. Hydrodynamics. vol 2. 270 - 271; Deighton, Bell and Co., Cambridge, 1888, or Lamb, op

cit: 602 - 604.

[15] Lamb, op. cit., pp.615,616.

[16] Knudsen M. Die Gesetze der Molekularströmung und der inneren Reibungsströmung der Gase durch Rohren. Ann. Physik, 1909, 28: 75 - 130.

[17] Knudsen M. Molekularstromung des Wasserstoffs durch Rohren und das Hitzdrahtmanometer. Ann. Physik, 1911, 35: 389 - 396.

[18] Gaede W. Die aussere Reibung der Gase. Ann. Physik, 1913, 41: 289 - 336.

[19] Ebert H. Das Strömen von Gasen bei niedrigen Drucken. Physikalische Zeitschrift, 1932, 33: 145 - 151.

[20] Ebert H. Darstellung der Stromungsvorgange von Gasen bei niedrigen Druchen mittels Reynoldsscher Zahlen. Zeitschrift fur Physik, 1933,85: 561 - 564.

[21] Millikan R A. The General Law of Fall of a Small Spherical Body Through a Gas, and its Bearing upon the Nature of Molecular Reflection from Surfaces. Physical Review, 1923, 22: 1 - 23.

[22] Smoluchowski M V. Zur kinetischen Theorie der Transpiration und Diffusion verdunnter Gase. Ann. Physik, 1910, 33: 1559 - 1570.

[23] Sänger E. Gaskinetik Sehr hoher Fluggeschweindigkeiten. Deutsche Luftfahrtforschung, Bericht 972, Berlin, 1938.

[24] Sänger E, Bredt J. Uber einer Raketenantrieb für Fernbomber. Deutsche Luftfahrtforschung, Untersuchungen u. Mitteilungen Nr. 3538: 141 - 173, Berlin, 1944.

[25] The error function is tabulated. See, for instance, Jahnke-Ende Tables of Functions. 3rd ed. 24; B.G. Teubner, Liepzig, 1938 // or Tables of Probability Functions, Vols. I and II; National Bureau of Standards, Washington, D.C., 1943.

Propagation of Plane Sound Waves in Rarefied Gases

Hsue-shen Tsien and Richard Schamberg
(*California Institute of Technology, Pasadena, California*)

If the density of gas is very small, the conventional Navier-Stokes equations are not accurate enough. The present investigation includes the effect of the so-called third approximation to the solution of Boltzmann-Maxwell equation as obtained by D. Burnett. The results of this more accurate calculation show that, even under extreme conditions, the velocity of propagation deviates from its usual value by only 2 percent.

1. Introduction

The problem of the propagation of plane sound waves in a viscous fluid was treated as early as 1845 by G. G. Stokes, who investigated the effect of viscous action but neglected heat conduction. The effect of viscosity was found to consist of a frequency dependent damping of the amplitudes of the sound waves whose velocity of propagation, to a first approximation, was equal to the adibatic propagation speed in a frictionless, compressible fluid. As was first pointed out by G. Kirchhoff in 1868, the influence of heat conduction is of the same order of magnitude as that of the viscosity, so that for a consistent solution both factors must be accounted for. This was done in the solution of H. Lamb,① and resulted in a more highly damped motion than that predicted by Stokes, whereas the speed of propagation remained approximately equal to that for the frictionless fluid.

The validity of the results is dependent, of course, on the validity of the fundamental Navier-Stokes equations which are the basis of these calculations. No question needs to be raised against the general laws of dynamics and kinematics and the only point of doubt is the correctness of the viscous stresses and of the heat flux used in the Navier-Stokes equations. In the Navier-Stokes equations, the viscous stresses are taken as the product of the viscosity coefficient and the linear combination of the first-order space derivatives of the velocity components, and the heat flux is taken as the product of the coefficient of heat conductivity and the gradient of the temperature of the fluid. Results of calculations using these equations agree very well with experimental observations. This fact may be used as the empirical justification

---

Received June 19, 1946.

Journal of the Acoustical Society of America, vol. 18, pp. 334 – 341, 1946.

① H. Lamb, *Hydrodynamics* (Cambridge University Press, London, England, 1932), sixth edition, pp. 647 – 650.

of the Navier-Stokes equations.

The kinetic theory of gases, as cultivated by S. Chapman and D. Enskog, shows, however, that the viscous stresses and the heat flux as used in the Navier-Stokes equation are only first-order approximations. If the number of gas molecules contained in a cube of dimension intrinsic to the problem, such as the wave-length of sound propagation, is small, then the first-order approximation is no longer sufficient. This means that if the wave-length is very small as in the case of ultrasonic waves or if the density of gas is very low, the Navier-Stokes equation is no longer valid. Since there are many assumptions introduced in the kinetic theory to make the calculation treatable, one may question the reliability of the theoretical results. However, the success of the kinetic theory in explaining many phenomena and the prediction of the first-order viscous stresses and heat flux which agrees with the observational data seem to indicate the reliability of such a theory.

In this paper the second approximation to the viscous stresses and the heat flux as given by Chapman and Cowling[①] is used to calculate the propagation of plane sound waves in rarefied gases or of plane sound waves of very small wave-lengths in gases of normal density. This investigation is prompted by the fact that the present knowledge of the state of the atmosphere at high altitudes is almost exclusively obtained through the measurement of the anomalous sound propagation. A knowledge of the effect of low density of the medium on the propagation velocity and the damping of waves will be, perhaps, useful in a critical examination of this method of obtaining data for high altitudes. The result of this investigation is most reassuring as it shows that, even under extreme conditions, the increase in the propagation velocity from the normal value at high density is less than 2 percent. In fact, the effect of the additional terms to the viscous stresses and the heat flux tends to maintain the constancy of the sound velocity with respect to the density of the medium.

2. Symbols

The following symbols are used in the succeeding sections:

b damping coefficient

c propagation speed of sound wave

c_0 adiabatic speed of sound in perfect fluid

c_p specific heat at constant pressure

K_2, K_3 numerical constants

L wave-length

p hydrostatic pressure

p_{xx} component of stress tensor

q_x x-component of heat flux vector

① S. Chapman and T. G. Cowling, *The Mathematical Theory of Non-Uniform Gases* (Cambridge University Press, London, England, 1939), pp. 265 – 269.

R Reynolds number, $R = \frac{\rho_0 c_0}{\mu} \cdot \frac{L}{2\pi}$

r_a amplitude ratio

t time

T absolute temperature

u velocity of fluid particles

x distance in direction of propagation

α dimensionless speed of sound $\alpha = c/c_0$

β reciprocal of Prandtl number $\beta = \lambda/(\mu c_p)$

γ ratio of specific heats

θ_2, θ_4 numerical constants

λ coefficient of thermal conductivity

ρ density

μ coefficient of absolute viscosity

ν frequency of the sound in cycles per second

$(\ )_0$ "undisturbed" value of variables

$(\ )'$ "perturbation" value of variable

$(\ )^*$ dimensionless form of variable

3. Basic Equations

The fundamental differential equations required to describe the propagation of plane sound waves express the conservation of mass, momentum, and energy of an element of a compressible fluid. If t is the time, x is the coordinate in the direction of propagation, ρ the density, and u the fluid velocity, then the equation of continuity is

$$\partial \rho/\partial t + \partial(\rho u)/\partial x = 0 \tag{1}$$

If p_{xx} is the stress in the x direction acting on a plane normal to the x axis, the momentum equation can be written as

$$\rho \frac{\partial u}{\partial t} + \rho u \frac{\partial u}{\partial x} = -\frac{\partial}{\partial x}(p_{xx}) \tag{2}$$

Furthermore, if p is the hydrostatic pressure, T the absolute temperature of the gas, q_x the heat flux in the x direction, and c_p the specific heat at constant pressure, the equation for the conservation of energy is

$$-\frac{\partial q_x}{\partial x} = c_p \rho \frac{\partial T}{\partial t} - \frac{\partial p}{\partial t} \tag{3}$$

For purposes of subsequent calculations, it is convenient to separate the dependent variables into their "undisturbed" and "perturbation" components according to Eq. (4) where the perturbations are denoted by a prime:

$$\begin{aligned} p &= p_0 + p', \quad \rho = \rho_0 + \rho', \\ T &= T_0 + T', \quad u = u' \end{aligned} \tag{4}$$

Since a sound wave is by definition a disturbance of infinitesimal amplitude, any quantities, involving squares or products of the perturbation variables or their derivatives can be neglected in comparison to these variables themselves. For instance, from the equation of state of a perfect gas

$$p/(\rho T) = \text{const} \tag{5}$$

one has

$$\frac{\partial \rho}{\partial t} = \frac{\partial \rho'}{\partial t} = \rho_0 \left[\frac{1}{p_0} \frac{\partial p'}{\partial t} - \frac{1}{T_0} \frac{\partial T'}{\partial t} \right] \tag{6}$$

The general form of the heat flux vector and the stress tensor are given by S. Chapman and T. Cowling[2] correct to terms of order μ^2. The following equations give the expressions for q_x and p_{xx} appropriate to the problem at hand by neglecting second-order quantities in the perturbation variables:

$$q_x = -\lambda \frac{\partial T}{\partial x} + \frac{2}{3}(\theta_2 + \theta_4) \frac{\mu^2}{\rho_0} \frac{\partial^2 u'}{\partial x^2} \tag{7}$$

$$p_{xx} = (p_0 + p') - \frac{4}{3}\mu \frac{\partial u'}{\partial x} - \frac{2}{3}\mu^2 \left[\frac{K_2}{p_0 \rho_0} \frac{\partial^2 p'}{\partial x^2} - \frac{K_3}{\rho_0 T_0} \frac{\partial^2 T'}{\partial x^2} \right] \tag{8}$$

where μ is the viscosity coefficient, λ is the heat conduction coefficient, θ_2, θ_4, K_2, K_3 are constants whose exact numerical values depend on the intramolecular structure of the molecules composing the gas.

Substitution of Eqs. (4) and (6) – (8) into the exact differential Eqs. (1) – (3), and subsequent omission of all second-order terms in the dependent variables, result in the linearized partial differential Eqs. (9) – (11).

$$\frac{1}{p_0} \frac{\partial p'}{\partial t} - \frac{1}{T_0} \frac{\partial T'}{\partial t} + \frac{\partial u'}{\partial x} = 0 \tag{9}$$

$$\rho_0 \frac{\partial u'}{\partial t} = -\frac{\partial p'}{\partial x} + \frac{4}{3}\mu \frac{\partial^2 u'}{\partial x^2} + \frac{2}{3}\mu^2 \left[\frac{K_2}{\rho_0 p_0} \frac{\partial^3 p'}{\partial x^3} - \frac{K_3}{\rho_0 T_0} \frac{\partial^3 T'}{\partial x^3} \right] \tag{10}$$

$$\rho_0 c_p \frac{\partial T'}{\partial t} = \frac{\partial p'}{\partial t} + \lambda \frac{\partial^2 T'}{\partial x^2} - \frac{2}{3}(\theta_2 + \theta_4) \frac{\mu^2}{\rho_0} \frac{\partial^3 u'}{\partial x^3} \tag{11}$$

It is convenient to reduce the above equations to dimensionless form by means of the non-dimensional parameters defined below:

$$p^* = p'/p_0, T^* = T'/T_0, u^* = u'/c_0 \tag{12}$$

$$x^* = x/L, t^* = t/(L/c_0) \tag{13}$$

where L is the wave-length and c_0 the adiabatic speed of sound propagation defined by

$$c_0 = [\gamma(p_0/\rho_0)]^{\frac{1}{2}} \tag{14}$$

γ here is the ratio of specific heats.

The physical constants of the gas may be expressed in terms of the two dimensionless parameters, β and R

$$\beta = \frac{\lambda}{\mu c_p}, R = \left[\frac{\rho_0 c_0}{\mu} \times \frac{L}{2\pi}\right] \tag{15}$$

$1/\beta = \mu c_p/\lambda$ is the Prandtl number and is a measure of the relative importance of the viscosity and the heat conductivity of the gas. From the point of view of the kinetic theory, viscosity is the result of the transfer of the momentum of the molecules, and heat conduction is the result of the transfer of the energy of the molecules. Both thus must be of the same order of magnitude. This is in agreement with experiment, because the Prandtl number is found to be of the order of unity. The Reynolds number is a measure of the relative importance of the inertia forces and the dissipative forces. If the dissipative forces are very small in comparison with the inertia forces, the Reynolds number will be very large. With increasing importance of the dissipative forces, the Reynolds number decreases. In the propagation of sound waves, the dissipative forces are measured by gradients of velocity and temperature and are inversely proportional to the wave-length L for any given amplitude. Therefore in this case, the Reynold's number R is directly proportional to the wave-length as shown by Eq. (15).

Substitution of Eqs. (12) to (15) into Eqs. (9) to (11) results in the set of three simultaneous partial differential Eqs. (16)

$$\left.\begin{aligned}
&\frac{\partial p^*}{\partial t^*} - \frac{\partial T^*}{\partial t^*} + \frac{\partial u^*}{\partial x^*} = 0 \\
&\left[\frac{\partial p^*}{\partial x^*} - \frac{2}{3}K_2 \frac{\gamma}{(2\pi R)^2}\frac{\partial^3 p^*}{\partial x^{*3}}\right] + \frac{2}{3}K_3 \frac{\gamma}{(2\pi R)^2}\frac{\partial^3 T^*}{\partial x^{*3}} + \left[\gamma \frac{\partial u^*}{\partial t^*} - \frac{4}{3}\frac{\gamma}{(2\pi R)}\frac{\partial^2 u^*}{\partial x^{*2}}\right] = 0 \\
&-\frac{\partial p^*}{\partial t^*} + \frac{\gamma}{\gamma - 1}\left[\frac{\partial T^*}{\partial t^*} - \frac{\beta}{(2\pi R)}\frac{\partial^2 T^*}{\partial x^{*2}}\right] + \frac{2}{3}(\theta_2 + \theta_4)\frac{\gamma}{(2\pi R)^2}\frac{\partial^3 u^*}{\partial x^{*3}} = 0
\end{aligned}\right\} \tag{16}$$

If the properties of the gas are known, the values of γ, β, K_2, K_3, θ_2, and θ_4 are fixed. Then the only remaining parameter of the problem is R which is really a measure of the wave-length. Therefore, the solutions of the problem should be expressed as a function of R.

4. Solution of the Differential Equations

The general solution of this set of linear partial differential equations (16) is evidently of the form

$$\left.\begin{aligned}
p^* &= A_1 \exp\{2\pi[(\mathrm{i} - b)x^* - \mathrm{i}\alpha t^*]\} \\
T^* &= A_2 \exp\{2\pi[(\mathrm{i} - b)x^* - \mathrm{i}\alpha t^*]\} \\
u^* &= A_3 \exp\{2\pi[(\mathrm{i} - b)x^* - \mathrm{i}\alpha t^*]\}
\end{aligned}\right\} \tag{17}$$

where the A's represent arbitrary constants. The coefficients b and α are both real quantities, to be determined as functions of the Reynolds number R. It follows from the definitions (13) and (14) that the damping coefficient per wave-length is $2\pi b$, and the physical velocity of propagation of the disturbance is given by $c = \alpha c_0$. By substituting Eqs. (17) into Eqs. (16), the following linear homogeneous equations for A_1, A_2, and A_3 are obtained:

$$[-\mathrm{i}\alpha]A_1 + [\mathrm{i}\alpha]A_2 + [\mathrm{i}-b]A_3 = 0$$

$$\left[(\mathrm{i}-b) - \frac{2}{3}K_2\frac{\gamma}{R^2}(\mathrm{i}-b)^3\right]A_1 + \left[\frac{2}{3}K_3\frac{\gamma}{R^2}(\mathrm{i}-b)^3\right]A_2 + \left[-\mathrm{i}\gamma\alpha - \frac{4}{3}\frac{\gamma}{R}(\mathrm{i}-b)^2\right]A_3 = 0$$

$$[+\mathrm{i}\alpha]A_1 + \left[\frac{\gamma}{\gamma-1}\left\{-\mathrm{i}\alpha - \frac{\beta}{R}(\mathrm{i}-b)^2\right\}\right]A_2 + \left[\frac{2}{3}(\theta_2+\theta_4)\frac{\gamma}{R^2}(\mathrm{i}-b)^3\right]A_3 = 0$$

In order that the values of A_1, A_2, and A_3 should not be simultaneously equal to zero and thus give a trivial solution, the determinant formed by the coefficients of A must be equal to zero. Thus

$$\begin{vmatrix} [-\mathrm{i}\alpha] & [\mathrm{i}\alpha] & [\mathrm{i}-b] \\ \left[(\mathrm{i}-b) - \frac{2}{3}\frac{K_2\gamma}{R^2}(\mathrm{i}-b)^3\right] & \left[\frac{2}{3}\frac{K_3\gamma}{R^2}(\mathrm{i}-b)^3\right] & \left[-\gamma\mathrm{i}\alpha - \frac{4}{3}\frac{\gamma}{R}(\mathrm{i}-b)^2\right] \\ [+\mathrm{i}\alpha] & \frac{\gamma}{\gamma-1}\left[-\mathrm{i}\alpha - \frac{\beta}{R}(\mathrm{i}-b)^2\right] & \left[\frac{2}{3}(\theta_2+\theta_4)\frac{\gamma}{R^2}(\mathrm{i}-b)^3\right] \end{vmatrix} = 0$$

By separating the real and imaginary parts of this determinant, one has the following two simultaneous Eqs. (18) and (19) from which b and α can be determined.

$$A\left[\frac{\alpha}{R^4}(1-b^2)(1-14b^2+b^4)\right] + B\left[\frac{\alpha}{R^2}(1-6b^2+b^4)\right] + C\left[\frac{2b}{R^3}(3-10b^2+3b^4)\right] + D\left[\frac{4b}{R}(1-b^2)\right] - E\left[\frac{2b}{R}\alpha^2\right] + \alpha(\alpha^2+b^2-1) = 0 \tag{18}$$

$$A\left[\frac{2\alpha b}{R^4}(3-10b^2+3b^4)\right] + B\left[\frac{4b\alpha}{R^2}(1-b^2)\right] - C\left[\frac{1}{R^3}(1-b^2)(1-14b^2+b^4)\right] - D\left[\frac{1}{R}(1-6b^2+b^4)\right] + E\left[\frac{\alpha^2}{R}(1-b^2)\right] - 2\alpha b = 0 \tag{19}$$

where

$$\left.\begin{aligned} A &= (4/9)\gamma(\gamma-1)(\theta_2+\theta_4)(K_2-K_3) \\ B &= \frac{2}{3}(\gamma-1)(\theta_2+\theta_4) - (4/3)\gamma\beta - \\ &\quad \frac{2}{3}\gamma(K_2-K_3) - \frac{2}{3}K_3 \\ C &= \frac{2}{3}\gamma\beta K_2, D = \beta, E = (4/3)+\gamma\beta \end{aligned}\right\} \tag{20}$$

For the complete solution of the problem it would be necessary to find the nine roots of Eqs. (18) and (19). However, from the nature of the classical solution of the problem it is known that the pair of roots of b and α which is of the greatest significance is the one with the least damping. In other words, for vanishing viscosity, or $R \to \infty$, $\alpha \to 1$, and $\beta \to 0$ so that the wave should be undamped and propagate with the normal speed c_0. Therefore, the appropriate forms for b and α are the following:

$$b = \frac{b_1}{R} + \frac{b_3}{R^3} + \frac{b_5}{R^5} + \ldots \tag{21}$$

$$\alpha = 1 + \frac{\alpha_2}{R^2} + \frac{\alpha_4}{R^4} + \frac{\alpha_6}{R^6} + \ldots \tag{22}$$

Equations (21) and (22) may be substituted into (18) and (19), and the resulting equations may be arranged in order of ascending powers of $1/R$. In order that the resulting equations are satisfied for arbitrary values of R, each of the coefficients of $1/R^n$ must be equal to zero. This results in a set of algebraic equations from which the coefficients b_1, b_3, ..., and α_2, α_4, ... are determined successively in the following order: b_1, α_2, b_3, α_4, ...

Table I

Medium	1	2	3
γ	$\frac{7}{5}$	1.393	$\frac{5}{3}$
$\beta = D$	1.333	1.610	$\frac{3}{2}$
A	−0.837 20	−0.818 44	−1.661 1
B	−1.436 9	−1.980 4	−0.678 67
C	2.523 7	3.032 2	3.380 0
E	3.200 0	3.576 1	3.833 3
b_1	0.933 3	0.983 0	1.166 7
b_3	−0.083 9	−0.247 7	2.425 3
b_5	3.019 5	2.515 6	22.434
α_2	0.780 7	0.857 0	0.631 0
α_4	−4.231 4	−5.701 5	−4.983 8
α_6	−2.834 4	−6.906 2	−2.543 8

Equations (23) and (24) give the final result of this solution. It should be noted that each coefficient is a function only of the previously determined coefficients and of the physical constants A, B, ..., E, of Eq. (20).

$$\left.\begin{aligned} b_1 &= \frac{1}{2}(E - D) \\ b_3 &= \left[-\frac{1}{2}C + 2Bb_1 + \left(3D - \frac{1}{2}E\right)b_1^2 + E\alpha_2 - b_1\alpha_2\right] \\ b_5 &= \Big[3Ab_1 + (15/2)Cb_1^2 - 2Bb_1^3 - \frac{1}{2}Db_1^4 + 2Bb_3 + (6D - E)b_1b_3 + \\ &\quad \frac{1}{2}E\alpha_2^2 + E\alpha_4 - Eb_1^2\alpha_2 + 2Bb_1\alpha_2 - b_1\alpha_4 - b_3\alpha_2\Big] \end{aligned}\right\} \tag{23}$$

$$\left.\begin{aligned}
\alpha_2 &= -\left[\frac{1}{2}B + (2D-E)b_1 + \frac{1}{2}b_1^2\right] \\
\alpha_4 &= -\left[\frac{1}{2}A + 3Cb_1 - 3Bb_1^2 - 2Db_1^3 + (2D-E)b_3 + b_1b_3 + \frac{1}{2}B\alpha_2 + \right. \\
&\quad \left. \frac{3}{2}\alpha_2^2 - 2Eb_1\alpha_2 + \frac{1}{2}\alpha_2 b_1^2\right] \\
\alpha_6 &= -\left[-(15/2)Ab_1^2 - 10Cb_1^3 + \frac{1}{2}Bb_1^4 + 3Cb_3 + \frac{1}{2}b_3^2 + (2D-E)b_5 - 6Bb_1b_3 - \right. \\
&\quad 6Db_1^2b_3 + b_1b_5 + \frac{1}{2}A\alpha_2 + \frac{1}{2}\alpha_2^3 + \frac{1}{2}B\alpha_4 + 3\alpha_4\alpha_2 - Eb_1\alpha_2^2 - 3Bb_1^2\alpha_2 - 2Eb_1\alpha_4 + \\
&\quad \left. \frac{1}{2}b^2\alpha_4 - 2Eb_3\alpha_2 + b_1b_2\alpha_2\right]
\end{aligned}\right\} \quad (24)$$

It is interesting to note that the first-order solution

$$b = b_1/R = (1/2R)[4/3 + (\gamma - 1)\beta], \quad \alpha = 1 \tag{25}$$

is identical with the solution given by H. Lamb[1].

5. Numerical Calculation

The constants A, B, ..., E depend on the physical constants of the gas, β, and γ, and on θ_2, θ_4, K_2, K_3. The latter depend on the nature of the force field which surrounds the molecule. In the kinetic theory, the molecule is generally represented by a mass point situated at the origin of a spherically symmetric, repulsive force field whose strength is assumed proportional to $1/r^s$ where r is the distance from the center of the molecule. Because of their relative mathematical simplicity, the two types of molecules most frequently considered are

1. The "Maxwell molecule," having $s = 5$;
2. The rigid elastic spherical molecule corresponding to $s = \infty$.

Calculations[3] based on the observed temperature dependence of the viscosity of real gases show that actually $5 < s < 15$. In particular, for air the exponent $s \cong 8.5$, and for the monatomic gases of helium and neon $s \cong 14$. Since the spherical molecule appears to be a closer approximation to reality than the Maxwellian molecule, the values of K_2 and K_3 given by D. Burnett[2] for spherical molecules are used throughout the present calculations. However, the only values available for θ_2 and θ_4 are those calculated by Chapman[2] for Maxwellian molecules. Nevertheless, the variation in the constants K is only of the order of 10 percent for the two extreme cases considered. The variation in θ must be also of the same order of magnitude.

In accordance with the above discussion, the numerical values for the molecular constants are then given by Eqs. (26).

$$\begin{aligned} \theta_2 &= 45/8, & \theta_4 &= 3, \\ K_2 &= 2.028, & K_3 &= 2.418 \end{aligned} \tag{26}$$

Equations (23), (24), and (20) will now be applied to calculate the damping coefficient, b,

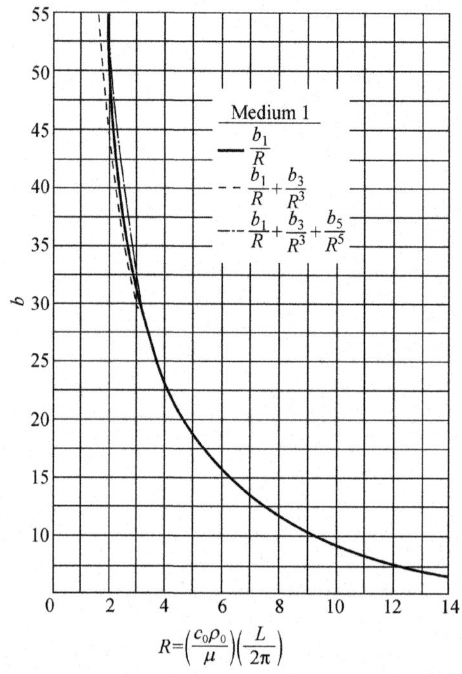

Fig. 1 Convergence of the expansion $b = (b_1/R) + (b_3/R^3) + (b_5/R^5) + \ldots$ as a function of the Reynold's number R for medium 1

and the dimensionless velocity of propagation, α, in the following three gaseous media, as characterized by the values of the physical constants γ and β:

1. Air at normal atmospheric temperature. This also corresponds closely to the theoretical diatomic gas.

2. Air at 400℃.

3. Theoretical monatomic gas. This corresponds very closely to real monatomic gases, such as helium, argon, and neon, at normal temperatures.

Table I gives the values of γ and β for each of the three media, as well as the values of the derived constants A, B, ..., E as calculated from (20) and (26), and the values of the coefficients b_1, b_3, b_5, and α_2, α_4, α_6 as calculated from Eqs. (23) and (24), respectively. The results of Table I are then used to calculate the damping coefficient b, and the propagation coefficient α as a function of the Reynolds number R, by means of Eqs. (21) and (22), respectively.

The nature of the convergence of the expansions (21) and (22) is shown for the typical case of medium 1 in Figs. 1 and 2. It should be noted that the solutions $b = (b_1/R) + (b_3/R^3)$, $\alpha = 1 + (\alpha_2/R^2) + (\alpha_4/R^4)$ are not necessarily accurate to terms of order $1/R^3$ and $1/R^4$, respectively, since the original differential Eqs. (16) are correct only to order $1/R^2$.

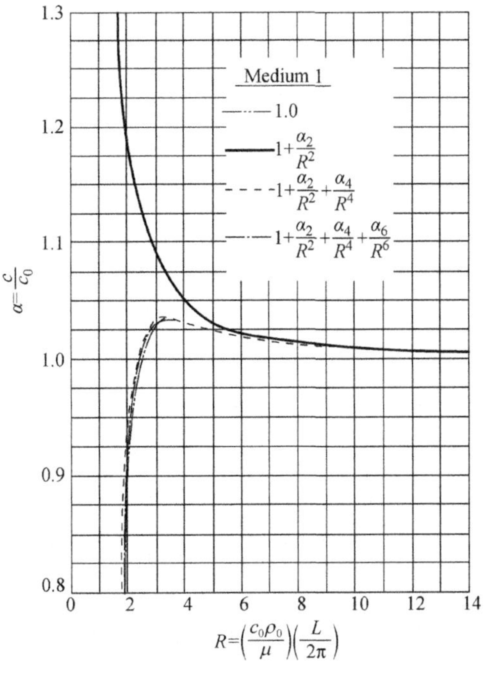

Fig. 2 Convergence of the expansion $\alpha = 1 + (\alpha_2/R^2) + (\alpha_4/R^4) + \ldots$ as a function of the Reynolds number R for medium 1

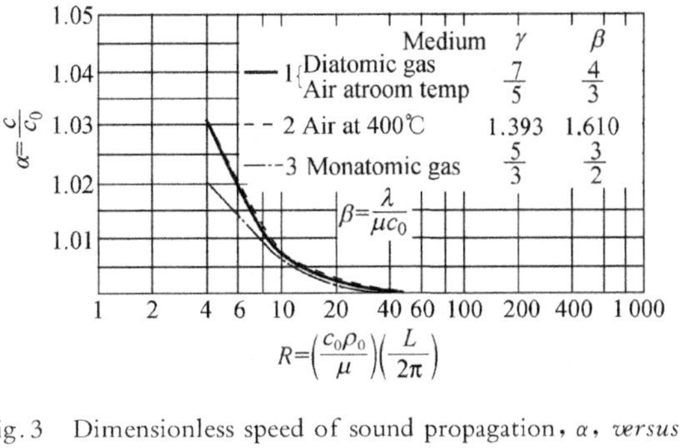

Fig. 3 Dimensionless speed of sound propagation, α, *versus* Reynolds number R for three media

6. Results

The values of α and b as functions of the Reynolds number for each of the three gaseous media considered are shown in Figs. 3 and 4. The effect of decreasing Reynolds number is to increase the damping coefficient in nearly inverse ratio, whereas the speed of propagation, αc_0, increases only slightly. Figures 3 and 4 also show that the molecular structure of the gas influences the magnitude of the propagation parameters to a much larger extent than does the mean temperature of the gas. This is apparently because of the fact that Eq. (20) depend more critically on the value of γ, which occurs also

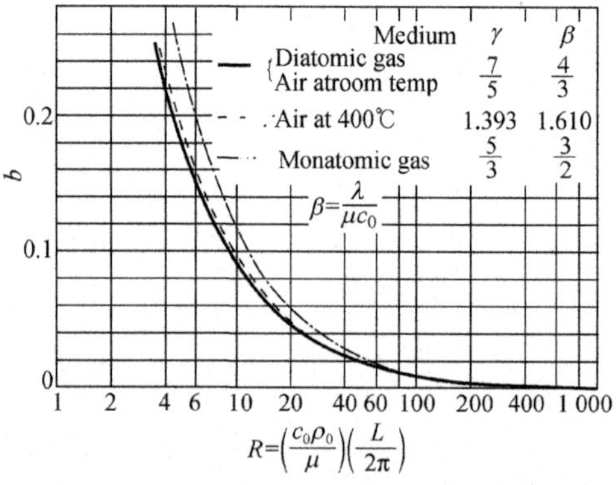

Fig. 4 Damping coefficient, b, versus Reynolds number R for three media

as ($\gamma - 1$), than on β. Both γ and β depend on the number of atoms which compose the molecule, whereas only β is significantly temperature dependent.

The magnitude of the damping effect is more readily ascertained in terms of the amplitude ratio, r_a, which is defined as the ratio of two successive maximum amplitudes of the sound wave. Hence $r_a = e^{-2\pi b} < 1$. Figure 5 shows the amplitude ratio as a function of the Reynolds number for the monatomic and the diatomic gas.

For practical applications of the results, one must first calculate the Reynolds number R corresponding to the wave-length L. Since for sound propagation problems the given physical parameter is the frequency ν instead of the wave-length L, R should be calculated in terms

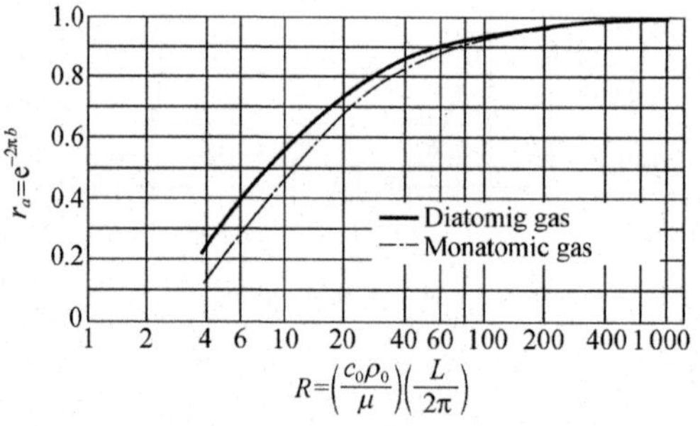

Fig. 5 Amplitude ratio, r_a, versus Reynolds number R for diatomic and monatomic gases

Table Ⅱ

Temperature/℃	$R_0 \times 10^{-8}$
0	1.321
20	1.248
40	1.186
60	1.130

(continued)

Temperature/℃	$R_0 \times 10^{-8}$
80	1.081
100	1.037
150	0.947
200	0.874
250	0.815
300	0.766
350	0.725
400	0.689
450	0.658
500	0.630

of ν in the following way:

$$R = (c/c_0 \times \gamma p_0/\mu) \times 1/(2\pi\nu) \tag{27}$$

However, $c/c_0 = \alpha$ is very close to unity, so that

$$R \cong (\gamma p_0/\mu) \times 1/(2\pi\nu) \tag{28}$$

To facilitate numerical calculations, Table II is constructed for R_0 which gives the value of R, at various temperatures, for air at the standard pressure of one atmosphere, for $\gamma = 1.4$, and for the frequency $\nu = 1\,000$ cycles per second. For any other values of the pressure p_0 in atmospheres, and of the frequency ν, the Reynolds number R is then given by

$$R = R_0\alpha\left(\frac{p_0}{1}\right)\left(\frac{1\,000}{\nu}\right) \cong R_0\left(\frac{p_0}{1}\right)\left(\frac{1\,000}{\nu}\right) \tag{29}$$

In general, the results of this investigation show that although the damping of the waves is greatly increased by the decrease in the Reynolds number R, the speed of propagation is practically unaltered. This justifies the procedure adopted in the analysis of anomalous sound propagation where the normal adiabatic propagation speed is used throughout. However, one must be aware of the fact that kinetic theory uses the smooth spherical model for the molecules and thus does not allow the interchange between translational kinetic energy with the vibrational and the rotational energies of the molecules. Because of the greatly decreased number of molecular collisions in rarefied conditions corresponding to small values of R, it will be difficult to excite the vibrational and the rotational degrees of freedom and the gas tends to behave more closely like a monatomic gas with a corresponding increase in the value of γ. This change in the properties of the gas is not directly taken into account in the calculations presented, as it is beyond the framework of the usual kinetic theory. On the other hand, this effect can be easily accounted for by an appropriate change in the value of γ as the Reynolds number decreases.

References

[1] Lamb H. Hydrodynamics. Cambridge University Press, London, England, 1932,6th: 647 - 650.
[2] Chapman S, Cowling T G. The Mathematicual Theory of Non-Uniform Gases. 265 - 269.
[3] Chapman S, Cowling T G. reference[2]: 221 - 223.

Similarity Laws of Hypersonic Flows

By Hsue-shen Tsien

Introduction. Hypersonic flows are flow fields where the fluid velocity is much larger than the velocity of propagation of small disturbances, the velocity of sound. Th. von Kármán[1] has pointed out that in many ways the dynamics of hypersonic flows is similar to Newton's corpuscular theory of aerodynamics. The pressure acting on an inclined surface is thus greater than the free stream pressure by a quantity which is approximately proportional to the square of the angle of inclination instead of the usual linear law for conventional supersonic flows. E. Sänger[2] has, in fact, used this concept to design the optimum wing and body shapes for hypersonic flight at extreme speeds.

Recently, von Kármán[3] has obtained the similarity laws for transonic flows where the fluid velocity is very near to the velocity of sound. He deduced these laws by using an affine transformation of the fluid field so that the differential equations of the flows are reduced to a single non-dimensional equation. In this paper, the same method is used to derive the similarity laws for hypersonic flows. These laws will be, perhaps, useful in correlating the experimental data to be obtained in the near future by hypersonic wind tunnels now under construction.

Differential Equation for Hypersonic Flows. If u, v are the components of velocity in the x, y directions and a is the local velocity of sound, the differential equations for irrotational two-dimensional motion are

$$\left(1-\frac{u^2}{a^2}\right)\frac{\partial u}{\partial x}-\frac{uv}{a^2}\left(\frac{\partial u}{\partial y}+\frac{\partial v}{\partial x}\right)+\left(1-\frac{v^2}{a^2}\right)\frac{\partial v}{\partial y}=0 \tag{1}$$

$$\frac{\partial v}{\partial x}-\frac{\partial u}{\partial y}=0 \tag{2}$$

Now if a slender body is present in an otherwise uniform stream of velocity V in the x-direction, equation (2) is satisfied by introducing the disturbance velocity potential φ defined as

$$u=V+\frac{\partial\varphi}{\partial x},\quad v=\frac{\partial\varphi}{\partial y} \tag{3}$$

If a_0 is the velocity of sound for the gas at rest and a^0 is the velocity of sound corresponding to the free stream velocity V, then there are the following relations:

Journal of Mathematics and Physics, vol. 25, pp. 247 – 251, 1946.

$$\left.\begin{aligned} a^2 &= a_0^2 - \frac{\gamma-1}{2}(u^2+v^2) = a_0^2 - \frac{\gamma-1}{2}\left[V^2 + 2V\frac{\partial\varphi}{\partial x} + \left(\frac{\partial\varphi}{\partial x}\right)^2 + \left(\frac{\partial\varphi}{\partial y}\right)^2\right] \\ a^{0^2} &= a_0^2 - \frac{\gamma-1}{2}V^2 \end{aligned}\right\} \tag{4}$$

where γ is the ratio of the specific heats.

For hypersonic flows over a slender body, both a^0 and $\frac{\partial\varphi}{\partial x}$, $\frac{\partial\varphi}{\partial y}$ are small in comparison with V. By substituting equations (3) and (4) into equation (1) and retaining terms up to second order, one has

$$\begin{aligned} &\left[1-(\gamma+1)M^0\frac{1}{a^0}\frac{\partial\varphi}{\partial x} - \frac{\gamma-1}{2}\frac{1}{a^{0^2}}\left(\frac{\partial\varphi}{\partial y}\right)^2 - M^{0^2}\right]\frac{\partial^2\varphi}{\partial x^2} - 2M^0\frac{1}{a^0}\frac{\partial\varphi}{\partial y}\frac{\partial^2\varphi}{\partial x\partial y} + \\ &\quad\left[1-(\gamma-1)M^0\frac{1}{a^0}\frac{\partial\varphi}{\partial x} - \frac{\gamma+1}{2}\frac{1}{a^{0^2}}\left(\frac{\partial\varphi}{\partial y}\right)^2\right]\frac{\partial^2\varphi}{\partial y^2} = 0 \end{aligned} \tag{5}$$

Here M^0 is the Mach number of the free stream of

$$M^0 = \frac{V}{a^0} \tag{6}$$

Similarity Laws in Two-Dimensional Flow. If $2b$ is the length or chord of the body and δ the thickness of the body, the non-dimensional coordinates ξ and η can be defined as

$$x = b\xi, y = b\left(\frac{\delta}{b}\right)^n\eta \tag{7}$$

where n is the exponent yet to be determined. von Kármán[1] has shown that for hypersonic flow over a slender body the variation of fluid velocity due to the presence of the body is limited within a narrow region close to the body, the hypersonic boundary layer. Therefore, in order to investigate this velocity variation, one must expand the coordinate normal to the surface of the body. This is similar to the case of ordinary viscous boundary layer, where Prandtl's simplified boundary layer equation is obtained from the exact Navier-Stokes equations by a coordinate expansion normal to the surface of the body. From this reasoning then, n must be positive so that η is much greater than $(y)/(b)$. This surmise is substantiated by the later calculations to be shown presently.

The appropriate non-dimensional form for the velocity potential φ is

$$\varphi = a^0 b\frac{1}{M^0}f(\xi,\eta) \tag{8}$$

By substituting equations (7) and (8) into equation (5), one has

$$\begin{aligned} &\left(\frac{\delta}{b}\right)^{2n}\left[1-(\gamma+1)\frac{\partial f}{\partial\xi} - \frac{\gamma-1}{2}\frac{1}{M^{0^2}(\delta/b)^{2n}}\left(\frac{\partial f}{\partial\eta}\right)^2\right]\frac{\partial^2 f}{\partial\xi^2} - M^{0^2}\left(\frac{\delta}{b}\right)^{2n}\frac{\partial^2 f}{\partial\xi^2} - 2\frac{\partial f}{\partial\eta}\frac{\partial^2 f}{\partial\xi\partial\eta} + \\ &\quad\left[1-(\gamma-1)\frac{\partial f}{\partial\xi} - \frac{\gamma+1}{2}\frac{1}{M^{0^2}(\delta/b)^{2n}}\left(\frac{\partial f}{\partial\eta}\right)^2\right]\frac{\partial^2 f}{\partial\eta^2} = 0 \end{aligned} \tag{9}$$

The boundary conditions at infinity require that the flow velocity be V. Thus

$$\frac{\partial f}{\partial \xi} = \frac{\partial f}{\partial \eta} = 0 \quad \text{at} \quad \infty \tag{10}$$

If the slender body is a symmetrical one, then the condition at the surface of the body can be written as

$$\left(\frac{\partial \varphi}{\partial y}\right)_{y=0} = a^0 M^0 \left(\frac{\delta}{b}\right) h(\xi) \quad \text{for} \quad -1 < \xi < 1 \tag{11}$$

where $h(\xi)$ for $-1 < \xi < 1$ is a given function describing the thickness distribution along the length of the body. Equation (11) can be converted into the following form by means of equations (7) and (8):

$$\left(\frac{\partial f}{\partial \eta}\right)_{\eta=0} = M^{0^2} \left(\frac{\delta}{b}\right)^{1+n} h(\xi) \tag{12}$$

Since the body is thin, δ/b is very small. Therefore the first group of terms in equation (9) is negligible in comparison with the rest. Then both the differential equation and the boundary conditions can be made to contain only a single parameter if one sets

$$n = 1 \tag{13}$$

That is, if

$$M^0 \frac{\delta}{b} = K$$

Then equation (9) becomes

$$\left[1 - (\gamma - 1)\frac{\partial f}{\partial \xi} - \frac{\gamma + 1}{2}\frac{1}{K^2}\left(\frac{\partial f}{\partial \eta}\right)^2\right]\frac{\partial^2 f}{\partial \eta^2} = K^2 \frac{\partial^2 f}{\partial \xi^2} + 2\frac{\partial f}{\partial \eta}\frac{\partial^2 f}{\partial \xi \partial \eta} \tag{14}$$

and the boundary conditions become

$$\frac{\partial f}{\partial \xi} = \frac{\partial f}{\partial \eta} = 0 \quad \text{at} \quad \infty \tag{15}$$

and

$$\left(\frac{\partial f}{\partial \eta}\right)_{n=0} = K^2 h(\xi) \quad \text{for} \quad -1 < \xi < 1$$

The meaning of this similarity law is the following: If a series of bodies having the same thickness distribution but different thickness ratios (δ/b) are put into flows of different Mach numbers M^0 such that the products of M^0 and (δ/b) remain constant and equal to K, then the flow patterns are similar in the sense that they are governed by the same function $f(\xi, \eta)$, determined by equations (14) and (15).

If p_0 is the stagnation pressure, p^0 the free stream pressure, and p the local pressure, then

$$p^0 = p_0\left[1+\frac{\gamma-1}{2}M^{0^2}\right]^{-(\gamma/\gamma-1)}$$

$$p = p_0\left[1+\frac{\gamma-1}{2}\frac{u^2+v^2}{a^2}\right]^{-(\gamma/\gamma-1)}$$

In notations introduced previously and by retaining terms of proper magnitudes, one can write the expression for the local pressure as

$$p = p^0\left[1-(\gamma-1)\frac{\partial f}{\partial \xi}-\frac{\gamma-1}{2}\frac{1}{K^2}\left(\frac{\partial f}{\partial \eta}\right)^2\right]^{\gamma/\gamma-1} \tag{16}$$

The drag D of the body can then be calculated. It is given by the following expression:

$$D = 2\int_{-b}^{b}(p)_{\eta=0}h(\xi)\left(\frac{\delta}{b}\right)\mathrm{d}x$$

$$= 2bp^0\left(\frac{\delta}{b}\right)\int_{-1}^{1}\left[1-(\gamma-1)\frac{\partial f}{\partial \xi}-\frac{\gamma-1}{2}\frac{1}{K^2}\left(\frac{\partial f}{\partial \eta}\right)^2\right]_{\eta=0}^{\gamma/\gamma-1}h(\xi)\mathrm{d}\xi$$

If one wishes to compute the drag coefficient C_D, then one uses

$$C_D = \frac{D}{\frac{1}{2}\rho^0V^2(2b)}$$

$$= \frac{1}{M^{0^2}}\cdot\left\{\frac{2}{\gamma}K\int_{-1}^{1}\left[1-(\gamma-1)\frac{\partial f}{\partial \xi}-\frac{\gamma-1}{2}\frac{1}{K^2}\left(\frac{\partial f}{\partial \eta}\right)^2\right]_{\eta=0}^{\gamma/\gamma-1}h(\xi)\mathrm{d}\xi\right\} \tag{17}$$

For a given thickness distribution, the quantity within the brackets is only a function of K, the similarity parameter. Therefore, one can write

$$C_D = \frac{1}{M^{0^2}}\Delta(K) = \frac{1}{M^{0^2}}\Delta\left(M^0\frac{\delta}{b}\right) \tag{18}$$

Similarly, one obtains for the lift coefficient C_L of the lift L the following law:

$$C_L = \frac{L}{\frac{1}{2}\rho^0V^2(2b)} = \frac{1}{M^{0^2}}\Lambda(K) = \frac{1}{M^{0^2}}\Lambda\left(M^0\frac{\delta}{b}\right) \tag{19}$$

These similarity laws show that for bodies of the same thickness distribution at angles of attack proportional to the thickness ratio (δ/b), the quantities ($C_DM^{0^2}$) and ($C_LM^{0^2}$) are functions of the single parameter $K = M^0(\delta/b)$.

Equations (18) and (19) agree with the results of the more limited linearized theory of Ackeret[4]. According to this theory, for similar bodies in the sense stated above the drag coefficient and the lift coefficient are given by

$$C_D \sim \frac{\left(\frac{\delta}{b}\right)^2}{\sqrt{M^{0^2}-1}}$$

$$C_L \sim \frac{\frac{\delta}{b}}{\sqrt{M^{0^2}-1}}$$

For hypersonic flows of very large values of M^0, these expressions reduce to

$$C_D \sim \left(\frac{\delta}{b}\right)^2 \Big/ M^0 \tag{20}$$

$$C_L \sim \left(\frac{\delta}{b}\right) \Big/ M^0 \tag{21}$$

Equations (20) and (21) agree with equations (18) and (19). Equations (18) and (19) are, however, more general and complete.

Axially Symmetrical Flows. For axially symmetrical flows, the ordinate y is the radial distance from the axis to the point concerned. Then a similar analysis leads to the following differential equation and boundary conditions:

$$\left[1-(\gamma-1)\frac{\partial f}{\partial \xi}-\frac{\gamma+1}{2}\frac{1}{K^2}\left(\frac{\partial f}{\partial \eta}\right)^2\right]\frac{\partial^2 f}{\partial \eta^2}+ \left[1-(\gamma-1)\frac{\partial f}{\partial \xi}-\frac{\gamma-1}{2}\frac{1}{K^2}\left(\frac{\partial f}{\partial \eta}\right)^2\right]\frac{1}{\eta}\frac{\partial f}{\partial \eta}=2\frac{\partial f}{\partial \eta}\frac{\partial^2 f}{\partial \xi \partial \eta}+K^2\frac{\partial^2 f}{\partial \xi^2} \tag{22}$$

$$\frac{\partial f}{\partial \xi}=\frac{\partial f}{\partial \eta}=0 \quad \text{at} \quad \infty \tag{23}$$

$$\left(\eta \frac{\partial f}{\partial \eta}\right)_{\eta=0}=K^2 h(\xi) \quad \text{for} \quad -1<\xi<1 \tag{24}$$

where $h(\xi)$ is the distribution function for cross-sectional areas along the length of the body and $K = M^0(\delta/b)$ as for two-dimensional flow.

The drag coefficient C_D referred to the maximum cross section of the body is then governed by the following similarity law:

$$C_D=\frac{1}{M^{0^2}}\Delta\left(M^0\frac{\delta}{b}\right) \tag{25}$$

California Institute of Technology.

References

[1] von Kármán Th. The Problem of Resistance in Compressible Fluids. Proceedings of the 5th Volta Congress, Rome 1936: 275 - 277.

[2] Sänger E. Gleitkörper für sehr hohe Fluggeshwindigkeiten. German Patent 411/42. Berlin, 1939.

[3] von Kármán Th. Similarity Laws of Transonic Flows, to be published soon.

[4] Durand Aerodynamic Theory. III, J. Springer, Berlin, 1935: 234 - 236.

One-Dimensional Flows of a Gas Characterized by van der Waals Equation of State*

By Hsue-shen Tsien

Introduction. It is generally assumed in the mathematical analysis of flows of a compressible fluid that the equation of state for the fluid is the equation for a perfect gas. This assumption, while sufficient for most cases of air flow under low pressures, does not allow an accurate enough representation of the properties of a gas near its point of condensation. For these flows of vapors, numerical and graphical methods are generally adopted. For instance, the isentropic expansion of steam can be traced on the Mollier diagram. A. Busemann[1] has developed a graphical method for determining the expansion around a corner of the supersonic flows of any real fluid.

However, the numerical or graphical methods are tedious and involve many laborious calculations. Furthermore, the diagrams constructed for one fluid cannot be adopted to a different fluid. The analytical method, on the other hand, is general and the results so established can be applied to any fluid with a mere change of the parameters. In this paper, an analytical method of calculating one-dimensional flows of a gas characterized by van der Waal's equation of state will be presented. The specific heat of the gas at constant volume is assumed to vary only with the temperature according to a quadratic law. Since for most applications envisaged the deviation of the results from the conventional ones for a perfect gas of constant specific heats is expected to be not more than ten percent, the method of small perturbation will be used so that terms of second order in the deviations from the perfect gas will be neglected. The particular problems solved are the isentropic expansion in a nozzle and the normal shock in supersonic flows. The calculation is prompted by the recent interest in hypersonic wind tunnels where extremely large expansion ratios are involved and the deviations from the perfect gas are not negligible.

The flow of a perfect gas with variable specific heats is treated by W. J. Walker[2]. It seems that he does not use the method of small perturbation explicitly and his results are thus somewhat awkward for flow calculations. The present analysis is believed to be more convenient to use.

Isentropic Expansion of a van der Waal Gas. If p is the pressure, T the temperature, ρ the

* Received February 18, 1947.
Journal of Mathematics and Physics, vol. 25, pp. 301 – 324, 1947.

density and R the gas constant, van der Waal's equation of state is

$$p = \frac{RT}{\rho^{-1} - b} - a\rho^2 \tag{1}$$

where a and b are constants, representing the effects of intermolecular attraction and molecular dimensions respectively. Let C_V be the specific heat of the gas at constant volume, assumed to be a function of temperature only. Then for isentropic expansion, the first law of thermodynamics requires that

$$C_V \mathrm{d}T + p\mathrm{d}\rho^{-1} = C_V \mathrm{d}T - p\mathrm{d}\rho/\rho^2 = 0 \tag{2}$$

By substituting equation (1) into equation (2), one has

$$C_V \mathrm{d}T - \{RT(\rho - b\rho^2)^{-1} - a\}\mathrm{d}\rho = 0 \tag{3}$$

Generally, the variation of the specific heat C_V can be quite accurately represented by the following expression:

$$C_V = \alpha + \beta T + \gamma T^2 \tag{4}$$

where α, β, γ are constants. Then equation (3) can be written as

$$(\alpha + \beta T + \gamma T^2)\mathrm{d}T - \{RT(\rho - b\rho^2)^{-1} - a\}\mathrm{d}\rho = 0 \tag{5}$$

This equation is the differential law for the variation of T with ρ in an isentropic expansion.

To integrate equation (5), the calculations are greatly simplified by the assumption that quantities in second order of a, b, β, γ can be neglected. This is justified on the ground that a, b, β, γ are the deviation from a perfect gas and these deviations are not expected to exceed ten per cent for most applications. With this assumption, the temperature T as a function of the density ρ can be expressed in the following form:

$$T(\rho) = T^{(0)}(\rho) + aT^{(1)}(\rho) + bT^{(2)}(\rho) + \beta T^{(3)}(\rho) + \gamma T^{(4)}(\rho) \tag{6}$$

By substituting equation (6) into equation (5) and by equating the quantities independent of a, b, β, γ to zero, one has

$$\alpha \mathrm{d}T^{(0)} - (R/\rho)T^{(0)}\mathrm{d}\rho = 0 \tag{7}$$

This is the equation connecting the temperature and density for a perfect gas undergoing isentropic expansion. The equations for the correction terms $T^{(1)}$, $T^{(2)}$, $T^{(3)}$ and $T^{(4)}$ are obtained by substituting equation (6) into equation (5) and then equating separately the coefficients of a, b, β, and γ in the resulting expression to zero. Thus,

$$\alpha \mathrm{d}T^{(1)} - \{(R/\rho)T^{(1)} - 1\}\mathrm{d}\rho = 0 \tag{8}$$

$$\alpha \mathrm{d}T^{(2)} - \{(R/\rho)(T^{(2)} + \rho T^{(0)})\}\mathrm{d}\rho = 0 \tag{9}$$

$$\alpha \mathrm{d}T^{(3)} + T^{(0)}\mathrm{d}T^{(0)} - (R/\rho)T^{(3)}\mathrm{d}\rho = 0 \tag{10}$$

$$\alpha \mathrm{d}T^{(4)} + T^{(0)^2}\mathrm{d}T^{(0)} - (R/\rho)T^{(4)}\mathrm{d}\rho = 0 \tag{11}$$

Equations (7) to (11) are now the linear first order differential equations for $T^{(0)}$, $T^{(1)}$, $T^{(2)}$,

$T^{(3)}$ and $T^{(4)}$ respectively. Their solutions are

$$T^{(0)} = C\rho^{R/\alpha} \tag{12}$$

$$T^{(1)} = \rho/(R-\alpha) \tag{13}$$

$$T^{(2)} = (RC/\alpha)\rho^{(R/\alpha)+1} \tag{14}$$

$$T^{(3)} = -(C^2/\alpha)\rho^{2R/\alpha} \tag{15}$$

$$T^{(4)} = -[C^3/(2\alpha)]\rho^{3R/\alpha} \tag{16}$$

The constants of integrations for $T^{(1)}$, $T^{(2)}$, $T^{(3)}$ and $T^{(4)}$ are absorbed into the constant C. The temperature T according to equation (6) can then be written as

$$T = C\rho^{R/\alpha} + a\frac{\rho}{R-\alpha} + b\frac{CR}{\alpha}\rho^{(R/\alpha)+1} - \beta\frac{C^2}{\alpha}\rho^{2R/\alpha} - \gamma\frac{C^3}{2\alpha}\rho^{3R/\alpha} \tag{17}$$

To eliminate the constant C, the quantities connected with the starting point of the expansion can be utilized. Let these quantities be denoted by the subscript 0. Then

$$T_0 = C\rho_0^{R/\alpha} + a\frac{\rho_0}{R-\alpha} + b\frac{CR}{\alpha}\rho_0^{(R/\alpha)+1} - \beta\frac{C^2}{\alpha}\rho_0^{2R/\alpha} - \gamma\frac{C^3}{2\alpha}\rho_0^{3R/\alpha} \tag{18}$$

The constant C can be eliminated between equations (17) and (18). Then the temperature ratio T/T_0 can be written as

$$\frac{T}{T_0} = \left(\frac{\rho}{\rho_0}\right)^{\lambda}\left[1 + \left(\frac{a\rho_0}{RT_0}\right)\frac{\lambda}{1-\lambda}\left\{1 - \left(\frac{\rho}{\rho_0}\right)^{1-\lambda}\right\} - (b\rho_0)\lambda\left\{1 - \frac{\rho}{\rho_0}\right\} + \left(\frac{\beta T_0}{\alpha}\right)\left\{1 - \left(\frac{\rho}{\rho_0}\right)^{\lambda}\right\} + \left(\frac{\lambda T_0^2}{\alpha}\right)\frac{1}{2}\left\{1 - \left(\frac{\rho}{\rho_0}\right)^{2\lambda}\right\}\right] \tag{19}$$

where λ is given by

$$\lambda = R/\alpha \tag{20}$$

If the specific heat C_V were constant, then $\beta = \gamma = 0$. If the gas were perfect, then $a = b = 0$. Then equation (19) would be reduced to the simple relation $T/T_0 = (\rho/\rho_0)^{\lambda}$, the well-known equation for perfect gas.

By inverting equation (19) and retaining only terms of proper order of magnitudes, one has

$$\frac{\rho}{\rho_0} = \left(\frac{T}{T_0}\right)^{1/\lambda}\left[1 - \left(\frac{a\rho_0}{RT_0}\right)\frac{1}{1-\lambda}\left\{1 - \left(\frac{T}{T_0}\right)^{(1/\lambda)-1}\right\} + (b\rho_0)\left\{1 - \left(\frac{T}{T_0}\right)^{1/\lambda}\right\} - \left(\frac{\beta T_0}{\alpha}\right)\frac{1}{\lambda}\left\{1 - \frac{T}{T_0}\right\} - \left(\frac{\gamma T_0^2}{\alpha}\right)\frac{1}{2\lambda}\left\{1 - \left(\frac{T}{T_0}\right)^2\right\}\right] \tag{21}$$

Other equations involving the pressure p can be easily obtained by using the equation of states (1). The results are

$$\frac{p}{p_0} = \left(\frac{\rho}{\rho_0}\right)^{\lambda+1}\left[1 + \left(\frac{a\rho_0}{RT_0}\right)\frac{1}{1-\lambda}\left\{1 - \left(\frac{\rho}{\rho_0}\right)^{1-\lambda}\right\} - (b\rho_0)(1+\lambda)\left\{1 - \left(\frac{\rho}{\rho_0}\right)\right\} + \left(\frac{\beta T_0}{\alpha}\right)\left\{1 - \left(\frac{\rho}{\rho_0}\right)^{\lambda}\right\} + \left(\frac{\gamma T_0^2}{\alpha}\right)\frac{1}{2}\left\{1 - \left(\frac{\rho}{\rho_0}\right)^{2\lambda}\right\}\right] \tag{22}$$

$$\frac{\rho}{\rho_0} = \left(\frac{p}{p_0}\right)^{1/(1+\lambda)} \left[1 - \left(\frac{a\rho_0}{RT_0}\right)\frac{1}{1-\lambda^2}\left\{1 - \left(\frac{p}{p_0}\right)^{(1-\lambda)/(1+\lambda)}\right\} + (b\rho_0)\left\{1 - \left(\frac{p}{p_0}\right)^{1/(1+\lambda)}\right\} - \left(\frac{\beta T_0}{\alpha}\right)\frac{1}{1+\lambda}\left\{1 - \left(\frac{p}{p_0}\right)^{\lambda/(1+\lambda)}\right\} - \left(\frac{\gamma T_0^2}{\alpha}\right)\frac{1}{2(1+\lambda)}\left\{1 - \left(\frac{p}{p_0}\right)^{2\lambda/(1+\lambda)}\right\}\right] \tag{23}$$

$$\frac{p}{p_0} = \left(\frac{T}{T_0}\right)^{1+(1/\lambda)} \left[1 - \left(\frac{a\rho_0}{RT_0}\right)\frac{\lambda}{1-\lambda}\left\{1 - \left(\frac{T}{T_0}\right)^{(1/\lambda)-1}\right\} - \left(\frac{\beta T_0}{\alpha}\right)\frac{1}{\lambda}\left\{1 - \frac{T}{T_0}\right\} - \left(\frac{\gamma T_0^2}{\alpha}\right)\frac{1}{2\lambda}\left\{1 - \left(\frac{T}{T_0}\right)^2\right\}\right] \tag{24}$$

$$\frac{T}{T_0} = \left(\frac{p}{p_0}\right)^{\lambda/(1+\lambda)} \left[1 + \left(\frac{a\rho_0}{RT_0}\right)\frac{\lambda^2}{1-\lambda^2}\left\{1 - \left(\frac{p}{p_0}\right)^{(1-\lambda)/(1+\lambda)}\right\} + \left(\frac{\beta T_0}{\alpha}\right)\frac{1}{1+\lambda}\left\{1 - \left(\frac{p}{p_0}\right)^{\lambda/(1+\lambda)}\right\} + \left(\frac{\gamma T_0^2}{\alpha}\right)\frac{1}{2(1+\lambda)}\left\{1 - \left(\frac{p}{p_0}\right)^{2\lambda/(1+\lambda)}\right\}\right] \tag{25}$$

Thus once the properties of the gas are known and the initial state of the gas specified, the "correction parameters" $a\rho_0/(RT_0)$, $(b\rho_0)$, $(\beta T_0/\alpha)$ and $(\gamma T_0^2/\alpha)$ can be immediately calculated. Then the subsequent states of expansion can be determined by equations (19) to (25). The functions of density ratios, pressure ratios and temperature ratios in these equations are universal for any gas and are independent of the initial conditions.

Expansion in a Nozzle. For isentropic expansion of the gas in a nozzle, one is interested in determining the velocity of flow. Let this velocity be denoted by v, then the dynamic equation of the flow is

$$\rho v \mathrm{d}v = -\mathrm{d}p \tag{26}$$

On the other hand, equation (2) can be rewritten as

$$C_V \mathrm{d}T + p/\rho - \mathrm{d}p/\rho = 0 \tag{27}$$

By combining equations (4), (26) and (27), the following differential equation is obtained:

$$(\alpha + \beta T + \gamma T^2)\mathrm{d}T + \mathrm{d}(p/\rho) + v\mathrm{d}v = 0 \tag{28}$$

If the conditions at the starting point of the expansion, or the stagnation point, are denoted by the subscript 0, the result of integrating equation (28) is

$$\frac{1}{2}v^2 = \alpha T_0\left\{1 - \frac{T}{T_0}\right\} + \frac{\beta T_0^2}{2}\left\{1 - \left(\frac{T}{T_0}\right)^2\right\} + \frac{\gamma T_0^3}{3}\left\{1 - \left(\frac{T}{T_0}\right)^3\right\} + \frac{p_0}{\rho_0}\left\{1 - \left(\frac{p}{p_0}\right)\left(\frac{\rho_0}{\rho}\right)\right\} \tag{29}$$

With the temperature ratio and pressure ratio determined by the equations given in the previous section, the velocity of the flow can be then calculated by equation (29).

For many applications such as wind tunnel design, the most convenient parameter for velocity is the Mach number M, the ratio of the velocity v and the velocity of sound c defined by

$$c^2 = \mathrm{d}p/\mathrm{d}\rho \tag{30}$$

where p and ρ are those in an isentropic process. This quantity can thus be calculated by differentiating equation (22). The result is

$$c^2 = (1+\lambda)\frac{p_0}{\rho}\left(\frac{\rho}{\rho_0}\right)^{\lambda}\left[1+\left(\frac{a\rho_0}{RT_0}\right)\frac{1}{1-\lambda}\left\{1-\frac{2}{1+\lambda}\left(\frac{\rho}{\rho_0}\right)^{1-\lambda}\right\}-(b\rho_0)(1+\lambda)\times\right.$$
$$\left.\left\{1-\frac{2+\lambda}{1+\lambda}\left(\frac{\rho}{\rho_0}\right)\right\}+\left(\frac{\beta T_0}{\alpha}\right)\left\{1-\frac{1+2\lambda}{1+\lambda}\left(\frac{\rho}{\rho_0}\right)^{\lambda}\right\}+\left(\frac{\lambda T_0^2}{\alpha}\right)\frac{1}{2}\left\{1-\frac{1+3\lambda}{1+\lambda}\left(\frac{\rho}{\rho_0}\right)^{2\lambda}\right\}\right] \tag{31}$$

Expressed in terms of local conditions p, T, and ρ, the velocity of sound is given by

$$c^2 = (1+\lambda)\frac{p}{\rho}\left[1-\left(\frac{a\rho}{RT}\right)\frac{1}{1+\lambda}+(b\rho)-\left(\frac{\beta T}{\alpha}\right)\frac{\lambda}{1+\lambda}-\left(\frac{\gamma T^2}{\alpha}\right)\frac{\lambda}{1+\lambda}\right] \tag{32}$$

Equation (32) can be obtained from equation (31) by the simple artifice of removing all the subscripts. This is the same as moving the reference point de-by the subscript 0 to the point concerned. Equation (32) is considerably different from the conventional formula for the velocity of sound,

$$c^2 = \left(1+\frac{R}{C_v}\right)\frac{p}{\rho} \tag{33}$$

It can be, however, easily shown by means of equation (4) that equation (32) can be written as

$$c^2 = \left(1+\frac{R}{C_v}\right)\frac{p}{\rho}\left[1-\left(\frac{a\rho}{RT}\right)\frac{1}{1+\lambda}+(b\rho)\right] \tag{34}$$

The difference between equations (33) and (34) is then purely due to the "thermal" imperfections of the gas.

In equation (29) the density ratio and the pressure ratio can be replaced as functions of the temperature ratio by equation (21) and (24). Then by replacing the velocity v by the Mach number M through equation (31), one obtains the relation between the temperature ratio T/T_0 and the Mach number:

$$\frac{T}{T_0} = \frac{1}{(1+\lambda M^2/2)}\left[1+\left(\frac{a\rho_0}{RT_0}\right)t_1-(b\rho_0)t_2+\left(\frac{\beta T_0}{\alpha}\right)t_3+\left(\frac{\gamma T_0^2}{\alpha}\right)t_4\right] \tag{35}$$

Here t_1, t_2, t_3 and t_4 are functions of M and λ given by the following equations:

$$\left.\begin{aligned}
t_1 &= \frac{\lambda}{\lambda+\lambda}\left\{M^2\left(1+\frac{\lambda}{2}M^2\right)^{-1/\lambda}+\left(1+\frac{\lambda}{2}M^2\right)^{1-(1/\lambda)}-1\right\}\\
t_2 &= \frac{1}{1+\lambda}\left\{(M^2-1)\left(1+\frac{\lambda}{2}M^2\right)^{-1-(1/\lambda)}+2\left(1+\frac{\lambda}{2}M^2\right)^{-1/\lambda}-1\right\}\\
t_3 &= \frac{1}{1+\lambda}\left\{\frac{1}{2}+\lambda\left(1+\frac{\lambda}{2}M^2\right)^{-1}-\left(\frac{1}{2}+\lambda\right)\left(1+\frac{\lambda}{2}M^2\right)^{-2}\right\}\\
t_4 &= \frac{1}{1+\lambda}\left\{\frac{1}{3}+\lambda\left(1+\frac{\lambda}{2}M^2\right)^{-2}-\left(\frac{1}{3}+\lambda\right)\left(1+\frac{\lambda}{2}M^2\right)^{-3}\right\}
\end{aligned}\right\} \tag{36}$$

It is easily recognized that the factor outside the bracket on the right side of equation (35) is the conventional value of the temperature ratio for a perfect gas. The functions t_1, t_2, t_3 and t_4 give then the corrections due to "thermal" and "calorical" imperfections.

By using the relations between the pressure ratio, the density ratio and the temperature ratio given in the previous section, the pressure ratio and the density ratio can be calculated in terms of the Mach number. These relations are the following:

$$\frac{p}{p_0} = \frac{1}{(1+\lambda M^2/2)^{1+(1/\lambda)}}\left[1+\left(\frac{a\rho_0}{RT_0}\right)P_1-(b\rho_0)P_2+\left(\frac{\beta T_0}{\alpha}\right)P_3+\left(\frac{\gamma T_0^2}{\alpha}\right)P_4\right] \tag{37}$$

$$\frac{\rho}{\rho_0} = \frac{1}{(1+\lambda M^2/2)^{1/\lambda}}\left[1+\left(\frac{a\rho_0}{RT_0}\right)r_1-(b\rho_0)r_2+\left(\frac{\beta T_0}{\alpha}\right)r_3+\left(\frac{\gamma T_0^2}{\alpha}\right)r_4\right] \tag{38}$$

where

$$\left.\begin{aligned}
P_1 &= M^2\left(1+\frac{\lambda}{2}M^2\right)^{-1/\lambda}-\frac{1}{1-\lambda}\left\{1-\left(1+\frac{\lambda}{2}M^2\right)^{1-(1/\lambda)}\right\}\\
P_2 &= (M^2-1)\left(1+\frac{\lambda}{2}M^2\right)^{-1-(1/\lambda)}+2\left(1+\frac{\lambda}{2}M^2\right)^{1/\lambda}-1\\
P_3 &= \left(\frac{1}{4}M^2-1\right)\left(1+\frac{\lambda}{2}M^2\right)^{-2}+\left(1+\frac{1}{2\lambda}\right)\left(1+\frac{\lambda}{2}M^2\right)^{-1}-\frac{1}{2\lambda}\\
P_4 &= \left(\frac{1}{6}M^2-1\right)\left(1+\frac{\lambda}{2}M^2\right)^{-3}+\left(1-\frac{1}{3\lambda}\right)\left(1+\frac{\lambda}{2}M^2\right)^{-2}+\frac{1}{2\lambda}\left(1+\frac{\lambda}{2}M^2\right)^{-1}-\frac{1}{6\lambda}
\end{aligned}\right\} \tag{39}$$

and

$$\left.\begin{aligned}
r_1 &= \frac{1}{1+\lambda}M^2\left(1+\frac{\lambda}{2}M^2\right)^{-1/\lambda}-\frac{2}{1-\lambda^2}\left\{1-\left(1+\frac{\lambda}{2}M^2\right)^{1-(1/\lambda)}\right\}\\
r_2 &= \frac{1}{1+\lambda}(M^2-1)\left(1+\frac{\lambda}{2}M^2\right)^{-1-(1/\lambda)}+\left(1+\frac{2}{1+\lambda}\right)\left(1+\frac{\lambda}{2}M^2\right)^{-1/\lambda}-\left(1+\frac{1}{1+\lambda}\right)\\
r_3 &= \frac{1}{1+\lambda}\left(\frac{M^2}{4}-1\right)\left(1+\frac{\lambda}{2}M^2\right)^{-2}+\frac{4\lambda+1}{2\lambda(1+\lambda)}\left(1+\frac{\lambda}{2}M^2\right)^{-1}-\frac{2\lambda+1}{2\lambda(1+\lambda)}\\
r_4 &= \frac{1}{1+\lambda}\left(\frac{M^2}{6}-1\right)\left(1+\frac{\lambda}{2}M^2\right)^{-3}+\frac{9\lambda+1}{6\lambda(1+\lambda)}\left(1+\frac{\lambda}{2}M^2\right)^{-2}-\frac{3\lambda+1}{6\lambda(1+\lambda)}
\end{aligned}\right\} \tag{40}$$

Here again the functions P_1, P_2, P_3, P_4 and r_1, r_2, r_3, r_4 are correction functions for the pressure ratio and the density ratio respectively.

To calculate the cross-sectional area A of the nozzle, one needs the equation of continuity,

$$\mathrm{d}(\rho v A) = 0 \tag{41}$$

From equation (41), the change of the area A can be expressed as follows

$$\frac{\mathrm{d}A}{A} = -\left(\frac{\mathrm{d}\rho}{\rho}+\frac{\mathrm{d}v}{v}\right) \tag{42}$$

But by using equation (26), the first term on the right of above equation is equal to

$$\frac{\mathrm{d}\rho}{\rho} = \frac{\mathrm{d}\rho}{\mathrm{d}p}\frac{\mathrm{d}p}{\rho} = -\frac{v^2}{c^2}\frac{\mathrm{d}v}{v} = -M^2\frac{\mathrm{d}v}{v}$$

Therefore, equation (42) can be written as

$$dA/A = -(1 - M^2)dv/v \tag{43}$$

This equation shows that at the throat of the nozzle where $dA = 0$, the local Mach number is unity for any equation of state. This is the well-known result generally deduced for flow of a perfect gas.

Equation (41) requires that $\rho v A$ = constant. Therefore, if quantities at the throat are denoted by an asterisk, then

$$\frac{A}{A^*} = \frac{\rho^* v^*}{\rho v} = \left(\frac{\rho^*}{\rho_0}\right)\left(\frac{\rho_0}{\rho}\right)\left(\frac{v^*}{\sqrt{p_0/\rho_0}}\right)\left(\frac{\sqrt{p_0/\rho_0}}{c}\right)\frac{1}{M} \tag{44}$$

All quantities entering into this equation can be expressed in terms of the Mach number M by using the formulae previously derived. The result of this calculation, retaining only first order terms of a, b, β, and γ, is

$$\frac{A}{A^*} = \frac{1}{M}\left(\frac{1+\lambda M^2/2}{1+(\lambda/2)}\right)^{\frac{1}{2}+(1/\lambda)} \cdot \left[1 + \left(\frac{a\rho_0}{RT_0}\right)f_1 - (b\rho_0)f_2 + \left(\frac{\beta T_0}{\alpha}\right)f_3 + \left(\frac{\gamma T_0^2}{\alpha}\right)f_4\right] \tag{45}$$

where

$$\left.\begin{aligned}
f_1 &= \frac{1+(\lambda/2)}{1+\lambda}\left[\left(1+\frac{\lambda}{2}\right)^{-1/\lambda} - M^2\left(1+\frac{\lambda}{2}M^2\right)^{-1/\lambda}\right] + \\
&\quad \frac{1}{1-\lambda}\left[\left(1+\frac{\lambda}{2}\right)^{1-(1/\lambda)} - \left(1+\frac{\lambda}{2}M^2\right)^{1-(1/\lambda)}\right] \\
f_2 &= \frac{1+(\lambda/2)}{1+\lambda}\left[(2+\lambda)\left(1+\frac{\lambda}{2}\right)^{-1-(1/\lambda)} - \{1+(1+\lambda)M^2\}\left(1+\frac{\lambda}{2}M^2\right)^{-1-(1/\lambda)}\right] \\
f_3 &= \frac{(1+(\lambda/2))\left(\frac{1}{4}+(\lambda/2)\right)}{1+\lambda}\left[\left(1+\frac{\lambda}{2}\right)^{-2} - M^2\left(1+\frac{\lambda}{2}M^2\right)^{-2}\right] + \\
&\quad \frac{1}{4}\frac{(1+(\lambda/2))}{(1+\lambda)}\left[\left(1+\frac{\lambda}{2}\right)^{-1} - M^2\left(1+\frac{\lambda}{2}M^2\right)^{-1}\right] + \\
&\quad \frac{1+\lambda-(\lambda^2/2)}{\lambda(1+\lambda)}\left[\left(1+\frac{\lambda}{2}\right)^{-1} - \left(1+\frac{\lambda}{2}M^2\right)^{-1}\right] \\
f_4 &= \frac{(1+(\lambda/2))\left(\frac{1}{6}+(\lambda/2)\right)}{1+\lambda}\left[\left(1+\frac{\lambda}{2}\right)^{-3} - M^2\left(1+\frac{\lambda}{2}M^2\right)^{-3}\right] + \\
&\quad \frac{1}{6}\frac{(1+(\lambda/2))}{(1+\lambda)}\left[\left(1+\frac{\lambda}{2}\right)^{-2} - M^2\left(1+\frac{\lambda}{2}M^2\right)^{-2}\right] + \\
&\quad \frac{1+\lambda-\lambda^2}{2\lambda(1+\lambda)}\left[\left(1+\frac{\lambda}{2}\right)^{-2} - \left(1+\frac{\lambda}{2}M^2\right)^{-2}\right] + \\
&\quad \frac{1}{6}\frac{(1+(\lambda/2))}{(1+\lambda)}\left[\left(1+\frac{\lambda}{2}\right)^{-1} - M^2\left(1+\frac{\lambda}{2}M^2\right)^{-1}\right]
\end{aligned}\right\} \tag{46}$$

All characteristics of nozzle flow are thus given by equations (35), (37), (38) and (45).

In these equations, the corrections introduced by the imperfections of the gas are separated into a series of terms, each of which is a product of two factors. One factor is a universal function of the Mach number M and the ratio $\lambda = R/\alpha$, but independent of the initial conditions of the expansion. The other factor is determined by these initial conditions only. This is a considerable simplification allowed by the small perturbation method adopted.

Shock Wave in Supersonic Flow. In addition to flow through a convergent-divergent nozzle, one dimensional flows of a gas have the important feature of the formation of normal shock waves in the supersonic region. If the Mach number before the shock is very large, the pressure ratio across the shock is also very large. Then one must expect considerable deviations from the imperfections of the gas. To determine these deviations, consider the shock as fixed in space and denote the conditions ahead of the shock by the subscript 1 and conditions after the shock by the subscript 2. The flow velocities ahead and after the shock, being normal to the shock front, are thus v_1 and v_2 respectively. The equation of continuity is then

$$\rho_1 v_1 = \rho_2 v_2 \tag{47}$$

The equation of momentum is

$$\rho_1 v_1^2 - \rho_2 v_2^2 = p_2 - p_1 \tag{48}$$

The equation for the conservation of energy is, according to equation (28),

$$\alpha T_1 + \frac{\beta}{2} T_1^2 + \frac{\gamma}{3} T_1^3 + \frac{p_1}{\rho_1} + \frac{1}{2} v_1^2 = \alpha T_2 + \frac{\beta}{2} T_2^2 + \frac{\gamma}{3} T_2^3 + \frac{p_2}{\rho_2} + \frac{1}{2} v_2^2 \tag{49}$$

The equation of states is, of course, the van der Waal equation

$$p_2 = \frac{RT_2\rho_2}{1 - b\rho_2} - a\rho_2^2 \tag{50}$$

Equations (47) to (50) form the system of four simultaneous equations for the four unknowns v_2, ρ_2, p_2, T_2.

For flow calculations, it is generally more convenient to express the velocities as Mach numbers. With the aid of equation (32), equations (47), (48) and (49) can be rewritten as the following:

$$\rho_1 M_1^2 (1+\lambda) p_1 \left[1 - \frac{1}{1+\lambda}\frac{a\rho_1}{RT_1} + b\rho_1 - \frac{\lambda}{1+\lambda}\frac{\beta T_1}{\alpha} - \frac{\lambda}{1+\lambda}\frac{\gamma T_1^2}{\alpha}\right]$$

$$= \rho_2 M_2^2 (1+\lambda) p_2 \left[1 - \frac{1}{1+\lambda}\frac{a\rho_2}{RT_2} + b\rho_2 - \frac{\lambda}{1+\lambda}\frac{\beta T_2}{\alpha} - \frac{\lambda}{1+\lambda}\frac{\gamma T_2^2}{\alpha}\right] \tag{51}$$

$$M_1^2 (1+\lambda) p_1 \left[1 - \frac{1}{1+\lambda}\frac{a\rho_1}{RT_1} + b\rho_1 - \frac{\lambda}{1+\lambda}\frac{\beta T_1}{\alpha} - \frac{\lambda}{1+\lambda}\frac{\gamma T_1^2}{\alpha}\right] -$$

$$M_2^2 (1+\lambda) p_2 \left[1 - \frac{1}{1+\lambda}\frac{a\rho_2}{RT_2} + b\rho_2 - \frac{\lambda}{1+\lambda}\frac{\beta T_2}{\alpha} - \frac{\lambda}{1+\lambda}\frac{\gamma T_2^2}{\alpha}\right] = p_2 - p_1 \tag{52}$$

$$\alpha T_1 + \frac{\beta}{2}T_1^2 + \frac{\gamma}{3}T_1^3 + \frac{p_1}{\rho_1} + \frac{1}{2}M_1^2(1+\lambda)\frac{p_1}{\rho_1} \times$$
$$\left[1 - \frac{1}{1+\lambda}\frac{a\rho_1}{RT_1} + b\rho_1 - \frac{\lambda}{1+\lambda}\frac{\beta T_1}{\alpha} - \frac{\lambda}{1+\lambda}\frac{\gamma T_1^2}{\alpha}\right]$$
$$= \alpha T_2 + \frac{\beta}{2}T_2^2 + \frac{\gamma}{3}T_2^3 + \frac{p_2}{\rho_2} + \frac{1}{2}M_2^2(1+\lambda)\frac{p_2}{\rho_2} \times$$
$$\left[1 - \frac{1}{1+\lambda}\frac{a\rho_2}{RT_2} + b\rho_2 - \frac{\lambda}{1+\lambda}\frac{\beta T_2}{\alpha} - \frac{\lambda}{1+\lambda}\frac{\gamma T_2^2}{\alpha}\right] \tag{53}$$

In the system of equations (50) to (53), one can, for instance, eliminate the variables ρ_2, T_2 and M_2 and obtain an equation for p_2. This process of elimination is somewhat tedious, but nevertheless is straightforward. The result, retaining only terms linear in a, b, β and γ, is

$$\left[\frac{2+\lambda}{2}\left(\frac{p_2}{p_1}\right)^2 - \{(1+\lambda)M_1^2+1\}\left(\frac{p_2}{p_1}\right) + \left\{(1+\lambda)M_1^2 - \frac{\lambda}{2}\right\}\right] +$$
$$\left(\frac{a\rho_1}{RT_1}\right)\left[(1+\lambda)M_1^2\left\{1 - \frac{1}{1 - \frac{1}{(1+\lambda)M_1^2}\left(\frac{p_2}{p_1}-1\right)}\right\} + M_1^2\left(\frac{p_2}{p_1}-1\right) + \right.$$
$$\left.\frac{\frac{2+\lambda}{2}\left(\frac{p_2}{p_1}\right)^2 - \{(1+\lambda)M_1^2+1\}\left(\frac{p_2}{p_1}\right) + \left\{(1+\lambda)M_1^2 - \frac{\lambda}{2}\right\}}{1+\lambda}\right] -$$
$$(b\rho_1)\left[\frac{2+\lambda}{2}\left(\frac{p_2}{p_1}\right)^2 - \{1+(1+\lambda)M_1^2\}\left(\frac{p_2}{p_1}\right) + \left\{(1+\lambda)M_1^2 - \frac{\lambda}{2}\right\}\right] +$$
$$\left(\frac{\beta T_1}{\alpha}\right)\left[(1+\lambda)M_1^2\left\{\frac{1}{2} - \frac{1}{2}\left(\frac{p_2}{p_1}\right)^2\left[1 - \frac{\frac{p_2}{p_1}-1}{(1+\lambda)M_1^2}\right]^2 + \frac{\lambda}{1+\lambda}\left(\frac{p_2}{p_1}-1\right)\right\} + \right.$$
$$\left.\frac{\frac{2+\lambda}{2}\left(\frac{p_2}{p_1}\right)^2 - \{(1+\lambda)M_1^2+1\}\left(\frac{p_2}{p_1}\right) + \left\{(1+\lambda)M_1^2 - \frac{\lambda}{2}\right\}}{(1+\lambda)/\lambda}\right] +$$
$$\left(\frac{\gamma T_1^2}{\alpha}\right)\left[(1+\lambda)M_1^2\left\{\frac{1}{3} - \frac{1}{3}\left(\frac{p_2}{p_1}\right)^3\left[1 - \frac{\frac{p_2}{p_1}-1}{(1+\lambda)M_1^2}\right]^3 + \frac{\lambda}{1+\lambda}\left(\frac{p_2}{p_1}-1\right)\right\} + \right.$$
$$\left.\frac{\frac{2+\lambda}{2}\left(\frac{p_2}{p_1}\right)^2 - \{(1+\lambda)M_1^2+1\}\left(\frac{p_2}{p_1}\right) + \left\{(1+\lambda)M_1^2 - \frac{\lambda}{2}\right\}}{(1+\lambda)\lambda}\right] = 0 \tag{54}$$

According to the principle of small perturbation method adopted here, the pressure ratio (p_2/p_1) can be expressed as

$$\left(\frac{p_2}{p_1}\right)=\left(\frac{p_2}{p_1}\right)^{(0)}+\left(\frac{a\rho_1}{RT_1}\right)\left(\frac{p_2}{p_1}\right)^{(1)}-(b\rho_1)\left(\frac{p_2}{p_1}\right)^{(2)}+$$
$$\left(\frac{\beta T_1}{\alpha}\right)\left(\frac{p_2}{p_1}\right)^{(3)}+\left(\frac{\gamma T_1^2}{\alpha}\right)\left(\frac{p_2}{p_1}\right)^{(4)} \quad (55)$$

where none of the functions $(p_2/p_1)^{(i)}$ $(i=0,1,2,3,4)$ contains the parameter a, b, β or γ. By substituting equation (55) into equation (54) and by equating the terms independent of a, b, β or γ to zero, one has the equation for the first approximation $(p_2/p_1)^{(0)}$ as

$$\frac{2+\lambda}{2}\left(\frac{p_2}{p_1}\right)^{(0)^2}-\{(1+\lambda)M_1^2+1\}\left(\frac{p_2}{p_1}\right)^{(0)}+\left\{(1+\lambda)M_1^2-\frac{\lambda}{2}\right\}=0 \quad (56)$$

The significant root of this quadratic equation is

$$\left(\frac{p_2}{p_1}\right)^{(0)}=\frac{1}{2+\lambda}[2(1+\lambda)M_1^2-\lambda] \quad (57)$$

This is the well-known result for shock wave of a perfect gas. The correction for intermolecular attraction can be determined by equating to zero the terms with the factor $[a\rho_1/(RT_1)]$ in the result of substituting (55) into equation (54). Thus

$$\left[(2+\lambda)\left(\frac{p_2}{p_1}\right)^{(0)}-\{(1+\lambda)M_1^2+1\}\right]\left(\frac{p_2}{p_1}\right)^{(1)}+(1+\lambda)M_1^2\times$$
$$\left\{1-\frac{1}{1-\dfrac{1}{(1+\lambda)M_1^2}\left[\left(\dfrac{p_2}{p_1}\right)^{(0)}-1\right]}\right\}+M_1^2\left[\left(\frac{p_2}{p_1}\right)^{(0)}-1\right]=0 \quad (58)$$

Similarly, one has

$$\left[(2+\lambda)\left(\frac{p_2}{p_1}\right)^{(0)}-\{(1+\lambda)M_1^2+1\}\right]\left(\frac{p_2}{p_1}\right)^{(2)}=0 \quad (59)$$

$$\left[(2+\lambda)\left(\frac{p_2}{p_1}\right)^{(0)}-\{(1+\lambda)M_1^2+1\}\right]\left(\frac{p_2}{p_1}\right)^{(3)}+(1+\lambda)M_1^2\times$$
$$\left\{\frac{1}{2}-\frac{1}{2}\left(\frac{p_2}{p_1}\right)^{(0)^2}\left[1-\frac{\left(\dfrac{p_2}{p_1}\right)^{(0)}-1}{(1+\lambda)M_1^2}\right]^2+\frac{\lambda}{1+\lambda}\left[\left(\frac{p_2}{p_1}\right)^{(0)}\right]\right\}=0 \quad (60)$$

$$\left[(2+\lambda)\left(\frac{p_2}{p_1}\right)^{(0)}-\{(1+\lambda)M_1^2+1\}\right]\left(\frac{p_2}{p_1}\right)^{(4)}+(1+\lambda)M_1^2\times$$
$$\left\{\frac{1}{3}-\frac{1}{3}\left(\frac{p_2}{p_1}\right)^{(0)^3}\left[1-\frac{\left(\dfrac{p_2}{p_1}\right)^{(0)}-1}{(1+\lambda)M_1^2}\right]^3+\frac{\lambda}{1+\lambda}\left[\left(\frac{p_2}{p_1}\right)^{(0)}-1\right]\right\}=0 \quad (61)$$

Since the value of $(p_2/p_1)^{(0)}$ is given by equation (57), these equations are linear equations for the functions $(p_2/p_1)^{(i)}$ $(i=1,2,3,4)$. They can be easily solved. The final result for (p_2/p_1) can be written in the following form:

$$\frac{p_2}{p_1}=\frac{2(1+\lambda)M_1^2-\lambda}{(2+\lambda)}\cdot\left[1+\left(\frac{a\rho_1}{RT_1}\right)\Pi_1-(b\rho_1)\Pi_2+\left(\frac{\beta T_1}{\alpha}\right)\Pi_3+\left(\frac{\gamma T_1^2}{\alpha}\right)\Pi_4\right] \tag{62}$$

where

$$\left.\begin{aligned}
\Pi_1&=-\frac{2\lambda M_1^2(M_1^2-1)}{(\lambda M_1^2+2)[2(1+\lambda)M_1^2-\lambda]},\quad \Pi_2=0\\
\Pi_3&=\frac{2\lambda M_1^2}{[2(1+\lambda)M_1^2-\lambda]}\left[\frac{1}{(2+\lambda)^3}\left\{(1+\lambda)+\frac{1}{M_1^2}\right\}\times\right.\\
&\quad\left.\left\{\lambda(1+\lambda)M_1^2+4(1+\lambda)-\frac{\lambda}{M_1^2}\right\}-1\right]\\
\Pi_4&=\frac{2\lambda M_1^2}{[2(1+\lambda)M_1^2-\lambda]}\left[\frac{1}{(2+\lambda)^5}\,\frac{1}{3}\left\{(1+\lambda)+\frac{1}{M_1^2}\right\}\left\{4\lambda^2(1+\lambda)^2M_1^4+\right.\right.\\
&\quad 2\lambda(1+\lambda)(12+12\lambda-\lambda^2)M_1^2+[(4+4\lambda-\lambda^2)^2+32(1+\lambda)^2]-\\
&\quad\left.\left.2\lambda(12+12\lambda-\lambda^2)\frac{1}{M_1^2}+\frac{4\lambda^2}{M_1^4}\right\}-1\right]
\end{aligned}\right\} \tag{63}$$

By using equation (62) and previously given relations (50) to (53), the density ratio ρ_2/ρ_1, the temperature ratio T_2/T_1 and the Mach number M_2 can be determined. This calculation gives the following relations:

$$\frac{\rho_2}{\rho_1}=\frac{(2+\lambda)M_1^2}{\lambda M_1^2+2}\left[1+\left(\frac{a\rho_1}{RT_1}\right)D_1-(b\rho_1)D_2+\left(\frac{\beta T_1}{\alpha}\right)D_3+\left(\frac{\gamma T_1^2}{\alpha}\right)D_4\right] \tag{64}$$

where

$$\left.\begin{aligned}
D_1&=\frac{2(1+\lambda)M_1^2-\lambda}{(1+\lambda)(\lambda M_1^2+2)}\Pi_1+\frac{2(M_1^2-1)}{(1+\lambda)(\lambda M_1^2+2)}\\
D_2&=\frac{2(M_1^2-1)}{(\lambda M_1^2+2)}\\
D_3&=\frac{2(1+\lambda)M_1^2-\lambda}{(1+\lambda)(\lambda M_1^2+2)}\Pi_3+\frac{2\lambda(M_1^2-1)}{(1+\lambda)(\lambda M_1^2+2)}\\
D_4&=\frac{2(1+\lambda)M_1^2-\lambda}{(1+\lambda)(\lambda M_1^2+2)}\Pi_4+\frac{2\lambda(M_1^2-1)}{(1+\lambda)(\lambda M_1^2+2)}
\end{aligned}\right\} \tag{65}$$

$$\frac{T_2}{T_1}=\frac{\{2(1+\lambda)M_1^2-\lambda\}(\lambda M_1^2+2)}{(2+\lambda)^2M_1^2}\cdot\left[1+\left(\frac{a\rho_1}{RT_1}\right)\tau_1-(b\rho_1)\tau_2+\left(\frac{\beta T_1}{\alpha}\right)\tau_3+\left(\frac{\gamma T_1^2}{\alpha}\right)\tau_4\right] \tag{66}$$

where

$$\left.\begin{aligned}\tau_1 &= \Pi_1 - D_1 + \frac{(2+\lambda)^2 M_1^4}{(\lambda M_1^2+2)^2}\frac{(2+\lambda)}{\{2(1+\lambda)M_1^2-\lambda\}} - 1\\ \tau_2 &= 0,\quad \tau_3 = \Pi_3 - D_3,\quad \tau_4 = \Pi_4 - D_4\end{aligned}\right\} \tag{67}$$

$$M_2 = \sqrt{\frac{\lambda M_1^2+2}{2(1+\lambda)M_1^2-\lambda}}\cdot\left[1+\left(\frac{a\rho_1}{RT_1}\right)m_1 - (b\rho_1)m_2 + \left(\frac{\beta T_1}{a}\right)m_3 + \left(\frac{\gamma T_1^2}{a}\right)m_4\right] \tag{68}$$

where

$$\left.\begin{aligned}
m_1 &= \frac{\left(1+\frac{\lambda}{2}\right)M_1^2}{(1+\lambda)(\lambda M_1^2+2)}\left[\frac{(2+\lambda)^3 M_1^4}{(\lambda M_1^2+2)^2\{2(1+\lambda)M_1^2-\lambda\}} - 1\right] - \frac{\left(1+\frac{\lambda}{2}\right)\left(M_1^2+\frac{1}{1+\lambda}\right)}{(\lambda M_1^2+2)}\Pi_1\\
m_2 &= \frac{\left(1+\frac{\lambda}{2}\right)M_1^2}{(\lambda M_1^2+2)}\left[\frac{(2+\lambda)M_1^2}{(\lambda M_1^2+2)} - 1\right]\\
m_3 &= \frac{\lambda\left(1+\frac{\lambda}{2}\right)M_1^2}{(1+\lambda)(\lambda M_1^2+2)}\left[\frac{(\lambda M_1^2+2)\{2(1+\lambda)M_1^2-\lambda\}}{(2+\lambda)^2 M_1^2} - 1\right] - \frac{\left(1+\frac{\lambda}{2}\right)\left(M_1^2+\frac{1}{1+\lambda}\right)}{(\lambda M_1^2+2)}\Pi_3\\
m_4 &= \frac{\lambda\left(1+\frac{\lambda}{2}\right)M_1^2}{(1+\lambda)(\lambda M_1^2+2)}\left[\frac{(\lambda M_1^2+2)^2\{2(1+\lambda)M_1^2-\lambda\}^2}{(2+\lambda)^4 M_1^4} - 1\right] - \frac{\left(1+\frac{\lambda}{2}\right)\left(M_1^2+\frac{1}{1+\lambda}\right)}{(\lambda M_1^2+2)}\Pi_4
\end{aligned}\right\} \tag{69}$$

Equations (62), (64), (66) and (68) show that all pertinent quantities for the shock can be calculated by applying a series of correction terms to the formulae for perfect gas. Similar to results obtained in the preceeding section, these correction terms are products of two factors. One factor is a universal function of the Mach number M and the ratio $\lambda = R/a$, but independent of the initial conditions of the shock. The other factor is determined by these initial conditions only.

Parameters for Constants *a* and *b*. In the analysis of the previous sections, the two parameters $(a\rho)/RT$ and $(b\rho)$ appear in all the formulae for computation. The constants a and b are connected with the critical pressure p_c and the critical temperature T_c in the following way[3]:

$$p_c = \frac{1}{27}\frac{a}{b^2},\quad T_c = \frac{8}{27}\frac{a}{Rb} \tag{70}$$

By solving these equations for a and b, one has

$$a = \frac{27}{64}\frac{(RT_c)^2}{p_c}, \quad b = \frac{1}{8}\frac{RT_c}{p_c} \tag{71}$$

Therefore, if only first order quantities in a, b are considered, then equation (71) and the equation of states give

$$\frac{a\rho}{RT} = \frac{27}{64}\left(\frac{T_c}{T}\right)^2\left(\frac{p}{p_c}\right) \tag{72}$$

and

$$b\rho = \frac{1}{8}\left(\frac{p}{p_c}\right)\left(\frac{T_c}{T}\right) \tag{73}$$

These forms of the parameters are believed to be more convenient in actual computations and they clearly bring out the non-dimensional character of the parameters.

Properties of Air. The experimentally determined values of a and b for air are given by J. Jeans[4]. From these constants, the "critical pressure" and the "critical temperature" can be computed by using equation (70). The values in engineering units are

$$p_c = 324.6 \text{ psia}, \quad T_c = 181.1\ {}^\circ\text{R} \tag{74}$$

Since air is a mixture of several gases, such critical pressure and critical temperature are not real quantities, but convenient substitutes for a and b in the van der Waal equation of state.

The specific heat of air at constant volume is given by J. H. Keenan and J. Kaye[5] as a function of temperature. By fitting the tabulated data from $T = 300\,^\circ\text{F}$ to $T = 1\,000\,^\circ\text{R}$ to equation (4) with the method of least square, the constants α, β and γ are determined as

$$\begin{aligned} \alpha &= 0.174\,57 \text{ B. t. u per lb. per } {}^\circ\text{F} \\ \beta &= -0.018\,77 \times 10^{-3} \text{ B. t. u per lb. per } ({}^\circ\text{F})^2 \\ \gamma &= 0.024\,41 \times 10^{-6} \text{ B. t. u per lb. per } ({}^\circ\text{F})^3 \end{aligned} \tag{75}$$

Since the gas constant R for air is 0.068 5 B. t. u per lb. per °F,

$$\lambda = R/\alpha = 0.392\,0 \tag{76}$$

Tabulation of Functions for Air. To facilitate the application of results obtained in the previous sections, the universal correction functions specified by equations (36), (39), (40), (46), (63), (65), (67) and (69) are tabulated for air in Tables 1 to 8. The values of factors outside the bracket of equations (35), (37), (38), (45), (62), (64), (66) and (68), which are the values of the different quantities for a perfect gas, are also given in the first columns of these tables. It is seen that the correction functions have appreciable magnitudes even for moderate Mach numbers. Thus even for a Mach number as low as 3, the deviations from the perfect gas cannot be neglected if the correction parameters are not extremely small.

Table I Functions for temperature ratio in expansion of air, cf. equations (35) and (36)

M	$\left(1+\frac{1}{2}\lambda M^2\right)^{-1}$	t_1	t_2	t_3	t_4
0	1.000 00	0	0	0	0
0.5	0.953 289	0.042 18	0.038 70	0.045 31	0.043 97
1.0	0.836 120	0.110 12	0.075 16	0.146 67	0.131 75
1.5	0.693 963	0.127 69	0.036 36	0.246 02	0.200 94
2.0	0.560 538	0.090 41	−0.044 82	0.315 71	0.236 17
2.5	0.449 438	0.028 66	−0.122 01	0.356 32	0.249 04
3.0	0.361 795	−0.033 96	0.178 58	0.377 20	0.251 65
3.5	0.294 031	−0.087 50	−0.215 78	0.386 60	0.250 56
4.0	0.241 780	−0.130 01	−0.239 25	0.389 82	0.248 56
4.5	0.201 248	−0.162 71	−0.253 92	0.389 92	0.246 62
5.0	0.169 492	−0.187 61	−0.263 15	0.388 52	0.245 02
5.5	0.144 321	−0.206 56	−0.269 05	0.386 49	0.243 76
6.0	0.124 131	−0.221 06	−0.272 89	0.384 28	0.242 81
6.5	0.107 747	−0.232 25	−0.275 44	0.382 10	0.242 08
7.0	0.094 304 0	−0.240 97	−0.277 16	0.380 05	0.241 53
7.5	0.083 160 1	−0.247 84	−0.278 35	0.378 18	0.241 11
8.0	0.073 833 4	−0.253 29	−0.279 18	0.376 49	0.240 79
8.5	0.065 958 7	−0.257 67	−0.279 77	0.374 98	0.240 54
9.0	0.059 255 7	−0.261 21	−0.280 20	0.373 63	0.240 34
9.5	0.053 507 4	−0.264 11	−0.280 52	0.372 43	0.240 19
10	0.048 543 7	−0.266 50	−0.280 76	0.371 36	0.240 07
11	0.040 459 6	−0.270 14	−0.281 07	0.369 54	0.239 89
12	0.034 218 5	−0.272 71	−0.281 26	0.368 08	0.239 77
13	0.029 304 9	−0.274 59	−0.281 37	0.366 90	0.239 69
14	0.025 370 4	−0.275 97	−0.281 44	0.365 93	0.239 64
15	0.022 172 9	−0.277 02	−0.281 49	0.365 12	0.239 60
16	0.019 540 4	−0.277 83	−0.381 52	0.364 45	0.239 57
17	0.017 347 9	−0.278 46	−0.281 55	0.363 89	0.239 55
18	0.015 502 9	−0.278 96	−0.281 56	0.363 41	0.239 53
19	0.013 936 1	−0.279 36	−0.281 57	0.363 00	0.239 52
20	0.012 595 6	−0.279 69	−0.281 58	0.362 64	0.239 51
∞	0	−0.281 61	−0.281 61	0.359 20	0.239 46

Table Ⅱ Functions for pressure ratio in expansion of air, cf. equations (37) and (39)

M	$\left(1+\frac{1}{2}\lambda M^2\right)^{-1-(1/\lambda)}$	P_1	P_2	P_3	P_4
0	1.000 000	0	0	0	0
0.5	0.843 775	0.103 67	0.137 41	0.041 75	0.096 55
1.0	0.529 633	0.234 75	0.266 88	0.102 77	0.258 53
1.5	0.273 263	0.174 52	0.129 12	0.092 92	0.323 18
2.0	0.128 024	−0.061 00	−0.159 14	0	0.278 12
2.5	0.058 428 5	−0.356 48	−0.433 24	−0.139 19	0.182 11
3.0	0.027 045 2	−0.632 13	−0.634 13	−0.288 62	0.079 57
3.5	0.012 949 5	−0.858 87	−0.766 24	−0.428 13	−0.010 71
4.0	0.006 464 13	−1.035 09	−0.849 57	−0.549 97	−0.084 47
4.5	0.003 369 26	−1.168 89	−0.901 66	−0.653 04	−0.143 06
5.0	0.001 831 00	−1.269 83	−0.934 45	−0.739 01	−0.189 26
5.5	0.001 034 58	−1.346 19	−0.955 41	−0.810 42	−0.225 82
6.0	0.000 605 831	−1.404 37	−0.969 04	−0.869 78	−0.254 97
6.5	0.000 366 481	−1.449 11	−0.978 08	−0.919 32	−0.278 44
7.0	0.000 228 316	−1.483 88	−0.984 20	−0.960 87	−0.297 54
7.5	0.000 146 082	−1.511 18	−0.988 42	−0.995 94	−0.313 25
8.0	0.000 095 751 1	−1.532 85	−0.991 37	−1.025 73	−0.326 29
8.5	0.000 064 152 2	−1.550 21	−0.993 48	−1.051 19	−0.337 22
9.0	0.000 043 847 1	−1.564 26	−0.995 01	−1.073 08	−0.346 46
9.5	0.000 030 519 1	−1.575 73	−0.996 14	−1.092 02	−0.354 34
10	0.000 021 598 9	−1.585 17	−0.996 97	−1.108 49	−0.361 11
11	0.000 011 311 1	−1.599 54	−0.998 08	−1.135 56	−0.372 05
12	0.000 006 239 21	−1.609 72	−0.998 74	−1.156 66	−0.380 43
13	0.000 003 598 10	−1.617 10	−0.999 15	−1.173 40	−0.386 98
14	0.000 002 156 43	−1.622 57	−0.999 41	−1.186 88	−0.392 20
15	0.000 001 336 55	−1.626 70	−0.999 58	−1.197 89	−0.396 42
16	0.000 000 853 240	−1.629 88	−0.999 70	−1.206 99	−0.399 88
17	0.000 000 559 150	−1.632 37	−0.999 78	−1.214 59	−0.402 75
18	0.000 000 375 078	−1.634 33	−0.999 83	−1.221 01	−0.405 16
19	0.000 000 256 927	−1.635 91	−0.999 87	−1.226 47	−0.407 21
20	0.000 000 179 350	−1.637 18	−0.999 90	−1.231 15	−0.408 95
∞	0	−1.644 74	−1.000 00	−1.275 51	−0.425 17

Table Ⅲ Functions for density ratio in expansion of air, cf. equations (38) and (40)

M	$\left(1+\frac{1}{2}\lambda M^2\right)^{-1/\lambda}$	r_1	r_2	r_3	r_4
0	1.000 000	0	0	0	0
0.5	0.885 120	−0.010 02	−0.016 17	−0.003 57	−0.004 22
1.0	0.633 441	−0.117 78	−0.174 83	−0.043 90	−0.047 70
1.5	0.393 772	−0.385 74	−0.513 57	−0.153 11	−0.148 65
2.0	0.228 395	−0.743 95	−0.885 93	−0.355 71	−0.272 26
2.5	0.130 003	−1.095 87	−1.181 24	−0.495 51	−0.382 55
3.0	0.074 752 8	−1.391 55	−1.380 80	−0.665 83	−0.466 59
3.5	0.044 041 4	−1.621 59	−1.506 42	−0.814 72	−0.526 04
4.0	0.026 735 7	−1.794 51	−1.583 59	−0.939 79	−0.566 86
4.5	0.016 741 9	−1.922 99	−1.631 00	−1.042 95	−0.594 71
5.0	0.010 802 9	−2.018 49	−1.660 50	−1.127 53	−0.613 83
5.5	0.007 168 57	−2.089 97	−1.679 19	−1.196 91	−0.627 10
6.0	0.004 880 57	−2.143 99	−1.691 27	−1.254 06	−0.636 45
6.5	0.003 401 31	−2.185 29	−1.699 24	−1.301 42	−0.643 15
7.0	0.002 421 06	−2.217 24	−1.704 62	−1.340 93	−0.648 02
7.5	0.001 756 63	−2.242 23	−1.708 31	−1.374 13	−0.651 61
8.0	0.001 296 85	−2.262 00	−1.710 90	−1.385 59	−0.654 30
8.5	0.000 972 612	−2.277 80	−1.712 74	−1.426 17	−0.656 34
9.0	0.000 739 965	−2.290 56	−1.714 07	−1.446 71	−0.657 91
9.5	0.000 570 372	−2.300 96	−1.715 04	−1.464 45	−0.659 13
10	0.000 444 937	−2.309 50	−1.715 77	−1.479 85	−0.660 09
11	0.000 279 566	−2.322 50	−1.716 74	−1.505 10	−0.661 46
12	0.000 182 334	−2.331 67	−1.717 31	−1.524 75	−0.662 35
13	0.000 122 782	−2.338 32	−1.717 66	−1.540 30	−0.662 96
14	0.000 084 997 8	−2.343 24	−1.717 88	−1.552 81	−0.663 37
15	0.000 060 278 5	−2.346 96	−1.718 03	−1.563 02	−0.663 67
16	0.000 043 665 4	−2.349 82	−1.718 13	−1.571 44	−0.663 88
17	0.000 032 231 5	−2.352 05	−1.718 20	−1.578 48	−0.664 04
18	0.000 024 194 0	−2.353 81	−1.718 25	−1.584 41	−0.664 16
19	0.000 018 436 1	−2.355 22	−1.718 28	−1.589 46	−0.664 25
20	0.000 014 240 3	−2.356 36	−1.718 31	−1.593 79	−0.664 32
∞	0	−2.363 13	−1.718 39	−1.634 71	−0.664 63

Table Ⅳ Functions for area ratio in expansion of air, cf. equations (45) and (46)

M	$\frac{1}{M}\left(\frac{1+\frac{1}{2}\lambda M^2}{1+\frac{1}{2}\lambda}\right)^{\frac{1}{2}+(1/\lambda)}$	f_1	f_2	f_3	f_4
0	∞	0.145 56	0.229 30	0.052 51	0.060 55
0.5	2.116 17	0.073 05	0.111 24	0.026 84	0.029 93
1.0	1.000 00	0	0	0	0
1.5	1.858 36	0.095 79	0.118 36	0.039 51	0.035 73
2.0	2.673 72	0.335 19	0.366 03	0.148 48	0.118 15
2.5	4.196 67	0.616 43	0.601 54	0.292 33	0.206 21
3.0	6.778 83	0.872 42	0.774 15	0.439 87	0.278 94
3.5	10.939 78	1.083 9	0.887 65	0.574 53	0.332 68
4.0	17.389 0	1.240 88	0.959 25	0.690 62	0.370 56
4.5	27.055 3	1.362 18	1.004 00	0.788 03	0.396 85
5.0	41.119 6	1.453 42	1.032 18	0.868 84	0.415 11
5.5	61.048 4	1.522 28	1.050 18	0.935 69	0.427 89
6.0	88.628 2	1.574 67	1.061 89	0.991 10	0.436 96
6.5	126.000 4	1.614 90	1.069 66	1.037 24	0.443 49
7.0	175.697	1.646 14	1.074 92	1.075 88	0.448 25
7.5	240.677	1.670 65	1.078 55	1.108 45	0.451 77
8.0	324.360	1.690 09	1.081 09	1.136 08	0.454 42
8.5	430.665	1.705 66	1.082 90	1.159 67	0.456 43
9.0	564.048	1.718 26	1.084 21	1.179 95	0.457 97
9.5	729.534	1.728 53	1.085 18	1.197 47	0.459 18
10	932.759	1.736 99	1.085 90	1.212 71	0.460 13
11	1 478.244	1.749 86	1.086 85	1.237 73	0.461 49
12	2 259.20	1.758 97	1.087 42	1.257 22	0.462 37
13	3 346.47	1.765 57	1.087 77	1.272 68	0.462 97
14	4 824.31	1.770 47	1.087 99	1.285 12	0.463 39
15	6 791.55	1.774 17	1.088 14	1.295 28	0.463 68
16	9 362.89	1.777 01	1.088 24	1.303 67	0.463 89
17	12 670.1	1.779 24	1.088 31	1.310 68	0.464 05
18	16 863.5	1.781 08	1.088 35	1.316 60	0.464 17
19	22 112.7	1.782 40	1.088 39	1.321 63	0.464 26
20	28 608.5	1.783 54	1.088 41	1.325 94	0.464 33
∞	∞	1.790 29	1.088 50	1.366 81	0.464 64

Table V Functions for pressure ratio across a shock wave in air, cf. equations (62) and (63)

M_1	$\frac{2(1+\lambda)M_1^2-\lambda}{2+\lambda}$	Π_1	Π_2	Π_3	Π_4
1.0	1.000 00	0	0	0	0
1.5	2.454 85	−0.130 30	0	−0.033 50	0.010 38
2.0	4.488 29	−0.245 60	0	−0.023 92	0.074 21
2.5	7.110 37	−0.339 89	0	0.003 18	0.185 19
3.0	10.316 04	−0.414 02	0	0.041 79	0.352 59
3.5	14.093 7	−0.471 18	0	0.089 97	0.591 20
4.0	18.458 2	−0.515 19	0	0.146 88	0.919 76
4.5	23.404 7	−0.549 30	0	0.212 14	1.360 32
5.0	28.933 1	−0.576 01	0	0.285 55	1.938 06
5.5	35.043 5	−0.597 17	0	0.366 99	2.681 19
6.0	41.735 8	−0.614 14	0	0.456 39	3.620 97
6.5	49.010 0	−0.627 91	0	0.553 70	4.791 63
7.0	56.866 2	−0.639 20	0	0.658 91	6.230 43
7.5	65.304 4	−0.648 56	0	0.771 99	7.977 57
8.0	74.324 4	−0.656 40	0	0.892 92	10.076 29
8.5	83.926 4	−0.663 01	0	1.021 70	12.572 8
9.0	94.110 4	−0.668 64	0	1.158 32	15.516 2
9.5	104.876	−0.673 47	0	1.302 78	18.958 7
10	116.224	−0.677 64	0	1.455 07	22.955 5
11	140.666	−0.684 43	0	1.783 12	32.847 3
12	167.435	−0.689 67	0	2.142 47	45.692 5
13	196.532	−0.693 79	0	2.533 10	62.039 4
14	227.957	−0.697 09	0	2.955 01	82.484 3
15	261.709	−0.699 78	0	3.408 19	107.671 0
16	297.789	−0.701 99	0	3.892 64	138.291
17	336.197	−0.703 83	0	4.408 35	175.084
18	376.933	−0.705 38	0	4.955 33	218.837
19	419.997	−0.706 69	0	5.533 57	270.384
20	465.388	−0.707 82	0	6.143 07	330.608

Table Ⅵ Functions for density ratio across shock wave in air, cf. equations (64) and (65)

M_1	$\frac{(2+\lambda)M_1^2}{\lambda M_1^2+2}$	D_1	D_2	D_3	D_4
1.0	1.000 00	0	0	0	0
1.5	1.867 45	0.432 46	0.867 45	0.195 25	0.259 48
2.0	2.681 61	0.677 16	1.681 61	0.421 86	0.633 96
2.5	3.359 55	0.761 83	2.359 55	0.673 19	1.172 96
3.0	3.894 36	0.752 27	2.894 36	0.949 03	1.945 19
3.5	4.307 85	0.698 72	3.307 85	1.215 85	3.036 47
4.0	4.626 69	0.629 93	3.626 69	1.584 51	4.548 07
4.5	4.874 02	0.560 09	3.874 02	1.949 48	6.596 10
5.0	5.067 80	0.495 30	4.067 80	2.348 66	9.311 40
5.5	5.221 39	0.437 67	4.221 39	2.783 48	12.839 6
6.0	5.344 59	0.387 43	4.344 59	3.254 96	17.341 3
6.5	5.444 56	0.344 03	4.444 56	3.763 86	22.991 9
7.0	5.526 59	0.306 66	4.526 59	4.310 74	29.982 2
7.5	5.594 60	0.274 49	4.594 60	4.896 01	38.602 5
8.0	5.651 51	0.246 72	4.651 51	5.519 98	48.819 1
8.5	5.699 56	0.222 68	4.699 56	6.182 89	61.122 4
9.0	5.740 46	0.201 80	4.740 46	6.884 92	75.678 8
9.5	5.775 54	0.183 57	4.775 54	7.626 20	92.754 5
10	5.805 83	0.167 60	4.805 83	8.406 85	112.631 0
11	5.855 16	0.141 12	4.855 16	10.086 57	161.988
12	5.893 24	0.120 29	4.893 24	11.924 6	226.306
13	5.923 22	0.103 65	4.923 22	13.921 2	308.382
14	5.947 23	0.090 17	4.947 23	16.076 7	411.260
15	5.966 74	0.079 11	4.966 74	18.391 2	538.224
16	5.982 80	0.069 95	4.982 80	20.864 8	692.803
17	5.996 18	0.062 27	4.996 18	23.497 6	878.770
18	6.007 44	0.055 77	5.007 44	26.289 7	1 100.14
19	6.017 00	0.050 23	5.017 00	29.241 0	1 361.17
20	6.025 19	0.045 47	5.025 19	32.351 7	1 666.36

Table Ⅶ Functions for temperature ratio across a shock wave in air, cf. equations (66) and (67)

M_1	$\frac{\{2(1+\lambda)M_1^2-\lambda\}(2+\lambda M_1^2)}{(2+\lambda)^2 M_1^2}$	τ_1	τ_2	τ_3	τ_4
1.0	1.000 00	0	0	0	0
1.5	1.314 54	−0.142 14	0	−0.228 75	−0.249 10
2.0	1.673 73	−0.320 58	0	−0.445 78	−0.559 76
2.5	2.116 46	−0.514 38	0	−0.670 02	−0.987 77
3.0	2.647 69	−0.695 44	0	−0.907 24	−1.592 60
3.5	3.271 62	−0.853 16	0	−1.161 88	−2.445 27
4.0	3.989 50	−0.985 40	0	−1.437 63	−3.628 31
4.5	4.801 93	−1.094 37	0	−1.737 34	−5.235 77
5.0	5.709 21	−1.183 65	0	−2.063 12	−7.373 34
5.5	6.711 53	−1.256 87	0	−2.416 49	−10.158 4
6.0	7.808 98	−1.317 15	0	−2.798 57	−13.720 3
6.5	9.001 65	−1.367 09	0	−3.210 16	−18.200 3
7.0	10.289 56	−1.408 76	0	−3.651 83	−23.751 7
7.5	11.672 8	−1.443 77	0	−4.124 02	−30.624 9
8.0	13.151 3	−1.473 39	0	−4.627 06	−38.742 8
8.5	14.725 1	−1.498 63	0	−5.161 19	−48.549 6
9.0	16.394 2	−1.520 29	0	−5.726 59	−60.162 6
9.5	18.158 7	−1.538 98	0	−6.323 42	−73.795 8
10	20.018 5	−1.555 21	0	−6.951 78	−89.675 5
11	24.024 2	−1.581 82	0	−8.303 45	−129.141
12	28.411 3	−1.602 53	0	−9.782 11	−180.613
13	33.179 9	−1.615 64	0	−11.388 1	−246.343
14	38.329 9	−1.632 10	0	−13.121 7	−328.776
15	43.861 3	−1.642 86	0	−14.983 0	−430.553
16	49.774 2	−1.651 74	0	−16.972 2	−554.512
17	56.068 6	−1.659 15	0	−19.089 3	−703.686
18	62.744 4	−1.665 40	0	−21.334 3	−881.301
19	69.801 7	−1.670 72	0	−23.707 4	−1 090.78
20	77.240 4	−1.675 28	0	−26.208 6	−1 335.75

Table Ⅷ Functions for Mach number after a shock wave in air, cf. equations (68) and (69)

M_1	$\left(\frac{\lambda M_1^2+2}{2(1+\lambda)M_1^2-\lambda}\right)^{\frac{1}{2}}$	m_1	m_2	m_3	m_4
1.0	1.000 000	0	0	0	0
1.5	0.700 575	0.442 64	0.809 96	0.123 97	0.178 64
2.0	0.576 491	0.968 28	2.254 72	0.292 20	0.562 86
2.5	0.511 511	1.345 34	3.963 52	0.522 18	1.299 06
3.0	0.473 427	1.529 16	5.635 83	0.815 62	2.554 33
3.5	0.449 186	1.564 50	7.124 87	1.172 74	4.537 73
4.0	0.432 843	1.510 75	8.389 80	1.592 50	7.493 99
4.5	0.421 326	1.412 42	9.441 02	2.073 87	11.705 7
5.0	0.412 922	1.296 93	10.307 4	2.616 01	17.493 3
5.5	0.406 600	1.179 64	11.020 8	3.218 25	25.215 4
6.0	0.401 739	1.068 07	11.610 0	3.880 11	35.268 3
6.5	0.397 915	0.965 61	12.099 4	4.601 25	48.086 3
7.0	0.394 859	0.873 30	12.508 3	5.381 40	64.141 7
7.5	0.392 379	0.790 99	12.852 5	6.220 35	83.944 5
8.0	0.390 340	0.718 00	13.144 0	7.117 97	108.042
8.5	0.388 643	0.653 39	13.392 7	8.074 13	137.021
9.0	0.387 213	0.596 22	13.606 2	9.088 75	171.504
9.5	0.386 003	0.545 58	13.790 6	10.161 77	212.153
10	0.384 964	0.500 64	13.950 9	11.293 1	259.667
11	0.383 294	0.425 04	14.213 8	13.730 7	378.274
12	0.382 016	0.364 58	14.418 5	16.401 3	533.674
13	0.381 021	0.315 69	14.580 7	19.304 6	732.822
14	0.380 230	0.275 72	14.711 2	22.440 7	983.273
15	0.379 589	0.242 69	14.817 6	25.809 3	1 293.19
16	0.379 065	0.215 13	14.905 6	29.410 6	1 671.35
17	0.378 630	0.191 92	14.979 0	33.244 4	2 127.11
18	0.378 266	0.172 21	15.041 0	37.310 6	2 670.46
19	0.377 957	0.155 34	15.093 7	41.609 4	3 311.98
20	0.377 692	0.140 80	15.138 9	46.140 6	4 062.84

Application of Results to a Hypersonic Wind Tunnel. As an example, the results of analysis will be applied presently to a hypersonic wind tunnel of Mach number 15 in its test section. Let the pressure at inlet to the nozzle be 600 psia, and the temperature 800°R. Then the values of correction parameters are

$$\frac{a\rho_0}{RT_0} = \frac{27}{64}\left(\frac{181.1}{800}\right)^2 \frac{600}{324.6} = 0.039\,6$$

$$b\rho_0 = \frac{1}{8}\times\frac{181.1}{800}\times\frac{600}{324.6} = 0.052\,30$$

$$\frac{\beta T_0}{\alpha} = -\frac{0.018\,77\times 0.800}{0.174\,57} = -0.086\,02$$

$$\frac{\gamma T_0^2}{\alpha} = \frac{0.024\,41\times 0.800^2}{0.174\,57} = 0.089\,49$$

With the aid of Tables Ⅰ to Ⅳ, the conditions of air at the test section can be calculated as follows:

$$T = 800\times 0.022\,172\,9\,(1-0.039\,96\times 0.277\,02+0.052\,30\times 0.281\,49-0.086\,02\times 0.365\,12+0.089\,49\times 0.239\,60) = 17.626\ ^\circ\mathrm{R}$$

$$p = 600\times 0.000\,001\,336\,55\,(1-0.039\,96\times 1.626\,70+0.052\,30\times 0.999\,58+0.086\,02\times 1.197\,89-0.089\,49\times 0.396\,42) = 0.000\,845\,91\ \mathrm{psia}$$

$$A/A^* = 6\,791.55\,(1+0.039\,96\times 1.774\,17-0.052\,30\times 1.088\,14-0.086\,02\times 1.295\,28+0.089\,49\times 0.463\,68) = 6\,411.6$$

Considerable deviations from the values for a perfect gas are thus present.

Now if there is a normal shock wave immediately after the test section, Tables 5 to 8 can be used to calculate conditions after the shock wave. Here the correction parameters are

$$\frac{a\rho_1}{RT_1} = \frac{27}{64}\left(\frac{181.1}{17.626}\right)^2 \frac{0.000\,845\,91}{324.6} = 0.000\,12$$

$$b\rho_1 = \frac{1}{8}\times\frac{181.1}{17.626}\times\frac{0.000\,845\,91}{324.6} = 0.000\,003\,4$$

$$\frac{\beta T_1}{\alpha} = -\frac{0.018\,77\times 0.017\,626}{0.174\,56} = -0.001\,895\,2$$

$$\frac{\gamma T_1^2}{\alpha} = \frac{0.024\,41\times 0.017\,626^2}{0.174\,57} = 0.000\,043\,44$$

Then the temperature and the pressure can be calculated as follows:

$$T_2 = 17.626\times 43.861\,3(1-0.000\,12\times 1.642\,86+0.001\,895\,2\times 14.983-0.000\,043\,44\times 430.55) = 780.44\ ^\circ\mathrm{R}$$

$$p_2 = 0.000\,845\,91\times 261.709(1-0.000\,12\times 0.699\,78-0.001\,895\,2\times 3.408\,2+0.000\,043\,44\times 107.67) = 0.220\,97\ \mathrm{psia}$$

$$M_2 = 0.379\,589(1+0.000\,12\times 0.242\,69-0.000\,003\,4\times 14.818-0.001\,895\,2\times 25.809+0.000\,043\,44\times 1\,293.2) = 0.382\,34$$

The deviations from the values for a perfect gas are much smaller here than for the expansion. The reason is, of course, the much lower pressure level for the shock process. If a system of compressors is used to recompress the air after the shock to the pressure p_0 at inlet to the nozzle so that the wind tunnel can be operated continuously, the overall compression ratio required is thus p_0/p_2, or approximately 2 700.

MASSACHUSETTS INSTITUTE OF TECHNOLOGY.

References

[1] Busemann A. Gasdynamik// Part 1, Vol. IV of Handbuch der Experimentalphysik, Akademische Verlag, Leipzig, 1931: 421 - 431.

[2] Walker W J. The Effect of Variable Specific Heats on the Discharge of Gases Through Orifices or Nozzles. Phil. Mag. 1922, 43(6): 589 - 592 // Specific Heat Variations in Relation to the Dynamic Action of Gases and Their Equation of State, Ibid., 50: 1244 - 1260 (1925).

[3] Epstein P S. Textbook of Thermodynamics. John Wiley & Sons, New York 1937: 12.

[4] Jeans J. An Introduction to the Kinetics Theory of Gases. Cambridge University Press, Cambridge, 1940: 81 - 83.

[5] Keenan J H, Kaye J. Thermodynamics Properties of Air. John Wiley & Sons, New York, 1945: 36.

Corrections on the Paper "One-Dimensional Flows of a Gas Characterized by van der Waals Equation of State"

By Hsue-shen Tsien

In the recent paper by the author①, a mistake② has been made in not including the variation of internal energy of the gas caused by change in volume. As a result, the formulae are incorrect in the terms connected with the parameter a of the Van der Waal's equation. The following corrected equations should be substituted for the corresponding equations in that paper.

$$C_v dT + T(\partial p/\partial T) d\rho^{-1} = 0 \tag{2}$$

$$C_v dT - \{RT(\rho - b\rho^2)^{-1}\} d\rho = 0 \tag{3}$$

$$(\alpha + \beta T + \gamma T^2) dT - \{RT(\rho - b\rho^2)^{-1}\} d\rho = 0 \tag{5}$$

$$dT^{(1)}(\rho) = 0 \ (8), \quad T^{(1)}(\rho) = 0 \tag{13}$$

$$T = C\rho^{R/\alpha} + b\frac{CR}{\alpha}\rho^{R/\alpha+1} - \beta\frac{C^2}{\alpha}\rho^{2R/\alpha} - \gamma\frac{C^3}{2\alpha}\rho^{3R/\alpha} \tag{17}$$

and similar correction for (18). The term involving $(a\rho_0/RT_0)$ in (20) and (21) should be

① Journal of Mathematics and Physics, *25*, pp. 301 - 324 (1947).

② The author is indebted to Professor Joseph H. Keenan for calling his attention to this mistake.

dropped. For (22), (23), (24), (25), (31), (32), and (34) the factor in connection with the parameter a should be corrected according to the following scheme:

Equation	Incorrect factor	Correct factor
(22)	$\frac{1}{1-\lambda}$	1
(23)	$-\frac{1}{1-\lambda^2}$	$-\frac{1}{1+\lambda}$
(24)	$-\frac{\lambda}{1-\lambda}$	1
(25)	$\frac{\lambda^2}{1-\lambda^2}$	$-\frac{\lambda}{1+\lambda}$
(31)	$\frac{1}{1-\lambda}$	1
(32)	$-\frac{1}{1+\lambda}$	$-\frac{1-\lambda}{1+\lambda}$
(34)	$-\frac{1}{1+\lambda}$	$-\frac{1-\lambda}{1+\lambda}$

In addition, the correct form of other eqtations are:

$$C_v \mathrm{d}T + \mathrm{d}(p/\rho) - a\mathrm{d}\rho - \mathrm{d}p/\rho = 0 \tag{27}$$

$$(\alpha + \beta T + \gamma T^2)\mathrm{d}T + \mathrm{d}(p/\rho) - a\mathrm{d}\rho + v\mathrm{d}v = 0 \tag{28}$$

$$\frac{1}{2}v^2 = \alpha T_0\left\{1 - \frac{T}{T_0}\right\} + \frac{\beta T_0^2}{2}\left\{1 - \frac{T^2}{T_0^2}\right\} + \frac{\gamma T_0^3}{3}\left\{1 - \frac{T^3}{T_0^3}\right\} + \frac{p_0}{\rho_0}\left\{1 - \left(\frac{p}{p_0}\right)\left(\frac{\rho_0}{\rho}\right)\right\} - a\rho_0\left(1 - \frac{\rho}{\rho_0}\right) \tag{29}$$

$$t_1 = \frac{\lambda}{1+\lambda}\left\{(M^2+2)\left(1 + \frac{\lambda}{2}M^2\right)^{-1/\lambda} - 2\right\} \tag{36}$$

$$P_1 = (M^2+2)\left(1 + \frac{\lambda}{2}M^2\right)^{-1/\lambda} - 1 + \left(1 + \frac{\lambda}{2}M^2\right)^{1-1/\lambda} \tag{39}$$

$$r_1 = \frac{1}{1+\lambda}\left\{(M^2+2)\left(1 + \frac{\lambda}{2}M^2\right)^{-1/\lambda} - 2\right\} \tag{40}$$

$$f_1 = \frac{1}{1+\lambda}\left[(2+\lambda)\left(1 + \frac{\lambda}{2}\right)^{-1/\lambda} - \{(1+\lambda) + M^2\}\left(1 + \frac{\lambda}{2}M^2\right)^{-1/\lambda}\right] - \tag{46}$$

$$a\rho_1 + \alpha T_1 + \frac{\beta}{2}T_1^2 + \frac{\gamma}{3}T_1^3 + \frac{p_1}{\rho_1} + \frac{1}{2}v_1^2 = -a\rho_2 + \alpha T_2 + \frac{\beta}{2}T_2^2 + \frac{\gamma}{3}T_2^3 + \frac{p_2}{\rho_2} + \frac{1}{2}v_2^2 \tag{49}$$

Π_1, as given by (63) should be multiplied by $(1-\lambda)$, and the second term for D_1 in (65) should be also multiplied by $(1-\lambda)$. The corrected expressions for m_1, m_2, m_3, m_4 in (69) are:

$$\left.\begin{aligned}
m_1 &= \frac{1-\lambda}{2(1+\lambda)}\left[\frac{(2+\lambda)^3 M_1^4}{(\lambda M_1^2+2)^2\{2(1+\lambda)M_1^2-\lambda\}}-1\right]-\frac{1}{2}(D_1+\Pi_1)\\
m_2 &= \frac{1}{2}\left[\frac{(2+\lambda)M_1^2}{\lambda M_1^2+2}-1\right]-\frac{1}{2}(D_2+\Pi_2)\\
m_3 &= \frac{\lambda}{2(1+\lambda)}\left[\frac{\{2(1+\lambda)M_1^2-\lambda\}(\lambda M_1^2+2)}{(2+\lambda)^2 M_1^2}-1\right]-\frac{1}{2}(D_3+\Pi_3)\\
m_4 &= \frac{\lambda}{2(1+\lambda)}\left[\frac{\{2(1+\lambda)M_1^2-\lambda\}^2(\lambda M_1^2+2)^2}{(2+\lambda)^4 M_1^4}-1\right]-\frac{1}{2}(D_4+\Pi_4)
\end{aligned}\right\} \tag{69}$$

The tabulated functions are correct except t_1, P_1, r_1, f_1, Π_1, D_1, τ_1, and the m's.

Massachusetts Institute of Technology.

Flow Conditions near the Intersection of a Shock Wave with Solid Boundary*

By Hsue-shen Tsien

1. Introduction. Recent investigations by J. Ackeret, F. Feldmann, and N. Rott[1] and H. - W. Liepmann[2] on the interactions of shock and boundary layer in transonic flows have shown many novel features hitherto unrecognized. One of these new phenomena is the presence of a strong expansion zone near the solid boundary immediately behind the shock. The same situation appears also in the results of the numerical calculations of transonic flows by H. W. Emmons[3]. Since Emmons' investigation is carried out without considering the viscosity and the heat conduction of the fluid, the occurrence of expansion zone after the shock must be independent of these fluid properties. A closer consideration on the order of magnitude of quantities which enter into the phenomenon shows that the effects of the small viscosity and heat conductance of the fluid is indeed small and negligible when compared with the very large pressure and velocity changes across a shock. In other words, although the shock in transonic flow may have been caused by the presence of the boundary layer, a result of the viscosity of the fluid, the final flow characteristics can be described without introducing viscosity and heat conductance. The situation is thus not unlike that of fully developed turbulent flow which can be described without the direct aid of viscosity, although the phenomenon itself is born from the effects of viscosity.

The purpose of the present note is to show that, neglecting the effects of viscosity and heat conductance, there is a simple relation between the Mach number near the intersection of the shock with the solid boundary and the curvature of the solid boundary. This same relation was obtained by Emmons[4] without considering the pressure gradients along the solid boundary. The present derivation is believed to be more general. The pressure gradients before and after the shock along the solid boundary are also simply related to the curvature of the boundary, with the gradient after the shock larger than the gradient before the shock.

2. Basic Equations. Let r and θ be the polar coordinates and u, v the components of velocity in the radial and circumferential direction. Further, let ρ, p be the density and pressure of the fluid. The equations of motion are then

$$\rho u \frac{\partial u}{\partial r} + \rho v \frac{\partial u}{r\partial\theta} - \frac{\rho v^2}{r} = -\frac{\partial p}{\partial r} \tag{1}$$

---

* Received December 12, 1946.

Journal of Mathematics and Physics, vol.26, pp.69 - 75, 1947.

$$\rho u \frac{\partial v}{\partial r} + \rho v \frac{\partial v}{r\partial\theta} + \frac{\rho u v}{r} = -\frac{1}{r}\frac{\partial p}{\partial\theta} \tag{2}$$

The equation of continuity is

$$\frac{\partial \rho u}{\partial r} + \frac{\rho u}{r} + \frac{1}{r}\frac{\partial \rho u}{\partial\theta} = 0 \tag{3}$$

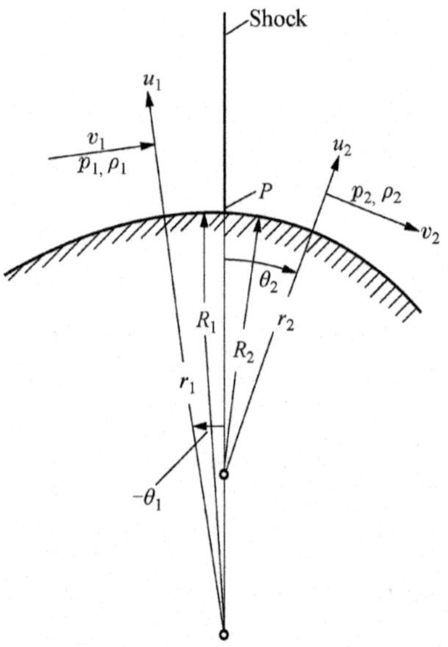

Fig. 1 Shock and solid boundary

These equations will be applied presently to the two sides of the shock (Fig. 1). The shock is necessarily normal to the boundary as the boundary is considered to have continuous slope and the flow immediately to the boundary should not change direction by passing through the shock. As a further simplification of the problem the shock will be assumed to be straight. Since the following calculation is limited to the immediate neighborhood of the point P of intersection of the shock and the solid boundary, the effect of any finite curvature of the shock will be of second order. Therefore the assumption of a straight shock really does not restrict the physical problem considered. Let the subscript 1 and 2 denote quantities corresponding to the upstream side and to the dowstream side respectively, and let the superscript 0 denote quantities at point P. Furthermore, let R_1 and R_2 be the radii of curvature of the solid boundary at left and at right of P. The investigation then allows for a change in the curvature of the boundary, even though the slope is continuous. Since the radial component of velocity is zero at the solid boundary, equation (1) gives

$$\frac{\rho_1^0 v_1^{0^2}}{R_1} = \left(\frac{\partial p_1}{\partial r_1}\right)^0$$

Since $\gamma p_1^0/\rho_1^0 = a_1^{0^2}$, γ being the ratio of specific heat, and $v_1^0/a_1^0 = M_1^0$ which is the Mach number of flow at the left of the point P, the above equation can be written as

$$\left[\frac{\partial(p_1/p_1^0)}{\partial r_1}\right]^0 = \frac{1}{R_1}\gamma M_1^{0^2} \tag{4}$$

Similarly,

$$\left[\frac{\partial(p_2/p_2^0)}{\partial r_2}\right]^0 = \frac{1}{R_2}\gamma M_2^{0^2} \tag{5}$$

By specializing equation (2) for the left of point P, one has

$$-\frac{1}{R_1}\left(\frac{\partial p_1}{\partial\theta_1}\right)^0 = \rho_1^0 v_1^0 \left(\frac{\partial v_1}{r_1 \partial_1 \theta}\right)^0 = v_1\left(\frac{\partial \rho_1 v_1}{r_1 \partial\theta_1}\right)^0 - v_1^{0^2}\left(\frac{\partial \rho_1}{r_1 \partial\theta_1}\right)^0 \tag{6}$$

But

$$\left(\frac{\partial\rho_1}{r_1\partial\theta_1}\right)^0 = \left(\frac{d\rho_1}{dp_1}\right)^0\left(\frac{\partial p_1}{r_1\partial\theta_1}\right) = \frac{1}{a^{0^2}}\left(\frac{\partial p_1}{r_1\partial\theta_1}\right)^0$$

Here $(dp_1/d\rho_1)^0$ is identified with $a_1^{0^2}$ because along any stream line in a non-viscous flow the fluid undergoes isentropic compression or expansion. Equation (3) gives

$$\left(\frac{\partial\rho_1 v_1}{r_1\partial\theta_1}\right)^0 = -\left(\frac{\partial\rho_1 u_1}{\partial r_1}\right)^0 = -\rho_1^0\left(\frac{\partial u_1}{\partial r_1}\right)^0$$

Therefore, equation (6) can be written as

$$(M_1^{0^2}-1)\frac{1}{R_1}\left(\frac{\partial p_1}{\partial\theta_1}\right)^0 = -\rho_1^0 v_1^0\left(\frac{\partial u_1}{\partial r_1}\right)^0 \tag{7}$$

Similarly for the right of the point P, one has

$$(1-M_2^{0^2})\frac{1}{R_2}\left(\frac{\partial p_2}{\partial\theta_2}\right)^0 = \rho_2^0 v_2^0\left(\frac{\partial u_2}{\partial r_2}\right)^0 \tag{8}$$

3. Relations of Quantities on the Two Sides of Shock. Now introduce the independent variable η as the normal distance from the solid boundary, i. e.,

$$\eta = r_1 - R_1 = r_2 - R_2 \tag{9}$$

Then equation (4) gives the pressure p_1 along the left of the shock:

$$p_1/p_1^0 = 1+\gamma M_1^{0^2}\eta/R_1 + \dots \tag{10}$$

Now the flow ahead of the shock can generally be considered as irrotational. Then the flow is not only isentropic along a given stream line but has the same entropy throughout the field. The temperature ratio T_1/T_1^0 is then given by

$$T_1/T_1^0 = (p_1/p_1^0)^{(\gamma-1)/\gamma} = 1+(\gamma-1)M_1^{0^2}\eta/R_1+\dots \tag{11}$$

However,

$$T_1/T_1^0 = \left(1+\frac{\gamma-1}{2}M_1^{0^2}\right)\Big/\left(1+\frac{\gamma-1}{2}M_1^2\right) \tag{12}$$

Equations (11) and (12) then give the Mach number M_1 as follows:

$$M_1^2 = M_1^{0^2}\left[1-2\left(1+\frac{\gamma-1}{2}M_1^{0^2}\right)\frac{\eta}{R_1}+\dots\right] \tag{13}$$

This is the Mach number along the left of the shock.

The incident velocity to the shock is not normal for points away from the solid boundary because of the presence of radial velocity u_1 at such points. However, a moment's reflection will show that this effect is of second order in calculating pressure p_2. Hence, for the accuracy adopted, the normal shock formula can be used. Equations (10) and (13) then give the pressure ratio as

$$\frac{p_2}{p_1^0}=\frac{p_1}{p_1^0}\left(\frac{2\gamma}{\gamma+1}M_1^2-\frac{\gamma-1}{\gamma+1}\right)=\left(\frac{2\gamma}{\gamma+1}M_1^{0^2}-\frac{\gamma-1}{\gamma+1}\right)\left[1+\frac{2\gamma M_1^{0^4}-\gamma(\gamma+3)M_1^{0^2}}{2\gamma M_1^{0^2}-(\gamma-1)}\frac{\eta}{R_1}+\ldots\right]$$

Therefore,

$$\frac{p_2}{p_2^0}=1+\frac{2\gamma M_1^{0^4}-\gamma(\gamma+3)M_1^{0^2}}{2\gamma M_1^{0^2}-(\gamma-1)}\frac{\eta}{R_1}+\ldots \tag{14}$$

On the other hand, equation (5) requires that

$$\left[\frac{\partial(p_2/p_2^0)}{\partial\eta}\right]^0=\frac{1}{R_2}\frac{2\gamma+\gamma(\gamma-1)M_1^{0^2}}{2\gamma M_1^{0^2}-(\gamma-1)} \tag{15}$$

By substituting equation (14) into equation (15), one has

$$[2+(\gamma-1)M_1^{0^2}]R_1/R_2=2M_1^{0^4}-(\gamma+3)M_1^{0^2}$$

or,

$$M_1^{0^4}-\frac{1}{2}\left\{(\gamma+3)+\frac{R_1}{R_2}(\gamma-1)\right\}M_1^{0^2}-\frac{R_1}{R_2}=0 \tag{16}$$

The appropriate root of this quadratic equation is

$$M_1^{0^2}=\frac{1}{4}\left\{(\gamma+3)+\frac{R_1}{R_2}(\gamma-1)\right\}+\sqrt{\frac{1}{16}\left\{(\gamma+3)+\frac{R_1}{R_2}(\gamma-1)\right\}^2+\frac{R_1}{R_2}} \tag{17}$$

Hence, the Mach numbers before and after the shock are not free but definitely determined by the ratio of the radii of curvature of the solid boundary. This is the same relation obtained by Emmons[4] without considering the pressure gradients along the solid boundary. Fig. 2 shows

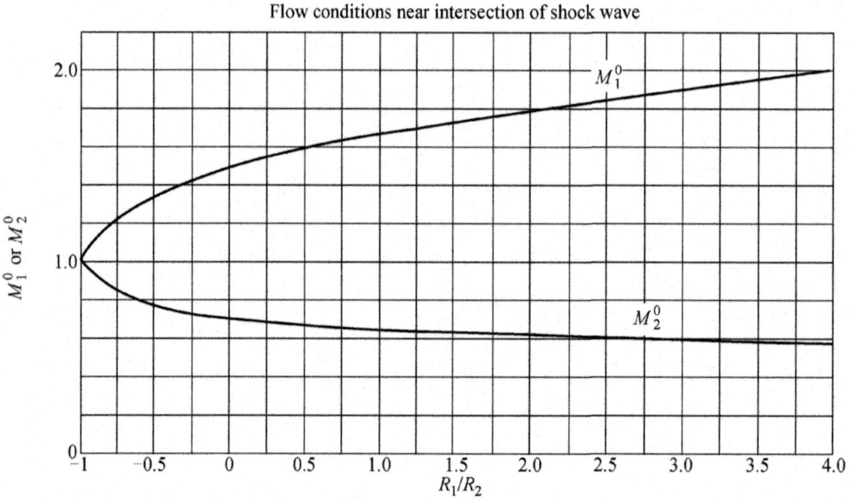

Fig. 2 Relation between the Mach number M_1^0 before the shock, the Mach number M_2^0 after the shock at the boundary and the ratio R_1/R_2 of the radii of curvature of the boundary

this relation by assuming $\gamma = 1.4$. It is seen that for large values of Mach number before the shock, the curvature of the boundary after the shock must be larger than that before the shock. If the curvature of the boundary is the same, then $M_1^0 = 1.661$.

The radial component of velocity u is not altered by going through the shock. Thus $u_1 = u_2$ along the shock. Furthermore, the condition of continuity requires that $\rho_1^0 v_1^0 = \rho_2^0 v_2^0$. Therefore, equations (7) and (8) yield immediately the relation

$$\left(\frac{\partial p_2}{\partial s}\right)^0 \Big/ \left(\frac{\partial p_1}{\partial s}\right)^0 = -\frac{M_1^{0^2} - 1}{1 - M_2^{0^2}} \tag{18}$$

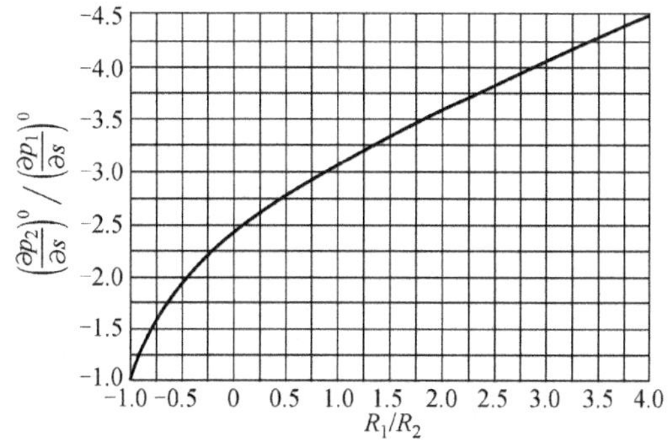

Fig. 3 Ratio $\left(\frac{\partial p_2}{\partial s}\right)^0 \Big/ \left(\frac{\partial p_1}{\partial s}\right)^0$ of the pressure gradients along the boundary after the shock and before the shock as function of the ratio R_1/R_2 of the radii of curvature of the boundary

where s is length along the solid boundary. Equation (18) then gives the ratio of the pressure gradient along the boundary before and after the shock. This ratio is plotted against the ratio of radii of curvature R_1/R_2 in Fig. 3 with $\gamma = 1.4$. It is interesting that the pressure gradients are of opposite sign. That is, if the flow before the shock is expanded in the flow direction, the flow after the shock will be compressed in the flow direction. If the flow before the shock is compressed in the flow direction, the flow after the shock will be expanded. Physically, this is easy to understand by observing the flow pattern near the point of intersection of the shock with the solid boundary as shown in Fig. 4. If the stream lines converge before the shock, they will converge even more after the shock. But converging stream lines in the supersonic region before the shock correspond to compression and converging stream lines in the subsonic region after the shock correspond to expansion. If the stream lines diverge, then the reverse is true. In both cases, however, the pressure gradients before and after the shock are of opposite sign. Therefore, an expansion zone after the shock is necessarily connected with a compression zone before the shock. This is in complete agreement with the experimental data[1] and Emmons' calculations[3].

Figure 3 also shows that the pressure gradient after the shock is always much steeper than the pressure gradient before the shock. This is also in accordance with the measurements of

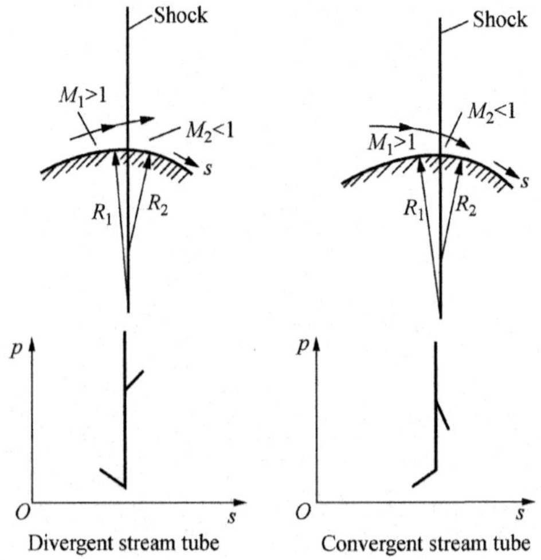

Fig. 4 Stream tube cross-section and the pressure gradient along the boundary

Ackeret, Feldmann and Rott[1]. The flow after the shock is, of course, rotational.

References

[1] Ackeret J, Feldmann F, Rott N. Untersuchungen an Verdichtungsstossen und Grenzschichten in Schnell bewegten Garen. Mitteilungen aus dem Institut fur Aerodynamik, ETH, No. 10, Zurich 1946.

[2] Liepmann H-W. Investigations of the Interaction of Boundary Layer and Shock Waves in Transonic Flow. Report to Air Materiel Command, Guggenheim Aeronautical Laboratory California Institute of Technology, Pasadena, 1946.

[3] Emmons H W. The Theoretical Flow of a Frictionless Adiabatic. Perfect Gas Past an NACA 0012 Airfoil, NACA Technical Note 1946.

[4] Emmons H W. The Theoretical Flow of a Frictionless. Adiabatic, Perfect Gas Inside of a Two-Dimensional Hyperbolic Nozzle, Appendix II, NACA Technical Note No. 1003, 1946.

Massachusetts Institute of Technology.

Lower Buckling Load in the Non-Linear Buckling Theory for Thin Shells

By Hsue-shen Tsien
(Massachusetts Institute of Technology)

For thin shells the relation between the load P and the deflection ε beyond the classical buckling load is very often non-linear. For instance, when a uniform thin circular cylinder is loaded in the axial direction, the load P when plotted against the end-shortening ε has the characteristic shown in Fig. 1. If the strain energy S and the total potential $\varphi = S - P\varepsilon$ are calculated, their behavior can be represented by the curves shown in Figs. 2 and 3. It can be demonstrated that the branches OC and AB corresponds to stable equilibrium configurations and the branch BC to unstable equilibrium configurations. The point B is then the point of transition from stable to unstable equilibrium configurations.

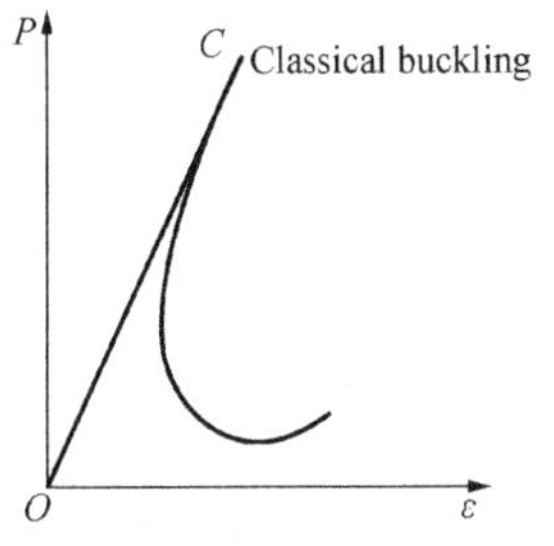

Fig. 1

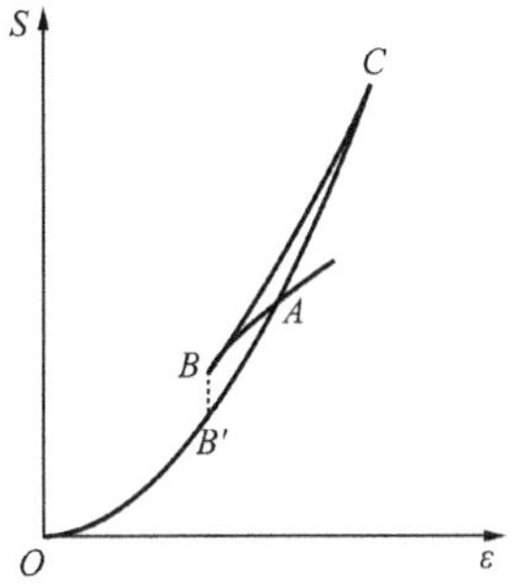

Fig. 2

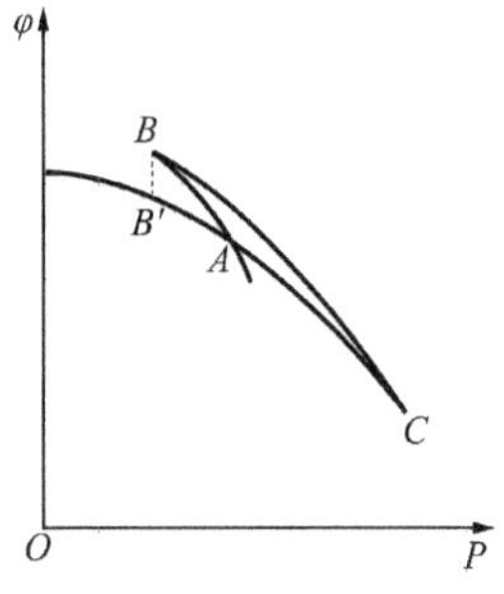

Fig. 3

It was proposed by the author in a previous paper[1] that the point A was the critical point for buckling of the structure under external disturbances, using the S, ε curve for "testing machine" loading and the φ, P curve for "deadweight" loading. The load P for the unbuckled configuration of the shell corresponding to the point A was called the lower buckling load of the shell. The energy represented by the vertical distance from the point A to the curve BC is then the minimum external excitation required to cause the buckling at point A.

However, if the external excitation is large, there is no reason why buckling cannot occur at the point B' directly under the point B. The minimum external excitation required is then

---

Received April 2, 1947.
Quarterly of Applied Mathematics, vol. 5, pp. 236 – 237, 1947.

[1] H. S. Tsien, *A theory for the buckling of thin shells*, J. Aero. Sciences 9, 373 – 384 (1942).

given by the energy represented by the distance $B'B$. This amount of energy is actually absorbed by the structure during buckling. Since the curve BA represents the final state of the structure after buckling, for buckling to happen between B' and A, energy is absorbed, and for buckling to happen between A and C, energy is released. But in any event, the lower limit of buckling load is definitely given by the point B', not the point A. Therefore the lower buckling load should be the load P corresponding to the point B'.

By referring to Figs. 11 and 13 of the aforementioned paper, and assuming a square wave pattern, we find the lower buckling stress σ of thin uniform cylindrical shells under axial load to be given by

$$\sigma = 0.42\, Et/R$$

for testing machine loading and

$$\sigma = 0.19\, Et/R$$

for deadweight loading. The corresponding values under the previously proposed criteria are $\sigma = 0.46 Et/R$ and $\sigma = 0.298\, Et/R$ for the two cases.

Rockets and Other Thermal Jets Using Nuclear Energy

With a General Discussion on the Use of Porous Pile Materials *

by Hsue-shen Tsien

1 Simple Theory of Space Rockets

To bring out the crucial problem in rocket propulsion, consider the motion of a rocket in a force-free space, i. e., a space without a gravitational field. Let m_1 denote the mass of the rocket at any time and dm_2 the mass ejected from the rocket in a small time interval. The absolute velocities of the rocket and the ejected mass are u_1 and $-u_2$ respectively. The relative velocity w of the ejected mass with respect to the rocket is then u_1+u_2 (Fig. 1).

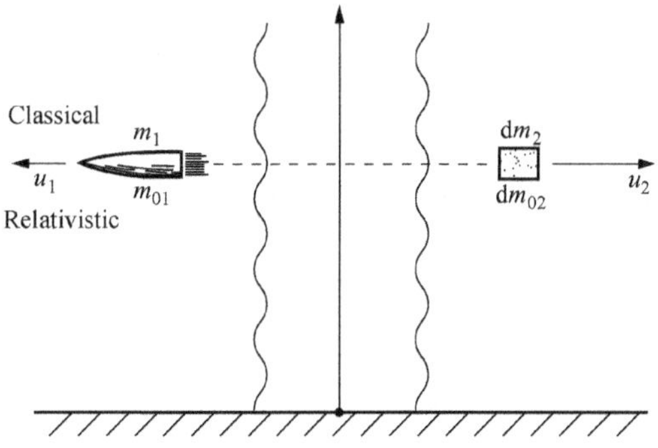

Fig. 1

The conservation of mass requires that

$$dm_2 = -dm_1 \tag{1}$$

The conservation of momentum requires that

$$d(m_1 u_1) = dm_2 \cdot u_2 \tag{2}$$

And finally

$$w = u_1 + u_2, \text{ or } u_2 = w - u_1 \tag{3}$$

A Chapter in Nuclear Science and Engineering, vol. 2, pp. 1 - 25, 1947.

* This paper was published as chapter 11 of The Science and Engineering of nuclear Power, volume II, Ed. by Clark Goodman, 1949, Addison-Wesley.

Equation (2) can be expanded into

$$u_1 \cdot dm_1 + m_1 \cdot du_1 = u_2 \cdot dm_2$$

By combining with (1) and (3), the above equation can be written as

$$m_1 du_1 = -w dm_1 \tag{4}$$

Equation (4) has the integrated form

$$\log m_1 = -\frac{u_1}{w} + \text{constant} : m_1 = Ce^{-\frac{u_1}{w}} \tag{5}$$

where C is a constant to be determined. Now let M_1 and M_2 be the mass of the rocket when $u_1 = 0$ and when $u_1 = U$ respectively. M_1 is thus the initial mass of the rocket and M_2 the final mass for the final velocity U. Then (5) gives the following equations:

$$M_1 = C \quad M_2 = Ce^{-\frac{U}{w}}$$

Or by denoting the ratio of mass M_2/M_1 by ν,

$$\nu = e^{-\frac{U}{w}}, \text{ or } \frac{U}{w} = \log(1/\nu) \tag{6}$$

(6) is the basic equation for the design of a rocket moving without gravitational force and air resistance. Since even the most energetic chemical propellant can hardly give an exhaust velocity, w, greater than 11 000 ft/sec, and for escape from the earth, U has to be at least 36 700 ft/sec, the mass ratio, ν, has to be smaller than 1/28. Therefore even for this optimistic condition of free space, in order to reach the moon, one has, for example, to build a rocket with an initial weight of 28 tons, containing 27 tons of propellant. With consideration of gravitational force and air resistance, the situation would be even worse*. Such demand on the design of a space rocket is definitely beyond engineering possibility. The situation can be improved somewhat, of course, by using the principle of the step rocket so that not all of the container for the propellant need be accelerated to the final velocity but most of it dropped as its function ceases. Nevertheless, the primary object of rocket engineering for intercontinental or interplanetary travel is to increase the effect exhaust velocity w of the propellant and so reduce the ratio of initial mass to final mass.

Before the advent of nuclear energy, any further increase of the exhaust velocity from 11 000 ft/sec seemed hardly possible. However, the nuclear fuels are much more energetic. Thus the question arises: How can nuclear fuels be used in a rocket engine and what is the probable performance of such a nuclear fuel rocket? It is the purpose of this chapter to point out the problems connected with the utilization of nuclear energy in rockets and to indicate the more probable directions of this development.

---

* Superiors in parentheses refers to bibliography at end of chapter.

2 Relativistic Theory of Space Rockets

In anticipation of the extreme velocity possible with nuclear propellants, the simple theory of the last section should be extended to take into account the relativistic effect on the motion. Such an extension of the theory has been made by J. Ackeret.* The following discussion follows the general outline of Ackeret's treatment.

The theory of special relativity which applies to the free space concerned here, states that if c is the velocity of light in a certain coordinate system in which a mass m travels with a velocity u, then the mass m is greater than the same mass at rest. The "rest mass" is m_0. The relation between m and m_0 is as follows:

$$m = \frac{m_0}{\sqrt{1 - \frac{u^2}{c^2}}} \tag{7}$$

The "total energy" associated with this mass m is

$$mc^2 \tag{8}$$

Since the "total energy" associated with the rest mass m_0 is

$$m_0 c^2$$

the kinetic energy of motion is given by the difference, or

$$(m - m_0)c^2 \tag{9}$$

Equation (7) shows that if the velocity u is only a very small fraction of the constant velocity of light c, the mass m is approximately equal to the rest mass m_0. The expression for kinetic energy given by (9) will also reduce to the familiar form of $\frac{1}{2}m_0 u^2$ under this restriction of small velocity.

If m_{0_1} is the rest mass of the rocket at a certain time in the coordinate system at rest with respect to the initial position of the rocket, dm_{0_2} is the rest mass of the mass ejected at a small time interval in the same coordinate system, then the conservation of mass and energy requires

$$d\left\{\frac{m_{0_1} c^2}{\sqrt{1 - \frac{u_1^2}{c^2}}}\right\} = -\frac{dm_{0_2} c^2}{\sqrt{1 - \frac{u_2^2}{c^2}}} \tag{10}$$

where u_1 and $-u_2$ are the velocity of the rocket and the ejected mass in the same coordinate system. The conservation of momentum then gives the following equation:

* J. Ackeret, Helvetica Physica Acta, 19, 103 (1946).

$$d\left\{\frac{m_{0_1}u_1}{\sqrt{1-\frac{u_1^2}{c^2}}}\right\}=\frac{d m_{0_2}u_2}{\sqrt{1-\frac{u_2^2}{c^2}}} \tag{11}$$

The final equation for the relative exhaust velocity w is different from (3) by the relativistic effect. It is now

$$u_2=\frac{w-u_1}{1-\frac{u_1 w}{c^2}} \tag{12}$$

If the velocities are small compared with the velocity of light, the denominator on the right of (12) is approximately unity; then (12) again reduces to the classical relation (3). The three equations (10), (11), and (12) determine completely the motion concerned.

Equations (10) and (11) can be expanded, keeping in mind that the velocity of light c is a constant. Then

$$\frac{d m_{0_1}}{\sqrt{1-\frac{u_1^2}{c^2}}}+m_{0_1}\frac{\frac{u_1^2}{c^2}}{\left(1-\frac{u_1^2}{c^2}\right)^{3/2}}\frac{du_1}{u_1}=-\frac{d m_{0_2}}{\sqrt{1-\frac{u_2^2}{c^2}}} \tag{13}$$

$$u_1\frac{d m_{0_1}}{\sqrt{1-\frac{u_1^2}{c^2}}}+m_{0_1}\frac{du_1}{\left(1-\frac{u_1^2}{c^2}\right)^{3/2}}=u_2\frac{d m_{0_2}}{\sqrt{1-\frac{u_2^2}{c^2}}} \tag{14}$$

By eliminating dm_{0_2} and u_2 with the aid of (12), the final relation between m_{0_1} and u_1 is

$$\frac{d m_{0_1}}{m_{0_1}}=-\frac{du_1}{w\left(1-\frac{u_1^2}{c^2}\right)}=-\frac{c}{2w}\left\{\frac{1}{1+\frac{u_1}{c}}+\frac{1}{1-\frac{u_1}{c}}\right\}d\left(\frac{u_1}{c}\right) \tag{15}$$

The integration of (15) then gives

$$\log m_{0_1}=-\frac{c}{2w}\log\frac{1+\frac{u_1}{c}}{1-\frac{u_1}{c}}+\text{constant}$$

If one denotes the initial rest mass by M_1 and the final rest mass by M_2 and final velocity by U, the mass ratio ν is given by

$$\nu=\frac{M_2}{M_1}=\left(\frac{1-\frac{U}{c}}{1+\frac{U}{c}}\right)^{c/2w} \tag{16}$$

or by inverting (16),

$$\frac{U}{c} = \frac{1 - \nu^{2w/c}}{1 + \nu^{2w/c}} \tag{17}$$

Figure 2 shows the relation between $\frac{U}{c}$, the ratio of the final velocity to the velocity of light, and the rest mass ratio ν for various ratios w/c of exhaust velocity to the velocity of light. Of course, the maximum velocity of motion of the rocket is the velocity of light in accordance with the principle of relativity. This is in contrast with the classical theory which imposes no limit on the velocity of the rocket as $\nu \to 0$, as shown by Eq. (6). Considerable deviation from the classical theory is present even for larger values of ν, particularly if the exhaust velocity is very high. This is shown in Fig. 3 in which the final velocity ratio U/c is plotted against w/c for a given mass ratio ν equal to 0.2. For the case $w = c$, when the exhaust velocity is equal to the velocity of light, the relativistic value of final velocity is nearly 0.6 of the classical value. In other words, if high velocity particles such as fission fragments or neutrons could be used as the propellant, the final velocity is substantially less than the classical value.

3 Idealized Optimum Design using Nuclear Energy

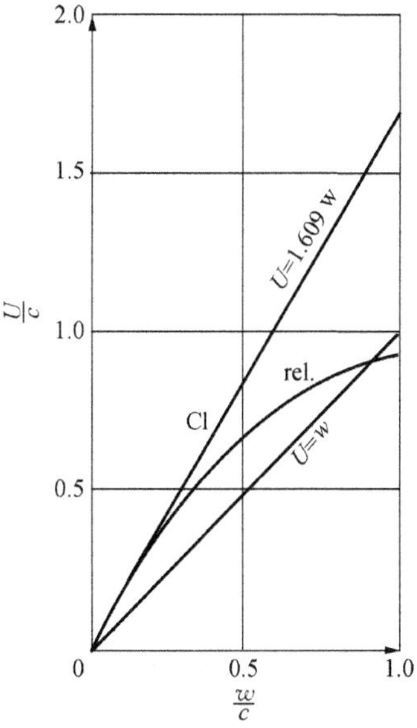

Fig. 2

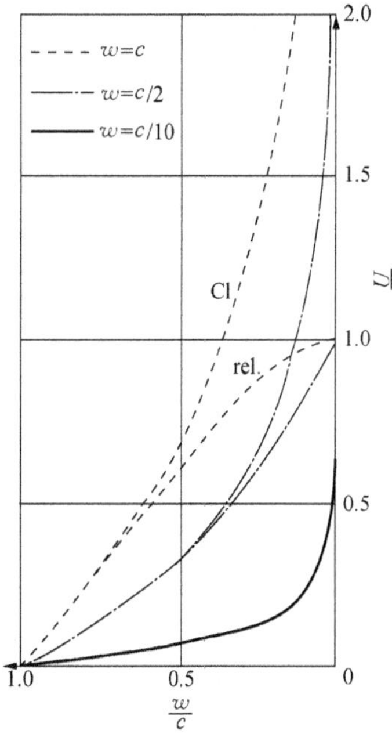

Fig. 3

According to Fig. 2, for a fixed mass ratio, the final velocity is a maximum if the exhaust velocity c is equal to the velocity of light. This would require the use of photons as the propellant. However, this can be easily shown to be impracticable. The alternative is to use material particles such as electrons, neutrons, protons, atoms, or molecules. This means that the exhaust velocity c will be a good deal smaller than the velocity of light and the mass ratio for achieving a given final velocity will be also less favorable. Then, to achieve the most favorable mass ratio for a given final velocity, we must unconditionally strive for the highest exhaust velocity. However, if the nuclear fuel is very expensive and, on the other hand, the mass ratio is not critical if within reason, then a logical question would be: for a given final mass M_2, and a given total amount of energy $\varepsilon M_2 c^2$, what should be the propellant weight bM_2 in order to maximize the final velocity U? It is seen that the total rest mass of the rocket at the start is

$$M_1 = M_2 + \varepsilon M_2 + bM_2 = M(1 + \varepsilon + b) \tag{18}$$

Of M_1, εM_2 goes into the energy for propulsion. The mass ratio ν is then

$$\nu = \frac{M_2}{M_1} = \frac{1}{1 + \varepsilon + b} \tag{19}$$

Now if $dm_0^{\prime}$ is the ejected rest mass at a certain time instant in a coordinate system moving with the rocket, its velocity is w in this coordinate system. Therefore, the kinetic energy of the ejected mass at every instant is

$$\frac{dm'_0 c^2}{\sqrt{1 - \frac{w^2}{c^2}}} - dm'_0 c^2 \tag{20}$$

This must come from the transformation of $(dm'_0)\frac{\varepsilon}{b}$ into energy. Thus

$$b\left\{\frac{1}{\sqrt{1 - \frac{w^2}{c^2}}} - 1\right\} = \varepsilon, \quad \text{or} \quad \frac{w}{c} = \sqrt{1 - \left(\frac{b}{b + \varepsilon}\right)^2} \tag{21}$$

By substituting into (17), the final velocity U is given by

$$\frac{U}{c} = \frac{1 - \left(\frac{1}{1 + \varepsilon + b}\right)^{2\sqrt{1 - \left(\frac{b}{b+\varepsilon}\right)^2}}}{1 + \left(\frac{1}{1 + \varepsilon + b}\right)^{2\sqrt{1 - \left(\frac{b}{b+\varepsilon}\right)^2}}} \tag{22}$$

For a given value of ε, one can then investigate the variation of $\frac{U}{c}$ with different values of b. Fig. 4 shows this variation for $\varepsilon = 0.02$. The final velocity has a flat maximum at $b \cong 4$. The maximum is 16.1% of the velocity of light or 1.585×10^8 ft/sec However, for such a large value of b and such a small value of ε, the velocities involved in the problem cannot be very large in comparison with the velocity of light. Then the classical theory should be a good approximation. In fact, for this approximation, the kinetic energy of the ejected mass at every instant is $dm(1/2)w^2$. This energy comes from $(dm)\frac{\varepsilon}{b}c^2$. Therefore,

$$\frac{w}{c} = \sqrt{\frac{2\varepsilon}{b}} \tag{23}$$

The mass ratio ν is, of course, $1/(b+1)$. Then Eq. (6) gives

$$\frac{U}{c}=\frac{w}{c}\log(1+b)=\sqrt{\frac{2\varepsilon}{b}}\log(1+b) \tag{24}$$

Figure 4 shows that the agreement between the relativistic theory and the classical theory is very close for the case concerned. Only for very small values of b is there an appreciable deviation. This is easily understood from the fact that as the propellant mass decreases, the exhaust velocity increases and relativistic effects are expected. For instance, as $b \to 0$, the exhaust velocity approaches c. Then the correct limiting value of U/c must be calculated from Eq. (22) and is $U = 0.002\,c$, while the classical theory of Eq. (24) gives $U = 0$.

As long as ε is small and b is large, Eq. (24) holds. Then it is seen that the optimum b for highest U is always approximately equal to 4. This means that the most efficient utilization of fuel for a space rocket is achieved by keeping ν approximately equal to 1/5. However, to design such a rocket is not always possible due to the temperature limitation imposed by available engineering materials. For instance, with $\varepsilon = 0.02$ and $b = 4$, the propellant contains energy equal to 0.5% of its mass. It is known that the fission products of U-235 contain an average energy equivalent to 1% of their mass, which is approximately a million times the energy released by ordinary chemical reaction. Therefore, for the optimum design, the temperature of the propellant in the rocket motor has to be approximately half a million times the ordinary combustion temperature. This is clearly an impossibility from the point of view of construction materials. Even if a tenfold increase in combustion temperature is allowed by special design, b would have to be 4×10^5 instead of 4. The final velocity according to Eq. (24) would be only 1/40 as large as the optimum case and even this is achieved at the enormous ratio of 4×10^5 for the initial to the final mass of the rocket.

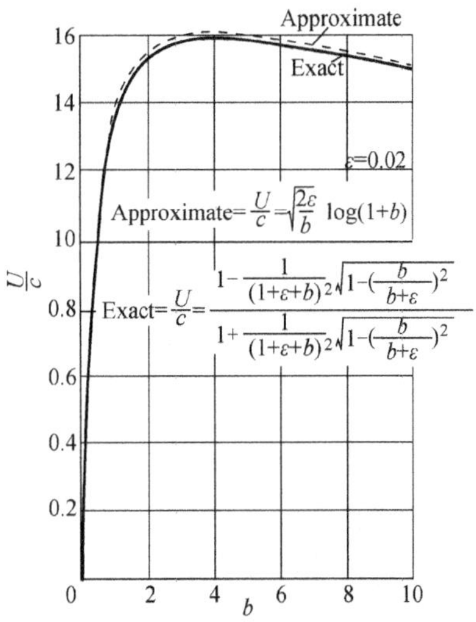

Fig. 4

4 Nuclear Energy Rocket

The difficulty of constructing a simple nuclear fuel rocket is, therefore, the enormous temperature developed in the combustion chamber which would then disintegrate in an instant. Since any engineering material is characterized by its molecular structure, the disintegration of the combustion chamber is the disintegration of the molecular structure. Thus, the problem to be faced in the design of a rocket using nuclear fuel is the problem of the essential difference in the order of magnitude of nuclear phenomena and molecular phenomena. A compromise solution is not at all easy, and if finally made to function, will be much less spectacular than the pure U-235 rocket, as the energy intensity must be scaled down towards the level of molecular phenomena or the level of conventional propellants.

An obvious possible solution is the utilization of the classical thermal neutron pile reactor. Here the fission rate, and thus the rate of heat generation, is regulated by controlling the density of neutrons with a neutron absorber, such as cadmium rods or sheets. All the fission material is present in the "combustion" chamber at one time, only the rate of "burning" is controlled. The heat generated is transmitted to the working fluid or the actual propellant of the rocket by conduction. A form of such a rocket is sketched in Fig. 5. The rocket has a central control rod. This control rod may be made of porous cadmium metal kept at a low temperature by forcing cold hydrogen through it. Hydrogen gas is used as the working fluid. The gas is heated to 6 000°R by passing through tapered tubes having porous walls of 1/8″ thickness made of a mixture of U-235, U-238, and carbon, the moderator (Fig. 6). The use of porous wall construction greatly reduced the difficulty of the heat transfer problem by its tremendous surface area. As the following calculation shows, the temperature difference between the hot active material and the gas is only of the order of 120°F, in spite of the extreme rapidity of the heating in a thickness of only 1/8″.

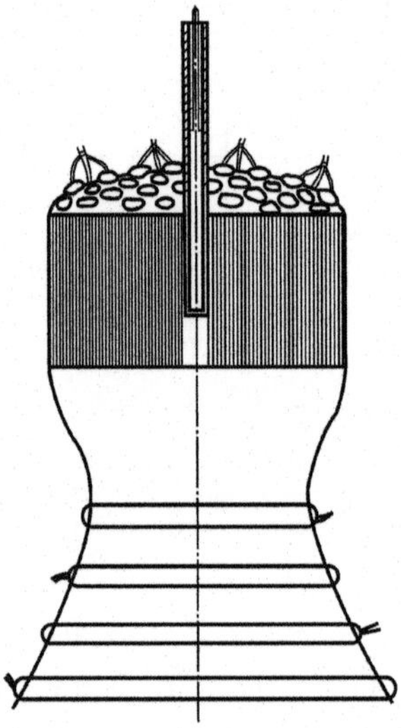

Fig. 5

Solid carbon contains 1.13×10^{23} atoms per cm^3. Let the presence of uranium atoms in the solid keep the number of all atoms the same. If θ is the fraction of the volume of the pile occupied by the hydrogen gas, the number of atoms of U-235, U-238, and carbon per cm^3 is then $(1-\theta)\ 1.13 \times 10^{23}$. Under standard conditions, the density of H_2 gas is 0.000 899 grams per cm^3. Assume that the average conditions of the hydrogen gas in the pile is 300 psi and 2 730K. The density is then 0.000 183 3 grams per cm^3. The number of H-atoms per cm^3 of the pile is therefore $0.001\,110 \times 10^{23}\theta$. Let N (235) be the number of U-235 nuclei per cm^3, and u = e similar notations for other nuclei. Then if

$$N(235)/N(238) = \delta$$
$$N(C)/N(238) = \alpha$$

we have

$$N(235) = 1.13 \times 10^{23}(1-\theta)\frac{\delta}{1+\delta+\alpha}$$

$$N(238) = 1.13 \times 10^{23}(1-\theta)\frac{1}{1+\delta+\alpha}$$

$$N(C) = 1.13 \times 10^{23}(1-\theta)\frac{\alpha}{1+\delta+\alpha}$$

$$N(H) = 0.001\,110 \times 10^{23}\theta$$

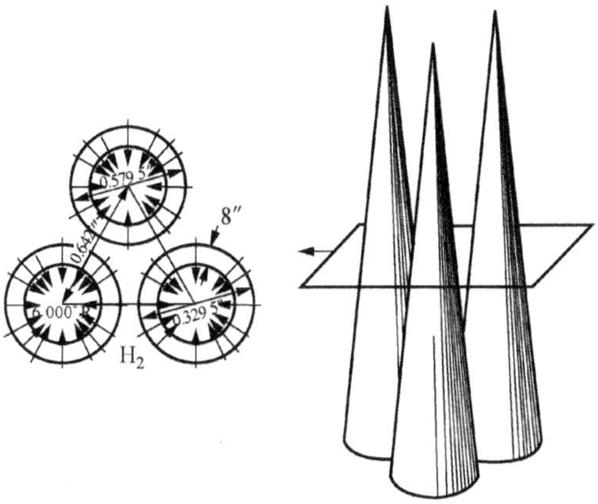

Fig. 6

Let $\sigma_f(235)$ be the fission cross section of U-235, $\sigma_c(238)$, $\sigma_c(C)$, $\sigma_c(H)$ the capture cross section of U-238, carbon, and hydrogen respectively, all at thermal neutron energy. Take

$$\sigma_f(235) = 500 \times 10^{-24}\,\text{cm}^2$$
$$\sigma_c(238) = 2 \times 10^{-24}\,\text{cm}^2$$
$$\sigma_c(C) = 0.004\,5 \times 10^{-24}\,\text{cm}^2$$
$$\sigma_c(H) = 0.31 \times 10^{-24}\,\text{cm}^2$$

Then f, the thermal utilization factor, is

$$\begin{aligned} f &= \frac{\sigma_f(235)N(235)}{\sigma_f(235)N(235) + \sigma_c(238)N(238) + \sigma_c(C)N(C) + \sigma_c(H)N(H)} \\ &= \frac{\delta}{\{0.004 + \delta + 9 \times 10^{-6}\alpha\} + 0.61 \times 10^{-6}\left(\frac{\theta}{1-\theta}\right)(1+\delta+\alpha)} \end{aligned} \tag{25}$$

The resonance escape probability during slowing down, p, is given by

$$-\log p = \int \frac{\sigma_r(E)N(238)}{\sigma_r(E)N(238) + \sigma_s(E)N_s} \frac{\text{d}E}{\xi E} \tag{26}$$

to be integrated from the thermal energy to the energy of fission neutrons. Here $\sigma_r(E)$ is the resonant capture cross-section of U-238 at energy E, $\sigma_s(E)$ is the scattering cross-section at Energy E, and N_s is the number of scattering nuclei per cm^3, ξ is the average logarithmic energy loss per collision. Since $\sigma_r(E)N(238) \ll \sigma_s(E)N_s$, and σ_s is nearly independent of E in the range concerned,

$$\int \frac{\sigma_r(E)N(238)}{\sigma_r(E)N(238) + \sigma_s(E)N_s} \frac{\text{d}E}{\xi E} = \left(\int \frac{(E)\text{d}E}{E}\right) \frac{N(238)}{\xi(C)\sigma_s(C)N(C) + \xi(H)\sigma_s(H)N(H)} \tag{27}$$

where $\xi(C)$ is the value of ξ for carbon and $\xi(H)$ is the value of ξ for hydrogen. The following values for the different quantities can be used:

$$\int \frac{\sigma_r(E)\mathrm{d}E}{E} = 88 \times 10^{-24}\,\mathrm{cm}^2$$

$$\sigma_s(\mathrm{C}) = 4.8 \times 10^{-24}\,\mathrm{cm}^2$$

$$\sigma_s(\mathrm{H}) = 20 \times 10^{-24}\,\mathrm{cm}^2$$

$$\xi(\mathrm{C}) = 0.15$$

$$\xi(\mathrm{H}) = 1$$

Therefore, by substituting values previously determined, one has

$$-\log p = \frac{88}{0.72\alpha + 0.02\left(\frac{\theta}{1-\theta}\right)(1+\delta+\alpha)} \tag{28}$$

The neutron reproduction factor k is then ηfp where η is the number of neutrons produced per fission. Take $\eta = 2.3$, then

$$k = 2.38\,\frac{\mathrm{e}^{-\left\{\frac{88}{0.02\left(\frac{\theta}{1-\theta}\right)(1+\delta)+\left[0.72+0.02\left(\frac{\theta}{1-\theta}\right)\right]\alpha}\right\}}}{0.004+\delta+0.61\times 10^{-6}\left(\frac{\theta}{1-\theta}\right)(1+\delta)+\left[9\times 10^{-6}+0.61\times 10^{-6}\left(\frac{\theta}{1-\theta}\right)\right]\alpha} \tag{29}$$

Now the optimum mixture α ratio can be determined by making k a maximum, or $\mathrm{d}k/\mathrm{d}\alpha = 0$. This value of α, after neglecting small quantities, is

$$\alpha_{\mathrm{opt.}} = \frac{44}{0.72+0.02\left(\frac{\theta}{1-\theta}\right)}\sqrt{4.5(4+1\,000\delta)\,\frac{0.72+0.02\left(\frac{\theta}{1-\theta}\right)}{0.9+0.061\left(\frac{\theta}{1-\theta}\right)}} \tag{30}$$

The corresponding optimum reproduction factor is then

$$k_{\mathrm{opt.}} = \frac{2.38}{22} \times \frac{\mathrm{e}^{-\frac{2}{\sqrt{4.5(4+1\,000\delta)\frac{0.72+0.02\left(\frac{\theta}{1-\theta}\right)}{0.9+0.061\left(\frac{\theta}{1-\theta}\right)}}}}}{0.000\,45(4+1\,000\delta)} \tag{31}$$

To calculate the critical size of the pile, the two diffusion lengths, L and L_f, have to be determined, the first for thermal neutrons and the second for the slowing down of neutrons. According to previous lectures by F. L. Friedman,

$$L^2 = \frac{1}{3[\sigma_s(\mathrm{C})N(\mathrm{C}) + \sigma_s(\mathrm{H})N(\mathrm{H})]} \times$$

$$\frac{1}{[\sigma_f(235)N(235) + \sigma_c(238)N(238) + \sigma_c(\mathrm{C})N(\mathrm{C}) + \sigma_c(\mathrm{H})N(\mathrm{H})]}$$

For the optimum case determined above,

$$L^2 \approx \frac{2 \times 10^4}{1.5 \times 1.13^2 (1-\theta)^2 \left\{0.9 + 0.061\left(\frac{\theta}{1-\theta}\right)\right\}} \times \frac{1}{\left\{4.8 + 0.02\left(\frac{\theta}{1-\theta}\right)\right\}\sqrt{4.5(4 + 1\,000\delta)\dfrac{0.72 + 0.02\left(\frac{\theta}{1-\theta}\right)}{0.9 + 0.061\left(\frac{\theta}{1-\theta}\right)}}} \tag{32}$$

Similarly,

$$L_f^2 = \frac{\log(E_{\text{fission}}/E_{\text{thermal}})}{3\{\sigma_s(\mathrm{N})N(\mathrm{C}) + \sigma_s(\mathrm{H})N(\mathrm{H})\}\{\xi(\mathrm{C})\sigma_s(\mathrm{C})N(\mathrm{C}) + \xi(\mathrm{H})\sigma_s(\mathrm{H})N(\mathrm{H})\}}$$

where E_{fission} is the energy of fission neutrons $\approx 10^6$ eV, and E_{thermal} is the energy of thermal neutrons, taken to be 1/40 eV. Thus,

$$L_f^2 = \frac{17.5}{3 \times 0.113^2 (1-\theta)^2 \left[4.8 + 0.02\left(\frac{\theta}{1-\theta}\right)\right]\left[0.72 + 0.02\left(\frac{\theta}{1-\theta}\right)\right]} \tag{33}$$

Both L^2 and L_f^2 are measured in cm^2.

The characteristic constant κ_0^2 is given by

$$\kappa_0^2 = \frac{1}{L_f^2}\log\frac{k}{1 + \kappa_0^2 L^2}$$

For a cylindrical pile of radius R and length l, the critical radius is then determined by

$$R_{\text{cri}} = \frac{\sqrt{2.404\,8^2 + \pi\dfrac{R^2}{l}}}{\kappa_0}$$

5 Specific Examples of Nuclear Energy Rocket

Now assume that the porous wall of the tubes is 0.40 porous, and that the porous material occupies one-half of the total volume of the pile, the other half being occupied by the gas flow space. Then, $\theta = 0.70$. Assume also that $\delta = 0.10$, i.e., there are ten U-238 atoms for every one U-235 atom. Then the previous formulae give the following values:

$$\alpha_{\text{opt.}} = 1\,070$$
$$k_{\text{opt.}} = 1.986$$
$$L^2 = 1\,235\ \text{cm}^2$$
$$L_f^2 = 1\,366\ \text{cm}^2$$
$$\kappa_0^2 = 0.282\,4 \times 10^{-3}\ \text{cm}^{-2}$$

Now let us further assume that the length of the pile is equal to its radius, or $R/l = 1$. Then

$$R_{\text{cri}} = \frac{\sqrt{2.404\,8^2 + \pi^2}}{\kappa_0} = 242.8\ \text{cm} = 95.7\ \text{in}$$

Assume a density of 2.26 grams per cm^3 for the solid material, the weight of the solid material in the pile is then $\pi R^2 l(1-\theta)$ 2.26 grams, or 33.6 tons. The amount of U-235 contained in this mass is 120 pounds.

The crucial problem of the nuclear energy rocket is the heat transfer problem. One must strive to take out heat at the highest possible rate so that the thrust of the rocket will be large and the corresponding unit weight of the pile will be small. This can be achieved by using porous construction. To estimate the heat transfer, the porous material can be visualized as composed of tubes of uniform radius and 1/8″ length. Let the radius of the tube be $\sqrt{10}\times 10^{-4}$ ft $= 0.003\,79''= 0.096\,2$ mm. Assume the average flow velocity through the tube to be 20 ft. per sec. The coefficient of viscosity of hydrogen at the prevailing temperature is approximately 1.76×10^{-5} lb/ft sec The density of the gas can be taken as before, 0.000 183 3 grams per cm^3, or 0.011 43 lb per ft^3. The Reynolds number of flow based upon the diameter of the tube is then 4. The flow is then certainly laminar.

The gas enters these capillary tubes at a very low temperature and is heated during its passage to a temperature of 6 000°R. Since the rate of energy generation by fission can be considered as constant within the dimensions of these capillary tubes, the temperature rise of the gas along the tube is approximately uniform. For the laminar flow through such capillary tubes, the difference ΔT between the temperature of the solid at each point along the capillary tube and the average temperature of the gas at each section of the capillary tube is given by*

$$\Delta T = \frac{11}{48}\frac{\rho}{\mu}\frac{1}{\gamma}u_m r^2\frac{T_M}{\mathrm{d}} \tag{34}$$

Here the Prandtl number of the gas is taken as unity. ρ is the density of the gas, u the coefficient of viscosity, γ the ratio of specific heats (1.4), u_m the average flow velocity, r the radius of the tube, T_M the maximum temperature of the gas, and d the length of the tube. For the case under consideration,

$$\Delta T = \frac{11}{48}\times\frac{0.011\,43}{1.76\times 10^{-5}}\times\frac{1}{1.4}\times 20\times 10^{-7}\times\frac{6\,000}{\frac{1}{8}\times\frac{1}{12}} = 122.3^\circ\mathrm{F}$$

The highest temperature of the solid is then 6 122°R = 3 400K = 3 127℃.

The pressure drop along the capillary tubes, or the pressure drop across the porous wall is

$$\Delta p = 8\mu\frac{u_m \mathrm{d}}{r^2} = 8\times 1.76\times 10^{-5}\,\frac{20\times\frac{1}{8}\times\frac{1}{12}}{10^{-7}}\text{lb per ft}^2 \tag{35}$$

$$= 0.063\,5\ \text{psi}(\text{l bs/in}^2)$$

This is a very reasonable pressure drop. Therefore, the scheme of using the porous wall is, by

* See S. Goldstein "Modern Development in Fluid Dynamics" Vol. II, p. 622, Oxford University Press (1938).

this rough estimate, feasible.

Now the performance of the rocket can be estimated:

The total sectional area of the pile is πR^2. It is assumed that one-half of this area is occupied by the porous material. Therefore, for the example considered, this area is 14 360 in^2. The thickness of the wall is 1/8″, the total length of the circumference of the porous tubes is thus $8 \times 14\,360$ in. Since the porosity of the wall is 0.40, the effective flow area for H_2 gas is $8 \times 14\,360 \times 95.7 \times 0.40\ in^2$. The weight flow of gas is then $8 \times 14\,360 \times 95.7 \times 0.40 \times \frac{1}{12^2} \times 20 \times 0.011\,43 = 6\,980$ lb per sec. Take the exhaust velocity to be 24 000 ft per sec, which is a reasonable value for the prevailing conditions.* Then the thrust of the rocket is

$$\text{Thrust} = \frac{6\,980}{32.2} \times 24\,000 \times \frac{1}{2\,000}\ \text{tons} = 2\,600\ \text{tons}$$

The dimensions of the porous tubes are shown in Fig. 6, with a hexagonal pattern. The percentage of flow area to total sectional area at the downstream end of the pile is 72.5%. The density of the heated hydrogen gas is 0.009 37 lb per ft^3. The flow velocity at the exit of the pile is thus 5 150 ft per sec. The velocity of sound of hydrogen at the prevailing conditions is 14 430 ft per sec. Hence no compressibility difficulty of the flow is expected.

At present, the most advanced design of long-range rockets is that of V-2. It has the following weight breakdown:

V-2 Rocket

	kg
Explosive charge (Amatol)	980
Fuselage	1 750
Pumping unit	450
Combustion chamber	550
Auxiliary devices	300
Alcohol + liquid oxygen	8 750
Auxiliary fuels for turbine	200
	12 980

It has a thrust of 27 200 kg. Therefore, the empty weight is 3 050 kg or 23.4% of the gross weight. The thrust is approximately twice the gross weight. Therefore, it is reasonable to assume that if the initial weight of the rocket is 0.60 of the thrust and the fuel weight is 80% of the initial rocket weight, the fuel load for the rocket concerned is 1 246 tons. The burning time, or the time of operation of the rocket is 358 seconds. The heat taken out is $1\,505 \times 10^8$ Btu per sec. or 1.59×10^8 kW. The total heat generated in the pile during the burning is thus

* See M. Summerfield and F. J. Malina: "The Problem of Escape from the Earth by Rocket," Journal of Aeronautical Sciences, Vol. 14 (1947).

5.39×10^{10} Btu, which is equivalent to the fission of 1.455 lb. of U-235. Therefore, only approximately 1% of the fissionable material in the pile is actually burned. By neglecting the small air resistance of such a large rocket, the maximum velocity of this rocket if fired vertically, starting at rest, is calculated to be 27 150 ft per sec. The maximum velocity of a V−2 rocket is only 5 000 ft per sec.

The results of the present calculation are summarized in the following table where another example with richer fissionable material is also included:

Summary of results

	Example I $N(235)/N(238) = \frac{1}{10}$	Example II $N(235)/N(238) = 1$
Diameter of pile, inches	191.4	150.4
Length of pile, inches	95.7	75.2
Weight of active material, tons	33.6	16.33
Weight of U-235, lbs	120	190.2
Weight flow of hydrogen, lbs/sec	6 980	3 360
Thrust, tons	2 600	1 251
Energy rate, Btu per sec. kW	1.505×10^8	0.726×10^8
	1.59×10^8	0.765×10^8
Initial Weight of Rocket, 60% of thrust, tons	1 560	751
Weight of hydrogen carried, 48% of thrust, tons	1 246	601
Duration of Burning, seconds	358	358
Total amount of heat generated, Btu	5.39×10^{10}	2.60×10^{10}
Maximum velocity, vertical Trajectory, ft/sec	27 150	27 150

It must be stated here that we have made the rather optimistic assumption of taking all the porous passages as effective capillary tubes. Actually, some of the porous passages may be plugged and are thus not useful for flow and heat transfer. This would tend to reduce the flow of hydrogen through the porous wall if the pressure difference across the porous wall is kept the same. This reduced flow would in turn decrease the thrust of the rocket. Therefore, it is of utmost importance to avoid the loss of flow capacity, i.e., the permeability of the material must be made high without increasing the porosity.

In this analysis, the energy production in the pile is so far assumed to be uniform. Actually, the energy production rate is much higher at the center and falls off to zero at the boundary of the pile. However, this does not necessarily mean that the gas temperature will be nonuniform. The rate of gas flow through the walls of the tubes can be easily adjusted by either one or a combination of the following parameters: the pressure drop, the porosity, and the grain size. Hence, by increasing the rate of gas flow in the central portion of the pile and decreasing the rate of gas flow in the outer portion of the pile, the final gas temperature can be made uniform. The overall performance will be approximately the same as calculated. Moreover, the fact that the rate of energy production at the boundary of the pile is low can be utilized to keep the temperature of the pile low near the walls of the reaction chamber and thus automatically solve the problem of proper cooling of the walls.

6 Possibilities of Reducing the Critical Size

The critical size of the rocket analyzed in the previous sections is relatively large, of the order of 1 000 tons. There are reasons to believe the desired size for the not too remote future is of the order of 100 tons. Therefore, one needs to consider the means by which the critical size can be reduced. It may be suggested that the critical size can be reduced by using a reflector. However, for an automotive nuclear power plant, such as a nuclear rocket, nuclear ramjet, and nuclear turbojet, one must be aware that the weight of the power plant is of utmost importance in the overall performance. One must reduce the weight to a minimum. Therefore, if the use of a reflector adds to the weight of the power plant per unit of energy generated, then its use is not desirable. A simple analysis shows that if the reflector is thin, say of the order of the absorption length of the reflector material for neutrons, then the decrease of the critical size is equal to the thickness of the reflector. In other words, the overall size of the pile including the reflector is approximately the same as the pile without the reflector. However, for the piles analyzed in the previous sections, the average density of the pile content is only 0.68 grams per cm^3. If beryllium is used as a reflector, the density of the reflector is 1.8 grams per cm^3, or nearly three times the density of the pile content. Hence, by using a reflector, while the actual size of the pile is reduced, the total weight of the pile will increase. This is undesirable. Therefore, the advantage of using a reflector for neutrons is generally very limited in such automotive power plants.

The critical dimensions of the pile are determined by the two lengths L and L_f, given by Eq.(32) and (33) respectively. By enriching the pile material, or by increasing the value of δ in Eq.(32), L can be reduced. However, L_f remains the same. In other words, the probability of neutron capture by fissionable nuclei can be evidently improved by enrichment to such an extent that the neutron, once slowed to the thermal region, does not have a good chance of leaking out before capture. But to slow the neutron down from the fission energy to thermal energy requires still the same number of collisions with the moderator, and during this slowing down process, the chances of leaking out of the pile are still large. Thus, in order to decrease this chance of leaking and to reduce the critical size, the obvious solution is to reduce the number of necessary collisions with the moderator. This can be accomplished by increasing the slowing down power through a better choice of moderator. For instance, beryllium would be better than carbon by a factor of almost 3.* (Cf. Table on page 315.) However, the rather low melting point of beryllium excludes its use in a nuclear energy rocket where high temperatures are of primary importance.

---

* See table 9-b on page 301, vol. I, Science and Engineeing of Nuclear Power.

The only remaining possibility of reducing the critical size is through the use of fast neutron or epithermal neutron fission. Here one reduces the necessary number of collisions during slowing down by simply not trying to reach low energy. For the mixture considered in the preceding calculations, if the energy at the end of slowing down is higher than 1/40 eV, then the value of L_f^2 is given as follows:

L_f^2 for carbon and hydrogen mixture of sec. 5

Final Energy/eV	L_f^2 /cm^2
0.025	1 366
0.25	593
2.5	258
25	112
250	48.7

It is then evident that the use of epithermal or fast neutron fission is certainly one of the most flexible methods to reduce the critical size. By using neutrons of about 7 eV with proper enrichment, the above table indicates that a pile of 1/2 the dimensions shown in the table on page 190 can be made critical. This means that nuclear rockets of approximately 100 tons initial weight could be built.

7 Application of Nuclear Fuel to other Thermal Jets

Although the calculations presented in the previous sections refer specifically to a nuclear rocket, they can be also applied to other power plants. For instance, by replacing the hydrogen gas by air at lower pressure, the whole pile design can be used for a ramjet or a turbojet. However, this means that the size of the "combustion chamber" with thermal neutrons will be of the order of 10 ft, which is enormous from the viewpoint of present aeronautical engineering. Take the case of ramjets. Due to the reduced pressure and temperature rise of the air in the pile compared with the rocket case, the energy production rate at low altitudes is probably 1/5 that of the rocket pile of the same size. Therefore, for Example II in the table on page 190, the energy production rate will be 14.52×10^6 Btu per second. For a ramjet at Mach number 2, one pound of thrust requires 20 Btu per second. Hence, the pile discussed can give 726 000 pounds of thrust at low altitude. This is approximately 1 000 times the thrust of ramjets currently considered. It is then apparent that here again one must utilize epithermal or fast neutron fission to reduce the critical size of the pile so that nuclear ramjets and nuclear turbojets can be brought within the realm of probability.

However, for any application of nuclear fuel to automotive power plants such as rockets, ramjets, and turbojets, one must keep in mind that the nuclear fuel can be "burned" only in relatively large mass. For instance, for a design like Example I on page 190, the weight of the pile, 34 tons, has to be carried even if we wish to operate the rocket for only ten seconds. Therefore, for short duration operation, the advantage of the high energy in nuclear fuels may be cancelled by the necessity of carrying large amounts of it. The situation is even clearer in the case of ramjets and turbojets which take the working fluid from the atmosphere, and only consume fuels. Now, if q is the heat generated by the pile in Btu per second per pound of pile material, and h is the heat value of the chemical fuel, for example, gasoline, with 19 000 Btu per pound, and t is the time of operation in seconds, then if $t > 19\,000/q$, the nuclear fuel will be lighter than the chemical fuel; if $t < 19\,000/q$, the nuclear fuel will be heavier than the chemical fuel. Figure 7 shows this line of demarkation. Thus, for a ramjet using the pile of Example II on page 190, with a rate of heat generation equal to 14.52×10^6 Btu per second, as calculated in the preceding paragraph, and a weight equal to 16.33 tons, q is equal to 0.445×10^3 Btu per second per pound. Therefore, the operating time of the nuclear ramjet must be greater than 44 seconds before there is a possibility of weight saving.

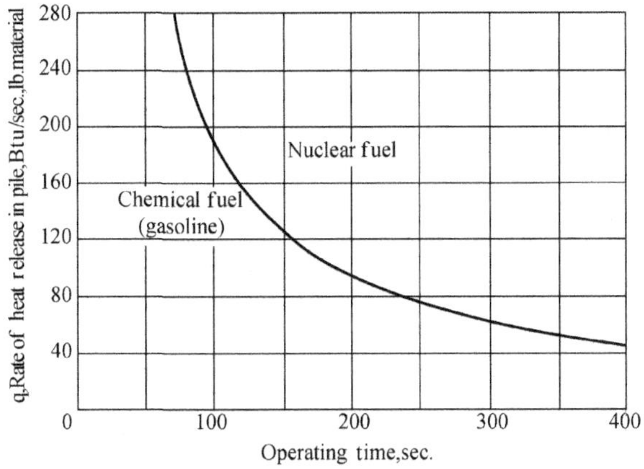

Fig. 7

8 Advantages of Using Porous Pile Material

Throughout these discussions, the use of porous pile material is assumed. It cannot be overemphasized that the method of heat removal from the pile by passing the working fluid through a porous reacting material has many basic advantages. In fact, it can even be claimed that this method is "intrinsic" to nuclear power engineering. The reason is as follows: The principle of efficient design in heat transfer is always to bring the heat-producing elements as close as possible to the heat-absorbing elements, so that the heat does not have to traverse a long route before reaching the absorber. This principle is clearly illustrated by Figure 8. This is a diagram prepared by the Escher Wyss Engineering Works of Zurich, Switzerland, for a heat exchanger of their closed cycle gas turbine power plant. The absolute values are not important for our consideration, but the relative values are very pertinent to the point concerned. It shows that by reducing the diameter of the tubes in the heat exchanger from 19 mm to 2.4 mm, the volume of the exchanger is reduced by a factor of 10, and the weight by a factor of 8. However, in the conventional engineering practice, it is difficult to carry the principle to the ultimate, and the most advanced design in this respect is the laminar flow radiator, with

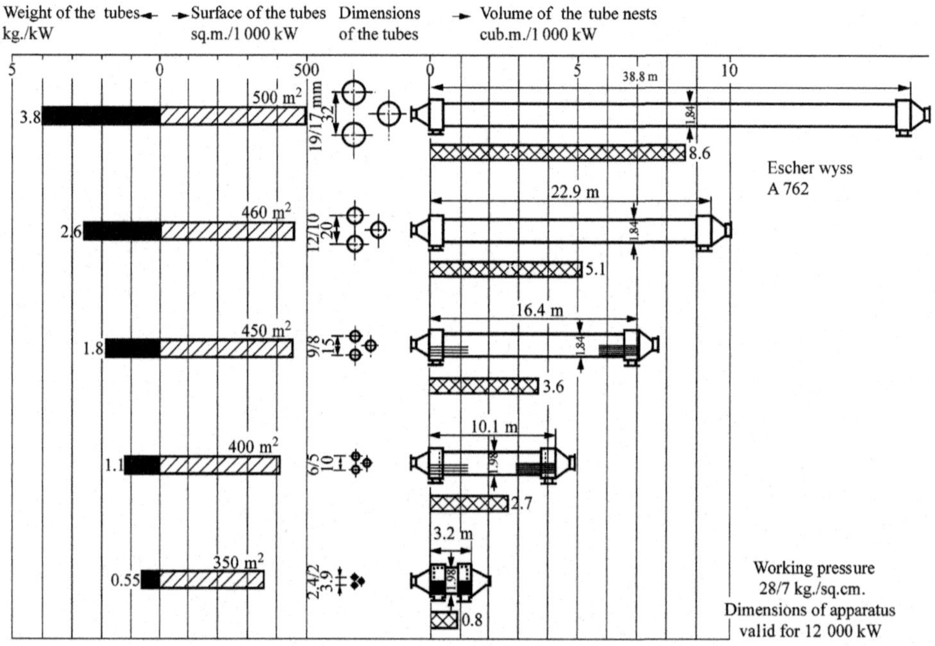

Fig. 8

finely divided fins and coolant passages. It is only in nuclear engineering that this principle of efficient heat transfer design can be carried to its logical conclusion: porous construction. Here the flow passages are of capillary size and there is a tremendous surface area in a unit volume. Besides the advantage of inherent large heat transfer capacity resulting from the large area of contact between the heat-producing and the heat-absorbing material, the heat is removed almost at the point of its production, thus avoiding unnecessary resistance to heat flow.

Unfortunately, the investigation of flow resistance and heat transfer problems of porous materials is only at its beginning and no extensive data are yet available. However, if the flow passages through the porous materials can be treated as capillary tubes, as assumed in section 5, then equations (34) and (35) give the following relation for gases:

$$\Delta T \cdot \Delta p \cong \frac{(\rho u_{\mathrm{m}})^2}{\rho} T_{\mathrm{M}}$$

where ΔT is the temperature difference between the solid material and the gas, Δp the pressure difference across the porous material, ρu_{m} the mass flow per unit area of passage per unit time, ρ the mean density of the gas, and T_{M} the final temperature of the gas after passing through the porous material. If θ is the porosity of the material, then

$$\rho u_{\mathrm{m}} = \frac{m}{\theta^{2/3} \cdot \varphi}$$

where m is the mass flow per unit area of porous material per unit time, and φ is the fraction of pore passages which are unplugged.

$$\Delta T \cdot \Delta p \cong \left(\frac{m}{\varphi \theta^{2/3}}\right)^2 \frac{T_M}{\rho} \tag{36}$$

Equation (36) can be used to calculate ΔT if other quantities are determined or specified. For instance, one can easily run a cold test to obtain a relation between Δp and m, and then equation (36) gives an estimate of ΔT with T_M specified.

References

[1] Malina F J, Smith A M O. Journal of Aero Sciences. 1938, 5: 199.

[2] Ackeret J. Helvetica Physica Acta, 1946, 19: 103.

[3] Goldstein S. Modern Development in Fluid Dynamics. Vol. II, Oxford University Press, 1938: 622.

[4] Summerfield M, Malina F J. The Problem of Escape from the Earth by Rocket. Jour. of Aeronautical Sciences, 1947, 14: 471 - 480.

Engineering and Engineering Sciences*

Hsue-shen Tsien**

Introduction

When one reviews the development of human society in the last half of century, one, is, certainly struck by the phenomenal growth of the importance of technical and scientific research as a determining factor in national and international affairs. It is quite clear that while technical and scientific research was pursued in an unplanned individualistic manner during the earlier days, such research is now carefully controlled in any major nation. Thus technical and scientific research has become a matter of state along with the age old matters such as the agriculture, financial policy, or the foreign relations. A closer examination for the reason of such growth of the importance of research would naturally yield the answer that research is now an integral part of modern industry and we cannot speak of a modern industry without mentioning research. Since industry is now the foundation of a nation's strength and welfare, technical and scientific research is then the key to a nation's strength and welfare.

But then one might argue that since the pioneering days of industrial age, technical and scientific research was related to the industrial development, then what is exactly the reason for making the research so important today? The answer to this question is the rate at which the modern industry is forced to develop due to national and international competition. At this rapid rate of development, research must be greatly intensified with the almost immediate application of basic scientific findings. Perhaps, nothing is so dramatic as an illustration as the wartime development of radar and nuclear energy. That the successful development of radar and nuclear energy contributed much to the victorious conclusion of the World War II in the side of Democracy is an established fact. Thus here intensive research has brought the findings of the basic science of physics through practical engineering and to successful applications to weapons of war in the short interval of a few years. Thus the distance between a pure scientific fact and industrial application is now very short. In other words, the difference between a long-haired pure scientist and a short-haired practical engineer is very small indeed, and a close cooperation between them is essential for the successful development of the

---

Journal of the Chinese Institute of Engineers, vol. 6, pp. 1 – 14, 1948.

* First given in lectures for the engineering students at the National Chekiang University, the Chiao-tung University and the National Tsing-Hua University, during the summer of 1947.

** Professor of Aerodynamics, Dept. of Aeronautical Engineering, Massachusetts Institute of Technology.

industry.

This need for close cooperation of the pure scientist and the practical engineer produced a new profession — the engineering research men or engineering scientist. They form the bridge between the pure science and the engineering. They are the men who apply the basic scientific knowledge to engineering problems. The purpose of the present article is to discuss what the engineering scientist can do, what is their job, in engineering, and then what kind of education and training he needs in order to do the job assigned to him.

Contributions of an Engineering Scientist to Engineering Development

The contributions of a engineering scientist to engineering development can be briefly stated as the economy in effort both in manpower and in money. This economy is achieved through a sound and general analysis of the problem on hand to point out 1) whether a proposed engineering scheme is at all possible 2) if feasible, what would be the best way of carrying out the proposal and finally 3) if a certain project failed, what is the cause of failure and what would be the remedy. It is needless to say that if an engineering scientist can fulfill these assignments then the cut and try in any research and development is to a large extent eliminated. All the effort and money can then be concentrated on the best approach or the few better methods of attacking the problem having the best chance of success.

It might be argued, nevertheless, that these three main problems assigned to the engineering scientist are really the three basic problems in engineering. What is that an engineering scientist can do which an engineer cannot do? The answer to this question is that as the engineering profession becomes more and more ocomplex, there is a need for specialization. The present requirement of knowledge for a satisfactory solution of its three problems stated includes a good training not only in engineering but also in mathematics, physics and chemistry, as will be discussed in greater detail in the subsequent paragraphs. Therefore the training of an engineering scientist is quite different from the conventional training of an engineer. In other words, he must be the specialist to solve just the three basic problems of engineering development.

The Feasibility of a Proposal — Long Range Rockets

To gain a better understanding of the way in which an engineering scientist solves the three basic problems of engineering development, a few illustrative examples will be described. The first example is the investigation of the possibility of extremely long range rocket. A rocket is propelled by the reaction of the exhaust jet obtained through the combustion of the propellent carried. The performance of the rocket motor is expressed as the specific propellant consumption in pounds of propellant per hour required to generate one pound of thrust. This figure is slightly modified by the change of the atmospheric pressure, but can be generally taken to be constant and equal to an average value. The specific propellant consumption then fixes the motor performance. The range of the rocket will evidently depend upon the amount of propellant carried, i. e. , the ratio of the propellant weight to the gross weight of the rocket. This is the propellant loading ratio. The rocket, during its flight, is opposed by the air resistance. We see thus that for an engineering scientist to solve the problem of the possibilities of long range rocket he has to have three kinds of basic information: the rocket motor performance, the structural efficiency and the aerodynamic forces at high speeds. For rocket motor performance he will depend upon the rocket engineer for test data. For the structural efficiency he will depend upon the stress man for data. For aerodynamic forces at high speeds, he will depend upon the high speed wind tunnel testing for data. The engineering scientist on the job must then synthesize these informations by good engineering judgment, by the application of the laws of dynamics and the skill of solving differential equations. The result is the calculated range of the rocket. If he uses the best rocket motor performance, the lowest practical value of specific fuel consumption, if he uses the best construction to achieve the highest propellant loading ratio and if he uses the best aerodynamic design of the shape of the rocket to minimize the air resistance, he will then obtain the largest range that can be achieved by a rocket.

This formulation of the problem of long range rocket assumes that the best motor performance, best structural efficiency and best aerodynamic shape are known to the analyst. But the real situation may not be so easy. An engineering scientist will find that while previous experience shows that the performance of the propellant can be predicted to within 10% accuracy by making a careful calculation of the combustion temperatures and the composition of the exhaust gas using chemical and thermodynamic equilibrium in the motor combustion chamber and by then calculating the exhaust velocity of the rocket using the dynamics of gaseous flow, the types of propellant actually tested are very few. In searching for the best possible motor performance, he may wish to know the probable specific propellant consumption for more energetic chemical reactions hitherto untried. This means the engineering scientist has made the theoretical estimate of these untried propellants. For instance, he may wish to calculate the performance of the liquid fluorine and liquid hydrogen rocket. If he does this kind of calculations, he will find two important facts about chemical rocket propellants. These are:

1) There is a strong tendency for the ordinary combustion products such as carbon dioxide, and water to dissociate at the extremely high temperature of the combustion chamber and these dissociations absorb heat. Therefore calculations on the propellant performance using low temperature calorimeter data is totally unreliable. In other words, thermodynamics and chemical equilibrium are matters of primary importance here.

2) There is no "wonder" propellant which will give a tenfold increase in the performance, i.e. , lower the specific consumption to one tenth of the value for present propellant. This is easily seen in the following table[1] which shows that the best propellant is the combustion fluorine and hydrogen which gives a specific consumption not less than one half of the more

conventional nitric acid and aniline combination.

Hence by this kind of investigation, the engineering scientist can achieve a broad orientation in an entire new field of engineering. He knows what to expect and is able to judge critically the validity of claims made by any inventors. Such ability of judgment generally requires years to achieve, if try and error is the only method used. Engineering science is then the useful tool to shorten this crucial process of "learning the trade".

Similarly, the engineering scientist may find the information and structural efficiency and aerodynamic forces quite incomplete. He is then forced to investigate a particular promising type of structure or to investigate a high speed flows over aerodynamic shapes to determine the probable air resistances at high speeds. In other words, to solve the problem of the possibility of long range rocket, the engineering scientist may be required not only to solve a very difficult problem in exterior ballistics but may also have to solve problems in thermodynamics and combustion, in elasticity and strength of materials and in fluid mechanics. His problem is thus not easy, but his reward is also large.

Rocket Propellants

Calculated Performance at Sea-Level, Chamber Pressure is 20 Atmospheres

Oxidizer	Fuel	Oxidizer/fuel weight ratio	Chamber temperature °Rankine	Specific consumption lbs per hr per lb thrust
Fluorine	Hydrazine	1.186	6 970	12.33
	Hydrazine	2.371	9 500	11.50
	Hydrogen	18.85	10 210	10.20
	Hydrogen	9.42	8 530	9.71
	Hydrogen	6.28	6 296	10.20
Oxygen ethyl alcohol		1.275	5 530	15.45
	(75% + water 25%)			
	Gasoline	2.62	5 930	14.95
	Hydrogen	3.80	5 500	10.20
Red fuming				
Nitric acid aniline		3.000	5 525	16.30

Best Method of Attack — Manufacture of Fissionable Material

Very often in engineering practice, one is confronted with the problem of choosing the best method of attack among a possible few. Here again the services of the engineering scientist is invaluable. Let us take the example of the manufacture of fissionable material. According to H. D. Smyth[2], the different possible methods are:

1) Manufacture of plutonium — 239 by the slow neutron pile using natural uranium and the chemical separation of plutonium.

2) Manufacture of uranium — 235 by electro-magnetic separation from the inert uranium — 238 in the natural uranium.

3) Manufacture of uranium — 235 by separation from uranium — 238 utilizing thermal diffusion.

4) Manufacture of uranium — 235 by isotope separation utilizing gaseous diffusion.

All methods except the first are involved in a physical process in that the materials to be separated have "identical" chemical properties. During the development of nuclear energy for the atomic bomb by the United States, all four methods were pursued. This way of attacking all possible methods simultaneously is certainly a wartime expedient when the time is very limited and the success of the project is a dire necessity. In normal times, engineering scientists should be called into service to analyze the four different processes to determine which one of them is the most economical. Of course, the engineering scientists will need much detailed information which must be obtained by either theoretical analysis or experimentations. For instance, in the first method, he would have to determine the fission cross-section, or fission probability, of uranium — 235, the resonant absorption cross-section of the moderator, etc. Then by the known principles of nuclear physics, he has to work out the process of the diffusion of neutrons in the pile, the distribution of neutron density in the pile, and finally the critical size of the pile. He has also to find the best method of constructing the pile by trying out in his calculations the different ways of placing the pieces of the uranium and the moderator. After all this investigation, the engineering scientist is able to say that the probable economy of the manufacture of plutonium — 239 is by the slow neutron pile method.

Now by a similar approach with laboratory experiment and theoretical calculation, the engineering scientist would be able to estimate the economy of all other proposed methods. Then the question of the best method to manufacture the fissionable material can be answered. It is quite evident now that if such an analysis of the relative economy of the different processes were possible, the plutonium process, or method 1), will be the chosen method. General Leslie R. Groves has revealed to the McMahan Committee that as of June 1945, the monthly operating cost of the processes were:

Hanford Plutonium Plant	\$3 500 000
Oak Ridge Diffusion Plant	6 000 000
Oak Ridge Electro-Magnetic Plant	12 000 000

Thus the Hanford Plutonium Plant is the most economical one in spite of the fact that it must be the one with the largest capacity for the fissionable materials.

Now suppose the preliminary analysis by the engineering scientist decides on the plutonium process, what would be the consequences? According to General Groves again, the investment spent and committed for plants and facilities as of June 30, 1945 is as follows:

Manufacturing facilities:

	Hanford Plant	\$350 000 000	
	Others	892 000 000	
Housing for workers:			\$1 242 000 000
	Hanford Plant	\$48 000 000	
	Others	114 500 000	162 500 000
Research			186 000 000
Workmen's compensation and medical care			4 500 000
Total			\$1 595 000 000

Therefore if the authority that directed the development of nuclear energy for the United States during wartime were able to decide on the plutonium process, then approximately one billion dollars could have been saved. In other words, two thirds of the investment could have been saved, if it were possible to use fully the services of engineering scientists.

Reason and Remedy for a Failure — the Tacoma Narrows Bridge

The third problem which may require the attention of an engineering scientist is the discovery of the reason and method of remedy for a failure. While the two previous problems the investigation of the feasibility and the best method of attacking a new development, are work to be done before starting the main part of the engineering, the third problem is, of course, something to be done afterwards. Let us take the example of the Tacoma Narrows Bridge. This bridge was first opened to traffic on July 1, 1940, and it was a suspension bridge of extremely narrow roadbed, as can be seen in the following table of dimensions:

Dimensions of the first Tacoma Narrows Bridge, Washington, U. S. A.

Center span	2 800 ft
Side span, west side	1 100 ft
Side span, east side	1 100 ft
West side, back stay	497 ft
East side, back stay	261.8 ft
Total length	4 759.2 ft
Width of roadway	26 ft
Total width including sidewalks	39 ft

After this bridge was built, it was found that the bridge was extremely flexible. During windy nights, a ghost effect often occurred as the head lights of approaching cars appear and then disappear caused by the lateral and longitudinal oscillation of the roadway. On 10:00 A. M., November 7, 1940 the bridge started to oscillate rather violently in torsion by the prevailing strong wind. This oscillation increased its amplitude and finally an hour later the bridge broke at approximately the mid-span. Of course, there is then developed among civil engineers great interest as to the cause of failure of the bridge, a kind never observed before. Civil engineers normally deal with static forces of rather large magnitudes. For instance, the stress in the bridge member is generally of the order of tens of tons per square inch. Now the air or wind forces on a surface is probably of the order of one fifth of a pound per square inch. It was rather difficult at first for the civil engineers to understand how such small wind forces could have broke the strong bridge.

The true mechanism of failure was finally explained by a Committee composed of O. H. Armann, Th. von Kármán and G. B. Woodruff[3]. The report was a typical example of investigation by engineering scientists. It consisted of model testing and theoretical computation. The true reason for the failure of the bridge was the resonant oscillation excited by the wind forces. A phenomena well-known to aeronautical engineers as the flutter, but quite outside the experience of a civil engineer. The wind forces although small have the same period or is always in phase with the oscillation of the roadbed and therefore can build up the amplitude of oscillation to ruinous magnitude. It is seen then that by incorporating damping and by stiffening the bridge to increase the natural frequency the failure can be avoided. This is the principle for the design of the new bridge.

Here again, the services of an engineering scientist is able to clear up a most perplexing engineering question, and can be used to avoid further mistakes in an engineering design.

Unification—Basic Research in Engineering Science

From the above discussion, it might be construed that the problems in engineering science are individual problems and the engineering scientist is to deal with particular case without much generalized scheme. This impression is however not correct. Among the multitude of problems in the current development of engineering there are phenomena which occur repeatedly in many branches of engineering. These phenomena can then be abstracted out of the direct routine problems which the engineering scientist has to solve and be formulated into individual fields of research. The results of such research will then not only benefit one field of engineering, but to all of them. This is the basic research in engineering sciences, through which the greatly diversified engineering activities are united.

Historically, such basic research in engineering sciences was started by the great mathematician F. Kelin in the Göttingen University, Germany, shortly before the World War I. His school has produced such eminent engineering scientists as Th. von Kármán and S. Timoshenko. At that time, the main fields of engineering activity had to deal with mechanics. It is thus natural that the basic research in engineering science was simply called "applied mechanics" (angewandte Mechanik). However the ever-widening fields of engineering now extend to subjects which are not treated in the applied mechanics as first conceived by the German school. Let us then divide the current topics of basic research in engineering sciences into three catagories: ① Research in the field which are not within the old boundary of applied mechanics; ② Research in field which are near the old boundary of applied mechanics, ③ Research in the fields which are within the old boundary of applied mechanics. To understand what is the character of basic research in engineering sciences, its ramifications and relations with different engineering problems, it will be profitable to examine these fields of research in greater detail:

1. Research in the Fields Which Are Not Within the Old Boundary of Applied Mechanics

a) Solid State of Matter

The engineering science of metallurgy has really progressed very little beyond the application of Gibb's phase rule. In fact, the present knowledge of materials is obtained through a tremendous amount of tedious laboratory tests. This large body of empirical data has practically no coordination or systematic interpretation. On the other hand, the physical theory of solid state, based upon quantum mechanics, is developed generally by physicists as a branch of pure science. In other words, there is a wide gap between the practical engineering and the scientific investigation. This gap has to be bridged. This effort of unitizing the physical theory with metallurgy will bring about not only a systematic interpretation of the accumulated empirical data but certainly will also indicate new possibilities in the field of development of materials. It is also certain that after a satisfactory engineering science of materials is developed, the search for an engineering material to satisfy a given specification will be greatly facilitated.

Another field of investigation is the ceramic materials. The present engineering materials are dominated by metals which consist of atomic crystals. There is no reason to believe that the other types of material, consisting of ionic crystals such as the ceramic materials, cannot be utilized as engineering materials for machine construction. In fact the recent demand for materials to withstand extremely high temperatures naturally points research in this direction.

b) Electronics

The electronics engineering can be divided into two main divisions: The division which deals with electronic tubes themselves and the division which treats the circuits and the radiation fields. The second division mainly involves an application of the classical Maxwell theory. The general character of the results is expected, in spite of the fact that such calculations may be very complicated and may require advance mathematical technique. The performance of tubes is, however, seldom comprehensively analyzed. The design of these tubes is generally worked out by numerous tests, guided by a few basic principles. However, the electronics engineering has now passed its heroic age of invention and creation and has entered the age of engineering development. The empirical approach may not be the most economical one in this new situation where detailed improvement of the various devices has to be carried out. This is especially true for very high frequency tubes where the electron inertia effect can no longer be neglected. It seems necessary to develop an engineering method of calculating such flow fields of electron cloud under the combined effect of rapidly varying external electric and magnetic fields. If this is done, then the characteristics of electronic tubes or other similar devices can be analyzed and the experimental data coordinated.

c) Nuclear Engineering

While the understanding of the nuclear structure is yet to be achieved, a general interpretation of the nuclear reaction seems to have developed to a satisfactory degree. In fact, the elementary processes of nuclei reactions such as collision, capture, excitation, and emission of new particles from the compound nuclei could be measured and studied individually. If these empirical data are available, then the overall microscopic performance of the reaction can be predicted by applying the methods of chemical kinetics. Impossible and undesirable processes can then be eliminated from further consideration and large scale tests. This approach to the utilization of atomic energy, or atomic engineering, seems to be able to lead to fruitful results without the danger of uncontrolled experiments. In other words, the stage seems to be set for the rapid development of utilization of nuclear reactions similar to the utilization of molecular reactions such as combustion.

2. Research in the Fields Which Are Near the Old Boundary of Applied Mechanics

a) Combustion

The theory of combustion has been studied by chemists mainly from the point of view of chemical kinetics. However, the problems which grow out of the recent development of jet propulsion generally involve very high speeds of flow. In such problems the effect of the inertia of the fluid elements certainly cannot be neglected. In fact, a study of a simple one dimensional problem has shown rather unexpected results due to this inertia effect. Then a complete and satisfactory solution of combustion problems must combine the science of fluid motion, i.e., hydrodynamics with the science of chemical kinetics. As an initial approach to this problem the effect of diffusion and turbulence on combustion must be studied.

b) Metal Forming by Plastic Deformation

The large number of metal forming processes is based upon the plastic deformation of the material. For instance, the widely used process of sheet metal forming is by pressing. This process, until recently, was practically carried out purely by experience. During the design of the dies for this process one has to use the cut and try method guided by a few empirical principles. This method is generally very uneconomical. It then seems necessary to develop a satisfactory theory so that such dies can be designed for each individual problem without resorting to numerous experiments. This new science of plastic forming, of course, will be based upon the methods of the theory of elasticity and a complete knowledge of the solid state of matter which is another topic of research as stated in the previous section.

3. Research in the Fields Which Are Within the Old Boundary of Applied Mechanics

a) Turbulence

During the last 15 years the problem of turbulence in fluid flows has been studied intensively and simple rules have been developed for satisfactory solutions of engineering problems in this field. However, the theory still lacks an explanation of the fundamentally important fact that the exchange coefficient in turbulent flows is enormous, as compared with that of laminar flows. The correct understanding of this phenomenon is the nub of the turbulence problem. It is believed that this understanding can only be achieved through a detailed survey of the turbulent fluid field together with theoretical analysis. Measurements on the turbulent velocities, the correlation coefficients and diffusion characteristics must be carried out.

Another possible field of investigation would be the application of the presently known knowledge of turbulence to the other fields of engineering such as combustion and the mixing problems in chemical engineering. Such applications are believed to be extremely useful.

b) Gasdynamics

The recent advance in aeronautics makes the science of gasdynamics one of the most important and urgently needed knowledges. The fundamental problems are connected with the interaction of viscosity and compressibility of the fluid. It was believed that the effects of viscosity, or Reynolds number, and the effect of compressibility, or Mach number, can be separated. However, it is now realized that such separation is impossible. On the other hand, the problem of the interaction is very complex, particularly due to the possible appearance of turbulence in the fluid. The detailed phenomenon must be studied simultaneously by both theoretical analysis and by experiment. In conjunction with this investigation, the effects of second viscosity coefficient and the relaxation time should be considered.

The possibility of flight at extreme speeds presents another very interesting problem of fluid dynamics at a very high Mach number. It is known that at such high Mach numbers, say, Mach number greater than 5, the fluid behaves very similarly to a stream of particles. In other words, the fluid reaction on a moving body will be very similar to that predicted by Newton on the assumption of no interaction between the particles of the fluid. The question of flying at extremely high altitudes leads to another interesting problem. This is the problem of fluid motion at extremely low density, i. e., the fluid flow in which the mean free path of the molecules is comparable to the dimension of the body moving in it. The solution of these problems is believed to be essential for the next assignment of aviation, the trans-oceanic flight at velocities faster than the velocity of sound.

From the above discussion on the different fields of basic research in engineering sciences, it would seem that the subjects are well within the general field of physics. But then why should they be called research topics in engineering science? The reasons are two-fold: Firstly, there is a fundamental difference between the point of view of a physicist and the point of view of an engineering scientist. The physicist's point of view is that of a pure scientist, interested essentially in simplifying the problem to such an extent that an "exact" solution can be made. The engineering scientist is more interested to solve the problem as given to him. It will be complicated, so only approximate solutions, though sufficiently accurate for engineering purposes, will be attempted. Thus physicist will give exact solutions of an over simplified problem while engineering scientist wants the approximate solutions of the realistic problem. The work of a physicist may be at times impractical, but that of an engineering scientist must always be practical. The second reason for separating the engineering science from the general field of physics is simply that physicist has no deep interest in engineering problems. Because of these two reasons, the engineering scientist is forced to pick up where the physicist has left it and develop the physical principles into tools of practical engineering.

Training of an Engineering Scientist

For the engineering scientist to solve the problems assigned to them and to carry out research in the basic engineering science he needs definitely an education quite different from the education of an engineer. Then what is exactly the necessary training of an engineering scientist? It would perhaps be best to first see that is the necessary tools for an engineering scientist. He must have these tools. These are:

1) Principles of engineering design and practice.

2) Scientific foundations of engineering.

3) Mathematical method of engineering analysis.

In the first group of tools, is the conventional engineering subjects such as mechanical drawing, drafting and machine design, engineering materials and processes, shop practice. In the second group of subjects, is the physics and the chemistry which is generally contained in a good engineering curriculum. But here the training of an engineer scientist would be different from the conventional engineer in that he must know much more about physics and chemistry. For instance, his knowledge about mechanics must not stop at the statics and dynamics of rigid bodies and the stress in simple beams and columns. He must learn the principles of the theory of elasticity and plasticity. His knowledge about fluid motion must not stop at the hydraulics. He must learn the principles of hydrodynamics and fluid mechanics. His knowledge in thermodynamics must not stop at the first law and the second law or calculation of the idealized Otto cycle or Diesel cycle. He must learn the physical meaning of entropy from the point of view of statistical mechanics and the broad concept of thermodynamical equilibrium. Then he must know the elementary structure of matter from the nucleus up to molecules. In other words, he must know many subjects which a physicist or a chemist has to learn.

The third group of subjects is mathematical methods and principles of mathematics which would help the understanding of the use of mathematical methods. Thus it includes subjects such as advanced calculus, functions of complex variable, principles of mathematical analysis, ordinary differential equations, partial differential equations. In other words, he must know most of the subject which an applied mathematician should know.

It is quite evident that the prospective engineering scientist cannot hope to cram all his learning into four years of college. In fact he has to be first trained in general engineering which may take three years in a good engineering school after high school, and then he has to spend approximately another three years to learn the science and the mathematics. Therefore it takes at least six years after high school to train an engineering scientist while the general practice now to train the conventional requires only four years. Engineering scientists are then definitely specialists who form only a few percent among the total personnel engaged in engineering and industry and which has to be trained from persons having the particular talent and inclination.

However as yet, only the necessary tools of the engineering scientists has been discussed. The fact that he is given these tools does not necessarily mean that he is trained in using these tools. How can he be trained to use these tools? Here the degree of training cannot be measured in the number of courses the student takes or the number of years he spent in the school. Learning how to use the tools effectively can be only achieved by experiencing the use of tools. Of course, the process can be accelerated with the aid of expert guidance. Therefore to complete the training of an engineering scientist after six years of schooling, the prospective specialist must spend one to two years working on a specific problem under the supervision of an experienced senior man. A good way of realizing this would be to study for a doctor's degree with an authoritative instructor in a well-equipped university. The unhurried academic atmosphere in an educational institution is certainly conductive to thinking which is, after all, the only way to gain wisdom. Wisdom gives insight to a complex problem and insight to a problem is the key to its successful solution.

Concluding Remarks

The training of a competent engineering scientist is a long process of seven to eight years, and the effort and ability required to complete such training is also proportionately great. Fortunately the reward is also large. From the discussion on the character of work performed by engineering scientist, it can be seen that they form the nucleus of men in any engineering development, they are the pioneers of new frontiers in industry. In fact the very essence of engineering science — the technique of transforming a basic scientific truth to practical means of human welfare really goes beyond the realm of present industry. Medicine is the application of chemistry, physics and physiology to cure and to prevent diseases. Agriculture is the application of chemistry, physics and plant physiology to produce food. Both are then engineering in the broad sense of the word and both will then benefit from the methods of engineering science. Hence it is appropriate to call the engineering scientists the most immediate and direct workers towards the goal of the pursuit of science, so aptly expressed by Professor Harold C. Urey: "We wish to abolish drudgery, discomfort, and want from the lives of men and bring them pleasure, leisure and beauty".

References

[1] Seifert J S, Mills M M, Summerfield M. The Physics of Rockets. American Journal of Physics. 1947, 15: 1-21, 121-140, 255-272.

[2] Smyth H D. Atomic Energy for Military Purposes. Princeton University Press, 1945.

[3] Armann O H, Th. von Kármán, Woodruff G B. The Failure of the Tacoma Narrows Bridge. Report to Federal Works Agency, March 28, 1941.

On Two-Dimensional Non-steady Motion of a Slender Body in a Compressible Fluid

C. C. Lin, E. Reissner and H. S. Tsien

1. Introduction. Many problems of aerodynamics involve flow around a slender body. It is known that in such cases a perturbation approach often leads to linear differential equations. Recently, von Karman[1] pointed out that for steady motions in the transonic range, certain non-linear terms have to be retained even in a perturbation theory. This fact was also discussed by C. Kaplan[2] together with a comparison with results of his earlier iteration calculation. Tsien[3] has carried the same point of view to hypersonic flow. Quite recently, I. E. Garrick[4], during his discussion of flutter in the transonic range, wrote down an equation for the non-steady disturbance potential which contains the same non-linear term as the equation derived by von Karman. Garrick, however, gave no discussion of the size of the different terms for different flow conditions and of the appropriate similarity laws. In this article, we shall present a simple unified treatment of the unsteady two-dimensional polytropic potential flow around a slender body. We shall give for different situations a systematic estimate of the size of the various terms in the differential equation for the velocity potential. This then includes all the known special cases and yields others not discussed by previous authors.

The general conclusions are as follows. Let δ be the larger of the two quantities: the thickness ratio of the body and the ratio of the amplitude of lateral motion of the body and its chord. Let M_∞ be the Mach number at infinity, and let $1/k$ be the characteristic period of time for transient motions; e.g., in the case of oscillating wings, k would be the dimensionless ratio of the product of the angular frequency and the chord length to the free stream velocity. These are the three important parameters.

A. If M_∞ is not large, there are the following possibilities:

If $|1-M_\infty| = O(\delta^{2/3})$ and $k = O(\delta^{2/3})$ (or smaller) we are in the transonic range, and the problem is non-linear. In this case, if $k = o(\delta^{2/3})$, the problem is quasi-stationary①. Notice that the comparison between k and $|1-M_\infty|$ does not enter the problem.

If either or both of the above conditions are not satisfied, that is, if $|1-M_\infty|$ or/and $k \gg \delta^{2/3}$ the general linearized theory holds, even though M_∞ may become unity. In such cases,

---

Received March 1, 1948; Revised July 26. 1948.

Journal of Mathematics and Physics, vol. 27, pp. 220 - 231, 1948.

① The notation $y = 0(x)$ and $y = o(x)$ are used in the following conventional sense. If $y = 0(x)$, $\lim_{x\to 0} y/x$ is bounded or zero. If $y = o(x)$, $\lim_{x\to 0} y/x = 0$.

if $k = o(|1 - M_\infty|)$ the problem is quasi-stationary.

B. If $M_\infty \gg 1$, there are the following possibilities:

If $M_\infty \ll 1/\delta$ and $1/(k\delta)$, the general linearized theory holds.

If either or both of the above conditions are not satisfied, the equations are non-linear and in the first of the two cases the flow is hypersonic. Moreover, we then have no longer a perturbation theory in the sense that the pressure perturbation is not small.

The method of attack is to introduce a parameter for each type of quantity to characterize its size. In this way, one can distinguish small quantities of various orders. One then neglects those quantities which are definitely of higher orders in comparison with other terms in the same equation.

To fix our ideas, we shall consider the periodic oscillations of a two-dimensional thin airfoil at small angles of attack. The scheme is obviously capable of including more general cases.

2. General formulation of the problem. The basic equations for unsteady two-dimensional polytropic potential motions are (in dimensionless form),

$$\left.\begin{aligned} &\phi_{tt} + 2\phi_x\phi_{xt} + 2\phi_y\phi_{yt} + \phi_x^2\phi_{xx} + 2\phi_x\phi_y\phi_{xy} + \phi_y^2\phi_{yy} = c^2(\phi_{xx} + \phi_{yy}) \\ &\phi_t + \frac{1}{2}(\phi_x^2 + \phi_y^2) + \frac{c^2}{\gamma - 1} = \frac{1}{2} + \frac{1}{(\gamma - 1)M_1^2} \end{aligned}\right\} \tag{1}$$

where ϕ is the (dimensionless) velocity potential, c is the (dimensionless) velocity of sound, and M_1 is the Mach number of the reference velocity U_1 which may be taken to be either the free stream velocity U_∞ or the critical velocity c_*. The reference length is the chord length b. The non-dimensional time variable t is constructed by multiplying the actual time by U_1 and dividing by b. On the airfoil

$$y = h(x, t) \tag{2}$$

the boundary condition

$$h_t + \phi_x h_x = \phi_y \tag{3}$$

must be satisfied. If the velocity potential is written as

$$\phi = x + \phi' \tag{4}$$

then, at infinity, we have the boundary conditions

$$1 + \phi'_x = \frac{U_\infty}{U_1} = \left\{1 + \frac{1}{(\gamma - 1)M_1^2}\right\}^{\frac{1}{2}} \Big/ \left\{1 + \frac{1}{(\gamma - 1)M_\infty^2}\right\}^{\frac{1}{2}}, \quad \phi'_x = 0 \tag{5}$$

If the reference velocity is taken to be the critical velocity c_* then

$$M_1 = 1 \tag{6}$$

If it is taken to be the free stream velocity then

$$M_1 = M_\infty \tag{7}$$

In the latter case,

$$\phi'_x = 0, \phi'_y = 0, \text{ at infinity} \tag{8}$$

We shall make the latter choice because it enables one to cover arbitrary free-stream Mach number. The former choice imposes an additional restriction that when the magnitude of ϕ'_x at infinity is small compared with unity it must at the same time be of the order of $|1-M_\infty|$. It is therefore limited to the transonic case. In fact, this additional restriction may be taken to be the criterion for transonic flow. Besides this additional restriction, and certain unessential changes in formulae, the results are not affected otherwise. Thus, it is very easy to include existing discussions of transonic flow in the present development. This will be carried through in some detail when the transonic case is brought up.

3. Introduction of parameters. Let us now introduce dimensionless quantities of the order of magnitude unity:

$$\left.\begin{aligned} &\xi = x, \eta = \lambda y, \tau = kt, \\ &g(\xi,\tau) = \delta^{-1}h(x,t), f(\xi,\eta,\tau) = \varepsilon^{-1}\phi'(x, y, t) \end{aligned}\right\} \tag{9}$$

where λ, k, δ, ε are parameters, which shall specify the orders of magnitude of the various quantities. In particular

$$k = \omega b/U_1 \tag{10}$$

is the dimensionless angular frequency, and δ is the thickness ratio of the airfoil.

The components of velocity and the velocity of sound are

$$\begin{aligned} &u = 1 + \varepsilon f_\xi, v = \varepsilon\lambda f_\eta \\ &c^2 = \frac{1}{M_1^2} - (\gamma - 1)\varepsilon\left[kf_\tau + f_\xi + \frac{1}{2}\varepsilon(f_\xi^2 + \lambda^2 f_\eta^2)\right] \end{aligned} \tag{11}$$

and the pressure coefficient is

$$C_p = \frac{p - p_1}{\frac{1}{2}\rho_1 U_1^2} = \frac{2}{\gamma M_1^2}\left\{\left[1 - (\gamma - 1)\varepsilon M_1^2\left(kf_\tau + f_\xi + \frac{1}{2}\varepsilon(f_\xi^2 + \lambda^2 f_\eta^2)\right)\right]^{\gamma/(\gamma-1)} - 1\right\} \tag{12}$$

4. Theory of small perturbations. The assumption of *small* disturbances demands that the departure of the velocity components *and* the pressure from the reference conditions is small. Thus, we have the conditions

$$\varepsilon \ll 1, \quad \varepsilon\lambda \ll 1 \tag{13}$$

from the expressions (11) for the velocity components, and the conditions

$$\varepsilon M_1^2 \ll 1, \quad (\varepsilon\lambda M_1)^2 \ll 1, \quad \varepsilon k M_1^2 \ll 1 \tag{13A}$$

from the expression (12) for the pressure coefficient.

The complete equation (1) is

$$k^2 f_{\tau\tau} + 2(1+\varepsilon f_\xi)kf_{\xi\tau} + 2(\varepsilon\lambda^2)kf_\eta f_{\eta\tau} + (1+2\varepsilon f_\xi + \varepsilon^2 f_\xi^2)f_{\xi\xi} + 2(\varepsilon\lambda^2)(1+\varepsilon f_\xi)f_{\xi\eta} +$$
$$(\varepsilon\lambda^2)^2 f_\eta^2 f_{\eta\eta} = \left\{\frac{1}{M_1^2} - (\gamma-1)\varepsilon\left[kf_\tau + f_\xi + \frac{1}{2}\varepsilon(f_\xi^2 + \lambda^2 f_\eta^2)\right]\right\}(f_{\xi\xi} + \lambda^2 f_{\eta\eta}) \tag{14}$$

Under the condition (13) of small disturbances, without any knowledge of the relative sizes of the parameters, the above equation simplifies to

$$k^2 f_{\tau\tau} + 2kf_{\xi\tau} + \left[1 - \frac{1}{M_1^2} + \varepsilon(\gamma+1)f_\xi\right]f_{\xi\xi} - \frac{\lambda^2}{M_1^2}f_{\eta\eta} = 0 \tag{15}$$

In this equation the orders of magnitude of the different terms are not necessarily the same and this allows further simplification of the equation, to be discussed subsequently. However all terms in (14) that are neglected are definitely small compared with some term retained in (15). For instance, the term $k\varepsilon(\gamma-1)f_\tau f_{\xi\xi}$ is small compared with $2kf_{\xi\tau}$ and is thus neglected.

Boundary conditions. The boundary condition (3) is

$$\delta[kg_\tau + (1+\varepsilon f_\xi)g_\xi] = \varepsilon\lambda f_\eta; \quad \eta = 0, \quad 0 < \xi < 1$$

which simplifies to

$$\eta = 0, \quad 0 < \xi < 1; \quad \varepsilon\lambda f_\eta = \delta[kg_\tau + g_\xi] \tag{16}$$

Equation (16) implies that the larger of the two quantities $k\delta$ and δ must be of the order of $\varepsilon\lambda$. Note that the possibility of satisfying the boundary condition at $\eta = 0$ rather than at the actual boundary depends on the fact that $\eta \ll 1$ at the actual boundary in all cases, as can be verified d posteriori. Besides this fact we must assume that the solution is regular near $\eta = 0$ in such a way that this simplification is permitted.

The boundary conditions at infinity do not impose any restriction when taken in the form (8). That is, when $M_1 = M_\infty$, the conditions at infinity are

$$f_\xi(-\infty, \eta) = f_\eta(-\infty, \eta) = 0 \tag{17}$$

The formula (12) for the pressure coefficient becomes

$$C_p = 2\varepsilon\left[kf_\tau + f_\xi + \frac{1}{2}\varepsilon\lambda^2 f_\eta^2\right] \tag{18}$$

Equation (15)-(18) are the basic equations for the small perturbation theory of a two-dimensional potential motion. Additional simplifications may be carried out for special cases. In fact, all the known perturbation theories are obtainable from them by such specialisations, as will be shown in the following sections.

Similarity of disturbance prevails when two cases have identical equations and boundary conditions.

5. Further simplification in special cases. We shall carry out some further reduction in the individual cases to bring out the important parameters involved. We shall take $M_1 = M_\infty$ in our discussions.

Our first step is the elimination of the parameter λ, as follows. We distinguish two cases by considering the boundary condition (16). We set

$$(A)\ \lambda = \delta/\varepsilon \quad \text{when } k = O(1) \tag{19}$$

$$(B)\ \lambda = \delta k/\varepsilon \quad \text{when } 1 \ll k \tag{20}$$

In the first case, (15), (16), and (19) give

$$f_{\eta\eta} = \frac{\varepsilon^2 M_\infty^2 k^2}{\delta^2} f_{\tau\tau} + \frac{2k\varepsilon^2 M_\infty^2}{\delta^2} f_{\xi\tau} + \frac{\varepsilon^2}{\delta^2}(M_\infty^2 - 1) f_{\xi\xi} + (\gamma+1)\frac{\varepsilon^2}{\delta^2} M_\infty^2 f_\xi f_{\xi\xi} \tag{21}$$

$$\eta = 0, \quad 0 < \xi < 1; \quad f_\eta = g_\xi + k g_\tau \tag{22}$$

and

$$C_p = 2\varepsilon\left[k f_\tau + f_\xi + \frac{1}{2}\frac{\delta^2}{\varepsilon} f_\eta^2\right] \tag{23}$$

The conditions (13) for small disturbances become

$$\varepsilon \ll 1, \quad \delta \ll 1, \quad \varepsilon M_\infty^2 \ll 1, \quad M_\infty \delta \ll 1 \tag{24}$$

In the second case, we have

$$f_{\eta\eta} = \frac{\varepsilon^2 M_\infty^2}{\delta^2} f_{\tau\tau} - \frac{\varepsilon^2}{k^2\delta^2} f_{\xi\xi} \tag{25}$$

$$\eta = 0, \quad 0 < \xi < 1; \quad f_\eta = g_\tau \tag{26}$$

$$C_p = 2\varepsilon k f_\tau + k^2\delta^2 f_\eta^2 \tag{27}$$

The conditions (13) become

$$\varepsilon \ll 1, \quad k\delta \ll 1, \quad k\delta M_\infty \ll 1, \quad \varepsilon k M_\infty^2 \ll 1 \tag{28}$$

To have a meaningful boundary value problem, we must retain the term with $f_{\eta\eta}$ in equations (21) and (25). That is why these equations are solved explicitly for $f_{\eta\eta}$. Further simplification depends on the order of magnitude of the coefficients. These will then appear as certain conditions on the parameters δ, k, and M_∞ of the problem:

Class A, $k = O(1)$. In this class of problems, there are the four parameters

$$\left.\begin{aligned} c_1 &= (\gamma+1)\frac{\varepsilon^3}{\delta^2} M_\infty^2, & c_2 &= \frac{\varepsilon^2}{\delta^2}(M_\infty^2 - 1), \\ c_3 &= \frac{2k\varepsilon^2 M_\infty^2}{\delta^2}, & c_4 &= \frac{\varepsilon^2 k^2 M_\infty^2}{\delta^2} \end{aligned}\right\} \tag{29}$$

But only the order of magnitude of three of them need be discussed, since $c_4 = O(c_3)$. The various cases may be distinguished further as follows:

$$\text{Case 1. } c_1 = 1, \quad c_2 = O(1), \quad c_3 = O(1) \tag{30}$$

$$\text{Case 2. } c_1 \ll 1, \quad |c_2| = 1, \quad c_3 = O(1) \tag{31}$$

$$\text{Case 3. } c_1 \ll 1, \quad |c_2| \ll 1, \quad c_3 = 1 \tag{32}$$

That is, c_1, c_2 and c_3 are successively taken to be the principal parameter. Notice that in Case 2 we impose the restriction $c_1 \ll 1$, rather than let $c_1 = 0(1)$, because the case $c_1 \sim 1$ is essentially included in Case 1. A similar remark applies to Case 3.

Let us work out Case 1 in some detail and then summarize the results. The other cases can be treated in a very similar manner. We have, from (29) and (30),

$$\left.\begin{aligned} \varepsilon &= (\gamma+1)^{-1/3}(\delta/M_\infty)^{2/3} \\ c_2 &= (M_\infty^2-1)[(\gamma+1)M_\infty\delta]^{-2/3} = O(1) \\ c_3 &= 2kM_\infty^2[(\gamma+1)M_\infty\delta]^{-2/3} = O(1) \\ c_4 &= k^2M_\infty^2[(\gamma+1)M_\infty\delta]^{-2/3} \end{aligned}\right\} \tag{33}$$

Thus, it is clear that one must have

$$M_\infty^2 - 1 = O[(\gamma+1)M_\infty\delta]^{2/3}, \quad k = O[(\gamma+1)M_\infty\delta]^{2/3}$$

or, since $M_\infty \sim 1$,

$$M_\infty - 1 = O(\delta^{2/3}), \quad k = O(\delta^{2/3}) \tag{34}$$

If one checks back with the condition (24), one sees that they are all satisfied when $\delta \ll 1$. The similarity parameters are essentially c_2 and c_3, because c_4 is negligible in view of (34).

The result may be summarized as follows:

Case 1. Transonic case. When $\delta \ll 1$ and the conditions (34) are satisfied, we have the transonic case with always a non-linear equation. The fundamental equations are

$$\left.\begin{aligned} &\frac{2k}{[(\gamma+1)\delta]^{2/3}} f_{\xi\tau} + \left\{\frac{M_\infty - 1}{[(\gamma+1)\delta]^{2/3}} + f_\xi\right\} f_{\xi\xi} - f_{\eta\eta} = 0 \\ &\eta = 0, \quad 0 < \xi < 1; \quad f_\eta = g_\xi \\ &C_p = 2\delta^{2/3}(\gamma+1)^{-1/3} f_\xi \end{aligned}\right\} \tag{35}$$

In this case, there are the two parameters

$$\frac{k}{[(\gamma+1)\delta]^{2/3}} \quad \text{and} \quad \frac{M_\infty - 1}{[(\gamma+1)\delta]^{2/3}} \tag{36}$$

If either or both of them are large compared with unity, we pass out of the present case. When $k \ll \delta^{2/3}$, the problem is quasi-stationary.

Case 2. Subsonic or supersonic case. The condition $|c_2| = 1$ leads to the following relation for the order of magnitude of the velocity disturbance

$$\varepsilon = \delta\,|1 - M_\infty^2|^{-\frac{1}{2}} \tag{37}$$

The conditions (24) and (31) become

$$\delta \ll 1, \quad \delta M_\infty \ll 1$$

and

$$k = O(1 - M_\infty^{-2}), \quad |1 - M_\infty^2| \gg \delta^{2/3} \tag{38}$$

When these are satisfied, we have the usual cases of supersonic or subsonic flow.

The fundamental equations are

$$\left.\begin{aligned}&k^2 f_{\tau\tau} + 2k f_{\xi\tau} + \left(1 - \frac{1}{M_\infty^2}\right)\left(f_{\xi\xi} + \frac{1 - M_\infty^2}{|1 - M_\infty^2|} f_{\eta\eta}\right) = 0\\&\eta = 0, \quad 0 < \xi < 1; \quad f_\eta = g_\xi + k g_\tau\\&C_p = 2\delta |1 - M_\infty^2|^{-1/2}(f_\xi + k f_\tau)\end{aligned}\right\} \tag{39}$$

The two parameters occurring in the differential equation are k and $1 - M_\infty^{-2}$. The pressure coefficient is further proportional to δ. When $k = 0$, the problem is stationary; when $k \ll |1 - M_\infty^2|$ the problem is quasi-stationary.

Case 3. Highly oscillatory case. The condition $c_3 = 1$ leads to the relation

$$\varepsilon = \delta / \sqrt{2kM_\infty} \tag{40}$$

The conditions (24) and (32) become

$$\delta \ll 1, \quad \delta M_\infty \ll 1$$

and

$$k \gg \delta^{2/3}, \quad k \gg \left|1 - \frac{1}{M_\infty^2}\right| \tag{41}$$

This is the case where the time rate of change predominates. The fundamental equations are

$$\left.\begin{aligned}&k f_{\tau\tau} + 2(f_{\xi\tau} - f_{\eta\eta}) = 0\\&\eta = 0, \quad 0 < \xi < 1; \quad f_\eta = g_\xi + k g_\tau\\&C_p = (2\delta / \sqrt{2kM_\infty})(f_\xi + k f_\tau)\end{aligned}\right\} \tag{42}$$

When $k \ll 1$, the terms in k in (42) can be omitted.

Class B, $k \gg 1$. In this class of problems, we have the two parameters

$$c_4 = \varepsilon^2 M_\infty^2 / \delta^2, \quad c_5 = \varepsilon^2 / k^2 \delta^2 \tag{43}$$

in Eq. (24). Consequently, there are two cases:

Case 4. $$c_4 = 1, \quad c_5 = O(1) \tag{44}$$

Case 5. $$c_4 \ll 1, \quad c_5 = 1 \tag{45}$$

Case 4. Highly oscillatory case. The condition $c_4 = 1$ leads to

$$\varepsilon = \delta / M_\infty$$

The other condition in (44) and the condition (28) lead to

$$1/k = O(M_\infty), \quad k\delta M_\infty \ll 1 \tag{46}$$

The fundamental equations are

$$\begin{aligned}&f_{\tau\tau} - f_{\eta\eta} - (1/k^2 M_\infty^2) f_{\xi\xi} = 0\\&\eta = 0, \quad 0 < \xi < 1; \quad f_\eta = g_\tau\end{aligned} \tag{47}$$

$$C_p = 2(\delta k / M_\infty) f_\tau$$

There is the only parameter kM_∞.

Case 5. Quasi-stationary nearly incompressible flow.

The condition $c_5 = 1$ leads to

$$\varepsilon = k\delta \tag{48}$$

The other condition in (42) and the condition (27) lead to

$$kM_\infty \ll 1; \quad k\delta \ll 1 \tag{49}$$

The fundamental equations are

$$\left.\begin{aligned} & f_{\xi\xi} + f_{\eta\eta} = 0 \\ & \eta = 0, \quad 0 < \xi < 1; \quad f_\eta = g_\tau \\ & C_p = 2\delta k^2 f_\tau \end{aligned}\right\} \tag{50}$$

The above results are summarized in Table 1.

For convenience of the discussion, the various items are numbered on the left. Item 1 gives the characterizing conditions in terms of the important parameters M_∞, k and δ. The equations which hold in the various cases are indicated in item 2. Item 3 gives the type of formula for the pressure coefficient. The functions F are not the same. Their arguments give the similarity parameters in the individual cases. In item 4, the physical nature of the various cases is described. Finally, it is indicated that, only in one case, is the theory non-linear; i.e., when

$$|1 - M_\infty| = O(\delta^{2/3}) \quad \text{and} \quad k = O(\delta^{2/3}) \tag{51}$$

But it should be emphasized that in this transonic case, the nonsteady effects appear even at relatively low reduced frequencies. Take for instance the case for which the thickness ratio and the amplitude ratio are of the same order of magnitude and where the value for thickness ratio is 5% or $\delta = 0.05$. Then non-steady effects are already important when $k \approx (0.05)^{2/3}$ according to (51). Consider particularly the case of harmonic oscillation, the period T of the corresponding motion is then

$$T = 2\pi b/(kU) = 46.5\, b/U$$

By taking $b = 5$ ft and $U = 1\,000$ ft/sec, $T = 0.237$ second. This is a relatively slow oscillation. In subsonic or supersonic flow, the non-steady effects of such slow oscillation are negligible as shown by the conditions of case 2 in Table I. But for transonic flow, the non-steady effects are not negligible. This fact must be kept in mind when the stability of aircraft in transonic flight is considered, as it is likely that a quasi-stationary calculation will not be sufficiently accurate.

Linearization. If either of the conditions (51) is not satisfied, the equation is linear. Note also that the boundary conditions and the formula for the pressure coefficient are always linear. Thus, if one includes all the linear terms in the equation, one does not have to distinguish between the cases 2 to 5. To be sure, some of the terms are negligible in one case and some in other. But it is clear that the complete linearized equations can be applied whenever either or both of the conditions (51) is not satisfied, since we are retaining all the necessary terms. The system of linear equations is the following,

$$\left.\begin{aligned} &\phi'_{tt}+2\phi'_{xt}+(1-M_1^{-2})\phi'_{xx}-M_1^{-2}\phi'_{yy}=0\\ &y=0,\quad 0<x<b;\quad \phi'_y=h_x+h_t\\ &C_p=2(\phi'_x+\phi'_t)\end{aligned}\right\}\tag{52}$$

Table 1 Classification of various cases. Arguments in function for pressure coefficient are the similarity parameters

<table>
<tr><td></td><td colspan="3">$k=O(1)$</td><td colspan="2">$1\ll k$</td></tr>
<tr><td></td><td>Case 1</td><td>Case 2</td><td>Case 3</td><td>Case 4</td><td>Case 5</td></tr>
<tr><td>1</td><td>$1-\frac{1}{M_\infty^2}=O(\delta^{2/3})$
$k=O(\delta^{2/3})$
$\delta\ll 1$</td><td>$\lvert 1-M_\infty\rvert\gg\delta^{2/3}$
$k=O\left(1-\frac{1}{M_\infty^2}\right)$
$\delta\ll 1,\ \delta M_\infty\ll 1$</td><td>$k\gg\delta^{2/3}$
$k\gg\left\lvert 1-\frac{1}{M_\infty^2}\right\rvert$
$\delta\ll 1,\ \delta M_\infty\ll 1$</td><td>$k^{-1}=O(M_\infty)$
$k\delta M_\infty\ll 1$</td><td>$kM_\infty\ll 1$
$k\delta\ll 1$</td></tr>
<tr><td>2</td><td>Eq. (35)</td><td>Eq. (39)</td><td>Eq. (42)</td><td>Eq. (47)</td><td>Eq. (50)</td></tr>
<tr><td>3</td><td>$C_p/[\delta^{2/3}(\gamma+1)^{1/3}]$
$=F\left[\frac{k}{[(\gamma+1)\delta]^{2/3}},\frac{M_\infty-1}{[(\gamma+1)\delta]^{2/3}}\right]$</td><td>$C_p\ \sqrt{\lvert 1-M_\infty^2\rvert}/\delta$
$=F\left(k,1-\frac{1}{M_\infty^2}\right)$</td><td>$\sqrt{M_\infty}C_p/\delta=F(k)$</td><td>$C_p/(\delta k^3)=$
$F(kM_\infty)$</td><td>$C_p/(\delta k^2)=$
$F(kM_\infty)$</td></tr>
<tr><td>4</td><td>transonic</td><td>subsonic
supersonic</td><td colspan="2">oscillating terms dominating</td><td>quasi-stationary
incompressible</td></tr>
<tr><td>5</td><td>non-linear</td><td colspan="4">linear</td></tr>
</table>

For example, when an airfoil is accelerated through the sonic speed with a sufficiently high acceleration, k can be made to be much larger than $\delta^{2/3}$ and the linear theory applies. The physical interpretation is the following. The breakdown of the linearization is due to the accumulation of disturbances in a steady flow at transonic speeds. On the other hand, if there is a high enough rate of time variation, such disturbances have no chance to build up, and the linearized treatment is justified.

6. Influence of wind tunnel walls. If the airfoil is placed in a two-dimensional wind tunnel, there are further conditions of vanishing vertical velocity at the walls $y=H_1$ and $y=H_2$ (reduced to non-dimensional form by dividing by the chord length). In order that one may have similarity, there are therefore the additional conditions of identical

$$\eta_1=\lambda H_1,\quad\text{and}\quad \eta_2=\lambda H_2$$

Referring to Eqs. (18) and (19) and the value of ε in the individual cases, one can easily show that the parameters for the five cases are respectively

$$\begin{array}{lll} \text{Case 1.} & [(\gamma+1)\delta]^{1/3}H_1 & [(\gamma+1)\delta]^{1/3}H_2 \\ \text{Case 2.} & |1-M_\infty^2|^{\frac{1}{2}}H_1 & |1-M_\infty^2|^{\frac{1}{2}}H_2 \\ \text{Case 3.} & \sqrt{2kM_\infty}H_1 & \sqrt{2kM_\infty}H_2 \\ \text{Case 4.} & kM_\infty H_1 & kM_\infty H_2 \\ \text{Case 5.} & H_1 & H_2 \end{array} \tag{53}$$

7. The Case of Large Pressure Disturbance. In all the above cases, the restrictions

$$M_\infty \ll 1/\delta \quad \text{and} \quad M_\infty \ll 1/(k\delta) \quad (k \gg 1) \tag{54}$$

are implied. Thus, if either or both of these conditions are not satisfied, the linearized theory breaks down. This can happen when $M_\infty \gg 1$, and is indeed the hypersonic range with non-linear equations. In such cases, even though the velocity perturbations may be small (i. e., condition (13) satisfied), the conditions (13A) for small pressure perturbation cannot be all satisfied. (For if they are, we will go back to one of the linearized cases discussed above); that is, this case cannot be treated properly by a theory of small perturbations. Thus, the expansion of the formula (12) for the pressure coefficient cannot be carried out. One can merely write

$$C_p = \frac{2}{\gamma M_1^2}\left\{\left[1-(\gamma-1)\varepsilon M_1^2\left(kf_\tau + f_\xi + \frac{1}{2}\varepsilon\lambda^2 f_\eta^2\right)\right]^{\gamma/(\gamma-1)} - 1\right\} \tag{55}$$

Indeed, the reduction of the fundamental equation (14) is very limited. One can show that it simplifies to

$$k^2 f_{\tau\tau} + 2kf_{\xi\tau} + f_{\xi\xi} = \varepsilon\lambda^2\left\{\left[\frac{1}{\varepsilon M_1^2} - (\gamma-1)(kf_\tau + f_\xi) - \frac{1}{2}(\gamma+1)\varepsilon\lambda^2 f_\eta^2\right]f_{\eta\eta} - 2f_\eta(kf_{\tau\eta} + f_{\xi\eta})\right\} \tag{56}$$

To clarify the problem further, we again distinguish between the cases

$$\text{(A) } k = O(1), \quad \text{(B) } k \gg 1$$

Case A. Hypersonic Flow. In this case, we have (cf. (19) and (54))

$$\varepsilon\lambda = \delta, \quad \text{and} \quad 1/(M_\infty\delta) = O(1) \tag{57}$$

It turns out that in terms of the parameters

$$\alpha = 1/(M_\infty\delta) = O(1) \quad \text{and} \quad \beta = \varepsilon\lambda^2 \tag{58}$$

one can rewrite (55) and (56) as

$$\begin{aligned} C_p &= \frac{2}{\gamma M_\infty^2}\left\{\left[1-(\gamma-1)\frac{\alpha^2}{\beta}\left(kf_\tau + f_\xi + \frac{\beta}{2}f_\eta^2\right)\right]^{\gamma/(\gamma-1)} - 1\right\} \\ k^2 f_{\tau\tau} + 2kf_{\xi\tau} + f_{\xi\xi} + 2\beta f_\eta(kf_{\eta\tau} + f_{\xi\eta}) &= \beta^2\left(\alpha^2 - \frac{1}{2}(\gamma+1)f_\eta^2\right) - (\gamma-1)\beta(kf_\tau + f_\xi) \end{aligned} \tag{59}$$

It is thus clear that β must be of the order of unity. Since it is at our disposal, one may put

$$\beta = 1 \tag{60}$$

Thus, the final equations for $k = O(1)$ are

$$\left.\begin{aligned}
&k^2 f_{\tau\tau} + 2k f_{\xi\tau} + f_{\xi\xi} + 2f_\eta (k f_{\eta\tau} + f_{\xi\eta}) = \left\{\left(\alpha^2 - \frac{1}{2}(\gamma+1) f_\eta^2\right) - (\gamma - 1)(k f_\tau + f_\xi)\right\} f_{\eta\eta} \\
&\eta = 0, \quad 0 < \xi < 1; \quad f_\eta = g_\xi + k g_\tau \\
&C_p = \frac{2}{\gamma M_\infty^2}\left\{\left[1 - (\gamma-1)\alpha\left(k f_\tau + f_\xi + \frac{1}{2} f_\eta^2\right)\right]^{\gamma/(\gamma-1)} - 1\right\}
\end{aligned}\right\} \tag{61}$$

Case B. Highly Oscillatory Flow. In the case $k \gg 1$, we have (cf. (20) and (54))

$$\varepsilon\lambda = k\delta, \quad \text{and} \quad 1/(M_\infty \delta k) = O(1) \tag{62}$$

A similar argument with

$$\alpha' = 1/(k\delta M_\infty) = O(1), \quad \beta' = \varepsilon\lambda^2 k^{-1} = 1 \tag{63}$$

leads to

$$\left.\begin{aligned}
&f_{\tau\tau} + 2 f_\eta f_{\eta\tau} = \left\{\alpha'^2 - \frac{1}{2}(\gamma+1) f_\eta^2 - (\gamma - 1) f_\tau\right\} f_{\eta\eta} \\
&\eta = 0, \quad 0 < \xi < 1, \quad f_\eta = g_\xi \\
&C_p = \frac{2}{\gamma M_\infty^2}\left\{\left[1 - (\gamma-1)\alpha'\left(f_\tau + \frac{1}{2} f_\eta^2\right)\right]^{\gamma/(\gamma-1)} - 1\right\}
\end{aligned}\right\} \tag{64}$$

Thus, when $k = 0(1)$, there are the similarity parameters

$$k, \alpha = 1/(M_\infty \delta), \quad \text{and} \quad \gamma$$

when $k \ll 1$, the problem is quasi-stationary, and the parameter k disappears. In the highly oscillating case, the similarity parameters are

$$\alpha' = 1/(M_\infty \delta k) \quad \text{and} \quad \gamma$$

8. The Transonic case with $M_l = 1$. If the reference velocity is taken to be the critical velocity, so that $M_1 = 1$, there is an apparent simplification, because the Mach number no longer appears explicitly in the equation (15). However, the boundary condition at infinity introduces it back again. In fact, at infinity,

$$\varepsilon f_\xi = 2(M_\infty - 1)/(\gamma + 1)$$

and this introduces a relation between the order of magnitude of ε and ($M_\infty - 1$). Arguments along the above lines lead to the following results.

$$\left.\begin{aligned}
&(2k/\delta^{2/3}) f_{\xi\tau} + f_\xi f_{\xi\xi} - f_{\eta\eta} = 0 \\
&\eta = 0, \quad 0 < \xi < 1; \quad f_\eta = g_\xi \\
&\xi = -\infty; \quad f_\xi = 2\frac{M_\infty - 1}{[(\gamma+1)\delta]^{2/3}}, \quad f_\eta = 0 \\
&C_p = 2\delta^{2/3}(\gamma+1)^{-1/3} f_\xi \\
&C_{p\infty} = \frac{p - p_\infty}{\frac{1}{2}\rho_\infty U_\infty^2} = 2\delta^{2/3} f_\xi - 4(1 - M_\infty)/(\gamma + 1)
\end{aligned}\right\} \tag{65}$$

and this is readily shown to be equivalent to the contents of (35).

Massachusetts Institute of Technology

References

[1] von Kármán Th. The Similarity Law of Transonic Flow. J. Math. Phys. 1947, 26: 182-190.

[2] Kaplan C. On Similarity Rules for Transonic Flows. NACA Tech. Note No. 1527, 1948: 16.

[3] Tsien H S. Similarity Laws of Hypersonic Flows. J. Math. Phys., 1946, 25: 247-251.

[4] Garrick I E. A Survey of Fluttor. Papers Presented at the NACA-University Conference in Aerodynamics, 1948, 291-292.

Wind-Tunnel Testing Problems in Superaerodynamics

Hsue-shen Tsien*
(*Massachusetts Institute of Technology*)

Summary

The problems in the experimentation of rarefied gas are discussed. First, the extremely large viscous effects in a wind-tunnel nozzle are shown. Then the difficulties of flow measurement are surveyed, pointing out particularly the unconventional behavior of the Pitot tube in rarefied gas. The performance of a hot-wire anemometer is then studied in some detail to show its feasibility. Finally, the rules for achieving complete flow similarity of rarefied gas flow are formulated.

Introduction

Wind tunnels are, perhaps, the most useful tool in aerodynamic investigations and certainly have contributed much in the modern development of fluid mechanics. It is thus natural, when one turns to a new field of aerodynamics, the aerodynamics of rarefied gases or superaerodynamics, that one should think of using the wind tunnel again. Only here it has to be adapted to entirely new circumstances, and many new problems, both in its design and in its operation, appear. It is the purpose of this paper to discuss some of these problems so as to gain an orientation in the new field of experimentation.

(1) Tunnel Design

To test models in the wind tunnel at its test section, it is of primary importance to obtain a uniform stream at the desired temperature, pressure, and velocity. For subsonic wind tunnels with ordinary pressures, this can be achieved without much difficulty. For supersonic wind tunnels at ordinary pressure, the expansion part of the tunnel before the test section or the nozzle is first designed to obtain a uniform stream at its exit without considering the effects of the viscosity of air. Then the boundary layer along the wall of the nozzle is calculated with the pressure gradient thus determined. Finally, the displacement thickness of the boundary layer, or the needed space for the boundary layer flow at lower velocity, is provided by making the nozzle larger than the dimensions first determined by the calculated amount[1]. This design

---

Presented at the Session on Aerodynamics of the Upper Atmosphere, Sixteenth Annual Meeting, I. A. S., New York, January 26 - 29, 1948.

Journal of the Aeronautical Sciences, vol. 15, pp. 573 - 580, 1948.

* Professor of Aerodynamics, Guggenheim Aeronautical Laboratory.

procedure is found to give satisfactory nozzles for supersonic wind tunnels.

However when one tries to apply the same design procedure to the superaerodynamics wind tunnel, one is immediately confronted with the difficulty of extremely large viscous effect. In other words, the boundary layer will be so thick as to occupy the main portion of the nozzle passage. To demonstrate this effect, let the length of the test section be L and the width of the square test section be b. Then the Reynolds Number based upon the conditions in the test section is $Re = UL/\nu$, where U is the velocity in the test section. If, as a rough estimate, we take the thickness of the boundary layer to be zero at the beginning of the test section and equal to a value δ calculated by the well-known Blasius formula for a flat plate at the end of the test section, then

$$\delta = 3.65\, L/\sqrt{Re} \tag{1}$$

Now, if this boundary layer actually occupies half the tunnel width $b/2$, then

$$\delta = b/2 = 3.65\, L/\sqrt{Re} \tag{2}$$

On the other hand, the ratio of the mean free path l and the boundary-layer thickness δ is known[2] to be

$$l/\delta = (1.255\sqrt{\gamma}/3.65)(M/\sqrt{Re}) \tag{3}$$

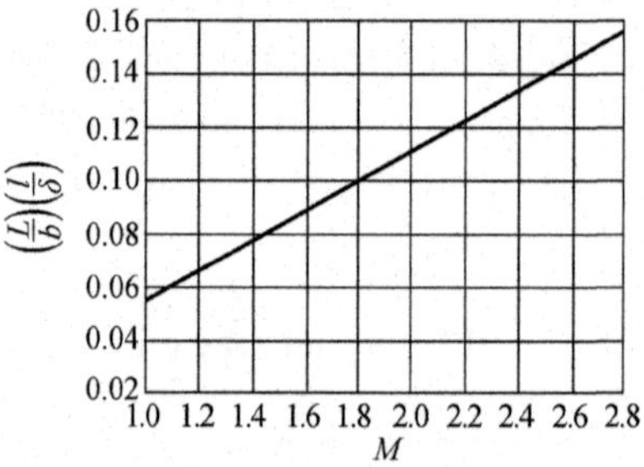

Fig. 1 l, mean free path; δ, boundary layer thickness at end of the test section of width b and length L; and M, Mach Number of the test section

where γ is the ratio of specific heats and can be taken as 1.4 and M is the Mach Number in the test section. By combining Eqs. (2) and (3), we have

$$(l/\delta)(L/b) = 0.055\,7M \tag{4}$$

This relation is shown in Fig. 1. Thus for a Mach Number M equal to 2 and $L/b = 2$, the boundary layer will completely fill up the test section, if the mean free path is equal to 5.6 per cent of the boundary-layer thickness or 2.8 per cent of the tunnel width. This means that the extremely strong viscous effect at low densities makes the ordinary concept of designing a wind tunnel totally inapplicable.

The extremely thick boundary layer where the velocity increases from a small value near the wall to some supersonic velocity at the center of the nozzle, also gives subsonic velocities in a rather large portion of the nozzle. Since pressure disturbances downstream can be transmitted upstream in subsonic flows, the flow in the test section of a low-pressure tunnel will be sensitive to changes in the diffuser even if the main stream velocity at the center of the nozzle is supersonic. This is, of course, a new phenomenon in the superaerodynamics tunnel not found in conventional supersonic wind tunnels.

The large viscous effects can also be demonstrated by calculating the ratio of frictional loss on the walls of the test section and the shock loss in the diffuser after the test section. Consider the diffuser to be a straight tube of approximately the same cross-sectional area as the test section, then the pressure loss due to friction, Δp_1, is

$$\begin{aligned}\Delta p_1 &= \text{frictional force}/b^2\\ &= (\rho U^2/2)4bLC_f(1/b^2)\end{aligned}$$

Taking C_f to be the Blasius value or $C_f = 1.328/\sqrt{Re}$, we have

$$\Delta p_1 = 2\rho U^2(L/b)(1.328/\sqrt{Re}) \tag{5}$$

Now, if p is the static pressure in the test section, then the pressure for ideal isentropic compression in the diffuser is

$$p\{1+[(\gamma-1)/2]M^2\}^{\gamma/(\gamma-1)}$$

If the actual pressure rise in the diffuser is estimated as that due to a normal shock without further recovery, then the actual pressure rise is

$$\{[2\gamma/(\gamma+1)]M^2-[(\gamma-1)/(\gamma+1)]\}p$$

Therefore, the pressure loss due to shock Δp_2 is

$$\Delta p_2 = \left[\left(1+\frac{\gamma-1}{2}M^2\right)^{\gamma/(\gamma-1)} - \left(\frac{2\gamma}{\gamma+1}M^2-\frac{\gamma-1}{\gamma+1}\right)\right]p \tag{6}$$

By combining Eqs. (5) and (6), the ratio of these two pressure losses is

$$\frac{\Delta p_1}{\Delta p_2} = \frac{2\gamma M^2(L/b)(1.328/\sqrt{Re})}{\left(1+\frac{\gamma-1}{2}M^2\right)^{\gamma/(\gamma-1)} - \left(\frac{2\gamma}{\gamma+1}M^2-\frac{\gamma-1}{\gamma+1}\right)} \tag{7}$$

Introducing the mean free path ratio given by Eq. (3), we have

$$\frac{\Delta p_1}{\Delta p_2} = \left(\frac{L}{b}\right)\left(\frac{l}{\delta}\right)\frac{6.528\gamma M}{\left(1+\frac{\gamma-1}{2}M^2\right)^{\gamma/(\gamma-1)} - \left(\frac{2\gamma}{\gamma+1}M^2-\frac{\gamma-1}{\gamma+1}\right)} \tag{8}$$

This relation is plotted in Fig. 2. Therefore, if the Mach Number M is 2 and $L/b = 2$ as before, then, when the ratio (l/δ) is 0.056, the ratio of frictional loss to shock loss is 0.628. Hence, the frictional loss and the shock loss are of the same order of magnitude.

These large viscous effects are fully confirmed by the recent tests on the 1 by 1-in, low-

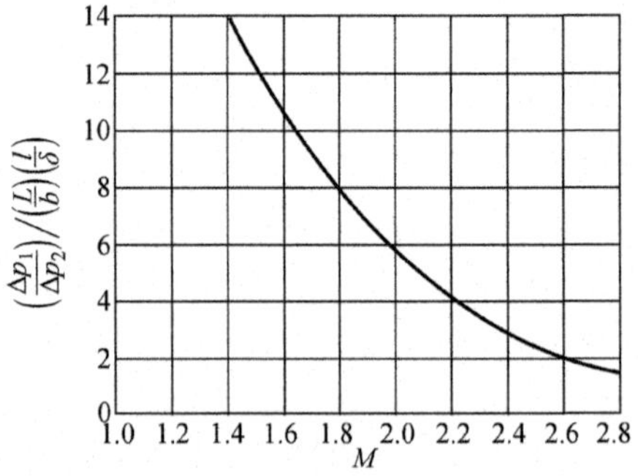

Fig. 2 l, mean free path; δ, boundary layer thickness at end of the test section of width b and length L; and M, Mach Number of the test section. Δp_1, pressure loss by friction; Δp_2, pressure loss by shock

pressure wind tunnel at the University of California.* The test nozzle (Fig. 3) was designed for Mach Number 4 without considering the viscous effect of the medium. During test, the static pressure on the wall at the exit of the nozzle is measured. This pressure is equal to 175 and 68 microns① for the two tests presented in Figs. 4 and 5. The apparent Mach Number M is the Mach Number calculated from the dynamic pressure measured by a Pitot tube by using the Rayleigh formula. Since there is the complication of large viscous effect in the Pitot tube reading as shown in the following section, this apparent Mach Number is only qualitative and cannot be taken as the exact value. However it is apparent from Figs. 4 and 5 that the boundary layer in the test section is indeed extremely thick and fills up the whole space. This large boundary-layer thickness makes the space available for the expansion of the central potential flow very small, if it exists at all. Therefore, the maximum Mach Number reached at the center of the nozzle is much smaller than the design Mach Number of 4. At the lower pressure, the influence of slip at the wall is also evident. This has the tendency to make the flow more uniform. However, the extremely low Mach Number at the test section indicates again the strong viscous effect in converting much of the pressure energy into heat energy.

These elementary calculations and preliminary test results make it clear that, for the design of the nozzle and test section for a superaerodynamics wind tunnel, it is no longer possible to separate the compressibility effect and the viscous effect. In fact, the concept of boundary layer is also of doubtful value because of the extremely small Reynolds Number encountered. Therefore, to design such a nozzle to obtain the nearest approximation to the ideal uniform flow, it will be necessary to use the exact Navier-Stokes equations instead of the approximate boundary-layer equations. Of course, it may be argued that for superaerodynamics the Navier-

* Experimental work done under contract with the Office of Naval Research. The author is deeply indebted to Profs. R. G. Folsom and E. D. Kane for permission to use their unpublished results.

① 1 000 μmHg = 1 mmHg; one atmosphere = 0.760×10^6 μmHg.

Fig. 3 Test section of University of California Low Pressure Wind Tunnel No. 2

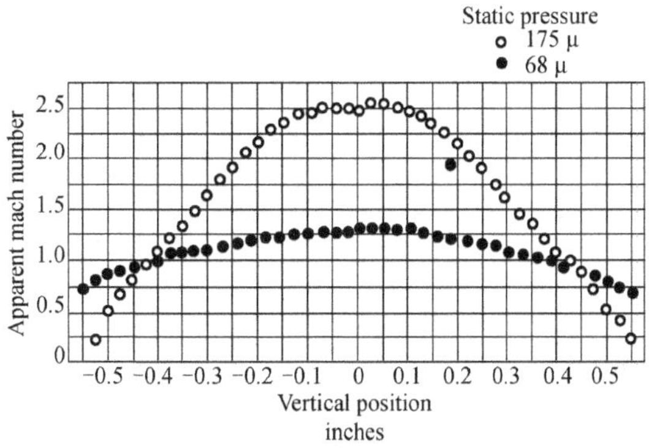

Fig. 4 Vertical velocity distribution in test section of University of California Low Pressure Wind Tunnel No. 2

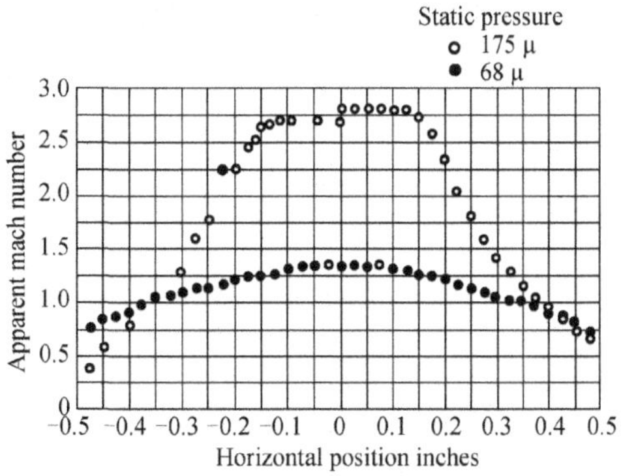

Fig. 5 Horizontal velocity distribution in test section of University of California Low Pressure Wind Tunnel No. 2

Stokes equations are no longer exact and that additional corrections must be added[2]. However, recent investigations by Schamberg[3] have shown that these additional corrections are small in the case of the slip flows concerned here and will not essentially alter the flow pattern. Hence, for a first approximation just like the nonviscous isentropic flow is a first approximation for ordinary supersonic nozzles, we can use the Navier-Stokes equations. The simplest case to be considered is certainly the axially symmetric nozzles. If x is the coordinate in the axial direction, r is the coordinate in the radial direction, and u and v are the corresponding velocity components, the equations are:

$$[\partial(\rho u)/\partial x] + (1/r)(\partial/\partial r)(\rho r v) = 0 \tag{9}$$

$$\rho(\mathrm{D}u/\mathrm{D}t) = -(\partial p/\partial x) + (\mathrm{Grad}\ \tau)_x \tag{10}$$

$$\rho(\mathrm{D}v/\mathrm{D}t) = -(\partial p/\partial_r) + (\mathrm{Grad}\ \tau)_r \tag{11}$$

$$\rho\frac{\mathrm{D}}{\mathrm{D}t}\left(\frac{u^2+v^2}{2}+c_pT\right) = \Phi - \{u(\mathrm{Grad}\ \tau)_x + v(\mathrm{Grad}\ \tau)_r\} + \frac{\partial}{\partial x}\left(\frac{\lambda\partial T}{\partial x}\right) + \frac{1}{r}\frac{\partial}{\partial r}\left(\lambda r\frac{\partial T}{\partial r}\right) \tag{12}$$

where

$$\mathrm{D}/\mathrm{D}t = u(\partial/\partial x) + v(\partial/\partial r)$$

$$\Phi = \text{dissipation function}$$

$$\tau = \text{stress tensor}$$

Eqs. (9), (10), (11), and (12), together with the equation of state,

$$p/\rho = RT \tag{13}$$

then determine the five unknowns u, v, ρ, p, and T. Of course, the actual process of making this calculation will be extremely tedious, and some approximate method of solution may have to be developed. One possibility would be to adapt the Kármán-Polhausen method for boundary layer to this case: We integrate the differential equations once with respect to r and thus only try to satisfy the equations "on the average" over the cross section of the nozzle. The "distribution" of u, v over the cross section will then be set in the form of a polynomial in r. Initial study by this procedure has already been made by Schaaf[4] at the suggestion of the author.

For ordinary supersonic diffusers, high efficiency of pressure recovery can generally be achieved by using a long diffuser. However, for a superaerodynamic wind tunnel, because of the extremely large loss through friction, long diffusers are undesirable. In fact, the pressure loss can be reduced by using the shortest possible diffuser.

(2) Flow Measurement

The quantities that determine the flow field are two of the three variables p, ρ, and T and the velocity components. The quantities p, ρ, T are related by the equation of state, and therefore only two are necessary for the determination of all three. Generally, for wind-tunnel work, the quantities actually measured are p, ρ, and q, the magnitude of the velocity.

For the measurement of pressure, a manometer is used. For ordinary pressures, one uses a fluid manometer filled with water, alcohol, or mercury. However for the extremely low pressure encountered in the superaerodynamic flow, some other form of manometer is necessary. One of the most successful types is the Pirani gage. The conventional form of Pirani gage has a pressure sensitivity of about 10^{-2} micron. It utilizes the change of temperature of a wire heated with constant energy caused by a change in the pressure of the gas surrounding it. The temperature change is measured by the change in the resistance of the wire. The wire is located in a small chamber which is connected to the point of measurement by a hole, flush with the gas stream if static pressure is to be measured. The question of best design of the connecting tube for quick response has been studied by Schaaf[5].

To measure the density ρ, the conventional method utilizes the difference in the velocity of light rays in mediums of different density. With different optical arrangements, we have the shadowgraph method, the schlieren method, and the interferometer method. However, if the density of the medium is low as in the case of superaerodynamic flow, the sensitivity of these methods becomes extremely poor. For instance in the case of the schlieren method, the percentage change in illumination I on passing through a region of thickness b is given by

$$\frac{\Delta I}{I} = k\frac{f}{e}\,0.000\,294\left(\frac{\rho}{\rho_0}\right)\left[\frac{b}{\rho}\frac{\Delta\rho}{\Delta n}\right] \tag{14}$$

where ρ_0 is the air density at 32°F. and 1 atmosphere pressure and where $\Delta\rho/\Delta n$ is the density gradient normal to the light ray. f and e are the focal length and the normal unobscured width of the light source image perpendicular to the knife edge. k is a factor of order 1, determined by the particular optical path used. Therefore the sensitivity of the schlieren method decreases in proportion to the factor (ρ/ρ_0). Some improvement can be made by altering the quantities f and e, but practical limitations and diffraction difficulties do not allow the increase of sensitivity to satisfactory values.

A new approach to this problem of density measurement is the method of absorption. It is found for instance, that oxygen at low pressures shows a strong absorption band at wave lengths around 1 470Å. or infrared light. * The percentage absorption is proportional to the number of molecules that meet the light ray and is, therefore, proportional to the density of the gas. The measurement is then similar to that of the interferometer method where the density is determined. A similar method is the utilization of the afterglow of nitrogen. These methods are now being studied by Evans[6].

The conventional method for the measurement of velocity is through the use of dynamic pressure rise in a Pitot tube. A straightforward application of this method is, however, difficult for rarefied gases, since the formula used is based upon the neglect of viscosity effects. But for rarefied gases the viscosity effect is of great importance, as pointed out in the previous section. Then the dynamic pressure would be different from that given by the usual formula. To estimate this effect, let us consider the case of low Mach Number so that compressibility effects can be neglected. Then, as a first approximation, take the flow field around the Pitot tube as

* Here "ultraviolet light" has been corrected as "intrared light". — Noted by editor.

that of a source of strength S in nonviscous flow of uniform velocity U. The "radius" of the tube a is

$$a = \sqrt{S/(\pi U)}$$

and the stagnation point is located at

$$R = \sqrt{S/(4\pi U)} = a/2 \tag{15}$$

The velocity introduced by the source is then

$$-U(1/4)(a^2/R^2)$$

By calculating the viscous stress from this approximate disturbance velocity, we have, for flow along the axis,

$$u\frac{\partial u}{\partial r} + \frac{1}{\rho}\frac{\partial p}{\partial r} = \nu U a^2 \frac{1}{2}\frac{1}{r^4} \tag{16}$$

Hence, if p_0 is the stagnation pressure and p° is the static pressure,

$$p_0 - p^\circ = \frac{1}{2}\rho U^2 + \mu U a^2 \frac{1}{2}\int_\infty^R \frac{dr}{r^4} = \frac{1}{2}\rho U^2 - \frac{1}{2}\mu U \frac{a^2}{R^3}$$

or

$$p_0 - p^\circ = \frac{1}{2}\rho U^2\left(1 - \frac{8}{9\sqrt{3}}\frac{\nu}{aU}\right) \tag{17}$$

For rarefied gases, the value of $\nu/(aU)$ or the reciprocal of the Reynolds Number of the Pitot tube could be of the order of unity. Then the dynamic pressure rise $p_0 - p^\circ$ is not the usual value $(1/2)\rho U^2$ but a value much less than that. In fact, previous investigations by Barker[7] and Homann[8] indicate that the Reynolds Number $a'U/\nu$, where a' is the radius of the mouth of tube, must exceed 30 in order to reach the usual dynamic pressure rise $(1/2)\rho U^2$.

Table 1 The functions F_1 and F_2 [cf. Eq. (21)]

U/c	F_1	F_2
0	1.772 45	0
0.2	1.737 51	0.070 20
0.4	1.638 80	0.272 69
0.6	1.492 48	0.585 60
0.8	1.320 21	0.978 43
1.0	1.143 28	1.420 53
1.2	0.978 25	1.885 55
1.4	0.834 80	2.354 92
1.6	0.716 28	2.818 12
1.8	0.621 53	3.271 17
2.0	0.546 83	3.713 56
2.2	0.487 90	4.146 72

(continued)

U/c	F_1	F_2
2.4	0.440 90	4.573 00
2.6	0.402 68	4.993 95
2.8	0.371 00	5.411 17
3.0	0.344 20	5.874 98

When the velocity of flow is high, we have an added complication due to the shock. The conventional Rayleigh formula for Pitot tubes in supersonic flow is based upon the assumption of a thin shock wave ahead of the Pitot tube. Now the thickness of the shock is proportional to the mean free path. Hence, in rarefied flows, the thickness of the shock will be so increased as to cause interference with flow in the neighborhood of the Pitot tube. This, together with the viscous effect mentioned in the previous paragraph, definitely shows the inapplicability of the Rayleigh formula for supersonic velocity of rarefied gases.

(3) Hot-Wire Anemometer

With the great complications in applying the conventional velocity-measuring device to superaerodynamic flows, one is naturally led to the thought of other avenues of approach. One possibility is the use of hot wires. If the wire diameter is of the order of 0.000 1 in and if the pressure of the gas stream is approximately 100 microns, the ratio of the mean free path to the wire diameter will be approximately 180. Therefore, the flow around the wire is definitely the free molecule flow[2]. Thus we have a simple physical situation, which is an improvement over the rather uncertain circumstances of mixed dynamic and viscous effects for the measurement of velocity by a Pitot tube. It therefore seems worth while to explore this possibility by a trial calculation of the performance of such a hot-wire anemometer.

If θ is the inclination of the solid surface to a gas stream that has a macroscopic velocity U and a Maxwellian molecular velocity distribution, the translational energy of molecules E_{itr} incident upon the unit area is

$$E_{itr} = \rho \frac{c}{2\sqrt{\pi}}\left\{ e^{-(U/c)^2 \sin^2\theta}\left(c^2 + \frac{1}{2}U^2\right) + \sqrt{\pi}\frac{U}{c}\sin\theta\left(\frac{5}{4}c^2 + \frac{1}{2}U^2\right)\left[1 + \operatorname{erf}\left(\frac{U}{c}\sin\theta\right)\right]\right\} \tag{18}$$

where $c^2 = 2RT$, T is the temperature of the gas stream, and erf is the error function. Now let r be the radius of the hot wire. Then the total energy E_i incident upon a unit length of the wire is the sum of translational energy and internal energy. If c_v is the specific heat at constant volume, this total energy per unit length of wire is

$$\begin{aligned} E_i = \rho \frac{c}{\sqrt{\pi}} r \Big\{ & \left[\frac{1}{2}U^2 + \left(\frac{1}{2}R + c_v\right)T\right]\int_{-\pi/2}^{\pi/2} e^{-(U/c)^2 \sin^2\theta} d\theta + \\ & \sqrt{r}\left[\frac{1}{2}U^2 + (R + c_v)T\right]\int_{-\pi/2}^{\pi/2} \frac{U}{c}\sin\theta\left[1 + \operatorname{erf}\left(\frac{U}{c}\sin\theta\right)\right] d\theta \Big\} \end{aligned} \tag{19}$$

The integrals in Eq. (19) can be expressed in terms of tabulated functions (see Appendix). Thus,

$$E_i = \rho c r\left\{\left[\frac{1}{2}U^2 + \left(\frac{1}{2}R + c_V\right)T\right]F_1\left(\frac{U}{c}\right) + \left[\frac{1}{2}U^2 + (R + c_V)T\right]F_2\left(\frac{U}{c}\right)\right\} \tag{20}$$

where

$$F_1\left(\frac{U}{c}\right) = \frac{1}{\sqrt{\pi}}\int_{-\pi/2}^{\pi/2} e^{-(U^2/c^2)\sin^2\theta}\,d\theta = \sqrt{\pi}\,e^{-(1/2)(U/c)^2}\,I_0\left(\frac{1}{2}\frac{U^2}{c^2}\right) \tag{21}$$

$$\begin{aligned} F_2\left(\frac{U}{c}\right) &= \int_{-\pi/2}^{\pi/2}\left(\frac{U}{c}\sin\theta\right)\left[1 + \operatorname{erf}\left(\frac{U}{c}\sin\theta\right)\right]d\theta \\ &= \sqrt{\pi}\left(\frac{U}{c}\right)^2 e^{-(1/2)(U/c)^2}\left[I_0\left(\frac{1}{2}\frac{U^2}{c^2}\right) + I_1\left(\frac{1}{2}\frac{U^2}{c^2}\right)\right] \end{aligned} \tag{22}$$

The I_0 and I_1 are the modified Bessel functions of the first kind of orders zero and one, respectively. The functions F_1 and F_2 are tabulated in Table 1.

If T_w is the wall temperature and α is the accommodation coefficient, the difference between the energy E_i incident upon the surface and the energy E_r carried by the molecules re-emitted from the surface is given by

$$E_i - E_r = \alpha(E_i - E_w)$$

where E_w is the energy that would be carried away by the molecules if the re-emission were at the temperature T_w of the wire. Therefore,

$$\begin{aligned} E_i - E_r = \alpha\rho c r\Big\{&\left[\frac{1}{2}U^2 + \left(\frac{1}{2}R + c_V\right)(T - T_w)\right]F_1\left(\frac{U}{c}\right) + \\ &\left[\frac{1}{2}U^2 + (R + c_V)(T - T_w)\right]F_2\left(\frac{U}{c}\right)\Big\} \end{aligned} \tag{23}$$

This difference of energy is then the net energy input to the wire per unit length of the wire by the air stream.

If i is the electric current heating the wire and Ω is the resistance of the wire per unit length at the wire temperature, the heat input per unit length of wire by the heating current is $i^2\Omega$. Heat is lost from the wire by radiation. If σ is Stefan-Boltzmann constant and ε is the emissivity of the wire surface, the radiation heat loss per unit length is $2\pi r\varepsilon\sigma T_w^4$. Therefore, if the wire has reached a steady condition, the heat balance requires

$$\begin{aligned} \alpha\rho\sqrt{2RT}r\Big\{&F_1\left(\frac{U}{c}\right)\left[\frac{1}{2}U^2 + \left(\frac{1}{2}R + c_V\right)(T - T_w)\right] + \\ &F_2\left(\frac{U}{c}\right)\left[\frac{1}{2}U^2 + (R + c_V)(T - T_w)\right]\Big\} + i^2\Omega = 2\pi r\varepsilon\sigma T_w^4 \end{aligned} \tag{24}$$

This equation can be put into somewhat simpler form by using the following relation:

$$R = c_p - c_V = c_V(\gamma - 1) \tag{25}$$

Furthermore, if we take T_0 as the reference temperature at which the resistance Ω is Ω_0 and the corresponding temperature coefficient of the resistance is β, the resistance Ω can be expressed as

$$\Omega = \Omega_0 [1 + \beta(T_w - T_0)] \tag{26}$$

Now let

$$\lambda = \Omega/\Omega_0 \tag{27}$$

Then, from Eq. (26),

$$T_w/T_0 = [(\lambda - 1)/(\beta T_0)] + 1 \tag{28}$$

Now introduce ρ_0 as the reference density and i_0 as the reference heating current; then Eq. (24) can be written as

$$\left(1+\frac{\lambda-1}{\beta T_1}\right)^4 = \left[\frac{\alpha\rho_0(RT_0)^{3/2}}{\pi\sqrt{2}\varepsilon\sigma T_0^4}\right]\left(\frac{\rho}{\rho_0}\right)\left(\frac{T}{T_0}\right)^{3/2}\left\{F_1\left(\frac{U}{c}\right)\left[\frac{U^2}{c^2}+\frac{1}{2}\frac{\gamma+1}{\gamma-1}\left(1-\frac{1+\frac{\lambda-1}{\beta T_0}}{T/T_0}\right)\right]+\right.$$
$$\left.F_2\left(\frac{U}{c}\right)\left[\frac{U^2}{c^2}+\frac{\gamma}{\gamma-1}\left(1-\frac{1+\frac{\lambda-1}{\beta T_0}}{T/T_0}\right)\right]\right\}+\left(\frac{i_0^2\Omega_0}{2\pi r\varepsilon\sigma T_0^4}\right)\lambda\left(\frac{i}{i_0}\right)^2 \tag{29}$$

The particular values of the reference temperature T_0, the reference density ρ_0, and the reference current i_0 are not yet fixed. We fix these quantities now by requiring that

$$\beta T_0 = 1 \tag{30}$$

$$\alpha\rho_0(RT_0)^{3/2}/(\pi\sqrt{2}\varepsilon\sigma T_0^4) = 1 \tag{31}$$

$$i_0^2\Omega_0/(2\pi r\varepsilon\sigma T_0^4) = 1 \tag{32}$$

Then Eq. (29) simplifies into

$$\lambda^4 = \left(\frac{p}{p_0}\right)\left(\frac{T}{T_0}\right)^{1/2}\times\left\{F_1\left(\frac{U}{c}\right)\left[\left(\frac{U}{c}\right)^2+\frac{1}{2}\frac{\gamma+1}{\gamma-1}\left(1-\frac{\lambda}{T/T_0}\right)\right]+\right.$$
$$\left.F_2\left(\frac{U}{c}\right)\left[\left(\frac{U}{c}\right)^2+\frac{\gamma}{\gamma-1}\left(1-\frac{\lambda}{T/T_0}\right)\right]\right\}+\lambda\left(\frac{i}{i_0}\right)^2 \tag{33}$$

This is then the performance equation of the hot wire in free molecular flow.

Now let us investigate in greater detail the case of a bright platinum wire. To satisfy Eq. (30), $T_0 = 492°\text{R}$. The value for ε and α can be taken to be 0.08 and 0.90, respectively. Then Eq. (31) gives the corresponding pressure p_0 for ρ_0 and T_0 as

$$p_0 = \pi\sqrt[4]{2}\,\sigma T_0^4/(\alpha\sqrt{RT_0}) = 3.37 \text{ microns}$$

Let the radius r of the wire be 0.000 1 in. Then Eq. (32) gives the reference heating current i_0 as

$$i_0 = \sqrt{2\pi r\varepsilon\sigma T_0^4/\Omega_0} = 0.274 \text{ milliamp}$$

where the resistivity of the platinum is taken as 10.96×10^{-6} ohm-cm. Therefore the order of magnitude of the different quantities is entirely satisfactory.

If the wire is used with a constant heating current, then Eq. (33) can be used to calculate the relation between the resistance ratio λ and the velocity ratio (U/c) at constant air-stream density and temperature. This is done for $p/p_0 = 1$, $T/T_0 = 1$, and $i/i_0 = 1$, * and the result is given in Fig. 6. It is seen that the sensitivity of the instrument is good. Of course, the behavior of the hot-wire anemometer will be actually determined by calibration for any experiment. Since the performance of the wire is strongly influenced by the accommodation coefficient α, as shown by Eq. (29), it will be necessary to find materials that can hold this coefficient constant for a considerable period of time so that no frequent calibration is required. However, the present analysis seems to indicate the feasibility of such an instrument for measurements in rarefied gases, and further research is definitely desirable.

(4) Parameters of Flow

The two parameters that are directly connected with the flow field are the Reynolds Number Re, defined as $Re = UL/\nu^\circ$ (where ν° is the kinematic viscosity, L is the typical linear dimension of the body), and the Mach Number M° of the free stream. This is true even for slip flows and free molecule flows because of the fact that the ratio of mean free path to the typical dimension can be also expressed in terms of the Reynolds Number and the Mach Number.

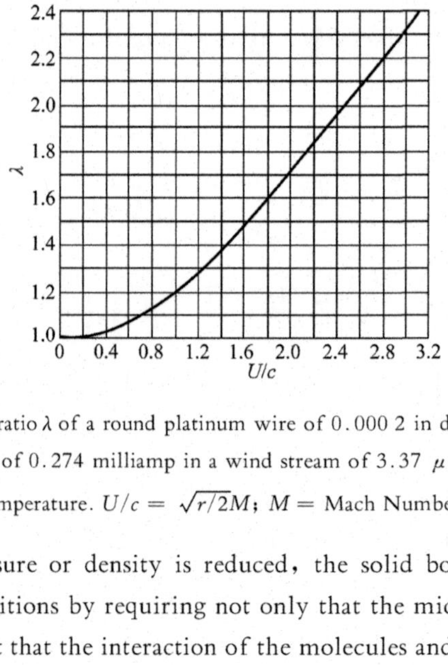

Fig. 6 Resistance ratio λ of a round platinum wire of 0.000 2 in diameter, heated by a constant current of 0.274 milliamp in a wind stream of 3.37 μ static pressure and 492°R static temperature. $U/c = \sqrt{r/2}M$; $M =$ Mach Number of air stream

However, as the pressure or density is reduced, the solid boundary of the flow enters actively into the flow conditions by requiring not only that the microscopic stream velocity be tangential to the surface but that the interaction of the molecules and the wall be considered and that the radiation of energy to and from the wall be taken into account. The interaction of the molecules with the wall is so far expressed through the fraction s of molecules that are diffusely

---

* The author is indebted to L. Mack for the numerical computations.

re-emitted from the wall and the accommodation coefficient α. It is known that both s and α are functions of the temperature of the wall, and there is reason to believe that they are also functions of the molecular velocity distribution. Therefore the interaction of the molecules with the wall is the same only if the wall temperature, the gas temperature, and the Mach Number of the gas above the wall are the same. These considerations seem to indicate then that for the model test to be similar to the prototype, the model must be made of same surface material as the prototype, the fluid must be the same, and, furthermore, the following parameters must be the same: (1) Reynolds Number, Re; (2) Mach Number, M°; and (3) free-stream temperature T°.

The radiation heat loss from the surface is equal to $\varepsilon\sigma T_w^4$ per unit area. However, if the model is surrounded by the walls of the test chamber, there is also a heat input due to radiation from walls of the test section to the model. Let us call this quantity q_c. Then the net heat loss per unit area of the surface of the model is $\varepsilon\sigma T_w^4 - q_c$. This quantity can be rendered nondimensional by dividing it by $\rho^\circ U c_p T^\circ$. Call this new parameter Λ_m; then

$$\Lambda_m = [(\varepsilon\sigma T_w^4 - q_c)/(\rho^\circ U c_p T^\circ)]_m \tag{34}$$

For the prototype, the heat from the walls of the test chamber is absent, but there may be solar radiation and the radiation from the earth and surrounding atmosphere[9]. Denoting this amount by q, the parameter Λ for the prototype is then

$$\Lambda = (\varepsilon\sigma T_w^4 - q)/(\rho^\circ U c_p T^\circ) \tag{35}$$

In order for the flow also to be same with respect to the radiation heat transfer,

$$\Lambda = \Lambda_m \tag{36}$$

Because of the previous conditions on the Reynolds Number and free-stream temperature, Eq. (36) is the same as

$$(\varepsilon^\circ T_w^4 - q)/(\varepsilon^\circ T_w^4 - q_c) = L_m/L \tag{37}$$

where L_m is the typical linear dimension of the model and L is the typical linear dimension of the prototype. This means that the wall temperature of the test chamber must be so controlled that q_c satisfies Eq. (37).

This set of rather strict similarity rules for model testing in superaerodynamic flows is certainly difficult to satisfy. In what way the rules can be relaxed is the problem of future research.

Appendix

Evaluation of the functions F_1 and F_2

For the function F_1,

$$F_1(z) = \frac{1}{\sqrt{\pi}}\int_{-\pi/2}^{\pi/2} e^{-z^2\sin^2\theta}\,d\theta = \frac{2}{\sqrt{\pi}}\int_0^{\pi/2} e^{-(z^2/2)(1-\cos 2\theta)}\,d\theta$$

$$= \frac{1}{\sqrt{\pi}} e^{-z^2/2}\int_0^{\pi} e^{(z^2/2)\cos\varphi}\,d\varphi = \sqrt{\pi}e^{-z^2/2}I_0\left(\frac{z^2}{2}\right)$$

where I_0 is the modified Bessel function of first kind and under zero. The last step is made possible by the substitution $2\theta = \varphi$.

For the function F_2,

$$F_2(z) = \int_{-\pi/2}^{\pi/2} (z\sin\theta)[1 + \mathrm{erf}(z\sin\theta)]\mathrm{d}\theta = \int_{-\pi/2}^{\pi/2} (z\sin\theta)\mathrm{erf}(z\sin\theta)\mathrm{d}\theta$$

By the definition of the error function, we have

$$F_2(z) = (z/\sqrt{\pi})\int_{-\pi/2}^{\pi/2} 2\sin\theta\left(\int_0^{z\sin\theta} \mathrm{e}^{-s^2}\mathrm{d}s\right)\mathrm{d}\theta$$

This form can be simplified by partial integration. Thus

$$\begin{aligned} F_2(z) &= (4/\sqrt{\pi})z^2\int_0^{\pi/2} \cos^2\theta\,\mathrm{e}^{-z^2\sin^2\theta}\mathrm{d}\theta \\ &= (2/\sqrt{\pi})z^2\int_0^{\pi/2} (1 + \cos 2\theta)\mathrm{e}^{-z^2\sin\theta}\mathrm{d}\theta \\ &= \sqrt{\pi}z^2\mathrm{e}^{-z^2/2} \times \left[\mathrm{I}_0\left(\frac{z^2}{2}\right) + \frac{1}{\pi}\int_0^{\pi}\cos\varphi\,\mathrm{e}^{+z^2/2\cos\varphi}\mathrm{d}\varphi\right] \end{aligned}$$

Therefore

$$F_2(z) = \sqrt{\pi}z^2\mathrm{e}^{-z^2/2}\left[\mathrm{I}_0\left(\frac{z^2}{2}\right) + \mathrm{I}_1\left(\frac{z^2}{2}\right)\right]$$

where I_1 is the modified Bessel function of first kind and order one.

References

[1] Puckett A E. Supersonic Nozzle Design. Journal of Applied Mechanics (A. S. M. E.), 1946, 13: A-266.

[2] Tsien H S. Superaerodynamics, Mechanics of Rarefied Gases. Journal of the Aeronautical Sciences, 1946, 13(12): 653.

[3] Schamberg R. The Fundamental Differential Equations and the Boundary Conditions for High Speed Slip-Flow, and Their Application to Several Specific Problems. Thesis, California Institute of Technology, 1947.

[4] Schaaf S A. Viscosity Effects in Wind Tunnel No. 2. University of California, Department of Engineering Report No. HE-150-16, 1947.

[5] Schaaf S A. The Theory of Minimum Response Time for Vacuum Gages. University of California, Department of Engineering Report No. HE-150-21, 1947.

[6] Evans R A. Flow Visualization at Low Pressures. University of California, Department of Engineering Report No. HE-150-21, 1947.

[7] Barker M. On the Use of Very Small Pitot-tubes for Measuring Wind Velocity. Proc. Royal Society (A), 1922, 101: 435 - 445.

[8] Homann F. Einfluss Grosser Zahigkeit bei Stromung um Zylinder. Forschung Ingews., 1936, 7: 1 - 10.

[9] Johnson H A, Possner L. A Design Manual for Determining the Thermal Characteristics of High Speed Aircraft. Chapt. 4, A. A. F. Technical Report No. 5632, 1947.

Airfoils in Slightly Supersonic Flow

Hsue-shen Tsien* and Judson R. Baron**

(Massachusetts Institute of Technology)

Summary

An investigation is made to determine the performance of simple thin airfoils in the slightly supersonic flow region with the aid of the nonlinear transonic theory first developed by von Kármán[1]. Expressions for the pressure coefficient across an oblique shock and a Prandtl-Meyer expansion are developed in terms of a transonic similarity parameter. Aerodynamic coefficients are calculated in similarity form for the flat plate and asymmetric wedge airfoils, and curves are plotted. Sample curves for a flat plate and a specific asymmetric wedge are plotted on the usual coordinate grid of C_l, C_d, and $C_{m_{c/4}}$ versus angle of attack and C_l versus Mach Number to illustrate the apparent features of nonlinear flow.

Introduction

Recently, theodore von Kármán[1] has derived a set of similarity rules for flows around slender bodies in a stream having close to sonic velocity. He called this set of rules the transonic similarity rules. Essentially, the results are as follows: If δ is the thickness ratio of the body, α the angle of attack of the airfoil, C_l, C_d, C_m the sectional lift, drag, and moment coefficients, respectively, then

$$C_l(\alpha/\delta) = (\delta^{2/3}/\Gamma^{1/3})L(K) \quad (1)$$

$$C_d(\alpha/\delta) = (\delta^{5/3}/\Gamma^{1/3})D(K) \quad (2)$$

and

$$C_m(\alpha/\delta) = (\delta^{2/3}/\Gamma^{1/3})M(K) \quad (3)$$

In these equations, K is the similarity parameter defined as

$$K = (1/2)[(M^{0^2} - 1)/(\Gamma\delta)^{2/3}] \quad (4)$$

Received September 18, 1948.

Journal of the Aeronautical Sciences, vol. 16, pp. 55 - 61, 1949.

This paper is based on a thesis by J. R. Baron, under the direction of the senior author, submitted to the Graduate School in partial fulfillment of the requirements for the degree of Master of Science in Aeronautical Engineering at Massachusetts Institute of Technology.

* Professor of Aerodynamics.

** Research Aerodynamicist, Supersonic Laboratory.

where M^0 is the free stream Mach Number and Γ is a parameter of the property of the fluid — i.e.,

$$\Gamma = (\gamma + 1)/2 \tag{5}$$

where γ is the ratio of the specific heats of the gas. Furthermore, if ξ is a nondimensional parameter, such as percentage of chord, designating a surface point on the airfoil, then the pressure coefficient, C_p, at that point is given by

$$C_p = \frac{p - p^0}{(1/2)\rho^0 U^2} = \frac{\delta^{2/3}}{\Gamma^{1/3}} p(K;\xi) \tag{6}$$

The appropriate differential equation for such transonic flows is, however, nonlinear and, in general, difficult to solve. For the special case of slightly supersonic flow, the problem is, on the other hand, much simpler, since the exact solutions associated with this problem, the oblique shock and Prandtl-Meyer flow, are known. The task is then simply to reduce the exact solutions to forms that are appropriate to transonic conditions involving the parameter K. This is indeed already indicated by von Kármán[1] himself. The purpose of this paper is to complete the calculations of von Kármán, to render them more systematic and then, finally, to apply these results to calculate the rather peculiar aerodynamic characteristics of simple thin airfoils in slightly supersonic flow with shocks attached to the leading edge.

Oblique Shock

Let β be the shock angle — i. e., the angle between the free stream and the shock front — and let θ be the angle through which the flow is deflected. The basic oblique shock relations[2] are then

$$\frac{p}{p^0} = \frac{2\gamma}{\gamma + 1} M^{0^2} \sin^2\beta - \frac{\gamma - 1}{\gamma + 1} \tag{7}$$

and

$$\frac{\tan(\beta - \theta)}{\tan\beta} = \frac{\gamma - 1}{\gamma + 1} + \frac{2}{\gamma + 1} \frac{1}{M^{0^2} \sin^2\beta} \tag{8}$$

On the other hand, the pressure coefficient C_p for a point on the surface is given by

$$C_p = [2/(\gamma M^{0^2})][(p/p^0) - 1] \tag{9}$$

In the three equations, Eqs. (7), (8), and (9), the two quantities β and (p/p^0) can be eliminated. The resultant relation is

$$\tan\theta\left(1 - \frac{C_p}{2}\right)\left(\frac{\Gamma C_p}{2} + \frac{1}{M^{0^2}}\right)^{1/2} = \frac{C_p}{2}\left[1 - \left(\frac{\Gamma C_p}{2} + \frac{1}{M^{0^2}}\right)\right]^{1/2} \tag{10}$$

Eq. (10) is the exact oblique shock relation. For the transonic conditions, θ is small and is equivalent to δ in von Kármán's formulas. Thus

$$K = (1/2)[(M^{0^2} - 1)/(\Gamma\theta)^{2/3}] \tag{11}$$

From this equation the free stream Mach Number, M^0, can be expressed in terms of K and θ. In fact,

$$1/M^{0^2} = 1 - 2K(\Gamma\theta)^{2/3} + \ldots \tag{12}$$

The appropriate form for the pressure coefficient is

$$C_p = (\theta^{2/3}/\Gamma^{1/3})p^{(1)}(K) \tag{13}$$

where $p^{(1)}$ is the pressure function associated with the flow after the oblique shock. Then Eqs. (12) and (13) can be substituted into the general relation, Eq. (10), and, by taking only the lowest order of terms, an expression for $p^{(1)}$ is obtained:

$$1 = (p^{(1)^2}/4)[2K - (p^{(1)}/2)] \tag{14}$$

Eq. (14) is a cubic equation for $p^{(1)}$. To solve this equation, it is convenient to put

$$j = 2/p^{(1)} \tag{15}$$

Then Eq. (14) is converted into the normal form of cubic equations,

$$f^3 - 2Kf + 1 = 0 \tag{16}$$

According to standard solutions[3], the character of the roots of this cubic equation is dependent upon whether

$$(1/4) - (8/27)K^3$$

is greater than, equal to, or smaller than zero. In the first case, there will be one real root and two complex conjugate roots. The complex roots have, of course, no physical significance and need not be considered further. However, the single root is negative, as can be seen from Eq. (16) by making K extremely small. Negative f means negative $p^{(1)}$. But physically the flow is compressed by the shock, and thus $p^{(1)}$ must be positive. Therefore, no useful solution exists for

$$(1/4) - [(8/27)K^3] > 0$$

When K is equal to the critical value K^* such that

$$(1/4) - [(8/27)(K^*)^3] = 0$$

or

$$K^* = (3/2)(1/4^{1/3}) = 0.945 \tag{17}$$

there are two distinct roots: $f = 1/2^{1/3}$ and $f = -2^{2/3}$. The second root is again without physical significance. Corresponding to the first root, the critical value $p^{(1)}$ is

$$p^{(1)}(K^*) = 2^{4/3} = 2.502 \tag{18}$$

K^* is actually then the lowest value for K, or the largest value for θ at a given M^0, for which the result has physical significance. This value of K must then correspond to the maximum flow deflection or critical condition for attachment at the leading edge. This checks with the calculation of von Kármán[1].

For larger values of K, the physically meaningful values[3] of $p^{(1)}$ are

$$p_1^{(1)} = 1\Big/\left[\sqrt{\frac{2}{3}}K\cos\left(\frac{\pi-\varphi}{3}\right)\right] \tag{19}$$

and

$$p_2^{(2)} = 1\Big/\left[\sqrt{\frac{2}{3}}K\cos\left(\frac{\pi+\varphi}{3}\right)\right] \tag{20}$$

where φ is given by

$$\cos\varphi = (1/4)\sqrt{(27/2)(1/K^3)} \tag{21}$$

Since $K > K^*$, φ is greater than zero but smaller than $(\pi/2)$. Under these conditions, $p_2^{(1)} > p_1^{(1)}$. $p_1^{(1)}$ then corresponds to the weak shock case, and $p_2^{(1)}$ corresponds to the strong shock case. Following the general practice, $p_1^{(1)}$ will be used for attached leadingedge shocks.

The numerical values of $p_1^{(1)}$ and $p_2^{(1)}$ are tabulated in Table 1, and the data are plotted in Fig. 1. $p_1^{(1)}$ can be expressed as an inverse power series of K. Calculation shows that

$$p_1^{(1)} = \frac{1}{K^{1/2}}\Big[1.414\,22 + \frac{0.250\,00}{K^{3/2}} + \frac{0.110\,48}{K^3} + \frac{0.062\,50}{K^{9/2}} + \frac{0.039\,87}{K^6} + \frac{0.027\,34}{K^{15/2}} + \frac{0.019\,66}{K^9} + \frac{0.014\,63}{K^{21/2}} + \frac{0.011\,19}{K^{12}} + \frac{0.008\,73}{K^{27/2}} + \ldots\Big] \tag{22}$$

The first two terms of Eq. (22) agree with von Kármán's calculation. He did not consider the additional terms.

Table 1 Pressure function for weak oblique shock, $p_1^{(1)}$, Strong oblique shock, $p_2^{(1)}$, prandtl-meyer expansion, $p^{(2)}$, and the ratio of $(M^2-1)/(M^{0^2}-1)$ for various values of K. [Eqs. (19), (20), (29), and (36)]

K	$p_1^{(1)}$	$p_2^{(1)}$	$\left(\frac{M^2-1}{M^{0^2}-1}\right)_1^{(1)}$	$-p^{(2)}$	$\left(\frac{M^2-1}{M^{0^2}-1}\right)^{(2)}$
0.00	...	...	...	2.080 1	∞
0.20	...	...	...	1.795 6	5.489 1
0.40	...	...	...	1.598 9	2.998 7
0.60	...	...	...	1.450 3	2.208 6
0.80	...	...	...	1.333 3	1.833 3
0.945	2.502 8	2.535 8	...	1.262 9	1.668 2
0.975	2.119 9	3.025 2	...	1.249 5	1.640 8
1.00	1.999 3	3.236 0	0.000 0	1.238 7	1.619 3
1.25	1.514 9	4.624 3	0.394 1	1.143 0	1.457 2
1.50	1.305 1	5.757 4	0.565 0	1.065 1	1.355 0

(continued)

K	$p_1^{(1)}$	$p_2^{(1)}$	$\left(\dfrac{M^2-1}{M^{0^2}-1}\right)_1^{(1)}$	$-p^{(2)}$	$\left(\dfrac{M^2-1}{M^{0^2}-1}\right)^{(2)}$
1.75	1.171 4	6.819 9	0.665 3	1.000 6	1.285 9
2.00	1.077 3	7.629 5	0.730 7	0.946 1	1.236 5
2.50	0.939 3	9.906 9	0.812 2	0.858 6	1.171 7
3.00	0.848 7	11.920 4	0.858 6	0.790 9	1.131 8
4.00	0.723 5	15.949 0	0.909 6	0.692 3	1.086 5
5.00	0.642 8	19.615 5	0.935 7	0.622 8	1.062 3
6.00	0.584 4	23.872 0	0.951 3	0.570 5	1.047 5
8.00	0.503 8	32.362 5	0.968 5	0.496 2	1.031 0
10.00	0.449 6	40.338 8	0.977 5	0.444 8	1.022 2
13.00	0.393 6	53.078 6	0.984 9	0.390 5	1.015 0
20.00	0.316 8	78.492 9	0.992 1	0.315 6	1.007 9
30.00	0.258 4	127.746 6	0.995 7	0.257 4	1.004 3
50.00	0.200 1	199.044 6	0.998 0	0.200 0	1.002 0
80.00	0.158 1	472.143 5	0.999 0	0.158 4	1.001 0
100.00	0.141 4	∞	0.999 3	0.142 0	1.000 7
∞	0.000 0	∞	1.000 0	0.000 0	1.000 0

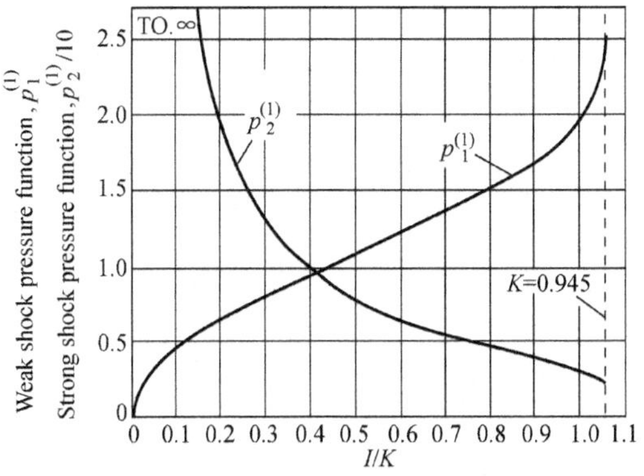

Fig. 1 Variation of compression shock pressure function with K[Eqs. (19) and (20)]

Let p^* denote the critical pressure corresponding to local sonic velocity. Then for isentropic flow,

$$\frac{p^*}{p^0} = \left[\frac{\dfrac{\gamma+1}{2} + \dfrac{\gamma-1}{2}(M^{0^2}-1)}{(\gamma+1)/2}\right]^{\gamma/(\gamma-1)}$$

The corresponding pressure coefficient, C_p^*, is then, according to Eq. (9),

$$C_p^* = \frac{2}{\gamma M^{0^2}}\left\{\left[1+\frac{\gamma-1}{\gamma+1}(M^{0^2}-1)\right]^{\gamma/(\gamma-1)}-1\right\}$$

When $M^0 \approx 1$ for transonic flows, the above equation simplifies to

$$C_p^* = (M^{0^2}-1)/\Gamma \tag{23}$$

The flow after the strong shock is always subsonic. Sonic velocity after shock can only be associated with the weak shock. But it is known that the entropy change of a weak shock is negligible, and therefore Eq. (23) can be used for the weak shock case.

In general, as shown by Eq. (13),

$$C_p = [(M^{0^2}-1)/(2\Gamma)](p_1^{(1)}/K) \tag{24}$$

By comparing Eqs. (23) and (24), the critical value of $(p_1^{(1)})^*$ corresponding to local sonic velocity after the shock is

$$p_1^{(1)*}/K_s = 2 \tag{25}$$

where K_s is the value of K corresponding to $(p_1^{(1)})^*$ By dividing Eq. (14) by K^3,

$$\frac{1}{K^3} = \frac{1}{4}\left(\frac{p_1^{(1)}}{K}\right)^2\left[2-\frac{1}{2}\left(\frac{p_1^{(1)}}{K}\right)\right] \tag{26}$$

and then substituting Eq. (25) into Eq. (26), the equation for K_s for sonic velocity after a shock is obtained:

$$K_s = 1 \tag{27}$$

Thus, for supersonic velocity after the leading edge shock, $K > 1$.

For airfoil calculations, one will also need the ratio $(M^2-1)/(M^{0^2}-1)$, where M is the Mach Number after the shock. This can easily be determined by noting that in general for isentropic flow

$$\frac{p}{p^0} = \left\{\frac{1+[(\gamma-1)/(\gamma+1)](M^{0^2}-1)}{1+[(\gamma-1)/(\gamma+1)](M^2-1)}\right\}^{\gamma/(\gamma-1)}$$

where both (M^2-1) and $(M^{0^2}-1)$ are small under transonic conditions. Then Eq. (9) gives

$$C_p = [(M^{0^2}-1)/\Gamma]\{1-[(M^2-1)/(M^{0^2}-1)]\} \tag{28}$$

By combining Eqs. (24) and (28), the required ratio is given as

$$(M^2-1)/(M^{0^2}-1) = 1-(1/2)(p/K) \tag{29}$$

This formula is, of course, generally true. If the value p is taken from Eq. (19), then the ratio will be that for the weak oblique shock. The numerical values are tabulated in Table 1, and the data are plotted in Fig. 2.

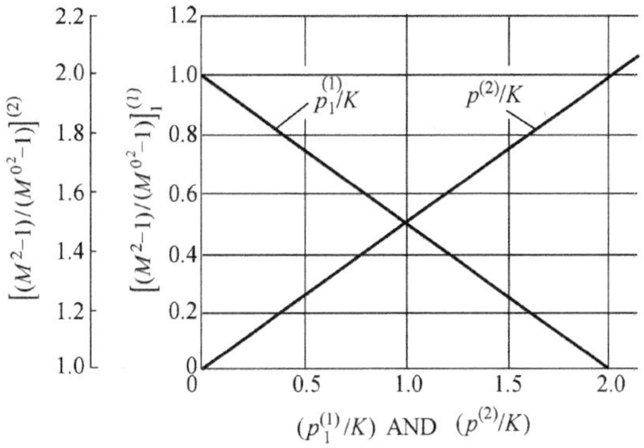

Fig. 2 Variation of $(M^2-1)/(M^{0^2}-1)$ with (p/K)[Eq. (29)]

Prandtl-Meyer Expansion

If q is the velocity and θ is the angle of turn during expansion, the Prandtl-Meyer expansion is determined by the relation[4]

$$dq = q\,d\theta/\sqrt{M^2-1}$$

However, in general,

$$dp = -\rho q\,dq$$

Thus, by eliminating dq from these two equations,

$$dp = -\rho q^2\,d\theta/\sqrt{M^2-1}$$

or

$$d\left(\frac{p}{p^0}\right) = -\left(\frac{\rho^0}{p^0}\right)\frac{(\rho/\rho^0)q^2\,d\theta}{\sqrt{M^2-1}} \tag{30}$$

Now since the expansion is adiabatic,

$$a^2 + \frac{\gamma-1}{2}q^2 = a^{0^2} + \frac{\gamma-1}{2}U^2$$

Thus by using the isentropic relations, q^2 can be calculated from the above equation as

$$q^2 = \frac{2}{\gamma-1}a^{0^2}\left[1 + \frac{\gamma-1}{2}M^{0^2} - \left(\frac{p}{p^0}\right)^{(\gamma-1)/\gamma}\right]$$

Furthermore

$$M^2 - 1 = \left(\frac{2}{\gamma-1} + M^{0^2}\right)\left(\frac{p^0}{p}\right)^{(\gamma-1)/\gamma} - \frac{\gamma+1}{\gamma-1}$$

These relations, together with Eq. (9), then convert Eq. (30) into

$$\frac{dC_p}{d\theta} = -\frac{2}{M^{0^2}} \frac{\left(1+\frac{\gamma M^{0^2}}{2}C_p\right)^{1/\gamma}\left[\frac{2}{\gamma-1}+M^{0^2}-\frac{2}{\gamma-1}\left(1+\frac{\gamma M^{0^2}}{2}C_p\right)^{(\gamma-1)/\gamma}\right]}{\left[\left(\frac{2}{\gamma-1}+M^{0^2}\right)\left(1+\frac{\gamma M^{0^2}}{2}C_p\right)^{-(\gamma-1)/\gamma}-\frac{\gamma+1}{\gamma-1}\right]^{1/2}} \tag{31}$$

This is the exact equation. It can be simplified by noting that for transonic conditions C_p is small and M^0 is close to 1. Then, expanding the terms on the right-hand side of Eq. (31) into a power series of C_p, one finds that the significant terms are

$$\frac{dC_p}{d\theta} = -2/(M^{0^2}-1)^{1/2}\left(1-\frac{\Gamma}{M^{0^2}-1}C_p\right)^{1/2} \tag{32}$$

Now, similar to Eq. (24), one can write

$$C_p = \frac{\theta^{2/3}}{\Gamma^{1/2}}p^{(2)}(K) = \frac{M^{0^2}-1}{2\Gamma}\left(\frac{p^{(2)}}{K}\right) \tag{33}$$

where $p^{(2)}$ is the pressure function associated with the Prandtl-Meyer expansion. Eq. (32) then be comes

$$\frac{1}{3}K^{5/2}\frac{d}{dK}\left(\frac{p^{(2)}}{K}\right) = \frac{1}{\sqrt{2-(p^{(2)}/K)}} \tag{34}$$

This can easily be integrated. The result is

$$[2-(p^{(2)}/K)]^{3/2} = (3/K^{3/2})+C \tag{35}$$

where C is the integration constant. To determine the constant, note that, when $\theta = 0$, no change is made in the free stream and $C_p = 0$. Then, when $K \to \infty$, $p^{(2)}/K \to 0$. Hence, the constant in Eq. (35) is equal to $(2)^{3/2}$. Then Eq. (35) can be solved for $p^{(2)}$, and the result is

$$p^{(2)} = \{2K - [(2K)^{3/2}+3]^{2/3}\} \tag{36}$$

Therefore, when $M^0 = 1$, and thus $K = 0$, the pressure coefficient is proportional to $\theta^{2/3}$, or

$$C_p = -[(3\theta)^{2/3}/\Gamma^{1/3}] \quad (M^0 = 1) \tag{37}$$

This is in agreement with von Kármán's calculation[1]. For other values of K, the pressure function, $p^{(2)}$, and the ratio $(M^2-1)/(M^{0^2}-1)$ are tabulated in Table 1 and plotted in Fig. 2 and Fig. 3.

Flat Plate Airfoil

The aerodynamic coefficients of an airfoil can be obtained by integrating the pressure coefficients over the surface of the airfoil. If the local surface inclination to the free stream is θ, the sectional coefficients are

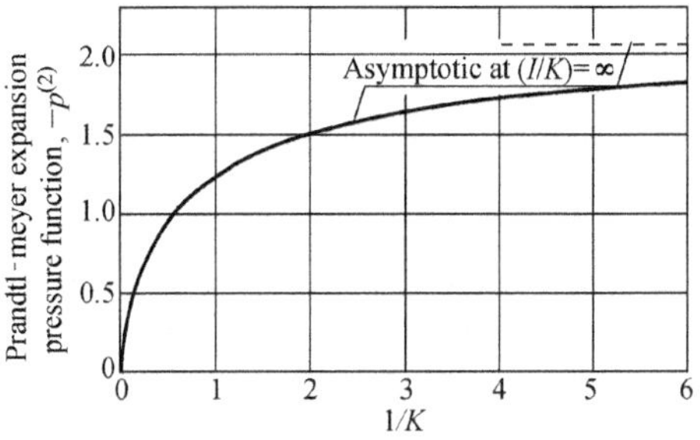

Fig. 3 Variation of Prandtl-Meyer expansion pressure function with K[Eq. (36)]

$$\left.\begin{aligned} C_l &= (1/c)\int_c (C_{p_{\text{lower}}} - C_{p_{\text{upper}}})\,\mathrm{d}x \\ C_d &= (1/c)\int_c (C_{p_{\text{lower}}} - C_{p_{\text{upper}}})(\sin\theta)\,\mathrm{d}x \\ -C_{m_{c/4}} &= (1/c)\int_c (C_{p_{\text{lower}}} - C_{p_{\text{upper}}})[C.\ P. - (c/4)]\,\mathrm{d}x \end{aligned}\right\} \tag{38}$$

where c is the airfoil chord and $C.P.$ is the center of pressure.

Consider now a flat plate inclined at an angle, α, to the flow such that an oblique shock and a Prandtl-Meyer expansion take place at the leading edge of the lower and upper surfaces, respectively. By integrating Eq. (38), noting that $\sin\alpha$ may be approximated by α, and utilizing Eqs. (13) and (33), one has

$$\left.\begin{aligned} C_l &= (\alpha^{2/3}/\Gamma^{1/3})L(K) \\ C_d &= (\alpha^{5/3}/\Gamma^{1/3})L(K) \\ -C_{m_{c/4}} &= (1/4)C_l \end{aligned}\right\} \tag{39}$$

where $L(K) = [p_1^{(1)} - p^{(2)}]$ and $K = (M^{0^2} - 1)/-2(\alpha\Gamma)^{2/3}$.

Rearranging Eq. (39), the aerodynamic coefficients may be plotted in a simple manner involving only the lift function $L(K)$. Thus,

$$L(K) = \frac{\Gamma^{1/3}}{\alpha^{2/3}}C_l = \frac{\Gamma^{1/3}}{\alpha^{5/3}}C_d = -4\frac{\Gamma^{1/3}}{\alpha^{2/3}}C_{m_{c/4}} \tag{40}$$

This has been carried out in Fig. 4. Curves of C_l versus M^0 and C_l versus α have been calculated from these data and are presented in Figs. 5 and 6. Drag and moment coefficient curves can be seen to be similar to the lift coefficient curve with only a change in the scale of the ordinate. Data obtained by the use of Ackeret's linearized theory have been included with the curves for comparison. Two features are immediately apparent from the figures: First, the lift versus angle of attack curve is not a straight line as predicted by the linear theory but has a definite upward curvature; secondly, the predicted lift is at all times greater than Ackeret's prediction, and this difference increases with increasing angle of attack.

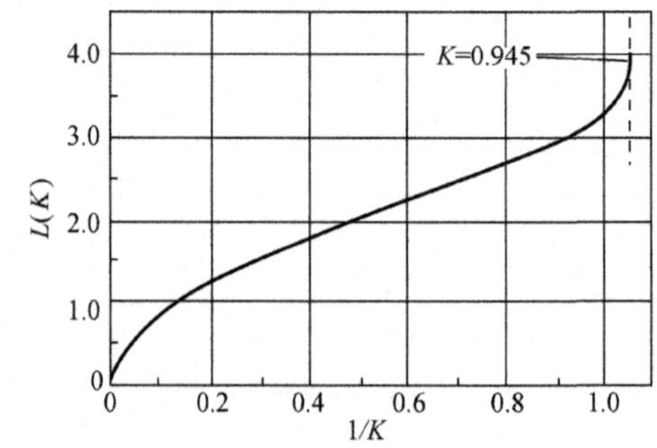

Fig. 4 Lift, drag, and moment similarity curve for flat plate airfoil [Eq. (40)]

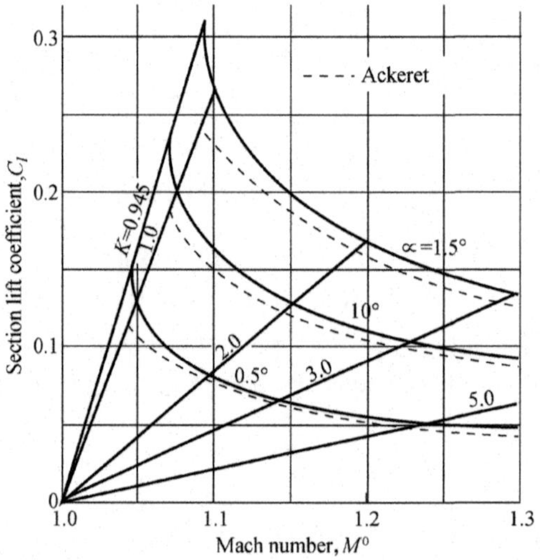

Fig. 5 Variation of section lift coefficient with Mach Number for the flat plate airfoil

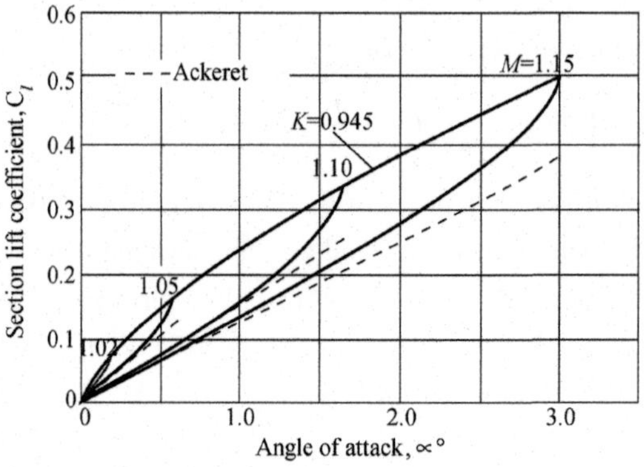

Fig. 6 Variation of section lift coefficient with angle or attack for the flat plate airfoil

Asymmetric Wedge Airfoil

As an example of finite thickness airfoils, consider now the asymmetric wedge depicted in Fig. 7. Denoting the vertex angle at the leading edge by θ, Table 2 lists the C_p and K for each region of the airfoil. It should be noted that the C_p for region (2) given in Table 2 is relative to region (1). Account may be taken of this by noting that

$$(C_p)_{2_{\text{relative to free stream}}} = \frac{p_2 - p^0}{q} = \frac{p_2 - p_1}{q} + \frac{p_1 - p^0}{q}$$
$$= (C_p)_{2_{\text{relative to region (1)}}} + (C_p)_1 \qquad (41)$$

to the order of magnitude of these calculations.

Table 2 Pressure coefficient and K for regions of asymmetric wedge airfoil

Region i	C_p	K_i
1	$\frac{(\theta-\alpha)^{2/3}}{\Gamma^{1/3}} p_1^{(1)}$	$\left(\frac{1}{2}\right)\frac{M^{0^2}-1}{[(\theta-\alpha)\Gamma]^{2/3}}$
2	$\frac{(2\theta)^{2/3}}{\Gamma^{1/3}} p^{(2)}$	$\left(\frac{1}{2}\right)\frac{M^{0^2}-1}{(2\theta\Gamma)^{2/3}}$
3	$\frac{\alpha^{2/3}}{\Gamma^{1/3}} p_1^{(1)}$	$\left(\frac{1}{2}\right)\frac{M^{0^2}-1}{(\alpha\Gamma)^{2/3}}$

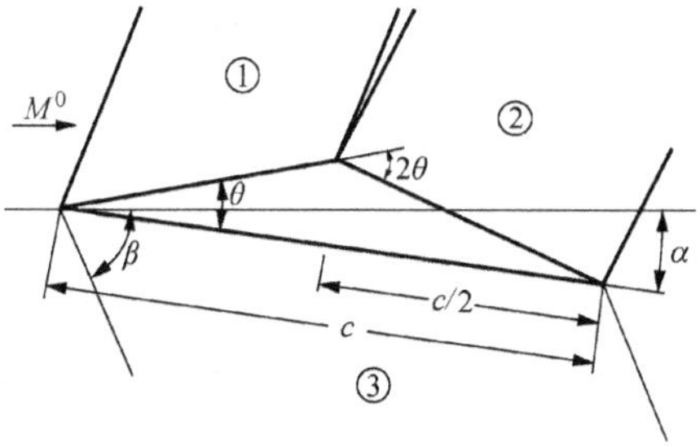

Fig. 7 Asymmetric wedge airfoil

Since all angles are small, the coefficients may be written in integrated form as

$$\left.\begin{aligned} C_l &= (C_p)_3 - (1/2)(C_p)_1 - (1/2)(C_p)_2 \\ C_d &= \alpha(C_p)_3 + (1/2)(\theta-\alpha)(C_p)_1 - (1/2)(\theta+\alpha)(C_p)_2 \\ -C_{m_{c/4}} &= (1/4)[(C_p)_3 - (C_p)_2] \end{aligned}\right\} \qquad (42)$$

In order to transform Eq. (42) into a form similar to Eq. (40) for the flat plate, two parameters, K_w and m are now defined such that

$$K_w = (1/2)[(M^{0^2} - 1)/(\theta\Gamma)^{2/3}] \tag{43}$$

and

$$m = \alpha/\theta \tag{44}$$

The expressions for K and C_p in terms of these parameters are tabulated in Table 3. Taking into account Eq. (41), Eq. (42) can now be written

Table 3 Pressure coefficient and K for regions of asymmetric wedge airfoil in terms of the similarity parameters K_w and m

Region i	C_p/R	K_i
1	$(1-m)^{2/3}$	$K_w/(1-m)^{2/3}$
2	$(2)^{2/3}$	$\frac{K_w}{(2)^{2/3}}\left[\frac{M_1^2-1}{M^{0^2}-1}\right]$
3	$(m)^{2/3}$	$K_w/m^{2/3}$

$R = (\theta^{2/3}/\Gamma^{1/3})p(K_i)$.

$$\left.\begin{aligned} C_l(\Gamma^{1/3}/\theta^{2/3}) &= m^{2/3}p_1^{(1)}(K_3) - (1-m)^{2/3}p_1^{(1)}(K_1) - (2)^{-1/3}p^{(2)}(K_2) \\ &= L(K,m) \\ C_d(\Gamma^{1/3}/\theta^{5/3}) &= m^{5/3}p_1^{(1)}(K_3) + (1-m)^{5/3}p_1^{(1)}(K_1) - (2)^{-1/3}(1+m)p^{(2)}(K_2) \\ &= D(K,m) \\ -C_{m_{c/4}}(\Gamma^{1/2}/\theta^{2/3}) &= (1/4)[m^{2/3}p_1^{(1)}(K_3) - (2)^{2/3}p^{(2)}(K_2)] = M(K,m) \end{aligned}\right\} \tag{45}$$

where the right-hand sides of Eq. (45) are functions only of K_w and m. These are the similarity relations for the asymmetric wedge airfoil. Calculations have been carried out and plotted in Figs. 8a, 8b, and 8c, for a range of values of $0 \leqslant m \leqslant 1$ and $0.632 \leqslant K_w \leqslant 1.90$. The chosen upper value of K_w was approximated by using a free stream with M^0 equal to 1.30 as a maximum and a wedge with $\theta = 3°$ as a minimum. The chosen lower value of K_w was determined by allowing a small range of values for m in which the present theory is applicable. The theoretical value of $K_w \to \infty$ — i.e., either a flat plate or a wedge in a stream of infinite M^0 — is seen to lie along the horizontal axis. Note, in Figs. 8a, 8b, and 8c, the limiting line for an attached shock ($K_3 = 0.945$) at the leading edge when approximately $0.55 \leqslant m \leqslant 1.0$, and note also the limiting line for sonic flow ($K_1 = 1.0$) in region ① of the airfoil when approximately $0 \leqslant m \leqslant 0.55$. The latter is a necessary condition for applying the Prandtl-Meyer flow in region ②.

As is to be expected, for the asymmetric wedge, zero lift occurs at a small angle of attack. It should be noticed that, as K_w increases (Fig. 8a), the lift versus angle of attack curves become linear. This can be explained by considering the two variables that determine K_w: M^0 and θ. If θ is considered constant, then K_w increases as M^0 increases, and thus, for

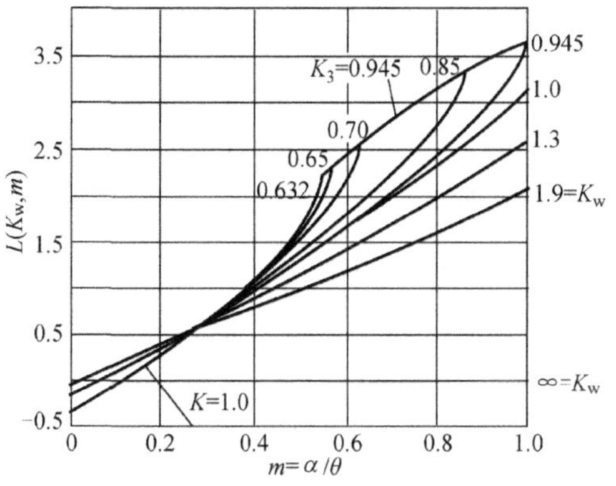

Fig. 8a Lift similarity curves for asymmetric wedge airfoil [Eq. (45)]

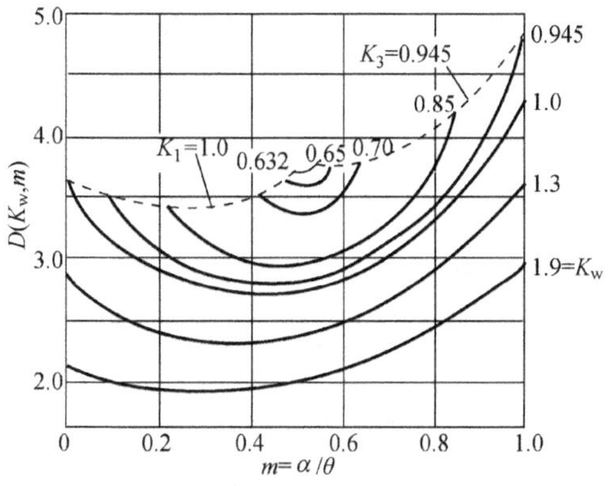

Fig. 8b Drag similarity curves for asymmetric wedge airfoil [Eq. (45)]

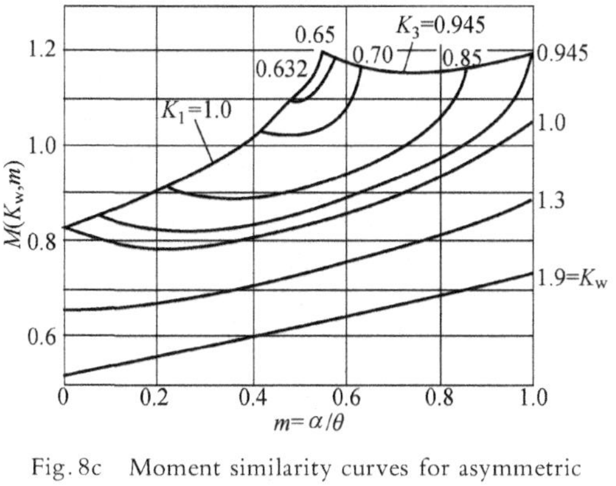

Fig. 8c Moment similarity curves for asymmetric wedge airfoil [Eq. (45)]

Mach Numbers much greater than 1.0, the lift curve is linear, as is well known. Considering, on the other hand, M^0 constant, then K_w increases as the wedge becomes thinner, and the ratio m represents a smaller variation of angle of attack. In the limit, when $K_w \to \infty$, the representation is that of a flat plate at zero angle of attack for all values of m.

To illustrate the nature of the coefficients in the usual form, calculations have been carried out for the case of $\theta = 3°$ and are plotted in Figs. 9 through 12.

Fig. 9 shows that, at small angles of attack giving positive lift, the lift coefficient actually increases with increase in Mach Number. This is a behavior not predicted by Ackeret's linear theory, which requires a decrease in lift coefficient with Mach Number according to the factor $1/(M^{0^2} - 1)^{1/2}$.

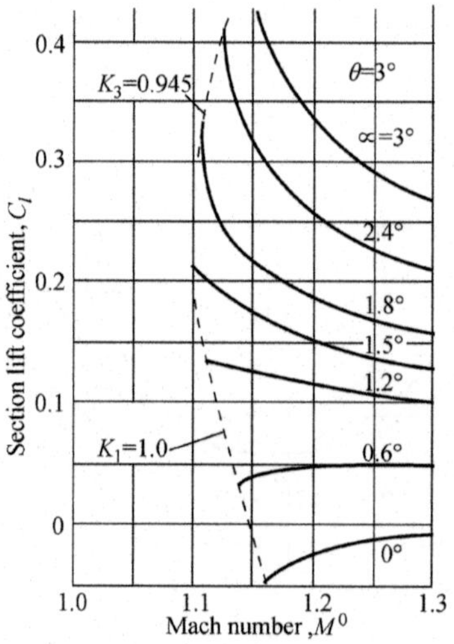

Fig. 9 Variation of section lift coefficient with Mach Number for the asymmetric wedge airfoil ($\theta = 3°$)

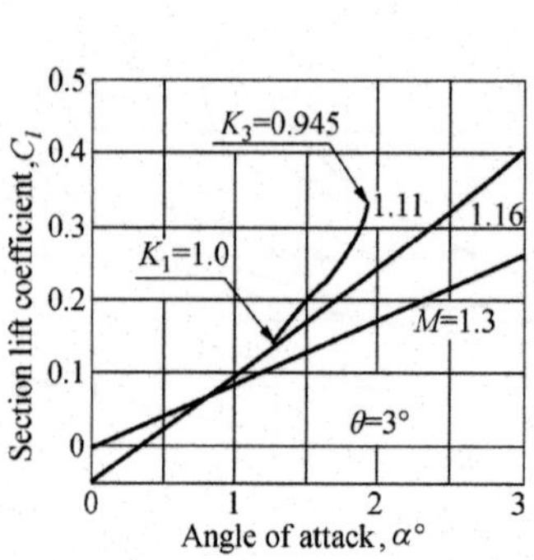

Fig. 10 Variation of section lift coefficient with angle of attack for the asymmetric wedge airfoil ($\theta = 3°$)

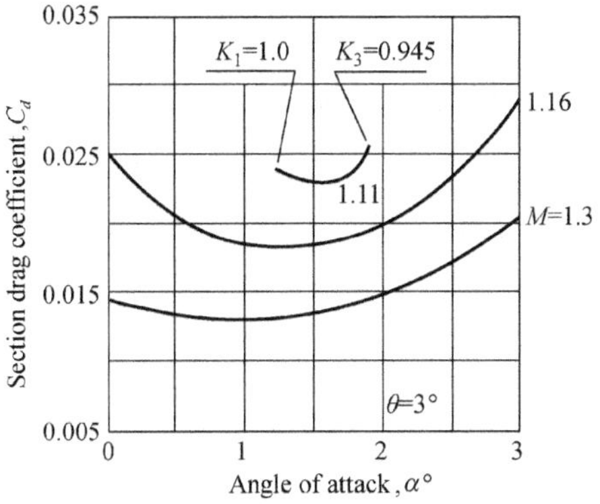

Fig. 11 Variation of section drag coefficient with angle of attack for the asymmetric wedge airfoil ($\theta = 3°$)

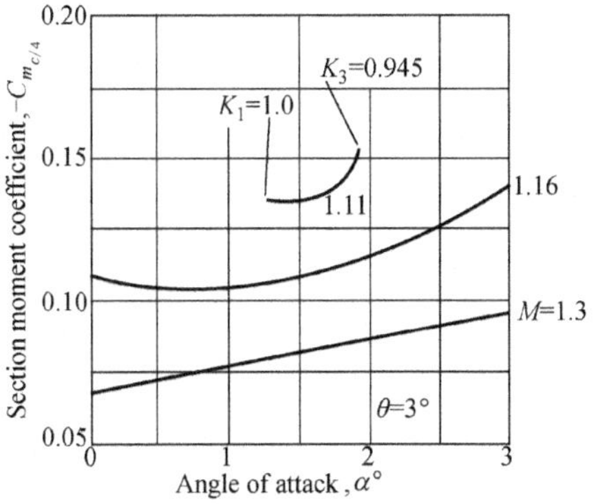

Fig. 12 Variation of section moment coefficient (about quarter-chord) with angle of attack for the asymmetric wedge airfoil ($\theta = 3°$)

References

[1] von Kármán Th. The similarity Law of Transonic Flow. Journal of Math. And Physics, 1947, 26(3): 182 - 190.

[2] Liepmann H W, Puckett A E. Introduction to Aerodynamics of a Compressible Fluid. John Wiley & Sons, New York, 1947: 51 - 58.

[3] Burington R S. Handbook of Mathematical Tables and Formulas. 2nd ed., Handbook Publishers, Sandusky Ohio, 1947: 7 - 9.

[4] Ibid., reference [2]: 212.

Interaction between Parallel Streams of Subsonic and Supersonic Velocities*

H. S. Tsien** and M. Finston***

(*Massachusetts Institute of Technology*)

Summary

An essential feature of the interaction phenomena between the compression shock and the boundary layer is the coexistence of parallel streams of subsonic and supersonic velocities. The compression disturbance of the shock is propagated "ahead" of the shock through the subsonic portion of the boundary layer, and "softening" of the shock takes place. It is the purpose of the present paper to demonstrate definitely this essential feature of boundary layer-shock interaction by using a simplified model. The boundary layer is replaced by a uniform subsonic stream of finite width, bounded on one side by the solid wall and on the other side by the interface with the uniform supersonic stream of infinite extent. The flow fields of two cases are analyzed in detail: (a) a compression wave in the supersonic stream incident upon the subsonic layer, and (b) outgoing compression waves generated by a sudden change in the slope of the solid wall. Both types of disturbances are assumed to be small, so that linearization of the differential equations is possible. A general result is that the distance of upstream propagation is directly proportional to the width of the subsonic layer. Therefore, when the boundary layer is turbulent and the subsonic layer is extremely thin, there is negligible softening of the shock.

Introduction

The phenomenon of interaction between the shock and the boundary layer in supersonic flow over a solid surface, as first explicitly demonstrated by Ackeret, Feldmann, and Rott[1], and then by Liepmann[2], has aroused great interest among the investigators of fluid mechanics. However, because of the great complexity of the phenomenon that involves both viscous heat-conducting effects and the effects of compressibility, it has not been possible to give a complete account based upon the first principles. The above-cited authors themselves have tried to elucidate some features of their experimental observations. A more conscientious effort on the qualitative relations in the interaction of shock and boundary layer is that of Lees[3].

Presented at the Fluid Mechanics Session, Seventeenth Annual Meeting, I. A. S., New York, January 24 - 27, 1949.

Journal of the Aeronautical Sciences, vol. 16, pp. 515 - 528, 1949.

* This paper is based on work supported in part by the Bureau of Ordnance, U. S. Navy.

** Professor of Aerodynamics.

*** Staff Member, Supersonic Laboratory

Lagerstrom[4] introduced the effects of viscosity and heat-conduction into the discussion of general compressible flow outside of the boundary layer. However, in order to make the mathematics involved tractable, he limits himself to the consideration of small disturbances and is thus able to linearize the differential equations. Under this assumption, the velocity at every point of the flow field has to be extremely close to the free-stream velocity — i.e., the flow is completely supersonic. On the other hand, it is the general belief of the investigators of the interaction phenomenon that one of the essential features is the coexistence of subsonic velocities in the boundary layer with the supersonic velocities in the main stream. This important fact is most clearly demonstrated by Howarth in a recent paper[5]. Howarth considers the problem of effects of a disturbance generated in a uniform supersonic stream that is bounded on one side by an interface with a uniform subsonic stream. Both regions are thus of semi-infinite extent. By assuming small disturbances and a perfect compressible fluid, he shows that the disturbances are propagated upstream through the subsonic region and that there is a region of expansion immediately behind the point of incidence of a compression wave in the interface.

Howarth's results are in general agreement with experimental observations on the shock and boundary-layer interaction. Thus the concept that at least one of the essential features of this phenomenon is the coexistence of subsonic flow with supersonic flow is proved quantitatively. The purpose of the present paper is to continue this line of attack and to improve Howarth's model for the real flow. The boundary layer is simulated by a uniform subsonic stream of unite width bounded on one side by a solid wall and on the other side by the interface with a uniform supersonic stream of semi-infinite extent. The fluid is assumed to be nonviscous and nonheat-conducting and the disturbances are assumed to be small. Two cases will be considered. First, the case of an incident compression wave from the supersonic stream on the interface will be studied. Second, the wall is assumed to have a break with a small constant slope after the break. For both cases, the pressure distribution at the interface and at the wall will be calculated. The results can then be used to interpret the experiments on the interaction of shock and the boundary layer.

Incident Wave

Let the x-axis be parallel to the undisturbed flows and coincide with the undisturbed interface (Fig. 1). For $y > 0$, the flow is supersonic, and quantities there will be denoted by the subscript 1. Let the line $y = -b$ represent the solid wall. For $-b < y < 0$, the flow is subsonic, and quantities there will be denoted by the subscript 2. Furthermore, the pressure, the density, and the Mach Number will be denoted by p, ρ, and M, respectively. Then $M_1 > 1$ and $M_2 < 1$. The linearized differential equation for the disturbance velocity potential φ_1 in the supersonic stream is

$$(M_1^2 - 1)(\partial^2 \varphi_1 / \partial x^2) - (\partial^2 \varphi_1 / \partial y^2) = 0 \qquad (1)$$

The general solution of this equation can be written as

$$\varphi_1 = f(x+\sqrt{M_1^2-1}y)+g(x-\sqrt{M_1^2-1}y) \tag{2}$$

where f and g are arbitrary functions. If the undisturbed flow is in the positive x- direction, then waves represented by the first term on the right of Eq. (2) are the incoming waves. The second term represents the outgoing waves originating from the interface.

To fix ideas, the form of the function f will now be specialized. Let

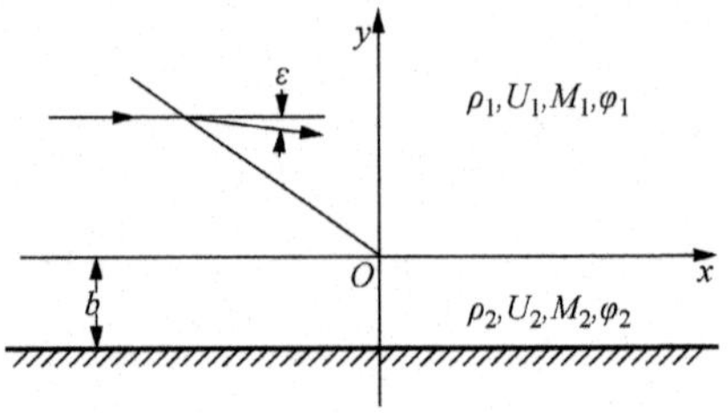

Fig. 1 Compression wave with deflection angle ε incident upon the interface of parallel subsonic and supersonic streams

$$\left.\begin{aligned} f(x+\sqrt{M_1^2-1}y) &= U_1\varepsilon\frac{1}{\beta\sqrt{M_1^2-1}}e^{-\beta(x+\sqrt{M_1^2-1}y)}, \quad \text{for} \quad x+\sqrt{M_1^2-1}y>0 \\ f(x+\sqrt{M_1^2-1}y) &= 0, \quad \text{for} \quad x+\sqrt{M_1^2-1}y<0 \end{aligned}\right\} \tag{3}$$

where U_1 is the undisturbed velocity in the supersonic stream, and ε, β are constants. The reason for choosing this form for f will be clear presently. By differentiating the potential φ_1 with respect to x and y, one obtains the x and y components u_1 and v_1 of the disturbance velocity in the supersonic stream. Therefore, the disturbance velocities of the incoming waves in the x and y directions are

$$-U_1\varepsilon(1/\sqrt{M_1^2-1})e^{-\beta(x+\sqrt{M_1^2-1}y)}$$

and

$$-U_1\varepsilon\, e^{-\beta(x+\sqrt{M_1^2-1}y)}, \quad \text{for} \quad x+\sqrt{M_1^2-1}y>0$$

For $x+\sqrt{M_1^2-1}y<0$ there is no disturbance. By making $\beta\to 0$, the incoming waves then degenerate into a single compression wave with a deflection angle equal to ε, and the origin of the x, y plane is the point of incidence of this compression wave with the interface of supersonic and subsonic streams.

For convenience of calculation, the discontinuous form of the disturbance velocities due to incoming waves can be written in the equivalent Fourier-integral form. This is possible by noting that

$$\left.\begin{array}{ll} e^{-\beta z}, & z>0 \\ 0, & z<0 \end{array}\right\} = \frac{1}{\pi}\int_0^\infty \left\{\frac{\beta}{\beta^2+\lambda^2}\cos\lambda z + \frac{\lambda}{\beta^2+\lambda^2}\sin\lambda z\right\} d\lambda \tag{4}$$

Then the potential of the outgoing wave should be also written in the form of a Fourier-integral. Thus,

$$g\left(x-\sqrt{M_1^2-1}y\right) = U_1\int_0^\infty \left\{A_1(\lambda)\sin\lambda\left(x-\sqrt{M_1^2-1}y\right)+B_1(\lambda)\cos\lambda\left(x-\sqrt{M_1^2-1}y\right)\right\} d\lambda \tag{5}$$

where $A_1(\lambda)$ and $B_1(\lambda)$ are the undetermined Fourier coefficients. As will be shown, the quantities of particular interest are the disturbance velocities at the interface. These can be obtained from Eqs. (3) and (5) with the aid of Eq. (4). Thus,

$$u_{1y=+0} = U_1\left[-\frac{\varepsilon}{\pi\sqrt{M_1^2-1}}\int_0^\infty \left\{\frac{\beta}{\beta^2+\lambda^2}\cos\lambda x + \frac{\lambda}{\beta^2+\lambda^2}\sin\lambda x\right\} d\lambda + \int_0^\infty \{\lambda A_1(\lambda)\cos\lambda x - \lambda B_1(\lambda)\sin\lambda x\} d\lambda\right] \tag{6}$$

$$v_{1y=+0} = U_1\left[-\frac{\varepsilon}{\pi}\int_0^\infty \left\{\frac{\beta}{\beta^2+\lambda^2}\cos\lambda x + \frac{\lambda}{\beta^2+\lambda^2}\sin\lambda x\right\} d\lambda - \sqrt{M_1^2-1}\int_0^\infty \{\lambda A_1(\lambda)\cos\lambda x - \lambda B_1(\lambda)\sin\lambda x\} d\lambda\right] \tag{7}$$

The linearized differential equation for the disturbance velocity potential φ_2 in the subsonic stream is

$$(1-M_2^2)(\partial^2\varphi_2/\partial x^2) + (\partial^2\varphi_2/\partial y^2) = 0 \tag{8}$$

The appropriate form of solution for the case concerned is

$$\varphi_2 = U_2\int_0^\infty \cosh\lambda\sqrt{1-M_2^2}(y+b)\{A_2(\lambda)\sin\lambda x + B_2(\lambda)\cos\lambda x\} d\lambda \tag{9}$$

where $A_2(\lambda)$ and $B_2(\lambda)$ are again the undetermined Fourier coefficients. That this form of φ_2 satisfies the differential equation can be easily verified. The boundary condition on the wall is

$$v_{2y=-b} = 0 \tag{10}$$

This condition is satisfied also by φ_2 specified by Eq. (9). At the interface, $y = 0$, the disturbance velocities of the subsonic side are thus

$$u_{2y=-0} = U_2\int_0^\infty \cosh\lambda\sqrt{1-M_2^2}b\{\lambda A_2(\lambda)\cos\lambda x - \lambda B_2(\lambda)\sin\lambda x\} d\lambda \tag{11}$$

and

$$v_{2y=-0} = U_2\sqrt{1-M_2^2}\int_0^\infty \sinh\lambda\sqrt{1-M_2^2}b\left\{\lambda A_2(\lambda)\sin\lambda x + \lambda B_2(\lambda)\cos\lambda x\right\} d\lambda \tag{12}$$

In the solutions for supersonic stream and subsonic stream, there are four unknowns A_1,

A_2, B_1, and B_2. These have to be determined by the conditions at the interface. Since the disturbances are assumed to be small, the values of quantities at the interface can be always taken at $y=0$, the undisturbed interface. The physical conditions to be satisfied at the interface are equal static pressure and equal inclination of the flow. Since under undisturbed conditions the static pressures in the supersonic stream must be equal to the static pressure in the subsonic stream, only the change in static pressure due to the disturbance velocities needs to be considered. The change in static pressure is equal to $-\rho U u$ under the assumption of small disturbances; thus

$$-\rho_1 U_1 u_{1_{y=+0}} = -\rho_2 U_2 u_{2_{y=-0}} \tag{13}$$

The condition for equal flow inclination at the interface is similarly,

$$v_{1_{y=+0}}/U_1 = v_{2_{y=-0}}/U_2 \tag{14}$$

Eqs. (13) and (14) are then the required conditions at the interface.

By substituting Eqs. (6), (7), (11), and (12) into Eqs. (13) and (14), one can then equate the corresponding Fourier coefficients on the two sides of these boundary equations. With some reductions, the resulting equations from Eq. (13) are

$$A_1(\lambda) - \frac{M_2^2}{M_1^2}\cosh\left(\lambda b\sqrt{1-M_2^2}\right)A_2(\lambda) = \frac{\varepsilon}{\pi\sqrt{M_1^2-1}}\left[\frac{\beta}{\lambda(\beta^2+\lambda^2)}\right] \tag{15}$$

$$B_1(\lambda) - \frac{M_2^2}{M_1^2}\cosh\left(\lambda b\sqrt{1-M_2^2}\right)B_2(\lambda) = -\frac{\varepsilon}{\pi\sqrt{M_1^2-1}}\left[\frac{1}{\beta^2+\lambda^2}\right] \tag{16}$$

Similarly, from Eq. (14) one has

$$A_1(\lambda) + \frac{\sqrt{1-M_2^2}}{\sqrt{M_1^2-1}}\sinh\left(\lambda b\sqrt{1-M_2^2}\right)B_2(\lambda) = -\frac{\varepsilon}{\pi\sqrt{M_1^2-1}}\left[\frac{\beta}{\lambda(\beta^2+\lambda^2)}\right] \tag{17}$$

$$B_1(\lambda) - \frac{\sqrt{1-M_2^2}}{\sqrt{M_1^2-1}}\sinh\left(\lambda b\sqrt{1-M_2^2}\right)A_2(\lambda) = \frac{\varepsilon}{\pi\sqrt{M_1^2-1}}\left(\frac{1}{\beta^2+\lambda^2}\right) \tag{18}$$

Eqs. (15) to (18) are the four equations for the four unknowns A_1, A_2, B_1, and B_2. Solution of these equations then determines the four quantities. For instance,

$$\left[\left(\frac{M_2^2}{M_1^2}\right)^2\cosh^2\left(\lambda b\sqrt{1-M_2^2}\right)+\left(\frac{1-M_2^2}{M_1^2-1}\right)\sinh^2\left(\lambda b\sqrt{1-M_2^2}\right)\right]A_2(\lambda)^{①}$$
$$= -\frac{2\varepsilon}{\pi\sqrt{M_1^2-1}(\beta^2+\lambda^2)}\times\left[\frac{\beta}{\lambda}\left(\frac{M_2^2}{M_1^2}\right)\cosh\left(\lambda b\sqrt{1-M_2^2}\right)+\right.$$
$$\left.\sqrt{\frac{1-M_2^2}{M_1^2-1}}\sinh\left(\lambda b\sqrt{1-M_2^2}\right)\right] \tag{19}$$

① "$A_2\lambda$" in the original paper has been corrected as "$A_2(\lambda)$". —Noted by editor.

$$\left[\left(\frac{M_2^2}{M_1^2}\right)^2\cosh^2(\lambda b\sqrt{1-M_2^2})+\left(\frac{1-M_2^2}{M_1^2-1}\right)\sinh^2(\lambda b\sqrt{1-M_2^2})\right]B_2(\lambda)$$

$$=\frac{2\varepsilon}{\pi\sqrt{1-M_2^2}(\beta^2+\lambda^2)}\times\left[\left(\frac{M_2^2}{M_1^2}\right)\cosh(\lambda b\sqrt{1-M_2^2})-\frac{\beta}{\lambda}\sqrt{\frac{1-M_2^2}{M_1^2-1}}\sinh(\lambda b\sqrt{1-M_2^2})\right] \quad (20)$$

There are similar expressions for A_1 and B_1. With the Fourier coefficients so determined, the solution of the problem is then complete.

However, the quantities that are of particular interest are the pressure distributions along the interface and along the solid wall. If Δp is the change in pressure from the undisturbed value, then $\Delta p=-\rho U u$. At the interface, let the change in pressure be denoted by $(\Delta p)_i$; then Eq. (11) gives

$$(\Delta p)_i=-\rho_2 U_2^2\int_0^\infty\cosh(\lambda\sqrt{1-M_2^2}b)\{\lambda A_2(\lambda)\cos\lambda x-\lambda B_2(\lambda)\sin\lambda x\}d\lambda \quad (21)$$

The integral can be evaluated after the A_2 and B_2 from Eqs. (19) and (20) are substituted into the integrand. To simplify the resultant expression, it is convenient to introduce the following notations:

$$\cos\theta=\left[\left(\frac{M_2^2}{M_1^2}\right)^2-\left(\frac{1-M_2^2}{M_1^2-1}\right)\right]\Big/\left[\left(\frac{M_2^2}{M_1^2}\right)^2+\left(\frac{1-M_2^2}{M_1^2-1}\right)\right] \quad (22)$$

with

$$\sin\frac{\theta}{2}=\sqrt{\left(\frac{1-M_2^2}{M_1^2-1}\right)\Big/\left[\left(\frac{M_2^2}{M_1^2}\right)^2+\frac{1-M_2^2}{M_1^2-1}\right]} \quad (23)$$

and

$$\cos\frac{\theta}{2}=\left(\frac{M_2^2}{M_1^2}\right)\Big/\sqrt{\left(\frac{M_2^2}{M_1^2}\right)^2+\frac{1-M_2^2}{M_1^2-1}} \quad (24)$$

Then Eq. (21) can be written as

$$\frac{(\Delta p)_i}{(1/2)\rho_1 U_1^2}=\frac{4\varepsilon}{\pi\sqrt{M_1^2-1}}\left[\cos^2\frac{\theta}{2}\int_0^\infty\left\{\frac{\beta}{\beta^2+\lambda^2}\cos\lambda x+\frac{\lambda}{\beta^2+\lambda^2}\sin\lambda x\right\}d\lambda+\frac{1}{2}\sin^2\theta\times\right.$$
$$\int_0^\infty\frac{(\beta\cos\lambda x+\lambda\sin\lambda x)d\lambda}{(\beta^2+\lambda^2)[\cosh(2b\sqrt{1-M_2^2}\lambda)+\cos\theta]}+\frac{1}{2}\sin\theta\times \quad (25)$$
$$\left.\int_0^\infty\frac{\{\lambda\sinh(2b\sqrt{1-M_2^2}\lambda)\cos\lambda x-\beta\sinh(2b\sqrt{1-M_2^2}\lambda)\sin\lambda x\}d\lambda}{(\beta^2+\lambda^2)[\cosh(2b\sqrt{1-M_2^2}\lambda)+\cos\theta]}\right]$$

Now the problem can be greatly simplified by making $\beta\to 0$ — i.e., the incoming waves are finally reduced to a simple compression wave with a deflection angle ε. At this limit, the first integral of Eq. (25) becomes the unit step function $l(x)$ — i.e.,

$$\left.\begin{aligned}l(x)&=1,\quad x>0\\&=0,\quad x<0\end{aligned}\right\} \quad (26)$$

By the substitution $\lambda/\beta=z$, the first part of the second integral is

$$\lim_{\beta\to 0}\int_0^{\infty}\frac{\beta\cos\lambda x\,d\lambda}{(\beta^2+\lambda^2)\left[\cosh(2b\sqrt{1-M_2^2}\lambda)+\cos\theta\right]}=\lim_{\beta\to 0}\int_0^{\infty}\frac{\cos(\beta xz)\,dz}{(z^2+1)\left[\cosh(2b\sqrt{1-M_2^2}\beta z)+\cos\theta\right]}$$

$$=\frac{1}{1+\cos\theta}\int_0^{\infty}\frac{dz}{z^2+1}=\frac{\pi}{2(1+\cos\theta)}$$

$$=\frac{\pi}{4\cos^2(\theta/2)}$$

Denoting the limiting value of $(\Delta p)_i$ by $(\Delta p)_i^*$,

$$\frac{(\Delta p)_i^*}{(1/2)\rho_1 U_1^2}=\frac{4\varepsilon}{\sqrt{M_1^2-1}}\left[\cos^2\frac{\theta}{2}l(\xi)+\frac{1}{2}\sin^2\frac{\theta}{2}+\frac{1}{2\pi}\sin^2\theta\int_0^{\infty}\frac{\sin(S\xi/\pi)\,dS}{S[\cosh S+\cos\theta]}+\frac{1}{2\pi}\sin\theta\int_0^{\infty}\frac{\sinh S\cos(S\xi/\pi)\,dS}{S[\cosh S+\cos\theta]}\right]\tag{27}$$

where

$$S=2b\lambda\sqrt{1-M_2^2}\tag{28}$$

$$\xi=\pi x/(2b\sqrt{1-M_2^2})\tag{29}$$

ξ is thus the characteristic nondimensional distance parameter. The first integral in Eq. (27) is an odd function of ξ, while the second integral is an even function of ξ. As shown in the Appendix, for positive values of ξ, these integrals have the following values:

$$\frac{1}{2\pi}\int_0^{\infty}\frac{\sin(S\xi/\pi)\,dS}{S[\cosh S+\cos\theta]}=\frac{1}{8\cos^2(\theta/2)}+\frac{1}{2\pi\sin\theta}\sum_{n=0}^{\infty}e^{-(2n+1)\xi}\left\{\frac{e^{-(\theta/\pi)\xi}}{2n+1+(\theta/\pi)}-\frac{e^{(\theta/\pi)\xi}}{2n+1-(\theta/\pi)}\right\}\tag{30}$$

and

$$\frac{1}{2\pi}\int_0^{\infty}\frac{\sinh S\cos(S\xi/\pi)\,dS}{S[\cosh S+\cos\theta]}=\frac{1}{2\pi}\sum_{n=0}^{\infty}e^{-(2n+1)\xi}\left\{\frac{e^{-(\theta/\pi)\xi}}{2n+1+(\theta/\pi)}+\frac{e^{(\theta/\pi)\xi}}{2n+1-(\theta/\pi)}\right\}\tag{31}$$

By substituting the values of these integrals into Eq. (27), the pressure distribution at the interface for positive ξ is, finally,

$$\frac{(\Delta p)_i^*}{(1/2)\rho_1 U_1^2}=\frac{4\varepsilon}{\sqrt{M_1^2-1}}[1+\sin\theta F_1(\xi;\theta)],\quad \xi>0\tag{32}$$

where the function $F_1(\xi;\theta)$ is

$$F_1(\xi;\theta)=\frac{1}{\pi}\sum_{n=0}^{\infty}\frac{e^{-[2n+1+(\theta/\pi)]\xi}}{2n+1+(\theta/\pi)}\tag{33}$$

By observing the proper symmetry of the integrals in Eq. (27), one can easily obtain the pressure distribution for negative values of ξ:

$$\frac{(\Delta p)_i^*}{(1/2)\rho_1 U_1^2}=\frac{4\varepsilon}{\sqrt{M_1^2-1}}\sin\theta F_2(\xi;\theta),\quad \xi<0\tag{34}$$

where

$$F_2(\xi;\theta) = \frac{1}{\pi}\sum_{n=0}^{\infty}\frac{e^{-[2n+1-(\theta/\pi)]|\xi|}}{2n+1-(\theta/\pi)} \tag{35}$$

To determine the pressure distribution along the wall, one obtains first, from Eq (9), the x-component of disturbance velocity u_2 at $y=-b$, the wall. Thus with values of A_2 and B_2 from Eqs. (19) and (20),

$$u_{2y=-b} = U_2\int_0^{\infty}\{\lambda A_2(\lambda)\cos\lambda x - \lambda B_2(\lambda)\sin\lambda x\}\,d\lambda$$

The pressure change at the wall $(\Delta p)_w$ can then be calculated as $-\rho_2 U_2 u_2(y=-b)$. Now let $\beta\to 0$ so as to obtain the value $(\Delta p)_w^*$ for a single compression wave of deflection ε. The procedure of calculation is similar to that for the pressure distribution along the interface. The final result is

$$\frac{(\Delta p)_w^*}{(1/2)\rho_1 U_1^2} = \frac{4\varepsilon}{\sqrt{M_1^2-1}}\left[\frac{1}{2}+\cos^2\frac{\theta}{2}\left(\frac{2}{\pi}\right)\int_0^{\infty}\frac{\cosh\frac{S}{2}\sin\frac{S\xi}{\pi}dS}{S[\cosh S+\cos\theta]}+\right.$$
$$\left.\sin\frac{\theta}{2}\cos\frac{\theta}{2}\left(\frac{2}{\pi}\right)\int_0^{\infty}\frac{\sinh\frac{S}{2}\cos\frac{S\xi}{\pi}dS}{S[\cosh S+\cos\theta]}\right] \tag{36}$$

The first integral in Eq. (36) is again an odd function of ξ and the second integral is an even function of ξ. As given in the Appendix, for positive values of ξ, these integrals can be evaluated as follows:

$$\frac{2}{\pi}\int_0^{\infty}\frac{\cosh\frac{S}{2}\sin\frac{S\xi}{\pi}dS}{S[\cosh S+\cos\theta]} = \frac{1}{2\cos^2\frac{\theta}{2}} -$$
$$\frac{1}{\pi\cos\frac{\theta}{2}}\sum_{n=0}^{\infty}(-1)^n e^{-(2n+1)\xi}\left\{\frac{e^{-(\theta/\pi)\xi}}{2n+1+\frac{\theta}{\pi}}+\frac{e^{(\theta/\pi)\xi}}{2n+1-\frac{\theta}{\pi}}\right\} \tag{37}$$

for $\xi>0$.

$$\frac{2}{\pi}\int_0^{\infty}\frac{\sinh\frac{S}{2}\cos\frac{S\xi}{2}dS}{S[\cosh S+\cos\theta]} = -\frac{1}{\pi\sin\frac{\theta}{2}}\sum_{n=0}^{\infty}(-1)^n e^{-(2n+1)\xi}\left\{\frac{e^{-(\theta/\pi)\xi}}{2n+1+\frac{\theta}{\pi}}-\frac{e^{(\theta/\pi)\xi}}{2n+1-\frac{\theta}{\pi}}\right\} \tag{38}$$

for $\xi>0$. Then, for positive values of ξ, the pressure distribution can be written as

$$\frac{(\Delta p)_w^*}{(1/2)\rho_1 U_1^2} = \frac{4\varepsilon}{\sqrt{M_1^2-1}}\left[1-2\cos\frac{\theta}{2}F_3(\xi;\theta)\right],\quad \xi>0 \tag{39}$$

where

$$F_3(\xi;\theta) = \frac{1}{\pi}\sum_{n=0}^{\infty}(-1)^n \frac{e^{-[2n+1+(\theta/\pi)]|\xi|}}{2n+1+(\theta/\pi)} \tag{40}$$

By utilizing the symmetry properties of the integrals in Eq. (36), one can immediately write down the expression for the pressure distribution along the wall for negative values of ξ:

$$\frac{(\Delta p)^*_w}{(1/2)\rho_1 U_1^2} = \frac{4\varepsilon}{\sqrt{M_1^2-1}} 2\cos\frac{\theta}{2} F_4(\xi;\theta), \quad \xi < 0 \tag{41}$$

where

$$F_4(\xi;\theta) = \frac{1}{\pi}\sum_{n=0}^{\infty}(-1)^n \frac{e^{-[2n+1-(\theta/\pi)]|\xi|}}{2n+1-(\theta/\pi)} \tag{42}$$

Eqs. (32), (34), (39), and (41) then give the formulas for computing the pressure distributions of interest.

For comparison with experimental observations on the shock-boundary layer interaction, it is of interest to determine the position of the displaced interface between the supersonic and the subsonic flows. Let the displacement from the undisturbed position be denoted by η. The slope $d\eta/dx$ of the disturbed interface is found by substituting Eqs. (19) and (20) into Eq. (12) so that

$$\frac{d\eta}{dx} = \frac{V_{2y=-0}}{U_2} = -\frac{2\varepsilon\sqrt{\dfrac{1-M_2^2}{M_1^2-1}}}{\pi\left[\left(\dfrac{M_2^2}{M_1^2}\right)^2 + \dfrac{1-M_2^2}{M_1^2-1}\right]} \times$$

$$\left[\int_0^{\infty} \frac{\left\{\beta\left(\dfrac{M_2^2}{M_1^2}\right)\sinh\left(2\lambda b\sqrt{1-M_2^2}\right) + \lambda\sqrt{\dfrac{1-M_2^2}{M_1^2-1}}\left(-\dfrac{1}{2}+\dfrac{1}{2}\cosh\left(2\lambda b\sqrt{1-M_2^2}\right)\right)\right\}\sin\lambda x\, d\lambda}{(\beta^2+\lambda^2)\left\{\cosh\left(2\lambda b\sqrt{1-M_2^2}\right)+\cos\theta\right\}} -\right.$$

$$\left.\int_0^{\infty} \frac{\left\{\lambda\left(\dfrac{M_2^2}{M_1^2}\right)\sinh\left(2\lambda b\sqrt{1-M_2^2}\right) - \beta\sqrt{\dfrac{1-M_2^2}{M_1^2-1}}\left[-\dfrac{1}{2}+\dfrac{1}{2}\cosh\left(2\lambda b\sqrt{1-M_2^2}\right)\right]\right\}\cos\lambda x\, d\lambda}{(\beta^2+\lambda^2)\left\{\cosh\left(2\lambda b\sqrt{1-M_2^2}\right)+\cos\theta\right\}}\right]$$

Introducing ξ, S and passing to the limit $\beta \to 0$, as above

$$\frac{d\eta}{dx} = -\frac{2\varepsilon}{\pi}\sin\frac{\theta}{2}\left[\sin\frac{\theta}{2}\left\{\int_0^{\infty}\frac{\cosh S\sin\dfrac{S\xi}{\pi}dS}{S[\cosh S+\cos\theta]} - \int_0^{\infty}\frac{\sin\dfrac{S\xi}{\pi}dS}{S[\cosh S+\cos\theta]}\right\} -\right.$$

$$\left.\cos\frac{\theta}{2}\int_0^{\infty}\frac{\sinh S\cos\dfrac{S\xi}{\pi}dS}{S[\cosh S+\cos\theta]}\right]$$

It can be shown that, for $\xi > 0$,

$$I(\xi) = \int_0^{\infty}\frac{\cosh S\sin(S\xi/\pi)dS}{S[\cosh S+\cos\theta]} = \frac{\pi}{2(1+\cos\theta)} - \pi\cot\theta[F_1(\xi;\theta) - F_2(\xi;\theta)]$$

With Eqs. (30) and (31) and $I(-\xi) = -I(\xi)$,

$$\begin{aligned} d\eta/dx &= 2\varepsilon \sin\theta F_1(\xi;\theta), \quad \xi > 0 \\ &= 2\varepsilon \sin\theta F_2(\xi;\theta), \quad \xi < 0 \end{aligned} \tag{43}$$

Noting that the displacement vanishes far ahead of the disturbance, integration, with respect to ξ, of these results gives

$$\begin{aligned} \frac{\pi}{2b\sqrt{1-M_2^2}}[\eta(\xi;\theta)] &= 2\varepsilon \sin\theta \frac{1}{\pi}\sum_{n=0}^{\infty} \frac{e^{-[2n+1-(\theta/\pi)]|\xi|}}{\left(2n+1-\frac{\theta}{\pi}\right)^2}, \quad \xi < 0 \\ &= 2\varepsilon \sin\theta \frac{1}{\pi}\sum_{n=0}^{\infty}\left\{\frac{1}{\left(2n+1-\frac{\theta}{\pi}\right)^2} + \frac{1}{\left(2n+1+\frac{\theta}{\pi}\right)^2} - \frac{e^{-[2n+1+(\theta/\pi)]|\xi|}}{\left(2n+1+\frac{\theta}{\pi}\right)^2}\right\}, \quad \xi > 0 \end{aligned} \tag{44}$$

Since F_1, $F_2 \to +\infty$ as $\xi \to 0$, the interface slope is infinite at the point of incidence of the compression wave in the interface. However, it is easy to show that, where δ is a positive quantity,

$$\left(\frac{d\eta}{dx}\right)_{\xi=+\delta} - \left(\frac{d\eta}{dx}\right)_{\xi=-\delta} = 4\varepsilon \sin\theta \frac{1}{\pi}\sum_{n=0}^{\infty} \frac{e^{-(2n+1)\delta}\left\{-(2n+1)\sinh\frac{\theta\delta}{\pi} - \frac{\theta}{\pi}\cosh\frac{\theta\delta}{\pi}\right\}}{(2n+1)^2 - (\theta/\pi)^2}$$

Then, the limit as $\delta \to 0$ is

$$\left(\frac{d\eta}{dx}\right)_{\xi=+0} - \left(\frac{d\eta}{dx}\right)_{\xi=-0} = -4\varepsilon \sin\theta\left(\frac{\theta}{\pi^2}\right)\sum_{n=0}^{\infty} \frac{1}{(2n+1)^2 - (\theta/\pi)^2} \tag{45}$$

This is a negative quantity. The significance of these calculations will be discussed in a later section, together with the numerical results.

Inclined Wall

In the problem of the inclined wall, the solid wall is assumed to have a small inclination ε for $x > 0$. For $x < 0$, the wall is parallel to the undisturbed streams of supersonic and subsonic velocities (Fig. 2). The x-axis is again the undisturbed interface between the semi-infinite uniform supersonic stream of Mach Number M_1 and the uniform subsonic stream of width b and Mach Number M_2. Since now the disturbances all originate from the inclined wall, there are no incoming waves from the supersonic stream. The appropriate disturbance velocity potential φ_1 for the supersonic region is then

$$\varphi_1 = U_1\int_0^{\infty}\{a_1(\lambda)\sin\lambda\left(x-\sqrt{M_1^2-1}\,y\right) + b_1(\lambda)\cos\lambda\left(x-\sqrt{M_1^2-1}\,y\right)\}d\lambda \tag{46}$$

where a_1 and b_1 are again the undetermined Fourier coefficients. For the disturbance velocity potential φ_2 in the subsonic region, one can take

$$\varphi_2 = U_2\left[\frac{\varepsilon}{\pi\sqrt{1-M_2^2}}\int_0^\infty \frac{\sinh\lambda\sqrt{1-M_2^2}(y+b)}{\lambda}\left\{\frac{\beta}{\beta^2+\lambda^2}\cos\lambda x+\frac{\lambda}{\beta^2+\lambda^2}\sin\lambda x\right\}d\lambda + \int_0^\infty \cosh\lambda\sqrt{1-M_2^2}(y+b)\{a_2(\lambda)\sin\lambda x+b_2(\lambda)\cos\lambda x\}d\lambda\right] \tag{47}$$

where a_2 and b_2 are the Fourier coefficients to be determined. The y-component v_2 of the disturbance velocity at the wall, obtained by differentiating the expression in Eq. (47) with respect to y and then setting $y=-b$, is solely due to the first integral. Then, if $\beta\to 0$, one has

$$\begin{aligned} v_{2y=-b} &= U_2\varepsilon, \quad x>0 \\ &= 0, \quad x<0 \end{aligned}$$

This means that for $x>0$ the wall is inclined at a small angle ε from the free stream. Therefore the boundary condition at the wall is satisfied by choosing φ_2 as given by Eq. (47).

To determine the Fourier coefficients a_1, a_2, b_1, and b_2, the physical conditions at the interface as expressed by Eqs. (13) and (14) again have to be used. In other words, the velocity components at the interface from the supersonic side and from the subsonic side must be first obtained from the expressions for φ_1 and φ_2, and then these values are substituted into Eqs. (13) and (14). By equating the corresponding Fourier terms, the following equations for the unknowns a_1, a_2, b_1, and b_2 are obtained:

$$a_1(\lambda)-\frac{M_2^2}{M_1^2}\cosh\left(\lambda\sqrt{1-M_2^2}b\right)[a_2(\lambda)]=\frac{M_2^2}{M_1^2}\frac{\varepsilon\sinh\left(\lambda\sqrt{1-M_2^2}b\right)}{\pi\sqrt{1-M_2^2}}\left(\frac{1}{\beta^2+\lambda^2}\right) \tag{48}$$

$$b_2(\lambda)-\frac{M_2^2}{M_1^2}\cosh\left(\lambda\sqrt{1-M_2^2}b\right)[b_2(\lambda)]=\frac{M_2^2}{M_1^2}\frac{\varepsilon\sinh\left(\lambda\sqrt{1-M_2^2}b\right)}{\pi\sqrt{1-M_2^2}}\left[\frac{\beta}{\lambda(\beta^2+\lambda^2)}\right] \tag{49}$$

$$a_1(\lambda)+\sqrt{\frac{1-M_2^2}{M_1^2-1}}\sinh\left(\lambda\sqrt{1-M_2^2}b\right)[b_2(\lambda)]=-\frac{\varepsilon\cosh\left(\lambda\sqrt{1-M_2^2}b\right)}{\pi\sqrt{M_1^2-1}}\left[\frac{\beta}{\lambda(\beta^2+\lambda^2)}\right] \tag{50}$$

$$b_1(\lambda)-\sqrt{\frac{1-M_2^2}{M_1^2-1}}\sinh\left(\lambda\sqrt{1-M_2^2}b\right)[a_2(\lambda)]=\frac{\varepsilon\cosh\left(\lambda\sqrt{1-M_2^2}b\right)}{\pi\sqrt{M_1^2-1}}\left(\frac{1}{\beta^2+\lambda^2}\right) \tag{51}$$

The solution of these equations yields the values for a_2 and b_2 given by the following expressions:

$$\left[\left(\frac{M_2^2}{M_1^2}\right)^2\cosh^2\left(\lambda b\sqrt{1-M_2^2}\right)+\frac{1-M_2^2}{M_1^2-1}\sinh^2\left(\lambda b\sqrt{1-M_2^2}\right)\right]a_2(\lambda)$$
$$=-\frac{\varepsilon}{\pi\sqrt{M_1^2-1}(\beta^2+\lambda^2)}\left[\frac{\beta}{\lambda}\left(\frac{M_2^2}{M_1^2}\right)+\frac{1}{2}\sqrt{\frac{M_1^2-1}{1-M_2^2}}\left\{\left(\frac{M_2^2}{M_1^2}\right)^2+\frac{1-M_2^2}{M_1^2-1}\right\}\sinh\left(2\lambda b\sqrt{1-M_2^2}\right)\right] \tag{52}$$

$$\left[\left(\frac{M_2^2}{M_1^2}\right)^2\cosh^2\left(\lambda b\sqrt{1-M_2^2}\right)+\frac{1-M_2^2}{M_1^2-1}\sinh^2\left(\lambda b\sqrt{1-M_2^2}\right)\right]b_2(\lambda)$$
$$=-\frac{\varepsilon}{\pi\sqrt{M_1^2-1}(\beta^2+\lambda^2)}\left[-\frac{M_2^2}{M_1^2}+\frac{1}{2}\sqrt{\frac{M_1^2-1}{1-M_2^2}}\left\{\left(\frac{M_2^2}{M_1^2}\right)^2+\frac{1-M_2^2}{M_1^2-1}\right\}\sinh\left(2\lambda b\sqrt{1-M_2^2}\right)\right] \tag{53}$$

Similar expressions for a_1 and b_1 can be also computed.

With the Fourier coefficients so determined, the desired pressure distributions along the interface and the wall can be easily calculated by following exactly the procedure outlined in the problem of an incident wave. No details of calculation will be given here but only the final results. For the pressure distribution along the interface ($\beta \to 0$),

$$\frac{(\Delta p)_i^*}{(1/2)\rho_1 U_1^2} = \frac{2\varepsilon}{\sqrt{M_1^2 - 1}} \left[1 - \frac{1}{2} \frac{\cot(\theta/2)}{\sin(\theta/2)} F_4(\xi;\theta) - \frac{1}{2} \cos\frac{\theta}{2} \left(3 - \cot^2\frac{\theta}{2}\right) F_3(\xi;\theta) \right], \quad \xi > 0 \tag{54}$$

and

$$\frac{(\Delta p)_i^*}{(1/2)\rho_1 U_1^2} = \frac{2\varepsilon}{\sqrt{M_1^2 - 1}} \left[\frac{1}{2} \cos\frac{\theta}{2} \left(3 - \cot^2\frac{\theta}{2}\right) F_4(\xi;\theta) + \frac{1}{2} \frac{\cot(\theta/2)}{\sin(\theta/2)} F_3(\xi;\theta) \right], \quad \xi < 0 \tag{55}$$

The pressure distribution along the wall is given by ($\beta \to 0$)

$$\frac{(\Delta p)_w^*}{(1/2)\rho_1 U_1^2} = \frac{2\varepsilon}{\sqrt{M_1^2 - 1}} \left[1 + 2\cot\frac{\theta}{2} F_1(\xi;\theta) \right], \quad \xi > 0 \tag{56}$$

and

$$\frac{(\Delta p)_w^*}{(1/2)\rho_1 U_1^2} = \frac{2\varepsilon}{\sqrt{M_1^2 - 1}} 2\cot\frac{\theta}{2} F_2(\xi;\theta), \quad \xi < 0 \tag{57}$$

In Eqs. (54) to (57), the functions F_1, F_2, F_3, and F_4 are functions defined by Eqs. (33), (35), (40), and (42). The variable θ is connected with the Mach Numbers M_1 and M_2 as specified by Eqs. (22) to (24). These equations then supply the desired information for the problem of an inclined wall. The numerical results and the discussions of the results will be presented in the next section.

Numerical Results and Discussion

The parameter θ occurs in all expressions for the pressure distributions. As shown by Eq. (22), it is connected with the Mach Numbers M_1 and M_2 and can be taken to have the range from 0 to π. When $0 < M_2 < 1$, the value of θ increases to π when the ratio M_2/M_1 tends to zero. When $M_1 > 1$ but finite, θ tends to 0 when M_2 tends to 1. The general relationships between θ and M_1, M_2 are given by Fig. 3. It is seen that, if the representative Mach Numbers for the boundary-layer flow and the free stream are taken to be 0.8 and 2, then the corresponding value of θ in the flow model is approximately $3\pi/4$. This fact should be kept in mind when the theoretical results are compared with the experiments on the shock and boundary-layer interactions.

The series for the functions F_1, F_2, F_3, and F_4 as given by Eqs. (33), (35), (40), and (42) are suitable for numerical computation when ξ is large, since the convergence is then extremely rapid. For small values of ξ, modified series as given in the Appendix is more appropriate. F_1 and F_2 have a logarithmic infinity at $\xi = 0$, but F_3 and F_4 are finite at $\xi = 0$. Table 1 gives the tabulated values for these functions.

The results of computation for the case of an incoming compression wave are shown in Figs. 4, 5, 6, and 7. In these graphs, the abscissas are the characteristic nondimensional length ξ along the streams defined by Eq. (29). The physical distance x for any value of ξ is thus proportional to $b\sqrt{1-M_2^2}$. The width b of the undisturbed subsonic stream is then the measure of length in the problems concerned. The ordinates of these graphs are

$$(\Delta p)^* / [(1/2)\rho_1 U_1^2 (2\varepsilon/\sqrt{M_1^2 - 1})]$$

Since, under the general assumption of small disturbances, a compression wave of ε deflection is associated with a pressure rise equal to

$$(1/2)\rho_1 U_1^2 (2\varepsilon/\sqrt{M_1^2 - 1})$$

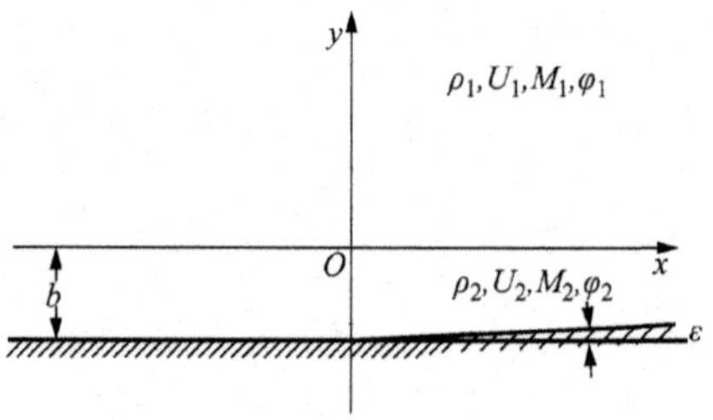

Fig. 2 Inclined wall with angle ε

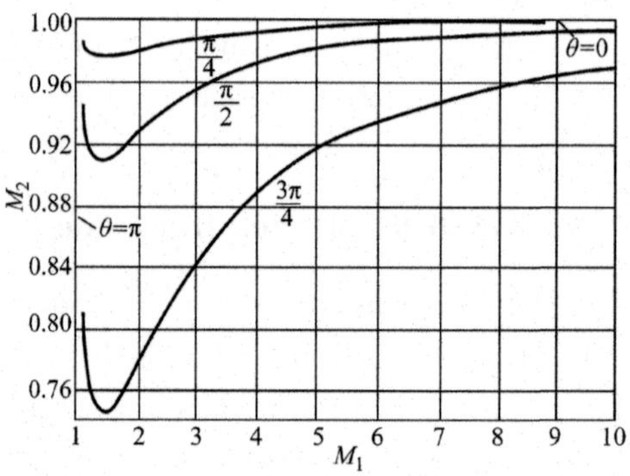

Fig. 3 The Mach Number parameter θ as function of Mach Numbers M_1 and M_2 in the supersonic and the subsonic streams, respectively

the ordinates of these graphs are then the ratio of pressure rise to the pressure rise caused by the simple incoming compression wave. It is then immediately seen that the pressure rise far downstream of the point of incidence of the wave on the interface is twice as large as the pressure rise for a simple compression wave. When an incoming compression wave is incident upon a solid wall, the compression wave is reflected as a compression wave, and the pressure rise after the reflection is twice as large as the pressure rise before the reflection. Therefore, except for the local interaction, the incoming compression wave is reflected as if by a solid wall without the subsonic stream.

Another feature of these graphs is that the pressure distribution along the interface has a peak at the point of incidence ($\xi = 0$) of the compression wave. Therefore before the point of

Table 1

	$\theta = 0$				$\theta = \pi/4$			
ξ	F_1	F_2	F_3	F_4	F_1	F_2	F_3	F_4
0	∞	∞	0.251 2	0.251 2	∞	∞	0.194 0	0.349 6
0.05	0.592 7	0.592 7	0.243 1	0.243 1	0.509 0	0.676 1	0.185 8	0.341 6
0.10	0.482 5	0.482 5	0.234 3	0.234 3	0.401 0	0.571 6	0.177 1	0.332 8
0.25	0.336 9	0.336 9	0.210 7	0.210 7	0.261 6	0.433 5	0.154 2	0.307 9
0.40	0.262 9	0.262 9	0.188 7	0.188 7	0.194 2	0.361 1	0.133 4	0.283 5
0.50	0.227 5	0.227 5	0.173 5	9.173 5	9.162 7	0.325 1	0.120 4	0.267 4
0.75	0.164 9	0.164 9	0.139 1	0.139 1	0.110 4	0.258 0	0.090 7	0.229 1
1.00	0.124 6	0.124 6	0.112 1	0.112 1	0.078 3	0.209 4	0.069 3	0.193 2
1.75	0.056 1	0.056 1	0.054 8	0.054 8	0.029 0	0.115 7	0.028 2	0.113 5
2.50	0.026 6	0.026 6	0.026 0	0.026 0	0.011 8	0.065 6	0.011 2	0.064 6
4.00	0.005 9	0.005 9			0.001 8	0.021 3		0.021 2
	$\theta = \pi/2$				$\theta = 3\pi/4$			
ξ	F_1	F_2	F_3	F_4	F_1	F_2	F_3	F_4
0	∞	∞	0.156 9	0.553 8	∞	∞	0.131 2	1.177 6
0.05	0.452 0	0.946 1	0.148 6	0.545 7	0.409 2	1.605 6	0.122 9	1.169 3
0.10	0.345 9	0.827 8	0.140 1	0.534 0	0.305 0	1.489 5	0.114 4	1.160 8
0.25	0.211 1	0.680 3	0.117 5	0.514 7	0.176 2	1.325 2	0.093 0	1.133 8
0.40	0.149 8	0.586 9	0.098 8	0.483 6	0.118 9	1.232 1	0.075 6	1.105 0
0.50	0.121 7	0.544 4	0.087 2	0.465 4	0.093 5	1.180 2	0.065 1	1.081 7
0.75	0.077 1	0.462 0	0.063 1	0.420 2	0.055 2	1.087 2	0.044 5	1.032 3
1.00	0.051 3	0.399 3	0.044 8	0.376 6	0.034 5	1.012 5	0.029 8	0.977 7
1.75	0.015 6	0.267 6	0.015 2	0.264 0	0.008 7	0.825 6	0.008 4	0.820 0
2.50	0.005 4	0.182 1	0.004 9	0.181 5	0.002 8	0.682 0	0.002 3	0.680 4
4.00	0.000 6	0.086 4		0.086 4	0.000 2	0.468 8		0.468 8

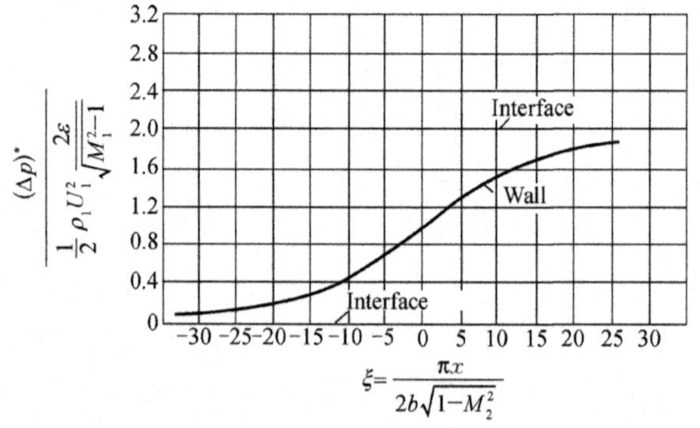

Fig. 4 Pressure increment $(\Delta p)^*$ for the case of incident wave plotted against the distance x from the point of incidence of the wave upon the interface $\theta = 0$

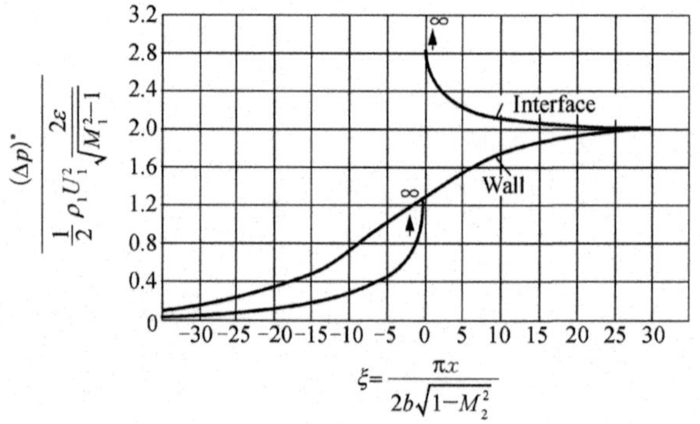

Fig. 5 Pressure increment $(\Delta p)^*$ for the case of incident wave plotted against the distance x from the point of incidence of the wave upon the interface $\theta = \pi/4$

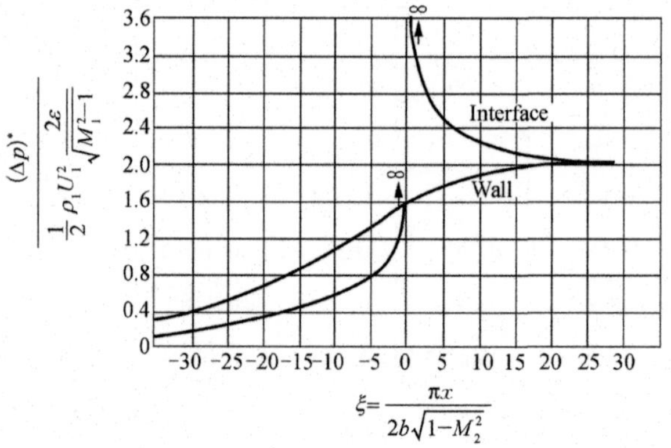

Fig. 6 Pressure increment $(\Delta p)^*$ for the case of incident wave plotted against the distance x from the point of incidence of the wave upon the interface $\theta = \pi/2$

incidence the flow is compressed, but after the point of incidence the flow is expanded. In the supersonic stream then, there is a series of compression wavelets sloping downstream, ahead of

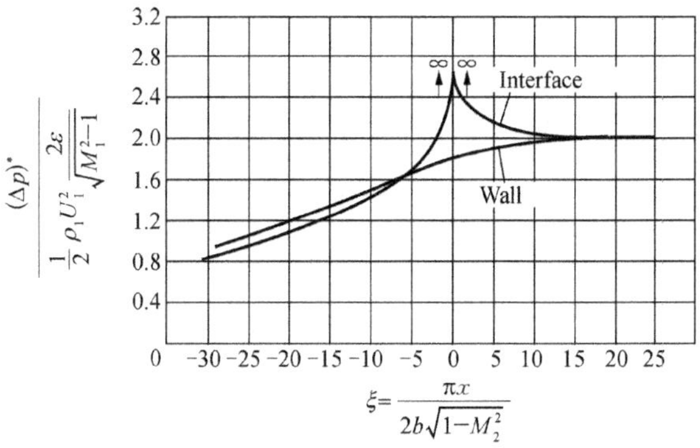

Fig. 7 Pressure increment $(\Delta p)^*$ for the case of incident wave plotted against the distance x from the point of incidence of the wave upon the interface $\theta = 3\pi/4$

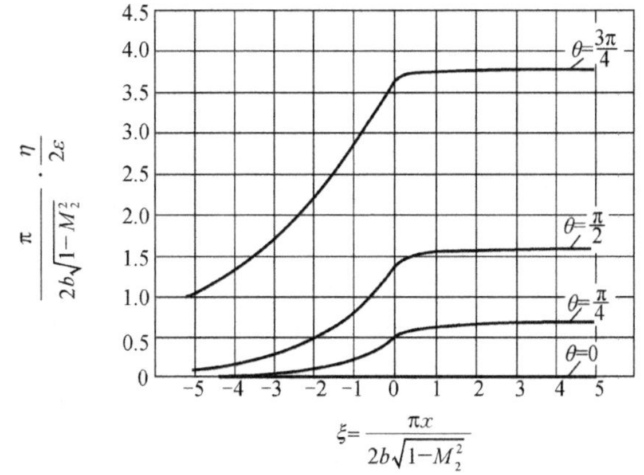

Fig. 8 Incident wave case, the displacement η of the interface (or increment of the thickness of the subsonic stream) plotted against the distance x from the point of incidence of wave upon the interface

the incoming compression wave. After the incident compression wave, there is an expansion region. The pressure distribution along the wall has no peak and is a smooth rising curve. Most of the pressure rise along the wall occurs, however, before the point of incidence of the compression wave. In fact for all points ahead of the point of incidence, the wall pressure is higher than the pressure at the interface. The region where there is strong interaction between the supersonic stream and the subsonic stream is of the order of tens of widths b of the subsonic stream. The length of this region is increased as θ approaches π.

Fig. 8 shows the results of computations for the interface displacement. A noticeable feature of these curves is the rapid change from an infinite slope at $\xi = 0$ to small values at $\xi \neq 0$. This is especially marked for $\xi > 0$. Eq. (45) shows that the slope for positive x is less than the slope for negative x. Therefore, as far as the subsonic flow is concerned, the corner at the point of incidence of the compression wave is a compression corner. This result agrees with computations for $(\Delta p)_i^*$, inasmuch as it implies compression upstream of $\xi = 0$, an infinite rise in pressure at $\xi = 0$, and expansion downstream of $\xi = 0$. It is also seen that the displacement increases as θ increases. Recalling that $\theta \to \pi$ as $M_2/M_1 \to 0$, for $0 < M_2 < 1$, this means the displacement increases as M_2 decreases — i.e., as the disturbance propagates upstream with greater strength (see Figs. 4 – 7), as would be expected.

One further comparison with the pressure calculations may be made. For $\xi < 0$, the wall pressure exceeds interface pressure except in an extremely small region about the singularity. The pressure gradient is such, then, as to be consistent with an increasing interface slope as $\xi \to 0$. For $\xi > 0$ the gradient is from the interface to the wall, consistent with a decreasing slope as ξ increases.

When a shock occurs over a laminar boundary layer, the case of the so-called λ-shock (Fig. 9), experiments show that there is a compression region in the free supersonic stream ahead of the shock and an expansion region after the shock. Furthermore, the pressure rise along the wall is known to be continuous. These features of the λ-shock are then completely demonstrated by the incident wave case discussed above. It is evident that the analysis presented is not a quantitative explanation of λ-shock, since the shock is a rather strong disturbance, while small disturbances and linearization of the differential equation are the basis of the present analysis. Furthermore, in the analysis, the viscous and the heat-conducting effects are completely neglected. Perhaps a more important omission is the velocity gradient in the boundary layer. The theoretical calculation assumes uniform supersonic and subsonic streams, and therefore the vorticity is concentrated at the interface instead of distributed throughout the subsonic layer, as is the case for the boundary layer. However, the close qualitative agreement between the theory and the experimental observations on the λ-shock indicates the importance of the effects of interaction between the supersonic and the subsonic region in the shock-boundary-layer phenomenon.

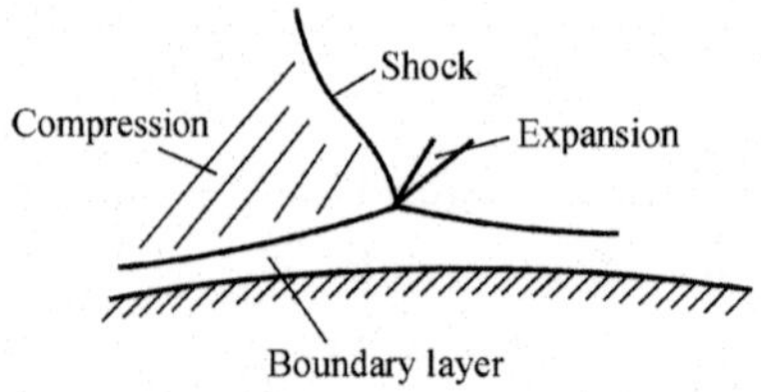

Fig. 9 Flow pattern of a λ-shock. Interaction of shock and laminar boundary layer

One might now ask: Why is the interaction between the shock and a turbulent boundary layer different from that of the λ-shock? In the case of a turbulent boundary layer, one does not observe the extended compression region before the shock and the expansion region after the shock. If the theory presented is of any real value, it must explain this case also. The authors believe the solution of this paradox lies in the thickness of the subsonic layer in case of turbulent boundary layer. Because of the much steeper velocity gradient resulting from the

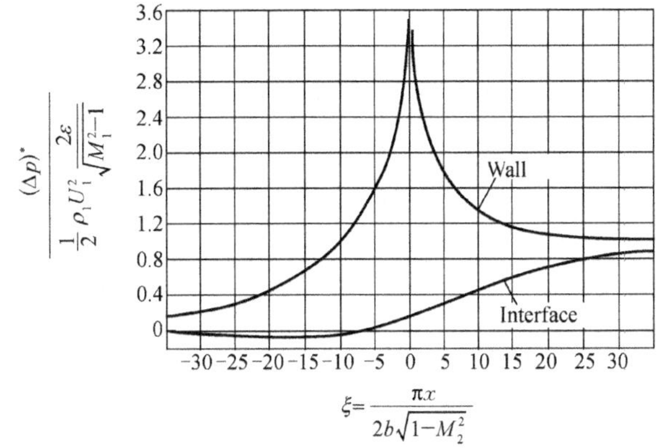

Fig. 10 Pressure increment $(\Delta p)^*$ for case of inclined wall plotted against the distance x from the break in the wall $\theta = \pi/4$

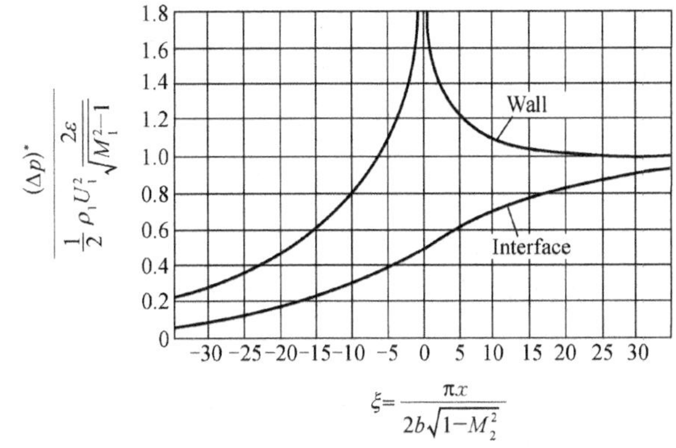

Fig. 11 Pressure increment $(\Delta p)^*$ for case of inclined wall plotted against the distance x from the break in the wall $\theta = \pi/2$

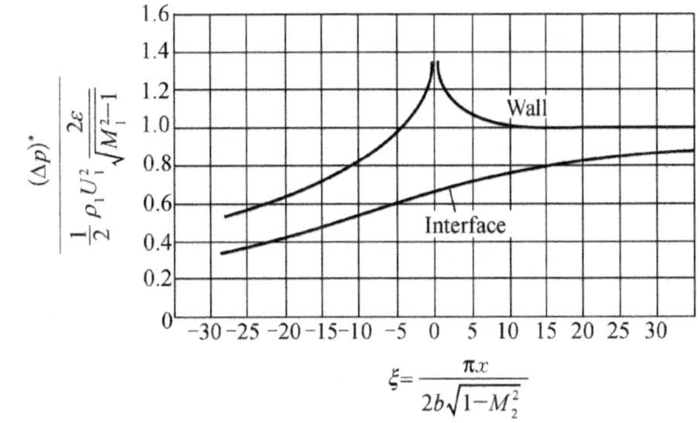

Fig. 12 Pressure increment $(\Delta p)^*$ for case of inclined wall plotted against the distance x from the break in the wall $\theta = 3\pi/4$

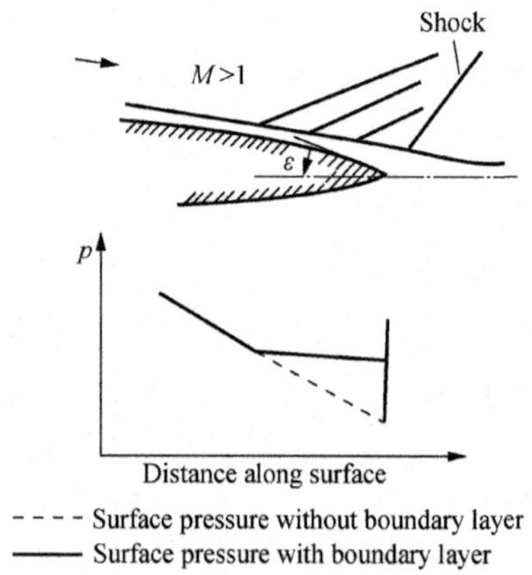

Fig. 13 Flow pattern near the trailing edge of a supersonic airfoil

extremely intensive turbulent flow exchange, the subsonic part of the turbulent boundary layer is generally extremely thin. For usual dimensions the subsonic layer in a turbulent boundary layer is only of the order of hundreths of an inch thick, while the subsonic layer in a laminar boundary layer is of the order of tenths of an inch thick. Therefore, the characteristic interaction effect between the subsonic and the supersonic flows, if it occurs, will be limited to a region of only a few tenths of an inch for the turbulent boundary layer. This would be difficult to observe, especially if, as is usually the case, the shock is not stationary but oscillates slightly in position. Furthermore, the large velocity gradient in the turbulent layer will certainly modify the flow considerably from that calculated theoretically, both because of the strong influence of vorticity and because of the effects of viscosity and heat-conductance. Therefore, while the theory presented cannot give a clear explanation for the interaction between shock and the turbulent boundary layer, the theory and the experiments are not contradictory.

The numerical results for the second problem of the inclined wall are presented in Figs. 10, 11, and 12. Here the general features of the pressure distributions are the same as for the problem of the incident wave. Only now the role of the wall and the interface is changed. The pressure distribution along the wall now has a peak at the starting point of the inclination, while the pressure at the interface rises steadily. At distances far downstream of the break in the wall, the pressure is equal to that due to a wedge of vertex ε, the angle of inclination of the wall. This is, of course, expected. But the important result is the rather large compression ahead of the change in flow direction caused by the inclination. One can compare these calculated results with the measured pressure distribution over the upper surface near the trailing edge of an airfoil in a supersonic stream (Fig. 13). At the trailing edge, the flow has to turn through an angle so that compatibility conditions for the flows along the upper surface and the flow along the lower surface can be satisfied. This change in the direction of flow is the same as that caused by the inclination of wall shown in Fig. 2. Therefore, from the results of the calculations, one can expect a rise of pressure over the surface of airfoil ahead of the trailing edge above that predicted by a theory without taking into account the effects of the boundary layer. This is actually observed[6]. Of course, when the boundary layer near the trailing edge is turbulent instead of laminar, the subsonic layer will be so thin as to cause negligible forward propagation of pressure. Then the pressure rise before the trailing edge will be greatly reduced, and the simple classical theory for pressure distribution over a supersonic airfoil will be more accurate.

References

[1] Ackeret J, Feldmann F, Rott N. Untersuchungan an Verdichtungsstosssen und Grenzschichten in schnell beweglen Gasen, Mitteilungen aus dem Institut fur Aerodynamik. E T. H., No. 10 (1946); or N. A. C. A. T. M. No. 1113, 1947.

[2] Liepmann H W. The Interaction Between Boundary Layer and Shock Waves in Transonic Flow. Journal of the Aeronautical Sciences, 1946, 13: 623 - 637.

[3] Lees L. Remarks on the Interaction Between Shock Wave and Boundary Layer in Transonic and Supersonic Flow. Report No. 120, Aeronautical Engineering Laboratory, Princeton University (Project Squid), 1947.

[4] Lagerstrom P A, Cole J D, Trilling L. Viscous Effects in Compressible Flow. Paper presented at the Institute on Heat Transfer and Fluid Mechanics, June 23, 1948.

[5] Howarth L. The Propagation of Steady Disturbances in a Supersonic Stream Bounded on One Side by a Parallel Subsonic Stream. Proc. Cambridge Philosophical Society, 1948, 44: 380 - 390.

[6] Ferri, A., Experimental Results with Airfoil Tested in the High Speed Tunnel at Guidonia. N. A. C. A. T. M. No. 946, 1940.

Appendix

(Ⅰ) Evaluation of Integrals

The infinite integrals in the expressions for the pressure distributions can be evaluated by contour integration. For instance,

$$\frac{1}{2\pi}\int_0^{\infty}\frac{\sin\frac{S\xi}{\pi}\mathrm{d}S}{S[\cos h S+\cos\theta]}=\frac{1}{4\pi}\int_{-\infty}^{\infty}\frac{\sin\frac{S\xi}{\pi}\mathrm{d}S}{S[\cos h S+\cos\theta]}$$

$$=\frac{1}{4\pi\mathrm{i}}\int_{-\infty}^{\infty}\frac{\mathrm{e}^{\mathrm{i}S\xi/\pi}\mathrm{d}S}{S[\cos h S+\cos\theta]}$$

Now consider S as a complex variable and take the contour as (1) the real axis indented at the origin, and (2) semicircle in the upper-half of the complex plane with infinite radius. It is easy to show that the contribution to the contour integral along the semicircle is zero. The contour encloses the following simple poles:

$$S=\mathrm{i}[(2n+1)\pi=\theta],\quad n=1,2,3,\ldots$$

Thus,

$$\frac{1}{2\pi}\int_0^\infty \frac{\sin\frac{S\xi}{\pi}\mathrm{d}S}{S[\cosh S+\cos\theta]} = \frac{1}{2}\Bigg[\frac{1}{2(1+\cos\theta)} + \sum_{n=0}^{\infty}\frac{\mathrm{e}^{-[2n+1+(\theta/\pi)]\xi}}{\mathrm{i}[(2n+1)\pi+\theta]\sinh \mathrm{i}[(2n+1)\pi+\theta]} +$$

$$\sum_{n=0}^{\infty}\frac{\mathrm{e}^{-[2n+1-(\theta/\pi)]\xi}}{\mathrm{i}[(2n+1)\pi-\theta]\sinh \mathrm{i}[(2n+1)\pi-\theta]}\Bigg]$$

$$=\text{(Formula continued at top of facing page)}\ \frac{1}{8\cos^2(\theta/2)} +$$

$$\frac{1}{2\pi\sin\theta}\sum_{n=0}^{\infty}\mathrm{e}^{-(2n+1)\xi}\times\left\{\frac{\mathrm{e}^{(\theta/\pi)\xi}}{2n+1+(\theta/\pi)} - \frac{\mathrm{e}^{-(\theta/\pi)\xi}}{2n+1-(\theta/\pi)}\right\},\quad \xi>0$$

This is the result given in Eq. (30).

The other integrals can be evaluated in a similar way. The results are already given in Eqs. (31), (37), and (38).

(Ⅱ) Continuity of Wall Pressure

From the computed results it appears that the wall pressure is continuous for the incident wave case and the interface pressure is continuous for the inclined wall case. From Eqs. (39), (41) and (50), (52), it is seen that this condition implies that

$$F_3(0;\theta)+F_4(0;\theta) = (1/2)\sec(\theta/2) \qquad \text{(A)}$$

This may be verified in several ways. The simplest seems to be the following:

$$\frac{\mathrm{d}F_3}{\mathrm{d}\xi} = -\frac{1}{\pi}\sum_{n=0}^{\infty}(-1)^n\mathrm{e}^{-[2n+1+(\theta/\pi)]\xi} = -\frac{1}{\pi}\frac{\mathrm{e}^{-[1+(\theta/\pi)]\xi}}{1+\mathrm{e}^{-2\xi}}$$

$$\frac{\mathrm{d}F_4}{\mathrm{d}\xi} = -\frac{1}{\pi}\frac{\mathrm{e}^{-[1-(\theta/\pi)]\xi}}{1+\mathrm{e}^{-2\xi}}$$

Now $F_3(\infty;\theta) = F_4(\infty;\theta) = 0$, so

$$F_3(\xi;\theta)+F_4(\xi;\theta) = \frac{1}{\pi}\int_\xi^\infty \frac{\mathrm{e}^{-[1+(\theta/\pi)]t}+\mathrm{e}^{-[1-(\theta/\pi)]t}}{1+\mathrm{e}^{-2t}}\mathrm{d}t$$

and

$$F_3(0;\theta)+F_4(0;\theta) = \frac{1}{\pi}\int_0^\infty \frac{\cosh(\theta/\pi)t}{\cosh t}\mathrm{d}t$$

Finally,

$$F_3(0;\theta)+F_4(0;\theta) = \frac{1}{2\pi}\int_{-\infty}^\infty \frac{\cosh(\theta/\pi)t}{\cosh t}\mathrm{d}t$$

Now consider t as a complex quantity and integrate $\cosh(\theta/\pi)t/\cosh t$ around a rectangular contour (1) along the real axis from $(-R,0)$ to $(R,0)$, (2) normal to the real axis along $(R,0)$ to (R,π) and $(-R,\pi)$ to $(-R,0)$, and (3) parallel to the real axis from (R,π) to $(-R,\pi)$. The only pole enclosed is at $(t+\mathrm{i}s) = \mathrm{i}(\pi/2)$ and the residue is $-\mathrm{i}\cos(\theta/2)$. By making $R\to\infty$, Eq. (A) can be shown to be true.

(Ⅲ) Series for Computing F-functions at Small Values of Argument

The series for F_1, F_2, F_3, and F_4 as given by Eqs. (33), (35), (40), and (42) are suitable for numerical computation only when $|\xi|$ is large. When $|\xi|$ is small, the convergence of these series is slow and computation tedious. On the other hand, the physical problem requires the values of these functions at small values of $|\xi|$. The following series for these functions are thus more convenient.

For $\xi > 0$

$$F_1(\xi;\theta) = \frac{1}{\pi}\sum_{n=0}^{\infty}\frac{e^{-[2n+1+(\theta/\pi)]\xi}}{2n+1+(\theta/\pi)}$$

Thus

$$\frac{dF_1}{d\xi} = -\frac{1}{\pi}\sum_{n=0}^{\infty}e^{-[2n+1+(\theta/\pi)]\xi} = -e^{-[1+(\theta/\pi)]\xi}\sum_{n=0}^{\infty}(e^{-2\xi})^n = -\frac{1}{\pi}\frac{e^{-[1+(\theta/\pi)]\xi}}{1-e^{-2\xi}}$$

Since $F_1(\infty;\theta) = 0$,

$$F_1(\xi;\theta) = \frac{1}{\pi}\int_{\xi}^{\infty}\frac{e^{-[1+(\theta/\pi)]t}}{1-e^{-2t}}dt$$

Now let

$$e^{-2t} = S$$

Then

$$F_1 = \frac{1}{2\pi}\int_{t}^{\eta}\frac{dS}{S^k(1-S)}$$

where $\eta = e^{-2\xi}$ and $k = (1/2) - (\theta/2\pi)$. For θ such that $0 \leqslant \theta \leqslant \pi$, $0 \leqslant k \leqslant 1/2$. Also, $0 \leqslant \eta \leqslant 1$.

Now

$$F_1 = \frac{1}{2\pi}\int_0^{\eta}\frac{[(1-S)+S]dS}{S^k(1-S)} = \frac{1}{2\pi}\int_0^{\eta}\frac{dS}{S^k} + \frac{1}{2\pi}\int_0^{\eta}\frac{S^{(1-k)}dS}{1-S}$$

Thus

$$2\pi F_1 = \frac{\eta^{(1-k)}}{1-k} - [S^{(1-k)}\log(1-S)]_0^{\eta} + (1-k)\int_0^{\eta}\frac{\log(1-S)}{S^k}dS$$

The remaining integration may be effected by a further change of variable. Let $1 - S = z$. Then one has to evaluate

$$\begin{aligned}\int_{(1-\eta)}^{1}\frac{\log z}{(1-z)^k}dz &= \sum_{n=0}^{\infty}(-1)^n\binom{-k}{n}\times\int_{1-\eta}^{1}z^n\log z\,dz \\ &= -\sum_{n=0}^{\infty}(-1)^n\binom{-k}{n}\times\left\{\frac{(1-\eta)^{n+1}}{n+1}\log(1-\eta) - \frac{1}{(n+1)^2}[(1-\eta)^{n+1}-1]\right\}\end{aligned}$$

where $\binom{-k}{n}$ is the binomial coefficient:

$$\binom{-k}{n} = \frac{(-k)(-k-1)\ldots(-k-n+1)}{n!}$$

Finally,

$$2\pi F_1 = \eta^{(1-k)}\left\{\frac{1}{1-k} - \log(1-\eta)\right\} -$$
$$\sum_{n=0}^{\infty}(-1)^n\binom{-k}{n}\left\{\frac{(1-\eta)^{n+1}}{n+1}\log(1\eta) - \frac{1}{(n+1)^2}[(1-\eta)^{n+1}-1]\right\}$$

a form that is convenient for evaluation at small values of ξ.

In similar fashion modified series for F_2, F_3, and F_4 may be determined. Then,

$$2\pi F_4 = \eta^{(1-k)}\left\{\frac{1}{1-k} - \log(1+\eta)\right\} + (1-k) \times \sum_{n=1}^{\infty}(-1)^{n+1}\frac{\eta^{(n+1-k)}}{n(n+1-k)}$$

where η, k are as above. $2\pi F_2$, $2\pi F_3$ have the same forms as $2\pi F_1$, $2\pi F_4$, respectively, except that $k = (1/2) + (\theta/2\pi)$.

Research in Rocket and Jet Propulsion*

Dr. Hsue-shen Tsien
(*Robert H. Goddard Professor,*
California Institute of Technology)

When considering the problems of basic research in rocket and jet propulsion, it is profitable to keep in mind the salient features of rocket-and jet-propulsion engineering. These are: short duration of operation of the powerplant and extreme intensity of reaction in the motor.

That the duration of operation of the powerplant is short stems from the high specific consumption of the propellant. On the other hand, the dry weight of the rocket engine is much lower than that of other engines of equal output. Therefore, the total installation weight (the sum of dry weight and propellant consumed) can be lower than other powerplants if the duration of operation is short[1].

Furthermore, the specific consumption of rocket engines at all speeds, and of ramjet engines at supersonic speeds, in terms of lb/(h · lbf) thrust, is essentially independent of flight speed. Therefore, the propulsive work done by the engine on the vehicle, per lb of fuel or propellant consumed, will be larger if the flight speed is larger. For this reason, it is advantageous to operate the rocket and the ramjet engine at large thrust and thus accelerate the vehicle to high speed.

The great kinetic energy of the vehicle at the end of the "burning time" of the powerplant is then utilized to achieve range by coasting. This form of dynamic trajectory is demonstrated to be superior to steady flight with long-drawn-out operation of the rocket and ramjet. Accordingly, all applications of these powerplants will involve intensive, but short-duration, operations of the engines.

The extreme intensity of reaction in the motor means high operating temperatures. To find materials which can withstand high stresses at high temperatures is the main material problem in rocket and jet-propulsion engineering. However, the problem here is different in one aspect from the material problem in turbojet and gas-turbine design. This is the short operating time of the unit. For expendable units, such as missiles, the operating time is generally of the order of minutes. Even for vehicles which are intended for repeated operation, it still is likely that the optimum performance will be obtained by a design which requires replacing the high-

---

Aero Digest, vol. 60, pp. 120 – 125, 1950.

* Condensed from a paper presented during the combined annual conventions of the American Rocket Society and the American Society of Mechanical Engineers, Dec. 1, 1949, in New York.

temperature and highly stressed parts after each operation.

By adopting this concept of designing for minutes instead of designing for thousands of hours as in the case of gas turbines, the material will be stressed for ultimate strength and not for creep. This difference is illustrated in Fig. 1, where the stresses are plotted against temperature. The lower curves are design curves for creep, and the upper curves are the ultimate stress, a stress which is practically independent of the rate of strain.

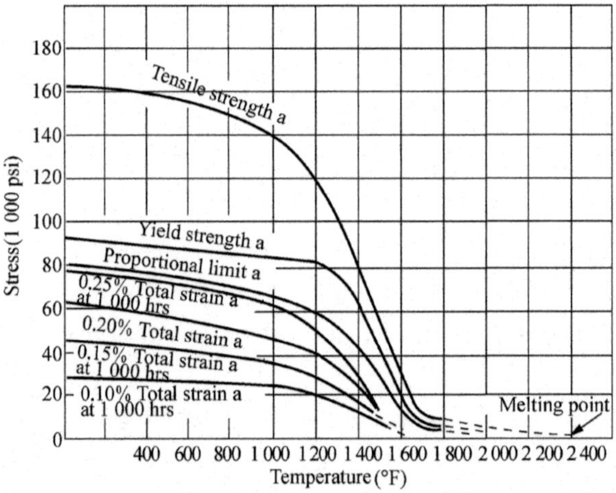

Strength vs temperature of inconel X heat treated

4 hrs at 2 100°F, 24 hrs at 1 550°F, and 20 hrs at 1 300°F

Fig. 1

For long operating time, the ultimate stress is not a design criterion, as the rate of strain near this stress is so large that the limiting strain will be reached long before the intended lifetime of the part, and the part will then fail. If the part is designed to have a life of only a few minutes, it can be stressed six times higher. This is a tremendous possibility in design and occurs only in rocket-and jetpropulsion engineering.

To explore this advantage leads, however, to complex problems in the stress and deflection analysis. The high rate of strain means constantly changing dimensions of the part, and its influence must be determined. The problem is not that of plasticity where the stress-strain relation is non-linear, nor that of elasticity, because now the material flows. In other words, the material must be considered as a visco-elastic medium.

As a first approximation, the stress-strain relation can be still considered as linear. To be specific, let σ_x, σ_y, σ_z, τ_{xy}, τ_{yz}, τ_{zx} be the six stress components, ε_x, ε_y, ε_z, γ_{xy}, γ_{yz}, γ_{zx} be the six strain components. The stress-strain relation for isotropic, visco-elastic media can be written as

$$P\sigma_x = \Phi(\lambda e + 2\mu\varepsilon_x)$$
$$P\sigma_y = \Phi(\lambda e + 2\mu\varepsilon_y)$$
$$P\sigma_z = \Phi(\lambda e + 2\mu\varepsilon_z)$$

$P\tau_{xy} = \Phi\mu\gamma_{xy}$, $P\tau_{yz} = \Phi\mu\gamma_{yz}$, $P\tau_{zx} = \Phi\mu\gamma_{zx}$ where λ and μ are constants and

$$e = \varepsilon_x + \varepsilon_y + \varepsilon_z$$

The operators P and Φ are linear time operators defined as:

$$P = \frac{\partial^m}{\partial t^m} + a_{m-1}\frac{\partial^{m-1}}{\partial t^{m-1}} + \ldots + a_0$$

$$\Phi = \frac{\partial^n}{\partial t^n} + b_{n-1}\frac{\partial^{n-1}}{\partial t^{n-1}} + \ldots + b_0$$

The a's and b's define the property of the material. They could be functions of time, but not functions of the space variables. Thus, a material with changing properties, caused by the drift toward thermodynamic and chemical equilibrium, also can be represented by these operators.

Variable Stress

An analysis[2,3] of the mechanics of such materials reveals that, if the load on the part is specified by a time factor $g(t)$, then the stress distribution at any instant can be calculated as if the material is purely elastic with the same instantaneous load. The deflection of the structure is, of course, different. But it is specified by a time factor $h(t)$ which is independent of the particular value and distribution of the load and is only dependent on $g(t)$ and is determined by

$$Qh(t) = Pg(t)$$

The $h(t)$ is thus a "universal" function in the sense that it is related only to $g(t)$ and the properties of the material. The other characteristics of the problem do not enter into its determination. In particular, the function $h(t)$ may be measured directly, experimentally, on a pure tension bar with the tension varied with time according to $g(t)$. This is then a considerable simplification of the mechanics of visco-elastic media and a useful tool in the application of the idea of design for short-time flow of material.

The extreme intensity of reaction in the rocket motor and in the combustion chambers of ramjets and pulsejets, and the high velocity of gas flow, lead to a very high rate of heat transfer to

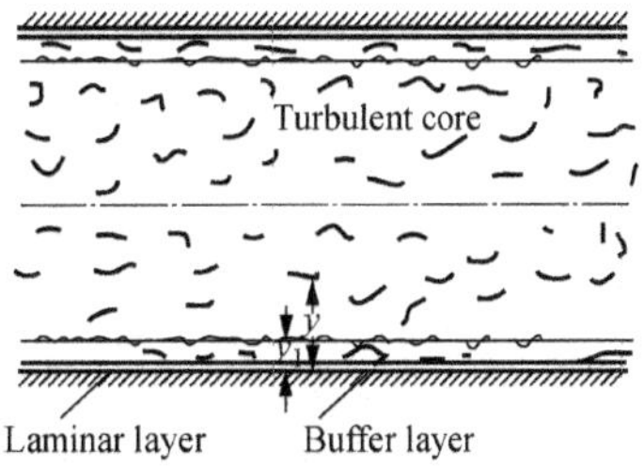

(1) Turbulent core

$$\frac{U}{U_1} = 1\,394 + 55\log_{10}(y/y_1)$$

(2) Buffer layer and laminar layer

$$\frac{\tau}{\rho} = \nu\frac{dU}{dy} - \overline{u'v'} + \frac{d\nu}{dT}T'\left(\frac{\partial v'}{\partial x} + \frac{\partial u'}{\partial y}\right)$$

Fig. 2

the walls. For instance, at the throat of a rocket nozzle, heat flux as high as 6 Btu per sec per sq in has been observed. Changed into conventional units in other branches of engineering, this is more than 3 million Btu per hour per sq ft. To cope with this high heat flux, designers have been forced to extrapolate the empirical laws of heat transfer to a cooling liquid and to seek other unconventional methods, such as surfaceboiling heat transfer.

Heat Transfer

To absorb the high heat flux by circulating a cooling liquid in a duct surrounding the hot chamber, one must use large differences between the wall temperature of the cooling liquid under turbulent-flow conditions. Here the problem is the lack of proper understanding of the basic mechanism. At present, the designer relies on empirical rules which are only safe to use within the range of variables of the test result.

To extrapolate without the guidance of a sound understanding of the phenomena is satisfactory. Of course, the problem of turbulent heat transfer has been attacked successfully by O. Reynolds, L. Prandtl, G. I. Taylor, Th. von Karmen and others. But their work is based upon the assumption that the temperature difference between the wall and the bulk of the liquid is small, so that the flow is essentially isothermal.

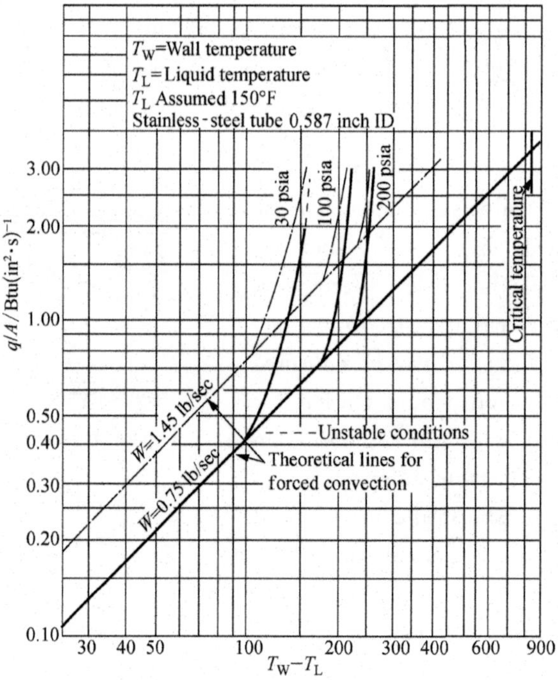

Fig. 3

Turbulent flow in a duct—say a circular pipe—can be divided into three regions (Fig. 2):

the turbulent central core where Reynolds' turbulent shearing stress dominates the molecular or viscous shearing stress, the laminar layer next to the wall where the viscous shearing stress dominates the turbulent shearing stress, and the buffer layer where both shearing stresses are important. For the turbulent central core, which occupies most of the pipe, previous experiments with isothermal flow indicate that the flow in general, and the velocity profile in particular, are controlled by the shear stress τ at the boundary of the turbulent core, the density ρ of the liquid and a linear dimension y_1. Since the boundary of the turbulent core is very close to the wall, τ is practically equal to the wall shearing stress τ_0. Together with ρ, τ_0 can define a velocity U_τ by

$$U_\tau^2 = \frac{\tau_0}{\rho}$$

Then, if U is the velocity at a point y from the wall, the non-dimensional equation for the velocity profile must be

$$\frac{U}{U_\tau} = f(y/y_1)$$

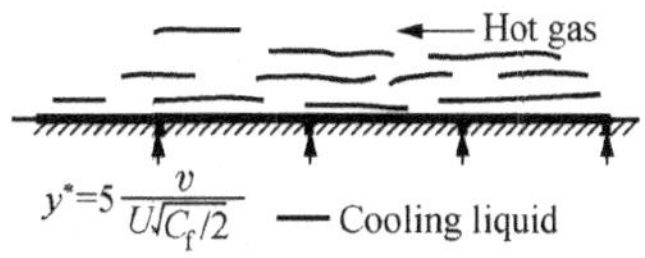

Fig. 4

Clearly, the only available linear dimension for flow near the boundary of the turbulent core is the distance of this boundary from the wall. Hence, for flow near the boundary of the turbulent core, y_1 must be the thickness of the laminar and buffer layer. From previous experiments with isothermal flow in smooth pipes[4], it is found that

$$\frac{U}{U_\tau} = 13.94 + 5.5 \log_{10} \frac{y}{y_1}$$

Since temperature differences in the liquid will change only the viscosity, and, according to experiments viscosity does not enter directly into the turbulent core flow, the velocity relation given above must hold also for non-isothermal flow.

The problem now is to determine the thickness y_1. This thickness will vary with the temperature conditions. The work of H. Reichardt[5] does not account for this variation and is therefore unsatisfactory. Thus, the main effect of higher temperature differences is in the buffer and the laminar layers. Here the variation of viscosity with temperature changes the flow. For instance, the effective shear stress τ is given by

$$\frac{\tau}{\rho} = \nu \frac{dU}{dy} - \overline{u'v'} + \overline{\frac{d\nu}{dT} T' \left(\frac{\partial v'}{\partial x} + \frac{\partial u'}{\partial y} \right)}$$

where v is the kinematic viscosity. T the temperature, $u'v'$ the instantaneous turbulent velocities in the directions parallel with the wall and normal to the wall, and T' is the temperature fluctuation. The bar over the second and third terms means averaging with respect to time.

The third term does not occur for isothermal flows. By its appearance in the equation for shear and the variable v in the first term means now the effects of heat conduction, and the effects of shear are now coupled. The solution is thus more difficult than the corresponding isothermal problem, but the difficulty is believed to be surmountable.

When the wall temperature is raised beyond that of the boiling of the liquid under prevailing pressure in the pipe, local vaporization takes place and bubbles are formed over the surface. But, since the main bulk of the liquid is still at a temperature below the boiling point, these bubbles cannot grow indefinitely. In fact, experiments by F. Kreith and M. Summerfield show that they contract again and have a life span of about 1/100 second. During its short life span, the bubble does not seem to move appreciably from the wall. The main consequence of the bubble formation and disappearance is the strong agitation of the fluid near the wall.

It is then understandable that the heat flux can be increased to many times that of the case without local boiling. This fact is shown clearly in Fig. 3, taken from the work of Kreith and Summerfield[6,7]. This means that a high rate of cooling can be achieved without high flow velocity in the cooling duct. The thusreduced pressure drop in the cooling duct will decrease the necessary pumping work of the coolant. Boiling-heat transfer then can be used to good advantage for many designers. The problem for research here, of course, is a closer understanding of the turbulent agitation due to bubble formation and thus better correlation of tests for different liquids and different test conditions.

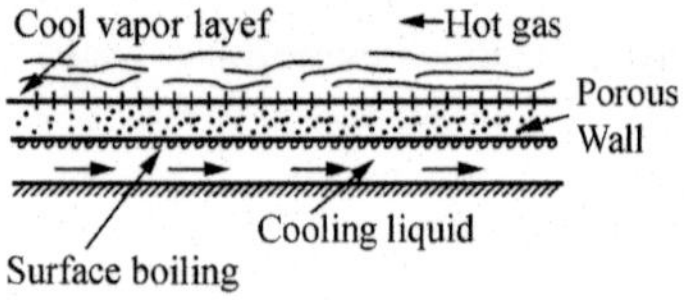

Fig. 5

If the wall temperature is increased beyond a critical value over the boiling point of the liquid, it has been found that a vapor envelope forms over the surface, and the heat flux is reduced by the insulating effect of the stagnant vapor. Therefore, with specified pressure and flow velocity, there is definitely a maximum value of heat-flux density even with local boiling at the surface. If still, high heat-flux density is desired, then other means of cooling have to be used.

Wall-Temperature Effects

However, even before reaching this intrinsic limit of boiling-heat transfer, the wall temperature at the inside surface of the rocket motor may be too high for the material strength, due to the necessary temperature gradient through the wall for the heat flux. For instance, if the heat flux is 6 Btu per sec per sq in, the wall thickness is 1/16 in, and the temperature at the cool side of the wall is 600°F, the temperature at the hot side of the wall will be 1 950°F if stainless steel is used. This temperature is certainly too high for good strength. New, powerful cooling methods for extremely high heat flux are sweat-cooling and film-cooling.

Film cooling (Fig. 4) is achieved by establishing a thin liquid film in contact with the hot gas over the surface to be cooled. Due to the shearing stress acting over the liquid-gas interface, the liquid flows in the downstream direction. Simultaneously, the heating of the film by the hot gas evaporates the liquid. It is seen that, so long as there is a liquid film, the wall temperature is kept below the boiling point of the liquid.

It is noted also that, to protect the wall from the hot gas, liquid film has to be reestablished by injection through holes in the wall when the film upstream injection is evaporated. Of course, the intervals of injection can be lengthened by injecting more liquid and establishing a thicker film each time. However, the difficulty here is the instability of the film against the turbulent flow in the gaseous boundary layer. The resultant partial breakaway of the liquid in the form of droplets constitutes a loss in effective cooling liquid. The problem here is then the determination of the relative cooling efficiency with respect to film thickness.

From experience on one-phase turbulent boundary layer, it is found that the laminar sublayer thickness y^* is determined. (Continued on page 124)

$$y^* = 5 \frac{\nu}{U\sqrt{\frac{C_f}{2}}}$$

where ν is the kinematic viscosity of the fluid, U is the freestream velocity and C_f is the local friction coefficient. If this relation holds also for two-phase turbulent boundary layers such as exist in film cooling, y^* is the limiting film thickness for perfect efficiency. If the film thickness is larger than y^*, instability of the film and breakaway of droplets is likely to occur. It is then seen that there is an advantage in having a higher kinematic viscosity v, as layer thickness is allowed.

If $U = 1\,000$ ft/sec, $C_f = 0.004$, and $v = 0.319 \times 10^{-5}$ ft/sec for water at 212°F, y^* is only 4.3×10^{-6} in. This result indicates that, for theoretical maximum cooling efficiency, the film should be very thin and reestablished frequently along the wall. The limiting case is sweat-cooling, where the coolant is forced through the porous wall and injection and evaporation occur at the same time.

Sweat-cooling is, however, not limited to the liquid coolant. The coolant may be gaseous. In fact, the most extensive experiments are made by P. Duwez and H. L. Wheeler[8] with gaseous coolants. However, it is shown by the above investigators that the coolant cannot be allowed to evaporate in the porous wall, as then the flow is essentially not stable, with wide fluctuations in the wall temperatures.

Generally, then, the most efficient sweat-cooling system, with least expenditures of the coolant, is one that evaporates the liquid coolant on the "outside" surtace of the porous wall before entering the porous material (Fig. 5). In a sense, this system is a combination of boiling-heat transfer and sweat-cooling. No extensive experiment on this method of cooling has yet been made.

It is evident that, with either film-cooling or sweat-cooling, there is no limit to the temperature of the combustion gas that can be handled effectively. Therefore, one need have no misgivings about the high-energy fuels and propellants for cooling difficulties. Furthermore, for rocket, ramjet and pulsejet, there is no contact of the combustion gas with a delicate moving part, such as turbine blades in a turbojet; and the combustion gas can be corrosive and can contain finely divided solid particles. These factors remove practically all restriction on the choice of fuels and propellants. Such strange combinations as liquid hydrogen and liquid fluorine, and diborane (B_2H_6) and air, are to be considered.

The more urgent problems of combustion in jet propulsion are those connected with fluid mechanical aspects. These are the auto-ignition of liquid jets, the evaporation of liquid droplets, the mixing of gaseous components, the mutual influence of combustion and turbulence of low combustion in the heterogeneous mixture etc. For ramjets, the most perplexing problem today is the problem of flame stabilization. This is a problem which confronts all ramjet designers. Worse still, the mechanism of flame stabilization is not yet understood. As a result, the flame-holder design for the combustion chamber is always done by *ad hoc* experimentation.

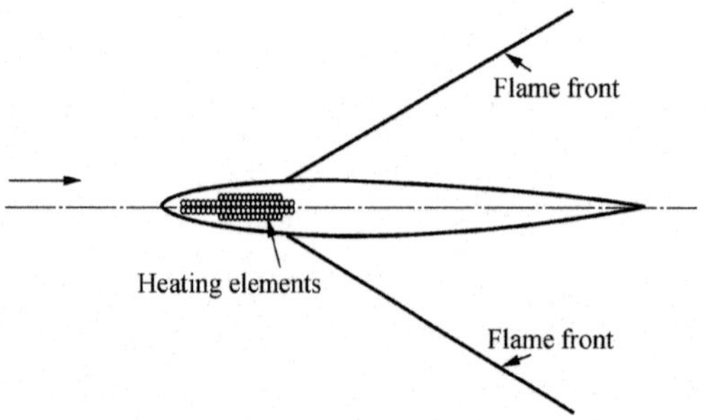

Fig. 6

Clearly, there is a need for experiments with the simplest physical conditions, so that the parameters can be controlled. The work of A. S. Scurlock[9] with homogeneous gas stream and controlled turbulence in the initial stream is the most noteworthy effort in this direction. However, for a true understanding of the mechanism, further detailed exploration of flow field is necessary.

It seems that one of the important aspects of flame stabilization is the interaction of the flame front and the boundary layer. To test this concept, a flame holder is the shape of a streamlined body may be tried (Fig. 6). To start the combustion, the airfoil has to be heated first by, say, an electric current, to a high temperature. Once the flame is started, the airfoil aft of the flame is heated by the hot gas. Heat is then conducted through the body of the airfoil to the front section, where heat is given up to the cold gas mixture through the boundary layer. The cold gas mixture in the boundary layer being heated by the body will increase the concentration of active carriers and finally ignite at the intersection of the flame front and the surface of the body.

It is evident that such a flame holder without turbulent and eddying loss has many practical applications despite the difficulty in starting. In fact, by increasing the length of the airfoil in the hot gas, the temperature at the front part can be increased, and thus the gas velocity can be raised without blowouts.

The ultimate aim of all this basic research is, of course, to improve the performance of rocket and jet-propelled vehicles. However, even when given the best powerplant, the designer still has to determine the best way of using it for optimum performance of the completed vehicle. For instance, what would be the optimum thrust programming for a sound rocket? What would be the gain possible by varying the thrust during ascent? Is this gain justified by the additional complication in the design?

The basic variational problem of thrust programming was studied by G. Hamel[10]. However he made no detailed calculations to allow the designer to weigh the importance of different aspects of the problem.

Long-Range Trajectory

But the fundamental question in the performance analysis is the trajectory, particularly the long-range trajectory. Earlier in this discussion, the reason for favoring the dynamic trajectory of varying velocity was given. But what particular dynamic trajectory? To avoid the penalty of high drag at high velocity in the dense atmosphere and yet to be able to accelerate the vehicle quickly, it is clear that the vehicle should be launched vertically.

Performance of the vertical trajectory of a rocket is well-known. But is a rocket the only powerplant capable of vertical trajectory? Certainly the ramjet—once boosted to a sufficiently high velocity—also can produce enough thrust to make an accelerated vertical flight. A ramjet would weigh more than a rocket per *lb* of thrust produced, but the fuel consumption is very much smaller. Preliminary estimates by Z. H. Schindel[11] show that the advantage of low fuel consumption overcomes the disadvantage of heavier dry weight. Therefore, there is a definite gain in substituting the lowest stage of a multi-stage rocket with a ramjet. Of course, to boost the ramjet to operating speed, it has to be operated as a ducted rocket in the first few seconds.

What can one say about the remainder of the trajectory? Since the high velocity of the vehicle is reached outside the atmosphere by vertical or near vertical ascent, the first part of the trajectory has to be liftless and this is elliptical. When the vehicle returns to the atmosphere at practically the same speed as it leaves the atmosphere, the lift of the body of the vehicle can be produced by putting the body into an angle of attack. The question here then is one of programming the angle of attack of the body so that the maximum range is obtained.

As an example of such a dynamic trajectory, the flight of a 3 000-mile rocket vehicle is studied under the assumption of steady glide after the initial elliptical path. The average lift-drag ratio in glide is taken to be 4. The result of this analysis is as follows:

Length	78.9 ft
Maximum diameter of body	8.86 ft
Gross weight	96 500 lbf
Fuel load	72 400 lbf
Weight after "Brennschluss"	24 100 lbf
Propellant loading fraction	0.750
Exhaust velocity	12 000 ft/sec
Propellant	liquid O_2 liquid H_2;
	liquid F_2 liquid H_2
Maximum velocity	9 140 mph

(continued)

Range at conclusion of elliptic path	1 200 miles
Range contributed by glide	1 800 miles
Altitude at beginning of glide	27 miles
Landing speed	150 mph
Landing angle	20°
Flight duration	Less than one hour

Thus it is seen that the requirements of a transcontinental rocket-liner are not at all beyond the grasp of present-day technology. The wings need not be large to achieve a reasonable landing speed, and the specifications on the structural weights are not impossible. When will such a rocket-liner be realized? That is a difficult question. But one thing is certain: the basic research as outlined in this discussion definitely will hasten the day of long-range rocket travel.

References

[1] Lowell A L. A Guide to Aircraft Powerplant Selection. Aeronautical Engineering Review, 1947, 6(4): 22-25.

[2] Alfrey T. Non-Homogeneous Stresses in Visco-elastic Media. Quarterly of Applied Mechanics, 1944, 2: 113-119.

[3] Tsien H S. A Generalization of Alfrey's Theorem for Visco-elastic Media. Quarterly of Applied Mechanics, 1950.

[4] Nikuradse J. Gesetzmässigkeiten der turbulenten Strömung in glatten Rohren. Ver Deutsch. Ing. Forschungsheft, 1932: 356.

[5] Reichardt H. Die Wärmeübertragung in turbulenten Reisbungschichten. Z. a. M. M., 1940, 20: 297-328.

[6] Kreith F, Summerfield M. Heat Transfer to Water at High Flux Densities with and without Surface Boiling. ASME Transactions. // Also: Investigations of Heat Transfer at High Heat-Flux Densities: Experimental Study with Water of Friction Drop and Forced Convection with and without Surface Boiling in Tubes. Progress Report No. 4-68, Jet Propulsion Laboratory, CIT, 1948.

[7] Kreith F, Summerfield M. Heat Transfer from an Electrically Heated Tube to Aniline at High Heat Flux. Progress Report No. 4-88, Jet Propulsion Laboratory, CIT, 1949.

[8] Duwez P, Wheeler H L. Experimental Study of Cooling by Injection of a Fluid through a Porous Material. J. of Aero. Sciences, 1948, 15: 509-521.

[9] Scurlock A S. Flame Stabilization and Propagation in High-Velocity Gas Streams. Meteor Report, No. 19, Mass. Institute of Technology, 1948.

[10] Hamel G. Uber eine mit dem Problem der Rakete zusammenhängende Aufgabe der Variationsrechnung. Z. a. M. M., 1927, 7: 451-452.

[11] Schindel Z H. Application of Ramjet to High Altitude Sounding Vehicles. M. S. Thesis. Dept. Aeronautical Engineering, Mass. Inst. of Technology, 1947.

A Generalization of Alfrey's Theorem for Visco-elastic Media

By H. S. Tsien

(California Institute of Technology)

1. Introduction. For the non-homogeneous stresses in isotropic incompressible viscoelastic media characterized by linear relations between the components of stress, strain and their derivatives with respect to time, T. Alfrey has shown[1] that in the case of the first boundary value problem, the stress distribution is identical with that in an incompressible elastic material under the same instantaneous surface forces. A similar result was obtained for the second boundary value problem where the displacements at the boundary are specified. It is the purpose of the present note to generalize this theorem to isotropic compressible media for problems involving body forces. Only the first boundary value problem will be discussed, as the corresponding theorem on the second boundary value problem is self-evident.

2. First boundary value problem. Let the displacements along the x, y, z directions be u, v, w. Then the typical expressions* of the six strain components can be written as:

$$\varepsilon_x = \frac{\partial u}{\partial x}, \qquad \gamma_{xy} = \frac{\partial u}{\partial y} + \frac{\partial v}{\partial x} \tag{1}$$

If the six stress components are denoted by σ_x, σ_y, σ_z, τ_{xy}, τ_{yz}, τ_{zx}, the components of body force by X, Y, Z, and the surface force per unit area by $\overline{X}$, $\overline{Y}$, $\overline{Z}$, the equations of equilibrium are

$$\frac{\partial \sigma_x}{\partial x} + \frac{\partial \tau_{xy}}{\partial y} + \frac{\partial \tau_{zx}}{\partial z} + X = 0 \tag{2}$$

Here the body forces X, Y, Z are the result of external field or agent and will not be identified with the inertia forces of the material. The inertia forces are here considered to be negligible as is actually the case for a wide class of problems. If l, m, n are the direction cosines of the normal to the surface, then the surface conditions are:

Received Sept. 7, 1949.

Quarterly of Applied Mathematics, vol. 8, pp. 104 – 106, 1950.

* Throughout this note, only typical expressions are explicitly given; other expressions can be readily obtained by cyclic permutations.

$$\overline{X} = l\sigma_x + m\tau_{xy} + n\tau_{zx} \tag{3}$$

To determine the stress distribution completely, there are in addition six equations of compatibility:

$$\frac{\partial^2 \varepsilon_x}{\partial y^2} + \frac{\partial^2 \varepsilon_y}{\partial x^2} = \frac{\partial^2 \gamma_{xy}}{\partial x \partial y}, \quad 2\frac{\partial^2 \varepsilon_x}{\partial y \partial z} = \frac{\partial}{\partial x}\left(-\frac{\partial \gamma_{yz}}{\partial x} + \frac{\partial \gamma_{zx}}{\partial y} + \frac{\partial \gamma_{xy}}{\partial z}\right) \tag{4}$$

It remains to specify the relations between the components of stress, strain and their time derivatives. These relations will be assumed to be linear, corresponding to problems of small strain. If in addition the material is assumed to be isotropic, then purely on the ground of invariance under space coordinate transformation it can be shown that the required relations have to be of the following form:

$$\begin{aligned} P\sigma_x &= Q(\lambda e + 2\mu\varepsilon_x) \\ P\tau_{xy} &= Q\mu\gamma_{xy} \end{aligned} \tag{5}$$

where μ and λ are constants, and

$$e = \varepsilon_x + \varepsilon_y + \varepsilon_z \tag{6}$$

The operators P and Q are time operators defined as:

$$\begin{aligned} P &= \frac{\partial^m}{\partial t^m} + a_{m-1}\frac{\partial^{m-1}}{\partial t^{m-1}} + \ldots + a_0 \\ Q &= \frac{\partial^n}{\partial t^n} + b_{n-1}\frac{\partial^{n-1}}{\partial t^{n-1}} + \ldots + b_0 \end{aligned} \tag{7}$$

The a's and b's define the characteristics of the material. They could be functions of time, but not functions of the space variables. Thus a material with changing properties, caused by the drift towards thermodynamic and chemical equilibrium, can be also represented by these operators.

By eliminating the strains between the compatibility equations (4) and the stress-strain relation (5), one has:

$$\begin{aligned} &P\left[\nabla^2\sigma_x + \frac{2\lambda + 2\mu}{3\lambda + 2\mu}\frac{\partial^2\theta}{\partial x^2} + \frac{\lambda}{\lambda + 2\mu}\left(\frac{\partial X}{\partial x} + \frac{\partial Y}{\partial y} + \frac{\partial Z}{\partial z}\right) + 2\frac{\partial X}{\partial x}\right] = 0 \\ &P\left[\nabla^2\tau_{xy} + \frac{2\lambda + 2\mu}{3\lambda + 2\mu}\frac{\partial^2\theta}{\partial x \partial y} + \left(\frac{\partial X}{\partial y} + \frac{\partial Y}{\partial x}\right)\right] = 0 \end{aligned} \tag{8}$$

where

$$\theta = \sigma_x + \sigma_y + \sigma_z \tag{9}$$

The equation (8) are sufficient to solve the first boundary value problem: The surface forces $\overline{X}$, $\overline{Y}$, $\overline{Z}$ and body forces X, Y, Z are specified for all values of t. For any given t, these forces must be in equilibrium. The problem is to determine the distribution of stresses fulfilling these boundary conditions.

Now let

$$\overline{X} = \overline{X}^* g(t), \quad \overline{Y} = \overline{Y}^* g(t), \quad \overline{Z} = \overline{Z}^* g(t) \tag{10}$$

The starred quantities are functions of space coordinates only. Then for equilibrium the body force must vary with time in a similar way. Thus:

$$X = X^* g(t), \quad Y = Y^* g(t), \quad Z = Z^* g(t) \tag{11}$$

The stress components can now be written in the same manner:

$$\sigma_x = \sigma_x^* g(t), \quad \tau_{xy} = \tau_{xy}^* g(t) \tag{12}$$

By substituting equations (11) and (12) into equation (8), it is easily shown that the starred quantities satisfy the stress equations for a purely elastic medium with Lamé's constants λ and μ. By substituting equations (10) and (12) into the boundary condition (3), it is seen that the starred quantities also satisfy their corresponding boundary conditions. Therefore, in the case of the first boundary value problem, the stress distribution is identical with that in purely elastic material under the same instantaneous surface forces and body forces.

To determine the displacements u, v and w, one introduces the unknown time function $h(t)$ such that:

$$u = u^* h(t), \quad v = v^* h(t), \quad w = w^* h(t) \tag{13}$$

where the starred quantities are again functions of space variables only. When equation (13) are substituted into equation (5), u^*, v^*, w^* are found to be the displacements of a purely elastic medium under the loading $\overline{X}^*$, $\overline{Y}^*$, $\overline{Z}^*$ and X^*, Y^*, Z^*. Furthermore, $h(t)$ is determined by:

$$\mathrm{Q}h(t) = \mathrm{P}g(t) \tag{14}$$

with the initial condition that at $t = 0$, h and its first $(n-1)$ derivatives vanish. The function $h(t)$ is thus universal in the sense that it depends only on $g(t)$ and the properties of the material. The other characteristics of the problem do not enter into its determination. In particular, the function $h(t)$ may be directly determined experimentally on a pure tension bar with the tension varied with time according to $g(t)$.

By superimposing solutions, the time dependence of the applied forces can be generalized as shown by Alfrey[1].

Reference

[1] Alfrey T. Non-homogeneous stresses in visco-elastic media. Q. Appl. Math. 1944, 2: 113-119.

Instruction and Research at the Daniel and Florence Guggenheim Jet Propulsion Center

By Hsue-shen Tsien
(*Member ARS, Robert H. Goddard Professor,*
Daniel and Florence Guggenheim Jet Propulsion Center,
California Institute of Technology, Pasadena, Calif.)

In "America Fledges Wings" R. M. Cleveland[①] stated: "Despite the importance of the role played by the Daniel Guggenheim Fund for the promotion of aeronautics as a herald, as an awakener, as a quickening spark in the manifold fields of practical aviation, its most important and probably most lasting contribution lay in its implementation and its creation of centers of research."

These centers of research are the well-known great schools of aeronautical engineering at the New York University, the Stanford University, the University of Michigan, the Massachusetts Institute of Technology, the California Institute of Technology, the University of Washington, and the Georgia School of Technology. It is a fact that a great majority of practicing aeronautical engineers today are either wholly educated in one of these centers or have had contact with one of these centers. Moreover, the strong influence of the Guggenheim Fund is not limited to this phase of aeronautical engineering. These centers of research contributed to a large extent to the fundamental knowledge of aeronautical science which forms the scientific basis of aeronautical engineering. Today we see an even more broadened effect exerted by the Guggenheim schools as men originally educated in these Guggenheim research centers establish new research laboratories and new schools of aeronautics in universities all over the world.

Jet-Propulsion Centers

The year 1930 marked the beginning of another phase of development instigated by the Guggenheim Fund. In that year Daniel Guggenheim made a special grant to the late Dr. Robert H. Goddard for liquid-propellant rocket research, to be carried on in Roswell, N. Mex. This work was continued under the Daniel and Florence Guggenheim Foundation until the death of

---

Presented at the Annual Convention of the AMERICAN ROCKET SOCIETY, Hotel Statler, New York, N. Y., Dec. 1, 1949.

Journal of the American Rocket Society, vol. 20, pp. 51 – 64, 1950.

① Numbers in parentheses refer to Bibliography on page 64.

Dr. Goddard. Research done by Dr. Goddard opened up the entirely new field of rocket engineering and heralded the dawn of the second epoch of aeronautics. This is the epoch of hyperaviation of flight with tremendous speeds at extremely high altitude. To propel a vehicle for hyperaviation, the conventional aircraft power plants are not adequate, and one must rely on radical propulsion systems such as rockets and ramjets. Dr. Goddard's work is among the first scientific experimentations in this field.

With this historical background of the Guggenheim Foundation in mind, it is then entirely fitting that the Foundation should decide during 1948 to establish two new centers of research to be called The Daniel and Florence Guggenheim Jet Propulsion Centers, one at Princeton University and the other at the California Institute of Technology. The purpose of the centers is threefold: (1) To train young engineers and scientists in the field of rocket and jet-propulsion technology on the postgraduate level, thus endeavoring to breed a new generation of pioneers to push the frontier of aviation to the next "higher" domain; (2) to instigate research and advanced thinking in rocket and jet propulsion, thus endeavoring to contribute to the basic knowledge necessary for the sound development of this new field; and (3) to promote peacetime commercial and scientific uses of rockets and jet propulsion. To carry out this program, there are the chairs of Robert H. Goddard professorships in honor of Dr. Goddard. Each Goddard professorship will be associated with a number of younger staff members and postgraduate fellows. The fellowships will be known as The Daniel and Florence Guggenheim Jet Propulsion Fellowships.

Instruction and Research of Jet Propulsion

Instruction and research in jet propulsion at the California Institute of Technology did not start however with the establishment of the Guggenheim Center; they started earlier.

During the academic year 1943 - 1944, the California Institute, at the request of the then Air Technical Service Command, Army Air Forces, initiated a course in rocket and jet propulsion limited to officer personnel assigned to the Institute for graduate study. Lectures were prepared by members of the Jet Propulsion Laboratory of the Institute and the GALCIT.① Several lectures on special topics were given by invited speakers. The course was planned by Theodore von Kármán, then Director of both the Jet Propulsion Laboratory and the GALCIT. The course covered in a comprehensive manner the basic principles of all jet-propulsion systems and the performance of jet-propelled vehicles. The lecture notes for the course, which was repeated during 1944 - 1945, were edited into the voluminous "Reference Text of Jet Propulsion"[2].

Research in rocket and jet propulsion at the California Institute started even earlier[3]. The so-called GALCIT Rocket Research Project was initiated more or less informally in 1936. Early phases of the research were financed by a gift from Weld Arnold, now Colonel Arnold (a

① Guggenheim Aeronautical Laboratory, California Institute of Technology.

tremendous sum of approximately \$1 000!) This modest start led to a rapid growth under the exigency of war from demands of the AAF Materiel Command and the ASF Ordnance Department. The result is the Jet Propulsion Laboratory with a staff now numbering more than 575 persons and facilities valued at approximately \$7 000 000.

Thus, it is fortunate that the newly established Guggenheim Jet Propulsion Center at the California Institute of Technology could obtain help and guidance from these two earlier developments in rocket and jet propulsion at the Institute, not to mention the inspiration it receives from the Guggenheim Graduate School of Aeronautics and the Guggenheim Aeronautical Laboratory. The center at the California Institute is a part of the Division of Engineering of the Institute. Its somewhat autonomous position is the result of the consideration that the solution of engineering problems in jet propulsion draws on knowledge and practice of older branches of engineering, particularly mechanical engineering and aeronautics. Thus, the program of instruction in jet propulsion should properly include material from both of these engineering fields. Furthermore, it is expected that, in general, students entering the course work in jet propulsion will have had their undergraduate preparation in mechanical engineering or aeronautics. Thus, the program of instruction in jet propulsion will have two separate options, allowing men from both aeronautics and mechanical engineering to follow their previous inclinations and developments. Both options lead to the degree of master of science upon the completion of the fifth-year program. For men in the aeronautics option, the degree of aeronautical engineer will be given upon the completion of a sixth-year program. Similarly, the degree of mechanical engineer will be given to men upon the completion of the sixth-year program of the mechanical-engineering option. More advanced study will lead to the degree of doctor of philosophy.

The actual courses of study in rocket and jet propulsion started September, 1949. Of course, the subjects of instruction and topics studied are in a state of flux. As experience is gained, there will be modifications in the materials covered and emphasis placed. It will take a number of years before the program of instruction can be stabilized.

The Guggenheim Jet Propulsion Genter is not provided with large-scale research facilities as these are available at the Government-sponsored Jet Propulsion Laboratory at the Institute. It is expected that if basic research in rocket and jet propulsion requiring expensive equipment coincides with the interest and program of the Jet Propulsion Laboratory, it could be carried out there.

Characteristics of Rocket and Jet-Propulsion Engineering

But what are the problems of basic research in rocket and jet propulsion? Before answering this question, it is profitable to keep in mind the salient features of rocket and jet-propulsion engineering. These are: (a) Short duration of operation of the power plant; and (b) extreme intensity of reaction in the motor. That the duration of operation of the power plant is short, stems from the fact of high specific consumption of propellant for the rocket engine. On the other hand, the dry weight of the rocket engine is much lower than that of other engines of equal output. Therefore the total installation weight of a rocket engine, which is the sum of dry weight and propellant consumed, can be lower than other power plants if the duration of operation is short[4]. Furthermore, the specific consumption of rocket engines at all speeds and of ramjet engines at supersonic speeds in terms of lb per hr per lb of thrust is essentially independent of flight speed. Therefore the propulsive work done by the engine on the vehicle per lb of fuel or propellant consumed will be larger if the flight speed is larger. For this reason, it is advantageous to operate the rocket and the ramjet engines at large thrust and thus accelerate the vehicle to high speed. The large kinetic energy of the vehicle at the end of the "burning time" of the power plant is then utilized to achieve range by coasting. This form of dynamic trajectory is demonstrated to be superior to steady flight with long drawn-out operation of the rocket and the ramjet. Therefore all applications of these power plants will involve intensive but short duration operations of the engines.

Material Problems

The extreme intensity of reaction in the motor means high operating temperature. To find materials which can withstand high stresses at high temperature is then the main material problem in rocket and jet-propulsion engineering. However, the problem here is in one aspect characteristically different from the material problem in turbojet and gas-turbine design. This is the short operating time of the unit. For expendable units such as missiles, the operating time is generally of the order of minutes. Even for vehicles which are intended for repeated operation, it is still likely that the optimum performance is obtained by a design which requires replacing the high-temperature and highly stressed parts after each operation.

By adopting this concept of designing for minutes instead of designing for thousands of hours, as in the case of turbojets and gas turbines, the material will be stressed for ultimate strength and not for creep. This difference is illustrated in Fig. 1 where the stresses are plotted against temperature. The lower curves are design curves for creep, and the upper curves are the ultimate stress, a stress which is practically independent of the rate of strain. For long operating time, the ultimate stress is not a design criterion, as the rate of strain near this stress is so large that the limiting strain will be reached long before the intended

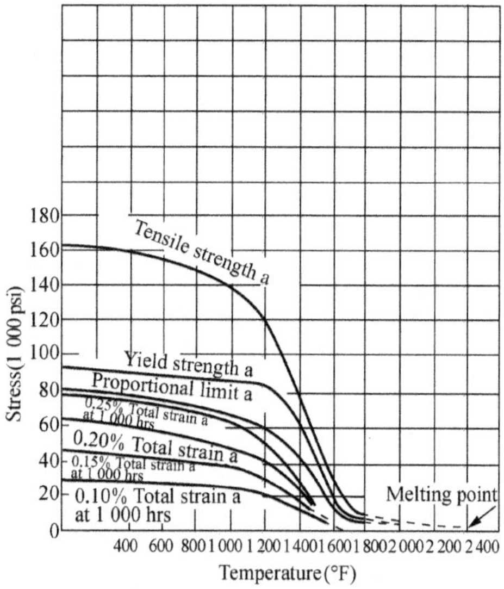

Fig. 1 Strength versus temperature of inconel X heat-treated 4 hr at 2 100 °F, 24 hr at 1 550 °F, and 20 hr at 1 300 °F

lifetime of the part, and the part will then fail. As seen from Fig. 1, if the part is designed to have a life of only a few minutes, it can be stressed six times higher. This is a tremendous possibility in design and occurs only in rocket and jet-propulsion engineering.

To explore this advantage leads however to complex problems in the stress and deflection analysis. The high rate of strain means constantly changing dimensions of the part, and its influence must be determined. The problem is not that of plasticity where the stress-strain relation is nonlinear, nor that of elasticity because now the material flows. In other words, the material must be considered as a viscoelastic medium. As a first approximation, the stress-strain relation can be still considered as linear. To be specific, let σ_x, σ_y, σ_z, τ_{xy}, τ_{yz}, τ_{zx} be the six stress components, ε_x, ε_y, ε_z, γ_{xy}, γ_{yz}, γ_{zx} be the six strain components. The stress-strain relation for isotropic viscoelastic media can be written as:

$$
\begin{aligned}
P\sigma_x &= Q(\lambda e + 2\mu\varepsilon_x) \\
P\sigma_y &= Q(\lambda e + 2\mu\varepsilon_y) \\
P\sigma_z &= Q(\lambda e + 2\mu\varepsilon_z) \\
P\tau_{xy} &= Q\mu\gamma_{xy}, \\
P\tau_{yz} &= Q\mu\gamma_{yz}, \\
P\tau_{zx} &= Q\mu\gamma_{zx}
\end{aligned}
$$

where λ and μ are constants and

$$e = \varepsilon_x + \varepsilon_y + \varepsilon_z$$

The operators P and Q are linear time operators defined as:

$$P = \frac{\partial^m}{\partial t^m} + a_{m-1}\frac{\partial^{m-1}}{\partial t^{m-1}} + \ldots + a_0$$

$$Q = \frac{\partial^n}{\partial t^n} + b_{n-1}\frac{\partial^{n-1}}{\partial t^{n-1}} + \ldots + b_0$$

The a's and b's define the property of the material. They could be functions of time, but not functions of the space variables. Thus a material with changing properties, caused by the drift toward thermodynamic and chemical equilibrium, can be also represented by these operators.

An analysis[5,6] of the mechanics of such materials reveals that if the load on the part is specified by a time factor $g(t)$, then the stress distribution at any instant can be calculated as if the material is purely elastic with the same instantaneous load. The deflection of the structure is, of course, different. But it is specified by a time factor $h(t)$ which is independent of the particular value and distribution of the load and is only dependent on $g(t)$ and is determined by

$$Qh(t) = Pg(t)$$

The $h(t)$ is thus a "universal" function in the sense that it is related only to $g(t)$ and the properties of the material. The other characteristics of the problem do not enter into its determination. In particular, the function $h(t)$ may be directly measured experimentally on a pure tension bar with the tension varied with time according to $g(t)$. This is then a considerable simplification of the mechanics of viscoelastic media and a useful tool in the application of the idea of design for short-time flow of material.

Heat Transfer

The extreme intensity of reaction in the rocket motor and in the combustion chambers of ramjet and pulsejet, and the high velocity of gas flow lead to a high rate of heat transfer to the walls, if the wall is to be kept at safe operating temperature. For instance, at the throat of a rocket nozzle, heat flux as high as 12 B.t.u. per sec per sq in. has been observed. Changed into conventional units in other branches of engineering, this is more than 6 million B.t.u. per hr per sq ft. To cope with this high heat flux, designers have been forced to extrapolate the empirical laws of heat transfer to a cooling liquid and to seek other unconventional methods, such as surface boiling heat transfer.

To absorb the high heat flux by circulating a cooling liquid in a duct surrounding the hot chamber, one must use large differences between the wall temperature of the cooling duct and the bulk temperature of the cooling liquid under turbulent-flow conditions. Here the problem is the lack of proper understanding of the basic mechanism. At present, the designer relies on empirical rules which are only safe to use within the range of variables of the test result. To extrapolate without the guidance of a sound understanding of the phenomena is not satisfactory. Of course the problem of turbulent heat transfer has been attacked successfully by O. Reynolds, L. Prandtl, G. I. Taylor, Theodore von Kármán, and others. But their work is based upon the assumption that the temperature difference between the wall and the bulk of the liquid is small so that the flow is essentially isothermal.

Turbulent flow in a duct, say a circular pipe, can be divided into three regions (Fig. 2): the turbulent central core, where Reynolds' turbulent shearing stress dominates the molecular or viscous shearing stress; the laminar layer next to the wall where the viscous shearing stress dominates the turbulent shearing stress; and the buffer layer where both shearing stresses are important. For the turbulent central core, which occupies most of the pipe, previous experiments with isothermal flow indicate that the flow in general, and the velocity profile in particular, are controlled by the shear stress τ at the boundary of the turbulent core, the density ρ of the liquid, and a linear dimension y_1. Since the boundary of the turbulent core is very close to the wall, τ is practically equal to the wall shearing stress τ_0. Together with ρ, τ_0 can define a velocity U_τ by

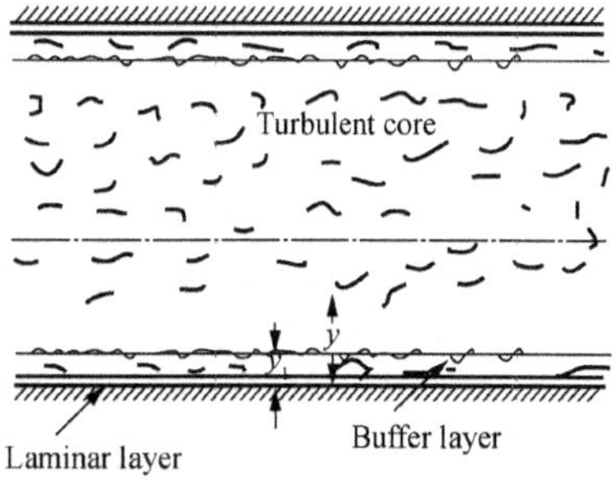

Fig. 2 The three regions which occur in turbulent flow in a duct turbulent core:

$$\frac{U}{U_\tau} = 13.94 + 5.5 \log_{10} \frac{y}{y_1}$$

Buffer layer and laminate layer:

$$\frac{\tau}{\rho} = \nu \frac{\mathrm{d}U}{\mathrm{d}y} - \overline{u'v'} + \frac{\mathrm{d}\nu}{\mathrm{d}T} \overline{T' \frac{\partial v'}{\partial x} + \frac{\partial u}{\partial y}}$$

$$U_\tau = \sqrt{\frac{\tau_0}{\rho}}$$

Then if U is the velocity at a point y from the wall, the nondimensional equation for the velocity profile must be

$$\frac{U}{U_\tau} = f(y/y_1)$$

Clearly, the only available linear dimension for flow near the boundary of the turbulent core is the distance of this boundary from the wall. Hence, for flow near the boundary of the turbulent core, y_1 must be the thickness of the laminar and buffer layer. From previous experiments with isothermal flow in smooth pipes[7], it is found that

$$\frac{U}{U_\tau} = 13.94 + 5.5 \log_{10} \frac{y}{y_1}$$

Since temperature differences in the liquid will change only the viscosity, and according to experiments viscosity does not enter directly into the turbulent core flow, the velocity relation given above must also hold for nonisothermal flow.

The problem now is to determine the thickness y_1. This thickness will vary with the temperature conditions. The work of H. Reichardt[8] does not account for this variation and is therefore unsatisfactory. The main effect of higher temperature differences therefore is in the buffer layer and the laminar layer. Here the variation of viscosity with temperature changes the flow. For instance, the effective shear stress τ is given by

$$\frac{\tau}{\rho} = \nu \frac{dU}{dy} - \overline{u'v'} + \overline{\frac{d\nu}{dT} T' \left(\frac{\partial v'}{\partial x} + \frac{\partial u'}{\partial y} \right)}$$

where ν is the kinematic viscosity, T the temperature, $u'v'$ the instantaneous turbulent velocities in the directions parallel to the wall and normal to the wall, and T' is the temperature fluctuation. The bar over the second and third terms means averaging with respect to time. The third term does not occur for isothermal flows. Its appearance and the appearance of the variable ν in the first term mean the effects of heat conduction and the effects of shear are now coupled. The solution is thus more difficult than the corresponding isothermal problem, but the difficulty is believed to be surmountable.

When the wall temperature is raised beyond that of the boiling of the liquid under prevailing pressure in the pipe, local vaporization takes place and bubbles are formed over the surface. But since the main bulk of the liquid is still at a temperature below the boiling point, these bubbles cannot grow indefinitely. In fact experiments by F. Kreith and M. Summerfield show that they contract again and have a life span of about 1/100 sec. During its short life span, the bubble does not seem to move appreciably from the wall. The main consequence of the bubble formation and disappearance is then the strong agitation of the fluid near the wall. It is then understandable that the heat flux can be increased to many times that of the case without local boiling. This fact is shown clearly in Fig. 3, taken from the work of Kreith and Summerfield[9,10]. This means that a high rate of cooling can be achieved without high flow velocity in the cooling duct. The thus reduced pressure drop in the cooling duct will decrease the necessary pumping work of the coolant. Boiling heat transfer then can be used to good advantage for many designers. The problem for research here, of course, is a closer understanding of the turbulent agitation due to bubble formation and thus better correlation of tests for different liquids and different test conditions.

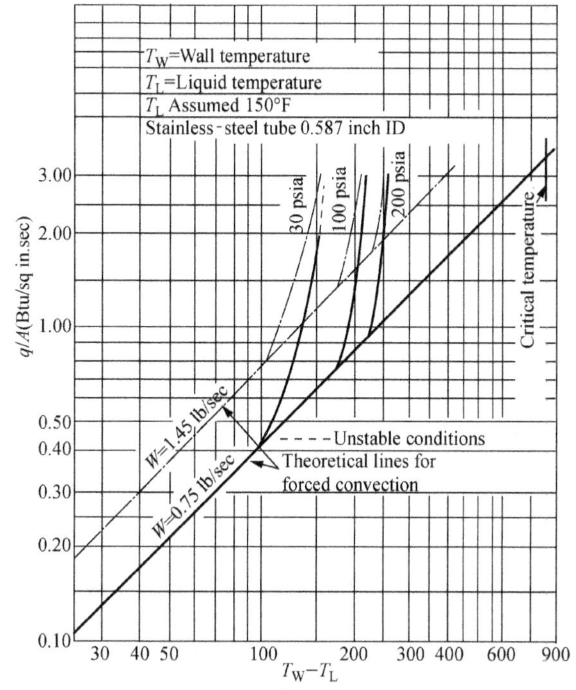

Fig. 3 Comparison of forced convection heat transfer with and without surface boiling

If the wall temperature is increased beyond a critical value over the boiling point of the liquid, it has been found that a vapor envelope forms over the surface and the heat flux is reduced by the insulating effect of the stagnant vapor. Therefore, with specified pressure and flow velocity, there is definitely a maximum value of heat-flux density even with local boiling at the surface. If still high heat-flux density is desired, then other means of cooling have to be used. However even before reaching this intrinsic limit of boiling heat transfer, the wall temperature at the inside surface of the rocket motor may be too high for the material strength, due to the necessary temperature gradient through the wall for the heat flux. For instance, if the heat flux is 6 Btu per sec per sq in., the wall thickness is 1/16 in., and the temperature at the cool side of the wall is 600 °F, the temperature at the hot side of the wall will be 1 950 °F if stainless steel is used. This temperature is certainly too high for good strength.

New powerful cooling methods for extremely high heat flux are sweat cooling and film cooling.

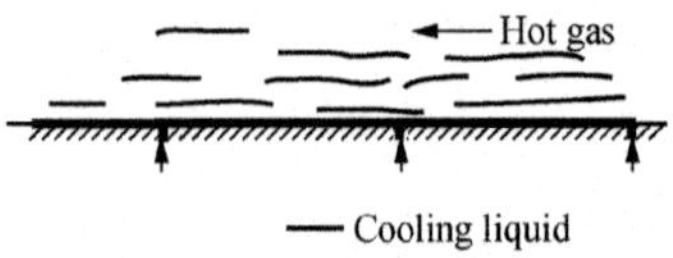

Fig. 4 Film cooling showing downstream flow of cooling unit

Film cooling (Fig. 4) is achieved by establishing a thin liquid film in contact with the hot gas over the surface to be cooled. Due to the shearing stress acting over the liquid-gas interface, the liquid flows in the down-stream direction. Simultaneously, the heating of the film by the hot gas evaporates the liquid. It is seen that so long as there is a liquid film, the wall temperature is kept below the boiling point of the liquid. It is also seen that to protect the wall from the hot gas, liquid film has to be reestablished by injection through holes in the wall when the film from upstream injection is evaporated. Of course, the intervals of injection can be lengthened by injecting more liquid and establishing a thicker film each time. However, the difficulty here is the instability of the film against the turbulent flow in the gaseous boundary layer. The resultant partial breakaway of the liquid in the form of droplets constitutes a loss in effective cooling liquid. The problem here is then the determination of the relative cooling efficiency with respect to film thickness.

From experiments on one-phase turbulent boundary layer, it is found that the laminar sublayer thickness y^* is determined by

$$y^* = 5\frac{\nu}{U\sqrt{C_f/2}}$$

where ν is the kinematic viscosity of the fluid, U is the free stream velocity, and C_f the local friction coefficient. If this relation also holds for two-phase turbulent boundary layers such as exist in film cooling, y^* is the limiting film thickness for perfect efficiency. If the film thickness is larger than y^*, instability of the film and breakaway of droplets is likely to occur. It is then seen that there is an advantage in having a higher kinematic viscosity v, as a thicker layer is then allowed. If $U = 1\,000$ ft per sec, $C_f = 0.004$, and $v = 0.319 \times 10^{-5}$ ft^2 per sec for water at 212 ℉, y^* is only 4.3×10^{-6} in. This result indicates that for theoretical maximum cooling efficiency, the film should be thin and re-established frequently along the wall. The limiting case is then sweat cooling where the coolant is forced through the porous wall, and injection and evaporation occur at the same time.

Sweat cooling is, however, not limited to the liquid coolant. The coolant may be gaseous. In fact, the most extensive experiments are made by P. Duwez and H. L. Wheeler[11] with gaseous coolants. It is shown by the above investigators however that the coolant cannot be allowed to evaporate in the porous wall, as then the flow is essentially not stable with wide fluctuations in the wall temperatures. Generally then, the most efficient sweat-cooling system with least expenditures of the coolant is one that evaporates the liquid coolant on the "outside" surface of the porous wall before entering the porous material, Fig. 5. In a sense, this system is a combination of boiling heat transfer and sweat cooling. No extensive experiment on this method of cooling has yet been made.

Combustion

It is evident that with either film cooling or sweat cooling, there is no limit to the temperature of the combustion gas that can be effectively handled. One need not have therefore any misgivings about high energy fuels and propellants for cooling difficulties. Furthermore, for rocket, ramjet, and pulsejet, there is no contact of the combustion gas with a delicate moving part such as turbine blades in a turbjoet; and the combustion gas can be corrosive and can contain finely divided solid particles. These factors practically remove all restriction in the choice of fuels and propellants. Such strange combinations as liquid hydrogen and liquid fluorine, and diborane (B_2H_6) and air are to be considered. To consider these combinations in combustion immediately raises two types of problems. The first type is problems in thermochemistry. What is the heat of combustion of these compounds? What are the thermodynamic properties of the combustion products? What are the equilibrium constants of the reactions? Information is particularly meager in connection with lithium, boron, fluorine, etc. The second type of problem concerns questions in chemical kinetics. How are the reactions of these combinations carried out? What are the individual component reactions? What are the rates of these component reactions? How can the reactions be speeded up or retarded?

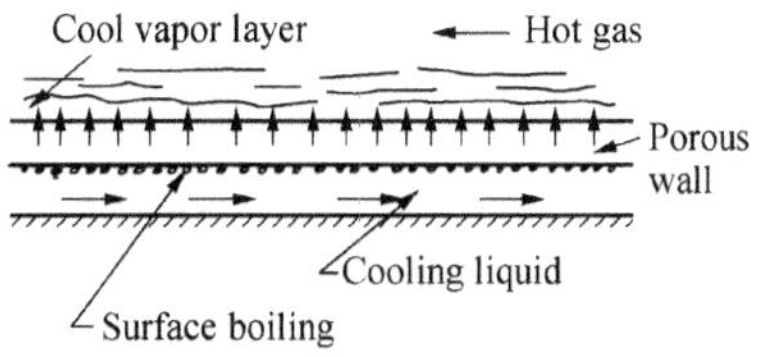

Fig. 5 Sweat cooling showing evapo-ration on the outside of porous wall before liquid coolant enters porous material

A first application of this thermochemical and chemical-kinetic information to combustion problems is the calculation of the absolute flame speed. At present, it is generally accepted that the old thermal ignition theory of flame speed, due originally to Mallard and Le Chatelier, is unsatisfactory. Due to the complexity of the problem, different investigators have emphasized different aspects of the problem. No simple general theory has yet evolved. One basic assumption of these theorise however, seems to need a clarification. This is the assumption that the rate of reaction calculated on the basis of homogeneous reaction can actually be used in determination of reaction in a flame front where the composition of the gas sometimes varies greatly in a distance of only a score of mean free paths of the molecules. In other words, the question is whether the nonuniformity involved can seriously change the reaction rate. What is needed then is a kinetic theory of nonuniform gas involving chemical reaction. For instance, to calculate the flame speed of a mixture of hydrogen gas and iodine vapor, one has to consider four species of molecules, H_2, I_2, HI, and $(HI)_2$ which is the activated complex of H_2 and I_2 after a successful collision. The Boltzmann differential-integral equation for the distribution functions of the molecules has to be solved by considering all possible collisions between these molecules. For some of the collisions the kinetic energy is not conserved.

More urgent problems of combustion in jet propulsion are those connected with fluid-mechanical aspects. These are the autoignition of liquid jets, the evaporation of liquid droplets, the mixing of gaseous components, the mutual influence of combustion and turbulence of flow, combustion in heterogeneous mixture, etc. For ramjets, the most perplexing problem today is the problem of flame stabilization. This is a problem which confronts all ramjet designers. Worse still, the mechanism of flame stabilization is not yet understood. As a result, the flame-holder design for the combustion chamber is always done by *ad hoc* experimentation. Clearly, there is a need for experiments with the simplest physical conditions so that the parameters can be accurately controlled. The work of A. S. Scurlock[12] with homogeneous gas stream and controlled turbulence in the initial stream is the most noteworthy effort in this direction. For a true understanding of the mechanism however, further detailed exploration of flow field is necessary.

It seems that one of the important aspects of flame stabilization is the interaction of the flame front and the boundary layer. To test this concept, a flame holder in the shape of a streamlined body may be tried (Fig. 6). To start the combustion, the airfoil has to be first heated by, say, electric current to a high temperature. Once the flame is started, the airfoil aft of the flame is heated by the hot gas. Heat is then conducted through the body of the airfoil to the front section where heat is given up to the cold gas mixture through the boundary layer. The cold gas mixture in the boundary layer being heated by the body will increase the concentration of active carriers and finally ignite at the intersection of the flame front and the surface of the body. It is evident that such a flame holder without turbulent and eddying loss has many practical applications in spite of the difficulty in starting. In fact, by increasing the length of the airfoil in the hot gas, the temperature at the front part can be increased, and thus, the gas velocity can be raised without blowouts.

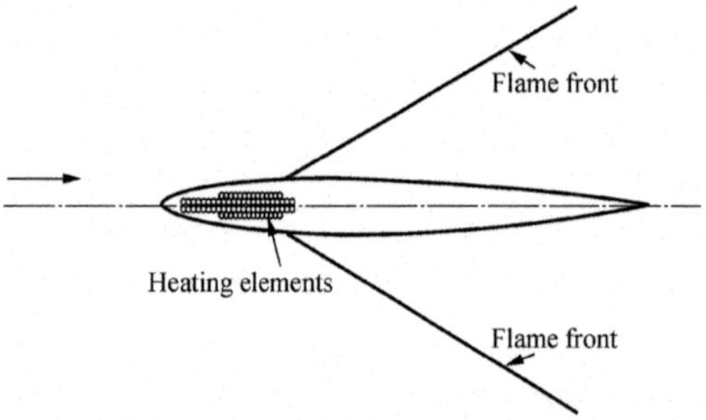

Fig. 6 Flame holder in shape of streamlined body

Performance of Rocket and Jet-Propelled Vehicles

The ultimate aim of all this basic research is of course to improve the performance of rocket and jet-propelled vehicles. However even when given the best power plant, the designer still has to determine the best way of using the power plant for optimum performance of the completed vehicle. For instance, what would be the optimum thrust programming for a sound rocket? What would be the gain possible by varying the thrust during ascent? Is this gain justified by the additional complication in the design? The basic variational problem of thrust programming was studied by G. Hamel[13]. However, he made no detailed calculations to allow the designer to weigh the importance of different aspects of the problem.

But the fundamental question in the performance analysis is the trajectory, particularly long-range trajectory. Earlier in this discussion the reason for favoring the dynamic trajectory of varying velocity was given. But what particular dynamic trajectory? To avoid the penalty of high drag at high velocity in the dense atmosphere and yet to be able to accelerate the vehicle quickly, it is clear that the vehicle should be launched in vertical position. Performance of the vertical trajectory of a rocket is well known. But is rocket the only power plant capable of vertical trajectory? Certainly the ramjet, once boosted to a sufficiently high velocity, can also produce enough thrust to make an accelerated vertical flight. A ramjet would weigh more than a rocket per lb of thrust produced, but the fuel consumption is much smaller. Preliminary estimate by L. H. Schindel[14] shows that the advantage of low fuel consumption overcomes the disadvantage of heavier dry weight. Therefore, there is a definite gain by substituting the lowest stage of a multistage rocket with a ramjet. Of course, to boost the ramjet to operating speed, it has to be operated as a ducted rocket in the first few seconds.

What can one say about the remainder of the trajectory? Since the high velocity of the vehicle is reached outside the atmosphere by vertical or near vertical ascent, the first part of the trajectory has to be liftless, and thus is elliptical. When the vehicle returns to the atmosphere at practically the same speed as it leaves the atmosphere, the lift of the body of the vehicle can be produced by putting the body into an angle of attack. The question here then is one of programming the angle of attack of the body so that maximum range is obtained. As an example of such a dynamic trajectory, the flight of a 3 000-mile rocket vehicle is studied under the assumption of steady glide after the initial elliptical path. The average lift-drag ratio in glide is taken to be 4. The result of this analysis is as follows:

Length	78.9 ft
Maximum diameter of body	8.86 ft (Fig.7)
Gross weight	96 500 lbf
Fuel load	72 400 lbf
Weight at end of burning	24 100 lbf
Propellant loading fraction	0.750
Exhaust velocity	12 000 ft/sec
Propellant	Liquid O_2 liquid H_2; liquid F_2 liquid H_2
Maximum velocity	9 140 mph
Range at conclusion of elliptic path	1 200 miles
Range contributed by glide	1 800 miles
Altitude at beginning of glide	27 miles
Landing speed	150 mph
Landing angle	20 degrees
Flight duration	Less than 1 hr

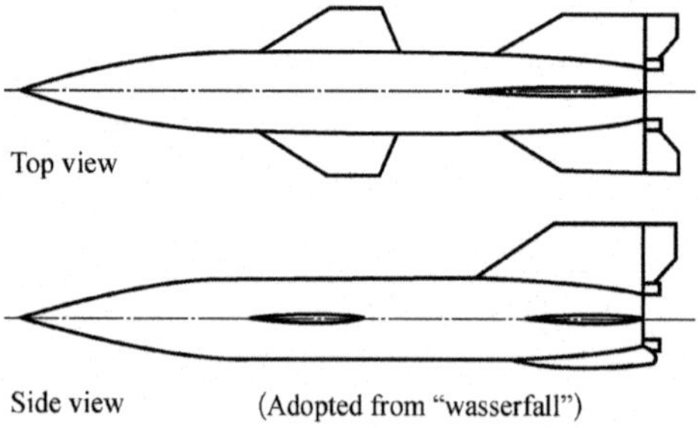

Fig. 7

It is thus seen that the requirements of a transcontinental rocket liner is not at all beyond the grasp of present-day technology. The wings need not be large to achieve a reasonable landing speed, and the specifications on the structural weight are not impossible. When will such a rocket liner be realized? That is a difficult question to answer. But one thing is certain: The basic research as outlined in this discussion will definitely hasten that day of long-range rocket travel.

References

[1] Reginald M. Cleveland. America Fledges Wings. New York, N. Y., Pitman Publishing Corporation, 1942.

[2] Prepared by the Staffs of the Guggenheim Aeronautical Laboratory and the Jet Propulsion Laboratory of the California Institute of Technology. Jet Propulsion. Air Material Command, 1946.

[3] Stanton R. Research and Development at the Jet Propulsion Laboratory. GALCIT. Engineering and Science Monthly, 1946, 9(7): 5 - 14.

[4] Lowell A L. A Guide to Aircraft Power-Plant Selection. Aeronautical Engineering Review, 1947, 6(4): 22 - 25.

[5] Alfrey T. Nonhomogeneous Stresses in Viscoelastic Media. Quarterly of Applied Mathematics, 1944, 2: 113 - 119.

[6] Tsien H S. A Generalization of Alfrey's Theorem for Viscoelastic Media. Quarterly of Applied Mathematics, 1950, Ⅷ(1): 104 - 106.

[7] Nikuradse J. Gesetzmässigkeiten der turbulenten Strömung in glatten Rohren. Verein Deutscher Ingenieure, Forschungsheft, 1932, No. 356.

[8] H. Reichardt. Die Wärmeübertragung in turbulenten Reisbungschichten. Zeitschrift für Angewandte Mathematik und Mechanik, 1940, 20: 297 - 328.

[9] Kreith F, Summerfield M. Heat Transfer to Water at High Flux Densities with and without Surface Boiling. ASME Transactions, 1949, 71: 805 - 815. // Also: Investigations of Heat Transfer at High Heat-Flux Densities: Experimental Study with Water of Friction Drop and Forced Convection with and without Surface Boiling in Tubes. Progress Report No. 4 -68, Jet Propulsion Laboratory, California Institute of Technology, 1948.

[10] Kreith F, Summerfield M. Heat Transfer from an Electrically Heated Tube to Aniline at High Heat Flux. Progress Report No. 4 - 88, Jet Propulsion Laboratory, California Institute of Technology, 1949.

[11] Duwez P, Wheeler H L. Experimental Study of Cooling Injection of a Fluid through a Porous Material. Journal of Aeronautical Sciences, 1948,15: 509 - 521.

[12] Scurlock A C. Flame Stabilization and Propagation in High-Velocity Gas Streams. Meteor Report, No. 19, Mass. Institute of Technology, 1948.

[13] Hamel G. Uber eine mit dem Problem der Rakete zusammenhängende Aufgabe der Variationsrechnung. Zeitschrift für Angewandte Mathematik und Mechanik, 1927, 7: 451 - 452.

[14] Schindel L H. Application of Ramjet to High Altitude Sounding Vehicles. Master of Science Thesis. Department of Aeronautical Engineering, Massachusetts Institute of Technology, 1947.

Influence of Flame Front on the Flow Field

By H. S. Tsien[①]
(*Pasadena, Calif.*)

Flame front is a region in the flow field where rapid change in the chemical composition of the fluid occurs with consequent release of chemical energy in the form of heat. In the majority of cases the phenomenon is a very complicated one involving the heat transfer by conduction and radiation, the changes in concentration of the different components by diffusion and chemical reaction. Owing to this and the difficult problem of chemical kinetics, only recently the complete theory of flame front has been formulated, particularly by the group under J. O. Hirschfelder[1,2]. Fortunately, as a result of the rapid rate of chemical reaction, the thickness of the flame front under ordinary conditions is generally very small, being less than 1 mm. Therefore, if one is interested in the influence of flame front on the flow field but not on the detailed structure of the flame, the flame can be assumed as infinitesimally thin, and only the final changes of the state of fluid due to combustion need be considered. This procedure is entirely analogous to that of treating the shock wave as having zero thickness in studying dynamics of compressible fluids. This simplification will be adopted for the present investigation.

Neglecting the change in specific heats of the gas by combustion and assuming perfect gas, very simple relations for quantities before the combustion and after the combustion can be obtained. This will be determined first. With these relations, the production of vorticity due to nonuniform condition before the flame front will be studied. After these preliminary results, the problem of flame width in a two-dimensional combustion chamber of constant width with a flame holder at the center will be solved approximately. This problem was first solved by A. C. Scurlock[3]. The present calculation is, however, very much simpler and is extended to take into account the compressibility of the gas. The effect of compressibility gives an anomalous spreading of the flame in the channel, and its significance in the efficiency of combustion and combustion-chamber design will be discussed.

Presented at the 1950 Heat Transfer and Fluid Mechanies Institute, Los Angeles, Calif., June 28 – 30, 1950.

ASME Transaction: Journal of Applied Mechanics, vol. 18, pp. 188 – 194, 1951.

Discussion of this paper should be addressed to the Secretary, ASME, 29 West 39th Street, New York, N. Y., and will be accepted until July 10, 1951, for publication at a later date. Difcussion received after the closing date will be returned.

NOTE: Statements and opinions advanced in papers are to be understood as individual expressions of their authors and not those of the Society. Manuscript received by the Applied Mechanics Division, October 23, 1950.

① Robert H. Goddard Professor of Jet Propulsion, Daniel and Florence Guggenheim Jet Propulsion Center, California Institute of Technology. Mem, ASME.

Flame Front

Consider the flame front to be stationary, and the unburned gas flows into it with a normal velocity S and leaves it with a velocity w_2; S is then the normal flame velocity. Let p, ρ, and γ be the pressure, density, and the ratio of specific heats, respectively. The subscript 1 will denote quantities before the combustion and the subscript 2 after the combustion. Then the equation of continuity is

$$\rho_1 S = \rho_2 w_2 \tag{1}$$

The momentum equation is

$$\rho_1 S^2 + p_1 = \rho_2 w_2^2 + p_2 \tag{2}$$

If λ is the ratio of the stagnation temperature after combustion to the stagnation temperature before the combustion, the energy equation is

$$\lambda\left[\frac{1}{2}S^2 + \frac{\gamma}{\gamma - 1}\frac{p_1}{\rho_1}\right] = \frac{1}{2}w_2^2 + \frac{\gamma}{\gamma - 1}\frac{p_2}{\rho_2} \tag{3}$$

Consider the quantities ρ_1, S, and λ as given by the detailed theory of flame or by experiment. Then Equations (1), (2), and (3) are three equations for the three unknowns w_2, p_2, and ρ_2. The solution corresponding to normal burning can be written as follows

$$\frac{w_2}{S} = \frac{\rho_1}{\rho_2} = \lambda + \frac{\gamma+1}{2}\lambda(\lambda - 1)M_1^2 + \frac{\gamma+1}{2}\lambda(\lambda - 1)\{1 + (\gamma+1)(\lambda - 1)\}M_1^4 + \ldots \tag{4}$$

$$\frac{p_2}{p_1} = 1 - \gamma(\lambda - 1)M_1^2 - \frac{\gamma+1}{2}\gamma\lambda(\lambda - 1)M_1^4 - \frac{\gamma+1}{2}\gamma\lambda(\lambda - 1)\{1 + (\gamma+1)(\lambda - 1)\}M_1^5 + \ldots \tag{5}$$

The temperature ratio T_2/T_1 is

$$\frac{T_2}{T_1} = \lambda - \frac{\gamma - 1}{2}\lambda(\lambda - 1)M_1^2 - \frac{\gamma^2 - 1}{2}\lambda^2(\lambda - 1)M_1^4 + \ldots \tag{6}$$

In these equations, the quantity M_1 is the ratio of the flame velocity S to the sound velocity a_1 in gas before the combustion, or the Mach number of the flame. Since, under ordinary conditions, S is of the order of 1 fps while a_1 is of the order of 1 000 fps, M_1 is very small and generally only the first terms in Equations (4), (5), and (6) are necessary.

Production of Vorticity by Flame

It is known that in a nonviscous and non-heat-conducting fluid, if the pressure is only a function of density, then the vorticity of any fluid element is a constant. These conditions of flow are satisfied approximately by real fluid without heat addition or combustion. Most fluid motions of practical interest originate from a uniform state, where vorticity is zero or the motion is irrotational. Then the motion will remain irrotational. This irrotationality of flow simplifies greatly the analysis of the field. Hence it is of interest to investigate the extent to which this condition is destroyed by flame front. In other words, the production of vorticity by flame should be calculated.

For simplicity, consider the two-dimensional flow. Let the gas be of uniform composition and having a constant sum of enthalpy and the kinetic energy, or isoenergetic, before combustion. In view of the generally irrotational flow without combustion, the flow before the flame front will be assumed to be irrotational and thus isentropic. The problem specifically is then to calculate the vorticity ω after the flame. Let σ be the specific entropy and ψ the stream function. Then it is known[4] that for steady flows

$$\omega = \rho_2 T_2 \frac{d\sigma_2}{d\psi_2} \tag{7}$$

where again the subscript 2 denotes quantities after the flame front. For perfect gas, Equation (7) can be written as

$$\omega = \frac{1}{\gamma - 1} p_2 \frac{d}{d\psi_2} \left\{ \log\left(\frac{p_2}{p_1}\right) - \gamma \log\left(\frac{\rho_2}{\rho_1}\right) + \log \frac{p_1}{\rho_1^2} \right\} \tag{8}$$

Before the flame, σ_1 is constant, so p_1/ρ_1^γ is constant. Since $d\psi_2 = d\psi_1$, Equation (8) simplifies to

$$\omega = \frac{1}{\gamma - 1} p_1 \left(\frac{p_2}{p_1}\right) \frac{d}{d\psi_1} \left\{ \log\left(\frac{p_2}{p_1}\right) + \gamma \log\left(\frac{\rho_1}{\rho_2}\right) \right\} \tag{9}$$

The pressure ratio p_2/p_1 and the density ratio ρ_2/ρ_1 are given by Equations (4) and (5). Therefore the production of vorticity is controlled by the variation of the flame Mach number M_1 or S/a_1 and the parameter λ along the flame front.

Perhaps due to very intense transport phenomena generated by the large temperature rise in the flame, the normal flame velocity S is observed to be only weakly dependent upon the local conditions before combustion. According to H. Sachsse[5], the normal flame velocity of methane-oxygen mixture is increased to 3 times the value at room temperature by preheating the mixture to 1 000 C. Later experiments by Sachsse and E. Bartholomé[6] indicated an increase of approximately 30 per cent in flame velocity by preheating various gas mixtures from 20 to 100 C. From this evidence, it seems that the normal flame velocity increases roughly as the absolute temperature of the "unburned" gas mixture. The experiments on the influence of pressure on the flame velocity do not seem to give conclusive results, but in any event, the influence is not large. Therefore, for the computation of the production of vorticity by flame front, two separate cases can be considered. For the first case, the flame speed S is taken to be a constant. For the second case, the flame speed S is to be proportional to the absolute temperature T_1.

Write

$$\left.\begin{aligned} \frac{p_2}{p_1} &= F \\ \frac{\rho_1}{\rho_2} &= G \end{aligned}\right\} \tag{10}$$

Then Equation (9) can be written as

$$\omega = \frac{1}{\gamma - 1} F\left[\left\{\frac{1}{F}\frac{\partial F}{\partial M_1^2} + \frac{\gamma}{G}\frac{\partial G}{\partial M_1^2}\right\}\left(p_1 \frac{\mathrm{d}M_1^2}{\mathrm{d}p_1}\right) + \left\{\frac{1}{F}\frac{\partial F}{\partial \lambda} + \frac{\gamma}{G}\frac{\partial G}{\partial \lambda}\right\}\left(p_1 \frac{\mathrm{d}\lambda}{\mathrm{d}p_1}\right)\right]\frac{\mathrm{d}p_1}{\mathrm{d}\psi_1} \tag{11}$$

If ΔH is the beat addition per unit mass of the gas due to chemical reaction, then from the definition of λ

$$\lambda = \frac{\Delta H}{\frac{1}{2}S^2 + \frac{\gamma}{\gamma - 1}\frac{p_1}{\rho_1}} + 1 \tag{12}$$

Therefore λ is not a constant, in spite of the fact that the heat released ΔH can be considered as a constant with good approximstion.

For the first case, S is a constant, and

$$M_1^2 = \frac{S^2}{a_0^2}\left(\frac{a_0}{a_1}\right)^2 = \frac{S^2}{a_0^2}\left(\frac{p_0}{p_1}\right)^{\frac{\gamma-1}{\gamma}}$$

where the subscript 0 refers to the stagnation condition before the flame. Then

$$p_1 \frac{\mathrm{d}M_1^2}{\mathrm{d}p_1} = -\frac{\gamma - 1}{\gamma}\frac{S^2}{a_1^2} = -\frac{\gamma - 1}{\gamma}M_1^2 \tag{13}$$

Similarly

$$p_1 \frac{\mathrm{d}\lambda}{\mathrm{d}p_1} = -\frac{\gamma - 1}{\gamma}(\lambda - 1)\frac{1}{1 + \frac{\gamma - 1}{2}M_1^2} \tag{14}$$

For the second case, S is proportional to the temperature T_1 or to a_1^2. Then

$$M_1^2 = \frac{S_0^2}{a_0^2}\left(\frac{a_1}{a_0}\right)^2 = \frac{S_0^2}{a_0^2}\left(\frac{p_1}{p_0}\right)^{\frac{\gamma-1}{\gamma}}$$

Therefore

$$p_1 \frac{\mathrm{d}M_1^2}{\mathrm{d}p_1} = \frac{\gamma - 1}{\gamma}\frac{S_0^2}{a_0^2}\left(\frac{a_1}{a_0}\right)^2 = \frac{\gamma - 1}{\gamma}M_1^2 \tag{15}$$

The corresponding derivative of λ is

$$p_1 \frac{\mathrm{d}\lambda}{\mathrm{d}p_1} = -\frac{\gamma - 1}{\gamma}(\lambda - 1)\frac{1 + (\gamma - 1)M_1^2}{1 + \frac{\gamma - 1}{2}M_1^2} \tag{16}$$

The derivative of pressure p_1 with respect to ψ_1 can be expressed in a more convenient form: If q_1 is the magnitude of velocity immediately ahead of the shock, n the normal distance from streamline to streamline, then

$$\mathrm{d}\psi_1 = \rho_1 q_1 \mathrm{d}n$$

Furthermore, the balance of centripetal forces by pressure requires

$$\frac{\mathrm{d}p_1}{\mathrm{d}n} = -\frac{\rho_1 q_1^2}{R_1}$$

where R_1 is the radius of curvature of the streamline immediately ahead of the shock, positive when the streamline is concave with respect to positive direction of q_1. From these two relations

$$\frac{\mathrm{d}p_1}{\mathrm{d}\psi_1} = -\frac{q_1}{R_1} \tag{17}$$

By substituting Equations (13), (14), and (17) into Equation (11), and by using Equations (4) and (5), the vorticity ω generated by the flame front for the case of constant flame speed S can be determined. By substituting Equations (15), (16), and (17) into Equation (11), the vorticity ω for the case of variable flame speed can be computed.

However, it is important to note that the value of the flame speed is generally so small as to make M_1 negligible compared with unity. Then $p_1(\mathrm{d}M_1^2/\mathrm{d}p_1)$ is negligibly small in comparison with p_1 $(\mathrm{d}\lambda/\mathrm{d}p_1)$ and the latter is approximately the same for both cases, i.e.

$$p_1 \frac{\mathrm{d}\lambda}{\mathrm{d}p_1} \cong -\frac{\gamma-1}{\gamma}(\lambda-1) \quad \text{for} \quad M_1^2 \ll 1 \tag{18}$$

By making the same approximation for the functions for F and G and their derivatives, the vorticity generated ω is simply

$$\omega \cong \frac{q_1}{R_1}\left(\frac{\lambda-1}{\lambda}\right) \quad \text{for} \quad M_1^2 \ll 1 \tag{19}$$

It is seen from Equation (19) that when $R_1 \to \infty$, $\omega \to 0$ as expected. Furthermore, when no combustion occurs, no heat is added, and $\lambda = 1$, then $\omega = 0$. But when there is combustion, the combustion will generate appreciable vorticity of the order of q_1/R_1.

Flame Width in a Uniform Channel

The problem of spreading of the flame in a homogeneous premixed combustible from an idealized point flame holder located at the axis of a two-dimensional uniform channel, Fig. 1, was first solved by A. C. Scurlock[3]. For interpreting the experimental data, he needs the relation between the flame width y_1 and the fraction of gas burned. He assumes for simplicity of calculation, that the fluid is a perfect incompressible fluid. The assumption of incompressible flow is justified on the ground that the velocity of flow is small compared with the speed of sound. This means then the flame Mach number M_1 is negligibly small. From

Equations (4), (5), and (6), it is seen that

$$\frac{\rho_1}{\rho_2} \approx \lambda, \quad \frac{p_2}{p_1} \cong 1 \tag{20}$$

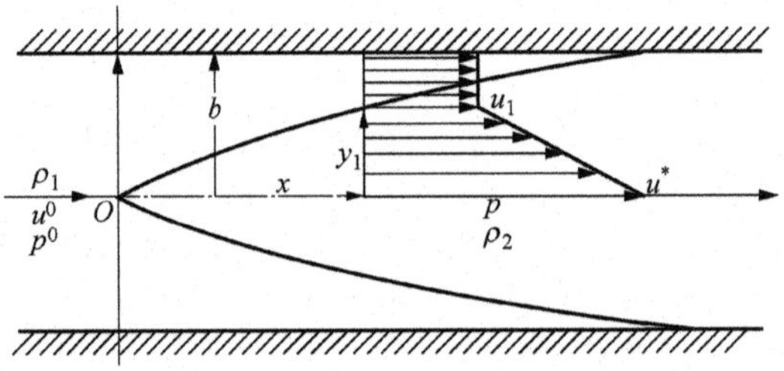

Fig. 1 Flame propagation in a two-dimensional combustion chamber of constant width from the point flame holder 0

Therefore under the assumption of incompressible flow, the effect of combustion is to change the density by a factor of λ (the ratio of stagnation temperatures) while the pressure remains constant. Scurlock then observed that since he is interested in cases where the flow velocities are very much larger than the normal flame velocity S, the flame fronts will be inclined at small angles from the channel axis. The result is that all streamlines are nearly parallel to the channel axis. Then as an approximation, the magnitude of velocity at any point is taken to be the magnitude of x-component u (parallel to the channel axis) of velocity, and neglect the effects of the curvature of the streamlines. If curvature of the streamlines is neglected, the pressure variation in the y-direction (normal to the channel axis) due to centrifugal forces must be also neglected. Equation (20) further shows that there is no pressure change by crossing the flame front; then it is evident that the pressure p in any cross section of the channel must be constant, whether in the unburned gas or the burned gas. This means the pressure is continuous in the whole field. Therefore the velocity u must also be continuous by crossing the flame front.

The entire problem is then reduced to a quasi-one-dimensional calculation: The fluid density is constant in respective regions of unburned and burned regions. The ratio of densities is λ. The unburned gas flows with constant uniform velocity u^0 and density ρ_1 until it reaches the section containing the flame holder 0, Fig. 1. The gas immediately after the flame holder has still the same velocity u^0 but a density $\rho_2 = \rho_1/\lambda$. At a section x downstream of the flame holder, the velocity at the channel axis is increased to u^* and the velocity in the unburned gas, uniform in the unburned region of the section, is increased from u^0 to u_1. The pressure p at x is however, smaller than the pressure p^0 of the approaching unburned gas. By using Bernoulli's theorem

$$\left.\begin{aligned} \frac{1}{2}\rho_1(u_1^2 - u^{0^2}) &= p^0 - p \\ \frac{1}{2}\rho_2(u^{*^2} - u^{0^2}) &= p^0 - p \end{aligned}\right\} \tag{21}$$

Therefore by eliminating p

$$\frac{u^*}{u^0} = \sqrt{1 + \lambda\left(\frac{u_1^2}{u^{0^2}} - 1\right)} \tag{22}$$

This equation shows that u^* is always larger than u_1.

At the section x, the velocity u in the burned region decreases from the value u^* at the axis to u_1 at the flame front $y = y_1$. Scurlock[4], using a laborious numerical method, has computed the velocity profile for various values of λ. Fig. 2 is taken from his paper. The accuracy of the result is, of course, predicated by the assumptions. That it cannot be exact is seen by using the result of the previous section. Along the axis, the curvature of streamline is zero. From Equation (19), the vorticity $(\partial v/\partial x - \partial u/\partial y)$ along the axis is then always zero. Furthermore the y-component of velocity v is by symmetry zero along the axis. Therefore $\partial v/\partial x$ is zero along the axis. The $\partial u/\partial y$ must be also zero along the axis. This is not so in Scurlock's result. This discrepancy must be, nevertheless, localized. In gross features then, Scurlock's results are accurate for the purpose of flamewidth determinations.

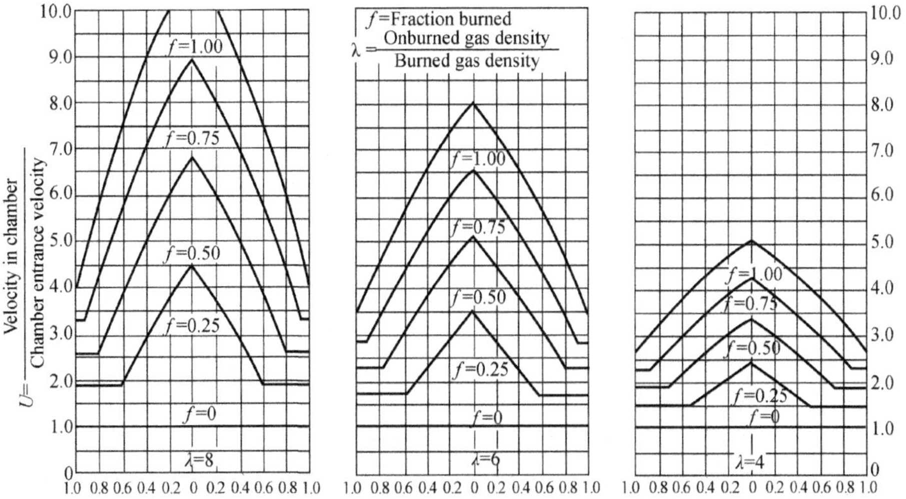

$$\eta = \frac{y}{b} = \frac{\text{Distance from center of chamber}}{\text{Distance from center of chamber to wall}}$$

Fig. 2 Velocity profile in combustion chamber with different flame width and at various density ratios (computed by A. C. Scurlock, reference 3)

On the other hand, if gross features are the only results that can be expected from the simplified quasi-one-dimensional calculation, the calculation could be made very much simpler: Take the velocity profile from $y = 0$ to $y = y_1$ to be linear. Then with Equation (22)

$$\frac{u}{u^0} = \sqrt{1 + \lambda\left(\frac{u_1^2}{u^{0^2}} - 1\right)} - \left\{\sqrt{1 + \lambda\left(\frac{u_1^2}{u^{0^2}} - 1\right)} - \frac{u_1}{u^0}\right\}\frac{y}{y_1} \tag{23}$$

The condition that the same mass must flow through each section then specifies

$$\rho_2 \int_0^{y_1} u\,dy + \rho_1 (b - y_1) u_1 = \rho_1 b u^0 \tag{24}$$

where b is the half width of the channel. By substituting Equation (23) into (24) and by observing $\rho_1/\rho_2 = \lambda$

$$\frac{1}{2\lambda}\left(\frac{y_1}{b}\right)\left[\sqrt{1+\lambda\left(\frac{u_1^2}{u^{0^2}}-1\right)}+\frac{u_1}{u^0}\right]+\frac{u_1}{u^0}\left(1-\frac{y_1}{b}\right)=1$$

By solving for (y_1/b), denoted by η, one has the simple relation

$$\eta = \frac{y_1}{b} = \frac{U-1}{\left(1-\frac{1}{2\lambda}\right)U - \frac{1}{2\lambda}\sqrt{1+\lambda(U^2-1)}} \tag{25}$$

where

$$U = u_1/u^0 \tag{26}$$

The fraction of the gas burned f is

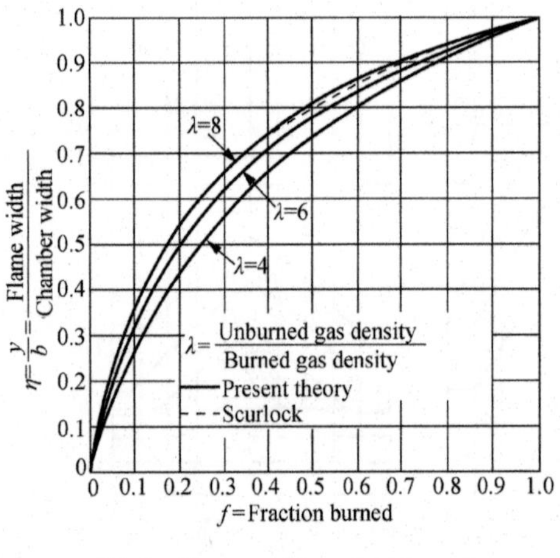

Fig. 3 Flame width as function of fraction burned at various density ratios λ, for incompressible fluid

$$f = 1 - \frac{\rho_1 (b - y_1) u_1}{\rho_1 b u^0} = 1 - U(1-\eta) \tag{27}$$

By combining Equation (25) with (26)

$$f = \frac{(U-1)\left[\sqrt{1+\lambda(U^2-1)}+U\right]}{(2\lambda-1)U - \sqrt{1+\lambda(U^2-1)}} \tag{28}$$

Equations (25) and (27) can be considered as the parametric representation of the relation between the nondimensional flame width η and the fraction burned f. Computations[①] using these equations have been carried out for $\lambda = 4$, 6, and 8. The results are compared with Scurlock's results in Fig. 3. The agreement is satisfactory. It seems then there is no need for the complicated numerical procedure of Scurlock[②].

Effect of Compressibility on Flame Width

Since the cases of interest are cases where the normal flame speed is small compared with the gas velocity, it is reasonable to speak of compressible flow of gas in general while still

① The author is indebted to Mr. D. Shonerd for carrying out the numerical computations in this paper.

② In the Appendix, a complete mathematical formulation of Scurlock's problem is given.

considering the flame Mach number M_1 in Equations (4), (5), and (6) to be negligibly small. Then the conclusions drawn is the previous section about changes of density and pressure in crossing the flame front still hold. In particular, the temperature of the gas is increased by a factor λ in crossing the flame holder. The λ at different points of the flame front is not the same as shown by Equation (12). However, λ will be taken to be a constant as an approximation. Only here Bernoulli's equation must be modified for the compressibility effects. Therefore, in place of Equation (21)

$$\left.\begin{aligned} \frac{1}{2}u^{*2} + c_p T^* &= \frac{1}{2}u^{0^2} + c_p \lambda T^0 \\ \frac{1}{2}u_1^2 + c_p T_1 &= \frac{1}{2}u^{0^2} + c_p T^0 \end{aligned}\right\} \tag{29}$$

where T^0, T_1, and T^* are the temperatures of the approaching unburned gas, the unburned gas at section x, and the burned gas at section x and the channel axis; c_p is the specific heat at constant pressure. Along each streamline, the entropy of gas is a constant in either the unburned region or the burned region. Therefore, in either region the corresponding isentropic relations hold. Equations (29) can then be modified to

$$\left.\begin{aligned} \frac{1}{2}u^{*2} + \frac{\gamma}{\gamma-1}\frac{p^0}{\rho^0}\lambda\left(\frac{p}{p^0}\right)^{\frac{\gamma-1}{\gamma}} &= \frac{1}{2}u^{0^2} + \frac{\gamma}{\gamma-1}\lambda\frac{p^0}{\rho^0} \\ \frac{1}{2}u_1^2 + \frac{\gamma}{\gamma-1}\frac{p^0}{\rho^0}\left(\frac{p}{p^0}\right)^{\frac{\gamma-1}{\gamma}} &= \frac{1}{2}u^{0^2} + \frac{\gamma}{\gamma-1}\frac{p^0}{\rho^0} \end{aligned}\right\} \tag{30}$$

By eliminating the pressure ratio p/p^0, Equation (22) is again obtained. Thus the relation between the burned velocity u^* at the axis and the unburned velocity u_1 is not modified by the compressibility.

If the linear velocity profile through the burned region is again assumed, Equation (23) remains true. However, now it is necessary to distinguish the density ρ_1 of the unburned gas at section x from the density ρ^0 of the approaching unburned gas. This ratio is easily obtained as

$$\frac{\rho_1}{\rho^0} = \left[1 - \frac{\gamma-1}{2}M^{0^2}(U^2-1)\right]^{\frac{1}{\gamma-1}} \tag{31}$$

where M^0 is the Mach number of the approaching unburned gas, u^0/a^0, and U is again u_1/u^0. By using the approximation that the density of fluid decreases by the constant factor $1/\lambda$ after crossing the flame, the continuity condition is now

$$\frac{\rho_1}{\lambda}\int_0^{y_1} u\,dy + \rho_1(b-y_1)u_1 = \rho^0 b u^0 \tag{32}$$

By using Equations (23) and (31), Equation (32) gives

$$\eta = \frac{y_1}{b} = \frac{U - \left[1 - \frac{\gamma-1}{2}M^{0^2}(U^2-1)\right]^{-\frac{1}{\gamma-1}}}{\left(1-\frac{1}{2\lambda}\right)U - \frac{1}{2\lambda}\sqrt{1+\lambda(U^2-1)}} \tag{33}$$

The fraction burned f is then

$$f = 1 - \frac{\rho_1}{\rho_0} U(1 - \eta) \tag{34}$$

Equations (33) and (34) together with Equation (31) are the parametric representation of the relation between the nondimensional flame width η and the fraction burned, f. The results of calculation are plotted in Figs. 4, 5, and 6. It is seen that the compressibility has little effect on the relation between the flame width η and the fraction burned f. The curves at different approach Mach numbers M^0 lie very close to the incompressible-flow curve calculated by using Equations (25) and (28). Therefore the procedure adopted by Scurlock in using the incompressible curve for all his computations is indeed justified. Figs. 4, 5, and 6, however, show another very important feature of the problem: Both the flame width η and the fraction burned f, have maximum values at higher M^0. For $M^0 = 0.4$, the maximum flame width is only 1/2 of the channel width and the maximum amount of gas burned is only 1/3 of the input. At higher values of M^0, these fractions are even smaller. This definitely shows that for a combustion chamber of constant width, as assumed in the present analysis, it will be difficult if not impossible to have complete combustion at high flow velocities, even with a good flame holder to initiate the flame.

What is the physical situation which causes this anomaly? By crossing the flame front the flow velocity is maintained, but the temperature of the gas is increased by λ times. Therefore the Mach number of the gas is reduced by combustion. In other words, the Mach number of the unburned gas is always higher than the Mach number of the burned gas. With equal reduction in pressure, the stream tube will contract less if the Mach number is higher. In fact, for supersonic flow, the stream tube will expand instead of contract when the pressure is reduced. Therefore the effect of compressibility of the gas is to make the width of the unburned gas relatively larger than the width for incompressible fluid. The effect is more prominent when the initial Mach number is higher. Therefore there will be one initial Mach number M_c^0, called the critical Mach number, for a given value of λ, at which

$$\frac{d\eta}{dU} = 0 \quad \text{at} \quad \eta = 1, \quad U = U_c \tag{35}$$

If the initial Mach number M^0 is greater than M_c^0, then $\frac{d\eta}{dU} = 0$ at $\eta < 1$. Then the flame width will not be able to increase beyond the value of η corresponding to that at $d\eta/dU = 0$. It seems that for complete combustion M^0 should be less than M_c^0.

Equations (33) and (35) give the following conditions for the critical state

$$\left(1 - \frac{1}{2\lambda}\right) - \frac{1}{2} \frac{U_c}{\sqrt{1 + \lambda(U_c^2 - 1)}} = 1 - M_c^{0^2} U_c \left\{1 - \frac{\gamma - 1}{2} M_c^{0^2} (U_c^2 - 1)\right\}^{-\frac{\gamma}{\gamma-1}} \tag{36}$$

$$\left(1-\frac{1}{2\lambda}\right)U_c-\frac{1}{2\lambda}\sqrt{1+\lambda(U_c^2-1)}=U_c-\left\{1-\frac{\gamma-1}{2}M_c^{0^2}(U_c^2-1)\right\}^{-\frac{1}{\gamma-1}} \tag{37}$$

From Equations (36) and (37)

$$\frac{\gamma-1}{2}M_c^{0^2}=\frac{\lambda U_c+\sqrt{1+\lambda(U_c^2-1)}}{\left(\frac{\gamma+1}{\gamma-1}U_c^2-1\right)\left\{\lambda U_c+\sqrt{1+\lambda(U_c^2-1)}\right\}-2\frac{\lambda-1}{\gamma-1}U_c} \tag{38}$$

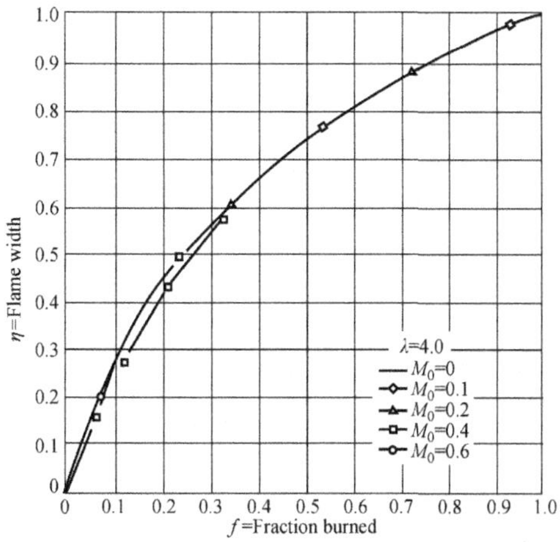

Fig. 4 Flame width as function of fraction burned at density ratio $\lambda = 4$ for various initial Mach Numbers M^0

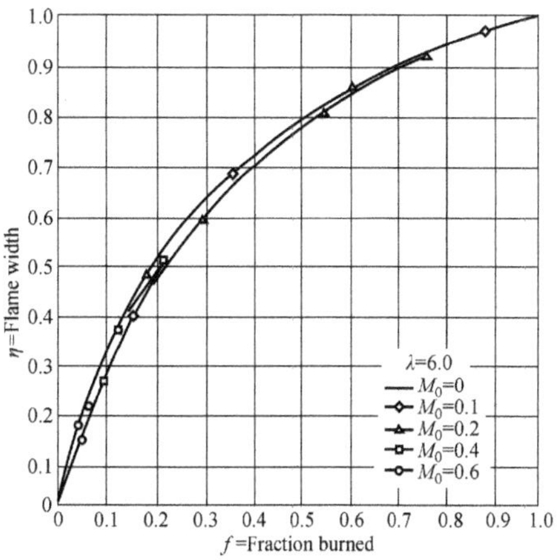

Fig. 5 Flame width as function of fraction burned at density ratio $\lambda = 6$ for various initial Mach Numbers M^0

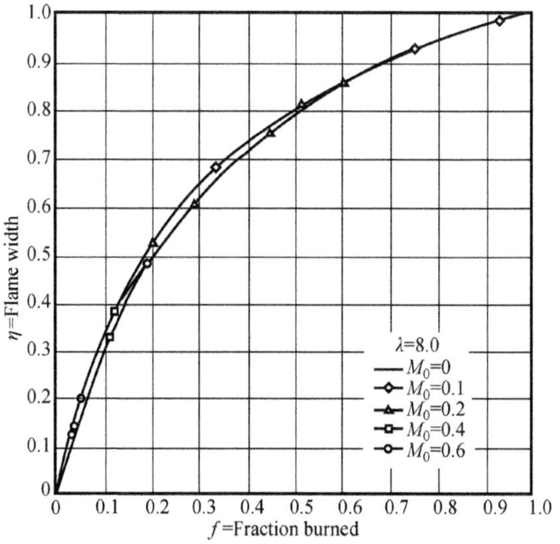

Fig. 6 Flame width as function of fraction burned at density ratio $\lambda = 8$ for various initial Mach Numbers M^0

This equation and Equation (36) can be used to determine the relation between M_c^0, λ, and U_c by numerical method. The Mach number M_c corresponding to U_c at the critical state is

$$M_c^2 = \frac{M_c^{0^2} U_c^2}{1 - \frac{\gamma - 1}{2} M_c^{0^2} (U_c^2 - 1)} \tag{39}$$

The results of calculation are given in Fig. 7 where M_c^0, M_c, U_c are plotted against λ. From the previous discussion it is seen that in order to have complete combustion, the approach Mach number M^0 must be kept below the critical value M_c^0. For a heating ratio $\lambda = 8$, M_c^0 is only 0.15. Then the velocity for complete combustion is below 200 fps. It is interesting to observe that the local Mach number M_c, for the unburned gas at the critical state, is very close to unity but slightly supersonic.

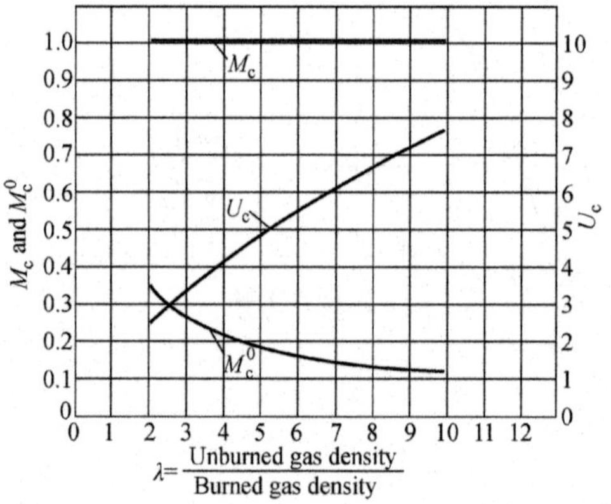

Fig. 7 Critical conditions for complete combustion at various density ratios λ

It is unfortunate that the results of the present theoretical analysis cannot be checked with Scurlock's experiments. The few critical cases recorded by Scurlock with high inlet velocity to the combustion chamber and small flame holder are made in a combustion chamber too short to ascertain whether the flame reached the chamber wall. This is, of course, natural as Scurlock concentrated his attention on the flame-holder performance, but not in the flame-spreading and combustion efficiency. From the present investigation it would seem that the problem of flame-spreading and combustion efficiency is a problem by itself, apart from the problem of flame holding, or flame initiation. In fact, one should not limit oneself only to combustion chambers of constant width, as it is evident that the way to achieve complete combustion at high approach Mach number is to use an expanding combustion chamber for reducing the flow Mach number as the combustion progresses. The present analysis is only a beginning in this problem. It serves to point out the general importance of mutual influence of flame front and the flow, and to show specifically the limitations of constant-width combustion chambers.

Appendix

Formulation of Scurlock's Problem as an Integral Equation

Let the flame width increase from ξ to $\xi + d\xi$ when distance downstream of the flame holder increases from t to dt, Fig. 8. The mass of unburned gas at t is $\rho_1 (b - \xi) v$ where v is the velocity of the unburned gas at t (i.e., at $x = t$, $u_1 = v$). The decrease in the mass flow of unburned gas, or the mass of gas burned dm between t and $t + dt$ is thus

$$dm = -\rho_1 d[(b - \xi)v] \tag{40}$$

At section x, this fraction of burned mass occupies the width dy with a velocity u and density (ρ_1/λ). From Bernoulli's equation applied between section t to x, one has, similar to Equation (22)

$$u = \sqrt{\lambda}\sqrt{u_1^2 - \left(1 - \frac{1}{\lambda}\right)v^2} \tag{41}$$

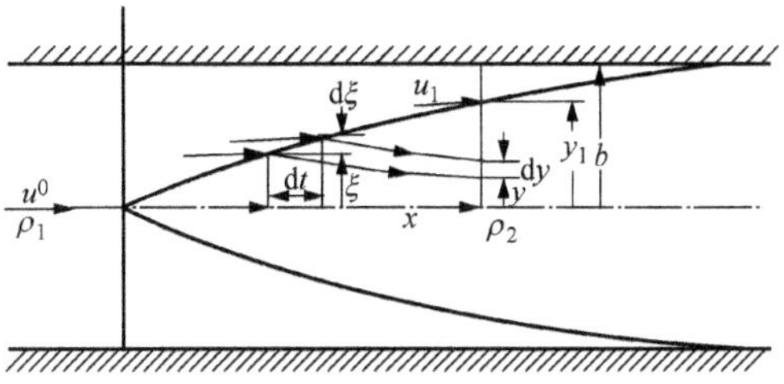

Fig. 8 Scurlock's problem

By using the relation

$$\frac{\rho_1}{\lambda} u\, dy = dm$$

and Equation (40)

$$dy = -\sqrt{\lambda}\frac{d[(b - \xi)v]}{\sqrt{u_1^2 - \left(1 - \frac{1}{\lambda}\right)v^2}} \tag{42}$$

The streamline passing through $y = 0$ corresponds to $v = u^0$ and the streamline passing through $y = y_1$ corresponds to $v = u_1$. Therefore, by integrating Equation (42)

$$y_1 = -\sqrt{\lambda}\int_{v=u^0}^{v=u_1}\frac{d[(b - \xi)v]}{\sqrt{u_1^2 - \left(1 - \frac{1}{\lambda}\right)v^2}} \tag{43}$$

Result of a partial integration of Equation (43) can be written in nondimensional variables as

$$\eta(U) = \frac{\lambda}{(\lambda - 1)^{3/2}}\left\{\sin^{-1}\sqrt{\frac{\lambda - 1}{\lambda}} - \sin^{-1}\frac{1}{U}\sqrt{\frac{\lambda - 1}{\lambda}}\right\} + \frac{1}{\sqrt{\lambda}}\int_1^U \frac{V^2}{\left[U^2 - \left(1 - \frac{1}{\lambda}\right)V^2\right]^{3/2}}\eta(V)\,dV \tag{44}$$

where $V = v/u^0$ and $\eta(V) = \xi/b$.

Equation (44) is an integral equation of the second kind for the unknown function $\eta(U)$. When λ is large, a very accurate value of η can be obtained by replacing the $\eta(V)$ in the integral of Equation (44) by the approximate value given by Equation (25).

References

[1] Hirschfelder J O, Curties C E. Theory of Propagation of Flames, Part Ⅰ. General Equations. Third Symposium on Combustion, Flames and Explosion Phenomenon, Williams and Wilkins, Baltimore, Md., 1949: 121 - 127.

[2] Theory of Flame Propagation. Journal of Chemical Physics, 1949, 17: 1076.

[3] Scurlock A C. Flame Stabilization and Propagation in High—Velocity Gas Streams. Meteor Report No. 19, Massachusetts Institute of Technology, 1948.

[4] Vazsonyi A. On Rotational Gas Flows. Quarterly of Applied Mathematics, 1945, 3: 29 - 37.

[5] Sachsse H. Über die Temperaturabhängigkeit der Flammengeschwindigkeit und das Temperaturgefälle in der Flammenfront. Zeitschrift für physikalische Chemie (A). 1937, 180: 305 - 313.

[6] Sachsse H, Bartholomé E. Beiträge zur Frage der Flammengeschwindigkeit. Zeitschrift für Elektrochemie und angewandte physikalische Chemie, 1949, 53: 183 - 190.

Optimum Thrust Programming for a Sounding Rocket

By H. S. Tsien① and Robert C. Evans②

The problem of optimum thrust programming for a sounding rocket of minimum starting weight to reach specified height with given final weight and propellant characteristics is first formulated as a problem in variational calculus. The general solution for arbitrary drag function is given. The solution is then applied to two special cases, one with quadratic drag dependence on velocity and the other with linear drag dependence on velocity. Complete numerical data are given. The results are then compared with the results of constant thrust to show the advantages of thrust programming. Thrust programming is shown to be able to increase appreciably the pay load of a high altitude sounding rocket.

For a rocket in vertical flight, the aerodynamic drag of the rocket body and the gravitational pull are in the same direction and directly opposite to the direction of the thrust force. The performance computation is thus relatively simple. It was indeed carried out by F. J. Malina and A. M. O. Smith[1]. They have shown that if the aerodynamic drag is absent, then the best way of using the propellant is to use it in the shortest possible time. Theoretically, a rocket of given weight fraction of propellant will reach the highest altitude if the thrust is applied as a single impulse and thus reaches maximum velocity immediately. This result can be understood, perhaps, by considering the other extreme of having the thrust equal to the weight of the rocket at every time instant. Then the rocket, having zero acceleration at all times, will not be able to leave the ground. Long drawn-out operation of the rocket, therefore, is definitely unprofitable if aerodynamic drag can be neglected.

When there is aerodynamic drag, the high velocity reached at low altitudes by an impulsive start will give a very high drag which tends to reduce the maximum altitude of the rocket. In fact, the calculations made by Malina and Smith, and more recently by Ivey, Bowen, and Oborny[2], show that, assuming constant thrust, the optimum initial acceleration is one to three g, depending on the ratio of drag and weight of the rocket. The more general

---

Received April 16, 1951.

Presented at the 1950 Annual Convention of the American Rocket Society, Hotel Statler, New York, N. Y., Dec. 1.

Journal of the American Rocket Society, vol. 21, pp. 99 – 107, 1951.

① Robert H. Goddard Professor, Daniel and Florence Guggenheim Jet Propulsion Center, California Institute of Technology, Pasadena, Calif.

② Research Assistant, Daniel and Florence Guggenheim Jet Propulsion Center, California Institute of Technology, Pasadena, Calif.

problem, however, is that of optimum thrust programming, i. e., to find the optimum time variation of the thrust for maximum altitude. This is the theoretical optimum design. Practical design is complicated by the added weight on the power plant to make the thrust variable and certainly cannot reach the theoretical optimum condition. The theoretical optimum gives a standard of comparison and shows how much could be expected by varying the thrust.

The problem of optimum thrust programming was studied quite early by G. Hamel[3]. He made the simplifying assumptions that the density of air decreases exponentially with altitude, and that the effective exhaust velocity c of the rocket motor and the gravitational constant g do not vary with altitude. Using variational calculus he gave the solution of the problem of optimum thrust programming. However, his paper is very brief and is not easy to understand. It is the purpose of the present authors to give a complete discussion of the problem together with more extensive numerical data.

Formulation of the Problem

Let M be the mass of the rocket and s the height at time instant t. Following Hamel, the effective exhaust velocity c of the rocket motor will be taken as a constant. The velocity of the rocket is $\mathrm{d}s/\mathrm{d}t$ and will be denoted as $\dot{s}$. For a specified rocket body, the aerodynamic drag D is a function of the altitude s, and the velocity $\dot{s}$. If g is the constant value of the gravitational force per unit mass, the equation of motion of the rocket during the powered flight is

$$\frac{\mathrm{d}M}{\mathrm{d}t}+\frac{M}{c}\left\{\frac{\mathrm{d}\dot{s}}{\mathrm{d}t}+g\right\}=-\frac{D(s,\dot{s})}{c} \tag{1}$$

The altitude s is measured from the point of launching. Therefore, at the beginning of powered flight, $t=0$, $s=0$, $M=M^0$, and $\dot{s}=\dot{s}_0$ where M^0 is the mass and $\dot{s}_0$ is the velocity after boosting. At the end of the powered flight, $t=t_1$, $M=M_1$, $s=s_1$, $\dot{s}=\dot{s}_1$. M_1 is then the final mass. Equation (1) together with these boundary conditions gives the following expression for mass M^0,

$$M^0=\exp\left(-\frac{\dot{s}_0}{c}\right)\times\left\{\int_0^{t_1}\frac{D(s,\dot{s})}{c}\exp\left(\frac{\dot{s}+gt}{c}\right)\mathrm{d}t+M_1\exp\left(\frac{\dot{s}_1+gt_1}{c}\right)\right\} \tag{2}$$

If M_0 is the initial mass of the rocket including the propellant for boosting by a single impulse to the velocity s_0, then

$$M_0=M^0\exp\left(\frac{\dot{s}_0}{c}\right) \tag{3}$$

Therefore, by combining (2) and (3),

$$M_0=\int_0^{t_1}\frac{D(s,\dot{s})}{c}\exp\left(\frac{\dot{s}+gt}{c}\right)\mathrm{d}t+M_1\exp\left(\frac{\dot{s}_1+gt_1}{c}\right) \tag{4}$$

Equation (4) does not explicitly contain the velocity after boosting, or initial velocity $\dot{s}_0$. If the initial velocity is zero, the acceleration of the rocket is smooth. If the initial velocity is not zero, the rocket starts with an impulse and then gradually accelerates. In any case, however, Equation (4) gives the total initial mass including the boosting charge.

Now let the problem of optimum thrust programming be formulated as follows: Given M_1, c, g, and the drag function $D(s, \dot{s})$, what should be the function $s(t)$, such that M_0 is a minimum? The auxiliary conditions are $s(0) = 0$ and that s_1 and $\dot{s}_1$ must be such as to reach the specified summit altitude S. To reach the given summit altitude with the specified M_1 and $D(s, \dot{s})$, s_1 and $\dot{s}_1$ are related, say

$$\dot{s}_1 = \phi(s_1) \tag{5}$$

where ϕ is a given function. For instance, at very high altitudes where the aerodynamic drag is negligible due to the low air density,

$$\dot{s}_1 \cong \sqrt{2g(S - s_1)} \tag{6}$$

To find the conditions for the solution of this variational problem, let the required s function be

$$s = s(t) \tag{7}$$

with $s(0) = 0$.

Now let there be an arbitrary function $\eta(t)$ such that

$$\eta(0) = 0 \tag{8}$$

but otherwise completely unspecified. Then the "neighboring" functions to $s(t)$ can be constructed as

$$\bar{s}(t) = s(t) + k(\varepsilon)\eta(t) \tag{9}$$

where k is a parameter but not a function of time. Because of Equation (8), $\bar{s}$ satisfies the initial condition $\bar{s}(0) = 0$. The duration of powered flight, or the burning time, for the optimum solution is t_1. For the neighboring solution, the burning time is $t_1 + \varepsilon$. Thus k is a function of ε. For the optimum solution, k and ε both vanish. Therefore,

$$\left.\begin{aligned} k(0) &= 0 \\ k(\varepsilon) &\cong \varepsilon k'(0) \end{aligned}\right\} \tag{10}$$

By considering only terms up to first order in ε,

$$\bar{s}_1 = \bar{s}(t_1 + \varepsilon) = s(t_1) + \varepsilon \dot{s}(t_1) + k'(0)\varepsilon\eta(t_1) \tag{11}$$

$$\dot{\bar{s}}_1 = \left(\frac{\mathrm{d}\bar{s}}{\mathrm{d}t}\right)_{t=t_1+\varepsilon} = \dot{\bar{s}}(t_1 + \varepsilon) = \dot{s}(t_1) + \varepsilon\ddot{s}(t_1) + k'(0)\varepsilon\dot{\eta}(t_1) \tag{12}$$

where $\dot{\eta} = \dfrac{\mathrm{d}\eta}{\mathrm{d}t}$. However $\bar{s}_1$ and $\dot{\bar{s}}_1$ must satisfy Equation (5) so that the neighboring solutions will represent rockets reaching the specified summit altitude, the stated auxiliary condition.

Therefore,

$$\left.\begin{aligned} \dot{\bar{s}}_1 &= \phi(s_1) + \left(\frac{d\phi}{ds}\right)_{s_1} \{\bar{s}_1 - s_1\} + \ldots \\ &= \dot{s}(t_1) + \left(\frac{d\phi}{ds}\right)_{s_1} \{\bar{s}_1 - s_1\} + \ldots \end{aligned}\right\} \tag{13}$$

$\bar{s}_1$ and $\dot{\bar{s}}_1$ from Equations (11) and (12) can be substituted into Equation (13). After some simplification, one has

$$\left[\left(\frac{d\phi}{ds}\right)_{s_1} \eta(t_1) - \dot{\eta}(t_1)\right] k'(0) = \ddot{s}(t_1) - \left(\frac{d\phi}{ds}\right)_{s_1} \dot{s}(t_1) \tag{14}$$

This equation then determines the value of $k'(0)$. With the $k'(0)$ so determined, the neighboring solutions will definitely satisfy all the auxiliary conditions.

With $\eta(t)$ specified, the total initial mass M_0 will be dependent upon ε. Let

$$F(s, \dot{s}, t) = D(s, \dot{s}) \exp\left(\frac{\dot{s} + gt}{c}\right) \tag{15}$$

Then by substituting Equations (9) and (12) into (4),

$$\begin{aligned} M_0(\varepsilon) = &\frac{1}{c}\int_0^{t_1+\varepsilon} F\{s + k(\varepsilon)\eta, \dot{s} + k(\varepsilon)\dot{\eta}, t\} dt + \\ &M_1 \exp\left\{\frac{\dot{s}(t_1) + \varepsilon \ddot{s}(t_1) + k(\varepsilon)\dot{\eta}(t_1) + gt_1 + g\varepsilon}{c}\right\} \end{aligned} \tag{16}$$

The condition for $s(t)$ to correspond to the optimum solution can now be stated simply as

$$\left(\frac{\partial M_0}{\partial \varepsilon}\right)_{\varepsilon=0} = 0 \tag{17}$$

By carrying out the required differentiation, one has

$$\begin{aligned} \left(\frac{\partial M_0}{\partial \varepsilon}\right)_{\varepsilon=0} = &\frac{1}{c}k'(0)\int_0^{t_1} \eta\left[\frac{\partial F}{\partial s} - \frac{d}{dt}\left(\frac{\partial F}{\partial \dot{s}}\right)\right] dt + \frac{1}{c}k'(0)\eta(t_1)\left(\frac{\partial F}{\partial \dot{s}}\right)_{t_1} + \\ &\frac{1}{c}F(s_1, \dot{s}_1, t_1) + \frac{1}{c}M_1[\ddot{s}_1 + g + k'(0)\dot{\eta}(t_1)] \exp\left(\frac{\dot{s}_1 + gt_1}{c}\right) \end{aligned}$$

But $\eta(t)$ is arbitrary other than the condition $\eta(0) = 0$. Therefore, in order for the above expression to vanish,

$$\frac{\partial F}{\partial s} - \frac{d}{dt}\left(\frac{\partial F}{\partial \dot{s}}\right) = 0 \tag{18}$$

and

$$k'(0)\eta(t_1)\left(\frac{\partial F}{\partial \dot{s}}\right)_{t_1} + F(s_1, \dot{s}_1, t_1) + M_1[\ddot{s}_1 + g + k'(0)\dot{\eta}(t_1)] \exp\left(\frac{\dot{s}_1 + gt_1}{c}\right) = 0 \tag{19}$$

Equation (18) is the familiar Euler-Lagrange differential equation. Equation (19) is the result of the auxiliary condition of the problem.

By eliminating $k'(0)$ between Equations (14) and (19), one has

$$0=\left[\ddot{s}_1-\left(\frac{d\phi}{ds}\right)_{s_1}\dot{s}_1\right]\eta(t_1)\left(\frac{\partial F}{\partial\dot{s}}\right)_{t_1}+\left[\left(\frac{d\phi}{ds}\right)_{s_1}\eta(t_1)-\dot{\eta}(t_1)\right]\cdot F(s_1,\dot{s}_1,t_1)+$$

$$M_1\left[\left\{\left(\frac{\partial\phi}{\partial s}\right)_{s_1}\eta(t_1)-\dot{\eta}(t_1)\right\}(\ddot{s}_1+g)+\dot{\eta}(t_1)\cdot\left\{\ddot{s}_1-\left(\frac{d\phi}{ds}\right)_{s_1}\dot{s}_1\right\}\right]\exp\left(\frac{\dot{s}_1+gt_1}{c}\right)$$

But $\eta(t)$ is arbitrary. Therefore, in order for the above equation to be true, the sums of quantities which multiply into $\eta(t_1)$ and $\dot{\eta}(t_1)$ must be zero separately. Therefore

$$\left\{\ddot{s}_1-\left(\frac{d\phi}{ds}\right)_{s_1}\dot{s}_1\right\}\left(\frac{\partial F}{\partial\dot{s}}\right)_{t_1}+\left(\frac{d\phi}{ds}\right)_{s_1}F(s_1,\dot{s}_1,t_1)+M_1\left(\frac{d\phi}{ds}\right)_{s_1}(\ddot{s}_1+g)\exp\left(\frac{\dot{s}_1+gt_1}{c}\right)=0 \quad (20)$$

and

$$F(s_1,\dot{s}_1,t_1)+M_1\left[g+\left(\frac{d\phi}{ds}\right)_{s_1}\dot{s}_1\right]\exp\left(\frac{\dot{s}_1+gt_1}{c}\right)=0 \quad (21)$$

Equations (18), (20), and (21) are now the complete answer to the variational problem. That is, the thrust programming must be such that Equation (18) is satisfied at every instant of the powered flight and, in addition, at the end of the powered flight the conditions given by Equations (20) and (21) must be fulfilled.

These conditions can be reduced to simpler forms if the relation in Equation (15) is reintroduced. Then Equation (18) becomes

$$\frac{\partial D}{\partial s}=\frac{\partial^2 D}{\partial s\partial\dot{s}}\dot{s}+\frac{\partial^2 D}{\partial\dot{s}^2}\ddot{s}+\frac{1}{c}\left\{\frac{\partial D}{\partial s}\dot{s}+\frac{\partial D}{\partial\dot{s}}(2\ddot{s}+g)+\frac{D}{c}(\ddot{s}+g)\right\} \quad (22)$$

When the drag is specified as a function of s and $\dot{s}$, Equation (22) gives the differential equation for the trajectory $s(t)$. Equations (20) and (21) now become

$$\left\{\ddot{s}_1-\left(\frac{d\phi}{ds}\right)_{s_1}\dot{s}_1\right\}\left\{\left(\frac{\partial D}{\partial\dot{s}}\right)_{s_1}+\frac{D(s_1,\dot{s}_1)}{c}\right\}+\left(\frac{d\phi}{ds}\right)_{s_1}D(s_1,\dot{s}_1)+M_1\left(\frac{d\phi}{ds}\right)_{s_1}(\ddot{s}_1+g)=0 \quad (23)$$

and

$$D(s_1,\dot{s}_1)+M_1\left[\left(\frac{d\phi}{ds}\right)_{s_1}\dot{s}_1+g\right]=0 \quad (24)$$

where the subscript $(\ )_1$ denotes the quantities evaluated at $t=t_1$. Equation (24) is, however, automatically satisfied if Equation (5) represents the relation at the beginning of coasting flight. The reason is as follows: During coasting, the rocket motor is stopped, and there is no expenditure of propellant; therefore, $dM/dt=0$. Then Equation (1) reduces to

$$\left[\left(\frac{d\phi}{dt}\right)_{t_1}+g\right]M_1+D(s_1,\dot{s}_1)=0 \quad (25)$$

This is the same as Equation (24).

By eliminating $\left(\frac{d\phi}{ds}\right)_{s_1}$ between Equations (23) and (24), the condition at the end of the powered flight is finally expressed as

$$\dot{s}_1(M_1\ddot{s}_1+D_1+M_1g)\left\{\left(\frac{\partial D}{\partial \dot{s}}\right)_1+\frac{D_1}{c}\right\}=(M_1\ddot{s}_1+D_1+M_1g)(D_1+M_1g)$$

where $D_1 = D(s_1, \dot{s}_1)$ is the drag at the end of burning. However, the factor $(M_1\ddot{s}_1+D_1+M_1g)$ is never zero, therefore,

$$\dot{s}_1\left\{\left(\frac{\partial D}{\partial \dot{s}}\right)_1+\frac{D_1}{c}\right\}=D_1+M_1g \tag{26}$$

The problem of optimum thrust programming can now be discussed in more concrete terms. Since the aerodynamic drag D enters linearly and homogeneously into Equation (22), that equation is actually a second-order differential equation for $s(t)$, independent of the size of the rocket body. However, being a second-order differential equation with only one initial condition $s(0) = 0$, the initial velocity or velocity after boosting $\dot{s}_0$ is yet free and undetermined. It is determined, however, by the condition, Equation (26), at the end of the powered flight. In other words, for a given size of the rocket and for a given final mass M_1, there is a corresponding optimum booster velocity $\dot{s}_0$ and subsequent optimum thrust programming for any specified summit altitude S. In general then, the optimum solution always involves an impulsive start from rest. This is a characteristic of the problem.

Quadratic Drag Law

To carry out the computation, the air density is assumed to decrease exponentially with respect to altitude. As a first example, the drag is taken to vary as the square of the velocity. Then the aerodynamic drag is given by

$$D = W_0\,\dot{s}^2 \exp(-\alpha s) \tag{27}$$

This corresponds to a constant drag coefficient. Substituting this expression into Equation (22) yields

$$\frac{\ddot{s}}{g}=\frac{v\{v^2+(1-\beta)v-2\beta\}}{\beta\{v^2+4v+2\}} \tag{28}$$

where

$$v=\frac{\dot{s}}{c},\quad \beta=\frac{g}{\alpha c^2} \tag{29}$$

both are nondimensional parameters. Integrating Equation (28) for $t(v)$ and $s(v)$,

$$\frac{gt}{c} = \ln\frac{v_0}{v} + \frac{\gamma}{2}\ln\frac{2v+(1-\beta)-\gamma}{2v+(1-\beta)+\gamma}\cdot\frac{2v_0+(1-\beta)+\gamma}{2v_0+(1-\beta)-\gamma} + \frac{\beta+1}{2}\ln\frac{v^2+(1-\beta)v-2\beta}{v_0^2+(1-\beta)v_0-2\beta} \tag{30}$$

$$\alpha s = v - v_0 + \frac{\gamma}{2}\ln\frac{2v+(1-\beta)-\gamma}{2v+(1-\beta)+\gamma}\cdot\frac{2v_0+(1-\beta)+\gamma}{2v_0+(1-\beta)-\gamma} + \frac{\beta+1}{2}\ln\frac{v^2+(1-\beta)v-2\beta}{v_0^2+(1-\beta)v_0-2\beta} \tag{31}$$

where

$$\gamma = \sqrt{(1-\beta)^2+4\beta} \tag{32}$$

and ln indicates the natural logarithm to the base e.

By using Equations (27), (28), (30), and (31), the mass M at any time instant is found to be as follows:

$$\frac{M}{M_1} = \exp\left\{-\left(v+\frac{gt}{c}\right)\right\}\left[\frac{W_0c^2}{M_1g}\cdot\beta\cdot v_0\cdot\exp(v_0)\times \{v_0^2+(1-\beta)v_0-2\beta\}\left\{\frac{v+2}{v^2+(1-\beta)v-2\beta} - \frac{v_1+2}{v_1^2+(1-\beta)v_1-2\beta}\right\} + \exp\left(v_1+\frac{gt_1}{c}\right)\right] \tag{33}$$

Setting $v = v_0$, $t = 0$ gives the mass M^0 after boosting:

$$\frac{M^0}{M_1} = \frac{W_0c^2}{M_1g}\cdot\beta\cdot v_0\left[(v_0+2)-(v_1+2)\frac{v_0^2+(1-\beta)v_0-2\beta}{v_1^2+(1-\beta)v_1-2\beta}\right] + \exp\left(v_1-v_0+\frac{gt_1}{c}\right) \tag{34}$$

The initial mass of the rocket including the booster change is given by Equation (3), or in terms of the nondimensional velocity v,

$$M_0 = M^0\exp(v_0)$$

Therefore,

$$\frac{M_0}{M_1} = \frac{M^0}{M_1}\exp(v_0) \tag{35}$$

The thrust at any instant can be obtained from Equation (1) noting that $F = c\dfrac{\mathrm{d}M}{\mathrm{d}t}$

$$\frac{F}{M_1g} = \frac{W_0c^2}{M_1g}v^2\exp(-\alpha s) + \frac{M}{M_1}\left(1+\frac{\ddot{s}}{g}\right) \tag{36}$$

where the acceleration $\ddot{s}/g$ can be computed from Equation (28).

The end conditions are given by Equations (5) and (26). Substituting Equation (27) into

Equation (26) gives one condition

$$\frac{W_0 c^2}{M_1 g} v_1^2 \exp(-\alpha s_1) = \frac{1}{1+v_1} \tag{37}$$

After the fuel is exhausted, the equation of motion is

$$(\ddot{s} + g)M_1 + W_0 \dot{s}^2 \exp(-\alpha s) = 0 \tag{38}$$

This equation can be easily integrated. By using the condition that at the summit, $s = S$, $\dot{s} = 0$, one has the following equation for the velocity v_1 at the beginning of the coasting flight

$$v_1^2 = -2\beta \exp\left(2\beta \frac{W_0 c^2}{M_1 g} \xi_1\right) \int_{\xi_1}^{\xi_2} \exp\left(-2\beta \frac{W_0 c^2}{M_1 g} x\right) \frac{\mathrm{d}x}{x} \tag{39}$$

where

$$\xi_1 = \exp(-\alpha s_1), \quad \xi_2 = \exp(-\alpha S) \tag{40}$$

The integral in Equation (39) can be evaluated using the series expansion

$$\int \exp(-\alpha x) \frac{\mathrm{d}x}{x} = \ln|x| - \frac{\alpha x}{1 \cdot 1!} + \frac{(\alpha x)^2}{2 \cdot 2!} - \dots \tag{41}$$

With any fixed value of S and the drag parameter $W_0 c^2 / M_1 g$, Equations (37) and (39) determine v_1 and αs_1. The parameter $W_0 c^2 / M_1 g$ is the nondimensional drag and weight ratio of the rocket.

Calculations were carried out for two sets of summit altitudes and exhaust velocities assuming $\alpha = 1/22\,000$ ft. One case was for a summit altitude of 500 000 ft and an exhaust velocity of 5 500 fps. The other case was for a summit altitude of 3 000 000 ft and an exhaust velocity of 8 000 fps. Actually, an iteration procedure was adopted to fit the end conditions. For any chosen value of the drag weight ratio $W_0 c^2 / M_1 g$, a value of v_1 was first assumed and substituted in Equation (37) to solve for αs_1. This result was put into Equation (39) which was solved for v_1. This process was repeated until the desired accuracy was obtained. A plot of $W_0 c^2 / M_1 g$ versus v_1, which was found to be almost a straight line on semilogarithmic graph paper, with v_1 plotted on the linear scale, simplified the process by giving an excellent first estimate of the velocity. The initial velocity ratio v_0 was then solved for by a trial and error procedure using Equation (31).

Values of $W_0 c^2 / M_1 g$ were assumed for the calculations, but the final results were put in terms of M_0. This was done because physically the results in terms of the initial mass are easier to visualize. The quantity $W_0 c^2$ is the drag of the rocket body at a flight velocity equal to the exhaust velocity c and at sea level. The parameter then expresses the ratio of this drag to initial weight. The results are shown in Figs. 1 to 8. The discussion of these results is postponed until a later section.

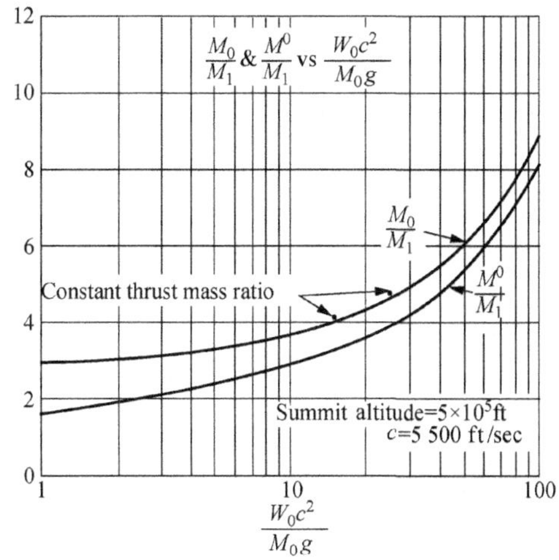

Fig. 1 Initial mass ratios versus the drag ratio W_0c^2/M_0g, the ratio of the sea level drag at a velocity equal to the exhaust velocity c, to the initial weight M_0g

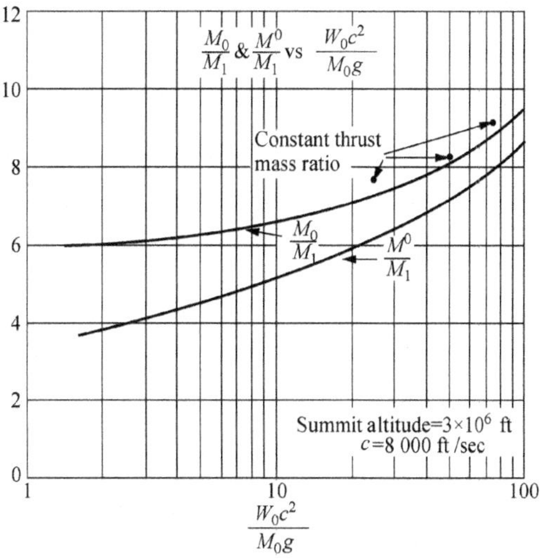

Fig. 2 Initial mass ratios versus the drag ratio W_0c^2/M_0g, the ratio of the sea level drag at a velocity equal to the exhaust velocity c, to the initial weight M_0g

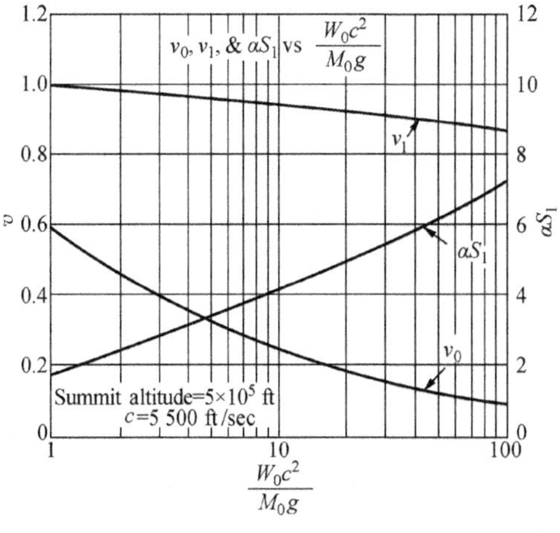

Fig. 3 Booster-velocity ratio $v_0 = \dot{s}_0/c$, maximum velocity ratio, $v_1 = \dot{s}_1/c$ and altitude parameter αs_1 at the end of burning versus W_0c^2/M_0g, the ratio of the sea level drag at a velocity equal to the exhaust velocity c, to the initial weight M_0g

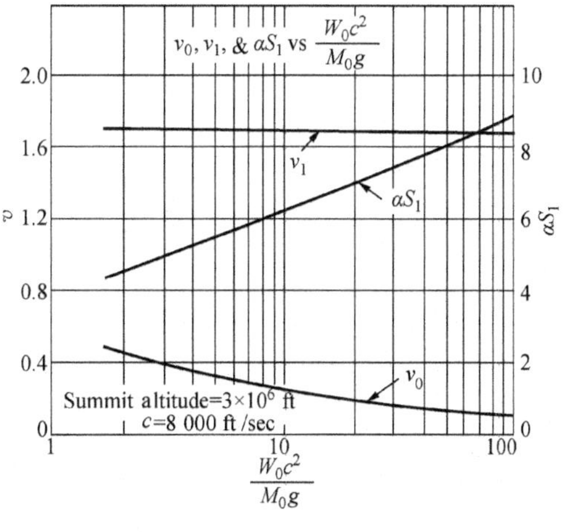

Fig. 4 Booster-velocity ratio $v_0 = \dot{s}/c$, maximum velocity ratio, $v_1 = \dot{s}/c$ and altitude parameter αs_1 at the end of burning versus W_0c^2/M_0g, the ratio of the sea level drag at a velocity equal to the exhaust velocity c, to the initial weight M_0g

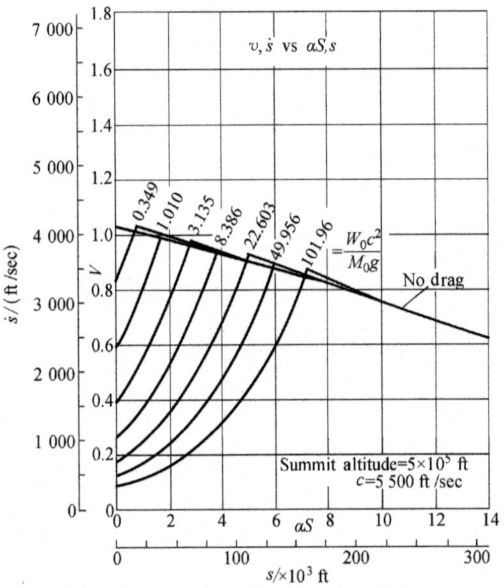

Fig. 5 Velocity $\dot{s}$ versus altitude s, W_0c^2/M_0g is the drag ratio

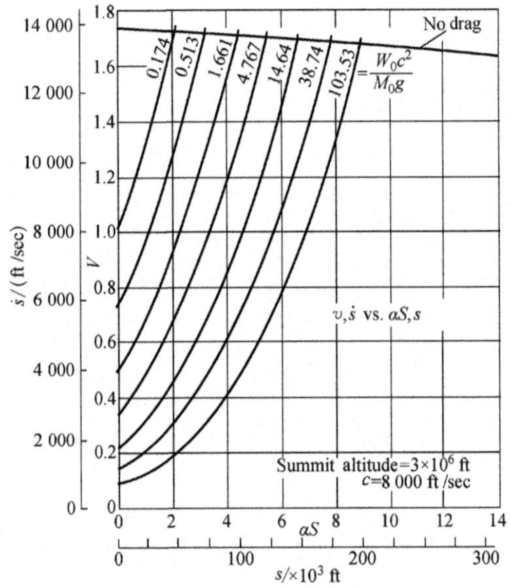

Fig. 6 Velocity $\dot{s}$ versus altitude s, W_0c^2/M_0g is the drag ratio

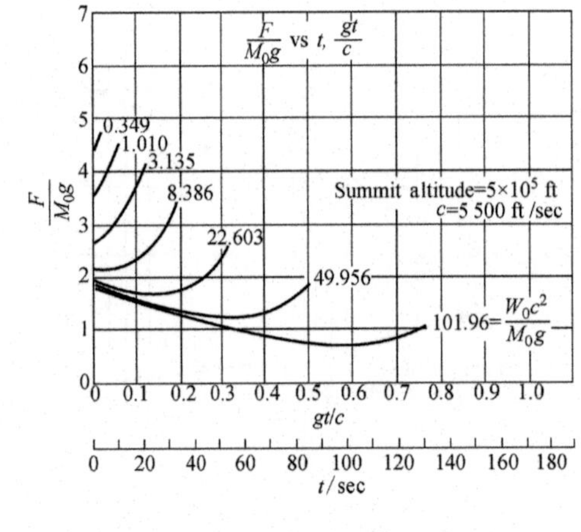

Fig. 7 F/M_0g, the ratio of the thrust to the initial weight versus time t. W_0c^2/M_0g is the drag ratio

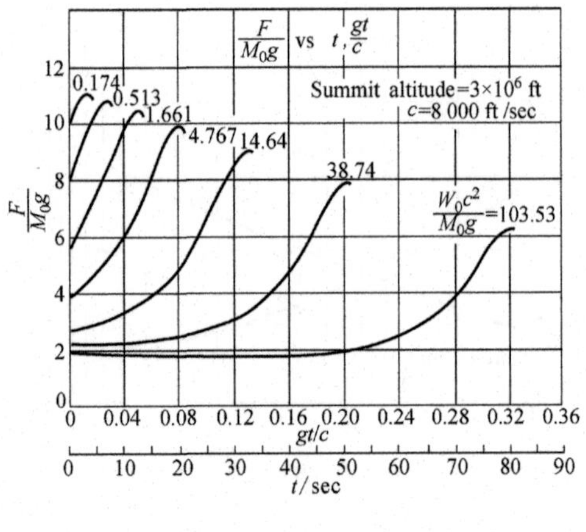

Fig. 8 F/M_0g, the ratio of the thrust to the initial weight versus time t. W_0c^2/M_0g is the drag ratio

Linear Drag Law

For supersonic flight of a rocket, perhaps, a better approximation of the drag is given by

$$D = (A\dot{s} + B)\exp(-\alpha s) \tag{42}$$

This again corresponds to an exponential air-density law, but the drag coefficient decreases as the velocity increases. Substituting Equation (42) into Equation (22) gives

$$\frac{\ddot{s}}{g} = \frac{v^2 + [B/(Ac) - \beta]v - [\beta + B/(Ac(1+\beta))]}{\beta[v + 2 + B/(Ac)]} \tag{43}$$

where v and β are the same as the previous section and are defined by Equation (29). Equation (43) can be integrated and gives

$$\begin{aligned}\frac{gt}{c} = {} & \frac{\beta}{2}\ln\frac{v^2 + v[B/(Ac) - \beta] - [\beta + B/(Ac(1+\beta))]}{v_0^2 + v_0(B/(Ac) - \beta) - [\beta + B/(Ac(1+\beta))]} + \\ & \frac{\beta[4 + \beta + B/(Ac)]}{2\lambda}\ln\frac{2v + (B/(Ac) - \beta) - \lambda}{2v + (B/(Ac) - \beta) + \lambda}\cdot\frac{2v_0 + (B/(Ac) - \beta) + \lambda}{2v_0 + (B/(Ac) - \beta) - \lambda}\end{aligned} \tag{44}$$

and

$$\begin{aligned}\alpha s = {} & v - v_0 + \left(1 + \frac{\beta}{2}\right)\ln\frac{v^2 + v(B/(Ac) - \beta) - [\beta + B/(Ac(1+\beta))]}{v_0^2 + v_0(B/(Ac) - \beta) - [\beta + B/(Ac(1+\beta))]} + \\ & \frac{\beta[4 + \beta + B/(Ac)]}{2\lambda}\ln\frac{2v + (B/(Ac) - \beta) - \lambda}{2v + (B/(Ac) - \beta) + \lambda}\cdot\frac{2v_0 + (B/(Ac) - \beta) + \lambda}{2v_0 + (B/(Ac) - \beta) - \lambda}\end{aligned} \tag{45}$$

where

$$\lambda = \sqrt{(B/(Ac) + \beta)^2 + 4(B/(Ac) + \beta)} \tag{46}$$

The mass M at any instant is found to be

$$\begin{aligned}\frac{M}{M_1} = {} & \left[\frac{Ac}{M_1 g}\cdot\beta(\exp v_0)\{v_0^2 + v_0(B/(Ac) - \beta) - (\beta + B/(Ac[1+\beta]))\}\times\right. \\ & \left\{\frac{v + B/(Ac) + 1}{v^2 + v(B/(Ac) - \beta) - (\beta + B/(Ac[1+\beta]))} - \right. \\ & \left.\frac{v_1 + B/(Ac) + 1}{v_1^2 + v_1(B/(Ac) - \beta) - (\beta + B/(Ac[1+\beta]))}\right\} + \\ & \left.\exp\left(v_1 + \frac{gt_1}{c}\right)\right]\exp\left\{-\left(v + \frac{gt}{c}\right)\right\}\end{aligned} \tag{47}$$

The initial mass M_0 including the boosting can be obtained from the above equation and Equation (35)

$$\begin{aligned}\frac{M_0}{M_1} = {} & \exp\left(v_1 + \frac{gt_1}{c}\right) + \frac{Ac}{M_1 g}\beta\left[\{v_0 + B/(Ac) + 1\} - \{v_1 + B/(Ac) + 1\}\times\right. \\ & \left.\frac{v_0^2 + v_0(B/(Ac) - \beta) - (\beta + B/(Ac[1+\beta]))}{v_1^2 + v_1(B/(Ac) - \beta) - (\beta + B/(Ac[1+\beta]))}\right]\cdot\exp(v_0)\end{aligned} \tag{48}$$

The thrust can be obtained from Equation (1). Thus

$$\frac{F}{M_1 g} = \frac{Ac}{M_1 g}(v + B/(Ac)) \exp(-\alpha s) + \frac{M}{M_1}\left(1 + \frac{\ddot{s}}{g}\right) \tag{49}$$

The acceleration s/g can be determined from Equation (43).

The end condition is given by Equations (26) and (5). Substituting Equations (42) and (43) into Equation (26), one has

$$\frac{Ac}{M_1 g}(v_1^2 + v_1 B/(Ac) - B/(Ac)) \exp(-\alpha s_1) = 1 \tag{50}$$

To supply the relation indicated by Equation (5), the coasting flight trajectory must be determined. By using Equation (42), the equation of motion for the coasting flight is

$$\left(\frac{\ddot{s}}{g} + 1\right) + \frac{Ac}{M_1 g}(v + B/(Ac)) \exp(-\alpha s) = 0 \tag{51}$$

This equation cannot be solved by simple quadrature. However, the effect of air drag is generally small. Then Equation (51) can be solved approximately by first neglecting the air drag and then making the necessary small correction. Let v^0 be the value of the velocity ratio without air drag and v' be the small correction on v. Then

$$v = v^0 + v' \tag{52}$$

The equation for v^0 is simply

$$\frac{d(v_0)^2}{ds} + \frac{2g}{c^2} = 0 \tag{53}$$

Now substitute Equation (52) into Equation (51) and retain only linear terms in v',

$$v^0 \frac{dv'}{ds} + v' \frac{dv^0}{ds} + \frac{A}{M_1 c}(v^0 + B/(Ac)) \exp(-\alpha s) = 0 \tag{54}$$

This is the differential equation for v', the correction term. Equation (53) has the solution

$$v^0 = \sqrt{2\beta(\alpha S - \alpha s)} \tag{55}$$

Substitute Equation (55) into Equation (54),

$$\frac{dv'}{d\zeta} + \frac{v'}{2\zeta} = \frac{Ac}{M_1 g}\beta\left\{1 + \frac{B/(Ac)}{\sqrt{2\beta\zeta}}\right\} \exp(\zeta - \alpha S)$$

where

$$\zeta = \alpha S - \alpha s$$

Solving for v'

$$v' = \frac{1}{\sqrt{\zeta}}\frac{Ac}{M_1 g}\beta \exp(-\alpha S)\int_0^\zeta \left\{\sqrt{x} + \frac{B/(Ac)}{\sqrt{2\beta}}\right\} \exp(x) dx \tag{56}$$

To obtain a suitable series for the computation of the integral in Equation (56), consider

$$f(\zeta) = \int_0^{\zeta} \sqrt{x} \exp(x) \mathrm{d}x$$

Let $x = \zeta(1-u)$, then

$$f(\zeta) \exp(-\zeta) = \zeta^{3/2} \int_0^1 \sqrt{1-u} \exp(-\zeta u) \mathrm{d}u$$

By expanding the radical in the integrand into a power series and integrating term by term, $f(\zeta)$ can be shown to be equal to the following series

$$f(\zeta) = \sqrt{\zeta}\left[\{\exp(\zeta) - 1\} - \frac{1}{2\zeta}\{\exp(\zeta) - (1+\zeta)\} - \frac{1}{4\zeta^2}\left\{\exp(\zeta) - \left(1+\zeta+\frac{\zeta^2}{2}\right)\right\} - \cdots \right]$$

Then

$$v' = \frac{1}{\sqrt{\zeta}} \cdot \frac{Ac}{M_1 g}\beta \left\{ f(\zeta) + B/(Ac) \frac{1}{\sqrt{2\beta}}(\exp(\zeta) - 1) \right\} \times \exp(-\alpha S) \tag{57}$$

Equations (50) and (57) must be solved simultaneously to determine the velocity and altitude at the end of burning for a given drag weight ratio, $Ac/M_1 g$, for the rocket. The quantity Ac is now the drag of the rocket body at a flight velocity equal to the exhaust velocity c and at sea level, if $B = 0$.

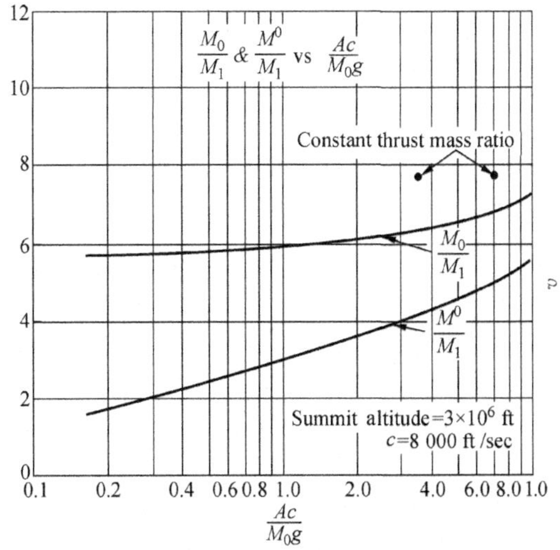

Fig. 9 Initial mass ratios versus the drag ratio $Ac/M_0 g$, the ratio of the sea level drag at a velocity equal to the exhaust velocity c, to the initial weight $M_0 g$

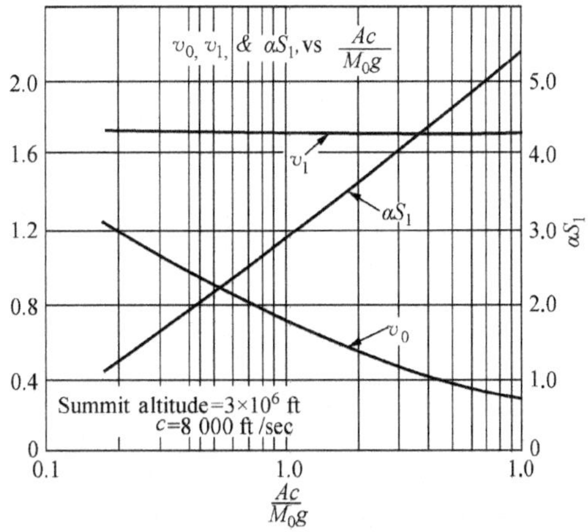

Fig. 10 Booster-velocity ratio $v_0 = \dot{s}_0/c$, maximum velocity ratio $v_1 = \dot{s}_1/c$ and altitude parameter αs_1, at the end of burning versus $Ac/M_0 g$, the ratio of the sea level drag at a velocity equal to the exhaust velocity c, to the initial weight $M_0 g$

Calculations were carried out for a summit altitude of 3 000 000 ft, an exhaust velocity of 8 000 fps, and no drag at zero velocity, assuming $\alpha = 1/22\ 000$ ft. An iteration procedure identical to that used in the previous example was used. The results are shown in Figs. 9 to 12. The parameter here is Ac/M_0g, the ratio of the drag of the rocket at velocity c and at sea level to the initial weight.

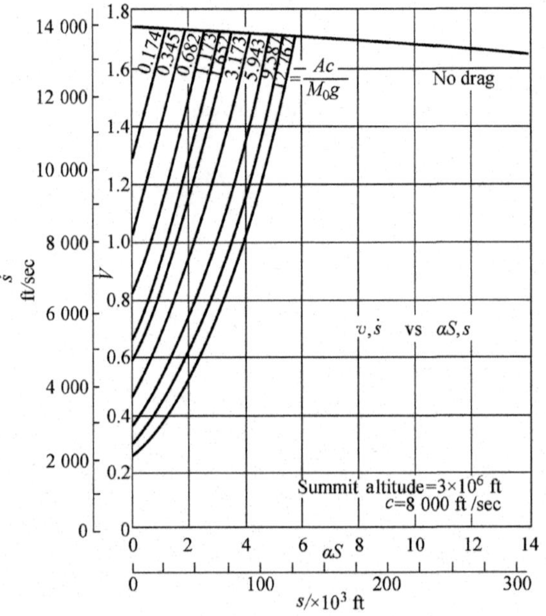

Fig. 11 Velocity $\dot{s}$ versus altitude s. Ac/M_0g is the drag ratio

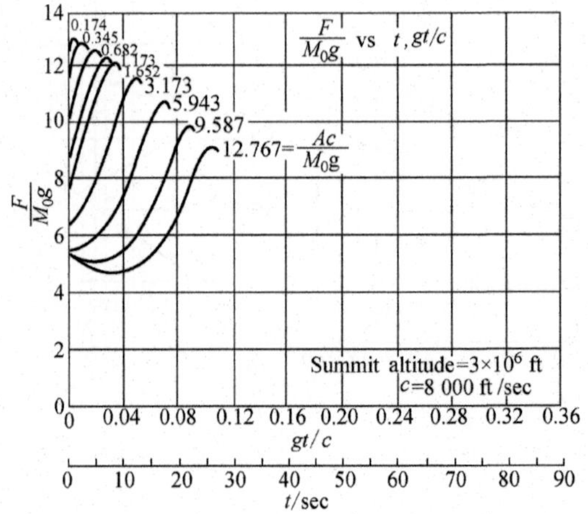

Fig. 12 F/M_0g, the ratio of the thrust to the initial weight versus time t. Ac/M_0g is the drag ratio

In order to determine the gain possible by using the optimum thrust programming as against the conventional constant thrust rocket, a few cases of constant thrust were computed using the same drag laws. It was assumed the constant thrust rocket had no initial velocity and the ratio of thrust to initial weight $F/M_0 g$ was 2.7. This was chosen to agree with the best initial acceleration given by Ivey, Bowen, and Oborny[2] for high-performance rockets. The comparisons were based on equal ratios of the drag at the exhaust velocity c and at sea level to the initial weight $M_0 g$ and summit altitude S. The mass ratio M_0/M_1 is now larger than the optimum value due to the fact that the final mass M_1 is now smaller. A step-by-step procedure was used in computing the mass ratio of final mass to initial mass which gave an error of approximately one per cent. This error was always nonconservative and would therefore favor the constant-thrust examples in the comparisons. The results are indicated by the dots in Figs. 1, 2, and 9.

Discussion of Results

Both the linear drag law and the quadratic drag law cases have certain common characteristics. As seen from Figs. 3, 4, 7, 8, 10, and 12, for low values of the drag parameter, the average thrust and initial velocity were high. Therefore, the rocket will burn its fuel rapidly and less energy will be expended against gravity. As the drag parameter increases, more energy is used to overcome the drag, and it becomes advantageous to have a lower velocity at low altitudes. Therefore, the initial velocity and thrust will be lower, and the altitude at the end of burning will be greater than in the case of low drag parameter. But it is important to note that there is always a finite initial velocity $\dot{s}_0$, so that the optimum thrust programming involves a boosting impulse. Another result the two drag law cases have in common is the effect of the drag parameters on the ratio of the initial mass to the final mass, M_0/M_1 (Figs. 1, 2, and 9). As the drag parameters increase, this ratio M_0/M_1 increases for a given summit altitude. As the drag parameters decrease, the ratio M_0/M_1 asymptotically approaches the no-drag value of exp $\dot{s}_0/c$ where $\dot{s}_0$ is such that the rocket has the kinetic energy at zero altitude equal to the potential energy at the summit altitude.

In general, for optimum programming, the thrust after the initial impulse should increase with altitude. Exceptions to this occur for portions of the powered flight when the values of the drag parameter are very high, or when the rocket has a very high performance. In these cases, for a portion of the burning time, the thrust can be seen to decrease. This is the result of either a very slowly increasing acceleration of the rocket or a very rapidly decreasing mass of the rocket. Since the accelerating force is the major fraction of the thrust and equal to the product of the instantaneous mass and acceleration, the thrust can decrease. In general, however, the thrust should increase. For a constant geometry rocket motor, this is accomplished to a limited extent by the decreasing atmospheric pressure acting on the nozzle exit. But for most cases, such an increase is only a fraction of the ideal, as determined by the present calculation.

The linear drag law case has a higher initial velocity and shorter burning time than has the quadratic drag law case for similar mass ratios M_0/M_1. This is true because the drag does not increase as rapidly with velocity in the linear drag law case; therefore, the linear drag law case will favor the higher velocities and shorter burning times. The drag in the quadratic law case increases as the square of the velocity; therefore, this case will favor the lower velocities and longer burning times.

To give an example of the magnitudes of the ratios W_0c^2/M_0q and Ac/M_0g for the drag and the initial weight, consider the case of the V-2 rocket. At a velocity of 6 000 fps the drag is roughly 100 000 lb. The initial weight of the V-2 is approximately 25 000 lb. This gives a ratio of drag at the exhaust velocity at sea level to the initial weight of roughly 4 if the exhaust velocity is 6 000 fps. Therefore, 4 might be an approximate value of the ratio Ac/M_0g. In the quadratic drag law case, a better estimate of W_0c^2/M_0g might be 10. This will account for the quadratic drag law giving a lower value of the drag than is actually the case during the major portion of the powered flight, especially at the lower altitudes. Since the summit altitude of the V-2 is approximately 100 miles, Fig. 3 can be used to estimate the optimum booster velocity to be roughly 1 500 fps, assuming $W_0c^2/M_0g = 10$. If the drag is relatively less, the booster velocity will increase. Therefore, the optimum thrust programming involves considerable boosting.

Several constant-thrust mass ratios were calculated for comparison with the ideal mass ratios. It was assumed the constant thrust rockets had no initial velocity and followed the same drag laws as the ideal cases. An initial acceleration of 1.7 g was chosen to agree with the best initial acceleration given by Ivey, Bowen, and Oborny[2] for high performance rockets. This value might not be the optimum for the constant thrust cases computed because it is dependent on the drag laws followed by the rocket. The values for the constant thrust rocket calculations can be seen in Figs. 1, 2, and 9. The advantages of the ideal rocket would be less if the constant thrust rocket had the best possible initial velocity and acceleration. In fact, it seems certain that the initial acceleration of 1.7 g is too small for the linear drag case of Fig. 9. The possible gain by using optimum thrust programming is thus more correctly represented by Fig. 2. Then it is seen that the gain is small in terms of the initial mass M_0. On the other hand, one must bear in mind that the pay load of a high altitude sounding rocket is only a very small fraction of the initial weight. Therefore, even if the gain by better thrust programming may be small in terms of the initial weight, the gain in terms of pay load could be appreciable. Of course, the numerical results obtained by the present calculation are only qualitative. For any practical design, besides consideration for the additional weight needed for thrust variation, the optimum-thrust programming has to be determined with a more factual drag versus velocity relation.

Acknowledgment

During his stay at the California Institute of Technology, Prof. Raymund Sänger of Eidgenössiche Technische Hochschule, Zürich, Switzerland, has kindly checked the mathematics of this calculation, and suggested a few important simplifications. For this, the authors are deeply grateful.

References

[1] Malina F J, Smith A M O. Flight Analysis of the Sounding Rocket. Journal of Aeronautical Sciences, 1938, 5: 199 - 202.

[2] Ivey H R, Bowen Jr. E N., Oborny L F. Introduction to the Problem of Rocket-Powered Aircraft Performance. NACA Technical Note No. 1401, 1947.

[3] Hamel G. Über eine mit dem Problem der Rakete zusammenhängende Aufgabe der Variationsrechnung. Zeitschrift für angewandte Mathematik und Mechanik, Berlin, 1927, 7: 451 - 452.

The Emission of Radiation from Diatomic Gases. Ⅲ. Numerical Emissivity Calculations for Carbon Monoxide for Low Optical Densities at 300 K and Atmospheric Pressure[①]

S. S. Penner, M. H. Ostrander[②] and H. S. Tsien

(Guggenheim Jet Propulsion Center, California Institute of Technology, Pasadena, California)

Numerical emissivity calculations at 300 K and atmospheric pressure for nonoverlapping rotational lines have been carried out for CO using a dispersion formula for the line-shape representation. Use of the best available experimental data on integrated absorption and rotational line-width leads to calculated emissivities which are in excellent agreement with extrapolated empirical data published by Hottel and Ullrich. In particular, the theoretical dependence of emissivity on optical density, for small optical densities at 300 K, has been shown to follow experimental observations with satisfactory precision.

For small optical densities the calculated emissivity is found to be proportional to the square root of the assumed rotational line-width, thus emphasizing the need for accurate line-width determinations at elevated temperatures. The limits of validity of the treatment utilizing nonoverlapping rotational lines are defined by examining overlapping between adjacent weak and strong rotational lines.

The calculation of emissivities can be simplified by the use of approximate treatments. Thus absolute values of the emissivity can be predicted within 10 percent by utilizing a treatment for nonoverlapping, equally spaced, and equally intense lines, together with empirically determined values for the equivalent mean integrated absorption of the rotational lines of CO. A better analytic solution, which does not involve the assumptions of equal spacing and equal intensity of the rotational lines, has been obtained by utilizing asymptotic relations for large values of modified Bessel functions.

Ⅰ. Introduction

THE general problem of emission and absorption of radiation involves two essentially distinct parts. One part of radiant heat transfer investigations is concerned with the

---

Received September 27, 1951.

Journal of Applied Physics, vol.23, pp.256 – 263, 1952.

① Supported, in part, by the ONR under contract Nonr-220 (O_3), NR 015 210, with the California Institute of Technology.

② This article uses, in part, the results of a thesis submitted by M. H. Ostrander in partial fulfillment of requirements for the degree of Aeronautical Engineer, California Institute of Technology, June, 1951.

determination of intensities if emissivities, absorptivities, and scattering coefficients are known. These studies are restricted to the problem of radiative transfer. For any defined geometric arrangement the radiative transfer problems can be formulated without difficulty. An exhaustive account of this work may be found in a recently published book by Chandrasekhar[1]. The other part of the general problem is the determination of emissivities, absorptivities, and scattering coefficients, which enter into the radiative transfer calculations. Until very recently, practical applications have been based almost entirely on the use of empirically determined emissivities. These emissivities are compiled, for instance, in McAdams' well-known treatise on heat transfer[2]. However, the problems involved in the theoretical calculation of emissivities were solved, in principle, a good many years ago. The results of these theoretical studies have been applied, to a limited extent, in the use of spectroscopic measurements for flame temperature determinations from line intensities[3] and for the analysis of absorption and emission of radiation in the atmosphere[4]. An important paper by Dennison[5], written more than twenty years ago, contains many of the basic theoretical relations which have been used for numerical emissivity calculations on CO in the present studies.

In preceding publications an approximate method for carrying out emissivity calculations at elevated total pressures, leading to complete overlapping of rotational lines, has been described[6~8]. Limited comparison of calculated emissivities with experimental measurements has shown that the approximate method is, as expected, not accurately valid at atmospheric pressure for a diatomic emitter such as CO. In particular, the calculated dependence of emissivity on optical density for nonoverlapping rotational lines is not predicted correctly by a method based on the use of average absorption coefficients for entire vibration-rotation bands[7]. It is the purpose of the present investigation to demonstrate that the observed discrepancies between the previously proposed theory and the experiments can be removed by a more satisfactory analysis of the emission of radiation from vibration-rotation bands with rotational lines which do not overlap appreciably.

In Section Ⅱ we present a summary of basic relations which are used for calculations on diatomic molecules with nonoverlapping rotational lines. Numerical emissivity calculations for CO at 300 K and a comparison with extrapolated experimental data are presented in Section Ⅲ. The limits of validity of a treatment for nonoverlapping rotational lines are discussed in Section Ⅳ. In Section Ⅴ approximate methods for performing emissivity calculations are considered under the conditions which are of interest for the present calculations.

Ⅱ. Summary of Theoretical Relations

For diatomic molecules which do not possess a Q branch but a rotational line shape which may be represented by a dispersion formula, the spectral absorptions coefficient P_ω at the wave number ω resulting from energy transitions corresponding to the fundamental vibration-rotation band is[5,9,10]

$$P_\omega = (b/\pi)\sum_j \{S_j[(\omega-\omega_j)^2+b^2]^{-1}+S_{j-1}[(\omega-\omega_{j-1})^2+b^2]^{-1}\} \tag{1}$$

where[①]

$$S_j = [N_T\pi\varepsilon^2/(3\mu cQ)][(\omega_j/\omega^*)j\times\{\exp[-E(0,j)/(kT)]\}]FG(\omega^*/\omega_e) \tag{2a}$$

$$S_{j-1} = [N_T\pi\varepsilon^2/(3\mu cQ)][(\omega_{j-1}/\omega^*)j\times\{\exp[-E(0,j-1)/(kT)]\}F'G'](\omega^*/\omega_e) \tag{2b}$$

Here N_T = total number of molecules per unit volume per unit of total pressure; μ = reduced mass of the diatomic molecule under discussion; c = velocity of light; ε = rate of change of electric moment with internuclear distance; b = rotational half-width in cm^{-1}; ω_e = wave number corresponding to an infinitesimal oscillation at the equilibrium interatomic distance; ω^* = wave number of the (forbidden) transition $j=0\rightarrow j=0$ and $n=0\rightarrow n=1$, where j and n denote, respectively, rotational and vibrational quantum numbers; ω_j = wave number corresponding to the transitions $j\rightarrow j-1$, $n=0\rightarrow n=1$; ω_{j-1} = wave number corresponding to the transitions $j-1\rightarrow j$, $n=0\rightarrow n=1$; $E(n,j)$ = energy of the nth vibrational and jth rotational level;

$$F = 1+8\gamma j[1+(5\gamma j/4)-(3\gamma/4)]$$

$$F' = 1-8\gamma j[1-(5\gamma j/4)-(3\gamma/4)]$$

$\gamma = h/8\pi^2 Ic\omega_e$ where I represents the equilibrium moment of inertia of the radiating molecule;

$$G = 1-\exp[-hc\omega_j/(kT)]$$

$$G' = 1-\exp[-hc\omega_{j-1}/(kT)]$$

Q = complete internal partition function referred to zero as reference state.

At elevated temperatures the actual value of P_ω must be obtained by adding a small number of series of the type given in Eq. (1) corresponding to vibrational transitions of the form $n\rightarrow n+1$. Furthermore, contributions from vibration-rotation bands for which $n\rightarrow n+2$, $n\rightarrow n+3$, etc., become important for large optical densities. Detailed applications are discussed in reference 9.

The total equilibrium intensity of radiation emitted per unit surface area per unit time into a solid angle of 2π steradians in the wave number interval between ω and $\omega+d\omega$ by heated gases distributed uniformly and of optical density X (X = product of partial pressure p of emitter and radiation path length L) is given by the well-known relation

$$R(\omega) = R^0(\omega)[1-\exp(-P_\omega X)] \tag{3}$$

① Equations (2a) and (2b) differ from the corresponding relations obtained from the treatment of B. L. Crawford, Jr., and H. L. Dinsmore (J. Chem. Phys. **18**, 983, 1682 (1950)) through the occurrence of the factors F and F', respectively. These factors are nearly equal to unity for the values of j which are important for the present studies.

Here $R^0(\omega)$ is the intensity of radiation emitted by a blackbody in the wave number interval between ω and $\omega + d\omega$ into a solid angle of 2π steradians per unit surface area per unit time. It is given in analytic form by the Planck blackbody distribution function. The engineering emissivity E is defined by the relation

$$E = \int_0^\infty R(\omega)\,d\omega / (\sigma T^4) \tag{4}$$

where σ is the Stefan-Boltzmann constant.

Subject to the approximations and assumptions involved in the use of Eq. (1), it is evident that the problem of emissivity calculations reduces to the evaluation of integrals of the type appearing in Eq. (4). This integration can be accomplished if the width of the rotational lines is so small that they may be treated as being completely separated. The results obtained for nonoverlapping rotational lines are useful at low pressures[4,9,11]. The integrals can also be evaluated, approximately, if the pressures are so high that very extensive overlapping of the rotational lines occurs. The results obtained in this case are useful for emissivity calculations at elevated pressures in rocket combustion chambers[6~8]. Emissivity calculations at intermediate pressures are more difficult to carry out.

We proceed by considering the problem of emissivity calculations for nonoverlapping rotational lines. In order to evaluate the infinite integral

$$\int_0^\infty R(\omega)\,d\omega = \int_0^\infty R^0(\omega)\,[1 - \exp(-P_\omega X)]\,d\omega \tag{5}$$

for nonoverlapping rotational lines, the approximation is first made that $R^0(\omega)$ may be treated as constant in the wave number interval for which $\exp(-P_\omega X)$ is sensibly different from unity. Replacing $R^0(\omega)$ by an appropriate average value introduces negligibly small errors into the emissivity calculations for nonoverlapping rotational lines at low pressures, since the intervals $\Delta\omega$ for which $P_\omega X$ is significantly different from zero are exceedingly small. For example, at atmospheric pressure and room temperature $\Delta\omega$ is of the order of 0.3 cm^{-1} for CO. The change of $R^0(\omega)$ in wave number intervals of this size is less than 1 percent.

At low pressures the rotational lines are so narrow that the only term contributing to $R^0(\omega)$, for example, in the vicinity of the rotational line centered at ω_j is the term $S_j[(\omega - \omega_j)^2 + b^2]^{-1}$. Hence we can break up the integration interval in such a way that each subinterval is centered about one of the wave numbers ω_j, ω_{j-1}[9]. Thus the following close approximation is obtained

$$\int_0^\infty R(\omega)\,d\omega = \sum_{\Delta\omega_i}\Big\{R^0(\omega_j)\int_{\Delta\omega_i}[1 - \exp(-P_j X)]\,d\omega + R^0(\omega_{j-1})\int_{\Delta\omega_i}[1 - \exp(-P_{j-1} X)]\,d\omega\Big\} \tag{6}$$

Table I Integrated absorption for the rotational lines of the fundamental of CO at 300 K.

Transition $j \to j-1$	$10^3 A_j$ (cm^{-2}-$atmos^{-1}$)	S_j (cm^{-2}-$atmos^{-1}$)	Transition $j-1 \to j$	$10^3 A_{j-1}$ (cm^{-2}-$atmos^{-1}$)	S_{j-1} (cm^{-2}-$atmos^{-1}$)
1→0	5.528	2.064	0→1	5.572	2.081
2→1	10.71	4.000	1→2	10.88	4.062
3→2	15.28	5.705	2→3	15.64	5.841
4→3	19.02	7.101	3→4	19.62	7.326
5→4	21.79	8.135	4→5	22.65	8.457
6→5	23.51	8.780	5→6	24.64	9.200
7→6	24.24	9.050	6→7	25.58	9.551
8→7	24.00	8.960	7→8	25.54	9.535
9→8	22.97	8.577	8→9	24.63	9.199
10→9	21.32	7.960	9→10	23.04	8.603
11→10	19.23	7.180	10→11	20.94	7.819
12→11	16.88	6.310	11→12	18.53	6.919
13→12	14.45	5.397	12→13	15.99	5.969
14→13	12.07	4.509	13→14	13.46	5.025
15→14	9.851	3.678	14→15	11.06	4.131
16→15	7.857	2.934	15→16	8.889	3.319
17→16	6.127	2.288	16→17	6.986	2.608
18→17	4.673	1.745	17→18	5.370	2.005
19→18	3.490	1.303	18→19	4.039	1.508
20→19	2.551	0.952 5	19→20	2.975	1.111
22→21	1.281	0.478 3	21→22	1.622	0.605 7
24→23	0.592 4	0.221 2	23→24	0.773	0.288 5
26→25	0.252 9	0.094 4	25→26	0.340	0.126 9
28→27	0.099 8	0.037 3	27→28	0.138	0.051 6
30→29	0.036 4	0.013 6	29→30	0.052	0.019 5
35→34	0.002 1	0.007 8	34→35	0.001	0.005 2
40→39	0.000 1	0.003 7	39→40	0.000	0.000 0

where $R^0(\omega_j)$ and $R^0(\omega_{j-1})$ denote, respectively, the spectral intensities of a blackbody radiator evaluated at the line centers corresponding to the indicated transitions. Similarly, P_j and P_{j-1} are the characteristic values of the spectral absorption coefficients valid if only the indicated transitions occurred. The subintervals $\Delta\omega_i$ are chosen conveniently to extend from a wave number midway between two line centers to the adjacent wave numbers which are located similarly. For nonoverlapping lines the error involved in replacing

$$\int_{\Delta\omega_i} [1-\exp(-P_j X)]\,d\omega \quad \text{by} \quad \int_{-\infty}^{\infty} [1-\exp(-P_j X)]\,d\omega$$

etc., is negligibly small. The magnitude of this error increases with increasing optical density X and can be estimated readily by numerical calculations. Equation (6) now becomes

$$\int_0^{\infty} R(\omega)\,d\omega = \sum_j \Big\{ R^0(\omega_j)\int_{-\infty}^{\infty} [1-\exp(-P_j X)]\,d\omega + R^0(\omega_{j-1})\int_{-\infty}^{\infty} [1-\exp(-P_{j-1} X)]\,d\omega \Big\} \qquad (7)$$

For fixed values of b the infinite integrals appearing in Eq. (7) are readily evaluated with the result[4,9,11]

$$\int_0^\infty R(\omega)\mathrm{d}\omega = 2\pi b\sum_{j=1}^{\infty}\{R^0(\omega_j)x_j \times \exp(-x_j)[\mathrm{I}_0(x_j)+\mathrm{I}_1(x_j)]+R^0(\omega_{j-1})x_{j-1}\times \exp(-x_{j-1})[\mathrm{I}_0(x_{j-1})+\mathrm{I}_1(x_{j-1})]\} \tag{8}$$

where

$$x_j = S_jX/(2\pi b),\quad x_{j-1} = S_{j-1}X/(2\pi b)$$

and I_0, I_1 denote modified Bessel functions of the first kind. Calculations of the emissivity E from Eqs. (4) and (8), which neglect contributions from upper harmonics, can now be carried out by utilizing intensity[12] and line-width data[13] obtained at room temperature. The contribution to the emissivity E from the first overtone is obtained by adding to Eq. (8) a series which is of the same form as Eq. (8), except that the integrated absorption S_j is replaced by S'_j and S_{j-1} by S'_{j-1} where the primed quantities identify appropriate values for the vibrational transition $0\to 2$.

Ⅲ. Representative Emissivity Calculations for CO at 300 K

For the present calculations the spectroscopic constants of Sponer were used[14]. Since these data have been

Table Ⅱ Integrated absorption for the rotational lines of the first overtone of CO at 300 K.

Transition $j\to j-1$	$10^2\ S_j$ ($\mathrm{cm^{-2}\text{-}atmos^{-1}}$)	Transition $j-1\to j$	$10^2\ S_{j-1}$ ($\mathrm{cm^{-2}\text{-}atmos^{-1}}$)
1→0	1.428	0→1	1.440
2→1	2.768	1→2	2.811
3→2	3.948	2→3	4.042
4→3	4.914	3→4	5.070
5→4	5.629	4→5	5.852
6→5	6.076	5→6	6.367
7→6	6.263	6→7	6.610
8→7	6.201	7→8	6.598
9→8	5.935	8→9	6.365
10→9	5.508	9→10	5.953
11→10	4.968	10→11	5.411
12→11	4.366	11→12	4.788
13→12	3.735	12→13	4.130
14→13	3.120	13→14	3.477
15→14	2.545	14→15	2.859
16→15	2.030	15→16	2.297
17→16	1.583	16→17	1.805
18→17	1.208	17→18	1.388
19→18	0.901 7	18→19	1.043 7

(continued)

Transition $j \to j-1$	$10^2\ S_j$ (cm^{-2}-$atmos^{-1}$)	Transition $j-1 \to j$	$10^2\ S_{j-1}$ (cm^{-2}-$atmos^{-1}$)
20→19	0.659 1	19→20	0.768 8
22→21	0.331 0	21→22	0.419 1
24→23	0.153 1	23→24	0.199 6
26→25	0.065 3	25→26	0.087 8
28→27	0.025 8	27→28	0.035 7
30→29	0.009 4	29→30	0.013 5
35→34	0.005 4	34→35	0.003 6
40→39	0.002 8	39→40	0.000 0

revised recently[15,16], all of the wave numbers of line centers, as well as the values of blackbody radiation from the line centers, will be somewhat in error. However, the over-all effect on the calculated values of emissivity is certainly negligibly small (less than 0. 1 percent).

Assuming the validity of the ideal gas law and using recently published integrated intensity data[12] for the fundamental vibration-rotation band of CO, it is found that

$$(\omega^* /\omega_e)[N_T \pi \varepsilon^2/(3\mu c Q)] = 371.9\ \text{cm}^{-2} \cdot \text{atmos}^{-1}\ ^*$$

The terms appearing in the large square brackets of Eqs. (2a) and (2b) are designated as A_j and A_{j-1}, respectively. Numerical values of A_j, A_{j-1}, and of S_j, S_{j-1} at 300 K for the fundamental vibration-rotation band of CO are listed in Table Ⅰ.

The integrated absorption for the rotational lines of the first overtone, S'_j and S'_{j-1}, are obtained, in close approximation, from S_j and S_{j-1}, respectively, by multiplying by the ratio of integrated absorption of the first overtone to the integrated absorption for the fundamental. The experimentally observed ratio[12] is $1.64/237 = 6.92 \times 10^{-3}$. Using this ratio, the results summarized in Table Ⅱ are obtained for S'_j and S'_{j-1} at 300 K.

By utilizing the intensity data listed in Tables Ⅰ and Ⅱ, the emissivity has been calculated[17] from Eqs. (4) and (8) as a function of optical density and rotational half-width.

Table Ⅲ Calculated and extrapolated emissivities for CO at 300 K as a function of optical density and rotational half-width.

$T = 300$ K $b = 0.06\ \text{cm}^{-1}$ $X = pL$ (cm-atmos)	E (Calculated)	E (Extrapolated from reference 18)
0.1	3.62×10^{-4}	4.1×10^{-4}
0.5	9.30×10^{-4}	1.05×10^{-3}
2.0	1.92×10^{-3}	2.2×10^{-3}
6.0	3.43×10^{-3}	3.35×10^{-3}

* c^2 appearing in this formula in the original paper has been corrected as c. — Noted by editor.

(continued)

$T = 300$ K $b = 0.07\ cm^{-1}$ $X = pL$ (cm-atmos)	E (Calculated)	E (Extrapolated from reference 18)
0.1	3.81×10^{-4}	4.1×10^{-4}
0.5	9.90×10^{-4}	1.05×10^{-3}
2.0	2.03×10^{-3}	2.20×10^{-3}
6.0	3.64×10^{-3}	3.35×10^{-3}
30.0	8.24×10^{-3}	5.8×10^{-3}
70.0	1.26×10^{-2}	7.9×10^{-3}

$T = 300$ K $b = 0.08\ cm^{-1}$ $X = pL$ (cm-atmos)	E (Calculated)	E (Extrapolated from reference 18)
0.1	4.13×10^{-4}	4.1×10^{-4}
0.5	1.08×10^{-3}	1.05×10^{-3}
2.0	2.21×10^{-3}	2.2×10^{-3}
6.0	3.90×10^{-3}	3.35×10^{-3}
30.0	8.84×10^{-3}	5.9×10^{-3}
70.0	1.35×10^{-2}	7.9×10^{-3}

The results of these calculations are summarized in Table Ⅲ and plotted in Fig. 1, where they are compared with extrapolated empirical data reported by Ullrich[18].

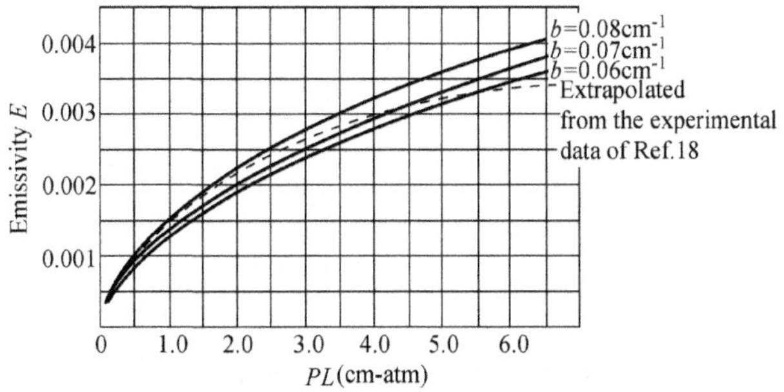

Fig. 1 Emissivity as a function of optical density at 300 K and atmospheric pressure

Reference to the data listed in Table Ⅲ and shown in Fig. 1 indicates that the calculated emissivity is proportional to the square root of the assumed value of the rotational half-width. Almost exact agreement between calculated and observed emissivities, for small optical densities, is obtained for $b = 0.076\ cm^{-1}$ at atmospheric pressure. This value is in fair agreement with infrared line-width measurements on CO obtained by standard techniques[13,19]. We therefore arrive at the important conclusion that emissivities calculated from spectroscopic data are in quantitative agreement with empirically determined results, well within the claimed limits of accuracy of the latter measurements. The fact that the calculated emissivities appear to be too large for optical densities in excess of 2 cm-atmos is the result of failure of the assumption that no overlapping occurs between rotational lines. This matter will be discussed in Section Ⅳ.

Reference to the data listed in Table Ⅲ and plotted in Fig. 1 shows that the emissivity E is very accurately a linear function of $(X)^{\frac{1}{2}}$. This result, together with the observation that E is proportional to the square root of the rotational half-width, is obviously in accord with Elsasser's well-known square-root law for nonoverlapping, equally intense, and equally spaced rotational lines[4]. Unfortunately, the simplified treatment[4] cannot be used to predict quantitatively (cf. Section VA for details) the slope of a plot of E *vs* $(X)^{\frac{1}{2}}$ in terms of the known basic spectroscopic constants①.

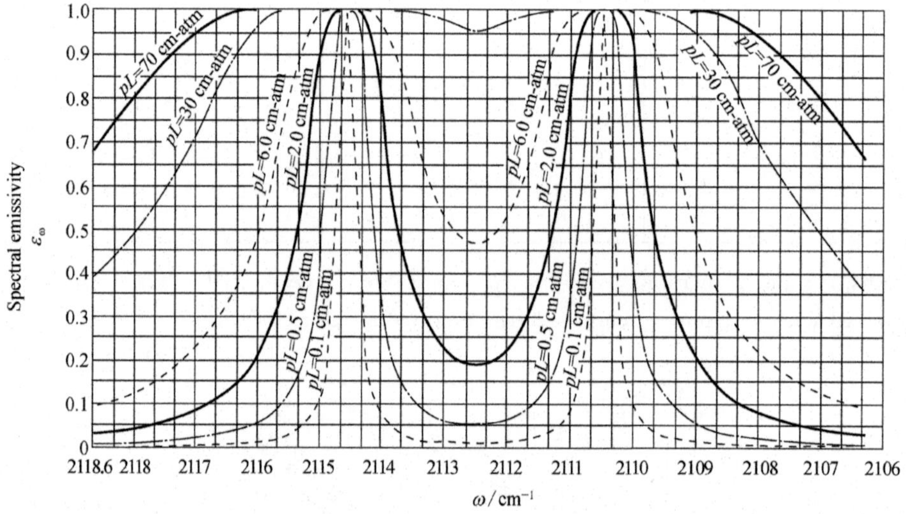

Fig. 2 Spectral emissivity ε_ω as a function of ω and optical density for two adjacent intense rotational lines ($n=0 \rightarrow n=1$, $j=8 \rightarrow j=7$ and $j=7 \rightarrow j=6$) at 300 K and atmospheric pressure

The results of the present calculations are of considerable practical importance. They clearly show that, for small optical densities, the emissivity is a sensitive function of rotational half-width. Thus we must conclude that not only is it not justified to extrapolate emissivities indiscriminately over temperature, pressure, or optical density intervals by using unsound simplified equations, but it is also essential to consider the problem with proper regard for the concentration of infrared-inert gases.

① The same difficulty arises also if attempts are made to calculate the rotational half-width from experimental measurements of transmission. Compare reference 13.

Ⅳ. Limits of Validity of the Treatment for Nonoverlapping Rotational Lines for CO[17]

To illustrate the effect of overlapping on P_ω, representative calculations involving the contributions to P_ω from two adjacent rotational lines have been carried out. In this case the correct expression for the spectral emissivity is evidently

$$\varepsilon_\omega = 1 - \exp\left[-(P_{\omega 1} + P_{\omega 2})X\right] \tag{9}$$

where $P_{\omega 1}$ and $P_{\omega 2}$ represent, respectively, the values of the spectral absorption coefficients arising from each of the two rotational lines separately. If the two rotational lines are separated sufficiently to permit neglect of overlapping, then the spectral emissivity is represented by the relation

$$\varepsilon_\omega' = 1 - \exp(-P_{\omega 1}X) + 1 - \exp(-P_{\omega 2}X) \tag{10}$$

By evaluating the differences $\varepsilon_\omega' - \varepsilon_\omega$ from Eqs. (9) and (10) we can determine the error incurred in making numerical emissivity calculations, if overlapping between two adjacent rotational lines is ignored. The resulting expression,

$$\varepsilon_\omega' - \varepsilon_\omega = (P_{\omega 1}P_{\omega 2})X^2 - (P_{\omega 1}^2 P_{\omega 2} + P_{\omega 1}P_{\omega 2}^2)X^3/2 + \ldots \tag{11}$$

is useful for small values of $P_{\omega 1}X$ and $P_{\omega 2}X$.

Emissivity calculations for two rotational lines based on the correct expression given in Eq. (9) can be used as a qualitative guide in ascertaining the limits of validity of theoretical calculations based on a treatment for nonoverlapping rotational lines. Representative values of ε_ω calculated from Eq. (9) are shown in Figs. 2 and 3 as a function of ω with X treated as a variable parameter. Reference to Fig. 2 indicates that the emissivity at a point midway between the rotational line centers which are characterized by the transitions $n = 0 \rightarrow n = 1$, $j = 8 \rightarrow j = 7$, and $j = 7 \rightarrow j = 6$, respectively, remains very small (i.e., $\varepsilon_\omega < 0.2$) for $X = 2$ cm-atmos. Since the indicated transitions correspond to the two most intense rotational lines, it appears justifiable to conclude that emissivity calculations based on the use of Eq. (8) will yield valid results at least for $X \leqslant 2$ cm-atmos. Calculations similar to those shown in Fig. 2 for intense adjacent rotational lines are shown in Fig. 3 for weak rotational lines, namely, for the lines corresponding to the transitions $n = 0 \rightarrow n = 1$, $j = 19 \rightarrow j = 18$, and $j = 18 \rightarrow j = 17$. As is to be expected, the emissivity ε_ω is represented more adequately by Eq. (10) for larger values of X than it was for the more intense rotational lines described in Fig. 2.

Although the precise evaluation of errors arising from the use of Eq. (8) is difficult to carry out, it is evident from the data described in Figs. 2 and 3 that the calculated values of the emissivity are reliable at least to $X = 2$ cm-atmos and probably to somewhat larger values. This conclusion is borne out by the comparison between calculated and observed emissivities described in Section Ⅲ.

Ⅴ. Approximate Emissivity Calculations for Diatomic Molecules with Nonoverlapping Rotational Lines

The numerical work involved in the emissivity calculations indicated in Section Ⅲ is very heavy. It is thus worthwhile to consider the possibility of developing approximate analytical expressions which avoid much of the tedious computations. In this section we shall present two such approximations, in the order of increasing accuracy.

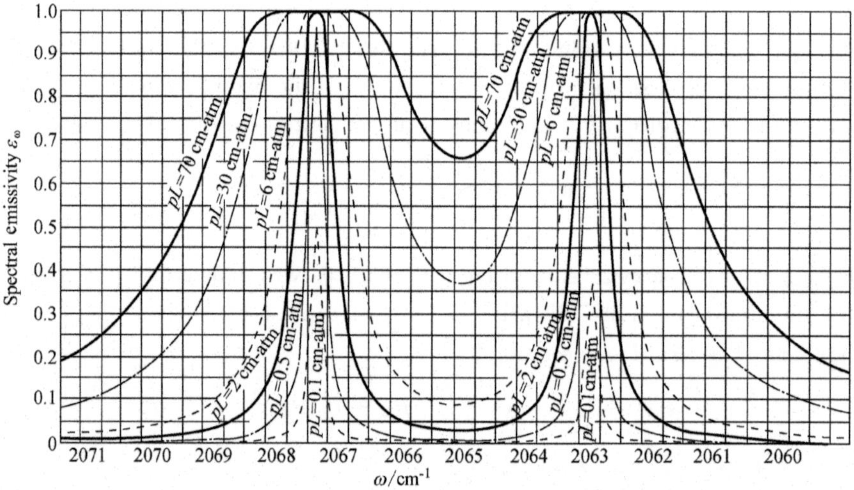

Fig. 3 Spectral emissivity ε_ω as a function of ω and optical density for two adjacent weak rotational lines ($n = 0 \rightarrow n = 1$, $j = 19 \rightarrow j = 18$ and $j = 18 \rightarrow j = 17$) at 300 K and atmospheric pressure

ⅤA. Emissivity Calculations for CO at 300 K for Nonoverlapping Rotational Lines, Assuming Equal Spacing and Intensity of Lines

From Eqs. (3) and (4) it is apparent that

$$E \simeq (\bar{R}^0/\sigma T^4)\int_0^\infty [1 - \exp(-P_\omega X)]\mathrm{d}\omega \tag{12}$$

where $\bar{R}^0$ represents an average value for $R^0(\omega)$ in the wave number interval for which P_ω is sensibly different from zero. The principal contribution to radiant heat transfer arises in the present problem from the fundamental vibration-rotation band. For moderate values of the optical density and nonoverlapping, equally spaced, and equally intense rotational lines it is well-known that[4]

$$\int_0^\infty [1 - \exp(-P_\omega X)]\mathrm{d}\omega = \Delta\omega_F \cdot 2(\bar{S}_F bX)^{\frac{1}{2}}/q \tag{13}$$

Here $\Delta\omega_F$ is the effective bandwidth[6], $\bar{S}_F$ represents the integrated absorption of the equally intense rotational lines[4,13] for the fundamental vibration-rotation band of CO, and q is the spacing of the equally spaced rotational lines. Combining Eqs. (12) and (13) leads to the result

$$E = 2[\bar{R}^0 \Delta\omega_F/(\sigma T^4)](\bar{S}_F bX)^{\frac{1}{2}}/q \tag{14}$$

As is known from the known limits of validity of the assumption that the lines are equally spaced and equally intense, Eq. (14) predicts accurately all of the qualitative features of an exact numerical solution for small values of X. Thus

$$\partial E/\partial(X)^{\frac{1}{2}} = \text{constant for fixed } b$$

and

$$\partial E/\partial(b)^{\frac{1}{2}} = \text{constant for fixed } X$$

Actually, we can do better than predict the functional form of E. Thus it has been shown[13] that for CO at 300 K

$$\bar{S}_F = 1.9 \text{ cm}^{-2}\text{atmos}^{-1} \tag{15}$$

Also $\Delta\omega_F \simeq 250 \text{ cm}^{-1}$ extending from 2 000 to 2 250 cm^{-1}, $q = 4 \text{ cm}^{-1}$, and $b = 0.08 \text{ cm}^{-1}$. Hence, Eq. (12) becomes at 300 K

$$E \simeq 1.48 \times 10^{-3}(X)^{\frac{1}{2}} \tag{16a}$$

where X is in cm-atmos, $b = 0.08$ cm.

The emissivity E calculated for CO at 300 K for a rotational half-width of 0.08 cm^{-1} (cf. Table Ⅲ) is plotted as a function of $(X)^{\frac{1}{2}}$ in Fig. 4. Reference to Fig. 4 shows that

$$E = 1.64 \times 10^{-3}(X)^{\frac{1}{2}} \quad (b = 0.08 \text{ cm}^{-1},\ 0.1 \leqslant X \leqslant 2 \text{ cm-atmos}) \tag{16b}$$

Comparison of Eqs. (16a) and (16b) shows that we can predict absolute emissivity values with an accuracy of about 10 percent without performing numerical calculations.

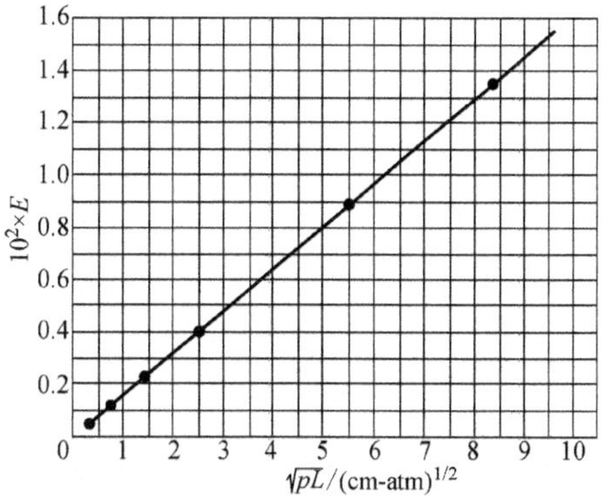

Fig. 4 Calculated emissivity E as a function of the square root of the optical density at 300 K and atmospheric pressure ($b = 0.08 \text{ cm}^{-1}$)

Considering that the objective of the present investigations is the theoretical calculation of emissivities from spectroscopic data, the following objections may be raised to the use of Eq. (14): (1) the absolute error involved in the use of Eq. (14) cannot be estimated without performing accurate numerical calculations; (2) Eq. (15) is an empirical relation whence it follows that the dependence of $\bar{S}_F$ on temperature and total pressure cannot be predicted accurately. Nevertheless, particularly in view of the fact that Elsasser[4] has obtained an integral representation for partial overlapping between equally spaced and equally intense rotational lines, the use of simplified treatments of the type given in Eq. (14) may be indicated in some cases. Thus we note that as the temperature is increased

$\bar{P}_F$ must approach the ratio of integrated absorption to bandwidth[8], a quantity which can be calculated[6] without difficulty.

V B. Emissivity Calculations for CO at 300 K for Nonoverlapping Rotational Lines Using Asymptotic Expressions for the Modified Bessel Functions

The emission functions

$$f(x) = x \exp(-x)[I_0(x) + I_1(x)] \tag{17}$$

appearing in Eq. (8) have been studied and evaluated by Elsasser[4] who has emphasized the fact that, even for moderate numerical values of x, asymptotic expressions for the modified Bessel functions yield very good approximations to the value of $f(x)$. However, this result has not heretofore been utilized in connection with accurate treatments on nonoverlapping rotational lines in which proper allowance is made for the change in intensity from one rotational line to another. It is the purpose of the present discussion to evaluate the integral appearing in Eq. (8) and to determine the value of the emissivity E from Eq. (4) under conditions where the asymptotic expressions for the modified Bessel functions can be used.

Before describing the details of the integration it may be in order to emphasize the fact that the results of the present approximate treatment must be used with caution since the analysis rests on two seemingly contradictory assumptions. Thus overlapping between rotational lines is minimized by small values of x (compare Section IV), whereas the asymptotic expressions for the modified Bessel functions apply most accurately for large values of x. Fortunately a range of values of x exists (particularly at low pressures where the rotational half-width is small) for diatomic molecules with relatively distant rotational spacing, where both restrictive conditions are satisfied closely. In particular, for the problem under discussion, it has already been shown that overlapping between rotational lines is unimportant for optical densities smaller than about 2 cm-atmos (compare Sections III and IV). Furthermore, the range of values of integrated absorption for the rotational lines of the fundamental vibration-rotation band (cf. Table I) is such that the error made by setting[4]

$$f(x) = 2[x/(2\pi)]^{\frac{1}{2}} \tag{18}$$

is small for most of the rotational lines, provided the optical density is greater than about 0.25 cm-atmos.① Hence, it is to be expected that a useful approximate solution will be

---

① By utilizing the known values of $f(x)$ and of S_j, S_{j-1} it is readily shown that the error involved in using asymptotic forms for the Bessel functions is

less than 5 percent for $1 \leqslant j \leqslant 18$ for $X = 2$ cm-atmos,

less than 5 percent for $2 \leqslant j \leqslant 16$ for $X = 1$ cm-atmos,

less than 15 percent for $2 \leqslant j \leqslant 14$ for $X = 0.25$ cm-atmos, etc.

Since most of the contributions to E result from the rotational lines with $1 \leqslant j \leqslant 16$, the preceding statement suggests the range of validity of the present treatment.

obtained at least for 0.25 cm-atmos $\leqslant pL \leqslant$ 2.0 cm-atmos. The approximate treatment utilizing asymptotic expressions for the modified Bessel functions is, of course, not applicable to calculations on the first overtone because of the small values of S'_j, S'_{j-1} (cf. Table Ⅱ). However, as was noted in Sections Ⅲ and ⅤA, the contribution to the emissivity E from the first overtone is negligibly small in the present problem.

Utilizing the approximation expressed by Eq.(18) in Eq.(8) it is found from Eq.(4) that

$$E = 30(u/\pi)^4 \sum_{j=1}^{\infty} \left\{ \frac{(\omega_j/\omega^*)^3}{[\exp(u\omega_j/\omega^*)] - 1} [bS_j X/(\omega^*)^2]^{\frac{1}{2}} + \frac{(\omega_j/\omega^*)^3}{[\exp(u\omega_{j-1}/\omega^*)] - 1} [bS_{j-1} X/(\omega^*)^2]^{\frac{1}{2}} \right\} \tag{19}$$

where explicit relations have been introduced for the Stefan-Boltzmann constant and, by using the well-known Planck distribution law, for the spectral radiation intensity emitted from a blackbody. Here we have adopted the notation of Mayer and Mayer[20] with

$$u = hc\omega^*/(kT) = h\nu^*/(kT)$$

In terms of the customary notation for the spectroscopic constants[21] we now introduce the following close approximations, neglecting terms involving $(\gamma j)^3$ and higher,

$$\omega_j/\omega^* = 1 - 2\gamma j - (\delta/\gamma)\gamma^2 j^2 \tag{20}$$

$$F^{\frac{1}{2}} = 1 + 4\gamma j - 3\gamma^2 j^2 \tag{21}$$

$$G^{\frac{1}{2}}[\exp(u\omega_j/\omega^*) - 1]^{-1} = e^{-u}(1-e^{-u})^{-\frac{1}{2}}\{1 + \gamma j u[2 + e^{-u}(1-e^{-u})^{-1}] + \gamma^2 j^2 [u(2u + \delta/\gamma) + u e^{-u}(1-e^{-u})^{-1}(3u + \delta/2\gamma) + (3/2)u^2 e^{-2u}(1-e^{-u})^{-2}]\} \tag{22}$$

$$\exp\{-[E(0,j) - E(0,0)]/(2kT)\} = [\exp(-\gamma j^2 u/2)][1 - (\gamma j u/2) + (\gamma^2 j^2 u^2/8)], \tag{23}$$

and[20]

$$Q \exp[E(0,0)/(kT)] = (\gamma u)^{-1}(1-e^{-u})^{-1}[1 + \gamma(u/3 + 8/u) + \delta(e^u - 1)^{-1} + 2x^* u(e^u - 1)^{-2}] \tag{24}$$

The relations corresponding to Eqs.(20) to (23) for the transitions $j-1 \rightarrow j$ are obtained by replacing j by $-j$. Because of this symmetry, the sum of terms from the transitions $j \rightarrow j-1$ and $j-1 \rightarrow j$ for every j is a function of j^2 only, and the linear terms in j will cancel. Hence the following result is obtained from Eq.(19), Eqs.(2a), (20) to (23), and Eq.(2b) together with the relations corresponding to Eqs.(20) to (23) for the transitions $j-1 \rightarrow j$:

$$E = 30u^5\pi^{-4}e^{-u}f(\gamma,\delta,x^*,u)[(\gamma b/\omega^*)(AX/\omega^*)]^{\frac{1}{2}} \sum_j 2j^{\frac{1}{2}} e^{-\gamma u j^2/2}[1 + \gamma^2 j^2 g(\gamma,\delta,u)] \tag{25}$$

where the temperature-independent quantity A is given by the relation

$$A = [N_T \pi \varepsilon^2/(3\mu c^2 u)](\omega^*/\omega_e) \tag{26}$$

The function $f(\gamma,\delta,x^*,u)$, which is obtained from Eq. (24),

$$f(\gamma,\delta,x^*,u) = 1 - \gamma(u/6 + 4/u) - \delta[2(e^u - 1)]^{-1} - x^* u(e^u - 1)^{-2} \tag{27}$$

is evidently close to unity. The function $g(\gamma,\delta,u)$ is

$$g(\gamma,\delta,u) = (3/2)(ue^{-u})^2(1-e^{-u})^{-2} + ue^{-u}(1-e^{-u})^{-1}[5u/2 + \delta/(2\gamma) - 3] + u(\delta/\gamma - 9/2 + 9u/8) + 27/2 - 7\delta/(2\gamma) \tag{28}$$

The Euler-Maclaurin summation formula[22] can now be employed to evaluate the summation over j appearing in Eq. (25). Details are given in the Appendix. The result is

$$E = 30u^5\pi^{-4}e^{-u}[(\gamma b/\omega^*)(AX/\omega^*)]^{\frac{1}{2}} \times f(\gamma,\delta,x^*,u)\{1.225\,[2/(\gamma u)]^{\frac{1}{4}} \times [1 + [3\gamma/(2u)]g(\gamma,\delta,u)] - 0.417\} \tag{29}$$

Since $f(\gamma,\delta,x^*,u)$ is close to unity and, at room temperature,

$$[3\gamma/(2u)]g(\gamma,\delta,u) \ll 1, \quad 0.417 \ll 1.255\,(2/\gamma u)^{\frac{1}{4}}$$

the following close approximation is obtained:

$$E = 0.634\,5u^5 e^{-u}(\gamma u)^{-\frac{3}{4}}[(\gamma b/\omega^*)(AX/\omega^*)]^{\frac{1}{2}} \tag{30}$$

Utilizing the spectroscopic constants of Sponer[14] for CO and setting[12]

$$A = 22.95 \text{ cm}^{-2}\text{atmos}^{-1}$$

it is found from Eq. (29) for $b = 0.08 \text{ cm}^{-1}$ that

$$E = 1.67 \times 10^{-3}(X)^{\frac{1}{2}} \quad (X \text{ in cm-atmos}, b = 0.08 \text{ cm}^{-1}) \tag{16c}$$

The expression for E given in Eq. (16c) is seen to be in excellent agreement with the result of numerical calculations given in Eq. (16c)[23]. Hence, we conclude that Eqs. (29) and (30) represent useful approximations for nonoverlapping rotational lines under appropriate conditions. It is evident from the previous discussion that Eq. (16c) cannot apply, even approximately, for values of X much smaller than 0. 25 cm-atmos. This conclusion is in agreement with the well-known fact that the absorption of radiation is not proportional to the square root of the optical density for very small values of the optical density.

The expression given for E in Eq. (29) represents a significant improvement over Eq. (14) (which was derived from Elsasser's treatment for equally spaced and equally intense rotational lines). Thus the parameters appearing in Eq. (29) are, in most cases, well-known molecular constants. However, Eq. (29) must be used with discretion, particularly as regards the calculation of emissivity as a function of temperature[24].

Appendix

Application of the Euler-Maclaurin summation formula[22] to the sum appearing in Eq.

(25) leads to the relation

$$2\sum_{j=1}^{\infty} j^{\frac{1}{2}} e^{-\gamma u j^2/2}[1+\gamma^2 j^2 g(\gamma,\delta,u)] \simeq 2\int_1^{\infty} j^{\frac{1}{2}} e^{-\gamma u j^2/2} \times [1+\gamma^2 j^2 g(\gamma,\delta,u)]\mathrm{d}j + (11/12) \quad \text{(A1)}$$

The integral from 1 to ∞ can be expressed as the difference between the integral from 0 to ∞ and from 0 to 1. The integral from 0 to 1 can be evaluated approximately as 4/3. Therefore,

$$2\sum_{j=1}^{\infty} j^{\frac{1}{2}} e^{-\gamma u j^2/2}[1+\gamma^2 j^2 g(\gamma,\delta,u)] \simeq 2\int_0^{\infty} j^{\frac{1}{2}} e^{-\gamma u j^2/2}[1+\gamma^2 j^2 g(\gamma,\delta,u)]\mathrm{d}j - (5/12)$$

$$= \Gamma(3/4)[2/(\gamma u)]^{\frac{3}{4}}[1+[3\gamma/(2u)]g(\gamma,\delta,u)] - (5/12) \quad \text{(A2)}$$

References

[1] Chandrasekhar S. Radiative Transfer. Clarendon Press, Oxford, 1950.

[2] Hottel H C. "Radiant heat transfer" in McAdam's Heat Transmission. McGraw-Hill Book Company, Inc., New York, 1942.

[3] For a review, containing numerous references to the original literature, see Am. J. Phys. 1949, **17**, 422, 491.

[4] Elsasser W M. Harvard Meteorological Studies No. 6. Blue Hill Observatory, Milton, Massachusetts, 1942.

[5] Dennison D M. Phys. Rev. 31, 503(1928).

[6] Penner S S. J. Appl. Phys. 21, 685(1950); Benitez L E, Penner S S. 21, 907(1950).

[7] Penner S S. J. Appl. Mech. 18, 53(1951).

[8] Penner S S, Weber D. J. Appl. Phys. 22, 1164(1951).

[9] Penner S S, Chem J. Phys. 19, 272, 1434(1951).

[10] Oppenheimer J R. Proc. Cambridge Phil. Soc. 23, 327(1926).

[11] Ladenburg R, Reiche F. Ann. Physik 42, 181(1913).

[12] Penner S S, Weber D. J. Chem. Phys. 19, 807, 817, 974(1951).

[13] Penner S S, Weber D. J. Chem. Phys. 19, 1351, 1361(1951).

[14] Sponer H. Molekülspektren. Verlag Julius Springer, Berlin, 1935.

[15] Rao K N. J. Chem. Phys. 18, 213(1950).

[16] Silverman, Plyler, and Benedict, Paper presented before the Symposium on Molecular Structure and Spectroscopy, Columbus, Ohio, 1951.

[17] Further details concerning this calculation may be found in the thesis of M. H. Ostrander.

[18] Ullrich W., Dr. Sci. thesis. M.I.T., Cambridge, Massachusetts, 1935.

[19] Matheson L A. Phys, Rev. 40, 813(1932).

[20] Mayer J E, Mayer M G. Statistical Mechanics. John Wiley and Sons, Inc., New York, 1940.

[21] A compilation of numerical values may be found on p. 468 of reference 20.

[22] See, for example, reference 20, p. 431.

[23] The fact that the present analysis leads to emissivities which are somewhat too large is a necessary consequence of the use of the asymptotic relation given in Eq. (18), which overestimates $f(x)$ for every finite value of x.

[24] Compare Section Ⅳ of reference 9.

The Transfer Functions of Rocket Nozzles

H. S. Tsien①

(*Daniel and Florence Guggenheim Jet Propulsion Center, California Institute of Technology, Pasadena, Calif.*)

The transfer function is defined as the fractional oscillating mass flow rate divided by the fractional sinusoidal pressure oscillation in the rocket combustion chamber. This is calculated as a function of the frequency of oscillation. For very small frequencies, the transfer function is approximately 1 with a small "lead component." For very large frequencies, the transfer function is considerably larger than 1, and is approximately $1+(\gamma M_1)^{-1}$ where γ is the ratio of specific heats of the gas, and M_1 is the Mach number at entrance to the nozzle.

Recently, the problem of combustion instability of rocket motor has been studied by several authors[1~3]. In these investigations, it is assumed that the percentage increase of the mass rate of flow through the nozzle is equal to the percentage increase of pressure in the rocket cylinder. It is, however, not certain whether this assumption is correct. Since this flow condition enters in a direct manner into the instability calculation, the relation between flow variations and the pressure variations should be determined more carefully. It is the purpose of this paper to do this. The result of the present study is expressed as the transfer function of the rocket nozzle, i.e., the ratio of the fractional increases in mass flow and the chamber pressure as function of the frequency of the oscillation. It indicates that the transfer function is rather a complex function of the nozzle geometry and the frequency of oscillation, and that the previous very simple assumption is only justified for very small frequencies of oscillation.

Flow Conditions

The flow in the nozzle will be considered as onedimensional, i.e., at each nozzle section the conditions are taken to be uniform and the independent variables of the problem are then the time t and the distance x along the nozzle axis in the direction of flow (Fig. 1). Let p be the pressure, ρ the density, and u the velocity. The primed quantities are the oscillating quantities; thus ρ' is the density oscillation. Similarly, the unprimed quantities are the steady-state or undisturbed quantities. Therefore p'/p is the oscillating pressure at x expressed as fraction of the steady-state pressure at x. These fractional quantities are assumed to be small so that only

Received January 2, 1952.

Journal of the American Rocket Society, vol. 22, pp. 139 - 143, 1952.

① Robert H. Goddard Professor of Jet Propulsion.

first order terms need to be considered. Since the mass rate of flow per unit area is equal to the product of density and velocity, the fractional increment of the mass rate of flow is $\left(\frac{\rho'}{\rho}+\frac{u'}{u}\right)$. Hence the purpose of this paper can now be stated as simply to compute the ratio $\left(\frac{\rho'}{\rho}+\frac{u'}{u}\right)\Big/\left(\frac{p'}{p}\right)$ at the entrance to the nozzle as a function of the oscillation frequency ω. This ratio will be called the transfer function of the rocket nozzle.

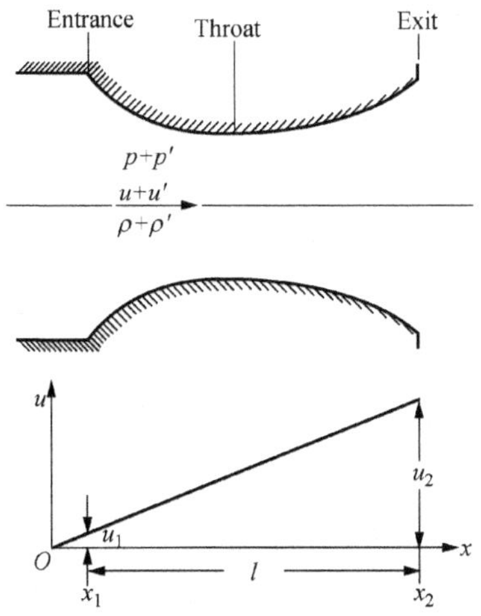

Fig. 1 Rocket nozzle with linear velocity distribution

The conditions at the entrance to the nozzle are fixed by the plausible assumption that the temperature of the combustion gas is not changed by variations in pressure. Let the gas be considered as a perfect gas, and let the subscript 1 denote the entrance to the nozzle; then

$$\left(\frac{p'}{p}\right)_1-\left(\frac{\rho'}{\rho}\right)_1=0 \tag{1}$$

Since the flow in the divergent part of the rocket nozzle is supersonic, the obvious additional condition for the complete determination of the flow is simply that the propagation of oscillations must be toward the downstream direction. For instance, if $U(x)$ is the velocity of the propagation of density oscillations at the section x, then for an observer who travels with the speed U, the density is stationary, i. e.,

$$\frac{\partial}{\partial t}\left(\frac{\rho'}{\rho}\right)+U\frac{\partial}{\partial x}\left(\frac{\rho'}{\rho}\right)=0 \tag{2}$$

This is the equation for determining the velocity of propagation. The condition of downstream propagation means positive U. Thus, in the nozzle the signs of $\partial\left(\frac{\rho'}{\rho}\right)\Big/\partial t$ and $\partial\left(\frac{\rho'}{\rho}\right)\Big/\partial x$ must be opposite. Similarly, the signs of $\partial\left(\frac{u'}{u}\right)\Big/\partial t$ and $\partial\left(\frac{u'}{u}\right)\Big/\partial x$ must also be opposite in the nozzle.

Formulation of the Problem in Nozzle

In the nozzle, if A is the cross section of the nozzle at x, then the continuity equation is

$$A\frac{\partial \rho'}{\partial t}+\frac{\partial}{\partial x}[A(\rho+\rho')(u+u')]=0 \tag{3}$$

The steady-state flow or the undisturbed flow satisfies, however, the simpler continuity equation

$$A\rho u = \text{const}$$

By eliminating A from Equation (3) by this equation, and by retaining only the first-order terms of the oscillating quantities, one obtains the linearized continuity equation as

$$\frac{\partial}{\partial t}\left(\frac{\rho'}{\rho}\right)+u\frac{\partial}{\partial x}\left(\frac{\rho'}{\rho}+\frac{u'}{u}\right)=0 \tag{4}$$

Similarly, the dynamic equation is

$$\frac{\partial}{\partial t}\left(\frac{u'}{u}\right)+\left(\frac{\rho'}{\rho}+2\frac{u'}{u}\right)\frac{du}{dx}+u\frac{\partial}{\partial x}\left(\frac{u'}{u}\right)=\left(\frac{p'}{p}\right)\frac{du}{dx}-\frac{p}{\rho u}\frac{\partial}{\partial x}\left(\frac{p'}{p}\right) \tag{5}$$

For any fluid mass, the entropy is maintained. Therefore, if γ is the ratio of specific heats of the gas,

$$\left(\frac{\partial}{\partial t}+u\frac{\partial}{\partial x}\right)\left[\left(\frac{p'}{p}\right)-\gamma\left(\frac{\rho'}{\rho}\right)\right]=0 \tag{6}$$

Equations (4), (5), and (6) are three equations for the three dependent variables (ρ'/ρ), (p'/p), and (u'/u).

It is convenient in the following calculations to introduce a specific nozzle shape such that the steady velocity in the nozzle increase simply linearly with x (Fig. 1). The simplest way to specify this is

$$u=\frac{u_1}{x_1}x \tag{7}$$

where the subscript 1 denotes again quantities at the entrance to the nozzle. x_1 is thus the x- coordinate at the entrance to the nozzle. Therefore, the origin of the x-axis is not generally at the entrance to the nozzle; it is there only if the steady velocity at entrance is equal to zero. With Equation (7), Equation (6) becomes

$$\left(\frac{\partial}{\partial t}+\frac{u_1}{x_1}\frac{\partial}{\partial \log x}\right)\left[\left(\frac{p'}{p}\right)-\gamma\left(\frac{\rho'}{\rho}\right)\right]=0 \tag{8}$$

Therefore, if the entropy oscillations at the entrance to the nozzle is $\varepsilon e^{i\omega t}$, i.e.,

$$\left(\frac{p'}{p}\right)_1-\gamma\left(\frac{\rho'}{\rho}\right)_1=\varepsilon e^{i\omega t} \tag{9}$$

then, according to Equation (8) in general

$$\left(\frac{p'}{p}\right)-\gamma\left(\frac{\rho'}{\rho}\right)=\varepsilon\exp\left[\mathrm{i}\omega\left(t-\frac{x_1}{u_1}\log\frac{x}{x_1}\right)\right] \tag{10}$$

By eliminating (p'/p) between Equations (5) and (10), the resultant equation together with Equation (4) constitute a system of two equations for the two unknowns (ρ'/ρ) and (u'/u).

Now let x_2 be the x-coordinate at the exit of the nozzle, or if l is the length of the nozzle,

$$x_2 = x_1 + l \tag{11}$$

Define a nondimensional independent variable ξ as

$$\xi = x/x_2 \tag{12}$$

Then

$$\xi_1 = x_1/x_2 = x_1/(x_1+l) = 1\Big/\left(1+\frac{l}{x_1}\right) \tag{13}$$

The oscillating quantities with frequency ω can now be written as

$$\left(\frac{\rho'}{\rho}\right)= f(\xi)\mathrm{e}^{\mathrm{i}\omega t} \tag{14}$$

$$\left(\frac{u'}{u}\right)= g(\xi)\mathrm{e}^{\mathrm{i}\omega t} \tag{15}$$

By introducing these expressions into the differential equations for ρ'/ρ and u'/u, equations for f and g are derived. The results can be, however, written more compactly if one uses the reduced frequency β, defined as

$$\beta = \frac{\omega x_1}{u_1} = \frac{\omega x_2}{u_2} = \frac{\omega l}{u_2}\frac{1}{1-\xi_1} \tag{16}$$

where u_2 is the exit velocity of the nozzle. β is then the ratio of the angular frequency ω and the velocity gradient u_1/x_1 along the nozzle axis. Then from Equations (4), (5), and (10) the equations for the f and g are obtained as

$$\xi\left[\frac{\mathrm{d}f}{\mathrm{d}\xi}+\frac{\mathrm{d}g}{\mathrm{d}\xi}\right]+\mathrm{i}\beta f = 0 \tag{17}$$

and

$$(2+\mathrm{i}\beta)g+(1-\mathrm{i}\beta)f-\xi\frac{\mathrm{d}f}{\mathrm{d}\xi} = \gamma f - \frac{1}{M^2}\xi\frac{\mathrm{d}f}{\mathrm{d}\xi}+\varepsilon\left(\frac{\xi}{\xi^1}\right)^{\mathrm{i}\beta}\left[1+\frac{\mathrm{i}\beta}{\gamma M^2}\right] \tag{18}$$

where M is the local Mach number of the undisturbed flow at x or ξ. In fact, for the assumed linear velocity distribution in the nozzle,

$$M^2 = \frac{M_1^2 \xi^2}{\xi_1^2 - \frac{\gamma - 1}{2} M_1^2 (\xi^2 - \xi_1^2)} = \frac{M_1^2 \left(\frac{x}{x_1}\right)^2}{1 - \frac{\gamma - 1}{2} M_1^2 \left(\frac{x^2}{x_1^2} - 1\right)} \tag{19}$$

M_1 is the Mach number at the entrance to the nozzle.

By eliminating $g(\xi)$ from Equations (17) and (18), a single second-order equation for $f(\xi)$ is obtained. The result can, however, be reduced to more convenient form by using a new independent variable z defined as

$$z = \frac{\frac{\gamma + 1}{2} M_1^2 (\xi / \xi_1)^2}{1 + \frac{\gamma - 1}{2} M_1^2} \tag{20}$$

It is easy to show that z is actually the square of the ratio of u to the so-called critical sound speed. Thus $z = 1$ at the throat of the nozzle. In terms of z, the differential equation for f is

$$z(1-z)\frac{d^2 f}{dz^2} - \left\{2 + \frac{2i\beta}{(\gamma+1)}\right\} z \frac{df}{dz} - \frac{i\beta(2+i\beta)}{2(\gamma+1)} f = -i\beta\varepsilon\left(\frac{z}{z_1}\right)^{-i(\beta/2)} \left[\frac{1 - i\beta\frac{\gamma-1}{2\gamma}}{2(\gamma+1)} + \frac{2+i\beta}{4\gamma}\frac{1}{z}\right] \tag{21}$$

The relation between $f(z)$ and $g(z)$ is

$$(2+i\beta)g(z) = [(\gamma-1)+i\beta]f(z) - (\gamma+1)(1-z)\frac{df}{dz} + \varepsilon\left(\frac{z}{z_1}\right)^{-i(\beta/2)}\left[1 - \frac{i\beta(\gamma-1)}{2\gamma} + \frac{i\beta(\gamma+1)}{2\gamma}\frac{1}{z}\right] \tag{22}$$

z_1 is, of course, the value of z corresponding to ξ_1 as given by Equation (13); i.e.,

$$z_1 = \frac{\frac{\gamma + 1}{2} M_1^2}{1 + \frac{\gamma - 1}{2} M_1^2} \tag{23}$$

The initial condition needed to solve Equation (21) is specified by Equations (1) and (9). That is, at $x = x_1$ or $z = z_1$,

$$f(z_1) = -\frac{\varepsilon}{\gamma - 1} \tag{24}$$

The complete solution requires the cognizance of the fact that the propagation of oscillations must be toward the downstream direction as discussed in the previous section.

When $f(z)$ is determined, one can compute $g(z)$ by Equation (22). Then the density oscillations and the velocity oscillations are determined as functions proportional to the amplitude of entropy oscillation ε. Since the point of interest is the ratio of the oscillations, the arbitrary ε does not really enter into the final result. In fact, according to Equation (1), the ratio of the fractional increase in mass rate of flow and that of pressure, or the transfer function $G(\beta)$ is

$$G(\beta)=\left[\left(\frac{\rho'}{\rho}\right)_1+\left(\frac{u'}{u}\right)_1\right]\Big/\left(\frac{p'}{p}\right)_1=\left[\left(\frac{\rho'}{\rho}\right)_1+\left(\frac{u'}{u}\right)_1\right]\left(\frac{\rho'}{\rho}\right)_1=1+\frac{g(z_1)}{f(z_1)} \tag{25}$$

Therefore $G(\beta)$ is unity as assumed by previous investigators of combustion instability only if there is no velocity oscillation at the entrance, or $(u'/u)_1=0$. It is perhaps worthwhile to point out the fact that the present definition of transfer function $G(\beta)$ corresponds to the practice of servomechanism analysis, if the pressure oscillation $(p'/p)_1$ is taken as the "input" and the mass flow oscillation as the "output". Hence the transfer function $G(\beta)$ will be useful also in the synthesis of the systems for servocontrol or servostabilization of the rocket motor, as suggested by W. Bollay[4].

Solution for Small Frequency

The basic differential equation with linear velocity distribution in the nozzle as given by Equation (21) can be reduced to the hypergeometric differential equation by the substitution $f(z)=z\cdot w(z)$. $w(z)$ is then the hypergeometric function. However, the calculation is rather difficult. For a first discussion, only cases of either very small frequency or very large frequency will be considered here.

If the frequency β is very small, the functions $f(z)$ and $g(z)$ can be expanded in terms of this parameter

$$f(z;\beta)=f^{(0)}(z)+\beta f^{(1)}(z)+\dots \tag{26}$$

$$g(z;\beta)=g^{(0)}(z)+\beta g^{(1)}(z)+\dots \tag{27}$$

These expressions are appropriate if the interval of z concerned does not include $z=0$. Fortunately, the physical problem generally specifies nonvanishing u at the entrance to the nozzle. Therefore, $\xi_1\neq 0$ and thus $z_1\neq 0$. The restriction on the validity of the expansions is thus generally complied. By substituting Equations (26) and (27) into the basic Equations (21) and (22), and then equating terms of same powers in β, one has

$$z(1-z)\frac{d^2f^{(0)}}{dz^2}-2z\frac{df^{(0)}}{dz}=0 \tag{28}$$

and

$$z(1-z)\frac{d^2f^{(1)}}{dz^2}-2z\frac{df^{(1)}}{dz}=i\left\{\frac{2}{\gamma+1}z\frac{df^{(0)}}{dz}+\frac{1}{\gamma+1}f^{(0)}-\frac{\varepsilon}{2}\left(\frac{1}{\gamma+1}+\frac{1}{\gamma z}\right)\right\} \tag{29}$$

Furthermore

$$2g^{(0)}(z)=(\gamma-1)f^{(0)}(z)-(\gamma+1)(1-z)\frac{df^{(0)}}{dz}+\varepsilon \tag{30}$$

$$2g^{(1)}(z) = (\gamma - 1)f^{(1)}(z) - (\gamma + 1)(1 - z)\frac{df^{(1)}}{dz} + \\ i[f^{(0)}(z) - g^{(0)}(z)] - \frac{i\varepsilon}{2}\left[\log\left(\frac{z}{z_1}\right) + \frac{\gamma - 1}{\gamma} - \frac{\gamma + 1}{\gamma z}\right] \tag{31}$$

The initial condition of Equation (24) now becomes

$$f^{(0)}(z_1) = -\frac{\varepsilon}{\gamma - 1} \tag{32}$$

$$f^{(1)}(z_1) = 0 \tag{33}$$

A straightforward calculation will give the following relation for the propagation velocity U for the density fluctuation defined by Equation (2),

$$\frac{U}{u} = -\frac{i\beta f(z)}{2z\frac{df}{dz}} = -\frac{1}{2z}\frac{i\beta[f^{(0)}(z) + \beta f^{(1)}(z) + \ldots]}{\frac{df^{(0)}}{dz} + \beta\frac{df^{(1)}}{dz} + \ldots} \tag{34}$$

where u is, of course, the undisturbed flow velocity at the section concerned. Therefore, in order that the propagation speed be finite even for small β, $f^{(0)}(z)$ must be a constant. Hence the appropriate solution of Equation (28) together with the condition of Equation (32) is

$$f^{(0)}(z) = -\frac{\varepsilon}{\gamma - 1} \tag{35}$$

Then Equation (30) gives immediately

$$g^{(0)}(z) = 0 \tag{36}$$

Up to this order of approximation then, $G = 1$, as shown by Equation (25). Therefore the assumption of the previous investigators of the combustion instability of rocket motor is indeed correct if the frequency is vanishingly small.

With $f^{(0)}(z)$ given by Equation (35), the equation for $f^{(1)}(z)$, Equation (29), can be written as

$$\frac{d}{dz}\left[(1 - z)^2\frac{df^{(1)}}{dz}\right] = -\frac{i\varepsilon}{2}\left[\frac{1}{\gamma - 1} + \frac{1}{\gamma z}\right]\frac{1 - z}{z}$$

Therefore the solution is

$$\frac{df^{(1)}}{dz} = -\frac{i\varepsilon}{2}\left[\frac{1}{\gamma(\gamma - 1)}\frac{\log z}{(1 - z)^2} + \frac{1}{(\gamma - 1)}\frac{1}{(1 - z)} - \frac{1}{\gamma}\frac{1}{z(1 - z)}\right] + \frac{C}{(1 - z)^2}$$

The quantity inside the square bracket is finite and positive for $z > 0$, while the last term is infinite at the throat of the nozzle where $z = 1$. Hence, according to Equation (34), the condition for finite positive propagation speed U requires $C = 0$. Therefore[①]

① Professor L. Crocco has very kindly pointed out that the condition for positive finite propagation speed U is equivalent to the suppression of all singularities at the sonic point. This observation may simplify the analysis for general values of β.

$$\frac{\mathrm{d}f^{(1)}}{\mathrm{d}z}=\frac{\mathrm{i}\varepsilon}{2}\left[-\frac{1}{\gamma(\gamma-1)}\frac{\log z}{(1-z)^2}-\frac{1}{(\gamma-1)}\frac{1}{(1-z)}+\frac{1}{\gamma}\frac{1}{z(1-z)}\right] \tag{37}$$

This equation together with Equations (31) and (33) then gives

$$g^{(1)}(z_1)=\frac{\mathrm{i}\varepsilon}{4\gamma}\left\{\frac{\gamma+1}{\gamma-1}\frac{\log z_1}{1-z_1}+1\right\} \tag{38}$$

Therefore, up to the first order in β, the transfer function $G(\beta)$ is

$$G(\beta)\cong 1+\frac{\beta g^{(1)}(z_1)}{f(z_1)}=1+\mathrm{i}\beta\left\{\frac{\gamma+1}{4\gamma}\frac{\log\left(\frac{1}{z_1}\right)}{1-z_1}-\frac{\gamma-1}{4\gamma}\right\},\quad \beta\ll 1 \tag{39}$$

Therefore, when the frequency β is not exactly zero, the transfer function has a small "lead component," proportional to the frequency.

Solution for Large Frequencies

If the value of β is very large, one must find the asymptotic solution of Equation (21). The dominating terms in Equation (21) are

$$z(1-z)\frac{\mathrm{d}^2 f}{\mathrm{d}z^2}-\frac{2\mathrm{i}\beta}{\gamma+1}z\frac{\mathrm{d}f}{\mathrm{d}z}+\frac{\beta^2}{2(\gamma+1)}f=\beta^2\varepsilon\left(\frac{z}{z_1}\right)^{-(\mathrm{i}\beta/2)}\frac{1}{4\gamma}\left[\frac{1}{z}-\frac{\gamma-1}{\gamma+1}\right] \tag{40}$$

For the particular integral f^*, take

$$f^*(z)=Z(z)\left(\frac{z}{z_1}\right)^{-(\mathrm{i}\beta/2)}$$

where $Z(z)$ is a function of z not involving β. Therefore, by retaining only the highest order term,

$$\frac{\mathrm{d}f^*}{\mathrm{d}z}\cong-\frac{\mathrm{i}\beta}{2}\frac{Z(z)}{z}\left(\frac{z}{z_1}\right)^{-(\mathrm{i}\beta/2)}$$

$$\frac{\mathrm{d}^2 f^*}{\mathrm{d}z^2}\cong-\frac{\beta^2}{4}\frac{Z(z)}{z^2}\left(\frac{z}{z_1}\right)^{-(\mathrm{i}\beta/2)}$$

By substituting these derivatives into Equation (40), it is found that

$$Z(z)=-\frac{\varepsilon}{\gamma}$$

thus

$$f^*(z)=-\frac{\varepsilon}{\gamma}\left(\frac{z}{z_1}\right)^{-(\mathrm{i}\beta/2)} \tag{41}$$

To find the complementary function, let

$$f(z)=\mathrm{e}^{\mathrm{i}\beta\lambda(z)}$$

then

$$\frac{\mathrm{d}f}{\mathrm{d}z} = \mathrm{i}\beta \mathrm{e}^{\mathrm{i}\beta\lambda(z)} \frac{\mathrm{d}\lambda}{\mathrm{d}z}$$

$$\frac{\mathrm{d}^2 f}{\mathrm{d}z^2} \cong -\beta^2 \mathrm{e}^{\mathrm{i}\beta\lambda(z)} \left(\frac{\mathrm{d}\lambda}{\mathrm{d}z}\right)^2$$

By substituting these into the homogeneous equation corresponding to Equation (40), one has

$$\frac{\mathrm{d}\lambda_{1,2}}{\mathrm{d}z} = \frac{1}{(\gamma+1)(1-z)}\left[1 \pm \sqrt{1+\frac{\gamma+1}{2}\frac{1-z}{z}}\right]$$

Thus

$$\lambda_1(z) = \frac{1}{\gamma+1}\int_{z_1} \frac{\mathrm{d}z}{1-z}\left[1+\sqrt{1+\frac{\gamma+1}{2}\frac{1-z}{z}}\right], \quad \lambda_1(z_1) = 0$$

and

$$\lambda_2(z) = \frac{1}{\gamma+1}\int_{z_1} \frac{\mathrm{d}z}{1-z}\left[1-\sqrt{1+\frac{\gamma+1}{2}\frac{1-z}{z}}\right], \quad \lambda_2(z_1) = 0 \tag{42}$$

The complete solution for large β is thus

$$f(z)\mathrm{e}^{\mathrm{i}\omega t} = -\frac{\varepsilon}{\gamma}\mathrm{e}^{\mathrm{i}\beta\left[\frac{u_2}{x_2}t-\frac{1}{2}\log\left(\frac{z}{z_1}\right)\right]} + B\mathrm{e}^{\mathrm{i}\beta\left[\frac{u_2}{x_2}t+\lambda_1(z)\right]} + D\mathrm{e}^{\mathrm{i}\beta\left[\frac{u_2}{x_2}t+\lambda_2(z)\right]} \tag{43}$$

The first term is the particular integral; it satisfies the condition of downstream propagation of disturbances. The same condition requires, however, that $\lambda(z)$ be negative and increasingly negative as z increases from z_1. $\lambda_1(z)$ does not satisfy this condition and hence has to be rejected. Therefore, $B = 0$. D is fixed by the initial condition of Equation (24), and is $-\varepsilon/\gamma(\gamma-1)$. Therefore, finally

$$f(z) = -\frac{\varepsilon}{\gamma}\left[\left(\frac{z}{z_1}\right)^{-(\mathrm{i}\beta/2)} + \frac{1}{\gamma-1}\mathrm{e}^{\mathrm{i}\beta\lambda_2(z)}\right] \tag{44}$$

where $\lambda_2(z)$ is given Equation (42).

By taking the dominating terms of Equation (22) for large β, one has

$$g(z) = f(z) - (\gamma+1)(1-z)\frac{1}{\mathrm{i}\beta}\frac{\mathrm{d}f}{\mathrm{d}z} + \varepsilon\left(\frac{z}{z_1}\right)^{-(\mathrm{i}\beta/2)}\left[\frac{\gamma+1}{2\gamma}\frac{1}{z} - \frac{\gamma-1}{2\gamma}\right]$$

with $f(z)$ as given Equation (44), then $g(z)$ is determined as

$$g(z) = -\frac{\varepsilon}{\gamma(\gamma-1)}\sqrt{1+\frac{\gamma+1}{2}\frac{1-z}{z}}\,\mathrm{e}^{\mathrm{i}\beta\lambda_2(z)}, \quad \beta \gg 1 \tag{45}$$

Therefore, the transfer function $G(\beta)$ for large β is

$$G(\beta) = 1 + \frac{1}{\gamma}\sqrt{1+\frac{\gamma+1}{2}\frac{1-z_1}{z_1}}, \quad \beta \gg 1$$

But z_1 is specified by Equation (23), therefore

$$G(\beta) = 1 + \frac{1}{\gamma M_1}, \quad \beta \gg 1 \tag{46}$$

Therefore, for large frequencies the transfer function is again real and positive, i.e., the mass flow oscillations and the pressure oscillations at the entrance to the nozzle are again in phase. However, the mass flow oscillations are considerably larger than the pressure oscillations; the ratio is given by Equation (46).

Concluding Remarks

In the previous section the calculation of the transfer function of the rocket nozzle is considered as a problem in nonsteady one-dimensional flow. For very small frequencies of oscillations, the transfer function is indeed approximately equal to unity, as assumed by previous investigators of combustion instability. However, the present calculation also shows that the transfer function is a complex number and has a lead component increasing with increase in frequency. For very large frequencies, the analyses give results indicating larger mass flow oscillations than supposed before. But the transfer function for infinite frequency β is again a positive real number. Therefore, the Nyquist diagram of the transfer function $G(\beta)$ is a curve, not a line. As an example, for $\gamma = 1.22$, and $M_1 = 0.2$, according to Equations (39) and (46)

$$\begin{aligned} G(\beta) &\cong 1 + 1.438\, i\beta, && \text{for } \beta \ll 1 \\ &\cong 5.1, && \text{for } \beta \gg 1 \end{aligned}$$

The accuracy of the present results is limited by the underlying assumptions. First of all, the nozzle shape is such as to give a linear velocity distribution in the nozzle. Then the gas is assumed to be perfect and nonreacting. In reality, viscosity, heat conduction, and residual chemical reactions in the exhaust gas will modify the result. At large frequencies or for very short nozzles, the wave length may be small in comparison with the nozzle length, then the one-dimensional flow assumption may introduce appreciable error. For these reasons, it is desirable to measure the transfer function of the rocket nozzle experimentally. Here, however, one must remember that the flow phenomenon in connection with the combustion stability is profoundly influenced by the entropy oscillations generated in the combustion processes. Direct stability measurements on rocket motor with artificially oscillated propellant flow will be necessary. The present analysis indicates, however, that the nozzle could be cut short with practically no divergent part as the transfer function is not influenced by the particular exit velocity so long as the exit velocity is supersonic. The propellant flow rate will then be smaller than a full-length nozzle, a desirable saving in experimentation. The present investigation also gives the characteristic nondimensional frequency as the angular frequency ω divided by the velocity gradient along the nozzle. The transfer functions for rocket nozzles of different sizes can be correlated through this nondimensional frequency or reduced frequency.

In conclusion then, it is certain that the present paper gives only a general outline of the problem of transfer function of rocket nozzles. The complete detail has yet to be filled in by further studies.

References

[1] Gunder D F, Friant D R. Stability of Flow in a Rocket Motor. Journal of Applied Mechanics, 1950,17(3): 327.

[2] Summerfield M. A Theory of Unstable Combustion in Liquid Propellant Rocket Systems. Journal of the American Rocket Society, 1951,21(5): 108.

[3] Crocco L. Aspects of Combustion Stability in Liquid Propellant Rocket Motors, Parts I. Journal of the American Rocket Society, 1951,21(6): 163.

[4] Bollay W. Aerodynamic Stability and Automatic Control. Journal of the Aeronautical Sciences, 1951,18(9): 569-623,605.

A Similarity Law for Stressing Rapidly Heated Thin-Walled Cylinders①

H. S. Tsien② and C. M. Cheng③

(*Daniel and Florence Guggenheim Jet Propulsion Center, California Institute of Technology, Pasadena, Calif.*)

When a thin cylindrical shell of uniform thickness is very rapidly heated by hot high-pressure gas flowing inside the shell, the temperature of material decreases steeply from a high temperature at the inside surface to ambient temperatures at the outside surface. Young's modulus of material thus varies. The purpose of the present paper is to reduce the problem of stress analysis of such a cylinder to an equivalent problem in conventional cylindrical shell without temperature gradient in the wall. The equivalence concept is expressed as a series of relations between the quantities for the hot cylinder and the quantities for the cold cylinder. These relations give the similarity law whereby strains for the hot cylinder can be simply deduced from measured strains on the cold cylinder and thus greatly simplify the problem of experimental stress analysis.

The cylinder of a solid propellant rocket is subjected to very rapid heating during its short duration of operation. The temperature distribution across the thin cylindrical wall, although approximately the same in every section, is not linear. This condition is most severe at the end of the combustion of the propellant grain. From the point of view of a materials engineer, this case is distinguished from others by the time rate of heating, which is so large as to not allow sufficient time for appreciable change in the structure of the material. The strength of the wall material under this operating condition is quite different from that under slow heating. This fact is clearly and conclusively shown by R. L. Noland in a recent paper[1]. From the point of view of a stress analyst, the rational design of a solid propellant rocket cylinder is thus complicated by the very large thermal stress and variable Young's modulus of the material across the wall as a result of the large temperature gradient. Furthermore, experimental stress determination under actual firing tests is rather difficult due to the short test time available and the high temperature.

It is believed that for these reasons the only case which has been analyzed by reliable rational method is the case of rocket cylinder under uniform internal pressure. The bending

---

Received February 14, 1952.

Journal of the American Rocket Society, vol. 22, pp. 144 - 149, 1952.

① This paper is based on part of a thesis submitted by the junior author for partial fulfillment of the requirements of Ph. D. in Mechanical Engineering, California Institute of Technology.

② Robert H. Goddard Professor of Jet Propulsion.

③ Graduate Assistant in Mechanical Engineering.

stresses due to canted nozzles, the stresses due to end enclosures, mounting lugs, etc., are only estimated by very rough methods. The purpose of this paper is to improve this situation by suggesting an approach which will reduce the general stress problem of hot cylinder to a problem of an equivalent cold cylinder. The equivalent problem can then be solved either analytically by the conventional method or directly by experimental stress determination. In either choice, the problem is believed to be greatly simplified. This law of equivalence between hot cylinder and cold cylinder can be called the similarity law.

Stresses and Strains of a Thin-Walled Cylinder

The fact that the thickness of the cylinder is small in comparison to its radius and length allows a great simplification in the strain analysis. To wit, the deformation of every point of the cylinder can be described sufficiently accurately by the displacements of the points on a single surface within the wall of the cylinder. This surface is the so-called median surface. The position of the median surface is so determined that a bending of the median surface will not induce net extensional forces in the plane of the median surface, across the thickness of the wall. When Young's modulus is a constant, as is the case for a cold cylinder, the median surface lies midway between the outer and the inner boundary surfaces of the cylinder. When Young's modulus is not a constant but decreases with increase in temperature, the median surface is displaced toward the cold side, as will be seen presently.

Let x, θ, z be the coordinate system with origin on the cylindrical median surface such that x points in the axial direction of the cylinder, θ is in the circumferential direction, measured on the median surface, and z is normal to the median surface, pointing toward the axis of the cylinder. Let U, V, and W be displacements of a point (x, θ) on the median surface in the directions x, θ, and z, respectively. They are thus functions of x and θ; but not of z. Then the above-mentioned fundamental simplification of thin shells can be stated as follows: if the direct strains in the x and θ directions are e_x and e_θ, and the shear strain $\gamma_{x\theta}$, then

$$\left.\begin{aligned} e_x &= \frac{\partial U}{\partial x} - z\frac{\partial^2 W}{\partial x^2} \\ e_\theta &= \frac{1}{R}\frac{\partial V}{\partial \theta} - \frac{W}{R} - \frac{z}{R^2}\left(\frac{\partial V}{\partial \theta} + \frac{\partial^2 W}{\partial \theta^2}\right) \\ \gamma_{x\theta} &= \frac{\partial V}{\partial x} + \frac{1}{R}\frac{\partial U}{\partial \theta} - 2\frac{z}{R}\left(\frac{\partial^2 W}{\partial x \partial \theta} + \frac{\partial V}{\partial x}\right) \end{aligned}\right\} \tag{1}$$

where R is the radius of the median cylindrical surface, or the "radius of the cylinder." This result is sometimes described as the Kirchhoff bending assumption: Plane normal to the median surface before bending remains so after bending.

The significant stresses in a thin shell are the direct stresses σ_x and σ_θ in x and θ, and the shear stress $\tau_{x\theta}$. All other stresses are small in comparison to these three. Now let T be the temperature of wall above a reference temperature, say the room temperature. T is assumed to be only a function of thickness ordinate z, but not of x and θ. Thus the heating of the cylinder is assumed to be uniform over the entire surface of the cylinder. This is generally very closely approximated in reality. If α is the coefficient of thermal expansion, the thermal expansion strain is then αT. By Hooke's law one has

$$\left.\begin{aligned} e_x &= \frac{1}{E}(\sigma_x - \nu\sigma_\theta) + \alpha T \\ e_\theta &= \frac{1}{E}(\sigma_\theta - \nu\sigma_x) + \alpha T \\ \gamma_{x\theta} &= \frac{2(1+\nu)}{E}\tau_{x\theta} \end{aligned}\right\} \tag{2}$$

where E is Young's modulus, and v is Poisson's ratio. E is, of course, a function of temperature or a function of z. v, however, will be assumed to be constant for lack of definite information. By solving for the stresses, one obtains from Equation (2),

$$\left.\begin{aligned} \sigma_x &= \frac{E(z)}{1-\nu^2}\{(e_x + \nu e_\theta) - (1+\nu)\alpha T(z)\} \\ \sigma_\theta &= \frac{E(z)}{1-\nu^2}\{(e_\theta + \nu e_x) - (1+\nu)\alpha T(z)\} \\ \tau_{x\theta} &= \frac{E(z)}{2(1+\nu)}\gamma_{x\theta} \end{aligned}\right\} \tag{3}$$

For thin-walled cylinders, the equilibrium equations are expressed in terms of "sectional averages" of the stresses given in Equation (3). That is, one speaks of the normal forces N_x and N_θ, the shearing force $N_{x\theta}$, the bending moments M_x, M_θ, and the twisting moment $M_{x\theta}$. They are related to σ_x, σ_θ, and $\tau_{x\theta}$ by the following equations

$$N_x = \int \sigma_x \mathrm{d}z, \quad N_\theta = \int \sigma_\theta \mathrm{d}z, \quad N_{x\theta} = \int \tau_{x\theta} \mathrm{d}z \tag{4}$$

$$M_x = -\int \sigma_x z \,\mathrm{d}z, \quad M_\theta = -\int \sigma_\theta z \,\mathrm{d}z, \quad M_{x\theta} = -\int \tau_{x\theta} z \,\mathrm{d}z \tag{5}$$

The integrals in the above equations all extend across the thickness of the wall. By substituting Equations (1) and (3) into Equations (4) and (5), one has for example

$$\begin{aligned} N_x = D_0 &\left(\frac{\partial U}{\partial x} + \frac{\nu}{R}\frac{\partial V}{\partial \theta} - \nu\frac{W}{R}\right) - \\ D_1 &\left(\frac{\partial^2 W}{\partial x^2} + \frac{\nu}{R^2}\frac{\partial V}{\partial \theta} + \frac{\nu}{R^2}\frac{\partial^2 W}{\partial \theta^2}\right) - N_T \end{aligned}$$

and

$$\begin{aligned} -M_x = D_1 &\left(\frac{\partial U}{\partial x} + \frac{\nu}{R}\frac{\partial V}{\partial \theta} - \nu\frac{W}{R}\right) - \\ D_2 &\left(\frac{\partial^2 W}{\partial x^2} + \frac{\nu}{R^2}\frac{\partial V}{\partial \theta} + \frac{\nu}{R^2}\frac{\partial^2 W}{\partial \theta^2}\right) - M_T \end{aligned}$$

where

$$\left.\begin{aligned} D_0 &= \frac{1}{1-\nu^2}\int E(z)\,dz \\ D_1 &= \frac{1}{1-\nu^2}\int E(z)\,z\,dz \end{aligned}\right\} \tag{6}$$

$$D_2 = \frac{1}{1-\nu^2}\int E(z)\,z^2\,dz \tag{7}$$

and

$$N_T = \frac{\alpha}{1-\nu}\int E(z)\,T(z)\,dz \tag{8}$$

$$M_T = \frac{\alpha}{1-\nu}\int E(z)\,T(z)\,z\,dz \tag{9}$$

The integrals again extend across the thickness of the wall. It is evident from the above expressions for N_x and M_x that considerable simplification can be achieved by choosing the median surface in such a way that

$$D_1 = \frac{1}{1-\nu^2}\int E(z)\,z\,dz = 0 \tag{10}$$

This is actually the condition to locate the median surface. Since Young's modulus E decreases with increase in temperature, it is seen from Equation (10) that the median surface is nearer to the cold boundary surface than to the hot boundary surface. For a rocket cylinder, hot inside but cold outside, the median surface is near to the outside surface. With this choice of the median surface, the forces and the moments are related to the displacements by the following simpler equations:

$$\left.\begin{aligned} N_x &= D_0\left(\frac{\partial U}{\partial x} + \frac{\nu}{R}\frac{\partial V}{\partial \theta} - \nu\frac{W}{R}\right) - N_T \\ N_\theta &= D_0\left(\frac{1}{R}\frac{\partial V}{\partial \theta} - \frac{W}{R} + \nu\frac{\partial U}{\partial x}\right) - N_T \\ N_{x\theta} &= \frac{1-\nu}{2}D_0\left(\frac{\partial V}{\partial x} + \frac{1}{R}\frac{\partial U}{\partial \theta}\right) \end{aligned}\right\} \tag{11}$$

$$\left.\begin{aligned} M_x &= D_2\left(\frac{\partial^2 W}{\partial x^2} + \frac{\nu}{R^2}\frac{\partial^2 W}{\partial \theta^2} + \frac{\nu}{R^2}\frac{\partial V}{\partial \theta}\right) + M_T \\ M_\theta &= D_2\left(\frac{1}{R^2}\frac{\partial^2 W}{\partial \theta^2} + \frac{1}{R^2}\frac{\partial V}{\partial \theta} + \nu\frac{\partial^2 W}{\partial x^2}\right) + M_T \\ M_{x\theta} &= (1-\nu)D_2\left(\frac{1}{R^2}\frac{\partial^2 W}{\partial x\partial \theta} + \frac{1}{R^2}\frac{\partial V}{\partial x}\right) \end{aligned}\right\} \tag{12}$$

It is worth while to point out two facts: Firstly, the choice of reference temperature is quite arbitrary. Changing the reference temperature will change the value of N_T, Equation

(8). But a corresponding adjustment in the normal strain will leave the normal forces N_x, N_θ given by Equation (11) unchanged. Therefore the physical problem is quite independent of the choice of reference temperature. M_T is independent of the reference temperature due to Equation (10). Secondly, the present system of equations reduces to those of the conventional thin shell theory if temperature gradient is absent, or if Young's modulus is independent of the temperature of material. In that case the median surface is midway between the boundary surfaces, and D_2 is simply the usual flexural rigidity of the shell.

Nondimensional Quantities and Equations of Equilibrium

It is useful to introduce nondimensional quantities by taking R as the reference length. Thus

$$\xi = x/R \tag{13}$$

$$u = U/R, \quad v = V/R, \quad w = W/R \tag{14}$$

$$n_\xi = N_x/D_0, \quad n_\theta = N_\theta/D_0, \quad n_{\xi\theta} = N_{x\theta}/D_0, \quad n_T = N_T/D_0 \tag{15}$$

and

$$m_\xi = M_x R/D_2, \quad m_\theta = M_\theta R/D_2, \quad m_{\xi\theta} = M_{x\theta} R/D_2, \quad M_T = M_T R/D_2 \tag{16}$$

Therefore Equations (11) and (12) become now

$$\left.\begin{aligned} n_\xi &= \frac{\partial u}{\partial \xi} + v\frac{\partial v}{\partial \theta} - vw - n_T, \quad n_\theta = \frac{\partial v}{\partial \theta} - w + v\frac{\partial u}{\partial \xi} - n_T \\ n_{\xi\theta} &= \frac{1-v}{2}\left(\frac{\partial v}{\partial \xi} + \frac{\partial u}{\partial \theta}\right) \end{aligned}\right\} \tag{17}$$

and

$$\left.\begin{aligned} m_\xi &= \frac{\partial^2 w}{\partial \xi^2} + v\frac{\partial^2 w}{\partial \theta^2} + v\frac{\partial v}{\partial \theta} + m_T, \quad m_\theta = \frac{\partial^2 w}{\partial \theta^2} + \frac{\partial v}{\partial \theta} + v\frac{\partial^2 w}{\partial \xi^2} + m_T \\ m_{\xi\theta} &= (1-v)\left(\frac{\partial^2 w}{\partial \xi \partial \theta} + \frac{\partial v}{\partial \xi}\right) \end{aligned}\right\} \tag{18}$$

The equations of equilibrium in terms of forces and moments are here exactly the same as the conventional theory[2]. The only innovation is to write them in nondimensional form. For this purpose, one has to define the nondimensional quantities q_ξ, q_θ, and p of the dimensional sectional shearing forces Q_x, Q_θ and normal pressure loading P, against z-direction as follows:

$$q_\xi = Q_x/D_0 \quad q_\theta = Q_\theta/D_0 \quad p = PR/D_0 \tag{19}$$

Then the equilibrium equations of forces are

$$\left.\begin{aligned} &\frac{\partial n_\xi}{\partial_\xi} + \frac{\partial n_{\xi\theta}}{\partial \theta} - q_\xi\frac{\partial^2 w}{\partial \xi^2} - q_\theta\left(\frac{\partial v}{\partial \xi} + \frac{\partial^2 w}{\partial \xi \partial \theta}\right) - n_{\xi\theta}\frac{\partial^2 v}{\partial \xi^2} - n_\theta\left(\frac{\partial^2 v}{\partial \xi \partial \theta} - \frac{\partial w}{\partial \xi}\right) = 0 \\ &\frac{\partial n_{\xi\theta}}{\partial \xi} + \frac{\partial n_\theta}{\partial \theta} - q_\xi\left(\frac{\partial v}{\partial \xi} + \frac{\partial^2 w}{\partial \xi \partial \theta}\right) - q_\theta\left(1 + \frac{\partial v}{\partial \theta} + \frac{\partial^2 w}{\partial \theta^2}\right) + n_\xi\frac{\partial^2 v}{\partial \xi^2} + n_{\xi\theta}\left(\frac{\partial^2 v}{\partial \xi \partial \theta} - \frac{\partial w}{\partial \xi}\right) = 0 \\ &\frac{\partial q_\xi}{\partial \xi} + \frac{\partial q_\theta}{\partial \theta} + 2n_{\xi\theta}\left(\frac{\partial v}{\partial \xi} + \frac{\partial^2 w}{\partial \xi \partial \theta}\right) + n_\xi\frac{\partial^2 w}{\partial \xi^2} + n_\theta\left(1 + \frac{\partial v}{\partial \theta} + \frac{\partial^2 w}{\partial \theta^2}\right) = p \end{aligned}\right\} \tag{20}$$

The equations for the equilibrium of moments are

$$\left.\begin{aligned}\frac{\partial m_{\xi\theta}}{\partial\xi}+\frac{\partial m_\theta}{\partial\theta}+m_\xi\frac{\partial^2 v}{\partial\xi^2}+m_{\xi\theta}\left(\frac{\partial^2 v}{\partial\xi\partial\theta}-\frac{\partial w}{\partial\xi}\right)+\beta q_\theta=0\\-\frac{\partial m_\xi}{\partial\xi}-\frac{\partial m_{\xi\theta}}{\partial\theta}+m_{\xi\theta}\frac{\partial^2 v}{\partial\xi^2}+m_\theta\left(\frac{\partial^2 v}{\partial\xi\partial\theta}-\frac{\partial w}{\partial\xi}\right)-\beta q_\xi=0\end{aligned}\right\}\tag{21}$$

where

$$\beta=R^2D_0/D_2\tag{22}$$

β is thus a quantity of the order of $(R/b)^2$, b the thickness of the wall. There are eleven individual equations in the system of Equations (17), (18), (20), and (21). With specified loading p, there are also eleven unknowns, u, v, w, n_ξ, n_θ, $n_{\xi\theta}$, m_ξ, m_θ, $m_{\xi\theta}$, q_ξ and q_θ. The system of equations is thus complete.

Infinite Cylinder Under Uniform Internal Pressure

The simplest special case in the present general problem is the case of very long cylinder under uniform internal pressure. If the rocket cylinder is long in comparison with its diameter, the actual stress system during operation is approximated by this idealized simple case. In this problem of infinitely long uniformly loaded cylinder, the forces n_ξ, n_θ, and moments m_ξ, m_θ are constants independent of ξ and θ. The shearing force $n_{\xi\theta}$ and the twisting moment $m_{\xi\theta}$ vanish. u is proportional to ξ or $\frac{\partial u}{\partial\xi}$ is a constant, say k_1. v vanishes. w is a constant, and is negative in the present coordinate system, say $-k_2$. Then Equations (17), (18), and (20) give

$$\left.\begin{aligned}n_\xi^0&=k_1+\nu k_2-n_T\\n_\theta^0&=\nu k_1+k_2-n_T\\m_\xi^0&=m_\theta^0=m_T\end{aligned}\right\}\tag{23}$$

where the superscript 0 denote the quantities in the present simple stress system. When the temperature distribution and the material properties are specified, Equation (23) gives the strains k_1 and k_2 in terms of the internal pressure p, and the axial load n_ξ. If the axial load is produced by the same internal pressure, then it can easily be shown that

$$n_\xi^0=p^0/2\tag{24}$$

It is of interest to note that the bending moment m_ξ and m_θ are equal to m_T and are independent of the conditions of loading.

By solving Equation (23) for k_1 and k_2, one has

$$\left.\begin{aligned}k_1&=\frac{1}{1-\nu^2}(n_\xi^0-\nu p^0)+\frac{1}{1+\nu}n_T\\k_2&=\frac{1}{1-\nu^2}(p^0-\nu n_\xi^0)+\frac{1}{1+\nu}n_T\end{aligned}\right\}\tag{25}$$

If the design condition is the maximum strain of the material, then Equation (25) gives the criterion directly from the pressure and temperature loading.

Linearized Theory for General Secondary Loading

As stated in the previous section, the actual stress system in a rocket chamber is approximately that of infinitely long cylinder under uniform internal pressure. This stress system can be called the primary stresses. The deviations from the primary stress system are results of bending due to canted nozzle, to end enclosures, to mounting lugs, etc. These additional stresses or the secondary stresses are, however, only a fraction of the primary stresses. Therefore it is justified to consider the second order terms of additional stresses and deformations as small in comparison to the first order terms, and thus negligible. In other words,

$$\left.\begin{array}{lll} u = k_1\xi + u', & v = v', & w = -k_2 + w' \\ n_\xi = n_\xi^0 + n'_\xi, & n_\theta = p^0 + n'_\theta, & n_{\xi\theta} = n'_{\xi\theta} \\ m_\xi = m_T + m'_\xi, & m_\theta = m_T + m'_\theta, & m_{\xi\theta} = m'_{\xi\theta} \\ \text{and} & & \\ q_\xi = q'_\xi, \; q_\theta = q'_\theta, \; p = p^0 + p' & & \end{array}\right\} \tag{26}$$

where k_1 and k_2 are given by Equation (25). The primed quantities are then the secondary deformations and the secondary stresses, they are considered to be small in comparison to the primary deformations and stresses. From Equations (17) and (18), one has the following relations between the deformations and the stresses,

$$\left.\begin{array}{l} n'_\xi = \dfrac{\partial u'}{\partial \xi} - \nu w' + \nu \dfrac{\partial v'}{\partial \theta}, \quad n'_\theta = \nu \dfrac{\partial u'}{\partial \xi} - w' + \dfrac{\partial v'}{\partial \xi}, \quad n'_{\xi\theta} = \dfrac{1-\nu}{2}\left(\dfrac{\partial v'}{\partial \xi} + \dfrac{\partial u'}{\partial \theta}\right) \\ \text{and} \\ m'_\xi = \dfrac{\partial^2 w'}{\partial \xi^2} + \nu \dfrac{\partial^2 w'}{\partial \theta^2} + \nu \dfrac{\partial v'}{\partial \theta}, \\ m'_\theta = \nu \dfrac{\partial^2 w'}{\partial \xi^2} + \dfrac{\partial^2 w'}{\partial \theta^2} + \dfrac{\partial v'}{\partial \theta}, \\ m'_{\xi\theta} = (1-\nu)\left(\dfrac{\partial^2 w'}{\partial \xi \partial \theta} + \dfrac{\partial v'}{\partial \xi}\right) \end{array}\right\} \tag{27}$$

By substituting Equation (26) into the equilibrium equations, Equations (20), and dropping out the second order terms of the primed quantities, a system of linearized equations is obtained. This system can be further reduced by substituting the q'_ξ and q'_θ obtained from the last two equations of bending moment equilibrium into the third equation. The final result is the following system of three equations: The first is the equilibrium of forces in the axial direction; the second is the equilibrium of forces in the circumferential direction; and the third is the equilibrium of forces in the radial direction:

$$\left.\begin{aligned}
&\frac{\partial^2 u'}{\partial \xi^2}+\frac{1-\nu}{2}\frac{\partial^2 u'}{\partial \theta^2}+\frac{1+\nu}{2}\frac{\partial^2 v'}{\partial \xi \partial \theta}-\nu\frac{\partial w'}{\partial \xi}=0\\
&\frac{\partial^2 v'}{\partial \theta^2}+\frac{1-\nu}{2}\frac{\partial^2 v'}{\partial \xi^2}+\frac{1+\nu}{2}\frac{\partial^2 u'}{\partial \xi \partial \theta}-\frac{\partial w'}{\partial \xi}=0\\
&\text{and}\\
&\frac{\partial^4 w'}{\partial \xi^4}+2\frac{\partial^4 w'}{\partial \xi^2 \partial \theta^2}+\frac{\partial^2 w'}{\partial \theta^4}-\beta\left(\nu\frac{\partial w'}{\partial \xi}-w'+\frac{\partial v'}{\partial \theta}\right)=-\beta p'+\beta\left(n_{\xi^0}-\frac{m_T}{\beta}\right)\frac{\partial^2 w'}{\partial \xi^2}+\beta p^0\frac{\partial^2 w'}{\partial \theta^2}
\end{aligned}\right\}\tag{28}$$

These equations have been simplified on the basis that β is a large quantity of the order of the square of the radius-thickness ratio.

In the latter equation, p' is the secondary load imposed on the cylinder expressed as a distributed pressure over the surface of the cylinder, directed radially outward. If the load is a concentrated load, then it has to be expanded into a product of Fourier series and Fourier intergral as done by S. W. Yuan[3] in his treatment of concentrated load on a cold cylindrical shell. Other types of loads can be similarly expanded. Then Equation (28) is a system of three equations for three unknowns u', v', and w'. The forces and moments are related to these displacements by Equation (27).

The problem of general secondary loading as expressed by Equations (27) and (28) is very similar to the problem of general loading on a cold cylindrical shell, and can thus be treated by the known methods developed for this conventional problem. In fact, the only difference between the hot cylinder and the cold cylinder is the appearance of the term m_T in Equation (28). However, even this difference is trivial: The reason is the very large magnitude of β as shown by Equations (6), (7), and (22). In fact, if, as is generally the case, N_T is of the same order of magnitude as N_x^0, then the above-cited equations show that the ratio of m_T/β and n_ξ^0 is at most of the order of b/R, where b is the thickness of the shell. Since the shell is considered to be thin, or $b/R \ll 1$, the terms involving m_T in Equation (28) can be dropped without impairing the accuracy of the present theory. When this is done, then in the system of equations given by Equations (27) and (28), the effects of thermal stresses and variable Young's modulus are not explicit. As far as the nondimensional equations are concerned, the problem of hot cylinder is identical to the problem of cold cylinder, and the basic equations are now essentially the same as that adopted by L. H. Donnell for his study of the stability of thin cylindrical shells[4]. This is the basis of the similarity law discussed in the following section.

Similarity Law for General Loading

If the problem of secondary loading is to be solved analytically, the results of the previous section show that it can be reduced to an equivalent problem of cold cylinder and solved accordingly. However, a more useful application of the equivalence concept lies in the possibility of experimentally determining the stress and the strain on the equivalent cold cylinder and then using the similarity law to determine the stress and the strain in the hot cylinder. There are mainly two advantages of this semi-experimental approach: (a) The experiments on a cold cylinder can be done more easily and much more accurately than possible on a hot cylinder. The test period can be as long as desired and not limited to the short burning time of the rocket. (b) The stresses induced by mounting lugs, etc., are very difficult to approximate by simple load systems amenable to theoretical calculations. For instance, the loads from a mounting lug are not really a concentrated force and a concentrated moment. To take them as a concentrated force and a concentrated moment would grossly overestimate the actual stress. Such difficulties disappear if the loading is done experimentally.

With such experimental stress determination in mind, it will be convenient to have the hot cylinder and the equivalent cold cylinder of same general sizes. Thus the radius R and the length L will be the same for both cylinders. In other that the nondimensional differential equations, Equation (28), be the same for the hot cylinder as for the cold cylinder, the parameters in these differential equations must be the same. That is, if a quantity of the cold cylinder is denoted by a bar over the quantity

$$\frac{1}{R^2}\beta = \frac{D_0}{D_2} = \frac{\overline{D}_0}{\overline{D}_2} \tag{29}$$

$$n_\xi^0 = \frac{N_x^0}{D_0} = \frac{\overline{N}_x^0}{\overline{D}_0}, \quad p^0 = \frac{P^0}{D_0} = \frac{\overline{P}^0}{\overline{D}_0} \tag{30}$$

The condition of Equation (29) can be satisfied by making the thickness $\overline{b}$ of the cold cylinder smaller than the thickness b of the hot cylinder. This is of course to be expected, since Young's modulus of the hot material is smaller and hence material is "softer" than the cold material. When the thickness $\overline{b}$ is determined from b by using Equation (29), $\overline{D}_0$ can be computed. Then Equation (30) gives the internal pressure $\overline{P}^0$ and the axial load $\overline{N}_x^0$ from the specified P^0 and N_x^0 for the hot cylinder. These steps then fix the geometry of the cold cylinder and the primary system of loads.

For the additional secondary loads, the fact that Equations (27) and (28) are linear equations can be utilized to introduce an added freedom in specifying the loads. Linear relations are not altered by multiplying the variable by a constant. Therefore for additional loads and additional displacements, the nondimensional quantities for the cold cylinder and the nondimensional quantities for the hot cylinder need not be identical, but differ by a factor ε. Thus

$$(\overline{u}', \overline{v}', \overline{w}') = \varepsilon(u', v', w')$$

and

$$(\overline{n}'_\xi, \overline{n}'_\theta, \overline{n}'_{\xi\theta}; \overline{m}'_\xi, \overline{m}'_\theta, \overline{m}'_{\xi\theta}) = \varepsilon(n'_\xi, n'_\theta, n'_{\xi\theta}; m'_\xi, m'_{\theta\xi}, m'_\theta) \tag{31}$$

Then

$$\overline{p}' = \varepsilon p' \tag{32}$$

But the nondimensional pressure loading p is related to the actual pressure loading by Equation

(19). Therefore the secondary pressure loadings $\bar{P}'$ for the cold cylinder and the secondary pressure loading P' for the hot cylinder are related through

$$\bar{P}' = \left(\frac{\bar{D}_0}{D_0}\varepsilon\right)P' \tag{33}$$

The ratio of the pressure load is then $\bar{D}_0\varepsilon/D_0$. Since the radius R and the length L of both cylinders are the same, other types of loads such as concentrated force, or moment for the cold cylinder and for the hot cylinder must also bear the same ratio. Needless to say, the loads for the cold cylinder must be applied at corresponding points for loads in the hot cylinder.

The additional forces N_x', $N_{x\theta}'$, the additional shear Q_x', and the additional moments N_x', $M_{x\theta}'$ at the ends of the cylinder are controlled by Equations (15), (16), and (19). It is easily seen that because of Equation (29), the ratio of these quantities for the cold cylinder and the hot cylinder is again $\bar{D}_0\varepsilon/D_0$.

Therefore, knowing the load system on the hot cylinder, one can find the corresponding load system for the cold cylinder. The factor ε for the secondary loads can be chosen at convenience of the experimenter. For instance, ε might be so chosen as to make the ratio $\bar{D}_0\varepsilon/D_0$ equal to unity. Then the secondary load system for the cold cylinder is exactly the same as the hot cylinder. When the proper load for the cold cylinder is selected and the corresponding strain on the cold cylinder determined by strain gages, the inverse equivalence problem is then to find the strain in the hot cylinder from the test data on cold cylinder.

Take for instance, the axial strain $e_x(z)$. For the cold cylinder, according to Equations (1) and (25),

$$\bar{e}_x(\bar{z}) = k_1 - \frac{1}{1+\nu}n_T + \left(\frac{\partial \bar{u}'}{\partial \xi} - \frac{\bar{z}}{R}\frac{\partial^2 \bar{w}'}{\partial \xi^2}\right) \tag{34}$$

where $\bar{z}$ is the value of z measured from the median surface of the cold cylinder, midway between the boundary surfaces. Now let $\bar{e}_x$ be the average of the measured axial strains on the outer surface and on the inner surface of the cold cylinder, and let $\Delta\bar{e}_x$ be the difference of the measured axial strains on the outer surface and on the inner surface of the cold cylinder. Then from Equation (34)

$$\bar{e}_x = k_1 - \frac{1}{1+\nu}n_T + \frac{\partial \bar{u}'}{\partial \xi} = k_1 - \frac{1}{1+\nu}n_T + \varepsilon\frac{\partial u'}{\partial \xi} \tag{35}$$

and

$$\Delta\bar{e}_x = \frac{\bar{b}}{R}\frac{\partial^2 \bar{w}'}{\partial \xi^2} = \frac{\bar{b}}{R}\varepsilon\frac{\partial^2 w'}{\partial \xi^2}$$

For the hot cylinder, the axial strain is given by

$$e_x(z) = k_1 + \frac{\partial u'}{\partial \xi} - \frac{z}{R}\frac{\partial^2 w'}{\partial \xi^2} \tag{36}$$

By eliminating $\frac{\partial u'}{\partial \xi}$ and $\frac{\partial^2 w'}{\partial \xi^2}$ from Equations (35) and (36), one has

$$e_x(z) = \left(1 - \frac{1}{\varepsilon}\right)k_1 + \frac{1}{(1+\nu)\varepsilon}n_T + \frac{1}{\varepsilon}\left(\bar{e}_x - \frac{z}{\bar{b}}\Delta\bar{e}_x\right) \tag{37}$$

The value of z is measured radially inward from the median surface of the hot cylinder and thus is larger in magnitude for the inside surface than for the outside surface.

Similarly,

$$e_\theta(z) = \left(1 - \frac{1}{\varepsilon}\right)k_2 + \frac{1}{(1+\nu)\varepsilon}n_T + \frac{1}{\varepsilon}\left(\bar{e}_\theta - \frac{z}{\bar{b}}\Delta\bar{e}_\theta\right) \tag{38}$$

and

$$\gamma_{x\theta}(z) = \frac{1}{\varepsilon}\left(\bar{\gamma}_{x\theta} - \frac{z}{\bar{b}}\Delta\bar{\gamma}_{x\theta}\right) \tag{39}$$

where $\bar{e}_\theta$ is the average of the measured circumferential strain on the outer surface and the inner surface of the cold cylinder, $\Delta\bar{e}_\theta$ is the difference of these strains; $\bar{\gamma}_{x\theta}$ and $\Delta\bar{\gamma}_{x\theta}$ are the corresponding quantities for the shearing strain. In Equations (37) and (38), k_1 and k_2, and n_T are the primary strains computed from Equations (25) and (8). Therefore these equations allow the calculation of the strains in the hot cylinder from test results from cold cylinder, and thus complete the desired similarity law.

For a stress analyst, the next step is perhaps the calculation of the principle strains at each value of z in the shell, and examine whether the larger of these principal strains exceeds the design limit of the material at the temperature prevailing at that point.

Example of Dimensioning the Equivalent Cold Cylinder

As an example of the procedure outline in the previous section, the data given by Noland[1] will be used to find the equivalent cold cylinder. The temperature distribution in the wall is taken from Fig. 2 of that paper and is reproduced as Fig. 1 here. The material is assumed to be 19-9 DL, and the variation of Young's modulus with temperature is plotted in Fig. 2, again using Noland's data. First, the position of the median surface will be determined, using Equation (10). This is found to be $0.588\,b$ from the inside surface. Next, by taking the cold cylinder to be at 100°F, the ratio of the thicknesses of the hot cylinder and the equivalent cold cylinder is computed by using Equations (6), (7), and (29). It is found that

$$\bar{b} = 0.936\,b$$

Thus the equivalent cold cylinder of the same material is 93.6 per cent as thick as the hot cylinder.

The ratio of the loads on the cold cylinder and the hot cylinder is controlled by $\bar{D}_0/D_0$. This is computed as

$$\bar{D}_0/D_0 = 1.29$$

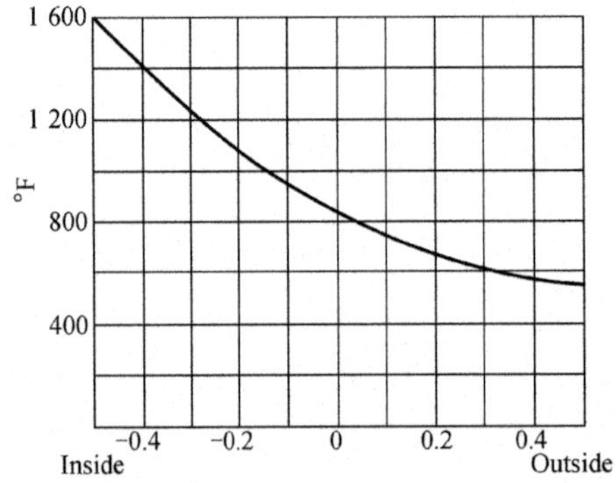

Fig. 1 Temperature distribution in the wall

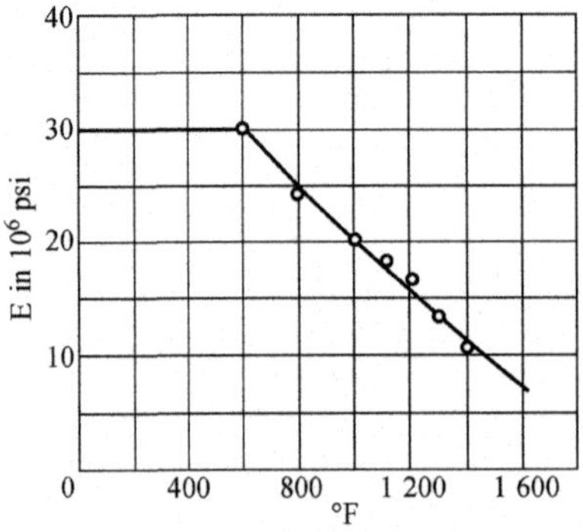

Fig. 2 Young's modulus as a function of temperature

By taking $\alpha = 10^{-5}$ /°F, and $b = 0.095$ in., the values of thermal stresses are

$$N_T = 22\,400 \text{ lb/in} \qquad n_T = 0.958 \times 10^{-2}$$

$$M_T = 192 \text{ lb.} \qquad m_T = -0.124$$

If the radius of the cylinder is 2.25 in, and if the internal pressure P^0 is 1 500 psi, then

$$p^0 = 1.44 \times 10^{-3}$$

If the axial tension N_x^0 is due to the same internal pressure, then

$$n_\xi^0 = \frac{p^0}{2} = 0.72 \times 10^{-3}$$

Now

$$\beta = 12\left(\frac{R}{b}\right)^2$$

Therefor the ratio of m_T/β and n_ξ^0 is

$$m_T/(\beta n_\xi^0) = -0.022\,4$$

This is indeed smaller than the thickness-radius ratio $b/R = 0.042\,2$. Hence the surmise that m_T/β is negligible against n_ξ^0 is now justified by numerical calculation.

Junction Stress Between Cylinder and Head

The previous formulation of the similarity law for stresses in the hot cylinder and the cold cylinder is based upon the assumption that the secondary load system is specified. This is not true for the junction stresses induced by fitting, say, a hemispherical head to the cylinder. Such stresses are determined by the equality of deformations of the head and the cylinder at the junction. If the semi-spherical shell has the same thickness as the cylindrical shell, then the temperature distribution in the spherical shell will be the same as in the cylindrical shell. An analysis shows that the "similar" cold test specimen can be made also of uniform thickness $\bar{b}$, determined by Equation (29). However, the similarity of both the primary loading and secondary loading now requires an additional restriction: ε must now be unity. When these conditions are fulfilled, the similarity of junction stresses in the hot cylinder and the cold cylinder will be assured, and the relations previously developed for computing the stresses in the hot cylinder from test data on the cold cylinder remain valid.

Ring Stiffener Around Cylinder

To strengthen the thin cylindrical shell against concentrated loads from the mounting lugs, a ring stiffener is often attached to the outside surface. It remains cold. Young's modulus of the ring material thus is that of cold material. To determine the dimensions of a "similar" ring for the cold cylinder, the conditions to be satisfied are those of Equations (31). In other words, for a ratio ε of deformation of the ring on the cold cylinder to that on the hot cylinder, the force required must bear the ratio $\varepsilon\bar{D}_0/D_0$. This means that the ratio of the stiffness of the ring for the cold cylinder to that on the hot cylinder is $\bar{D}_0/D_0$. If the ring is rectangular in section, the correct ratio of the width of the two rings for complete similarity is then $\bar{D}_0/D_0$. The ring on the cold cylinder is thus wider than on the hot cylinder. When this condition on the ring dimension is satisfied, the simple similarity relations for the stresses in the shell are again correct.

References

[1] Noland R L. Strengths of Several Steels for Rocket Chambers Subjected to High Rates of Heating. Journal of American Rocket Society, 1951, 21(6): 154-162.

[2] Timoshenko S. Theory of Plates and Shells. McGraw-Hill Book Company, New York, 1940: 389.

[3] Yuan S W. The Cylindrical Shells Subjected to Concentrated Loads. Quarterly of Applied Mathematics, 1946,4: 13-26.

[4] Donnell L H. Stability of Thin-Walled Tubes Under Torsion. NACA Technical Report 473, 1933.

On the Determination of Rotational Line Half-Widths of Diatomic Molecules*

S. S. Penner and H. S. Tsien
(Guggenheim Jet Propulsion Center, California Institute of Technology, Pasadena, California)

A simple closed-form expression is obtained for the fractional intensity of radiation absorbed by vibration-rotation bands with collision-broadened spectral lines. The resulting expressions greatly reduced the labor involved in obtaining apparent rotational half-widths from experimental measurements.

I. Introduction

At room temperature the integrated absorption of rotational lines is given, in adequate approximation, by the relations[1,2]

$$S_{j\to j-1}{}^{0\to1} = \alpha_F(\omega_{j\to j-1}{}^{0\to1}/\omega^*)[(j-\lambda)(j+\lambda)/j]\times\exp[-E(0,j)/(kT)]Q^{-1} \quad (1a)$$

$$S_{j-1\to j}{}^{0\to1} = \alpha_F(\omega_{j-1\to j}{}^{0\to1}/\omega^*)[(j-\lambda)(j+\lambda)/j]\times\exp[-E(0,j-1)/(kT)]Q^{-1} \quad (1b)$$

$$S_{j\to j}{}^{0\to1} = \alpha_F(\omega_{j\to j}{}^{0\to1}/\omega^*)[\lambda^2(2j+1)/j(j+1)]\times\exp[-E(0,j)/(kT)]Q^{-1} \quad (1c)$$

$$S_{j\to j'}{}^{0\to2}/S_{j\to j'}{}^{0\to1} = \alpha_0/\alpha_F \quad (2)$$

Here α_F and α_0 represent, respectively, the integrated absorption of the fundamental and first overtone; $\omega_{j\to j'}{}^{0\to1}$ is the wave number corresponding to the rotational transition $j\to j'$ and the vibrational transition $0\to1$; ω^* = wave number for the transition $j=0\to j=0$, $n=0\to n=1$; λ = quantum number measuring the component of electronic angular momentum about the internuclear axis; $E(0, j)$ = energy of the zeroth vibrational and j'th rotational level; k = Boltzmann constant; T = absolute temperature; Q = complete partition function.

In an absorption experiment with a source of spectral radiant intensity $R_{s\omega}$, the fractional absorbed intensity $R_{a\omega}/R_{s\omega}$, integrated over an entire vibration-rotation band of width $\Delta\omega_B$, is

$$\int_{\Delta\omega_B}(R_{a\omega}/R_{s\omega})d\omega = \int_{\Delta\omega_B}[1-\exp(-P_\omega X)]d\omega \quad (3)$$

where P_ω represents the spectral absorption coefficient and X is the optical density. For collision-

---

Received January 21, 1952.
The Journal of Chemical Physics, vol. 20, pp. 827 - 828, 1952.
* Supported, in part, by the ONR under Contract Nonr-220(03), NR 015 210.

broadened spectral lines which do not overlap[3~5]

$$\int_{\Delta\omega_B} (R_{a\omega}/R_{s\omega})\mathrm{d}\omega = \sum_j 2\pi b_{j\to j'}{}^{0\to 1}[f(x_{j\to j'}{}^{0\to 1})] \tag{4}$$

where

$$2\pi b_{j\to j'}{}^{0\to 1}[f(x_{j\to j'}{}^{0\to 1})] \simeq 2(S_{j\to j'}{}^{0\to 1} b_{j\to j'}{}^{0\to 1} X)^{\frac{1}{2}} \text{ for large values of } x_{j\to j'}{}^{0\to 1} \tag{5a}$$

and

$$2\pi b_{j\to j'}{}^{0\to 1}[f(x_{j\to j'}{}^{0\to 1})] \simeq S_{j\to j'}{}^{0\to 1} X \text{ for small values of } x_{j\to j'}{}^{0\to 1} \tag{5b}$$

The symbol $b_{j\to j'}{}^{0\to 1}$ denotes one-half of the spectral half-width for the indicated transition. Equations (5a) and (5b) represent asymptotic limiting forms in which the right-hand sides exceed the value of $2\pi b_{j\to j'}{}^{0\to 1} \times [f(x_{j\to j'}{}^{0\to 1})]$ for every value of $x_{j\to j'}{}^{0\to 1}$. Equation (5a) constitutes a better approximation for $x_{j\to j'}{}^{0\to 1} > (2/\pi)$, whereas Eq. (5b) is preferable for $x_{j\to j'}{}^{0\to 1} < (2/\pi)$. Since, at room temperature, $\alpha_F = \sum_j S_{j\to j'}{}^{0\to 1}$, it is evident from Eqs. (4) and (5b) that

$$\int_{\Delta\omega_B} (R_{a\omega}/R_{s\omega})\mathrm{d}\omega = \alpha_F X, \text{ for small values of } x_{j\to j'}{}^{0\to 1} \tag{6}$$

Equation (6) is an illustration of the well-known result that accurate measurements of α_F can be obtained for very small values of the optical density. It is evident that under these conditions the experimental results are independent of the line width.

Ⅱ. Calculation of Rotational Half-Widths from Experimental Data

In order to obtain a useful procedure for the calculation of rotational half-widths, it is necessary to assume that $b_{j\to j'}{}^{0\to 1} = b_F$, $b_{j\to j'}{}^{0\to 2} = b_0$ and to evaluate the right-hand side of Eq. (4) by using Eqs. (5a) and (1a) to (2). We shall outline the procedure for the fundamental vibration-rotation hand.

Equations (1a) to (1c), (4), and (5a) lead to the relation

$$\begin{aligned}\int_{\Delta\omega_F} (R_{a\omega}/R_{s\omega})\mathrm{d}\omega = {} & 2\{(\alpha_F b_F X/Q)\exp[E(0,0)/(kT)]\}^{\frac{1}{2}} \times \\ & \sum_j [(j-\lambda)(j+\lambda)/j]^{\frac{1}{2}}\{(\omega_{j\to j-1}{}^{0\to 1}/\omega^*)^{\frac{1}{2}} \times \\ & \exp\{-[E(0,j)-E(0,0)]/2kT\} + (\omega_{j-1\to j}{}^{0\to 1}/\omega^*)^{\frac{1}{2}} \times \\ & \exp\{-[E(0,j-1)-E(0,0)]/(2kT)\} + \\ & \lambda[(2j+1)/j(j+1)]^{\frac{1}{2}}(\omega_{j\to j}{}^{0\to 1}/\omega^*)^{\frac{1}{2}} \times \\ & \exp\{-[E(0,j)-E(0,0)]/(2kT)\}\end{aligned} \tag{7}$$

Equation (7) is the basic equation from which useful results can be obtained for two special cases.

A. Diatomic Molecules Without Q-Branch ($\lambda=0$)

For diatomic molecules without Q-branch ($\lambda = 0$) Eq. (7) becomes

$$\int_{\Delta\omega_F}(R_{a\omega}/R_{s\omega})d\omega = 2\{(\alpha_F b_F X/Q)\exp[E(0,0)/(kT)]\}^{\frac{1}{2}}\times$$
$$\sum_j j^{\frac{1}{2}}\{(\omega_{j\to j-1}{}^{0\to 1}/\omega^*)^{\frac{1}{2}}\times$$
$$\exp\{-[E(0,j)-E(0,0)]/(2kT)\}+(\omega_{j-1\to j}{}^{0\to 1}/\omega^*)^{\frac{1}{2}}\times$$
$$\exp\{-[E(0,j-1)-E(0,0)]/(2kT)\} \quad (7a)$$

Utilizing customary spectroscopic notation[6] and expanding the various terms in Eq. (7a) in powers of γj, it is found that, correct to powers of $\gamma^2 j^2$ [7],

$$\int_{\Delta\omega_F}(R_{a\omega}/R_{s\omega})d\omega = 2\{(\alpha_F b_F X/Q)\exp[E(0,0)/(kT)]\}^{\frac{1}{2}}\times$$
$$\sum_j 2j^{\frac{1}{2}}\exp[-(\gamma u j^2/2)][1+\gamma^2 j^2 h(\gamma,\delta,u)]$$

where

$$h(\gamma,\delta,u) = (u^2/8)+(u/2)-[(\delta/\gamma)+1]/2$$

The sum over j can be evaluated readily by use of the Euler-Maclaurin summation formula[7]. In this manner it is found that, in good approximation,

$$\int_{\Delta\omega_F}(R_{a\omega}/R_{s\omega})d\omega = 4.1\,(2)(\alpha_F b_F X)^{\frac{1}{2}}(\gamma u)^{-\frac{1}{4}} \quad (8)$$

where a suitable expansion has been introduced[6] for $\{Q\exp[E(0,0)/(kT)]\}^{\frac{1}{2}}$. The error in Eq. (8) is estimated to be of the order of $(\gamma u)^{\frac{3}{4}}$. It is evident that Eq. (8) holds for the first overtone with α_F and b_F replaced by α_0 and b_0, respectively.

B. Diatomic Molecules with Q-Branch ($\lambda\neq 0$)

In the general case $\lambda \neq 0$ we must proceed from Eq. (7) directly. After suitable expansion of various quantities appearing in Eq. (7) it is found that

$$\int_{\Delta\omega_F}(R_{a\omega}/R_{s\omega})d\omega = 2(\alpha_F b_F X)^{\frac{1}{2}}(\gamma u)^{+\frac{1}{2}}\times$$
$$\sum_j\{2[(j-\lambda)(j+\lambda)/j]^{\frac{1}{2}}\times$$
$$[\exp(-\gamma u j^2/2)][1+\gamma^2 j^2 h(\gamma,\delta,u)]\}+$$
$$\{\lambda[(2j+1)/j(j+1)]^{\frac{1}{2}}[\exp(-\gamma u j^2/2)]\times$$
$$[1-(u\gamma j/2)+(u^2\gamma^2 j^2/8)]\} \quad (7b)$$

Since λ is generally a small integer and since the contributions made to the value of the sum by rotational lines with small values of j are relatively small, we may approximate $[(j-\lambda)(j+\lambda)/j]^{\frac{1}{2}}$ by $j^{\frac{1}{2}}$, whence it follows that the first brace of the sum in Eq. (7b) again leads to the result given in Eq. (8). Thus it is only necessary to estimate the total contributions made by the Q-branch. An adequate approximation is obtained by utilizing again the Euler-Maclaurin summation formula. The result is

$$\int_{\Delta\omega_F}(R_{a\omega}/R_{s\omega})\,d\omega = 4.1\,(2)(\alpha_F b_F X)^{\frac{1}{2}}(\gamma u)^{-\frac{1}{4}}[1+1.4\,(8)\lambda(\gamma u)^{\frac{1}{2}}] \tag{9}$$

The error in Eq. (9) is also of the order of $(\gamma u)^{\frac{3}{4}}$. Equation (9) applies to the first overtone if α_F and b_F are replaced by α_0 and b_0, respectively.

Because of very close spacing of the rotational lines in the Q-branch, the right-hand side of Eq. (9) will generally be too large. For this reason it may be indicated to use Eq. (9) in the form

$$\int_{\Delta\omega_F}(R_{a\omega}/R_{s\omega})\,d\omega = 4.1\,(2)(\alpha_F b_F X)^{\frac{1}{2}}(\gamma u)^{-\frac{1}{4}}[1+1.4\,(8)\lambda\beta(\gamma u)^{\frac{1}{2}}] \tag{9a}$$

where β is an empirically determined correction factor which is less than or equal to unity. The contribution of the Q-branch is usually so small that it may be neglected in the treatment of experimental data.

In conclusion, we wish to emphasize again that the results which we have obtained are useful only if the Lorentz dispersion formula is adequate to describe the experimental measurements. This condition seems to apply to simple diatomic molecules such as CO and NO but not to HCl and HBr[3,8]. If the empirical data obtained cannot be correlated quantitatively by expressions of the form given in Eqs. (8) and (9), then it is generally safe to assume that an oversimplified description of spectral line-shape has been used[9].

References

[1] Penner S S. J. Chem. Phys. 19, 272, 1434(1951).

[2] Crawford, Jr. B L, Dinsmore HL. J. Chem. Phys. 18, 983, 1682(1950).

[3] Penner S S, Weber D. J. Chem. Phys. 19, 1351, 1361(1951).

[4] Ladenburg R, Reiche F. Ann. Physik 42, 181(1913).

[5] Elsasser W M. Harvard Meteorological Studies No. 6. Blue Hill Observatory, Milton, Massachusetts, 1942.

[6] Mayer J E, Mayer M G. Statistical Mechanics. John Wiley and Sons, Inc. New York, 1940.

[7] Penner, Ostrander, Tsien. For details, describing a similar evaluation. J. Appl. Phys. 1952, 23: 256.

[8] Penner S S, Weber D. paper presented before the Symposium on Molecular Structure and Spectroscopy. Ohio State University, Columbus, Ohio, 1951.

[9] In this connection reference should be made to the extensive literature on pressure broadening of spectral lines, for example,

Lorentz H A. Proc. Amst. Akad. Sci. 8, 59(1906);

Lenz W. Physik Z. 80, 423(1933);

Weisskopf V F. Physik Z. 34, 1(1933);

Lindholm E. Arkiv. Mat. Astron. Fysik. 32, 17(1945);

van Vleck J H. Weisskopf V F. Revs. Modern Phys. 17, 227(1945);

van Vleck J H. Margenau H. Phys. Rev. 76, 1211(1949);

Anderson P W. Phys. Rev. 76, 647(1949);

Margenau H. Phys. Rev. 82, 156(1951).

Automatic Navigation of a Long Range Rocket Vehicle

H. S. Tsien①, T. C. Adamson② and E. L. Knuth②
(Daniel and Florence Guggenheim Jet Propulsion Center, California Institute of Technology, Pasadena, Calif.)

The flight of a rocket vehicle in the equatorial plane of a rotating earth is considered with possible disturbances in the atmosphere due to changes in density, in temperature, and in wind speed. These atmospheric disturbances together with possible deviations in weight and in moment of inertia of the vehicle tend to change the flight path away from the normal flight path. The paper gives the condition for the proper cut-off time for the rocket power, and the proper corrections in the elevator angle so that the vehicle will land at the chosen destination in spite of such disturbances. A scheme of tracking and automatic navigation involving a high-speed computer and elevator servo is suggested for this purpose.

The behavior of a vehicle flying through air is closely dependent upon the aerodynamic forces acting upon the vehicle. If, during one period of oscillation of the vehicle, there is appreciable variation of the response of aerodynamic forces to the attitude of the vehicle through variations in speed, in aerodynamic coefficients, in air density, etc., then the behavior of the disturbed flight path cannot be described by a linear differential equation of constant coefficients. In fact, the basic differential equation actually has coefficients that are specified functions of time. A very simple example of such motion is that of an artillery rocket during burning of the propellant grain. As shown by J. B. Rosser, R. R. Newton, and G. L. Gross[1], the basic differential equation for this particular case can be written as Bessel's differential equation for the order 1/2. The general character of the solutions of such differential equations is quite different from the character of solutions of differential equations with constant coefficients. For instance, while for equations with constant coefficients the stability of solutions for the homogeneous equation is generally sufficient to insure the stability of solutions with reasonable forcing functions, this simple state of affairs no longer prevails for equations with variable coefficients. The present theory of control and stability is built almost

---

Received April 28, 1952.

Journal of the American Rocket Society, vol. 22, pp. 192 - 199, 1952.

① Robert H. Goddard Professor of Jet Propulsion.

② Daniel and Florence Guggenheim Jet Propulsion Fellows. The computations involved in this paper were carried out by R. C. Evans, Daniel and Florence Guggenheim Jet Propulsion Fellow, and F. W. Hartwig, Captain U. S. A. F., in addition to the junior authors.

exclusively upon the theory of differential equations with constant coefficients. Therefore to study the disturbed motion of rockets, new methods have to be used.

R. Drenick in a recent paper[2] demonstrated the usefulness of ballistic disturbance theory in solving the control and guidance problem of ballistic trajectories described by equations with time-varying coefficients. His theory is based upon the method of adjoint functions, first introduced by G. A. Bliss[3] during World War I. The purpose of the present paper is to make Drenick's theory more complete and definite and to apply it to the problem of automatic navigation of a long-range winged rocket vehicle. A system of control involving a fast computer is suggested, whereby the range errors due to changing atmospheric conditions and deviation from the standard weight of the vehicle are automatically corrected. The main objective here is not, however, to give the final design of such an automatic navigation system, but rather to show the power of the ballistic disturbance theory for solving such problems and the various elements necessary for such a navigational system.

Equations of Motion

In order not to complicate matters, the vehicle is assumed to move in the equatorial plane of the rotating earth (Fig. 1). The planar motion is possible due to the absence of cross Coriolis force in the equatorial plane. The co-ordinate system is fixed with respect to the rotating earth, i. e., actually it rotates with the angular velocity Ω, the speed of earth rotation. The value of Ω is as follows:

$$\Omega = 7.292\,1 \times 10^{-5}\,\text{rad/sec} \tag{1}$$

In the equatorial plane, the position of the vehicle at any time instant t is specified by the radius r and angle θ from the starting point of the vehicle. r_0 is the mean earth radius, its value is

$$r_0 = 20.88 \times 10^6\ \text{ft} \tag{2}$$

If g is the gravitational constant at the surface of the earth without the centrifugal force due to

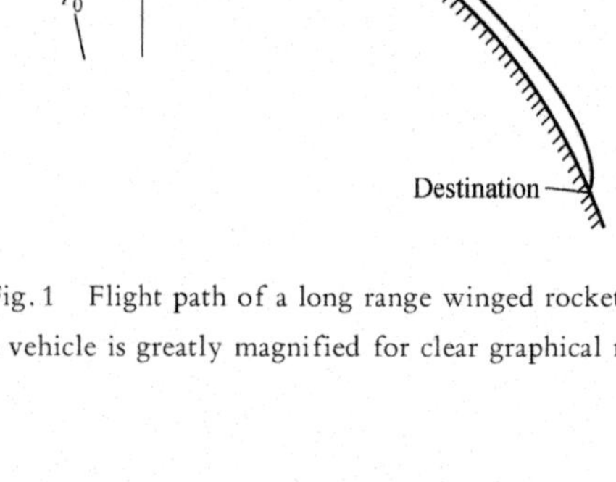

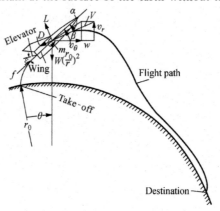

Fig. 1 Flight path of a long range winged rocket
(The size of the vehicle is greatly magnified for clear graphical representation.)

rotation, then

$$g = 32.2577 \text{ ft/sec}^2 \tag{3}$$

Let R and Θ be the force per unit mass acting on the vehicle in the radial and the circumferential directions, respectively. Then the equations of motion of the center of gravity of the vehicle are

$$\left.\begin{aligned} \frac{dr}{dt} &= \dot{r} \\ \frac{d\theta}{dt} &= \dot{\theta} \\ \frac{d\dot{r}}{dt} &= R + r(\dot{\theta} \pm \Omega)^2 - g\left(\frac{r_0}{r}\right)^2 \\ r\frac{d\dot{\theta}}{dt} &= \Theta - 2\dot{r}(\dot{\theta} \pm \Omega) \end{aligned}\right\} \tag{4}$$

where the plus sign in the second terms to the right will be valid for flights toward the east, and the minus sign for flights toward the west.

The forces acting on the center of gravity of the vehicle are the thrust f, the lift L, and the drag D (Fig. 1). Let W be the instantaneous weight of the vehicle with respect to g, and V the magnitude of air velocity relative to the vehicle. Then it is convenient to introduce the parameters Ψ, Λ, and Δ as follows:

$$\Psi = \frac{fg}{W}, \quad \Lambda = \frac{Lg}{WV}, \quad \Delta = \frac{Dg}{WV} \tag{5}$$

It will be assumed that the natural wind velocity w is in the horizontal direction, positive if it is a head wind. w is considered as a function of altitude r. If v_r is the radial velocity and v_θ the circumferential velocity, i.e.,

$$\left.\begin{aligned} v_r &= \dot{r} \\ v_\theta &= r\dot{\theta} \end{aligned}\right\} \tag{6}$$

then relative air velocity V is computed as

$$V^2 = v_r^2 + (v_\theta + w)^2 \tag{7}$$

If β is the angle between the thrust line and the horizontal direction, then the components of forces R and Θ per unit mass are

$$\left.\begin{aligned} R &= \Psi\sin\beta + (v_\theta + w)\Lambda - v_r\Delta \\ \Theta &= \Psi\cos\beta - v_r\Lambda - (v_\theta + w)\Delta \end{aligned}\right\} \tag{8}$$

If N is the moment of forces about the center of gravity, divided by the moment of inertia of the vehicle, the equation for the angular acceleration is

$$\frac{d\dot{\beta}}{dt} = \frac{d\dot{\theta}}{dt} + N \tag{9}$$

To completely specify the motion of the vehicle, the lift L, the drag D, and the moment m about the center of gravity have to be given as functions of time. According to the aerodynamic convention, the L and D will be expressed in terms of the lift coefficient C_L and the drag coefficient C_D as follows:

$$\left.\begin{aligned} L &= \frac{1}{2}\rho V^2 S C_L \\ D &= \frac{1}{2}\rho V^2 S C_D \end{aligned}\right\} \tag{10}$$

where ρ is the air density, a function of the altitude r, and S is a fixed reference area, say the wing area of the vehicle. In the present problem, since the motion of the vehicle is restricted to the equatorial plane, the attitude of the vehicle essential for aerodynamic calculations is determined by the angle of attack[①] α, i.e., the angle between the thrust line, or body axis and the relative air velocity vector (Fig. 1). The control on the motion of the vehicle is affected, however, through the elevator angle ε. The parameters which will affect C_L and C_D are thus α and ε. In addition, the aerodynamic coefficients are functions of the Reynolds number Re and the Mach number M. Thus

$$\left.\begin{aligned} C_L &= C_L(\alpha, \varepsilon, M, Re) \\ C_D &= C_D(\alpha, \varepsilon, M, Re) \end{aligned}\right\} \tag{11}$$

It will be assumed that the thrust line passes through the center of gravity of the vehicle; thus the thrust gives no moment. Since the angular motion of the vehicle during the powered flight is expected to be slow, the jet damping moment of the rocket is negligible. The only moment acting on the vehicle is then the aerodynamic moment m. m can be also expressed as a coefficient C_M as follows:

$$m = \frac{1}{2}\rho V^2 S l C_M \tag{12}$$

where l is a reference length. The moment coefficient C_M is again a function of the four parameters α, ε, M, and Re, or

$$C_M = C_M(\alpha, \varepsilon, M, Re) \tag{13}$$

If I is the moment of inertia of the vehicle, then the magnitude of N in Equation (9) is

$$N = m/I \tag{14}$$

With the rotations defined above, the system of equations of motion is as follows:

---

① Drenick (Ref. 2) seems to take the sense of the angle of attack α opposite to the convention of aerodynamicists.

$$\left.\begin{aligned}
\frac{dr}{dt} &= v_r \\
\frac{d\theta}{dt} &= v_\theta / r \\
\frac{d\beta}{dt} &= \dot{\beta} \\
\frac{dv_r}{dt} &= \Psi \sin\beta + (v_\theta + w)\Lambda - v_r\Delta + r\left(\frac{v_\theta}{r} \pm \Omega\right)^2 - g\left(\frac{r_0}{r}\right)^2 = F \\
\frac{dv_\theta}{dt} &= \Psi \cos\beta - v_r\Lambda - (v_\theta + w)\Delta - 2v_r\left(\frac{v_\theta}{r} + \Omega\right) + \frac{v_\theta v_r}{r} = G \\
\frac{d\dot{\beta}}{dt} &= \frac{1}{r}\{\Psi\cos\beta - v_r\Lambda - (v_\theta + w)\Delta\} - 2\frac{v_r}{r}\left(\frac{v_\theta}{r} + \Omega\right) + N = H
\end{aligned}\right\} \qquad (15)$$

This system of equations is a set of first order equations for the six unknowns r, θ, β, v_r, v_θ, and $\dot{\beta}$. To solve it, the six initial values at the start, $t = 0$, for the unknowns must be specified. In addition, the thrust f, the weight W, the moment of inertia I must be given for every time instant. To determine the aerodynamic forces, the elevator angle ε must be specified as a function of time. The properties of the atmosphere must be known; i.e., wind velocity w, air density ρ, air viscosity and air velocity of sound must be given as functions of the altitude r. The angle of attack α of the vehicle cannot be specified; it is a quantity to be computed from the angle β and the relative air velocity vector V.

Normal Flight Path

Let the properties of the atmosphere be standardized and known as the properties of the normal atmosphere. The average characteristics of the vehicle and its power plant can be taken to be representative. Then if the elevator angle ε is given as a function of time, the flight path of the vehicle is determined and can be calculated by integrating the system of Equation (15). The actual execution of this computation will be probably done on an electro-mechanical computer. This flight path of a standardized vehicle in normal atmosphere can be called the normal flight path.

The dominating feature of the normal flight path is its range. This range is the distance between the take-off point and the landing point. The problem of navigation is then to calculate the proper time for cut-off of the rocket and the proper variation of the elevator angle during flight so that the range is that desired. This problem of navigation for the standardized vehicle in normal atmosphere can be solved mathematically before the actual take-off of the vehicle, since all information for the normal flight path is known or specified beforehand.

Disturbance Equations

Natural atmospheric characteristics do not, of course, coincide with those assumed for the atmosphere. The wind velocity at each altitude changes according to the weather conditions;

the temperature T is also a varying quantity. Therefore one should expect variations from the normal flight path due to changes in atmospheric conditions. The actual vehicle also may be somewhat different from the standardized vehicle, in weight, rocket performance, etc. Therefore actual flight path will be different from the normal flight path if the same elevator angle programming is used. The problem of navigation of an actual vehicle is that of correcting the elevator angle programming so that the range of the actual flight will be the same as the normal flight path and the destination is reached without error. Due to the rapidity of flight, this navigational problem cannot be solved by conventional method; but should be solved by an automatic computing system, which responds to every deviation from the normal conditions with a speed approaching instant action.

The general problem of automatic navigation is very difficult indeed. However, the deviations from the normal conditions are expected to be small, since the normal flight path is, after all, a good representation of the average situation. This fact immediately suggests that only first order quantities in deviations need be considered. This "linearization" is the basis of the ballistic disturbance theory and the present theory of automatic navigation.

Let quantities of the normal flight path be denoted by a bar over them and deviations by the δ sign. Thus for the actual flight path,

$$\left.\begin{array}{lll} r=\bar{r}+\delta r, & \theta=\bar{\theta}+\delta\theta, & \beta=\bar{\beta}+\delta\beta \\ v_r=\bar{v}_r+\delta v_r, & v_\theta=\bar{v}_\theta+\delta v_\theta, & \dot{\beta}=\dot{\bar{\beta}}+\delta\dot{\beta} \end{array}\right\} \tag{16}$$

The deviations of the actual atmosphere from the normal atmosphere are expressed as the deviation of density $\delta\rho$, the deviation of temperature δT, and the deviation of wind velocity δw ; thus

$$\rho=\bar{\rho}+\delta\rho, \quad T=\bar{T}+\delta T, \quad w=\bar{w}+\delta w \tag{17}$$

The deviation of the actual vehicle from the normal vehicle is assumed to be limited only to deviation of weight δW and the moment of inertia δI. That is

$$W=\bar{W}+\delta W, \quad I=\bar{I}+\delta I \tag{18}$$

The rate of propellant flow and the effective exhaust velocity of the rocket is assumed to be standard. The wing area S and the aerodynamic characteristics of the vehicle as expressed by Equations (11) and (13) are also assumed to be invariant.

By substituting Equations (16), (17), and (18) into the equations of motion, Equation (15), and retaining only first term deviations, one has

$$\left.\begin{array}{l} \dfrac{d\delta r}{dt}=+\delta v_r \\ \dfrac{d\delta\theta}{dt}=-\dfrac{\bar{v}_\theta}{\bar{r}^2}\delta r+\dfrac{1}{\bar{r}}\delta v_\theta \\ \dfrac{d\delta\beta}{dt}=+\delta\dot{\beta} \end{array}\right\} \tag{19}$$

$$\left.\begin{aligned}
\frac{\mathrm{d}\delta v_r}{\mathrm{d}t} &= a_1\delta r + a_2\delta\beta + a_3\delta v_r + a_4\delta v_\theta + a_5\delta\varepsilon + a_6\delta\rho + a_7\delta T + a_8\delta w + a_9\delta W \\
\frac{\mathrm{d}\delta v_\theta}{\mathrm{d}t} &= b_1\delta r + b_2\delta\beta + b_3\delta v_r + b_4\delta v_\theta + b_5\delta\varepsilon + b_6\delta\rho + b_7\delta T + b_8\delta w + b_9\delta W \\
\frac{\mathrm{d}\delta\dot{\beta}}{\mathrm{d}t} &= c_1\delta r + c_2\delta\beta + c_3\delta v_r + c_4\delta v_\theta + c_5\delta\varepsilon + c_6\delta\rho + c_7\delta T + c_8\delta w + c_9\delta W + c_{10}\delta I
\end{aligned}\right\} \tag{20}$$

The coefficients a's, b's, and c's are partial derivatives of F, G, and H defined by Equations (15), evaluated on the normal flight path. That is, for example,

$$\left.\begin{aligned}
a_1 &= \left(\frac{\overline{\partial F}}{\partial r}\right), & a_2 &= \left(\frac{\overline{\partial F}}{\partial \beta}\right), & a_3 &= \left(\frac{\overline{\partial F}}{\partial v_r}\right), \\
a_4 &= \left(\frac{\overline{\partial F}}{\partial v_\theta}\right), & a_5 &= \left(\frac{\overline{\partial F}}{\partial \varepsilon}\right), & a_6 &= \left(\frac{\overline{\partial F}}{\partial \rho}\right), \\
a_7 &= \left(\frac{\overline{\partial F}}{\partial T}\right), & a_8 &= \left(\frac{\overline{\partial F}}{\partial w}\right), & a_9 &= \left(\frac{\overline{\partial F}}{\partial W}\right)
\end{aligned}\right\} \tag{21}$$

These coefficients are calculated in detail and given in the Appendix.

The system of Equations (19) and (20) is the system of disturbance equations. They are linear. The coefficients when evaluated on the normal flight path are finally specified functions of time. If the deviations of the atmospheric properties $\delta\rho$, δT, and δw are known, and if $\delta\varepsilon$, δW, and δI are specified, then this system of differential equations determines the deviations δr, $\delta\theta$, $\delta\beta$, δv_r, δv_θ, and $\delta\dot{\beta}$ from the normal trajectory. This is the direct problem in the ballistic disturbance theory. The problem of automatic navigation is however different from this. What is required is the functions $\delta\varepsilon$, correction to the elevator angle, such that the range error is zero. As suggested by Drenick[2], this navigational problem can best be solved by the method of adjoint functions of Bliss.

Adjoint Functions for Range Correction

The principle of the method of adjoint functions is as follows[3]: Let $y_i(t)$, $i = 1, \ldots, n$, be determined by a system of linear equations

$$\frac{\mathrm{d}y_i}{\mathrm{d}t} = \sum_{j=1}^{n} a_{ij} y_j + Y_i \tag{22}$$

where a_{ij} are given coefficients which may be functions of the time t, and Y_i are specified functions of time. Now introduce a new set of functions $\lambda_i(t)$, called the adjoint functions to $y_i(t)$, which satisfy the following system of differential equations

$$\frac{\mathrm{d}\lambda_i}{\mathrm{d}t} = -\sum_{j=1}^{n} a_{ji}\lambda_j \tag{23}$$

By multiplying Equation (22) by λ_i and Equation (23) by y_i, and then adding the resultant equations, it can be shown that

$$\frac{\mathrm{d}}{\mathrm{d}t}\sum_{i=1}^{n}\lambda_i y_i = \sum_{i=1}^{n}\lambda_i Y_i \tag{24}$$

Equation (24) can be integrated from $t = t_1$ to $t = t_2$,

$$\sum_{i=1}^{n}\lambda_i y_i\Big|_{t=t_2} = \sum_{i=1}^{n}\lambda_i y_i\Big|_{t=t_1} + \int_{t=t_1}^{t=t_2}\Big(\sum_{i=1}^{n}\lambda_i Y_i\Big)\mathrm{d}t \tag{25}$$

Bliss named Equation (25) the "Fundamental Formula."

For the present problem, the y_i are the disturbance quantities, i.e.,

$$\left.\begin{aligned} y_1 &= \delta r, & y_2 &= \delta\theta, & y_3 &= \delta\beta \\ y_4 &= \delta v_r, & y_5 &= \delta v_\theta, & y_6 &= \delta\dot{\beta} \end{aligned}\right\} \tag{26}$$

Then, according to Equation (20), the adjoint functions λ_i satisfy the following differential equations

$$\left.\begin{aligned} -\frac{\mathrm{d}\lambda_1}{\mathrm{d}t} &= -\frac{\bar{v}_\theta}{\bar{r}^2}\lambda_2 + a_1\lambda_4 + b_1\lambda_5 + c_1\lambda_6 \\ -\frac{\mathrm{d}\lambda_2}{\mathrm{d}t} &= 0 \\ -\frac{\mathrm{d}\lambda_3}{\mathrm{d}t} &= a_2\lambda_4 + b_2\lambda_5 + c_2\lambda_6 \\ -\frac{\mathrm{d}\lambda_4}{\mathrm{d}t} &= \lambda_1 + a_3\lambda_4 + b_3\lambda_5 + c_3\lambda_6 \\ -\frac{\mathrm{d}\lambda_5}{\mathrm{d}t} &= \frac{1}{\bar{r}}\lambda_2 + a_4\lambda_4 + b_4\lambda_5 + c_4\lambda_6 \\ -\frac{\mathrm{d}\lambda_6}{\mathrm{d}t} &= \lambda_3 \end{aligned}\right\} \tag{27}$$

The Y_i are then

$$\left.\begin{aligned} Y_1 &= Y_2 = Y_3 = 0 \\ Y_4 &= a_5\delta\varepsilon + a_6\delta\rho + a_7\delta T + a_8\delta w + a_9\delta W \\ Y_5 &= b_5\delta\varepsilon + b_6\delta\rho + b_7\delta T + b_8\delta w + b_9\delta W \\ Y_6 &= c_5\delta\varepsilon + c_6\delta\rho + c_7\delta T + c_8\delta w + c_9\delta W + c_{10}\delta I \end{aligned}\right\} \tag{28}$$

The Equations (27) do not determine the λ-functions completely. To do that, a set of values for λ must be specified at a certain instant. For the problem of automatic navigation, the requirement is vanishing range error; then the required boundary conditions for Equations (27) can be determined as follows: If t_2 is the time instant of landing of the actual vehicle, $\bar{t}_2$ the time instant of landing of the normal flight path, then

$$t_2 = \bar{t}_2 + \delta t_2 \tag{29}$$

Similarly with the subscript 2 denoting the quantities at the landing instant

$$\left.\begin{aligned} r_2 &= \bar{r}_2 + \delta r_2 \\ \theta_2 &= \bar{\theta}_2 + \delta\theta_2 \end{aligned}\right\} \tag{30}$$

But

$$\left.\begin{aligned} \delta r_2 &= (\bar{v}_r)_{t=\bar{t}_2}\delta t_2 + (\delta r)_{t=\bar{t}_2} \\ \delta\theta_2 &= \frac{1}{r_0}(\bar{v}_\theta)_{t=\bar{t}_2}\delta t_2 + (\delta\theta)_{t=\bar{t}_2} \end{aligned}\right\} \tag{31}$$

δr_2 is, however, zero by definition, because landing means contact with the surface of earth, $\bar{r}_2 = r_2 = r_0$. By eliminating δt_2 from the Equations (31),

$$\delta\theta_2 = \left[-\frac{1}{\bar{r}}\left(\frac{\bar{v}_\theta}{\bar{v}_r}\right)\delta r + \delta\theta\right]_{t=\bar{t}_2} \tag{32}$$

Therefore if the magnitudes of λ_i at the landing instant $t = \bar{t}_2$ are specified as

$$\left.\begin{aligned} \lambda_1 &= -\frac{1}{\bar{r}}\left(\frac{\bar{v}_\theta}{\bar{v}_r}\right) \\ \lambda_2 &= 1 \\ \lambda_3 &= \lambda_4 = \lambda_5 = \lambda_6 = 0 \end{aligned}\right\} t = \bar{t}_2 \tag{33}$$

then the error in range is given by

$$\delta\theta_2 = \sum_{i=1}^{n}\lambda_i y_i\Big|_{t=\bar{t}_2} = [\lambda_1\delta r + \lambda_2\delta\theta + \lambda_3\delta\beta + \lambda_4\delta v_r + \lambda_5\delta v_\theta + \lambda_6\delta\dot{\beta}]_{t=\bar{t}_2} \tag{34}$$

When the normal flight path is determined, the coefficients in Equation (27) are specified as functions of time. These equations together with the end conditions of Equation (33) then determine the adjoint functions λ_i. The integration has to be performed "backwards" for $t < \bar{t}_2$, by perhaps an electro-mechanical computer. With the adjoint functions so determined, one can use the Fundamental Formula of Equation (25) to modify the equation for the range error given by Equation (34): Let $\bar{t}_1$ denote the time instant for the power cut-off. Then the condition for the error $\delta\theta_2$ in range to be zero can be expressed as

$$\left.\begin{aligned} \delta\theta_2 = 0 = [\lambda_1\delta r + \lambda_2\delta\theta + \lambda_3\delta\beta + \lambda_4\delta v_r + \lambda_5\delta v_\theta + \lambda_6\delta\dot{\beta}]_{t=\bar{t}_1} + \\ \int_{t_1}^{\bar{t}_2}[\lambda_4 Y_4 + \lambda_5 Y_5 + \lambda_6 Y_6]\,dt \end{aligned}\right\} \tag{35}$$

This is the basic equation for automatic navigation. It will be exploited presently.

Cut-Off Condition

The condition of Equation (35) for arbitrary disturbances can be broken down into two parts: The sum and the integral to be set to zero separately. Therefore the condition to be satisfied at the normal cut-off instant $\bar{t}_1$ is

$$[\lambda_1\delta r + \lambda_2\delta\theta + \lambda_3\delta\beta + \lambda_4\delta v_r + \lambda_5\delta v_\theta + \lambda_6\delta\dot{\beta}]_{t=\bar{t}_1} = 0 \tag{36}$$

Since the normal cut-off instant $\bar{t}_1$ is a standard time instant, but not necessarily the actual cut-off instant t_1 , i.e.,

$$t_1 = \bar{t}_1 + \delta t_1 \tag{37}$$

Equation (36) should be converted into a more useful form involving the quantities at the actual cut-off instant. This is easily done, because up to the first order quantities, according to Equation (16)

$$(\delta r)_{t=\bar{t}_1} = (r)_{t=t_1} - \left(\frac{\overline{dr}}{dt}\right)_{t=\bar{t}_1} \delta t_1 - (\bar{r})_{t=\bar{t}_1}$$

Or

$$(\delta r)_{t=\bar{t}_1} = (r)_{t=t_1} - (\bar{r})_{t=\bar{t}_1} - (\bar{v}_r)_{t=\bar{t}_1} \delta t_1$$

Similarly,

$$(\delta\theta)_{t=\bar{t}_1} = (\theta)_{t=t_1} - (\bar{\theta})_{t=\bar{t}_1} - \left(\frac{1}{\bar{r}}\bar{v}_\theta\right)_{t=\bar{t}_1} \delta t_1$$

$$(\delta\beta)_{t=\bar{t}_1} = (\beta)_{t=t_1} - (\bar{\beta})_{t=\bar{t}_1} - (\dot{\bar{\beta}})_{t=\bar{t}_1} \delta t_1$$

$$(\delta v_r)_{t=\bar{t}_1} = (v_r)_{t=t_1} - (\bar{v}_r)_{t=\bar{t}_1} - (\bar{F})_{t=\bar{t}_1} \delta t_1$$

$$(\delta v_\theta)_{t=\bar{t}_1} = (v_\theta)_{t=t_1} - (\bar{v}_\theta)_{t=\bar{t}_1} - (\bar{G})_{t=\bar{t}_1} \delta t_1$$

$$(\delta\dot{\beta})_{t=\bar{t}_1} = (\dot{\beta})_{t=t_1} - (\dot{\bar{\beta}})_{t=\bar{t}_1} - (\bar{H})_{t=\bar{t}_1} \delta t_1$$

where $\bar{F}$, $\bar{G}$, and $\bar{H}$, are the values of these quantities given by Equation (15) evaluated on the normal flight path. In fact, they should be evaluated at an instant just before the normal cut-off time $\bar{t}_1$ so that the accelerating force of the rocket is included and the rates of change of velocities are those of a powered flight. Now define J and $\bar{J}$ as follows,

$$J = [\lambda_1^* r + \lambda_2^* \theta + \lambda_3^* \beta + \lambda_4^* v_r + \lambda_5^* v_\theta + \lambda_6^* \dot{\beta}]_{t=t_1} \tag{38}$$

and

$$\bar{J} = [\lambda_1^* \bar{r} + \lambda_2^* \bar{\theta} + \lambda_3^* \bar{\beta} + \lambda_4^* \bar{v}_r + \lambda_5^* \bar{v}_\theta + \lambda_6^* \dot{\bar{\beta}}]_{t=\bar{t}_1}$$

where λ_i^* are the values of λ_i evaluated at the normal cut-off time t_1. Then the condition to be satisfied at the actual cut-off instant t_1 is

$$J = \bar{J} + \left\{\lambda_1^* \bar{v}_r + \lambda_2^* \frac{\bar{v}_\theta}{\bar{r}} + \lambda_3^* \dot{\bar{\beta}} + \lambda_4^* \bar{F} + \lambda_5^* \bar{G} + \lambda_6^* \bar{H}\right\}_{t=t_1} (t_1 - \bar{t}_1) \tag{39}$$

This is the equation to determine the proper instant of power cut-off.

When the normal flight path is known, $\bar{J}$ and the quantity within the bracket to the right of Equation (39) are fixed. Then the whole right-hand side of Equation (39) can be considered as a linearly increasing function of time t, if t is substituted for t_1. Simultaneously J can be computed at every instant before cut-off by using the predetermined λ_i^* and values of position and velocity of the actual vehicle obtained by tracking stations. The magnitudes of the quantities on the two sides of Equation (39) can then be continuously compared. When they are equal to each other, condition (39) is satisfied. Then the power cut-off signal is given and the rocket power is shut off①.

Condition for Automatic Navigation

When the rocket power is shut off earlier or later than the normal cut-off instant $\bar{t}_1$, the propellant left in the tank, if not dumped, will alter the weight W and the moment of inertia I of the vehicle. It is also possible that the pay load of the vehicle is not that specified for the standard vehicle. Then after power-off, there is a fixed δW and δI, fixed in the sense that they do not change with time, and are known once the power cut-off is affected. Of different character are the deviations $\delta\rho$, δT, δw of the actual atmosphere from the standard atmosphere. These are not known unless they are measured. In the following, it is proposed to use the vehicle itself as a measuring instrument, and proceed as follows:

After the cut-off condition is satisfied, the condition for zero range error is that the integral in Equation (35) should vanish. Now since the Y_i in that integrand involving arbitrary disturbances $\delta\rho$, δT, and δw is not known beforehand, this condition can be satisfied only if the integrand itself vanishes. That is, according to Equations (28),

$$\begin{aligned}&(\lambda_4 a_5 + \lambda_5 b_5 + \lambda_6 c_5)\delta\varepsilon + (\lambda_4 a_6 + \lambda_5 b_6 + \lambda_6 c_6)\delta\rho + \\ &(\lambda_4 a_7 + \lambda_5 b_7 + \lambda_6 c_7)\delta T + (\lambda_4 a_8 + \lambda_5 b_8 + \lambda_6 c_8)\delta w + \\ &(\lambda_4 a_9 + \lambda_5 b_9 + \lambda_6 c_9)\delta W + \lambda_6 c_{10}\delta I = 0\end{aligned}$$

Or, with the following notation

$$\left.\begin{aligned} d_5 &= \lambda_4 a_5 + \lambda_5 b_5 + \lambda_6 c_5 \\ d_6 &= \lambda_4 a_6 + \lambda_5 b_6 + \lambda_6 c_6 \\ d_7 &= \lambda_4 a_7 + \lambda_5 b_7 + \lambda_6 c_7 \\ d_8 &= \lambda_4 a_8 + \lambda_5 b_8 + \lambda_6 c_8 \\ D &= -(\lambda_4 a_9 + \lambda_5 b_9 + \lambda_6 c_9)\delta W - \lambda_6 c_{10}\delta I \end{aligned}\right\} \tag{40}$$

this condition can be written as

$$d_5\delta\varepsilon + d_6\delta\rho + d_7\delta T + d_8\delta w = D \tag{41}$$

Equation (20) can be rewritten as

$$\left.\begin{aligned} a_5\delta\varepsilon + a_6\delta\rho + a_7\delta T + a_8\delta w &= A \\ b_5\delta\varepsilon + b_6\delta\rho + b_7\delta T + b_8\delta w &= B \\ c_5\delta\varepsilon + c_6\delta\rho + c_7\delta T + c_8\delta w &= C \end{aligned}\right\} \tag{42}$$

---

① Drenick's cut-off condition (Ref. 2) differs from that of Equation (39) in that the second terms to the right is not present. It seems that this neglect is not justified and will introduce first order range error.

where

$$\left.\begin{aligned} A &= \frac{d\delta v_r}{dt} - a_1\delta r - a_2\delta\beta - a_3\delta v_r - a_4\delta v_\theta - a_7\delta W \\ B &= \frac{d\delta v_\theta}{dt} - b_1\delta r - b_2\delta\beta - b_3\delta v_r - b_4\delta v_\theta - b_7\delta W \\ C &= \frac{d\delta\dot{\beta}}{dt} - c_1\delta r - c_2\delta\beta - c_3\delta v_r - c_4\delta v_\theta - c_9\delta W - c_{10}\delta I \end{aligned}\right\} \tag{43}$$

If the tracking stations for the vehicle will measure the quantities A, B, and C, then the atmospheric disturbances $\delta\rho$, δT, and δw can be determined by solving for these variations using Equation (42). This is essentially using the vehicle itself as a measuring instrument for $\delta\rho$, δT, and δw. When $\delta\rho$, δT, and δw are known, Equation (41) gives the proper elevator angle correction $\delta\varepsilon$.

A mathematically equivalent way to calculate $\delta\varepsilon$ would be to directly solve for $\delta\varepsilon$ using the system of Equations (41) and (42). Thus

$$\begin{vmatrix} a_5 & a_6 & a_7 & a_8 \\ b_5 & b_6 & b_7 & b_8 \\ c_5 & c_6 & c_7 & c_8 \\ d_5 & d_6 & d_7 & d_8 \end{vmatrix} \delta\varepsilon = \begin{vmatrix} A & a_6 & a_7 & a_8 \\ B & b_6 & b_7 & b_8 \\ C & c_6 & c_7 & c_8 \\ D & d_6 & d_7 & d_8 \end{vmatrix} \tag{44}$$

This equation specifies the necessary change in the elevator angle at every instant to be calculated from the quantities a's, b's, c's, and A, B, C, D at the same instant. These quantities consist partly of predetermined information from the normal flight path, and partly of measured information on the position and the velocities of vehicle obtained by tracking the vehicle. At high altitudes where the air density is very small, the aerodynamic forces will be almost negligible in comparison with the gravitational and inertia forces. Then the quantities A, B, and C of Equation (43) will be the small difference of large magnitudes. These are then the quantities most difficult to determine accurately. If the actual elevator angle is made to conform with the one calculated by Equation (44), then in conjunction with the proper power cut-off as specified in the last section the vehicle will be navigated to the chosen landing point in spite of the atmospheric disturbances.

Discussion

When the general character of the flight path is chosen from the over-all engineering considerations, the first step is the calculation of the normal flight path using the properties of the standard atmosphere and the expected performance of the vehicle with normal weight. The knowledge of the normal flight path then determines the a's, b's, and c's. The Equation (27) together with the end conditions of Equation (33) allows the calculation of the adjoint functions λ_i. All this information should be on hand before the actual flight of the vehicle, and may be called the "stored data."

Before the power cut-off, the elevator angle may be programmed according to that for the normal flight path, and the stability of the vehicle is supplied by the jet vanes or by the auxiliary rockets. The tracking stations, however, go immediately into action and supply the vehicle with information on its positions and velocities. This information goes first into the cut-off computer which, using the stored information, continuously compares the magnitudes of quantities on the two sides of Equation (39), the cut-off condition. When that condition is satisfied, the power cut-off is affected.

At the instant of power cut-off, the tracking information is switched to the computer for automatic navigation. The instant of power cut-off also fixes the amount of propellant in the tank and thus determines the variations of weight δW and inertia moment δI from the standard. This information together with the stored data on the normal flight path then allows the computer to generate the elevator correction angle $\delta\varepsilon$ according to Equations (40), (43), and (44). Theoretically the value of $\delta\varepsilon$ must be obtained without time delay from the instant when the information is received, because Equation (44) is a condition of equality of two quantities evaluated at identical time instants. Practically there will be time delay due to the finite computing time. However, it is now clear that this computing time must be made very short in order to satisfy the condition of automatic navigation as accurately as possible. The computed correction $\delta\varepsilon$ combined with the elevator angle $\bar{\varepsilon}$ determined for the normal flight path then gives the actual elevator angle setting ε. The design of the control mechanism for the elevator from here on follows the conventional feed-back servomechanism, with the usual criteria of quick action, stability, and accuracy. The general scheme of the automatic navigation of the vehicle then can be represented by the sketch in Fig. 2.

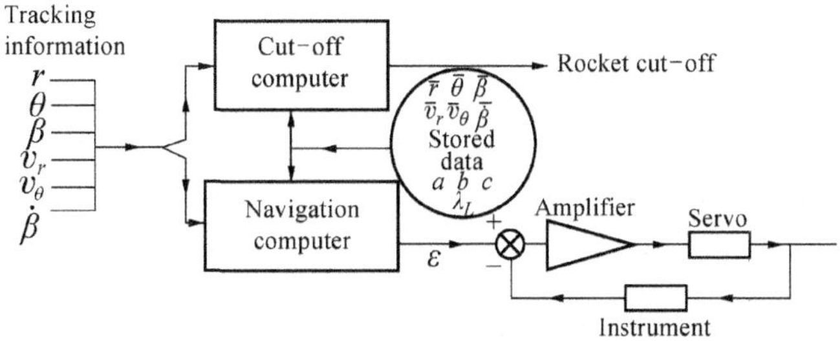

Fig. 2 System of automatic navigation

The primary purpose or the "duty" of the computer is then to properly digest the dynamic and the aerodynamic information of the vehicle and thus to decide the right flight path correction that will insure landing at the chosen destination. The purpose or the duty of the elevator servo is now simply to follow the command of the computer. If the servo is considered to be internal to the vehicle, then the computer can be considered to be a mechanism to account for the external conditions of the vehicle. The separation of the internal from the external is not necessary for the control of conventional aircraft. For conventional aircraft, the basic disturbance dynamic equations have constant coefficients and the computer can be merged with the servo through a single "amplifier" of essentially RC-circuits.

The computers envisaged here are carried in the vehicle and receive the information on positions and velocities from the fixed ground tracking stations[4]. It will be beyond the scope of the present paper to discuss their design. But the required accuracy and speed to generate proper signal almost instantly would indicate that they should be of the electronic digital type. If this is indeed the proper type to be used, then the separation of cut-off computer and the navigation computer is not necessary. All could be done in one computer by proper programming.

Appendix

Calculation of Coefficients

The quantities F, G, and H are defined by Equation (15). They contain the parameters Ψ, Λ, Δ, and N. According to their definition as given by Equations (5) and (14), they can be written as follows:

$$\left.\begin{aligned} \Psi &= \frac{fg}{W} \\ \Lambda &= \frac{g}{W}\frac{1}{2}\rho S C_L \sqrt{v_r^2 + (v_\theta + w)^2} \\ \Delta &= \frac{g}{W}\frac{1}{2}\rho S C_D \sqrt{v_r^2 + (v_\theta + w)^2} \\ N &= \frac{1}{I}\frac{1}{2}\rho S C_M \{v_r^2 + (v_\theta + w)^2\} \end{aligned}\right\} \tag{45}$$

where the aerodynamic coefficients C_L, C_D, C_M are functions of angle of attack α, elevator angle ε, Mach number M, and the Reynolds number Re. These aerodynamic parameters are related to quantities immediately connected with the flight path as follows:

$$\alpha = \beta - \tan^{-1}\frac{v_r}{v_\theta + w}, \quad M = V/a(r), \quad Re = \rho V l/\mu(r) \tag{46}$$

where $a(r)$ is the sound velocity in atmosphere, and $\mu(r)$ is the coefficient of viscosity of air, both functions of altitude r. In the following calculation, the thrust f will be considered to be a function of altitude only. It is also assumed that the composition of atmosphere at different altitudes remains the same as that of the standard atmosphere, only the density ρ and the temperature T change. Thus the variations of a and μ at any altitude are variations due to temperature T.

For Ψ

$$\frac{\partial \Psi}{\partial r} = \frac{g}{W}\frac{\partial f}{\partial r}, \quad \frac{\partial \Psi}{\partial W} = -\frac{\Psi}{W} \tag{47}$$

All other partial derivatives are zero.

For Λ

$$\left.\begin{aligned}
\frac{\partial \Lambda}{\partial r} &= \Lambda\left[-\frac{1}{\rho}\frac{\mathrm{d}\rho}{\mathrm{d}r}\left\{1+\frac{Re}{C_L}\frac{\partial C_L}{\partial Re}\right\}+\frac{1}{V^2}\frac{\mathrm{d}w}{\mathrm{d}r}\times\left\{\left(\frac{M}{C_L}\frac{\partial C_L}{\partial M}+\frac{Re}{C_L}\frac{\partial C_L}{\partial Re}+1\right)\times(v_\theta+w)+\right.\right.\\
&\qquad \left.\left.\frac{1}{C_L}\frac{\partial C_L}{\partial \alpha}v_r\right\}-\frac{M}{C_L}\frac{\partial C_L}{\partial M}\frac{1}{a}\frac{\mathrm{d}a}{\mathrm{d}r}-\frac{Re}{C_L}\frac{\partial C_L}{\partial Re}\frac{1}{\mu}\frac{\mathrm{d}\mu}{\mathrm{d}r}\right]\\
\frac{\partial \Lambda}{\partial v_r} &= \Lambda\frac{v_r}{V^2}\left[\frac{M}{C_L}\frac{\partial C_L}{\partial M}+\frac{Re}{C_L}\frac{\partial C_L}{\partial Re}+1-\frac{1}{C_L}\frac{\partial C_L}{\partial \alpha}\frac{v_\theta+w}{v_r}\right]\\
\frac{\partial \Lambda}{\partial v_\theta} &= \Lambda\frac{v_\theta+w}{V^2}\left[\frac{M}{C_L}\frac{\partial C_L}{\partial M}+\frac{Re}{C_L}\frac{\partial C_L}{\partial Re}+1+\frac{1}{C_L}\frac{\partial C_L}{\partial \alpha}\frac{v_r}{v_\theta+w}\right]\\
\frac{\partial \Lambda}{\partial \beta} &= \Lambda\frac{1}{C_L}\frac{\partial C_L}{\partial \alpha}\\
\frac{\partial \Lambda}{\partial \varepsilon} &= \Lambda\frac{1}{C_L}\frac{\partial C_L}{\partial \varepsilon}\\
\frac{\partial \Lambda}{\partial \rho} &= \Lambda\frac{1}{\rho}\left[1+\frac{Re}{C_L}\frac{\partial C_L}{\partial Re}\right]\\
\frac{\partial \Lambda}{\partial T} &= -\Lambda\left[\frac{M}{C_L}\frac{\partial C_L}{\partial M}\frac{1}{2T}+\frac{Re}{C_L}\frac{\partial C_L}{\partial Re}\frac{1}{\mu}\frac{\partial \mu}{\partial T}\right]\\
\frac{\partial \Lambda}{\partial w} &= \Lambda\frac{v_\theta+w}{V^2}\left[\frac{M}{C_L}\frac{\partial C_L}{\partial M}+\frac{Re}{C_L}\frac{\partial C_L}{\partial Re}+1+\frac{1}{C_L}\frac{\partial C_L}{\partial \alpha}\frac{v_r}{v_\theta+w}\right]=\frac{\partial \Lambda}{\partial v_\theta}\\
\frac{\partial \Lambda}{\Lambda W} &= -\frac{\Lambda}{W}
\end{aligned}\right\}\quad(48)$$

For Δ, the partial derivatives are obtained from the above equations by substituting Δ for Λ, and C_D for C_L.

For N,

$$\left.\begin{aligned}
\frac{\partial N}{\partial r} &= N\left[-\frac{1}{\rho}\frac{\mathrm{d}\rho}{\mathrm{d}r}\left\{1+\frac{Re}{C_M}\frac{\partial C_M}{\partial Re}\right\}+\frac{1}{V^2}\frac{\mathrm{d}w}{\mathrm{d}r}\times\left\{\left(\frac{M}{C_M}\frac{\partial C_M}{\partial M}+\frac{Re}{C_M}\frac{\partial C_M}{\partial Re}+2\right)\times(v_\theta+w)+\right.\right.\\
&\qquad \left.\left.\frac{1}{C_M}\frac{\partial C_M}{\partial \alpha}v_r\right\}-\frac{M}{C_M}\frac{\partial C_M}{\partial M}\frac{1}{a}\frac{\mathrm{d}a}{\mathrm{d}r}-\frac{Re}{C_M}\frac{\partial C_M}{\partial Re}\frac{1}{\mu}\frac{\mathrm{d}\mu}{\mathrm{d}r}\right]\\
\frac{\partial N}{\partial v_r} &= N\frac{v_r}{V^2}\left[\frac{M}{C_M}\frac{\partial C_M}{\partial M}+\frac{Re}{C_M}\frac{\partial C_M}{\partial Re}+2-\frac{1}{C_M}\frac{\partial C_M}{\partial \alpha}\frac{v_\theta+w}{v_r}\right]\\
\frac{\partial N}{\partial v_\theta} &= N\frac{(v_\theta+w)}{V^2}\left[\frac{M}{C_M}\frac{\partial C_M}{\partial M}+\frac{Re}{C_M}\frac{\partial C_M}{\partial Re}+2+\frac{1}{C_M}\frac{\partial C_M}{\partial \alpha}\frac{v_r}{v_\theta+w}\right]\\
\frac{\partial N}{\partial \beta} &= N\frac{1}{C_M}\frac{\partial C_M}{\partial \alpha}\\
\frac{\partial N}{\partial \varepsilon} &= N\frac{1}{C_M}\frac{\partial C_M}{\partial \varepsilon}\\
\frac{\partial N}{\partial \rho} &= N\frac{1}{\rho}\left[1+\frac{Re}{C_M}\frac{\partial C_M}{\partial Re}\right]\\
\frac{\partial N}{\partial T} &= -N\left[\frac{M}{C_M}\frac{\partial C_M}{\partial M}\frac{1}{2T}+\frac{Re}{C_M}\frac{\partial C_M}{\partial Re}\frac{1}{\mu}\frac{\partial \mu}{\partial T}\right]\\
\frac{\partial N}{\partial w} &= \frac{\partial N}{\partial v_\theta}\\
\frac{\partial N}{\partial I} &= -\frac{N}{I}
\end{aligned}\right\}\quad(49)$$

With these partial derivatives, the coefficients a's, b's, and c's can be easily calculated:

$$\left.\begin{aligned}
a_1 &= \frac{\partial F}{\partial r} = \frac{\partial \Psi}{\partial r}\sin\beta + \frac{\mathrm{d}w}{\mathrm{d}r}\Lambda + (v_\theta + w)\frac{\partial \Lambda}{\partial r} - v_r\frac{\partial \Delta}{\partial r} + \\
&\quad \left(\frac{v_\theta}{r} \pm \Omega\right)^2 - 2\frac{v_\theta}{r}\left(\frac{v_\theta}{r} \pm \Omega\right) + 2\frac{g}{r}\left(\frac{r_0}{r}\right)^2 \\
a_2 &= \frac{\partial F}{\partial \beta} = \Psi\cos\beta + (v_\theta + w)\frac{\partial \Lambda}{\partial \beta} - v_r\frac{\partial \Delta}{\partial \beta} \\
a_3 &= \frac{\partial F}{\partial v_r} = (v_\theta + w)\frac{\partial \Lambda}{\partial v_r} - \Delta - v_r\frac{\partial \Delta}{\partial v_r} \\
a_4 &= \frac{\partial F}{\partial v_\theta} = \Lambda + (v_\theta + w)\frac{\partial \Lambda}{\partial v_\theta} - v_r\frac{\partial \Delta}{\partial v_\theta} + 2\left(\frac{v_\theta}{r} \pm \Omega\right) \\
a_5 &= \frac{\partial F}{\partial \varepsilon} = (v_\theta + w)\frac{\partial \Lambda}{\partial \varepsilon} - v_r\frac{\partial \Delta}{\partial \varepsilon} \\
a_6 &= \frac{\partial F}{\partial \rho} = (v_\theta + w)\frac{\partial \Lambda}{\partial \rho} - v_r\frac{\partial \Delta}{\partial \rho} \\
a_7 &= \frac{\partial F}{\partial T} = (v_\theta + w)\frac{\partial \Lambda}{\partial T} - v_r\frac{\partial \Delta}{\partial T} \\
a_8 &= \frac{\partial F}{\partial w} = \Lambda + (v_\theta + w)\frac{\partial \Lambda}{\partial w} - v_r\frac{\partial \Delta}{\partial w} \\
a_9 &= \frac{\partial F}{\partial W} = \frac{\partial \Psi}{\partial W}\sin\beta + (v_\theta + w)\frac{\partial \Lambda}{\partial W} - v_r\frac{\partial \Delta}{\partial W}
\end{aligned}\right\} \tag{50}$$

$$\left.\begin{aligned}
b_1 &= \frac{\partial G}{\partial r} = \frac{\partial \Psi}{\partial r}\cos\beta - v_r\frac{\partial \Lambda}{\partial r} - \frac{\mathrm{d}w}{\mathrm{d}r}\Delta - (v_\theta + w)\frac{\partial \Delta}{\partial r} + \frac{v_r v_\theta}{r^2} \\
b_2 &= \frac{\partial G}{\partial \beta} = -\Psi\sin\beta - v_r\frac{\partial \Lambda}{\partial \beta} - (v_\theta + w)\frac{\partial \Delta}{\partial \beta} \\
b_3 &= \frac{\partial G}{\partial v_r} = -\Lambda - v_r\frac{\partial \Lambda}{\partial v_r} - (v_\theta + w)\frac{\partial \Delta}{\partial v_r} - 2\left(\frac{1}{2}\frac{v_\theta}{r} \pm \Omega\right) \\
b_4 &= \frac{\partial G}{\partial v_\theta} = -v_r\frac{\partial \Lambda}{\partial v_\theta} - \Delta - (v_\theta + w)\frac{\partial \Delta}{\partial v_\theta} - \frac{v_r}{r} \\
b_5 &= \frac{\partial G}{\partial \varepsilon} = -v_r\frac{\partial \Lambda}{\partial \varepsilon} - (v_\theta + w)\frac{\partial \Delta}{\partial \varepsilon} \\
b_6 &= \frac{\partial G}{\partial \rho} = -v_r\frac{\partial \Lambda}{\partial \rho} - (v_\theta + w)\frac{\partial \Delta}{\partial \rho} \\
b_7 &= \frac{\partial G}{\partial T} = -v_r\frac{\partial \Lambda}{\partial T} - (v_\theta + w)\frac{\partial \Delta}{\partial T} \\
b_8 &= \frac{\partial G}{\partial w} = -v_r\frac{\partial \Lambda}{\partial w} - \Delta - (v_\theta + w)\frac{\partial \Delta}{\partial w} \\
b_9 &= \frac{\partial G}{\partial W} = \frac{\partial \Psi}{\partial W}\cos\beta - v_r\frac{\partial \Lambda}{\partial W} - (v_\theta + w)\frac{\partial \Delta}{\partial W}
\end{aligned}\right\} \tag{51}$$

$$
\left.\begin{aligned}
c_1 &= \frac{\partial H}{\partial r} = -\frac{1}{r^2}\left\{\Psi\cos\beta - v_r\Lambda - (v_\theta + w)\Delta - 2v_r\left(\frac{v_\theta}{r} \pm \Omega\right)\right\} + \\
&\quad \frac{1}{r}\left\{\frac{\partial\Psi}{\partial r}\cos\beta - v_r\frac{\partial\Lambda}{\partial r} - (v_\theta + w)\frac{\partial\Delta}{\partial r} - \frac{\mathrm{d}w}{\mathrm{d}r}\Delta + 2\frac{v_r v_\theta}{r^2}\right\} + \frac{\partial N}{\partial r} \\
c_2 &= \frac{\partial H}{\partial\beta} = \frac{1}{r}\left\{-\Psi\sin\beta - v_r\frac{\partial\Lambda}{\partial\beta} - (v_\theta + w)\frac{\partial\Delta}{\partial\beta}\right\} + \frac{\partial N}{\partial\beta} \\
c_3 &= \frac{\partial H}{\partial v_r} = \frac{1}{r}\left\{-\Lambda - v_r\frac{\partial\Lambda}{\partial v_r} - (v_\theta + w)\frac{\partial\Delta}{\partial v_r} - 2\left(\frac{v_\theta}{r} \pm \Omega\right)\right\} + \frac{\partial N}{\partial v_r} \\
c_4 &= \frac{\partial H}{\partial v_\theta} = \frac{1}{r}\left\{-v_r\frac{\partial\Lambda}{\partial v_\theta} - \Delta - (v_\theta + w)\frac{\partial\Delta}{\partial v_\theta} - 2\frac{v_r}{r}\right\} + \frac{\partial N}{\partial v_\theta} \\
c_5 &= \frac{\partial H}{\partial\varepsilon} = \frac{1}{r}\left\{-v_r\frac{\partial\Lambda}{\partial\varepsilon} - (v_\theta + w)\frac{\partial\Delta}{\partial\varepsilon}\right\} + \frac{\partial N}{\partial\varepsilon} \\
c_6 &= \frac{\partial H}{\partial\rho} = \frac{1}{r}\left\{-v_r\frac{\partial\Lambda}{\partial\rho} - (v_\theta + w)\frac{\partial\Delta}{\partial\rho}\right\} + \frac{\partial N}{\partial\rho} \\
c_7 &= \frac{\partial H}{\partial T} = \frac{1}{r}\left\{-v_r\frac{\partial\Lambda}{\partial T} - (v_\theta + w)\frac{\partial\Delta}{\partial T}\right\} + \frac{\partial N}{\partial T} \\
c_8 &= \frac{\partial H}{\partial w} = \frac{1}{r}\left\{-v_r\frac{\partial\Lambda}{\partial w} - \Delta - (v_\theta + w)\frac{\partial\Delta}{\partial w}\right\} + \frac{\partial N}{\partial w} \\
c_9 &= \frac{\partial H}{\partial W} = \frac{1}{r}\left\{\frac{\partial\Psi}{\partial W}\cos\beta - v_r\frac{\partial\Lambda}{\partial W} - (v_\theta + w)\frac{\partial\Delta}{\partial W}\right\} \\
c_{10} &= \frac{\partial H}{\partial I} = \frac{\partial N}{\partial I}
\end{aligned}\right\} \tag{52}
$$

After the power cut-off, the thrust f vanishes. Thus for $t > \bar{t}_1$, Ψ and its derivatives are zero.

References

[1] Rosser J B, Newton R R, Gross G L. Mathematical Theory of Rocket Flight. McGrow Hill Book Co., Inc., New York, 1947, particularly chapter Ⅲ.

[2] Drenick R. The Perturbation Calculus in Missile Ballistics. Journal of the Franklin Institute, 1951, 251: 423 - 436.

[3] Bliss G A. Mathematics for Exterior Ballistics. John Wiley & Sons, New York, 1944.

[4] Tuska C D. For possible tracking system. see "Pictorial Radio", Journal of the Franklin Institute, 1952, 253: 1 - 20; 95 - 124.

A Method for Comparing the Performance of Power Plants for Vertical Flight

H. S. Tsien①

(*Daniel and Florence Guggenheim Jet Propulsion Center, California Institute of Technology, Pasadena, Calif.*)

A new method of power plant selection for vertical flight is proposed. It can be used to determine whether the performance of a rocket design can be improved by substituting for the rocket motor a different power plant such as a ramjet. Calculations indicate that there are advantages in using the ramjet provided the power plant can be made to operate under rapid acceleration and at high altitudes.

A problem constantly facing engineers who use jet propulsion power plants is to determine the best power plant among a multitude of possible power plants for a particular design application. A very general method of power plant selection was proposed, perhaps for the first time, by W. Bollay and E. Redding, based upon the concept of lowest total installation weight. The total installation weight for a specified thrust at given altitude and speed of flight is the sum of dry power plant weight plus the weight of fuel and fuel tank for a given duration of flight at that altitude and speed. The fuel and fuel tank weight increase with the increase in flight duration. Thus, a light power plant with large fuel consumption, such as a rocket, is competitive with a heavier power plant with smaller fuel consumption only at short flight durations. This concept was extensively developed by Th. von Kármán in his general analysis of jet propulsion power plants[1]. This method of power plant selection was also described by A. L. Lowell[2].

However, no actual vehicle will fly at constant speed and constant altitude. There is always a definite flight plan describing the speed and altitude as functions of time. Hence, true power plant selection must depend upon a sort of weighted average of different speeds, and altitudes according to the particular flight plan of the vehicle under consideration. If the speed and altitude of the vehicle are rapidly varying, as in the case of accelerated vertical flight, then the selection based upon total installation weight at a fixed speed and a fixed altitude would be quite wrong. The purpose of the present note is to give a different method of power plant selection for vehicles in vertical flight. A meteorological sounding vehicle is a direct example of such an application. For other types of vehicle, the vertical powered flight is often a good

---

Received January 3, 1952.

Journal of the American Rocket Society, vol. 22, pp. 200 - 203, 1952.

① Robert H. Goddard Professor of Jet Propulsion.

approximation to the true flight trajectory during the application of propulsive power. Therefore the proposed method is believed to have a wider range of usefulness than is, perhaps, first apparent.

General Relation

It will be assumed in this analysis that the gravitational acceleration and the effective exhaust velocity of the power plant are constants. For moderate altitudes, the decrease of the gravitational acceleration from its sea-level value is very small and thus negligible. The effective exhaust velocity of the power plant, defined as ratio of thrust generated by the power plant to the mass rate of consumption of fuel or propellant carried within the vehicle, is of course a variable, not a constant, even for a given power plant using a given fuel or propellant. For a rocket, this variation is caused by the change in atmospheric pressure with altitude, and is usually small enough to be neglected. For a ramjet, the effective exhaust velocity increases very rapidly with speed of the vehicle in the subsonic range, but in the supersonic range the exhaust velocity is again almost constant with respect to the speed of the vehicle, within the useful speed range of the engine[3]. For other power plants, the variation of the effective exhaust velocity may be more complicated; but as a first approximation, it is generally possible to use an average value as the assumed constant exhaust velocity.

For a rocket vehicle, the air drag is proportional to the cross-sectional area of the body, but the mass of the vehicle is proportional to the volume of the body. Therefore for vehicles of similar design, the air drag is proportional to the square of the body diameter, while the mass of the vehicle is proportional to the cube of the body diameter. For large rocket vehicles the drag force is then negligible with respect to the gravitational force. Calculations seem to show that for a high performance rocket of 50 tons gross weight, the air drag reduces the velocity at the end of powered flight by only 5 per cent. Hence for a large rocket vehicle, the effects of air drag on the performance can be neglected in the approximate analysis attempted here. When another power plant is used, the design cannot perhaps be made as compact as the rocket. For instance, the ramjet requires a rather large duct to produce a sizable thrust. Then the air drag may not be negligible even for a large vehicle. However, in such cases the air drag of the power plant installation can be charged against the thrust produced. As an example, the air drag of the outside surfaces of the ramjet duct can be deducted from the thrust, produced by the ramjet, and the power plant is considered to produce a smaller "net thrust" while the body drag is considered to remain at the same magnitude without the ramjet duct. Therefore if one takes the effective, exhaust velocity as that based upon the net thrust, then the same argument for a large rocket applies and the drag of the body can be neglected for large vehicles.

For vehicles that are not so large, neglect of the air drag will certainly introduce an error. But the emphasis here is the comparison of the performance obtainable from different power plants rather than the absolute value of performance. Therefore the error made in this way is believed to be not large, and the method given below is useful even for moderate-sized vehicles.

Let m be the mass of the vehicle at time instant t when the vertical position and vertical velocity of the vehicle are $y(t)$ and $\dot{y}(t)$, respectively. Denote by c and g the constant exhaust velocity and the gravitational acceleration. Then the balance of inertia force and gravitational against the thrust (Fig. 1) gives

$$c\frac{\mathrm{d}m}{\mathrm{d}t}+m\left(\frac{\mathrm{d}\dot{y}}{\mathrm{d}t}+g\right)=0 \tag{1}$$

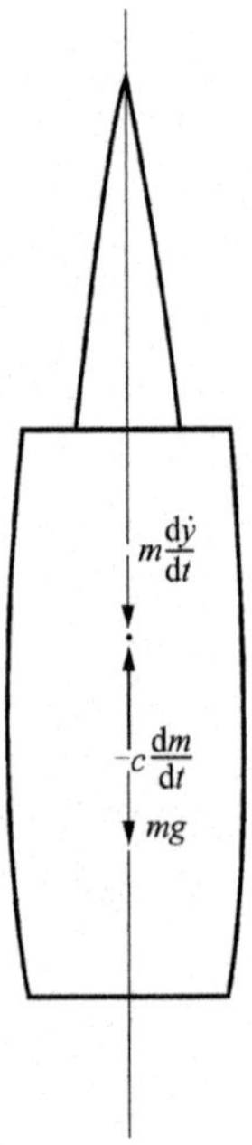

Fig. 1 Balance of forces on a vehicle in vertical asecnt

If the initial velocity at $t = 0$ is zero, as is usually the case, Equation (1) integrates to

$$\log\frac{m}{m_0}=-\frac{1}{c}(\dot{y}+gt) \tag{2}$$

If the subscript 1 denotes conditions at the end of the powered flight, then Equation (2) gives

$$\log\frac{m_0}{m_1}=\frac{1}{c}(\dot{y}_1+gt_1) \tag{3}$$

This is the fundamental performance equation for vertical flight. It has been derived previously by many authors[4], but the present deviation clearly shows that it is quite general and independent of the particular way the thrust is programmed during the powered flight.

For a vehicle intended for obtaining long range but having a vertical powered trajectory, the performance is essentially determined by the velocity at the end of powered flight or $\dot{y}_1$. For an atmospheric sounding vehicle, the summit altitude is determined by y_1 and $\dot{y}_1$. The two vehicles having the similar thrust programming during powered flight and the same $\dot{y}_1$ and t_1 will have the same y_1 and $\dot{y}_1$. To simplify the calculation, the performances of two vehicles with different power plants are made to be the same by specifying that the values of y_1, $\dot{y}_1$, and t_1 at end of the powered flight are the same for the two vehicles. This condition will be satisfied if the acceleration programs of the two vehicles are the same. The thrust programs of the two vehicles are different, however, due to the different fuel consumption of the two power plants. Therefore, one of the criteria for equal performance is the value of $\dot{y}_1+gt_1$, occurring on the right of Equation (3). According to Equation (3), then, to have the same performance, $c\log(m_0/m_1)$ must be the same for vehicles. This conditions will be used to compare the performance of different power plants.

Now let w_0, w_s, w_e, and w_1 denote the gross weight, the weight of structures and pay load, the engine weight, and the weight of the vehicle, respectively, at the end of powered flight. Then

$$w_1 = w_s + w_e \tag{4}$$

Equation (3) can be written as

$$c \log \frac{w_0}{w_s + w_e} = \dot{y}_1 + g t_1 \tag{5}$$

Consider now two vehicles, both having the same gross weight w_0; one is a rocket with $c = c^*$, $w_s = w_s^*$, $w_e = w_e^*$, and

$$\frac{w_0^*}{w_s^* + w_e^*} = \frac{1}{1 - \zeta} \tag{6}$$

where ζ is the so-called propellant loading ratio or the fraction of propellant weight in the gross weight. The other vehicle with $w_0 = w_0^*$, $w_s = w_s^*$, but with c and w_e different from c^* and w_e^*. This means that the structural weight and pay load of the vehicle with the alternate power plant remain the same as the rocket, but the engine weight is different. To compare the performance of the power plants, the maximum allowable engine weight w_e for equal performance should be calculated. If the actual engine weight is less than this calculated maximum, then the pay load can be increased over that possible for the rocket. In other words, the rocket motor can be substituted by the new power plant with a net gain in performance. If the actual engine weight is higher than the calculated maximum, then the alternate power plant will give a poorer performance than the rocket engine.

The conditions to determine the maximum allowable engine weight w_e are

$$\begin{aligned} c^* \log \frac{1}{1-\zeta} &= c^* \log \frac{w_0^*}{w_s^* + w_e^*} = \dot{y}_1 + g t_1 \\ c \log \frac{w_0}{w_s + w_e} &= c \log \frac{w_0^*}{w_s^* + w_e} = \dot{y}_1 + g t_1 \end{aligned} \tag{7}$$

Therefore by eliminating $\dot{y}_1$ and t_1, one has

$$\frac{w_e}{w_e^*} - 1 = \left(\frac{w_1^*}{w_e^*}\right)[(1-\zeta)^{-(1-(c^*/c))} - 1] \tag{8}$$

If instead of the effective exhaust velocity, the engine consumption is specified by the specific consumption s in lb of fuel per lb of thrust per unit time, then, since s is inversely proportional to c, Equation (8) can be written as

$$\frac{w_e}{w_e^*} - 1 = \left(\frac{w_1^*}{w_e^*}\right)[(1-\zeta)^{-(1-(s/s^*))} - 1] \tag{9}$$

where s is the specific consumption of the engine under investigation and s^* the specific consumption of the comparison rocket engine. The left sides of Equations (8) and (9) are the maximum allowed increase in engine weight over the rocket engine weight divided by the rocket engine weight. The first factors on the right sides of Equations (8) and (9) are the ratio for the rocket vehicle of the weight at end of powered flight or burnout and the engine weight. The second factor on the right sides of Equations (8) and (9) are plotted in Fig. 2.

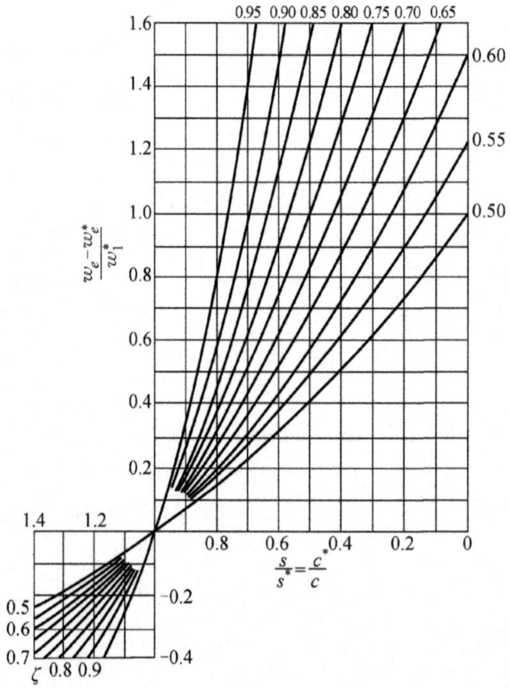

Fig. 2 Engine weight factor as functions of rocket propellant loading ratio ζ and specific fuel consumption s. cf. Equations (8) and (9)

Fig. 2 shows that as the propellant loading ratio ζ for the rocket vehicle increases, the allowable increase in engine weight for a fixed reduction in specific consumption s is extended. Thus there is wider latitude in the choice of power plant for high performance vehicles than there is for a low-performance vehicle. This is certainly an encouraging fact for design engineers. It also points out the fact that the choice of power plant cannot be made independent of the performance of the vehicle but rather is intimately related to the performance of the vehicle.

Applications

As a first example, consider the possibility of using a ramjet as the propulsive power plant for a vehicle of the performance of a V-2 rocket. The weight break-down of a V-2 is given in Table 1[5]:

Therefore $\zeta = 8\,950/12\,980 = 0.69$. The ratio $w_1^* / w_e^* = 4\,030/1\,000 = 4.03$. The effect exhaust velocity including the fuel for turbine drive can be taken as $c^* = 7\,000$ ft/sec or $s^* = 16.56$ lb per lb thrust per hr.

Table 1 V-2 rocket

Pay load:			
Explosive charge	980 kg		
Auxiliary devices	300 kg		
		1 280 kg	
Structural Engine		1 750 kg	
Pumping unit	450 kg		
Rocket motor	550 kg		
		1 000 kg	
Weight at burn-out			4 030 kg
Propellant and Fuel:			
Propellant		8 750 kg	
Fuel for turbine		200 kg	
			8 950 kg
Gross weight			12 980 kg

For a ramjet, the fuel consumption based upon net thrust can be taken as $s = 4$ lb per lb thrust per hr[3]. Then, by using Fig. 2, the maximum allowable increase in engine weight is given by

$$\frac{w_e}{w_e^*} = 6.77$$

This means that the substitute ramjet engine could weigh as much as 14 900 lb without impairing the performance of the V-2. The V-2 rocket has a thrust of 55 000 lb. This is also the average thrust to be produced by the ramjet. Then the maximum allowable ramjet weight is 0.271 lb per lb of thrust. The actual ramjet weight is probably less than this value. Therefore if the ramjet can be made to operate properly under conditions of rapid vertical acceleration and altitude variation up to high altitude, then there is advantage in substituting the ramjet for the rocket engine. This possibility has motivated more accurate calculations of a ramjet in vertical flights by L. H. Schindel[6] and J. V. Rowny[7]. A ramjet as the power plant for the first stage of a high performance seems particularly attractive.

There are studies[8, 9] on the possibility of using the nuclear reactor as an energy source for rocket propulsion. In such a nuclear rocket the reactor may take the form of porous material through which the working fluid, say hydrogen, is passed and heated. A recent investigation by L. Green[10] indicated the feasibility of such a scheme. Now one of the difficulties of using nuclear reactor is the heavy weight of the reactor, particularly when a radiation shield is required. Counteracting the weight is the higher effective exhaust velocity possible. For instance, if hydrogen could be heated to 6 000 °F, then the data given by F. J. Malina and M. Summerfield[11] point to a value of c equal to 26 000 ft/sec or $s = 4.46$ lb per lb of thrust per hr. This aspect of the problem of a nuclear rocket can, however, be analyzed by using the present method. For example, if the high performance rocket with energetic chemical propellant has the following specifications.

$$\zeta = 0.80$$
$$s^* = 10 \text{ lb per lb of thrust per hr}$$
$$w_1^* / w_e^* = 2.7$$

Then according to Equation (9), a nuclear rocket engine giving performance equal to the chemical rocket can have an engine weight 4.89 times the chemical rocket motor. In other words, if the nuclear reactor together with its necessary auxiliary construction increases the engine weight by more than 389 per cent, then the nuclear rocket is not feasible for the performance studied. If the increase in weight is not so much, then the nuclear rocket is worth while.

Detail Improvement of Rocket Engine

A rocket designer is often confronted with the problem of improving a given design at the expense of increasing the engine weight. For example, the effective exhaust velocity of a propellant can be generally increased by increasing the combustion chamber pressure. But increasing the chamber pressure would require heavier construction and an increase in the feed pressure for the propellant. Increasing the feed pressure would in turn increase the weight of the feed system. Therefore the improvement in propellant consumption is to be achieved only with higher engine weight. The present analysis can be used to determine whether such a change will or will not improve the over-all performance of the vehicle.

For small changes in effective exhaust velocity or specific consumption, Equations (8) and (9) can be simplified to relate the allowable fractional increase in engine weight $\Delta w_e / w_e^*$ to fractional increase in effective exhaust velocity $\Delta c / c^*$ or specific consumption $\Delta s / s^*$. Thus, taking only the first order terms, one has from Equations (8) and (9).

$$\frac{\Delta w_e}{w_e^*} = \frac{\Delta c}{c^*}\left(\frac{w_1^*}{w_e^*}\right)\log\left(\frac{1}{1-\zeta}\right) \tag{10}$$

and

$$\frac{\Delta w_e}{w_e^*} = \frac{\Delta s}{s^*}\left(\frac{w_1^*}{w_e^*}\right)\log(1-\zeta) \tag{11}$$

These equations give the allowable fractional increase in engine weight for equal performance. If the actual increase in engine weight is less, then the modification in design improves performance, and is thus desirable. If the actual increase is more, the modification is not practical.

By taking the V-2 as an example, it can be easily calculated that 1 per cent improvement in consumption is worth 4.72 per cent increase in engine weight.

References

[1] von Kármán Th. Comparative Study of Jet Propulsion Systems as Applied to Missiles and Transonic Aircraft.

Memorandum No. JPL-2, 1944, Jet Propulsion Laboratory, California Institute of Technology.
[2] A Guide to Aircraft Power Plant Selection. Aeronautical Engineering Review, April 1947: 22 - 25.
[3] The theoretical performance of ramjet as calculated by many authors; e.g., J. Reid. The Gas Dynamic Theory of the Ramjet. Aeronautical Research Council, London, Report & Memorandum, No. 2370, 1950.
[4] Malina F J, Smith A M O. Flight Analysis of the Sounding Rocket. Journal of the Aeronautical Sciences, 1932, 5: 199 - 202.
[5] Kooy J M J, Uytenbogart J W H. Ballistics of the Future. New York: McGraw-Hill Book Co., Inc., 1946, p.297.
[6] Schindel L H. Application of Ramjet to High Altitude Sounding Vehicle. M. S. thesis (Aeronautical Engineer), Massachusetts Institute of Technology, 1948.
[7] Rowny J V. Application of Ramjet to Vertical Ascent. thesis (Aeronautical Engineer), California Institute of Technology, 1949.
[8] Shepherd J R, Cleaver A V. The Atomic Rocket. Journal of the British Interplanetary Society. 1949, 7: 185 - 194, 234 - 241.
[9] Tsien H S. Rockets and Other Thermal Jets Using Nuclear Energy. in: The Science and Engineering of Nuclear Power, vol. II, Cambridge, Mass.: Addison-Wesley Press, Inc., 1949: 124.
[10] Green Jr. L. Gas Cooling of a Porous Heat Source. Journal of Applied Mechanics. 1952, 19: 173 - 178.
[11] Malina F J, Summerfield M. The Problem of Escape from the Earth by Rocket. Journal of the Aeronautical Sciences, 1947, 14: 471 - 480, particularly Table 2.

Servo-Stabilization of Combustion in Rocket Motors

H. S. Tsien①

(*Daniel and Florence Guggenheim Jet Propulsion Center*, *California Institute of Technology*, *Pasadena*, *Calif.*)

This paper shows that the combustion in the rocket motor can be stabilized against any value of time lag in combustion by a feedback servo link from a chamber pressure pickup, through an appropriately designed amplifier, to a control capacitance on the propellant feed line. The technique of stability analysis is based upon a combination of the Satche diagram and the Nyquist diagram. For simplicity of calculation, only low-frequency oscillations in monopropellant rocket motors are considered. However, the concept of servo-stabilization and method of analysis are believed to be generally applicable to other cases.

The phenomenon of rough burning in liquid-propellant rocket motor has been interpreted as the instability of the coupled system of propellant feed and combustion chamber by D. F. Gunder and D. R. Friant[1], M. Yachter[2], M. Summerfield[3], and L. Crocco[4]. The essential feature of these theories is the time lag between the instant of injection of the propellant and the instant when the propellant is burned into hot gas. Crocco has further improved on this concept by considering the time lag as an integrated effect of consecutive stages, each of which is controlled by the prevailing pressure in the combustion chamber. As a result of this new concept, Crocco showed the possibility of intrinsic instability with constant injection rate not influenced by the chamber pressure.

The present paper will first give a slightly more general formulation of Crocco's concept of time lag, allowing arbitrary pressure dependence of lag. Then the problem of intrinsic stability is discussed by applying a method suggested by M. Satche[5]. This method is based upon a modification of the Nyquist diagram and is particularly useful for systems having time lag. For easy reference, this new diagram will be called the Satche diagram. The later sections of the paper will show the possibility of stabilizing the combustion by means of a feedback servo for all values of time lag. Such possibility of servo-stabilization was first mentioned by W. Bollay in his admirable paper[6] on the application of servomechanisms to aeronautics. The present study definitely shows the power of this idea.

Time Lag in Combustion

Let $\dot{m}_b(t)$ be the mass rate of generation of hot gas by combustion at time instant t.

Received February 22, 1952.

Journal of the American Rocket Society, vol. 22, pp. 256 – 262, 1952.

① Robert H. Goddard Professor of Jet Propulsion.

Consider, for simplicity, a monopropellant motor. Then the mass rate of injection at t can be denoted by $\dot{m}_i(t)$. Let $\tau(t)$ be the time lag for that parcel of propellant which is burned at the instant t. Then the mass burned during the interval from t to $t + dt$ must be equal to the mass injected during the time from $t-\tau$ to $t-\tau+d(t-\tau)$. Thus

$$\dot{m}_b(t)dt = \dot{m}_i(t-\tau)d(t-\tau) \tag{1}$$

The mass of hot gas generated is either used to fill the combustion chamber by raising its pressure $p(t)$, or is discharged through the rocket nozzle. If the frequency of the possible oscillations in the chamber is small, then the pressure in the chamber can be considered as uniform, and as a first approximation[7] the rate of flow through the nozzle can be taken as proportional to the instantaneous chamber pressure $p(t)$. Thus if $\overline{\dot{m}}$ is the steady mass rate flow through the system, $\overline{M}_g$ is the average mass of hot gas in the chamber, and if the volume occupied by the unburned liquid propellant is neglected

$$\dot{m}_b dt = \overline{\dot{m}}\left(\frac{p}{\bar{p}}\right)dt + d\left(\overline{M}_g\frac{p}{\bar{p}}\right) \tag{2}$$

where $\bar{p}$ is the steady state pressure in the combustion chamber.

By following Crocco, the nondimensional variables for the chamber pressure and the rate of injection are defined as

$$\varphi = \frac{p-\bar{p}}{\bar{p}}, \quad \mu = \frac{\dot{m}_i - \overline{\dot{m}}}{\overline{\dot{m}}} \tag{3}$$

φ and μ are then the fractional deviation of pressure and injection rate from the average. With Equation (3), $\dot{m}_b$ can be eliminated from Equations (1) and (2), and

$$\frac{\overline{M}_g}{\overline{\dot{m}}}\frac{d\varphi}{dt} + \varphi + 1 = \left(1 - \frac{d\tau}{dt}\right)[\mu(t-\tau)+1] \tag{4}$$

To calculate the quantity $d\tau/dt$, Crocco's concept of pressure dependence of time lag has to be introduced. If the rate at which the liquid propellant is prepared for the final rapid transformation into hot gas is a function $f(p)$, then the lag τ is determined by

$$\int_{t-\tau}^{t} f(p)dt = \text{const} \tag{5}$$

By differentiating Equation (5) with respect to t,

$$[f(p)]_t - [f(p)]_{t-\tau}\left(1 - \frac{d\tau}{dt}\right) = 0$$

The concept of small perturbation from the steady state will now be explicitly introduced: Assume that the deviation of the pressure p from the steady state value $\bar{p}$ is small. Then $f(p)$ at the instant t and $f(p)$ at the instant $t-\tau$ can be expanded as Taylor's series around $\bar{p}$. By taking only the first order terms,

$$[f(p)]_t = f(\bar{p}) + \bar{p}\left(\frac{df}{dp}\right)_{p=\bar{p}} \varphi(t)$$

$$[f(p)]_{t-\tau} = f(\bar{p}) + \bar{p}\left(\frac{df}{dp}\right)_{p=\bar{p}} \varphi(t-\tau)$$

Here τ is the lag at the average pressure $\bar{p}$, a constant now. Then

$$1 - \frac{d\tau}{dt} = 1 + \left(\frac{d\log f}{d\log p}\right)_{p=\bar{p}} [\varphi(t) - \varphi(t-\tau)] \tag{6}$$

By combining Equations (4) and (6), the following equation is obtained

$$\frac{d\varphi}{dz} + \varphi = \mu(z-\delta) + n[\varphi(z) - \varphi(z-\delta)] \tag{7}$$

where

$$n = \left(\frac{d\log f}{d\log p}\right)_{p=\bar{p}} \tag{8}$$

and

$$z = t/\theta_g, \quad \theta_g = \overline{M}_g / \overline{\dot{m}} \tag{9}$$

If n is a constant independent of $\bar{p}$, then $f(p)$ is proportional to p^n. This is the form of $f(p)$ assumed by Crocco. The present formulation of the problem is slightly more general in that $f(p)$ is arbitrary and the value of n is to be computed by using Equation (8), and is a function of $\bar{p}$. θ_g is, of course, the gas transit time.

Intrinsic Instability

Crocco called the instability of combustion with constant rate of injection the intrinsic instability. If the injection rate is constant and not influenced by the chamber pressure p, then $\mu \equiv 0$. Therefore the stability problem is controlled by the following simple equation obtained from Equation (7),

$$\frac{d\varphi}{dz} + (1-n)\varphi(z) + n\varphi(z-\delta) = 0 \tag{10}$$

Now let

$$\varphi(z) \sim e^{sz}$$

Then

$$s + (1-n) + n e^{-\delta s} = 0 \tag{11}$$

This is the equation for the exponent s.

Crocco determined the value of the complex number s by studying the set of two equations for the real and the imaginary parts of Equation (11). However, if the point of interest is whether the system is stable or not, one can use the well-known Cauchy theorem with advantage. Let

$$G(s) = e^{-\delta s} - \left[-\frac{1-n}{n} - \frac{s}{n} \right] \tag{12}$$

Then the question of stability is determined by whether $G(s)$ has zeros in the right half of the complex s-plane. This question itself can be in turn answered by watching the argument of $G(s)$ when s traces a contour enclosing the right half s-plane. Specifically, let s trace clockwise the contour consisting of the imaginary axis and a large half circle to the right of the imaginary axis (Fig. 1). If the vector $G(s)$ makes a number of complete clockwise revolutions, then that number is, according to Cauchy's theorem, the difference between the number of zeros and the number of poles of $G(s)$ in the right half s-plane. Since $G(s)$ evidently has no poles in the s-plane, the number of revolutions of $G(s)$ is the number of zeros. Hence for stability, the vector $G(s)$ must not make any complete revolutions, as s traces the specified contour. Therefore the stability question can be answered by plotting graphically $G(s)$ on the complex plane. This graph is, of course, the well-known Nyquist diagram.

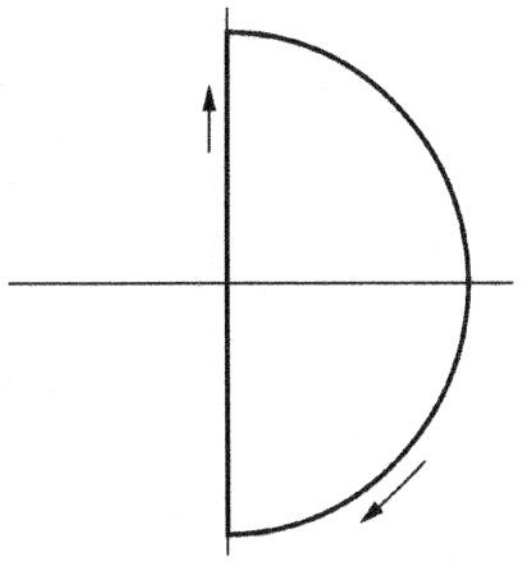

Fig. 1 Contour traced by the variable s for the satche diagram or the nyquist diagram

A direct application of this method to $G(s)$ given by Equation (12) is, however, inconvenient for the complication caused by lag term $e^{-\delta s}$ [8]. M. Satche[5], however, proposed a very elegant and ingenious method of treating such a system with time lag: Instead of $G(s)$, break it into two parts,

$$G(s) = g_1(s) - g_2(s) \tag{13}$$

where

$$g_1(s) = e^{-\delta s}$$

$$g_2(s) = -\frac{1-n}{n} - \frac{s}{n} \tag{14}$$

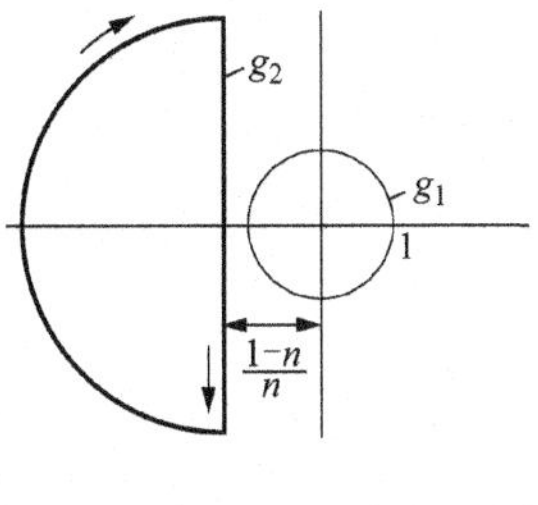

Fig. 2 Stable satche diagram for intrinsic oscillations; $0 < n < 1/2$

The vector $G(s)$ is thus a vector with vertex in $g_1(s)$ and its tail on $g_2(s)$. The graph of $g_1(s)$ is the unit circle for s on the imaginary axis. For s on the large half circle, $g_1(s)$ is within the unit circle. The graph of $g_2(s)$ is the straight line (Fig. 2) paralleled to the imaginary axis when s is on the imaginary axis. When s is on the large half circle, $g_2(s)$ is a half of a large circle closing the contour on the left. A moment's reflection will show that in order for the vector $G(s)$ not to make complete revolutions for any value of δ, the $g_2(s)$ contour must lie completely out of the $g_1(s)$ contour. That is, for unconditional intrinsic stability

$$\frac{1-n}{n} > 1 \quad \text{or} \quad \frac{1}{2} > n > 0 \tag{15}$$

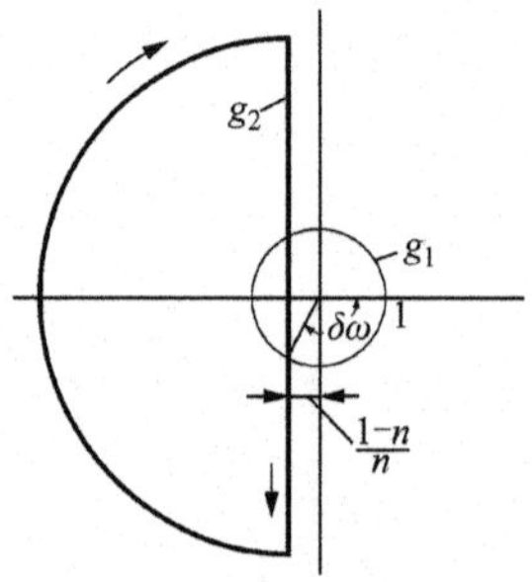

Fig. 3 Unstable satche diagram for intrinsic oscillations; $n > 1/2$

When $n > 1/2$, the $g_1(s)$ contour and the $g_2(s)$ contour intersect. Stability is still possible, however, if for $g_2(s)$ within the unit circle (Fig. 3), $g_1(s)$ is to the right of $g_2(s)$. This condition is satisfied if

$$\cos(\delta\sqrt{2n-1}) > -\frac{1-n}{n}$$

Or if

$$\delta < \delta^*$$

where

$$\delta^* = \frac{1}{\sqrt{2n-1}}\cos^{-1}\left(-\frac{1-n}{n}\right) = \frac{1}{\sqrt{2n-1}}\left(\pi - \cos^{-1}\frac{1-n}{n}\right) \tag{16}$$

When $\delta = \delta^*$, then with

$$\omega^* = \sqrt{2n-1} \tag{17}$$

$G(i\omega^*) = 0$. Therefore when $\delta = \delta^*$, φ has the oscillatory solution with the angular frequency ω^*.

These results on intrinsic stability were obtained by Crocco. The present discussion with the Satche diagram, however, seems to be simpler. For the more complicated stability problem treated below with feed system and servo control, the solution is hardly practical without the Satche diagram.

System Dynamics with Servo Control

Consider now a system including the propellant feed and a servo control represented by Fig. 4. In order to approximate the elasticity of the feed line, a spring load capacitance is put at the midway point between the propellant pump and the injector. The spring constant is to be computed from the feed-line dimensions.① Near the injector there is another capacitance controlled by the servo. The servo receives its signal from the chamber pressure pickup through an amplifier. If the feed system and the motor design are fixed by the designer, the question is whether it is possible to design an appropriate amplifier so that the whole system will be stable. Because there is no accurate information on the time lag of combustion, a practical design should specify unconditional stability, i.e., stability for any value of δ.

① See the Appendix for details.

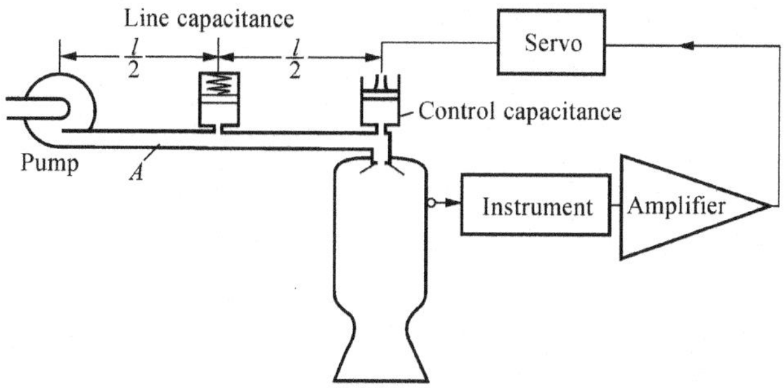

Fig. 4 Servo-controlled liquid monopropellant rocket motor

Let $\dot{m}_0$ be the instantaneous mass flow rate out of the propellant pump, and p_0 be the instantaneous pressure at the outlet of pump. The average flow rate must be $\overline{\dot{m}}$. The average pressure is $\bar{p}_0$. The pump characteristics can be represented by the following equation,

$$\frac{p_0 - \bar{p}_0}{\bar{p}_0} = -\alpha \frac{\dot{m}_0 - \overline{\dot{m}}}{\dot{m}} \tag{18}$$

If the time rate of change of mass flow is small, α is simply related to the slope of the head-volume curve of the pump at constant speed near the steady-state operating point. For constant pressure pump or the simple pressure feed, α is zero. For conventional centrifugal pumps, α is approximately 1. For displacement pumps, α is very large.

Let $\dot{m}_1$ be the instantaneous mass rate of flow after the spring loaded capacitance, χ the spring constant of the capacitance, and p_1 the instantaneous pressure at the capacitance. Then

$$\dot{m}_0 - \dot{m}_1 = \rho\chi \frac{\mathrm{d}p_1}{\mathrm{d}t} \tag{19}$$

where ρ is the density of the propellant, a constant.

In the following calculation, the pressure drop in the line by frictional forces will be neglected. Then the pressure difference $p_0 - p_1$ is due to the acceleration of the flow only. That is

$$p_0 - p_1 = \frac{l}{2A} \frac{\mathrm{d}\dot{m}_0}{\mathrm{d}t} \tag{20}$$

where A is the cross-sectional area of the feed line, a constant, and l is the total length of the feed line. Similarly, if p_2 is the instantaneous pressure at the control capacitance.

$$p_1 - p_2 = \frac{l}{2A} \frac{\mathrm{d}\dot{m}_1}{\mathrm{d}t} \tag{21}$$

If the mass capacity of the control capacitance is C, then

$$\dot{m}_1 - \dot{m}_i = \frac{\mathrm{d}C}{\mathrm{d}t} \tag{22}$$

Since the control capacitance is very close to the injector, the inertia of the mass of propellant between the control capacitance and the injector is negligible. Then

$$p_2 - p = \frac{1}{2}\frac{\dot{m}_i^2}{\rho A_i^2} \tag{23}$$

where A_i is the effective orifice area of the injector. A_i can be eliminated from the calculation by noting that at steady state, the difference of pressures $\bar{p}_0$ and $\bar{p}$, or $\Delta\bar{p}$ is

$$\bar{p}_0 - \bar{p} = \Delta\bar{p} = \frac{1}{2}\frac{\bar{\dot{m}}_i^2}{\rho A_i^2} \tag{24}$$

Equations (18) to (24) describe the dynamics of the feed system. By a straightforward process of elimination of variables, a relation between $\dot{m}_i$, p, and C is obtained. To express this relation in nondimensional form, the following quantities are introduced, following the notation of Crocco:

$$P = \frac{\bar{p}}{2\Delta\bar{p}}, \quad E = \frac{2\Delta\bar{p}}{\bar{\dot{m}}\theta_g}\rho\chi, \quad J = \frac{l\dot{m}}{2\Delta\bar{p}A\theta_g} \tag{25}$$

and

$$\kappa = C/(\bar{\dot{m}}\,\theta_g) \tag{26}$$

where θ_g is the gas transit time given by Equation (9). Then the nondimensional equation relating φ, μ, and κ is

$$P\left\{1 + E\left(P + \frac{1}{2}\right)\frac{d}{dz} + \frac{JE}{2}\frac{d^2}{dz^2}\right\}\varphi + \left[\left\{1 + \alpha\left(P + \frac{1}{2}\right)\right\} + \left\{\alpha E\left(P + \frac{1}{2}\right) + J\right\}\frac{d}{dz} + \left\{\frac{\alpha JE}{2}\left(P + \frac{1}{2}\right) + \frac{JE}{2}\right\}\frac{d^2}{dz^2} + \frac{J^2E}{4}\frac{d^3}{dz^3}\right]\mu + \left[\alpha\left(P + \frac{1}{2}\right)\frac{d}{dz} + J\frac{d^2}{dz^2} + \frac{\alpha JE}{2}\left(P + \frac{1}{2}\right)\frac{d^3}{dz^3} + \frac{J^2E}{4}\frac{d^4}{dz^4}\right]\kappa = 0 \tag{27}$$

where z is the nondimensional time variable defined by Equation (9).

The dynamics of the servo control are specified by the composite of the instrument characteristics of the pressure pickup, the response of the amplifier, and the properties of the servo. Since it is not the purpose of the present paper to discuss the detailed design of the servo control, the over-all dynamics of the servo control are represented by the following operator equation:

$$F\left(\frac{d}{dz}\right)\varphi = \kappa \tag{28}$$

where F is the ratio of two polynomials with the denominator of higher order than the numerator.

Equations (7), (27), and (28) are the three equations for the three variables φ, μ, and κ. Since they are equations with constant coefficients, the appropriate forms for the variables are

$$\varphi = a\,e^{sz}, \quad \mu = b\,e^{sz}, \quad \kappa = c\,e^{sz} \tag{29}$$

By substituting Equation (29) into Equations (7), (27), and (28), three homogeneous equations for a, b, and c are obtained. In order for a, b, c to be nonzero, the determinant formed by their coefficients must vanish. This condition can be written as follows:

$$\begin{aligned}&[s+(1-n)]\left[\frac{J^2E}{4}s^3+\frac{JE}{2}\left\{1+\alpha\left(P+\frac{1}{2}\right)\right\}s^2+\right.\\&\left.\left\{\alpha E\left(P+\frac{1}{2}\right)+J\right\}s+\left\{1+\alpha\left(P+\frac{1}{2}\right)\right\}\right]+\\&e^{-\delta s}\left\{\frac{nJ^2E}{4}s^3+\left[\frac{nJE}{2}\left\{1+\alpha\left(P+\frac{1}{2}\right)\right\}+\frac{JEP}{2}\right]s^2+\right.\\&\left[n\left\{\alpha E\left(P+\frac{1}{2}\right)+J\right\}+\alpha EP\left(P+\frac{1}{2}\right)\right]s+\\&\left[n\left\{n+\alpha\left(P+\frac{1}{2}\right)\right\}+P\right]+\\&\left.sF(s)\left[\frac{J^2E}{4}s^3+\frac{\alpha JE}{2}\left(P+\frac{1}{2}\right)s^2+Js+\alpha\left(P+\frac{1}{2}\right)\right]\right\}=0\end{aligned} \tag{30}$$

This is the equation for determining the exponent s. $F(s)$ is now recognized as the over-all transfer function of the servo-control link. The complete system stability depends upon whether Equation (30) gives roots that have positive real parts.

Instability Without Servo Control

The system characteristics without the servo control can be simply obtained from the basic Equation (30) by setting $F(s)=0$. Let it be assumed that the polynomial multiplied into $e^{-\delta s}$ has no zero in the positive half s-plane, as is usually the case. Then Equation (30) can be divided by that polynomial without introducing poles in the positive half s-plane into the resultant function. That is, for the Satche diagram, one has again

$$G(s)=g_1(s)-g_2(s),\quad g_1(s)=e^{-\delta s}$$

$g_1(s)$ is thus again the "unit circle." $g_2(s)$ is now much more complicated:

$$g_2(s)=-\left[\frac{s}{n}+\frac{1-n}{n}\right]\times\frac{\frac{J^2E}{4}s^3+\frac{JE}{2}\left\{1+\alpha\left(P+\frac{1}{2}\right)\right\}s^2+\left\{\alpha E\left(P+\frac{1}{2}\right)+J\right\}s+\left\{1+\alpha\left(P+\frac{1}{2}\right)\right\}}{\frac{J^2E}{4}s^3+\frac{JE}{2}\left\{1+\alpha\left(P+\frac{1}{2}\right)+\frac{P}{n}\right\}s^2+\left\{\alpha E\left(P+\frac{1}{2}\right)\left(1+\frac{P}{n}\right)+J\right\}s+\left\{1+\alpha\left(P+\frac{1}{2}\right)+\frac{P}{n}\right\}} \tag{31}$$

The intercept of $g_2(s)$, when s is pure imaginary, is given by setting $s=0$ in Equation (31), i.e.,

$$g_2(0)=-\frac{1-n}{n}\,\frac{1+\alpha\left(P+\frac{1}{2}\right)}{1+\alpha\left(P+\frac{1}{2}\right)+\frac{P}{n}} \tag{32}$$

Since all the parameters n, α, P are positive, the magnitude of $g_2(0)$ is now smaller than the magnitude of $g_2(0)$ given by Equation (14) for the intrinsic stability problem. Thus the effect of the feed system is to move the $g_2(s)$ curve toward the unit circle of $g_1(s)$ in the Satche diagram. For instance, for $n = 1/2$, $g_2(s)$ is just tangent to the unit circle for the intrinsic system without considering the propellant feed. But with the propellant feed system, $g_2(s)$ contour will intersect the unit circle and the system will become unstable, for time lag δ exceeds a certain finite value. The influence of the feed system is thus always destabilizing. This is further confirmed by considering the asymptote of $g_2(s)$ for large imaginary s, obtained from Equation (31). That is

$$g_2(s) \sim -\left[\frac{s}{n} + \left(\frac{1-n}{n} - \frac{2P}{Jn^2}\right) + \ldots\right], \quad |s| \gg 1 \tag{33}$$

Therefore, for large imaginary s, $g_2(s)$ approaches asymptotically a line parallel to the imaginary axis at a distance

$$\frac{1-n}{n} - \frac{2P}{Jn^2}$$

to the left of the imaginary axis. The effect of the feed system is again to move $g_2(s)$ toward the unit circle.

It is thus evident that for the parameter n near 1/2 or larger than 1/2, it would be impossible to design the system for unconditional stability. In the Satche diagram, $g_1(s)$ contour and $g_2(s)$ will always intersect without a servo control.

Complete Stability with Servo Control

If the polynomial $H(s)$

$$\begin{aligned} H(s) = {} & \frac{J^2E}{4}s^3 + \left[\frac{JE}{2}\left\{1 + \alpha\left(P + \frac{1}{2}\right)\right\} + \frac{JEP}{2n}\right]s^2 + \\ & \left[\alpha E\left(P + \frac{1}{2}\right) + \frac{\alpha EP}{n}\left(P + \frac{1}{2}\right)\right]s + \\ & \left[1 + \alpha\left(P + \frac{1}{2}\right) + \frac{P}{n}\right] + \\ & \frac{1}{n}sF(s)\left[\frac{J^2E}{4}s^3 + \frac{\alpha JE}{2}\left(P + \frac{1}{2}\right)s^2 + Js + \alpha\left(P + \frac{1}{2}\right)\right] \end{aligned} \tag{34}$$

which multiplies into $e^{-\delta s}$ in Equation (30), has no poles and zeros in the right half s-plane, then the occurrence zeros of the expression in Equation (30) in the right half s-plane can be determined from the Satche diagram with

$$g_1(s) = e^{-\delta s}$$

and

$$g_2(s) = -\left[\frac{s}{n} + \frac{1-n}{n}\right]\left\{\frac{J^2E}{4}s^3 + \frac{JE}{2}\left[1+\alpha\left(P+\frac{1}{2}\right)\right]s^2 + \left[\alpha E\left(P+\frac{1}{2}\right)+J\right]s + \left[1+\alpha\left(P+\frac{1}{2}\right)\right]\right\}\Big/H(s) \tag{35}$$

As s traces the contour of Fig. 1, $g_1(s)$ is again a unit circle. Therefore, if simultaneously the $g_2(s)$ contour is completely outside the unit circle, there can be no root of Equation (30) in the right half s-plane. In other words, if the transfer function $F(s)$ of the servo-control link is so designed as to place the $g_2(s)$ contour completely out of the unit circle (Fig. 5), then the system is stabilized for all time lags.

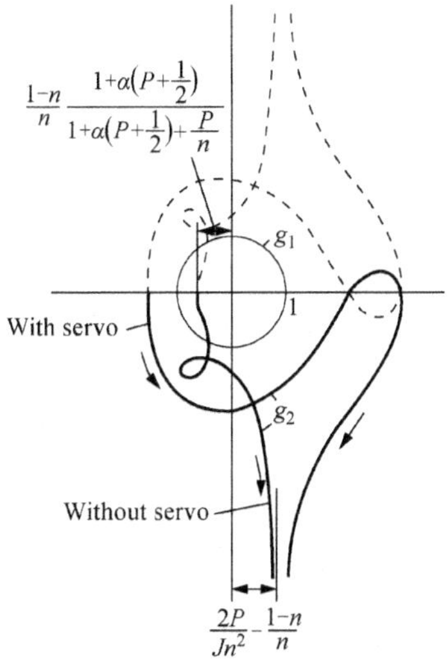

Fig. 5 Satche diagram for the original and for the servo-stabilized system

As an example, take

$$n = \frac{1}{2}, \quad P = \frac{3}{2}, \quad J = 4, \quad E = \frac{1}{4}, \quad \alpha = 1$$

Then without the servo control, the $g_2(s)$ is

$$g_2(s) = -\frac{1}{2}\frac{(2s+1)(2s^3+3s^2+9s+6)}{s^3+3s^2+6s+6}$$

Of primary interest is the behavior of $g_2(s)$ when s is a pure imaginary number $i\omega$, ω real. Thus

$$g_2(i\omega) = -\frac{1}{2}\frac{(6-21\omega^2+4\omega^4)(6-3\omega^2)+\omega^2(21-8\omega^2)(6-\omega^2)}{(6-3\omega^2)^2+\omega^2(6-\omega^2)^2} - \frac{1}{2}i\omega\frac{(21-8\omega^2)(6-3\omega^2)-(6-21\omega^2+4\omega^4)(6-\omega^2)}{(6-3\omega^2)^2+\omega^2(6-\omega^2)^2}$$

This contour for $\omega \geqslant 0$ is plotted in Fig. 6. It is evident that for sufficiently large values of time lag, the system will be unstable. On the other hand, if the $g_2(s)$ contour can be changed by the servo control to, say,

$$g_2(s) = -2\frac{(s+2)(s+3)}{(s+6)}$$

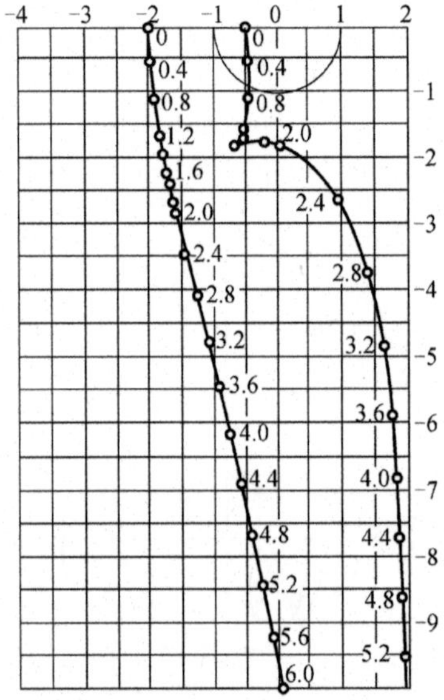

Fig. 6 Satche diagram for the original and for the servo-stabilized system
$P = 3/2, J = 4, E = 1/4, \alpha = 1$
($g_2(i\omega)$ without servo intersects the unit circle;
$g_2(i\omega)$ with servo is outside the unit circle. Numbers beside points are the value of ω)

Then, as plotted in Fig. 6, the new g_2 contour is completely outside of the unit circle of $g_1(s)$. Therefore the system is now unconditionally stable. A straight-forward calculation from Equations (31) and (35) shows that the required transfer function $F(s)$ for the servo link is

$$F(s) = -4.875\frac{(s+1.0528)(s^2+0.7164s+2.6304)}{s(s+2)(s+3)(s+0.5332)(s^2+0.4668s+3.7511)}$$

The servo link has thus the character of an integrating circuit. If, with given response of the chamber pressure pickup and of the servo for the control capacitance, an amplifier could be designed to give an over-all transfer function close to that specified above, the combustion can be stabilized by such a servo control.

As the second example, take

$$n = \frac{1}{2}, \quad P = \frac{3}{2}, \quad J = 4, \quad E = \frac{1}{4}, \quad \alpha = 0$$

Since $\alpha = 0$, the feed pressure p_0 is thus constant with even variable flow of propellant. The case then corresponds to that of a simple pressure feed. Without the servo control,

$$g_2(s) = -\frac{1}{2}\frac{(2s+1)(2s^3+s^2+8s+2)}{s^3+2s^2+4s+4}$$

When s is pure imaginary,

$$g_2(i\omega) = -\frac{1}{2}\frac{(4-2\omega^2)(2-17\omega^2+4\omega^4)+\omega^2(4-\omega^2)(12-4\omega^2)}{(4-2\omega^2)^2+\omega^2(4-\omega^2)^2} - \frac{1}{2}i\omega\frac{(4-2\omega^2)(12-4\omega^2)-(4-\omega^2)(2-17\omega^2+4\omega^4)}{(4-2\omega^2)^2+\omega^2(4-\omega^2)^2}$$

This contour of g_2 is plotted in Fig. 7. It is evident that without servo control the combustion will be unstable for sufficiently long time lag. In fact, the system is even less stable than the system considered in the first example: It will become unstable at shorter time lag. The part of the g_2 contour near $\omega = 2$ is of special interest. Near $\omega = 2$, the contour comes so close to the unit circle of g_1 that if the value of time lag δ is such as to make g_1 and g_2 for $\omega \sim 2$ very close to each other, then an almost undamped oscillation at $\omega \sim 2$ can occur. This critical value of δ is evidently smaller than the critical δ determined from the true intersection of g_2 with the unit circle at $\omega \sim 0.65$. Such near instability at smaller values of time lag can be easily overlooked in the analytic treatment of the stability condition by Crocco, and yet such possible instability should not be dismissed. This, perhaps, indicates the superiority of the present graphical method.

For unconditional stability, g_2 should be displaced out of the unit circle, to, say, the same

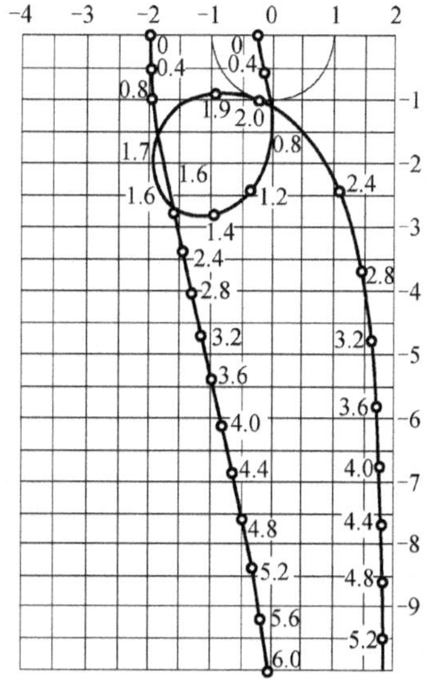

Fig. 7 Satche diagram for the original and for the servo-stabilized system

$P = 3/2, J = 4, E = 1/4, \alpha = 0$

($g_2(i\omega)$ without servo intersects the unit circle;

$g_2(i\omega)$ with servo is outside the unit circle. Numbers beside points are the value of ω)

"stable" contour as in the first example. The required transfer function $F(s)$ is calculated to be

$$F(s) = -4.875 \frac{(s+0.8126)(s^2-0.04337s+2.6506)}{s^2(s+2)(s+3)(s^2+4)}$$

The required servo link must then have the character of double integrating circuit. Furthermore, the transfer function has two purely imaginary poles at $\pm 2i$. This unrealistic requirement on the amplifier comes from the original feed-system dynamics and is due to the neglect of frictional damping in the feed line. In any actual system, the frictional damping in the feed line will remove these purely imaginary poles of the required transfer function $F(s)$ and replace them by two complex conjugate poles.

Stability Criteria

In the preceding discussion of servo-stabilization, it is assumed that the polynomial $H(s)$, Equation (34), has no pole or zero in the right half s-plane. This is, however, not necessarily the case. In general then, one should first investigate the number of zeros and poles of $H(s)$ in the right half s-plane. To do this, it should be recognized that the polynomial in Equation (34) before the factor $F(s)$ usually does not have zeros in the right half s-plane. Therefore instead of studying $H(s)$, one can study the ratio of $H(s)$ and that polynomial. That is, the number of zeros and poles of $H(s)$ in the right half s-plane is the same as the number of zeros and poles of the following function

$$1+K(s) = \frac{H(s)}{\frac{J^2E}{4}s^3+\left[\frac{JE}{2}\left\{1+\alpha\left(P+\frac{1}{2}\right)\right\}+\frac{JEP}{2n}\right]s^2+\left[\alpha E\left(P+\frac{1}{2}\right)+\frac{\alpha EP}{n}\left(P+\frac{1}{2}\right)\right]s+\left[1+\alpha\left(P+\frac{1}{2}\right)+\frac{P}{n}\right]} \tag{36}$$

where

$$K(s) = \frac{\frac{1}{n}sF(s)\left[\frac{J^2E}{4}s^3+\frac{\alpha JE}{2}\left(P+\frac{1}{2}\right)s^2+Js+\alpha\left(P+\frac{1}{2}\right)\right]}{\frac{J^2E}{4}s^3+\left[\frac{JE}{2}\left\{1+\alpha\left(P+\frac{1}{2}\right)\right\}+\frac{JEP}{2n}\right]s^2+\left[\alpha E\left(P+\frac{1}{2}\right)+\frac{\alpha EP}{n}\left(P+\frac{1}{2}\right)\right]s+\left[1+\alpha\left(P+\frac{1}{2}\right)+\frac{P}{n}\right]} \tag{37}$$

According to the Nyquist criterion, the number of poles and zeros for $1+K(s)$ in the right half s-plane can be found by plotting the Nyquist diagram of $1+K(s)$ with s tracing the contour of Fig. 1. In fact, if $1+K(s)$ or $H(s)$ has r zeros and q poles in right half s-plane then $K(s)$ will carry out $r-q$ clockwise revolutions around the point -1, as s traces the contour of Fig. 1. Hence the necessary information on $H(s)$ can be obtained by plotting the Nyquist diagram of $K(s)$.

When one divides the Equation (30) by $H(s)$ in order to obtain $g_1(s)$ and $g_2(s)$ as given by Equation (35), g zeros and r poles are introduced in the right half s-plane. The g poles of $K(s)$ must come from $F(s)$, since the polynomial in the denominator of Equation (37) has no zero in the right half s-plane. Therefore the original expression in Equation (30) also has g poles in the right half s-plane. Hence in order for the original expression in Equation (30) to have no zero in the right half s-plane, $g_2(s)$ must make $-q+(q-r)=-r$ clockwise revolutions around the unit circle. In order for stability to be unconditional, i.e., stable for all time lag, the $g_2(s)$ contour should never intersect the unit circle. Therefore the general unconditional stability criteria are, first, $g_2(s)$ contour completely outside of the unit circle; and, second, $g_2(s)$ making r counterclockwise revolutions around the unit circle as s traces the conventional contour enclosing the right half s-plane. These are the criteria for stability with the Satche diagram. To determine r, one has to use the Nyquist diagram of $K[s]$, Equation (37). Thus the stability problem for the general case requires both the Satche diagram and the Nyquist diagram (Fig. 8).

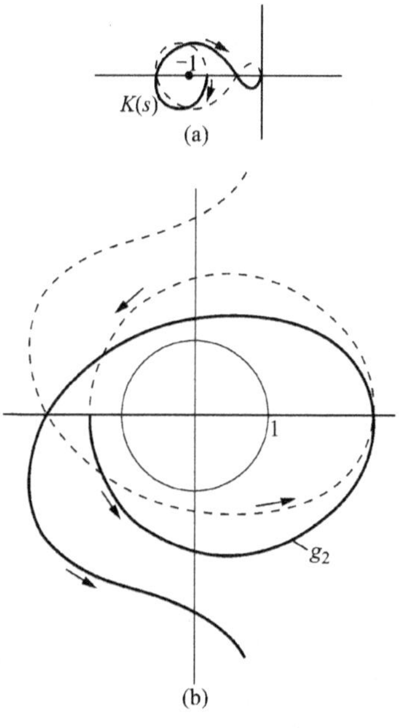

Fig. 8 Full curve for positive ω; dotted curve for negative ω

(a) Nyquist diagram for $K(s)$, with two zeros for $1+K(s)$ in right half s-plane.

(b) Corresponding stable Satche diagram

Concluding Remarks

In the previous sections of this paper, the theoretical possibility of completely stabilizing the combustion for any value of time lag by servo control is demonstrated. The great flexibility of electronic amplifier seems to indicate that this theoretical possibility can be always realized.

On the other hand, without the servo link, unconditional stability is shown to be generally impossible. Therefore the concept of feedback servo is indeed a powerful tool in controlling the behavior of a time-lag system. It is to be realized, of course, that the proposed scheme is but one among many. No attempt is made here to give an exhaustive treatment of all possible schemes. The best scheme is certainly to be determined by detailed considerations on all aspects of the engineering problem, such as the possibility of high-frequency acoustic oscillations which are not considered here. The main purpose here is to give a general discussion of the concept together with a suggested general method of analyzing the stability by the Satche diagram.

It is of interest to point out that stabilization by servo control is only one phase of the general concept of feedback link. The opposite case of destabilization could be of importance also. For instance, consider the so-called valveless pulsejet. It is not always possible to operate the engine with the desired pulsation. With a feedback servo linking the combustion chamber pressure pickup through an amplifier to the fuel line, the system can be destabilized at the desired operating frequency and thus operate the engine at that frequency of pulsation. This application of servodestabilization gives the valveless pulsejet a new flexibility and an extended range of operation. Therefore it seems worth while to explore carefully all possible applications of feedback control to systems with time lag.

Appendix

Calculation of Parameters J and E

If L^* and c^* are the characteristic length and the characteristic velocity of the motor, and if T_c is the chamber temperature, R the gas constant, the transit time θ_g is $\theta_g = L^* c^* / RT_c$.

To calculate J and E defined by Equation (25), it is more convenient to use the average propellant velocity v in the feed line. Thus $\bar{m} = \rho A v$.

Thus, according to Equation (25)

$$J = \frac{1}{2}\rho v\left(\frac{l}{\theta_g}\right)\Big/\Delta\bar{p}$$

A consistent set of units would be ρ in slugs per cubic foot, v in feet per second, l in feet, θ_g in seconds, and $\Delta\bar{p}$ in pounds per square foot.

If d is the diameter of the feed line, h its thickness, and E' Young's modulus of the tube material, then χ, the change in volume of the feed line per unit rise in pressure, is

$$\chi = l\pi\left(\frac{\mathrm{d}}{2}\right)^2 d\Big/(E'h)$$

Therefore Equation (25) gives

$$E = \frac{2\Delta\bar{p}}{E'}\left(\frac{d}{h}\right)\frac{l/\theta_g}{v}$$

A consistent set of units would be $\Delta \bar{p}$ in pounds per square inch, E' in pounds per square inch, l in feet, θ_g in seconds, and v in feet per second.

References

[1] Gunder D F, Friant D R. Stability of Flow in a Rocket Motor. Journal of Applied Mechanics, 1950, 17: 327 - 333.

[2] Yachter M. Discussion of above paper. Journal of Applied Mechanics, 1951, 18: 114 - 115.

[3] Summerfield M. A Theory of Unstable Combustion in Liquid Propellant Rocket Systems. Journal of the American Rocket Society, 1951, 21: 108 - 114.

[4] Crocco L. Aspects of Combustion Stability in Liquid Propellant Rocket Motors, Parts I and II. Journal of the American Rocket Society, 1951, 21: 163 - 178; 1952, 22: 7 - 16.

[5] Satche M, Discussion on "Stability of Linear Oscillating Systems with Constant Time Lag (by Ansoff H. I.)". Journal of Applied Mechanics, 1949, 16: 419 - 420.

[6] Bollay W. Aerodynamic Stability and Automatic Control. Journal of Aeronautical Sciences, 1951, 18: 569 - 623, 605.

[7] Tsien H S. The Transfer Functions of Rocket Nozzles. Journal of the American Rocket Society, 1952, 22: 139 - 143.

[8] Ansoff H I. Stability of Linear Oscillating Systems with Constant Time Lag. Journal of Applied Mechanics, 1949, 16: 158 - 164.

Physical Mechanics, A New Field in Engineering Science

H. S. Tsien①

(*Daniel and Florence Guggenheim Jet Propulsion Center, California Institute of Technology, Pasadena, Calif.*)

The purpose of physical mechanics is to predict the engineering behavior of matter in bulk from the microscopic properties of its molecular and atomic constituents. The constants and basic concepts of this new engineering science, of particular importance to rocket and jet propulsion, are discussed in this paper.

The term physical mechanics has been used in the past to indicate a course in classical mechanics for sophomores and juniors in college[1]. Physical mechanics here is used to designate a new field in engineering science, the purpose of which is to predict the macroscopic behavior of matter, of interest to engineers, from the known microscopic properties of the constituents of matter. The need for such a branch of science arises originally from the advanced engineering problems in jet propulsion, aeronautics, and atomic power, but impact of this new discipline is inevitable on all fields of engineering. This article will discuss the contents of physical mechanics and its point of view. Above all, the purpose here is to attract the attention of scientists and engineers to this new and fruitful field.

Basic Concepts

The constituents of matter are molecules and atoms. The atom, in turn, consists of a central nucleus and an electron cloud surrounding the nucleus. According to the prevailing view, the nucleus is ultimately made up of protons and neutrons. This relentless drive of physicist to the "heart" of the matter was perhaps motivated by the desire to interpret all nature phenomena by a unified theory from the elementary structures. During this development of physical science in the past century, there was a continuous interplay between two phases of the study. One phase was the investigation of the structure of molecules and atoms by analytical procedures such as x-ray analysis and electron diffraction, molecular and atomic spectroscopy. The other phase was the explanation of the bulk properties of matter such as heat capacity, the

---

Received September 12, 1952.

Journal of the American Rocket Society, vol.23, pp.14-16, 1953.

① Robert H. Goddard Professor of Jet Propulsion. The author is indebted to Dr. S. S. Penner, Assistant Professor of Jet Propulsion, for many stimulating discussions during the formulation of concepts in this paper.

pressure of fluids, from the molecular and atomic structure. The second phase of the investigation, developed by the use of the statistical mechanics and the kinetic theory, was of particular importance to the physicist and the chemists in that the earlier pictures of the molecular and atomic structures were quite uncertain and vague. Therefore the physical scientists needed the comfort of seeing their theory verified by "every-day" experience with matter in bulk.

The present knowledge of the molecular and atomic structures is, however, very complete and well founded. To the physical scientists then, the interpretation of the macroscopic behavior of matter from the atomic theory is only of side interest. A physicist's main interest is one step deeper: The structure of atomic nuclei and the properties of their constituent elementary particles. This situation in physics and chemistry leads naturally to a reversal of the procedure. The statistical mechanics and the kinetic theory are not used to verify the atomic theory through the bulk properties of matter, but rather should be used to predict the bulk behaviors of matter from the known properties of the molecules and atoms. Since engineers always deal with matter in bulk, the prediction of bulk properties is then of vital interest to engineers, and is logically a branch of engineering science.

However, it might be argued that the mere fact of the engineer's interest in bulk properties of matter does not necessarily mean the need for physical mechanics. The bulk properties can be measured directly; then the theoretical calculations of physical mechanics will not be needed. This has been the situation till very recently. When the engineer needs the thermodynamic properties of steam or ammonia, he measures them. When the engineer needs the viscosity of water, he again measures it. If such direct measurements can be made easily with the required accuracy, there is no incentive to introduce new methods. Recently, however, particularly with the advent of rocket and jet propulsion engineering and nuclear engineering, bulk properties of matter at unusual conditions are required. For instance, thermodynamic properties at very high temperatures, say, 4 000 K, enter into engineering calculations. The experimental measurement of thermodynamic properties at such temperatures is certainly very difficult, if not impossible. On the other hand, the thermodynamic properties of gases at high temperatures can be calculated by the method of statistical mechanics with ease and certainty, once the properties of the constituent molecules and atoms are known. This circumstance can be easily understood by the observation that although the bulk temperature of the gas may be very high in the conventional engineering sense, the average energy of a single molecule or atom is quite moderate and within the range of certain knowledge of a physicist or a chemist. For instance, the average or representative kinetic energy of molecule or atom at a temperature T in the absolute scale is $3/2\, kT$, k being Boltzmann's constant. The value of k is given by

$$\begin{aligned} k &= 1.380 \times 10^{-16} \text{ erg per K} \\ &= 0.861 \times 10^{-4} \text{ electron-volt per K} \end{aligned}$$

Therefore even at the "fantastic" temperature of 10 000 K, the average kinetic energy of the atoms is still only 1.292 electron-volt, an energy at which the behavior of atoms is known with accuracy.

Physical Mechanics as an Engineering Science

The problems in physical mechanics can be classified into two categories: Problems in thermodynamic properties of matter at equilibrium, and problems in transport properties of matter not at equilibrium. Although the powerful methods of statistical mechanics are equally as applicable to matter at thermodynamic equilibrium as to matter not at equilibrium, concrete useful results are obtained easily only for the first case. For the transport properties, such as viscosity, heat conduction, and diffusion, the methods of the kinetic theory of matter give quantitative answers.

Examples of the first category of problems are the thermodynamic functions of gases, solids and liquids, the equation of states for imperfect or highly compressed gases, chemical equilibrium constants, and the thermodynamic behavior of electrically charged gases, such as gas containing ions and electrons. The previous investigations and results in this field are summarized, for example, in the two excellent books by Sir R. Fowler[2] and by Fowler and E. A. Guggenheim[3].

Examples of the second category of problems are, besides those already mentioned, the neutron diffusion in a nuclear reactor, and the moderation of fast neutrons from the fission nuclei to thermal neutrons by the reactor materials. Therefore many problems in nuclear engineering dealing with the macroscopic effects of elementary nuclear processes belong logically to the science of physical mechanics. The theory of transport properties of gases is given in the famous book by S. Chapman and T. G. Cowling[4]. Besides those problems of transport properties, the radiative emissivity of gases at high temperatures and the spectra obtained from regions of active combustion are also important problems of the second category.

It is evident that there has been certainly no lack of effort by physicists and physical chemists in the general field of what is here called physical mechanics. Physical chemists were particularly diligent in carrying out many extremely ted ous calculations[5]. Then, is the task of establishing physical mechanics as an engineering science that of simply disseminating this work of physicists and chemists among the engineering profession? Unfortunately, the task is not as simple as that. To establish physical mechanics as an engineering science, one must follow the example of other branches of engineering science, such as fluid mechanics, by introducing the successful guiding principles of the socalled "angewandte Mechanik." These principles of engineering science were first formulated and applied by a group of brilliant "applied mechanicists" of Göttingen University at the turn of this century[6]. The following sections will be devoted to a discussion of these principles as applied to physical mechanics.

Use of Approximate Models

One of the principles of engineering science is the approximate solution of complex problem by using a simplified model which gives a satisfactory representation of reality. Because of the very fact that the model is simple, it cannot have all the properties of the real system, and it can be only designed to emphasize the most important feature of the real system under the particular physical situation concerned. Under a different physical situation, the real system may very well reveal a different property which is important. Then a different model has to be adopted. Therefore the success of the choice of models rests on the clear understanding of the physical circumstances. A "physical mechanicist" can have, however, assistance in the difficult task in two ways. He can always study the experimental observations connected with the phenomenon and thus gain insight to the problem. Then he is helped by knowing the logical requirement, that the models representing the same physical system, although they may be different under different situations, should nevertheless be compatible with each other and should not be contradictory to each other. The following is an example for the point under discussion.

From diffraction experiments and spectroscopic studies, the structure of molecules can be determined in terms of interatomic distances and bond angles. Such data, together with the van der Waals sizes of atoms, then give a definite model of a molecule as a structure of spheres (atoms) properly fused together. At ordinary temperatures, the rotation of molecules in a gas is fully excited. Thus the odd-shaped molecule is really under rapid rotation, and for molecules that are not excessively elongated the angular asymmetry is averaged out. Therefore, when one is considering the interaction of molecules, as, for example, in the calculation of the second virial coefficient and the transport properties of gases, one may well consider the molecules as spheres with a diameter r^* equal to twice the maximum distance between the center of mass of the molecule and the boundary of the molecular model. This is shown to be actually the case by S. D. Hamann[7] and by Hamann and J. F. Pearse[8] for nonpolar molecules with the Lennard-Jones potential given by

$$\varepsilon(r) = \varepsilon^* \left[\left(\frac{r^*}{r} \right)^{12} - \left(\frac{r^*}{r} \right)^{6} \right] \tag{1}$$

where r is the distance between the centers of mass of the molecules, ε the interaction energy, and ε^* the interaction energy at the distance r^*, the equilibrium distance. This is a very satisfactory state of affairs, as the chosen model of molecular interaction is entirely compatible with a wide range of other physical phenomena and concepts.

For polar molecules with permanent dipole moment μ, W. H. Stockmayer[9] proposed an interaction potential which is a hybrid between the Lennard-Jones potential of Equation (1) and the dipole interaction potential,

$$\varepsilon = \varepsilon^* \left[\left(\frac{r^*}{r} \right)^{12} - \left(\frac{r^*}{r} \right)^{6} \right] - \frac{\mu^2}{r^3} [2\cos\theta_1 \cos\theta_2 - \sin\theta_1 \sin\theta_2 \cos\varphi] \tag{2}$$

where θ_1, θ_2, and φ are angles specifying the orientation of the interacting dipole axes. Following the work of W. H. Keesom[10], Stockmayer calculated the second virial coefficient

$B(T)$, where T is the temperature, as

$$B(T) = \frac{1}{4}N\int_0^{\infty} r^2 \mathrm{d}r \int_0^{\pi} \sin\theta_1 \,\mathrm{d}\theta_1 \int_0^{\pi} \sin\theta_2 \,\mathrm{d}\theta_2 \times \int_0^{2\pi} [1 - \mathrm{e}^{-\varepsilon/kT}]\,\mathrm{d}\varphi \tag{3}$$

where N is Avogadro's number, ε is the interaction potential given by Equation (2), and k is Boltzmann's constant. The fitting of the experimentally determined second virial coefficients to Stockmayer's formula was carried out by J. O. Hirschfelder, F. T. McClure, and I. F. Weeks[11], and also by Hamann and Pearse[8] for methyl chloride and methyl fluoride. The results are, however, anomalous. For instance, the "size" r^* for steam and ammonia turns out to be approximately the van der Waals diameter of the oxygen atom and nitrogen atom, respectively, *alone*, without any room allowed for the hydrogen atoms in these molecules. It thus seems that Stockmayer's model of polar molecules is not entirely unquestionable.

An explanation of this difficulty can be obtained by observing the fact that if the molecule is under almost free rotation, then the dipole attraction and the dipole repulsion at any distance r between molecules average out by the prevailing random orientation. Consequently, when the rotation of molecules is fully excited, the dipole moment of the molecule makes no contribution to the interaction potential. In other words, as far as the calculations of the second virial coefficient and transport properties are concerned, it may be more realistic to neglect the difference between a polar molecule and a nonpolar molecule. This assumption will give a great simplification for the calculation of transport properties of polar gases. By fitting the second virial coefficient data for steam and ammonia to the formula for nonpolar molecules, satisfactory sizes r^* for these molecules are obtained. Of course, the validity of this point of view has yet to be proved by a critical examination of the theory of virial coefficients.

A Question of Methodology

In recent years, perhaps because of the influx of mathematicians to the field of applied mechanics, the level of logical organization and of mathematical argument of research papers in this field is generally quite high. This modern trend, by itself, contributes to the elegance of presentation and facilitates the general understanding of the work. However, clarity of thought and the use of advanced mathematical tools have more than just this to contribute. In many problems of engineering science, the very complexity of the problems demands the most efficient and powerful tools for their solution. It is even to be expected that in many cases the problem will not be solved by any means other than the most efficient and powerful method. Therefore, to establish physical mechanics as an engineering science, it is necessary to emphasize this point of methodology.

As an example of lack of clarity of organization, one may take a recent paper[12] on the thermodynamic properties of completely ionized hydrogen. The criticism here is not that the results are incorrect, but rather that the same results could be obtained very simply and logically in a direct way. For this problem, one should first recognize that all thermodynamic properties of matter are contained in the partition function, or alternately, the free energy expression. Therefore the logical first step would be to establish the free energy of such ionized hydrogen, composed of equal numbers of positively charged protons and negatively charged electrons. But then the problem is exactly the same as the problem of solutions of electrolytes which dissociate into positive and negative ions. For solutions of electrolytes, there is the well-known Debye-Hückel theory①. The approximations involved and the validity of the assumptions involved in this theory are now clearly understood. By applying the Debye-Hückel theory to the problem of ionized hydrogen, one is at once clear about the power and the limitations of the solution. This alone is a worth-while saving in effort.

To show how easily the thermodynamics of completely ionized hydrogen can be determined with the rational method, one notes that if F is the free energy of the assembly, E the internal energy, V the volume, P the pressure, and T the temperature, then according to the general laws of thermodynamics

$$P = -\frac{\partial F}{\partial V} \tag{4}$$

$$E = -T^2 \frac{\partial (F/T)}{\partial T} \tag{5}$$

According to the Debye-Hückel theory①, the first approximation to the free energy of Coulomb interaction F^{el} of equal numbers of positively and negatively charged particles with $\pm ze$ charge on each particle is

$$F^{el} = -\frac{2}{3}\sqrt{\pi} NkT\left(\frac{N^{1/3}z^2e^2}{V^{1/3}kT}\right)^{3/2} \tag{6}$$

where N is the total number of particles. Therefore, with Equations (4) and (5), the deviation ΔP of pressure and the deviation ΔE of internal energy from an assembly without Coulomb interaction are

$$\Delta P = -\frac{1}{3}\sqrt{\pi}\frac{NkT}{V}\left(\frac{N^{1/3}z^2e^2}{V^{1/3}kT}\right)^{3/2} \tag{7}$$

$$\Delta E = -\sqrt{\pi} NkT\left(\frac{N^{1/3}z^2e^2}{V^{1/3}kT}\right)^{3/2} \tag{8}$$

For ionized hydrogen, $z = 1$, and if H is the atomic weight of hydrogen, and ρ the density of the mixture

$$N/V = 2\rho/H \tag{9}$$

Therefore for ionized hydrogen

$$\Delta P = -\frac{\sqrt{\pi}}{3}\left(\frac{2\rho}{H}\right)^{3/2}\frac{e^3}{\sqrt{kT}} \tag{10}$$

① See, for instance, Ref. 3, Chapter IX.

and

$$\Delta E = -\sqrt{\pi}\left(\frac{2\rho}{H}\right)^{3/2} \frac{Ve^3}{\sqrt{kT}} \tag{11}$$

These results are the same as those in Reference [12] and are obtained here with very little calculation.

As an example of the advantages of mathematical technique in shortening the calculation, one may consider the second virial coefficient $B(T)$ for the Lennard-Jones potential of Equation (1). It is known[①] that, with N as Avogardo's number, and

$$r_0 = 2^{-1/6} r^* \tag{12}$$

$$\frac{B(T)}{\frac{2\pi}{3} N r_0^3} = -\frac{1}{4}\left(\frac{4\varepsilon^*}{kT}\right)^{1/4} \sum_{n=0}^{\infty} \frac{\Gamma\left(\frac{n}{2}-\frac{1}{4}\right)}{n!}\left(\frac{4\varepsilon^*}{kT}\right)^{n/2} \tag{13}$$

This series, although convergent for all values of T, is nevertheless inconvenient for very small values of temperature. For instance, it is stated (5) that for $kT/\varepsilon^* = 0.3$, about thirty terms are necessary to obtain an accuracy of five significant figures. The situation evidently calls for asymptotic expansion of the function $B(T)$ instead of the Taylor series of Equation (13). The asymptotic series can be obtained easily as

$$\frac{B(T)}{\frac{2\pi}{3} N r_0^3} \sim -\sqrt{\frac{\pi kT}{2\varepsilon^*}}\, e^{\varepsilon^*/(kT)} \times \sum_{n=0}^{\infty} \frac{\Gamma\left(n+\frac{3}{4}\right)\Gamma\left(n+\frac{5}{4}\right)}{\Gamma\left(\frac{3}{4}\right)\Gamma\left(\frac{5}{4}\right) n!}\left(\frac{kT}{\varepsilon^*}\right)^n \tag{14}$$

This expression has not only the advantage of being easier to use at small values of T, but also the advantage of clearly demonstrating the exponential behavior of the function at low temperatures. Such definite indication of functional behavior is often a help in understanding the interactions of the different elements of a problem.

Concluding Remarks

In the preceding discussion, the subject matter and the basic concepts of the new engineering science, physical mechanics, are outlined. It is an engineering science mainly because its foremost purpose is to help solve the engineering problems. And since it is an engineering science, physical mechanics should be a subject of training for any research or development engineers, of equal importance with fluid mechanics and solid mechanics. Because of its close relation to jet propulsion and rocket development, physical mechanics is now taught as a graduate course at the California Institute of Technology in the Daniel and

---

① See, for instance, Ref. 3, p. 280.

Florence Guggenheim Jet Propulsion Center. However, the course is open to students of other fields of engineering with the proper preparation in mathematics, physics, and chemistry.

To skeptical purists among physicists and physical chemists, this discussion may appear to be overly optimistic or even immodest. For them, the author can only point to the unquestionable success of fluid mechanics and solid mechanics in modern engineering. There is no reason to expect the future of physical mechanics to be radically different.

References

[1] Lindsay R B. Physical Mechanics — An Intermediate Test for Students of the Physical Sciences. D. Van Nostrand Co., New York: 1950.

[2] Fowler R H. Statistical Mechanics. Cambridge University Press, 1936.

[3] Fowler R H, Guggenheim E A. Statistical Thermodynamics. Cambridge University Press, 1949.

[4] Chapman S, Cowling T G. The Mathematical Theory of Non-Uniform Gases. Cambridge University Press, 1939.

[5] Hirschfelder J O, Curtiss C F, Bird R B, Spotz E L. Properties of Gases. The manuscript of this book was issued by authors as reports of the Naval Research Laboratory, Department of Chemistry, University of Wisconsin.

[6] Millikan C B. For this interesting anecdote of the birth of modern engineering science. see the chapter on "Aeronautics" in "Physics in Industry". American Institute of Physics, 1937.

[7] Hamann S D. The Interpretation of Intermolecular Force Constants. Journal of Chemical Physics, 1951, 19: 655.

[8] Hamann S D, Pearse J F. The Second Virial Coefficients of Some Organic Molecules. Transactions of the Faraday Society, 1952, 48(2): 101 - 106.

[9] Stockmayer W H. Second Virial Coefficients of Polar Gases. Journal of Chemical Physics, 1941, 9: 398 - 402.

[10] Keesom W H . On the Deduction from Boltzmann's Entropy Principle of the Second Virial Coefficient. Comm. Phys. Lab. Leiden, Suppl. 24B, § 6, 1912.

[11] Hirschfelder J O, McClure F T, Weeks I F. Second Virial Coefficients and the Force between Complex Molecules. Journal of Chemical Physics, 1942, 10: 201 - 211.

[12] Williamson R E. On the Equation of State of Ionized Hydrogen. Astrophysics Journal. 1946, 103: 139.

The Properties of Pure Liquids

H. S. Tsien[1]

(*Daniel and Florence Guggenheim Jet Propulsion Center*,
California Institute of Technology, *Pasadena*, *Calif.*)

By a semiempirical approach, a method is found to calculate the specific heat of a normal pure liquid at constant pressure from the specific heat of the gaseous state at the same temperature. It is also found that the coefficient of thermal expansion, the compressibility, and the velocity of sound of the liquid can be calculated accurately if the density, the molecular weight, and the normal boiling temperature of the liquid at atmospheric pressure are known. Finally, a method of computing the thermal conductivity of all liquids, except liquid metals, from compressibility and density is developed. For normal liquids, the thermal conductivity can again be determined if only the normal boiling temperature, the density, and the molecular weight are known.

In rocket and jet propulsion engineering, because of the necessity of considering a very wide range of possible fuels and propellants, one often meets the situation that the physical properties, such as heat capacity and thermal conductivity, of liquid of interest are not listed in the handbooks. Naturally the question to ask is whether such physical properties can be estimated from known simple quantities such as the boiling point, the molecular weight, and the density of the liquid. It is clear that such a correlation of liquid properties must come from the theory of the liquid state. Although there was no lack of work by physicists and physical chemists in this field, the agreement in the numerical values of theoretical prediction and of experimental observation is usually very poor. Therefore it may be justified to say that there are only very few useful results in the engineering sense. The difficulty here is evidently due to the rather indefinite structure of the liquid state in comparison to other states of matter: For the gaseous state, the interaction between the molecules can be almost neglected, and the predominating feature of the state is the translational and internal motions of the individual molecules. For the solid state, the reverse is true. The predominating feature is the interaction between the molecules or the atoms. For the liquid state of matter, the molecular interaction and the molecular motion are of equal importance. This fact leads to great complexities, and any theory of liquids, necessarily based upon a simplified model, is incomplete and is predicated upon many assumptions. The divergences between the theoretical predictions and the experimental observations are thus to be expected.

Received October 14, 1952.

Journal of the American Rocket Society, vol. 23, pp. 17 – 24, 1953.

① Robert H. Goddard Professor of Jet Propulsion.

In this paper a somewhat new approach to the subject will be attempted. The theory will not be used to predict the physical properties from the molecular and the atomic characteristics, but rather used as a framework to fit the experimental data. In other words, the theory only gives the parameters that will enter into a relation, while the exact form of the functional relation is to be determined by the experimental data. Thus the approach here is that of "dimensional analysis" so successful in the older fields of engineering science, such as fluid mechanics and solid mechanics[1]. The two specific useful results of this investigation are a method to calculate the specific heats and a method to calculate the thermal conductivity of liquids. These methods are generally applicable to the so-called normal liquids, but a more general form for the thermal conductivity is available to include all liquids, normal or otherwise.

Lennard-Jones and Devonshire Theory of Liquids

One of the fairly successful theories of normal liquids is that given by Lennard-Jones and Devonshire[2]. It is a theory of the "free volume" type in that the liquid molecule is assumed to move within a cage formed by the neighboring molecules. In this theory, the properties of the cage are determined by smoothing the bimolecular interactions of a face-centered cubic lattice where the nearest distance between molecules is a. The bimolecular interaction potential ε is taken from the theory of gaseous states and is expressed as

$$\varepsilon(r) = \varepsilon_m\left[\left(\frac{r^*}{r}\right)^{12} - 2\left(\frac{r^*}{r}\right)^6\right] \tag{1}$$

where r is the distance between the molecules, r^* the equilibrium distance, and ε_m is the magnitude of the potential at $r = r^*$.

The free energy F for an assembly of N molecules is found in this way to be

$$\frac{F}{N} = -kT\log\frac{(2\pi mkT)^{3/2}}{h^3} - kT\log j(T) - kT - \Lambda^*\left\{1.2\left(\frac{V^*}{V}\right)^2 - 0.5\left(\frac{V^*}{V}\right)^4\right\} - kT\log(2\pi\gamma gV) \tag{2}$$

In this equation, k is the Boltzmann constant, T the absolute temperature, $j(T)$ the internal partition function of the molecule. And furthermore

$$\Lambda^* = z\varepsilon_m \tag{3}$$

where z is the coordination number and is 12 for face-centered cubic lattice, V is the volume per molecule, and V^* is the characteristic volume. They are related to a and r^* as follows

$$V = \frac{1}{\gamma}a^3,\quad V^* = \frac{1}{\gamma}r^{*3},\quad \gamma = \frac{1}{\sqrt{2}} \tag{4}$$

g is the following complicated integral

$$g = \int_0^{1/4} y^{1/2}\exp\left\{-\frac{\Lambda^*}{kT}\left(\frac{V^*}{V}\right)^4 l(y) + 2\frac{\Lambda^*}{kT}\left(\frac{V^*}{V}\right)^2 m(y)\right\}\mathrm{d}y \tag{5}$$

where

$$l(y)=\frac{1+12y+25.2y^2+12y^3+y^4}{(1-y)^{10}}-1,\quad l(0)=0 \tag{6}$$

$$m(y)=\frac{1+y}{(1-y)^4}-1,\quad m(0)=0 \tag{7}$$

Once the free energy is calculated, the thermodynamic properties of the liquid can be obtained by simple differentiations. For instance, the energy per molecule E/N is

$$\frac{E}{N}=-T^2\frac{\partial}{\partial T}\left(\frac{F/N}{T}\right) \tag{8}$$

The pressure P is

$$P=-\frac{\partial}{\partial V}\left(\frac{F}{N}\right) \tag{9}$$

Mathematically then, the most difficult task of the theory is the evaluation of the integral g in Equation (5). Recently, R. H. Wentorf, R. J. Buehler, J. O. Hirschfelder, and C. F. Curtiss[3] have carried out this tedious integration by numerical method, together with some inconsequential improvements in its formulation. Unfortunately, their extensive numerical tables are quite unsuitable for the present purpose of discovering the analytical functional relationships between the different quantities. However, for liquid states at low or atmospheric pressures, the ratio V/V^* is very nearly unity, while the ratio Λ^*/kT is of the order of 20. It is thus appropriate to seek the asymptotic expansion of the function g for large values of Λ^*/kT. This can be done as follows:

Let

$$s=\frac{\Lambda^*}{kT}\left(\frac{V^*}{V}\right)^4 \tag{10}$$

Then by expanding the functions $l(y)$ and $m(y)$ in power series of y, one has

$$-\frac{\Lambda^*}{kT}\left(\frac{V^*}{V}\right)^4 l(y)+2\frac{\Lambda^*}{kT}\left(\frac{V^*}{V}\right)^2 m(y)=-s\cdot\eta$$

$$=-s\left[\left\{22-10\left(\frac{V}{V^*}\right)^2\right\}y+\left\{200.2-28\left(\frac{V}{V^*}\right)^2\right\}y^2+\left\{1\,144-60\left(\frac{V}{V^*}\right)^2\right\}y^3+\ldots\right] \tag{11}$$①

η is thus the power series within the square bracket of Equation (11). By inverting this power

① In 1962, the author corrected the first number appearing on the left side of Eq. (11) from 24 to 22, followed by suecessive corrections below, though the corrections had no essential influence on the main conclusions in this paper. In editing this bork, the above-mentioned corrections were accepted. — Noted by editor.

series, it is found that

$$y = a_1\eta + a_2\eta^2 + a_3\eta^3 + \dots \tag{12}$$

where

$$a_1 = \frac{1}{22 - 10\left(\frac{V}{V^*}\right)^2}$$

$$a_2 = -\frac{200.2 - 28\left(\frac{V}{V^*}\right)^2}{\left[22 - 10\left(\frac{V}{V^*}\right)^2\right]^3}$$

and

$$a_3 = \frac{2\left[200.2 - 28\left(\frac{V}{V^*}\right)^2\right]^2 - \left[1\,144 - 60\left(\frac{V}{V^*}\right)^2\right]\left[22 - 10\left(\frac{V}{V^*}\right)^2\right]}{\left[22 - 10\left(\frac{V}{V^*}\right)^2\right]^5} \tag{13}$$

At the lower limit of integration for g, $y = 0$, so $\eta = 0$. At the upper limit of integration, $y = 1/4$, so $\eta = l(1/4) - 2(V/V^*)^2 m(1/4)$. $l(1/4)$ is, however, approximately 90, $m(1/4)$ is 2.95, and s is very large. Therefore it is correct to set the upper limit of integration in η as ∞. Then

$$g \sim a_1^{3/2}\int_0^\infty e^{-s\cdot\eta}\left[\eta^{1/2} + \frac{5}{2}\frac{a_2}{a_1}\eta^{3/2} + \frac{7}{2}\left\{\frac{a_3}{a_1} + \frac{1}{4}\left(\frac{a_2}{a_1}\right)^2\right\}\eta^{5/2} + \dots\right]d\eta$$

$$= \frac{\sqrt{\pi}}{2}\left(\frac{a_1}{s}\right)^{3/2} \times \left[1 + \frac{3\times 5}{2\times 2}\frac{a_2}{a_1}\frac{1}{s} + \frac{3\times 5\times 7}{2\times 2\times 2}\left\{\frac{a_2}{a_1} + \frac{1}{4}\left(\frac{a_3}{a_1}\right)^2\right\}\frac{1}{s^2} + \dots\right] \tag{14}$$

By substituting the value of s from Equation (10), and the value of a's from Equation (13), the following expression for $\log g$ is obtained

$$\log g \sim \log\frac{\sqrt{\pi}}{2} + \frac{3}{2}\log\frac{kT\left(\frac{V}{V^*}\right)^4}{\Lambda^*\left[22 - 10\left(\frac{V}{V^*}\right)^2\right]} - \frac{3\times 5}{2\times 2}\left(\frac{V}{V^*}\right)^8\frac{kT}{\Lambda^*}\frac{200.2 - 28\left(\frac{V}{V^*}\right)^2}{\left[22 - 10\left(\frac{V}{V^*}\right)^2\right]} +$$

$$\frac{3\times 5\times 7}{2\times 2\times 2}\left(\frac{V}{V^*}\right)^8\left(\frac{kT}{\Lambda^*}\right)^2 \times$$

$$\frac{12\left[200.2 - 28\left(\frac{V}{V^*}\right)^2\right]^2 - 7\left[1\,144 - 60\left(\frac{V}{V^*}\right)^2\right]\left[22 - 10\left(\frac{V}{V^*}\right)^2\right]}{7\left[22 - 10\left(\frac{V}{V^*}\right)^2\right]^4} \tag{15}$$

It is evident that the expansion of Equation (15) is the appropriate one for large values of Λ^*/kT. By combining Equation (15) with Equation (2), the free energy of the liquid state can be obtained. Differentiation according to Equations (8) and (9) then gives the other thermodynamic quantities.

Specific Heat at Constant Volume

If $\frac{E^{\text{int}}}{N}$ is the internal energy per molecule, i.e.,

$$\frac{E^{\text{int}}}{N} = kT^2 \frac{\partial}{\partial T}\{\log j(T)\} \tag{16}$$

then Equation (8) gives

$$\frac{E}{N} = \frac{E^{\text{int}}}{N} + 3kT - \Lambda^* \left\{1.2\left(\frac{V^*}{V}\right)^2 - 0.5\left(\frac{V^*}{V}\right)^4\right\} - \frac{3\times 5}{2\times 2}\frac{200.2 - 28\left(\frac{V}{V^*}\right)^2}{\left\{22 - 10\left(\frac{V}{V^*}\right)^2\right\}^2}\left(\frac{V}{V^*}\right)^4 \frac{kT}{\Lambda^*}kT \tag{17}$$

where terms of third power in T and higher are dropped. By differentiating Equation (17) once more with respect to T, the specific heat at constant volume can be determined. Let C_V^{l} be the molar specific heat of the liquid state at constant volume, and C^{int} be the molar specific heat of the internal energy alone, i.e., by taking N to be Avogadro's number

$$C^{\text{int}} = \frac{\partial}{\partial T}E^{\text{int}} \tag{18}$$

Then

$$C_V^{l} = C^{\text{int}} + 3R - \frac{3\times 5}{2\times 2}\frac{200.2 - 28\left(\frac{V}{V^*}\right)^2}{\left\{22 - 10\left(\frac{V}{V^*}\right)^2\right\}^2}\left(\frac{V}{V^*}\right)^4 \frac{kT}{\Lambda^*}R \tag{19}$$

where R is the universal gas constant or $R = Nk$.

Equation (19) demonstrates the gratifying result that aside from the small correction of the third term, the specific heats at constant volume for the liquid state and the solid state are the same. This is in agreement with the concept that for pressures and temperatures below the critical pressure and the critical temperature, there are more points of similarity between the liquid state and the solid state than there are between the liquid state and the gaseous state. The full classical value of $3R$ means the absence of quantal effects. This is of course generally true, as will be discussed in more detail in the appendix.

The molar specific heat at constant pressure C_p^{g} for the gaseous state can be calculated as

$$C_p^{g} = C^{\text{int}} + \frac{5}{2}R \tag{20}$$

For molecules that are not excessively elongated as to restrict their rotational freedom in the liquid state, the molecular energy of the internal degrees of freedom must be the same in the liquid state as in the gaseous state. Therefore the values of C^{int} in Equations (19) and (20) must be the same. Then Equation (19) can be also written as

$$C_V^{\text{l}} = C_p^{\text{g}} + R\left[0.5 - \frac{15}{2\times 2}\frac{200.2 - 28\left(\frac{V}{V^*}\right)^2}{\left\{22 - 10\left(\frac{V}{V^*}\right)^2\right\}^2}\left(\frac{V}{V^*}\right)^4\frac{kT}{\Lambda^*}\right] \tag{21}$$

Now since the third term to the right of Equation (21) is a small correction term, it would be all right to use the following approximations, true for liquid state at low or atmospheric pressures, i.e., at pressures much lower than the critical pressure. Furthermore

$$\left.\begin{aligned} V/V^* &\sim 1 \\ \frac{kT_{\text{b}}}{\Lambda^*} &\sim 16.5^{-1} \end{aligned}\right\} \tag{22}$$

where T_{b} is the normal boiling temperature of the liquid at atmospheric pressure. Then Equation (21) becomes

$$C_V^{\text{l}} = C_p^{\text{g}} + R\left[0.5 - 0.5\frac{T}{T_{\text{b}}}\right] \tag{23}$$

Therefore the specific heat of liquids at constant volume can be easily calculated once the specific heat of the gaseous state and the boiling point are known.

Thermal Expansion and Compressibility

The equation of state for the liquid can be determined by using Equation (9). By using only two terms of the expansion for g as given by Equation (15), one has

$$P = \frac{\Lambda^*}{V}\left\{2\left(\frac{V^*}{V}\right)^4 - 2.4\left(\frac{V^*}{V}\right)^2\right\} + \frac{kT}{V}\frac{77 - 20\left(\frac{V}{V^*}\right)^2}{11 - 5\left(\frac{V}{V^*}\right)^2} \tag{24}$$

Now the coefficient of thermal expansion α is defined as

$$\alpha = \frac{1}{V}\left(\frac{\partial V}{\partial T}\right)_p \tag{25}$$

Then by differentiating Equation (24) with respect to T, keeping the pressure P constant, one has

$$\alpha\left[\frac{\Lambda^*}{V}\left\{10\left(\frac{V^*}{V}\right)^4 - 7.2\left(\frac{V^*}{V}\right)^2\right\} + \frac{kT}{V}\left\{\frac{77 - 20\left(\frac{V}{V^*}\right)^2}{11 - 5\left(\frac{V}{V^*}\right)^2} - \frac{330\left(\frac{V}{V^*}\right)^2}{\left[11 - 5\left(\frac{V}{V^*}\right)^2\right]^2}\right\}\right] = \frac{k}{V}\frac{77 - 20\left(\frac{V}{V^*}\right)^2}{11 - 5\left(\frac{V}{V^*}\right)^2}$$

However, the pressure P is very small in comparison with the size of both terms in the right of Equation (24), i.e.,

$$\frac{\Lambda^*}{V}\left\{2\left(\frac{V^*}{V}\right)^4 - 2.4\left(\frac{V^*}{V}\right)^2\right\} + \frac{kT}{V}\,\frac{77-20\left(\dfrac{V}{V^*}\right)^2}{11-5\left(\dfrac{V}{V^*}\right)^2} \cong 0$$

Therefore the equation for the thermal expansion can be simplified to

$$\alpha\left[\Lambda^*\left\{8\left(\frac{V^*}{V}\right)^4 - 4.8\left(\frac{V^*}{V}\right)^2\right\} - kT\,\frac{330\left(\dfrac{V}{V^*}\right)^2}{\left\{11-5\left(\dfrac{V}{V^*}\right)^2\right\}^2}\right] = k\,\frac{77-20\left(\dfrac{V}{V^*}\right)^2}{11-5\left(\dfrac{V}{V^*}\right)^2}$$

By again using the approximations of Equation (22), one has finally

$$\alpha T_b = \frac{0.576}{3.2 - 0.556\,\dfrac{T}{T_b}} \tag{26}$$

The compressibility β of the liquid is defined as

$$\beta = -\frac{1}{V}\left(\frac{\partial V}{\partial P}\right)_T \tag{27}$$

By following a similar procedure as outlined in the preceding paragraph for α, the compressibility is found to be

$$\beta\frac{RT_b}{V_l} = \frac{1}{52.8 - 9.17\left(\dfrac{T}{T_b}\right)} \tag{28}$$

where R is again the universal gas constant, V_l is the volume per mole of the liquid.

Equations (26) and (28) give the coefficient of thermal expansion α and the compressibility β according to the Lennard-Jones and Devonshire theory of liquids. Other theories of liquids give different formulas. For instance, the free volume theory of H. Eyring and J. O. Hirschfelder[4] gives

$$\alpha T_b = \frac{3}{9.4 - 4\left(\dfrac{T}{T_b}\right)} \tag{29}$$

and

$$\beta\frac{RT_b}{V_l} = \frac{1}{3.13\left(9.4\,\dfrac{T_b}{T} - 4\right)} \tag{30}$$

Such discrepancies between the theories probably indicate that neither theory is really accurate enough for calculations of α and β. But both theories give αT_b and $\beta R T_b / V_l$ as functions of the temperature ratio T/T_b. Therefore it seems justified to consider such functions as unknown theoretically, but to be determined by experimental data. Once determined, these functions are then universal and applicable to all normal liquids.

Table 1 Coefficient of thermal expansion of liquids

Liquid	Formula	T_b/K	T/K	$\alpha \times 10^3$/K^{-1}	T/T_b	αT_b
Acetone	$(CH_3)_2CO$	329.7	293.2	1.071	0.890	0.353
Aniline	$C_6H_5NH_2$	457.6	293.2	0.855	0.641	0.382
Arsenic trichloride	$AsCl_3$	403.4	293.2	1.029	0.728	0.415
Benzene	C_6H_6	353.2	293.2	1.237	0.830	0.437
Bromine	Br_2	332.0	293.2	1.132	0.883	0.376
Carbon disulfide	CS_2	319.5	293.2	1.218	0.918	0.390
Carbon tetrachloride	CCl_4	350.0	293.2	1.236	0.838	0.433
Chloroform	$CHCl_3$	334.5	293.2	1.273	0.877	0.426
Ethyl ether	$(C_2H_5)_2O$	307.8	293.2	1.656	0.953	0.510
Ethyl iodide	C_2H_5I	345.4	293.2	1.179	0.848	0.407
Trimethyl ethane	C_5H_{12}	301.2	293.2	1.598	0.973	0.481
Phosphorus tribromide	PBr_3	446.1	293.2	0.868	0.657	0.387
Phosphorus trichloride	PCl_3	348.7	293.2	1.154	0.841	0.402
	...	...	...	1.211	...	0.422
Phosphorus oxychloride	$POCl_3$	378.5	293.2	1.116	0.775	0.423
Pentane	C_5H_{12}	309.4	293.2	1.608	0.948	0.498
i-Propyl chloride	C_3H_7Cl	308.6	293.2	1.591	0.950	0.491
Isoprene	C_5H_8	307.2	293.2	1.567	0.955	0.481
Silicon bromide	$SiBr_4$	426.2	293.2	0.983	0.688	0.419
Silicon chloride	$SiCl_4$	330.8	293.2	1.430	0.886	0.473
Stannic chloride	$SnCl_4$	387.3	293.2	1.178	0.757	0.456
Titanium tetrachloride	$TiCl_4$	409.6	293.2	0.998	0.715	0.409
o-Toluidine	$C_7H_7NH_2$	473.0	293.2	0.847	0.620	0.401

Table 1 lists the values of the coefficient of thermal expansion at 20℃ taken from the Landolt-Börnstein Tabellen for various normal liquids at atmospheric pressure. The nondimensional quantity αT_b is then plotted against T/T_b in Fig. 1, where the relations specified by Equations (26) and (29) are also plotted. Although the theories do predict the increasing thermal expansion with temperature, a behavior contrary to that of the gaseous state, the experimental data lie between the theoretical curves and are grouped definitely around a different curve. This empirical curve is drawn as a heavy line in Fig. 1. Therefore the surmise that αT_b is a function of T/T_b is now justified. By using this empirical curve, the thermal expansion of normal liquids can be calculated to 10 per cent accuracy once the boiling point T_b is known.

Table 2 lists the values of the compressibility of normal liquids at approximately atmospheric pressure taken again from the Landolt-Börnstein Tabellen. The nondimensional quantity $\beta RT_b/V_l$ is then plotted against the temperature ratio T/T_b in Fig. 2. The theoretical curves specified by Equations (28) and (30) are also plotted. It is seen that a similar situation as for α exists. In fact, the empirical curve can be very closely represented by

$$\beta\frac{RT_b}{V_l} = \frac{1}{101.6 - 82.4\left(\dfrac{T}{T_b}\right)} \tag{31}$$

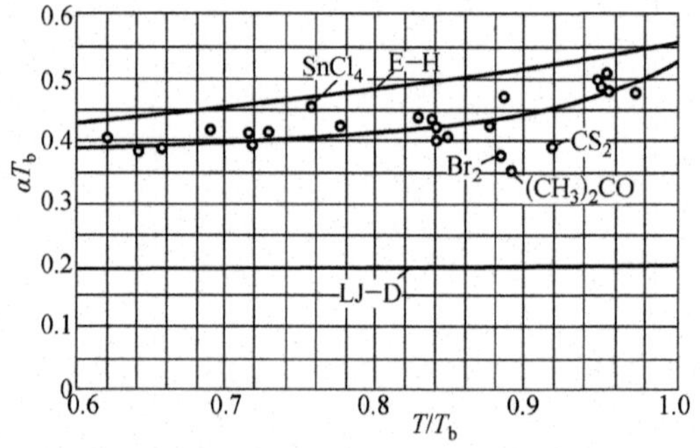

Fig. 1 α, Coefficient of thermal expansion; T, temperature; T_b, normal boiling temperature. E-H, relation given by eyring and hirschfelder (Ref. 4), equation (29). LJ-D, relation deduced from Lennard-Jones and devonshire theory of liquids, equation (26)

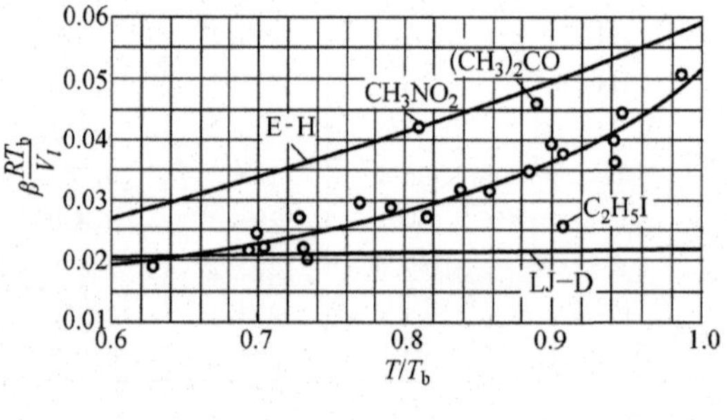

Fig. 2 β, Compressibility; T, temperature; T_b, normal boiling temperature; V_l, molar volume of liquid; R, universal gas constant E-H, relation given by eyring and hirschfelder (Ref. 4), equation (30). LJ-D, relation deduced from Lennard-Jones and devonshire theory of liquids, equation (28)

Therefore the compressibility of any normal liquid can be calculated with 10 per cent accuracy if the boiling point T_b, the liquid density, and the molecular weight are known.

Specific Heat at Constant Pressure

For many engineering calculations, what is desired is not the specific heat at constant volume C_v^l but rather the specific heat at constant pressure C_p^l. According to the general thermodynamic laws, the difference between C_p^l and C_v^l is given by

$$C_p^l - C_v^l = \frac{\alpha^2}{\beta} V_l T \tag{32}$$

where α and β are the thermal expansion and compressibility, respectively. By combining Equations (23) and (32), the molar specific heat at constant pressure for the liquid state is related to the molar specific heat at constant pressure for the gaseous state as follows

$$C_p^l - C_p^g = R\left\{0.5 - 0.5\frac{T}{T_b} + \frac{(\alpha T_b)^2}{\left(\beta\frac{RT_b}{V_l}\right)}\frac{T}{T_b}\right\} \tag{33}$$

The relation given by Equation (33) can be compared with the equation given by S. W. Benson[5] for the difference of molar specific heats at constant pressure for the saturated liquid and saturated vapor. Benson's relation is

Table 2 Compressibility of liquids

Liquid	Formula	Molecular weight	T_b/K	T/ K	Density, gr/cc	$\beta \times 10^6$/ atm^{-1}	T/T_b	$\frac{\beta R T_b}{V_l}$
Acetone	$(CH_3)_2CO$	58.08	329.7	293.2	0.792	125.6	0.890	0.046 4
Benzene	C_6H_6	78.11	353.3	303.2	0.868	98.5	0.858	0.031 8
	...	...	...	333.2	0.836	116.4	0.944	0.036 2
Carbon disulfide	CS_2	76.13	319.7	303.2	1.261	102.0	0.948	0.044 4
Carbon tetrachloride	CCl_4	153.84	350.0	293.2	1.595	105.8	0.838	0.031 6
Chlorobenzene	C_6H_5Cl	112.56	405.2	283.2	1.107	72	0.698	0.022 6
Chloroform	$CHCl_3$	119.39	334.5	303.2	1.49	109.5	0.907	0.037 5
Ethyl ether	$(C_2H_5)_2O$	74.12	307.8	303.2	0.713	210	0.986	0.051 0
Ethyl bromide	C_2H_5Br	108.98	311.2	293.2	1.430	120	0.942	0.040 2
Ethyl iodide	C_2H_5I	155.98	345.4	313.2	1.91	74	0.907	0.025 7
n-Heptane	C_7H_{16}	100.20	371.7	303.2	0.684	134	0.815	0.027 4
n-Hexane	C_6H_{14}	86.17	342.2	303.2	0.66	159	0.885	0.034 3
Nitrobenzene	$C_6H_5NO_2$	123.11	484.1	303.2	1.198 7	49	0.627	0.019 0
Nitromethane	CH_3NO_2	61.04	374.2	303.2	1.13	73.6	0.810	0.041 9
n-Octane	C_8H_{18}	114.23	399.0	293.2	0.704	101.6	0.735	0.020 5
Paraldehyde	$C_6H_{12}O_3$	132.16	397.6	291.2	0.994	88.2	0.733	0.021 6
Silicon bromide	$SiBr_4$	347.72	426.2	298.2	2.814	86.6	0.700	0.024 5
Silicon chloride	$SiCl_4$	169.89	330.8	298.2	1.483	165.2	0.902	0.039 1
Titanium tetrachloride	$TiCl_4$	189.73	409.6	298.2	1.726	89.8	0.728	0.027 5
Stannic chloride	$SnCl_4$	260.53	387.3	298.2	2.232	108.9	0.770	0.029 6
Toluene	$C_6H_5CH_3$	92.13	384.0	303.2	0.862	96.5	0.790	0.028 5
o-Xylene	$C_6H_4(CH_3)_2$	106.16	417.2	293.2	0.875	79.7	0.703	0.022 4

$$C_p^l - C_p^g = n\Delta E_{vap}\alpha + R$$

where n is approximately 5/3 for a large number of substances, ΔE_{vap} is the molar energy of evaporation, and α the thermal expansion coefficient defined by Equation (25). However this relation, which requires a knowledge of n and ΔE_{vap}, seems to be less convenient to use than Equation (33) together with the semi-empirical information on the thermal expansion α and the compressibility β.

With the empirically determined relations for the coefficient of thermal expansion α and the compressibility β, the right side of Equation (33) can be calculated as a function of the temperature ratio T/T_b. This is carried out in Table 3. It is seen that except possibly near the boiling point, the difference between the molar specific heats of liquid and gas at constant pressure is very nearly $5R$, or

$$C_p^l - C_p^g = 10 \text{ cal/K mole} \tag{34}$$

This is indeed a remarkably simple result.

Table 3 Difference of specific heats of liquid and gas

T/T_b	αT_b	$\beta \frac{RT_b}{V_l}$	$\frac{(\alpha T_b)^2}{\left(\beta \frac{RT_b}{V_l}\right)} \frac{T}{T_b}$	$0.5 - 0.5 \frac{T}{T_b}$	$\frac{C_p^l - C_p^g}{R}$
0.6	0.392	0.019 2	4.80	0.20	5.00
0.7	0.398	0.023 2	4.78	0.15	4.93
0.8	0.410	0.028 0	4.80	0.10	4.90
0.9	0.445	0.036 5	4.88	0.05	4.93
1.0	0.525	0.052 0	5.30	0	5.30

The comparison of the calculated specific heat of liquid using Equation (34) with the experimental data is exhibited in Table 4. Since the theory is developed for normal liquids, molecules containing hydroxyl group or amino group, and molecules that are excessively elongated should be excluded. The experimental data are taken also from the Landolt-Börnstein Tabellen. For diatomic molecules, the theoretical value of C_p^g without vibrational heat is used. This value of 7 cal/℃ is correct for the prevailing low temperature. For silver chloride, this value of C_p^g may be too low. The first half of the table shows excellent agreement between the calculated and the experimental molar specific heat of liquid at constant pressure, with differences well within the experimental error. The only exception is carbon disulfide. This success of the theory is notable for the very wide range of temperature covered, from 120 K for nitric oxide to 763 K for silver chloride.

The second half of Table 4 indicates, however, considerable discrepancy between the calculated specific heat and the experimental specific heat. The calculated values are too large by approximately 4 cal/℃. A similar discrepancy for Trouton's ratio, the ratio of heat of evaporation and the boiling temperature T_b, also occurs for this group of liquids[6]. Such differences are well outside the probable experimental error. Furthermore, the temperatures concerned, although low, are not low enough for the occurrence of quantal effects (see appendix). Nor is there any likelihood of a different molecular interaction than the first group of liquids in the table. This is shown by their similar transport properties in the gaseous state. The only possible explanation seems to be the association effects. For instance, oxygen tends to associate to O_4 molecule in liquid state. If so, the molar specific heat of the liquid will be twice as large as listed in the table, or 25 cal/℃. The fictitious O_4 molecule in gaseous state then should have a C_p^g of 15 cal/℃. This is an entirely reasonable value.

From the foregoing discussion, it seems justified to use Equation (34) for normal liquids at room temperature or higher when association and dissociation are absent. For normal liquids, the accuracy of present method is very much higher than the method, suggested by R. R. Wenner[7], based upon counting individual atoms in the liquid molecule. Of course it may be argued that when it is necessary to predict the specific heat of liquid, the specific heat for the gaseous state is generally also not available, and then Equation (34) is of no practical utility. Fortunately, however, it is not necessary to depend on direct experimental determination of the specific heat of gaseous state. C_p^g can be very accurately calculated from the fundamental frequencies of the molecule determined by spectroscopy. Or C_p^g can be calculated with sufficient accuracy from the averaged frequencies for each type of chemical bonds, a method recently rendered more complete by R. V. Meghreblian[8]. As an example of such a situation, one may consider the molecule $CHCl_2Br$, bromodichloromethane. This molecule is not even listed in the well-known Handbook of Chemistry and Physics. But Meghreblian has calculated its molar specific heat C_p^g at 27℃ to be 16.2 cal/℃. Then according to Equation (34), the molar specific heat of the liquid C_p^l at the same temperature is 26.2 cal/℃. Since the molecular weight of this compound is 163.85, the specific heat at constant pressure of the liquid is 0.159 8 cal/gr ℃, or 0.159 8 Btu/lb°F.

Table 4 Specific heat of liquids at constant pressure

Liquid	Formula	Molecular weight	T_b/K	T/K	C_p^g/ (cal/K), exp	C_p^l/ (cal/K), calc'd	C_p^l/ (cal/K), exp
Ammonia	NH_3	17.032	239.8	213.2	8.0	18.0	17.9
	...	...	...	273.2	8.7	18.7	18.7
Acetone	$(CH_3)_2CO$	58.08	329.7	313.2	20.1	30.1	30.8
Benzene	C_6H_6	78.11	353.3	293.2	21.8	31.8	32.5
	...	...	...	323.2	23.3	33.3	34.3
Bromine	Br_2	159.83	332.0	270.0	7.0[a]	17.0	17.1
Carbon disulfide	CS_2	76.13	319.7	290.7	12.0	22.0	18.4
Carbon tetrachloride	CCl_4	153.84	350.0	273.2	21.5	31.5	30.9
	...	...	...	293.2	20.7	30.7	31.8
Chloroform	$CHCl_3$	119.39	334.5	313.2	17.2	27.2	27.9
Ethyl ether	$(C_2H_5)_2O$	74.12	307.8	303.2	31	41	40.5
Nitric oxide	NO	30.01	121.4	120	7.0[a]	17.0	17.3
Silver chloride	AgCl	143.34	1823	763	$<7.0, >9.0$	$<17.0, >19.0$	18.5
Stannic chloride	$SnCl_4$	260.53	387.3	287 − 371	24.4	34.4	38.5
Sulphur dioxide	SO_2	64.06	263.2	273.2	9.9	19.9	20.4
Argon	A	39.944	87.4	85.0	5.0	15.0	10.5
Carbon monoxide	CO	28.01	81.1	69.4	7.0[a]	17.0	14.27

(continued)

Liquid	Formula	Molecular weight	T_b/K	T/K	C_p^g/ (cal/K), exp	C_p^l/ (cal/K), calc'd	C_p^l/ (cal/K), exp
Methane	CH_4	16.04	111.7	100	8.0	18.0	13.01
Nitrogen	N_2	28.016	77.3	64.7	7.0[a]	17.0	13.15
	...	...	...	72.8	7.0[a]	17.0	13.33
Oxygen	O_2	32.00	90.1	73.2	7.0[a]	17.0	12.60

[a] Theoretical value for diatomic molecules without vibrational heat.

Liquid Metals

Pure metal atoms do not associate into molecules. Therefore the specific heat of pure liquid metals should be correlated on the basis of one gram atomic weight. When this is done, pure liquid metals give a specific heat at constant pressure from 6.4 cal/℃ to 8 cal/℃. Therefore it is reasonable to take as a first approximation

$$C_p^l = 7 \text{ cal/K} \tag{35}$$

per gram atomic weight. Since the specific heat at constant volume should be close to 3 R or 6 cal/℃, Equation (35) shows that the difference between C_p and C_v for liquid metals is only 1 cal/℃. This is very much less than the corresponding value found for normal liquids as discussed in the previous section. For the particular case of liquid mercury, a more detailed discussion has been given by J. F. Kincaid and H. Eyring[9]. They have also pointed out that the difference in behavior of the liquid metals from that of normal liquids is due to the difference in the interaction potential of the constituent particles. Since the interaction in normal liquids is the interaction between molecules, the interaction in liquid metals is the interaction between the metallic atoms. This difference is certainly expected.

Velocity of Sound

The velocity of propagation of small disturbances is generally called the velocity of sound c and is a very important quantity in fluid dynamics. The general formula for computing this quantity is

$$c^2 = -\frac{v^2 \mathrm{d}P}{\mathrm{d}v} = -\frac{1}{\rho}\left(\frac{V\mathrm{d}P}{\mathrm{d}V}\right) \tag{36}$$

where v is the volume for unit mass of the liquid, ρ is the density or $1/v$, and the derivative is to be computed according to the adiabatic process.

If the frequency of the sound wave is sufficiently low, or if the characteristic time of the small disturbance is longer than the relaxation time for reaching thermodynamic equilibrium, then it can be easily shown that

$$c^2 = \frac{1}{\rho\beta}\frac{C_p^l}{C_V^l} \tag{37}$$

where β is the compressibility defined by Equation (27). For normal liquids, β is given by Equation (31), and according to Table 3, the difference between C_p and C_V is approximately 9.6 cal/℃. Let M be the molecular weight. Then

$$c^2 = \frac{RT_b}{M}\left(101.6 - 82.4\frac{T}{T_b}\right)\frac{C_p^l}{C_p^l - 9.6} \tag{38}$$

The comparison between the calculated and the experimental values of the velocity of sound is shown in Table 5. It is seen that with the exception of carbon disulfide, Equation (38) predicts the velocity of sound to within one per cent of the experimental value. This accuracy is perhaps expected, since Equation (38) already contains the empirically deduced relations for $C_p - C_v$ and β. Nevertheless, the agreement is satisfying in that it indicates the inner consistency of the theory. The case of CS_2 may be partially explained by the exceptionally low experimental value of C_p. Table 4 shows that the theoretical value for C_p^l of carbon disulfide is considerably higher. If so, the calculated velocity of sound for CS_2 will be lowered and the agreement with the experimental data will be better. Therefore it is justified to say that Equation (38) gives a satisfactory prediction of the velocity of sound for pure normal liquids. Since C_p^l increases with temperature, Equation (38) shows that the velocity of sound decreases with temperature. This behavior of normal liquid is contrary to that of gaseous state, for which the velocity of sound increases with temperature.

Transport Properties

The properties of liquids discussed in the preceding sections are all properties at thermodynamic equilibrium. Transport properties are properties of matter not at equilibrium, and the theory of transport properties is very much more complex than the theory of equilibrium properties. A "basic" approach to the theory of transport properties of liquids was made by J. G. Kirkwood and by M. Born and H. S. Green. However, no useful result has yet been obtained by their theories. Here the method of simple model followed by fitting the theoretical relation to empirical data will be used. An example of this method applied to the transport properties of liquids is Eyring's theory of viscosity of liquids[10].

According to Eyring's theory, the viscosity μ of a liquid at temperature T is related to the energy of vaporization ΔE_{vap} per mole of the liquid as follows.

$$\mu = \frac{hN}{V_l}\exp\left[\Delta E_{vap}/(2.45\,RT)\right] \tag{39}$$

where h is Planck's constant, N Avogadro's number, and V_l the molar volume of the liquid. For normal liquids, energy of vaporization is related to the normal boiling temperature T_b by Trouton's rule. That is

$$\Delta E_{vap} \cong 9.4\,RT_b \tag{40}$$

By combining Equations (39) and (40), the viscosity in centipoises is given by

$$\mu \cong \frac{0.3990}{V_1} \exp(3.83\, T_b/T) \tag{41}$$

where V_1 is in cc per mole.

Thermal Conductivity

The elementary theory of heat conduction in gas gives[11] the thermal conductivity λ of a gas as

$$\lambda = \frac{1}{3} c l C_v \tag{42}$$

where c is the velocity of sound, C_v the specific heat of the gas at constant volume per unit volume, and l is the mean free path of the molecules in the gas. In other words, l is the mean distance for which the molecules will maintain their individual velocities. I. Estermann and J. E. Zimmerman[12] noted that if l is interpreted as the distance, a lattice wave in a solid will travel before scattering, then the thermal conductivity of solid due to lattice oscillations calculated by R. E. B. Makinson[13] can be easily obtained from the relation of Equation (42). This observation clearly shows the fundamental character of Equation (42). It should then be true for all three states of matter, if the different quantities entering into it are properly identified. P. W. Bridgman[4], in fact, suggested such a correlation as early as 1931. However, the following theory differs from that of Bridgman in important details, as will be explained presently.

Table 5 Velocity of Sound

Liquid	Formula	Molecular weight	T_b/K	T/K	C_p^l /(cal/K)	$c_{calc} \times 10^{-5}$ /(cm/sec)	$c_{exp}^{a} \times 10^{-5}$ /(cm/sec)
Benzene	C_6H_6	78.11	353.3	290.2	39.8	1.176	1.166
Carbon disulfide	CS_2	76.13	319.7	288.2	18.2	1.350	1.161
Chloroform	$CHCl_3$	119.39	334.5	288.2	27.9	0.967	0.983
Ethyl ether	$(C_2H_5)_2O$	74.12	307.8	288.2	39.8	1.022	1.032

[a] Taken from Smithsonian tables.

For liquids other than liquid metals, the heat must be conducted as oscillations of the molecules from its mean position, similar to the heat conduction in solid by "lattice waves." Therefore if Equation (42) will be used to calculate the thermal conductivity of a liquid, the c is the velocity of sound in the liquid. This is exactly what Bridgman proposed. It will not be correct, however, to compute c by Equation (37), because the frequencies of the lattice waves are generally so high as to not allow the thermodynamic equilibrium assumed for Equation (37). Therefore there are some questions in determining c. For lack of better information, let c be tentatively taken simply as

$$c \approx 1/\sqrt{\rho\beta} \tag{43}$$

Thus the uncertain ratio of specific heats in Equation (37) is dropped.

The appropriate mean free path l for the lattice waves in liquid must be the size of the local organization of the liquid molecules. This size of local organization is a few times the molecular spacing a. Therefore according to Equation (4)

$$l \approx V^{1/3} \tag{44}$$

where V is the liquid volume per molecule. l is thus of the order of a few ångströms.

Bridgman proposed to identify c_v as the total heat capacity including the internal degrees of freedom. However, with l only of the order of 10^{-7} cm, and with c as large as 10^5 cm per sec, the characteristic time for the lattice waves must be of the order of 10^{-12} sec. This is much shorter than the known relaxation time for the internal degrees of freedom of molecules. Then it is reasonable to suppose that the internal degrees of freedom of the molecule do not participate in the conduction of heat. Therefore for computing the thermal conductivity of liquids, the specific heat should be that of external degrees of freedom of the molecules only. The specific heat per mole is then approximately $3R$. Or

$$c_v \approx \frac{3R}{V_l} \tag{45}$$

By combining Equations (43), (44), and (45), the thermal conduction of liquid can be written as

$$\lambda \approx \frac{1}{\sqrt{\rho\beta}} V^{1/3} \frac{3R}{V_l}$$

The factor of proportionality of the above relation remains undetermined. To determine it, one has to introduce the experimental data. This is done in Table 6 and Fig. 3. The compressibility data were taken from the Landolt-Börnstein Tabellen and the thermal conductivity from the Appendix of McAdam's book on heat transfer[15]. These data include such "abnormal" liquids as water and alcohols. It thus seems that the great majority of experimental thermal conductivity falls within 20 per cent of the calculated value if one simply takes

$$\begin{aligned} \lambda &= \frac{1}{\sqrt{\rho\beta}} V^{1/3} \frac{3R}{V_l} \\ &= \frac{1}{\sqrt{\rho\beta}} \left(\frac{M}{N\rho}\right)^{1/3} \frac{3R\rho}{M} \end{aligned} \tag{46}$$

Therefore the factor of proportionality is just unity. By comparing Equations (42) and (44), it is seen that mean free path l of the lattice waves is approximately 3 times the intermolecular distance. The local structure of the liquid may be pictured as a cubic lattice having nearly 12 neighbors to the central molecule. This deduction is entirely in agreement with the adopted concept of the liquid state. Equation (46) shows that the thermal conductivity of all liquids, normal or otherwise, can be calculated satisfactorily if the molecular weight, the density, and the compressibility of the liquid are known.

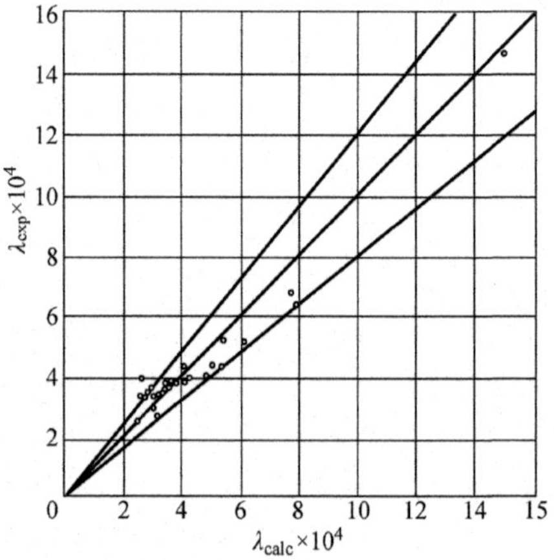

Fig. 3 Comparison of calculated, λ_{calc}, and experimental, λ_{exp}, values of the thermal conductivity of liquids. the unit of λ is Cal/cm sec ℃

Table 6 Thermal conductivity

Liquid	Formula	Molecular weight	Temperature /℃	Density /gr/cc	$\beta\times 10^{-5}$ /atm^{-1}	$(\rho\beta)^{-1/2}$ $\times 10^{-5}$ /cm/sec	$V^{1/3}\times$ 10^8/cm	$\frac{V^{1/3}}{\sqrt{\rho\beta}}\frac{3R}{V_l}$ $\times 10^6$	$\lambda_{exp}\times 10^6$ cal/cm /℃ sec	$\frac{\lambda_{exp}}{\frac{V^{1/3}}{\sqrt{\rho\beta}}\frac{3R}{V_l}}$
Acetic acid	CH_3COOH	60.05	20	1.049	90.6	1.033	4.56	0.000 491	0.000 409	0.833
Acetone	$(CH_3)_2CO$	58.08	20	0.792	125.6	1.010	4.96	0.000 407	0.000 430	1.056
Allyl alcohol	$C_2H_3CH_2OH$	58.08	20~30	0.855	75	1.257	4.83	0.000 533	0.000 430	0.807
Amyl alcohol	$C_5H_{11}OH$	88.15	17.7	0.814	90.5	1.172	5.65	0.000 365	0.000 392	1.075
Aniline	$C_6H_5NH_2$	93.12	20	1.022	36.1	1.657	5.32	0.000 577	0.000 413	0.715
Benzene	C_6H_6	78.11	30	0.868	98.5	1.089	5.30	0.000 382	0.000 380	0.995
	...	...	60	0.836	116.4	1.020	5.37	0.000 350	0.000 359	1.026
Carbon dioxide	CO_2	44.01	13.3[a]	0.960	624.4	0.412	4.23	0.000 246	0.000 245	0.995
Carbon disulfide	CS_2	76.13	30	1.261	102.0	0.888	4.64	0.000 407	0.000 384	0.945
Carbon tetrachloride	CCl_4	153.84	20	1.595	105.8	0.775	5.43	0.000 260	0.000 392	1.510
Chlorobenzene	C_6H_5Cl	112.56	10	1.107	72	1.128	5.53	0.000 336	0.000 343	1.020
Chloroform	$CHCl_3$	119.39	30	1.49	109.5	0.788	5.11	0.000 299	0.000 330	1.104
Ethyl alcohol	C_2H_5OH	46.07	20	0.789	112	1.071	4.59	0.000 502	0.000 434	0.865
Ethyl bromide	C_2H_5Br	108.98	20	1.430	120	0.769	5.01	0.000 301	0.000 289	0.962
Ethyl ether	$(C_2H_5)_2O$	74.12	30	0.713	210	0.808	5.56	0.000 258	0.000 330	1.280
Ethyl iodide	C_2H_5I	155.98	40	1.91	74	0.847	5.15	0.000 318	0.000 264	0.830
Ethylene glycol	$(CH_2OH)_2$	62.07	0	1.12	34	1.63	4.51	0.000 790	0.000 632	0.800
Glyecrine	$(CH_2OH)_2$-CHOH	92.09	20	1.260	22	1.912	4.94	0.000 773	0.000 678	0.877
n-Heptane	C_7H_{16}	100.20	30	0.684	134	1.051	6.25	0.000 268	0.000 334	1.246
n-Hexane	C_6H_{14}	86.17	30	0.66	159	0.983	6.01	0.000 270	0.000 330	1.220
Hexyl alcohol	$C_6H_{12}OH$	102.17	30	0.818	60	1.437	5.91	0.000 405	0.000 384	0.948

(continued)

Liquid	Formula	Molecular weight	Temperature /℃	Density /gr/cc	$\beta\times10^{-5}$ /atm^{-1}	$(\rho\beta)^{-1/2}\times10^{-5}$ /cm/sec	$V^{1/3}\times10^{8}$/cm	$\frac{V^{1/3}}{\sqrt{\rho\beta}}\frac{3R}{V_l}\times10^6$	$\lambda_{exp}\times10^6$ cal/cm /℃ sec	$\lambda_{exp}\Big/\frac{V^{1/3}}{\sqrt{\rho\beta}}\frac{3R}{V_l}$
Methyl alcohol	CH_3OH	32.04	20	0.7928	123.5	1.016	4.07	0.000 609	0.000 512	0.841
Nitrobenzene	$C_6H_5NO_2$	123.11	30	1.199	49	1.313	5.54	0.000 422	0.000 393	0.930
Nitromethane	CH_3NO_2	61.04	30	1.13	73.6	1.104	4.47	0.000 544	0.000 517	0.950
n-Octane	C_8H_{18}	114.23	20	0.704	101.6	1.190	6.45	0.000 282	0.000 347	1.230
Paraldehyde	$C_6H_{12}O_3$	132.16	18	0.994	88.2	1.075	6.04	0.000 291	0.000 355	1.220
Toluene	$C_6H_5CH_3$	92.13	30	0.862	96.5	1.103	5.62	0.000 347	0.000 355	1.023
Water	H_2O	18.02	30	0.996	47.9	1.457	3.104	0.001 490	0.001 470	0.987
o-Xylene	$C_6H_4(CH_3)_2$	106.16	20	0.875	79.7	1.207	5.87	0.000 347	0.000 372	1.070

[a] At 87 atm.

As a further check on Equation (46), the variation of the thermal conductivity of water with temperature is computed and compared with the experiments. Since the ratio of specific heats for liquid water is very nearly equal to one, the velocity $1/\sqrt{\rho\beta}$ for water is in fact the velocity of sound c. Thus for water, the ratio of thermal conductivity at two temperatures designated by subscripts 1 and 2, is given by

$$\frac{\lambda_2}{\lambda_1}=\frac{c_2}{c_1}\left(\frac{\rho_2}{\rho_1}\right)^{2/3}$$

This relation is compared with experimental data on λ in Table 7 where 13℃ is taken as the lower comparison temperature. The ratios of conductivity at two higher temperatures of 19 and 31℃ check closely with the test values. Therefore Equation (46) can also predict satisfactorily correct temperature variation of the thermal conductivity.

Table 7 Thermal conductivity of water

Temperature, ℃	$c\times10^{-1}$ /cm/sec[a]	ρ /gr/cc	λ_{exp} /cal/sec /℃ cm	$\left(\frac{\lambda_2}{\lambda_1}\right)_{calc}$	$\left(\frac{\lambda_2}{\lambda_1}\right)_{exp}$
13	1.441	0.999 4	0.001 410	1.000	1.000
19	1.461	0.998 4	0.001 431	1.012	1.015
31	1.505	0.995 4	0.001 475	1.040	1.046

[a] Taken from Smithsonian tables.

For normal liquids, the compressibility β is given by Equation (31); then Equation (46) reduces to

$$\lambda=\frac{3R}{N^{1/3}}\left\{\frac{RT_b}{M}\left(101.6-82.4\frac{T}{T_b}\right)\right\}^{1/2}\left(\frac{\rho}{M}\right)^{2/3} \tag{47}$$

Therefore the thermal conductivity of normal liquid can be calculated once the molecular weight M, the boiling temperature T_b, and the density are known. Since density decreases with temperature, Equation (47) shows that the conductivity of normal liquids also decreases with temperature. The increasing conductivity of water with temperature shown in Table 7 indeed indicates the abnormality of water as a liquid.

Concluding Remarks

The properties of pure liquids studied in the preceding sections are properties at low pressures, i.e., at pressures below the critical pressure of the substance. Then the pressure of the liquid is not a parameter in the calculations. In fact, for normal liquids, the only essential parameter is the normal boiling temperature of the liquid. This parameter, together with the density and the molecular weight of the liquid, then determines the coefficient of thermal expansion, the compressibility, the specific heats, the velocity of sound, the viscosity, and the thermal conductivity. The required information for calculating the properties is thus generally available in the handbook.

From the point of view of physical mechanics of predicting macroscopic behavior of matter in bulk from microscopic behavior of the molecules, the remaining task is then the determination of the liquid density and the normal boiling temperature from the knowledge of the molecular structure. Since both of these quantities are related to the bimolecular interaction, the missing link is then a method of calculating the molecular interaction from the structural formula of the liquid molecules. The well-known theory of molecular interaction due to F. London and W. Heitler and others has unfortunately not yet been developed into a form useful for this purpose. Here clearly is a field for future research.

Appendix

Quantal Effects at Low Temperatures

S. D. Hamann[16] has investigated the quantal correction to the Lennard-Jones and Devonshire theory of liquids by approximating the spherical cage of the theory by a cube with a square-well potential. The size of the cube is determined by the diameter of the sphere within which the Lennard-Jones-Devonshire potential is negative. Let the ratio of this size to the molecular distance a be x. When $V \simeq V^*$, as is the case for liquid at low pressures, $x \simeq 0.2$. Hamann showed that the temperature T for appreciable quantal effects is given by the equation

$$\frac{2\pi mkT}{h^2}\left(\frac{4\pi}{3}\sqrt{2}V\right)^{2/3} x^2 \simeq 1$$

where m is the mass of a molecule, k the Boltzmann constant, h the Planck constant, and V the volume of liquid per molecule. For N_2 molecule, if the density of the liquid is taken to be 0.8, this temperature for appreciable quantal effects is only 5.5 K. Therefore it is certain that none of the anomalies in Table 5 can be due to quantum mechanical reasons.

References

[1] Tsien H S. Physical Mechanics, a New Field in Engineering Science. ARS Journal, 1953: 14-16.

[2] Lennard-Jones J E, Devonshire A F. Critical Phenomena in Gases. Proc. Royal Soc. London (A), 1937, 163: 53 - 70; 1938 , 163: 1 - 11.//Fowler R H, Guggenheim E A. Statistical Mechanics. Cambridge University Press, 1949, § 808: 336.

[3] Wentorf, Jr. R H, Buehler R J, Hirschfelder J O, Curtiss C F. Lennard-Jones and Devonshire Equation of State of Compressed Gases and Liquids. Journal of Chemical Physics, 1950, 18: 1484 - 1500.

[4] Eyring H. Hirschfelder J O. The Theory of the Liquid State. Journal of Physics Chemistry, 1937, 41: 249 - 257.

[5] Benson S W. Heat Capacities of Liquids and Vapors. Journal of Chemical Physics, 1947, 15: 866 - 867.

[6] Barclay I M, Butler J. A. V. The entropy of Solution. Transactions Faraday Society, 1938, 34(2): 1445 - 1454.

[7] Wenner R R. Thermochemical Calculations. New York: McGraw-Hill Book Co. , Inc. , 1941: 16.

[8] Meghreblian R V. Approximate Calculations of Special Heats for Polyatomic Gases. Journal of the American Rocket Spciety, 1951, 21: 127 - 131.

[9] Kincaid J F. Eyring H. A Partition Function for Liquid Mercury. Journal of Chemical Physics, 1937, 5: 587 - 596.

[10] Hirschfelder J O, Curtiss C F, Bird R B, Spotz E L. The Transport Properties at High Densities, chapter in the book: The Properties of Gases. issued as report of the Naval Research Laboratory, Department of Chemistry, University of Wiscon*sin* , 1951.

[11] Loeb L. Kinetic Theory of Gases. New York: McGraw-Hill Book Co. , Inc,. 1927, p.240.

[12] Estermann I. Zimmerman J E. Heat Conduction in Alloys at Low Temperatures. Journal of Applied Physics, 1952, 23: 578 - 588.

[13] Makinson R. E. B. The Thermal Conductivity of Metals. Proceedings of the Cambridge Philosophical Society, 1938, 34: 474 - 477.

[14] Bridgman P W. Physics of High Pressures. Bel G. and Sons. London 1931.//Lawson A W. On Heat Conductivity in Liquids. Journal of Chemical Physics, 1950, 18: 1421.

[15] McAdams W H. Heat Transmission. New York: McGraw-Hill Book Co. , Inc. , 2nd Ed. , 1942, 389.

[16] Hamann S D. A Quantum Correction to the Lennard-Jones and Devonshire Equation of States. Transactions Faraday Society, 1952, 48(2): 303 - 307.

Similarity Laws for Stressing Heated Wings

H. S. Tsien*
(*California Institute of Technology*)

Summary

It will be shown that the differential equations for a heated plate with large temperature gradient and for a similar plate at constant temperature can be made the same by a proper modification of the thickness and the loading for the isothermal plate. This fact leads to the result that the stresses in the heated plate can be calculated from measured strains on the unheated plate by a series of relations, called the "similarity laws." The application of this analog theory to solid wings under aerodynamic heating is discussed in detail. The loading on the unheated analog wing is, however, complicated and involves the novel concept of feedback and "body force" loading. The problem of stressing a heated box-wing structure can be solved by the same analog method and is briefly discussed.

Introduction

The high stagnation temperature for flight of aircraft at supersonic speeds results in severe aerodynamic heating of the surfaces of aircraft. For instance, in a recent paper by Kaye[1], the transient temperature distribution in a wedge-shaped solid wing was calculated for accelerated flight and was found to be rapidly varying both with respect to the space points in the wing and with respect to time instants. When there are large temperature gradients in the material, there are generally large thermal stresses due to the uneven thermal expansion of the material. Therefore, concurrent with the severe aerodynamic heating, there is the problem of determining the thermal stresses in the wing. The purpose of this paper is to suggest a method of stressing such heated wings.

The starting point of the stress analysis is the temperature distribution in the wing. Because the rate of change of temperature is small in comparison to the speed of sound propagation in the material, the stress calculation can be considered as a quasi-steady problem. That is, the stress in the wing at each time instant can be calculated from the temperature distribution at that time instant without the inertia effects from the time variation of the material displacements required by the changing stresses and the thermal expansions. This problem in general theory of elasticity is greatly simplified by the fact that the wing is thin,

---

Received June 10, 1952.

Journal of the Aeronautical Sciences, vol. 20, pp. 1 - 11, 1953.

* Robert H. Goddard Professor of Jet Propulsion, Daniel and Florence Guggenheim Jet Propulsion Center.

and therefore the Kirchhoff hypothesis for bending of thin plates holds. In fact, the problem of thermal stresses in an elastic plate has been treated some time ago by Nádai[2]. The present study generalizes the earlier theory in two directions: Nádai's assumption of linear temperature profile across the plate is no longer necessary. The temperature profile is now arbitrary. Secondly, Young's modulus E of the material is now allowed to vary as a function of temperature. Thus, the effects of decrease in Young's modulus with increase in temperature can be taken into account in the present theory.

However, the main purpose of the present paper is not just to construct a theory for heated plates. The main purpose is to utilize the theory to formulate a law of similarity, whereby the problem of stressing a heated wing can be solved by performing a set of experiments on a properly proportioned and properly loaded wing at room temperature without heating. This concept of similarity law has been explored before by the present author[3] for thin-walled cylinders. It is believed that this approach to the problem of stressing a heated wing has many practical advantages over a pure analytical solution, as will be discussed later in the paper.

Basic Equations for Heated Plate

Let the plane of the plate be the x-y plane, and its variable thickness be $b(x, y)$. The z- axis is normal to the plate and pointing downward. The temperature above the ambient is specified by $T(x, y; z)$. Actually, the distance z is measured from the "median surface," the position of which will be specified presently. Therefore, it is implied here that the median surface, although not exactly coinciding with the x-y plane, is nevertheless sufficiently flat to be considered as the x-y plane. Let u, v, and w be the displacement of a point (x, y) on the median surface in the x, y, and z directions, respectively, due to elastic strain and thermal expansion. Then, according to Kirchhoff's bending hypothesis, the total strains of a point (x, y, z) in the plate are given by

$$\left.\begin{aligned} \varepsilon_x &= (\partial u/\partial x) - z(\partial^2 w/\partial x^2) \\ \varepsilon_y &= (\partial v/\partial y) - z(\partial^2 w/\partial y^2) \\ \gamma_{xy} &= (\partial u/\partial y) + (\partial v/\partial x) - 2z(\partial^2 w/\partial x\partial y) \end{aligned}\right\} \tag{1}$$

where ε_x is the direct strain in x, ε_y is the direct strain in y, and γ_{xy} is the shear strain in the x-y plane. All other strains are small and negligible for thin plates.

Let $E(T) = E(x, y; z)$ be the variable Young's modulus, ν the constant Poisson's ratio, and σ_x, σ_y, τ_{xy} the significant components of stresses. Then the strains can be computed from these quantities as follows:

$$\left.\begin{aligned} \varepsilon_x &= [(\sigma_x - \nu\sigma_y)/E(x, y; z)] + aT(x, y; z) \\ \varepsilon_y &= [(\sigma_y - \nu\sigma_x)/E(x, y; z)] + aT(x, y; z) \\ \gamma_{xy} &= 2(1+\nu)T_{xy}/E(x, y; z) \end{aligned}\right\} \tag{2}$$

where a is the coefficient of thermal expansion. By solving Eq. (2) for stresses, one has

$$\left.\begin{aligned}\sigma_x &= \frac{E(x, y; z)}{1-\nu^2}[(\varepsilon_x+\nu\varepsilon_y)-(1+\nu)aT(x, y; z)]\\ \sigma_y &= \frac{E(x, y; z)}{1-\nu^2}[(\varepsilon_y+\nu\varepsilon_x)-(1+\nu)aT(x, y; z)]\\ \tau_{xy} &= [E(x, y; z)/2(1+\nu)]\gamma_{xy}\end{aligned}\right\} \tag{3}$$

In the theory of thin plates, the important quantities are not the stresses but are rather the sectional forces and the sectional moments derived from the stresses. These sectional quantities are defined as

$$N_x = \int \sigma_x \mathrm{d}z, \quad N_y = \int \sigma_y \mathrm{d}z, \quad N_{xy} = \int \tau_{xy} \mathrm{d}z \tag{4}$$

$$M_x = \int \sigma_x z \mathrm{d}z, \quad M_y = \int \sigma_y z \mathrm{d}z, \quad M_{xy} = \int \tau_{xy} z \mathrm{d}z \tag{5}$$

where all integrations extend across the whole thickness of the plate. N_x and N_y are then sectional normal forces, N_{xy} is the sectional shear, M_x and M_y are the sectional bending moments, and M_{xy} is the sectional twisting moment. By substituting Eq. (1) into Eq. (3) and then into Eqs. (4) and (5), one has, for example,

$$N_x = D_0\left(\frac{\partial u}{\partial x}+\nu\frac{\partial v}{\partial y}\right)-D_1\left(\frac{\partial^2 w}{\partial x^2}+\nu\frac{\partial^2 w}{\partial y^2}\right)-N_T$$

$$M_x = D_1\left(\frac{\partial u}{\partial x}+\nu\frac{\partial v}{\partial y}\right)-D_2\left(\frac{\partial^2 w}{\partial x^2}+\nu\frac{\partial^2 w}{\partial y^2}\right)-M_T$$

where

$$D_0(x, y) = \frac{1}{1-\nu^2}\int E(x, y; z)\mathrm{d}z \tag{6}$$

$$D_1(x, y) = \frac{1}{1-\nu^2}\int E(x, y; z)z\mathrm{d}z \tag{7}$$

$$D_2(x, y) = \frac{1}{1-\nu^2}\int E(x, y; z)z^2\mathrm{d}z \tag{8}$$

$$N_T = \frac{a}{1-\nu}\int E(x, y; z)T(x, y; z)\mathrm{d}z \tag{9}$$

$$M_T = \frac{a}{1-\nu}\int E(x, y; z)T(x, y; z)z\mathrm{d}z \tag{10}$$

N_T and M_T being due to thermal expansion of the material can thus be called thermal sectional normal force and thermal bending moment, respectively. From the above typical expressions for N_x and M_x, it is seen that a simplification results if the median surface is so chosen that

$$D_1 = 0 \tag{11}$$

In fact, this is the condition for fixing the position of the median surface. With the median surface so determined, the sectional quantities can be calculated as

$$\left.\begin{aligned} N_x &= D_0\left(\frac{\partial u}{\partial x}+\nu\frac{\partial v}{\partial y}\right)-N_T \\ N_y &= D_0\left(\frac{\partial v}{\partial y}+\nu\frac{\partial u}{\partial x}\right)-N_T \\ N_{xy} &= \frac{1-\nu}{2}D_0\left(\frac{\partial v}{\partial y}+\frac{\partial u}{\partial y}\right) \end{aligned}\right\} \tag{12}$$

$$\left.\begin{aligned} M_x &=- D_2\left(\frac{\partial^2 w}{\partial x^2}+\nu\frac{\partial^2 w}{\partial y^2}\right)-M_T \\ M_y &=- D_2\left(\frac{\partial^2 w}{\partial y^2}+\nu\frac{\partial^2 w}{\partial x^2}\right)-M_T \\ M_{xy} &= (1-\nu)D_2\frac{\partial^2 w}{\partial x\partial y} \end{aligned}\right\} \tag{13}$$

The equilibrium of forces in the medium surface requires

$$\left.\begin{aligned} (\partial N_x/\partial x)+(\partial N_{xy}/\partial y) &= 0 \\ (\partial N_{xy}/\partial x)+(\partial N_y/\partial y) &= 0 \end{aligned}\right\} \tag{14}$$

By substituting Eq. (12) into Eq. (14), the following two equations for u and v are obtained:

$$\left.\begin{aligned} \frac{\partial}{\partial x}\left[D_0\left(\frac{\partial u}{\partial x}+\nu\frac{\partial v}{\partial y}\right)\right]+\frac{1-\nu}{2}\frac{\partial}{\partial y}\left[D_0\left(\frac{\partial v}{\partial x}+\frac{\partial u}{\partial y}\right)\right] &= \frac{\partial N_T}{\partial x} \\ \frac{1-\nu}{2}\frac{\partial}{\partial x}\left[D_0\left(\frac{\partial v}{\partial x}+\frac{\partial u}{\partial y}\right)\right]+\frac{\partial}{\partial y}\left[D_0\left(\frac{\partial v}{\partial y}+\nu\frac{\partial u}{\partial x}\right)\right] &= \frac{\partial N_T}{\partial y} \end{aligned}\right\} \tag{15}$$

When the temperature distribution is specified, the right sides of these equations are known. Then Eq. (15) is a set of simultaneous partial differential equations for u and v. If the boundary conditions of the problem are specified in the displacements, Eq. (15) is the proper basis of solution.

If the boundary conditions of the problem are specified in terms of forces instead of displacements, it would be more convenient to use the stress function $\varphi(x, y)$ as the basis of solution. $\sum$ is defined through the following relations and satisfies Eq. (14) automatically:

$$N_x = \frac{\partial^2 \varphi}{\partial y^2}, \quad -N_{xy} = \frac{\partial^2 \varphi}{\partial x\partial y}, \quad N_y = \frac{\partial^2 \varphi}{\partial x^2} \tag{16}$$

By substituting these relations into Eq. (12), one has

$$\left.\begin{aligned} \frac{\partial u}{\partial x} &= \frac{1}{(1-\nu^2)D_0}\left[\left(\frac{\partial^2\varphi}{\partial y^2}-\nu\frac{\partial^2\varphi}{\partial x^2}\right)+(1-\nu)N_T\right] \\ \frac{\partial v}{\partial y} &= \frac{1}{(1-\nu^2)D_0}\left[\left(\frac{\partial^2\varphi}{\partial x^2}-\nu\frac{\partial^2\varphi}{\partial y^2}\right)+(1-\nu)N_T\right]- \\ \left(\frac{\partial v}{\partial x}+\frac{\partial u}{\partial y}\right) &= \frac{2}{(1-\nu)D_0}\frac{\partial^2\varphi}{\partial x\partial y} \end{aligned}\right\} \tag{17}$$

When the first equation of Eqs. (17) is differentiated twice with respect to y, the second equation is differentiated twice with respect to x, the third equation is differentiated with respect to x and y, and the results are added, a single equation for φ is obtained:

$$\frac{1}{(1-\nu^2)}\frac{\partial^2}{\partial x^2}\left[\frac{1}{D_0}\left(\frac{\partial^2\varphi}{\partial x^2}-\nu\frac{\partial^2\varphi}{\partial y^2}\right)\right]+\frac{2}{(1-\nu)}\frac{\partial^2}{\partial x\partial y}\left(\frac{1}{D_0}\frac{\partial^2\varphi}{\partial x\partial y}\right)+$$
$$\frac{1}{1-\nu^2}\frac{\partial^2}{\partial y^2}\left[\frac{1}{D_0}\left(\frac{\partial^2\varphi}{\partial y^2}-\nu\frac{\partial^2\varphi}{\partial x^2}\right)\right]+\frac{1}{1+\nu}\nabla^2\left(\frac{N_T}{D_0}\right)=0 \tag{18}$$

where ∇^2 is the Laplacian operator. At the boundary of the plate (Fig. 1a), if l, m are the direction cosines of the "outside" normal n to the boundary direction s, then sectional normal force N_n to the boundary and the sectional shearing force N_{ns} along the boundary are given as follows:

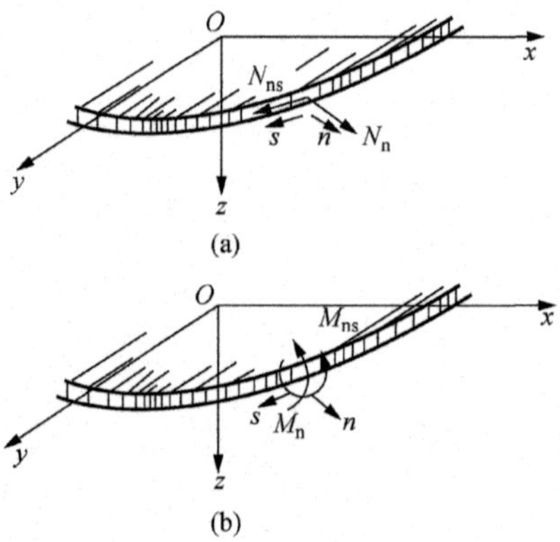

Fig. 1 Boundary forces and moments

$$\begin{aligned}N_n &= \frac{N_x+N_y}{2}+\frac{N_x-N_y}{2}(l^2-m^2)+N_{xy}(2lm)\\ &= \frac{1}{2}\nabla^2\varphi+\frac{1}{2}(l^2-m^2)\left(\frac{\partial^2\varphi}{\partial y^2}-\frac{\partial^2\varphi}{\partial x^2}\right)-2lm\frac{\partial^2\varphi}{\partial x\partial y}\end{aligned} \tag{19}$$

$$\begin{aligned}N_{ns} &= -(N_x-N_y)lm+(l^2-m^2)N_{xy}\\ &= -lm\left(\frac{\partial^2\varphi}{\partial y^2}-\frac{\partial^2\varphi}{\partial x^2}\right)-(l^2-m^2)\frac{\partial^2\varphi}{\partial x\partial y}\end{aligned} \tag{20}$$

With the boundary described and the boundary forces specified, Eqs. (18), (19), and (20) then give the proper formulation of the problem.

If p is the downward pressure loading on the plate, then the equilibrium of forces normal to the plane of the plate requires

$$\frac{\partial^2 M_x}{\partial x^2}-2\frac{\partial^2 M_{xy}}{\partial x\partial y}+\frac{\partial^2 M_y}{\partial y^2}=-p-N_x\frac{\partial^2 w}{\partial x^2}-2N_{xy}\frac{\partial^2 w}{\partial x\partial y}-N_y\frac{\partial^2 w}{\partial y^2}$$

By substituting the sectional moments obtained from Eq. (13) into the above equilibrium equation, an equation for the lateral deflection w results:

$$\frac{\partial^2}{\partial x^2}\left[D_2\left(\frac{\partial^2 w}{\partial x^2}+\nu\frac{\partial^2 w}{\partial y^2}\right)\right]+2(1-\nu)\frac{\partial^2}{\partial x\partial y}\left(D_2\frac{\partial^2 w}{\partial x\partial y}\right)+\frac{\partial^2}{\partial y^2}\left[D_2\left(\frac{\partial^2 w}{\partial y^2}+\nu\frac{\partial^2 w}{\partial x^2}\right)\right]$$
$$=-\nabla^2 M_T+p+N_x\frac{\partial^2 w}{\partial x^2}+2N_{xy}\frac{\partial^2 w}{\partial x\partial y}+N_y\frac{\partial^2 w}{\partial y^2} \tag{21}$$

Part of the boundary conditions will be expressed in terms of w directly. For instance, if the boundary s with outside normal n is a "fixed edge," then on that boundary

$$w=0,\quad \partial w/\partial n=0 \tag{22}$$

The other part of the boundary conditions is described in terms of sectional moments. If l, m are the direction cosines of the outside normal, the sectional bending moments M_n, M_s and the sectional twisting moment M_{ns} on the boundary (Fig. 1b) are, then,

$$\left.\begin{aligned}
M_n &= \frac{M_x+M_y}{2}+\frac{M_x-M_y}{2}(l^2-m^2)-M_{xy}(2lm)\\
&=-M_T-\frac{1+\nu}{2}D_2\nabla^2 w-\frac{1-\nu}{2}D_2(l^2-m^2)\left(\frac{\partial^2 w}{\partial x^2}-\frac{\partial^2 w}{\partial y^2}\right)-2(l-\nu)D_2 lm\frac{\partial^2 w}{\partial x\partial y}\\
M_s &= \frac{M_x+M_y}{2}-\frac{M_x-M_y}{2}(l^2-m^2)+M_{xy}(2lm)\\
&=-M_T-\frac{1+\nu}{2}D_2\nabla^2 w+\frac{1-\nu}{2}D_2(l^2-m^2)\left(\frac{\partial^2 w}{\partial x^2}-\frac{\partial^2 w}{\partial y^2}\right)+2(1-\nu)D_2 lm\frac{\partial^2 w}{\partial x\partial y}
\end{aligned}\right\} \tag{23}$$

$$\begin{aligned}
M_{ns} &= (M_x-M_y)lm+(l^2-m^2)M_{xy}\\
&=-(1-\nu)D_2 lm\left(\frac{\partial^2 w}{\partial x^2}-\frac{\partial^2 w}{\partial y^2}\right)+(1-\nu)D_2(l^2-m^2)\frac{\partial^2 w}{\partial x\partial y}
\end{aligned} \tag{24}$$

For a "free edge," the Kirchhoff boundary conditions are, then,

$$M_n=0,\quad \frac{\partial M_n}{\partial n}+\frac{M_n-M_s}{r}+\frac{\partial M_{ns}}{\partial s}=0 \tag{25}$$

where r is the radius of curvature of the boundary s, or

$$1/r=(1/m)(dl/ds) \tag{26}$$

Plate at Constant Temperature

For a plate at the constant reference room temperature, all the temperature terms in the equations developed in the previous section vanish. By using a bar over a quantity to differentiate the present case from the previous case, one has for strains

$$\left.\begin{aligned}
\bar{\varepsilon}_x &= (\partial\bar{u}/\partial x)-\bar{z}(\partial^2\bar{w}/\partial x^2)\\
\bar{\varepsilon}_y &= (\partial\bar{v}/\partial y)-\bar{z}(\partial^2\bar{w}/\partial y^2)\\
\bar{\gamma}_{xy} &= (\partial\bar{v}/\partial x)+(\partial\bar{u}/\partial y)-2\bar{z}(\partial^2\bar{w}/\partial x\partial y)
\end{aligned}\right\} \tag{27}$$

Let $\bar{E}$ be Young's modulus, now a constant, then

$$\left.\begin{aligned}\bar{\sigma}_x &= [\bar{E}/(1-\nu^2)](\bar{\varepsilon}_x+\nu\bar{\varepsilon}_y)\\ \bar{\sigma}_y &= [\bar{E}/(1-\nu^2)](\bar{\varepsilon}_y+\nu\bar{\varepsilon}_x)\\ \bar{\tau}_{xy} &= [\bar{E}/2(1+\nu)]\bar{\gamma}_{xy}\end{aligned}\right\} \tag{28}$$

The sectional forces and the sectional moments defined as Eqs. (4) and (5) are

$$\left.\begin{aligned}\bar{N}_x &= \bar{D}_0\left(\frac{\partial\bar{u}}{\partial x}+\nu\frac{\partial\bar{v}}{\partial y}\right)\\ \bar{N}_y &= \bar{D}_0\left(\frac{\partial\bar{v}}{\partial y}+\nu\frac{\partial\bar{u}}{\partial x}\right)\\ \bar{N}_{xy} &= \frac{1-\nu}{2}\bar{D}_0\left(\frac{\partial\bar{v}}{\partial x}+\frac{\partial\bar{u}}{\partial y}\right)\end{aligned}\right\} \tag{29}$$

$$\left.\begin{aligned}\bar{M}_x &= -\bar{D}_2\left(\frac{\partial^2\bar{w}}{\partial x^2}+\nu\frac{\partial^2\bar{w}}{\partial y^2}\right)\\ \bar{M}_y &= -\bar{D}_2\left(\frac{\partial^2\bar{w}}{\partial y^2}+\nu\frac{\partial^2\bar{w}}{\partial x^2}\right)\\ \bar{M}_{xy} &= (1-\nu)\bar{D}_2\frac{\partial^2\bar{w}}{\partial x\partial y}\end{aligned}\right\} \tag{30}$$

where, with $\bar{b}(x, y)$ as the thickness of the plate at the point (x, y),

$$\bar{D}_0(x, y) = \frac{\bar{E}}{1-\nu^2}\bar{b}(x, y), \quad \bar{D}_2(x, y) = \frac{\bar{E}}{1-\nu^2}\frac{\bar{b}^3}{12} \tag{31}$$

The median surface now lies midway between the upper and the lower surfaces of the plate.

Now introduce a potential $F(x, y)$ for the body forces X and Y, which are the forces per unit area of the median surface in the x and y directions:

$$X = -\partial F/\partial x, \quad Y = -\partial F/\partial y \tag{32}$$

Then the equations of equilibrium of forces in the medium surface are

$$\left.\begin{aligned}(\partial\bar{N}_x/\partial x)+(\partial\bar{N}_{xy}/\partial y)-(\partial F/\partial x) &= 0\\ (\partial\bar{N}_{xy}/\partial x)+(\partial\bar{N}_y/\partial y)-(\partial F/\partial y) &= 0\end{aligned}\right\} \tag{33}$$

By substituting Eq. (29) into the above equations, a set of equations for $\bar{u}$ and $\bar{v}$ is obtained:

$$\left.\begin{aligned}\frac{\partial}{\partial x}\left[\bar{D}_0\left(\frac{\partial\bar{u}}{\partial x}+\nu\frac{\partial\bar{v}}{\partial y}\right)\right]+\frac{1-\nu}{2}\frac{\partial}{\partial y}\left[\bar{D}_0\left(\frac{\partial\bar{v}}{\partial x}+\frac{\partial\bar{u}}{\partial y}\right)\right] &= \frac{\partial F}{\partial x}\\ \frac{1-\nu}{2}\frac{\partial}{\partial x}\left[\bar{D}_0\left(\frac{\partial\bar{v}}{\partial x}+\frac{\partial\bar{u}}{\partial y}\right)\right]+\frac{\partial}{\partial y}\left[\bar{D}_0\left(\frac{\partial\bar{v}}{\partial y}+\nu\frac{\partial\bar{u}}{\partial x}\right)\right] &= \frac{\partial F}{\partial y}\end{aligned}\right\} \tag{34}$$

Eq. (33) will be automatically satisfied, if a new stress function $\bar{\varphi}(x, y)$ is introduced such that

$$\bar{N}_x = \frac{\partial^2\bar{\varphi}}{\partial y^2}+F, \quad -\bar{N}_{xy} = \frac{\partial^2\bar{\varphi}}{\partial x\partial y}, \quad \bar{N}_y = \frac{\partial^2\bar{\varphi}}{\partial x^2}+F \tag{35}$$

The equation for $\bar{\varphi}$ is now

$$\frac{1}{1-\nu^2}\frac{\partial^2}{\partial x^2}\left[\frac{1}{D_0}\left(\frac{\partial^2\bar{\varphi}}{\partial x^2}-\nu\frac{\partial^2\bar{\varphi}}{\partial y^2}\right)\right]+\frac{2}{1-\nu}\frac{\partial^2}{\partial x\partial y}\left(\frac{1}{D_0}\frac{\partial^2\bar{\varphi}}{\partial x\partial y}\right)+$$
$$\frac{1}{1-\nu^2}\frac{\partial^2}{\partial y^2}\left[\frac{1}{D_0}\left(\frac{\partial^2\bar{\varphi}}{\partial y^2}-\nu\frac{\partial^2\bar{\varphi}}{\partial x^2}\right)\right]+\frac{1}{1+\nu}\nabla^2\left(\frac{F}{\bar{D}_0}\right)=0 \quad (36)$$

At the boundary of the plate,

$$\bar{N}_n=\frac{\bar{N}_x+\bar{N}_y}{2}+\frac{\bar{N}_x-\bar{N}_y}{2}(l^2-m^2)+\bar{N}_{xy}2lm$$
$$=F+\frac{1}{2}\nabla^2\bar{\varphi}+\frac{1}{2}(l^2-m^2)\left(\frac{\partial^2\bar{\varphi}}{\partial y^2}-\frac{\partial^2\bar{\varphi}}{\partial x^2}\right)-2lm\frac{\partial^2\bar{\varphi}}{\partial x\partial y} \quad (37)$$

$$\bar{N}_{ns}=-(\bar{N}_x-\bar{N}_y)lm+(l^2-m^2)\bar{N}_{xy}$$
$$=-lm\left(\frac{\partial^2\bar{\varphi}}{\partial y^2}-\frac{\partial^2\bar{\varphi}}{\partial x^2}\right)-(l^2-m^2)\frac{\partial^2\bar{\varphi}}{\partial x\partial y} \quad (38)$$

If $\bar{p}(x, y)$ is the downward pressure loading of the plate at the point (x, y), the equilibrium of lateral forces gives the following equation for $\bar{w}$:

$$\frac{\partial^2}{\partial x^2}\left[\bar{D}_2\left(\frac{\partial^2\bar{w}}{\partial x^2}+\nu\frac{\partial^2\bar{w}}{\partial y^2}\right)\right]+2(1-\nu)\frac{\partial^2}{\partial x\partial y}\left(\bar{D}_2\frac{\partial^2\bar{w}}{\partial x\partial y}\right)+\frac{\partial^2}{\partial y^2}\left[\bar{D}_2\left(\frac{\partial^2\bar{w}}{\partial y^2}+\nu\frac{\partial^2\bar{w}}{\partial x^2}\right)\right]$$
$$=\bar{p}+\bar{N}_x\frac{\partial^2\bar{w}}{\partial x^2}+2\bar{N}_{xy}\frac{\partial^2\bar{w}}{\partial x\partial y}+\bar{N}_y\frac{\partial^2\bar{w}}{\partial y^2} \quad (39)$$

The sectional bending moments $\bar{M}_n$, $\bar{M}_s$ and the sectional twisting moment $\bar{M}_{ns}$ are, then,

$$\left.\begin{aligned}\bar{M}_n&=-\frac{1+\nu}{2}\bar{D}_2\nabla^2\bar{w}-\frac{1-\nu}{2}\bar{D}_2(l^2-m^2)\left(\frac{\partial^2\bar{w}}{\partial x^2}-\frac{\partial^2\bar{w}}{\partial y^2}\right)-2(1-\nu)\bar{D}_2lm\frac{\partial^2\bar{w}}{\partial x\partial y}\\ \bar{M}_s&=-\frac{1+\nu}{2}\bar{D}_2\nabla^2\bar{w}+\frac{1-\nu}{2}\bar{D}_2(l^2-m^2)\left(\frac{\partial^2\bar{w}}{\partial x^2}-\frac{\partial^2\bar{w}}{\partial y^2}\right)+2(1-\nu)\bar{D}_2lm\frac{\partial^2\bar{w}}{\partial x\partial y}\end{aligned}\right\} \quad (40)$$

$$\bar{M}_{ns}=-(1-\nu)\bar{D}_2lm\left(\frac{\partial^2\bar{w}}{\partial x^2}-\frac{\partial^2\bar{w}}{\partial y^2}\right)+(1-\nu)\bar{D}_2(l^2-m^2)\frac{\partial^2\bar{w}}{\partial x\partial y} \quad (41)$$

Similarity Laws for Solid Thin Wings

By comparing the corresponding equations for the heated plate and for the plate at room temperature, their similarity is evident. The question is then whether it is possible to find a corresponding plate at room temperature such that, with proper loading, the plate will give solutions that are similar to that of the heated plate. To fix ideas, let the heated plate and the unheated plate both have the same plane form. The temperature distribution of the heated plate is specified so that N_T and M_T are known functions of the coordinates (x, y). The heated plate is built in at one end but free at all other edges (Fig. 2). The lateral pressure loading p for the heated plate is also specified. The problem is to find the thickness distribution $\bar{b}(x, y)$ and the loading of the unheated plate, such that $\bar{\varphi}(x, y)$ and $\bar{w}(x, y)$ are the same as $\varphi(x, y)$ and

$w(x, y)$, respectively, except a proportionality factor.

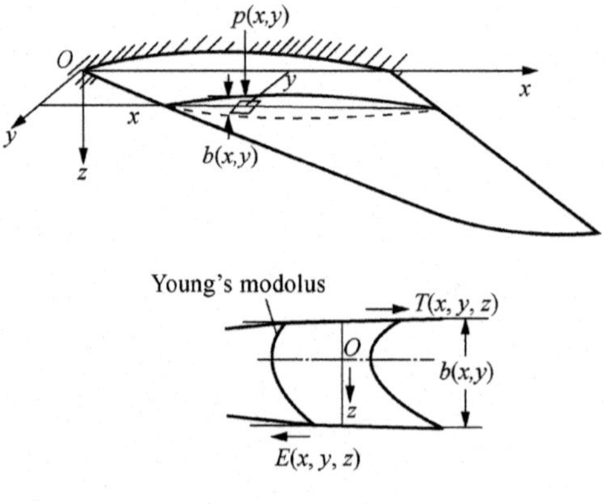

Fig. 2 Loads and boundary conditions of a solid heated wing

Let α and β be two constants; then it is clear that the Eqs. (18) and (36) for $\varphi(x, y)$ and $\bar{\varphi}(x, y)$ will be exactly the same if

$$\bar{D}_0 = \alpha D_0 \tag{42}$$

$$\bar{\varphi} = \beta\varphi \tag{43}$$

$$F = \beta N_T \tag{44}$$

Then, because of Eqs. (16) and (35).

$$\bar{N}_x = \beta(N_x + N_T), \quad \bar{N}_{xy} = \beta N_{xy}, \quad \bar{N}_y = \beta(N_\beta + N_T) \tag{45}$$

Now, if λ and μ are two additional constants, and

$$\bar{D}_2 = \lambda D_2 \tag{46}$$

$$\bar{w} = \mu w \tag{47}$$

then the equation for $\bar{w}$, Eq. (39), can be rewritten as

$$\frac{\partial^2}{\partial x^2}\left[D_2\left(\frac{\partial^2 w}{\partial x^2} + \nu\frac{\partial^2 w}{\partial y^2}\right)\right] + 2(1-\nu)\frac{\partial^2}{\partial x \partial y}\left(D_2\frac{\partial^2 w}{\partial x \partial y}\right) + \frac{\partial^2}{\partial y^2}\left[\left(\frac{\partial^2 w}{\partial y^2} + \nu\frac{\partial^2 w}{\partial x^2}\right)\right]$$
$$= \frac{1}{\lambda\mu}p + \frac{\beta}{\lambda\mu}N_T\,\nabla^2\bar{w} + \frac{\beta}{\lambda}\left(N_x\frac{\partial^2 w}{\partial x^2} + 2N_{xy}\frac{\partial^2 w}{\partial x \partial y} + N_y\frac{\partial^2 w}{\partial y^2}\right)$$

Therefore it can be deduced, by comparing the above equation with Eq. (21), that, in order for w and $\bar{w}$ to satisfy the same equation,

$$\lambda = \beta \tag{48}$$

$$\bar{p} = -\beta N_T\,\nabla^2\bar{w} - \beta\mu\,\nabla^2 M_T + \beta\mu p \tag{49}$$

On the boundary of the heated plate, the forces N_n and N_{ns} vanish. By comparing Eqs. (19) and (20) with Eqs. (37) and (38) and by noting Eqs. (43) and (44),

$$\bar{N}_n = \beta N_T, \quad \bar{N}_{ns} = 0 \tag{50}$$

Similarly, by comparing Eqs. (23) and (24) with Eqs. (37) and (38) and by noting Eqs. (46) and (47), the boundary conditions of Eq. (25) for the free edge of the heated plate reduces to

$$\left.\begin{aligned} \bar{M}_n &= \beta\mu M_T \\ \frac{\partial \bar{M}_n}{\partial n} + \frac{\bar{M}_n - \bar{M}_s}{r} + \frac{\partial \bar{M}_{ns}}{\partial s} &= \beta\mu \frac{\partial M_T}{\partial n} \end{aligned}\right\} \tag{51}$$

The boundary conditions for the built-in edge, Eqs. (22), give

$$\bar{w} = 0, \quad \partial\bar{w}/\partial n = 0 \tag{52}$$

As a first application of the theory developed, consider the simple case of a thin wing of solid section: The purpose is to find the stresses in the heated wing by performing experiments on an unheated "corresponding" or analog wing. The first step is to understand clearly the meaning of Eqs. (42) and (46). These equations, in fact, determine the thickness $\bar{b}$ of the unheated wing. Now let g be ratio $E/\bar{E}$ and η the nondimensional thickness variable measured from the upper surface of the plate such that $\eta = -1$ at upper surface and $\eta = 1$ at the lower surface. Let η_0 be the value of η at the median surface. Then,

$$D_0(x, y) = \frac{\bar{E}b}{2(1-\nu^2)}\int_{-1}^{1} g(x, y; \eta)\mathrm{d}\eta$$

$$D_1(x, y) = \frac{\bar{E}b^2}{4(1-\nu^2)}\int_{-1}^{1} g(x, y; \eta)(\eta - \eta_0)\mathrm{d}\eta$$

$$D_2(x, y) = \frac{\bar{E}b^3}{8(1-\nu^2)}\int_{-1}^{1} g(x, y; \eta)(\eta - \eta_0)^2\mathrm{d}\eta$$

By using Eq. (31) and the above relations, the conditions of Eqs. (11), (42), and (46) can be written as

$$\int_{-1}^{1} g(x, y; \eta)(\eta - \eta_0)\mathrm{d}\eta = 0 \tag{53}$$

$$\frac{\alpha}{2}b(x, y)\int_{-1}^{1} g(x, y; \eta)\mathrm{d}\eta = \bar{b}(x, y) \tag{54}$$

$$\frac{\lambda}{8}b^3(x, y)\int_{-1}^{1} g(x, y; \eta)(\eta - \eta_0)^2\mathrm{d}\eta = \frac{1}{12}\bar{b}^3(x, y) \tag{55}$$

By eliminating b, $\bar{b}$, and η_0 from these equations, one obtains

$$\left(\frac{1}{12}\frac{\alpha^3}{\lambda}\right)\left(\int_{-1}^{1} g\mathrm{d}\eta\right)^4 = \left(\int_{-1}^{1} g\eta^2\mathrm{d}\eta\right)\times\left(\int_{-1}^{1} g\mathrm{d}\eta\right) - \left(\int_{-1}^{1} g\eta\,\mathrm{d}\eta\right)^2 \tag{56}$$

Equation (56) shows that the "Young's modulus profiles" $g(\eta)$ for various points (x, y) of the heated plate are not entirely arbitrary but must satisfy that relation so that the "similarity" between the heated wing and the isothermal analog wing is possible. If g is a fixed function—i. e., the temperature profiles across the plate are similar—or Young's modulus is constant, then Eq. (56) is certainly satisfied. * For any given problem of a heated wing then, the first step is to compute the value of α^3/λ or α^3/β for various points of the wing by Eq. (56). If these values do not differ very much, then, as an approximation, the average of the computed α^3/β can be used. Then the similarity procedure is possible. But in any event, the value of α^3/β is fixed by the problem, not at the free choice of the stress analyst. Thus, if β is chosen, then α is fixed. Furthermore, λ is equal to β according to Eq. (48). Therefore, out of the four constants, α, β, λ, and μ, only two (β and μ) can be chosen arbitrarily. When this is done, Eq. (54) determines the appropriate thickness $\bar{b}(x, y)$ for the unheated analog wing.

With the geometry of the analog wing determined, the next step is to specify the loads on it. Eq. (50) shows that the unheated wing must be loaded with a sectional tension force βN_T at the boundary but no sectional shearing force. The first of Eqs. (51) shows that, at the free edge, the analog wing should have a bending moment equal to $\beta\mu M_T$. The numerical value of μ is arbitrary and is at the disposal of the experimenter. The second of Eqs. (51) can be interpreted as an upward support force $-\beta\mu(\partial M_T/\partial n)$ per unit length of boundary but no twisting moment at the boundary. Eq. (52) shows that a built-in edge of the heated wing corresponds to a built-in edge of the analog wing.

On the unheated wing, Eqs. (32) and (44) specify a body force

$$X = -\beta\frac{\partial N_T}{\partial x}, \quad Y = -\beta\frac{\partial N_T}{\partial y} \tag{57}$$

These are forces in the x and y direction per unit area of the wing surface and are loads perhaps new to structures testing. The lateral pressure loading p on the analog wing is specified by Eq. (49). If $\Delta\bar{\varepsilon}_x$ and $\Delta\bar{\varepsilon}_y$ denote the differences of x and y components of direct strains measured on the isothermal wing on the top and the bottom surfaces—i. e., according to Eq. (28)

$$\left.\begin{aligned}
\Delta\bar{\varepsilon}_x(x, y) &= \bar{\varepsilon}_x\left(x, y; -\frac{\bar{b}}{2}\right) - \bar{\varepsilon}_x\left(x, y; \frac{\bar{b}}{2}\right) = \bar{b}(x, y)\frac{\partial^2 \bar{w}}{\partial x^2} \\
\Delta\bar{\varepsilon}_y(x, y) &= \bar{\varepsilon}_y\left(x, y; -\frac{\bar{b}}{2}\right) - \bar{\varepsilon}_y\left(x, y; \frac{\bar{b}}{2}\right) = \bar{b}(x, y)\frac{\partial^2 \bar{w}}{\partial y^2}
\end{aligned}\right\} \tag{58}$$

then Eq. (49) can be written as

$$\bar{p} = (-\beta N_T/\bar{b})(\Delta\bar{\varepsilon}_x + \Delta\bar{\varepsilon}_y) - \beta\mu\nabla^2 M_T + \beta\mu p \tag{59}$$

Needless to say, if there are concentrated lateral loads, the ratio of these loads on the test wing and on the original wing is also equal to $\beta\mu$.

---

* Mathematically speaking, Eq. (56) is a functional equation. The solution of it gives the explicit required character of g. This problem is discussed in the Appendix.

The physical significance of Eq. (59) is different from that of Eq. (57) in that the body force loading is completely specified before the test because N_T is a known quantity, while the first term of $\bar{p}$ itself depends upon the experimentally determined $\Delta\bar{\varepsilon}_x$ and $\Delta\bar{\varepsilon}_y$. Therefore, if the loads are considered as input to the wing structure and the experimentally measured strains as output, then Eq. (57) shows that the output also partly determines the input. In other words, there is a "feedback link" in the experimental setup of the analog wing.

With the loading on the unheated wing so specified, the relation between $\bar{\varphi}$ and φ and between $\bar{w}$ and w are given by Eqs. (43) and (47). The stresses in the original heated wing can then be computed from the measured strains on the test wing. For instance, from Eqs. (1) and (3),

$$\sigma_x(x, y; z) = \frac{E(x, y; z)}{1-\nu^2}\left[\left(\frac{\partial u}{\partial x}+\nu\frac{\partial u}{\partial y}\right)-z\left(\frac{\partial^2 w}{\partial x^2}+\nu\frac{\partial^2 w}{\partial y^2}\right)-(1+\nu)aT(x, y; z)\right]$$

By using Eq. (17), the above equation is reduced to

$$\sigma_x(x, y; z) = \frac{E(x, y; z)}{1-\nu^2}\left\{\frac{N_T}{D_0}+\frac{1}{D_0}\frac{\partial^2\varphi}{\partial y^2}-z\left(\frac{\partial^2 w}{\partial x^2}+\nu\frac{\partial^2 w}{\partial y^2}\right)-(1+\nu)aT(x, y; z)\right\}$$

But according to Eqs. (42), (43), (29), and (35),

$$\frac{1}{D_0}\frac{\partial^2\varphi}{\partial y^2}=\frac{\alpha}{\beta}\frac{1}{\bar{D}_0}\frac{\partial^2\bar{\varphi}}{\partial y^2}=\frac{\alpha}{\beta}\frac{1}{\bar{D}_0}[\bar{N}_x-\beta N_T]=\frac{\alpha}{\beta}\left(\frac{\partial\bar{u}}{\partial x}+\nu\frac{\partial\bar{v}}{\partial y}\right)-\frac{N_T}{D_0}$$

However,

$$\frac{\partial\bar{u}}{\partial x}=\frac{1}{2}\left[\bar{\varepsilon}_x\left(x, y; -\frac{\bar{b}}{2}\right)+\bar{\varepsilon}_x\left(x, y; \frac{\bar{b}}{2}\right)\right]=\bar{\varepsilon}_x(x, y) \tag{60}$$

$$\frac{\partial\bar{v}}{\partial y}=\frac{1}{2}\left[\bar{\varepsilon}_y\left(x, y; -\frac{\bar{b}}{2}\right)+\bar{\varepsilon}_y\left(x, y; \frac{\bar{b}}{2}\right)\right]=\bar{\varepsilon}_y(x, y) \tag{61}$$

where $\bar{\varepsilon}_x(x, y)$ is the average of the x-strain at the top and the bottom surfaces of the analog wing at the point (x, y) and $\bar{\varepsilon}_y(x, y)$ is the average of the y-strain. Thus, together with Eq. (58), the above expressions give

$$\sigma(x, y; z) = \frac{E(x, y; z)}{1-\nu^2}\left\{\frac{\alpha}{\beta}(\bar{\varepsilon}_x+\nu\bar{\varepsilon}_y)-\frac{z}{\mu b}(\Delta\bar{\varepsilon}_x+\nu\Delta\bar{\varepsilon}_y)-(1+\nu)aT(x, y; z)\right\} \tag{62}$$

It may be proper to point out again that the quantity z in the above equation and in the subsequent equations is the distance from the median surface, positive if the point is below the median surface. Similar to Eq. (62), one has

$$\sigma_y(x, y; z) = \frac{E(x, y; z)}{1-\nu^2}\left[\frac{\alpha}{\beta}(\bar{\varepsilon}_y+\nu_x\bar{\varepsilon})-\frac{z}{\mu b}(\Delta\bar{\varepsilon}_y+\nu\Delta\bar{\varepsilon}_x)-(1+\nu)aT(x, y; z)\right] \tag{63}$$

If the difference $\Delta\bar{\gamma}_{xy}$ of the shear strain and the average shear strain $\bar{\bar{\gamma}}_{xy}$ are defined as

$$\Delta\bar{\gamma}_{xy}(x, y) = \bar{\gamma}_{xy}\left(x, y; -\frac{\bar{b}}{2}\right) - \bar{\gamma}_{xy}\left(x, y; \frac{\bar{b}}{2}\right) \tag{64}$$

$$\bar{\bar{\gamma}}_{xy}(x, y) = \frac{1}{2}\left[\bar{\gamma}_{xy}\left(x, y; -\frac{\bar{b}}{2}\right) + \bar{\gamma}_{xy}\left(x, y; \frac{\bar{b}}{2}\right)\right] \tag{65}$$

the shear stress in the original heated wing is calculated as

$$\tau_{xy}(x, y; z) = \frac{E(x, y; z)}{2(1+\nu)}\left[\frac{\alpha}{\beta}\bar{\bar{\gamma}}_{xy}(x, y) - \frac{z}{\mu\bar{b}}\Delta\bar{\gamma}_{xy}(x, y)\right] \tag{66}$$

Eqs. (62), (63), and (66) allow the calculation of the stresses in the heated wing from test data of the unheated wing. They also show the advantage having the constants (β/α) and μ smaller than unity, so that the strains of the unheated wing are magnified in computing the stresses in the heated wing. Then, for a specrned load on the heated wing, the test load on the unheated wing is reduced and the analog wing will not be over-strained. Fig. 3 gives a summary of the loading of the unheated test wing, together with the equation numbers of relations that specify the quantities in terms of the quantities of the original heated wing.

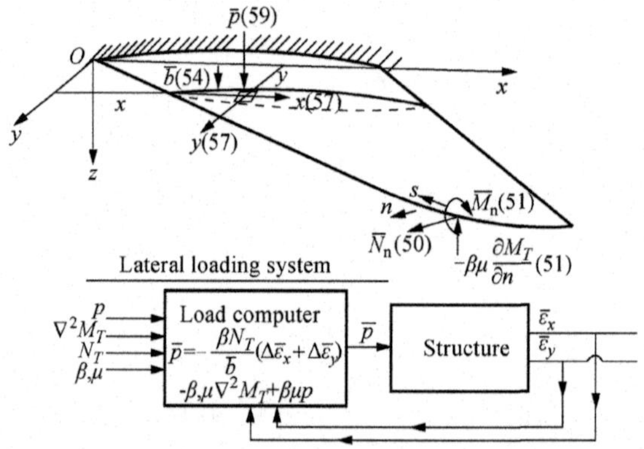

Fig. 3 Loads and boundary conditions of the analog test wing
(Numbers in parentheses correspond to numbered equations in the text where the appropriate relation with quantities of the original heated wing can be found)

Alternate Test Procedure for Thin Solid Wings

In the previous section, the simulated tests on the isothermal wing are proposed with simultaneous loading both in the plane of the median surface and in lateral pressure. This results in a rather complicated load system at the free boundary of the wing. To simplify the load system at the boundary, the loading in the plane of the median surface and the lateral loading can be separated—that is, two separate tests, properly correlated, are made on the analog wing, and the stresses in the heated wing are obtained by a synthesis of the test results.

To determine first the sectional forces N_x, N_y, and N_{xy}, the unheated wing is dimensioned to satisfy Eq. (42). It is then loaded with body forces specified by Eq. (57), but no lateral load is applied. At the "free edge" of the wing, it is loaded according to Eq. (50). Then the magnitude of the sectional forces for the heated wing can be computed from Eq. (45) if $\bar{N}_x$, $\bar{N}_y$, and $\bar{N}_{xy}$ are determined. These sectional forces can be calculated from the measured strains on the wing. Although theoretically there should be no bending of the wing, actually bending will be present because of deviations of the real wing from the idealized wing assumed in the theory. The effects of such spurious bending can be eliminated by taking the average of the strains of the top and the bottom surfaces. Let the superscript 1 denote the quantities induced by this loading in the plane of the median surface. Introduce the following notations for the average strains:

$$\left.\begin{aligned} \bar{\bar{\varepsilon}}_x^{(1)}(x, y) &= \frac{1}{2}\left[\bar{\varepsilon}_x^{(1)}\left(x, y; -\frac{\bar{b}}{2}\right)+\bar{\varepsilon}_x^{(1)}\left(x, y; \frac{\bar{b}}{2}\right)\right] \\ \bar{\bar{\varepsilon}}_y^{(1)}(x, y) &= \frac{1}{2}\left[\bar{\varepsilon}_y^{(1)}\left(x, y; -\frac{\bar{b}}{2}\right)+\bar{\varepsilon}_y^{(1)}\left(x, y; \frac{\bar{b}}{2}\right)\right] \\ \bar{\bar{\gamma}}_{xy}^{(1)}(x, y) &= \frac{1}{2}\left[\bar{\gamma}_{xy}^{(1)}\left(x, y; -\frac{\bar{b}}{2}\right)+\bar{\gamma}_{xy}^{(1)}\left(x, y; \frac{\bar{b}}{2}\right)\right] \end{aligned}\right\} \tag{67}$$

Then the sectional forces N_x, N_y, and N_{xy} for the heated wing can be calculated as

$$\left.\begin{aligned} N_x(x, y) &= \frac{\alpha}{\beta}D_0\,[\bar{\bar{\varepsilon}}_x^{(1)} + \nu\bar{\bar{\varepsilon}}_y^{(1)}] - N_T \\ N_y(x, y) &= \frac{\alpha}{\beta}D_0\,[\bar{\bar{\varepsilon}}_y^{(1)} + \nu\bar{\bar{\varepsilon}}_x^{(1)}] - N_T \\ N_{xy} &= \frac{\alpha}{\beta}\,\frac{1-\nu}{2}D_0\bar{\bar{\gamma}}_{xy}^{(1)} \end{aligned}\right\} \tag{68}$$

When the average strains of Eq. (67) are measured, the loading in the plane of the median surface, the body forces, and the boundary forces can be removed. The second step of the test is to load the unheated wing laterally. Let Eqs. (46) and (47) be satisfied. Then in order that $\bar{w}$ and w satisfy the same differential equation, the lateral loading $\bar{p}$ on the test wing must be specified as

$$\bar{p} = -\lambda\mu\,\nabla^2 M_T + \lambda\mu p + \lambda\left[N_x\frac{\partial^2\bar{w}}{\partial x^2} - 2N_{xy}\frac{\partial^2\bar{w}}{\partial x\partial y} - N_y\frac{\partial^2\bar{w}}{\partial y^2}\right] \tag{69}$$

On the "free edges" of the wing, there is a bending moment equal to $\lambda\mu M_T$ and an upward support force $-\lambda\mu(\partial M_T/\partial n)$ per unit length of the boundary.

Let the superscript 2 denote the quantities induced by bending; then

$$\left.\begin{aligned} \Delta\bar{\varepsilon}_x^{(2)}(x, y) &= \bar{\varepsilon}_x^{(2)}\left(x, y; -\frac{\bar{b}}{2}\right) - \bar{\varepsilon}_x^{(2)}\left(x, y; \frac{\bar{b}}{2}\right) = \bar{b}(x, y)\frac{\partial^2\bar{w}}{\partial x^2} \\ \Delta\bar{\varepsilon}_y^{(2)}(x, y) &= \bar{\varepsilon}_y^{(2)}\left(x, y; -\frac{\bar{b}}{2}\right) - \bar{\varepsilon}_y^{(2)}\left(x, y; \frac{\bar{b}}{2}\right) = \bar{b}(x, y)\frac{\partial^2\bar{w}}{\partial y^2} \\ \Delta\bar{\gamma}_{xy}^{(2)}(x, y) &= \bar{\gamma}_{xy}^{(2)}\left(x, y; -\frac{\bar{b}}{2}\right) - \bar{\gamma}_{xy}^{(22)}\left(x, y; \frac{\bar{b}}{2}\right) = 2\bar{b}(x, y)\frac{\partial^2\bar{w}}{\partial x\partial y} \end{aligned}\right\} \tag{70}$$

With these strains differences between the top and the bottom surfaces, the loading Eq. (69) can be written as

$$\bar{p}(x, y) = -\lambda\mu \nabla^2 M_T + \lambda\mu p(x, y) + \lambda\Big[N_x \frac{\Delta\bar{\varepsilon}_x^{(2)}}{b} + N_{xy} \frac{\Delta\bar{\gamma}_{xy}^{(2)}}{b} + N_y \frac{\Delta\bar{\varepsilon}_y^{(2)}}{b}\Big] \tag{71}$$

where the sectional forces N_x, N_y, and N_{xy} are computed by using Eq. (68) and the strain data obtained during the first part of the test. This equation again demonstrates the feedback character of the lateral loading. Fig. 4 summarizes the load systems for the two successive tests.

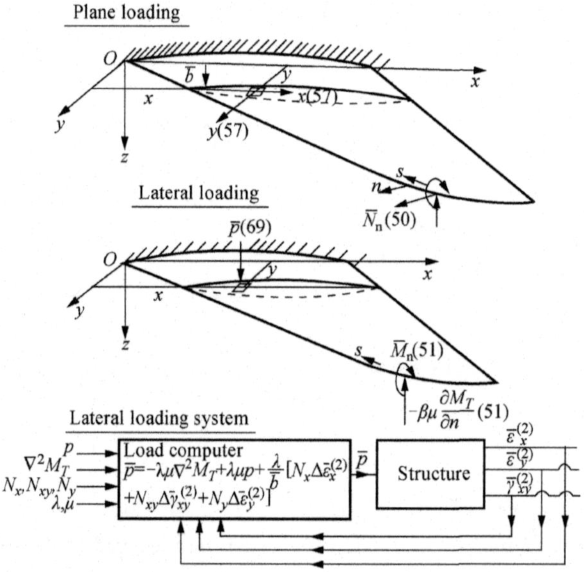

Fig. 4 Loads and boundary conditions of the analog test wing, alternate procedure
(Numbers in parentheses correspond to numbered equations in the next where the appropriate relation with quantities of the original heated wing can be found)

When the strain averages of Eq. (67) and the strain differences of Eq. (70) are measured by the two successive tests on the isothermal analog wing, the stresses in the original heated wing can be computed by using equations similar to Eqs. (62), (63), and (65), replacing $\bar{\varepsilon}$, $\bar{\gamma}$ by $\bar{\varepsilon}^{(1)}$, $\bar{\gamma}^{(1)}$ and $\Delta\bar{\varepsilon}$, $\Delta\bar{\gamma}$ by $\Delta\bar{\varepsilon}^{(2)}$, $\Delta\bar{\gamma}^{(2)}$ of Eqs. (67) and (70), respectively. However, there is one important difference between the est procedure of the present section and the test procedure of the previous one: α can be chosen independent of β, because now β is not necessarily equal to λ. But α and λ are connected because of the condition of Eq. (56). Out of the four constants α, β, λ, and μ, three can be chosen arbitrarily. This greater degree of freedom and simpler load system in each stage of tests than the test procedure described in the previous section may prove to be advantageous in actual application.

Box Wing

In the previous sections, the similarity laws for thin solid wings are formulated for obtaining the stresses in a heated wing by testing an analog unheated wing. A majority of actual wing structures is, however, much more complicated than a solid wing. For example, the main load-carrying member of the wing may be a box structure with top and bottom panels, ribs, and deep beams joining the top and bottom panels. Since only the top and bottom panels of such a box wing are exposed to the air stream, the aerodynamic heating of the structure is limited to these panels.

To analyze such a box wing, the different structure elements can be broken down, and each individual part, such as the top panel, can be considered as a plate. To each individual part then, the theory developed in the earlier sections of this paper can be applied. The thickness $\bar{b}$ for the analog unheated plate for each part of the structure can be specified by the method described previously. The only restriction is that the four constants α, β, λ, and μ must be the same for all structure elements. The loads on the individual unheated plates corresponding to the different structure elements can be determined according to equations given in the earlier sections. At the boundary or the junction of the different unheated analog panels, there are now two types of load. One type of load comes from the similarity law. This load system is specified previously as the load system for "free edges." The other type of load comes from the junction condition that the deformations of the different parts of the structure must fit together. If the analog panels are put together, then the junction loads will be automatically supplied by the structure. Therefore the procedure to use the method of similarity is as follows:

The first step is to determine the thickness $\bar{b}$ of the analog isothermal box wing by Eqs. (54) and (56). In a box structure, there is appreciable influence on the sectional forces in the median surface of the plates because of bending deflection. Therefore the separation of bending from the "extensional loads" in the plane of the plates is no longer appropriate. The complete load system for the analog wing must be applied simultaneously. This consists of the body force loads in the plane of the plates, the lateral pressure $\bar{p}$, and the "free edge" loads at the junction of the different parts of the structure. The stresses in the original heated box wing can then be calculated by using Eqs. (12), (63), and (65) with the measured strains.

Discussion

In the previous sections, it is shown that the problem of stressing a heated wing can be solved by tests on an analog unheated wing through the similarity laws. The temperature distribution in the wing is assumed to be given by a theoretical calculation using aerodynamic heating data and the theory of heat transfer in solid materials[1]. The loading system required for the analog wing is indeed complicated. Furthermore, the lateral pressure load $\bar{p}$ specified by either Eq. (59) or Eq. (71) involves the concept of feedback in that $\bar{p}$ is partially dependent upon the measured strains and cannot be predetermined. This is, perhaps, a novel concept in structure testing. The possible justification for this great complication in structural testing is the serious difficulty of simulating aerodynamic heating in a structures laboratory and of measuring strains at high temperatures.

The basic principle of the theory of similarity law can be traced to the well-known fact that the general three-dimensional thermal stress problem can be reduced to a problem in isothermal material by introducing appropriate fictitious three-dimensional body forces and surface forces on the boundary of the body. The general three-dimensional formulation of the "analogy" is, however, not useful for structure testing purposes, because there is no possible method of applying three-dimensional body forces. If one dimension of the body is small, such as a thin plate or a thin shell, then the body force is two-dimensional. The loading for the analog isothermal structure can be done, although not easily, as shown in the previous discussion. From this general argument, it is clear then that the present "similarity theory" for thermal stresses can certainly be extended to any thin elastic shells, although the actual execution of the analog testing may be more difficult than the flat plate case studied here.

To avoid the task of testing a heated wing, a test wing at room temperature is proposed as an analog. In a sense then, the whole concept of the method of similarity is that of analog-machine computing. There is, however, an advantage of the present method over the recognized machine computer in that the main physical member of the problem, the elastic plate, remains and is not replaced by an approximate system such as electric network. The unheated analog wing is thus the closest analog the aerodynamically heated wing can have and is the most accurate analog. The price of this accuracy and detailed reproduction of the original problem is the complicated test setup required. However, an enterprising structures test engineer will probably welcome such challenge to his ingenuity and demand on his technical skill.

Appendix—Young's Modulus Profile

By writing $f(\eta)$ as

$$f(\eta) = \sqrt{(1/12)(\alpha^3/\lambda)}\, g(\eta) \tag{72}$$

Eq. (56) can be written as

$$\left[\int_{-1}^{1} f(\eta)\mathrm{d}\eta\right]^4 = \left[\int_{-1}^{1} f(\eta)\eta^2\mathrm{d}\eta\right]\left[\int_{-1}^{1} f(\eta)\mathrm{d}\eta\right] - \left[\int_{-1}^{1} f(\eta)\eta\mathrm{d}\eta\right]^2 \tag{73}$$

Now any continuous function $f(\eta)$ for $-1 \leqslant \eta \leqslant 1$ can be expanded into a series of Legendre polynomials $P_n(\eta)$ * — i.e.,

* The author is deeply indebted to Prof. A. Erdélyi, of the California Institute of Technology, for suggesting this method of solving the problem.

$$f(\eta) = \sum_{n=0}^{\infty} a_n P_n(\eta) \tag{74}$$

where a_n's are constant coefficients. Then[4]

$$\int_{-1}^{1} f(\eta)\mathrm{d}\eta = \int_{-1}^{1} f(\eta) P_0(\eta)\mathrm{d}\eta = 2a_0$$

$$\int_{-1}^{1} f(\eta)\eta\mathrm{d}\eta = \int_{-1}^{1} f(\eta) P_1(\eta)\mathrm{d}\eta = \frac{2}{3}a_1$$

$$\int_{-1}^{1} f(\eta)\eta^2\mathrm{d}\eta = \frac{2}{3}\int_{-1}^{1} f(\eta)\left[P_2(\eta) + \frac{1}{2}P_0(\eta)\right]\mathrm{d}\eta = \frac{2}{3}\left(\frac{2}{5}a_2 + a_0\right)$$

Therefore Eq. (73) can be written as

$$12a_0^2 = 1 + \frac{2}{5}\left(\frac{a_2}{a_0}\right) - \frac{1}{3}\left(\frac{a_1}{a_0}\right)^2 \tag{75}$$

The restriction of Eq. (73) is thus simply a relation among the first three coefficients a_0, a_1, and a_2 of the expansion Eq. (74). The later coefficients, a_n, $n \geqslant 3$, are entirely free. Therefore the $f(\eta)$, although not entirely arbitrary, has a wide degree of freedom. Hence, it is probable that the Young's modulus profile will conform to the condition of Eq. (56) for the similarity theory up to a high order.

References

[1] Kaye J. The Transient Temperature Distribution in a Wing Plying at Supersonic Speeds. Journal of the Aeronautical Sciences, 1950, 17, (12): 787-807,816.

[2] Nádai A. Elastische Platten. Julius Springer, Berlin,1925:274.

[3] Tsien H S, Cheng C M. Similarity Law for Stressing Rapidly Heated Thin-Walled Cylinders. Journal of the American Rocket Society, 1952, 22: 144.

[4] Whittaker E T, Watson G N. Modern Analysis. Chapt. 15; Cambridge University Press, London, England, 1920.

Take-Off from Satellite Orbit

H. S. Tsien①

(*Daniel and Florence Guggenheim Jet Propulsion Center, California Institute of Technology, Pasadena, Calif.*)

The mass ratio or the characteristic velocity for the take-off of a space ship from the satellite orbit is computed for two cases: the radial thrust, and the circumferential thrust. The circumferential thrust is much more efficient in that the required mass ratio is much less than for the radial thrust. Both cases show, however, an increase of the required mass ratio and the characteristic velocity with a reduction in acceleration. With circumferential thrust, the characteristic velocity increases by a factor of two, when the acceleration is reduced from 1/2 g to 1/3 000 g.

For take-off of a rocket from the earth surface, it is convenient to have the initial trajectory in the vertical direction, and then the thrust should be considerably larger than the initial weight of the rocket to overcome the gravity and to give an appropriate acceleration. Depending upon the relative magnitudes of the aerodynamic drag and the weight, the initial ratio of the thrust and the weight should be between 2 and 3 for minimum expenditure of the propellant. The situation is quite different for a space ship taking off from the satellite orbit: In a satellite orbit, the gravitational attraction is completely balanced by the centrifugal force, and the vehicle is effectively in a weightless state. This fact has led many fanciers of interplanetary travel to conclude that take-off from satellite orbit requires only a very minute thrust. For instance, L. Spitzer[1] proposed a nuclear power plant for a space ship to be accelerated at only 1/3 000 g. Another example is the extensive discussion of interorbital transport techniques by H. Preston-Thomas[2], based upon the assumption of equally small acceleration. On the other hand, W. von Braun[3] seems to prefer a very much larger acceleration of approximately 1/2 g for take-off from the satellite orbit.

The magnitude of the acceleration has a strong bearing on the optimum type of power plant to be used: The ion-beam rocket is only feasible for very small acceleration, while for moderate acceleration, chemical rocket is required. Therefore the question of the magnitude of acceleration is an important one for interplanetary flight. The purpose of this note is to compute the relation between the acceleration and the mass ratio required for escape from the earth's gravitational field, starting from the satellite orbit. It is hoped that the present investigation will give the future generation of astronautical engineers a rational basis for

---

Received November 19, 1952.

Journal of the American Rocket Society, vol. 23, pp. 233 – 236, 1953.

① Robert H. Goddard Professor of Jet Propulsion.

designing space ships.

Basic Equations

The problem considered is the motion of a space ship under the influence of the rocket thrust and the gravitational attraction of a single massive body, say the earth. Then if the rocket thrust is in the plane of trajectory, the trajectory of the space ship will remain in a plane. Let the position of the ship at any time instant t be given by the polar co-ordinates r and θ (r is the distance from the center of attraction, and θ the angular position). If the components of the rocket thrust per unit mass of the vehicle are R in the radial direction and Θ in the circumferential direction, and if g is the magnitude of gravitational attraction at the starting satellite orbit $r = r_0$ (Fig. 1), then the equations of motion of the space ship are

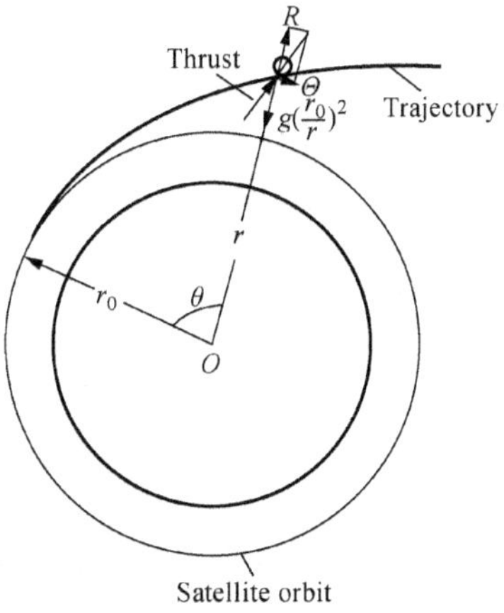

Fig. 1 Take-off from the satellite orbit with thrust in the plane of satellite orbit

$$\frac{d^2 r}{dt^2} = R + r\left(\frac{d\theta}{dt}\right)^2 - g\left(\frac{r_0}{r}\right)^2 \tag{1}$$

and

$$\frac{d}{dt}\left(r^2 \frac{d\theta}{dt}\right) = r\Theta \tag{2}$$

By using the subscript 0 to indicate quantities at the starting instant $t = 0$, the equilibrium condition of the satellite orbit is given by

$$r_0\left(\frac{d\theta}{dt}\right)_0^2 = g \tag{3}$$

Initially, the radial velocity is zero, i.e.,

$$\left(\frac{dr}{dt}\right)_0 = 0 \tag{4}$$

These are the initial conditions.

For the space ship to have sufficient energy to escape the earth gravitational field at the end of the powered flight, the sum of the kinetic energy and potential energy must vanish at the end of the accelerating period. Let that instant be denoted by the subscript 1. Thus, at $t = t_1$

$$\frac{1}{2}\left[\left(\frac{\mathrm{d}r}{\mathrm{d}t}\right)_1^2 + \left(r\frac{\mathrm{d}\theta}{\mathrm{d}t}\right)_1^2\right] - g\frac{r_0^2}{r_1} = 0 \tag{5}$$

With any specified variation of the thrust forces R and Θ as functions of time, the above system of equations determine completely the take-off trajectory of the space ship. In the following sections, two special cases of practical significance will be discussed in detail: the case R = const, $\Theta = 0$, purely radial thrust; and the case $R = 0$, Θ = const, purely circumferential thrust.

Radial Thrust

If the thrust is always radial and is proportional to the instantaneous mass of the vehicle, a nondimensional thrust factor μ can be introduced as

$$R = \mu g \tag{6}$$

Furthermore, let

$$\rho = \frac{r}{r_0}, \quad \tau = \sqrt{\frac{g}{r_0}}t \tag{7}$$

ρ is thus the nondimensional radial distance, and τ is the nondimensional time. Then Equations (1) and (2) can be written in the nondimensional form as

$$\frac{\mathrm{d}^2\rho}{\mathrm{d}\tau^2} = \mu + \rho\left(\frac{\mathrm{d}\theta}{\mathrm{d}\tau}\right)^2 - \frac{1}{\rho^2} \tag{8}$$

and

$$\frac{\mathrm{d}}{\mathrm{d}\tau}\left(\rho^2\frac{\mathrm{d}\theta}{\mathrm{d}\tau}\right) = 0 \tag{9}$$

Equation (9) can be immediately integrated and by using the initial condition of Equation (3), the result of integration is

$$\frac{\mathrm{d}\theta}{\mathrm{d}\tau} = \frac{1}{\rho^2} \tag{10}$$

By substituting this equation into Equation (8), the final equation for ρ is

$$\frac{\mathrm{d}^2\rho}{\mathrm{d}\tau^2} = \mu + \frac{1}{\rho^3} - \frac{1}{\rho^2} \tag{11}$$

The nondimensional radial velocity is $\mathrm{d}\rho/\mathrm{d}\tau$. This is related to the physical radial velocity $\mathrm{d}r/\mathrm{d}t$ as follows

$$\frac{dr}{dt} = \sqrt{gr_0}\,\frac{d\rho}{d\tau} \tag{12}$$

Equation (11) can be rewritten as

$$\frac{1}{2}\,\frac{d}{d\rho}\left(\frac{d\rho}{d\tau}\right)^2 = \mu + \frac{1}{\rho^3} - \frac{1}{\rho^2}$$

Since $d\rho/d\tau = 0$, when $\tau = 0$ and $\rho = 1$ according to Equation (4), the result of integrating the above equation is

$$\left(\frac{d\rho}{d\tau}\right)^2 = 2\mu(\rho - 1) + \left(1 - \frac{1}{\rho^2}\right) - 2\left(1 - \frac{1}{\rho}\right) \tag{13}$$

Therefore the nondimensional time τ can be calculated as a function of the radius ρ as follows

$$\tau = \int_1^{\rho} \frac{\rho\, d\rho}{\sqrt{(\rho - 1)(2\mu\rho^2 - \rho + 1)}} \tag{14}$$

With Equations (10) and (13), the end condition of Equation (5) can be written as

$$\frac{1}{2}\left[\left\{2\mu(\rho_1 - 1) + \left(1 - \frac{1}{\rho_1^2}\right) - 2\left(1 - \frac{1}{\rho_1}\right)\right\} + \frac{1}{\rho_1^2}\right] - \frac{1}{\rho_1} = 0$$

Or simply

$$\rho_1 = 1 + \frac{1}{2\mu} \tag{15}$$

Then the velocities at the end of acceleration period are

and

$$\left.\begin{aligned} \left(\frac{dr}{dt}\right)_1 &= \sqrt{gr_0}\,\frac{\sqrt{1 + (1/\mu)}}{1 + (1/(2\mu))} \\ \left(r\,\frac{d\theta}{dt}\right)_1 &= \sqrt{gr_0}\,\frac{1}{1 + (1/(2\mu))} \end{aligned}\right\} \tag{16}$$

The time τ_1 for the powered flight can be obtained from Equation (14) by setting the upper limit of integration to ρ_1. The result of this integration is①

$$\tau_1 = \sqrt{\frac{2}{\mu}}\left[\frac{\sqrt{2(\mu + 1)}}{2\mu + 1} + F\left(\frac{1}{\sqrt{8\mu}},\ \cos^{-1}\frac{2\mu - 1}{2\mu + 1}\right) + E\left(\frac{1}{\sqrt{8\mu}},\ \cos^{-1}\frac{2\mu - 1}{2\mu + 1}\right)\right] \tag{17}$$

where F and E are the elliptical integrals of first kind and second kind, respectively.

If $M(t)$ is the instantaneous mass of the space ship, and c the effective exhaust velocity of the rocket, then

$$RM = \mu g M = -c\,\frac{dM}{dt} = -c\sqrt{\frac{g}{r_0}}\,\frac{dM}{d\tau}$$

---

① The author is indebted to Dr. Y. T. Wu who kindly supplied the relation of Equation (17).

Therefore the mass ratio M_0/M_1 can be calculated as follows:

$$\log_e(M_0/M_1) = \frac{\sqrt{gr_0}}{c}\mu\tau_1$$

By using the result of Equation (17)

$$\frac{c}{\sqrt{gr_0}}\log_e(M_0/M) = \frac{2\sqrt{\mu(\mu+1)}}{2\mu+1} + \sqrt{2\mu}\left\{F\left(\frac{1}{\sqrt{8\mu}}, \cos^{-1}\frac{2\mu-1}{2\mu+1}\right) + E\left(\frac{1}{\sqrt{8\mu}}, \cos^{-1}\frac{2\mu-1}{2\mu+1}\right)\right\} \tag{18}$$

When the acceleration is very large, $\mu \gg 1$, the integrand in Equation (14) can be expanded in terms of this parameter. Then the mass ratio is calculated as

$$\frac{c}{\sqrt{gr_0}}\log_e(M_0/M_1) = 1 + \frac{1}{24\mu^2} - \frac{1}{40\mu^3} + \ldots \tag{19}$$

The relation of Equations (18) and (19) is plotted in Fig. 2. For $\mu = 1/8$, the mass ratio becomes infinite. The reason is that at this value of acceleration, there is a radial position where the thrust force is equal to the gravitational attraction and no further increase in the energy of the vehicle can occur. Therefore the radial thrust per unit mass, if maintained constant throughout the powered flight, should be larger than 1/8 g. With increasing thrust, the required mass ratio for escape from the earth's gravitational field decreases. This strong

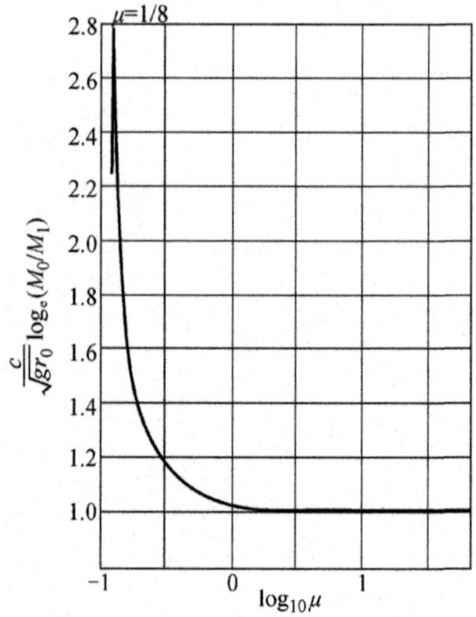

Fig. 2 Mass ratio factor ($c/\sqrt{gr_0}$) $\log_e(M_0/M_1)$ against acceleration factor μ for radial thrust. c, effective exhaust velocity; g, gravity at the satellite orbit of radius r_0; M_0, initial mass; M_1, final mass; μ, the ratio of instantaneous thrust per unit mass and g for radial thrust

dependence of the mass ratio upon the acceleration factor is contrary to opinion that for take-off from satellite orbit only very small thrust is required. The asymptotic value of $\log_e(M_0/M_1)$ is $\sqrt{gr_0}/c$. However, there is no appreciable improvement in going to higher thrust than 1 g.

Equation (16) shows that at very large values of the acceleration factor μ, the acceleration is accomplished in so short an interval that the circumferential velocity at the end of the acceleration remains at the initial value of $\sqrt{gr_0}$. The radial velocity increases from nothing at the initial instant to the final value of $\sqrt{gr_0}$. The total kinetic energy is thus gr_0 at the end of acceleration and this is equal to the negative of potential energy at that instant, since the radial position r must be practically the initial value r_0 under very large thrust. The work of the rocket is to produce the radial velocity $\sqrt{gr_0}$. Thus it is evident that the value of $c \log_e(M_0/M_1)$ must be $\sqrt{gr_0}$, as the calculation shows.

Circumferential Thrust

If the thrust is always circumferential and proportional to the mass of the vehicle, then a new thrust factor ν can be introduced such that

$$\Theta = \nu g \tag{20}$$

By using the same nondimensional variables as defined in Equation (6), the equations of motion are

$$\frac{d^2\rho}{d\tau^2} = \rho\left(\frac{d\theta}{d\tau}\right)^2 - \frac{1}{\rho^2} \tag{21}$$

$$\frac{d}{d\tau}\left(\rho^2 \frac{d\theta}{d\tau}\right) = \nu\rho \tag{22}$$

The initial conditions of Equation (3) and (4) are

$$\left(\frac{d\theta}{d\tau}\right)_0 = 1, \quad \left(\frac{d\rho}{d\tau}\right)_0 = 0 \quad \text{at} \quad \rho = 1, \quad \tau = 0 \tag{23}$$

Therefore, Equation (21) gives another initial condition that

$$\left(\frac{d^2\rho}{d\tau^2}\right)_0 = 0 \tag{24}$$

By eliminating θ from Equations (21) and (22)

$$\frac{d}{d\tau}\left(\rho^3 \frac{d^2\rho}{d\tau^2} + \rho\right)^{1/2} = \nu\rho \tag{25}$$

This is a third-order differential equation with three initia conditions specified by Equations (23) and (24). No simple general solution can, however, be obtained. The following discussion will be concerned with the approximations that are valid for large values of ν or for small values of ν.

For very large values of ν, the acceleration period is expected to be short and the change of the radial position to be small. Then the value of ρ must be very close to the initial value of unity. By taking ρ to be unity, Equation (25) becomes

$$\frac{\mathrm{d}}{\mathrm{d}\tau}\left(\frac{\mathrm{d}^2\rho}{\mathrm{d}\tau^2}+1\right)^{1/2} = \nu$$

Then

$$\frac{\mathrm{d}^2\rho}{\mathrm{d}\tau^2}+1 = C^2 + 2C\nu\tau + \nu^2\tau^2$$

where C is the integration constant. C, however, must be 1 because of the initial condition of Equation (24). The appropriate approximate solution for ρ for very large ν is thus

$$\rho \cong 1 + \frac{1}{3}\nu\tau^3 + \frac{1}{12}\nu^2\tau^4 \tag{26}$$

To obtain higher terms in this power series, the usual series substitution method may be used. The calculation is some-what lengthy and therefore will not be reproduced there. The result is

$$\rho = 1 + \frac{1}{3}\nu\tau^3 + \frac{1}{12}\nu^2\tau^4 - \frac{\nu}{60}\tau^5 - \frac{23\nu^2}{360}\tau^6 + \ldots \tag{27}$$

By using the result of Equation (27), the radial velocity is obtained by differentiation. Then Equation (21) gives the circumferential velocity. The end condition of Equation (5) can be modified into the following more convenient form by mulitiplying it by $2r^2$

$$0 = \left[\left(\rho\frac{\mathrm{d}\rho}{\mathrm{d}\tau}\right)^2 + \left(\rho^2\frac{\mathrm{d}\theta}{\mathrm{d}\tau}\right)^2 - 2\rho\right]$$

By substituting the solution of Equation (27) into this condition, an equation for determining τ_1 is obtained

$$0 = -1 + 2\nu\tau_1 + \nu^2\tau_1^2 - \frac{2}{3}\nu\tau_1^3 + \nu^2\tau_1^4 + \frac{\nu}{30}(1+26\nu^2)\tau_1^5 - \frac{\nu^2}{90}(4-13\nu^2)\tau_1^6 + \ldots \tag{28}$$

The mass ratio M_0/M_1 can be calculated in the same way as in the previous section and can be determined through the new parameter x defined as follows

$$\frac{c}{\sqrt{gr_0}}\log_e(M_0/M_1) = \nu\tau_1 = x \tag{29}$$

Equation (28) then can be written as

$$0 = -1 + 2x + x^2 - \frac{2}{3}\frac{x^3}{\nu^2} + \frac{x^4}{\nu^2} + \frac{x^5}{30\nu^4} + \frac{13}{15}\frac{x^5}{\nu^2} - \frac{2}{45}\frac{x^6}{\nu^4} + \frac{13}{90}\frac{x^6}{\nu^2} + \ldots \tag{30}$$

Since the calculation is designed for large values of ν, the appropriate expansion of x should be a series in inverse powers ν. Equation (30) suggests specifically

$$x(\nu) = x^{(0)} + \frac{x^{(1)}}{\nu^2} + \frac{x^{(2)}}{\nu^4} + \ldots \tag{31}$$

where $x^{(0)}$, $x^{(1)}$, and $x^{(2)}$ are constants independent of ν. By substituting Equation (31) into Equation (30) and equating equal powers of ν, the following set of equations results.

$$x^{(0)^2} + 2x^{(0)} - 1 = 0 \tag{32}$$

$$x^{(1)} = \frac{1}{2(1 + x^{(0)})}\left[\frac{2}{3}x^{(0)^3} - x^{(0)^4} - \frac{13}{15}x^{(0)^5} - \frac{13}{90}x^{(0)^6}\right] \tag{33}$$

$$x^{(2)} = \frac{1}{2(1 + x^{(0)})}\left[-x^{(1)^2} + 2x^{(0)^2}x^{(1)} - 4x^{(0)^3}x^{(1)} - \frac{1}{30}x^{(0)^6} - \frac{13}{3}x^{(0)^4}x^{(1)} + \frac{2}{45}x^{(0)^6} - \frac{13}{15}x^{(0)^5}x^{(1)}\right] \tag{34}$$

The explicit numerical solutions are then

$$\begin{aligned} x^{(0)} &= \sqrt{2} - 1 = 0.414\,21 \\ x^{(1)} &= 0.002\,349 \\ x^{(2)} &= -0.000\,047\,91 \end{aligned} \tag{35}$$

This completes the calculation of mass ratio for large values of the acceleration factor ν.

For the other extreme case of very small values of ν, it is to be expected that the acceleration will be very small, and in Equation (25) the term $\rho^3 \mathrm{d}^2\rho/\mathrm{d}\tau^2$ will be very much smaller than ρ. Therefore a good approximation of Equation (25) at small ν is

$$\frac{\mathrm{d}}{\mathrm{d}\tau}\rho^{1/2} = \nu\rho \quad \text{or} \quad \frac{1}{2}\frac{\mathrm{d}\rho}{\rho^{3/2}} = \nu \mathrm{d}\tau$$

The solution of this equation with the initial condition of $\rho = 1$ at $\tau = 0$ is

$$\rho = \frac{1}{(1 - \nu\tau)^2} \tag{36}$$

Therefore

$$\frac{\mathrm{d}\rho}{\mathrm{d}\tau} = \frac{2\nu}{(1 - \nu\tau)^3}, \quad \frac{\mathrm{d}^2\rho}{\mathrm{d}\tau^2} = \frac{6\nu^2}{(1 - \nu\tau)^4} \tag{37}$$

At $\tau = 0$, the radial velocity and the radial acceleration are thus not zero, as required by the initial conditions of Equations (23) and (24). They are, however, very small, because ν is very small. Therefore the solution of Equation (36) is a good approximation to the exact solution.

To the same approximation, Equation (20) becomes

$$\rho\frac{\mathrm{d}\theta}{\mathrm{d}\tau} = \frac{1}{\rho^{1/2}} = 1 - \nu\tau \tag{38}$$

This means that at every instant, because of the extremely small acceleration, the centrifugal force per unit mass $r(\mathrm{d}\theta/\mathrm{d}t)^2$ practically balances the gravitational attraction. The end condition of Equation (5) can then be written as

$$\frac{4\nu^2}{(1-x)^6} - (1-x)^2 = 0 \tag{39}$$

where x is again $\nu\tau_1$. The appropriate solution for x is then

$$x = 1 - (2\nu)^{1/4} \tag{40}$$

Since the mass ratio, M_0/M_1, is related to x by Equation (29), Equation (40) actually gives the mass ratio for escaping the gravitational field with very small acceleration.

The parameter x is plotted against ν in Fig. 3, using Equation (31) with both Equations (35) and (40). When ν approaches zero, x approaches 1. When ν is very large, x approaches $\sqrt{2}-1$. As ν increases, x and hence the mass ratio, M_0/M_1, decrease monotonically. Therefore, same as the result for purely radial thrust, there is a strong influence of the magnitude of acceleration on the required mass ratio. However, as far as decreasing the mass ratio is concerned, there is no appreciable advantage in using ν greater than 1/2.

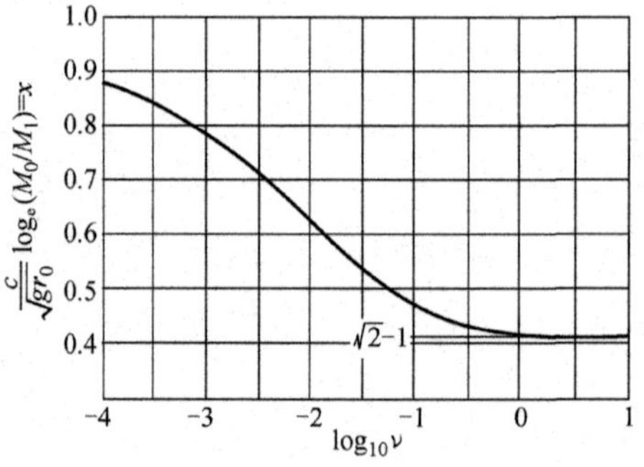

Fig. 3 Mass ratio factor ($c/\sqrt{gr_0}$) $\log_e(M_0/M_1)$ against acceleration factor ν for circumferential thrust. c, effective exhaust velocity; g, gravity at the satellite orbit of radius r_0; M_0 initial mass; M_1, final mass; ν, the ratio of instantaneous thrust per unit mass and g for circumferential thrust

When the acceleration factor ν is very large, the thrust force acts like an impulse. Since the thrust is in the circumferential direction, the rocket action only produces an increase in the circumferential velocity with practically no change in the radial position. The initial circumferential velocity is $\sqrt{gr_0}$ the required circumferential velocity for escape is $\sqrt{2gr_0}$. Thus the increase of velocity produced by the rocket action is $(\sqrt{2}-1)\sqrt{gr_0}$. This explains the asymptotic value of x for very large ν.

Discussion

By comparing Fig. 2 with Fig. 3, it is apparent that the radial thrust is much less efficient than the circumferential thrust for take-off from the satellite orbit. For large thrusts, the value of $\log(M_0/M_1)$ for radial thrust is more than twice that for circumferential thrust. Furthermore, in case of radial thrust, the ratio of thrust to the instantaneous mass, if maintained constant, must be larger than $g/8$. In case of circumferential thrust, no such limit exists. Therefore, circumferential thrust is definitely preferred.

The quantity $c \log_e(M_0/M_1)$ is a measure of the performance or the capability of the vehicle. It has the dimension of a velocity and is actually the increase of velocity which the vehicle is capable of in a space without gravitation. This quantity is conveniently called the *characteristic velocity* of the vehicle. Let this be denoted by V. Then for the case of circumferential thrust, Equation (29) gives

$$V = c\log_e(M_0/M_1) = \sqrt{gr_0}x = \frac{S}{\sqrt{2\lambda}}x \tag{41}$$

where S is the "escape velocity" from the surface of the earth, and λ is the ratio of the radii of the satellite orbit and the earth. S is equal to 11.2 km/sec. Fig.3 then shows that by decreasing the acceleration from 1/2 to 1/3 000 g, x, hence the required characteristic velocity V, will increase by a factor of two. This is a very important point for the designers of space ships.

References

[1] Spitzer L. Jr. Interplanetary Travel Between Satellite Orbits. Journal of the American Rocket Society, March-April 1952, 22: 92-96.

[2] Preston-Thomas H. Interorbital Transport Techniques. Journal of the British Interplanetary Society 1952, 11: 173-193.

[3] von Braun W. Man on the Moon, the Journey. Collie's, 1952, 18: 52.

Analysis of Peak-Holding Optimalizing Control

H. S. Tsien* and S. Serdengecti†

(*California Institute of Technology*)

Summary

The peak-holding optimalizing control is analyzed under the assumption of first-order input linear group and output linear group. Design charts are constructed for determining the required input drive speed and the consequent hunting loss with specified time constants of the input and output linear groups, the hunting period, and the critical indicated difference for input drive reversal.

Introduction

Optimalizing control was invented by C. S. Draper, Y. T. Li, and H. Laning, Jr[1,2]. Their basic idea can be summarized as follows: In almost all engineering systems, within the restrictions of operation, there is an optimum state of the system for performance. For instance, in an internal combustion engine, within the restriction of producing the load torque at the specified speed, there are optimum settings for the manifold pressure and the ignition timing for minimum fuel consumption. Another example is an airplane under cruising condition; then under the restriction of engine cruising r. p. m. and assigned altitude, there is an optimum combination of trim setting and engine throttle for maximum fuel economy or maximum miles per gallon of fuel. But more important than the existence of an optimum operating state is the fact that the optimum operating state cannot be exactly predicted in advance because of the natural changes in the environment of the engineering system: In the case of the internal combustion engine, it is the changes in the temperature and the humidity of the air; in the case of the airplane, it is unavoidable changes in the aerodynamic properties of the airplane and the engine performance with age. Therefore if the purpose is to operate always near the optimum state in spite of the "drift" of the system, then the control device for the engineering system must be so designed as to search out automatically the optimum state of operation and to confine the operation close to this state. This is the basic idea of optimalizing control.

The application of Draper's optimalizing control to the general cruise control of airplanes

---

Received April 30, 1954.

Journal of the Aeronautical Sciences, vol. 22, pp. 561 – 570, 1955.

* Robert H. Goddard Professor of Jet Propulsion, Daniel and Florence Guggenheim Jet Propulsion Center.

† Daniel and Florence Guggenheim Jet Propulsion Fellow.

was discussed by Shull[3]. Shull emphasized the possible elimination of extensive flight testing of new airplanes for performance determination, because the optimalizing control will automatically measure the performance whenever the airplane is flown. This in itself would constitute a great saving. But moreover, in critical circumstances such as flight through icing atmosphere, the ability of the optimalizing control to extract the best performance of a radically changed system (through ice deposition on the airplane) could be of utmost importance.

There are two fundamental problems in the theory of optimalizing control. One of the problems is the dynamic effects of the controlled system on the performance of the control. The other problem is the elimination of the noise interference. The two problems are somewhat interrelated, because if large deviations from the optimum state or the optimum operating point and hence large loss can be tolerated, then the noise interference will not be critical. The basic design aim of optimalizing control is to have the smallest loss or to operate as close to the optimum state as possible without the danger of having the control misled by the noise interference. Both of these problems were considered by the original inventors of optimalizing control. The noise problem is essentially the problem of detection of a sinusoidal variation under heavy random interference, a subject of much current research. The purpose of the present paper is to solve completely the first problem of dynamic effects under the assumption that the dynamic properties of the controlled system can be approximated by a first-order linear system. We shall begin with the brief review of the operating principles of an optimalizing control of the peak-holding type— a type least affected by the noise interference[1,2].

Principle of Operation

The heart of an optimalizing control system is the nonlinear component that characterizes the optimum operating condition of the controlled system. For simplicity of discussion, it is assumed that this basic component has a single input and a single output. For the time being the dynamic effects will be neglected and the output is assumed to be determined by the instantaneous value of the input. Since there is an optimum point, output as a function of input has a maximum at the output y_0 at the input x_0, as shown in Fig. 1. It is convenient to refer the output and the input to the optimum point and put the physical input as $x+x_0$ and the physical output as y^*+y_0. The optimum point is then the point $x=y^*=0$. The purpose of an optimalizing control is then to search out this optimum point and to keep the system in the immediate neighborhood of this point. In this neighborhood, the relation between x and y^* can be represented as

$$y^* = -kx^2 \tag{1}$$

where k is a characteristic constant of the controlled system.

The operation of a peak-holding optimalizing control, neglecting the dynamic effects, then would be as follows: Say the input x is below the optimum value and is thus negative. The input drive is then set to increase the input at a constant rate. At the time instant 1 (Fig. 2)

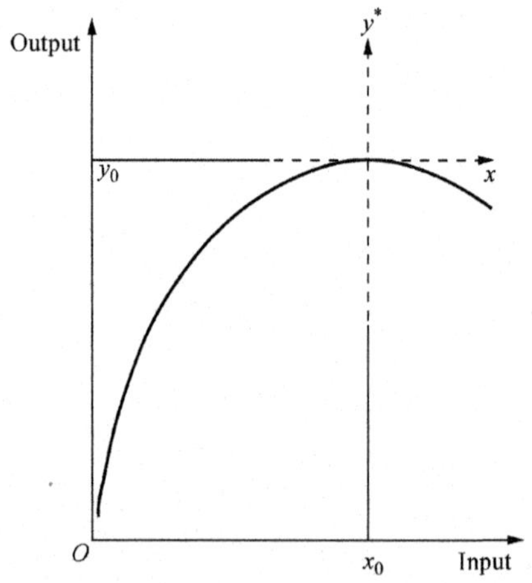

Fig. 1 Input-output characteristic of controlled system

the input changes from negative to positive and passes through the optimum point. The output y^* is thus maximum at the time instant 1 and is decreasing after the instant 1. Now if an output sensing instrument is so designed as to follow the output exactly when the output is increasing, but hold to the maximum value after the maximum is passed and the output starts to decrease; then there will be a difference between the reading of this output sensing instrument and the output itself after the time instant 1. This difference is shown in the lower graph of Fig. 2. When this difference is built up to a critical value c at the time instant 2, the input drive is tripped and the direction of the input drive is reversed, but still at the same constant rate as before. After the instant 2 then, the input decreases and the output increases till a maximum in output is again reached at the time instant 3. At time instant 3, the input, of course, again passes from positive to negative, and the indicated difference between the output sensing instrument and the output itself again builds up. At the time instant 4, the difference reaches the critical value c again, and the input drive direction is again reversed. At the time instant 5, the input x becomes zero again and another maximum of the output is reached. The period of input variation is thus the time interval from the instant 1 to the instant 5, and the input, when plotted as a function of time, consists of a series of straight line segments forming a saw-tooth variation. The period of output variation is the time interval from the instant 1 to the instant 3, and the output, when plotted as a function of time, consists of a series of parabolic arcs. The periodic variations of input and output are called the hunting of the system, and the period of output variation is called the hunting period T. The period of input variation is thus $2T$.

The extreme variation of output Δ (Fig. 2) is called the hunting zone. If a is the amplitude of the sawtooth variation of the input (Fig. 2), then due to Eq. (1),

$$\Delta = ka^2 \tag{2}$$

The difference between the maximum output and the average output of the hunting system is called

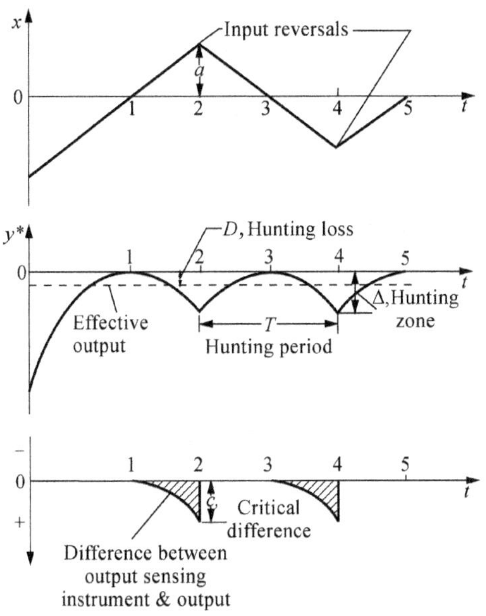

Fig. 2 Typical performance diagram for an ideal peak-holding optimalizing control system

the hunting loss D (Fig. 2). Because of the fact that the output is a series of parabolic arcs,

$$D = (1/3)\Delta = (1/3)ka^2 \tag{3}$$

For this idealized case, the critical indicated difference c between the output sensing instrument and the output itself is equal to Δ, the hunting zone. It is then clear from this discussion that in order to reduce the hunting loss for better efficiency of the system, one must try to reduce the hunting zone or the amplitude of input variation. Unfortunately the critical indicated difference is also reduced by such modification, and a limit is set by the noise interference on the proper tripping operation of the input drive.

The dynamic effects are so far neglected. But in any physical system, this is not possible because of the ever present inertial and damping forces. The output y^* given by Eq. (1) has to be considered then as the fictitious "potential output" but not the actual output y measured by the output indicating and sensing instrument. y^* is equal to y only when the period T of hunting becomes extremely long. The relation between y^* and y is determined by the dynamical effects. For the conventional engineering systems, these dynamical effects are determined by a linear relation. For instance, in the case of an internal combustion engine, the potential output is essentially the corrected effective pressure generated in the engine cylinders, while the actual output is the brake mean effective pressure of the engine. The dynamical effects are here mainly due to the inertia of the piston, the crankshaft, and other moving parts of the engine. For small changes in the operating conditions of the engine, such dynamical effects can be represented as a linear differential equation with constant coefficients.

Since the reference level of input and output is taken to be the optimum input x_0 and the optimum output y_0, the physical potential output is $y^* + y_0$ and the physical actual output is $y + y_0$. Thus the relation between the physical potential output and the physical actual output can be written as an operator equation

$$y + y_0 = F_0(\mathrm{d}/\mathrm{d}t)(y^* + y_0) \tag{4}$$

where F_0 is generally the quotient of two polynomials in the time differential operator $\mathrm{d}/\mathrm{d}t$. In the language of the Laplace transform then $F_0(s)$ is the transfer function. Let the linear system which transforms the potential output to actual output be called the output linear group. Then $F_0(s)$ is, specifically, the transfer function of the output linear group. By implication however, when the dynamical effects are negligible or when $s = 0$, the potential output is equal to the actual output. Therefore

$$F_0(0) = 1 \tag{5}$$

Since the optimum output y_0 only varies extremely slowly by the drift of the controlled system, during a time interval of many hunting periods y_0 can be taken as a constant. Then the condition of Eq. (5) simplifies Eq. (4) to

$$y = F_0(\mathrm{d}/\mathrm{d}t)y^* \tag{6}$$

In a similar manner, let x^* be the "potential input" that is actually the forcing function generated by the optimalizing control system but not the actual input x. It is x^* that has the saw-tooth form shown in Fig. 2, but not x. The relation between x^* and x is determined by the inertial and dynamical effects of the input drive system. This input drive system can be called the "input linear group" of the optimalizing control. The operator equation between the potential input x^* and the actual input x is

$$x = F_i(\mathrm{d}/\mathrm{d}t)x^* \tag{7}$$

$F_i(s)$ is thus the transfer function of the input linear group. Similar to Eq. (5), the meaning of potential and actual inputs implies

$$F_i(0) = 1 \tag{8}$$

Thus a simple representative block diagram of the complete optimalizing control system can be drawn as shown in Fig. 3. The nonlinear components of the system are thus the

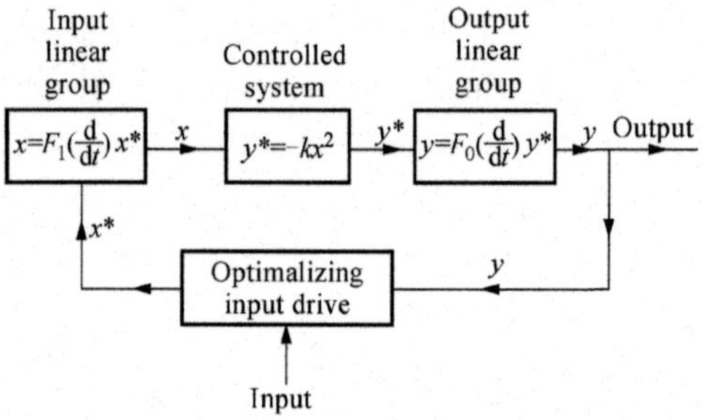

Fig. 3 Block diagram of a complete peak-holding optimalizing control system

optimalizing input drive and the controlled system itself.

Formulation of the Mathematical Problem

The general relation between the input x and the output y is determined by the system of Eqs. (1), (6), and (7), with the potential input x^* specified as a sawtooth curve with period $2T$ and amplitude a. Let ω_0 be the hunting frequency defined by

$$\omega_0 = 2\pi/T \tag{9}$$

then x^* can be expanded into a Fourier series,

$$\begin{aligned} x^* &= \frac{8a}{\pi^2}\sum_{n=0}^{\infty}\frac{(-1)^n}{(2n+1)^2}\sin(2n+1)\frac{\omega_0 t}{2} \\ &= \frac{8a}{\pi^2}\sum_{n=0}^{\infty}\frac{(-1)^n}{(2n+1)^2}\frac{1}{2i}\left(e^{[(2n+1)/2]i\omega_0 t} - e^{-[(2n+1)/2]i\omega_0 t}\right) \end{aligned} \tag{10}$$

Therefore by using Eq. (7), the actual input x is given by

$$\begin{aligned} x = {} & \frac{8a}{\pi^2}\sum_{n=0}^{\infty}\frac{(-1)^n}{(2n+1)^2(2i)}\times \\ & \left[F_i\left(\frac{2n+1}{2}i\omega_0\right)e^{[(2n+1)/2]i\omega_0 t} - F_i\left(-\frac{2n+1}{2}i\omega_0\right)e^{-[(2n+1)/2]i\omega_0 t}\right] \end{aligned} \tag{11}$$

By using Eqs. (11) and (16), the actual output y is given by

$$\begin{aligned} y = {} & \frac{16a^2k}{\pi^4}\sum_{n=0}^{\infty}\sum_{m=0}^{\infty}\frac{(-1)^{n+m}}{(2n+1)^2(2m+1)^2}\times\Big\{F_0[(n+m+1)i\omega_0]F_i\left(\frac{2n+1}{2}i\omega_0\right)\times \\ & F_i\left(\frac{2m+1}{2}i\omega_0\right)e^{(n+m+1)i\omega_0 t} - F_0[(n-m)i\omega_0]\times \\ & F_i\left(\frac{2n+1}{2}i\omega_0\right)F_i\left(-\frac{2m+1}{2}i\omega_0\right)e^{(n-m)i\omega_0 t} - \\ & F_0[-(n-m)i\omega_0]F_i\left(-\frac{2n+1}{2}i\omega_0\right)\times \\ & F_i\left(\frac{2m+1}{2}i\omega_0\right)e^{-(n-m)i\omega_0 t} + F_0[-(n+m+1)i\omega_0]\times \\ & F_i\left(-\frac{2n+1}{2}i\omega_0\right)F_i\left(-\frac{2m+1}{2}i\omega_0\right)\times e^{-(n+m+1)i\omega_0 t}\Big\} \end{aligned} \tag{12}$$

By comparing Eqs. (11) and (12), it is seen that the input has half the frequency of the output. This is, of course, to be expected from the basic parabolic relation of input and output as specified by Eq. (1).

The average of the actual output y with respect to time t, being here referred to the optimum output y_0, gives directly the hunting loss D. Equation (12) shows that this average value is the sum of terms with $n = m$ from the second and the third terms of that equation. Therefore, using Eq. (5),

$$D = \frac{32a^2k}{\pi^4}\sum_{n=0}^{\infty}\frac{1}{(2n+1)^4}F_i\left(\frac{2n+1}{2}i\omega_0\right)\times F_i\left(-\frac{2n+1}{2}i\omega_0\right) \tag{13}$$

This equation can be easily checked by observing that when the dynamic effects are absent, $F_i \equiv 1$, then the series can be easily summed and $D = (1/3)a^2k$ as required by Eq. (3). Equation (13) also shows that the average output and hence the hunting loss are independent of the output linear group. This agrees with the one's physical understanding: Only detailed time variation of the output is modified by the dynamics of the output linear group. In the case of an internal combustion engine, the average output specifies the power of the engine. The dynamics of the output linear group is determined by the inertia of the moving parts. The power of the engine is certainly independent of the inertia of the moving parts.

Equations (11) to (13) fully determine the performance of the optimalizing control system once the values of a, k, and ω_0 are specified and the transfer functions $F_i(s)$ and $F_0(s)$ of the input linear group and the output linear group are given. The following sections give the detailed calculations and results for the case of first-order input and output groups.

First-Order Input and Output Groups

The frequency ω_0 of the optimalizing control is usually low, and the important dynamic effects come from the inertia in the input and the output linear groups. Then these linear groups can be closely approximated by first-order systems. In other words, their transfer functions are

$$F_i(i\omega) = 1/(1+i\omega\tau_i) \tag{14}$$

$$F_0(i\omega) = 1/(1+i\omega\tau_0) \tag{15}$$

where τ_i and τ_0 are the characteristic time constants of the input linear group and the output linear group, respectively. It is evident that these transfer functions satisfy the conditions of Eqs. (5) and (8).

By substituting Eq. (14) into Eq. (11), the actual output x is given by

$$x = \frac{8a}{\pi^2}\sum_{n=0}^{\infty}\frac{(-1)^n}{2i(2n+1)^2}\left[\frac{e^{[(2n+1)/2]i\omega_0 t}}{1+(2n+1)i(\omega_0\tau_i/2)} - \frac{e^{-[(2n+1)/2]i\omega_0 t}}{1-(2n+1)i(\omega_0\tau_i/2)}\right] \tag{16}$$

When the summation is carried out, Eq. (16) yields the following equations for the input x:

$$x = NT\left[\frac{t}{T} - \frac{\tau_i}{T} + \frac{\tau_i}{T}\frac{e^{-[(t/T)/(\tau_i/T)]}}{\cosh(T/2\tau_i)}\right] \quad \text{for} \quad -\frac{1}{2}\leqslant\frac{t}{T}\leqslant\frac{1}{2} \tag{17a}$$

and

$$x = -NT\left[\frac{t}{T} - \left(1+\frac{\tau_i}{T}\right) + \frac{\tau_i}{T}\frac{e^{(1-t/T)/(\tau_i/T)}}{\cosh(T/2\tau_i)}\right] \quad \text{for} \quad \frac{1}{2}\leqslant\frac{t}{T}\leqslant\frac{3}{2} \tag{17b}$$

where N is the constant input drive speed — i.e.,

$$N = 2a/T \tag{18}$$

By using these equations, the variation of actual input x with respect to time can be calculated for any specified data. Examples of such calculations are shown in Figs. 4 and 5 for $\tau_i/T = 0.1$ and $\tau_i/T = 0.4$, respectively. Both show the expected effect of rounding-off of the sharp corners of the saw-tooth curve and a time delay. It is of interest to note that while the delay is almost equal to τ_i itself for small τ_i/T, the delay is less than τ_i for larger τ_i/T.

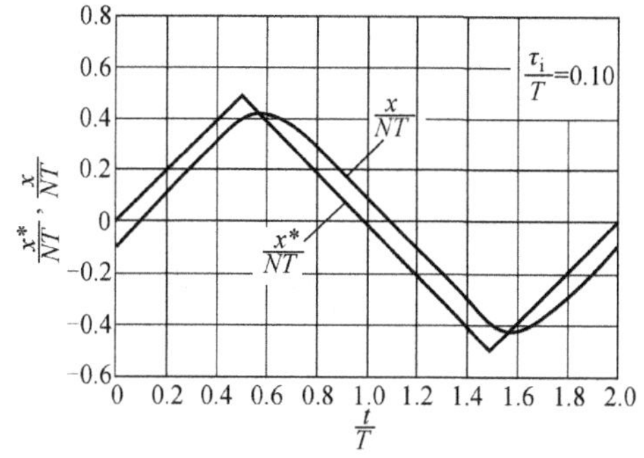

Fig. 4 "Potential" input and actual input for value of $\tau_i/T = 0.1$

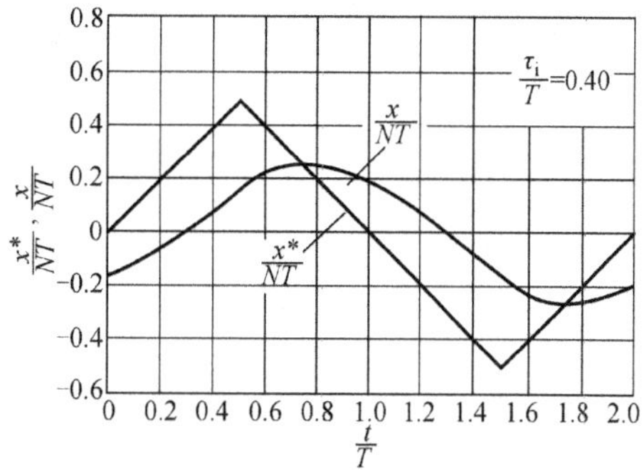

Fig. 5 "Potential" input and actual input for value of $\tau_i/T = 0.4$

With the first-order transfer function of Eq. (14), the hunting loss given by Eq. (13) becomes

$$D = \frac{32a^2k}{\pi^4}\sum_{n=0}^{\infty}\frac{1}{(2n+1)^4\{1+[(2n+1)/2]^2\omega_0^2\tau_i^2\}} \tag{19}$$

By carrying out the summation, Eq. (19) gives the hunting loss as

$$D = (N^2T^2k/12)[1-12(\tau_i/T)^2+24(\tau_i/T)^3\tanh(T/2\tau_i)] \tag{20}$$

Fig. 6 shows a dimensionless plot of this equation.

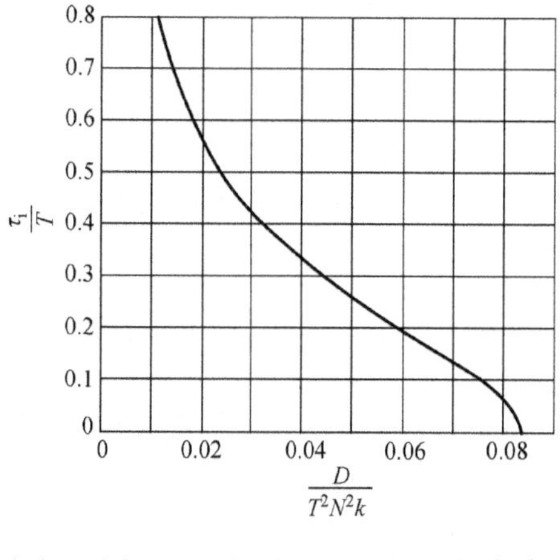

Fig. 6 Variation of dimensionless hunting loss, $D/(T^2N^2k)$ with τ_i/T

To calculate the actual output y, both Eqs. (14) and (15) have to be substituted into Eq. (12) — i.e.,

$$y = \frac{4T^2N^2k}{\pi^4}\sum_{n=0}^{\infty}\sum_{m=0}^{\infty}\frac{(-1)^{n+m}}{(2n+1)^2(2m+1)^2}\times$$
$$\left\{\frac{e^{i(n+m+1)\omega_0 t}}{[1+(n+m+1)i\omega_0\tau_0][1+(2n+1)i(\omega_0\tau_i/2)][1+(2m+1)i(\omega_0\tau_i/2)]}-\right.$$
$$\frac{e^{i(n-m)\omega_0 t}}{[1+(n-m)i\omega_0\tau_0][1+(2n+1)i(\omega_0\tau_i/2)][1-(2m+1)i(\omega_0\tau_i/2)]}-\tag{21}$$
$$\frac{e^{-i(n-m)\omega_0 t}}{[1-(n-m)i\omega_0\tau_0][1-(2n+1)i(\omega_0\tau_i/2)][1+(2m+1)i(\omega_0\tau_i/2)]}+$$
$$\left.\frac{e^{-i(n+m+1)\omega_0 t}}{[1-(n+m+1)i\omega_0\tau_0][1-(2n+1)i(\omega_0\tau_i/2)][1-(2m+1)i(\omega_0\tau_i/2)]}\right\}$$

By changing the summation indices, Eq. (21) can also be written as

$$y = \frac{4T^2N^2k}{\pi^4}\left(\sum_{s=-\infty}^{\infty}\frac{(-1)^{s-1}e^{is\omega_0 t}}{(1+is\omega_0\tau_0)}\times\right.$$
$$\sum_{n=0}^{\infty}\frac{1}{(2n+1)^2[(2n+1)-2s]^2[1+(2n+1)i(\omega_0\tau_i/2)]\{1-[(2n+1)-2s]i(\omega_0\tau_i/2)\}}+$$
$$\sum_{s=-\infty}^{\infty}\frac{(-1)^{s-1}e^{-is\omega_0 t}}{(1-is\omega_0\tau_0)}\times$$
$$\left.\sum_{n=0}^{\infty}\frac{1}{(2n+1)^2[(2n+1)-2s]^2[1-(2n+1)i(\omega_0\tau_i/2)]\{1+[(2n+1)-2s]i(\omega_0\tau_i/2)\}}\right)$$

or

$$y = \frac{8T^2N^2k}{\pi^4}\left\{-\sum_{n=0}^{\infty}\frac{1}{(2n+1)^4[1+(2n+1)^2(\omega_0\tau_i/2)^2]}+\sum_{s=1}^{\infty}\frac{(-1)^{s-1}e^{is\omega_0 t}}{(1+is\omega_0\tau_0)}\times\right.$$
$$\sum_{n=0}^{\infty}\frac{[(2n+1)^2+4s^2][(1+i\omega_0\tau_i s)+(\omega_0\tau_i/2)^2(2n+1)^2]+8(\omega_0\tau_i/2)^2s^2(2n+1)^2}{(2n+1)^2[(2n+1)^2-4s^2]^2[1+(\omega_0\tau_i/2)^2(2n+1)^2][(1+i\omega_0\tau_i s)^2+(\omega_0\tau_i/2)^2(2n+1)^2]}+$$
$$\sum_{s=1}^{\infty}\frac{(-1)^{s-1}e^{-is\omega_0 t}}{(1-is\omega_0\tau_0)}\times$$
$$\left.\sum_{n=0}^{\infty}\frac{[(2n+1)^2+4s^2][(1-i\omega_0\tau_i s)+(\omega_0\tau_i/2)^2(2n+1)^2]+8(\omega_0\tau_i/2)^2s^2(2n+1)^2}{(2n+1)^2[(2n+1)^2-4s^2]^2[1+(\omega_0\tau_i/2)^2(2n+1)^2][(1-i\omega_0\tau_i s)^2+(\omega_0\tau_i/2)^2(2n+1)^2]}\right\}\tag{22}$$

The last two summations in Eq. (22) are complex conjugate of each other, thus

$$y = \frac{8T^2N^2k}{\pi^4}\left\{-\sum_{n=0}^{\infty}\frac{1}{(2n+1)^4[1+(2n+1)^2(\omega_0\tau_i/2)^2]}+2\mathrm{Re}\sum_{s=1}^{\infty}\frac{(-1)^{s-1}e^{is\omega_0 t}}{(1+is\omega_0\tau_0)}\times\right.$$
$$\left.\sum_{n=0}^{\infty}\frac{[(2n+1)^2+4s^2][(1+i\omega_0\tau_i s)+(\omega_0\tau_i/2)^2(2n+1)^2]+8(\omega_0\tau_i/2)^2s^2(2n+1)^2}{(2n+1)^2[(2n+1)^2-4s^2]^2[1+(\omega_0\tau_i/2)^2(2n+1)^2][(1+i\omega_0\tau_i s)^2+(\omega_0\tau_i/2)^2(2n+1)^2]}\right\}\tag{23}$$

where Re means the real part of the expression following it. In order to carry out the summation with respect to the index n, Eq. (23) is resolved into the following partial fraction form:

$$y=\frac{8T^2N^2k}{\pi^4}\Bigg(-\sum_{n=0}^{\infty}\frac{1}{(2n+1)^4[1+(2n+1)^2(\omega_0\tau_i/2)^2]}+2\mathrm{Re}\sum_{s=1}^{\infty}\frac{(-1)^{s-1}e^{is\omega_0 t}}{(1+is\omega_0\tau_0)}\times$$
$$\Bigg\{\frac{1}{4s^2(1+is\omega_0\tau_i)}\sum_{n=0}^{\infty}\frac{1}{(2n+1)^2}+$$
$$\frac{(\omega_0\tau_i/2)^4}{2(1+is\omega_0\tau_i)^2[1+is(\omega_0\tau_i/2)]}\sum_{n=0}^{\infty}\frac{1}{[1+(\omega_0\tau_i/2)^2(2n+1)^2]}+$$
$$\frac{(\omega_0\tau_i/2)^4}{2(1+is\omega_0\tau_i)[1+is(\omega_0\tau_i/2)]}\sum_{n=0}^{\infty}\frac{1}{[(1+i\omega_0\tau_i s)^2+(\omega_0\tau_i/2)^2(2n+1)^2]}-\tag{24}$$
$$\frac{[1+is\omega_0\tau_i+4(\omega_0\tau_i/2)^2s^2]}{4s^2(1+is\omega_0\tau_i)^2}\sum_{n=0}^{\infty}\frac{1}{[(2n+1)^2+(i2s)^2]}+$$
$$\frac{2}{(1+is\omega_0\tau_i)}\sum_{n=0}^{\infty}\frac{1}{[(2n+1)^2+(i2s)^2]^2}\Bigg\}\Bigg)$$

By using the summation formulas given in the Appendix, the sums with respect to n can be evaluated and the result is, noting that $\tan\pi s=0$ for integer values of s,

$$y=\frac{8T^2N^2k}{\pi^4}\Bigg(-\left[\frac{\pi^4}{96}-\frac{\pi^2}{8}\left(\frac{\omega_0\tau_i}{2}\right)^2+\frac{\pi}{4}\left(\frac{\omega_0\tau_i}{2}\right)^3\tanh\frac{\pi}{\omega_0\tau_i}\right]+$$
$$2\mathrm{Re}\sum_{s=1}^{\infty}\frac{(-1)^{s-1}e^{is\omega_0 t}}{(1+is\omega_0\tau_0)}\Bigg\{\frac{\pi^2}{4}\frac{1}{(4s^2)(1+is\omega_0\tau_i)}+\tag{25}$$
$$\frac{\pi}{4}\frac{(\omega_0\tau_i/2)^3\tanh[\pi/(\omega_0\tau_i)]}{(1+is\omega_0\tau_i)^2[1+is(\omega_0\tau_i/2)]}\Bigg\}\Bigg)$$

Eq. (25) is again resolved into partial fractions in order to carry out the summation with respect to s, viz.,

$$y=\frac{8T^2N^2k}{\pi^4}\Bigg[-\left[\frac{\pi^4}{96}-\frac{\pi^2}{8}\left(\frac{\omega_0\tau_i}{2}\right)^2+\frac{\pi}{4}\left(\frac{\omega_0\tau_i}{2}\right)^3\tanh\frac{\pi}{\omega_0\tau_i}\right]+$$
$$\frac{\pi}{2}\Bigg(\frac{(\omega_0\tau_0/2)^3}{[(\omega_0\tau_0/2)-(\omega_0\tau_i/2)]}\Bigg\{\frac{2(\omega_0\tau_i/2)^3\tanh[\pi/(\omega_0\tau_i)]}{[(\omega_0\tau_0/2)-(\omega_0\tau_i/2)][\omega_0\tau_0-(\omega_0\tau_i/2)]}-\pi\Bigg\}\mathrm{Re}\sum_{s=1}^{\infty}\frac{(-1)^{s-1}e^{is\omega_0 t}}{(1+is\omega_0\tau_0)}+$$
$$\frac{(\omega_0\tau_i/2)^3}{[(\omega_0\tau_0/2)-(\omega_0\tau_i/2)]}\Bigg\{\pi-\frac{2(\omega_0\tau_i/2)^2\tanh[\pi/(\omega_0\tau_i)]}{[(\omega_0\tau_0/2)-(\omega_0\tau_i/2)]}\Bigg\}\mathrm{Re}\sum_{s=1}^{\infty}\frac{(-1)^{s-1}e^{is\omega_0 t}}{(1+is\omega_0\tau_i)}-\frac{\pi}{2}\left(\frac{\omega_0\tau_0}{2}+\frac{\omega_0\tau_i}{2}\right)\times$$
$$\mathrm{Re}\sum_{s=1}^{\infty}\frac{(-1)^{s-1}ie^{is\omega_0 t}}{s}+\frac{\pi}{4}\mathrm{Re}\sum_{s=1}^{\infty}\frac{(-1)^{s-1}e^{is\omega_0 t}}{s^2}-$$
$$\frac{(\omega_0\tau_i/2)^4\tanh[\pi/(\omega_0\tau_i)]}{[2(\omega_0\tau_0/2)-(\omega_0\tau_i/2)]}\mathrm{Re}\sum_{s=1}^{\infty}\frac{(-1)^{s-1}e^{is\omega_0 t}}{[1+is(\omega_0\tau_i/2)]}-$$
$$\frac{2(\omega_0\tau_i/2)^4\tanh[\pi/(\omega_0\tau_i)]}{[(\omega_0\tau_0/2)-(\omega_0\tau_i/2)]}\mathrm{Re}\sum_{s=1}^{\infty}\frac{(-1)^{s-1}e^{is\omega_0 t}}{(1+is\omega_0\tau_i)^2}\Bigg)\Bigg]\tag{26}$$

The result of carrying out the summations in Eq. (26) and simplifying the expressions is,

$$y = 2T^2N^2k\Big[-\Big\{\frac{1}{2}\Big(\frac{t}{T}\Big)^2 - \Big(\frac{\tau_i}{T}+\frac{\tau_0}{T}\Big)\Big(\frac{t}{T}\Big) + \Big[\frac{1}{2}\Big(\frac{\tau_i}{T}\Big)^2 + \frac{\tau_i\tau_0}{T^2} + \Big(\frac{\tau_0}{T}\Big)^2\Big]\Big\} +$$
$$\frac{1}{2}\Big(-\frac{(\tau_0/T)^2}{(\tau_0/T-\tau_i/T)}\Big\{\frac{2(\tau_i/T)^3\tanh(T/2\tau_i)}{[(\tau_0/T)-(\tau_i/T)][2(\tau_0/T)-(\tau_i/T)]} - 1\Big\}\frac{e^{-(t/T)/(\tau_0/T)}}{\sinh[T/(2\tau_0)]} +$$
$$\Big\{\frac{t}{T} + \frac{(\tau_i/T)^2}{[(\tau_0/T)-(\tau_i/T)]}\Big\}\frac{2(\tau_i/T)^2 e^{-(t/T)/(\tau_i/T)}}{[(\tau_0/T)-(\tau_i/T)]\cosh[T/(2\tau_i)]} +$$
$$\frac{(\tau_i/T)^3}{[2(\tau_0/T)-(\tau_i/T)]}\frac{e^{-(2t/T)/(\tau_i/T)}}{\cosh^2[T/(2\tau_i)]}\Big)\Big] \tag{27a}$$

for $-(1/2) \leqslant t/T \leqslant 1/2$ and

$$y = 2T^2N^2k\Big[-\Big\{\frac{1}{2}\Big(\frac{t}{T}\Big)^2 - \Big(\frac{\tau_0}{T}+\frac{\tau_i}{T}+1\Big)\Big(\frac{t}{T}\Big) + \Big[\frac{1}{2}\Big(\frac{\tau_i}{T}\Big)^2 + \frac{\tau_i\tau_0}{T^2} + \Big(\frac{\tau_0}{T}\Big)^2 + \frac{\tau_0}{T} + \frac{\tau_i}{T} + \frac{1}{2}\Big]\Big\} +$$
$$\frac{1}{2}\Big(-\frac{(\tau_0/T)^2}{[(\tau_0/T)-(\tau_i/T)]}\Big\{\frac{2(\tau_i/T)^3\tanh(T/2\tau_i)}{[(\tau_0/T)-(\tau_i/T)][2(\tau_0/T)-(\tau_i/T)]} - 1\Big\}\frac{e^{(1-t/T)/(\tau_0/T)}}{\sinh(T/2\tau_0)} +$$
$$\Big[\frac{t}{T} + \frac{(\tau_i/T)^2}{(\tau_0/T)-(\tau_i/T)} - 1\Big]\frac{2(\tau_i/T)^2 e^{(1-t/T)/(\tau_i/T)}}{[(\tau_0/T)-(\tau_i/T)]\cosh[T/(2\tau_i)]} +$$
$$\frac{(\tau_i/T)^3 e^{2(1-t/T)/(\tau_i/T)}}{[2(\tau_0/T)-(\tau_i/T)]\cosh^2[T/(2\tau_i)]}\Big)\Big] \tag{27b}$$

for $1/2 \leqslant t/T \leqslant 3/2$.

In Eqs. (27a) and (27b), there are apparent singularities whenever $\tau_0/T = \tau_i/T$ and $2\tau_0/T = \tau_i/T$; that is, the value of output y seemingly cannot be determined for these values of time constants. However this is deceptive. By using a simple limit procedure or by direct evaluation of Eq. (25) for these two cases, it can be shown that this is not the case. For example, for $\tau_i/T = \tau_0/T$

$$y = 2T^2N^2k\Big\{-\frac{1}{2}\Big(\frac{t}{T}\Big)^2 + 2\Big(\frac{\tau_i}{T}\Big)\Big(\frac{t}{T}\Big) - \frac{5}{2}\Big(\frac{\tau_i}{T}\Big)^2 + \Big[-\Big(\frac{\tau_i}{T}\Big)^2 - \frac{1}{2}\Big(\frac{t}{T}\Big)^2 +$$
$$\Big(\frac{\tau_i}{T}\Big)\Big(\frac{t}{T}\Big) + \frac{1}{8}\Big]e^{-[(t/T)/(\tau_i/T)]}\operatorname{sech}\frac{T}{2\tau_i} + \tag{28}$$
$$\frac{3}{2}\Big(\frac{\tau_i}{T}\Big)e^{-(t/T)/(\tau_i/T)}\operatorname{csch}\frac{T}{2\tau_i} + \frac{1}{2}\Big(\frac{\tau_i}{T}\Big)^2 e^{-(2t/T)/(\tau_i/T)}\operatorname{sech}^2\frac{T}{2\tau_i}\Big\}$$

for $-(1/2) \leqslant t/T \leqslant 1/2$, and, for $2\tau_0/T = \tau_i/T$,

$$y = -2T^2N^2k\Big\{\frac{1}{2}\Big(\frac{t}{T}\Big)^2 - \frac{3}{2}\Big(\frac{\tau_i}{T}\Big)\Big(\frac{t}{T}\Big) + \frac{5}{4}\Big(\frac{\tau_i}{T}\Big)^2 +$$
$$\Big[\Big(\frac{\tau_i}{T}\Big)\Big(\frac{t}{T}\Big) + 2\Big(\frac{\tau_i}{T}\Big)^2 + \frac{1}{2}\Big(\frac{\tau_i}{T}\Big)\coth\frac{T}{\tau_i} + \frac{1}{8}\Big(\frac{\tau_i}{T}\Big)\coth\frac{T}{2\tau_i}\Big]\frac{e^{-[(2t/T)/(\tau_i/T)]}}{\cosh^2 T/2\tau_i} + \tag{29}$$
$$\Big(\frac{\tau_i}{T}\Big)\Big(2\frac{t}{T} - 4\frac{\tau_i}{T}\Big)\frac{e^{-[(t/T)/(\tau_i/T)]}}{\cosh T/2\tau_i}\Big\}$$

for $-(1/2) \leqslant t/T \leqslant 1/2$.

An analysis for the continuity of Eqs. (27a) and (27b) at $t/T = 1/2$ shows that the values of y and its derivative with respect to t are the same at $t/T = 1/2$ whether they are computed from Eq. (27a) or from Eq. (27b).

Now the computation of the potential output y^* can be accomplished by letting $\tau_0/T = 0$ in Eqs. (27a) and (27b). Thus,

$$y^* = 2T^2N^2k\left[-\frac{1}{2}\left(\frac{t}{T}\right)^2 + \left(\frac{\tau_i}{T}\right)\left(\frac{t}{T}\right) - \frac{1}{2}\left(\frac{\tau_i}{T}\right)^2 - \left(\frac{t}{T} - \frac{\tau_i}{T}\right)\left(\frac{\tau_i}{T}\right)\frac{e^{-[(t/T)/(\tau_i/T)]}}{\cosh(T/2\tau_i)} - \frac{1}{2}\left(\frac{\tau_i}{T}\right)^2 \frac{e^{-[(2t/T)/(\tau_i/T)]}}{\cosh^2(T/2\tau_i)}\right] \tag{30a}$$

for $-(1/2) \leqslant t/T \leqslant 1/2$ and

$$y^* = 2T^2N^2k\left\{-\frac{1}{2}\left(\frac{t}{T}\right)^2 + \left(1 + \frac{\tau_i}{T}\right)\left(\frac{t}{T}\right) - \frac{1}{2}\left(1 + \frac{\tau_i}{T}\right)^2 - \left(\frac{\tau_i}{T}\right)\left[\frac{t}{T} - \left(\frac{\tau_i}{T} + 1\right)\right]\frac{e^{(1-t/T)/(\tau_i/T)}}{\cosh(T/2\tau_i)} - \frac{1}{2}\left(\frac{\tau_i}{T}\right)^2 \frac{e^{[2(1-t/T)]/(\tau_i/T)}}{\cosh^2(T/2\tau_i)}\right\} \tag{30b}$$

for $1/2 \leqslant t/T \leqslant 3/2$.

These expressions check with the result of direct calculation of y^* by Eqs. (1) and (17).

Figures 7 and 8 show the dimensionless plots of actual output y and potential output y^* for the particular values of τ_0/T and τ_i/T. In these figures it is clearly seen that the dynamic effects not only decrease the output of the system but also introduce a time lag and lower the maximum output of the system. Figure 8 with $\tau_i/T = 0.4$, $\tau_0/T = 0.6$, has the maximum value of y almost at the very instants of input drive reversal points, $t/T = n + (1/2)$. This is indeed an extreme case.

Design Charts

From the principle of operation of the peak-holding optimalizing control, it is seen that the most important quantity to be specified for its design is the critical indicated difference c between the reading of the special output sensing instrument and the output itself. By definition, c is the difference of the maximum of the actual output y and the value of y at the tripping instant of the input drive. The instant of reversing the input drive is typified by $t/T = 1/2$. If the corresponding instant of maximum y is t^*, then the critical indicated difference c is calculated as

$$c = y(t^*/T) - y(1/2) \tag{31}$$

by using any one of Eqs. (27), (28), or (29). Since the instant of input drive reversal must come after the instant of maximum output, $t^*/T < 1/2$.

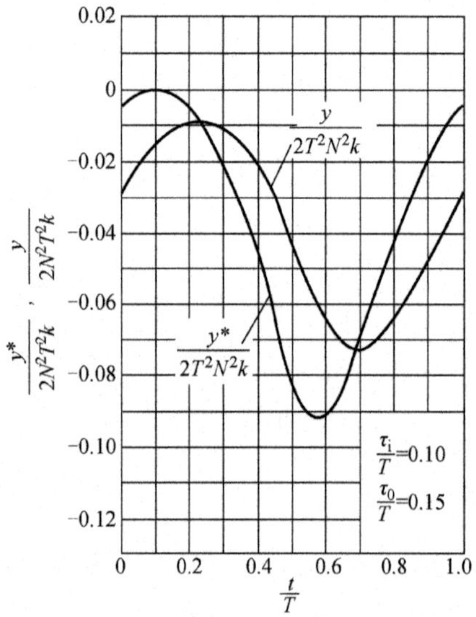

Fig. 7 "Potential" output and indicated output for values of $\tau_i/T = 0.1$ and $\tau_0/T = 0.15$

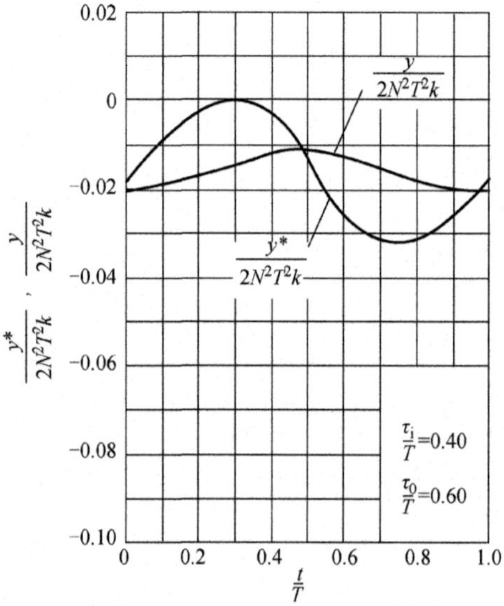

Fig. 8 "Potential" output and indicated output for values of $\tau_i/T = 0.4$ and $\tau_0/T = 0.6$

To determine t^*, one may use the condition of zero slope—i. e., $dy/dt = 0$. Then Eq. (27a) gives

$$-\left[\frac{t^*}{T}-\left(\frac{\tau_0}{T}+\frac{\tau_i}{T}\right)\right]+\frac{(\tau_0/T)}{2[(\tau_0/T)-(\tau_i/T)]}\left\{\frac{2(\tau_i/T)^3\tanh(T/2\tau_i)}{[(\tau_0/T)-(\tau_i/T)][2(\tau_0/T)-(\tau_i/T)]}-1\right\}\frac{e^{-[(t^*/T)/(\tau_0/T)]}}{\sinh(T/2\tau_0)}+\left\{1-\left(\frac{t^*}{T}\right)\left(\frac{T}{\tau_i}\right)-\frac{(\tau_i/T)}{[(\tau_0/T)-(\tau_i/T)]}\right\}\frac{(\tau_i/T)^2 e^{-(t^*/T)/(\tau_i/T)}}{[(\tau_0/T)-(\tau_i/T)]\cosh(T/2\tau_i)}-\frac{(\tau_i/T)^2 e^{-[(2t^*/T)/(\tau_i/T)]}}{[2(\tau_0/T)-(\tau_i/T)]\cosh^2(T/2\tau_i)}=0 \tag{32}$$

This transcendental equation for t^*/T may be solved by iteration. For instance, for small τ_0/T and τ_i/T, only terms within the first brackets are of importance, then $t^*/T \simeq (\tau_0+\tau_i)/T$. This is already recognized by Draper and co-workers[1,2]. The complete results of calculation are shown in Fig. 9, which shows that t^*/T is almost only a function of $(\tau_0+\tau_i)/T$ with minor modifications from the parameter τ_0/τ_i, the ratio of characteristic times of the output linear group and the input linear group. Values of t^*/T beyond 1/2 are not shown, as clearly then the maxima of the output will occur after the corresponding input drive reversal points and proper operation of the control will be difficult if not impossible.

With t^*/T determined, Eq. (31) gives c by substituting Eq. (27a). However the specified quantities of an optimalizing control are k, the characteristics of the controlled system, and τ_i, τ_0, the characteristics of the linear group. From considerations on the noise interference, the designer can make an appropriate choice of the period T and the critical indicated difference c

for input drive reversal. Therefore the quantities that the designer wishes to know, after he has the values of k, τ_i, τ_0, T, and c, are N, the input drive speed, and D, the hunting loss. Thus the result of calculation with Eq. (31) should be written as follows:

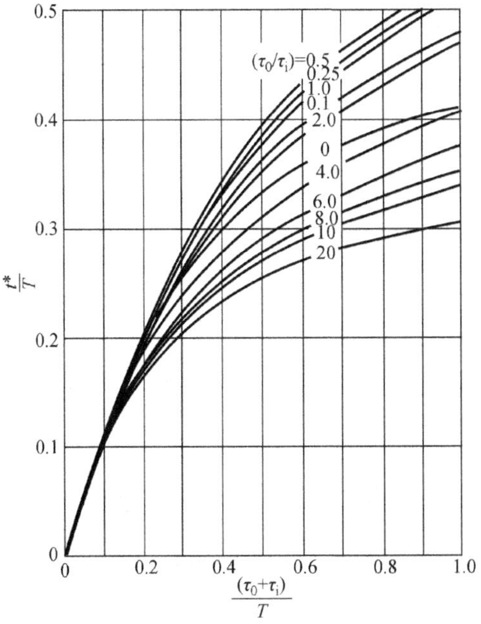

Fig. 9 Maximum output occurrence instant, t^*/T in interval $(0 \leqslant t/T \leqslant 1/2)$ versus $(\tau_0+\tau_i)/T$ with τ_0/τ_i as parameter

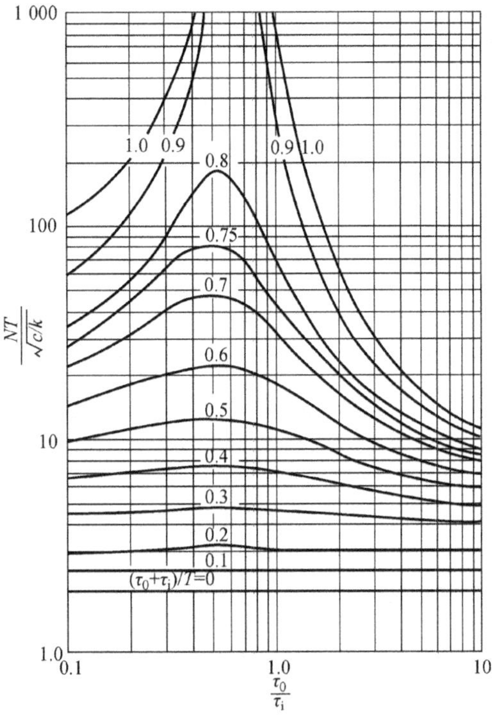

Fig. 10 Critical indicated difference parameter, $TN/\sqrt{c/k}$ versus τ_0/τ_i with $(\tau_0+\tau_i)/T$ as parameter

$$
\begin{aligned}
\frac{TN}{\sqrt{c/k}} = {} & \left(\left[\frac{1}{4}-\left(\frac{t^*}{T}\right)^2\right]+2\left(\frac{\tau_i}{T}\right)\left(\frac{\tau_0}{\tau_i}+1\right)\left(\frac{t^*}{T}-\frac{1}{2}\right)-\frac{(\tau_0/\tau_i)^2(\tau_i/T)}{[(\tau_0/\tau_i)-1]\sinh(\tau_i/\tau_0)(T/2\tau_i)}\times\right. \\
& \left\{\frac{2(\tau_i/T)\tanh(T/2\tau_i)}{[(\tau_0/\tau_i)-1][2(\tau_0/\tau_i)-1]}-1\right\}\left(e^{-[(t^*/T)/(\tau_0/\tau_i)(\tau_i/T)]}-e^{-\{1/[2(\tau_0/\tau_i)(\tau_i/T)]\}}\right)+ \\
& \frac{2(\tau_i/T)}{[(\tau_0/\tau_i)-1]\cosh(T/2\tau_i)}\left\{\left[\frac{t^*}{T}+\frac{(\tau_i/T)}{(\tau_0/\tau_i)-1}\right]e^{-[(t^*/T)/(\tau_i/T)]}-\right. \\
& \left.\left[\frac{1}{2}+\frac{(\tau_i/T)}{(\tau_0/\tau_i)-1}\right]e^{-(T/2\tau_i)}\right\}+ \\
& \left.\frac{(\tau_i/T)^2\left(e^{-[(2t^*/T)/(\tau_i/T)]}-e^{-(T/\tau_i)}\right)}{[2(\tau_0/\tau_i)-1]\cosh^2(T/2\tau_i)}\right)^{-(1/2)}
\end{aligned} \tag{33}
$$

When N is determined, Eq. (20) then gives the hunting loss D.

Figures 10 and 11 are the design charts for peakholding optimalizing control computed from the equations of the preceding analysis. Figure 10 gives $TN/\sqrt{c/k}$ as a function of τ_0/τ_i with $(\tau_0+\tau_i)/T$ as parameter. Figure 11 gives relative hunting loss D/c again as a function of τ_0/τ_i with $(\tau_0+\tau_i)/T$ as parameter. The peaks of curves near $\tau_0/\tau_i=1$ indicate a sort of resonant effect between the input linear group and output linear group. The hunting loss for fixed $(\tau_i+\tau_0)/T$ and c is smaller for τ_0/τ_i away from unity. For fixed τ_i, τ_0, and c, clearly the way to reduce the hunting loss is by increasing the period T.

Concluding Remarks

The present analysis gives the necessary input drive speed N and the hunting loss D for any specified hunting period T, time constants τ_i and τ_0 for the input linear group and the output linear group, and the chosen critical indicated difference c. T and c are fixed by considerations on the noise interference. The analysis shows that whenever the hunting period is relatively short with respect to the time constants τ_i, τ_0, or whenever $(\tau_i+\tau_0)/T$ is relatively large, the hunting loss will be large, especially when τ_i and τ_0 are nearly equal. To avoid such unfavorable condition, the designer should improve his input drive system so as to reduce the constant τ_i. τ_0 is, however, a constant of the intrinsic characteristic of the controlled system, due to, say, the inertia of the moving parts of the system. τ_0 is thus not at the disposal of the designer of the control system. However, suppose there is a compensating circuit between the output y and optimalizing input drive unit (Fig. 3), such that the effects of the output linear group is completely compensated. Then the effective signal for input drive reversal is not the actual output y, but the potential output y^*. In other words, the value of τ_0 is made to be effectively zero. Even if complete compensation is not achieved, the effective value of τ_0 can still be greatly reduced. For difficult cases then, such a compensating unit should certainly be added to reduce the hunting loss. This will be just a minor complication when compared with the additional equipment required for satisfactory noise filtering.

Appendix

Typical Summation Formulas

Re and Im mean, respectively, the "real part of" and the "imaginary part of" the expression following it.

(1) $$\sum_{n=0}^{\infty}\frac{1}{(2n+1)^2}=\frac{\pi^2}{8}$$

(2) $$\sum_{n=0}^{\infty}\frac{1}{(2n+1)^4}=\frac{\pi^4}{96}$$

(3) $$\sum_{n=0}^{\infty}\frac{1}{[1+(2n+1)^2z^2]}=\frac{\pi}{4z}\tanh\frac{\pi^2}{2z}$$

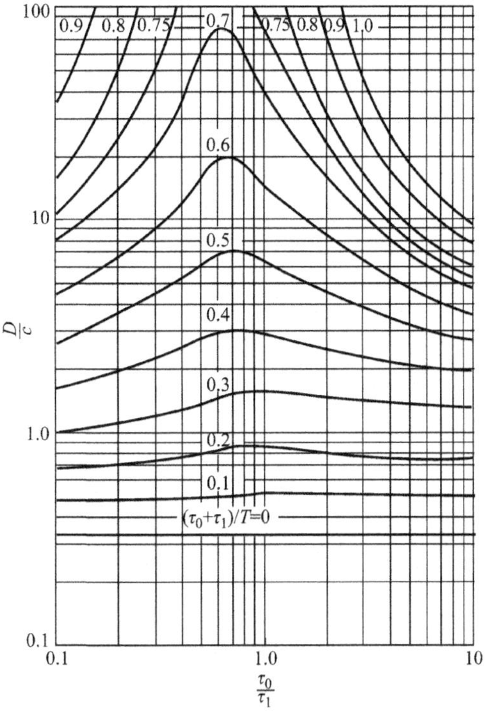

Fig. 11 Relative hunting loss, D/c versus τ_0/τ_i with $(\tau_0+\tau_i)/T$ as parameter

(4) $\mathrm{Re}\sum_{s=1}^{\infty}\frac{(-1)^{s-1}e^{is\omega_0 t}}{1+i2sb}=\frac{1}{2}-\frac{\pi}{4b}\frac{e^{-[(\omega_0 t)/(2b)]}}{\sinh\pi/2b}$ when $-\pi<\omega_0 t<\pi$

$=\frac{1}{2}-\frac{\pi}{4b}\frac{e^{\pi/b}}{\sinh\pi/2b}\times e^{-[(\omega_0 t)/(2b)]}$ when $\pi<\omega_0 t<3\pi$, etc.

(5) $\mathrm{Im}\sum_{s=1}^{\infty}\frac{(-1)^{s-1}e^{is\omega_0 t}}{s}=\frac{\omega_0 t}{2}$ when $-\pi<\omega_0 t<\pi$

$=\frac{\omega_0 t}{2}-\pi$ when $\pi<\omega_0 t<3\pi$, etc.

(6) $\mathrm{Re}\sum_{s=1}^{\infty}\frac{(-1)^{s-1}e^{is\omega_0 t}}{s^2}=\frac{\pi^2}{12}-\left(\frac{\omega_0 t}{2}\right)^2$ when $-\pi<\omega_0 t<\pi$

$=2\pi\left(\frac{\omega_0 t}{2}\right)-\left(\frac{\omega_0 t}{2}\right)^2-\frac{11}{12}\pi^2$ when $\pi<\omega_0 t<3\pi$, etc.

(7) $\mathrm{Re}\sum_{s=1}^{\infty}\frac{(-1)^{s-1}e^{is\omega_0 t}}{(1+i2sa)^2}=\frac{1}{2}-\frac{\pi}{4a^2}\frac{e^{-[(\omega_0 t)/(2a)]}}{\sinh^2(\pi/2a)}\times\left(\frac{\omega_0 t}{2}\sinh\frac{\pi}{2a}+\frac{\pi}{2}\cosh\frac{\pi}{2a}\right)$ when $-\pi<\omega_0 t<\pi$

$=\frac{1}{2}-\frac{\pi}{4a^2}\frac{e^{\pi/a}e^{-[(\omega_0 t)/(2a)]}}{\sinh^2(\pi/2a)}\times\left[\left(\frac{\omega_0 t}{2}-\pi\right)\sinh\frac{\pi}{2a}+\frac{\pi}{2}\cosh\frac{\pi}{2a}\right]$ when $\pi<\omega_0 t<3\pi$, etc.

References

[1] Draper C S., Li Y T, Principles of Optimalizing Control Systems and an Application to the Internal Combustion Engine. American Society of Mechanical Engineers Publication, September, 1951.

[2] Li Y T. Optimalizing System for Process Control Instruments. 1952, 25. 72 - 77 , 1952: 190 - 193, 228, 1952: 324 - 327, 350 - 352.

[3] Shull R, Jr. An Automatic Cruise Control Computer for Long Range Aircraft. Trans. I. R. E., Professional Group on Electronic Computers, 1952:47 - 51.

The Poincaré-Lighthill-Kuo Method

H. S. Tsien

(Daniel and Florence Guggenheim Jet Propulsion Center, California Institute of Technology, Pasadena, California)

Contents

Advances in Applied Mechanics, vol. 4, Academic Press, pp. 281–349, 1955.

Ⅰ. Introduction

1. Historical Development

In his famous book, "Les méthodes nouvelles de la mécanique céleste"[1], Poincaré devised a method for finding the periodic solution of a system of first order equations,

$$\frac{dx_i}{dt} = X_i(x_1, x_2, \dots, x_i, \dots, x_n; \varepsilon) \quad (i = 1, 2, \dots, n) \tag{1.1}$$

where t is the time variable, and ε is a small parameter representing the perturbation influences. The equations with $\varepsilon = 0$, corresponding to the unperturbed system, are particularly simple, and a periodic solution with period $T^{(0)}$ can be easily found. The essence of Poincaré's method is the expansion of the perturbed solution in the parameter ε. Not only the variables

$$x_i = x_i^{(0)} + \varepsilon x_i^{(1)} + \varepsilon^2 x_i^{(2)} + \dots \tag{1.2a}$$

are expanded, but also the period T:

$$T = T^{(0)} + \varepsilon T^{(1)} + \varepsilon^2 T^{(2)} + \dots \tag{1.2b}$$

In recent years, this method has found many applications in the theory of nonlinear oscillations (nonlinear mechanics) where the same equations (1.1) prevail. However, for nearly sixty years no extension of the principle of this method was made, and the full potentiality of Poincaré's invention remained unexploited.

On May 19, 1949, Lighthill[2] spoke before the London Mathematical Society about a technique for rendering approximate solutions to physical problems uniformly valid and introduced a very important extension of Poincaré's method. Lighthill's objective was to improve the well-known method of perturbation for calculating the approximate solution of a physical problem. The perturbation method is based upon the concept of expanding the exact solution in a power series of the small parameter ε, the zeroth order solution being independent of ε, the first order solution proportional to ε, etc. This method, elementary in principle and straightforward in execution, is very effective and yields useful results for a large class of problems. Nevertheless there are problems, not at all infrequent, where the zeroth order solution contains a singularity at a point or on a line within the domain of interest. Then not only will the singularity again appear at the same location in the higher order solutions, but it will become progressively more severe as the order of the solution increases. The power series expansion in ε breaks down near such singularities, and the classical perturbation fails to give a usable solution near the singular points.

Lighthill's method is designed to eliminate such difficulties and to render the expansion uniformly valid, or of uniform accuracy, over the whole domain of interest. The principle is to expand not only the dependent variable u, but also the independent variables x and y, say, in power series of ε. Then

$$u = u^{(0)}(\xi, \eta) + \varepsilon u^{(1)}(\xi, \eta) + \varepsilon^2 u^{(2)}(\xi, \eta) + \ldots \tag{1.3}$$

$$\begin{cases} x = \xi + \varepsilon x^{(1)}(\xi, \eta) + \varepsilon^2 x^{(2)}(\xi, \eta) + \ldots \\ y = \eta + \varepsilon y^{(1)}(\xi, \eta) + \varepsilon^2 y^{(2)}(\xi, \eta) + \ldots \end{cases} \tag{1.4}$$

where ξ, η take the place of the original independent variables x, y. Evidently $u^{(0)}(\xi, \eta)$ is simply the zeroth order solution of the classical perturbation method with ξ, η replacing x, y. If we neglect the higher order terms in u of (1.3), then the approximate solution is simply the zeroth order perturbation solution with the coordinates stretched or distorted by the transformation (1.4). This fact has led several authors to call Lighthill's method the method of coordinate perturbation.

Lighthill applied his method to problems involving partial differential equations when the zeroth order solution is obtained from a reduced linear equation of equal order as the exact equation. It soon becomes apparent, however, that Lighthill's original purpose of uniform validity throughout the domain of interest cannot always be realized. In many problems a good zeroth order approximation can be obtained only if a "boundary layer" solution is used. Kuo[3] first recognized this necessity in his elegant solution of the problem of the laminar incompressible boundary layer on a flat plate. This and his later work[4] on the supersonic laminar boundary layer constitute a further extension of Poincaré's original concept.

The elements of the method of Poincaré, Lighthill, and Kuo are no doubt used by many workers in applied mathematics, but the generality of the concept was perhaps never sufficiently emphasized. Thus, if we wish to recognize the importance of originality and bold exploration, we may give the credit for developing this very powerful method of engineering mathematics to the three authors cited and call it the PLK method.

2. Simple Example

To illustrate the principle of the PLK method, let us consider the following first order ordinary differential equation:

$$(x + \varepsilon u)\frac{\mathrm{d}u}{\mathrm{d}x} + u = 0 \tag{1.5}$$

Dividing the equation by $\mathrm{d}u/\mathrm{d}x$, we interchange the roles of dependent and independent variables and obtain

$$u\frac{\mathrm{d}x}{\mathrm{d}u} + x = -\varepsilon u$$

or

$$\frac{\mathrm{d}}{\mathrm{d}u}(xu) = -\varepsilon u \tag{1.6}$$

and obtain by integration

$$xu = -\frac{\varepsilon}{2}u^2 + C_0$$

where C_0 is an integration constant. If we impose the boundary condition

$$u(1) = 1 \tag{1.7}$$

then the exact solution of the differential equation (1.5) is

$$u = -\frac{x}{\varepsilon} + \sqrt{\left(\frac{x}{\varepsilon}\right)^2 + \frac{2}{\varepsilon} + 1} \tag{1.8}$$

Let us now try the classical perturbation method for the equation (1.5), i.e., expand u in powers of ε:

$$u(x) = u^{(0)}(x) + \varepsilon u^{(1)}(x) + \varepsilon^2 u^{(2)}(x) + \ldots \tag{1.9}$$

Substituting (1.9) into (1.5), and then equating equal powers of ε, we have

$$x\frac{\mathrm{d}u^{(0)}}{\mathrm{d}x} + u^{(0)} = 0 \tag{1.10}$$

$$x\frac{\mathrm{d}u^{(1)}}{\mathrm{d}x} + u^{(1)} = -u^{(0)}\frac{\mathrm{d}u^{(0)}}{\mathrm{d}x} \tag{1.11}$$

$$x\frac{\mathrm{d}u^{(2)}}{\mathrm{d}x} + u^{(2)} = -u^{(0)}\frac{\mathrm{d}u^{(1)}}{\mathrm{d}x} - u^{(1)}\frac{\mathrm{d}u^{(0)}}{\mathrm{d}x} \tag{1.12}$$

......

If we make $u^{(0)}(x)$ to satisfy the boundary condition (1.7), then

$$u^{(0)}(x) = \frac{1}{x} \tag{1.13}$$

With $u^{(0)}$ so determined, (1.12) gives

$$u^{(1)}(x) = -\frac{1}{2}\frac{1}{x^3} + \frac{C_1}{x}$$

But now the boundary condition (1.7) requires

$$u^{(1)}(1) = 0 \tag{1.14}$$

Hence the integration constant C_1 can be determined, and

$$u^{(1)}(x) = \frac{1}{2x}\left(1 - \frac{1}{x^2}\right) \tag{1.15}$$

Similarly we have

$$u^{(2)}(x) = \frac{1}{2x}\left(1 - \frac{1}{x^2}\right) - \frac{1}{2x}\left(1 - \frac{1}{x^4}\right) \tag{1.16}$$

The function $u^{(0)}(x)$ has a singularity at $x = 0$, and (1.15) and (1.16) show that this singularity becomes worse as the order of the perturbation solution is increased. The solution so obtained is thus worthless near $x = 0$. In fact, away from the singular point $x = 0$, the fractional error of the solution can be estimated simply by the order of ε neglected. Thus if, as we have done, we calculate up to the ε^2-term, then the fractional error is $O(\varepsilon^3)$. But near the singular point this estimate breaks down, and the error is very much larger than $O(\varepsilon^3)$. The solution obtained has nonuniform accuracy in the region of interest near $x = 0$, i.e., the solution is not *uniformly valid*. Now let us try a different procedure, and expand both u and x in powers of ε as required by the PLK method:

$$\left.\begin{aligned} u &= u^{(0)}(\xi) + \varepsilon u^{(1)}(\xi) + \dots \\ x &= \xi + \varepsilon x^{(1)}(\xi) + \dots \end{aligned}\right\} \tag{1.17}$$

The original differential equation (1.5) can now be written as

$$(x + \varepsilon u)\frac{du}{d\xi} + u\frac{dx}{d\xi} = 0 \tag{1.18}$$

By substituting (1.17) into (1.18) and equating equal powers of ε, we obtain

$$\xi\frac{du^{(0)}}{d\xi} + u^{(0)} = 0 \tag{1.19}$$

$$\xi\frac{du^{(1)}}{d\xi} + u^{(1)} = -(x^{(1)} + u^{(0)})\frac{du^{(0)}}{d\xi} - u^{(0)}\frac{dx^{(1)}}{d\xi} \tag{1.20}$$

Now (1.19) gives simply

$$u^{(0)}(\xi) = \frac{k_0}{\xi}$$

If we impose the condition

$$x^{(1)}(1) = 0 \tag{1.21}$$

such that $x = 1$ for $\xi = 1$, then $k_0 = 1$ is required by the boundary condition (1.7) for $u^{(0)}$. Thus

$$u^{(0)}(\xi) = \frac{1}{\xi} \tag{1.22}$$

With this solution for $u^{(0)}$, (1.20) becomes

$$\frac{d}{d\xi}(\xi u^{(1)}) = -\frac{1}{\xi}\frac{dx^{(1)}}{d\xi} + \frac{1}{\xi^2}x^{(1)} + \frac{1}{\xi^3}$$

In order to avoid having a $u^{(1)}$ with higher order singularity than $u^{(0)}$, we take advantage of the additional freedom in the choice of $x^{(1)}$ by setting

$$\frac{1}{\xi}\frac{dx^{(1)}}{d\xi} - \frac{1}{\xi^2}x^{(1)} = \frac{1}{\xi^3} \tag{1.23}$$

This is then the differential equation for $x^{(1)}(\xi)$. The solution with the condition (1.21) is

$$x^{(1)} = \frac{\xi}{2}\left(1 - \frac{1}{\xi^2}\right) \tag{1.24}$$

With $x^{(1)}$ so determined, $u^{(1)} \equiv 0$ in order to satisfy the original boundary condition in u. Then, up to this order of approximation, we have

$$\left.\begin{aligned} u &= \frac{1}{\xi} \\ x &= \xi + \varepsilon\frac{\xi}{2}\left(1 - \frac{1}{\xi^2}\right) \end{aligned}\right\} \tag{1.25}$$

Now the amazing fact is that by eliminating ξ from the pair (1.25), we have exactly the solution for u as given by (1.8). Therefore in this case the PLK method not only removes the difficulty of the singularity at $x = 0$, but yields a solution which is so good as to be, in fact, the exact solution.

3. Essential Features of the PLK Method

We see that the appropriate expansion of the exact solution of (1.5) in ε is different for different regions of x. The expansion (1.9) is only appropriate for large values of x away from the origin. Near the origin, an entirely different expansion is valid. Such changes in the character of the solution make the conventional perturbation method almost useless. In fact, to use the conventional perturbation method in such cases would require the most ingenious guessing. The PLK method, on the other hand, is a foolproof method treating all problems on a uniform basis; the intricacies of the problem will appear automatically and correctly without the need of foresight on the part of the investigator. In this respect, the PLK method is not unlike the method of the Laplace transform. For engineers then, this characteristic of the PLK method is of particular importance. We shall appreciate this point better when the examples in the subsequent chapters are discussed.

Another feature of the PLK method is the great flexibility in application. Although Lighthill originally emphasized the uniform validity of the solution, this cannot always be achieved, as will be seen in the following discussion. The point is that by introducing "stretched" coordinates we gain additional freedom in the process of solution, which is used to increase the accuracy of the zeroth order solution, originally poor near the singular point or singular line. How far we can succeed depends upon the problem. Unfortunately at the present time, the mathematical theory of the PLK method has not been sufficiently investigated to allow one to predict the degree of success of failure to be expected from the given differential equation and auxiliary conditions. But from what has already been achieved it seems certain that the PLK method will be a useful tool in engineering analysis, even if it may not yield results that are as good as can be desired.

With these points in mind, our exposition of the PLK method will be oriented towards applications rather than towards the mathematical foundations of the method. In fact, the mathematical justification of the method has as yet been examined in but a few cases. In addition, such mathematical analysis requires a codification of the method, while, as mentioned before, the important feature of the PLK method is the flexibility of approach. From the point of view of application, it is better to discuss the method by examples rather than by a general theory. This will be the spirit of the following exposition.

The author wishes to take this opportunity to acknowledge the instructive discussions he has had with and critical comments he received from his colleagues at the California Institute of Technology, particularly from Professor A. Erdélyi. The author's interest in this new mathematical method first arose from the imaginative work of Professor Y. H. Kuo at Cornell University. Professor C. C. Lin at the Massachusetts Institute of Technology has, however, cautioned the author about the limitations of the method. To all of them the author's thanks are due.

Ⅱ. Ordinary Differential Equations

1. Equations of First Order

To begin our discussion of the PLK method, let us study first its application to ordinary differential equations. Even for this relatively simple case, a complete mathematical proof of the convergence of the solution has only recently been supplied by Wasow[5], and this only for a very simple type of ordinary differential equation of the first order, i.e.,

$$(x+\varepsilon u)\frac{\mathrm{d}u}{\mathrm{d}x}+q(x)u=r(x) \tag{2.1}$$

where $q(x)$ and $r(x)$ are assumed to be *regular* near the origin $x = 0$. We are usually interested in obtaining a solution for $x \geqslant 0$. This equation was first investigated in the present context by Lighthill himself[2]. In the following discussion we shall follow Lighthill in giving only heuristic arguments. For mathematical proofs the reader is referred to Wasow's work.

Equation (2.1) serves as the typical case because the classical perturbation method fails to give a usable solution near $x = 0$. The difficulty here is of course that, on the line $x+\varepsilon u = 0$ in the u, x-plane, the differential equation is singular. We shall assume that ε is positive. (If ε is negative, we can reduce the equation to that of positive ε by taking $-u$ as the dependent variable). Then the line of singularity is that indicated in Fig. 1. On this line, since $-q(x)u+r(x)$ is generally not zero, $\mathrm{d}u/\mathrm{d}x$ according to (2.1) is infinitely large. If we use the classical perturbation method, we drop the term εu in the coefficient of $\mathrm{d}u/\mathrm{d}x$ for the zeroth approximation, and the singularity is moved to the u-axis. On the u-axis, the classical perturbation theory will give infinite slope of $u(x)$. Thus if u^* is the result of the classical perturbation theory, u^* must differ greatly from u near $x = 0$, as shown in Fig. 1, and this difference cannot be improved by higher order perturbations.

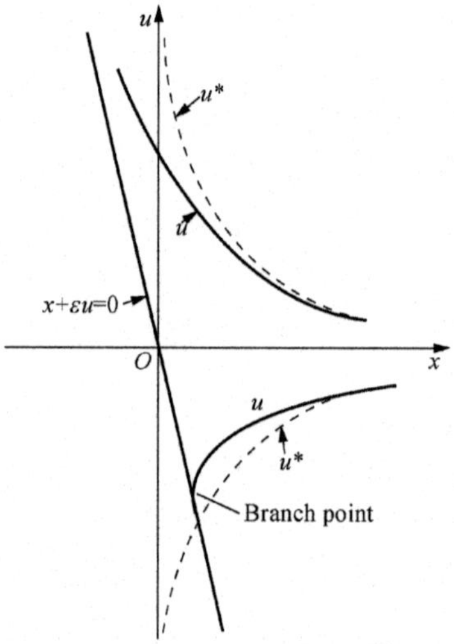

Fig. 1 Representation of solutions of $(x+\varepsilon u)\dfrac{\mathrm{d}u}{\mathrm{d}x}+q(x)u = r(x)$ in the x, u plane

Now let us see what can be done if we use the PLK method and put

$$\left.\begin{aligned} u &= u^{(0)}(\xi)+\varepsilon u^{(1)}(\xi)+\varepsilon^2 u^{(2)}(\xi)+\ldots \\ x &= \xi+\varepsilon x^{(1)}(\xi)+\varepsilon^2 x^{(2)}(\xi)+\ldots \end{aligned}\right\} \tag{2.2}$$

By substituting (2.2) into (2.1) and expanding every term in powers of ε, we have

$$\begin{aligned} &(\xi+\varepsilon x^{(1)}+\varepsilon^2 x^{(2)}+\ldots+\varepsilon u^{(0)}+\varepsilon^2 u^{(1)}+\ldots)(u^{(0)\prime}+\varepsilon u^{(1)\prime}+\varepsilon^2 u^{(2)\prime}+\ldots) \\ &= (1+\varepsilon x^{(1)\prime}+\varepsilon^2 x^{(2)\prime}+\ldots)\Big[r+\varepsilon x^{(1)}r'+\varepsilon^2 x^{(2)}r'+\frac{1}{2}\varepsilon^2 x^{(1)2}r''+\ldots- \\ &\left(q+\varepsilon x^{(1)}q'+\varepsilon^2 x^{(2)}q'+\frac{1}{2}\varepsilon^2 x^{(1)2}q''+\ldots\right)(u^{(0)}+\varepsilon u^{(1)}+\varepsilon^2 u^{(2)}+\ldots)\Big] \end{aligned} \tag{2.3}$$

where the prime indicates differentiation with respect to ξ, and all quantities are functions of ξ. [For instance, $q = q(\xi)$.] By equating the terms in (2.3) which are independent of ε we obtain the zeroth order equation

$$\xi u^{(0)\prime}(\xi)+q(\xi)u^{(0)}(\xi) = r(\xi) \tag{2.4}$$

The ratio of the coefficients of $u^{(0)}$ and $u^{(0)\prime}$ as $\xi\to 0$ is thus $O(1/\xi)$. Therefore, according to the theory of differential equations, $\xi = 0$ is a *regular singularity* of the differential equation. The solution of (2.4) is

$$u^{(0)}(\xi) = \exp\left\{-\int \frac{q\mathrm{d}\xi}{\xi}\right\}\left[\int \frac{r}{\xi}\left(\exp\int \frac{q\mathrm{d}\xi}{\xi}\right)\mathrm{d}\xi + C\right] \tag{2.5}$$

where C is an integration constant. Now since $q(\xi)$ is regular near $\xi = 0$, the value of q at $\xi = 0$ is finite and equal, say, to q_0. Then

$$\exp\int \frac{q\mathrm{d}\xi}{\xi} = \xi^{q_0} R(\xi) \tag{2.6}$$

where $R(\xi)$ means a function regular at $\xi = 0$. But $r(\xi)$ is also regular near $\xi = 0$, hence the first term within the square bracket on the right-hand side of (2.5) is $\xi^{q_0} R(\xi)$. Thus it is seen that, as $\xi \to 0$, we have

$$u^{(0)}(\xi) = R(\xi) + O(\xi^{-q_0}) \tag{2.7}$$

where the second term to the right indicates the order of magnitude of the singular part of the solution. This conclusion will be somewhat modified if q_0 is a nonpositive integer. In such cases, the first term within the square bracket on the right-hand side of (2.5) will contain a term $\log \xi$. Then

$$u^{(0)}(\xi) = R(\xi) + O(\xi^{-q_0}\log \xi), \quad q_0 = 0, -1, -2, \ldots \tag{2.8}$$

The coefficients of ε in (2.3) lead to the following first order equation

$$\frac{\mathrm{d}}{\mathrm{d}\xi}\left[u^{(1)}(\xi)\exp\int \frac{q\mathrm{d}\xi}{\xi}\right] = \frac{1}{\xi}[(r - qu^{(0)})x^{(1)\prime} +$$

$$(r' - q'u^{(0)} - u^{(0)\prime})x^{(1)} - u^{(0)}u^{(0)\prime}]\exp\int \frac{q\mathrm{d}\xi}{\xi} \tag{2.9}$$

If the classical perturbation method were followed, then $x^{(1)} \equiv 0$, and as $\xi \to 0$ (2.9) would give, by (2.7),

$$u^{(1)}(\xi) = O(\xi^{-q_0}) + O(u^{(0)}u^{(0)\prime}) = R(\xi) + O(\xi^{-q_0-1}) + O(\xi^{-2q_0-1}) \tag{2.10}$$

Again, when q_0 is a nonpositive integer, this conclusion should be modified by using (2.8),

$$u^{(1)}(\xi) = R(\xi) + O(\xi^{-q_0-1}\log \xi), \quad q_0 = 0, -1, -2, \ldots \tag{2.11}$$

We shall now see that the effects of these singularities are bad indeed. Take first the case of positive q_0. Then the worst singularity in $u^{(1)}$, according to (2.10), is ξ^{-2q_0-1}; and near the origin $\xi = x = 0$, the ratio of the second order term $\varepsilon u^{(1)}$ to the first order term $u^{(0)}$ is $\varepsilon\xi^{-2q_0-1}/\xi^{-q_0} = \varepsilon\xi^{-q_0-1}$. For small but finite ε, this ratio will diverge to infinity as $\xi \to 0$. This state of affairs will persist for higher order perturbations, and the series for u will diverge for given ε when $x = \xi$ is small enough. For nonpositive q_0, the term $O(\xi^{-q_0-1})$ in (2.10) is worse than the term $O(\xi^{-2q_0-1})$. Then either (2.10) or (2.11) will show that the ratio of $\varepsilon u^{(1)}$ to $u^{(0)}$ is $\varepsilon\xi^{-1}$. Higher order perturbations show similar behavior, and the perturbation series for u again diverges near the origin.

The point of the PLK method is of course to utilize the additional freedom of $x^{(1)}$ to control the singularity of $u^{(1)}$. We see that it is only necessary to make the square bracket on the right-hand side of (2.9) equal to $R(\xi) + O(\xi^{-q_0+1})$. This can evidently be done by a proper choice of $x^{(1)}$. In fact, we can even make the factor vanish as was indeed done in the simple example given in Section Ⅰ.2. However, it is sufficient to use only a finite number of terms of the expansion of $x^{(1)}$ in ascending powers of ξ. The cases $q_0 < 0$ and $q_0 > 0$ are somewhat different. Let us treat the case of positive q_0 first.

2. Case of $q_0 > 0$

This is the case represented in Fig. 1. As already shown by (2.7), the singular term of $u^{(0)}(\xi)$ is $A\xi^{-q_0}$, where A is a constant determined by the boundary condition of the solution. Therefore the most damaging terms in the square bracket to the right of (2.9) are

$$-q_0 A\xi^{-q_0} x^{(1)\prime} + q_0 A\xi^{-q_0-1} x^{(1)} + A^2 q_0 \xi^{-2q_0-1}$$

This group of terms can be eliminated if we require that, for small ξ,

$$x^{(1)} \sim -A\xi^{-q_0}/(q_0+1) \tag{2.12}$$

When this is done, $u^{(1)}(\xi) = O(\xi^{-2q_0})$ for small ξ. Analysis of higher order perturbations[2] produces similar results, and we have for $\xi \to 0$

$$\left.\begin{aligned} u &= A\xi^{-q_0} + \varepsilon O(\xi^{-2q_0}) + \varepsilon^2 O(\xi^{-3q_0}) + \varepsilon^3 O(\xi^{-4q_0}) + \ldots \\ x &= \xi + \varepsilon\left(-\frac{A\xi^{-q_0}}{q_0+1} + \ldots\right) + \varepsilon^2 O(\xi^{-2q_0}) + \varepsilon^3 O(\xi^{-3q_0}) + \ldots \end{aligned}\right\} \tag{2.13}$$

Let us take for the moment $A > 0$, the type of solution shown in Fig. 1. Then, according to (2.13), the point $x = 0$, the trouble point of the classical perturbation theory, corresponds to $\xi \cong \varepsilon^{1/(q_0+1)}$. Therefore the successive terms of the series for u and x near $x = 0$ have the ratio $\varepsilon^{1/(q_0+1)}$. Thus the series for both u and x are convergent near $x = 0$ if ε is small enough, and the difficulties of the classical perturbation method are now resolved. This is achieved by controlling the singularities in the $u^{(j)}(\xi)$'s and the $x^{(j)}(\xi)$'s in such a way that they have similar singularities for each j, as shown by (2.13).

Of course when $A < 0$, also shown in Fig. 1, the solution has a branch point before $x = 0$. This can be determined by setting $x + \varepsilon u = 0$. To a first approximation, then,

$$\xi - \varepsilon\frac{A\xi^{-q_0}}{q_0+1} + \varepsilon A\xi^{-q_0} \cong 0$$

In the case of a branch point, therefore,

$$\xi \cong \left(-\frac{\varepsilon A q_0}{q_0+1}\right)^{1/(q_0+1)}, \quad x \cong \left(1+\frac{1}{q_0}\right)\left(-\frac{\varepsilon A q_0}{q_0+1}\right)^{1/(q_0+1)} \tag{2.14}$$

By the argument given in the preceding paragraph, the solution by the PLK method is certainly good up to the branch point. When $A > 0$, there is no real branch point.

To see the details of the method, let us consider the following equation studied by Lighthill[2]:

$$(x+\varepsilon u)\frac{\mathrm{d}u}{\mathrm{d}x}+(2+x)u=0 \tag{2.15}$$

with the boundary condition

$$u=\mathrm{e}^{-1} \quad \text{at} \quad x=1 \tag{2.16}$$

Let ξ_1 be the value of ξ for $x=1$. ξ_1 is thus determined by

$$1=\xi_1+\varepsilon x^{(1)}(\xi_1)+\varepsilon^2 x^{(2)}(\xi_1)+\ldots$$

or

$$\xi_1=1-\varepsilon x^{(1)}(\xi_1)-\varepsilon^2 x^{(2)}(\xi_1)+\ldots$$

Now we can substitute in the functions $x^{(1)}(\xi_1)$ and $x^{(2)}(\xi_1)$ the value of ξ_1, given by the above expression, and then expand the result in powers of ε. Therefore we have

$$\begin{aligned}\xi_1&=1-\varepsilon[x^{(1)}(1)-\varepsilon x^{(1)}(1)x^{(1)\prime}(1)]-\varepsilon^2 x^{(2)}(1)+\ldots\\&=1-\varepsilon x^{(1)}(1)-\varepsilon^2[x^{(2)}(1)-x^{(1)}(1)x^{(1)\prime}(1)]+\ldots\end{aligned}$$

The boundary condition (2.16) can now be written as

$$\mathrm{e}^{-1}=u^{(0)}(1)-u^{(0)\prime}(1)\varepsilon x^{(1)}(1)+\varepsilon u^{(1)}(1)+\ldots$$

or

$$\left.\begin{aligned}&u^{(0)}(1)=\mathrm{e}^{-1}\\&u^{(1)}(1)=u^{(0)\prime}(1)x^{(1)}(1)\\&\cdots\cdots\end{aligned}\right\} \tag{2.17}$$

Equation (2.17) is the converted boundary conditions in ξ.

Now the zeroth order equation is

$$\xi\frac{\mathrm{d}u^{(0)}}{\mathrm{d}\xi}+(2+\xi)u^{(0)}=0$$

Thus with the boundary condition (2.17) we have

$$u^{(0)}(\xi)=\mathrm{e}^{-\xi}\xi^{-2} \tag{2.18}$$

According to (2.9), the first order equation is

$$\begin{aligned}\frac{\mathrm{d}}{\mathrm{d}\xi}[u^{(1)}(\xi)\mathrm{e}^{\xi}\xi^2]&=\frac{1}{\xi}[-(2+\xi)\mathrm{e}^{-\xi}\xi^{-2}x^{(1)\prime}-\{\mathrm{e}^{-\xi}\xi^{-2}-\mathrm{e}^{-\xi}\xi^{-2}-2\mathrm{e}^{-\xi}\xi^{3}\}x^{(1)}-\\&\qquad \mathrm{e}^{-\xi}\xi^{-2}\{-\mathrm{e}^{-\xi}\xi^{-2}-2\mathrm{e}^{-\xi}\xi^{-3}\}]\mathrm{e}^{\xi}\xi^2\\&=\frac{1}{\xi}\left[-(2+\xi)x^{(1)\prime}+\frac{2}{\xi}x^{(1)}+\mathrm{e}^{-\xi}\xi^{-2}\left(1+\frac{2}{\xi}\right)\right]\end{aligned}$$

It is seen that the worst singularity in $u^{(1)}$ will be eliminated if we put

$$x^{(1)\prime}-\frac{1}{\xi}x^{(1)}=\frac{1}{\xi^3}$$

or

$$x^{(1)}(\xi)=-\frac{1}{3\xi^2} \tag{2.19}$$

With this $x^{(1)}$, the equation for $u^{(1)}$ is reduced to

$$\begin{aligned}\frac{\mathrm{d}}{\mathrm{d}\xi}[u^{(1)}(\xi)\mathrm{e}^{\xi}\xi^2]&=\frac{1}{\xi}\left[-\xi x^{(1)\prime}-\frac{2}{\xi^3}+\mathrm{e}^{-\xi}\xi^{-2}\left(1+\frac{2}{\xi}\right)\right]\\&=-\frac{2}{3}\frac{1}{\xi^3}-\frac{2}{\xi^4}+\mathrm{e}^{-\xi}\left(\frac{1}{\xi^3}+\frac{2}{\xi^4}\right)\end{aligned}$$

To satisfy the boundary condition (2.17), we have then

$$u^{(1)}(\xi)=\mathrm{e}^{-\xi}\xi^{-2}\left[\frac{2}{3}\frac{1}{\xi^3}+\frac{1}{3}\frac{1}{\xi^2}-\int_\xi^1\mathrm{e}^{-\xi}\left(\frac{2}{\xi^4}+\frac{1}{\xi^3}\right)\mathrm{d}\xi\right] \tag{2.20}$$

The second order equation is

$$\begin{aligned}\xi u^{(2)\prime}+(2+\xi)u^{(2)}=&-(x^{(1)}+u^{(0)})u^{(1)\prime}-(x^{(2)}+u^{(1)})u^{(0)\prime}-x^{(2)}u^{(0)}-x^{(1)}u^{(1)}-\\&x^{(1)\prime}\{(2+\xi)u^{(1)}+x^{(1)}u^{(0)}\}-x^{(2)\prime}(2+\xi)u^{(0)}\end{aligned}$$

To eliminate the worst singularity in $u^{(2)}$, we set

$$\xi\frac{\mathrm{d}}{\mathrm{d}\xi}\left(\frac{x^{(2)}}{\xi}\right)=\frac{3}{2}\frac{1}{\xi^5}\quad\text{or}\quad x^{(2)}(\xi)=-\frac{3}{10}\frac{1}{\xi^4} \tag{2.21}$$

Collecting the results so far obtained, we have

$$\left.\begin{aligned}u&=\mathrm{e}^{-\xi}\xi^{-2}+\varepsilon\left[\mathrm{e}^{-\xi}\xi^{-2}\left\{\frac{2}{3\xi^3}+\frac{1}{3\xi^2}-\int_\xi^1\mathrm{e}^{-\xi}\left(\frac{2}{\xi^4}+\frac{1}{\xi^3}\right)\mathrm{d}\xi\right\}\right]+O\left(\frac{\varepsilon^2}{\xi^6}\right)\\x&=\xi-\frac{\varepsilon}{3\xi^2}-\frac{3\varepsilon^2}{10\xi^4}+O\left(\frac{\varepsilon^2}{\xi^6}\right)\end{aligned}\right\} \tag{2.22}$$

Now a straightforward calculation will show that at $x=0$

$$\xi=\left(\frac{\varepsilon}{3}\right)^{1/3}+\frac{9}{10}\left(\frac{\varepsilon}{3}\right)^{2/3}+O(\varepsilon)\quad\text{and}\quad u=\left(\frac{3}{\varepsilon}\right)^{2/3}-27\left(\frac{3}{\varepsilon}\right)^{1/3}+O(1) \tag{2.23}$$

For the present equation $u^{(0)}\sim 1/\xi^2$ as $\xi\to 0$ according to (2.18); A is thus positive and no real branch point appears.

In some cases, the first few $x^{(j)}(\xi)$'s can be set identically zero without impairing in any way the validity of the solution: For instance, if $A=\lim\limits_{\xi\to 0}\xi^{q_0}u^{(0)}(\xi)$ happens to be zero due to the boundary condition of the problem, then the worst possible singularity in $u^{(1)}$ is automatically eliminated, and there is no need to introduce $x^{(1)}$. In general if, for $j<i$, $\lim\limits_{\xi\to 0}\xi^{q_0}u^{(j)}(\xi)=0$,

then the first nonvanishing $x^{(j)}(\xi)$ will be $x^{(i)}(\xi)$.

3. Case of $q_0 = 0$

In this case $u^{(0)}(\xi)$ has a logarithmic singularity, as $\xi \to 0$:

$$u^{(0)}(\xi) = r_0 \log \xi + B + O(\xi \log \xi) \tag{2.24}$$

By setting the most serious terms in the square bracket on the right-hand side of (2.9) equal to zero, we have

$$x^{(1)\prime} - \frac{1}{\xi} x^{(1)} = \xi \frac{\mathrm{d}}{\mathrm{d}\xi}\left(\frac{x^{(1)}}{\xi}\right) = \frac{1}{\xi}[r_0 \log \xi + B]$$

Thus

$$x^{(1)}(\xi) = - r_0 \log \xi - (r_0 + B) \tag{2.25}$$

In general, we find when $\xi \to 0$ (see[2])

$$u^{(j)} = O(\log^{2j+1} \xi) \quad \text{and} \quad x^{(j)} = O(\log^{2j-1} \xi) \tag{2.26}$$

The ratio of successive terms in both series for u and x is thus equal to $\varepsilon \log^2 \xi$. But (2.25) shows that for a first approximation

$$x \cong \xi - \varepsilon [r_0 \log \xi + r_0 + B]$$

Thus $x = 0$ corresponds to $\xi \cong \varepsilon \log \varepsilon$. Hence the radius of convergence of the series given by the PLK method for this case is roughly $\varepsilon^{-1} \log^{-2} \varepsilon$ and is thus very much larger than ε for small ε. The branch point of the solution occurs for $x + \varepsilon u = 0$. This occurs when $\xi = \varepsilon r_0 + O(\varepsilon^2 \log^2 \varepsilon)$ or

$$x = - \varepsilon r_0 \log(\varepsilon r_0) - \varepsilon B + O(\varepsilon^3 \log^3 \varepsilon) \tag{2.27}$$

which is real and positive for small ε, if $r_0 > 0$. For $r_0 < 0$, no real branch point occurs.

4. Case of $q_0 < 0$

When q_0 is negative, the zeroth order solution $u^{(0)}$ is actually finite at $\xi = 0$. We may be led to think that this case is not so bad as the case $q_0 \geqslant 0$ studied in the previous sections. However, it turns out that this is quite deceptive, and the PLK method need not give a convergent solution down to the origin $x = 0$. Let us discuss the simpler case of $q_0 \leqslant -1$ first. We have already shown in Section Ⅱ.1 that in the classical perturbation theory the troublesome term in $u^{(1)}$ is $O(\xi^{-q_0-1})$ in (2.10), evolving from $u^{(0)}(0)u^{(0)\prime}$ in the square bracket on the right-hand side of (2.9). This term can be eliminated by taking $x^{(1)}$ to be a constant, different from zero and determined by

$$- x^{(1)} - u^{(0)}(0) = 0 \tag{2.28}$$

In fact all subsequent $x^{(j)}$ can be taken to be constant. Thus x differs from ξ by a constant depending on ε. To show this, it will be convenient to start with a transformation

$$x = \xi + \alpha, \quad u = v + \beta \tag{2.29}$$

where α and β are constants to be determined presently. The equation in v and ξ is

$$\{\xi + \varepsilon v + (\alpha + \varepsilon\beta)\}\frac{dv}{d\xi} = -q(\xi + \alpha)(v + \beta) + r(\xi + \alpha)$$

We shall now fix α and β by requiring that the coefficient of $dv/d\xi$ as well as the right-hand side vanish for $\xi = v = 0$. The point $\xi = v = 0$ is thus actually a nodal point. Therefore

$$\alpha + \varepsilon\beta = 0, \quad q(\alpha)\beta = r(\alpha) \tag{2.30}$$

This set of equations determines α and β as functions of ε. Moerover, because q and r are assumed to be regular, α and β can be expanded in powers of ε.

Now the transformed equation is of the type

$$(\xi + \varepsilon v)\frac{dv}{d\xi} + q(\xi)v = r(\xi), \quad r_0 = r(0) = 0 \tag{2.31}$$

Let $q_0 = q(0) \leqslant -1$. Equation (2.31) can be solved by the classical perturbation method, and

$$v = v^{(0)}(\xi) + \varepsilon v^{(1)}(\xi) + \varepsilon^2 v^{(2)}(\xi) + \ldots \tag{2.32}$$

The expansion of α in powers of ε furnishes all the $x^{(j)}$'s. The zeroth order solution is

$$v^{(0)}(\xi) = \exp\left(-\int \frac{q}{\xi}d\xi\right)\int \frac{r}{\xi}\exp\left(\int \frac{q}{\xi}d\xi\right)d\xi = \xi R(\xi) + O(\xi^{-q_0}\log^{\mu}\xi) \tag{2.33}$$

where μ is 1 when q_0 is a negative integer, and otherwise zero. The analysis by Lighthill[2] shows that the higher order perturbations are given by

$$v^{(j)}(\xi) = \xi R(\xi) + O(\xi^{-q_0}\log^{\nu j + \mu}\xi) \tag{2.34}$$

where $\nu = 2$ when $q_0 = -1$, and $\nu = 1$ otherwise. Thus the ratio of successive terms in the series (2.32) for v is $O(\varepsilon\log^{\nu}\xi)$ as $\xi \to 0$. This seems to indicate that the radius of convergence in ξ of the series is $\exp(-\text{const.}/\varepsilon^{1/\nu})$. In other words, the smallest ξ for which our solution is valid is $O[\exp(-\text{const.}/\varepsilon^{1/\nu})]$. Although this is smaller than any power of ε, it still means that the point $\xi = 0$, the nodal singularity, cannot be reached. Worse still, x exceeds ξ by α according to (2.29). If α is positive, our solution fails in a range of x extending from $x = 0$ to a value somewhat larger than α. Of course, if it turns out that the required solution of the physical problem does not include that troublesome range, then the PLK method is perfectly successful.

The case of $-1 < q_0 < 0$ has the characteristic of both cases $q_0 \geqslant 0$ and $q_0 \leqslant -1$. For our analysis, we find it convenient to apply again the transformation (2.29) together with (2.30). Therefore we shall consider the equation to be of the form (2.1) with $r_0 = 0$, and $-1 < q_0 < 0$. The point $u = x = 0$ is thus a nodal singularity. Here the full double expansion (2.2) is required. Lighthill has shown[2] that as $\xi \to 0$

$$x^{(j)}(\xi) = O(\xi^{-q_0}\log^{j-1}\xi) \quad \text{and} \quad u^{(j)}(\xi) = O(\xi^{-q_0}\log^{j}\xi) \tag{2.35}$$

Again, the essence of our method is to control the singularities of $x^{(j)}(\xi)$ and $u^{(j)}(\xi)$ in such a way that both functions have similar singularities for each j. As in the case of $q_0 \leqslant -1$, the radius of convergence of the series in ξ may again be of order $O[\exp(-\text{const.}/\varepsilon)]$, and the point $\xi = 0$ has to be excluded. Usually, however, a branch point of the solution appears before $\xi = 0$ is reached; the physical problem then does not require knowledge of the solution near this troublesome point, and the PLK method gives satisfactory results.

As an example for this case, let us consider the equation studied by Lighthill[2]

$$(y + \varepsilon w)\frac{dw}{dy} - \frac{1}{2}w = 1 + y^2 \tag{2.36}$$

under the condition

$$w(1) = -1 \tag{2.37}$$

The required preliminary transformation (2.29) can be written as

$$y = x + \alpha, \quad w = u - \left(\frac{\alpha}{\varepsilon}\right) \tag{2.38}$$

according to the first equation (2.30). Then the second equation (2.30) gives

$$\alpha^2 - \frac{1}{2\varepsilon}\alpha + 1 = 0$$

or, since $\alpha \to 0$ as $\varepsilon \to 0$,

$$\alpha = \frac{1}{4\varepsilon}[1 - \sqrt{1 - 16\varepsilon^2}] = 2\varepsilon + 8\varepsilon^3 + \ldots \tag{2.39}$$

The resultant equation is

$$(x + \varepsilon u)\frac{du}{dx} - \frac{1}{2}u = 2\alpha x + x^2 \tag{2.40}$$

The boundary condition (2.37) now becomes

$$u = \frac{\alpha}{\varepsilon} - 1 \quad \text{at} \quad x = 1 - \alpha$$

Using the expansion (2.2), we obtain

$$\begin{aligned} 1 + 8\varepsilon^2 + \ldots = {} & u^{(0)}(1) + \varepsilon\{-u^{(0)\prime}(1)[2 + x^{(1)}(1)] + u^{(1)}(1)\} + \\ & \varepsilon^2\Big\{u^{(0)\prime}(1)[2x^{(1)\prime}(1) - x^{(2)}(1)] + \frac{1}{2}u^{(0)\prime\prime}(1)[2 + x^{(1)}(1)]^2 - \\ & u^{(1)\prime}(1)[2 + x^{(1)}(1)] + u^{(2)}(1)\} + \ldots \end{aligned}$$

The converted boundary condition is thus

$$
\begin{aligned}
&u^{(0)}(1) = 1 \\
&u^{(1)}(1) = u^{(0)\prime}(1)[2+x^{(1)}(1)] \\
&u^{(2)}(1) = u^{(1)\prime}(1)[2+x^{(1)}(1)] - \frac{1}{2}u^{(0)\prime\prime}(1)[2+x^{(1)}(1)]^2 - \\
&\qquad u^{(0)\prime}(1)[2x^{(1)\prime}(1) - x^{(2)}(1)] \\
&\cdots\cdots
\end{aligned}
\tag{2.41}
$$

The zeroth order equation is

$$
\xi\frac{du^{(0)}}{d\xi} - \frac{1}{2}u^{(0)} = \xi^2 \tag{2.42}
$$

The solution of this equation with the first of the boundary conditions given by (2.41) is

$$
u^{(0)}(\xi) = \frac{1}{3}\xi^{1/2} + \frac{2}{3}\xi^2 \tag{2.43}
$$

The first order equation is

$$
\xi\frac{du^{(1)}}{d\xi} - \frac{1}{2}u^{(1)} = \left(\xi^2 + \frac{1}{2}u^{(0)}\right)x^{(1)\prime} + (2\xi - u^{(0)\prime})x^{(1)} + (4\xi - u^{(0)}u^{(0)\prime}) \tag{2.44}
$$

To remove here the most serious term to the right we put

$$
\xi^{1/2}x^{(1)\prime} - \frac{1}{\xi^{1/2}}x^{(1)} = \frac{1}{3}
$$

or

$$
x^{(1)} = -\frac{2}{3}\xi^{1/2} \tag{2.45}
$$

The complete solution turns out to be

$$
u = \frac{1}{3}\xi^{1/2} + \frac{2}{3}\xi^2 + \varepsilon\left(-\frac{21}{5}\xi^{1/2} + 8\xi - \frac{13}{9}\xi^{3/2} - \frac{16}{45}\xi^3\right) + O(\varepsilon^2\xi^{1/2}\log\xi)
$$

$$
x = \xi - \frac{2}{3}\varepsilon\xi^{1/2} + \frac{42}{5}\varepsilon^2\xi^{1/2} + O(\varepsilon^3\xi^{1/2}\log\xi)
$$

We can then use (2.38) and (2.39) to write the solution in terms of w and y :

$$
\left.
\begin{aligned}
w &= -2 + \frac{1}{3}\xi^{1/2} + \frac{2}{3}\xi^2 + \varepsilon\left(-\frac{21}{5}\xi^{1/2} + 8\xi - \frac{13}{9}\xi^{3/2} - \frac{16}{45}\xi^3\right) + O(\varepsilon^2) \\
y &= \xi + \varepsilon\left(2 - \frac{2}{3}\xi^{1/2}\right) + \frac{42}{5}\varepsilon^2\xi^{1/2} + O(\varepsilon^3)
\end{aligned}
\right\}
\tag{2.46}
$$

This solution can of course be obtained without first applying the transformation (2.38). Direct substitution of the double expansion for w and y will yield the same result. Nevertheless it is often more convenient to use the preliminary transformation.

At the point where $du/d\xi = 0$, the relation between y and ξ starts to double up, and we have the branch point of the solution. Using this condition, we have at the branch point

$$\left.\begin{aligned} \xi &= \frac{1}{9}\varepsilon^2 - \frac{14}{5}\varepsilon^3 + O(\varepsilon^4) \\ y &= 2\varepsilon - \frac{1}{9}\varepsilon^2 + \frac{14}{5}\varepsilon^3 + O(\varepsilon^4) \end{aligned}\right\} \tag{2.47}$$

The branch point thus occurs before the breakdown of the solution at the point $\xi = 0$.

5. Equations Requiring the Boundary Layer Method

It will be recalled that in the previous discussions concerning the differential equation (2.1) we have imposed the condition that $q(x)$ and $r(x)$ are regular functions at $x = 0$. Because $q(x)$ and $r(x)$ are regular, we can expand them in uniformly valid power series in ε, when x is replaced according to the second equation (2.2). Since such expansions are required to form the equations of successive order as shown by (2.3), the regularity of $q(x)$ and $r(x)$ in (2.1) is a very essential condition for the success of the PLK method. If $q(x)$ or $r(x)$, or both, are not regular at $x = 0$, then the equation cannot be solved by the PLK method, but must be treated by some different method such as the boundary layer method.

This observation is supported by the investigations of Carrier[6,7] who found for instance that the following equation cannot be solved by the PLK method:

$$(z^2 + \varepsilon u)\frac{du}{dz} + u = (2z^3 + z^2) \tag{2.48}$$

That this must be expected as a consequence of our regularity conditions, can be seen immediately if we make the transformation $z^2 = x$. Then (2.48) becomes

$$(x + \varepsilon u)\frac{du}{dx} + \frac{1}{2\sqrt{x}}u = \left(x + \frac{1}{2}\sqrt{x}\right) \tag{2.49}$$

The equation is now in our standard form, but $q(x)$ and $r(x)$ are not regular at $x = 0$, and the PLK method must fail in the present instance. Of course, it might be argued that we can expand z in the original equation (2.48) instead of x in the transformed equation (2.49). Then the PLK method seemingly can be carried through. But that is only an illusion, because there is really no difference between the power series expansion of z or of x. If the method fails in the form with x, it must fail in the form with z.

6. Second Order Equations

Equations of higher order, with a regular singularity such that the zeroth order solution has an algebraic or logarithmic singularity at the critical point like the equations discussed in the previous sections, can be treated by similar methods. It will be convenient to express such higher order equations as a set of simultaneous first order equations. For instance, the equation

$$\left(x+\varepsilon\frac{\mathrm{d}v}{\mathrm{d}x}+\varepsilon av\right)\frac{\mathrm{d}^2v}{\mathrm{d}x^2}+q(x)\frac{\mathrm{d}v}{\mathrm{d}x}+S(x)v=r(x) \tag{2.50}$$

should be rewritten as

$$(x+\varepsilon u+\varepsilon av)\frac{\mathrm{d}u}{\mathrm{d}x}+q(x)u+S(x)v=r(x), \quad \frac{\mathrm{d}v}{\mathrm{d}x}=u \tag{2.51}$$

Now the singularity of the equation lies on the plane $x+\varepsilon u+\varepsilon av=0$ in the (x, u, v) space. To treat the equation by the PLK method, we substitute for x, u, v the following expansions:

$$\left.\begin{aligned} x &= \xi+\varepsilon x^{(1)}(\xi)+\varepsilon^2 x^{(2)}(\xi)+\ldots \\ u &= u^{(0)}(\xi)+\varepsilon u^{(1)}(\xi)+\varepsilon^2 u^{(2)}(\xi)+\ldots \\ v &= v^{(0)}(\xi)+\varepsilon v^{(1)}(\xi)+\varepsilon^2 v^{(2)}(\xi)+\ldots \end{aligned}\right\} \tag{2.52}$$

Then the second equation (2.51) gives

$$v^{(0)\prime}+\varepsilon v^{(1)\prime}+\varepsilon^2 v^{(2)\prime}+\ldots=[1+\varepsilon x^{(1)\prime}+\varepsilon^2 x^{(2)\prime}+\ldots]\times[u^{(0)}+\varepsilon u^{(1)}+\varepsilon^2 u^{(2)}+\ldots] \tag{2.53}$$

The zeroth order equation is thus

$$\xi\frac{\mathrm{d}u^{(0)}}{\mathrm{d}\xi}+q(\xi)u^{(0)}+S(\xi)v^{(0)}=r(\xi), \quad \frac{\mathrm{d}v^{(0)}}{\mathrm{d}\xi}=u^{(0)} \tag{2.54}$$

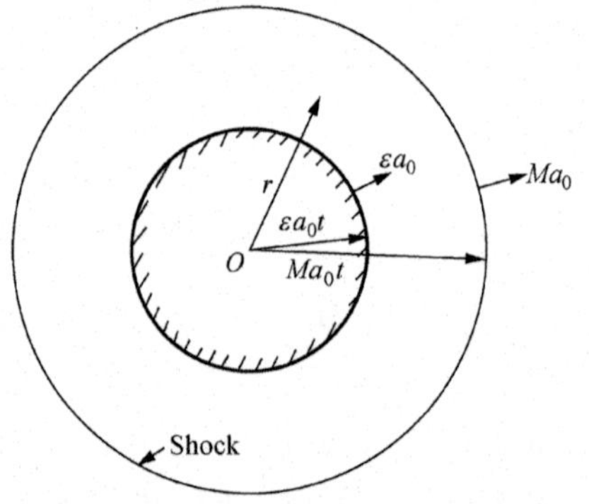

Fig. 2 Cylindrical shock due to expanding cylinder

When $q_0>0$, $v^{(0)}$ can be neglected in comparison with $u^{(0)}$ for small ξ, and $u^{(0)}\sim A\xi^{-q_0}$ as $\xi\to 0$. The quantity $v^{(0)}$ is obtained from the second equation (2.54) by integration in accordance with some boundary condition. When $q_0\leqslant -1$, the $x^{(j)}$'s may still be taken as constants as in Section Ⅱ.4. But now conditions similar to (2.30) will not determine $x^{(j)}$, because we have only two conditions for three shift constants associated with u, v and x, and $x^{(j)}$ must be determined from the differential equation and the boundary conditions. The following general rule holds: After $u^{(0)}(\xi)$, $u^{(1)}(\xi)$, ... $u^{(k-1)}(\xi)$; $v^{(0)}(\xi)$, $v^{(1)}(\xi)$, ... $v^{(k-1)}(\xi)$; $x^{(1)}$, ... $x^{(k-1)}(\xi)$ have been determined, $x^{(k)}$ is fixed by requiring the coefficient of ε^k in $x+\varepsilon u+\varepsilon av$ to vanish at $\xi=0$. Then the successive increase in the order of singularity of $u^{(k)}$ and $v^{(k)}$ caused by the factor ξ^{-1} is prevented. When $-1 < q_0 < 0$, similar to the case discussed in Section Ⅱ. 4, a combination of constant shift and expansion of the independent variable will be necessary.

As an example of applying the PLK method to second order equations, let us solve the problem of a cylindrical shock produced in air by a solid cylindrical surface expanding uniformly from zero radius; the air is assumed to be a perfect nonviscous nonconducting gas. This example was also discussed by Lighthill[2]. We denote the distance from the center of the cylinder by r and count the time t from the instant when the radius of the cylinder was zero (Fig. 2). Since there is no fundamental length or time involved, all velocities and pressures must depend on the parameter r/t. If a_0 is the sound velocity in the still air, we may write the velocity potential ϕ as

$$\phi = a_0^2 t f(x) \tag{2.55}$$

where x is the nondimensional parameter

$$x = r/(a_0 t) \tag{2.56}$$

Let the velocity of the expanding cylindrical surface be εa_0, and let the surface of the cylinder be defined by $x = \varepsilon$. Let the shock be at $Ma_0 t$. Then the region of flow is $\varepsilon \leqslant x \leqslant M$. Since the shock wave is of uniform strength, the flow behind it is adiabatic and Bernoulli's equation holds, i.e., the local sound velocity a is given by

$$a^2 = a_0^2\left[1-(\gamma-1)\left(f - xf' + \frac{1}{2}f'^2\right)\right] \tag{2.57}$$

where the prime denotes differentiation with respect to x, and γ is the ratio of the specific heats of air.

The equation for the velocity potential ϕ is

$$a^2 \nabla^2\phi = \frac{\partial^2\phi}{\partial t^2} + 2\frac{\partial\phi}{\partial r}\frac{\partial^2\phi}{\partial r\partial t} + \left(\frac{\partial\phi}{\partial r}\right)^2\frac{\partial^2\phi}{\partial r^2}$$

In terms of f, we have

$$\left[1-(\gamma-1)\left(f - xf' + \frac{1}{2}f^2\right)\right]\left[f'' + \frac{1}{x}f'\right] = (x - f')^2 f'' \tag{2.58}$$

The boundary conditions are specified by the requirements that (i) at the surface of the cylinder the velocity of the air is equal to the surface velocity of the cylinder; (ii) at the shock, ϕ is continuous, but in the still air $\phi = 0$, thus $\phi = 0$ at the shock; (iii) the relation between the velocity of the shock and the fluid speed behind it must satisfy the Rankine-Hugoniot relations. Expressed in terms of f, we have the following three conditions, respectively:

$$\left.\begin{aligned} f'(\varepsilon) &= \varepsilon \\ f(M) &= 0 \\ f'(M) &= 2\left(M - \frac{1}{M}\right)\bigg/(\gamma+1) \end{aligned}\right\} \tag{2.59}$$

Since we now have three boundary conditions for an equation of second order, we should be able to find a relation between ε and M, or a relation between the velocity of expansion and the shock velocity.

To put the problem in more familiar form, we set

$$f' = u, \quad f = v \tag{2.60}$$

Then (2.58) becomes

$$\left.\begin{aligned} &\left[1 - x^2 + (\gamma+1)xu - (\gamma-1)v - \frac{1}{2}(\gamma+1)u^2\right]\frac{du}{dx} + \\ &\frac{u}{x}\left[1 + (\gamma-1)\left(xu - v - \frac{1}{2}u^2\right)\right] = 0, \\ &\frac{dv}{dx} = u \end{aligned}\right\} \tag{2.61}$$

The boundary conditions are

$$u(\varepsilon) = \varepsilon, \quad u(M) = \frac{2}{(\gamma+1)}\left(M - \frac{1}{M}\right), \quad v(M) = 0 \tag{2.62}$$

When ε is very small, u and v will be small quantities; then (2.61) is approximated by

$$\left.\begin{aligned} &(1 - x^2)\frac{du}{dx} + \frac{u}{x} = 0 \\ &\frac{dv}{dx} = u \end{aligned}\right\} \tag{2.63}$$

The solution with the boundary conditions (2.62) is thus

$$\left.\begin{aligned} &u = \varepsilon^2\sqrt{\frac{1}{x^2} - 1}, \quad M = 1 \\ &v = \int_1^x u\,dx \end{aligned}\right\} \tag{2.64}$$

with a singularity at $x = 1$ corresponding to $q_0 = -\frac{1}{2}$. The problem thus belongs to the case $-1 < q_0 < 0$. The solution (2.64) does not exist for $x > 1$, while M, the ratio of the shock velocity to the sound velocity, must be greater than 1 for the exact solution. Thus the PLK method is clearly indicated. The zeroth approximation suggests that we adopt the following expansions,

$$\left.\begin{aligned} u &= \varepsilon^2 u^{(0)}(\xi) + \varepsilon^4 u^{(1)}(\xi) + \ldots \\ v &= \varepsilon^2 v^{(0)}(\xi) + \varepsilon^4 v^{(1)}(\xi) + \ldots \\ x &= \xi + \varepsilon^2 x^{(1)}(\xi) + \varepsilon^4 x^{(2)}(\xi) + \ldots \\ M &= 1 + \varepsilon^2 M^{(1)} + \varepsilon^4 M^{(2)} + \ldots \end{aligned}\right\} \tag{2.65}$$

where $u^{(0)}(\xi)$ and $v^{(0)}(\xi)$ are of course given by (2.64), or

$$\left.\begin{aligned} u^{(0)}(\xi) &= \sqrt{\frac{1}{\xi^2} - 1} \\ v^{(0)}(\xi) &= \int_1^{\xi} u^{(0)}(\xi)\,d\xi \end{aligned}\right\} \tag{2.66}$$

Now since the singularity is at $\xi = 1$, $(1-\xi)$ takes the role of ξ in previous sections. The most important terms in $x^{(1)}$, $x^{(2)}$, ..., according to the general theory for $-1 < q_0 < 0$, will be constants, although terms in $(1-\xi)^{1/2}$ multiplied by various powers of $\log(1-\xi)$ will also be required, as shown by (2.35). The coefficient of ε^2 in the factor of du/dx in (2.61) is

$$-2\xi x^{(1)} + (\gamma+1)\xi u^{(0)} - (\gamma-1)v^{(0)}$$

According to the general rule stated at the beginning of the present section, this sum should be zero at the singular point $\xi = 1$. But (2.66) shows that $u^{(0)}$ and $v^{(0)}$ are zero at $\xi = 1$. Therefore we take

$$x^{(1)} = 0 \tag{2.67}$$

With this value of $x^{(1)}$ the equation for $u^{(1)}$ is

$$\begin{aligned} &(1-\xi^2)u^{(1)\prime} + \frac{1}{\xi}u^{(1)} + [(\gamma+1)\xi u^{(0)} - (\gamma-1)v^{(0)}]u^{(0)\prime} + \\ &\frac{1}{\xi}u^{(0)}[(\gamma-1)\xi u^{(0)} - (\gamma-1)v^{(0)}] = 0 \end{aligned} \tag{2.68}$$

If $\xi \to 1$ in this equation, we obtain

$$u^{(1)}(1) = -(\gamma+1)\lim_{\xi\to 1}[u^{(0)}u^{(0)\prime}] = \gamma + 1 \tag{2.69}$$

To calculate $v^{(1)}(1)$ we have to use the boundary conditions. Let ξ_1 be the value of ξ corresponding to the shock; then, according to (2.62) and (2.65), we have on account of (2.67)

$$\left.\begin{aligned} &\xi_1 + \varepsilon^4 x^{(2)}(\xi_1) + \ldots = M \\ &\varepsilon^2 v^{(0)}(\xi_1) + \varepsilon^4 v^{(1)}(\xi_1) + \ldots = 0 \\ &\varepsilon^2 u^{(0)}(\xi_1) + \varepsilon^4 u^{(1)}(\xi_1) + \ldots = \frac{2}{(\gamma+1)}\left(M - \frac{1}{M}\right) \end{aligned}\right. \tag{2.70}$$

But we know from (2.66) and (2.69) that $u^{(0)} \cong \sqrt{2(1-\xi)}$, $u^{(1)} \cong \gamma + 1$, as $\xi \to 1$. Then the first and the third equation (2.70) give

$$M = \xi_1 + O(\varepsilon^4)$$

$$\frac{2}{(\gamma+1)}\frac{M+1}{M}(M-1) = \varepsilon^2\sqrt{2(1-\xi_1)} + \varepsilon^4(\gamma+1)$$

Therefore

$$\xi_1 = 1 - O(\varepsilon^4) \tag{2.71}$$

and

$$M-1=O(\varepsilon^4) \tag{2.72}$$

Then according to $v^{(0)}$ given by (2.66),

$$v^{(0)}(\xi_1)=O(\varepsilon^6) \tag{2.73}$$

Hence the second of the boundary condition (2.70) gives

$$v^{(1)}(1)=-\lim_{\varepsilon\to 0}[\varepsilon^{-2}v^{(0)}(\xi_1)]=\lim_{\varepsilon\to 0}[\varepsilon^{-2}O(\varepsilon^6)]=0 \tag{2.74}$$

With the values already determined, the coefficient of ε^4 in the coefficient of du/dx in (2.61) is at $\xi=1$

$$\lim_{\varepsilon\to 1}\left[-2\xi x^{(2)}+(\gamma+1)\xi u^{(1)}-(\gamma-1)v^{(1)}-\frac{1}{2}(\gamma+1)(u^{(0)})^2\right]=-2x^{(2)}+(\gamma+1)^2$$

According to the general principle this sum must be zero. Therefore

$$x^{(2)}=\frac{1}{2}(\gamma+1)^2+O(\sqrt{1-\xi}\log(1-\xi)) \tag{2.75}$$

Hence, by combining the first and third of the boundary conditions (2.70), we have

$$\varepsilon^2\sqrt{2(1-M)+\varepsilon^4(\gamma+1)^2}+\varepsilon^4(\gamma+1)+O(\varepsilon^6\log\varepsilon)=\frac{4}{\gamma+1}(M-1)+O[(M-1)^2]$$

By solving for $M-1$, we obtain finally

$$M=1+\frac{3}{8}(\gamma+1)^2\varepsilon^4+O(\varepsilon^6\log\varepsilon) \tag{2.76}$$

Equation (2.76) gives the desired relation between M and ε. It will be noted that this result has been obtained here with very little actual calculation, while Lighthill himself obtained the same result previously[8] by a different method, but only after some tedious computation. The power of the new method is thus clearly demonstrated.

7. Irregular Singularity

We now consider the differential equation

$$\frac{d^2u}{dx^2}+u=\varepsilon f\left(u,\frac{du}{dx}\right) \tag{2.77}$$

If we identify x with the time, this equation represents electrical or mechanical systems with small nonlinear terms. Such systems often show self-excited oscillations which differ considerably from the simple harmonic oscillations of period 2π when $\varepsilon=0$. The self-excited periodic solution is called the limit cycle of the system: it actually represents the problem of Poincaré. If we use the classical perturbation method, we substitute for u

$$u=u^{(0)}(x)+\varepsilon u^{(1)}(x)+\varepsilon^2u^{(2)}(x)+\ldots$$

The equation of order zero is

$$\frac{d^2u^{(0)}}{dx^2} + u^{(0)} = 0$$

The point $x = \infty$ is thus an irregular singularity of the differential equation. We can take the zeroth order solution to be

$$u^{(0)} = A\sin x$$

Then the first order equation is

$$u^{(1)''} + u^{(1)} = f(A\sin x, A\cos x)$$

Making a Fourier analysis of $f(A\sin x,\ A\cos x)$, we have

$$f(A\sin x, A\cos x) = \frac{1}{2}a_0 + \sum_1^{\infty}(a_n\cos nx + b_n\sin nx)$$

We now can easily determine $u^{(1)}$ as

$$u^{(1)}(x) = \frac{a_0}{2} + \frac{a_1}{2}x\sin x - \frac{b_1}{2}x\cos x +$$
$$\sum_2^{\infty}\left(\frac{a_n}{1-n^2}\cos nx + \frac{b_n}{1-n^2}\sin nx\right) + B\sin x + C\cos x$$

As $x \to \infty$, $u^{(1)}$ behaves like xe^{ix}. The point $x = \infty$ is thus a singularity of the perturbation equation. Higher order solutions have the same general character and the perturbation series diverges as $x \to \infty$.

To treat this problem by PLK method, we substitute (2.2) into (2.77). Then

$$\frac{u^{(0)''} + \varepsilon u^{(1)''} + \varepsilon^2 u^{(2)''} + \dots}{(1 + \varepsilon x^{(1)'} + \varepsilon^2 x^{(2)'} + \dots)^2} - \frac{(u^{(0)'} + \varepsilon u^{(1)'} + \varepsilon^2 u^{(2)'} + \dots)(\varepsilon x^{(1)''} + \varepsilon^2 x^{(2)''} + \dots)}{(1 + \varepsilon x^{(1)'} + \varepsilon^2 x^{(2)'} + \dots)^3} +$$
$$u^{(0)} + \varepsilon u^{(1)} + \varepsilon^2 u^{(2)} + \dots = \varepsilon f\left(u^{(0)} + \varepsilon u^{(1)} + \dots, \frac{u^{(0)'} + \varepsilon u^{(1)'} + \dots}{1 + \varepsilon x^{(1)'} + \dots}\right) \tag{2.78}$$

The zeroth order solution is the same as that for the classical perturbation method except that x is replaced by ξ. Thus

$$u^{(0)}(\xi) = A\sin\xi \tag{2.79}$$

But the first order equation is now

$$u^{(1)''} + u^{(1)} = \frac{1}{2}a_0 + a_1\cos\xi + b_1\sin\xi - 2Ax^{(1)'}\sin\xi +$$
$$Ax^{(1)''}\cos\xi + \sum_2^{\infty}(a_n\cos n\xi + b_n\sin n\xi) \tag{2.80}$$

The troublesome terms on the right-hand side of (2.80) are those connected with sin ξ and

$\cos\xi$. The $\sin\xi$ term can be eliminated if we put

$$x^{(1)} = \frac{b_1}{2A}\xi, \quad x^{(1)''} = 0 \tag{2.81}$$

As no help can be obtained from $x^{(1)}$, we set $a_1 = 0$ in order to eliminate $\cos\xi$. In other words,

$$\int_0^{2\pi} f(A\sin\xi, A\cos\xi)\cos\xi d\xi = 0 \tag{2.82}$$

This equation actually determines the amplitude A of the oscillation. The period of this self-excited oscillation or limit cycle is the change in x when ξ changes by 2π, since now $u^{(1)}$ as well as $u^{(0)}$ is periodic with the period 2π. Thus, according to (2.81), the period is given by

$$2\pi\left[1+\frac{\varepsilon b_1}{2A}+O(\varepsilon^2)\right] = 2\pi + \frac{\varepsilon}{A}\int_0^{2\pi} f(A\sin\xi, A\cos\xi)\sin\xi d\xi + O(\varepsilon^2) \tag{2.83}$$

Poincaré has shown[1] that the process can be extended to higher orders with $x^{(j)}(\xi)$ each proportional to ξ. The period of the limit cycle can then be calculated as a power series in ε as indicated by (1.2).

8. Combined Method; Sink Flow of a Viscous Gas

In all previous sections except Section Ⅱ.5, we have shown how the PLK method can be used to give a uniformly valid solution for cases where the classical perturbation method fails. The types of differential equations we have so far considered are, however, quite restricted. There are equations for which the restriction of Section Ⅱ.5 does not, yet the PLK method will not yield the valid solution over the whole domain of interest. Where the method fails, we must again resort to other methods of solution. It usually turns out that the "method of boundary layer" will provide the correct solution in the difficult region when new variables of the form $\varepsilon^{\mu}u$ and $\varepsilon^{\nu}x$ are introduced. Outside of the difficult region, the PLK method is still effective. For this type of problems, then, the complete solution requires a combination of methods. We shall demonstrate this technique of the combined method by studying the problem of sink flow of a viscous heat-conducting gas. Our discussion follows the work of Wu[9].

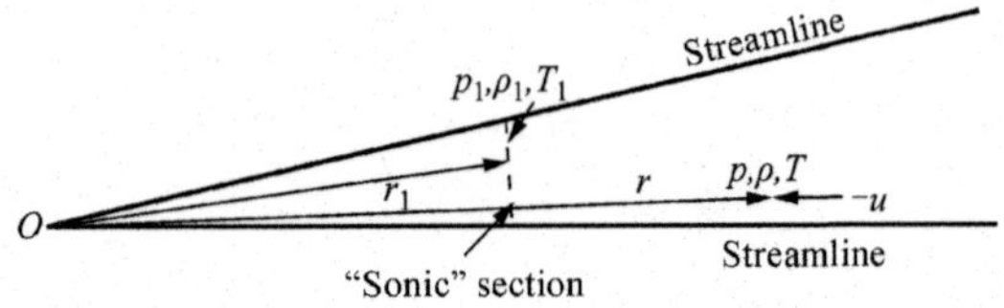

Fig. 3 Compressible sink flow

Here we are concerned with steady two dimensional flow with cylindrical symmetry (Fig. 3). The only independent variable is the radial distance r from the origin, and the radial velocity u is the only velocity component. For the sink flow u is always negative. Let the velocity far away from the origin be subsonic and vanish as $r \to \infty$. We are interested in the region of the flow where the local Mach number is nearly unity, because there the velocity gradient is large, and the viscous effects of the fluid can no longer be neglected.

Let p, ρ, T, μ, μ', λ, R, C_p, C_v denote, respectively, the pressure, density, absolute temperature, coefficients of shear and bulk viscosity, heat conductivity, gas constant, and the specific heats at constant pressure and volume. Then the momentum equation is

$$\rho u \frac{\mathrm{d}u}{\mathrm{d}r} = -\frac{\mathrm{d}p}{\mathrm{d}r} + \frac{\mathrm{d}}{\mathrm{d}r}\left[2\mu \frac{\mathrm{d}u}{\mathrm{d}r} + \frac{2}{3}(\mu' - \mu)\frac{1}{r}\frac{\mathrm{d}}{\mathrm{d}r}(ru)\right] + 2\mu \frac{\mathrm{d}}{\mathrm{d}r}\left(\frac{u}{r}\right) \tag{2.84}$$

and the energy equation is

$$\rho u r \frac{\mathrm{d}}{\mathrm{d}r}\left(\frac{u^2}{2} + C_p T\right) = \frac{\mathrm{d}}{\mathrm{d}r}\left\{r\left[\lambda \frac{\mathrm{d}T}{\mathrm{d}r} + \mu \frac{\mathrm{d}u^2}{\mathrm{d}r} + \frac{2}{3}(\mu' - \mu)\left(\frac{1}{2}\frac{\mathrm{d}u^2}{\mathrm{d}r} + \frac{u^2}{r}\right)\right]\right\} \tag{2.85}$$

If m is the sink strength, the continuity equation is simply

$$2\pi \rho u r = -m \tag{2.86}$$

The equation of state is assumed to be that of a perfect gas, or

$$p = R\rho T \tag{2.87}$$

It would be convenient to use nondimensional variables defined as follows

$$\begin{gathered} \bar{r} = r/r_1, \quad w = -u/a_1, \quad \theta = T/T_1 = (a/a_1)^2 \\ \bar{p} = p/p_1, \quad \bar{\rho} = \rho/\rho_1, \quad \bar{\mu} = \mu/\mu_1, \quad \bar{\mu}' = \mu'/\mu'_1 \end{gathered} \tag{2.88}$$

where quantities with the subscript 1 are fictitious quantities, which correspond to conditions at the local Mach number unity for an inviscid gas with vanishing heat conductivity. The ratio of the specific heats, γ, is assumed to be constant throughout; the sonic speed a_1 at $r = r_1$ is given by

$$a_1^2 = \gamma p_1/\rho_1 \quad \text{and} \quad 2\pi \rho_1 a_1 r_1 = m \tag{2.89}$$

The continuity equation becomes therefore

$$\bar{\rho} w \bar{r} = 1 \tag{2.90}$$

Here w is positive for sink flow. The equation of state (2.87) is

$$\bar{p} = \bar{\rho}\,\theta \tag{2.91}$$

We now introduce the parameter k by

$$\mu' - \mu = 3k\mu \tag{2.92}$$

and the Reynolds number

$$Re = \frac{m}{2\pi\mu_1} \tag{2.93}$$

The nondimensional form of the momentum equation (2.84) reduces to

$$\frac{1}{\bar{r}}\frac{\mathrm{d}w}{\mathrm{d}\bar{r}}=-\frac{1}{\gamma}\frac{\mathrm{d}\bar{p}}{\mathrm{d}\bar{r}}-\frac{2}{Re}\left\{\frac{\mathrm{d}}{\mathrm{d}\bar{r}}\left[\bar{\mu}\frac{\mathrm{d}w}{\mathrm{d}\bar{r}}+k\bar{\mu}\frac{1}{\bar{r}}\frac{\mathrm{d}}{\mathrm{d}\bar{r}}(\bar{r}w)\right]+\bar{\mu}\frac{\mathrm{d}}{\mathrm{d}\bar{r}}\left(\frac{w}{\bar{r}}\right)\right\} \tag{2.94}$$

By using (2.90), the energy equation can be integrated once. The integration constant will be chosen so that at $\bar{r}=\infty$ the limiting case of vanishing viscosities and heat conduction reduces to that of isoenergetic flow. Thus

$$\frac{w^2}{2}+\frac{\theta}{\gamma-1}+\frac{1}{Re}\bar{r}\bar{\mu}\left[\frac{1}{\sigma}\frac{\mathrm{d}}{\mathrm{d}\bar{r}}\left(\frac{\theta}{\gamma-1}\right)+(1+k)\frac{\mathrm{d}w^2}{\mathrm{d}\bar{r}}+2k\frac{w^2}{\bar{r}}\right]=\frac{\gamma+1}{2(\gamma-1)} \tag{2.95}$$

where σ is the Prandtl number

$$\sigma=\frac{C_p\mu}{\lambda} \tag{2.96}$$

The pressure $\bar{p}$ can now be eliminated by using (2.90) and (2.91). The result is conveniently expressed by a new independent variable η,

$$\eta=\log\bar{r} \tag{2.97}$$

The final equations for the two unknowns w and θ are then

$$\frac{\mathrm{d}w}{\mathrm{d}\eta}+\frac{1}{\gamma}\left[\frac{\mathrm{d}}{\mathrm{d}\eta}\left(\frac{\theta}{w}\right)-\frac{\theta}{w}\right]=-\frac{2}{Re}\left\{\bar{\mu}(1+k)\left(\frac{\mathrm{d}^2w}{\mathrm{d}\eta^2}-w\right)+\left[(1+k)\frac{\mathrm{d}w}{\mathrm{d}\eta}+kw\right]\frac{\mathrm{d}\bar{\mu}}{\mathrm{d}\eta}\right\} \tag{2.98}$$

and

$$\frac{w^2}{2}+\frac{\theta}{\gamma-1}+\frac{1}{Re}\bar{\mu}\left[\frac{\sigma^{-1}}{\gamma-1}\frac{\mathrm{d}\theta}{\mathrm{d}\eta}+(1+k)\frac{\mathrm{d}w^2}{\mathrm{d}\eta}+2kw^2\right]=\frac{\gamma+1}{2(\gamma-1)} \tag{2.99}$$

The solution we wish to obtain is one for small viscosity or large Re, tending to the inviscid subsonic solution at large radial distances. Thus at $\eta\to\infty$, $w=0$, $\theta=(\gamma+1)/2$. The critical point of the equation is $\eta=0$, where the inviscid solution has local Mach number unity.

To avoid unduly complicated calculations, we assume the viscosity coefficient to be temperature independent, hence

$$\bar{\mu}=1 \tag{2.100}$$

We can now introduce the small parameter ε of our problem as

$$\varepsilon=\frac{4\gamma}{\gamma+1}(1+k)\frac{1}{Re} \tag{2.101}$$

Then the basic system of differential equation becomes

$$w^2\frac{\mathrm{d}w}{\mathrm{d}\eta}+\frac{1}{\gamma}\left[w\frac{\mathrm{d}\theta}{\mathrm{d}\eta}-\theta\frac{\mathrm{d}w}{\mathrm{d}\eta}-\theta w\right]=-\frac{\gamma+1}{2\gamma}\varepsilon\left(\frac{\mathrm{d}^2w}{\mathrm{d}\eta^2}-w\right)w^2 \tag{2.102}$$

and

$$\theta+\frac{\gamma-1}{2}(1+b\varepsilon)w^2+\frac{\gamma+1}{4\gamma}\varepsilon\left[\frac{1}{\sigma(1+k)}\frac{\mathrm{d}\theta}{\mathrm{d}\eta}+(\gamma-1)\frac{\mathrm{d}w^2}{\mathrm{d}\eta}\right]=\frac{\gamma+1}{2} \tag{2.103}$$

where b is the constant

$$b = \frac{\gamma+1}{\gamma}\frac{k}{1+k} \tag{2.104}$$

We shall find it convenient to use w as the independent variable. Then, according to the PLK method, the expansions are

$$\begin{aligned} w &= \xi + \varepsilon w^{(1)}(\xi) + \varepsilon^2 w^{(2)}(\xi) + \ldots \\ \eta &= \eta^{(0)}(\xi) + \varepsilon \eta^{(1)}(\xi) + \varepsilon^2 \eta^{(2)}(\xi) + \ldots \\ \theta &= \theta^{(0)}(\xi) + \varepsilon \theta^{(1)}(\xi) + \varepsilon^2 \theta^{(2)}(\xi) + \ldots \end{aligned} \tag{2.105}$$

By substituting (2.105) in (2.102) and (2.103), we find the equations of order zero in the form

$$\begin{aligned} &\left[\xi^2 + \frac{\xi^2}{\gamma}\frac{d}{d\xi}\left(\frac{\theta^{(0)}}{\xi}\right) - \frac{1}{\gamma}\xi\theta^{(0)}\eta^{(0)\prime}\right](\eta^{(0)\prime})^2 = 0 \\ &\left[\theta^{(0)} - \left(\frac{\gamma+1}{2} - \frac{\gamma-1}{2}\xi^2\right)\right]\eta^{(0)\prime} = 0 \end{aligned} \tag{2.106}$$

where primes again indicate differentiation with respect to ξ. The quantity $\eta^{(0)\prime}$ is generally not zero, thus (2.106) gives the zeroth order solutions

$$\begin{aligned} \theta^{(0)} &= \frac{\gamma+1}{2} - \frac{\gamma-1}{2}\xi^2 \quad \text{and} \\ \eta^{(0)} &= -\log\xi - \frac{1}{\gamma-1}\log\left|\frac{\gamma+1}{2} - \frac{\gamma-1}{2}\xi^2\right| \end{aligned} \tag{2.107}$$

where the integration constant is chosen such that $\theta^{(0)}$ and $\eta^{(0)}$ represent the inviscid solution, when we put $\xi = w$.

The first order equations are now

$$\theta^{(1)} + (\gamma-1)\xi w^{(1)} = -\frac{\gamma-1}{2}b\xi^2 + \frac{\gamma+1}{2}\alpha\frac{\xi^2(1-\beta\xi^2)}{1-\xi^2} \tag{2.108}$$

and

$$\begin{aligned} &(\eta^{(0)\prime})^2\left\{\left(2\xi + \frac{\theta^{(0)}}{\gamma}\right)w^{(1)} + \left(\xi^2 - \frac{\theta^{(0)}}{\gamma}\right)w^{(1)\prime} + \frac{1}{\gamma}\xi^2\frac{d}{d\xi}\left(\frac{\theta^{(1)}}{\xi}\right) - \right. \\ &\left.\frac{1}{\gamma}[\eta^{(0)\prime}(\xi\theta^{(1)} + w^{(1)}\theta^{(0)}) + \xi\theta^{(0)}\eta^{(1)\prime}]\right\} \\ &= \frac{\gamma+1}{2\gamma}\xi^2[\eta^{(0)\prime\prime} + \xi(\eta^{(0)\prime})^3] \end{aligned} \tag{2.109}$$

where α and β are constants defined by

$$\alpha = \frac{\gamma-1}{\gamma}\left[1 - \frac{1}{2\sigma(1+k)}\right], \quad \beta = \frac{\gamma-1}{\gamma+1} \tag{2.110}$$

Equations (2.108) and (2.109) are two equations for three unknowns $\theta^{(1)}$, $w^{(1)}$, and $\eta^{(1)}$. This degree of freedom can be used to control the singularity of the solution. In fact, by substituting the zeroth order solution (2.107) in (2.109), we find that the latter equation is reduced to

$$\frac{d}{d\xi}[\eta^{(0)\prime}w^{(1)}-\eta^{(1)}]=(1-\alpha)\left[\frac{2\xi}{(1-\xi^2)^2}+\gamma\frac{\xi}{\xi^2-1}\right]+\frac{(1-\beta b)(1-\beta)\xi}{\beta(\beta\xi^2-1)^2}+\left[\frac{1-\beta b}{\beta}-(\beta b-\alpha)-\frac{(1+\alpha)2\beta}{1-\beta}\right]\frac{\xi}{\beta\xi^2-1} \tag{2.111}$$

The principle of the PLK method is to choose $w^{(1)}$ such that the resultant $w^{(1)}$ and $\eta^{(1)}$ will have similar singularities. This requirement gives the proper split of (2.111) as follows:

$$\frac{d}{d\xi}[\eta^{(0)\prime}w^{(1)}]=\frac{(1-\beta b)(1-\beta)}{\beta}\frac{\xi}{(\beta\xi^2-1)^2}+\left[\frac{1-\beta b}{\beta}-(\beta b-\alpha)-\frac{(1+\alpha)2\beta}{1-\beta}\right]\frac{\xi}{\beta\xi^2-1} \tag{2.112}$$

and

$$\frac{d\eta^{(1)}}{d\xi}=-(1-\alpha)\left[\frac{2\xi}{(1-\xi^2)^2}+\frac{\gamma\xi}{\xi^2-1}\right] \tag{2.113}$$

The solutions of these equations are

$$w^{(1)}(\xi)=-A\frac{\xi}{1-\xi^2}-B\frac{\xi(1-\beta\xi^2)}{1-\xi^2}\log\left|\frac{\gamma+1}{2}-\frac{\gamma-1}{2}\xi^2\right| \tag{2.114}$$

and

$$\eta^{(1)}(\xi)=-(1-\alpha)\left[\frac{1}{1-\xi^2}+\frac{\gamma}{2}\log|1-\xi^2|\right] \tag{2.115}$$

where

$$A=\frac{(1-\beta b)(1-\beta)}{2\beta^2}\quad\text{and}$$
$$B=\frac{1}{2\beta}\left[\frac{(1-\beta b)}{\beta}-(\beta b-\alpha)-\frac{(1+\alpha)2\beta}{1-\beta}\right] \tag{2.116}$$

The results of (2.114) and (2.115) are obtained by dropping the integration constants. We can, of course, keep these two integration constants, and they can then be considered as two free parameters of the solution. These parameters can be fixed by imposing some boundary conditions. Here the natural boundary conditions at $r\to\infty$ are all satisfied. However, there remain the conditions for pressure, stress, and heat flux rate on an "inside" boundary, $r=r_0$, say. These inside boundary conditions will determine the two integration constants. By dropping the integration constants, we fix the solution to a particular one among the many possible ones.

Further calculations by Wu[9] give the following final solutions:

$$w(\xi) = \xi - \varepsilon\left[A\frac{\xi}{1-\xi^2} + B\frac{\xi(1-\beta\xi^2)}{1-\xi^2}\times \log\left|\frac{\gamma+1}{2}-\frac{\gamma-1}{2}\xi^2\right|\right] + O\left(\frac{\varepsilon^2}{1-\xi^2}\right) \tag{2.117}$$

$$\eta(\xi) = -\left[\log\xi + \frac{1}{\gamma-1}\log\left|\frac{\gamma+1}{2}-\frac{\gamma-1}{2}\xi^2\right|\right] - \varepsilon(1-\alpha)\left[\frac{1}{1-\xi^2}+\frac{\gamma}{2}\log|1-\xi^2|\right] + \varepsilon^2\frac{2(1-\beta)(1-\alpha)}{(1-\xi^2)^4} + O\left(\frac{\varepsilon^2}{(1-\xi^2)^3}, \frac{\varepsilon^3}{(1-\xi^2)^7}\right) \tag{2.118}$$

$$\theta(\xi) = \left(\frac{\gamma+1}{2}-\frac{\gamma-1}{2}\xi^2\right) + \varepsilon(\gamma-1)\left[A\frac{\xi^2}{1-\xi^2} + B\frac{\xi^2(1-\beta\xi^2)}{1-\xi^2}\log\left|\frac{\gamma+1}{2}-\frac{\gamma-1}{2}\xi^2\right| - \frac{b}{2}\xi^2 + \frac{1}{2\beta}\alpha\frac{\xi^2(1-\beta\xi^2)}{1-\xi^2}\right] + O\left(\frac{\varepsilon^2}{(1-\xi^2)^2}\right) \tag{2.119}$$

As $\xi \to 0$, we have $w = 0$, $\eta \to \infty$, $\theta = (\gamma+1)/2$. Therefore (2.118) and (2.119) represent the correct solution for our subsonic sink flow.

Equation (2.118) shows that for $\xi = 1 - \kappa\varepsilon^{1/3}$, where κ a numerical constant of order 1, the magnitudes of the succeeding group of terms are *all the same*, namely $O(\varepsilon^{2/3})$. If ξ is pushed further towards 1, the higher order terms become more important than the lower order terms, and the series for η actually diverges. Therefore, in spite of the PLK method, we fail to obtain a uniformly valid solution beyond $\xi = 1 - \kappa\varepsilon^{1/3}$. Other ways of splitting the original equation (2.111) will not alter this natural limit in the admissible value of ξ. In fact, we can abandon the PLK method and try the classical perturbation method, i.e.,

$$\left.\begin{aligned}\eta &= \eta^{(0)}(w) + \varepsilon\eta^{(1)}(w) + \varepsilon^2\eta^{(2)}(w) + \ldots\\ \theta &= \theta^{(0)}(w) + \varepsilon\theta^{(1)}(w) + \varepsilon^2\theta^{(2)}(w) + \ldots\end{aligned}\right\} \tag{2.120}$$

The limited range of the solution now again appears in the series for η and is effectively the same as previously. One objection to the expansion of (2.120) is that there are now some spurious singularities in $\eta^{(1)}(w)$ at $w = \beta^{-1/2}$, corresponding to supersonic flow speed. Therefore from the point of view of generality of solution, the PLK method is definitely preferable. Furthermore, if the solution is to be pursued to a higher order, it will be safer to use a method which allows a certain control of the singularities.

To continue the solution beyond the limit $\xi = 1 - \kappa\varepsilon^{1/3}$, we have to use the "boundary layer method." The solution so far obtained gives however the necessary junction conditions. Taking $\kappa = 2/(\gamma+1)^{1/3}$, we find that the series for η is rapidly convergent, and the terms explicitly shown in (2.118) are sufficient for numerical calculations. In fact, for $k = -\dfrac{1}{3}$, $\left(\text{corresponding to } \mu' = 0 \text{ and } \sigma = \dfrac{3}{4} \text{ and implying } \alpha = 0\right)$, we have

$$\left.\begin{aligned} \eta &= 1.766\,(\gamma+1)^{1/3}\varepsilon^{2/3} \\ \frac{dw}{d\eta} &= -0.478\,\frac{\varepsilon^{-1/3}}{(\gamma+1)^{2/3}}+0.17 \\ \theta &= 1+O(\varepsilon^{1/3}) \end{aligned}\right\} \quad \text{at} \quad w = 1-2\left(\frac{\varepsilon}{\gamma+1}\right)^{1/3}+1.75\,(\gamma+1)^{1/3}\varepsilon^{2/3} \tag{2.121}$$

The boundary layer method requires a modification of the independent variable by a factor depending on ε together with expansions of the dependent variables. The results of the PLK method as given by (2.121) naturally suggest that the new independent variable ζ be defined by

$$\eta = \varepsilon^{2/3}\zeta \tag{2.122}$$

Accordingly, w and θ are expanded as

$$\left.\begin{aligned} w &= 1+\varepsilon^{1/3}w^{(1)}(\zeta)+\varepsilon^{2/3}w^{(2)}(\zeta)+\varepsilon w^{(3)}(\zeta)+\dots \\ \theta &= 1+\varepsilon^{1/3}\theta^{(1)}(\zeta)+\varepsilon^{2/3}\theta^{(2)}(\zeta)+\varepsilon\theta^{(3)}(\zeta)+\dots \end{aligned}\right\} \tag{2.123}$$

By setting $\alpha = 0$, in accordance with (2.121), and by substituting (2.122) and (2.123) in our original set of differential equations (2.102) and (2.103), we have as first order equations

$$\frac{d^2\omega^{(1)}}{d\zeta^2}+2\omega^{(1)}\frac{d\omega^{(1)}}{d\zeta} = 1-\beta \tag{2.124}$$

and

$$\theta^{(1)}(\zeta) = -(\gamma-1)\omega^{(1)}(\zeta)$$

The second order equations are

$$\left.\begin{aligned} \frac{d^2\omega^{(2)}}{d\zeta^2}+2\frac{d}{d\zeta}(\omega^{(1)}\omega^{(2)}) &= \frac{d}{d\zeta}(\omega^{(1)})^3-(1+\beta)\omega^{(1)} \\ \theta^{(2)}(\zeta) &= -(\gamma-1)\left[\omega^{(2)}(\zeta)+\frac{1}{2}\omega^{(1)^2}(\zeta)\right] \end{aligned}\right\} \tag{2.125}$$

The boundary conditions for these equations can be deduced from the junction conditions (2.121). Thus, at $\zeta = 1.766\,(\gamma+1)^{1/3}$,

$$\left.\begin{aligned} \omega^{(1)} &= -2/(\gamma+1)^{1/3}, \quad & \frac{d\omega^{(1)}}{d\zeta} &= -0.478/(\gamma+1)^{2/3} \\ \omega^{(2)} &= 1.75\,(\gamma+1)^{1/3}, \quad & \frac{d\omega^{(2)}}{d\zeta} &= 0.17 \end{aligned}\right\} \tag{2.126}$$

We have now formulated the boundary layer problem. Wu has carried out the numerical calculation for $\omega^{(1)}$. We shall not pursue these details here. But what has been discussed here serves to show the way of combining the PLK method with the boundary layer method in a physical problem, where the PLK method alone is insufficient.

Ⅲ. Hyperbolic Partial Differential Equations

1. Generalization to Hyperbolic Equations

In this section we shall see that the procedures of Sections Ⅱ. 1 to Ⅱ. 4 developed in connection with ordinary differential equations can be easily generalized to hyperbolic partial differential equations in two independent variables. Previously we were concerned with the solution near the regular singularity ($x=0$) of the zeroth order equation. The purpose of the PLK method is to make the perturbation solution convergent down to this singular point. For hyperbolic partial differential equations, the singular point is replaced by a whole line, the singular characteristic, near which the classical perturbation method fails to give a useful solution. That the line must be a characteristic can be seen as follows: Let us introduce a system x, y of curvilinear coordinates, such that the line of singularity is represented by $x=0$. At this line, the zeroth order solution $v^{(0)}$ has an algebraic or logarithmic singularity. This situation, quite analogous to that discussed previously, means that the coefficient of $\partial^2 v^{(0)}/\partial x^2$ in the zeroth order equation must tend to zero as $x \to 0$, while the coefficients of other second order derivatives remain finite. In other words, near $x=0$, the zeroth order differential equation is as follows:

$$x\frac{\partial^2 v^{(0)}}{\partial x^2}+B\frac{\partial^2 v^{(0)}}{\partial x\partial y}+C\frac{\partial^2 v^{(0)}}{\partial y^2}=\text{terms with first derivatives} \tag{3.1}$$

where B and C are different from zero at $x=0$. The variations $\mathrm{d}x$ and $\mathrm{d}y$ along a characteristic are given by

$$x(\mathrm{d}y)^2-B(\mathrm{d}x)(\mathrm{d}y)+C(\mathrm{d}x)^2=0 \tag{3.2}$$

At $x=0$, (3.2) gives $\mathrm{d}x=0$. The characteristic curve is thus indeed the line $x=0$.

Moreover, we shall now indicate that any hyperbolic equation will give a zeroth order equation of the form (3.1), and thus the classical perturbation method may run into difficulties at a characteristic corresponding to $x=0$. Let us write the zeroth order equation in the normal form with characteristic coordinates μ and ν, namely,

$$\frac{\partial^2 v^{(0)}}{\partial\mu\partial\nu}=\text{terms independent of second order derivatives} \tag{3.3}$$

If now we introduce the coordinate transformation

$$x=\mu\nu,\quad y=y(\mu,\nu) \tag{3.4}$$

we have

$$\begin{aligned}\frac{\partial^2 v^{(0)}}{\partial\mu\partial\nu}&=\frac{\partial}{\partial\mu}\left[\frac{\partial v^{(0)}}{\partial\nu}\right]=\frac{\partial}{\partial\mu}\left[\frac{\partial x}{\partial\nu}\frac{\partial v^{(0)}}{\partial x}+\frac{\partial y}{\partial\nu}\frac{\partial v^{(0)}}{\partial y}\right]\\&=\left(\frac{\partial x}{\partial\mu}\frac{\partial x}{\partial\nu}\right)\frac{\partial^2 v^{(0)}}{\partial x^2}+\left(\frac{\partial x}{\partial\nu}\frac{\partial y}{\partial\mu}\right)\frac{\partial^2 v^{(0)}}{\partial x\partial y}+\left(\frac{\partial y}{\partial\mu}\frac{\partial y}{\partial\nu}\right)\frac{\partial^2 v^{(0)}}{\partial y^2}+\frac{\partial^2 x}{\partial\mu\partial\nu}\frac{\partial v^{(0)}}{\partial x}+\frac{\partial^2 y}{\partial\mu\partial\nu}\frac{\partial v^{(0)}}{\partial y}\end{aligned}$$

$$= x\frac{\partial^2 v^{(0)}}{\partial x^2} + \mu\frac{\partial y}{\partial \mu}\frac{\partial^2 v^{(0)}}{\partial x \partial y} + \left(\frac{\partial y}{\partial \mu}\frac{\partial y}{\partial \nu}\right)\frac{\partial^2 v^{(0)}}{\partial y^2} + \frac{\partial v^{(0)}}{\partial x} + \frac{\partial^2 y}{\partial \mu \partial \nu}\frac{\partial v^{(0)}}{\partial y}$$

Thus, by writing equation (3.3) in independent variables x and y, we obtain an equation of the form of (3.1). The original exact equation (before the application of the perturbation method) must therefore be of the form

$$\left.\begin{aligned} &\frac{\partial v}{\partial x} = u \\ &\left[x + \varepsilon p_1\left(x, y, v, u, \frac{\partial v}{\partial y}, \frac{\partial u}{\partial y}, \frac{\partial^2 v}{\partial y^2}\right) + \ldots\right]\frac{\partial u}{\partial x} = \text{terms in } \varepsilon, x, y, u, v, \frac{\partial v}{\partial y}, \frac{\partial u}{\partial y}, \frac{\partial^2 v}{\partial y^2} \end{aligned}\right\} \tag{3.5}$$

The right-hand side of the second equation (3.5) is linear in x, y and the derivatives when $\varepsilon = 0$. This is now in a form very similar to our basic equation (2.1) of Section II.1. We are thus encouraged to use the same method. To treat (3.5) by that method we introduce

$$\left.\begin{aligned} u &= u^{(0)}(\xi, \eta) + \varepsilon u^{(1)}(\xi, \eta) + \ldots \\ v &= v^{(0)}(\xi, \eta) + \varepsilon v^{(1)}(\xi, \eta) + \ldots \\ x &= \xi + \varepsilon x^{(1)}(\xi, \eta) + \varepsilon^2 x^{(2)}(\xi, \eta) + \ldots \\ y &= \eta \end{aligned}\right\} \tag{3.6}$$

The variable y is not expanded, because our difficulty with the perturbation solution is associated with the variable x, but not with y. We shall see how the PLK method works in the present instance through an example.

Let us consider the following equation:

$$\left.\begin{aligned} &\frac{\partial u}{\partial y} = \varepsilon\left(u + \frac{\partial v}{\partial y}\right)\frac{\partial u}{\partial x} \\ &\frac{\partial v}{\partial x} = u \end{aligned}\right\} \tag{3.7}$$

To use the expansions (3.6), we first have to compute the derivatives $\partial/\partial x$ and $\partial/\partial y$. Now

$$\begin{aligned} \frac{\partial}{\partial x} &= \frac{\partial \eta}{\partial x}\frac{\partial}{\partial \eta} + \frac{\partial \xi}{\partial x}\frac{\partial}{\partial \xi}, \\ \frac{\partial}{\partial y} &= \frac{\partial \eta}{\partial y}\frac{\partial}{\partial \eta} + \frac{\partial \xi}{\partial y}\frac{\partial}{\partial \xi} \end{aligned} \tag{3.8}$$

But from the last equation (3.6) we have clearly

$$\frac{\partial \eta}{\partial x} = 0, \quad \frac{\partial \eta}{\partial y} = 1 \tag{3.9}$$

By differentiating the x-expansion with respect to y and x, we have then, using (3.9),

$$\left.\begin{aligned} 0 &= \frac{\partial \xi}{\partial y}\left[1+\varepsilon\frac{\partial x^{(1)}}{\partial \xi}+\varepsilon^2\frac{\partial x^{(2)}}{\partial \xi}+\ldots\right]+\left[\varepsilon\frac{\partial x^{(1)}}{\partial \eta}+\varepsilon^2\frac{\partial x^{(2)}}{\partial \eta}+\ldots\right] \\ 1 &= \frac{\partial \xi}{\partial x}\left[1+\varepsilon\frac{\partial x^{(1)}}{\partial \xi}+\varepsilon^2\frac{\partial x^{(2)}}{\partial \xi}+\ldots\right] \end{aligned}\right\} \tag{3.10}$$

Solving (3.10) for $\partial\xi/\partial x$ and $\partial\xi/\partial y$ and substituting the result in (3.8), we have

$$\left.\begin{aligned} \frac{\partial}{\partial x} &= \frac{1}{1+\varepsilon\dfrac{\partial x^{(1)}}{\partial \xi}+\varepsilon^2\dfrac{\partial x^{(2)}}{\partial \xi}+\ldots}\frac{\partial}{\partial \xi} \\ \frac{\partial}{\partial y} &= \frac{\partial}{\partial \eta}-\frac{\varepsilon\dfrac{\partial x^{(1)}}{\partial \eta}+\varepsilon^2\dfrac{\partial x^{(2)}}{\partial \eta}+\ldots}{1+\varepsilon\dfrac{\partial x^{(1)}}{\partial \xi}+\varepsilon^2\dfrac{\partial x^{(2)}}{\partial \xi}+\ldots}\frac{\partial}{\partial \xi} \end{aligned}\right\} \tag{3.11}$$

Then the original equation (3.7) can be written as

$$\begin{aligned} &\left[1+\varepsilon\frac{\partial x^{(1)}}{\partial \xi}+\varepsilon^2\frac{\partial x^{(2)}}{\partial \xi}+\ldots\right]\left[\frac{\partial u^{(0)}}{\partial \eta}+\varepsilon\frac{\partial u^{(1)}}{\partial \eta}+\ldots\right]- \\ &\left[\varepsilon\frac{\partial x^{(1)}}{\partial \eta}+\varepsilon^2\frac{\partial x^{(2)}}{\partial \eta}+\ldots\right]\left[\frac{\partial u^{(0)}}{\partial \xi}+\varepsilon\frac{\partial u^{(1)}}{\partial \xi}+\ldots\right] \\ &=\left[\varepsilon\frac{\partial u^{(0)}}{\partial \xi}+\varepsilon^2\frac{\partial u^{(1)}}{\partial \xi}+\ldots\right]\times \\ &\left[u^{(0)}+\varepsilon u^{(1)}+\ldots+\frac{\partial v^{(0)}}{\partial \eta}+\varepsilon\frac{\partial v^{(1)}}{\partial \eta}+\ldots-\right. \\ &\left.\frac{\left(\varepsilon\dfrac{\partial x^{(1)}}{\partial \eta}+\varepsilon^2\dfrac{\partial x^{(2)}}{\partial \eta}+\ldots\right)\left(\dfrac{\partial v^{(0)}}{\partial \xi}+\varepsilon\dfrac{\partial v^{(1)}}{\partial \xi}+\ldots\right)}{1+\varepsilon\dfrac{\partial x^{(1)}}{\partial \xi}+\varepsilon^2\dfrac{\partial x^{(2)}}{\partial \xi}+\ldots}\right] \end{aligned} \tag{3.12}$$

and

$$\left[\frac{\partial v^{(0)}}{\partial \xi}+\varepsilon\frac{\partial v^{(1)}}{\partial \xi}+\ldots\right]=\left[1+\varepsilon\frac{\partial x^{(1)}}{\partial \xi}+\varepsilon^2\frac{\partial x^{(2)}}{\partial \xi}+\ldots\right][u^{(0)}+\varepsilon u^{(1)}+\ldots] \tag{3.13}$$

The zeroth order equations are then

$$\left.\begin{aligned} \frac{\partial u^{(0)}}{\partial \eta} &= 0 \\ \frac{\partial v^{(0)}}{\partial \xi} &= u^{(0)} \end{aligned}\right\} \tag{3.14}$$

The solution is

$$u^{(0)}=u^{(0)}(\xi),\quad v^{(0)}=\int u^{(0)}(\xi)\mathrm{d}\xi+F(\eta) \tag{3.15}$$

By using (3.14), the first order equations are found as

$$\left.\begin{aligned}\frac{\partial u^{(1)}}{\partial \eta} &= \frac{\partial u^{(0)}}{\partial \xi}\left[\frac{\partial x^{(1)}}{\partial \eta} + u^{(0)} + \frac{\partial v^{(0)}}{\partial \eta}\right] \\ \frac{\partial v^{(1)}}{\partial \xi} &= \frac{\partial x^{(1)}}{\partial \xi} u^{(0)} + u^{(1)}\end{aligned}\right\} \tag{3.16}$$

Now if the initial conditions of the problem are such as to require

$$u^{(0)} \sim A\xi^{-q_0} \tag{3.17}$$

with $q_0 > 0$, the worst term in the bracket on the right hand side of the first equation (3.16) is $u^{(0)}$. This term can be cancelled, however, if for $\xi \to 0$,

$$\frac{\partial x^{(1)}}{\partial \eta} = - A\xi^{-q_0} \quad \text{or} \quad x^{(1)} = - A\eta\xi^{-q_0} \tag{3.18}$$

When this is done, $u^{(1)}$ for $\xi \to 0$ is no worse than $O(\xi^{-2q_0})$. The second equation (3.16) then shows that $v^{(1)}$ is also $O(\xi^{-2q_0})$. Therefore the behavior of the expansions here is very much the same as in the case of ordinary differential equations discussed in Section Ⅱ.2. For a fixed η, the series have much the same character as (2.13). Thus just as for the ordinary differential equations, the PLK method is sufficient to produce a valid perturbation solution down to $x = 0$. A first approximation to the solution near $x = 0$ is then

$$u = u^{(0)}(\xi), \quad x = \xi - \varepsilon A\eta\xi^{-q_0}, \quad v = \int u^{(0)}(\xi)\,d\xi + F(\eta) \tag{3.19}$$

Thus if $A\eta < 0$, there is a real line of branch points at $\partial x/\partial \xi = 0$ or

$$\xi = (-\varepsilon A\eta q_0)^{1/1+q_0}, \quad x = \left(1 + \frac{1}{q_0}\right)(-\varepsilon A\eta q_0)^{1/(1+q_0)} \tag{3.20}$$

If the initial conditions of the problem are such that near $\xi = 0$,

$$u^{(0)}(\xi) = u^{(0)}(0) + A\xi^{-q_0} \tag{3.21}$$

with $q_0 \leqslant -1$, then the value of $u^{(0)}$ near $\xi = 0$ is represented by $u^{(0)}(0)$ and $v^{(0)}$ by $v^{(0)}(0, \eta)$. In order to make the order of the singularity of $u^{(1)}$ the same as of $u^{(0)}$, we have to put

$$x^{(1)} = - \eta u^{(0)}(0) - v^{(0)}(0, \eta) \tag{3.22}$$

But if q_0 in (3.21) is such that $-1 < q < 0$, then we achieve the same purpose by requiring

$$x^{(1)} = - \eta[u^{(0)}(0) + A\xi^{-q_0}] - v^{(0)}(0, \eta) \tag{3.23}$$

The above example shows that the technique used here and the results obtained are quite similar to our previous treatment of the ordinary differential equation of Section Ⅱ.1. However, there is one essential difference: While the point of singularity for the ordinary differential equation is fixed at the point $x = 0$, and the value of q_0 is explicitly given by the differential equation itself, our equation (3.7) gives no such explicit information. In fact the so-called singularity can occur at any x, and q_0 is only specified by the initial condition, and not known otherwise. The zeroth order solution can have any one of the properties discussed above at different values of ξ. Therefore our treatment, with the discussion centered on individual points of ξ, although useful for the understanding of how the PLK method can be extended to hyperbolic equations, is not effective in obtaining an over-all picture of the solution. For an insight into the situation, we observe first that the nonuniformity of the classical perturbation series is caused by the terms on the right-hand side of the first equation (3.7). In fact, the most detrimental term is $\partial u/\partial x$ or $\partial^2 v/\partial x^2$ as seen from (3.16). This term only appears because the independent variables x, y in our equation (3.7) are the characteristic variables of the zeroth order equation, but not that of the exact full equation. If ξ, η are the exact characteristic coordinates, then the normal form of the equation is

$$\frac{\partial^2 v}{\partial \xi \partial \eta} = \text{terms independent of the second derivatives} \tag{3.24}$$

and the classical perturbation procedure will work. Thus the point is to use the exact characteristics instead of the characteristics of the zeroth order equation. We shall see that this is really what the PLK method tries to accomplish in the present case.

The system (3.7) can be written as a single second order equation

$$\frac{\partial^2 v}{\partial x \partial y} - \varepsilon\left(\frac{\partial v}{\partial x} + \frac{\partial v}{\partial y}\right)\frac{\partial^2 v}{\partial x^2} = 0 \tag{3.25}$$

Therefore the variations of dx and dy along an exact characteristic are related as follows:

$$-(dx)(dy) - \varepsilon\left(\frac{\partial v}{\partial x} + \frac{\partial v}{\partial y}\right)(dy)^2 = 0$$

Hence, if we indicate the exact characteristic variables by ξ and η, then

$$d\xi = dx + \varepsilon\left(\frac{\partial v}{\partial x} + \frac{\partial v}{\partial y}\right)dy$$

$$d\eta = dy$$

or

$$\left.\begin{aligned} \xi &= x + \varepsilon\int\left(\frac{\partial v}{\partial x} + \frac{\partial v}{\partial y}\right)dy \\ \eta &= y \end{aligned}\right\} \tag{3.26}$$

where the integration is to be carried out along lines of constant ξ. Equation (3.26) can be written in a form that conforms to (3.6). For instance, we have up to first order accuracy,

$$x = \xi - \varepsilon\int\left(\frac{\partial v^{(0)}}{\partial \xi} + \frac{\partial v^{(0)}}{\partial \eta}\right)d\eta, \quad y = \eta \tag{3.27}$$

This is exactly what our previous results in (3.18), (3.22), and (3.23) specify. For the present problem then, the independent variables ξ and η in (3.6), introduced for the PLK method, are nothing but the exact characteristic parameters of the hyperbolic equation. The mathematical implications of this fact will be discussed in Section Ⅲ.5.

2. Progressive Wave Far from Source

In this section we shall discuss another difficulty of the classical perturbation method when applied to hyperbolic partial differential equations, the physical character of which is quite different from that of the preceding section. It has to do with the natural spreading of a progressive wave, sometimes also called a simple wave, after it has travelled a large distance compared with its width. Such progressive waves are described by a hyperbolic equation which is approximately linear, but its exact form is quasilinear, the coefficients of the second order derivatives being functions of the lower derivatives of the unknown variable. Therefore, if (x, y) are the characteristics of the linearized approximate equation, the full equation can be written as follows:

$$\frac{\partial^2 v}{\partial x \partial y} + F = A\frac{\partial^2 v}{\partial x^2} + B\frac{\partial^2 v}{\partial x \partial y} + C\frac{\partial^2 v}{\partial y^2} + D \tag{3.28}$$

where F is linear in v, $\partial v/\partial x$ and $\partial v/\partial y$, and A, B, C are at least linear in v, $\partial v/\partial x$, $\partial v/\partial y$, but may be of higher order; D is at least of second order. When the waves are weak, the right-hand side of (3.28) can be neglected and we have the approximate equation

$$\frac{\partial^2 v}{\partial x \partial y} + F = 0 \tag{3.29}$$

showing that x, y are indeed the characteristic variables of the linearized equation. For wave propagation in the r-direction with constant velocity a_0, x and y are $a_0 t - r$ and $a_0 t + r$.

Now if $F \to 0$ as $x \to \infty$, one would conclude from (3.29) at first sight that v is propagated unchanged along the characteristic $y =$ constant. However, this conclusion is really erroneous. For, along $y =$ constant, some of the terms on the right-hand side of (3.28), so far neglected, may have constant sign and, on integration along y to large values of x, will produce an accumulated effect far more important than the effects of F which we take into account. The nonlinear terms, though negligible for small x, are thus essential for the correct description of the physical phenomena at large x. Such accumulative influences of waves far from the source have been stressed by Hayes[10] and made the basis of his pseudo-transonic similarity rule.

To see this effect in more detail, let us suppose that, as $x \to \infty$ while $y = O(1)$ corresponding to progressive waves, F is given by

$$F = \frac{\partial v}{\partial y}\left[\frac{n}{x} + O\left(\frac{1}{x^2}\right)\right] + \frac{\partial v}{\partial x} O\left(\frac{1}{x}\right) + vO\left(\frac{1}{x^2}\right) \tag{3.30}$$

where $n \geqslant 0$. For the particular cases of plane, cylindrical, and spherical waves, n is equal to 0, $\frac{1}{2}$, and 1, respectively. The coefficients in D are $O(1/x)$, but A, B, C are $O(1)$. The linearized equation is then approximately

$$0=\frac{\partial^2 v}{\partial x\partial y}+\frac{n}{x}\frac{\partial v}{\partial y}=x^{-n}\frac{\partial^2}{\partial x\partial y}(x^n v)$$

or

$$v\sim\frac{v^{(0)}(y)}{x^n} \tag{3.31}$$

This is the leading term in the solution of the linearized equation. $x^n v^{(0)}(y)$ will then be propagated unchanged along the characteristic $y=$ const. To improve this solution we substitute

$$v=\frac{v^{(0)}(y)}{x^n}+\frac{v^{(1)}(y)}{x^{n+1}}+\frac{v^{(2)}(y)}{x^{n+2}}+\dots \tag{3.32}$$

By substituting (3. 32) in (3. 29) with F given by (3. 30), we can determine $v^{(1)}(y)$, $v^{(2)}(y)$, etc. In other words, the linearized equation will produce a solution in descending powers of x, starting with $v^{(0)}(y)/x^n$. But this conclusion is seriously changed by the nonlinear terms on the right-hand side of (3.29). The most serious term is $C\partial^2 v/\partial y^2$. Since C contains v, $\partial v/\partial x$, $\partial v/\partial y$, the term $C\partial^2 v/\partial y^2$ may be $O(1/x^{2n})$. Clearly then, at large distances, the nonlinear terms are equally important as the linear terms, and, if $n<1$ the series solution (3.32) is inappropriate. Such changes in the relative importance of the nonlinear terms from small x to large x make the classical perturbation solution break down at $x\to\infty$. The application of the PLK method is again indicated.

3. Solution for Progressive Waves

In order to facilitate our discussion, (3.28) is written in the following form so that the most important derivatives are of first order:

$$\left.\begin{aligned}&\frac{\partial u}{\partial x}+F=A\frac{\partial^2 v}{\partial x^2}+B\frac{\partial u}{\partial x}+C\frac{\partial u}{\partial y}+D\\&\frac{\partial v}{\partial y}=u\end{aligned}\right\} \tag{3.33}$$

As shown in the previous section, the linearized solution is

$$u\sim v^{(0)\prime}(y)x^{-n}=u^{(0)}(y)x^{-n} \tag{3.34}$$

Now let us determine the line along which the quantity $x^n u$ is propagated unchanged. On such a line

$$\mathrm{d}(x^n u)=0=\left(nx^{n-1}u+x^n\frac{\partial u}{\partial x}\right)\mathrm{d}x+x^n\frac{\partial u}{\partial y}\mathrm{d}y$$

Thus the slope of such a line is

$$\frac{\mathrm{d}y}{\mathrm{d}x}=-\left(\frac{\partial u}{\partial x}+\frac{nu}{x}\right)\Big/\frac{\partial u}{\partial y} \tag{3.35}$$

According to the linearized equations (3.29) and (3.30), $\mathrm{d}y/\mathrm{d}x$ is zero. But actually $\mathrm{d}y/\mathrm{d}x$ is small but not zero. This means that $x^n u$ changes along $y =$ constant, but remains unchanged on some line which is slightly inclined relative to $y =$ constant. As x changes from a small value to a very large value, i. e., the waves have propagated far from their source, this line may deviate far from the initial line $y =$ constant. But since a constant value of $x^n u$ is carried by this line, the nonlinear terms will alter the solution drastically as $x \to \infty$.

But what are the lines with constant values of $x^n u$? For hyperbolic partial differential equations, such lines must be characteristics. In fact, this can also be seen by a different consideration. We note that the trouble at $x = \infty$ is caused by the term $C\partial^2 v/\partial y^2$ in (3.28). But the very existence of this term in (3.28) means that the coordinate system is not the true characteristic coordinate, as indeed x, y is the characteristic coordinate of the linearized equation. If we use the true characteristic coordinates, we will not have this difficulty. The variations of $\mathrm{d}x$ and $\mathrm{d}y$ along a true characteristic are given by

$$-(\mathrm{d}x)(\mathrm{d}y) = A(\mathrm{d}y)^2 - B(\mathrm{d}x)(\mathrm{d}y) + C(\mathrm{d}x)^2 \tag{3.36}$$

where A, B, C are by definition small in magnitude. The characteristic lines thus have slopes specified by

$$\frac{\mathrm{d}y}{\mathrm{d}x} = \frac{1}{2A}\left[-1 + B - \sqrt{(1-B)^2 - 4AC}\right] \cong -\frac{1}{0}$$

and

$$\frac{\mathrm{d}y}{\mathrm{d}x} = \frac{1}{2A}\left[-1 + B + \sqrt{(1-B)^2 - 4AC}\right] \cong -0$$

Thus to a first approximation, with emphasis on characteristic lines of small slope $\mathrm{d}y/\mathrm{d}x$, we can use the characteristic coordinates defined by

$$\xi = x, \quad \eta = y + \int C\mathrm{d}\xi \tag{3.37}$$

where the integration is along $\eta =$ constant. The correct independent variables to use are thus ξ and η, defined above. In fact, the principle of the solution is as follows: If, in the linearized problem, u, which is the derivative of v across the characteristic, can be expanded in descending powers of x with coefficients constant on each approximate characteristic $y =$ constant, seek a similar expansion with coefficients constant on each exact characteristic $\eta =$ constant, finding the latter curve by a second similar expansion of x. We see that in the present case the variable to be modified is y, and not x as in Section Ⅲ.1.

To a first approximation, then, we have according to (3.37)

$$\begin{aligned} x &= \xi \\ y &= \eta - \int C\mathrm{d}\xi \end{aligned} \tag{3.38}$$

Since in general $C = O(\xi^{-n})$, this means that $y = \eta + O(\xi^{1-n})$ or $\eta + O(\log \xi)$ if $n = 1$. The difference $y - \eta$ is indefinite as $\xi \to \infty$. For the case of plane waves, $n = 0$; the characteristics are straight lines fanning out. For the case of cylindrical waves, $n = \frac{1}{2}$; the characteristics are parabolas. For the case of spherical waves, $n = 1$; the characteristics spread out at a logarithmic rate. It is interesting to note that the behavior of v at large x is now seriously altered: According to the second equation of (3.33)

$$\frac{\partial v}{\partial \eta} = u \frac{\partial y}{\partial \eta} \sim u^{(0)}(\eta)\xi^{-n}\left(-\frac{\partial}{\partial \eta}\int C \mathrm{d}\xi\right) = O(\xi^{1-2n}) \tag{3.39}$$

while $u = O(\xi^{-n})$. The reason for this is of course that now the distance between the curves of constant η is $O(\xi^{1-n})$ or $O(\log \xi)$, $n = 1$ and thus greatly increases the magnitude of v for $x \to \infty$.

As an example of the application of the method explained in the preceding paragraphs, let us consider the problem of propagation of a spherical blast treated by Whitham[11]. Since the problem concerns mainly the motion at large distances from the center of blast, the motion is weak and can be calculated by assuming isentropic flow. The equation of spherically symmetrical isentropic motion is

$$a^2 \nabla^2 \phi = \frac{\partial^2 \phi}{\partial t^2} + 2\frac{\partial \phi}{\partial r}\frac{\partial^2 \phi}{\partial r \partial t} + \left(\frac{\partial \phi}{\partial r}\right)^2 \frac{\partial^2 \phi}{\partial r^2} \tag{3.40}$$

where ϕ is the velocity potential, r the radial distance, t the time, and a is the local velocity of sound given by the Bernoulli equation

$$a^2 = a_0^2 - (\gamma - 1)\left[\frac{\partial \phi}{\partial t} + \frac{1}{2}\left(\frac{\partial \phi}{\partial r}\right)^2\right] \tag{3.41}$$

Here a_0 is the velocity of sound in the undisturbed air. With $u = \partial\phi/\partial t$, $v = \partial\phi/\partial r$, the equations of motion may be written as

$$\left.\begin{aligned}
&\frac{\partial v}{\partial t} - \frac{\partial u}{\partial r} = 0\\
&\frac{\partial v}{\partial r}\left[a_0^2 - (\gamma-1)u + \frac{1}{2}(\gamma+1)v^2\right] - \frac{\partial u}{\partial t} - 2v\frac{\partial v}{\partial t} + \\
&\frac{2v}{r}\left[a_0^2 - (\gamma-1)u - \frac{1}{2}(\gamma-1)v^2\right] = 0
\end{aligned}\right\} \tag{3.42}$$

In the linearized theory, $\phi = f_0(a_0 t - r)/r$ for outgoing waves of the present problem; hence u and v are of the form

$$\left.\begin{aligned}
u &= \frac{f_1(a_0 t - r)}{r}\\
v &= -\frac{u}{a_0} + \frac{f_2(a_0 t - r)}{r^2}
\end{aligned}\right\} \tag{3.43}$$

that is, u and v are expanded in negative powers of r with coefficients constant on each approximate characteristic $a_0 t - r =$ constant. Thus expansions for u and v of a similar form are sought with coefficients constant on each *exact* characteristic $\eta =$ constant, where η is a function of r and t to be determined in the process. Hence u and v are assumed to be of the form

$$\left.\begin{aligned} u &= a_0^2\,[f(\eta)r^{-1} + g(\eta)r^{-2} + \ldots] \\ v &= -\frac{u}{a_0} + a_0\,[b(\eta)r^{-2} + c(\eta)r^{-3} + \ldots] \end{aligned}\right\} \tag{3.44}$$

Substitution of these in the condition expressing that $\eta =$ constant is a characteristic suggests a similar expansion for

$$a_0 t = r - \eta \log r - h(\eta) - m(\eta) r^{-1} \tag{3.45}$$

In fact, the first two terms of this expansion are just what we expected from our general theory. However, it is found that (3. 44) requires a modification with a corresponding modification of (3. 45) so that the equations of motion can be satisfied. The modification consists in replacing g, b, c, and m by $g_1(\eta)\log r + g_2(\eta)$, $b_1(\eta)\log r + b_2(\eta)$, $c_1(\eta)\log r + c_2(\eta)$, and $m_1(\eta)\log r + m_2(\eta)$, respectively. When this is done, equations (3. 42) are satisfied.

The only additional condition is that $\eta =$ constant is a characteristic of the system (3.42), i.e., along $\eta =$ constant, $(\mathrm{d}t/\mathrm{d}r)$ must satisfy the condition

$$\left(\frac{\mathrm{d}t}{\mathrm{d}r}\right)_\eta^2\left[a_0^2 - (\gamma - 1)u - \frac{\gamma+1}{2}v^2\right] + 2v\left(\frac{\mathrm{d}t}{\mathrm{d}r}\right)_\eta - 1 = 0 \tag{3.46}$$

By substituting the modified series for u, v, t into the last expression, and equating equal powers of r and $\log r$, we find that $g_1(\eta)$ is identically zero and all unknown functions of η can be expressed in terms of $h(\eta)$ with certain constants. The solution is

$$u = a_0^2\left[-\frac{k\eta}{r} + \frac{\kappa_1\eta^2 + \frac{1}{2}B_1}{r^2} + \ldots\right] \tag{3.47}$$

$$v = -\frac{u}{a_0} - a_0\left[\frac{\left(\frac{1}{2}k\eta^2 + B_1\right)\log r + \frac{1}{2}k\eta^2 + k\int_0^\eta \xi h'(\xi)\,\mathrm{d}\xi + B_2}{r^2}\right] + \ldots \tag{3.48}$$

$$a_0 t = r - \eta\log r - h(\eta) - \frac{\left(\frac{1}{2}k\eta^2 + B_1\right)\log r + \kappa_2\eta^2 + \frac{1}{4}(\gamma+5)B_1 + \int_0^\eta \xi h'(\xi)\,\mathrm{d}\xi + B_2}{r} + \ldots \tag{3.49}$$

where B_1 and B_2 are arbitrary constants so far undetermined, and

$$k = \frac{2}{\gamma + 1}, \quad \kappa_1 = k^2 - \frac{k}{4}, \quad \kappa_2 = \frac{5}{4} + \frac{3}{2}k \tag{3.50}$$

To determine B_1 and B_2 we have to use the condition at the leading shock S. There is another trailing shock S_1. In the (r, t) plane, the configuration of the shocks is as shown in Fig. 4. Due to the interaction with wavelets in the region between S and S_1, S is retarded and S_1 is accelerated as the shocks progress. As $r \to \infty$, both shocks degenerate to zero strength and are finally propagated with sound velocity a_0.

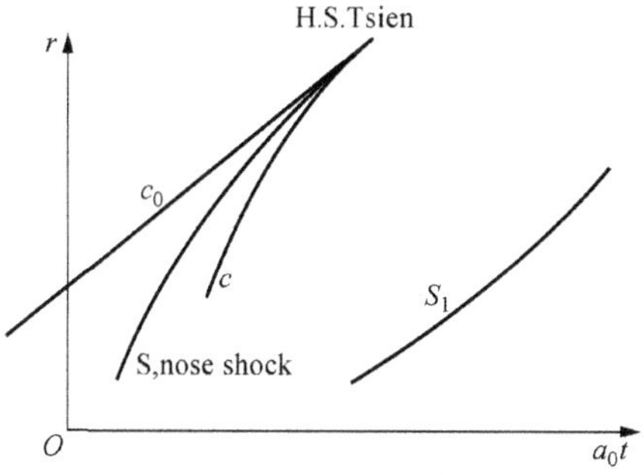

Fig. 4 Representation of spherical blast wave in the t, r plane

There are two boundary conditions to be satisfied at a shock and the most convenient forms for these are: (i) the "angle property" which says that, to the first order in the strength of the shock, the angles that the shock makes in the (r, a_0t) plane with the characteristics on each side of it are equal; and (ii) ϕ is continuous across the shock, hence $\partial\phi/\partial r + (\partial\phi/\partial t)/U = v + u/U$ takes the same value on each side, where U is the velocity of the shock.

Let C_0 be a characteristic of the undisturbed region ahead of S, and let C be a characteristic of the region between the shocks; then the equation of C is given by (3.49):

$$a_0 t = r - \eta \log r - h(\eta) + O(r^{-1}\log r), \quad \eta = \text{const on } C \tag{3.51}$$

For any fixed r, the value of t is bounded, because the characteristic must lie between the two shocks S and S_1, and hence η and $h(\eta)$ are bounded in this region. Let the equation of S be

$$a_0 t = r - f(r) \quad \text{on} \quad S \tag{3.52}$$

then from the angle property (i) and the fact that the equation for C_0 is

$$a_0 t = r + \text{const}$$

we have

$$f'(r) = \frac{1}{2}\eta r^{-1} + O(r^{-2}\log r) \tag{3.53}$$

At the shock we have, by eliminating $a_0t - r$ from (3.51) and (3.52),

$$f(r) = \eta \log r + h(\eta) + O(r^{-1}\log r) \tag{3.54}$$

We can then use η as a parameter to describe the shock, i.e., at the shock both r and t are functions of η. Therefore, by differentiating (3.54) with respect to η and then substituting $f'(r)$ from (3.53), we have

$$[\eta r^{-1}+O(r^{-2}\log r)]\frac{\mathrm{d}r}{\mathrm{d}\eta}+2\log r=-2h'(\eta)$$

or

$$\mathrm{d}[\eta^2\log r+O(\eta r^{-1}\log r)]=-2\eta h'(\eta)\mathrm{d}\eta$$

Integrating, we obtain

$$\eta^2\log r+O(\eta r^{-1}\log r)=-2\int\eta h'(\eta)\mathrm{d}\eta=-2\eta h(\eta)+2h_1(\eta)+b^2 \tag{3.55}$$

where

$$h_1(\eta)=\int_0^\eta h(\xi)\mathrm{d}\xi \tag{3.56}$$

and b is an arbitrary constant. Equation (3.55) can be solved for $\log r$:

$$\log r=\frac{b^2}{\eta^2}-\frac{2h(\eta)}{\eta}+\frac{2h_1(\eta)}{\eta^2}+O(r^{-1}\log^{3/2}r) \tag{3.57}$$

Then (3.54) gives

$$f=\frac{b^2}{\eta}-h(\eta)+\frac{2h_1(\eta)}{\eta}+O(r^{-1}\log r) \tag{3.58}$$

For a given η, (3.57) gives r, and (3.58) gives f. Then (3.52) gives a_0t. Therefore the system (3.52), (3.57), and (3.58) is the system of parametric equations for the shock S. When $h(\eta)$ is not known, it may be expanded as $h(0)+\eta h'(0)+O(\eta^2)$, since it is bounded. Then we can determine η as a function of $\log r$ from (3.57), and the equation of shock becomes

$$a_0t=r-b\log^{1/2}r-h(0)-\frac{1}{2}bh'(0)\log^{-1/2}r+O(\log^{-1}r) \tag{3.59}$$

The error is here rather large, and the parametric representation for the shock is definitely to be preferred if we know $h(\eta)$. The constant b cannot be determined within the framework of the problem, since we cannot fix the shock position absolutely, unless we specify the wave motion at small r and t.

The velocity of the shock U is the slope of S in the (r, t) plane, hence

$$\frac{1}{U}=\frac{\mathrm{d}t}{\mathrm{d}r}=\frac{1-f'(r)}{a_0}$$

Thus the shock condition (ii) when applied to S, gives because ϕ is identically zero ahead of S,

$$a_0v+u-uf'(r)=0$$

immediately behind S. Hence from (3.47), (3.48), and (3.53), we have

$$B_1 = 0, \quad B_2 = -\frac{1}{2}kb^2 \tag{3.60}$$

This completes the solution of the problem as far as it can be determined from the given data. Our example also shows that for actual calculations there is really no need to use the characteristic parameters of the linearized equation as independent variables. The radius r is more convenient to use for this particular problem and thus it is used in place of the characteristic parameter $a_0 t + r$.

4. Uniformly Valid Solution with Initial Condition

The problem discussed in the two preceding sections, although dealing with small disturbances at large distance from the source, can have a large disturbance at the source if $n > 0$, because then the initial disturbances will eventually be weakened like x^{-n} by the process of propagation. But since the initial disturbances are not definitely specified, there is some degree of arbitrariness in the solution, as is clearly shown in the problem of spherical blast in Section Ⅲ.3. If the initial conditions are specified, and if the initial disturbances are weak, then a consistent scheme of constructing a uniformly valid solution can be developed with the results already obtained. Then u and v are definitely of order ε, the small parameter estimating the size of the disturbances. Thus

$$\left.\begin{aligned} u &= \varepsilon u^{(0)}(\xi, \eta) + \varepsilon^2 u^{(1)}(\xi, \eta) + \ldots \\ v &= \varepsilon v^{(0)}(\xi, \eta) + \varepsilon^2 v^{(1)}(\xi, \eta) + \ldots \end{aligned}\right\} \tag{3.61}$$

The coordinates ξ, η approximate the characteristic coordinates x, y of the linearized equation. But in order to account for the anomalous effect of the nonlinear terms in the exact differential equation at large x, we have to distinguish between η and y. In fact, according to our previous discussion, ξ and η are related to x and y by (3.37) or (3.38).

To construct a uniformly valid first approximation, we take the first order solution of the linearized equation, $u^{(0)}(x, y)$, and replace x and y by ξ and η, where ξ and η are given by

$$x = \xi, \quad y = \eta - \varepsilon \int C^*(\xi, \eta)\,d\xi + O(\varepsilon^2) \tag{3.62}$$

Here C^* is the asymptotic form of the coefficient C for $\xi \to \infty$ in the nonlinear equation (3.33), when $u^{(0)}(\xi, \eta)$ is substituted. We are permitted to substitute $u^{(0)}$ for u in C, because we are here only interested in the first order correction in y, and we can use the asymptotic form of the coefficient, because, at small ξ, the difference between y and η is entirely unimportant. Where this difference is important, ξ must be so large that the asymptotic form is accurate. This particular procedure was actually discovered by Whitham in collaboration with Lighthill who later generalized it to the theory discussed in the preceding sections. Whitham applied the method to solve the problem of flow around a supersonic projectile[12,13] and to the problem of propagation of weak spherical shocks in stars[14]. We shall not discuss these very interesting examples here; although the principles have already been outlined in preceding paragraphs, the details are too lengthy to be presented in this exposition.

Instead, we shall give below a complete solution of a somewhat artificially simplified equation to demonstrate the technique of the method. This equation, which was also studied by Lighthill[2], is

$$\left.\begin{aligned} \frac{\partial u}{\partial x}+\frac{n}{x+y}u &= u\left(\frac{\partial^2 v}{\partial x^2}+\frac{\partial u}{\partial y}\right) \\ \frac{\partial v}{\partial y} &= u \end{aligned}\right\} \tag{3.63}$$

with the condition that

$$u = v = 0, \text{ on } \quad y = 0 \tag{3.64}$$

and

$$u = \varepsilon U(y) y^{-n}, \quad \text{on} \quad x = 0, \quad U(0) = 0 \tag{3.65}$$

ε is small and $0 < n < 1$. The linearized solution is

$$u = \varepsilon U(y)(x+y)^{-n}$$

Hence, according to the principle stated above, the uniformly valid solution of first order is

$$u = \varepsilon U(\eta)(\xi+\eta)^{-n}, \quad x = \xi, \quad y = \eta - \varepsilon U(\eta)\xi^{1-n}/(1-n) \tag{3.66}$$

In order to construct a solution uniformly valid up to ε^2, we first make the transformation

$$x = \xi, \quad y = y(\xi, \eta) \tag{3.67}$$

Then (3.63) becomes

$$\left.\begin{aligned} \frac{\partial u}{\partial \xi}-\frac{\dfrac{\partial y}{\partial \xi}}{\dfrac{\partial y}{\partial \eta}}\frac{\partial u}{\partial \eta}+\frac{nu}{\xi+y} &= u\left[\left(\frac{\partial}{\partial \xi}-\frac{\dfrac{\partial y}{\partial \xi}}{\dfrac{\partial y}{\partial \eta}}\frac{\partial}{\partial \eta}\right)\left(\frac{\partial v}{\partial \xi}-\frac{\partial y}{\partial \xi}u\right)+\frac{\dfrac{\partial u}{\partial \eta}}{\dfrac{\partial y}{\partial \eta}}\right] \\ \frac{\partial v}{\partial \eta} &= u\frac{\partial y}{\partial \eta} \end{aligned}\right\} \tag{3.68}$$

We wish, however, to have η as one of the exact characteristics, such that the coefficient of $\partial u/\partial \eta$ in the first equation of (3.68) vanishes. Thus the equation for y is

$$\frac{\partial y}{\partial \xi} = -u\left[\left(\frac{\partial y}{\partial \xi}\right)^2+1\right] \tag{3.69}$$

Then (3.68) simplifies to

$$\left.\begin{aligned} \frac{\partial u}{\partial \xi}+\frac{nu}{\xi+y} &= u\left(\frac{\partial^2 v}{\partial \xi^2}-\frac{\partial^2 y}{\partial \xi^2}u-2\frac{\partial y}{\partial \xi}\frac{\partial u}{\partial \xi}\right) \\ \frac{\partial v}{\partial \eta} &= u\frac{\partial y}{\partial \eta} \end{aligned}\right\} \tag{3.70}$$

Now let us substitute (3.61) and

$$x=\xi,\quad y=\eta+\varepsilon y^{(1)}(\xi,\eta)+\varepsilon^2 y^{(2)}(\xi,\eta)+\dots \tag{3.71}$$

in (3.69) and (3.70). The solution of order ε is

$$\left.\begin{aligned} u^{(0)}(\xi,\eta)&=\frac{U(\eta)}{(\xi+\eta)^n},\quad y^{(1)}(\xi,\eta)=-\frac{U(\eta)(\xi+\eta)^{1-n}}{1-n}\\ v^{(0)}(\xi,\eta)&=\int_0^\eta\frac{U(t)}{(\xi+t)^n}\mathrm{d}t\end{aligned}\right\} \tag{3.72}$$

The second order equations are

$$\frac{\partial u^{(1)}}{\partial\xi}+\frac{nu^{(1)}}{\xi+\eta}=\frac{ny^{(1)}u^{(0)}}{(\xi+\eta)^2}+u^{(0)}\frac{\partial^2 v^{(0)}}{\partial\xi^2},\quad \frac{\partial y^{(2)}}{\partial\xi}=-u^{(1)} \tag{3.73}$$

By substituting the first order solution (3.72) in (3.73), we have

$$\frac{\partial}{\partial\xi}[(\xi+\eta)^n u^{(1)}]=-\frac{nU^2(\eta)}{(1-n)(\xi+\eta)^{1+n}}+U(\eta)\int_0^\eta\frac{n(n+1)U(t)}{(\xi+t)^{n+2}}\mathrm{d}t$$

Now we have to rewrite the initial condition (3.65). It can be written as

$$u=\varepsilon U(y)y^{-n}=\varepsilon u^{(0)}(0,y)\quad\text{at}\quad x=0$$

or, by replacing the variable y by η in the above equation and introducing the expansion (3.61), as

$$\varepsilon u^{(0)}(0,\eta+\varepsilon y^{(1)}+\dots)+\varepsilon^2 u^{(1)}(0,\eta)+\dots=\varepsilon u^{(0)}(0,\eta)+\dots$$

Therefore, we obtain the initial condition for $u^{(1)}$ by equating the terms of order ε^2 in the form

$$u^{(1)}(0,\eta)=-\left[\frac{\mathrm{d}u^{(0)}(0,\eta)}{\mathrm{d}\eta}\right]y^{(1)}(0,\eta)=\left[\frac{\mathrm{d}}{\mathrm{d}\eta}(U(\eta)\eta^{-2})\right]\frac{U(\eta)\eta^{1-n}}{1-n}$$

With this condition, the differential equation for $u^{(1)}$ can be integrated to give

$$\begin{aligned} u^{(1)}&=\frac{U^2(\eta)}{1-n}\left[\frac{1}{(\xi+\eta)^{2n}}-\frac{1}{\eta^n(\xi+\eta)^n}\right]+\frac{U(\eta)\eta^{1-n}}{(1-n)(\xi+\eta)^n}\frac{\mathrm{d}}{\mathrm{d}\eta}(U(\eta)\eta^{-n})+\\ &\quad\frac{nU(\eta)}{(\xi+\eta)^n}\int_0^\eta\left\{\frac{1}{t^{n+1}}-\frac{1}{(\xi+t)^{n+1}}\right\}U(t)\mathrm{d}t \end{aligned} \tag{3.74}$$

For large x or ξ we have from (3.74)

$$u^{(1)}(\xi,\eta)\sim F(\eta)\xi^{-n} \tag{3.75}$$

where

$$F(\eta)=-\frac{U^2(\eta)}{(1-n)\eta^n}+\frac{U(\eta)\eta^{1-n}}{(1-n)}\frac{\mathrm{d}}{\mathrm{d}\eta}(U(\eta)\eta^{-2})+nU(\eta)\int_0^\eta\frac{U(t)\mathrm{d}t}{t^{n+1}} \tag{3.76}$$

Also, for large x or ξ, u and $y^{(2)}$ are given as

$$u \cong \frac{\varepsilon U(\eta) + \varepsilon^2 F(\eta) + \ldots}{x^n} \tag{3.77}$$

$$y^{(2)} \cong -\frac{F(\eta)\xi^{1-n}}{1-n} \tag{3.78}$$

To make the solution uniformly valid up to the order ε^2, it is only necessary to include the asymptotic form of $y^{(2)}$ given by (3.78). Thus

$$\left.\begin{aligned} u &= \varepsilon u^{(0)}(\xi, \eta) + \varepsilon^2 u^{(1)}(\xi, \eta) + \ldots \\ x &= \xi \\ y &= \eta + \varepsilon y^{(1)}(\xi, \eta) - \frac{\varepsilon^2 F(\eta)\xi^{1-n}}{1-n} \end{aligned}\right\} \tag{3.79}$$

Equation (3.77) clearly shows that, for any ξ and η, the series for y is convergent. The third equation (3.79) also indicates that, although the initial correction to y, $\varepsilon y^{(1)}$, may be quite large for large ξ, the ratio of the subsequent terms in y is always of order ε. Therefore the series for y is convergent for any finite ξ and η. The term $\varepsilon^2 y^{(2)}$ in y is quite important, however, for obtaining a solution u accurate to ε^2. To see this, let us calculate the change in η for fixed x and y due to this term. This change is evidently

$$\frac{\varepsilon^2 y^{(2)}}{\dfrac{\partial y}{\partial \eta}} = \frac{\varepsilon^2 y^{(2)}}{1 + O(\varepsilon \xi^{1-n})}$$

When x or y is large, $y^{(2)}$ is $O(\xi^{1-n})$. Thus this change in η due to $\varepsilon^2 y^{(2)}$ can be of order ε and the modification of u due to this change is $O(\varepsilon^2)$. For large x, it is therefore important to retain the term $\varepsilon^2 y^{(2)}$ in y.

5. Perturbation by Using Exact Characteristics

The application of the PLK method to hyperbolic partial differential equations explained in the preceding sections, although very effective in removing the difficulties of the classical perturbation method, has not yet been shown to be mathematically sound. The question is: Are the perturbation series really convergent, or do they only seem to be so? To answer this mathematical question, we must first recast the method so as to be more formal and thus amenable to mathematical argument. We have already seen in Section Ⅲ.1 that the key of the procedure is to change the coordinates to the exact characteristic coordinates. Again in Sections Ⅲ.3 and Ⅲ.4 we introduced one exact characteristic coordinate η, but left the x coordinate unchanged. This, however, was due to the fact that in the problem of progressive waves or simple waves there is no necessity of changing x. To formalize the method and to include all the cases treated, we then simply have to make use of both exact characteristic variables ξ and η. Then the scheme of the method is first to convert the hyperbolic partial differential equation to its normal form in characteristic variables ξ, η and then to expand the solution in powers of ε as follows:

$$
\left.\begin{aligned}
u &= \varepsilon u^{(0)}(\xi,\ \eta) + \varepsilon^2 u^{(1)}(\xi,\ \eta) + \varepsilon^3 u^{(2)}(\xi,\ \eta) + \ldots \\
v &= \varepsilon v^{(0)}(\xi,\ \eta) + \varepsilon^2 v^{(1)}(\xi,\ \eta) + \varepsilon^3 v^{(2)}(\xi,\ \eta) + \ldots \\
x &= \xi + \varepsilon x^{(1)}(\xi,\ \eta) + \varepsilon^2 x^{(2)}(\xi,\ \eta) + \ldots \\
y &= \eta + \varepsilon y^{(1)}(\xi,\ \eta) + \varepsilon^2 y^{(2)}(\xi,\ \eta) + \ldots
\end{aligned}\right\} \tag{3.80}
$$

Here x, y are the characteristic coordinates of the linearized equation. This is then the *classical* perturbation procedure with the exact characteristic variables ξ, η as independent variables. It was suggested by various authors, and the convergence of the series seemed to be implied by the general theory of hyperbolic partial differential equations. But it was Lin[15] and Fox[16] who for simple cases gave convergence proofs of the expansions (3.80) for all values of ξ and η, and sufficiently small ε. Lin[15] has also applied the procedure to several very interesting problems of plane supersonic flows. We shall not pursue these matters here and refer the interested reader to the original papers.

The importance of the work of Lin and Fox is the support which it gives to the PLK method by showing that the procedure described in the previous sections of the present chapter is mathematically sound. For solving engineering problems, the PLK method is preferred to the perturbation method based on exact characteristics. The first reason is that the PLK method has the advantage of economy of means. For instance, in the problems of progressive waves, only the η-characteristic is used, while x is left unchanged, because this is sufficient. For the problems of Section Ⅲ.1, the trouble is with the $\partial^2 v/\partial x^2$ term; there only the ξ-characteristic need be used, and y may be left unchanged. If the characteristic perturbation method is used, both x and y must be modified, and the work of calculation will be very much greater. This is particularly true when only the lowest order solution up to ε is required, as is usually the case for engineering applications. The PLK method is also more flexible and more general in that the principle can be applied to hyperbolic partial differential equations in more than two independent variables and of order higher than two. We just introduce sufficient distortion in the coordinate to remove the difficulty in the linearized equation.

Ⅳ. Elliptic Partial Differential Equations

1. Failure of the PLK Method in the Thin Airfoil Problem

Singularities of elliptic partial differential equations differ from those of hyperbolic equations in that they are point singularities. One well-known singularity of this type is the singularity at the nose of a thin airfoil in an incompressible perfect fluid when the solution is obtained by using the first order boundary condition—the classical thin airfoil theory. In fact, if the solution is pushed to the ε^3-term, with ε equal to the thickness-chord ratio of the airfoil, the singularity at the nose becomes worse than those of the first order and the second order solution. This seems to be a natural problem on which to try the PLK method. Lighthill[17] himself has investigated the question and obtained very useful results for round nosed airfoils.

Let us consider the simpler problem of flow around a symmetric airfoil at zero angle of attack, with the fluid velocity far from the airfoil normalized to unity. The stream function ψ has to satisfy Laplace's equation

$$\frac{\partial^2 \psi}{\partial x^2} + \frac{\partial^2 \psi}{\partial y^2} = 0 \tag{4.1}$$

The boundary condition can then be stated in the form $\psi = 0$ on the surface of the airfoil. Let the shape of the airfoil be represented by

$$y^* = \pm \varepsilon \sqrt{x}(F_0 + F_1 x + \ldots) \tag{4.2}$$

where F_0, F_1 are numerical constants and y^* is y on the surface of the section, and the subsequent terms are such as to give a cusp at the trailing edge $x = c$. We now introduce the following expansions according to the PLK method:

$$\left.\begin{aligned} \psi &= \eta + \varepsilon\psi^{(1)}(\xi, \eta) + \varepsilon^2\psi^{(2)}(\xi, \eta) + \varepsilon^3\psi^{(3)}(\xi, \eta) + \ldots \\ x &= \xi + \varepsilon x^{(1)}(\xi, \eta) + \varepsilon^2 x^{(2)}(\xi, \eta) + \varepsilon^3 x^{(3)}(\xi, \eta) + \ldots \\ y &= \eta \end{aligned}\right\} \tag{4.3}$$

Lighthill found that

$$x^{(1)} = 0 \tag{4.4}$$

and that in order to make the singularity of $\psi^{(3)}$ at the nose $x = 0$ not worse than the singularity of $\psi^{(1)}$ and $\psi^{(2)}$, we must have

$$x^{(2)} = \frac{1}{4}F_0^2 \tag{4.5}$$

which is a constant. The term $\varepsilon^2 x^{(2)}$ can be shown to be equal to one half of the radius of curvature of the nose of the airfoil. Thus up to the ε^2-term, the solution of the PLK method is equal to that of the corresponding classical perturbation solution with the x-coordinate shifted downstream to a distance equal to half the radius of the nose. By doing so, the nose singularity of the perturbation solution is "absorbed" into the airfoil and the solution outside of the section is actually free of singularities. The resultant solution is shown to be correct by comparison with the exact solution, which fortunately is available for such a simple case.

Difficulties appear, however, if the solution is pushed to higher order terms. Fox[16] has shown that the higher order solution is indeed not uniformly valid, and that all "higher order" terms give at the nose the same order of magnitude $O(\varepsilon^2)$. Hence no improvement of the solution beyond the second order can be obtained. This aspect of the result is very similar to that of the compressible sink flow discussed in Section II.8, where we also find that the high order terms cannot be used to reduce the error of the solution. Therefore we might be encouraged to guess that here as in the sink flow, the nose singularity can only be dealt with by a boundary layer solution. That is, we have to give up the idea of a single solution good for the entire field, and seek a solution good locally near the nose. The complete solution is then obtained by joining the solution at the nose with the solution obtained from the classical perturbation method or the PLK method. In fact, Van Dyke[18] has developed just such a theory, which, although not yet fully established mathematically, appears nevertheless to be correct from physical reasoning and by comparison with the exact solution. Van Dyke's investigation actually is more general and includes the case of subsonic compressible flow.

2. Probable Source of Difficulty

Is there then any warning that we might notice in applying the PLK method to the thin airfoil problem to indicate that the method might fail? It seems that there is such a warning: When we substitute the expansion (4.3) in the equation of the airfoil (4.2) in preparation for satisfying the boundary condition, we have to expand $\sqrt{x}$, among other things, in terms of ξ and $x^{(n)}(\xi, \eta)$, for x near zero. Since $\sqrt{x}$ is not regular at $x=0$, such an expansion clearly cannot be uniformly valid. The same difficulty was discussed in Section Ⅱ.5 in connection with the ordinary differential equations. If we try to make a formal expansion as Fox did, then the PLK method refuses to yield a uniformly valid solution. The present author has tried to avoid the necessity of expanding $\sqrt{x}$ in (4.2) by squaring (4.2) first. But then the boundary condition would require a power series expansion of $\psi^{(n)}$ for small ξ and η. But $\psi^{(n)}$ is not regular at $\xi=\eta=0$, therefore for this approach also no uniformly valid expansion can be obtained. The failure of the PLK method in this problem is thus perhaps to be expected.

From the above reasoning it is evident that the PLK method when applied to the airfoil problem will still fail, even if we introduce another expansion for y and write

$$\left.\begin{aligned} x &= \xi+\varepsilon x^{(1)}(\xi, \eta)+\varepsilon^2 x^{(2)}(\xi, \eta)+\dots \\ y &= \eta+\varepsilon y^{(1)}(\xi, \eta)+\varepsilon^2 y^{(2)}(\xi, \eta)+\dots \end{aligned}\right\} \tag{4.6}$$

This approach was indeed explored by Fox and found to be of no avail. Thus our surmise about the source of the difficulty is further strengthened. We may perhaps even venture to say that any singularity in the solution of an elliptic partial differential equation of the same general type as discussed here cannot be removed by the PLK method. That is, a solution uniformly valid to *all orders* is not possible. Of course, a solution uniformly valid to a finite order of ε might still be possible, as in the case of thin airfolis. Engineers may well be satisfied with such solutions of limited but uniform accuracy.

Ⅴ. Applications to Fluid Boundary Layer Problems

1. Boundary Layer Along a Flat Plate

The theory of the boundary layer along the surface of a solid body in a moving fluid of low viscosity originated with Prandtl. It is the prototype of all boundary layer problems in applied mechanics and mathematics. Prandtl's boundary layer theory is essentially the first order solution for very small viscosity. In recent years, various investigators have tried to improve the original theory to include high order terms. But here the difficulty is the occurrence of a singularity of Prandtl's solution at the nose of the body. Higher order perturbations will only make the singularity worse and actually produce infinite total viscous shearing force over the surface instead of a finite shearing force. The solution is thus entirely unacceptable.

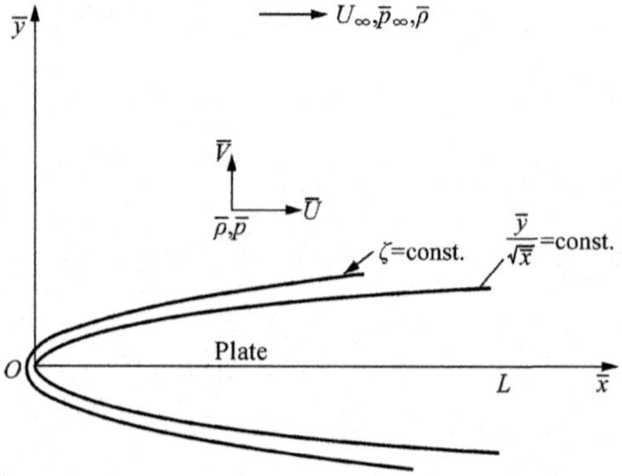

Fig. 5 Incompressible boundary layer over a flat plate

Kuo[3] realized that a satisfactory solution can be obtained by introducing a coordinate distortion within the framework of boundary layer theory, i. e., the physical coordinates are transformed twice. First the boundary layer transformation is introduced, which also requires a modified dependent variable, and then the boundary layer coordinate is again distorted. One might say that while the combined method discussed in Section Ⅱ. 8 is the boundary layer method "plus" the coordinate perturbation method, Kuo's contribution is the boundary layer method "multiplied" by the coordinate perturbation method. This very imaginative extension of the concepts of Poincaré and Lighthill was indeed very effective in removing the nonuniform validity of Prandtl's theory. Because of the importance of this work and the beauty of the results obtained, we shall give in the following sections a considerably detailed representation of a particular problem, the simplest of the group of problems studied by Kuo: the incompressible fluid boundary layer over a flat plate.

Let $\bar{x}$, $\bar{y}$ be rectangular Cartesian coordinates. The plate then occupies the strip $0 \leqslant \bar{x} \leqslant L$ with $\bar{y} = 0$ (Fig. 5), L being the length of the plate. Referred to this system, the $\bar{x}$ and the $\bar{y}$ components of the velocity are $\bar{u}$ and $\bar{v}$, respectively and the pressure is $\bar{p}$. The density $\bar{\rho}$ and the kinematic viscosity $\bar{\nu}$ are constants in this problem. The differential equations of the problem are the Navier-Stokes equations and the equation of continuity. We shall write these equations immediately in terms of nondimensional boundary layer variables instead of physical variables defined above. These boundary layer variables are defined in terms of the Reynolds number Re,

$$Re = U_\infty L / \bar{\nu} \tag{5.1}$$

where U_∞ is the fluid velocity very far from the plate. For fluids of very small viscosity, then, Re is very large. We shall introduce the small parameter ε of this problem by

$$\varepsilon^2 = \frac{1}{Re} = \frac{\bar{\nu}}{U_\infty L} \tag{5.2}$$

The boundary layer variables, according to Prandtl, are then

$$\left.\begin{aligned} u &= \bar{u}/U_\infty, \quad v = \frac{1}{\varepsilon}(\bar{v}/U_\infty) = \frac{1}{\varepsilon}V \\ x &= \bar{x}/L, \quad y = \frac{1}{\varepsilon}(\bar{y}/L) = \frac{1}{\varepsilon}Y \end{aligned}\right\} \tag{5.3}$$

thus v and y are modified by the parameter ε. We shall also introduce a nondimensional pressure p by

$$p = (\bar{p} - \bar{p}_\infty)/\bar{\rho}\, U_\infty^2 \tag{5.4}$$

To satisfy the continuity equation, we introduce the stream function ψ defined by

$$u = \psi_y, \quad v = -\psi_x \tag{5.5}$$

where, as hereafter, we have used the suffix notation of partial differentiation. In terms of the variables introduced, the Navier-Stokes equations are

$$\left.\begin{aligned} &\psi_y\psi_{xy} - \psi_x\psi_{yy} - \psi_{yyy} = -p_x + \varepsilon^2\psi_{xxy} \\ &p_y + \varepsilon^2(\psi_x\psi_{xy} - \psi_y\psi_{xx} + \psi_{xyy}) = -\varepsilon^4\psi_{xxx} \end{aligned}\right\} \tag{5.6}$$

These are two equations for the two unknowns ψ and p.

To solve this problem by the "boundary layer method," we write

$$\left.\begin{aligned} \psi(x, y) &= \psi^{(0)}(x, y) + \varepsilon\psi^{(1)}(x, y) + \varepsilon^2\psi^{(2)}(x, y) + \ldots \\ p(x, y) &= \qquad\qquad\quad \varepsilon p^{(1)}(x, y) + \varepsilon^2 p^{(2)}(x, y) + \ldots \end{aligned}\right\} \tag{5.7}$$

where the zeroth order pressure is missing because we deal with the flat plate problem. The quantities x, y are the undistorted boundary layer variables. We shall see that to go to the ε^2-terms, the x, y coordinate system has to be distorted. But we shall delay doing this until the necessity arises. The zeroth order stream function $\psi^{(0)}$ is now determined by substituting (5.7) in (5.6) and equating the part independent of ε. Thus

$$\psi_y^{(0)}\psi_{xy}^{(0)} - \psi_x^{(0)}\psi_{yy}^{(0)} - \psi_{yyy}^{(0)} = 0 \tag{5.8}$$

The first order equations are obtained by equating the coefficients of ε,

$$\left.\begin{aligned} &\psi_y^{(0)}\psi_{xy}^{(1)} + \psi_{xy}^{(0)}\psi_y^{(1)} - \psi_x^{(0)}\psi_{yy}^{(1)} - \psi_{yy}^{(0)}\psi_x^{(1)} = -p_x^{(1)} + \psi_{yyy}^{(1)} \\ &0 = -p_y^{(1)} \end{aligned}\right\} \tag{5.9}$$

The second equation (5.9) indicates that the well-known conclusion in Prandtl's boundary layer theory, namely, the constancy of pressure in any section of the layer, still holds. In fact, if we limit ourselves to terms up to ε, the phenomenon can be adequately described by the simpler boundary layer equations of Prandtl instead of the full Navier-Stokes equations given by (5.6).

The zeroth order equation (5.8) is the well-known Blasius equation. Using the substitution,

$$\psi^{(0)} = \sqrt{x} f_0(\zeta) \tag{5.10}$$

$$\zeta = \frac{y}{\sqrt{x}} \tag{5.11}$$

and denoting differentiation with respect to ζ by a prime, we get from (5.8)

$$2f'''_0 + f_0 f''_0 = 0 \tag{5.12}$$

The velocity components deduced from $\psi^{(0)}$ are then

$$u^{(0)} = \psi_y^{(0)} = f'_0(\zeta), \quad v^{(0)} = -\psi_x^{(0)} = \frac{1}{2\sqrt{x}}[\zeta f'_0(\zeta) - f_0(\zeta)] \tag{5.13}$$

Since the boundary conditions are $u = v = 0$ at $y = 0$ and $u = 1$ at $y = \infty$ the boundary conditions for f_0 can be obtained by using (5.13):

$$\left.\begin{aligned} f_0 = f'_0 = 0 \quad & \text{at} \quad \zeta = 0 \\ f'_0 = 1 \quad & \text{at} \quad \zeta = \infty \end{aligned}\right\} \tag{5.14}$$

The solution of (5.12) under these boundary conditions is given by the following power series and the asymptotic formula for large ζ:

$$f_0(\zeta) = \frac{\sigma}{2!}\zeta^2 + \left(-\frac{1}{2}\right)\frac{\sigma^2}{5!}\zeta^5 + \left(-\frac{1}{2}\right)^2 \frac{11\sigma^3}{8!}\zeta^8 + \left(-\frac{1}{2}\right)^3 \frac{375\sigma^4}{11!}\zeta^{11} + \ldots$$

$$\sigma \cong 0.332 \tag{5.15}$$

$$f_0(\zeta) \cong \zeta - 1.73 + 0.231 \int_\infty^\zeta d\zeta' \int_\infty^{\zeta'} e^{-(1/4)(\zeta''-1.73)^2} d\zeta'' \tag{5.16}$$

From (5.16) we can compute the velocity V_e at the edge of the boundary layer ($\zeta \to \infty$), using (5.13):

$$V_e = v_0 \varepsilon \frac{1}{\sqrt{x}} \tag{5.17}$$

where

$$v_0 \cong \frac{1.73}{2} = 0.865 \tag{5.18}$$

To determine $p^{(1)}$, necessary for the determination of the first order stream function $\psi^{(1)}$ from (5.9), we have to calculate the velocity field u and V outside the boundary layer, induced by the boundary layer. The effect of the boundary layer is represented approximately by the vertical velocity V_e, given by (5.17) for $0 \leqslant x \leqslant 1$. For $x < 0$, V_e vanishes. In the wake, $1 < x$, we can approximately take $V_e = 0$. Of course, V_e cannot be really zero at the boundary of the wake, but the difference is perhaps not important. Summarizing then, we have

$$V_e = \begin{cases} 0 & \text{for} \quad x < 0 \\ \varepsilon \dfrac{v_0}{\sqrt{x}} & \text{for} \quad 0 \leqslant x \leqslant 1 \\ 0 & \text{for} \quad 1 < x \end{cases} \tag{5.19}$$

Now let the potential field outside the boundary layer be expanded as a power series in ε, i.e.,

$$\left.\begin{aligned} u &= 1+\varepsilon U^{(1)}\left(\frac{\bar{x}}{L},\frac{\bar{y}}{L}\right)+\dots \\ V &= \varepsilon V^{(1)}\left(\frac{\bar{x}}{L},\frac{\bar{y}}{L}\right)+\dots \end{aligned}\right\} \tag{5.20}$$

Since the edge of the boundary layer is very close to the plate, i.e., the thickness of the boundary layer is of the order ε, the $U^{(1)}$ and $V^{(1)}$ can be found by the same method as the first order solution of the thin airfoil theory. Thus, we have

$$U^{(1)} - \mathrm{i}V^{(1)} = \frac{v_0}{\pi}\int_0^1 \frac{\mathrm{d}w}{\sqrt{w}(z-w)} = -\frac{\mathrm{i}v_0}{\sqrt{z}} + \frac{v_0}{\pi\sqrt{z}}\log\frac{1+\sqrt{z}}{1-\sqrt{z}} \tag{5.21}$$

where z denotes $(x+\mathrm{i}Y)$. At the edge of the boundary layer, the induced $U_e^{(1)}$ is given to this order of approximation by $U^{(1)}(x, 0)$. Thus

$$\begin{aligned} U_e^{(1)} = U^{(1)}(x,0) &= \frac{v_0}{\pi\sqrt{x}}\log\frac{1+\sqrt{x}}{1-\sqrt{x}} \\ &= \frac{2v_0}{\pi}\left[1+\frac{x}{3}+\frac{x^2}{5}+\dots+\frac{x^n}{2n+1}+\dots\right] \end{aligned} \tag{5.22}$$

Then by using the Bernoulli equation, the pressure $p^{(1)}$ at the edge of the boundary layer is

$$p^{(1)}(x) = -U_e^{(1)}(x) \tag{5.23}$$

Therefore the first order equation (5.9) becomes

$$\begin{aligned} &\psi_y^{(0)}\psi_{xy}^{(1)} + \psi_{xy}^{(0)}\psi_y^{(1)} - \psi_x^{(0)}\psi_{yy}^{(1)} - \psi_{yy}^{(0)}\psi_x^{(1)} - \psi_{yyy}^{(1)} = \frac{\mathrm{d}U_e^{(1)}}{\mathrm{d}x} \\ &= \frac{2v_0}{\pi}\left[\frac{1}{3}+\frac{2}{5}x+\dots+\frac{n-1}{2n-1}x^{n-2}+\dots\right] \end{aligned} \tag{5.24}$$

To solve (5.24), we expand $\psi^{(1)}$ as follows

$$\psi^{(1)} = \frac{2v_0}{\pi}\left[x^{1/2}f_1(\zeta)+\frac{1}{3}x^{3/2}f_2(\zeta)+\dots+\frac{x^{n-\frac{1}{2}}}{2n-1}f_n(\zeta)+\dots\right] \tag{5.25}$$

where the f's are functions of ζ only, and ζ is defined by (5.17). By substituting (5.10) and (5.25) in (5.24), we obtain, after equating the coefficients of the powers of x to zero, the differential equations for f_n:

$$2f'''_n + f_0 f''_n - 2(n-1)f'_0 f'_n + (2n-1)f''_0 f_n = -2(n-1) \tag{5.26}$$

for $n = 1,2,3\dots$ Now the x-component of the velocity deduced from $\psi^{(1)}$ is

$$u^{(1)} = \psi_y^{(1)} = \frac{2v_0}{\pi}\left[f'_1+\frac{x}{3}f'_2+\dots+\frac{x^n}{2n+1}f'_{n+1}+\dots\right] \tag{5.27}$$

At the edge of the boundary layer $\zeta \to \infty$, and $u^{(1)}$ should coincide with $U_e^{(1)}$ of (5.22). At the surface of the plate, where $\zeta = 0$, $u^{(1)} = v^{(1)} = 0$. Thus the boundary conditions for the third order differential equations (5.26) are

$$\left.\begin{aligned} f_n = f'_n = 0 \quad & \text{at} \quad \zeta = 0 \\ f'_n = 1 \quad & \text{at} \quad \zeta = \infty \end{aligned}\right\} \tag{5.28}$$

Since the coefficients of the equation (5.26) cannot be expressed analytically by simple functions for the whole range of ζ, an analytical solution valid for ζ and n is generally not possible. The problem is on the whole a numerical one. Kuo has found, by combining the numerical solutions of Howarth and Tani for low n with the asymptotic solution for large n, the following important formula for the skin friction coefficient C_f, which is the total shear force over one side of the plate divided by $\frac{1}{2}\bar{\rho}\, U_\infty^2 L$:

$$C_f = \frac{1.328}{\sqrt{Re}} + \frac{4.12}{Re} \tag{5.29}$$

Here Re is the Reynolds number defined by (5.1). This equation has been compared with experimental data of Janour[19] and found to be good for Re as low as 10. Below this value of the Reynolds number, Oseen's method of solution will be more appropriate.

2. Second Order Solution

If we follow the indicated procedure and write down the differential equation for $\psi^{(2)}$, we see that, unlike $\psi^{(1)}$, $\psi^{(2)}$ will have a leading edge singularity worse than $\psi^{(0)}$ and $\psi^{(1)}$, and that the additional shear stress will, in fact, not be integrable. To remedy this situation, Kuo considered the various functions in the expansions (5.7) to be functions of ξ and η, and introduced a further distortion of the boundary layer variables. Evidently the distortion need only be of $O(\varepsilon^2)$, as the trouble is with $\psi^{(2)}$. Thus

$$x = \xi + \varepsilon^2 x^{(2)}(\xi, \eta) + \dots, \quad y = \eta \tag{5.30}$$

The y remains unchanged; that this is sufficient will be seen presently. Then, the differentiations with respect to x and y become

$$\left.\begin{aligned} \frac{\partial}{\partial x} &\cong \frac{\partial}{\partial \xi} - \varepsilon^2 x_\xi^{(2)} \frac{\partial}{\partial \xi} \\ \frac{\partial}{\partial y} &\cong \frac{\partial}{\partial \eta} - \varepsilon^2 x_\eta^{(2)} \frac{\partial}{\partial \xi} \end{aligned}\right\} \tag{5.31}$$

Clearly, the changes effected by the new variables are in the second order terms. The zeroth order solution and the first order solution remain unchanged, except for the substitution x, $y \to \xi$, η. The equations for $\psi^{(2)}$ and $p^{(2)}$ are now

$$\psi_\eta^{(0)}\psi_{\xi\eta}^{(2)} + \psi_{\xi\eta}^{(0)}\psi_\eta^{(2)} - \psi_\xi^{(0)}\psi_{\eta\eta}^{(2)} - \psi_{\eta\eta}^{(0)}\psi_\xi^{(2)} - \psi_{\eta\eta\eta}^{(2)} = [\psi_\xi^{(1)}\psi_{\eta\eta}^{(1)} - \psi_\eta^{(1)}\psi_{\xi\eta}^{(1)}] +$$
$$[-p_\xi^{(2)} + \psi_{\xi\xi\eta}^{(0)} + \psi_{\eta\eta\eta}^{(0)}x_\xi^{(2)} + (p_\eta^{(2)} - 2\psi_{\xi\eta\eta}^{(0)})x_\eta^{(2)} + \psi_\eta^{(0)}\psi_\xi^{(0)}x_{\xi\eta}^{(2)} -$$
$$(\psi_\xi^{(0)^2} + 3\psi_{\xi\eta}^{(0)})x_{\eta\eta}^{(2)} - \psi_\xi^{(0)}x_{\eta\eta\eta}^{(2)}] \tag{5.32}$$

and

$$p_\eta^{(2)} + \psi_\xi^{(0)}\psi_{\xi\eta}^{(0)} - \psi_\eta^{(0)}\psi_{\xi\xi}^{(0)} + \psi_{\eta\eta\eta}^{(0)} = 0 \tag{5.33}$$

As before, let us introduce the similarity variable ζ by

$$\zeta = \frac{\eta}{\sqrt{\xi}} \tag{5.34}$$

Then it can be seen that the singularity of the first group of terms on the right-hand side of (5.32) is $O(\xi^{-1})$ when $\xi \to 0$ while that of the second group of terms, due to terms like $\psi_{\xi\xi\eta}^{(0)}$ is $O(\xi^{-2})$, when $\xi \to 0$. The second group of terms is the troublesome group. Thus $x^{(2)}$ should be such as to make the content of the second square bracket on the right-hand side of (5.32) harmless. To do this, we first observe that according to (5.33)

$$p^{(2)} = p_2(\xi) + P^{(2)}(\xi, \eta) \tag{5.35}$$

where

$$P^{(2)}(\xi, \eta) = \int_0 (\psi_\eta^{(0)}\psi_{\xi\xi}^{(0)} - \psi_\xi^{(0)}\psi_{\xi\eta}^{(0)} - \psi_{\xi\eta\eta}^{(0)})\,d\eta \tag{5.36}$$

Then $p_2(\xi)$ should be determined just like $p^{(1)}(\xi)$ from the second order potential flow. But because of the character of $\psi^{(1)}$, the vertical velocity $V_e^{(2)}$ at the edge of the boundary layer has the same singularity as $V_e^{(1)}$. From our experience with the first order theory, we know that the gradient $dp_2/d\xi$ is quite harmless. On the other hand, $P_\xi^{(2)}$ has a strong singularity. Thus the correct choice of $x^{(2)}$ must take into account $P_\xi^{(2)}$, but not necessarily $dp_2/d\xi$. Hence the equation for $x^{(2)}$ is

$$\psi_{\eta\eta\eta}^{(0)}x_\xi^{(2)} + (P_\eta^{(2)} - 2\psi_{\xi\eta\eta}^{(0)})x_\eta^{(2)} + \psi_\eta^{(0)}\psi_\xi^{(0)}x_{\xi\eta}^{(2)} -$$
$$(\psi_\xi^{(0)^2} + 3\psi_{\xi\eta}^{(0)})x_{\eta\eta}^{(2)} - \psi_\xi^{(0)}x_{\eta\eta\eta}^{(2)} = P_\xi^{(2)} - \psi_{\xi\xi\eta}^{(0)} \tag{5.37}$$

This is the all important equation for determining the coordinate distortion, and it is considerably more complicated than the equation for the distortion function in the earlier chapters.

Equations (5.10) and (5.11) now give $\psi^{(0)} = \sqrt{\xi} f_0(\zeta)$. By studying the structure of (5.37), it is seen that $x^{(2)}$ is a function of ζ only, i.e.

$$x^{(2)} = g_2(\zeta) \tag{5.38}$$

Now (5.37) can be written as

$$2(f_0-\zeta f_0')g_2'''+(f_0^2-\zeta f_0 f_0-6\zeta f_0'')g_2''+$$
$$[2f_0'(f_0-\zeta f_0')-6(f_0''+1/3\zeta f_0''')]g_2'$$
$$=6\zeta f_0''+\zeta^2 f_0'''+\frac{1}{2}\zeta f_0 f_0'-f_0^2+\frac{1}{2}\zeta^2 f_0'^2$$

This equation has the integrating factor $f_0(\zeta)$. Thus on multiplying both sides by $f_0(\zeta)$, an integration yields

$$2f_0(f_0-\zeta f_0')g_2''-[2f_0'(f_0-\zeta f_0')+4\zeta f_0 f_0''-f_0^3+\zeta f_0^2 f_0']g_2'=G(\zeta)+\text{const} \tag{5.39}$$

where

$$G(\zeta)=\int_0^\zeta f_0\left[6\zeta f_0''+\zeta^2 f_0'''+\frac{1}{2}\zeta f_0 f_0'-f_0^2+\frac{1}{2}\zeta f_0'^2\right]d\zeta$$
$$=\frac{3\sigma^2}{4}\zeta^4-\frac{3\sigma^3}{140}\zeta^7+\frac{999\sigma^4}{20\times(8!)}\zeta^{10}-\ldots \tag{5.40}$$

This series form of $G(\zeta)$ is obtained by using (5.15). Near $\zeta=0$, (5.39) is approximated by

$$\frac{d}{d\zeta}(\zeta^2 g_2')=-\frac{3}{2}\zeta^2+\frac{\text{const}}{\zeta^2} \tag{5.41}$$

or

$$g_2=-\frac{1}{4}\zeta^2+\frac{\text{const}}{\zeta^2}+\frac{\text{const}}{\zeta}+\text{const} \tag{5.42}$$

If we therefore impose the condition

$$x^{(2)}=g_2=0 \quad \text{at} \quad \eta=\zeta=0 \tag{5.43}$$

so that the plate is not "moved" by the coordinate distortion, all three integration constants in $g_2(\zeta)$ must be set equal to zero. Thus the single condition (5.43) determines $g_2(\zeta)$ completely. A more complete calculation gives

$$g_2(\zeta)=-\left[\frac{1}{2\times 2!}\zeta^2-\frac{\sigma}{14\times 5!}\zeta^5+\frac{7\sigma^2}{30\times 8!}\zeta^8-\ldots\right] \tag{5.44}$$

On the other hand, for large ζ, f_0 is approximated by $\zeta-1.73$ according to (5.16). Then (5.39) is reduced to

$$g_2'''+\frac{1}{2}(\zeta-1.73)g_2''+g_2'=-\frac{1}{2}\left(\frac{3}{2}\zeta-1.73\right) \tag{5.45}$$

By the substitutions

$$g_2'=-\frac{1}{2}\left(\zeta-\frac{1.73}{2}\right)+g'(t), \quad t=\frac{1}{\sqrt{2}}(\zeta-1.73) \tag{5.46}$$

we obtain

$$g''' + tg'' + 2g' = 0$$

This equation can be integrated twice, and the result is

$$g' + tg = C_1 t + C_2 \tag{5.47}$$

Also this equation can be easily integrated. Finally we obtain the following asymptotic formula for g_2, valid for large ζ :

$$\begin{aligned} g_2(\zeta) \sim -\frac{1}{4}\left(\zeta - \frac{1.73}{2}\right)^2 + C_1 + C_2 e^{-(1/4)(\zeta - 1.73)^2} + \\ C_3 e^{-(1/4)(\zeta-1.73)^2} \int_{\zeta_1}^{(1/\sqrt{2})(\zeta-1.73)} e^{t^2/4} dt \end{aligned} \tag{5.48}$$

By joining the two solutions given by (5.44) and (5.48) at $\zeta_1 = 3$, the values of C_1, C_2 and C_3 are obtained as 1.901, 1.264, and 0.431, respectively. Furthermore, we observe that

$$e^{-t^2} \int_{t_1}^{t} e^{t^2} dt$$

tends to zero as $t \to \infty$. Therefore $g_2(\zeta)$ tends to negative infinity for large ζ as $-\frac{1}{4}(\zeta - 1.73/2)^2$. Detailed calculation shows that $g_2(\zeta)$ is a smoothly monotone function of ζ, beginning on the parabola $-\frac{1}{4}\zeta^2$ and ending on the parabola $-\frac{1}{4}(\zeta - 1.73/2)^2$.

We have thus determined the stretching function completely by the principle of the PLK method. Here, however, we should note that the zeroth order approximation and the first order approximation are solutions of the boundary layer equation which is a parabolic partial differential equation, while the stretching function and the second order approximation are computed by using the full Navier-Stokes equations, which is an elliptic system of partial differential equation. Thus there is a change in the type of equation in going from the low order approximation to the high order approximation. We then expect that the uniformly valid solution with the stretched coordinates will reproduce the character of the exact Navier-Stokes equations, even if we use just $\psi^{(0)}$. We shall see this in the following section. In fact, as far as engineering applications are concerned, there is no need to enter into the calculation of $\psi^{(2)}$ itself.

3. Improvement of the Zeroth Order Solution by Coordinate Stretching

It will be recalled that the flow field of the Blasius solution is confined to the first quadrant, the plate coinciding with the positive ξ-axis, and both variables ξ and η being positive. Without coordinate distortion this is a highly unsatisfactory representation of the true flow field near the nose of the plate. From the definition of ζ (5.34), $\eta = 0$ corresponds to $\zeta = 0$ if $\xi \neq 0$ and it follows that $x = \xi$ as $g_2(0) = 0$. That is, the positive ξ-axis is transformed into the positive x-axis. On the other hand, if $\xi = 0$, but $\eta > 0$, then $\zeta \to \infty$ and $g_2 \to -\infty$; thus $x \to -\infty$. But when ξ and η vanish simultaneously such that ζ is arbitrary, the whole negative x-axis is swept by the equation $x = \varepsilon^2 g_2(\zeta)$. Thus the origin of the ξ, η-plane is mapped on the whole negative x-axis. For values of ξ different from zero, it can readily be shown that every line $\xi =$ const is mapped on a curve in the x, y-plane, which begins at a point on the positive x-axis and tends to negative infinity when η increases indefinitely. Consequently, the Blasius domain of the first quadrant is mapped onto the whole upper x, y-plane.

The curves of constant ζ-value are of interest. In the ξ, η-plane, these curves are parabolas with vertices at the origin. But from (5.30), (5.34), and (5.38), the curves $\zeta =$ constant in the x, y-plane are defined by

$$y^2 = \zeta^2 [x - \varepsilon^2 g_2(\zeta)] \tag{5.49}$$

Thus they are again parabolas; but now the vertices are separated and are located at $x = \varepsilon^2 g_2(\zeta)$; as ζ increases, they move along the negative x-axis to $-\infty$. If we pick $\zeta = 5.2$, say, as the bounding curve for the viscous region, then we see that the viscous effects diffuse out in front of the leading edge of the plate to a distance of order ε^2, or a physical distance of order $\bar{v}/U_\infty$. Therefore our coordinate distortion alone gives already a very much more reasonable picture of the flow than the classical Blasius solution.

By using equation (3.11) and noting that $x^{(1)} \equiv 0$, the velocity components are

$$\left.\begin{aligned} u = \frac{\bar{u}}{U} &= \frac{\partial \psi^{(0)}}{\partial y} = \frac{\partial \psi^{(0)}}{\partial \eta} - \frac{\varepsilon^2 \dfrac{\partial x^{(2)}}{\partial \eta}}{1 + \varepsilon^2 \dfrac{\partial x^{(2)}}{\partial \xi}} \frac{\partial \psi^{(0)}}{\partial \xi} = f_0'(\zeta) + \frac{\varepsilon^2 g_2'(\zeta)}{2\xi - \varepsilon^2 \zeta g_2'(\zeta)} (\zeta f_0' - f_0) \\ V = \frac{\bar{v}}{U} &= -\varepsilon \frac{\partial \psi^{(0)}}{\partial x} = -\frac{\varepsilon}{1 + \varepsilon^2 \dfrac{\partial x^{(2)}}{\partial \xi}} \frac{\partial \psi^{(0)}}{\partial \xi} = \frac{\varepsilon \sqrt{\xi}}{2\xi - \varepsilon^2 \zeta g_2'(\zeta)} (\zeta f_0' - f_0) \end{aligned}\right\} \tag{5.50}$$

Since the denominator $2\xi - \varepsilon^2 \zeta g_2'(\zeta)$ in these expressions vanishes only at the leading edge, the only singularity of the solution occurs at the leading edge. V is now finite everywhere and it is zero on the negative x-axis because $\xi = 0$ there, and $\zeta \neq 0$. The effect on u is represented by an extra term which is of second order practically everywhere in the boundary layer. Compared with the classical Blasius solution, where u was unity on the negative x-axis and on the y-axis, the essential improvement is that u now varies with ζ according to (5.50) for $\xi = 0$.

In order to bring out the exact nature of the singularity at the leading edge, the function g_2 may be explicity approximated for small ζ by the leading term, namely, by $-\zeta^2/4$ according to (5.44). With this form of g_2, (5.30) gives

$$x = \xi - \frac{\varepsilon^2}{4} \frac{y^2}{\xi} = \xi - \frac{1}{4\xi} Y^2$$

Thus

and

$$\left.\begin{aligned} 2\xi &= x+\sqrt{x^2+Y^2} \\ \frac{1}{2}\varepsilon^2\zeta^2 &= \sqrt{x^2+Y^2}-x \end{aligned}\right\} \tag{5.51}$$

Thus, if we approximate $\psi^{(0)}$ by the leading term $\sqrt{\xi}\,(\sigma\zeta^2/2)$, then near the leading edge

$$\begin{aligned} \psi^{(0)} &\cong \frac{\sigma}{\varepsilon^2}\frac{1}{\sqrt{2}}\{x+\sqrt{x^2+Y^2}\}^{\frac{1}{2}}\{\sqrt{x^2+Y^2}-x\} \\ &= \frac{\sigma}{\varepsilon^2}\frac{1}{\sqrt{2}}Y\{\sqrt{x^2+Y^2}-x\}^{\frac{1}{2}} \end{aligned} \tag{5.52}$$

If we write $x+\mathrm{i}Y=z$, $x-\mathrm{i}Y=\bar{z}$, then the above expression for $\psi^{(0)}$ is proportional to the real part of $z^{3/2}-\bar{z}^{1/2}z$. Thus, with the stretched coordinate, $\psi^{(0)}$ is a biharmonic function near the leading edge and exhibits the character of Stokes' approximation. It may further be noted that the velocities u and V, which were of different order of magnitude in the boundary layer theory, are now of the same order. This is another feature of the improved Blasius solution.

For large ζ, $g_2(\zeta)$ can again be roughly approximated as $-\frac{1}{2}\zeta^2$ according to (5.48). Then (5.50) is reduced to

$$\left.\begin{aligned} u &= 1-\frac{\varepsilon v_0}{\sqrt{2}}\frac{\sqrt{r-x}}{r}, \quad r^2=x^2+Y^2 \\ v &= \frac{\varepsilon v_0}{\sqrt{2}}\frac{\sqrt{r+x}}{r}, \end{aligned}\right\} \quad \zeta\to\infty \tag{5.53}$$

At the edge of the boundary layer, $Y=O(\varepsilon)$. Hence the velocity components at the edge of the boundary layer can be obtained by setting $Y=0$, and are thus equal to 1 and $\varepsilon v_0/\sqrt{x}$, in agreement with the Blasius solution. But the disturbance terms in (5.53) are actually the real and imaginary parts of $-\mathrm{i}\varepsilon 2v_0 z^{-1/2}$ where $z=x+\mathrm{i}Y$. Thus the disturbance velocities are indeed those of a potential flow, vanishing at distances far from the plate, and the unrealistic picture of the classical boundary layer theory is completely corrected by the stretching of coordinates.

One might expect a change in the shear stress at the plate due to the coordinate stretching. However, a detailed calculation shows that all changes in shear stress vanish at the plate where $\zeta=0$. Then the friction computed by the boundary layer theory is still correct, and our previous result, (5.29), still holds. Further improvement in the friction calculation can only come from a computation which uses $\psi^{(2)}$ and $x^{(3)}$. But this is hardly profitable in view of the excellent results already obtained.

4. Boundary Layer in Supersonic Flow

The success of the combined methods of boundary layer transformation and coordinate distortion in the Blasius problem of the incompressible boundary layer led to Kuo's investigation of the more difficult problems in supersonic boundary layer flows by the same method. Here the main complication results from the intimate interaction between the viscous boundary layer and the supersonic inviscid flow bounded by the nose shock just outside of the boundary layer. Because of the viscous forces, the gas in the boundary layer is slowed down and heated; the "thickness" of the layer then continuously increases along the length of the plate. This in turn deflects the outside stream and produces a pressure gradient along the plate. The difference between the supersonic flow and the incompressible flow discussed in the preceding sections lies in the fact that for the supersonic flow this "induced" pressure gradient is very much stronger than that in the incompressible flow. This is the more so if the Mach number is large. In fact, Kuo has found that the distortion of the boundary layer coordinate has to be introduced already in the first order in ε, i.e.,

$$x = \xi + \varepsilon x^{(1)}(\xi, \eta) + \varepsilon^2 x^{(2)}(\xi, \eta) + \dots, \quad y = \eta \tag{5.54}$$

Compared with the corresponding problem in compressible flow as indicated by (5.30), the complication arises at an earlier stage of the solution.

Kuo studied two problems of this type: one is the supersonic boundary layer on a flat plate[4], and the other is the supersonic boundary layer on a plate at an angle with the parallel stream far from the plate (boundary layer on a supersonic wedge). For both problems, however, the calculations involved are considerable and too complicated to be presented here. The interested reader is referred to Kuo's original work. It is hoped however that by discussing the simpler problem of the incompressible boundary layer in detail in the preceding sections, the technique and the power of this latest application of the PLK method have been amply clarified.

Ⅵ. Concluding Remarks

In the preceding sections we have given a somewhat lengthy exposition of the principles and technique of solving physical problems involving a small parameter ε by the method developed by Poincaré, Lighthill, and Kuo. We have used a number of examples, some of them quite complicated, to illustrate the method, but we have not given a general mathematical theory of it. This is really forced upon us, because there is as yet no general mathematical theory available. It is hoped, however, that the reader is not completely left to his own devices to decide whether the PLK method could help him to obtain a useful solution for his particular problem: We have throughout the discussion, shown also problems for which the PLK method has failed; in such instances, we have always tried to point out why it failed. (Sections Ⅱ.5, Ⅱ.8, Ⅳ.2).

The reasons we have given for the failure of the PLK method in the various problems were necessarily vague and heuristic. Here, then, is a problem worth the attention of our colleagues in mathematics. Could they be persuaded to study the question and then tell us just exactly for what type of problem the PLK method will work, and for what type of problem the PLK method will not work? For problems where the PLK method will not work, i. e., will not give a uniformly valid solution to all orders, or a solution of uniform arbitrary accuracy, the method may still give solutions that are uniformly valid to a finite order of ε, as in the thin airfoil problem. Can this be known by just looking at how the problem is formulated?

In the absence of answers to the above questions, the engineer really need not despair. For him, the best guide in estimating the correctness of his calculation is still his understanding of the physical problem. If a mathematical solution does not give the expected answer, he naturally has to question the validity of the mathematical solution. Therefore the fact that he does not fully "understand" the PLK method should not prevent him from trying to use the method to solve his problem. He may well keep in mind what Heaviside said when his intuitive operational calculus was questioned: "Shall I refuse my dinner because I do not fully understand the process of digestion?"

References

[1] Poincaré H. Les Méthodes Nouvelles de la Méchanique Céleste. vol. 1, Chap. Ⅲ, Paris, 1892.

[2] Lighthill M J. A technique for rendering approximate solutions to physical problems uniformly valid. Phil. Mag. 1949,40(7); 1179.

[3] Kuo Y H. On the flow of an incompressible viscous fluid past a flat plate at moderate Reynolds number. J. Math. and Phys. 1953, 32; 83.

[4] Kuo Y H. Viscous flow along a flat plate moving at high supersonic speeds. J. Aeron. Sci. (1956), 23: 125.

[5] Wasow W A. On the convergence of an approximation method of Lighthill. Abstract No. 40, Bull. Am. Math. Soc. 1955,61: 48; J. Rational Mech. Anal. 1955, 4: 751.

[6] Carrier G F. Boundary layer problems in applied mechanics. Advances in Appl. Mech. 1953, 3: 1.

[7] Carrier G F. Boundary layer problems in applied mechanics. Comm. Pure and Appl. Math. 1954, 7: 11.

[8] Lighthill M J. The position of the shock-wave in certain aerodynamic problems. Quart. J. Mech. and Appl. Math. 1948, 1: 309.

[9] Wu Y T. Two dimensional sink flow of a viscous, heat-conducting compressible fluid; cylindrical shock waves. Quart. Appl. Math.

[10] Hayes W D. Pseudotransonic similitude and first-order wave structure. J. Aeron. Sci. 1954, 21: 721.

[11] Whitham G B. The propagation of spherical blast. Proc. Roy. Soc. 1950, A203: 571.

[12] Whitham G B. The behavior of supersonic past a body of revolution. far from the axis, Proc. Roy. Soc. 1950, A201: 80.

[13] Whitham G B. The flow pattern of a supersonic projectile. Comm. Pure and Appl. Math. 1952, 5: 301.

[14] Whitham G B. The propagation of weak spherical shocks in stars. Comm. Pure and Appl. Math. 1953, 6: 397.

[15] Lin C C. On a perturbation theory based on the method of characteristics. J. Math. and Phys. 1954, 33: 117.

[16] Fox P A. On the use of coordinate perturbation in the solution of physical problems. Tech. Rept. No. 1,

Project for Machine Method of Computation and Numerical Analysis, Mass. Inst. Technol., Cambridge, Mass., 1953.

[17] Lighthill M J. A new approach to thin airfoil theory. Aeronaut. Quart, 1951, 3: 193.

[18] van Dyke M D. Subsonic edges in thin-wing and slender-body theory. Natl. Advisory Comm. Aeronaut., Tech. Note No. 3343 (1954).

[19] Janour Z. Resistance of a plate in parallel flow at low Reynolds number. Natl. Advisory Comm. Aeronaut., Tech. Mem. No. 1316 (1951).

Thermodynamic Properties of Gas at High Temperatures and Pressures

H. S. Tsien[1]

(*Daniel and Florence Guggenheim Jet Propulsion Center, California Institute of Technology, Pasadena, Calif.*)

1 Equation of States of Dense Gas

When the density of gas is high, it is well known that the simple equation of states for a perfect gas can no longer be expected to be valid. The most crude approximation to the equation of states for a dense gas is that of Van der Waal. If P is the pressure, v the volume per molecule, T the temperature, and k the Boltzmann constant, then the Van der Waal equation is

$$\left(P+\frac{a}{v^2}\right)(v-b)=kT \tag{1}$$

where a and b are two constants, small in magnitude. The constant b is usually simply identified as four times the volume of a molecule. If the molecules are assumed to be spheres of diameter D, then

$$b=\frac{2\pi}{3}D^3 \tag{2}$$

At high temperatures, the density of gas can be large only if the pressure is very high. Then the term a/v^2 is not important in comparison with P, and Equation (1) can be simplified into the so-called covolume equation of states

$$P(v-b)=kT$$

Or we can write

$$\frac{Pv}{kT}=1+\frac{1}{\dfrac{3}{2\pi}\dfrac{v}{v^*}-1} \tag{3}$$

where v^* is a volume defined by

$$v^*=D^3 \tag{4}$$

Received June 20, 1954.

Jet Propulsion, vol. 25, pp. 471 - 472, 1955.

① Director, Guggenheim Jet Propulsion Center.

However, because of the crude approximation in the Van der Waal equation of states, neither Equation (1) nor Equation (2) can be expected to be sufficiently accurate for gas at very high temperatures and high pressures. An example of such state of matter is the gaseous products of detonation of condensed explosives, where temperatures of several thousand degrees Kelvin and densities of the order of solids occur. Other more elaborate equations of states designed to cover the whole range of pressures and temperatures suffer from the same defect.

For products of detonation of condensed explosives, a more accurate equation is the Halford-Kistiakowsky-Wilson equation[1]

$$\frac{Pv}{kT} = 1 + KT^{-1/4} \exp(0.3\,K/T^{1/4}) \tag{5}$$

where

$$K = \sum_i n_i K_i \tag{6}$$

n_i is the number of moles of the ith molecular species per unit volume, and K_i is the empirical constant for the ith species. For water and ammonia, the K_i's are 108 and 164 cm^3 per mole, respectively. It is suggested that when K_i is not known, it can be calculated as

$$K_i = 5.5\left(\frac{2\pi}{3}\right) N D_i^3 \tag{7}$$

where N is Avogadro's number, and D_i is the low energy collision diameter of molecules of the ith species. K_i is thus equal to 22 times the volume of one mole of spherical molecules of diameter D_i. If there is only one species of molecules, then Equations (6) and (7) give

$$K = 5.5\left(\frac{2\pi}{3}\right) \Big/ \left(\frac{v}{v^*}\right) \tag{8}$$

where v^* is a volume defined by Equation (4). Thus for a single gas, the Halford-Kistiakowsky-Wilson equation can be written as

$$\frac{Pv}{kT} = 1 + \frac{11.51}{\left(\frac{v}{v^*}\right) T^{1/4}} \exp\left(\frac{3.453}{\frac{v}{v^*} T^{1/4}}\right) \tag{9}$$

It is easily seen from Equations (5) and (9) that they are an improvement over the covolume equation (3); now the "compressibility," $Pv/(kT)$, is unity as $T \to \infty$ even with v/v^* finite. This result is to be expected on the general ground that molecules are never rigid spheres, but "squeezy," i.e., closer approach is possible if two molecules collide with greater kinetic energy. At very high temperatures, the kinetic energy of the molecules is very high, then the effective size of the molecules in collision must be very small. Now the effects of gas imperfection is proportional to the molecular size. Therefore as $T \to \infty$, the effective molecular size vanishes and the gas becomes a perfect gas, even if the volume ratio is finite. Hence really satisfactory equation of states must have this property. This is so for the Halford-Kistia-kowsky-Wilson equation.

But even when the constant K is related to the low energy collision diameter of the molecules through Equations (7) and (8), still no proper account is made for the strength of molecular interaction. If we represent the interaction between a pair of molecules by the Lennard-Jones potential $\varepsilon(r)$, a function of the intermolecular distance r

$$\varepsilon(r) = 4\varepsilon^* \left[\left(\frac{D}{r} \right)^{12} - \left(\frac{D}{r} \right)^{6} \right] \tag{10}$$

then besides the collision diameter D, there is the parameter of equilibrium potential $-\varepsilon^*$, the potential when the pair of molecules are at their equilibrium distance $2^{1/6}D$. A convenient parameter for ε^* is the "characteristic temperature of interaction," Θ_1, defined as

$$\Theta_1 = \frac{\varepsilon^*}{k} \tag{11}$$

Then we expect the equation of states to be of the form

$$\frac{Pv}{kT} = 1 + f(T/\Theta_1, v/v^*) \tag{12}$$

i.e., the compressibility should be a function of temperature ratio T/Θ_1, but not temperature itself as is in the Halford-Kistiakowsky-Wilson equation. Furthermore, according to our argument on vanishing imperfection at $T \to \infty$

$$f(\infty, v/v^*) = 0 \tag{13}$$

With the general concept that the equation of states for dense gas must satisfy the conditions embodied in Equations (12) and (13), many of the equations of states proposed by various authors can be ruled out as unreliable. For instance, Cottrell and Paterson suggested[2] that

$$\frac{Pv}{kT} = 3 + \frac{\text{const}}{(T/\Theta_1)^{1/2}\left(\frac{v}{v^*} \right)} \tag{14}$$

Since the condition of Equation (13) is not satisfied, this equation cannot be reliable for very high temperatures. In a similar manner, the theoretical equation of states proposed by Zwansig[3] is also unacceptable, because he makes the untenable assumption that at very high temperatures the molecules interact as rigid spheres.

2 Lennard-Jones and Devonshire Theory

If there were sufficient experimental data for gas at very high temperatures and pressures, we can try to fit the data to the nondimensional equation of states of Equation (12) and determine the function $f(T/\Theta_1, v/v^*)$. This is, in effect, the application of the principle of corresponding states. Unfortunately, except for two old measurements on hydrogen and helium by Bridgman[4], no experimental data are available at high enough temperatures and pressures for this to be possible. We must then turn to theory to determine the proper equation of states. If we are interested in the region of density where the nondimensional volume v/v^* is near unity, i.e., densities of the order of liquid density at low pressure, then the Lennard-Jones and Devonshire theory of liquids and dense gas is a very good approximation to the true physical situation. Here the main defect of not allowing for empty lattice sites in the theory is not important. Furthermore, Wentorf, Buhler, Hirschfelder, and Curtiss[5] have carried out very extensive and accurate calculations for the Lennard-Jones and Devonshire theory. We can use their tabulated values directly. But for practical calculations, it would be more convenient to put their result in a simple analytic form for easy interpolation and extrapolation. This is the main purpose of the present note.

Of course theoretically the equation of states can be directly obtained from the formulation of Lennard-Jones and Devonshire theory without having to use the numerical results. But the interested range of parameters T/Θ_1 and v/v^* is such that no simple analytic result can be obtained. This is quite different from the case of liquid state where T/Θ_1 is small enough for the simple analytic treatment[6] to be successful.

By using the tabulated values of Pv/kT given by Wentorf and collaborators, we can plot T/Θ_1 against v/v^* for constant Pv/kT. The result of this preliminary investigation indicates that instead of two variables T/Θ_1 and v/v^* in f of Equation (12), a single variable η is sufficient with

$$\eta = (T/\Theta_1)^{1/6}(v/v^*) \tag{15}$$

Then, according to Equations (12) and (13)

$$\frac{Pv}{kT} = 1 + f(\eta), \quad f(\infty) = 0 \tag{16}$$

In Fig. 1, $1/f = (Pv/kT - 1)^{-1}$ is plotted against η. We see that the points are grouped around a straight line for $T/\Theta_1 = 10$, 20, 50, 100, and 400 and v/v^* near unity. In fact, we find the approximate equation of states for gas at very high temperatures and pressures as

$$\frac{Pv}{kT} = 1 + \frac{1}{0.278\eta - 0.177} \tag{17}$$

The condition of Equation (13) is thus satisfied. This result is accurate to about 10 per cent.

Although the equation of states, Equation (17), is obtained for dense gas at high temperature, it may be interesting to see how well it behaves for dilute gas. For dilute gas, v/v^* is very large, then η is very large. Therefore for dilute gas at high temperatures, we have

$$\frac{Pv}{kT} \cong 1 + \frac{1}{0.278\eta} = 1 + 3.594\,(T/\Theta_1)^{-1/6}(v/v^*)^{-1} \tag{18}$$

On the other hand, the exact equation of states for dilute gas at very high temperatures is given by the virial equation, retaining only the leading term in the expansion of the second virial coefficient[7]. Thus

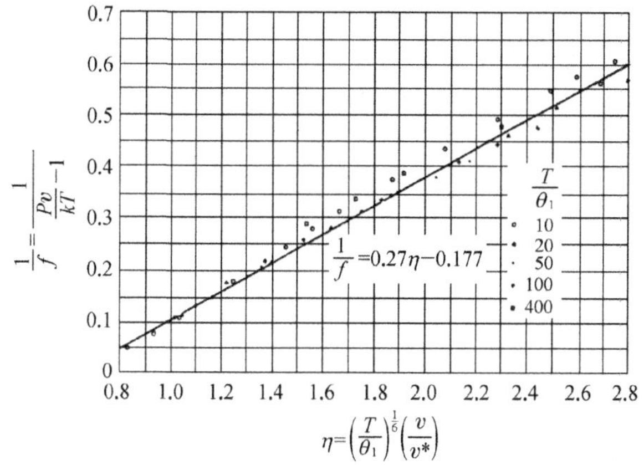

Fig. 1

$$\frac{Pv}{kT} \cong 1+\frac{2\sqrt{2}\pi}{3}\Gamma\left(\frac{3}{4}\right)(T/\Theta_1)^{-1/4}(v/v^*)^{-1} = 1+3.630\,(T/\Theta_1)^{-1/4}(v/v^*)^{-1} \tag{19}$$

By comparing Equations (18) and (19), it is seen that aside from the difference in the exponent of T/Θ_1, our approximate equation of dense gas at high temperatures even reproduces some of the characteristics of a dilute gas. Therefore for rough approximations, our equation of states, Equation (17), can be even used for the complete range of v/v^*. Of course, Equation (17) can be true only for high temperatures, say $T/\Theta_1 > 10$.

3 Other Thermodynamic Functions

Since the proposed equation of states is approximately true for even dilute gas, we may use it to calculate other thermodynamic functions by integrating to $v/v^* \to \infty$. For instance, if E is the energy per molecule, then the general thermodynamic law states that

$$\left(\frac{\partial E}{\partial v}\right)_T = T\left(\frac{\partial P}{\partial T}\right)_v - P$$

Thus, by differentiating Equation (17), using Equation (15)

$$\left(\frac{\partial E}{\partial v}\right)_T = -\frac{kT}{v}\frac{0.278\eta}{(0.278\eta-0.177)^2} \tag{20}$$

Thus, if E_∞ is the energy per molecule when $v/v^* \to \infty$, i.e., E_∞ is the energy per molecule calculated for a perfect gas, then the energy per molecule due to imperfection of the gas is

$$(E-E_\infty)_T = -\int_v^\infty \left(\frac{\partial E}{\partial v}\right)_T \mathrm{d}v \tag{21}$$

By substituting Equation (20) into Equation (21), and by considering $(E-E_\infty)/kT$ to be a function of η only, we obtain

$$(E-E_\infty)_T = \frac{kT}{6}\frac{1}{0.278\eta-0.177} \tag{22}$$

We observe that although the result involves the use of our approximate equation of states for large v/v^*, the contribution to $E - E_\infty$ at large v/v^* is small, and the error made is not as serious as one might conclude at first sight. In fact, by comparing the computed value using Equation (22) with the exact numerical value of the same function tabulated by Wentorf and collaborators, the difference is of the order of 10 per cent. Ten per cent accuracy is all we claim for the equation of state.

Similarly, we have the following formulas for molecular enthalpy H, molecular heat capacities, c_v and c_p, molecular entropy s

$$(H - H_\infty)_T = 7(E - E_\infty)_T \tag{23}$$

$$(c_V - c_{V\infty})_T = \frac{k}{6}\frac{0.231\,4\eta - 0.177}{(0.278\eta - 0.177)^2} \tag{24}$$

$$(c_p - c_{p\infty})_T = 7(c_V - c_{V\infty})_T\left[1 - \frac{0.278\eta}{0.231\,4\eta - 0.177}\cdot \frac{(0.278\eta - 0.177)^2 + (0.231\,4\eta - 0.177)}{(0.278\eta - 0.177)^2 + (0.556\eta - 0.177)}\right] \tag{25*}$$

$$(s - s_\infty)_T = \frac{k}{6}\frac{1}{0.278\eta - 0.177} - \frac{k}{0.177}\log\frac{0.278\eta}{0.278\eta - 0.177} \tag{26}$$

The above formulas for heat capacities and entropy are less accurate than for energy and enthalpy because of the differentiation used in their derivations. But these equations are at least consistent among themselves.

References

[1] Hirschfelder J O, Curtiss C F, Bird R B. Molecular Theory of Gases and Liquids. John Wiley, 1954: 263.

[2] Cottrell T L. Paterson S S. An Equation of State Applicable to Gases at Densities near That of the Solid and Temperatures far above the Critical. Proc. Roy. Soc. (A) 1952, 213: 214.

[3] Zwansig R W. High-Temperatures Equation of State by a Perturbation Method. I. Nonpolar Gases. J. Chem. Phys., 1954, 22; 1420.

[4] Bridgman P W., The Physics of High Pressures. G. Bell and Sons;1931;108.

[5] Wentorf R H., Buhler R J., Hirschfelder J O, Curtiss C F. Lennard-Jones and Devonshire Equation of State of Compressed Gases and Liquids. J. Chem. Phys., 1950, 18: 1484.

[6] Tsien H S. The Properties of Pure Liquids. Journal of the American Rocket Society, January-February 1953.

[7] Ref. 1, p. 163.

* Eq. (25) has been changed according to the corrections made by the anthor in 1962. — Noted by edifor.

Thermonuclear Power Plants

Hsue-shen Tsien①
(*Daniel and Florence Guggenheim Jet Propulsion Center, California Institute of Technology, Pasadena, Calif.*)

Some of the unique features of thermonuclear power plants and the essential problems in the technical design of such plants are discussed in this paper. The thermonuclear reaction rate for the fusion of deuterium is calculated on the basis of a similar analysis published by Gamow and Teller. The pressure, temperature, and minimum dimensions of the necessary reaction chamber are determined largely by consideration of reaction quenching and energy loss near the walls. Results are presented for the power output and the efficiency of a power station utilizing the deuterium fusion reaction. The comment by Greenstein that follows this paper deals particularly with the difficult problem of calculating the reaction quenching and energy loss rates at the walls.

1 Introduction

Agreat majority of the present discussions and plans on the utilization of nuclear energy for power production is related to the fission reaction. Although the nuclear fission power plants have many distinct advantages over the conventional power plants, the limited world reserve of uranium and thorium that can be practically mined makes the long term prospect of such power plants somewhat uncertain. On the other hand, the thermonuclear fusion reaction, particularly the "burning" of deuterium into helium, utilizes a very abundant fuel. Therefore, if the fusion reaction can be made to generate energy for power plants, the prospect of world energy supply will be very much brighter.

But can the fusion reaction be utilized in terrestrial power plants? This question has been examined by E. Sänger and his collaborators [1,2]. However, a critical reading of Sänger's work will show that part of his analysis is not valid because he has not gone deep enough into the subject. It is the purpose of this note to point out some of the unique features of the thermonuclear power plants and the essential problems in the technical design of such power plants. It will be seen that such engineering projects are truly of stupendous proportions and are a challenge to one's imagination. However, the reward to the welfare of the human race by a successful development of thermonuclear power plants is so great as to make the careful

---

Received Oct. 24, 1955.
Jet Propulsion, vol. 26, pp. 559–564, 1956.

① Formerly: Robert H. Goddard Professor of Jet Propulsion.

examination of this problem a very worthwhile research project.

2 Thermonuclear Reaction Rate

Thermonuclear reactions are reactions between charged nuclei. Because of the electric charge, necessarily positive, the approach of the nuclei to each other has to overcome the Coulomb repulsion between the nuclei. Therefore, only if the relative kinetic energy of the nuclei is high, can a close enough approach be obtained and a reaction take place. This required kinetic energy is so large that even at temperatures as high as 10^8 K only nuclei in the high energy end of the Maxwellian velocity distribution can achieve reaction. Therefore only a very small fraction of the nuclei participate in the reaction. In other words, the reaction rate is quite small. This observation leads to a great simplification of the calculation: The nuclear distribution can be considered as quasisteady; i.e., the Maxwellian distribution can in fact be used in spite of the slight lack of thermodynamic equilibrium during the reaction. Gamow and Teller[3] have developed such a theory of thermonuclear reactions. The following is a slightly modified form of the theory suitable for the present purpose.

Considering the colliding particles as rigid spheres, it is well known[4] that the number dN of collisions per unit volume per unit time between particles of type 1 and 2 with kinetic energy of collision between ε and $\varepsilon + d\varepsilon$ is

$$dN = \frac{n_1 n_2}{s}\left(\frac{2\pi kT}{\mu}\right)^{1/2} e^{-\varepsilon/kT} 2D_{12}^2 \frac{\varepsilon}{kT}\frac{d\varepsilon}{kT} \tag{1}$$

where n_1, n_2 are the number densities of particles of type 1 and 2, respectively; s is the symmetry number, equal to 2 if type 1 and type 2 are the same, and equal to 1 if not; μ is the reduced mass of type 1 and type 2 particles. D_{12} is the average diameter of the two types of particles; i.e., if D_1 and D_2 are the diameters of type 1 and 2 particles, then

$$D_{12} = \frac{1}{2}(D_1 + D_2) \tag{2}$$

To put this general formula into a form more useful for the present computation, molar fractions ν_1 and ν_2 are introduced

$$\nu_1 = \frac{n_1}{n},\ \nu_2 = \frac{n_2}{n},\ n = \Sigma_{ni} \tag{3}$$

where n is the total number of particles per unit volume; the particles may include electrons besides the nuclei. Furthermore, if M_1 and M_2 are the mass of particles of type 1 and 2, and A_1 and A_2 are the corresponding quantities expressed in terms of atomic mass units ("atomic weights" of the nuclei), and M is the mass of one atomic mass unit, then

$$\mu = \frac{M_1 M_2}{M_1 + M_2} = \frac{A_1 A_2}{A_1 + A_2} M = AM \tag{4}$$

Thus A is the reduced mass expressed in terms of atomic mass units. If V is the relative velocity of the two colliding particles, then ε, the relative translational energy, is defined as

$$\varepsilon = \frac{1}{2}\mu V^2 \tag{5}$$

If P is the thermodynamic pressure, then

$$n = P/(kT) \tag{6}$$

Equation (6) is true only if the assembly of particles is at thermodynamic equilibrium and if the particles essentially do not interact, i. e., the assembly is a perfect gas. At the extremely low gas density that will be considered, this is true to a high degree of accuracy. Of course, if the sum of the particles is not at thermodynamic equilibrium, e. g., fusion product neutrons which hardly collide a sufficient number of times with other particles to have a Maxwellian distribution of velocity within the dimension of region considered, then such particles must not be considered in calculating the "total" particle density n. The "pressure" produced by such particles on the wall of the containing vessel has to be treated separately. On the other hand, photons, if any, that are almost at thermodynamic equilibrium must be included in the particle density n.

The most important modification of (1) is the following: Nuclear reactions cannot be considered as collisions between rigid spheres but are expressed through an effective cross section σ. The effective cross section for collision of rigid spheres is a circle of radius D_{12}. Therefore

$$D_{12}^2 = \sigma/\pi \tag{7}$$

By substituting (3), (4), (6), and (7) into (1), the number dN is

$$\mathrm{d}N = \frac{4\nu_1\nu_2\sigma P^2}{s\sqrt{2\pi AM}}\frac{\mathrm{e}^{-\varepsilon/kT}}{(kT)^{7/2}}\varepsilon \mathrm{d}\varepsilon \tag{8}$$

It is important to note that in general the effective cross section is not a constant but function of energy ε. In particular, according to the theory of nuclear reaction, the quantal penetration of Coulomb barrier gives the effective cross section σ as

$$\sigma = \frac{\Lambda^2}{4\pi}\frac{\Gamma AMR^2}{\hbar^2}\times \exp\left[-\frac{2\pi \mathrm{e}^2 Z_1 Z_2\sqrt{AM}}{\hbar\sqrt{2\varepsilon}} + \frac{4\mathrm{e}\sqrt{2AMZ_1Z_2R}}{\hbar}\right] \tag{9}$$

where Λ is the de Broglie wavelength

$$\Lambda = \frac{2\pi\hbar}{\sqrt{2AM\varepsilon}} \tag{10}$$

R is the radius of the compound nucleus formed during reaction, and can be estimated as

$$R = [1.7 + 1.22(A_1 + A_2)^{1/3}]\times 10^{-13}\,\mathrm{cm} \tag{11}$$

Γ is the half-width of the nuclear resonance level. Z_1 and Z are the nuclear charges in units of electronic charge e.

By substituting (9) and (10) into (8), it is seen that the variation of $dN/d\varepsilon$ with respect to ε is due to the exponential factor

$$\exp\left\{-\left[\frac{\varepsilon}{kT}+\frac{2\pi e^2 Z_1 Z_2\sqrt{AM}}{\hbar\sqrt{2\varepsilon}}\right]\right\} \tag{12}$$

There is a minimum of the quantity within the square bracket with respect to ε which corresponds to the maximum of $dN/d\varepsilon$. If this ε is denoted by ε^*, then

$$\frac{1}{kT}=\frac{\pi e^2 Z_1 Z_2\sqrt{AM}}{\sqrt{2}\hbar\,\varepsilon^{*3/2}}$$

or

$$\varepsilon^*=\left\{\frac{\pi e^2 Z_1 Z_2\sqrt{AM}kT}{\sqrt{2}\hbar}\right\}^{2/3} \tag{13}$$

Near this value of ε^*, the expression in the square bracket of (12) can be approximated by

$$\begin{aligned}\frac{\varepsilon}{kT}+\frac{2\pi e^2 Z_1 Z_2\sqrt{AM}}{\hbar\sqrt{2\varepsilon}} &\cong \frac{\varepsilon^*}{kT}+\frac{2\pi e^2 Z_1 Z_2\sqrt{AM}}{\sqrt{2}\hbar\sqrt{\varepsilon^*}}+\frac{1}{2}\cdot\frac{3}{2}\times\frac{\pi e^2 Z_1 Z_2\sqrt{AM}}{\sqrt{2}\hbar\,\varepsilon^{*5/2}}(\varepsilon-\varepsilon^*)^2\\ &=3\frac{\varepsilon^*}{kT}+\frac{3}{4}\frac{1}{kT\varepsilon^*}(\varepsilon-\varepsilon^*)^2\end{aligned} \tag{14}$$

Thus, under this approximation, dN can be computed as

$$\begin{aligned}dN=&\frac{4\nu_1\nu_2\Gamma P^2}{s\sqrt{2\pi AM}(kT)^{7/2}}\frac{\pi R^2}{2}\exp\left[\frac{4e\sqrt{2AMZ_1Z_2R}}{\hbar}-\frac{3\varepsilon^*}{kT}\right]\times\\ &\exp\left[-\frac{3}{4}\frac{1}{kT\varepsilon^*}(\varepsilon-\varepsilon^*)^2\right]d\varepsilon\end{aligned} \tag{15}$$

By integrating over all ε, we have the number N of effective binary collisions per unit vol per unit time as

$$N=\frac{4\pi\nu_1\nu_2R^2}{s\sqrt{3}}\left(\frac{P}{kT}\right)^2\left(\frac{\Gamma}{kT}\right)\sqrt{\frac{kT}{2AM}}\sqrt{\frac{\varepsilon^*}{kT}}\times\exp\left[\frac{4e\sqrt{2AMZ_1Z_2R}}{\hbar}-3\frac{\varepsilon^*}{kT}\right] \tag{16}$$

From (13)

$$\frac{\varepsilon^*}{kT}=\left(\frac{\pi^2 e^4 Z_1^2 Z_2^2 AM}{2\hbar^2 kT}\right)^{1/3} \tag{17}$$

Equations (16) and (17) together determine the reaction rate. There is only one important difference between (16) and the original formula due to Gamow and Teller[3]; Gamow and Teller have not included the symmetry number s and thus may be wrong in some cases by a factor of 2.

If we denote by x the quantity $(kT)^{1/3}$, then the temperature dependent part of (16) can be written as

$$x^8 \exp\left[-3\left(\frac{\pi^2 e^4 Z_1^2 Z_2^2}{2\hbar^2}\right)^{1/3} x\right] \tag{18}$$

This quantity clearly has a maximum at some value of x, say x_0; x_0 is determined by

$$8 - 3\left(\frac{\pi^2 e^4 AMZ_1^2 Z_2^2}{2\hbar^2}\right)^{1/3} x_0 = 0 \tag{19}$$

Equation (19) gives the optimum reaction temperature T_0 for maximum reaction rate at constant pressure as

$$T_0 = \left(\frac{3}{8}\right)^3 \frac{\pi^2 e^4 AZ_1^2 Z_2^2}{2k\hbar^2} \tag{20}$$

By putting in the numerical values of physical constants

$$T_0 = 1.442 \times 10^8 AZ_1^2 Z_2^2 \quad \text{K} \tag{21}$$

Equations (20) and (21) show that the optimum reaction temperature depends only on the reduced mass and charges of the nuclei and is independent of the details of the reaction. T_0 is the smallest for proton-proton reaction, ($A = 1/2$, $Z_1 = Z_2 = 1$), for which $T_0 = 0.721 \times 10^8$ K.

The important parameter in the expression of reaction rate is the level width Γ. This has to be determined experimentally. However the experimental reaction cross sections are usually expressed as

$$\sigma = \frac{B}{\varepsilon} e^{-C/\sqrt{\varepsilon}} \tag{22}$$

where B and C are two empirically determined constants for any one reaction. Same as the preceding paragraphs, (8) and (22) can be combined to give the formulas

$$\frac{\varepsilon^*}{kT} = \left(\frac{C^2}{4kT}\right)^{1/3} \tag{23}$$

$$N = \frac{4\nu_1\nu_2}{s}\left(\frac{B}{kT}\right)\left(\frac{P}{kT}\right)^2 \sqrt{\frac{2kT}{3AM}}\sqrt{\frac{\varepsilon^*}{kT}}\, e^{-3(\varepsilon^*/kT)} \tag{24}$$

and

$$T_0 = \left(\frac{3}{8}\right)^3 \frac{C^2}{4k} \tag{25}$$

If the optimum temperature T_0 is used as a reference temperature

$$\frac{\varepsilon^*}{kT} = \frac{8}{3}\left(\frac{T_0}{T}\right)^{1/3} \tag{26}$$

For numerical computation, P is usually given in atmospheres. Thus

$$\frac{P}{kT} = 7.34 \times 10^{21} \frac{P}{T} \text{ cm}^{-3} \tag{27}$$

B is usually given in units of barns-kilovolts, thus

$$\frac{B}{kT} = 1.160 \times 10^{-17} \frac{B}{T} \text{ cm}^2 \tag{28}$$

C is usually given in units of $\text{kV}^{1/2}$, thus

$$T_0 = 1.528 \times 10^5 C^2 \quad \text{K} \tag{29}$$

If E is the energy production of a single binary reaction, then the rate of energy production Q per unit vol per unit time is obviously

$$Q = EN \tag{30}$$

3 Example: Deuterium Reaction

Because of the abundance of deuterium as a naturally occurring stable isotope of hydrogen, it is of interest to consider the burning of deuterium. The accurate reaction data were given recently by Arnold et al[5].

$$\sigma = \frac{B'}{\varepsilon'} e^{-C'/\sqrt{\varepsilon'}} \tag{31}$$

where ε' is the deuteron energy in kilovolts in the usual laboratory coordinate system, and $B' = 288$ barns-kV, $C' = 45.7\ \text{kV}^{1/2}$. Since the ε in (22) is the relative kinetic energy defined by[5] and the ratio of deuterium mass and the reduced mass is 2 in a deuteron-deuteron reaction,

$$\varepsilon' = 2\varepsilon \tag{32}$$

Therefore, by using Arnold's data. the reaction constants B and C are

$$B = B'/2 = 144 \text{ barnes-kV} \tag{33}$$

$$C = C'/\sqrt{2} = 32.36\ \text{kV}^{1/2} \tag{34}$$

Equation (29) immediately gives the optimum temperature T_0 for deuteron-deuteron reaction as

$$T_0 = 1.600 \times 10^8 \text{ K} \tag{35}$$

According to Arnold et al., the deuteron-deuteron reaction branches, with almost equal probability, into two reactions

$${}_1H^2 + {}_1H^2 \rightarrow {}_1H^3 + {}_1p^1 \tag{36}$$

$${}_1H^2 + {}_1H^2 \rightarrow {}_2He^3 + {}_0n^1 \tag{37}$$

Since the masses of the atomic species are given as follows

$$\left.\begin{aligned} A({}_1H^2) &= 2.014\,735 \\ A({}_1H^3) &= 3.016\,997 \\ A({}_2He^3) &= 3.016\,977 \\ A({}_1p^1) &= 1.008\,142 \\ A({}_0n^1) &= 1.008\,982 \end{aligned}\right\} \tag{38}$$

The reaction (36) then produces

$$2\times 2.014\,735-(3.016\,997+1.008\,142)=0.043\,31\text{ amu}=4.03\text{ MeV}=6.46\times 10^{-6}\text{ ergs} \tag{39}$$

The reaction (37) produces

$$2\times 2.0147\,35-(1.008\,982+3.016\,177)=0.003\,511\text{ amu}=3.27\text{ MeV}=5.24\times 10^{-6}\text{ ergs} \tag{40}$$

However, in a thermonuclear reaction chamber, (36) and (37) do not represent the end of reactions. The reaction products are immediately thermalized by elastic collisions with other particles; then the following reactions are possible

$${}_2He^3+{}_0n^1\rightarrow{}_1H^3+{}_1p^1 \tag{41}$$

$${}_1H^3+{}_1H^2\rightarrow{}_2He^4+{}_0n^1 \tag{42}$$

$${}_2He^3+{}_1H^2\rightarrow{}_2He^4+{}_1p^1 \tag{43}$$

At first sight it seems that the reaction (41) depending upon both components of the reaction products of low concentration will be very much less frequent than (42) and (43) depending upon only one component of reaction product. However at thermal energies corresponding to temperatures of the order of 10^8 K, the cross section of (41) is of the order of 10 barns, while according to Arnold[5], the reaction cross sections of (42) and (43) are very much smaller. Therefore (41) is in fact the only reaction of importance while (42) and (43) can be neglected. This being the case, then eventually all of the deuteron-deuteron reaction really results in the production of tritium according to (36). Therefore the average energy production E for every deuteron-deuteron reaction is according to (39)

$$E=6.46\times 10^{-6}\text{ergs} \tag{44}$$

The energy produced by burning a unit mass of deuterium is thus

$$\frac{6.46\times 10^{-6}\times 2.388\times 10^{-8}\times 0.605\times 10^{24}}{4.029}\text{cal/gr}=2.31\times 10^{10}\text{cal/gr} \tag{45}$$

Now assume $T=T_0=1.600\times 10^8$ K, $P=100$ atm. The feed gas D_2 will be completely ionized at this temperature, and reacting mixture will be composed initially of equal numbers of deuterons and electrons. Thus $\nu_1=\nu_2=1/2$, $s=2$. Then using (24), (26), (30) the energy production Q per unit vol per unit time is

$$Q = 6.46 \times 10^{-6} \times \frac{1}{2} \times \frac{1.160 \times 10^{-17} \times 144}{1.600 \times 10^{8}} \times$$

$$\left(\frac{7.34 \times 10^{23}}{1.600 \times 10^{8}}\right)^{2} \sqrt{\frac{8}{3} \times \frac{2 \times 8.316 \times 10^{7} \times 1.6 \times 10^{8}}{3 \times 1.007}} \times e^{-8} \text{ergs/cm}^{3}\,\text{sec} \tag{46}$$

$$= 0.365 \times 10^{8} \text{ergs/cm}^{3}\,\text{sec} = 3.65 \text{ watts/cm}^{3} = 0.874 \text{ cal/cm}^{3}\,\text{sec}$$

It is interesting to note that this volume rate of energy generation is only 1/10 of the rate of generation as in a modern aircraft gas turbine combustion chamber using hydrocarbon fuel. Therefore in spite of the extremely high temperature, thermonuclear reaction using deuterium is a relatively slow reaction. The reason for this anomaly is the extremely low density of the hot gas: There simply are not enough deuterons in a unit volume to give high reaction rate. However, as Sänger[1] has shown, other nuclei generally give even lower rates of energy production.

4 Thermonuclear Reaction Chamber

The moderate volume rate of energy production together with the extremely high gas temperature naturally call ones attention to the problem of quenching of the "flame" by excessive cooling. This problem is in fact the central problem of thermonuclear reaction chamber. There is certainly a critical size, say a critical diameter, of the reaction chamber below which the reaction cannot be maintained. As a very rough first estimate, one may take the chemical combustion as a model, and use the mean free path as the sizing length. Because of the relatively slow thermonuclear reaction, the chemical model should be one of poor reactivity. Thus the quenching diameter at atmospheric pressure can be taken as 1 cm. The pressure effect on quenching can be thought of as a Reynolds number effect. Then the quenching diameter at 100 atm will be 1/100 cm. To translate this value to thermonuclear reaction chamber, one notes the fact that the ratio of mean free path for the two cases is approximately 10^{6}. Thus the rough estimate of the critical diameter D_c of the reaction chamber is

$$D_c = \frac{1}{100} \times 10^{6} \text{cm} = 100 \text{ meters} \tag{47}$$

If the length is to be ten times the diameter, then the thermonuclear reaction chamber is a vessel of 100 meters diam and 1 000 meters long built to withstand a pressure of 100 atmospheres!

To examine the quenching problem in some detail, one must first estimate the mean free path of the fully ionized mixture of deuteron and electron. If n is the particle density, and if σ_s is the scattering cross section of the particles, then the general equation for the mean free path l is

$$l = \frac{0.177}{n\sigma_s} = \frac{0.177kT}{P\sigma_s} = 2.41 \times 10^{-23} \frac{T}{P\sigma_s} \tag{48}$$

For fully ionized particles, the cross section σ_s can be computed approximately according to Lin, Resler, and Kantrowitz[6] as

$$\sigma_s = 8.10\left(\frac{e^2}{3kT}\right)^2 \log\left(\frac{kT}{e\sqrt{4\pi P}}\Big/\frac{e^2}{3kT}\right) \tag{49}$$

By taking $T = 10^8$ K, σ_s from (49) is equal to $4.00 \times 10^{-20}\,\text{cm}^2$. Then (48) gives a mean free path of $l = 603$ cm. With such a large free path, the transfer of energy by collisions is extremely slow and inefficient. To improve the chances of collision, some particles of larger size must be introduced, e.g., atoms of heavier elements. The heavier atoms can have their outer electrons stripped (ionized) at the prevailing high temperature, but since some electrons remain attached to the nucleus, the size of the partially ionized atom can still be of the order of Å. Then such particles will be a scattering cross section of the order of $10^{-16}\,\text{cm}^2$. Even with only one per cent of such heavier elements in the mixture, the mean free path will be brought down to a few centimeters. This is indeed the mean free path used in the size estimate of the preceding paragraph. Needless to say, the heavier atoms introduced must not capture neutrons appreciably so as not to interfere with the very important energy producing reaction of (41).

However, even with the presence of heavy partially ionized atoms, the mixture will be still practically transparent to high energy neutrons generated by reaction (37). The energies carried by them cannot then be "kept" in the gas by collision, but rather are received directly by the walls of the reaction chamber. This is a direct energy leak and makes the quenching problem very much more difficult. In fact, out of the reaction (37), only the kinetic energy of ${}_2He^3$ is kept within the gaseous mixture. This energy is only 1/4 of the total given by (40), or 1.31×10^{-6} ergs. The energy produced by the reaction (36) is of course retained in the gaseous mixture and is equal to the difference of energies given by (39) and (40) or 1.22×10^{-6} ergs. Hence 50 per cent of the deuteron-deuteron reactions have an effective energy production of only

$$(1.31 + 1.22) \times 10^{-6} = 2.53 \times 10^{-6}\,\text{ergs} \tag{50}$$

The average of the reaction energy kept in the mixture is thus, using (39)

$$\frac{1}{2}[2.53 + 6.46] \times 10^{-6}\,\text{ergs} = 4.50 \times 10^{-6}\,\text{ergs} \tag{51}$$

Compared with gross energy production given by (44), this is only 69.6 per cent; 30.4 per cent of energy produced is delivered directly to the solid walls of the chamber.

Out of the energy kept in the reacting mixture, given by (51) for one single binary reaction or

$$1.606 \times 10^{10}\,\text{cal/gr} \tag{52}$$

a good fraction will be absorbed by the reacting deuterium in entering the flame: The deuterium gas is heated, dissociated, and finally ionized to reach the full flame temperature of say 1.600×10^8 K. According to Sänger[1], to heat up to this temperature, the deuterium takes up approximately

$$10^9\,\text{cal/gr} = 0.1 \times 10^{10}\,\text{cal/gr} \tag{53}$$

Now the crucial question is: how many grams of deuterium have to be heated in order that one gram of deuterium will be burned to completion? In other words, what is the combustion efficiency of the flame in the reaction chamber? By comparing (52) with (53), it is seen that if 16.06 grams of deuterium have to be heated to flame temperature to get one gram of deuterium burned, then there will be no heat left to be conducted and radiated to the wall *through* the gaseous mixture. But there must be heat conducted and radiated to the wall because the wall, being necessarily of solid material, must be at a temperature, say 2 000 K, which is very much lower than the flame temperature of 1.600×10^8 K. This shows that the ratio of mass to be heated and actually burned must be less than 16.06.

For lack of more accurate information, consider a combustion efficiency of the flame zone to be 1/6. That is, six grams of deuterium have to be heated to have one gram actually burned. Then the energy available for conduction and radiation to the wall per gram of deuterium burned is, according to (52) and (53)

$$(1.606 - 6 \times 0.1) \times 10^{10}\,\text{cal/gr} = 1.006 \times 10^{10}\,\text{cal/gr} \tag{54}$$

Therefore, by comparing with (45), only less than one-half of the gross energy production is available for "cooling" loss. In fact, with (46), the "cooling" loss energy Q_c produced per unit vol of flame per unit time is

$$Q_c = \frac{1.006}{2.31} \times 0.874\,\text{cal/cm}^3\,\text{sec} = 0.382\,\text{cal/cm}^3\,\text{sec} \tag{55}$$

Now let it be assumed that the flame in the 100 m diam reaction chamber be a cylindrical volume of some 60 m diam and 120 m long. Then within this 120 m of flame, the wall will receive by conduction and radiation *through* the mixture, a heat flux density q_c equal to

$$q_c = \frac{\frac{\pi}{4} \times 6\,000^2 \times 0.382}{\pi \times 10\,000} = 343\,\text{cal/cm}^2\,\text{sec} = 8.75\,\text{Btu/in}^2\,\text{sec} \tag{56}$$

This corresponds to a black body radiation at 3 990 K.

The question is, of course, whether the heat flux q_c actually equals that given by (56). For the specified conditions in the reaction chamber, if the actual q_c is larger than (56), then the critical reaction chamber diameter must be larger than the assumed 100 m. If less, then the critical diameter can be smaller. Therefore, one of the basic problems of thermonuclear reaction chamber design is the calculation of q_c or radiation heat flux through a gas layer of variable composition and variable temperature. The technical complication here is, of course, the fact that here the radiation mean free path is large in comparison with the physical dimensions and therefore the simple method developed by astrophysicists for the interior of stars is not applicable. On the other hand, all essential basic information required for the calculation is now available. The problem is thus only complicated but not insurmountable. But, in any event, the flame is almost transparent due to the low density and almost complete ionization. In fact, within the flame, radiation will come almost only from the specially introduced heavy atoms which are, however, of very, very low density. Therefore, the flame, although of extremely high temperature, is a relatively weak radiator. Hence the comparatively low effective black body temperature of 3 990 K may not be far from being correct.

5 Thermonuclear Power Station

The part of energy directly transmitted to the wall by fast neutrons is the difference between (45) and (52); thus the total energy flux to the wall q is

$$q = \frac{\frac{\pi}{4} \times 6\,000^2 \times \left[0.382 + 0.874 \times \frac{2.31 - 1.606}{2.31}\right]}{\pi \times 10\,000} \tag{57}$$

$$= 582 \text{ cal/cm}^2\text{sec} = 14.89 \text{ Btu/in}^2\text{sec}$$

Although nearly one-half of this energy is realized by slowing down the fast neutrons and is thus distributed in a layer of wall material, not merely delivered at the surface of the wall, nevertheless the tremendous heat density poses a cooling problem which cannot be solved by conventional cooling methods. It seems that the only feasible method is that of transpiration cooling. That is, the wall is made of porous material, say porous carbon or graphite, and cold deuterium gas is forced by pressure through the wall into the reaction chamber. Heat in the wall is picked up by the coolant gas and returned to the reaction chamber. By using a large enough quantity of coolant gas, the wall temperature can be kept at the desired low temperature of, say, 2 000 K. In fact, the application of transpiration cooling to nuclear reactors has already been considered by Kaeppeler[7]. He, however, has not included the important "spacing heating" effects of neutron slowing-down.

Behind the section of reaction chamber occupied by the flame zone, the heat flux due to neutrons is greatly reduced; then the coolant gas forced into the reaction chamber merely serves to lower the temperature of the gas from the flame zone (exhaust gas). At the end of the reaction chamber, the temperature across the chamber cross section should be fairly uniform and at, say, 1 000 K. The discharge pressure of this body of hot gas is of course essentially the chamber pressure which is taken to be 100 atm in the above discussion. The high pressure hot gas can be used to generate power through a gas turbine. It is perhaps worth while to note that the product gas is expected to contain only the weakly radioactive $_1H^3$ and thus should give no difficulty for the power generating machinery. The exhaust from the turbine after being cooled by heat exchanger will pass through the waste extraction system for removing the nuclear "ash." The purified gas will contain of course mainly D_2, but also has a very small concentration of H_2 and T_2 produced by reaction (36) and (41). This gas together with small amount of make-up D_2 to replace the deuterium burned is then compressed to high pressure and fed through the porous wall back to the reaction chamber. This then completes the cycle of the power plant. Fig. 1 is a diagram representing the components and the process of the system.

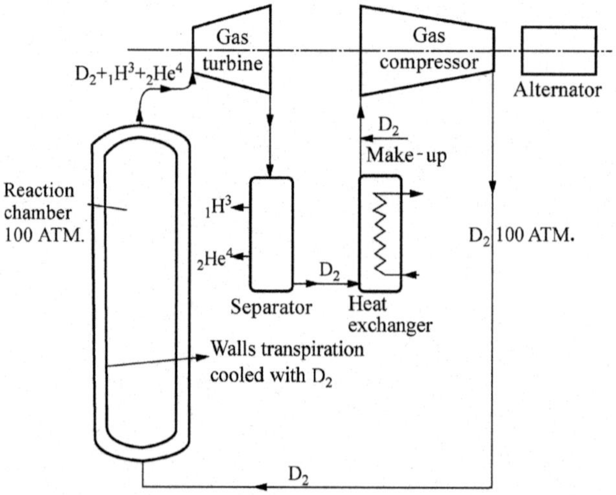

Fig. 1 Schematic diagram of thermonuclear power station July 1956

Of course, by cycling the gaseous mixture repeatedly through the reaction chamber, the concentration of H_2 and T_2 will build up and eventually participate importantly in the energy production through reaction (42), and other reactions. These reactions actually produce the nuclear ash ${}_2He^4$. Finally the composition of feed gas to the reaction chamber will be stabilized at a fixed ratio of H_2, D_2, and T_2 with the production of H and T by reactions (36) and (41) balanced by consumption of H and T through reactions (42) and others. Then the over-all result is that of feeding in deuterium D_2 and taking out ${}_2He^4$. Thus the reaction chamber effectively converts deuterons to helium according to

$$ {}_1H^2 + {}_1H^2 \rightarrow {}_2He^4 + \text{energy} \tag{58} $$

and the energy produced per gram of deuterium burned is very much larger than given by (45). Furthermore, in this final stage the reaction scheme is considerably more complicated than that discussed in Section 3 and the volume rate of energy generation and heat flux must be somewhat different from that calculated previously. However, these calculations will not be attempted here since the purpose of this study is merely to give a general outline of the problem.

6 Ignition

According to the studies made by Sänger[1], the ignition temperature, i. e., the temperature at which the rate of energy production is just balanced by the rate of energy lost, mainly through radiation, with the deuteron-deuteron reaction is approximately 10^7 K. Naturally the question is how can the thermonuclear reaction be initiated by heating the gaseous mixture to this very high temperature. Before the advent of nuclear fission and the fission bomb, such high temperatures seemed unapproachable. But now this is definitely not so. It may even be possible to obtain ignition without using the fission reaction. But at the moment, one can only say that ignition of the thermonuclear reaction is certainly possible; no detailed scheme can, however, be suggested.

7 Thermonuclear Power Industry

The rate of energy production Q according to (46) is 0.003 65 kW/cm^3 in the flame zone. If, as assumed previously, the flame zone is a cylinder of 60 m diam and 120 m long, the total energy production is

$$0.003\,65 \times (\pi/4) \times 6\,000^2 \times 12\,000 = 1.238 \times 10^9 \text{ kW} \tag{59}$$

If the thermodynamic efficiency of the power plant cycle is 25 per cent, the power of the station is

$$0.25 \times 1.238 \times 10^9 \text{ kW} = 0.309 \times 10^9 \text{ kW} \tag{60}$$

Thus continuous operation of the plant will product annually electric energy of the amount

$$0.309 \times 10^9 \times 24 \times 365 = 2.71 \times 10^{12} \text{ kW} \cdot \text{hr} \tag{61}$$

In 1954, the annual electric energy production in the United States was approximately 0.5×10^{12} kW hr. Thus in one thermonuclear power plant, perhaps one of minimum size, the capacity is over five times the total effective capacity of the United States! This points to the extreme importance of determining the critical quenching size of the thermonuclear reaction chamber accurately. The speculations in the preceding sections are based upon an assumed diameter of 100 m for the reaction chamber. Smaller size and lower flame temperature will naturally reduce the scale of the power plant. However for the deuteron-deuteron reaction, the ignition temperature is roughly 10^7 K. For steady burning, the flame temperature cannot be below this temperature.

To conclude this discussion, comparison of thermonuclear energy and other energy sources will be made: According to (45), the fusion energy of deuterium is

$$2.31 \times 10^{10} \times 1.8 \text{ Btu/lb} = 4.16 \times 10^{10} \text{ Btu/lb of } D_2 \tag{62}$$

The fission of one pound of U-235 gives 3.14×10^{10} Btu. Therefore fusion energy is almost 4/3 times as large as fission energy. Since the natural isotope concentration of deuterium in hydrogen is 1 : 7 000, in terms of natural hydrogen, the fusion energy is

$$(4.16 \times 10^{10})/7\,000 = 5.94 \times 10^6 \text{ Btu/lb of hydrogen} \tag{63}$$

or, referred to water

$$(5.94 \times 10)/9 = 6.60 \times 10^5 \text{Btu/lb of } H_2O \tag{64}$$

If the average chemical energy of coal is taken as 11,000 Btu/lb, one pound of water is potentially equivalent to sixty pounds of coal! But even all this is based upon only partial burning of deuteron to triton and proton. With complete burning into ${}_2He^4$, the thermonuclear energy of deuterium will be still larger. Therefore, if thermonuclear power plants can actually be constructed, then the source of fusion energy far exceeds the other terrestial energy resources, chemical or fission.

References

[1] Sänger E. Astrinautica Acta. 1955, 1: 61-88.

[2] Kaeppeler H J. J. Astronautics. 1955, 2: 50-56.

[3] Gamow G, Teller E. Physical Review. 1938, 53: 608. // Gamow G, Critchfield C I. Theory of Atomic Nucleus and Nuclear Energy Sources. Oxford, 1949, Chap. 10.

[4] Fowler R H. Guggenheim, E. A., Statistical Thermodynamics. Cambridge, 1939: 493.

[5] Arnold W R, Philips J A, Sawyer G A, Stovall E J. Jr., Tuck J L. Physical Review. 1954, 93: 483.

[6] Lin S-C, Resler E L, Kantrowitz A. Journal of Applied Physics. 1955, 26: 95.

[7] Kaeppeler H J. Jet Propulsion. 1954, 24: 316.

Index

Printed in the United States
By Bookmasters